Vitamin

E in
Health
and
Disease

Vitamin E in Health and Disease

edited by

Lester Packer
*University of California
Berkeley, California*

Jürgen Fuchs
*Johann Wolfgang Goethe University
Frankfurt, Germany*

CRC Press
Taylor & Francis Group
Boca Raton London New York

CRC Press is an imprint of the
Taylor & Francis Group, an **informa** business

CRC Press
Taylor & Francis Group
6000 Broken Sound Parkway NW, Suite 300
Boca Raton, FL 33487-2742

First issued in paperback 2019

© 1993 by Taylor & Francis Group, LLC
CRC Press is an imprint of Taylor & Francis Group, an Informa business

No claim to original U.S. Government works

ISBN-13: 978-0-8247-8692-2 (hbk)
ISBN-13: 978-0-367-40263-1 (pbk)

Library of Congress Cataloging-in-Publication Data

Vitamin E in health and disease / edited by Lester Packer, Jürgen
  Fuchs.
      p.   cm.
    Includes bibliographical references and index.
    ISBN 0-8247-8692-0 (alk. paper)
    1. Vitamin E--Therapeutic use. 2. Vitamin E--Physiological
effect.   I. Packer, Lester. II. Fuchs, J. (Jürgen).
    [DNLM: 1. Vitamin E--chemistry. 2. Vitamin E--pharmacology.
3. Vitamin E--therapeutic use.   QU 179 V8388]
RM666.T65V58   1992
612.3'99--dc20
DNLM/DLC
for Library of Congress   92-49976
                                                              CIP

**Visit the Taylor & Francis Web site at
http://www.taylorandfrancis.com**

**and the CRC Press Web site at
http://www.crcpress.com**

Preface

This volume presents the current knowledge and state-of-the-art information that will enhance the important and well-recognized field of vitamin E research. Not since the excellent books of B. Lubin and L.J. Machlin appeared more than a decade ago has the subject of vitamin E been treated in a comprehensive manner. During this time new information on the biological activity of vitamin E and its various isomers have become available. Many future applications can be anticipated for all these new findings. In view of this we were motivated to prepare a volume that contains contributions by the leading vitamin E researchers throughout the world and would be of interest to biochemists, physiologists, pharmacologists, and public health researchers.

It is generally accepted that free radicals are formed in biological systems through metabolism and that exogenous free radicals from environmental sources (irradiation, food, drugs, and so forth) also contribute significantly to free radical production in biological systems. Free radicals, being reactive species, are short-lived and do not travel very far; in fact, they react with molecular targets very close to where they are formed. Thus, free radical concentrations in biological systems are extremely low and difficult to detect by electron spin resonance (ESR) spectroscopy. Even indirect methods of reactions of radicals with specific biomolecules are not sensitive enough to quantitate their presence. However, the response of the antioxidant defenses that react with radical species serves as an indirect measure that free radicals have been formed. Thus, redox-based antioxidants change their oxidation state and the antioxidants become free radicals themselves. However, such major antioxidants as vitamin E and vitamin C have aromatic ring structures that delocalize free radicals. Hence, they are long-lasting persistent radicals which react only slowly and can frequently be generated in biological systems. However, they are partially destroyed by radical reactions and thus must be replenished.

Vitamin E, the most important lipophilic radical quencher, is surveyed in this volume. Considered in a total of six sections consisting of a wide range of chapters, its molecular characterization and mechanism of action in its role in clinical medicine are discussed. The rationale for developing biochemical and physiological characteristics, as well as medical and epidemiological studies, stems from the great importance that is now being attached to the usefulness of vitamin E, alone and in combination with other antioxidants, in slowing the course of chronic and degenerative disease and aging. Also, considerable evidence indicates that patients with acute clinical conditions such as ischemia-reperfusion injury (e.g., following myocardial infarction or cerebral ischemia) can benefit from prior administration of antioxidants. Epidemiological studies point strongly to vitamin E as perhaps the most important substance in preventing chronic disease and premature aging. In particular, many human intervention trials for cancer, cardiovascular diseases, and stroke have been conducted and many more are currently underway.

The importance of vitamin E in drug therapy is based on the principle that redox drugs generate reactive oxygen species (ROS). These pharmacological interventions with ROS-generating drugs often damage surrounding normal cells and tissues. Their use is widespread in anticancer therapy and in treatment of parasitic diseases, and it has been demonstrated that some of the toxic side reactions can be reduced with vitamin E therapy.

As a result of clinical interventions, it is clear that vitamin E therapy is essential in specific clinical conditions in addition to vitamin E deficiency. This has most frequently been observed in infants and children in whom genetic disease (abetalipoproteinemia) or chronic cholestatic liver disease results in an inability to properly absorb vitamin E, as well as in premature babies exposed to oxygen inhalation therapy. Despite much positive circumstantial evidence, the importance of vitamin E is still not conclusively proven for prevention or delay of chronic diseases and premature aging. However, it is frequently recommended that adults maintain an adequate vitamin E level, which depends on the person's lifestyle and the oxidative stress to which individuals with different genetic potentials are exposed.

The distinction between the physiological and pharmacological usefulness of vitamin E for health benefits and treatment of disorders related to free radical generation in cells and tissues is not completely clear. However, sufficient studies have been made providing evidence, albeit circumstantial, to indicate that progress will be swift in these areas.

The editors are grateful to many who offered advice and support, including Sharon Landvik of the Vitamin E Research and Information Service and Pharma Stroschein.

Lester Packer

Jürgen Fuchs

Contents

Contributors

Bruce N. Ames, Ph.D. Department of Molecular and Cell Biology, University of California, Berkeley, California

Erkki Antila, Ph.D., M.D. Department of Anatomy, University of Helsinki, Helsinki, Finland

Francisco J. Aranda, Ph.D. Department of Biochemistry and Molecular Biology, Facultad de Veterinaria, Universidad de Murcia, Murcia, Spain

Faik Atroshi, Ph.D. College of Veterinary Medicine, University of Helsinki, Helsinki, Finland

Pietro Avogaro, M.D. Regional Center for Atherosclerosis, Regional General Hospital, Venice, Italy

Angelo Manfredo Azzi, Ph.D., M.D. Institute for Biochemistry and Molecular Biology, University of Bern, Bern, Switzerland

R. A. Bakalova Medical Academy, Sofia, Bulgaria

Gianna Bartoli, M.D. Institute for Biochemistry and Molecular Biology, University of Bern, Bern, Switzerland

Adrianne Bendich, Ph.D. Hoffmann-La Roche, Inc., Nutley, New Jersey

Gabriele Bittolo-Bon, M.D. Regional Center for Atherosclerosis, Regional General Hospital, Venice, Italy

Shirley R. Blakely, Ph.D. Experimental Nutrition Branch, Division of Nutrition, Department of Health and Human Services, United States Food and Drug Administration, Washington, D.C.

Gunter Blankenhorn, Ph.D. Department of Biochemistry, R. P. Scherer, GmbH, Eberbach, Germany

Carmia Borek, Ph.D. Division of Radiation and Cancer Biology, Department of Radiation Oncology, New England Medical Center and Tufts University School of Medicine, Boston, Massachusetts

Wolf Bors, Ph.D. Institut für Strahlenbiologie, GSF-Forschungszentrum für Umwelt und Gesundheit, Neuherberg, Germany

Daniel Boscoboinik, M.D. Institute for Biochemistry and Molecular Biology, University of Bern, Bern, Switzerland

Graham W. Burton, Ph.D. Steacie Institute for Molecular Sciences, National Research Council of Canada, Ottawa, Ontario, Canada

Giuseppe Cazzolato, Tech. Regional Center for Atherosclerosis, Regional General Hospital, Venice, Italy

Malcolm Chiswick, M.D. (FRCP DHC) Neonatal Medical Unit, North Western Regional Perinatal Centre, St. Mary's Hospital, Manchester, England

Ching K. Chow, Ph.D. Department of Nutrition and Food Science, University of Kentucky, Lexington, Kentucky

Sabine Clewing, Ph.D. Department of Medical Information, Brenner-Efeka Pharma GmbH, Münster, Germany

William Cohn, Ph.D. Department of Vitamin and Nutrition Research, F. Hoffmann-La Roche Ltd., Basel, Switzerland

David G. Cornwell, Ph.D. Department of Medical Biochemistry, College of Medicine, Ohio State University, Columbus, Ohio

Bertil Diamant, M.D. Department of Clinical Pharmacology, Copenhagen University, Copenhagen, Denmark

Martina Dieber-Rotheneder Institute of Biochemistry, University of Graz, Graz, Austria

Ellen S. Dierenfeld, Ph.D. Animal Health Center, New York Zoological Society, Bronx, New York

Paolo Di Mascio, Ph.D.* Department of Physiological Chemistry, University of Düsseldorf, Düsseldorf, Germany

Harold H. Draper, Ph.D. Department of Nutritional Sciences, University of Guelph, Guelph, Ontario, Canada

Garry Graeme Duthie, Ph.D. Biochemistry Division, Rowett Research Institute, Aberdeen, Scotland

Nabil M. Elsayed, Ph.D. Department of Respiratory Research, Division of Medicine, Walter Reed Army Institute of Research, Washington, D.C., and Department of Environmental Health, School of Public Health, University of California, Los Angeles, California

---

**Present affiliation*: Instituto de Química Bio Sup, University of São Paulo, São Paulo, Brazil

Charles E. Elson, Ph.D.   Department of Nutritional Sciences, University of Wisconsin, Madison, Wisconsin

Hermann Esterbauer, Ph.D.   Institute of Biochemistry, University of Graz, Graz, Austria

Karl Folkers, M.D., Ph.D.   Laboratory for Biomedical Research, University of Texas at Austin, Austin, Texas

Balz B. Frei, Ph.D.*   Department of Molecular and Cell Biology, University of California, Berkeley, California

Hans-Joachim Freisleben, M.Pharm., Ph.D., D.Sci. (Med)   Gustav-Embden-Zentrum der Biologischen Chemie, University Clinics, Frankfurt, Germany

Jean-Charles Fruchart, Ph.D.   Department of Lipoproteins and Atherosclerosis, Pasteur Institute, Lille, France

Jürgen Fuchs, Ph.D., M.D.   Department of Dermatology and Venerology, University Clinic, Johann Wolfgang Goethe University, Frankfurt, Germany

K. Fred Gey, M.D.   Institute for Biochemistry and Molecular Biology, University of Bern, Bern, Switzerland

Lawrence I. Golbe, M.D.   Department of Neurology, University of Medicine and Dentistry of New Jersey—Robert Wood Johnson Medical School, New Brunswick, New Jersey

Juan C. Gómez-Fernández, Ph.D.   Department of Biochemistry and Molecular Biology, Facultad de Veterinaria, Universidad de Murcia, Murcia, Spain

Sam Goth   University of California, Berkeley, California

Erich Grundel, B.S.   Experimental Nutrition Branch, Division of Nutrition, United States Food and Drug Administration, Department of Health and Human Services, Washington, D.C.

Carmel Hensey   Institute for Biochemistry and Molecular Biology, University of Bern, Bern, Switzerland

Keith U. Ingold, D.Phil.   Steacie Institute for Molecular Sciences, National Research Council of Canada, Ottawa, Ontario, Canada

Mamie Leah Young Jenkins, Ph.D.   Experimental Nutrition Branch, Division of Nutrition, United States Food and Drug Administration, Department of Health and Human Services, Washington, D.C.

**Present affiliation*: Department of Nutrition, Harvard School of Public Health, Boston, Massachusetts

Kurt Johansen, B.S.   Kabi-Pharmacia, Stockholm, Sweden

Hideaki Kabuto   Department of Neurochemistry, Institute for Neurobiology, Okayama University Medical School, Okayama, Japan

Valerian E. Kagan, Ph.D., D.Sci.   Department of Molecular and Cell Biology, University of California, Berkeley, California

Stephan Kaiser, M.D.   Department of Physiological Chemistry, University of Düsseldorf, Düsseldorf, Germany

Jan Karlsson, Ph.D.   Department of Clinical Physiology, Karolinska Hospital, Stockholm, Sweden

Marguerite M. B. Kay   Department of Microbiology and Immunology, University of Arizona College of Medicine, Tucson, Arizona

Paul Knekt, Dr. P.H.   Research Institute for Social Security, Social Insurance Institution, Helsinki, Finland

Eckhard Koepcke, M.D.   Departments of Medicine and Obstetrics and Gynecology, Klinikum Südstadt, Rostock, Germany

Kanki Komiyama, Ph.D.   Research Division, The Kitasato Institute, Shirokane, Minato-ku, Tokyo, Japan

Motoharu Kondo, M.D., Ph.D.   First Department of Medicine, Kyoto Prefectural University of Medicine, Kyoto, Japan

Angelica Krebs   Institute of Biochemistry, University of Graz, Graz, Austria

Bodo Kuklinski   Department of Medicine, Klinikum Südstadt, Rostock, Germany

Sharon V. Landvik, M.S., R.D.   Vitamin E Research and Information Service, Edina, Minnesota

J. William Langston, M.D.   California Parkinson's Foundation, San Jose, California

Daniel C. Liebler, Ph.D.   Department of Pharmacology and Toxicology, College of Pharmacy, University of Arizona, Tucson, Arizona

Silvio Lippa   Department of Human Physiology-Clinical Chemistry, Catholic University of the Sacred Heart, Rome, Italy

Susana Llesuy, Ph.D.   Departments of Analytical Chemistry and Physical Chemistry, School of Pharmacy and Biochemistry, University of Buenos Aires, Buenos Aires, Argentina

Gérald Luc, M.D.   Department of Lipoproteins and Atherosclerosis, Pasteur Institute, Lille, France

Lawrence J. Machlin Human Nutrition Research, Hoffmann-La Roche, Inc., Nutley, New Jersey

Mitsuyoshi Matsuo, Ph.D. Tokyo Metropolitan Institute of Gerontology, Tokyo, Japan

Edgar Mentrup, M.Pharm., Ph.D. Pharma-Galenik, Hoechst AG, Frankfurt, Germany

Simin Nikbin Meydani, D.V.M., Ph.D. Nutritional Immunology Laboratory, United States Department of Agriculture—Human Nutrition Research Center on Aging at Tufts University, Boston, Massachusetts

Christa Michel Institüt für Strahlenbiologie, GSF-Forschungszentrum für Umwelt und Gesundheit, Neuherberg, Germany

D. A. G. Mickle, M.D., M.Sc., F.R.C.P.(C) Department of Clinical Biochemistry, Toronto General Hospital, Toronto, Ontario, Canada

José Milei, M.D. Department of Cardiology, Juan A. Fernández Hospital, Buenos Aires, Argentina

Makoto Mino Department of Pediatrics, Osaka Medical College, Takatsuki Osaka, Japan

Geraldine V. Mitchell Experimental Nutrition Branch, Division of Nutrition, United States Food and Drug Administration, Department of Health and Human Services, Washington, D.C.

Kenneth Patrick Mitton, B.Sc. (Ph.D. Student) Department of Biochemistry, University of Western Ontario, London, Ontario, Canada

Akitane Mori, M.D., Ph.D. Department of Neurochemistry, Institute for Neurobiology, Okayama University Medical School, Okayama, Japan

Kazuo Mukai Department of Chemistry, Faculty of Science, Ehime University, Matsuyama, Japan

David P. R. Muller, Ph.D. Division of Biochemistry and Metabolism, Institute of Child Health, University of London, London, England

Michael E. Murphy, Ph.D.* Department of Physiological Chemistry, University of Düsseldorf, Düsseldorf, Germany

Yuji Naito, M.D. First Department of Medicine, Kyoto Prefectural University of Medicine, Kyoto, Japan

Etsuo Niki, Ph.D. Research Center for Advanced Science and Technology, University of Tokyo, Tokyo, Japan

---

Present affiliation: University of Vermont, Burlington, Vermont

Roger Nordmann Department of Biomedical Research on Alcoholism, Université René-Descartes, Paris, France

Augustine S. H. Ong, Ph.D. Scientific and Technical Services, Malaysian Palm Oil Promotion Council, Kuala Lumpur, Malaysia

Lester Packer, Ph.D. Department of Molecular and Cell Biology, University of California, Berkeley, California

Rao V. Panganamala, Ph.D. Department of Medical Biochemistry, College of Medicine, Ohio State University, Columbus, Ohio

Jean A. T. Pennington, Ph.D., R.D. Division of Nutrition, Center for Food Safety and Applied Nutrition, United States Food and Drug Administration, Department of Health and Human Services, Washington, D.C.

Giuseppina Petrelli, M.D. Department of Dermatology, Facolta di Medicina e Chirurgia, Catholic University of the Sacred Heart, Rome, Italy

William A. Pryor, Ph.D. Biodynamics Institute, Louisiana State University, Baton Rouge, Louisiana

Herbert Puhl, Ph.D. Institute of Biochemistry, University of Graz, Graz, Austria

Asaf A. Qureshi, Ph.D. Advanced Medical Research, Madison, Wisconsin

Nilofer Qureshi, Ph.D. Mycobacteriology Research Laboratory, William S. Middleton Memorial Veterans Hospital and Department of Bacteriology, College of Agricultural and Life Science, University of Wisconsin, Madison, Wisconsin

Bruce Rappaport Department of Morphological Sciences, The Bruce Rappaport Faculty of Medicine and Rappaport Institute for Medical Research, Technion-Israel Institute of Technology, Haifa, Israel

Donald J. Reed, Ph.D. Department of Biochemistry and Biophysics, Oregon State University, Corvallis, Oregon

Abraham Z. Reznick, Ph.D. Department of Morphological Sciences, The Bruce Rappaport Faculty of Medicine and Rappaport Institute for Medical Research, Technion-Israel Institute of Technology, Haifa, Israel

S. R. Ribarov Medical Academy, Sofia, Bulgaria

Peter D. Richardson Division of Engineering and Department of Medicine, Brown University, Providence, Rhode Island

James McD. Robertson, D.V.M., M.Sc. (Med.) Department of Epidemiology and Biostatistics, University of Western Ontario, London, Ontario, Canada

Hélène Rouach Department of Biomedical Research on Alcoholism, Université René-Descartes, Paris, France

Luigi Rusciani, M.D. Department of Dermatology, Facolta di Medicina e Chirurgia, Catholic University of the Sacred Heart, Rome, Italy

Manfred Saran, Dipl. Phys. Institut für Strahlenbiologie, GSF-Forschungszentrum für Umwelt und Gesundheit, Neuherberg, Germany

Barbara Scheer Gustav-Embden-Zentrum der Biologischen Chemie, University of Frankfurt, Frankfurt, Germany

Joel L. Schwartz, D.M.D., D.M.Sc. Department of Oral Pathology, Harvard School of Dental Medicine, Harvard University, Boston, Massachusetts

Elena A. Serbinova, Ph.D. Department of Molecular and Cell Biology, University of California, Berkeley, California

Alan J. Sheppard, Ph.D. Experimental Methods Research Section, Division of Nutrition, Center for Food Safety and Applied Nutrition, United States Food and Drug Administration, Department of Health and Human Services, Washington, D.C.

Gerald Shklar, D.D.S., M.S. Department of Oral Medicine and Oral Pathology, Harvard School of Dental Medicine, Harvard University, Boston, Massachusetts

Helmut Sies, M.D. Department of Physiological Chemistry, University of Düsseldorf, Düsseldorf, Germany

Irene Simon-Schnass, M.D. Department of Medicine, Hermes Arzneimittel GmbH, Grosshesselohe, Munich, Germany

Sunil Sinha Northern Western Regional Perinatal Centre, Saint Mary's Hospital, Manchester, England

Ronald J. Sokol, M.D. Section of Gastroenterology/Nutrition, Department of Pediatrics, University of Colorado School of Medicine; Children's Hospital, Denver, Colorado

Manfred Steiner, M.D., Ph.D. Division of Hematology, Brown University, Providence, and Division of Hematology/Oncology, Memorial Hospital of Rhode Island, Pawtucket, Rhode Island

Adam Szewczyk Institute for Biochemistry and Molecular Biology, University of Bern, Bern, Switzerland

Hirohisa Takano, M.D. First Department of Medicine, Kyoto Prefectural University of Medicine, Kyoto, Japan

Al Tappel, Ph.D. Department of Food Science and Technology, University of California, Davis, California

Robert P. Tengerdy, Ph.D. Department of Microbiology, Colorado State University, Fort Collins, Colorado

Henning Theorell, M.D. General Practitioner, Stockholm, Sweden

Thomas Thürich Gustav-Embden-Zentrum der Biologischen Chemie, University of Frankfurt, Frankfurt, Germany

Maret G. Traber, Ph.D. Department of Medicine, New York University School of Medicine, New York, New York

John R. Trevithick, Ph.D. Department of Biochemistry, Faculty of Medicine, University of Western Ontario, London, Ontario, Canada

Masahiko Tsuchiya University of California, Berkeley, California

Shiro Urano, Ph.D. Department of Biochemistry and Isotopes, Tokyo Metropolitan Institute of Gerontology, Tokyo, Japan

José Villalaín Department of Biochemistry and Molecular Biology, Facultad de Veterinaria, Universidad de Murcia, Murcia, Spain

George Waeg Institute of Biochemistry, University of Graz, Graz, Austria

John L. Weihrauch Human Nutrition Information Service, United States Department of Agriculture, Hyattsville, Maryland

Tuomas Westermarck, M.D. Helsinki Central Institute for the Mentally Retarded, University of Helsinki, Helsinki, Finland

Eric Witt, Ph.D. Department of Molecular and Cell Biology, University of California, Berkeley, California

Masakazu Yamaoka, Ph.D. Biological Chemistry Division, National Chemical Laboratory for Industry, Tsukuba, Ibaraki, Japan

Isao Yokoi, M.D., Ph.D. Department of Neurochemistry, Institute for Neurobiology, Okayama University Medical School, Okayama, Japan

Toshikazu Yoshikawa, M.D., Ph.D. First Department of Medicine, Kyoto Prefectural University of Medicine, Kyoto, Japan

Guido Zimmer, M.D., Ph.D. Membrane Structure Group, Gustav-Embden-Zentrum der Biologischen Chemie, University of Frankfurt, Frankfurt, Germany

Z. Z. Zhelev Medical Academy, Sofia, Bulgaria

I
NATURAL OCCURRENCE AND TISSUE LEVELS

A. Plants

1

Natural Sources of Tocotrienols

Augustine S. H. Ong

Malaysian Palm Oil Promotion Council, Kuala Lumpur, Malaysia

INTRODUCTION

The tocotrienols, which are similar to tocopherols in molecular structure except that there are three double bonds in the isoprenoid side chain, are found widely distributed in the plant kingdom. In general they are found in such cereals as wheat, barley, rye, and rice and in some oils, such as palm oil and rice bran oil. Work has also shown that tocotrienols are present in rubber (*Hevea brasiliensis*) latex.

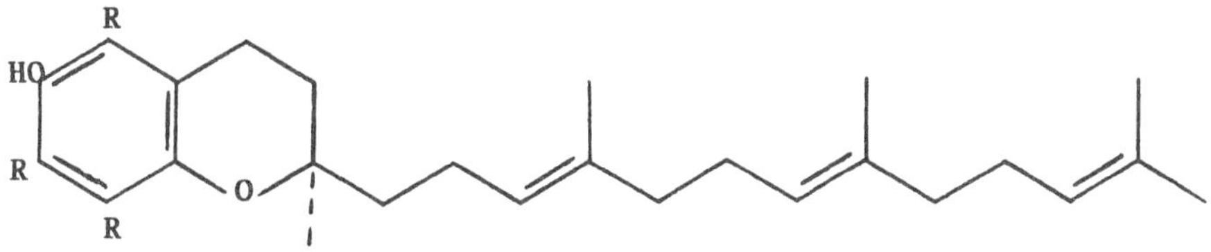

TOCOTRIENOLS

In the survey conducted in Finland on the presence of tocotrienols and tocopherols in Finnish foods, it was found that the richest sources of tocotrienols and tocopherols are wheat germ, bran, meal, and flour of ash content 1.2–1.4%, rye meal, and oat products (1).

Tocotrienols were also found in barley. Purification of the oily, nonpolar fraction of high-protein barley (*Hordeum vulgare* L.) flour by high-performance liquid chromatography yielded 10 major components, one of which was identified as *d*-α-tocotrienol. Of interest, this component was found to inhibit hepatic cholesterogenesis in chicks (2). The group in Finland surveyed various foods for the presence of tocotrienols and tocopherols, such as vegetables, fruits, and berries, in which α- and γ-tocotrienols were present in addition to α-, β-, γ-, and δ-tocopherols (3); in meat and meat products (4); dairy products and eggs (5); and finally in oils and fats, in which the total tocopherols content varies from 4 (coconut oil) to 242 mg per 100 g (wheat germ oil) (6).

In the literature survey of tocotrienols and tocopherols in oils and fats, it is found that tocotrienols are present in only a few of these oils, as shown in Table 1 (7–11). The data from Table 1 indicate that palm oil is the richest source of tocotrienols among the oils, and this is treated in greater detail because it is also the most practical source of tocotrienols currently.

TABLE 1 Tocopherols (T) and Tocotrienols (T₃) Determined in Vegetable Oils (mg/kg)

Oil	α-T	β-T	γ-T	δ-T	α-T$_3$	γ-T$_3$	δ-T$_3$
Castor	28	29	111	310	—	—	—
Cocoa butter	11	—	170	17	2	—	—
Coconut[a]	—	—	—	3	20	—	—
Corn[a]	134	18	412	39	—	—	—
Cottonseed	573	40	317	10	—	—	—
Groundnut	169	5	144	13	—	—	—
Linseed	—	—	—	—	—	—	—
Mustard	75	—	494	31	—	—	—
Olive	93	—	7	—	—	—	—
Palm	150	—	—	—	117	297	80
Palm[a]	133	—	—	—	130	204	45
Palm oil from fibers	1662	—	—	—	456	485	142
Rape[a]	70	16	178	7	—	—	—
Rice bran	324	18	53	—	236	349	—
Rice bran[a]	249	15	47	—	167	141	—
Safflower	477	—	44	10	—	—	—
Sesame[a]	12	6	244	32	—	—	—
Soybean[a]	116	34	737	275	—	—	—
Sunflower[a]	608	17	11	—	—	—	—
Wheat germ	1179	398	493	118	Tr	—	—

[a]Refined oils.
Source: From Refs. 7–11.

Besides these sources, it was found that large amounts of free and esterified α-, γ-, and δ-tocotrienols are present in rubber latex at a concentration of 0.82% of the lipid fraction (12) and also in bleached Thai rubberseed oil (13). These nonrubber ingredients in the natural rubber have an effect on aging properties (14). Tocotrienols are found in various by-products of the palm oil industry, such as palm fatty acid distillate (PFAD), a by-product of the physical refining of palm oil, and in the fibers. Palm oil can be processed by both chemical and physical refining. It is mainly processed by physical refining, however, because it is more economical and environment friendly. The steps involved in physical refining are given in Figure 1. During this process crude palm oil is degummed with phosphoric acid and treated with bleaching earth before subjected to a vacuum of 2–5 torr (mmHg) at a temperature of 260–270°C. Under these conditions the volatiles are condensed as PFAD. The composition of the PFAD is given in Table 2. The concentration of PFAD tocotrienols and tocopherols is about 0.48%. This content may vary from 0.16 to 1%, however, depending on the raw materials for the refining process as given in Table 3. The potential availability of tocotrienols and tocopherols from PFAD is given in Table 4. For 1990 the potential availability of tocotrienols and tocopherols from PFAD was 1092 t (15).

The extraction of tocotrienols and tocopherols from PFAD has been developed from laboratory to pilot-plant scale (16), and the process is given in Figure 2. Basically the process involves converting the unwanted free fatty acids and partial glycerides into methyl esters, which are then removed by distillation, and the residue purified of squalene, sterols, and other impurities. A pilot plant established at the Palm Oil Research Institute of Malaysia

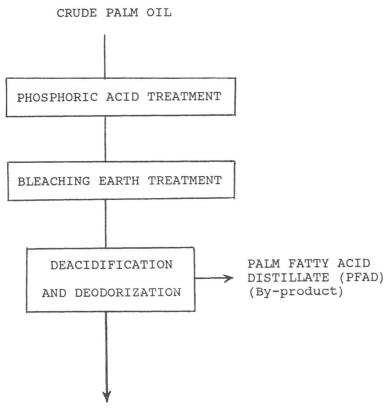

FIGURE 1 Physical refining process of palm oil.

TABLE 2 Composition (%) of Palm Fatty Acid Distillate (PFAD)

Free fatty acids		81.7
Glycerides		14.4
Triglycerides	4.1	
Diglycerides	7.6	
Monoglycerides	2.7	
Sterols		0.34
Cholesterol	0.004	
Campesterol	0.092	
Stigmasterol	0.032	
β-Sitosterol	0.212	
Hydrocarbons		1.47
Squalene	0.76	
Others	0.71	
Tocopherols + tocotrienols		0.48
Others		1.6

TABLE 3 Total Tocopherols (T) and Tocotrienols (T₃) in Palm Fatty
Acid Distillate (PFAD)

Feedstock of refining	Total T and T₃ in PFAD (ppm)		
	Average	Range	No. of samples[a]
Palm oil	3973	744–8191	26
Palm olein	3391	1018–7172	19
Palm stearin	1379	162–2408	4

[a]Samples were collected from 14 refineries.

(PORIM) is able to achieve an overall yield of 65% with a purity of 95–99%. The rate of production is 5–6 kg per batch. Work thus far has shown that the process is technically feasible, and an economical evaluation for commercialization of the production of tocotrienols from palm oil is being undertaken (17). An alternative process for the extraction of tocotrienols and tocopherols from PFAD has also been reported (18).

Several samples of tocotrienols obtained from the pilot plant have been sent for various nutrition and chemical studies in different parts of the world, and some of the results are provided here.

The tocotrienols from palm oil have been converted into free radicals, and the electron spin resonance spectra of the radicals derived from both the tocotrienols and their dimers have been determined (19). The role of these tocotrienols as antioxidants at the cellular level has been reported (20). Nutrition studies have shown in several animal models, including chickens, quails, and swine, and in humans that the tocotrienols have exhibited a hypocholesterolemic effect by reducing the activity of β-hydroxy-β-methylglutaryl coenzyme A reductase, thus suppressing the synthesis of cholesterol in the liver (21–24). Based on in vitro experiments by Holub et al., it has been shown that tocotrienols, particularly δ-tocotrienol, have a significant antiaggregation effect on blood platelets (25). In terms of carcinogenesis, evidence that tocotrienols exhibit anticancer properties has been obtained (26).

Thus it is indicated that tocotrienols have significant physiological properties and that palm oil is likely to be an important commercial source of tocotrienols.

TABLE 4 Potential Availability of Vitamin E from Palm Fatty Acid
Distillate (PFAD)

Year	CPO production[a] (1000 t)	PFAD (1000 t)	Vitamin E (t)
1986	4536	223	892
1987	4533	183	732
1988	5033	197	788
1989	6055	262	1048
1990	6311	273[b]	1092

[a]Crude palm oil.
[b]Estimate.

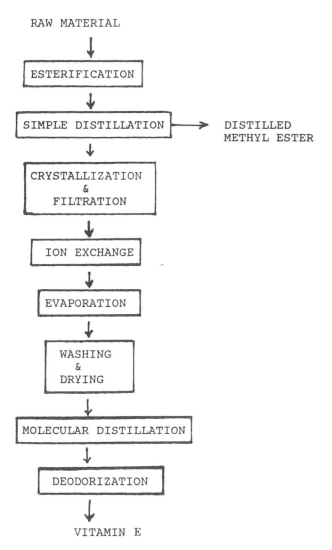

FIGURE 2 Vitamin E pilot-plant production process.

REFERENCES

1. Piironen V, Syvaoja E-L, Varo P, Salminen K, Koivistoinen P. Cereal Chem 1986; 63(2):78.
2. Qureshi AA, Burger WC, Peterson DM, Elson CE. J Biol Chem 1986; 261(23):10544.
3. Piironen V, Syvaoja E-L, Varo P, Salminen K, Koivistoinen P. J Agr Food Chem 1986; 34:742.
4. Piironen V, Syvaoja E-L, Varo P, Salminen K, Koivistoinen P. J Agr Food Chem 1985; 33:1215.
5. Syvaoja E-L, Piironen V, Varo P, Koivistoinen P, Salminen K. Milchwissenschaft 1985; 40:467.
6. Syvaoja E-L, Piironen VV, Varo P, Koivistoinen P, Salminen K. J Am Oil Chem Soc 1986; 63:328.
7. Carpenter AP. J Am Oil Chem Soc 1979; 56:668.
8. Muller-Mulot W. J Am Oil Chem Soc 1976; 53:732.
9. Gapor A, Berger KG, Hashimoto T, et al. Palm oil product technology in the eighties. Pushparajah E, Rajadurai M, eds. Publisher: Incorporated Society of Planter, 1983; 145.
10. Choo YM, et al. Unpublished results.

11. Nakasato S. Newsletter Malaysian Inst Chem 1989; 33:17.

12. Dunphy PJ, Whittle KJ, Pennock JC, Morton RA. Nature 1965; 207:521.

13. Yamaoka M, Chandrawong P, Tanaka A. Yukagaku 1988; 37(9):736.

14. Morimoto M. Proc Int Rubber Conf 1985; 2:61.

15. Gapor A, Kato A, Ong ASH. Proc Int Oil Palm/Palm Oil Conf 1987; 124.

16. Gapor A, Leong WL, Ong ASH, Kawada T, Watanabe H, Tsuchiya N. Australian Patent Pending, PI 7273/88, 1988.

17. Ong ASH, Halim Hassan A, Basiron Y, Choo YM, Gapor A. Symp Proc New Dev Palm Oil, Soc Chem Ind (SCI) Oils Fats Group, 1990; 45.

18. Goh SH, Lai FL. Malaysian Patent Pending, PI 8703215, 1988.

19. Goh SH, Hew NF, Ong ASH, Choo YM, Brumby S. J Am Oil Chem Soc 1990; 67(4):250.

20. Servinova E, Kegan V, Han D, Packer L. Free Radical Biol Med (In Press).

21. Ong ASH, Qureshi N, Atil O, Din ZZ, Bitgood JJ, Qureshi AA. Faseb J 1988; 2(5):A1540.

22. Qureshi AA, Qureshi N, Ong ASH, Gapor A, deWitt GF, Chong YH. Suppression of cholesterol biosynthesis and hypocholesterolemic effects of tocotrienols from palm oil in the chicken model. Presented at the National Conference on Oil Palm/Palm Oil, October 11–15, 1988, Kuala Lumpur, Malaysia.

23. Qureshi AA, Weber FE, Qureshi N, et al. Dietary tocotrienols reduce levels of plasma cholesterol, apolipoprotein B, thromboxane B_2 and platelet factor 4 in pigs with inherited hyperlipidemias. Presented at PORIM International Palm Oil Development Conference, September 5–9, 1989, Kuala Lumpur, Malaysia.

24. Qureshi N, Wright JJK, Pearce BC, et al. Suppression of apolipoprotein B, thromboxane B_2 and platelet factor 4 by palm oil and its tocotrienols inn genetically hypercholesterolemic quail. Presented at the X International Symposium on Drugs Affecting Lipid Metabolism (abstract), November 8–11, 1989, Houston, Texas.

25. Holub BJ, Sicilia F, Mahadevappa VG. Effect of tocotrienol derivatives on collegen and ADP-induced human platelet aggregation. Presented at the PORIM International Palm Oil Development Conference, September 5–9, 1989, Kuala Lumpur, Malaysia.

26. Komiyama K, Iijuka K, Yamaoka M, Watanabe H, Tsuchiya N, Umezawa I. Chem Pharm Bull 1989; 37(5):1369.

2

Analysis and Distribution of Vitamin E in Vegetable Oils and Foods

Alan J. Sheppard and Jean A. T. Pennington

United States Food and Drug Administration, Washington, D.C.

John L. Weihrauch

United States Department of Agriculture, Hyattsville, Maryland

INTRODUCTION

Vitamin E was first isolated from wheat germ in 1936 by Evans et al. (1). Shortly thereafter, Fernholz (2) characterized the structure of α-tocopherol. A number of workers were successful in synthesizing α-tocopherol, the most notable being the work of von Karrer et al. in 1938 (3,4). After the discovery, isolation, and characterization of α-tocopherol, vitamin E was usually described as α-tocopherol until recently.

NOMENCLATURE, TERMINOLOGY, AND DEFINITIONS

The nomenclature, terminology, and definitions of vitamin E or compounds exhibiting vitamin E activity vary with the audience being addressed. Therefore, it is necessary to examine and to appreciate the variability in nomenclature, terminology, and definitions used by different scientific compendia and organizations. The discussion that follows covers most of the current usage.

United States Pharmacopeia XXI (USP)

Currently, vitamin E is described as a form of α-tocopherol by the USP (5). This definition includes *d-* or *d,l-*α-tocopherol, *d-* or *d,l-*α-tocopheryl acetate, and *d-* or *d,l-*α-tocopheryl succinate. This compendium recognizes that the International Unit (IU) of vitamin E has been discontinued but that the IU of vitamin E, which is numerically equal to the USP unit of vitamin E, continues to be used in labeling some vitamin E products. USP nomenclature applies to vitamin E preparations and vitamin E capsules formulated and labeled per the USP Official Monographs. The relationship between α-tocopherol and its derivatives are summarized in Table 1.

American Institute of Nutrition (AIN)

A nomenclature policy for vitamins and related compounds has been established by AIN and published in the *Journal of Nutrition* (6) based on the nomenclature naming procedures

TABLE 1 Relative Vitamin E Activity According to the USP/NF and IUPAC Convention[a]

Isomer	Activity (IU/mg)
d,l-α-Tocopherol	1.1
d-α-Tocopherol	1.49
d-α-Tocopheryl acetate	1.36
d-α-Tocopheryl acid succinate	1.21
d,l-α-Tocopheryl acetate	1.0

[a]United States Pharmacopeia/National Formulary and the International Union of Pure and Applied Chemistry.

developed by the International Union of Nutrition Sciences Committee on Nomenclature (IUNSCN). This nomenclature policy is consistent with the Recommendations of the Commission on Biochemical Nomenclature (IUPAC-IUB; Table 2). In this system, the summation of the contributions to the dietary intake of different forms of vitamin E are made in milligrams of *RRR*-α-tocopherol equivalents, even though the alcohol is greatly inferior, with respect to its instability and variable biological activity, to *RRR*-α-tocopheryl acetate as the standard for biological tests.

Association of Official Analytical Chemists (AOAC)

The AOAC (7) adopted a set of nomenclature rules in 1980 based on the IUNSCN as outlined in the *Journal of Nutrition* (6). The AOAC states that the "term vitamin E should be used as generic descriptor for all tocol- and tocotrienol derivatives" that exhibit vitamin E activity. Consequently the term "tocopherols" is not only synonymous with the term vitamin E but is also a descriptor for all Me tocols. Under this set of nomenclature rules, α-tocopherol is a trivial name without stereochemical designation (Table 2). The compound isolated from natural

TABLE 2 Nomenclature for Vitamin E

Trivial name	IUPAC-IUB-AIN[a]	
	Name	Designated name
α-Tocopherol	5,7,8-Trimethyltocol	
d-α-Tocopherol	2*R*,4′*R*,8′*R*-α-tocopherol	*RRR*-α-tocopherol
l-α-Tocopherol	2*S*,4′*R*,8′*R*-α-tocopherol	2-Epi-α-tocopherol
d,l-α-Tocopherol (totally synthetic)	2DL,4′DL,8′DL-α-tocopherol	All-rac-α-tocopherol
d,l-α-Tocopherol[b] synthesis from natural phytol		2-Ambo-α-tocopherol
β-Tocopherol[c]		5,8-Dimethyltocol
γ-Tocopherol[c]		7,8-Dimethyltocol
δ-Tocopherol[c]		8-Methyltocol
Tocotrienol	2-Methyl-2-(4′,8′,12′-trimethylthrideca-3′,7′,11′-tnenyl) ehroman-6-ol	

[a]International Union or Pure and Applied Chemistry, International Union of Biology, and American Institute of Nutrition.
[b]Formerly the international standard for vitamin E.
[c]Preferred usage.

sources, $2R,4'R,8'R$-α-tocopherol, formerly known as α-tocopherol, should be designated as all-rac-α-tocopherol according to the AOAC. This designation is not consistent with that of the IUNSCN (Table 2).

National Research Council (NRC)

The NRC has issued a new edition of *Recommended Dietary Allowances* (8), in which the occurrence and biological activity of vitamin E are summarized in detail. It is emphasized that there are two groups of compounds found in plant materials that have vitamin E biological activity, that is, tocopherols and tocotrienols. These active compounds differ in the number and position of the methyl groups in the chroman ring system and in the structure of the side chain. The tocopherol group includes α-, β-, γ-, and δ-tocopherols. The most biologically active form of vitamin E is α-tocopherol. The biological activity of the vitamin E analogs has been determined by a number of workers with animal bioassays. These biological activity values for the various vitamin E analogs have been applied to humans with no species adjustment. The relative biological values of the vitamin E analogs are as follows: α-tocopherol, 100; β-tocopherol, 25–50; γ-tocopherol, 10–15; and α-tocotrienol, 30. The range or variability of the relative biological value estimates is due to the various types of biological assays contributing to the estimate. The NRC has adopted the IUNSCN nomenclature in which the natural α-tocopherol should be designated *RRR*-α-tocopherol (formerly *d*-α-tocopherol) and the synthetic compound should be designated all-rac-α-tocopherol (formerly *d,l*-α-tocopherol). The NRC has stipulated, for dietary purposes, that vitamin E activity be expressed as *RRR*-α-tocopherol equivalents (α-TE). One α-TE is defined as the activity of 1 mg of *RRR*-α-tocopherol. The total α-TE of a natural source of mixed forms is estimated with the following conversion factors: milligrams of β-tocopherol × 0.5; milligrams of γ-tocopherol × 0.1; and milligrams of α-tocotrienol × 0.3. If the synthetic all-rac-α-tocopherol is present, the milligrams of the compound present should be multiplied by 0.74.

Food Chemicals Codex (FCC III)

The FCC III (9) sets standards for chemicals added to foods, that is, food-grade chemicals, among which is vitamin E. In this compendium, the various vitamin E analogs are identified by their trivial names. The unit of measurement is milligrams of the vitamin E analog or milligrams of *d*-α-tocopherol per gram of total tocopherols, whichever is appropriate for the specific monograph being followed.

METHODOLOGY

The methodology for determining vitamin E has undergone several major changes. The original methods were biological fertility tests. Chemical methods were developed very quickly, followed by paper and thin-layer chromatography (TLC). Modern methods for the analysis of the tocopherols (both tocols and tocotrienols) are based either on gas-liquid chromatography (GLC) or high-performance liquid chromatography (HPLC).

Extraction

The extraction process is a very critical step in the analysis. It must be quantitative and nondestructive. This is especially true for the vitamin E analogs, which must be protected from oxidation and light, both of which lead to an underestimation of the actual vitamin E content. The extracting agents include ether, hexane, acetone, chloroform, ethanol, propanol, and, in some cases, combinations of solvents. Various other features can be added, such as mechanical shakers, heat, freeze-drying, Soxhlet, and saponification, depending on the nature of the matrix, the experience of the investigator(s), and the analytical goal.

Many of the early workers used hot absolute alcohol to extract vitamin E (10–12) from animal and plant tissues. Pudelkiewicz and Matterson (11) extracted wet homogenized chick livers for 20 h with hot absolute ethanol as described by Quaife and Harris (10). The extraction used by these workers is based on weighing the homogenized tissues or ground, dehydrated food or tissue from a Wiley mill into a Soxhlet thimble, which is then placed into an extraction cup. The cup is inserted into the Soxhlet extraction unit. These units have a heat source and cold water condenser and constitute a closed system. Other common systems are the Bailey-Walker and Goldfish units, and other extraction apparatus are readily available from most laboratory supply houses, Bieri (13) emphasized that the extraction of saponification mixtures is commonly accomplished with diethyl ether, which must be peroxide free. Bieri pointed ut that hexane and petroleum ether are often used as the extracting agents, but some workers believe that these two solvents do not extract as much of the tocopherols as the more polar ether or alcohol. This is a point not universally agreed on. A good outline of the alcohol-Soxhlet procedure applicable to foods is provided by Sheppard et al. (12), along with extract cleanup procedures that include column chromatography and sublimation, procedures discussed later in this chapter. Laidman et al. (14) emphasized the extraction of the tocopherols from plant materials. They demonstrated that wheat grain extraction with hot propan-2-ol and chloroform is superior to extraction with acetone. In this procedure, the grain is homogenized in the extracting solvent before transfer to a separatory funnel for water washing of the extracts and final collection of the cleaned extract for GLC analysis. Cohen and Lapointe (15) used a solution of dioxane-isooctane (20:80 vol/vol) and a wrist-action shaker to extract vitamin E from feeds. A good overall review of sampling, saponification, and extraction techniques and their associated problems as of 1980 has been done by Parrish (16). The USP XX and AOAC methods for pharmaceuticals use *n*-hexane as the extracting solvent based on the studies conducted by Pillsbury et al. (17).

Biological Methods

The biological assay methods using the rodent were developed in the 1920s and early 1930s (18,19). Female rats are usually preferred for the biological determination of vitamin E because they can be essentially healed several times from avitaminosis, whereas the damage done in male rats by the absence of vitamin E is irreparable. Later work by Bacharach and coworkers (20,21) demonstrated that female virgin rats (about 30–40 days after weaning) are preferable because the initial storage of the vitamin in the young can be controlled through placental and mammary transfer. There were two types of tests: prophylactic and curative. The latter method is considered the more reliable of the two. The criterion of proof is the occurrence of gestation and fetus resorption. The bouchon vaginale is the proof of conception. Implantation of the ova is verified by the finding of blood in the vagina between days 9 and 13 after mating. The percentage of fertility is measured in this test. An improved method is the so-called uterine index determination. This method is based on a graded response of the animals at day 9 of pregnancy. The evaluation consists of the determination of uterine weight before and after fetus removal, including living fetuses, dead fetuses, and resorption sites. The IU is defined as 1 mg synthetic racemic α-tocopherol acetate, which is the average amount of the standard preparation that, when administered orally, prevents resorption-gestation in rats deprived of vitamin E (22). There were some attempts to use other animals, such as the rabbit (23) and the transparent crustacea *Daphnia magna* (24,25). However, these test species never generated any popular following and the rat remained the primary vitamin E bioassay test species. The expense, time, facilities, degree of physiological expertise, and excessive variability quickly prompted extensive research activity for an alternative to the biological assay approach for routine analysis.

Chemical Methods

Several chemical methods were used during the very late 1930s through the early 1940s. Adaptations made to these methods during the 1950s to 1960s are still used for specialized applications. Among these methods were the gold chloride, nitric acid, cerimetric titration, and ferric chloride–dipyridyl colorimetric methods.

Gold Chloride Method

This method is based on the studies of von Karrer and coworkers in Zurich (26,27). Vitamin E is oxidized with gold chloride at about 50°C, with the progress of the reaction monitored electrochemically. The accuracy of the method is affected by other reducing substances, including the carotenoids. The method is not as sensitive as the ferric chloride method, which is discussed later. The gold chloride method never became a major analytical procedure for the measurement of vitamin E.

Nitric Acid Method

This method is also known as "tocopherol red." Vitamin E yields a characteristic red color when heated with alcoholic nitric acid (28). Some other naturally occurring substances give a similar red or yellow color when heated with alcoholic nitric acid. A filter limited to the waveband 450–520 μm helps eliminate these interferences when vitamin E is measured with the photometer. One disadvantage of this method is that the simple biologically inactive oxidation product of the tocopherols gives the same color. Ungnade and Smith (29) found "that when appropriately used, the method was very convenient and rapid." However, the method did not achieve widespread use.

Cerimetric Titration Method

For this method, the *d*- or *d,l*-α-tocopherol is dissolved in alcoholic sulfuric acid. Diphenylamine indicator is added to the tocopherol solution, followed by titration with ceric sulfate to a blue end point. The method is applicable to pharmaceutical preparations and was used as the official monograph method of the National Formulary (NF) through the 13th edition (30). If mixed tocopherol concentrates from natural sources or tocopherol derivatives constitute a test sample, the test portion must undergo saponification before the color development. The weakness of this method is that it is nonspecific and should be used for pharmaceutical preparations that are free of interfering substances. This method is no longer used as the monograph method of the NF, having been replaced in 1980 by a gas chromatographic method that is discussed subsequently.

Ferric Chloride–Dipyridyl Method

This method is also known as the Emmerie-Engel method, named for the workers who developed it (31). This method is based on the oxidation of vitamin E in alcoholic solution by ferric chloride. The addition of α,α'-dipyridyl develops a characteristic red color from the formation of ferrous chloride. This method is not very specific, because any reducing compound produces the red color. It is necessary to work in subdued light, because sunlight tends to cause the development of color that must be corrected for by the use of a blank determination of color background. It is possible to estimate the vitamin E content of some nonsaponified oils. However, it is better to saponify the test portion and remove the other reducing compounds. Emmerie and Engel (32) found that interfering substances could be adsorbed on Floridin (Fuller's earth) activated with hydrochloric acid. Hydrolysis with 2 N

KOH-methanol solution for 10 minutes was reported by Emmerie (33) as the best method to eliminate interfering reducing substances. Saponification may destroy some of the vitamin E under some circumstances (34). The absorbance of the vitamin E is determined at 520 μm with a suitable spectrophotometer. Emmerie (35) successfully used the method to determine the vitamin E content of butter. The ferric chloride–dipyridyl was the official monograph method of the USP through the 19th edition (36) for the assay of the α-tocopherol content of decavitamin capsules and tablets. Ferric chloride–dipyridyl chemistry is currently used as the AOAC official method for measuring α-tocopherol and α-tocopheryl acetate in foods and feeds, for identification of *RRR*- or all-rac-α-tocopherol in drugs and food or feed supplements, and for measuring supplemental α-tocopheryl acetate in foods and feeds (37,38).

Column, Paper, and Thin-Layer Chromatography

We view column, paper, and TLC as preparative procedures usually used in conjunction with a chemical method of measurement. However, we recognize that not all investigators are in full agreement with this approach. These preparative procedures are especially useful for removing a large proportion of the impurities that interfere with the assay of vitamin E. This situation is especially true when the vitamin E is isolated and purified from such natural products as foods, tissues, and organs. The test portion or its extract (usually in ether or petroleum ether) is saponified, which may remove the majority of the interfering impurities. Unfortunately, reducing substances accompany α-tocopherol in the unsaponifiable fraction, that is, vitamin A, carotenoids, sterols, and other tocopherols are present in naturally occurring α-tocopherol sources. Consequently, these reducing substances must be removed before α-tocopherol can be accurately determined by chemical methods, most of which are not compound specific (see Chemical Methods).

Column Chromatography

Column chromatography may be a useful step under certain conditions, as when preliminary purification of the unsaponifiable extract is needed before the individual tocols and/or tocotrienols can be separated. A wide variety of absorbants have been used to remove impurities from crude extracts before vitamin E analysis, including Celite 545-digitonin (12), Florex (11,39), Florisil and silicic acid (40), hydroxyalkoxypropyl Sephadex (41), deactivated alumina (42,43), secondary magnesium phosphate (44,45), Fuller's earth (45), and Floridin earth (46). The application of column chromatography in vitamin E methodology has been extensively reviewed by Desai (47) and Strohecker and Henning (48). In the latter reference, detailed step-by-step procedures are provided. We consider these two references to be outstanding, and there is little that can be added to them. Currently, a column chromatographic extract cleanup procedure is included in the AOAC official methods for the identification of all-rac-α-tocopherol (49) and the measurement of α-tocopheryl acetate supplement in foods and feeds by the colorimetric method (50).

Paper Chromatography

Brown (39) found that paper coated with Vaseline reliably identified the presence of naturally occurring tocopherols. A mobile phase of 75% ethanol in water (vol/vol) gave the best separations. Brown (51) was unable to quantitatively recover tocopherols from paper chromatograms.

Green et al. (52,53) successfully used both one- and two-dimensional paper chromatography to do structural studies on tocopherols isolated from rice. McHale et al. (54) were able to enhance the color of the tocopherol bands by impregnating the paper with a solution of zinc carbonate containing a small amount of sodium fluorescein before impregnation with liquid paraffin. The mobile phase was aqueous alcohol. Dilley and Crane (55) used paraffin-impregnated paper and a 3:1 vol/vol ethanol-water mobile phase to assay the tocopherols in plant tissues. Dasilva and Jensen (56) measured the α-tocopherol content of some blue-green algae using alumina-impregnateed paper. Other workers who contributed to the one-dimensional system development were Lambertsen and Braekkau (57) and Eggitt and Ward (58).

Evolving out of the experience of many investigators, a simpler one-dimensional paper chromatographic method has been developed that appears to be easier to use and as effective in separating the tocopherols as the two-dimensional paper system. In this method, the paper is impregnated with USP paraffin. The mobile phase is a mixture of absolute ethanol and water (80:20 vol/vol). The developing agent is a spray consisting of a mixture of 2,2'-bipyridyl and ferric chloride dissolved in absolute ethanol. This latter method is described in great detail by Strohecker and Henning (48).

Green et al. (59) and Marcinkiewicz and Green (60) developed an excellent paper chromatographic method for isolating and assaying all the individual tocopherols in natural products. It is a two-dimensional system that requires experience and gives good results when performed by a skilled chromatographer. This method was instrumental in the discovery of the various tocopherols that occur in nature. These early papers in two-dimensional paper chromatography contributed heavily toward the development of the official methodology adopted in England.

The Vitamin E Panel of the Society of Analytical Chemists in England (46) adopted an official standard method for the separation and identification of the individual tocopherols. The system is two-dimensional, with zinc carbonate–impregnated paper and a mobile-phase mixture. Benzene in cyclohexane (30:70 vol/vol) is used in the first direction and ethanol in water (75:25 vol/vol) in the second. The spots can be located by one of two procedures: viewed under an ultraviolet (UV) lamp after application of a fluorescein spray or with a template overlay on the unsprayed paper. The template is prepared by going through the mobile-phase steps, developing the color spots for the individual tocopherols with a dipyridyl–ferric chloride spray, and cutting out the individual tocopherol spots.

Edwin et al. (61) used a modification of the vitamin E panel method for determining the tocopherols in animal tissues. The modification consisted of a substitution of 80% ethanol for 75% ethanol as the mobile phase during the second-dimensional run. They reported superior separations with this minor modification. Booth (62) found that two-dimensional paper chromatography had the advantage of providing a more certain identification than other methods of tocopherol purification, such as Floridin/Florex XXS columns. Another advantage of their paper system was the elimination of artifact introduction during tocopherol determinations. By the mid-1960s, paper chromatography was generally replaced by TLC.

Thin-Layer Chromatography

It is generally accepted that TLC has several advantages over paper chromatography. These advantages include a shorter development time, superior separation of the individual tocopherols, larger loading capacity, and better quantitative elution of compounds from the absorbent.

Much of the early developmental work with both paper and TLC was thoroughly reviewed by Kofler et al. (63). Desai (47) provided a more recent review of the use of TLC for the separation and measurement of the tocopherols. Desai provides an illustration of a silica G separation of the tocopherols and tocopherol-related compounds found in animal tissues. Strohecker and Henning (48) discuss thin-layer techniques in depth and provide color photographs of TLC separations on silica gel G. The thin-layer methods for the tocopherols fall into two distinct groups, one-dimensional and two-dimensional. The two-dimensional methods developed as part of the effort to resolve β- and γ-tocopherol from each other.

Later improvements by Lovelady (64) in the one-dimensional system led to successful separations of β- and γ-tocopherols. Lovelady obtained resolution factors (R_f) of 0.255 and 0.287, respectively, for γ- and β-tocopherol on silica gel G with a mobile-phase mixture of cyclohexane, n-hexane, isopropyl ether, and ammonium hydroxide (40:40:20:2 vol/vol/vol/vol). Stowe (65) previously reported achieving a one-dimensional separation of γ- and β-tocopherol, but the R_f values were not reported. The γ- and β-tocopherol were not successfully separated on silica gel G plates with benzene-methanol or with cyclohexane-diethyl ether solvent phases (55,66). Kofler et al. (63) had difficulty in achieving complete γ- and β-tocopherol separations. Bolliger (66) provides an excellent discussion of the separation of the tocopherols on a variety of immobile phases and solvent systems. Tables are provided of R_f values and reagent colors.

The two-dimensional technique is discussed in depth in an excellent review by Desai (47). Generally the tocopherols can be separated from their corresponding tocotrienols and from other interfering compounds, such as the quinones and retinol. Whittle and Pennock (67) used a two-dimensional system to separate the tocopherols and tocotrienols with a reproducible recovery rate of about 92%. They used chloroform initially and a diisopropyl ether–petroleum ether mixture in the second direction. A satisfactory separation of γ- and β-tocopherols was achieved. Ames (68) conducted a collaborative study of the determination of vitamin E in foods and feeds for the AOAC. Incorporated in the overall methodology was a two-dimensional TLC procedure with an alumina absorbent. The method was adopted in 1971 as the official AOAC method for determining α-tocopherol in foods and feeds (37).

Molecular Distillation and Microsublimation

Distillation under high vacuum as a purification step in the isolation and determination of the tocopherols was first given emphasis by Dam et al. in 1941 (69) as an alternative to saponification and subsequent solvent extraction procedures. The technique has been used by a number of workers with a variety of equipment and procedures (11,69–75). Emmerson et al. (71) were the first to suggest that molecular distillation was a good possibility for the purification of vitamin E. Quaife and Harris (73) described a molecular semimicrostill that they used for concentrating the tocopherols from food fats. They used the system in their survey of the vitamin E content of foods (10). The distillation technique can effectively separate tocopherols from most other associated compounds, such as glycerides, chlorophylls, xanthophylls, carotenoids, quinone, ubiquinol, ubichromanol, and sterols. The "cold finger" microapparatus is cheaper to purchase and easier to operate but has a lower mass sample capacity than the semimicrostill. Sheppard and Ford (76) redesigned the sublimation apparatus (cold finger), recovering 98% of ^{14}C-labeled tocopheryl acetate added to lard (75). Up to 20 mg of tocopherols could be recovered quantitatively when the unit was operated at 160°C and 0.020 mmHg for 60 minutes. The technique is suitable for the extraction of tocopherols from small test portions, especially so when saponification may induce undesirable physiochemical changes. The use of the sublimation apparatus requires considerable operator skill, but the resultant product is of excellent quality and is suitable for subsequent analysis.

Gas Chromatography

This section addresses the analysis of the tocols and tocotrienols in foods, the analysis of vitamin E in concentrates and bulks, and the analysis of vitamin E in pharmaceuticals. A number of excellent reviews on the application of GLC for the analysis of vitamin E had appeared in the literature by 1972 (12,13,43,63,77,78).

Tocols and Tocotrienols in Foods

The separation and quantitation of the tocols and the tocotrienols in a single gas chromatography (GC) analysis is a goal that has been very difficult to achieve. The separation of the pair β-tocopherol and γ-tocopherol has been exceptionally challenging. Silicones have been the most satisfactory stationary phases evaluated (77,79–85). The majority of laboratories use SE-30 for the determination of α-, (β + γ), and δ-tocopherols. Typically, the packed columns were 5–7 feet in length and generally glass with a 5–6 mm inner diameter (ID). Column temperatures were usually 210–250°C. Other stationary phases that have been used include SE-52, QF-1, XE-60, and Apiezon N. The capacity was generally 0.5–2.0 mg. Nair and Turner (80) were successful in separating the dimethyltocopherols as their *p*-quinones, with a binary mixture of silicone rubber, SE-30, and XE-60, coated on 80–100 mesh Gas Chrom P. Attempts to separate the tocols and tocotrienols in a single GC analysis shifted from the wide-bore columns of the 5–7 feet length to longer and smaller ID columns usually packed with a lower percentage of stationary-phase material. The work of Slover et al. (83) is an excellent example. The columns were 1/8 inch outer diameter (OD) × 0.08 inch ID × 15 feet packed with either 0.5% Apiezon L or 2% SE-30 coated on 110–120 mesh Anakrom, acid and base washed, and silanized. The column temperature was 235°C, and the helium carrier gas flow rate was 15–20 ml/minute. The resolution of the β- and γ-tocopherol pair was better on the Apiezon L than on the SE-30 stationary phase, but separation was not complete. The Apiezon L stationary phase yielded a superior separation of the mixture of seven tocols and tocotrienols tested. The methodology was successfully used to measure the tocols and tocotrienols in a number of vegetable oils. Slover et al. (86) extended their stationary phase to include OV-17. The OV-17 gave a separation of the tocols and tocotrienols similar to that obtained with SE-30 and Apiezon L.

More recently the trend in the analysis of the tocols and tocotrienols has been the replacement of the packed-column technology with capillary technology. Slover and coworkers have been the major contributors in the application of capillary technology for the GC analysis of the tocols and tocotrienols (87–89). A typical application of a capillary column has been a 50 m × 0.25 mm capillary column coated with Dexsil 400 operated at 260°C, split ratio of 1:20, 1 ml/minute helium carrier gas, and nitrogen makeup gas at 40 ml/minute. The Slover group used the trimethylsilyl derivatives in their work. They were highly successful in the analysis of α-, γ-, and δ-tocopherol, as well as with the major sterols (campesterol, stigmasterol, and sitosterol) in a single analysis. However, the separation of the (β + γ)-tocopherol remained elusive. Nair and coworkers (82) used the *p*-quinones most successfully in the resolution of β- and γ-tocopherol, ironically with the older packed-column technology. These derivatives of the dimethyltocols would be expected to be resolved quite well from each other on SE-30 or Apiezon L capillary columns.

Vitamin E in Commercial Bulks and Concentrates

The FCC III GC method for mixed concentrates was collaboratively studied by Sheppard et al. (90). The method has been adopted as the official method of the USP and the AOAC for the analysis of α-, (β + γ)-, and δ-tocopherols in commercial bulks and concentrates. The tocopherols are determined as the propionate derivatives in the FCC method on a 2 m × 4 mm borosilicate column packed with 2–5% methylpolysiloxane gum on an 80–100 mesh, silanized diatomaceous earth support that has been acid-base washed. The column is maintained isothermally between 240 and 260°C. The flow rate of the carrier gas is adjusted to obtain a hexadecyl hexadecanoate (the internal standard) peak approximately 18–20 minutes after analyte introduction when a 2% column is used or 30–32 minutes when a 5% column is used. The R_f between the major peaks, which occur at retention times of about 0.50 (δ-tocopheryl propionate) and 0.63 [(β + γ)-tocopheryl propionates], relative to the internal standard at 1.00, is not less than 2.5. A total of 13 collaborators completed the study as directed. When a comparison of the collaborative mean value to the known value of the control was calculated, the percentage relative errors were 1.20, 4.25, 2.25, and 2.37 for α-tocopherol, (β + γ)-tocopherol, δ-tocopherol, and total tocopherols, respectively.

Vitamin E in Pharmaceuticals

Sheppard et al. (91) collaboratively studied a GLC method for determining the vitamin E content of tablets and capsules, and the method was adopted by the AOAC as an official method. This study demonstrated the superiority of GLC over the older colorimetric method. An external calibration procedure was used for calculating the quantity of specific vitamin E isomers in pharmaceuticals. Rudy et al. (92) developed additional data on the analysis of pharmaceuticals by an extensive collaborative study sponsored by the Pharmaceutical Manufacturers Association (PMA). The study verified the usefulness of GLC for the assay of α-tocopherol, α-tocopheryl acetate, and α-tocopheryl acid succinate. This method used dotri-acontane as an internal standard. The mean coefficient of variation obtained from the data of 13 collaborators was about ±5%.

In a preliminary study, they found no significant differences between the results obtained with the external standard and the internal standard calibration procedures. Based on these results, the more simple internal standard calibration procedure was incorporated into the method collaboratively studied. However, there was considerable debate and discussion within the PMA, NF, and the AOAC regarding the GLC method. Sheppard and Hubbard (93) conducted a collaborative study sponsored by the NF and the AOAC to resolve any differences and preconceived objections to adopting the GLC methodology as a replacement to the chemical methodology for official compendium purposes. The study had 11 collaborators and 11 analytical samples. The internal standard method of calibration was used, but hexadecyl hexadecanoate was substituted for doctriacontane based on a consensus of the NF, AOAC, and PMA representatives. The coefficients of variation of the reproducibility and repeatability were 3.4 and 1.6% for all laboratories and test samples, respectively. The coefficients of variation of reproducibility and repeatability for α-tocopheryl acid succinate were 2.1 and 1.5%, respectively. All the values fell within the 5% limit requirements of the NF. The NF and the AOAC subsequently adopted the method for their respective compendium. The AOAC methodology permits use of internal or external systems. A 2.4 m × 4 mm ID (uniform bore) Pyrex column packed with 5% SE-30 coated on 100–120 mesh Gas Chrom Q and operated at 270–285°C is used in this method. The injection system and detector temperatures are 295°C. The flow of the inert carrier gas, nitrogen or argon, is adjusted so that the α-tocopheryl acetate peak appears 23–27 min after analyte injection. Relative to hexadecyl palmitate and dotri-acontane, the internal standards, retention times are as follows: α-tocopherol, 0.53 and 0.75; α-tocopheryl hydrogen succinate, 0.54 and 0.76; and α-tocopheryl acetate, 0.62 and 0.86, respectively. With the external calibration system, the relative retention times are α-tocopheryl acetate, 1.0; α-tocopherol, 0.9; and α-tocopheryl hydrogen succinate, 0.9. To distinguish α-tocopherol from α-tocopheryl hydrogen succinate, one must obtain an analysis of the test extract and then combine it with a portion of the extract with acetic anhydride-pyridine (2:1 vol/vol) and rerun the GLC analysis. If the peak shifts to the α-tocopheryl acetate position in the second analysis, the test sample contains the alcohol; if it does not shift, the test sample contains α-tocopheryl hydrogen succinate. The NF does not use the external calibration procedure.

Liquid Chromatography

The emergence of HPLC for the analysis of vitamin E has accelerated since the publication of an HPLC method for vitamin E by Williams et al. in 1972 (94). Wiggins (95) and Saxby (96) published HPLC reviews with brief mention of the application of the method for the analysis of vitamin E.

Parrish (16), in an 1980 review of vitamin E analysis, briefly discussed HPLC. In 1973, Van Niekerk (97) reported an HPLC system for the direct determination of vitamin E in plant oils with a 150 cm long stainless steel 2 mm ID tube packed with Corasil II coupled to a 30 μl quartz flow cell. The mobile phase consisted of diisopropyl ether and *n*-hexane mixture (5:95 vol/vol) that was pumped through the column at 1.5 ml/minute. The analytical measurement was made with a fluorometer with the excitation wavelength set at 295 nm and the emission wavelength at 340 nm. A good separation of α, β-, γ-, and δ-tocopherols was obtained. The β- and γ-tocopherols were separated with a resolution index of 1.0. Cohen and Lapointe in 1978 (15) reported an HPLC method that effectively resolved α-tocopherol and α-tocopheryl acetate from vitamins D and A during the liposoluble vitamin analysis of feeds. Their system included a Bondapak C_{18} column, a methanol-water (95:5 vol/vol) solvent mixture, and a UV detector operated at 280 nm. During 1979, Thompson and Hatina (98) reported determining the tocopherols and tocotrienols in foods and tissue with a 25 cm ID of 5 μm Li Chromsorb Si60 silica column, a mobile phase of 0.2% isopropanol or 5% diethyl ether in moist hexane, and a fluorometer set at 290 mm excitation wavelength and 330 mm emission wavelength. The four tocopherols (α, β, γ, and δ) and three tocotrienols (α, β, and γ) were successfully separated and measured; 4 mg tocopherol produced a measurable peak. A total of 32 commercial infant formulas were analyzed by Landen (99) for their α-tocopheryl acetate on a Zorbax ODS column and a methylene chloride–acetonitrile-methanol (30:70:0.2 vol/vol/vol) mobile phase. Syvaoja and Salminen (100) reported in 1985 the successful determination of the tocopherols and tocotrienols in Finnish fish and fish products. They used a 25 × 0.4 cm ID 5 μm Li Chromsorb Si60 silica column, a 2.5 ml/minute flow rate of varying percentages of diisopropylether in hexane, and a fluorescence detector operated at an excitation wavelength of 290 nm and an emission wavelength of 325 nm.

During 1986, two important HPLC papers appeared (101,102). Lehmann et al. (101) successfully determined the vitamin E content of high and low linoleic acid diets using the HPLC method of Lehmann and Martin (103). The extract residue was dissolved in methanol and analyzed directly with no further cleanup. They used a reversed-phase column [as used by Marshall et al. (104)], a mobile phase of 100% methanol (2 ml/minute flow), and a fluorescence detector operated at an excitation wavelength of 292 nm and an emission wavelength of 335 nm. Recoveries of standard materials were 93.9 ± 7.4 and 93.9 ± 3.4% for α- and γ-tocopherol, respectively, when added to six different foods. Shen and Sheppard (102) demonstrated that baseline separations could be achieved for the four tocopherols (α, β, γ, and δ) on four normal-phased HPLC columns by using a hexane-based mobile phase with 2-propanol as the polar modifier. The separations were achieved on Porasel 10 μm silica (300 × 3.9 mm), Lichrosorb SI-100 5 μm silica (250 × 4.6 mm), and Spherisorb 5 μm silica (100 × 4.6 mm). The mobile-phase flow rate was 2 ml/minute. A UV-visible detector was operated at 295 nm. The columns were maintained at 30°C. The short column of Spherisorb 3 μm silica (100 × 4.6 mm) was equally effective in separating the tocopherols, but the analytical time and the amount of solvent required were substantially reduced compared to the more conventional columns. The method was successfully used to determine the vitamin E content of mixed tocopherol concentrates.

During 1989, Hogarty et al. (105) reported analyzing 40 foods for their tocopherol content using HPLC. They used a Zorbax-ODS (Mac-mod. 5 μm, 25 cm × 4.6 mm) column and a mobile phase of acetonitrile, methylene chloride, and 0.001% triethylamine and methanol

(700:300:50 vol/vol/vol) at a flow rate of 1 ml/minute. The column was connected to a fluorescence detector set at an excitation wavelength of 290 nm and an emission wavelength of 330 nm as previously used by Thompson and Hatina (98).

SIGNIFICANT SOURCES OF VITAMIN E

The foods listed in Table 3 accounted for 50% of the total vitamin E consumed by participants in the 1987–1988 Nationwide Food Consumption Survey (NFCS) conducted by the U.S. Department of Agriculture (106). The data show that significant sources of vitamin E include regular margarines (both hard and soft), regular mayonnaise and salad dressings, vitamin E-fortified breakfast cereals, vegetable shortenings used for home cooking, peanut butter, eggs, soybean salad and cooking oils, potato chips, whole milk, tomato products, and apples. Milk, eggs, and apples cannot be considered good sources of the vitamin; however, the large amounts of these foods that are eaten makes them significant sources. Breakfast cereals normally have low amounts of vitamin E, but because some are fortified to contain 100% of the U.S. Recommended Daily Allowances (U.S. RDA) of vitamin E per serving, they appear in the list as good sources. The listing in Table 3 should be interpreted with caution because ingredient oils contribute the vitamin E to such foods as margarines, mayonnnaise, salad dressings, vegetable shortenings, and potato chips. Because soybean oil is often the major oil used

TABLE 3 Significant Sources of Vitamin E in the American Diet[a]

Food	Food consumed[b] (kg)	Vitamin E consumed as α-TE[c] (mg)	% Total	% Cumulative total
Margarines, regular stick and tub with salt	5,185	660,127	13	13
Mayonnaise, regular, made with soybean oil	1,960	521,312	10	23
Breakfast cereals, vitamin E fortified	299	316,297	6	29
Shortening, vegetable, household, partially hydrogenated soy and cottonseed oil	1,952	282,988	5	34
Salad dressings, regular, French, and Italian	1,823	154,718	3	37
Peanut butter with salt	1,684	136,422	3	40
Eggs, chicken, whole, raw	12,147	109,813	2	42
Soybean oils and partially hydrogenated soybean oils, salad and cooking	645	108,126	2	44
Potato chips, salted	2,158	105,303	2	46
Milk, cow, whole, 3.3% fat raw or pasteurized, fluid	99,128	89,215	2	48
Tomato sauce, canned	4,510	58,623	1	49
Apples, raw, with peel	9,511	56,113	1	49

[a]Information based on the USDA Nationwide Food Consumption Survey (NFCS) 1987–88 (106).
[b]Food consumed by participants in the 1987–88 NFCS during 3 days of data collection.
[c]α-Tocopherol equivalents.

in these products, it deserves to be at the head of the list. The soybean oil in the table reflects liquid salad and cooking oil for home use.

Vitamin E Activity

The eight known tocopherols have different biological activities. The naturally occurring *RRR*-tocopherol has been assigned an activity of 1 mg α-TE per milligram. The relative activities of other tocopherols are listed in Table 4. The vitamin E activity of a food may be calculated by taking the sum of the values obtained by multiplying the number of milligrams of each component tocopherol by the appropriate factor given in Table 4. In many foods α- and γ-tocopherol contribute all the activity. Some cereals and related products contain significant amounts of β-tocopherol. Palm oil has relatively high amounts of the tocotrienols, which should be included when estimating the total vitamin E activity.

Tocopherol Content and Vitamin E Activity of Foods

Data on the vitamin E content of foods in Tables 5 through 9 were obtained from the published literature. The concentration of the tocopherols is expressed in milligrams per 100 g food. The total vitamin E activity was calculated from the concentration of the individual tocopherols and is reported as milligrams α-TE per 100 g food. These units permit comparisons with the Recommended Dietary Allowances published by the Food and Nutrition Board of the National Academy of Sciences (8).

Tocopherols in Oils, Fats, and Dressings

Most vegetable oils are among the richest sources of tocopherols (Table 6). For many oils, α-tocopherol is the major contributor to the total vitamin E activity; however, notable exceptions are soybean oil and corn oil, which have relatively high concentrations of γ-tocopherol. Most oils have either low concentrations of tocotrienols or none at all. An exception is palm oil, which has about 14 mg α-tocotrienol, 28 mg γ-tocotrienol, and nearly 7 mg δ-tocotrienol per 100 g oil. Among the oils, coconut oil has the lowest vitamin E activity at 0.7 mg α-TE, and wheat germ oil has the highest at 173 mg of α-TE per 100 g of oil. Wheat germ oil also has the highest β-tocopherol content at 71 mg per 100 g. Fish liver oils have relatively high α-tocopherol concentrations (e.g., cod liver oil). Fish body oils, like herring and menhaden oils, have moderate amounts of α-tocopherol. Much of this may be lost when fish oils are processed

TABLE 4 Vitamin E Activity of the Tocopherols and Tocotrienols[a]

Tocopherol	Activity as α-TE[b] (mg/mg compound)
RRR-α-tocopherol	1.0
RRR-β-tocopherol	0.5
RRR-γ-tocopherol	0.1
RRR-δ-tocopherol	0.03
RRR-α-tocotrienol	0.3
RRR-β-tocotrienol	0.05
Synthetic α-tocopheryl acetate	0.74

[a]The activities of γ-tocotrienol and δ-tocotrienol are unknown.
[b]α-Tocopherol equivalents.
Source: Adapted from Refs. 8 and 107.

TABLE 5 Tocopherol Content of Meat, Poultry, Fish, Dairy, and Egg Products (per 100 g Edible Portion)

Food	Tocopherol α (mg)	Tocopherol γ (mg)	Vitamin E activity as α-TE[a] (mg)
Meats			
Beef, round, lean, broiled	0.45	0.06	0.46
Beef, ground, 21% fat, broiled	0.37	0.03	0.37
Pork, ham, fresh, lean, roasted	0.09	0.02	0.09
Veal, rump, 21% fat, broiled	0.42	—[b]	0.42
Poultry			
Chicken, breast, roasted	0.20	0.10	0.20
Turkey, light meat, roasted	0.10	0.03	0.10
Fish			
Fillets, cooked			
Flounder, baked	1.90	—	1.90
Haddock, baked	0.40	—	0.40
Canned			
Salmon, in water, drained	1.00	—	1.00
Sardines, in tomato sauce[c]	5.40	0.03	5.40
Tuna, in water, drained	0.53	0.03	0.53
Tuna, in oil, drained[c]	1.00	4.80	1.48
Dairy and egg products			
Milk, cow's, fluid			
Whole, 3.3% fat	0.04	0.01	0.04
Low fat, 2% fat	0.03	0.02	0.03
Cream, fluid, light, coffee	0.15	0.02	0.15
Cheese			
American processed	0.40	—	0.40
Cheddar	0.30	—	0.30
Muenster	0.50	—	0.50
Swiss	0.60	—	0.60
Eggs, chicken			
Hard boiled	0.99	0.69	1.06
Scrambled	0.79	0.52	0.84

[a]α-Tocopherol equivalents.
[b]Dashes denote no value or trace.
[c]Contains 0.1 mg δ-tocopherol.
Source: Adapted from References 101 and 105.

to make them suitable for use in foods. McLaughlin and Weihrauch (109) observed that nine studies reported losses of vitamin E because of refining. These losses ranged from 0 to 79%. During hydrogenation, little or no vitamin E was lost. Shortenings, margarines, salad dressings, fried foods, and baked products tend to reflect the tocopherol pattern of the ingredient oil, which is predominantly soybean oil in many commercial products in the United States.

Tocopherols in Fruits, Vegetables, and Legumes

Vegetables have little α- and γ-tocopherol (Table 7). Potato chips were included to illustrate the effect of the absorbed frying fat, which contributes most of the vitamin E in this food. With

TABLE 6   Tocopherol Content and Vitamin E Activity of Oils, Fats, and Dressings (per 100 g Product)

Product	Tocopherols				Tocotrienols		Vitamin E activity as α-TE[a] (mg)	Reference
	α (mg)	β (mg)	γ (mg)	δ (mg)	α (mg)	β (mg)		
Oils								
Vegetable								
Canola	21.0	0.1	4.2	0.04	0.04	—[b]	21.5	108
Coconut[c]	0.5	—	—	0.6	0.5	0.1	0.7	84
Corn	11.2	5.0	60.2	1.8	—	—	19.8	84
Cottonseed	38.9	—	38.7	—	—	—	42.8	84
Olive	11.9	—	0.7	—	—	—	12.0	109
Palm[d]	25.6	—	31.6	7.0	14.6	3.2	33.5	84
Palm kernel	6.2	—	—	—	—	—	6.2	109
Peanut	13.0	—	21.4	2.1	—	—	15.2	84
Safflower	34.2	—	7.1	—	—	—	34.9	86
Sesame	13.6	—	29.0	—	—	—	16.5	84
Soybean[e]	7.5	1.5	79.7	26.6	0.2	0.1	17.1	108
Sunflower	48.7	—	5.1	0.8	—	—	49.2	84
Walnut	56.3	—	59.5	45.0	—	—	63.6	84
Wheat germ	133.0	71.0	26.0	27.1	2.6	18.1	173.6	84
Fish								
Cod liver	22.0	—	—	—	—	—	22.0	109
Herring	9.2	—	—	—	—	—	9.2	109
Menhaden	7.5	—	—	—	—	—	7.5	109
Fats								
Lard	1.2	—	0.7	—	0.7	—	1.5	110
Tallow	2.7	—	—	—	—	—	2.7	109
Vegetable shorten-ing, home	14.0	—	76.1	7.7	—	—	21.8	109
Margarine								
Stick								
Soybean	5.1	—	35.1	13.7	—	—	9.0	89
Corn	15.8	—	49.8	4.4	—	—	20.9	89
Sunflower	24.3	—	13.8	5.9	—	—	25.9	89
Soft								
Soybean	5.2	—	34.1	10.0	—	—	8.9	89
Corn	5.1	—	46.0	—	—	—	9.7	89
Sunflower	5.6	—	9.8	2.5	—	—	6.7	89
Dressings								
Blue cheese	4.7	—	57.5	25.4	—	—	11.2	105
French	3.1	—	61.9	18.2	—	—	9.8	105
Italian	4.0	—	62.7	17.8	—	—	10.8	105
Mayonnaise	3.2	—	42.1	—	—	—	7.4	101

[a]α-Tocopherol equivalents.
[b]Dashes denote no value or trace.
[c]Contains 1.9 mg γ-tocotrienol.
[d]Contains 28.6 mg γ-tocotrienol and 6.9 mg δ-tocotrienol.
[e]Contains 0.03 mg δ-tocotrienol.

TABLE 7 Tocopherol Content and Vitamin E Activity of Fruits, Vegetables, and Legumes (per 100 g Edible Portion)

Food	Tocopherols α (mg)	β (mg)	Vitamin E activity as α-TE[a] (mg)	Reference
Fruits				
Apples, unpeeled, raw	0.4	0.03	0.4	101
Apricots, dried[b]	6.2	0.2	6.2	111
Avocado	1.2	0.1	1.2	101
Bananas	0.2	—[c]	0.2	111
Grapes[d]	0.6	0.1	0.6	111
Oranges	0.4	—	0.4	111
Peaches	1.0	0.1	1.0	111
Pears	0.1	0.01	0.1	111
Plums	0.7	0.1	0.7	105
Raisins	0.4	0.03	0.4	101
Strawberries	0.6	0.2	0.6	111
Vegetables				
Asparagus, canned, drained	0.9	0.1	0.9	101
Beets, red, pickled	0.1	—	0.1	111
Broccoli	0.5	0.2	0.5	110
Cabbage	0.02	—	0.02	101
Carrots	0.4	0.02	0.4	101
Corn, yellow, frozen	0.1	0.3	0.1	101
Cucumber, raw, not peeled	0.1	0.1	0.1	101
Lettuce	0.6	0.3	0.6	111
Peas, green, frozen	—	1.6	0.2	101
Peppers, sweet, green	0.6	0.02	0.6	101
Potatoes, white	0.05	—	0.05	101
Potato chips[e]	3.4	16.1	5.4	112
Spinach frozen	1.8	0.1	0.5	101
Sweet potatoes, canned, drained	0.5	0.02	0.5	101
Legumes				
Beans, lima, dry[f]	—	7.2	0.7	108
Peas, dry[g]	0.3	6.4	0.9	108

[a]α-Tocopherol equivalents.
[b]Contains 0.1 mg α-tocotrienol.
[c]Contains 0.3 mg β-tocopherol and 7.9 mg δ-tocopherol.
[d]Contains 0.1 mg α-tocotrienol and 0.1 mg γ-tocotrienol.
[e]Contains 0.3 mg β-tocopherol and 7.9 mg δ-tocopherol.
[f]Contains 0.5 mg δ-tocopherol.
[g]Contains 0.6 mg δ-tocopherol.

the exception of frozen spinach, none of the other vegetables exceeds 1 mg α-TE per 100 g. Likewise, most fruits have vitamin E activities well below 1 mg α-TE per 100 g. An exception is dried apricots, which have 6.2 TE per 100 g. Other dried fruits may also have higher vitamin E content than the comparable fresh fruit. The vitamin E content of stone fruits, such as peaches and plums, appears to be somewhat higher than that of pome fruit. Few data were found on dry mature legumes. The data for lima beans and peas indicate that they are good sources of γ-tocopherol.

Tocopherols in Tomatoes and Tomato Products

Tomatoes have low amounts of α- and γ-tocopherol; these vitamins are present in higher concentration in tomato products with higher solids contents (Table 8). Condensed tomato products may be desirable as low-fat sources of the vitamin in the designing of fat-controlled diets. The higher concentrations of vitamin E in the condensed tomato products may be due, in some cases, to small amounts of vegetable oil added during processing.

Tocopherols in Nuts, Seeds, Cereal Grains, and Related Products

Many of the common nuts and seeds are good sources of vitamin E (Table 9). Those having the highest α-TE activities also have the highest α-tocopherol concentrations. Examples include almonds, Brazil nuts, filberts, and peanuts. Macadamia nuts are unique in that they contain only α-tocotrienol. Although cereal grains have relatively low amounts of tocopherols, they contain some of the tocotrienols in higher amounts than other foods. During the milling process, the vitamin E content of the whole grain is concentrated in certain milling fractions (e.g., wheat bran and wheat germ). The vitamin E content of baked products made with vegetable oil or fat tends to reflect the vitamin E content of the added fat (see the data for the chocolate chip cookies and sugar wafers in Table 9). Grain products that have no added fat retain the tocopherol pattern of the cereal flour from which they were made (e.g., macaroni, white bread, and rye bread).

Observations About the Data

Data on the tocopherols in vegetable oils, nuts, seeds, and other unprocessed foods tend to have predictable quantitative and qualitative patterns; these patterns are not discernible in processed and highly refined foods. Tocopherol deterioration is accelerated by heat, light, and aeration. Machlin (117) reported that in most foods, added α-tocopheryl acetate is completely stable. The users of these tables are cautioned that because of many variables, large

TABLE 8 Tocopherol Content, Vitamin E Activity, and Solids Content of Tomatoes and Tomato Products (per 100 g Edible Portion)

Food	Tocopherols α (mg)	Tocopherols β (mg)	Vitamin E activity as α-TE[a] (mg)	Solids[b] (g)
Tomatoes, fresh, raw	0.5	0.02	0.5	6
Tomatoes, stewed	0.9	0.2	0.9	9
Tomato sauce	1.4	0.1	1.4	11
Tomato soup[c]	0.7	0.3	0.7	19
Barbecue sauce	1.0	0.8	1.1	19
Tomato paste	4.1	0.3	4.1	26
Tomato catsup	1.1	0.1	1.1	31
Tomato chili sauce	2.7	0.2	2.7	32

[a] α-Tocopherol equivalents.
[b] Solids data adapted from References 113 and 114.
[c] Contains 0.1 mg δ-tocopherol.
Source: Adapted from References 101 and 105.

TABLE 9 Tocopherol Content and Vitamin E Activity of Nuts, Seeds, Cereal Grains, and Related Products (per 100 g Edible Portion)

Food	Tocopherols α (mg)	β (mg)	γ (mg)	δ (mg)	Tocotrienols α (mg)	β (mg)	Vitamin E activity as α-TE[a] (mg)	Reference
Nuts								
Almonds	45.2	0.3	1.9	0.1	0.2	—[b]	45.6	108
Brazil nuts	11.0	—	5.1	2.6	—	—	11.6	105
Cashews[c]	1.2	0.05	5.8	0.4	—	0.2	1.8	108
Coconuts	0.7	—	0.3	—	—	—	0.7	84
Filberts[d]	21.5	—	0.1	0.01	—	—	21.5	105
Macadamia	—	—	—	—	1.38	—	0.4	110
Peanuts	11.4	—	8.4	—	—	—	12.2	110
Peanut butter	6.2	—	9.4	—	—	—	7.1	110
Pecans	1.2	0.2	24.3	0.2	—	—	3.7	115
Pistachios[e]	3.1	0.2	30.2	0.8	0.5	—	6.5	108
Walnuts	1.4	—	9.2	0.6	—	—	2.3	105
Seeds								
Poppy	1.8	—	9.2	—	—	—	2.7	110
Sesame	—	—	22.7	—	—	—	2.3	84
Sunflower	49.5	2.7	—	—	—	—	5.9	110
Cereal grains								
Barley[f]	0.2	0.04	0.03	0.01	1.1	0.3	0.6	86
Corn[g]	0.6	—	4.5	—	0.3	—	1.1	84
Oats	0.5	0.1	—	—	1.1	0.2	1.0	86
Rice, white[h]	0.3	—	0.3	0.04	0.5	—	0.5	86
Rye	1.6	0.4	—	—	1.5	0.8	2.3	86
Wheat	1.0	0.7	—	—	0.4	2.8	1.6	86
Related products								
Bread, rye	0.1	—	0.9	—	—	—	0.2	101
Bread, white	0.03	—	0.2	—	—	—	0.05	101
Breakfast cereal[i]	104.5	—	—	—	—	—	77.4	105
Chocolate chip cookies	1.3	—	12.2	—	—	—	2.5	101
Oat bran	0.5	0.1	0.03	0.01	1.6	—	1.0	116
Macaroni	0.03	0.02	—	0.01	0.04	0.2	0.6	116
Sugar wafers	1.4	—	13.1	—	—	—	2.7	101
Wheat bran	16.3	10.1	—	—	11.0	53.7	27.3	86
Wheat flour	0.03	0.04	—	0.01	0.01	0.2	0.06	116
Wheat germ	115.3	66.0	—	—	2.6	8.1	149.5	86

[a]α-Tocopherol equivalents.
[b]Dashes denote no value or trace.
[c]Contains 0.4 mg γ-tocotrienol.
[d]Filberts are the same as hazelnuts.
[e]Contains 4.5 mg γ-tocotrienol and 0.03 mg δ-tocotrienol.
[f]Contains 0.2 mg γ-tocotrienol.
[g]Contains 0.5 mg γ-tocotrienol.
[h]Contains 0.5 mg γ-tocotrienol.
[i]Fortified with α-tocopheryl acetate to 100% of the U.S. RDA per serving.

deviations from the listed values are possible. Those in need of more precise vitamin E data are advised to obtain analyses of the foods of interest in the form consumed, for example, cooked or raw. The data for the individual tocopherols may be useful for scientists investigating tocopherols as antioxidants, alone or with other food components, and their possible involvement in aging, cancer, arthritis, circulatory problems, cataracts, and defense against air pollution, as suggested by Diplock et al. (118). Appropriate references to the data are given in the tables.

REFERENCES

1. Evans HM, Emerson OH, Emerson GA. The isolation from wheat germ oil of an alcohol, alpha-tocopherol, having the properties of vitamin E. J Biol Chem 1936; 113:319-32.
2. Fernholz E. On the constitution of alpha-tocopherol. J Am Chem Soc 1938; 60:700-5.
3. von Karrer P, Fritzsche H, Ringier BH, Salomon H. Synthese des alpha-Tocohperols. Helv Chim Acta 1938; 21:820-5.
4. von Karrer P, Fritzsche H, Ringier BH, Salomon H. Alpha-tocopherol. Helv Chim Acta 1938; 21:520-5.
5. United States Pharmacopeia (USP XXII), 20th rev. U.S. Pharmacopeia Convention. Rockville, MD, 1985; 1118-20.
6. Nomenclature policy: generic descriptors and trivial names for vitamins and related compounds. J Nutr 1979; 109:8-15.
7. Nomenclature rules for vitamin E (972.31). Official methods of analysis. Assoc Off Anal Chem, 15th ed. 1990; 2:1070-1.
8. Food and Nutrition Board. Recommended dietary allowances, 10th rev. ed. Washington, DC: National Academy of Sciences, 1989.
9. Food Chemicals Codex, 3rd ed. Washington, DC: National Academy Press, 1981.
10. Quaife ML, Harris PL. Chemical assay of foods for vitamin E content. Anal Chem 1943; 20:1221-4.
11. Pudelkiewicz WJ, Matterson LD. Effect of coenzyme Q_{10} on the determination of tocopherol in animal tissue. J Biol Chem 1960; 235:496-8.
12. Sheppard AJ, Prosser AR, Hubbard WD. Gas chromatography of vitamin E. In: McCormick DB, Wright LD, eds. Methods in enzymology, vitamins and coenzymes, part C. New York: Academic Press, 1971; 18:356-65.
13. Bieri JG. Chromatography of tocopherols. In: Marinetti GV, ed. Lipid chromatographic analysis. New York: Marcel Dekker, 1969; 2:459-78.
14. Laidman DL, Gaunt JK, Hall GS, Broad CT. Extraction of tocopherols from plant tissues. In: McCormick DB, Wright LD, eds. Advances in enzymology, Part C. 1971; 18:366-9.
15. Cohen H, Lapointe M. Method for the extraction and cleanup of animal feed for the determination of liposoluble vitamins D, A, and E by high-pressure liquid chromatography. J Agr Food Chem 1978; 26:1210-3.
16. Parrish DB. Determination of vitamin E in foods—a review. CRC Crit Rev Food Sci Nutr 1980; 13:161-87.
17. Pillsbury HC, Sheppard AJ, Libby DA. Gas-liquid chromatographic method for the determination of fat-soluble vitamins. V. Application to pharmaceuticals containing vitamin E. J Assoc Off Anal Chem 1967; 50:809-14.
18. Evans HM, Burr GO. The anti-sterility vitamine fat soluble E. Proc Natl Acad Sci USA 1925; 11:334-41.
19. Olcott HS, Matill HA. Vitamin E. I. Some chemical and physiological properties. J Biol Chem 1934; 104:423-34.
20. Bacharach AL, Allchorne E, Glynn HE. Investigations into the method of estimating vitamin E. I. The influence of vitamin E deficiency on implantation. Biochem J 1937; 31:2287-92.
21. Bacharach AL, Allchorne E. Investigations into the method of estimating vitamin E. II. Further observations on vitamin E deficiency and implantation. Biochem J 1938; 32:1298-300.

22. Hume EM. Standardization of vitamin A. Nature 1941; 148:472-3.

23. Mackenzie CG, McCollum EV. The cure of nutritional muscular dystrophy in the rabbit by alpha-tocopherol and its effect on creatine metabolism. J Nutr 1940; 19:345-62.

24. Vichoever A, Cohen I. The responses of *Daphnia magna* to vitamin E. Am J Pharm 1938; 110: 297-315.

25. Vichoever A. Report on *Daphnia* methods. J Assoc Off Anal Chem 1939; 22:715-8.

26. von Karrer P, Escher R, Fritzsche H, Keller H, Ringier BH, Salomon H. Konstitution und bestimmung des alpha-tocopherols und einiger ahnlicher verbindungen. Helv Chim Acta 1938; 21:939-53.

27. von Karrer P, Keller H.Quantitative bestimmung der tocopherols in verschiedenen ausgangsmaterialien. Helv Chim Acta 1938; 21:1161-9.

28. von Furter M, Meyer RE. Eine quantitative photometrische bestimmung von vitamin E. Helv Chim Acta 1939; 22:240-9.

29. Ungnade HE, Smith LI. The chemistry of vitamin E. XV. An extension of the analytical method of Furter and Meyer. J Agr Chem 1939; 4:397-400.

30. Vitamin E. National Formulary XIII, 13th ed. 1970; 758-61.

31. Emmerie A, Engel C. Colorimetric determination of alpha-tocopherol (vitamin E). Rec Trav Chim 1938; 57:1351-5.

32. Emmerie A, Engel C. Colorimetric determination of tocopherol (vitamin E). II. Adsorption experiments. Rec Trav Chim 1939; 58:283-9.

33. Emmerie A. Colorimetric determination of tocopherol (vitamin E). IV. The quantitative determination of tocopherol in oils after saponification. Rec Trav Chim 1940; 59:246-8.

34. Emmerie A, Engel C. Colorimetric determination of tocopherol (vitamin E). III. Estimation of tocopherol in blood-serum. Rec Trav Chim 1939; 59:895-902.

35. Emmerie A. Colorimetric determination of tocopherol (vitamin E). V. The estimation of tocopherol in butter. Rec Trav Chim 1941; 60:104-5.

36. Alpha tocopherol assay. United States Pharmacopeia, 19th rev. 1975; 630-1.

37. Alpha-tocopherol and alpha-tocopheryl acetate in foods and feeds. Colorimetric method (971.30). Official methods of analysis. Assoc Off Anal Chem, 15th ed. 1990; 2:1071-4.

38. Alpha-tocopheryl acetate (supplemental) in foods and feeds. Colorimetric method (948.26). Official methods of analysis. Assoc Off Anal Chem, 15th ed. 1990; 2:1075-6.

39. Brown F. The estimation of vitamin E. 2. Quantitative analysis of tocopherol mixtures by paper chromatography. Biochem J 1952; 52:523-6.

40. Dicks-Bushnell MW. Column chromatography in the determination of tocopherol: florisil, silicic acid and secondary magnesium phosphate. J Chromatogr 1967; 27:96-103.

41. Thompson JN, Erdody P, Maxwell WB. Chromatographic separation and spectrophotofluorometric determination of tocopherols using hydroxyalkoxypropyl Sephadex. Anal Biochem 1972; 50:267-80.

42. Burnell RH. Vitamin E assay by chemical methods. In: Gyorgy P, Pearson WN, eds. The vitamins, 2nd ed. New York: Academic Press, 1967; 2:261-304.

43. Laidman DL, Hall GS. Adsorption column chromatography of tocopherols. In: McCormick DB, Wright LD, eds. Advances in enzymology, Part C. 1971; 18:348-57.

44. Bro-Rasmussen F, Hjarde W. Determination of alpha-tocopherol by chromatography on secondary magnesium phosphate (with collaborative tests in four laboratories). Acta Chem Scand 1957; 11:34-43.

45. Kjolhede KT. The elimination of the "vitamin A and carotenoid error" in the chemical determination of tocopherol. Z Vitaminforsch 1942; 12:138-45.

46. Analytical Methods Committee. Report prepared by the vitamin-E panel. The determination of tocopherols in oils, foods and feeding stuffs. Analyst 1959; 84:356-72.

47. Desai ID. Assay methods. In: Macklin LJ, ed. Vitamin E. A. Comprehensive treatise, New York: Marcel Dekkker, 1980; 67-98.

48. Strohecker R, Henning HM. Vitamin E (alpha-tocopherol). In: Vitamin assay tested methods. Weinheim: Verlag Chemie, 1965; 283-6.

49. Identification of RRR- or all-rac-alpha-tocopherol in drugs and food or feed supplements. Polarimetric method (975.43). Official methods of analysis, 15th ed. Assoc Off Anal Chem 1990; 2:1074-5.

50. Alpha-tocopheryl acetate (supplemental) in foods and feeds. Colorimetric method (948.26). Official methods of analysis. Assoc Off Anal Chem, 15th ed. 1990; 2:1075-6.

51. Brown F. The estimation of vitamin E. 1. Separation of tocopherol mixtures occurring in natural products by paper chromatography. Biochem J 1952; 51:237-9.

52. Green J, McHale D, Marcinkiewicz S, Mamalis P, Watt PR. Tocoperhols. Part V. Structural studies on epsilon- and delta-tocopherol. J Chem Soc (Lond) 1959; 3362-73.

53. Green J, Mamalis P, Marcinkiewicz S, McHale D. Structure of epsilon-tocopherol. Chem Industry (Lond) 1960; 73-74.

54. McHale D, Green J, Marcinkiewicz S. Determination of multiple bonds by reversed phase paper chromatography. Chem Ind (Lond) 1961; 555-6.

55. Dilley RA, Crane FL. A specific assay for tocopherols in plant tissue. Anal Biochem 1963; 5:531-41.

56. Dasilva EJ, Jensen A. Content of alpha-tocopherol in some blue-green algae. Biochim Biophys Acta 1971; 239:345-7.

57. Lambertsen G, Braekkau OR. The spectrophotometric determination of alpha-tocopherol. Analyst 1959; 4:706-11.

58. Eggitt PWR, Ward LD. Chemical estimation of vitamin E activity in cereal products. I. Tocopherol pattern of wheat-germ oil. J Sci Food Agr 1953; 4:569-79.

59. Green J, Marcinkiewica S, Watt PR. The determination of tocopherols by paper chromatography. J Sci Food Agr 1955; 6:274-82.

60. Marcinkiewicz S, Green J. The complete analysis of tocopherol mixtures. Part II. The separation of nitroso-tocopherols by paper chromatography and their determination. Analyst 1959; 84:304-12.

61. Edwin EE, Diplock AT, Bunyan J, Green J. Studies on vitamin E. 1. The determination of tocopherols in animal tissues. Biochem J 1960; 75:450-6.

62. Booth VH. Spurious recovery tests in tocopherol determinations. Anal Chem 1961; 33:1224-6.

63. Kofler M, Sommer PF, Bolliger HR, Schmidli B, Vecchi M. Physicochemical properties and assay of the tocopherols. In: Harris RS, Wool IG, eds. Vitamins and hormones. New York: Academic Press 1962; 20:407-39.

64. Lovelady HG. Separation of beta- and gamma-tocopherols in the presence of alpha- and delta-tocopherols and vitamin A acetate. J Chromatogr 1973; 78:449-52.

65. Stowe HD. Separation of beta- and gamma-tocopherol. Arch Biochem Biophys 1963; 103:42-4.

66. Bolliger HR. Vitamins. In: Stahl E, ed. Thin-layer chromatography. A laboratory handbook. New York: Springer-Verlag, 1965; 210-48.

67. Whittle KJ, Pennock JF. The examination of tocopherols by two-dimensional thin-layer chromatography and subsequent colorimetric determination. Analyst 1967; 92:423-30.

68. Ames SR. Vitamins and other nutrients. Determination of vitamin E in foods and feeds—a collaborative study. J Assoc Off Anal Chem 1971; 54:1-12.

69. Dam H, Glavind J, Prange I, Ottensen J. Kgl Danske Videnskab Selskab Biol Meddel 1941; 16:1-39.

70. Baxter JG, Robeson CD, Taylor JD, Lehman RW. Natural alpha-, beta- and gamma-tocopherols and certain esters of physiological interest. J Am Chem Soc 1943; 65:918-24.

71. Emmerson OH, Emerson GA, Mohammad A, Evans HM. The chemistry of vitamin E. Tocopherols from various sources. J Biol Chem 1937; 122:99-107.

72. Glavind J, Heslet H, Prange I. The use of molecular distillation at the chemical determination of tocopherols (vitamin E) for the elimination of other reducing substances. Z Vitaminforsch 1943; 13:266-74.

73. Quaife ML, Harris PL. Molecular distillation as a step in the chemical determination of total and gamma-tocopherols. Ind Eng Chem Anal Ed 1946; 18:707-8.

74. Robeson CD, Baxter JG. Alpha-tocopherol, a natural antioxidant in a fish liver oil. J Am Chem Soc 1943; 65:940-43.

75. Sheppard AJ, Ford LA, Boehne JW, Libby DA. Gas-liquid chromatographic method for determination of fat-soluble vitamins. III. Evaluation of a vacuum distillation procedure for vitamin E isolation, using C-14 labelled alpha-tocopheryl acetate. J Assoc Off Anal Chem 1965; 48:977-80.

76. Sheppard AJ, Ford LA. Evaluation of an improved sublimation apparatus utilizing carbon-14 labeled methyl esters of fatty acids. J Assoc Off Anal Chem 1963; 46:947-9.

77. Wilson PW, Kodicek E, Booth VH. Separation of tocopherols by gas-liquid chromatography. Biochem J 1962; 84:524-31.

78. Sheppard AJ, Prosser AR, Hubbard WD. Gas chromatography of the fat-soluble vitamins: a review. J Am Oil Chem Soc 1972; 11:619-33.

79. Bieri JG, Andrews EL. The determination of alpha-tocopherol in animal tissues by gas-liquid chromatography. Iowa State J Sci 1963; 38:3-12.

80. Nair PP, Turner DA. The application of gas-liquid chromatography to the determination of vitamins E and K. J Am Oil Chem Soc 1963; 40:353-6.

81. Libby DA, Sheppard AJ. Gas-liquid chromatographic method for the determination of fat-soluble vitamins. I. Application to vitamin E. J Assoc Off Anal Chem 1964; 47:371-6.

82. Nair PP, Sarlos I, Machiz J. Microquantitative separation of isomeric dimethyltocols by gas-liquid chromatography. Arch Biochem Biophys 1966; 114:488-93.

83. Slover HT, Shelley LM, Burks TL. Identification and estimation of tocopherols by gas-liquid chromatography. J Am Oil Chem Soc 1967; 44:161-6.

84. Slover HT. Tocopherols in foods and fats. Lipids 1971; 6:291-6.

85. Slover HT, Lanza E. Quantitative analysis of food fatty acids by capillary gas chromatography. J Am Oil Chem Soc 1979; 56:933-43.

86. Slover HT, Lehmann J, Valis RJ. Vitamin E in foods: determination of tocols and tocotrienols. J Am Oil Chem Soc 1969; 46:417-20.

87. Slover HT, Thompson RH. Chromatographic separation of the stereoisomers of alpha-tocopherol. Lipids 1981; 16:268-75.

88. Slover HT, Thompson RH, Merola GV. Determination of tocopherols and sterols by capillary gas chromatography. J Am Oil Chem Soc 1983; 60:1524-8.

89. Slover HT, Thompson RH, Davis CS, Merola GV. Lipids in margarines and margarine-like foods. J Am Oil Chem Soc 1985; 62:775-86.

90. Sheppard AJ, Shen C-SJ, Rudolf TS, Butler SW, Atkinson JC. Food chemicals codex gas chromatographic method for mixed tocopherols concentrate—a collaborative study. U.S. Pharmacopeial Convention, 1987; 13:2155-62.

91. Sheppard AJ, Hubbard WD, Prosser AR. Collaborative study comparing gas-liquid chromatographic and chemical methods for quantitatively determining vitamin E content of pharmaceutical products. J Assoc Off Anal Chem 1969; 52:442-8.

92. Rudy BC, Mahn FP, Senkowski BZ, Sheppard AJ, Hubbard WD. Collaborative study of the gas-liquid chromatographic assay for vitamin E. J Assoc Off Anal Chem 1972; 55:1211-8.

93. Sheppard AJ, Hubbard WD. Collaborative study of a GLC method for vitamin E. J Pharm Sci 1979; 68:98-100.

94. Williams RC, Schmit JA, Henry RA. Quantitative analysis of the fat-soluble vitamins by high-speed liquid chromatography. J Chromatogr Sci 1972; 10:494-501.

95. Wiggins RA. Chemical analysis of vitamins A, D and E. Proc Anal Div Chem Soc 1976; 13:133-7.

96. Saxby MJ. Applications of high-pressure liquid chromatography in food analysis. In: King RD, ed. Developments in food analysis techniques. London: Applied Science Publishers, 1978; 1: 125-53.

97. Van Niekerk PJ. The direct determination of free tocopherols in plant oils by liquid-solid chromatography. Anal Biochem 1973; 52:533-7.

98. Thompson NJ, Hatina G. Determinationn of tocopherols and tocotrienols in foods and tissues by high performance liquid chromatography. J Liquid Chromatogr 1979; 2:327-44.

99. Landen WO. Application of gel permeation chromatography and nonaqueous reverse phase chromatography to high performance liquid chromatographic determination of retinyl palmitate and alpha-tocopheryl acetate in infant formulas. J Assoc Off Anal Chem 1982; 65:810-6.

100. Syvaoja EL, Salminen K. Tocopherols and tocotrienols in Finnish foods: fish and fish products. J Am Oil Chem Soc 1985; 62:1245-8.

101. Lehmann J, Martin HL, Lashley EL, Marshall MW, Judd JT. Vitamin E in foods from high and low linoleic acid diets. J Am Diet Assoc 1986; 86:1208-16.

102. Shen C-SJ, Sheppard AJ. A rapid high-performance liquid chromatographic method for separating tocopherols. J Micronutrient Anal 1986; 2:43-53.

103. Lehmann J, Martin HL. Liquid-chromatographic determination of alpha- and gamma-tocopherols in erythrocytes, with fluorescence detection. Clin Chem 1983; 29:1840-2.
104. Marshall MW, Iacono JM, Young CW, Washington VA, Slover HT, Leapley PM. Composition of diets containing 25 and 35 percent calories from fat. Analyzed vs. calculated values. J Am Diet Assoc 1975; 66:470-81.
105. Hogarty CJ, Ang C, Eitenmiller RR. Tocopherol content of selected foods by HPLC/fluorescence quantitation. J Food Comp Anal 1989; 2:200-9.
106. US Department of Agriculture, Human Nutrition Information Service. Nationwide Food Consumption Survey, 1987–88. Individual intakes, 3 days. Computer tape PB 90-504044. Springfield, VA: National Technical Information Service, 1990.
107. Horwitt M. Vitamin E abstracts. La Grange IL: VERIS, 1989.
108. Eitenmiller RR. Personal communication. University of Georgia, November 1990.
109. McLaughlin PJ, Weihrauch JL. Vitamin E content of foods. J Am Diet Assoc 1979; 79:647-65.
110. Bauernfeind J. Tocopherols in foods. In: Machlin LJ, ed. Vitamin E: a comprehensive treatise. New York: Marcel Dekker, 1977; 99-167.
111. Pilronen V, Syvaoja EL, Varo P, Salminen K, Koivistoinen P. Tocopherols and tocotrienols in Finnish foods: vegetables, fruits and berries. J Agr Food Chem 1986; 34:742-6.
112. Eitenmiller RR. Unpublished data. November 1990.
113. Watt BK, Merrill AL. Composition of foods: raw, processed, prepared. U.S. Department of Agriculture Handbook No. 8 (rev.), 1963.
114. Haytowitz DB, Matthews RH. Composition of foods: raw processed, prepared. Vegetables and vegetable products. U.S. Department of Agriculture Handbook No. 8-11, 1984.
115. Yao F. HPLC quantification of tocopherols in pecans and the relationship of tocopherol levels during storage to pecan kernel quality. Ph.D. thesis, Athens, GA: University of Georgia, 1990.
116. Padmalayam I, Dial S, Weihrauch J, Eitenmiller R. Tocopherol and tocotrienol determination in cereals and cereal products by normal phase HPLC. Talk prepared for IFT annual meeting, 1991.
117. Machlin LJ. Handbook of vitamins. New York: Marcel Dekker, 1984.
118. Diplock AT, Machlin LJ, Packer L, Pryor WA. Vitamin E: biochemistry and health implications. New York: New York Academy of Sciences, 1989.

B. Animals and Humans

3

Absorption, Transport and Delivery to Tissues

Maret G. Traber

New York University School of Medicine, New York, New York

William Cohn

F. Hoffmann-La Roche, Ltd., Basel, Switzerland

David P. R. Muller

Institute of Child Health, University of London, London, England

INTRODUCTION

Vitamin E is a generic term for tocopherols and tocotrienols, which have saturated and unsaturated phytyl tails, respectively. The α-, β-, δ-, and γ-tocopherols and tocotrienols differ in the number and position of the methyl groups on the chroman ring. The tocopherols can exist in a number of stereoisomeric forms depending on the chirality of the phytyl tail. Of these compounds *RRR*-α-tocopherol has the greatest biological activity and accounts for approximately 90% of the vitamin E found in tissues. This chapter, therefore principally considers the absorption and transport of *RRR*-α-tocopherol (α-tocopherol) from ingestion to uptake by tissues. The absorption and transport of γ-tocopherol, which is the predominant form in human diets and vitamin E supplements (α-tocopheryl acetate and the synthetic isomers of α-tocopherol), are also discussed.

ABSORPTION

Both free tocopherol and its acetate ester are water-insoluble, nonswelling amphiphiles, as are the triglycerides and cholesterol (1). Thus, many of the factors and processes necessary for the absorption of dietary lipids are also required for absorption of tocopherols. These factors include efficient emulsification, solubilization within mixed bile salt micelles, uptake by the small intestinal cell (enterocyte), packaging within lipoprotein particles (chylomicrons), and secretion into the circulation via the lymphatic system. These processes are considered in turn after considering the overall efficiency of tocopherol absorption.

Efficiency

As reviewed by Gallo-Torres (2), the efficiency of absorption of vitamin E remains largely unknown because of the number of uncontrolled variables involved in these measurements. In general, two experimental approaches have been used to calculate the efficiency of absorption of physiological amounts of vitamin E. In one approach, physiological doses of radio-labeled vitamin E have been administered and net absorption inferred by measurement of fecal (i.e., unabsorbed) radioactivity. Alternatively, a more direct approach has been used that measures the appearance of radiolabel in mesenteric or thoracic lymph following cannulation of the appropriate vessels. The use of these techniques in both humans and experimental animals has led to a consensus that the absorption of vitamin E is incomplete, although the reported percentage absorption has varied.

By measuring fecal radioactivity in normal human subjects, mean values of tocopherol absorption of 69% (3) and 72% (4) have been reported. However, Blomstrand and Forsgren (5) cannulated the thoracic duct of two subjects with a gastric carcinoma and lymphatic leukemia and reported a mean absorption of only 21 and 29% of radiolabel from meals containing tocopherol and tocopheryl acetate, respectively.

Variable percentage recoveries of tocopherol in lymph have also been reported following cannulation of the thoracic duct in rats. Gallo-Torres (6) reported a 10% recovery of radiolabel in 12 h after giving rats a bolus emulsion containing a total of 2 mg α-tocopheryl acetate by stomach intubation. In rats given α-tocopherol intraduodenally, a mean recovery of 15% was obtained in mesenteric lymph over a 24 h period compared with a recovery of 45% of oleic acid (7). Bjørneboe et al. point out that under physiological conditions, the absorption of oleic acid is virtually complete; this may therefore indicate that in vivo the absorption of tocopherol is of the order of 30%. However, Traber et al. (8) found that 65% of α-tocopherol appeared in the lymph after a slow infusion (0.12 mg/h) into the duodenum. They also found that the efficiency of absorption decreased as the amount of α-tocopherol given to the animals was increased (8). It is therefore likely that the reported differences in efficiency of absorption result from such variables as the method of administering the tocopherol (bolus compared to slow infusion) and the amount given, as well as from experimental details, such as the length of time allowed for recovery from the anesthetic.

The efficiency of absorption of tocopherol is also influenced by other dietary lipids, as reviewed in detail previously (2,9). In summary, monolein and triolein have been shown to stimulate the absorption of tocopherol to a greater extent than an equivalent amount of unsaturated fatty acids. Medium-chain triglycerides appear to enhance absorption compared to long-chain triglycerides, but retinoic acid and long-chain polyunsaturated fatty acids appear to reduce absorption. The influence of polyunsaturated fatty acids is complicated since the apparent reduction in absorption of tocopherol may result, at least in part, from its oxidation in vivo.

Luminal Events

A number of different approaches have been used to understand the luminal factors and events important for the efficient absorption of α-tocopherol. These include in vitro and in vivo studies in the experimental animal, studies in humans that make use of specific defects in fat absorption, and in vitro studies using human intestinal cells in culture. Essentially similar results and conclusions have been obtained using these different approaches.

Lipids, including tocopherol, must be emulsified and solubilized before their absorption across the brush-border membrane of the enterocyte. Emulsification begins in the stomach by predominantly mechanical forces that break up large emulsion particles into smaller particles. Within the small intestine, chyme mixes with pancreatic and biliary secretions, which are necessary for the efficient absorption of tocopherol. Pancreatic lipase is necessary for the hydrolysis of triglyceride in the small intestine to monoglycerides and fatty acids. These lipolytic products, together with bile salts (which are synthesized by the liver and secreted into the small intestine by the gallbladder), form molecular aggregates known as mixed micelles, which are able in turn to solubilize more hydrophobic molecules, such as tocopherol. These mixed micelles have a radius of 15-50 Å compared to emulsion particles with a radius of 1000-25,000 Å (1) and are able to transport tocopherol across the unstirred water layer to the brush-border membrane of the enterocyte.

The relative importance of the individual pancreatic and biliary factors necessary for the efficient absorption of tocopherol has been studied by investigating serum concentrations of vitamin E in patients with specific defects of fat absorption (10). Many studies have shown that an adequate luminal concentration of bile salts is necessary for the absorption of tocopherol. Thus, very low serum concentrations of tocopherol have been consistently reported in patients with liver disease and biliary obstruction (11-16). In such patients, luminal bile salt concentrations are generally less than the concentration necessary for micelle formation (critical micellar concentration, CMC), and therefore tocopherol cannot be solubilized in the small intestine. If, however, such patients are given bile salts together with vitamin E supplements, absorption can occur (14). The importance of bile salts for the absorption of tocopherol has also been shown in experimental animals both in vivo and in vitro. When bile was diverted away from the small intestine in rats, the absorption of tocopherol was greatly reduced (6,17). Similarly, perfusion studies in hamsters (18) and in vitro experiments using rat small intestinal slices (19) have also shown the importance of bile salts for the efficient absorption of tocopherol.

Patients in whom intestinal bile salt concentrations are below the CMC are unable to absorb conventional oral preparations of vitamin E (11-16). The only oral preparation of vitamin E that appears to be absorbed by such patients is tocopheryl polyethylene glycol-1000 succinate (TPGS) (8,16). TPGS is soluble in water up to a concentration of 20%, is amphipathic, and forms micelles at low concentrations in an aqueous medium (8). Following administration of TPGS, only free tocopherol is found in lymph (8), indicating that hydrolysis takes place before secretion from the enterocyte. This was also demonstrated in an in vitro system (20).

Patients with pancreatic insufficiency, as in cystic fibrosis, secrete greatly reduced amounts of pancreatic enzymes into the lumen of the small intestine. Such patients also show impaired absorption of tocopherol, presumably caused by a failure to hydrolyze triglyceride, thereby resulting in a failure to form mixed micelles (21-24). The deficiency of vitamin E can be overcome by giving such patients pancreatic enzyme replacement therapy and modest supplements of α-tocopheryl acetate (10 mg/kg/day) (22,24). The importance of pancreatic enzymes for the efficient absorption of tocopherol has also been shown in the rat by Gallo-Torres (6), who found that absorption was greatly impaired after pancreatic juice was diverted from the small intestine. Traber et al. (25) came to essentially similar conclusions regarding the importance of bile salts and pancreatic enzymes for the efficient absorption of tocopherol using a human intestinal cell line (Caco-2 cells) in tissue culture. They found that high concentrations of bile salts and fatty acids in the incubation medium was necessary for the uptake of tocopherol by cells. In vivo the fatty acids would be formed by the action of pancreatic lipase on dietary triglyceride.

The Enterocyte and Intracellular Events

Once tocopherol has been solubilized within bile salt micelles and transported across the unstirred water layer, the micelle comes into contact with the absorptive brush-border membrane of the enterocyte. Results of both in vitro (26) and in vivo (27) studies suggest that the uptake of tocopherol by the enterocyte takes place by passive diffusion. Thus, the process is nonsaturable, noncarrier-mediated, and unaffected by metabolic inhibitors and does not require energy. In the rat, maximal absorption takes place at the junction between the upper and middle thirds of the small intestine (2,26). It is therefore not surprising that absorption is impaired, and serum concentrations of tocopherol are reduced in patients with celiac disease, who have abnormal architecture of the small intestine with a reduced absorptive surface area as a result of a gluten enteropathy (10). The proximal small intestine is predominantly affected in patients with celiac disease, and they can therefore be used to study the site of tocopherol absorption in humans. When patients are placed on a gluten-free diet without added vitamin E supplements, serum vitamin E concentrations generally return to normal within a few weeks. This suggests that the proximal small intestine is the principal site for human tocopherol absorption (10).

Within the enterocyte, tocopherol is incorporated into chylomicrons and secreted into the intracellular spaces and lymphatics and thus into the bloodstream (7,28). This process does not appear to require a transfer protein (29). The necessity of chylomicron formation for the absorption of tocopherol was demonstrated in studies of patients with abetalipoproteinemia. These patients have an inborn error of lipoprotein metabolism and are unable to secrete apolipoprotein B (apo B)-containing lipoproteins, which include chylomicrons. Thus, tocopherol is not absorbed normally by these patients, and this results in virtually undetectable serum concentrations of tocopherol from birth (10,30-32). If extremely large oral doses of α-tocopheryl acetate (100 mg/kg/day) are given, a small proportion is absorbed as (1) plasma concentrations increase and may reach a maximum of one-tenth of normal (10), (2) the abnormal in vitro hemolysis is corrected (10,33), and (3) tocopherol concentrations in adipose tissue reach normal levels (34,35). Absorption of tocopherol in these patients may take place by either the portal system, an alternative pathway suggested from animal studies (36), or by the lymphatics bound to the abnormal high-density lipoproteins synthesized in the enterocytes of such patients (37).

As discussed, tocopherol is normally carried via the lymphatics to the bloodstream. It is therefore not surprising that absorption is reduced in such situations as intestinal lymphangiectasia, in which lymph flow is impaired (10).

Absorption of α-Tocopheryl Acetate

Supplements of vitamin E are generally given in the form of α-tocopheryl acetate in which the reactive hydroxyl group of α-tocopherol is esterified, rendering the molecule more stable than the free form. Before absorption α-tocopheryl acetate is hydrolyzed in the small intestinal lumen by pancreatic esterase (38), which is also known as cholesterol esterase, cholesterol ester hydrolase, sterol ester hydrolase, or carboxylic ester hydrolase. This enzyme, unlike pancreatic lipase, requires its substrate to be in a solubilized form. Bile salts and the lipolytic products of triglyceride are therefore necessary to solubilize the tocopheryl acetate before it can be hydrolyzed. In addition it has been shown that bile salts are also essential cofactors for this enzyme (38).

There is some confusion in the literature about whether there is a significant difference in the overall efficiency of absorption between tocopherol and its acetate ester and whether this varies from species to species. The situation has been clarified by Burton et al. (39), however, in both humans and rats using free and esterified α-tocopherols labeled with different amounts of deuterium. When an equimolar mixture of deuterated free α-tocopherol and its acetate ester was given with a meal to five normal subjects, no significant differences in the absorption of the two compounds were found (39). These authors obtained different results in the rat, however, depending on the way in which the vitamin was presented for absorption. If tocopherol and tocopheryl acetate were given together with vegetable oil (as done by Weiser et al., 40), the plasma concentration of deuterated α-tocopherol from the free form was only 50% of that from the ester. If, however, the two forms of the vitamin were administered together with the standard animal diet, the plasma concentrations of the deuterated tocopherols were virtually identical. A likely explanation for these observations is that increased oxidation and therefore destruction of free tocopherol take place in the intestine in the presence of the unsaturated fatty acids found in corn oil, and as a result less tocopherol is available for absorption.

Absorption of γ-Tocopherol

γ-Tocopherol is the major tocopherol in nature and therefore predominates in the diet (41-43). There are conflicting data in the literature, however, as to its efficiency of absorption compared to α-tocopherol. A number of studies have concluded that the absorption of γ-tocopherol is less efficient than that of α-tocopherol (42,44-46). Others have found no difference in the absorption of the two tocopherols (47,48), however, but concluded that γ-tocopherol disappears more rapidly from tissues than α-tocopherol (47). Furthermore, direct measurements of lymph from thoracic duct-cannulated rats demonstrated that the absorption of γ-tocopherol was not impaired by the presence of a 50-fold excess of α-tocopherol (20). The controversy appears to have been resolved by a study of Traber and Kayden (49), who fed a mixture of 1 g of α- and γ-tocopherols with a meal to normal humans. Both tocopherols increased similarly in plasma up to 12 h after the oral load, but by 24 h γ-tocopherol concentrations had decreased dramatically, whereas the concentrations of α-tocopherol remained essentially unchanged. These results suggest that both tocopherols are absorbed by the enterocyte and secreted into the bloodstream within chylomicrons to a similar extent but that thereafter their handling is different. Meydani et al. (50) came to similar conclusions in humans using rather smaller doses of α- and γ-tocopherols. Studies in cynomolgus monkeys using a single oral dose of deuterated α- and γ-tocopherols demonstrated that the liver secretes nascent very low density lipoprotein (VLDL) preferentially enriched in *RRR*-α-tocopherol (51). From these studies it appears that the liver, but not the intestine, is capable of discriminating between these two tocopherols, as discussed subsequently.

PLASMA TRANSPORT OF VITAMIN E

Vitamin E is transported in the blood by the plasma lipoproteins (52-55) and erythrocytes (56-58). There is no evidence for the existence of a specific vitamin E plasma carrier protein (9,59,60). This is in contrast to the transport of vitamin A by retinol binding protein, which apparently mediates the delivery of retinol to target tissues by a receptor pathway (61,62).

There are a number of important consequences resulting from the transport of vitamin E in lipoproteins. First, circulating polyunsaturated fatty acids (PUFAs) and other lipids are protected from free-radical attack by vitamin E (63). Second, plasma vitamin E concentrations do not entirely depend on dietary intake but vary with those of the lipoproteins. This is manifested by high correlations between plasma concentrations of tocopherol and total lipids or cholesterol (64-66). Thus, when lipoprotein concentrations are raised, as in familial hypercholesterolemia, vitamin E concentrations are also increased (67). Because of the relationship between plasma tocopherol and lipid concentrations, vitamin E status is generally expressed in relation to circulating lipids. Such ratios are particularly valuable in pathological states in which altered lipoprotein concentrations and vitamin E deficiency may be expected. For example, in chronic liver disease, in which plasma lipids and lipoproteins may be elevated, vitamin E deficiency could be missed if the concentrations of plasma lipids are not taken into account (15,68). Alternatively, in fat malabsorptive states, such as cystic fibrosis, plasma lipids and lipoprotein concentrations are frequently reduced and therefore vitamin E deficiency would be overestimated (24).

Normal plasma vitamin E concentrations in humans range from 11 to 37 μmol/L (5-16 mg/L) (69-71). When plasma lipids are taken into account the lower limits of normal are 1.6 μmol α-tocopherol/mmol lipid (0.8 mg/g) or 2.5 μmol α-tocopherol/mmol cholesterol (2.8 mg/g) (64).

Both α- and γ-tocopherols are present in the blood of humans and experimental animals. Plasma α-tocopherol is usually about 5- to 10-fold higher than γ-tocopherol, despite the fact that most diets are many fold richer in γ-tocopherol (45,72,73). Because absorption of the two homologs is similar, the discrimination results from postabsorptive events (49), as discussed subsequently.

Distribution Within Plasma Lipoproteins

There are four main lipoprotein classes: (1) chylomicrons, which are synthesized in the enterocyte and transport exogenous lipids, (2) very low density lipoproteins, which primarily transport the triglycerides from the liver, (3) low-density lipoproteins (LDL), and (4) high-density lipoproteins (HDL); the last two lipoprotein classes are involved in cholesterol transport. There are two additional lipoprotein species, the chylomicron remnants and the intermediate-density lipoproteins (IDL), which are metabolic products of the intravascular action of lipoprotein lipase on chylomicrons and VLDL, respectively. The amphiphilic proteins of the lipoproteins are called apolipoproteins (apos). Chylomicron remnants, IDL, and LDL are taken up by hepatocytes via receptor-mediated endocytosis. Receptors for LDL are also present on cells of peripheral tissues.

Although extensive homology exists between human and mammalian lipoproteins, the plasma concentrations of the various lipoprotein classes differ widely between various species (74). Furthermore, lipoprotein profiles are subject to marked changes as a function of nutritional state (75), developmental stage (76,77), and disease (78).

All the lipoprotein fractions contain α-tocopherol, and its distribution parallels the distribution of lipids in lipoproteins (52,58), as shown in Figure 1. As a consequence of the distribution of tocopherols in lipoproteins, all the lipoproteins are similarly protected against lipid peroxidation by free radicals. The highest concentrations of vitamin E in the rat are found in the HDL, followed by VLDL (47,79), which represent the most abundant liproprotein

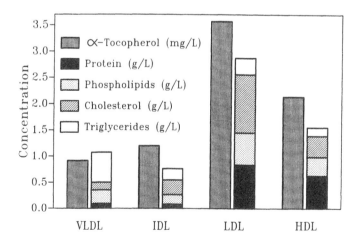

FIGURE 1 Distribution of α-tocopherol among the human plasma lipoproteins. Plasma was obtained from a normolipidemic fasting human subject. Lipoproteins were separated by gradient ultracentrifugation (112). The plasma concentrations of the major lipid and protein components in the lipoprotein fractions are shown.

classes in this species. In humans, however, LDL and HDL are the major carriers of tocopherols with no differences in total plasma vitamin E concentrations between the sexes (53). However, in males more tocopherol is found in LDL than in HDL, whereas in females the opposite is true. Although the levels of γ-tocopherol are considerably lower than α-tocopherol, the relative lipoprotein distributions of the two tocopherols are similar (47,73,80).

The distribution of α-tocopherol in lipoproteins does not parallel individual lipid classes, such as triglycerides, cholesterol, or phospholipids (Fig. 1), indicating that the vitamin does not share the same metabolic fate as any one of these lipids. Furthermore, the lipoprotein distribution of α-tocopherol does not reflect the relative importance of individual lipoprotein classes for the transport of vitamin E. For example, tocopherol enters the systemic circulation within chylomicrons, which in terms of turnover are the major carriers of vitamin E. As the secretion of newly synthesized chylomicrons fluctuates with the intake of dietary lipids, however, and clearance of this lipoprotein is fast, plasma concentrations of vitamin E associated with the chylomicron fraction are usually low. Thus, the distribution of vitamin E in the plasma lipoproteins represents a steady state that is adjusted by various metabolic processes, as discussed subsequently.

Spontaneous Transfer and Partitioning Between Lipoproteins

Little is known about the localization of vitamin E within lipoprotein particles. It is present in both the amphiphilic surface monolayer and the hydrophobic core of the lipoproteins. It has been estimated that at most one-third of the tocopherol in VLDL is in the amphiphilic phospholipid, cholesterol, and protein surface monolayer (81), whereas in HDL the majority of tocopherol is in the surface monolayers. These conclusions regarding HDL are based on the relative ease with which tocopherol can exchange between lipoproteins (56,81,82). Using artificial lipid monolayers, α-tocopherol has been shown to be aligned with the hydrophilic chroman moiety near the lipid/water interface and the phytyl tail aligned with the fatty acid tails of the phospholipids (83,84).

α-Tocopherol has been shown to exchange spontaneously between different lipoproteins (56,81), between lipoproteins and liposomes (82), and between lipoproteins and erythrocytes (56,81), as discussed later. The transfer of tocopherol between lipoproteins is not assisted by the neutral lipid transfer protein, which promotes the exchange of cholesteryl ester for triglyceride between HDL and the triglyceride-rich lipoproteins, that is, chylomicrons and VLDL (59).

Redistribution During Lipoprotein Metabolism

Following its absorption, vitamin E is secreted from the intestine incorporated in chylomicrons. These are then catabolized by the action of lipoprotein lipase, which is bound to the surface of the endothelial lining of the capillary walls (85). This results in the formation of chylomicron remnants, which are taken up by the liver. Once in the liver, a number of different in vitro and in vivo studies have shown that α-tocopherol is secreted into the bloodstream within nascent VLDL (28,51,86-89). Since the delivery of dietary α-tocopherol by chylomicrons fluctuates with the load of absorbed vitamin E, hepatic VLDL are important for maintaining α-tocopherol plasma concentrations (86-88).

During intravascular hydrolysis of triglycerides by lipoprotein lipase, some of the core material from the chylomicrons or VLDL is removed; consequently, less surface material consisting of unesterified cholesterol, phospholipids, and apolipoproteins is required (90). This redundant material moves along a surface pathway to HDL, and apolipoprotein E moves from HDL to chylomicrons or VLDL (90). The remnant particles contain the remaining core lipids (including cholesteryl esters, triglycerides, and vitamin E) and are cleared from the circulation by the liver via the apo E receptor pathway (91). VLDL are cleared from the circulation similarly; however, the IDL and LDL produced can be taken up both by the liver and by peripheral tissues via LDL (apolipoprotein B/E) receptors (78).

Vitamin E also partitions between lipoproteins during the catabolism of chylomicrons and VLDL (28,92). Some tocopherol is probably transferred with the excess surface material from chylomicrons and VLDL to HDL during the hydrolysis of triglycerides. Transfer of tocopherol from HDL to other lipoproteins could then occur, enriching all the lipoprotein fractions with the vitamin. In studies using deuterated α-tocopherol in humans, it was observed that during the catabolism of chylomicrons and particularly VLDL, enrichment of LDL and HDL with deuterated tocopherol took place (28,87). Ultimately, LDL and HDL bore the largest proportion of the deuterated tocopherol in the plasma (28,87).

Discrimination Between Tocopherols During Transport

It has long been recognized that not all stereoisomers of α-tocopherol have the same biopotency: *SRR*-α-tocopherol has about one-third the biopotency of *RRR*-(93). Studies of the discrimination between *RRR*- and *SRR*-α-tocopheryl acetates, compounds that differ only in the stereochemistry at the junction of the chroman ring and the phytyl tail, have been made possible with the use of deuterated tocopherols (94,95) or, alternatively, by quantitation of α-tocopherol stereoisomers applying a combination of chiral high-performance liquid chromatography and capillary gas chromatography (96). These stereoisomers are important because *RRR*-α-tocopherol is the naturally occurring form of the vitamin, and synthetic vitamin E (*all-rac*-α-tocopherol) contains eight stereoisomers, with half the stereoisomers having the S conformation at the 2 position. Ingold et al. (95), in a long-term study in rats, showed that *RRR*-α-tocopherol, rather than *SRR*-, accumulated in plasma, erythrocytes, and brain.

One mechanism for the enhanced biopotency of *RRR*-α-tocopherol has been elucidated using deuterated tocopherols. When humans consumed an equimolar mixture of *SRR*- and *RRR*-α-tocopheryl acetates labeled with different amounts of deuterium, the chylomicrons contained equal concentrations of both, but by 9 h the VLDL, LDL, and HDL fractions were preferentially enriched in *RRR*-α-tocopherol (87). The concentration of *SRR*-α-tocopherol in the plasma and the lipoprotein fractions decreased twice as fast as that of *RRR*- (Fig. 2) (87). Thus, discrimination between these stereoisomers did not occur during absorption and secretion in chylomicrons, but is explained by discrimination during secretion of VLDL. A direct demonstration that the liver secretes nascent VLDL preferentially enriched in *RRR*-α-tocopherol was carried out using a liver perfusion system in cynomolgus monkeys (51). Nascent VLDL contained about five times as much *RRR*-α-tocopherol as either *SRR*-α- or *RRR*-γ-tocopherols (51). Thus, there is good evidence that the liver is able to discriminate between tocopherol homologs and stereoisomers. This specificity has been attributed to an hepatic α-tocopherol binding protein (51,86,87,95).

Hepatic Tocopherol Binding Protein

A tocopherol binding protein (31 kD) has been described in rat liver (29,97) but not in intestine (29). It has been suggested that the binding protein is capable of transferring α-tocopherol between membranes (97,98). A similar molecular weight tocopherol binding protein has been described in human liver (99). The binding of tocopherol analogs to the rat protein has demonstrated that [³H]α-tocopherol could be displaced by α-tocopherol but not by trolox, which lacks the phytyl tail of α-tocopherol. γ-Tocopherol was only partially effective in displacing [³H]α-tocopherol, and α-tocopheryl acetate and the quinone were completely ineffective (29). The results of these competition studies suggest that the binding site of the protein recognizes the hydroxyl group and the methyl groups on the chroman ring, as well as the

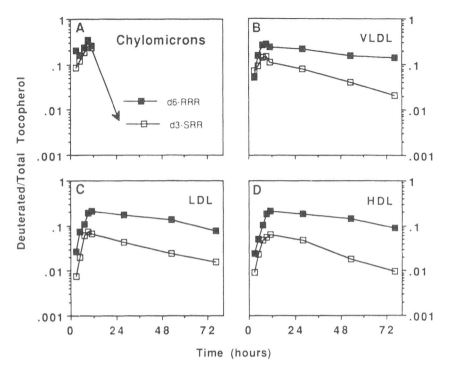

FIGURE 2　Mean deuterated/total α-tocopherol ratios in human lipoproteins. Four normal human subjects consumed a single oral dose containing *RRR-* or *SRR-α*-tocopheryl acetates labeled with either six (d_6-*RRR-*) or three (d_3-*SRR*) deuterium atoms. Blood samples were collected at the indicated intervals; the plasma was separated by centrifugation and the lipoproteins isolated at the following density intervals: VLDL, $d < 1.006$; LDL, $1.006 < d < 1.063$; and HDL, $d > 1.063$. The tocopherol contents were analyzed by gas chromatography-mass spectrometry. The chylomicrons contained similar ratios of each deuterated/total α-tocopherol at all time points. VLDL contained similar ratios of both during the first 9 h, and thereafter the *RRR-* to total ratio was significantly greater at all time points ($p < 0.05$). The LDL and HDL also contained similar ratios of both during the initial portion of the study, then the *RRR-* to total ratio was statistically greater during the 9-76 h period ($p < 0.05$). (From Ref. 87. Copied with permission from the *Journal of Lipid Research*.)

stereochemistry of the 2 position, where the phytyl tail joins the ring. The binding characteristics of this protein correlate with the in vivo discrimination of tocopherol (49,51,87). It is therefore likely that the tocopherol binding protein discriminates between the forms of vitamin E in the liver in vivo and that this protein is involved in the incorporation of *RRR-α*-tocopherol into nascent VLDL. Confirmation of this hypothesis awaits the complete purification of the protein.

Genetic Defects in Vitamin E Transport

A genetically linked abnormality of vitamin E transport in humans may result from a lack or a defect in the hepatic tocopherol binding protein. Nine patients worldwide have been described who exhibit neurological abnormalities characteristic of vitamin E deficiency but have no evidence of fat malabsorption (100-106). The absorption and transport of deuterated tocopherol has been studied in four of these patients—three siblings and one unrelated patient (86). In normal control subjects plasma deuterated tocopherol increased during chylomicron *and* VLDL catabolism, whereas in patients it increased *only* during chylomicron catabolism. This resulted in a premature and faster decline in the plasma tocopherol concentrations in the patients. This defect in transport of α-tocopherol from the liver can be alleviated by supplementation with oral doses of vitamin E. Improvement in neurological function, or at least prevention of further deterioration, has been reported following treatment with the vitamin (101-103,106).

TRANSFER TO TISSUES

Because tocopherols are carried by the plasma lipoproteins, mechanisms that provide tissues with lipids from the lipoproteins represent obvious pathways for the delivery of α-tocopherol (107). A variety of mechanisms have been identified for the movement of lipids from lipoproteins to tissues, including lipases, uptake by lipoprotein receptor-mediated endocytosis, and receptor-independent uptake, as well as spontaneous transfer and exchange reactions.

Exchange Mechanisms

Exchange of tocopherol between lipoproteins and membranes has long been recognized as a process by which vitamin E can be transferred to cells. Spontaneous transfer and exchange of α-tocopherol between lipoproteins and either red blood cells (56,81) or liposomal membranes (82) has been demonstrated in vitro. HDL tocopherol, but not VLDL tocopherol, equilibrated completely with rat erythrocytes (81). Because the erythrocyte lacks both lipoprotein receptors and lipoprotein lipase activity, these cells are likely to depend on transfer and exchange mechanisms for the adjustment of their tocopherol concentration in vivo.

Transfer During Lipolysis

Lipoprotein lipase, an enzyme bound to the capillary endothelial surface (85), has been shown in vitro to promote the transfer of tocopherol from chylomicrons to fibroblasts during triglyceride hydrolysis and fatty acid uptake (108). Release of lipoprotein lipase from the cell surface by heparin prevented the delivery of tocopherols and free fatty acids. Thus, the presence of the products of lipid hydrolysis or the changes in conformation of the remnants alone were not sufficient for the transfer of tocopherol to take place.

Tissues capable of synthesizing and secreting lipoprotein lipase, such as adipose tissue and muscle, may therefore obtain tocopherol by the lipoprotein lipase mechanism. There are some clinical data in support of a role of lipoprotein lipase in mediating the transfer of dietary tocopherol to tissues in vivo (108). The tocopherol content of adipose tissue from an 11-year-old child with lipoprotein lipase deficiency was reduced compared with that of normal subjects, and those in adult patients were low but within the normal range. These patients do not manifest any symptoms of vitamin E deficiency, so the relative importance of the lipoprotein lipase mechanism for maintaining tissue α-tocopherol concentrations is uncertain.

Tocopherol Delivery by the LDL (Apolipoprotein B/E) Receptor Mechanism

The high-affinity receptor pathway for LDL has been shown to function as a mechanism for α-tocopherol delivery to fibroblasts in tissue culture (109,110). Consistent with endocytosis of intact LDL, both α-tocopherol and apo B, the apolipoprotein of LDL, were taken together. When LDL were incubated with receptor-negative fibroblasts from a patient with homozygous familial hypercholesterolemia, delivery of α-tocopherol was reduced. Uptake of both [³H]α-tocopherol and [¹²⁵I]LDL by normal fibroblasts was greatly reduced when the lipoprotein was modified by methylation (110), which prevented binding of LDL to its receptor (111). The residual uptake of α-tocopherol from either methylated LDL by normal cells or from normal LDL by LDL receptor-deficient cells suggests that the LDL receptor mechanism is not the only pathway for delivery of tocopherols from LDL to cells. This is consistent with the observation that patients with the homozygous form of familial hypercholesterolemia do not manifest any biochemical or clinical evidence of vitamin E deficiency (109).

The role of the LDL receptor pathway in the maintenance of α-tocopherol concentrations in tissues of intact animals has been further investigated in the Watanabe heritable hyperlipidemic (WHHL) rabbit (112). These animals have a defect in the intracellular glycosylation of the immature LDL receptor, which causes decreased transport of the receptor to the cell surface and a virtual absence of LDL receptor activity (113). Thus, WHHL rabbits provide an animal model for homozygous familial hypercholesterolemia. Plasma concentrations of α-tocopherol in WHHL rabbits fed a control diet for 200 days were increased 10-fold compared to control rabbits, and this was associated with increased concentrations of VLDL, IDL, and LDL. Despite the LDL receptor deficiency in the WHHL animals, tissue concentrations of α-tocopherol, with the exception of the adrenal gland, were not reduced. The adrenal gland appears to rely on the LDL receptor pathway for obtaining a significant proportion of its α-tocopherol since the adrenal glands from the WHHL rabbits contained half as much α-tocopherol as control adrenals. Following the injection of LDL labeled with [³H]α-tocopherol into both normal and WHHL rabbits, there was evidence for the transfer of [³H]α-tocopherol to tissues independent of LDL uptake. In addition [³H]α-tocopherol was found to move between the various lipoprotein fractions. The results also indicate that the high concentrations of VLDL, IDL, and LDL in WHHL rabbits allow clearance by receptor-independent mechanisms to become more efficient. Taken together these data show that LDL α-tocopherol may be delivered to tissues by a number of different mechanisms, including uptake of α-tocopherol with LDL by receptor-dependent and receptor-independent pathways, and uptake of α-tocopherol without concomitant LDL uptake. Moreover, α-tocopherol can exchange from LDL to other lipoproteins, which in turn can mediate the transfer of α-tocopherol to tissues.

HDL-Mediated Delivery of Tocopherol

Little is known about HDL-mediated tissue uptake of vitamin E. HDL binding sites recognizing apolipoproteins A-I and A-II have been identified on cell surfaces. These binding sites are involved in cholesterol efflux from cells to HDL (114-116). HDL appears to play a central role in reverse cholesterol transport from peripheral tissues to the liver. Current studies suggest that HDL may also play a role in the uptake of tocopherol by tissues (Cohn, unpublished results) and could also contribute to reverse tocopherol transport from peripheral tissues to the liver.

HDL may also have a role in delivery of tocopherol to the nervous systsem, which is at risk in vitamin E deficiency (117,118). HDL particles enriched in apo E (119) have been detected in cerebrospinal fluid from humans and dogs, and LDL (apolipoprotein B/E) receptors that also recognize apo E-HDL have been demonstrated in brain from rat and monkeys (119). Recently α-tocopherol was detected in human cerebrospinal fluid at a concentration of 29.2 $\pm$ 9.5 nmol/L (12.6 $\pm$ 4.1 mg/L) (120), and it has been shown that greater than 80% of the tocopherol was carried within lipoproteins (Traber MG, Kayden HJ, unpublished observations). Thus, vitamin E could be delivered to the central and peripheral nervous system by HDL particles enriched in apo E (91).

Regulation of Vitamin E Uptake by Tissues

Mechanisms that effect the cotransfer of both lipids and vitamin E to tissues are advantageous in that they ensure the provision of peroxidizable lipids together with their protecting antioxidant (vitamin E). They have the disadvantage, however, that the tissues cannot regulate uptake of tocopherol based solely on tissue tocopherol requirements. Such mechanisms as the lipoprotein lipase-catalyzed transfer of fatty acids to tissues, or the LDL receptor pathway, are designed to perform specific functions in triglyceride and cholesterol metabolism, respectively. Thus, delivery of tocopherol by lipoprotein lipase is regulated by energy requirements, and the LDL receptor mechanism is regulated by the cholesterol requirements of the tissues. Similarly, maintenance of tocopherol tissue concentrations by spontaneous exchange processes from tocopherol-rich donors to tocopherol-poor acceptors (lipoproteins or membranes) are not regulated mechanisms. This lack of specificity may be compensated by the existence of multiple pathways for tocopherol delivery to the same tissue, and this apparent redundancy could account for the observation that patients with defective lipoprotein lipase or LDL receptors do not become vitamin E deficient.

CONCLUSIONS

Since the last comprehensive treatise on vitamin E in 1980 (121), there have been no major advances in our knowledge of the absorption of vitamin E, but our understanding of its transport has substantially increased. Much, however, remains to be learned regarding, for example, the regulation of plasma tocopherol concentrations and the determinants of its distribution within the different lipoprotein fractions. The mechanisms involved in the uptake of vitamin E by cells in vivo are poorly understood. This area is of great importance, and of potential clinical relevance, as it is now recognized that vitamin E is essential for normal neurological structure and function (117,118), and yet there is virtually no information regarding its uptake by neural tissues. Cellular uptake may depend on intracellular processes, including the transfer of tocopherol by binding proteins between organelles and other carriers. More work is clearly required to elucidate and understand these processes.

REFERENCES

1. Carey M, Small D. The characteristics of mixed micellar solutions with particular references to bile. Am J Med 1970; 49:590-608.
2. Gallo-Torres HE. Absorption. In: Machlin LJ, ed. Vitamin E: a comprehensive treatise. New York: Marcel Dekker, 1980; 170-92.
3. Kelleher J, Losowsky MS. The absorption of α-tocopherol in man. Br J Nutr 1970; 24:1033-47.
4. MacMahon MT, Neale G. The absorption of α-tocopherol in control subjects and in patients with intestinal malabsorption. Clin Sci 1970; 38:197-210.
5. Blomstrand R, Forsgren L. Labelled tocopherols in man. Int J Vitan Nutr Res 1968; 38:328-44.
6. Gallo-Torres H. Obligatory role of bile for the intestinal absorption of vitamin E. Lipids 1970; 5: 379-84.
7. Bjørneboe A, Bjørneboe G-EA, Bodd E, Hagen BF, Kveseth N, Drevon CA. Transport and distribution of α-tocopherol in lymph, serum and liver cells in rats. Biochim Biophys Acta 1986; 889: 310-5.
8. Traber MG, Kayden HJ, Green JB, Green MH. Absorption of water miscible forms of vitamin E in a patient with cholestasis and in rats. Am J Clin Nutr 1986; 44:914-23.
9. Bjørneboe A, Bjørneboe GA, Drevon CA. Absorption, transport and distribution of vitamin E. J Nutr 1990; 120:233-42.
10. Muller D, Harris J, Lloyd J. The relative importance of the factors involved in the absorption of vitamin E in children. Gut 1974; 15:966-71.

11. Harries J, Muller D. Absorption of vitamin E in children with biliary obstruction. Gut 1971; 12: 579-84.

12. Rosenblum JL, Keating JP, Prensky AL, Nelson JS. A progressive neurologic syndrome in children with chronic liver disease. N Engl J Med 1981; 304:503-8.

13. Guggenheim M, Ringel S, Silverman A, Grabert B. Progressive neuromuscular disease in children with chronic cholestasis and vitamin E deficiency: diagnosis and treatment with alpha tocopherol. J Pediat 1982; 100:51-8.

14. Sokol RJ, Heubi JE, Iannaccone S, Bove KE, Harris RE, Balistreri WF. The mechanism causing vitamin E deficiency during chronic childhood cholestasis. Gastroenterology 1983; 85:1172-82.

15. Jeffrey G, Muller D, Burroughs A, et al. Vitamin E deficiency and its clinical significance in adults with primary biliary cirrhosis and other forms of chronic liver disease. J Hepatol 1987; 4:307-17.

16. Sokol RJ, Heubi JE, Butler-Simon N, McClung HJ, Lilly JR, Silverman A. Treatment of vitamin E deficiency during chronic childhood cholestasis with oral d-α-tocopheryl polyethylene glycol 1000 succinate (TPGS). I. Intestinal absorption, efficacy and safety. Gastroenterology 1987; 93: 975-85.

17. MacMahon M, Thompson G. Comparison of the absorption of a polar lipid, oleic acid and a nonpolar lipid α-tocopherol from mixed micellar solutions and emulsions. Eur J Clin Invest 1970; 1: 161-6.

18. Akerib M, Steiner W. Inhibition of vitamin E absorption by a lipid fraction. Int J Vitan Nutr Res 1971; 41:42-3.

19. Pearson C, Legge A. Uptake of vitamin E by rat small intestinal slices. Biochim Biophys Acta 1972; 288:404-12.

20. Traber MG, Thellman CA, Rindler MJ, Kayden HJ. Uptake of intact TPGS (d-α-tocopheryl polyethylene glycol 1000 succinate) a water miscible form of vitamin E by human cells in vitro. Am J Clin Nutr 1988; 48:605-11.

21. Bennett M, Medwadowski B. Vitamin A, vitamin E and lipids in serum of children with cystic fibrosis or congenital heart defects compared with normal children. Am J Clin Nutr 1967; 20:415-21.

22. Harries J, Muller D. Absorption of different doses of fat soluble and water miscible preparations of vitamin E in children with cystic fibrosis. Arch Dis Child 1971; 46:341-4.

23. Farrell P, Bieri J, Fratantoni J, Wood R, Di Sant 'Agnese P. The occurrence and effects of human vitamin E deficiency. J Clin Invest 1977; 60:233-41.

24. Stead RJ, Muller DPR, Matthews S, Hodson ME, Batten JC. Effect of abnormal liver function on vitamin E status and supplementation in adults with cystic fibrosis. Gut 1986; 27:714-8.

25. Traber MG, Goldberg I, Davidson E, Lagmay N, Kayden HJ. Vitamin E uptake by human intestinal cells during lipolysis in vitro. Gastroenterology 1990; 98:96-103.

26. Hollander D, Rim E, Muralidhara K. Mechanism and site of small intestinal absorption of α-tocopherol in the rat. Gastroenterology 1975; 68:1492-9.

27. Muralidhara K, Hollander D. Intestinal absorption of α-tocopherol in the unanaesthetized rat. The influence of luminal constituents on the absorptive process. J Lab Clin Med 1977; 90:85-91.

28. Traber MG, Ingold KU, Burton GW, Kayden HJ. Absorption and transport of deuterium-substituted 2R,4$'R$,8$'R$-α-tocopherol in human lipoproteins. Lipids 1988; 23:791-7.

29. Catignani GL, Bieri JG. Rat liver α-tocopherol binding protein. Biochim Biophys Acta 1977; 497: 349-57.

30. Kayden HJ. Vitamin E deficiency in patients with abetalipoproteinemia. In: von Kress HF, Blum KU, eds. Vitamine A, E und K. Klinische und physiologisch-chemische probleme. Stuttgart: F.K. Schattauer Verlag, 1967; 301-8.

31. Kayden H. Abetalipoproteinemia. Annu Rev Med 1972; 23:285-96.

32. Kane J, Havel R. Disorders of the biogenesis and secretion of lipoproteins containing the B-apolipoproteins. In: Scriver R, Beaudet A, Sly W, Valle D, ed. The metabolic basis of inherited disease. New York: McGraw-Hill, 1989; 1139-64.

33. Kayden HJ, Silber R. The role of vitamin E deficiency in the abnormal autohemolysis of acanthocytosis. Trans Assoc Am Phys 1965; 78:334-41.

34. Kayden HJ, Hatam LJ, Traber MG. The measurement of nanograms of tocopherol from needle aspiration biopsies of adipose tissue: normal and abetalipoproteinemic subjects. J Lipid Res 1983; 24:652-6.

35. Kayden HJ. Tocopherol content of adipose tissue from vitamin E deficient humans. In: Porter R, Whelan J, eds. Biology of vitamin E. London: Pittman Books, 1983; 101:70-91.

36. MacMahon M, Neale G, Thompson G. Lymphatic and portal venous transport of α-tocopherol and cholesterol. Eur J Clin Invest 1971; 1:288-94.

37. Deckelbaum R, Eisenberg S, Oschry Y, Cooper M, Blum C. Abnormal high density lipoproteins of abetalipoproteinemia: relevance to normal HDL metabolism. J Lipid Res 1982; 23:1274-82.

38. Muller D, Manning J, Mathias P, Harries J. Studies on the hydrolysis of tocopheryl esters. Int J Vitan Nutr Res 1976; 46:207-10.

39. Burton GW, Ingold KU, Foster DO, et al. Comparison of free α-tocopherol and α-tocopheryl acetate as sources of vitamin E in rats and humans. Lipids 1988; 23:834-40.

40. Weiser H, Vecchi M, Schlachter M. Stereoisomers of α-tocopheryl acetate. IV. USP units and α-tocopherol equivalents of *all-rac-*, *2-ambo-* and *RRR-α*-tocopherol evaluated by simultaneous determination of resorption-gestation, myopathy and liver storage capacity in rats. Int J. Vitan Nutr Res 1986; 56:45-56.

41. Bieri JG, Evarts RP. Tocopherols and fatty acids in American diets. J Am Diet Assoc 1973; 62: 147-51.

42. Quaife M, Swanson W, Dju M, Harris P. Vitamin E in foods and tissues. Ann N Y Acad Sci 1949; 52:300-5.

43. Parker RS. Dietary and biochemical aspects of vitamin E. Adv Food Nutr Res 1989; 33:157-232.

44. Pearson C, Barnes M. Absorption and distribution of the naturally occurring tocochromanols in the rat. Br J Nutr 1970; 24:581-7.

45. Handelman GJ, Machlin LJ, Fitch K, Weiter JJ, Dratz EA. Oral α-tocopherol supplements decrease plasma γ-tocopherol levels in humans. J Nutr 1985; 115:807-13.

46. Baker H, Handelman GJ, Short S, et al. Comparison of plasma α- and γ-tocopherol levels following chronic oral administration of either all-rac-α-tocopheryl acetate or *RRR*-α-tocopheryl acetate in normal adult male subjects. Am J Clin Nutr 1986; 43:382-7.

47. Peake IR, Bieri JG. α- and γ-Tocopherols in the rat: in vitro and in vivo tissue uptake and metabolism. J Nutr 1977; 101:1615-22.

48. Bieri JG, Evarts RP. γ-Tocopherol: metabolism, biological activity and significance in human nutrition. Am J Clin Nutr 1974; 27:980-6.

49. Traber MG, Kayden HJ. Preferential incorporation of α-tocopherol vs γ-tocopherol in human lipoproteins. Am J Clin Nutr 1989; 49:517-26.

50. Meydani M, Cohn JS, Macauley JB, McNamara JR, Blumberg JB, Schaefer EJ. Postprandial changes in the plasma concentration of α- and γ-tocopherol in human subjects fed a fat-rich meal supplemented with fat-soluble vitamins. J Nutr 1989; 119:1252-8.

51. Traber M, Rudel L, Burton G, Hughes L, Ingold K, Kayden H. Nascent VLDL from liver perfusions of cynomolgus monkeys are preferentially enriched in *RRR*- compared with *SRR*-α tocopherol. Studies using deuterated tocopherols. J Lipid Res 1990; 31:687-94.

52. Bjornson LK, Kayden HJ, Miller E, Moshell AN. The transport of α-tocopherol and β-carotene in human blood. J Lipid Res 1976; 17:343-51.

53. Behrens WA, Thompson JN, Madere R. Distribution of alpha tocopherol in human plasma lipoproteins. Am J Clin Nutr 1982; 35:691-6.

54. Haga P, Ek J, Kran S. Plasma tocopherol levels and vitamin E/beta lipoprotein relationships during pregnancy and in cord blood. Am J Clin Nutr 1982; 36:1200-4.

55. Chow CK. Distribution of tocopherols in human plasma and red blood cells. Am J Clin Nutr 1975; 28:756-60.

56. Kayden HJ, Bjornson LK. The dynamics of vitamin E transport in the human erythrocyte. Ann N Y Acad Sci 1972; 203:127-40.

57. Machlin L. Vitamin E. In: Machlin L, ed. Handbook of vitamins. New York: Marcel Dekker, 1984; 99-145.

58. Gallo-Torres HE. Transport and metabolism. In: Machlin LJ, ed. Vitamin E: a comprehensive treatise. New York: Marcel Dekker, 1980; 193-267.

59. Granot E, Tamir I, Deckelbaum RJ. Neutral lipid transfer protein does not regulate α-tocopherol transfer between human plasma lipoproteins. Lipids 1988; 23:17-21.

60. Burton GW, Traber MG. Vitamin E: antioxidant activity, biokinetics and bioavailability. Annu Rev Nutr 1990; 10:357-82.

61. Rask L, Anundi H, Bohme J, et al. The retinol-binding protein. Scand J Clin Lab Invest 1980; 40 (Suppl.154):45-61.

62. Blomhoff R, Green M, Berg T, Norum K. Transport and storage of vitamin A. Science 1990; 250: 399-404.

63. Esterbauer H, Jurgens G, Quehenberger O, Koller E. Autoxidation of human low density lipoprotein: loss of polyunsaturated fatty acids and vitamin E and generation of aldehydes. J Lipid Res 1987; 28:495-509.

64. Horwitt MK, Harvey CC, Dahm DH, Searcy MT. Relationship between tocopherol and serum lipid levels for determination of nutritional adequacy. Ann N Y Acad Sci 1972; 203:223-36.

65. Widhalm K, Holzl M, Brubacher G. Lipids, lipoproteins and alpha-tocopherol: relationship and changes during adolescence. A longitudinal study. Ann Nutr Metab 1985; 29:12-8.

66. Davies T, Kelleher J, Losowsky M. Interrelation of serum lipoprotein and tocopherol levels. Clin Chim Acta 1969; 21:431-6.

67. Lambert D, Mourot J. Vitamin E and lipoproteins in hyperlipoproteinemia. Atherosclerosis 1984; 53:327-30.

68. Sokol R, Heubi J, Iannaccone S, Bove K, Balistreri W. Vitamin E deficiency with normal serum vitamin E concentrations in children with chronic cholestasis. N Engl J Med 1984; 310:1209-12.

69. Farrell P. Deficiency states, pharmacological effects, and nutrient requirements. In: Machlin L, ed. Vitamin E: a comprehensive treatise. New York: Marcel Dekker, 1980; 520-620.

70. Farrell PM. Vitamin E. In: Shils ME, Young VR, eds. Modern nutrition in health and disease. Philadelphia: Lea & Febiger, 1988; 340-54.

71. Horwitt MK, Elliott WH, Kanjananggulpan P, Fitch CD. Serum concentrations of α-tocopherol after ingestion of various vitamin E preparations. Am J Clin Nutr 1984; 40:240-5.

72. Jansson L, Nilsson B, Lindgren R. Quantitation of serum tocopherols by high performance liquid chromatography with fluorescence detection. J Chromatogr 1980; 181:242-7.

73. Hatam LJ, Kayden HJ. The failure of α-tocopherol supplementation to alter the distribution of lipoprotein cholesterol in normal and hyperlipoproteinemic persons. Am J Clin Pathol 1981; 76: 122-4.

74. Chapman M. Comparative analysis of mammalian plasma lipoproteins. In: Segrest J, Albers J, eds. Plasma lipoproteins. Part A. Preparation, structure and molecular biology. Orlando, FL: Academic Press, 1986; 128:70-143.

75. Knipping G, Kostner G, Holasek A. Studies on the composition of pig serum lipoproteins isolation and characterization of different apoproteins. Biochim Biophys Acta 1975; 393:88-99.

76. Fernando-Warnakulasuriya G, Eckerson M, Clark W, Wells M. Lipoprotein metabolism in the suckling rat: characterization of plasma and lymphatic lipoproteins. J Lipid Res 1983; 24:1626-38.

77. Van Lenten B, Roheim P. Changes in the concentrations and distributions of apolipoproteins in the aging rat. J Lipid Res 1982; 23:1187-95.

78. Gotto AM, Pownall HJ, Havel RJ. Introduction to the plasma lipoproteins. In: Segrest J, Albers J, eds. Methods in enzymology. Orlando, FL: Academic Press, 1986; 128:3-41.

79. Peake I, Windmueller H, Bieri J. A comparison of the intestinal absorption, lymph and plasma transport, and tissue uptake of α- and γ-tocopherols in the rat. Biochim Biophys Acta 1972; 260: 679-88.

80. Behrens WA, Madere R. Transport of α- and γ-tocopherol in human plasma lipoproteins. Nutr Res 1985; 5:167-74.

81. Bjornson LK, Gniewkowski C, Kayden HJ. A comparison of the exchange of α-tocopherol and of free cholesterol between rat plasma lipoproteins and erythrocytes. J Lipid Res 1975; 16:39-53.

82. Massey JB. Kinetics of transfer of α-tocopherol between model and native plasma lipoproteins. Biochim Biophys Acta 1984; 793:387-92.

83. Perly B, Smith ICP, Hughes L, Burton GW, Ingold KU. Estimation of the location of natural α-tocopherol in lipid bilayers by ^{13}C-NMR spectroscopy. Biochim Biophys Acta 1985; 819:131-5.

84. Ekiel IH, Hughes L, Burton GW, Jovall PA, Ingold KU, Smith ICP. Structure and dynamics of α-tocopherol in model membranes and in solution: a broad-line and high-resolution NMR study. Biochemistry 1988; 27:1432-40.

85. Nelsson-Ehle P, Garfinkel AS, Schotz MC. Lipolytic enzymes and plasma lipoprotein metabolism. Annu Rev Biochem 1980; 49:667-93.

86. Traber MG, Sokol RJ, Burton GW, et al. Impaired ability of patients with familial isolated vitamin E deficiency to incorporate α-tocopherol into lipoproteins secreted by the liver. J Clin Invest 1990; 85:397-407.

87. Traber MG, Burton GW, Ingold KU, Kayden HJ. *RRR*- and *SRR*-α-tocopherols are secreted without discrimination in human chylomicrons, but *RRR*-α-tocopherol is preferentially secreted in very low density lipoproteins. J Lipid Res 1990; 31:675-85.

88. Cohn W, Loechleiter F, Weber F. *a*-Tocopherol is secreted from rat liver in very low density lipoproteins. J Lipid Res 1988; 29:1359-66.

89. Bjørneboe A, Bjørneboe G-EA, Hagen BF, Nossen JO, Drevon CA. Secretion of α-tocopherol from cultured rat hepatocytes. Biochim Biophys Acta 1987; 922:199-205.

90. Havel RJ. Origin, metabolic fate, and function of plasma lipoproteins. In: Steinberg D, Olefsky JM, eds. Hypercholesterolemia and atherosclerosis, pathogenesis and prevention. New York: Churchill Livingston, 1987; 117-41.

91. Mahley RW. Apolipoprotein E: cholesterol transport protein with expanding role in cell biology. Science 1988; 240:622-30.

92. Bjørneboe A, Bjørneboe G-EA, Drevon CA. Serum half-life, distribution, hepatic uptake and biliary excretion of α-tocopherol in rats. Biochim Biophys Acta 1987; 921:175-81.

93. Weiser H, Vecchi M. Stereoisomers of α-tocopheryl acetate. II. Biopotencies of all eight stereoisomers, individually or in mixtures, as determined by rat resorption-gestation tests. Internat J Vitam Nutr Res 1982; 52:351-70.

94. Ingold KU, Hughes L, Slaby M, Burton GW. Synthesis of 2*R*,4 '*R*,8 '*R*-α-tocopherols selectively labelled with deuterium. J Labeled Comp Radiopharm 1987; 24:817-31.

95. Ingold KU, Burton GW, Foster DO, Hughes L, Lindsay DA, Webb A. Biokinetics of and discrimination between dietary *RRR*- and *SRR*-α-tocopherols in the male rat. Lipids 1987; 22:163-72.

96. Vecchi M, Walther W, Glinz E, et al. Chromatographische trennung und quantitative bestimmung aller acht stereoisomeren von α-tocopherol. Helv Chim Acta 1990; 73:782-9.

97. Behrens WA, Madere R. Transfer of α-tocopherol to microsomes mediated by a partially purified liver α-tocopherol binding protein. Nutr Res 1982; 2:611-8.

98. Murphy DJ, Mavis RD. Membrane transfer of α-tocopherol. J Biol Chem 1981; 256:10464-8.

99. Kaplowitz N, Yoshida H, Kuhlenkamp J, Slitsky B, Ren I, Stolz A. Tocopherol-binding proteins of hepatic cytosol. Ann N Y Acad Sci 1989; 570:85-94.

100. Burck U, Goebel HH, Kuhlendahl HD, Meier C, Goebel KM. Neuromyopathy and vitamin E deficiency in man. Neuropediatrics 1981; 12:267-78.

101. Kohlschutter A, Hubner C, Jansen W, Lindner SG. A treatable familial neuromyopathy with vitamin E deficiency, normal absorption, and evidence of increased consumption of vitamin E. J Inher Metab Dis 1988; 11:149-52.

102. Laplante P, Vanasse M, Michaud J, Geoffroy G, Brochu P. A progressive neurological syndrome associated with an isolated vitamin E deficiency. Can J Neurol Sci 1984; 11:561-4.

103. Harding AE, Matthews S, Jones S, Ellis CJK, Booth IW, Muller DPR. Spinocerebellar degeneration associated with a selective defect of vitamin E absorption. N Engl J Med 1985; 313:32-5.

104. Krendel DA, Gilchrest JM, Johnson AO, Bossen EH. Isolated deficiency of vitamin E with progressive neurologic deterioration. Neurology 1987; 37:538-40.

105. Yokota T, Wada Y, Furukawa T, Tsukagoshi H, Uchihara T, Watabiki S. Adult-onset spinocerebellar syndrome with idiopathic vitamin E deficiency. Ann Neurol 1987; 22:84-7.

106. Sokol RJ, Kayden HJ, Bettis DB, et al. Isolated vitamin E deficiency in the absence of fat malabsorption—familial and sporadic cases: characterization and investigation of causes. J Lab Clin Med 1988; 111:548-59.

107. Gonzales M. Serum concentrations and cellular uptake of vitamin E. Med Hypoth 1990; 32:107-10.

108. Traber MG, Olivecrona T, Kayden HJ. Bovine milk lipoprotein lipase transfers tocopherol to human fibroblasts during triglyceride hydrolysis in vitro. J Clin Invest 1985; 75:1729-34.

109. Traber MG, Kayden HJ. Vitamin E is delivered to cells via the high affinity receptor for low density lipoprotein. Am J Clin Nutr 1984; 40:747-51.

110. Thellman C, Shireman R. In vitro uptake of [³H]-α-tocopherol from low density lipoprotein by cultured human fibroblasts. J Nutr 1985; 115:1673-9.

111. Weisgraber K, Innerarity T, Mahley R. Role of lysine residues of lipoproteins in high affinity binding to cell surface receptors on human fibroblasts. J Biol Chem 1978; 253:9053-62.

112. Cohn W, Kuhn H. The role of the low density lipoprotein receptor for α-tocopherol delivery to tissues. Ann N Y Acad Sci 1989; 570:61-71.

113. Yamamoto T, Bishop R, Brown M, Goldstein J, Russel D. Deletion in cysteine-rich region of LDL receptor impedes transport to cell surface in WHHL rabbit. Science 1986; 232:1230-7.

114. Oram J, Brinton E, Bierman E. Regulation of high density lipoprotein receptor activity in cultured human skin fibroblasts and human arterial smooth muscle cells. J Clin Invest 1983; 72:1611-21.

115. Aviram M, Bierman E, Oram J. High density lipoprotein stimulates sterol translocation between intracellular and plasma membrane pools in human monocyte-derived macrophages. J Lipid Res 1989; 30:65-76.

116. Tozuka M, Fidge N. Purification and characterization of two high density lipoprotein binding proteins from rat and human liver. Biochem J 1989; 261:239-44.

117. Sokol RJ. Vitamin E deficiency and neurologic disease. Annu Rev Nutr 1988; 8:351-73.

118. Muller D. Vitamin E—its role in neurological function. Postgrad Med J 1986; 62:107-12.

119. Pitas RE, Boyles JK, Lee SH, Hui D, Weisgraber KH. Lipoproteins and their receptors in the central nervous system. J Biol Chem 1987; 262:14352-60.

120. Vatassery G, Nelson M, Maletta G, Kuskowski M. Vitamin E (tocopherols) in human cerebrospinal fluid. Am J Clin Nutr 1991; 53:95-9.

121. Machlin LJ, ed. Vitamin E: a comprehensive treatise. Basic and clinical nutrition. New York: Marcel Dekker, 1980.

4

Interrelationships of Vitamin E with Other Nutrients

Harold H. Draper

University of Guelph, Guelph, Ontario, Canada

INTRODUCTION

Vitamin E is unusual among the vitamins, although not unique, in the extent to which its dietary requirement depends upon the intake of other nutrients. The main determinant of its requirement is the intake of polyunsaturated fatty acids (PUFA). Depending upon the concentration and composition of these acids in the diet of experimental animals, their vitamin E requirement can vary by up to two orders of magnitude.

Vitamin E is also unusual in that its essentiality in the diet lacks absolute specificity. The classic diseases of vitamin E deficiency, including sterility in female rats and muscular dystrophy in growing laboratory and domestic animals, can be prevented and treated by administration of certain synthetic structurally unrelated lipid antioxidants, among which the rubber antioxidant N,N^1-phenylenediamine (DPPD) is the most active evaluated thus far.

These observations, together with the fact that the in vivo oxidation products of α-tocopherol are biologically inactive compounds analogous to those produced by its oxidation in peroxidizing unsaturated oils and by mild chemical oxidants, have established that the requirement for vitamin E reflects a need for a biologically active lipid antioxidant. This requirement is somewhat analogous to the requirement for a source of methyl groups that can be filled by one of several different methyl donors, including choline, betaine, and methionine. Despite its nonspecificity, vitamin E in the form of α-tocopherol is the most effective biological antioxidant from the standpoint of overall efficacy and safety.

In addition to its interrelationship with PUFA, vitamin E has important nutritional and metabolic interrelationships with other dietary ingredients, including selenium, sulfur amino acids (SAA), β-carotene, synthetic food antioxidants, and iron and other transition elements. These interrelationships have been reviewed by the author elsewhere (1-3), as well as by other investigators (4,5), and are discussed further elsewhere in this volume. An updated summary emphasizing the nutritional implications of these interrelationships is provided in the following sections.

DIETARY PUFA

As in animals, the amount of vitamin E required in the diet of humans is determined mainly by the mass and composition of PUFA in the tissues. Over protracted periods of time, the composition of PUFA in the tissues approaches that of PUFA in the habitual diet. PUFA are concentrated in cell membranes, where, by a mechanism that is still obscure, they have the capacity to sequester vitamin E in amounts commensurate with their own concentration and degree of unsaturation. In hepatocytes, for example, vitamin E is found predominantly in the PUFA-rich mitochondria and endoplasmic reticulum. In contrast, the synthetic anti-oxidant DPPD is passively distributed among cell compartments according to their total lipid content.

A natural association between PUFA and vitamin E is evident in plant oils. The genetic basis for this association is illustrated by the fact that it extends to different varieties of the same species; for example, selection of corn for a high PUFA content results in a concomitant increase in vitamin E (6). This interrelationship is of pivotal importance in the prevention of vitamin E deficiency on high-PUFA diets.

Attempts to define the vitamin E requirement as a function of PUFA intake (in a manner similar to the vitamin B_6 requirement as a function of protein intake) have not been successful. This is attributable to several factors, including the fact that the amount of vitamin E required to stabilize PUFA in the tissues is influenced to a greater extent by their degree of unsaturation than by their mass and that PUFA are not deposited in the tissues in the same proportions that they occur in the diet. Dietary PUFA are modified by elongation and desaturation and are catabolized to various degrees depending upon energy status. Studies on experimental animals have indicated that the relationship between the vitamin E requirement and PUFA intake is not linear. Jager (7) concluded that when the membranes are replete with PUFA, the intake of linoleic acid (the main PUFA in the diet) has little effect on the vitamin E requirement. A ratio of 0.4 mg α-tocopherol per g PUFA consumed was found to be characteristic of the U.S. mixed diet in 1973 (8). Allowance for the biological activity contributed by γ-tocopherol (the most prevalent form of vitamin E in the diet) and other vitamers raised the ratio to 0.5 mg α-tocopherol equivalents per g PUFA. Thompson and associates (9) found an α-tocopherol-PUFA ratio of 0.52 for a composite Canadian diet but concluded that this ratio is not a useful indicator of the adequacy of foods and diets.

Because expressing recommended dietary intakes of vitamin E on the basis of PUFA intake is infeasible, recommendations have been formulated to meet the requirements of those individuals in the population with the highest PUFA intakes. For example, the recommended dietary allowance (RDA) for U.S. adult males is 10 mg α-tocopherol equivalents per day (10). Despite the fact that the vitamin E intake of large numbers of U.S. men falls below the RDA, there is no significant evidence of vitamin E deficiency in this segment of the population. This is attributable to the fact that individuals with a vitamin E intake substantially below the RDA also have a lower intake of PUFA and therefore a lower vitamin E requirement.

All indigenous diets, regardless of their PUFA content, appear to contain sufficient vitamin E to prevent biochemical and clinical signs of deficiency. A recent survey of 451 healthy male and female Canadian adults yielded a mean plasma vitamin E level of 1.02 mg/dl, a value typical of adults consuming the mixed diet (11). γ-Tocopherol comprised 13% of vitamin intake and the rest was almost entirely α-tocopherol. There is no history of epidemic vitamin E diseases among populations whose indigenous diet culture was disrupted by the introduction of lower quality processed foods, as occurred in the case of thiamin, niacin, and vitamin C. In the only experimental study on the vitamin E requirement of human adults, up to 2 years was required for the first deficiency symptoms to appear in subjects fed a vitamin E-deficient diet, even when the diet was high in PUFA (12). The vitamin E content of animal fats is less than that of vegetable oils, but their lower PUFA content generates a lower requirement for vitamin E. Further, the vitamin E in animal fats consists almost entirely of α-tocopherol, its most active form, whereas the main vitamer in edible oils is γ-tocopherol, which has only 10% as much biological activity. Non-α-tocopherol vitamers are estimated to contribute about 20% of the total vitamin E activity in the mixed diet (10).

The oils of fish and marine mammals represent an exception to the general rule that the concentration of vitamin E in oils and fats is commensurate with the amount required to maintain the oxidative stability of the fatty acids they contain. The propensity of these highly unsaturated oils to become rancid when exposed to oxygen is well known, and indeed, the feeding of such oils to experimental animals is a common means of inducing vitamin E deficiency. When administered in the fresh, unoxidized state, however, fish oils are capable of furnishing enough vitamin E to prevent a deficiency in rats. The oxidative stability of their PUFA in the tissues of aquatic animals may be due to the lower temperature and oxygen tension that prevail in these species. Arctic native populations who consume large quantities of fish, seal, and whale as staple foods have plasma vitamin E levels comparable to those of consumers of the mixed North American diet (13). This is explainable by the fact that much of the oil in fish is removed during the drying process and most of the blubber of marine mammals is rendered to obtain oil for heat and light or is fed to dogs. Consequently, the intake of fat and PUFA by circumpolar aboriginal peoples is lower than commonly believed. Pharmaceutical preparations of *n* - 3 fish oil fatty acids prescribed for the prevention of heart attacks are fortified with vitamin E and other antioxidants and encapsulated to prevent oxidative deterioration.

TISSUE PUFA

The peroxidation of PUFA (RH) in the tissues is dealt with in detail elsewhere in this volume and is discussed here only to document the involvement of vitamin E and other nutrients. The prevailing assumptions are made that peroxidation is initiated mainly by hydroxyl (HO·), peroxy (ROO·), and alkoxy (RO·) radicals at positions allylic to the carbon-carbon double bonds of PUFA, where the C-H bond energy is weakest, and that uptake of oxygen by the resulting fatty acyl radical (R·) generates a peroxy radical (ROO·). This species, as well as the RO· radical formed by the metal-catalyzed decomposition of preformed hydroxyperoxides, is capable of denaturing a range of macromolecules and propagating a chain reaction that generates more acyl radicals. The resulting membrane dysfunction is responsible for many of the symptoms seen in vitamin E deficiency, among which the most obvious are the hemolytic anemia caused by peroxidation of red blood cell membranes in several species and the capillary leakage that occurs in exudative diathesis in birds.

An inverse relationship between the plasma α-tocopherol level of normal adults and their exhalation of pentane, a product of the oxidative decomposition of *n* - 6 fatty acids, has been reported by Lemoyne and coworkers (14). Their data (see Fig. 1) suggest that lipid peroxidation in human tissues is related to the intake of vitamin E within the normal range, and imply that peroxidation would be further decreased by vitamin E supplementation. Whether differences in pentane exhalation within the normal range have any implications for human health is unknown.

Although the molecular orientation of α-tocopherol in the lipid bilayer of membranes is still obscure, vitamin E is believed to function as a chain-breaking antioxidant by reducing fatty acyl hydroperoxy radicals (ROO·) to form more stable hydroperoxides (ROOH). These compounds are reduced to their corresponding hydroxy acids by glutathione-dependent peroxidases. The tocopheroxy radical (E-O·) generated in the process undergoes two transformations, one involving an intramolecular electron shift to form a carbanion at the 5-methyl position that self-reacts to yield a dihydroxy dimer. In the other transformation, the tocopheroxy ion undergoes further oxidation involving opening of the chroman ring and formation of α-tocopheryl *p*-quinone. The quinone is the most frequently reported oxidation product of α-tocopherol in biological materials, but oxidation of the vitamin using mild chemical oxidants, such as ferricyanide, or autoxidation in unsaturated oils yields a predominance of

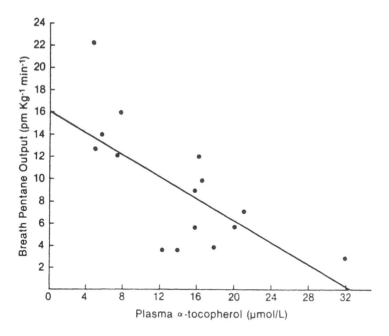

FIGURE 1 Relationship between the plasma α-tocopherol levels of adults and their exhalation of pentane.

dimer (15). The dimer is less easily detected than the *p*-quinone because its absorption spectrum and polarity are similar to those of α-tocopherol, whereas the *p*-quinone is readily detected from its bicuspid absorption peaks at 260 and 270 nm and high extinction coefficient. although large doses of the *p*-quinone are capable of relieving vitamin E deficiency by virtue of its enzymatic reduction to the hydroquinone, which has antioxidant properties, none of the stable oxidation products of vitamin E contribute significantly to its biological activity under physiological conditions.

Whether the α-tocopheroxy radical is reduced in vivo, resulting in recycling of the vitamin, remains a question of interest. Its reduction by various biological agents, including ascorbic acid, has been demonstrated by several groups of investigators (16-19), but whether these reactions occur under physiological conditions is uncertain. Recently, reduction of the tocopheroxy ion by mitochondrial and microsomal preparations was reported (20). The reduction was dependent on the presence of ubiquinone and NADH or NADPH, but not on vitamin C. The proposed reduction of the α-tocopheroxy radical by vitamin C in vivo is discussed later.

SELENIUM

The important relationship between vitamin E and Se was discovered as a result of the observation that the alkaline ash of certain biological materials could prevent two diseases, necrotic liver degeneration in rats and exudative diathesis in chicks, which previously were attributed exclusively to vitamin E deficiency. Following identification of the active ingredient in the ash as Se, it became apparent that there was an association between the epidemiology of nutritional muscular dystrophy in farm animals and the Se content of the soils upon which their feeds were grown. However, most field cases of Se-responsive diseases in animals also respond to the administration of vitamin E or a biologically active synthetic antioxidant. In the case of liver necrosis and exudative diathesis, a requirement for both nutrients has yet to be demonstrated; hence the occurrence of these diseases signifies a deficiency of both nutrients. So-called Se deficiency in domestic animals appears to be primarily the result of an increased vitamin E requirement in the presence of a low intake of selenium, often aggravated by deterioration of the vitamin E content of feeds during storage.

The perplexing metabolic interrelationship between vitamin E and selenium appeared to be explained by discovery of the Se-containing enzyme glutathione peroxidase (Se-GPx), which is capable of removing fatty acid hydroperoxides formed in the tissues by reduction to their corresponding hydroxy acids (ROH). This reduction prevents the decomposition of ROOH to form alkoxy radicals that initiate further peroxidation.

$$\text{ROOH} + 2\text{GSH} \xrightarrow{\text{Se-GPx}} \text{ROH} + \text{GSSG} + \text{H}_2\text{O}$$

However, further work indicated that the reduction of fatty acyl hydroperoxides is catalyzed mainly by a non-Se-dependent glutathione peroxidase and that the most important antioxidant reaction catalyzed by the Se-containing enzyme is probably the reduction of hydrogen peroxide, which can generate highly reactive $\text{HO}^{\cdot}$ radicals in the Fenton reaction (see later). From a functional standpoint, Se-GPx therefore has more in common with catalase, another hydrogen peroxide decomposing enzyme, than with the non-Se-dependent GPx.

Selenium is ingested by humans and animals primarily in the form of selenomethionine and selenocysteine present in cereal proteins. Substantial differences in the Se content of soils and consequently in the Se content of national food supplies has evoked an interest in the possible occurrence of Se deficiency in human populations. The intake of selenium by the residents of New Zealand has been estimated at 28-56 μg/day and the whole blood level at about 70 ng/ml, both values less than half those found for residents of the United States and Canada (21). Plasma glutathione peroxidase levels are also somewhat lower. Nevertheless, extensive clinical studies of the New Zealand population have failed to reveal any health problems attributable to Se deficiency. Radioisotope studies on New Zealand women indicated that 20-30 μg/day was adequate to maintain Se balance but that a somewhat higher intake was necessary to reach a plateau in red blood cell glutathione peroxidase activity.

Paradoxically, the specific essentiality of Se was established by the discovery that a severe cardiomyopathy among vitamin E-replete children in the Keshan district of China was caused by an extremely low concentration of Se in the local soils and in the foods grown on these soils (22). Children with Keshan disease have been estimated to require at least 20 μg Se per day (22). There is some evidence of adaptability to a low-Se intake that may explain the difference in the apparent amount of Se required to maintain Se balance in New Zealand women and the amount required to maintain normal plasma levels and body stores in U.S. adults (54 μg/day) (23).

In industrialized countries, differences in the Se content of foods grown in different localities are overridden by the widespread distribution of commercial food products. Small differences in blood levels of glutathione peroxidase may have no significance for human health when the level of vitamin E is adequate, as appears to be the case in human populations. Supplementation with Se is constrained by a narrow margin of safety for this element. A prospective study is underway on the possible value of raising Se intake from the lower to the higher end of the normal range of U.S. intakes in cancer prevention (24).

IRON

Sequestration of iron and other transition metals in the tissues is of central importance in preventing iron-catalyzed generation of $\text{HO}^{\cdot}$ radicals from hydrogen peroxide formed by dismutation of superoxide:

$$O_2^- + O_2^- + 2H^+ \rightarrow H_2O_2 + O_2$$

$$H_2O_2 + Fe^{2+} \rightarrow HO\cdot + OH^- + Fe^{3+}$$

Free iron is exquisitely toxic in biological systems, and most attempts to detect it in the blood plasma and tissues of animals and humans have been unsuccessful. Intravenous administration of trace quantities of inorganic iron salts results in hemolysis caused by oxidative decomposition of red blood cell membrane lipids. Dietary iron, which exists mainly in the ferric state but is absorbed in the ferrous state, is sequestered by transferrin, a glycoprotein that tightly binds this element. There is a large excess of transferrin in the plasma that represents reserve iron binding capacity. Lactoferrin, a similar protein, is found in milk and body fluids. Plasma iron is transferred to the cells by internalization of transferrin, and iron released in the cytoplasma is efficiently bound to various constituents, including citrate, ATP, and other phosphate esters. Iron not used in synthesis of iron proteins is stored in the protein ferritin. Although various heme proteins have been reported to catalyze lipid peroxidation, it now appears that such observations were due to release of free iron or to contamination of reagents with trace quantities of this element.

Excess iron absorption occurs in certain pathological conditions, in accidental or prescribed overdosing with iron supplements, and as the result of extraordinary cultural practices, the best known of which is the Bantu custom of drinking beer fermented in iron pots. Accumulation of iron in the liver and other organs gives rise to a pathological condition termed iron overload. The genetic disease idiopathic hemochromatosis causes a similar condition as the result of overabsorption of dietary iron.

To some extent, the effects of iron overload can be ameliorated by large doses of vitamin E. A more effective approach, however, is the administration of chelators that form a complex with excess iron, increasing its excretion in the urine and bile. One of the most effective chelators is the bacterial compound desferrioxamine.

Transition metals, particularly iron, can cause membrane damage by catalyzing the decomposition of lipid hydroperoxides to yield alkoxy radicals, which are capable of reacting with various cellular constituents as well as initiating further lipid peroxidation.

$$ROH + Fe^{2+} \rightarrow RO\cdot + OH^- + Fe^{3+}$$

As for lipid peroxidation initiated by $HO\cdot$ radicals in the Fenton reaction, peroxidation caused by Fe-catalyzed formation of $RO\cdot$ radicals consumes vitamin E.

$$RO\cdot + E\text{-}OH \rightarrow ROH + E\text{-}O\cdot$$

"Off-flavors" caused by aldehydes, ketones, and other products of the metal-catalyzed decomposition of fatty acid hydroperoxides are a major problem in the food industry. This problem is combated by reducing, as much as possible, contact between food and iron, copper, cobalt, and other metals during processing, using such iron chelators as EDTA and citrate, and synthetic food antioxidants, such as BHA or BHT, to scavenge alkoxy radicals.

It has been reported that administration of a large dose of vitamin E preoperatively is beneficial in reducing tissue damage caused by surgery (25) and other forms of trauma. Release of catalytic iron from heme and other iron binding proteins resulting in $HO\cdot$ generation and consequent peroxidation of cell membranes may contribute to tissue damage.

SULFUR AMINO ACIDS

Vitamin E has a nutritional interrelationship with sulfur amino acids arising from a requirement for these acids in the synthesis of glutathione needed for glutathione peroxidase reactions. There is also evidence that glutathione can reduce the tocopheroxy radical (26). The intake of SAA exceeds requirements on all human diets that are adequate in protein but can be a determinant of the type of clinical pathology that develops in experimental animals in vitamin E and/or Se deficiency. For example, simple vitamin E deficiency in the growing chick leads to the development of encephephalomalacia, a disease marked by necrotic degeneration of the cerebellar cortex that is expressed in convulsive seizures. Removal of both vitamin E and Se from the diet results in the development of exudative diathesis, a condition marked by leakage of plasma from the capillaries. Muscular dystrophy, the condition that occurs in most animal species fed a vitamin E-deficient diet, appears in the chick only when fed a diet limiting in sulfur amino acids as well as in vitamin E and Se. Otherwise, the animals succumb to one of the other two deficiency diseases.

Most of the glutathione in the tissues is maintained in the reduced form by glutathione reductases that catalyze the recycling of oxidized glutathione by NADPH generated in the pentose shunt.

$$GSSG + NADPH^+ + H^+ \rightarrow 2GSH + NADP^+$$

VITAMIN C

The question of whether a metabolic interrelationship exists between vitamin C, a water-soluble antioxidant, and vitamin E, a fat-soluble antioxidant, has been the subject of extensive speculation. Ascorbic acid can reduce various metal ions, SH compounds, quinones, and free radicals. The best known involvement of vitamin C in the redox state of metals in vivo is the reduction of ferric iron in the diet, which results in an increased absorption of dietary iron. The capacity of ascorbic acid to reduce ferric iron, thereby generating $HO^{\cdot}$ radicals in the Fenton reaction, is exploited in studies on microsomal lipid peroxidation. Ascorbic acid is toxic to iron overload patients. On the other hand, ascorbate can act as an antioxidant because of its ability to scavenge O_2^- and $HO^{\cdot}$ radicals as well as singlet oxygen ($^1O_2$).

Despite strong evidence that vitamin C is capable of regenerating vitamin E from its tocopheroxy free-radical state under laboratory conditions (16–19), evidence for this reaction derived from animal experiments is uniformly negative. Administration of high levels of vitamin C to rats has little if any effect on the development of vitamin E deficiency, and conversely, vitamin E deficiency has no significant effect on vitamin C status. Behrens and Madere (27) concluded that "the interactions that readily take place in vitro between these two vitamins do not occur in vivo." Experiments by Burton et al. (28) on guinea pigs chronically labeled with deuterated α-tocopherol acetate showed that the turnover of vitamin E in the tissues was the same in animals fed diets containing 50, 250, or 5000 mg ascorbic acid per kg diet. The authors concluded that "the long postulated 'sparing' action of vitamin C on vitamin E, which is well documented in vitro, is of negligible importance in vivo." Liebler and coworkers (19) deduced that reduction of the tocopheroxy ion by ascorbic acid would require catalysis to proceed in biological membranes. Although NADH- and NADPH-dependent recycling of tocopherols by ubiquinones in mitochondria and microsomes has been reported by Kagan et al. (20), recycling was not dependent on vitamin C. There is no evidence that vitamin C-deficient animals require increased amounts of vitamin E. Administration of megadoses of vitamin C (1 g/day) to human adults fails to influence their urinary excretionn of malondialdehyde, an indicator of in vivo lipid peroxidation (Draper, unpublished). These findings indicate that vitamin C has neither a prooxidant nor an antioxidant effect on lipids in vivo. Further, it is unlikely that vitamin E status is dependent upon a nutrient whose intake from natural sources varies over nearly a 10-fold range.

β-CAROTENE

β-Carotene and vitamin A are cooxidized with PUFA in foods during cooking and storage, and hence they contribute significantly, although not extensively, to the requirement for a dietary antioxidant. The antioxidants BHA and BHT are effective stabilizers of these and other lipids in foods. β-Carotene can also act as an antioxidant by virtue of its ability to quench $^1O_2$, a nonradical initiator of lipid peroxidation. β-Carotene can be a significant source of lipid antioxidant activity in plasma (29) but it is not an essential nutrient, and followers of a carnivorous diet culture have survived without it for generations. There is evidence of an association between β-carotene intake and the incidence of some cancers, but it is possible that β-carotene serves as a marker for a diet culture high in fruits and vegetables, low in fat, and high in fiber that affords some protection against cancer or a generally healthier life-style on the part of persons who consume such a diet.

REFERENCES

1. Draper HH. Nutrient interrelationships. In: Machlin LJ, ed. Vitamin E. A comprehensive treatise. New York: Marcel Dekker, 1980; 272-88.
2. Draper HH. Lipid-vitamin-mineral interactions in the diet and in the tissues. In: Bodwell CE, Erdman JW, Jr, eds. Nutrient interactions. New York: Marcel Dekker, 1988; 187-203.
3. Draper HH. Nutritional modulation of oxygen radical pathology. Adv Nutr Res 1990; 8:119-45.
4. Machlin LJ. Vitamin E. In: Machlin LJ, ed. Handbook of vitamins. Nutritional, biochemical and clinical aspects. New York: Marcel Dekker, 1984; 99-145.
5. Diplock AT, ed. Vitamin E. Biochemistry and health implications. Ann N Y Acad Sci 1989; 570: 1-555.
6. Levy RD. Genetics of vitamin E content of corn grain. Ph.D. Thesis, University of Illinois at Urbana-Champaign, 1973; 1-95.
7. Jager JC. Linoleic acid intake and vitamin E requirement. Unilever Research, Vlaardmigen, The Netherlands, 1973.
8. Bieri JG, Evarts RP. Tocopherols and fatty acids in American diets. The recommended allowance for vitamin E. J Am Diet Assoc 1973; 62:147-51.
9. Thompson JN, Beare-Rogers JL, Erddy P, Smith PC. Appraisal of the human vitamin E requirement based on examination of individual and composite Canadian diets. Am J Clin Nutr 1973; 26:1349-54.
10. Food and Nutrition Board. Recommended dietary allowances, 8th ed. Washington, DC: US Natl Acad Sci Publ 2216, 1974.
11. Behrens BA, Madère R. Alpha- and gamma-tocopherol concentrations in human serum. J Am Coll Nutr 1986; 5:91-6.
12. Horwitt MK. Interrelationships between vitamin E and polyunsaturated fatty acids in adult men. Vitam Horm 1962; 20:541-58.
13. Wo CKW, Draper HH. Vitamin E status of Alaskan Eskimos. Am J Clin Nutr 1975; 28:808-13.
14. Lemoyne M, Von Gossum A, Kurian R, Ostro M, Axler J, Jeejeebhoy KN. Breath pentane analysis as an index of lipid peroxidation: a functional test of vitamin E status. Am J Clin Nutr 1987; 46:267-72.
15. Csallany AS, Draper HH. Oxidation products of α-tocopherol formed in autoxidizing methyl linoleate. Lipids 1970; 5:1-8.
16. Packer JE, Slater TF, Willson RL. Direct observation of a free radical interaction between vitamin E and vitamin C. Nature 1979; 278:737-8.
17. Niki E, Tsuchiya J, Tanimura R, Kamiya Y. Regeneration of vitamin E from α-chromanoxyl radical by glutathione and vitamin C. Chem Lett (Jpn) 1982; 789-92.
18. Scarpa M, Rigo A, Majorino M, Ursini F, Gregolin C. Formation of α-tocopherol by ascorbate during peroxidation of phosphatidylcholine liposomes. Biochim Biophys Acta 1984; 801:215-9.
19. Liebler DC, Kaysen KL, Kennedy TA. Redox cycles of vitamin E: hydrolysis and ascorbic acid dependent reduction of 8a-(alkyldioxy) tocopherones. Biochem 1989; 28:9772-7.

20. Kagan V, Serbinova E, Packer L. Antioxidant effects of ubiquinones in microsomes and mitochondria are mediated by tocopherol recycling. Biochem Biophys Res Commun 1990; 169:851-7.

21. Thompson CD, Robinson MF. Selenium in human health and disease with emphasis on those aspects peculiar to New Zealand. Am J Clin Nutr 1980; 33:303-23.

22. Young G, Chang J, Wen Z, et al. The role of selenium in Keshan disease. Adv Nutr Res 1984; 6: 203-31.

23. Levander OA, Sutherland B, Morris VC, King JC. Selenium balance in young men during selenium depletion and repletion. Am J Clin Nutr 1981; 34:2662-9.

24. Clark LC, Combs GF Jr. Selenium compounds and the prevention of cancer: research needs and public health implications. J Nutr 1986; 116:170-3.

25. Cararochi NC, England MD, O'Brien JF, et al. Superoxide generation during cardiopulmonary bypass: a role for vitamin E? J Surg Res 1988; 30:519-21.

26. Leedle RA, Aust SD. The effect of glutathione on the vitamin E requirement for inhibition of liver lipid peroxidation. Lipids 1990; 25:241-5.

27. Behrens WA, Madere R. Ascorbic and dehydroascorbic acids status in rats fed diets varying in vitamin E levels. Int J Vitam Nutr Res 1989; 59:360-54.

28. Burton GW, Wronska V, Stone L, Foster DO, Ingold KV. Biokinetics of dietary *RRR*-α-tocopherol in the male guinea pig at three dietary levels of vitamin C and two levels of vitamin E. Evidence that vitamin C does not "spare" vitamin E in vivo. Lipids 1990; 25:199-210.

29. Burton GW, Ingold KV. Beta carotene: an unusual type of lipid antioxidant. Science 1984; 244: 569-71.

5

Effects of β-Carotene and Related Carotenoids on Vitamin E

Shirley R. Blakely, Geraldine V. Mitchell, Mamie Leah Young Jenkins, and Erich Grundel

United States Food and Drug Administration, Washington, D.C.

INTRODUCTION

Vitamin E (α-tocopherol) and β-carotene and related carotenoids are important nutritional components in the human diet. The two classes of compounds share many similarities, including occurrence in plant foods, lipid solubility, deposition in fatty portions of cells, such as membranes, and antioxidant properties. β-carotene and related carotenoids have the capacity to quench singlet oxygen and reduce the concentration of chain-carrying peroxy radicals at low partial pressures of oxygen, a property that complements the chain-breaking action of vitamin E at high oxygen concentrations (1).

The focus of this chapter is on studies in animals and humans that involve interactions of β-carotene and related carotenoids with vitamin E. β-Carotene, because of its conversion to vitamin A, may exhibit an indirect effect on vitamin E by enhancing the level of vitamin A in the system. These and other factors that influence the effects of carotenoids on vitamin E are discussed.

β-CAROTENE AND RELATED CAROTENOIDS

The antioxidant properties of β-carotene make it a prime test compound in the chemoprevention of cancer (2). β-Carotene has been reported to be inversely associated with lung (3) and breast (4) cancer in humans. It has also been found to have chemopreventive properties in a number of animal tumor models (5).

Adaptation to β-carotene in the diet of rodents occurs rapidly; both the uptake and the rate of conversion to vitamin A increase (6). Rapid conversion of β-carotene to vitamin A can confound the results of studies of the effects of this carotenoid. Additionally, the bioavailability of β-carotene in humans varies widely (7). Therefore, to observe increases in plasma and tissue levels of β-carotene, relatively large doses must be administered.

Canthaxanthin (4,4′-diketo-β-carotene) and apocarotenal (β-apo-8′-carotenal) are two carotenoids related to β-carotene that are approved for use in the human diet. Because of its lack of vitamin A activity, canthaxanthin is frequently used in animal studies to observe a carotenoid effect. It has been reported to be as effective as β-carotene in quenching singlet oxygen and trapping free radicals (8). With at least nine conjugated double bonds in their chemical structures, both canthaxanthin and apocarotenal meet the criterion for carotenoid efficacy as antioxidants (8).

TABLE 1 Plasma and Liver β-Carotene and Canthaxanthin from Adult Rats Fed β-Carotene or Canthaxanthin for 8 Weeks[a]

Diet groups	Level (%)	Plasma (μmol/L)	Liver (nmol/g)
β-Carotene	0.048	0.057 ± 0.004[2]	147 ± 18[1]
β-Carotene	0.200	0.118 ± 0.021[2]	197 ± 13[2]
Canthaxanthin	0.048	0.30 ± 0.02[3]	589 ± 55[3]
Canthaxanthin	0.200	0.35 ± 0.03[3]	1304 ± 123[4]

[a]Values are means $\pm$ SEM (standard error of the mean). $N = 10$. For the plasma or lipid column, means with a common superscript number are not significantly different at $P < 0.05$.

When β-carotene and canthaxanthin were fed to rats at either 0.048 or 0.2% of the diet for 8 weeks (Table 1), the subsequent rise in plasma and tissue levels of these compounds was three- to sixfold greater for canthaxanthin than for β-carotene. A dose-related rise in plasma β-carotene and liver canthaxanthin occurred in which levels in the high-treatment groups were more than 50% higher than those in the low-treatment groups. Therefore, the rat, which is generally considered a poor animal model for β-carotene studies, becomes an improved model when canthaxanthin is the test carotenoid.

EFFECT OF β-CAROTENE AND RELATED CAROTENOIDS ON VITAMIN E

Animal Studies

The administration of β-carotene (0.048%, wt/wt) to weanling rats for 8 weeks (9) resulted in a significant reduction in liver vitamin E, but plasma vitamin E was reduced only when β-carotene and excess vitamin A (400,000 IU/kg diet) were administered in combination (Table 2). Bendich and Shapiro (10) reported that feeding either β-carotene or canthaxanthin to rats for 66 weeks led to reductions in plasma vitamin E of approximately 47 and 55% in β-carotene and canthaxanthin groups, respectively. In this study, vitamin E was administered at 100 mg/kg diet and each of the carotenoids was administered at 2000 mg/kg, a ratio of vitamin E to carotenoid of 1:20.

TABLE 2 Liver and Plasma α-Tocopherol Levels in Young Rats Fed β-Carotene and Vitamin A Palmitate (VA) for 8 Weeks[a]

Diet	Level (%)	Plasma (μmol/L)	Liver[b] (nmol/g)
Control	0	10 ± 1.7[1]	93 ± 16[1]
Vitamin A_1	0.002	10 ± 1.7[1]	58 ± 9[2]
Vitamin A_2	0.022	7 ± 1.0[1]	60 ± 5[2]
β-Carotene	0.048	7 ± 1.0[1]	63 ± 7[2]
β-Carotene + VA_1	Same	5 ± 1.3[2]	53 ± 7[2]
β-Carotene + VA_2	Same	2 ± 0.8[2]	60 ± 9[2]

[a]Values are means $\pm$ SEM. $N = 7$–10. For each column, means with a common superscript number are not significantly different at $P < 0.05$.
[b]Significant vitamin A $\times$ β-carotene interaction for liver α-tocopherol levels.

TABLE 3 Plasma and Liver Tocopherol from Adult Rats Fed Canthaxanthin and Vitamin A Palmitate (VA) for 8 Weeks[a]

Diet groups	Level (%)	Plasma (μmol/L)	Liver[b] (nmol/g)
Control	0	22.6 ± 2.0[1]	309 ± 33[1]
Vitamin A	0.022	20.9 ± 1.9[1]	309 ± 16[1]
Canthaxanthin 1	0.048	22.8 ± 0.9[1]	244 ± 12[2]
Canthaxanthin 2	0.2	20.4 ± 1.3[1]	211 ± 14[2]
Canthaxanthin 1 + VA	Same	16.3 ± 1.0[2]	302 ± 16[1]
Canthaxanthin 2 + VA	Same	14.8 ± 1.0[2]	265 ± 19[1,2]

[a]Values are means ± SEM. $N = 10$. For each column, means with a common superscript number are not significantly different at $P < 0.05$.

In adult retired breeder rats fed two different levels of canthaxanthin (0.048 and 0.2% wt/wt) for 8 weeks with or without vitamin A, liver vitamin E levels were significantly reduced (Table 3). As observed in the study reported in Table 2, plasma vitamin E levels were reduced only in groups fed canthaxanthin plus excess vitamin A. The results of a study in which either 0.1 or 0.2% canthaxanthin or apocarotenal (with some provitamin A activity) was administered to a choline-deficient animal model for 12 weeks are shown in Table 4. Reductions in liver vitamin E levels of approximately 67 and 78% were observed in apocarotenal- and canthaxanthin-treated groups, respectively. Although there were significant liver stores of these antioxidant carotenoids, they appeared to have no effect on liver lipid peroxidation compared to the choline-deficient control group, which may have been a consequence of the reductions in liver vitamin E caused by these treatments.

Influence of Lipids

β-Carotene has greater bioavailability when administered with fat (7). Alam et al. (11) demonstrated that β-carotene concentrations in blood and tissues of rats are related to the degree of saturation of the fat in the diet. They reported that rats had higher liver β-carotene concentrations when fed a diet containing 0.2% β-carotene with 10% corn oil (11), safflower oil, or olive oil (12) than with 10% lard or coconut oil.

TABLE 4 Plasma and Liver Tocopherol from Choline-Deficient (CD) F344 Rats Fed Canthaxanthin (CX) or Apocarotenal (AC) for 12 Weeks[a]

Diet groups	Level (%)	Plasma (μmol/L)	Liver (μmol/g)	Liver (μmol/g fat)
CS[b]	0	50.9 ± 2.7[1]	0.51 ± 0.048[1,4]	13.0 ± 1.40[1]
CD	0	18.3 ± 1.5[2]	1.94 ± 0.120[2]	10.3 ± 0.69[2]
CD + CX	0.1	31.6 ± 2.5[3]	0.46 ± 0.026[1,4]	2.6 ± 0.16[3]
CD + CX	0.2	31.8 ± 1.9[3]	0.36 ± 0.034[1]	2.0 ± 0.17[3]
CD + AC	0.1	23.9 ± 1.4[2]	0.70 ± 0.044[3]	3.7 ± 0.28[3]
CD + AC	0.2	29.5 ± 3.0[3]	0.61 ± 0.054[3]	2.6 ± 0.22[3]

[a]Values are means ± SEM. $N = 10$. For each column, means with a common superscript number are not significantly different at $P < 0.05$.
[b]Choline sufficient.

Vitamin E levels in plasma were lower in rats fed β-carotene with 16% corn oil than in rats fed β-carotene with 5% corn oil in a recent 8 week experiment conducted in our laboratory (Table 5). Although plasma vitamin E decreased in the high corn oil group, the most dramatic decrease was observed when a combination of high corn oil, β-carotene, and vitamin A was administered. As with the studies by Alam et al. (11), liver β-carotene increased in the high corn oil group.

Influence of Vitamin A

Some of the first reports of antagonism toward vitamin E concerned vitamin A. Studies have shown that excess vitamin A decreases the uptake and deposition of vitamin E (13) as well as carotenoids (14). Therefore, in addition to the separate effect of carotenoids on vitamin E, β-carotene may have an indirect effect on vitamin E in rodents through its capacity to increase vitamin A in the serum and liver, thereby promoting the antagonism of vitamin A toward vitamin E (9).

Human Studies

Because of its potential to prevent the development of cancer at certain sites (3–5,15), β-carotene is being used as the test compound in several ongoing chemoprevention and intervention trials (2). In one such trial, data have been obtained on the effects of β-carotene on serum vitamin E. Levels of α-tocopherol were reduced 40% when β-carotene was administered daily to normal subjects for 8 months (16). In another clinical intervention trial in which 15 mg β-carotene and 25,000 IU vitamin A are administered daily, vitamin E and β-carotene are among the substances being monitored in the blood of subjects (17). Findings from this study will be of interest in view of the study with humans and animal studies in which β-carotene and vitamin A in combination antagonized vitamin E.

TABLE 5   Plasma α-Tocopherol and Plasma and Liver β-Carotene Levels in Rats Fed Corn Oil or Mixed Fat, 0.048% β-Carotene, and 0.022% Vitamin A Palmitate for 8 Weeks[a]

Diets[b]	Fat level (%)	α-Tocopherol (μmol/L)	Plasma β-carotene (μmol/L)	Liver β-carotene (μmol/g)
Corn oil				
Control	5	35 ± 5.3[1]	0.87 ± 0.13[1]	4.38 ± 0.35[1]
Control	16	22 ± 1.9[2]	0.87 ± 0.17[1]	9.83 ± 1.06[2]
Vitamin A	5	14 ± 1.2[3]	0.58 ± 0.06[2]	2.34 ± 0.18[3]
Vitamin A	16	12 ± 1.5[3]	0.49 ± 0.08[2]	4.28 ± 0.29[1]
Mixed fat[b]				
Control	5	29 ± 2.1[1]	1.03 ± 0.13[1]	4.35 ± 0.29[1]
Control	16	30 ± 2.1[1]	1.21 ± 0.17[1]	6.43 ± 0.69[1,2]
Vitamin A	5	21 ± 2.3[2]	0.46 ± 0.05[2]	2.16 ± 0.14[3]
Vitamin A	16	19 ± 1.9[2]	0.60 ± 0.13[2]	4.30 ± 1.14[1]

[a]Means ± SEM. N = 10. For each column, means with a common superscript number are not significantly different at $P < 0.05$.
[b]Corn oil is polyunsaturated fat. Mixed fat consisted of lard, beef tallow, hydrogenated coconut oil, and corn oil (mostly saturated, P/S ratio 0.26). Significant fat level × fat type × vitamin A interaction for plasma α-tocopherol and liver β-carotene.

Evidence also indicates that excess vitamin E decreases the levels of β-carotene in the serum of humans (18) and the plasma and liver of rats (12,19). In the human study (18), 30 mg β-carotene and 728 mg vitamin E were administered daily, a ratio of β-carotene to vitamin E of 1:24. In a clinical trial of smokers for the chemoprevention of lung cancer underway in Finland (20), the combined effects of β-carotene and vitamin E (at 20 and 50 mg/day, respectively), β-carotene alone (20 mg/day), or vitamin E alone (50 mg/day) on the development of lung cancer in 29,000 subjects are being examined. Results of the pilot study suggest a slight, nonsignificant reduction of serum β-carotene in the group receiving the combination capsule (20). The information obtained from these trials will be useful in determining the long-term effects on the bioavailability and efficacy of β-carotene administered singly or in combination with vitamin E.

CONCLUSIONS

β-Carotene and vitamin E are nutrient antioxidants that have been shown to have mutual and complementary effects in biological systems. Experiments in which excessive amounts of β-carotene and related carotenoids are administered can lead to reduced vitamin E levels, especially when administered on a long-term basis. Increasing tissue β-carotene at the expense of vitamin E may compromise the overall protection afforded to the cell by dietary antioxidants.

The concept of the "total antioxidant pool" has recently been raised in an editorial about the lack of positive results of a clinical trial in which β-carotene was the sole test compound administered (21). In view of the studies showing competition between β-carotene and vitamin E, experimental designs of future clinical trials and of trials using animal tumor models may focus on achieving a higher "total antioxidant pool" rather than increasing a single antioxidant.

REFERENCES

1. Burton GW, Ingold KU. β-carotene: an unusual type of lipid antioxidant. Science 1984; 224:569-73.
2. Boone CW, Kelloff GJ, Malone WE. Identification of candidate cancer chemopreventive agents and their evaluation in animal models and human clinical trials: a review. Cancer Res 1990; 50:2-9.
3. Menkes MS, Comstock GW, Vuilleumier JP, Helsing KJ, Rider AA, Brookmeyer R. Serum beta-carotene, vitamins A and E, selenium, and the risk of lung cancer. N Engl J Med 1986; 315:1250-4.
4. Wald NJ, Boreham J, Hayward JL, Bulbrook RD. Plasma retinol, beta-carotene, and vitamin E in relation to the future risk of breast cancer. Br J Cancer 1984; 49:321-4.
5. Santamaria L, Bianchi A, Arnaboldi A, et al. Chemoprevention of indirect and direct chemical carcinogenesis by carotenoids and oxygen radical quenchers. Ann N Y Acad Sci 1988; 534:584-96.
6. Lakshman MR, Asher KA, Attlesey MG, Satchithanandam S, Mychkovsky I, Coutlakis PJ. Absorption, storage, and distribution of beta-carotene in normal and beta-carotene fed rats: roles of parenchymal and stellate cells. J Lipid Res 1989; 30:1545-50.
7. Dimitrov NV, Meyer C, Ullrey DE, et al. Bioavailability of β-carotene in humans. Am J Clin Nutr 1988; 48:298-304.
8. Krinsky NI. Membrane antioxidants. Ann N Y Acad Sci 1988; 551:17-33.
9. Blakely SR, Grundel E, Jenkins MY, Mitchell GV. Alterations in beta-carotene and vitamin E status in rats fed beta-carotene and excess vitamin A. Nutr Res 1990; 10:1035-44.
10. Bendich A, Shapiro SS. Effect of β-carotene and canthaxanthin on the immune responses of the rat. J Nutr 1986; 2254-62.
11. Alam BS, Alam SQ, Bendich A, Shapiro SS. Effect of dietary lipids on hepatic and plasma β-carotene and vitamin A levels in rats fed β-carotene. Nutr Cancer 1988; 11:233-41.
12. Alam BS, Brown LR, Alam SQ. Influence of dietary fats and vitamin E on plasma and hepatic vitamin A and β-carotene levels in rats fed excess β-carotene. Nutr Cancer 1990; 14:111-6.

13. Combs GF Jr. Differential effects of high dietary levels of vitamin A on the vitamin E-selenium nutrition of young and adult chickens. J Nutr 1976; 106:967-75.

14. Dua PN, Day EJ, Tipton HC, Hill JE. Influence of dietary vitamin A on carotenoid utilization, nitrogen retention and energy utilization by the chick. J Nutr 1966; 90:117-22.

15. Zeigler RG. A review of epidemiologic evidence that carotenoids reduce the risk of cancer. J Nutr 1989; 119:116-22.

16. Xu MJ, Peng YM, Liu Y, Alberts DC, Plezia PM, Sayers SM. Effects of chronic oral administration of β-carotene on plasma α-tocopherol concentrations in normal subjects. Proc Am Assoc Cancer Res 1990; 31:126.

17. Omenn GS, Goodman G, Rosenstock L, et al. Cancer chemoprevention with vitamin A and beta-carotene in populations at high-risk for lung cancer. In: Cerutti PA, Nygaard OF, Simic MG, eds. Anticarcinogenesis and radiation protection. New York: Plenum Press, 1987; 279-83.

18. Willett WC, Stampfer MJ, Underwood BA, Taylor JO, Hennnekins CH. Vitamins A, E, and carotene: effects of supplementation on their plasma levels. Am J Clin Nutr 1983; 38:559-66.

19. Arnrich L, Arthur VA. Interactions of fat-soluble vitamins in hypervitaminoses. Ann N Y Acad Sci 1980; 355:109-18.

20. Albanes D, Virtamo J, Rautalahti M, et al. Pilot study: the U.S.-Finland Lung Cancer Prevention Trial. J Nutr Growth Cancer 1986; 3:207-14.

21. Meyskens FL. Coming of age—the chemoprevention of cancer. N Engl J Med 1990; 323:825-8.

II
CHEMISTRY, BIOCHEMISTRY, AND CELL BIOLOGY

A. Chemical Reactions, Antioxidant Action, and Biomembranes

6

Pulse Radiolysis as a Tool in Vitamin E Research

Wolf Bors, Manfred Saran, and Christa Michel

Institut für Strahlenbiologie, GSF-Forschungszentrum für Umwelt und Gesundheit, Neuherberg, Germany

INTRODUCTION

The action of vitamin E as antioxidant depends on its ability to break radical-propagated chain reactions (1,2). Thus, the odd-electron derivative of vitamin E, the phenoxyl radical, is an intrinsic intermediate in any anti-oxidative process. Its formation has repeatedly been demonstrated using electron spin resonance (ESR) spectroscopy, mainly in organic solvents (3–9), but also in micellar systems (10,11) and in microsomal and mitochondrial preparations (11, 12). It is therefore of interest to study this phenoxyl radical under controlled conditions and to determine its properties in detail. Although quite a variety of chemical, biochemical, and enzymatic processes exist for the generation of radicals, only radiolytic processes are specific enough to ensure selective generation methods of individual radical species. As opposed to steady-state radiolysis, the rapid technique of pulse radiolysis furthermore allows the direct monitoring of build-up and decay of radicals.

In the following, the principles of pulse radiolysis are outlined briefly before examples of their application to vitamin E research are discussed.

PRINCIPLES OF PULSE RADIOLYSIS

Extensive reviews of radiation chemistry and ancillary techniques are readily available (13–15). However, a short introduction to the basic chemistry and instrumentation is necessary to understand the subsequent application to vitamin E and its model compounds.

Chemistry

Radiation chemistry of aqueous solutions is the most thoroughly studied and the most versatile method to generate different types of radicals. Vitamin E, however, is insoluble in water and can therefore only be examined in organic solvents or micellar systems. To study the behavior of the phenoxyl radical itself, for example, its formation by various types of radicals as an expression of the antioxidative potential of the chromanol moiety, model compounds lacking the phytyl side chain must be investigated in aqueous solution.

Radiolysis of Water

The initial deposition of radiation energy in aqueous solution leads to ionization and excitation of the bulk water. Within several nanoseconds, these physical processes give rise to the so-called primary radicals: hydrated electrons (e_{aq}^-), hydroxyl radicals ($\cdot OH$), and hydrogen atoms ($\cdot H$), as well as to some dissolved hydrogen (H_2) and hydrogen peroxide (H_2O_2) (16–18):

$$H_2O \rightarrow E_{aq}^-, \ \cdot OH, \ \cdot H, \ H_2O_2, \ H_2 \tag{1}$$

The radiolytic yields (*G* values) of the radicals, that is, the number of molecules produced per 100 eV absorbed energy, in neat water and with sparsely ionizing radiation, such as x-rays, γ-rays, or electron beams (β-irradiation), are 2.7 (e_{aq}^-), 2.7 ($\cdot OH$), and 0.6 ($\cdot H$) (or 0.28, 0.28, and 0.06 in the new SI dimension of $\mu mol/J$); thus about 45% each of e_{7aq}^- and $\cdot OH$ and 10% of $\cdot H$ are formed (17).

It is important to recall that $\cdot OH$ radicals are the most strongly oxidizing chemical species, whereas e_{aq}^- (and less so $H\cdot$) are reducing. To simplify studies of radical reactions, it is necessary to limit the radical spectrum to either oxidizing or reducing species (the lower reactivity and low concentration of $H\cdot$ atoms in most cases allows us to neglect their presence). Conversion reactions are also essential to generate various other radicals. In the following, only the most important of these are listed (for a more comprehensive review, see Ref. 19).

Oxidizing Radicals.   Hydroxyl radicals alone (and at a doubled yield) are obtained when e_{aq}^- is converted to $\cdot OH$ by N_2O:

$$e_{aq}^- + N_2O + H_2O \rightarrow \cdot OH + OH^- + N_2 \tag{2}$$

Adding inorganic salts, whose anions readily scavenge $\cdot OH$ radicals, results in the formation of other electrophilic radicals, which are less reactive and thus more selective than $\cdot OH$ radicals:

$$\cdot OH + X^- \rightarrow X \cdot + OH^- \tag{3}$$

Examples for $X \cdot$ are $\cdot N_3$ (20) or $SO_x^{\overline{\cdot}}$ (21) radicals, whereas in the case of (pseudo)halides an initially formed $\cdot OH$ adduct,

$$\cdot OH + Y^- \rightarrow [YOH]^{\overline{\cdot}} \tag{4}$$

reacts immediately with excess anion to form the dimer radical (22),

$$YOH^- + Y^- \rightarrow Y_2^{\overline{\cdot}} + OH^- \tag{5}$$

Alkoxyl radicals, as organic equivalents of $\cdot OH$, can best be generated by reduction of organic hydroperoxides with e_{aq}^- (23):

$$e_{aq}^- + ROOH \rightarrow RO \cdot + OH^- \tag{6}$$

Peroxyl radicals are preferentially produced in solutions saturated with a mixture of N_2O and O_2 and containing organic solvents (24): Equation (2) is followed by

$$\cdot OH + RH \rightarrow R \cdot + H_2O \tag{7}$$

and

$$R \cdot + O_2 \rightarrow ROO \cdot \tag{8}$$

This latter reaction is generally diffusion controlled (24) owing to the biradical character of ground-state oxygen ($^3O_2$).

The inorganic equivalent of $ROO \cdot$, the hydroperoxyl radical ($HO_2 \cdot$) or its conjugate base, the superoxide anion ($O_2^{\overline{\cdot}}$), is formed in oxygen-saturated solutions containing sodium formate (25):

$$e_{aq}^- + O_2 \rightarrow O_2^{\bar{\cdot}} \tag{9}$$

$$H^{\cdot} + O_2 \rightarrow HO_2^{\cdot} \rightleftharpoons O_2^{\bar{\cdot}} + H^+ \tag{10}$$

$$^{\cdot}OH + HCOO^- \rightarrow CO_2^{\bar{\cdot}} + H_2O \tag{11}$$

$$CO_2^{\bar{\cdot}} + O_2 \rightarrow O_2^{\bar{\cdot}} + CO_2 \tag{12}$$

In these solutions, therefore, all three primary radicals are rapidly converted into one single radical species. A simply oxygenated solution contains both $O_2^{\bar{\cdot}}$ from Equations (9) and (10) and $^{\cdot}OH$ radicals. According to Equations (7) + (8), this would convert any organic solute into peroxyl radicals that absorb in the same wavelength region as $O_2^{\bar{\cdot}}$ itself (26) but may exhibit different reactivities.

Reducing Radicals. To study e_{aq}^- alone, $^{\cdot}OH$ is preferentially scavenged by *tert*-butanol, the alkyl radical formed being rather unreactive:

$$^{\cdot}OH + CH_3\!-\!\underset{\underset{\displaystyle CH_3}{|}}{\overset{\overset{\displaystyle CH_3}{|}}{C}}\!-\!OH \rightarrow {}^{\cdot}CH_2\!-\!\underset{\underset{\displaystyle CH_3}{|}}{\overset{\overset{\displaystyle CH_3}{|}}{C}}\!-\!OH + H_2O \tag{13}$$

The strongly reducing but more selectively reacting formate radical, $CO_2^{\bar{\cdot}}$, is formed via Equation (11). In the presence of O_2 it is only an intermediate, however, and formate-containing solutions must be purged with N_2O [Eq. (2)] to obtain $CO_2^{\bar{\cdot}}$ exclusively.

Radiolysis of Organic Solvents

As opposed to aqueous solutions, this area has been investigated less thoroughly (27–29). Basically, the same physical processes of ionization and excitation take place. However, because of the properties of the different solvents (including bond energies, polarity, and viscosity) the yield of *solvated* electrons varies considerably and no hydroxyl radicals are formed. Instead, more or less specific organic radicals or radical ions are produced, which, as alkyl radicals, are also reducing species. Radical cations (R^+) may be considered electron holes and radical anions ($R^{\bar{\cdot}}$) electron traps, in analogy to radiation effects in solid material (30).

Instrumentation for Pulse Radiolysis Studies

Two principal devices are necessary to run pulse radiolysis experiments: (1) an accelerator to generate radiation pulses of sufficient energy (to penetrate the cuvette wall and the solvent layer homogeneously) (31), and (2) monitoring devices to record radical buildup and decay. Ancillary devices for solvent delivery, dose reduction, and computer control (32) may vary from laboratory to laboratory and are not discussed here. Extensive documentation exists in the Radiation Chemistry Data Center at the Radiation Laboratory, Notre Dame University (33,34), with continuous updates of lists of rate constants (35) and other pertinent information (36).

The different types of accelerators and their specifications have been listed before (37, 38). In Table 1 the different monitoring techniques are compiled, of which optical spectroscopy is by far the most important and most commonly used technique. The kinetic behavior of a radical can easily be monitored at a single wavelength, but different instruments have been developed to rapidly record the full spectra of unstable radical intermediates:

1. Computer-controlled monochromators, which automatically record spectra by stepwise advance of the wavelength setting and subsequent triggering of individual pulses (38).

TABLE 1 Monitoring Devices for Pulse Radiolysis Studies[a]

Method	Type	Instrument	References
Spectroscopy	Optical absorption	Monochromator/PM	39
		Spectrograph/PDA	40
		Spectrograph/PMA	41
	Fluorescence spectroscopy		42
	Raleigh (light) scattering		43
	Resonance raman spectroscopy		38
Magnetic resonance	ESR	Time-resolved CIDNP	38, 44
	NMR		38
Electrochemistry	Conductivity		38, 45
	Polarography		38, 46

[a]PM, photomultiplier; PDA, photodiode array; PMA, photomultiplier array; ESR, electron spin resonance; NMR, nuclear magnetic resonance; CIDNP, chemically induced dynamic nuclear polarization.

2. Smooth spectrograph recordings of broad-wavelength regions can be done with diode arrays (40), which unfortunately lack a high sensitivity in the ultraviolet (UV) region.
3. Taking advantage of the high UV sensitivity of photomultipliers, we developed a spectrograph-based optical recording system that obtains spectra between 200 and 800 nm at a spectral resolution of 9 nm and, at present, a time resolution of 500 ns (41).

PULSE RADIOLYSIS OF VITAMIN E MODEL COMPOUNDS

Model compounds of vitamin E that lack the phytyl side chain are sufficiently water soluble to enable scavenging studies with different types of radicals. Since the 6-hydroxy group of the chromane ring in each case forms the phenoxyl radical, probably not too much information is lost if one compares the properties of the model phenoxyl radicals with that of the actual vitamin E.

In this respect, the two most commonly used substances are Trolox c® (2-carboxyl-2.5.7.8-tetramethyl-6-hydroxy-chromane) and PMC (2.2.5.7.8-pentamethyl-6-hydroxy-chromane).

It is evident from the methyl substitution pattern in Scheme 1 that both substances are modeled after α-tocopherol; no pulse radiolysis studies of the β-, γ- and δ-tocopherols or their model compounds have ever been performed.

Transient Spectra

Spectra of the phenoxyl radical of Trolox c have been recorded repeatedly after generation by ˙OH (47,48), bromide (Br_2^{-}) (48–51), or azide (˙N_3) radicals (49,52). Figure 1 reveals that there is very little difference in the spectrum of the phenoxyl radical of either Trolox c or PMC

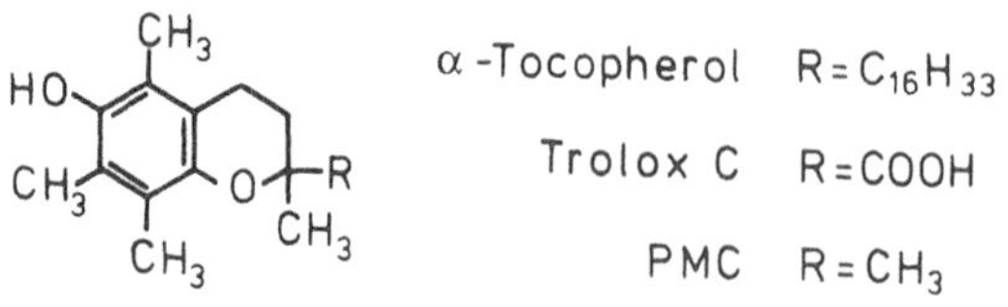

α-Tocopherol R = $C_{16}H_{33}$

Trolox C R = COOH

PMC R = CH_3

SCHEME 1

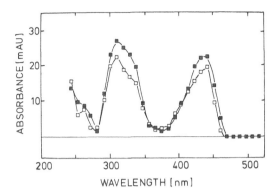

FIGURE 1 Transient spectra of phenoxyl radicals of vitamin E model compounds. N_2O-saturated solutions containing 10 mM NaN_3, pH 8.3–8.4 (unbuffered). Dose-normalized spectra observed 35 μs after the pulse: (solid squares) Trolox c, concentration 50 μM, average pulse dose 10.6 Gy; (open squares) pentamethylchroman, concentration 100 μM, pulse dose 12.0 Gy.

generated by $^\cdot N_3$ radicals or of the Trolox carbinol after $Br_2^{\cdot-}$ attack (51). Incidentally, only the spectra shown in Figure 1 and those by Bielski and coworkers (50,51) extend below 300 nm, whereas all others are more or less truncated with cutoffs above 300 nm (for PMC this is the first recorded radical, obtained after enhancing the solubility of PMC in 200 mM acetonitrile) (52).

Kinetic Data

These limited spectral data make it difficult to compare kinetic data, in particular since we originally observed different kinetic behavior of the three absorption bands (47). In Table 2 the known rate constants for the generation of $PheO^\cdot$ are collated; they demonstrate that Trolox c is indeed a very effective radical scavenger [for PMC only one rate constant with $^\cdot N_3$ is known, $(1.2 \times 10^9 \ M^{-1} \ s^{-1})$ (52), which is somewhat lower than that for Trolox c]. We have at present no explanation for the considerably lower values for $k_{\cdot OH}$ and $k_{\cdot N3}$ determined recently in the most exhaustive study on scavenging rate constants for Trolox c (48). As indicated in Table 2, the pH also has a strong influence, with some radicals only reacting with the dissociated phenolate (pK 11.7 or 11.9, Ref. 48 or 55).

The low univalent oxidation potential of Trolox c of 190 mV (55) enables this substance to be oxidized not only by strongly electrophilic inorganic and peroxyl radicals (see Table 2) but also by organic radicals (Table 3). In the case of the amino acid radicals, this was considered a "repair" process (49), whereas drug radicals were thought to act as prooxidants (56,57).

To distinguish from this repair reaction, we propose the term "annealing" for the reduction of the Trolox c (or α-tocopherol) phenoxyl radical back to the original phenol. In fact, a very important aspect of the antioxidative capacity of vitamin E is the stability of its phenoxyl radical, which does not react with O_2 (58) but is readily reduced by ascorbate (59, 60). The bimolecular decay rates of the phenoxyl radicals of the model substances are similarly slow to allow the determination of such annealing rate constants with reducing substances (Table 4). In the case of Trolox c we see a clear distinction between rapid reduction by the radicals ($O_2^{\cdot-}$ and the disulfide radical anions of dihydrolipoic acid and glutathione), intermediate values for ascorbate (and NADH), and complex kinetic behavior with the two thiols (52).

TABLE 2 Scavenging Rate Constants of Trolox C

Radical	pH	Rate constant $(M^{-1} s^{-1})$	Reference
Reactant PheOH			
$\cdot$OH	8.2	3.5×10^{10}	47
	7.0	2.5×10^{10}	49
	7.0	2.2×10^{9}	48[a]
$\cdot N_3$	7.0	5.0×10^{8}	48
	8.3	3.0×10^{9}	52
	10.0	3.6×10^{9}	52
$Br_2^{\cdot -}$	0.2–12	6.0×10^{8}	51
	7.0	3.8×10^{8}	49
	10.0	4.3×10^{8}	48
$(SCN)_2^{\cdot -}$	6.5	2.2×10^{8}	48
$SO_5^{\cdot -}$	9.0	1.2×10^{7}	53
$HO_2^{\cdot}$	2.0	2.0×10^{5}	50
$O_2^{\cdot -}$	10.0	1.0×10^{-1}	50
t-BuO$\cdot$	8.5	1.1×10^{9}	54
$CCl_3OO\cdot$	7.0	3.7×10^{8}	48
$CHCl_2OO\cdot$	7.0	1.1×10^{8}	48
$CH_2ClOO\cdot$	7.0	1.6×10^{7}	48
Reactant PheO$^-$			
$\cdot$OH	11.0	3.8×10^{9}	48[a]
$NO_2^{\cdot}$	11.5	5.0×10^{8}	49
$SO_3^{\cdot -}$	11.6	1.5×10^{8}	53
$\cdot CH_2$-CHO	13.5	1.8×10^{9}	55

[a]Value of $6.8 \times 10^9 \, M^{-1} \, s^{-1}$ was obtained for both pH 7.0 and 11.0 by competition kinetics (48).

TABLE 3 Repair of Organic Radicals by Trolox C

Substrate	pH	Rate constant $(M^{-1} s^{-1})$	Reference
Phenoxyl radicals			
Phenol	7.0	3.0×10^{8}	48
Salicylate	7.0	4.1×10^{8}	48
m-Cresol	7.0	2.8×10^{8}	48
p-Cresol	7.0	9.5×10^{7}	48
o-Cresol	7.0	$<10^{5}$	48
Tyrosine	7.0	3.2×10^{8}	48
	7.0	3.8×10^{8}	49
Acetaminophen	7.0	1.4×10^{8}	56
Different types of radicals			
Tryptophan	1.65	1.8×10^{9}	49
	7.0	5.2×10^{7}	49
Histidine	7.2	8.0×10^{8}	49
Methionine	7.0	7.1×10^{8}	49
Cysteine	7.0	1.0×10^{8}	48
	12.0	8.0×10^{8}	48
Aminopyrine	7.0	8.0×10^{8}	57
	13.0	6.1×10^{8}	57
Antipyrine	7.0	2.1×10^{9}	57

TABLE 4 Rate Constants for "Annealing" of Phenoxyl Radical of Vitamin E Model Compounds

PheO·	Reductant	Rate constant $(M^{-1} s^{-1})$	Reference
Trolox c	GSSG^{-a}	1.3×10^9	52
	LS$_2$$^{-b}$	2.5×10^8	52
	O$_2$$^-$	4.5×10^8	61
	Ascorbate	1.2×10^7	52
		8.3×10^6	48
	NADH	$<10^6$	62
	L(SH)$_2$^{c}	2.7×10^3	52
PMC	GSSG$^-$	3.7×10^7	52
	LS$_2$$^-$	4.7×10^7	52
	Ascorbate	1.9×10^7	52
	L(SH)$_2$)	1.3×10^4	52

[a]Disulfide radical anion of GSH.
[b]Cyclic disulfide radical anion of lipoic acid.
[c]Dihydrolipoic acid (dithiol).

Product Analysis

A detailed product analysis after pulse radiolytic oxidation to the phenoxyl radical has only been carried out for Trolox c (51), but several product studies exist for PMC after oxidation with chemically produced O$_2$$^-$ (63,64) or t-BuOOH/t-BuOO· (65,66). We have considerable doubt that the products identified after O$_2$$^-$ attack are specific for this radical, which actually reacts very slowly with Trolox c (50).

Also, the conjecture (51) should be qualified that, in view of the influence of the carboxyl group on the product pattern, Trolox c would not be an ideal model for vitamin E. This is certainly true for product studies, but the few rate constants known for PMC reveal that there are only small differences in the kinetic properties leading to the formation and decay of the phenoxyl radicals of either compound.

PULSE RADIOLYSIS OF VITAMIN E

As mentioned already, the insolubility of vitamin E in water allows pulse radiolysis studies only in organic solvents. In cyclohexane, where the cyclohexane peroxyl radical is the oxidizing species (67), the absorption maxima at 340 and 420 nm lie close to those depicted in Figure 1 for the model compounds in aqueous solution (310 and 440 nm). A slight shift to 425 nm is observed in ethanol (68) or in tetradecyltrimethylammonium bromide micelles (69).

More important than these spectroscopic and product analysis studies are the kinetic data obtained for vitamin E and its phenoxyl radical, respectively. The rate constant for the annealing of the vitamin E radical by ascorbate of 1.6×10^6 M^{-1}s^{-1} (59) compares with 8.3×10^6 or 1.2×10^7 M^{-1}s^{-1} for Trolox c (see Table 3). Scavenging rates for vitamin E, obtained by pulse radiolysis in a variety of solvents, are listed in Table 5.

Two rate constants stand out as far below average. The low rate constant for methyl linoleate peroxyl radicals may not be too unexpected since it was obtained in a micellar system (8.0×10^4 M^{-1}s^{-1}) (71); see, however, the high values for the inorganic radicals ·N$_3$ and Br$_2$$^-$ in two other types of micelles (69). Almost certainly, the value for ethanol peroxyl radicals of 9.4×10^4 M^{-1}s^{-1} (68) is off by several orders of magnitude, as the viscosity of ethanol

TABLE 5 Scavenging Rate Constants of α-Tocopherol Obtained by Pulse Radiolysis in Different Solvents[a]

Radical	Solvent	Rate constant (M^{-1} s^{-1})	Reference
$\cdot N_3$	H_2O/SDS-micelles	1.8×10^9	69
	H_2O/TTAB-micelles	2.2×10^9	69
$Br_2^{\cdot-}$	H_2O/TTAB-micelles	7.2×10^8	69
$CCl_3COO^\cdot$	H_2O/2-propanol/acetone	5.0×10^8	59
$CF_3CHClOO^\cdot$	H_2O + 10% t-BuOH	9.2×10^7	70
Solvent peroxyl radical			
$HROO^\cdot$	Cyclohexane	2.3×10^7	67
	n-Dodecan	1.5×10^7	67
	Trimethylpentane	1.4×10^7	67
	n-Octanoic acid	2.8×10^6	67
	Oleic acid	2.5×10^6	67
	Ethanol	9.4×10^4	68
	Methyl linoleate/micelles	8.0×10^4	71
Phenothiazine radical cations			
Chlorpromazine	H_2O/2-propanol/acetone	4.4×10^8	72
Metiazinic acid	H_2O/2-propanol/acetone	1.3×10^8	72
Promethazine	H_2O/2-propanol/acetone	1.1×10^8	72

[a]SDS, sodium dodecyl sulfate; TTAB, tetradecyl trimethyl ammonium bromide.

at room temperature is almost that of water (73); the data of Reference 67 are listed in Table 5 according to the increase in viscosity. Evidently, these pulse radiolysis experiments of ethanolic solutions of vitamin E have little relevance to the other studies.

CORRELATION OF PULSE RADIOLYSIS WITH OTHER DATA ON VITAMIN E PHENOXYL RADICAL

Pulse radiolytic investigations are generally done under clearly defined in vitro conditions. It is thus imperative to correlate these data with results obtained by other methods.

Kinetic Data

The basis for such a correlation may be the data in Table 6, that is, the rate constants of vitamin E with different types of radicals obtained in chemical systems. It is of interest that the rate constant for the annealing of the vitamin E phenoxyl radical by ascorbate (1.6×10^6 $M^{-1}s^{-1}$, obtained by pulse radiolysis in aqueous solutions containing molar concentrations of isopropanol and acetone) (59) drops only to 2×10^5 $M^{-1}s^{-1}$ in phosphatidylcholine liposomes (determined by ESR) (10) but to 5.5×10^2 M^{-1} s^{-1} in benzene-ethanol (2:1 vol/vol), determined by stop-flow spectrophotometry) (82). Rate constants were also reported after stop-flow spectrophotometry for the reduction of α-tocopheroxyl radicals by ubiquinol-10 in benzene or ethanol at 25°C (3.74×10^5 and 2.15×10^5 M^{-1} s^{-1}, respectively (83).

Contrary to an early report (84), it has not been possible to confirm the direct reduction of α-tocopheroxyl radicals by GSH during lipid peroxidation (85,86). We now find (52) that in analogy to the reaction of thiyl radicals ($RS^\cdot$) with fatty acids and alcohols (87), the $GS^\cdot$ thiol radical is capable of oxidizing the model compounds Trolox c and PMC to their phenoxyl radicals in aqueous solution with rate constants of 2.2×10^7 and 2.0×10^7 M^{-1} s^{-1},

TABLE 6 Scavenging Rate Constants of α-Tocopherol Obtained by Other Methods[a]

Radical	Solvent	Method	Rate constant ($M^{-1} s^{-1}$)	Reference
Stable radicals				
GO	Ethanol/25°C	Kinetic ESR	1.0×10^5	74
DBMPO	Ethanol/25°C	Stop-flow spectro- photometry	5.1×10^3	75
DPPH	Benzene	Spectrophotometry	1.6×10^3	76
	Methanol	Spectrophotometry	1.9×10^3	77
	Ethanol/25°C	Spectrophotometry	5.3×10^2	77
Peroxyl radicals				
(Poly)styrene	Benzene/30°C	LFP/ESR	2.6×10^6	78
	Benzene/30°C	LP/IAO	3.2×10^6	78
Methyl linoleate	t-BuOH/37°C	LP/IAO	5.1×10^5	79
Linoleic acid	Methanol/37°C	LP/IAO	4.9×10^5	80
Ethylbenzene[b]	Ethylbenzene/25°C	CL	1.5×10^5	81

[a]GO, galvinoxyl; DBMPO, 2.6-di-*tert*-butyl-4(4-methoxyphenyl)phenoxyl; DPPH, 1.1-diphenyl-2-pikrylhydrazyl; LFP/ESR, laser flash photolysis and ESR; LP/IAO, lag phase/inhibition of autoxidation; CL, chemiluminescence.
[b]For other rate constants with this peroxyl radical, determined at higher temperatures and with other methods, see Table 3 in Reference 79.

respectively: a value of $4.4 \times 10^5 \, M^{-1} s^{-1}$ was reported for the reaction of the penicillamine thiyl radical with vitamin E in isopropanol and water (1:1, vol/vol) (88). This would in effect reverse the repair reaction

$$PheO^{\cdot} + GSH \rightarrow GS^{\cdot} + PheOH \tag{14}$$

and therefore strongly suggest that under in vivo conditions vitamin E radicals can only be reduced by GSH in an enzymatic process. Such GSH-dependent free-radical reductases have in fact been described repeatedly (89–93).

Product Studies

Detailed analytical studies have been done for the reaction of vitamin E with both alkoxyl (94) and peroxyl radicals (95–98). Fatty acid alkoxyl radicals seem to very rapidly rearrange to epoxy allyl radicals, which are then trapped by both vitamin E (94) and its model compounds Trolox c (94) and PMC (99). Peroxyl radicals are stable enough to react as such. They have been shown to interact with phenolic antioxidants via a 2:1 stoichiometry; that is, a second peroxyl radical adds to a mesomeric structure of the phenoxyl radical (preferentially in *para* position) to form a quinone methide adduct that eventually hydrolyzes to the *p*-quinone (100). An early report on vitamin E products after reaction with peroxyl radicals, although confirming the 2:1 stoichiometry, rather suggested *ortho* peroxy adducts (95). In contrast, detailed recent studies identified the expected 8a-alkylperoxy adducts (96,98) and 4a,5- and 7,8-epoxy-8a-hydroperoxy tocopherones formed in a different pathway by the peroxyl radicals (97).

ESR Spectroscopy

It was mentioned in the Introduction that ESR spectroscopy provided the first evidence that vitamin E phenoxyl radical are indeed formed under in vivo conditions (11,12). The lower time resolution of ESR spectroscopy as opposed to pulse radiolysis ($f \simeq 1000$) is of minor importance, as long as the stability of the observed radical is sufficient. This is the case for α-tocopheroxyl radical in organic solvent but barely so for the model phenoxyl radicals in aqueous solutions. Nevertheless, Trolox c phenoxyl radicals have already been observed using ESR spectroscopy in aqueous solution (48,101).

CONCLUSION

The method of pulse radiolysis has been shown to be a versatile method to generate almost any type of radical in aqueous solutions. For studies of the water-insoluble vitamin E, model compounds lacking the hydrophobic phytyl side chain are fortunately available (Trolox c and pentamethylchromane). They all share the 6-hydroxy group of the chromane ring, which is the site of the phenoxyl radical. As expected, many pulse radiolytic investigations have used these model compounds to determine transient spectra and kinetic data for the formation and decay of the phenoxyl radical. Comparable studies with vitamin E proper were done in organic solvents and showed some spectral shifts and differences in kinetic behavior. Pulse radiolysis thus provided useful information on the behavior of the vitamin E phenoxyl radical, which is the obligatory intermediate in the chain-breaking processes of this most important lipid antioxidant.

REFERENCES

1. Burton GW, Joyce A, Ingold KU. Is vitamin E the only lipid-soluble, chain-breaking antioxidant in human blood plasma and erythrocyte membranes? Arch Biochem Biophys 1983; 221:281-90.
2. Ingold KU, Webb AC, Witter D, Burton GW, Metcalfe TA, Muller DPR. Vitamin E remains the major lipid soluble chain breaking antioxidant in human plasma even in individuals suffering severe Vitamin E deficiency. Arch Biochem Biophys 1987; 259:224-5.
3. Boguth W, Niemann H. Über die Reaktion von α-Tocopherol mit 1,1-Diphenyl-2-pikrylhydrazyl. IV. Das α-Tocopherolradikal. Int Z Vitam Forsch 1969; 39:429-37.
4. Ozawa T, Hanaki A, Matsumoto S, Matsuo M. ESR studies of radicals obtained by the reaction of α-tocopherol and its model compound with superoxide ion. Biochim Biophys Acta 1978; 531: 72-8.
5. Parteshko VG, Staren'kii AG, Koshechko VG. Investigation of free-radical oxidation of tocopherol. Proc Acad Sci USSR 1979; 245:165-71.
6. Doba T, Burton GW, Ingold KU. EPR spectra of some α-tocopherol model compounds. Polar and conformational effects and their relation to antioxidant activities. J Am Chem Soc 1983; 105: 6505-6.
7. Matsuo M, Matsumoto S. ESR spectra of the chromanoxyl radicals derived from tocopherols (vitamin E) and their related compounds. Lipids 1983; 18:81-6.
8. Tsuchiya J, Niki E, Kamiya Y. Oxidation of lipids. IV. Formation and reaction of chromanoxyl radicals as studied by ESR. Bull Chem Soc Jpn 1983; 56:229-32.
9. Rousseau-Richard C, Richard C, Martin R. Kinetics of bimolecular decay of α-tocopherol free radicals studied by ESR. FEBS Lett 1988; 233:307-10.
10. Scarpa M, Rigo A, Maiorino M, Ursini F, Gregolin C. Formation of α-tocopherol radical and re-cycling of α-tocopherol by ascorbate during peroxidation of phosphatidylcholine liposomes. An EPR study. Biochim Biophys Acta 1984; 801:215-9.
11. Mehlhorn RJ, Sumida S, Packer L. Tocopheroxyl radical persistence and tocopherol consumption in liposomes and in vitamin E enriched rat liver mitochondria and microsomes. J Biol Chem 1989; 264:13448-52.
12. Maguire J, Wilson DS, Packer L. Mitochondrial electron transport-linked tocopheroxyl radical reduction. J Biol Chem 1989; 264:21462-5.
13. Wardman P. Application of pulse radiolysis methods to study the reactions and structure of biomolecules. Rep Prog Phys 1978; 41:259-302.

14. Freeman GR. Adsorption of energy from ionizing radiation. In: Baxendale JH, Busi F, eds. The studies of fast processes and transient species by electron pulse radiolysis. NATO Adv. Study Inst Series, Vol. 86. Dordrecht: Reidel, 1982; 1-17.

15. Chatterjee A. Interaction of ionizing radiation with matter. In: Farhataziz, Rodgers MAJ, eds. Radiation chemistry. Principles and applications. Weinheim: VCH Verlag, 1987; 1-28.

16. Bielski BHJ, Gebicki JM. Application of radiation chemistry to biology. In: Pryor, WA, ed. Free radicals in biology, Vol. III. New York: Academic Press 1977; 1-51.

17. Buxton GV. Radiation chemistry of the liquid state. 1. Water and homogeneous aqueous solutions. In: Farhataziz, Rodgers MAJ, eds. Radiation chemistry. Principles and applications. Weinheim: VCH Verlag 1987; 321-49.

18. Jonah CD, Bartels DM, Chernovitz AC. Primary processes in the radiation chemistry of water. Radiat Phys Chem 1989; 34:145-56.

19. Bors W, Saran M, Michel C, Tait D. Formation and reactivities of oxygen free radicals. In: Breccia A, Greenstock CL, Tamba M, eds. Advances on oxygen radicals and radioprotectors. Bologna: Lo Scarabeo, 1984; 13-27.

20. Alfassi ZB, Schuler RH. Reaction of azide radicals with aromatic compounds. Azide as a selective oxidant. J Phys Chem 1985; 89:3359-63.

21. Huie RE, Neta P. Chemical behaviour of SO_3^- and SO_5^- radicals in aqueous solutions. J Phys Chem 1984; 88:5665-9.

22. Hunt JW, Wolff RK, Chenery S. Fast oxidative reactions following irradiation in aqueous solutions of Cl^-, Br^- CNS^-. In: Adams GE, Fielden EM, Michael BD, eds. Fast processes in radiation chemistry and biology. New York: Wiley & Sons, 1973; 109-22.

23. Bors W, Tait D, Michel C, Saran M, Erben-Russ M. Reactions of alkoxyl radicals in aqueous solutions. Isr J Chem 1984; 24:17-24.

24. Adams GE, Willson RL. Pulse radiolysis studies on the oxidation of organic radicals in aqueous solution. Trans Faraday Soc 1969; 65:2981-7.

25. Czapski G. Radiation chemistry of oxygenated aqueous solutions. Annu Rev Phys Chem 1971; 22:171-208.

26. Bors W, Michel C, Saran M. Superoxide anions do not react with hydroperoxides. FEBS Lett 1979; 107:403-6.

27. Busi F. Labile species and fast processes in liquid alkanes. In: Baxendale JH, Busi F, eds. The studies of fast processes and transient species by electron pulse radiolysis. NATO Adv. Study Inst. Series, Vol. 86. Dordrecht: Reidel, 1982; 417-31.

28. Warman JM. The dynamics of electrons and ions in nonpolar liquids. In: Baxendale JH, Busi F, eds. The studies of fast processes and transient species by electron pulse radiolysis. NATO Adv. Study Inst. Series, Vol. 86. Dordrecht: Reidel, 1982; 433-533.

29. Swallow AJ. Radiation chemistry of the liquid state. 2. Organic liquids. In: Farhataziz, Rodgers MAJ, eds. Radiation chemistry. Principles and applications. Weinheim: VCH Verlag, 1987; 351-75.

30. Willard JE. The radiation chemistry of organic solids. In: Farhataziz, Rodgers MAJ, eds. Radiation chemistry. Principles and applications. Weinheim: VCH Verlag, 1987; 395-434.

31. Charbonnier FM, Barbour JP, Brewster JL. A new high intensity nanosecond electron source for pulse radiolysis. In: Adams GE, Fielden EM, Michael BD, eds. Fast processes in radiation chemistry and biology. New York: Wiley & Sons, 1973; 3-15.

32. Foyt DC. Data acquisition and analysis in pulse radiolysis. I. Control, digitization, and analysis. In: Baxendale JH, Busi F, eds. The studies of fast processes and transient species by electron pulse radiolysis. NATO Adv. Study Inst. Series, Vol. 86. Dordrecht: Reidel, 1982; 199-212.

33. Helman WP, Ross AB. Radiation chemistry data center: information sevices produced from bibliographic data base. Radiat Phys Chem 1980; 16:425-30.

34. Helman WP, Ross AB. Bibliographies on radiation chemistry. XII. Rate constants for reactions of nonmetallic inorganic radicals in aqueous solution. Radiat Phys Chem 1990; 36:845-57.

35. Buxton GV, Greenstock CL, Helman WP, Ross AB. Critical review of rate constants for reactions of hydrated electrons, hydrogen atoms and hydroxyl radicals ($^{\cdot}OH/O^-$) in aqueous solution. J Phys Chem Ref Data 1988; 17:513-904.

36. Hug GL. Optical spectra of nonmetallic inorganic transient species in aqueous solution. NSRDS-NBS Report 69. Washington, DC, National Bureau of Standards, U.S. Department of Commerce, 1981.

37. Sauer MC. Sources of pulsed radiation. In: Baxendale JH, Busi F, eds. The studies of fast processes and transient species by electron pulse radiolysis. NATO Adv. Study Inst. Series, Vol. 86. Dordrecht: Reidel, 1982; 35-47.

38. Patterson LK. Instrumentation for measurement of transient behavior in radiation chemistry. In: Farhataziz, Rodgers MAJ, eds. Radiation chemistry. Principles and applications. Weinheim: VCH Verlag, 1987; 65-96.

39. Roffi G. Optical monitoring techniques. In: Baxendale JH, Busi F, eds. The studies of fast processes and transient species by electron pulse radiolysis. NATO Adv. Study Inst. Series, Vol. 86. Dordrecht: Reidel, 1982; 63-89.

40. Simic MG, Hunter EPL. The reactivities of organic oxygen (oxy) radicals. In: Bors W, Saran M, Tait D, eds. Oxygen radicals in chemistry and biology. Berlin: de Gruyter, 1984; 109-21.

41. Saran M, Vetter G, Erben-Russ M, et al. Pulse radiolysis equipment: a setup for simultaneous multiwavelength kinetic spectroscopy. Rev Sci Instrum 1987; 58:363-8.

42. Hodgson BW, Keene JP, Land EJ, Swallow AJ. Light-induced fluorescence of short-lived species produced by a pulse of radiation: the benzophenone ketyl radical. J Chem Phys 1975; 63:7671-2.

43. Rodgers MAJ. Light scattering techniques for investigation of transients produced in electron pulse radiolysis: In: Baxendale JH, Busi F, eds. The studies of fast processes and transient species by electron pulse radiolysis. NATO Adv. Study Inst. Series, Vol. 86. Dordrecht: Reidel, 1982; 189-97.

44. Trifunac AD. EPR and NMR detection of transient radicals and reaction products. In: Baxendale JH, Busi F, eds. The studies of fast processes and transient species by electron pulse radiolysis. NATO Adv. Study Inst. Series, Vol. 86. Dordrecht: Reidel, 1982; 163-178.

45. Asmus KD, Janata E. Conductivity monitoring techniques. In: Baxendale JH, Busi F, eds. The studies of fast processes and transient species by electron pulse radiolysis. NATO Adv. Study Inst. Series, Vol. 86. Dordrecht: Reidel, 1982; 91-113.

46. Asmus KD, Janata E. Polarography monitoring techniques. In: Baxendale JH, Busi F, eds. The studies of fast processes and transient species by electron pulse radiolysis. NATO Adv. Study Inst. Series, Vol. 86. Dordrecht: Reidel, 1982; 115-28.

47. Bors W, Michel C, Erben-Russ M, Kreileder B, Tait D, Saran M. Rate constants of sparingly water-soluble phenolic antioxidants with OH radicals. In: Bors W, Saran M, Tait D, eds. Oxygen radicals in chemistry and biology. Berlin: de Gruyter, 1984; 95-9.

48. Davies MJ, Forni LG, Willson RL. Vitamin E analogue Trolox c. ESR and pulse radiolysis studies of free radical reactions. Biochem J 1988; 255:513-22.

49. Bisby RH, Ahmed S, Cundall RB. Repair of amino acid radicals by a vitamin E analogue. Biochem Biophys Res Commun 1984; 119:245-51.

50. Cabelli DE, Bielski BHJ. Studies of the reactivity of Trolox with Mn^{3+}/Fe^{3+} complexes by pulse radiolysis. J Free Rad Biol Med 1986; 2:71-5.

51. Thomas MJ, Bielski BHJ. Oxidation and reaction of Trolox c, a tocopherol analogue, in aqueous solution. A pulse radiolysis study. J Am Chem Soc 1989; 111:3315-9.

52. Bors W, Michel C, Saran M. Manuscript in preparation.

53. Huie RE, Neta P. Oxidation of ascorbate and a tocopherol analogue by the sulfite-derived radicals SO_3^- and SO_5^-. Chem Biol Interact 1985; 53:233-8.

54. Erben-Russ M, Michel C, Bors W, Saran M. Absolute rate constants of alkoxyl radical reactions in aqueous solutions. J Phys Chem 1987; 91:2362-5.

55. Steenken S, Neta P. One-electron redox potentials of phenols. Hydroxy- and aminophenols and related compounds of biological interest. J Phys Chem 1982; 86:3661-7.

56. Bisby RH, Cundall RB, Tabassum N. Formation and reactivity of free radicals of acetaminophen formed by one electron oxidation. Life Chem Rep 1985; 3:29-34.

57. Forni LG, Mora-Arellano VO, Packer JE, Willson RL. Aminopyrine and antipyrine free radical cations: pulse radiolytic studies of one-electron transfer reactions. J Chem Soc Perkin II 1988; 1579-84.

58. Doba T, Burton GW, Ingold KU, Matsuo M. α-Tocopheroxyl decay: lack of effect of oxygen. J Chem Soc Chem Commun 1984; 461-2.

59. Packer JE, Slater TF, Willson RL. Direct observation of a free radical interaction between vitamin E and vitamin C. Nature 1979; 278:737-8.

60. Chen LH. Interaction of vitamin E and ascorbic acid. In Vivo 1989; 3:199-210.

61. Cadenas E, Merenyi G, Lind J. Pulse radiolysis study on the reactivity of Trolox c phenoxyl radical with superoxide anion. FEBS Lett 1989; 253:235-8.

62. Forni LG, Willson RL. Thiyl and phenoxyl free radical and NADH. Direct observation of one-electron oxidation. Biochem J 1986; 240:897-903.

63. Matsuo M, Matsumoto S, Iitaka Y, Hanaki A, Ozawa T. Epoxidation of an α-tocopherol model compound, 2,2,5,7,8-pentamethylchroman-6-ol, with potassium superoxide; x-ray crystal structure of 4a,5;7,8-diepoxy-4,7,8,8a-tetrahydro-8a-hydroxy-2,2,5,7,8-pentamethylchroman-6(5H)-one. J Chem Soc Chem Commun 1979; 105-6.

64. Matsuo M, Matsumoto S, Iitaka Y. Oxygenations of vitamin E and its model compound 2,2,5,7,8-pentamethylchroman-6-ol in the presence of the superoxide radical solubilized in aprotic solvents: unique epoxidations and recyclizations. J Org Chem 1987; 52:3514-20.

65. Suarna C, Craig DC, Cross KJ, Southwell-Keely PT. Oxidations of vitamin E (α-tocopherol) and its model compound 2,2,5,7-pentamethyl-6-hydroxychroman. A new dimer. J Org Chem 1988; 53:1281-4.

66. Matsuo M, Matsumoto S, Iitaka Y, Niki E. Radical scavenging reactions of vitamin E and its model compound 2,2,5,7,8-pentamethylchroman-6-ol, in a *t*-butyl-peroxyl radical generating system. J Am Chem Soc 1989; 111:7179-85.

67. Simic MG. Kinetic and mechanistic studies of peroxy, vitamin E and anti-oxidant free radicals by pulse radiolysis. In: Simic MG, Karel M, eds. Autoxidation in food and biological systems. New York: Plenum Press, 1980; 17-26.

68. Jore D, Patterson LK, Ferradini C. Pulse radiolysis study of α-tocopherol radical mechanisms in ethanolic solution. J Free Rad Biol Med 1986; 2:405-10.

69. Bisby RH, Ahmed S, Cundall RB, Thomas EW. Free radical reactions with α-tocopherol and *n*-stearoyl tryptophan methyl ester in micellar solution. Free Rad Res Commun 1986; 1:251-61.

70. Mönig J, Asmus KD, Schäffer M, Slater TF, Willson RL. Electron transfer reactions of halothane-derived peroxyl free radicals, $CF_3CHClOO$: measurement of absolute rate constants by pulse radiolysis. J Chem Soc Perkin II 1983; 1133-7.

71. Patterson LK. Studies of radiation-induced peroxidation in fatty acid micelles. In: Rodgers MAJ, Powers EL, eds. Oxygen and oxy-radicals in chemistry and biology. New York: Academic Press, 1981; 89-95.

72. Bahnemann D, Asmus KD, Willson RL. Phenothiazine radical-cations: electron transfer equilibria with iodide ions and the determination of one-electron redox potentials by pulse radiolysis. J Chem Soc Perkin II 1983; 1669-73.

73. Weast RC, ed. CRC Handbook of chemistry and physics, 67th ed. Boca Raton, FL: CRC Press, 1986; F39.

74. Niki E, Tsuchiya J, Yoshikawa Y, Yamamoto Y, Kamiya Y. Oxidation of lipids. XIII. Antioxidant activities of α-, β-, γ- and δ-tocopherols. Bull Chem Soc Jpn 1986; 59:497-501.

75. Mukai K, Watanabe Y, Uemoto Y, Ishizu K. Stopped-flow investigation of antioxidant activity of tocopherols. Bull Chem Soc Jpn 1986; 59:3113-6.

76. Boguth W, Repges R, Pracht I. Über die Reaktion von α-Tocopherol mit 1,1-Diphenyl-2-pikryl-hydrazyl. V. Kinetische und thermodynamische Daten. Int Z Vitam Forsch 1969; 39:439-45.

77. Rao AM, Singh UC, Rao CNR. Interaction of α-tocopherol with diphenyl picrylhydrazyl. A means to determine the polarity of the environment around α-tocopherol and its binding with lipids. Biochim Biophys Acta 1982; 711:134-7.

78. Burton GW, Doba T, Gabe EJ, et al. Autoxidation of biological molecules. 4. Maximizing the antioxidant activity of phenols. J Am Chem Soc 1985; 107:7053-65.

79. Niki E, Saito T, Kawakami A, Kamiya Y. Inhibition of oxidation of methyl linoleate in solution by vitamin E and vitamin C. J Biol Chem 1984; 259:4177-82.

80. Braughler JM, Pregenzer JF. The 21-aminosteroid inhibitors of lipid peroxidation: reactions with lipid peroxyl and phenoxy radicals. Free Rad Biol Med 1989; 7:125-30.

81. Niki E, Tanimura R, Kamiya Y. Oxidation of lipids. II. Rate of inhibition of oxidation by α-tocopherol and hindered phenols measured by chemiluminescence. Bull Chem Soc Jpn 1982; 55:1551-5.

82. Mukai K, Fukuda K, Ishizu K, Kitamura Y. Stopped flow investigation of the reaction between vitamin E radical and vitamin C in solution. Biochem Biophys Res Commun 1987; 146:134-9.

83. Mukai K, Kikuchi S, Urano S. Stopped-flow kinetic study of the regeneration reaction of tocopherol radical by reduced ubiquinone-10 in solution. Biochim Biophys Acta 1990; 1035:77-82.

84. Niki E, Tsuchiya J, Tanimura R, Kamiya Y. Regeneration of vitamin E from α-chromanoxyl radical by GSH and vitamin C. Chem Lett 1982; 789-92.

85. Barclay LRC. The cooperative antioxidant role of GSH with a lipid-soluble and a water-soluble antioxidant during peroxidation of liposomes initiated in the aqueous phase and in the lipid phase. J Biol Chem 1988; 263:16138-42.

86. Leedle RA, Aust SD. The effect of GSH on the vitamin E requirement for inhibition of liver microsomal lipid peroxidation. Lipids 1990; 25:241-5.

87. Schöneich C, Asmus KD. Reaction of thiyl radicals with alcohols, ethers and polyunsaturated fatty acids: a possible role of thiyl free radicals in thiol mutagenesis? Radiat Environ Biophys 1990; 29:263-71.

88. Schöneich C, Bonifacic M, Asmus KD. Reversible H-atom abstraction from alcohols by thiyl radicals: determination of absolute rate constants by pulse radiolysis. Free Rad Res Commun 1990; 6:393-405.

89. Haenen GRMM, Tai Tin Tsoi JNL, Verrmeulen NPE, Timmerman H, Bast A. 4-Hydroxy-2,3-trans-nonenal stimulates microsomal lipid peroxidation by reducing the GSH-dependent protection. Arch Biochem Biophys 1987; 259:449-56.

90. McCay PB, Lai EK, Brueggemann G, Powell SR. A biological antioxidant function for vitamin E: electron shuttling for a membrane-bound "free radical reductase." In: Galli C, Fedeli E, eds. Fat production and consumption. NATO Adv. Study Inst. Series A, Vol. 131. New York: Plenum Press, 1987; 145-56.

91. Schallreuter KU, Hordinsky MK, Wood JM. Thioredoxin reductase. Role in free radical reduction in different hypopigmentation disorders. Arch Dermatol 1987; 123:615-9.

92. Packer L, Maguire JJ, Mehlhorn RJ, Serbinova E, Kagan VE. Mitochondria and microsomal membranes have a free radical reductase activity that prevents chromanoxyl radical accumulation. Biochem Biophys Res Commun 1989; 159:229-35.

93. Fuchs J, Mehlhorn RJ, Packer L. Assay for free radical reductase activity in biological tissue by ESR spectroscopy. In: Packer L, Glazer AN, eds. Methods in enzymology, Vol. 186. Oxygen radicals in biological systems. Part B. Oxygen radicals and antioxidants. New York: Academic Press, 1990; 670-4.

94. Gardner HW, Eskins K, Grams GW, Inglett GE. Radical addition of linoleic hydroperoxides to α-tocopherol or the analogous hydroxy chroman. Lipids 1972; 7:324-4.

95. Winterle J, Dulin D, Mill T. Products and stoichiometry of reaction of vitamin E with alkylperoxy radicals. J Org Chem 1984; 49:491-5.

96. Liebler DC, Kaysen KL, Kennedy TA. Redox cycles of vitamin E: hydrolysis and ascorbic acid dependent reduction of 8a-(alkydioxy)tocopherones. Biochemistry 1989; 28:9772-7.

97. Liebler DC, Baker PF, Kaysen KL. Oxidation of vitamin E: evidence for competing autoxidation and peroxyl radical trapping reactions of the tocopheroxyl radicals. J Am Chem Soc 1990; 112: 6995-7000.

98. Yamauchi R, Matsui T, Kato K, Ueno Y. Reaction products of α-tocopherol with methyl linoleateperoxyl radicals. Lipids 1990; 25:152-8.

99. Kaneko T, Matsuo M. The radical-scavenging reactions of a vitamin E model compound, 2,2,5,7,8-pentamethylchroman-6-ol, with radicals from the Fe(II)-induced decomposition of a linoleic acid hydroperoxide, (9Z-11E)-13-hydroperoxy-9,11-octadecadienoic acid. Chem Pharm Bull 1985; 33: 1899-905.

100. Boozer CE, Hammond GS, Hamilton CE, Sen JN. Air oxidation of hydrocarbons. II. The stoichiometry and fate of inhibitors in benzene and chlorobenzene. J Am Chem Soc 1985; 77:3233-7.

101. Nakamura M. One electron oxidation of trolox c (a vitamin E analogue) by peroxidases. J Biochem 1990; 108:245-9.

7

Peroxyl Radical Trapping Reactions of α-Tocopherol in Biomimetic Systems

Daniel C. Liebler

University of Arizona, Tucson, Arizona

INTRODUCTION

α-Tocopherol (vitamin E [1]), the principal lipid-soluble antioxidant in biological membranes, reacts readily with many radical and nonradical oxidants. In membranes, α-tocopherol inhibits lipid peroxidation primarily by trapping peroxyl radicals, which propagate the radical chains (1):

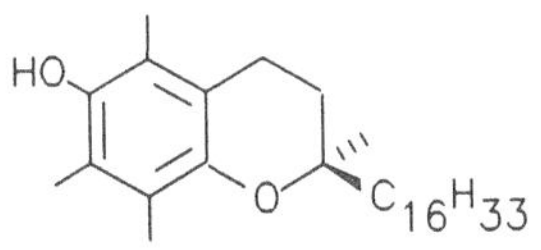

This reaction produces lipid hydroperoxides and the relatively stable tocopheroxyl radical [2], which does not propagate radical chains [Eq. (1)] (1).

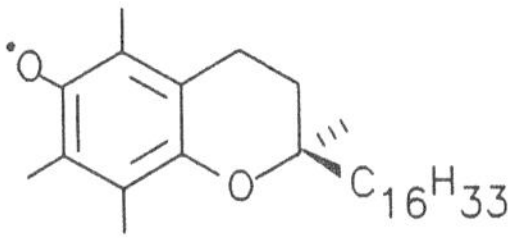

$$\text{LOO}^{\cdot} + \alpha\text{-TH} \rightarrow \text{LOOH} + \alpha\text{-T}^{\cdot} \tag{1}$$

Since the rate constant for this reaction is two to three orders of magnitude higher than that for peroxyl radical propagation (1,2), α-tocopherol protects membranes efficiently at levels typically as low as 1 α-tocopherol per 1000 phospholipids (3,4). Sustained antioxidant protection has been postulated to depend on regeneration of α-tocopherol from the tocopheroxyl radical by cellular reductants, including ascorbic acid [Eq. (2)] (5–10).

$$\alpha\text{-T}^{\cdot} + \text{electron donor}_{red} + \text{H}^{+} \rightarrow \alpha\text{-TH} + \text{electron donor}_{ox} \tag{2}$$

Tocopheroxyl radicals that do not complete a one-electron redox cycle may react instead with peroxyl radicals to form other products. Indeed, kinetic studies in homogeneous solution indicate that each α-tocopherol molecule can react with two peroxyl radicals according to Equations (1) and (3) (11–14).

$$\text{LOO}^{\cdot} + \alpha\text{-T}^{\cdot} \rightarrow \text{nonradical product(s)} \tag{3}$$

Until recently, the nonradical α-tocopherol oxidation products formed in reaction (3) had not been identified. This is surprising since products of α-tocopherol oxidation by other oxidants have been extensively investigated and characterized (15–20).

The oxidative fate of the tocopheroxyl radical in reaction (3) is of interest for three reasons. First, reaction (3) consumes the tocopheroxyl radical, terminates its antioxidant action, and leads to α-tocopherol turnover, where turnover here is defined as loss of the antioxidant molecule from the system. Second, reaction (3) may form products that, although devoid of antioxidant activity, could be reduced back to α-tocopherol to complete a two-electron redox cycle analogous to that proposed for the tocopheroxyl radical [Eqs. (1) and (2)]. Such products are *reversibly* oxidized if they can be reduced to α-tocopherol under conditions that either occur or can be created in biological systems. Third, reaction (3) may form products that have no antioxidant properties and cannot easily be reduced to α-tocopherol. These products are *irreversibly* oxidized and could serve as biochemical markers for α-tocopherol oxidation.

This chapter describes α-tocopherol oxidation by peroxyl radicals both in homogeneous solution and in lipid bilayers and the properties of the products in these model systems. These model systems are biomimetic in that they simulate physiological membrane environments. The models are nevertheless chemically well defined and free of other redox-active components that could confuse interpretation of the chemistry involved. The objectives of this work were (1) to quantitatively account for the disposition of α-tocopherol in these models, (2) to determine the mechanisms by which the products are formed, and (3) to assess whether α-tocopherol products could participate in biochemical redox cycles.

REACTIONS OF α-TOCOPHEROL WITH PEROXYL RADICALS IN HOMOGENOUS SOLUTION

Oxidation of α-Tocopherol by Peroxyl Radical Generating Systems

To investigate the reactions of α-tocopherol with peroxyl radicals, oxidants that specifically generate peroxyl radicals are employed. Peroxyl radicals may be generated conveniently in homogeneous solution by the decomposition of azo compounds, such as azobis-(2,4-dimethylvaleronitrile) (AMVN), which generate peroxyl radicals by thermolysis and subsequent oxygen addition [Eqs. (4) and (5)] (21).

$$[(CH_3)_2CHCH_2C(CN)(CH_3)N]_2 \rightarrow 2\ (CH_3)_2CHCH_2C(CN)(CH_3)C^{\cdot} + N_2 \qquad (4)$$

$$(CH_3)_2CHCH_2C(CN)(CH_3)C^{\cdot} + O_2 \rightarrow (CH_3)_2CHCH_2C(CN)(CH_3)COO^{\cdot} \qquad (5)$$

In two previous studies, Winterle et al. (14) and Yamauchi et al. (22) identified 8a-(alkyldioxy)tocopherones [3a] as products of α-tocopherol oxidation by peroxyl radicals generated from AMVN and analogous azo compounds.

3a   R = −OO−alkyl
3b   R = −OOH
3c   R = −OOC(CH₃)(CN)CH₂CH(CH₃)₂
3d   R = −OH

The alkyl portion of the 8a-alkyldioxy group was derived from the azo compound used. 8a-(Alkylkdioxy)tocopherone formation was proposed to result from radical addition of the alkylperoxyl radical to the tocopheroxyl radical 8a position.

Matsuo et al. (23) generated *tert*-butyl peroxyl radicals by thermolysis of di-*tert*-butyl-peroxyoxalate in *tert*-butyl hydroperoxide. α-Tocopherol oxidation did not yield 8a-(*tert*-butyl-dioxy)tocopherone but instead produced 4a,5-epoxy-(8a-hydroperoxy)tocopherones [4], which accounted for 28% of the products.

4   5

Yamauchi et al. (22) isolated analogs of epoxytocopherones [4] and [5] with an 8a-ethoxy group from AMVN-initiated oxidations in ethanol. The ethoxy group was apparently derived from the solvent. 8a-Hydroperoxytocopherone [3b] was also reported as an α-tocopherol oxidation product in both systems (22,23).

These studies demonstrate that peroxyl radicals oxidize α-tocopherol to two principal product groups: the 8a-substituted tocopherones and the epoxytocopherones. Depending on the peroxyl radical involved and the reaction solvent, either 8a-alkyldioxytocopherones or epoxytocopherones or both were formed. Whether these products accounted for all of the α-tocopherol oxidized in these systems is not clear. In addition, the relative yields of various products were not quantified. Another important question is how epoxytocopherone products are formed. When epoxytocopherones have been identified, they were proposed to arise via epoxidation of the corresponding tocopherone precursors.

Products of Peroxyl Radical Oxidation of α-Tocopherol in Acetonitrile

These studies were undertaken to quantitatively account for the products formed when peroxyl radicals react with α-tocopherol in a chemically defined model system (24). Acetonitrile was chosen in part for its miscibility with aqueous buffers, which permitted the effects of water on product distribution to be studied (see later). To monitor product distribution, [5-methyl-^{14}C]α-tocopherol ([^{14}C]α-tocopherol) (24,25) was incubated at 50°C with excess AMVN in air-saturated acetonitrile and the products were analyzed by reversed-phase high-performance liquid chromatography (HPLC; Fig. 1). Two groups of radiolabeled products were formed as [^{14}C]α-tocopherol was consumed. The polar product fraction eluting at 8 minutes was identified by ultraviolet-visible (UV-Vis) and nuclear magnetic resonance (NMR) spectroscopy and mass spectrometry as a mixture of diastereomeric epoxytocopherones [4]. Although not identified directly, the epoxide regioisomer [5] also appeared to be present since the corresponding epoxyquinone hydrolysis product was formed when this product fraction was exposed to aqueous acid (see later). The nonpolar fraction eluting at 19 minutes was identified as a mixture of four diastereomeric 8a-(alkyldioxy)tocopherones [3c] in which the alkyl group was derived from the AMVN molecule. In addition to the two major product groups, small amounts of 8a-hydroperoxytocopherone [3b] were produced. The product distributions for α-tocopherol oxidation in this system are summarized in Table 1. The products listed accounted for at least 95% of the α-tocopherol consumed.

When acetonitrile and water (6:4 vol/vol) was used as the reaction solvent, the hydrolysis products α-tocopherylquinone-2,3-oxide [6] and α-tocopherylquinone-5,6-oxide [7] were recovered instead of their respective epoxytocopherone precursors (Table 1).

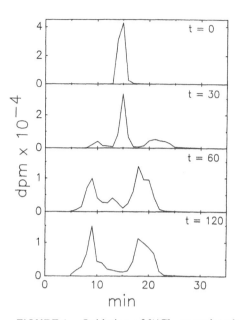

FIGURE 1   Oxidation of [^{14}C]α-tocopherol by AMVN in acetonitrile. [^{14}C]α-tocopherol (0.235 μmol; 0.238 μCi) and 4.36 μmol AMVN were incubated at 50°C under air in 0.7 ml acetonitrile. Products were analyzed by reversed-phase HPLC. (Reproduced with permission from Reference 24.)

TABLE 1   Product Distributions from Oxidation of α-Tocopherol by AMVN-Derived Peroxyl Radicals in Lipid Bilayers and Homogeneous Solution[a]

Experimental system	Reaction time (minutes)	Recovery of 1 (%)	Yield (%)					% Epoxides[b]
			3b	**3b**	**4/5**	**6/7**	**8**	
Liposomes	2400	29	6	11	23	20	7	64
CH$_3$CN/H$_2$O (6:4 vol/vol)	60	40	1	32	6	13	2	35
CH$_3$CN	60	20	3	51	22	ND[c]	ND	29

[a]Soy phosphatidylcholine liposomes containing 0.1 mol per 100 mol α-tocopherol and 20 mol per 100 mol AMVN were incubated at 37°C. Homogeneous solutions contained 2.35 μmol α-tocopherol and 43.6 μmol AMVN in 7 ml and were incubated at 50°C.
[b]Percentage of total products present as **4/5** and **6/7**.
[c]Not Detected.
Source: Adapted with permission from Reference 24.

                6   7

The four 8a-(alkyldioxy)tocopherone diastereomers [3c] and their hydrolysis product, α-tocopherolquinone [8], comprised the balance of the products. The 8a-(alkyldioxy)tocopherones thus appear to be more resistant to hydrolysis than the epoxytocopherone products. Treatment of the α-tocopherol product mixtures with dilute mineral acid quantitatively converted the epoxytocopherones [4/5] exclusively to epoxyquinones [6/7] and tocopherones [3b/3c] exclusively to tocopherolquinone [8]. Analysis of these quinone and epoxyquinone hydrolysis products thus provides a convenient method for determining tocopherone-epoxytocopherone product ratios.

8

Mechanism of Epoxide Product Formation

Our observation that peroxyl radical oxidation of α-tocopherol forms epoxytocopherones [4/5] agreed with those of Matsuo et al. (23) and Yamauchi et al. (22). These authors postulated that epoxytocopherones arose via epoxidation of tocopherone precursors. This straightforward hypothesis implies that epoxidation is a secondary event in α-tocopherol oxidation in which primary products are further oxidized. Since this hypothesis also implies that epoxide formation would lag behind tocopherone formation, we first studied the kinetics of ^{14}C-labeled product formation from [^{14}C]α-tocopherol in our model (24). Both epoxytocopherones [4/5] and 8a-(alkyldioxy)tocopherones [3c] were formed together at similar rates, and no lag in epoxide formation was evident. Once the [^{14}C]α-tocopherol was consumed, however, continued oxidation converted 8a-(alkyldioxy)tocopherones [3c] to products that eluted with the epoxide fraction. It thus appeared that further oxidation of 8a-(alkyldioxy)tocopherones could produce epoxytocopherones.

This possibility was further examined by studying the oxidation of 8a-(alkyldioxy)tocopherone [3c] and 8a-hydroperoxytocopherone [3b] by AMVN-derived peroxyl radicals in acetonitrile (24). When incubated alone with AMVN, both tocopherones were oxidized by the AMVN-derived peroxyl radicals to epoxytocopherones [4/5] and epoxyquinones [6/7]. When the oxidation mixture contained tocopherone [3b/3c] and α-tocopherol in equimolar amounts, the tocopherone was not oxidized until all of the α-tocopherol was first consumed. This indicated that tocopherones [3b/3c] could not serve as epoxide precursors when α-tocopherol was present because α-tocopherol prevented their oxidation. Epoxytocopherones [4/5] formed during α-tocopherol oxidation therefore must be formed by a mechanism that does not involve tocopherone intermediates.

To explain these results, we postulated that two competing reactions between the tocopheroxyl radical [2] and peroxyl radicals lead to the observed products (24). In the first reaction, peroxyl radicals add to the 8a position of [2] to form an 8a-(alkyldioxy)tocopherone product [Eq. (6)].

$$2 + ROO^{\cdot} \rightarrow 3a \qquad\qquad (6)$$

In the second, a peroxyl radical epoxidizes [2] to form an epoxytocopheroxyl radical (epoxy-[2]), which then adds oxygen to the 8a position and abstracts a hydrogen atom from a hydrogen donor to form epoxytocopherones [4/5] [Eqs. (7) through (9) or Fig. 2]. Although radical-radical reactions usually do not proceed with great efficiency, reaction (7) may be favored by the unusual stability of [2].

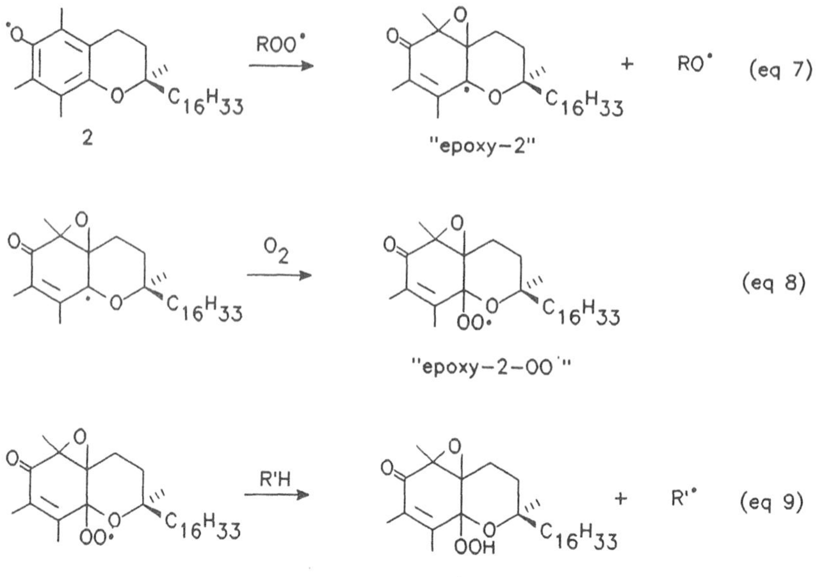

FIGURE 2 Proposed mechanism of epoxytocopherone [4] formation. An analogous mechanism involving epoxidation at the 7,8-position of [2] was proposed to lead to epoxytocophrone [5]. See text for discussion. (From Ref. 24.)

Whether these proposed reactions affect the antioxidant action of α-tocopherol is not clear. Unlike reaction (6), this three-reaction sequence results in no net radical trapping and generates a reactive peroxyl radical [epoxy-[2]-OO$\cdot$; Eq. (8)] as an intermediate. Were α-tocopherol to serve as the hydrogen donor R'H, as it probably would under the conditions of our experiments, reactions (7) through (9) would lead to α-tocopherol autoxidation. On the other hand, this reaction sequence could contribute to the antioxidant action of α-tocopherol if reaction (7) involved the formation of a stable, nonradical intermediate, which then slowly decomposed to epoxy-[2] and RO$\cdot$. No such metastable intermediate has yet been identified.

AUTOXIDATION AND PEROXYL RADICAL TRAPPING REACTIONS OF α-TOCOPHEROL IN LIPID BILAYERS

Products of α-Tocopherol Oxidation by Peroxyl Radicals in Lipid Bilayers

To determine whether α-tocopherol reacts with peroxyl radicals in phospholipid membranes as in homogeneous solution, we studied the α-tocopherol oxidation by AMVN-derived peroxyl radicals in a liposome system (26). The liposome system affords a reaction environment similar to biological membranes, where α-tocopherol resides with its isoprenoid chain buried in the bilayer and the chromanoxy group may interact with the aqueous phase (27). Unlike biological membranes, the chemically defined liposome system lacks other redox-active components that could obscure some reactions of α-tocopherol and its products. In liposomes containing AMVN, [^{14}C]α-tocopherol was depleted with near-linear kinetics over a 12 h period. As in homogeneous solution, the products included epoxytocopherones [4/5], their epoxyquinone hydrolysis products [6/7], 8a-(alkyldioxy)tocopherone [3c] and its quinone hydrolysis product [8], and a small amount of 8a-hydroperoxytocopherone [3b] (Table 1).

No tocopherone product containing a phospholipid moiety was found, which suggests that the α-tocopherol quantitatively scavenged the AMVN-derived peroxyl radicals and completely prevented their reaction with liposomal lipids. This was verified by the observation that significant accumulation of lipid-conjugated dienes occurred only after complete α-tocopherol oxidation. In contrast to homogeneous solution, where nonepoxide products [3] and [8] predominated, epoxide products [4] through [7] predominated in the liposomes (Table 1).

Stability of α-Tocopherol Oxidation Products in the Lipid Bilayers and Mechanism of Epoxide Product Formation

8a-(Alkyldioxy)tocopherone [3c] decomposed slowly in liposomes to tocopherolquinone [8] and epoxides [4] through [7]. At a concentration of 0.1 mol per 100 mol, about 70% of the initial tocopherone remained intact after 6 h at 37°C. Tocopherolquinone [8], the primary product, was formed in approximately 25-fold excess over the epoxides. 8a-Hydroperoxy-tocopherone [3b] decomposition yielded tocopherolquinone [8] and epoxides [4] through [7] in approximately a 3:1 ratio. The distribution of tocopherone [3c] decomposition products was not affected by including either AMVN or α-tocopherol in the lipid bilayer. Were epoxide formation from [3c] to depend on peroxyl radical attack, α-tocopherol would have inhibited epoxide formation whereas AMVN would have stimulated epoxide formation. Peroxyl radicals thus do not further oxidize tocopherone [3c] in the lipid bilayer. This would be expected were tocopherone [3c] no more reactive toward radicals that the surrounding phospholipid, which is present in great excess over [3c]. Tocopherone decomposition therefore appears to be a unimolecular reaction, possibly initiated by homolysis of the peroxide moiety.

Epoxide products [4] through [7] formed during α-tocopherol oxidation are not formed via tocopherones [3b] and [3c]. Epoxides are clearly the predominant products during α-tocopherol oxidation in the lipid bilayer, yet they are only minor products of tocopherone decomposition. Moreover, tocopherone [3c] decomposes too slowly under the incubation conditions to account for the levels of epoxides formed. These results in lipid bilayers collectively demonstrate that, as in homogeneous solution, 8a-(alkyldioxy)tocopherones and epoxides arise from two competing reactions of the tocopheroxyl radical [2] with peroxyl radicals.

HYDROLYSIS AND REDUCTION OF 8a-(ALKYLDIOXY)TOCOPHERONES: FEASIBILITY OF A TWO-ELECTRON α-TOCOPHEROL REDOX CYCLE

Previous studies indicated that tocopherones hydrolyze easily to tocopherolquinone [8] under acid conditions and that they react with reductants to regenerate α-tocopherol (14,19,28,29). Since 8a-(alkyldioxy)tocopherones appear likely to result from α-tocopherol oxidation in biological membranes (see earlier), these products have been postulated to participate in redox cycles of α-tocopherol (19,30). To explore this possibility further, we studied the hydrolysis and reduction reactions of 8a-(alkyldioxy)tocopherone [3c] in homogeneous solution and in lipid bilayers.

Hydrolysis of 8a-(Alkyldioxy)tocopherone to Tocopherolquinone

When incubated in acetonitrile-buffer mixtures, 8a-(alkyldioxy)tocopherone [3c] hydrolyzed to tocopherolquinone [8], which was monitored spectroscopically at 266 nm (31). The overall hydrolysis reaction displayed a kinetic profile consistent with an initial acid-catalyzed dissociation of 8a-(alkyldioxy)tocopherone [3c] to the tocopherone cation T^+, followed by the rapid pseudo–first-order hydrolysis of T^+ to 8a-hydroxytocopherone [3d] and acid-catalyzed rearrangement of [3d] to tocopherolquinone [8] [Eqs. (10 through (12)]:

$$3c \rightarrow T^+ \tag{10}$$

$$T^+ + H_2O \rightarrow 3d \tag{11}$$

$$3d \rightarrow 8 \tag{12}$$

Reaction (10) is actually an acid-catalyzed dissociation that proceeds in two steps (13) and (14):

$$\text{TOOR } (3c) + H^+ \rightarrow T(H^+)OOR \tag{13}$$

$$T(H^+)OOR \rightarrow T^+ + ROOH \tag{14}$$

Only when the reaction medium is acidic enough to drive reactions (13) and (14) (approximately pH < 5) does the overall conversion of [3c] to [8] proceed at a measurable rate. In further support of this scheme, authentic [3d] rearranged to quinone [8] in a manner kinetically consistent with its proposed intermediary role in 8a-(alkyldioxy)tocopherone hydrolysis (31). Moreover, ^{18}O-labeled quinone [8] was recovered when [3c] hydrolyzed in $H_2^{18}O$. This confirms the incorporation of water into the product in reaction (11).

Reduction of 8a-(Alkyldioxy)tocopherone to α-Tocopherol by Ascorbic Acid

Treatment of either 8a-(alkyldioxy)tocopherone [3c] or 8a-hydroxytocopherone [3d] with ascorbic acid reduced the tocopherones to α-tocopherol (31). Reduction of 8a-(alkyldioxy)tocopherone [3c] was accelerated at pH < 5, which corresponds to the pH requirement for [3c] hydrolysis to quinone [8]. This observation suggested that [3c] reduction to α-tocopherol and hydrolysis to quinone [8] involve a common intermediate, T^+. Consistent with this conclusion is the observation that 8a-hydroxytocopherone [3d] also was reduced most efficiently at low pH, which apparently raises the equilibrium concentration of T^+ in reaction (11). Additional support for T^+ as the ascorbic acid-reducible intermediate comes from the earlier work of Marcus and Hawley, who found that the corresponding tocopherone cation of the α-tocopherol model compound 2,2,5,7,8-pentamethyl-6-hydroxychroman, but not the 8a-hydroxy derivative, was electrochemically reducible (32). These observations support the combined scheme for hydrolysis and reduction of 8a-(alkyldioxy)tocopherones depicted in Figure 3. Both hydrolysis and reduction of [3c] proceed via T^+, which is formed by the acid-catalyzed loss of the 8a-alkyldioxy moiety. T^+ either may reversibly hydrolyze to 8a-hydroxytocopherone [3d] or be reduced by ascorbic acid. 8a-Hydroxytocopherone [3d] may rearrange to tocopherolquinone [8] but can only be reduced to α-tocopherol via T^+. Tocopherolquinone [8] was not reduced to α-tocopherol in any of the systems studied (31).

Reduction of 8a-(Alkyldioxy)tocopherones in Lipid Bilayers

Cellular electron donors, such as ascorbic acid, could theoretically reduce tocopherones to α-tocopherol to complete a two-electron redox cycle (19,30). Whether tocopherones [3c] or [3d] can be reduced efficiently in biological membranes remains unclear. Reduction of 8a-(alkyldioxy)tocopherone [3c] in liposomes incubated with ascorbic acid was slow and inefficient (31). When liposomes containing 3 mol per 100 mol [3c] were incubated with 5 mM ascorbate, only about 14% of the [3c] was consumed in 1 h. Moreover, only 10% of the [3c] consumed was reduced to α-tocopherol. The remaining 90% appeared as tocopherolquinone [8]. The overall reduction of 8a-(alkyldioxy)tocopherone [3c] is slow because there is too little acid available at pH 7 to rapidly convert [3c] to the reducible intermediate, T^+. Once formed, T^+ may be reduced inefficiently because ascorbic acid-dependent reduction cannot effectively compete with T^+ hydrolysis, even at high ascorbic acid concentrations (e.g., 5 mM). Whether this is due to a reduced ascorbic acid concentration at the bilayer surface or a reduced rate constant for the reduction is not known.

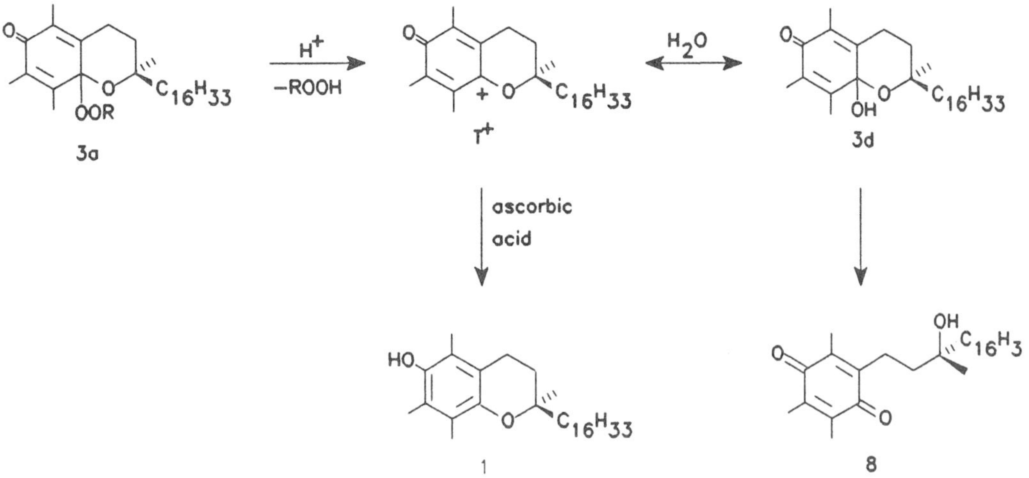

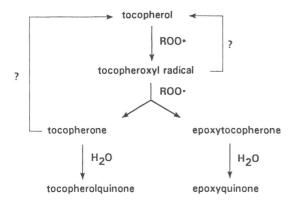

FIGURE 3 Proposed mechanisms for hydrolysis and reduction of 8a-(alkyldioxy)tocopherones. See text for discussion.

The foregoing discussion indicates that the lipid bilayer environment at physiological pH certainly does not favor a nonenzymic reduction of 8a-(alkyldioxy)tocopherones, such as [3c]. Efficient reduction clearly would require enzymic catalysis. 8a-Hydroxytocopherone [3d] reduction would not necessarily require acid catalysis, since [3d] exists in equilibrium with the reducible intermediate T^+. Nevertheless, the poor efficiency of nonenzymic T^+ reducton (see earlier) suggests that enzyme catalysis also would be required to reduce [3d] in biological membranes.

CONCLUSIONS

The peroxyl radical trapping reactions of α-tocopherol in biomimetic systems form distinctive primary products, 8a-(alkyldioxy)tocopherones and epoxytocopherones. These two prod-

FIGURE 4 Competing peroxyl radical-dependent oxidation pathways and putative redox cycles for α-tocopherol. See text for discussion.

uct classes arise via competing reactions of the tocopheroxyl radical (Fig. 4). Whereas 8a-(alkyldioxy)tocopherone formation results from a simple radical-radical termination, epoxide formation results from a more complex reaction sequence that is apparently initiated by epoxidation of the tocopheroxyl radical. Although the 8a-(alkyldioxy)tocopherone-forming pathway clearly results in radical trapping, the epoxide-forming pathway may actually cause α-tocopherol autoxidation. Further clarification of the epoxidation chemistry is necessary to assess its impact on the antioxidant performance of α-tocopherol. 8a-(Alkyldioxy)tocopherones may undergo reduction to complete a two-electron α-tocopherol redox cycle, but this process would require enzymic catalysis in biological membranes. Whether this actually occurs physiologically is under investigation. Epoxide products, on the other hand, appear to be irreversibly oxidized, since complete reduction of the epoxides represents a formidable biochemical problem. Are epoxide products formed in vivo during α-tocopherol antioxidant actions, and could they serve as biochemical markers for α-tocopherol antioxidant function? Moreover, if epoxides are formed in proportion to numbers of α-tocopherol redox cycles, could epoxides act as markers for redox cycles? Extension of these investigations from biomimetic systems to biological membranes should provide answers to these questions and will be the focus of our continuing studies of physiological α-tocopherol turnover.

ACKNOWLEDGMENTS

The author acknowledges the contributions of Todd Kennedy, Jeanne Burr, Kathryn Kaysen, and Peter Baker to this work. Work from the author's laboratory was supported in part by USPHS Grant CA47943 and by American Cancer Society Institutional Research Grant IN110.

REFERENCES

1. Burton GW, Ingold KU. Vitamin E: application of the principles of physical organic chemistry to the exploration of its structure and function. Acc Chem Res 1986; 19:194-201.
2. Howard JA, Ingold KU. Absolute rate constants for hydrocarbon autoxidation. VI. Alkyl aromatic and olefinic hydrocarbons. Can J Chem 1967; 45:793-802.
3. Kornbrust DJ, Mavis RD. Relative susceptibility of microsomes from lung, heart, liver, kidney, brain and testes to lipid peroxidation: correlation with vitamin E content. Lipids 1980; 15:315-22.
4. Sevanian A, Hacker AD, Elsayed N. Influence of vitamin E and nitrogen dioxide on lipid peroxidation in rat lung and liver microsomes. Lipids 1982; 17:269-77.
5. Packer JE, Slater TF, Willson RL. Direct observation of a free radical interaction between vitamin E and vitamin C. Nature 1979; 278:737-8.
6. Tappel AL, Brown WD, Zalkin H, Maier VP. Unsaturated lipid peroxidation catalyzed by hematin compounds and its inhibition by vitamin E. J Am Oil Chem Soc 1961; 38:5-9.
7. Niki E, Kawakami Y, Yamamoto Y, Kamiya Y. Oxidation of lipids. VIII. Synergistic inhibition of oxidation of phosphatidylcholine liposome in aqueous dispersion by vitamin E and vitamin C. Bull Chem Soc Jpn 1985; 58:1971-5.
8. Mehlhorn RJ, Sumida S, Packer L. Tocopheroxyl radical persistence and tocopherol consumption in liposomes and in vitamin E-enriched rat liver mitochondria and microsomes. J Biol Chem 1989; 264:13448-52.
9. Reddy CC, Scholz RW, Thomas CE, Massaro EJ. Vitamin E dependent reduced glutathione inhibition of rat liver microsomal lipid peroxidation. Life Sci 1982; 31:571-6.
10. Haenen GRMM, Bast A. Protection against lipid peroxidation by a microsomal glutathione dependent labile factor. FEBS Lett 1983; 159:24-8.
11. Pryor WA, Strickland T, Church DF. Comparison of the efficiencies of several natural and synthetic antioxidants in aqueous sodium dodecyl sulfate micelle solutions. J Am Chem Soc 1988; 110:2224-9.

12. Burton GW, Ingold KU. Autoxidation of biological molecules. 1. The antioxidant activity of vitamin E and related chain-breaking phenolic antioxidants in vitro. J Am Chem Soc 1981; 103:6472-7.

13. Niki E, Tanimura R, Kamiya Y. Oxidation of lipids. II. Rate of inhibition of oxidation by alpha-tocopherol and hindered phenols measured by chemiluminescence. Bull Chem Soc Jpn 1982; 55: 1551-5.

14. Winterle J, Dulin D, Mill T. Products and stoichiometry of reaction of vitamin E with alkylperoxy radicals. J Org Chem 1984; 49:491-5.

15. Clough RL, Yee BG, Foote CS, Chemistry of singlet oxygen. 30. The unstable primary product of tocopherol photooxidation. J Am Chem Soc 1979; 101:683-6.

16. Skinner WA, Alaupovic P. Vitamin E oxidation with alkaline ferrricyanide. Science 1963; 140: 803-5.

17. Matsuo M, Matsumoto S, Iitaka Y. Oxygenations of vitamin E (alpha-tocopherol) and its model compound 2,2,5,7,8-pentamethylchroman-6-ol in the presence of the superoxide radical solubilized in aprotic solvents: unique epoxidations and recyclizations. J Org Chem 1987; 52:3514-20.

18. Urano S, Yamanoi S, Hattori Y, Matsuo M. Radical scavenging reactions of alpha-tocopherol. II. The reaction with some alkyl radicals. Lipid 1977; 12:105-8.

19. Goodhue CT, Risley HA. Reactions of vitamin E with peroxides. II. Reaction of benzoyl peroxide with d-alpha-tocopherol in alcohols. Biochemistry 1965; 4:854-8.

20. Gardner HW, Eskins K, Grams GW, Inglett GE. Radical addition of linoleic acid hydroperoxides to alpha-tocopherol or the analogous hydroxychroman. Lipids 1972; 7:324-33.

21. Niki E. Free radical initiators as source of water- or lipid-soluble peroxyl radicals. Methods Enzymol 1990; 186:100-8.

22. Yamauchi R, Matsui T, Satake Y, Kato K, Ueno Y. Reaction products of alpha-tocopherol with a free radical initiator, 2,2'-azobis(2,4-dimethylvaleronitrile). Lipids 1989; 24:204-9.

23. Matsuo M, Matsumoto S, Iitaka Y, Niki E. Radical scavenging reactions of vitamin E and its model compound, 2,2,5,7,8-pentamethylchroman-6-ol, in a *tert*-butylperoxyl radical generating system. J Am Chem Soc 1989; 111:7179-85.

24. Liebler DC, Baker PF, Kaysen KL. Oxidation of vitamin E: evidence for competing autoxidation and peroxyl radical trapping reactions of the tocopheroxyl radical. J Am Chem Soc 1990; 112: 6995-7000.

25. Urano S, Hattori Y, Yamanoi S, Matsuo M. ^{13}C nuclear magnetic resonance studies on alpha-tocopherol and its derivatives. Chem Pharm Bull 1980; 28:1992-8.

26. Liebler DC, Kaysen KL, Burr JA. Peroxyl radical trapping and autoxidation reactions of alpha-tocopherol in lipid bilayers. Chem Res Toxicol 1991; 4:89-93.

27. Perly B, Smith ICP, Hughes L, Burton GW, Ingold KU. Estimation of the location of natural alpha-tocopherol in lipid bilayers by ^{13}C-NMR spectroscopy. Biochim Biophys Acta 1985; 819: 131-5.

28. Boyer PD. The preparation of a reversible oxidation product of alpha-tocopherol, alpha-tocopheroxide and of related oxides. J Am Chem Soc 1951; 73:733-40.

29. Durckheimer W, Cohen LA. The chemistry of 9-hydroxy-alpha-tocopherone, a quinone hemiacetal. J Am Chem Soc 1964; 86:4388-93.

30. McCay PB. Vitamin E: interactions with free radicals and ascorbate. Annu Rev Nutr 1985; 5:323-40.

31. Liebler DC, Kaysen KL, Kennedy TA. Redox cycles of vitamin E: hydrolysis and ascorbic acid dependent reduction of 8a-(alkyldioxy)tocopherones. Biochemistry 1989; 28:9772-7.

32. Marcus MF, Hawley MD. Electrochemical studies of the redox behavior of alpha-tocopherol. Biochim Biophys Acta 1970; 201:1-8.

8

Synthesis and Kinetic Study of Antioxidant and Prooxidant Actions of Vitamin E Derivatives

Kazuo Mukai

Ehime University, Matsuyama, Japan

INTRODUCTION

Vitamin E compounds (α-, β-, γ-, and δ-tocopherols) are well known as scavengers of active free radicals (LOO$^{\cdot}$, LO$^{\cdot}$, and HO$^{\cdot}$) generated in biological systems. The antioxidant actions of the tocopherols (TocH) have been ascribed to hydrogen abstraction from the OH group in the tocopherols by a peroxyl radical (LOO$^{\cdot}$). The hydrogen abstraction produces a tocopheroxyl radical (Toc$^{\cdot}$), which combines with another peroxyl radical [reactions (1) and (2)] (1,2):

$$LOO^{\cdot} + TocH \xrightarrow{k_1} LOOH + Toc^{\cdot} \tag{1}$$

$$LOO^{\cdot} + Toc^{\cdot} \xrightarrow{k_2} LOO\text{-}Toc \tag{2}$$

On the other hand, hydrophilic vitamin C (ascorbic acid, AsH$_2$) is, by itself, a very poor antioxidant (3,4), but it enhances the antioxidant activity of tocopherols by regenerating the tocopheroxyl to tocopherol (reaction 3) (5,6):

$$Toc^{\cdot} + Vit\ C \xrightarrow{k_3} TocH + Vit\ C^{\cdot} \tag{3}$$

Further, tocopherols can also act as an efficient scavenger of singlet oxygen ($^1O_2$) (7–9). It was shown that α-tocopherol scavenges $^1O_2$ by a combination of physical quenching (k_q) and chemical reaction (chemical quenching) (k_r). Because $k_q \gg k_r$, the quenching process is almost entirely "physical"; that is, α-tocopherol deactivates about 120 $^1O_2$ molecules before being destroyed by chemical reaction [reaction (4)] (7):

$$^1O_2 + TocH \xrightarrow{k_Q} \text{physical quenching + chemical reaction} \tag{4}$$

On the other hand, several investigators demonstrated that α-tocopherol at high concentrations acts as a prooxidant during the autoxidation of polyunsaturated fatty acids (LH) in an aqueous medium and bulk phase (10–12). This prooxidant effect of α-tocopherol leads to an increase in the level of hydroperoxides (LOOH) having a conjugated diene structure. It has been proposed that tocopheroxyl radicals participate in this prooxidant effect through reactions (5) and (6) (10,12):

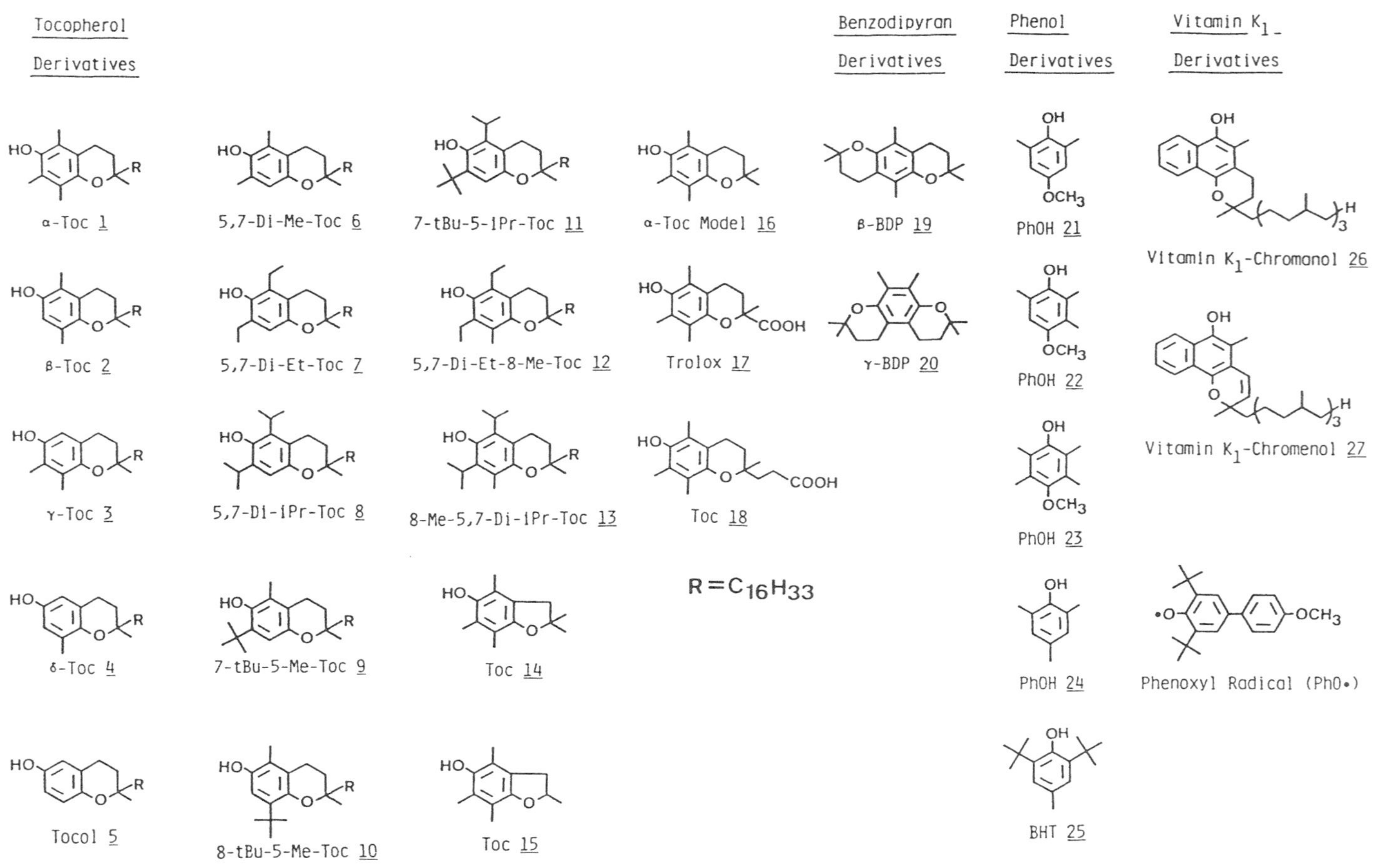

FIGURE 1   Molecular structures of tocopherols **1** through **18**, benzodipyrans **19** and **20**, phenols **21** through **25**, vitamin K₁-chromanol **26** and K₁-chromenol **27**, and substituted phenoxyl radical (PhO·).

$$\text{Toc}^{\cdot} + \text{LH} \xrightarrow{k_5} \text{TocH} + \text{L}^{\cdot} \tag{5}$$

$$\text{Toc}^{\cdot} + \text{LOOH} \xrightarrow{k_{-1}} \text{TocH} + \text{LOO}^{\cdot} \tag{6}$$

where (5) is the chain transfer reaction and (6) is the reverse reaction of (1). However, the kinetic studies of reactions (5) and (6) have not been undertaken, as far as we know.

In recent years, to obtain tocopherol compounds with a higher antioxidant activity than α-tocopherol, several investigators have prepared many tocopherol compounds (1,13–19). It was found that a few tocopherols have a higher antioxidant activity than α-tocopherol (1,17, 19). However, the number of the tocopherol compounds with a phytyl side chain at position 2 are very limited (1,19). Although α-tocopherol and an α-tocopherol model in which the phytyl side chain is replaced by a methyl group have similar antioxidant activities, the latter compound generally shows little or no vitamin E activity in vivo. The phytyl side chain appears to be necessary for the phospholipid penetration.

Therefore, to obtain tocopherol compounds with higher in vivo biological and/or in vitro antioxidant activity than α-tocopherol, we prepared many new tocopherol derivatives that have two or three alkyl substituents at the aromatic ring and the phytyl side chain at position 2 (see Fig. 1). Further, to clarify the structure-activity relationship in the antioxidant and prooxidant actions of tocopherols, systematic kinetic studies of reactions (1) and (3) through (6) have been performed using α-, β-, γ-, and δ-tocopherols, tocopherol derivatives [5] through [20], and related phenols [21] through [25].

FREE RADICAL SCAVENGING ACTIVITY OF TOCOPHEROL DERIVATIVES AND RELATED PHENOLS IN SOLUTION

Second-Order Rate Constants k_7 for the Reaction of Substituted Phenoxyl (PhO$^{\cdot}$) with Tocopherols and Related Phenols

As natural vitamin E compounds, α-, β-, γ-, and δ-tocopherols are well known. These tocopherols have different numbers of methyl substituents at the aromatic ring and show different biological activities (20). Burton and Ingold reported the second-order rate constants k_1 for the reaction of α-, β-, γ-, and δ-tocopherols with the poly(peroxystyryl)peroxyl radical using an O_2 consumption method (see Table 1) (17,21). We measured the reaction rates k_7 between the tocopherols and the substituted phenoxyl radical (PhO$^{\cdot}$) (see Fig. 1) in ethanol solution [reaction (7)], using stop-flow spectrophotometry as a model reaction of tocopherol with LOO$^{\cdot}$, LO$^{\cdot}$, and HO$^{\cdot}$ radicals in biological systems (22,23).

$$\text{PhO}^{\cdot} + \text{TocH} \xrightarrow{k_7} \text{PhOH} + \text{Toc}^{\cdot} \tag{7}$$

By reacting the PhO$^{\cdot}$ radical with α-tocopherol under pseudo–first-order conditions, the absorption of PhO$^{\cdot}$ at 376 nm decreases rapidly and the absorption of tocopheroxyl radical at 428 nm increases (see Fig. 2). The rate was measured by following the decrease in absorbance at 376 nm of the phenoxyl radical. The details of these experiments were reported in a previous paper (22). The second-order rate constants k_7 obtained are 5.12×10^3 (α-TocH), 2.24×10^3 (β-TocH), 2.42×10^3 (γ-TocH), and 1.00×10^3 (δ-TocH) M^{-1} s^{-1} in ethanol at 25°C. The relative k_7 values ($\alpha/\beta/\gamma/\delta$, 100:44:47:20) obtained by the stop-flow technique are in good agreement with the k_1 values (100:41:44:14) obtained by the O_2 consumption method, although the absolute values are about 600 times smaller than those for the reaction of tocopherols with the poly(peroxystyryl)peroxyl radical in chlorobenzene. The result suggests that

TABLE 1 Second-Order Rate Constants k_7, k_1, and k_Q and Half-Peak Oxidation Potentials $E_{p/2}$

Compounds	$10^{-3}k_7$ $(M^{-1}s^{-1})$	$10^{-6}k_1$ $(M^{-1}s^{-1})$	$10^{-8}k_Q$ $(M^{-1}s^{-1})$	$E_{p/2}$ (mV versus SCE)
Tocopherol derivatives				
α-Tocopherol **1**	5.12	3.20	2.06	860
β-Tocopherol **2**	2.24	1.30	1.53	920
γ-Tocopherol **3**	2.42	1.40	1.38	930
δ-Tocopherol **4**	1.00	—	0.53	990
Tocol **5**	0.56	1.80	0.28	1050
5,7-Di-Me-Toc **6**	2.39	—	0.97	890
5,7-Di-Et-Toc **7**	2.47	—	0.95	890
5,7-Di-iPr-Toc **8**	2.51	—	1.13	890
7-tBu-5-Me-Toc **9**	2.97	—	1.22	880
8-tBu-5-Me-Toc **10**	3.62	—	1.63	970
7-tBu-5-iPr-Toc **11**	2.16	—	1.02	910
5,7-Di-Et-8-Me-Toc **12**	3.64	—	2.11	840
8-Me-5,7-Di-iPr-Toc **13**	4.43	—	2.83	830
Toc **14**	6.99	5.70	3.57	810
Toc **15**	6.17	5.40	2.50	840
α-Toc Model **16**	4.20	3.80	1.96	870
Trolox **17**	2.23	1.10	0.81	890
Toc **18**	3.10	3.70	1.80	850
Benzodipyran derivatives				
β-BDP **9**	—	—	3.35	880
γ-BDP **20**	—	—	2.75	890
Phenol derivatives				
PhOH **21**	0.48	0.94	0.25	970
PhOH **22**	0.96	1.30	0.68	940
PhOH **23**	0.035	0.39	0.14	1070
PhOH **24**	0.017	0.085	0.038	1260
BHT **25**	0.035	0.014	0.034	1240
Vitamin K_1 derivatives				
Vitamin K_1-Chromanol **26**	35.4	—	—	780
Vitamin K_1-Chromenol **27**	24.8	—	—	750

the relative reactivities, that is, the relative antioxidant activities of tocopherols in homogeneous solution, do not depend on the kinds of radicals (substituted phenoxyl and peroxyl radicals) used, and the absolute rates are considerably different from each other.

Similarly,, the reaction rates k_7 of PhO˙ with tocopherols [5] through [20] and related phenols [21] through [25] have been observed (24,25). As listed in Table 1, 5,7-dimethyl-tocol [6] has quite similar rate constants to those of β- and γ-tocopherols, whereas δ-tocopherol is only approximately 20% as reactive as α-tocopherol and tocol is approximately 11% as reactive as α-tocopherol. This result indicates that the antioxidant activity of these tocopherols varies depending on the number of methyl substitutions. Consequently, we can expect that the antioxidant activity of these tocopherol compounds relates to the total electron-donating character of the methyl group substituted on the aromatic ring.

The values of log k_7 for tocopherols have been plotted against both the sum of the Hammett's σ constants ($\Sigma\sigma$) and Brown's σ^+ constants ($\Sigma\sigma^+$) for all the substituents on the tocopherols. The logarithms of rate constants k_7 were found to roughly correlate with $\Sigma\sigma$ or $\Sigma\sigma^+$ substituent constants, but the two cases could not be distinguished (18,23,24). Howard and

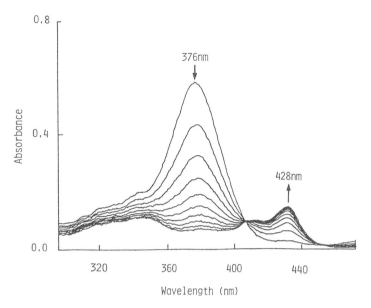

FIGURE 2 Change in electronic absorption spectrum of substituted phenoxyl radical (PhO˙) during reaction of PhO˙ with α-tocopherol in ethanol at 25°C. $[PhO˙]_{t=0} = 0.10$ mM, and $[\alpha\text{-tocH}]_{t=0} = 1.00$ mM. The spectra were recorded at 120 ms intervals. The arrow indicates the decrease in absorbance maximum with time.

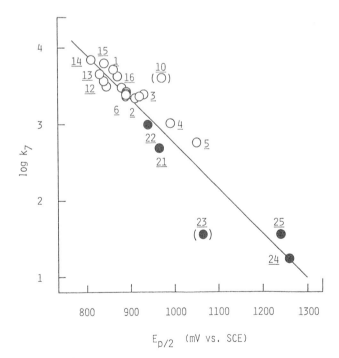

FIGURE 3 Plot of log k_7 versus $E_{p/2}$ for tocopherols **1** through **18** (open circles) and phenols **21** through **25** (solid circles).

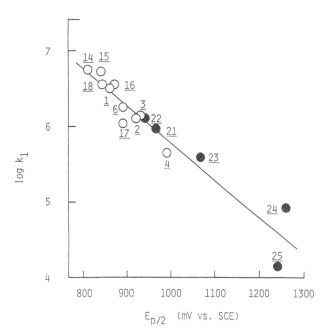

FIGURE 4   Plot of log k_1 versus $E_{p/2}$ for tocopherols (open circles) and phenols (solid circles).

Ingold (26) measured the rate constants k_1 for the reaction of ortho-alkyl phenols with poly-(peroxystyryl)peroxyl radicals. They reported that the steric effect due to ortho-alkyl substituents is important. However, the present result indicates that the steric effect on the reaction rate is not remarkable for the reactions of tocopherols with a substituted phenoxyl radical in ethanol. A similar trend was observed for the reaction of tocopherols with the poly(peroxystyryl)peroxyl radical (17,27).

Correlation Between Log k_1 (Log k_7) and $E_{p/2}$

Measurement of half-peak oxidation potentials $E_{p/2}$ of tocopherols has been performed (23, 24). The observed values are listed in Table 1. The values of log k_7 for tocopherols have been plotted against $E_{p/2}$. As shown in Figure 3, a plot of log k_7 versus $E_{p/2}$ is linear over most of the range with a slope of -5.5 V^{-1} (correlation coefficient = 0.98), except for [10] and [23]. The same correlation is given for the reaction of tocopherols with peroxyl radical, showing a slope of -5.0 V^{-1} (correlation coefficient = 0.96) (see Fig. 4). The tocopherols that have smaller $E_{p/2}$ values show higher reactivities. The excellent correlations between the log k_1 of the peroxyl system and the half-peak oxidation potentials $E_{p/2}$ and a similar good relationship for the present substituted phenoxyl system suggest a similar transition state for both reactions; that is, the transition state involves the charge-transfer intermediate in these free-radical scavenging reactions by tocopherols, as described later.

FINDING NEW TOCOPHEROL DERIVATIVES WITH THE HIGHEST ANTIOXIDANT ACTIVITY AMONG PHENOLIC ANTIOXIDANTS

In recent years, several investigators, including the present authors, have prepared many tocopherol compounds (1,13–19). However, the numbers of the tocopherol compound with higher antioxidant activities than α-tocopherol are very limited (1,17,19).

Burton et al. (17,21) reported the second-order rate constants k_1 for the reaction of α-tocopherol and some related phenols (for example, 4-methoxy-2,3,5,6-tetramethylphenol [23]) with poly(peroxystyryl)peroxyl radicals (LOO$^{\cdot}$). By comparing the k_1 value for α-tocopherol with those found for structurally related phenols that lacked the six-membered heterocyclic ring, they suggested that the structure of the ring was largely responsible for the high reactivity of α-tocopherol. That is, the high reactivity of α-tocopherol has been attributed to a strong orbital overlap between the $2p$ type of lone pair on the para oxygen atom and the aromatic π system. Further, they found that a better tocopherol compound is 2,3-dihydro-5-hydroxy-2,2,4,6,7-pentamethylbenzofuran (tocopherol [14]), which has a five-membered heterocyclic ring instead of the six-membered ring of α-tocopherol. They reported that the reaction rate of tocopherol [14] is 1.8 times higher than that of α-tocopherol. The high reactivity of tocopherol [14] has been attributed to stereoelectric factors relating to the orientation of the p-type lone pair on the oxygen in position 1 with respect to the aromatic ring.

As described in a previous section, we prepared many tocopherol compounds and studied the structure-activity relationship in antioxidant action of tocopherols (18,19,22–25). The result indicates that the k_1 and k_7 values increase as the total electron-donating capacity of the methyl substituents on the aromatic ring increases. Log k_1 and log k_7 obtained for tocopherols were found to correlate with their half-peak oxidation potentials $E_{p/2}$.

Vitamin K$_1$-chromanol [26] and K$_1$-chromenol [27] (see Fig. 1) have a structure similar to that of vitamin E chromanol (tocopherol). Compounds [26] and [27] have larger π-electron system than α-tocopherol, and thus we can expect that the half-peak oxidation potentials of [26] and [27] are smaller than that of α-tocopherol. In fact, as listed in Table 1, vitamin K$_1$-chromanol [26] and K$_1$-chromenol [27] show smaller $E_{p/2}$ values and were found to be 6.9 and 4.5 times more active than the α-tocopherol, respectively (25). Thus, these compounds have the highest antioxidant activity among phenolic antioxidants, including natural tocopherols, tocopherol derivatives, and related phenols in solution.

As listed in Table 1, tocopherols [14] and [15], with a five-membered heterocyclic ring, are 1.4 and 1.2 times as reactive as α-tocopherol in homogeneous solution, respectively. However, these tocopherols have no phytyl side chain, $-C_{16}H_{33}$, which is necessary for membrane penetration. On the other hand, both vitamin K$_1$-chromanol [26] and K$_1$-chromenol [27] have higher antioxidant activities than α-tocopherol and have a phytyl side chain. Therefore, compounds [26] and [27] may function as better antioxidants in vivo than α-tocopherol. Recently, Burton et al. (1) found a tocopherol derivative having 1.5–1.9 times higher biological activity than α-tocopherol. Here, the tocopherol is a derivative of tocopherol [14] in which one of the methyl groups at the 2 position has been replaced by a phytyl group, $-C_{16}H_{33}$. It will be interesting to study the biological activity of vitamin K$_1$-chromanol [26] and K$_1$-chromenol [27] in detail.

QUENCHING REACTION OF SINGLET OXYGEN BY TOCOPHEROL DERIVATIVES AND RELATED PHENOLS IN SOLUTION

Second-Order Rate Constants k_Q for the Reaction of $^1O_2$ with Tocopherols and Related Phenols

The rate of quenching of $^1O_2$ by 20 kinds of tocopherol derivatives, including α-, β-, γ-, and δ-tocopherols, and 5 structurally related phenols has been measured spectrophotometrically in ethanol at 35°C. Singlet oxygen was generated by the thermal decomposition of the naphthalene 1,4-endoperoxide (EP) (28). The overall rate constants k_Q ($= k_q + k_r$) for the reacction of $^1O_2$ with tocopherols and related phenols [reaction (4)] were determined in ethanol by Equation (8) derived from the steady-state treatment of Scheme 1 (28).

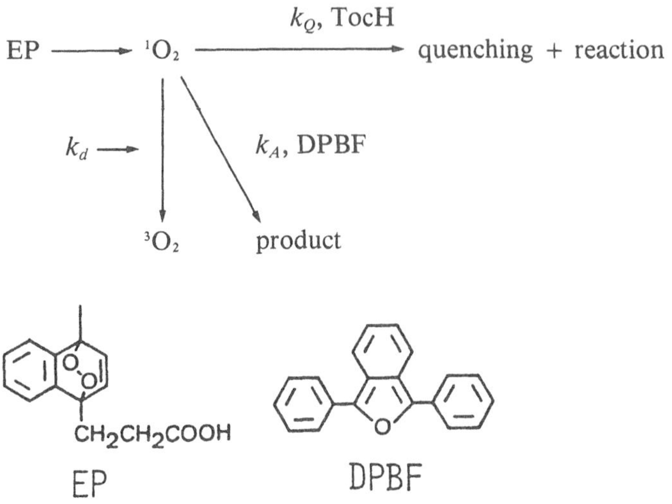

SCHEME 1

$$\frac{S_0}{S_s} = 1 + \frac{k_q + k_r}{k_d} \text{[TocH]} \tag{8}$$

where S_0 and S_s are slopes of the first-order plots of disappearance of the $^1O_2$ acceptor DPBF in the absence and presence of tocopherol, respectively. k_d is the rate of deactivation of $^1O_2$ in ethanol. The overall rate constants (k_Q) were calculated by using the value of k_d in ethanol ($k_d = 8.3 \times 10^4 \text{ s}^{-1}$), reported by Merkel and Kearns (29). The k_Q values obtained are summarized in Table 1 (30).

Yamauchi and Matsushita (31) have reported absolute second-order rate constants for the physical quenching (k_q) and chemical reaction (k_r) of α-, γ-, and δ-tocopherols with $^1O_2$, produced by methylene blue photosensitization in ethanol [reaction (4)]. The $^1O_2$ quenching rate constants k_q for α-, γ-, and δ-tocopherols in ethanol were estimated as 2.6×10^8, 1.8×10^8, and $1.0 \times 10^8 \text{ M}^{-1} \text{ s}^{-1}$, respectively. The rate constants k_r for the chemical reaction between each tocopherol and $^1O_2$ were 6.6×10^6, 2.6×10^6, and $0.7 \times 10^6 \text{ M}^{-1} \text{ s}^{-1}$ for α-, γ-, and δ-tocopherols, respectively, in ethanol at 20°C. Subsequently, the overall rate constants, k_Q ($k_Q = k_q + k_r$), obtained are 2.67×10^8 (α-TocH), 1.83×10^8 (γ-TocH), and 1.01×10^8 (δ-TocH) $\text{M}^{-1} \text{ s}^{-1}$. The relative k_Q values ($\alpha/\gamma/\delta = 100:67:26$) of α-, γ-, and δ-tocopherols obtained by the present work are in agreement with those (100:69:38) reported by Yamauchi and Matsushita, although the absolute k_Q values are 23–47% smaller than those for the reaction of corresponding tocopherols with the $^1O_2$ produced by methylene blue photosensitization. Kaiser et al. (32) also recently reported the rate constants k_q and k_r for the reaction of α-, β-, γ-, and δ-tocopherols with $^1O_2$ produced by the thermal decomposition of the endoperoxide of 3,3'-(1,4-naphthylidene)dipropionate. The relative k_Q ($= k_q + k_r$) values were found to decrease in the order $\alpha > \beta > \gamma > \delta$-tocopherols.

By comparing the rate constant k_Q observed for 5,7-dimethyltocol [6] with that of α-tocopherol, the former is only approximately 47% as reactive as the lattter. Further, β- and γ-tocopherols are 74 and 67% as reactive as α-tocopherol, respectively, whereas δ-tocopherol is only approximately 26% as reactive as α-tocopherol and tocol [5] is only approximately 14% as reactive as α-tocopherol. This result indicates that the quenching rate of these tocopherol compounds relates to the total electron-donating character of the alkyl group substituents on the aromatic ring.

The results listed in Table 1 demonstrate that two tocopherols [14] and [15] with a five-membered heterocyclic ring are 1.73 and 1.21 times as reactive as α-tocopherol, which has the highest antioxidant activity among natural tocopherols. Two benzodipyran compounds [19] and [20] having no OH group were also found to be 1.63 and 1.33 times more active than the α-tocopherol, respectively. However, tocopherols [17] and [18] and phenol derivatives [21] through [25] showed less reactivity than α-tocopherol. Therefore, tocopherol [14] with a five-membered heterocyclic ring has the highest antioxidant activity among phenolic antioxidants, including natural tocopherols, tocopherol derivatives, and related phenols in solution.

Correlation Between Log k_Q and $E_{p/2}$

The values of log k_Q for tocopherols [1] through [18] and related compounds [19] through [25] have been plotted against $E_{p/2}$. As shown in Figure 5, a plot of log k_Q versus $E_{p/2}$ is linear over most of the range, with a slope of -4.5 V^{-1} (correlation coefficient = -0.95). The tocopherols that have smaller $E_{p/2}$ values show higher reactivities.

Thomas and Foote (33) have extensively studied the reaction of $^1O_2$ with alkyl-substituted phenols. They have found that para-substituted 2,6-di-tert-butylphenols show a linear correlation between the log of the total rate of $^1O_2$ removal and their half-peak oxidation potentials $E_{p/2}$. Saito et al. (28) and Thomas and Foote (33) suggested that the most plausible scheme is a charge-transfer reaction between $^1O_2$ and phenol, leading to products by means of superoxide and phenol radical cation. As described earlier, the logarithm of the second-order rate constants k_Q obtained for tocopherols and related phenols was also found to correlate

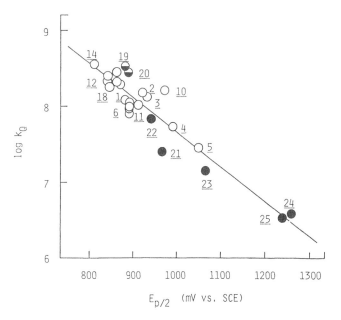

FIGURE 5   Plot of log k_Q versus $E_{p/2}$ for tocopherols **1** through **18** (open circles), benzodipyrans **19** and **20** (half-open), and phenols **21** through **25** (solid).

with their half-peak oxidation potentials $E_{p/2}$. These facts suggest that the transition state in the $^1O_2$ quenching reaction by tocopherols has the property of the charge-transfer intermediate.

Finding a Linear Correlation Between the Rates of Quenching of Singlet Oxygen and Scavenging of Peroxyl and Phenoxyl Radicals in Solution

As described in a previous section, log k_1 and log k_7 of tocopherols and related phenols were found to correlate with their half-peak oxidation potentials $E_{p/2}$, respectively. Similarly, log k_Q was found to correlate with $E_{p/2}$. Therefore, the values of k_Q were plotted against k_1 and k_7, respectively. As shown in Figures 6 and 7, the k_Q values were found to correlate linearly with the k_1 values (correlation coefficient = 0.94) and the k_7 values (correlation coefficient = 0.96), respectively. The ratios of k_Q to k_1 and k_Q to k_7 were estimated as 56 and 4.9×10^4 from the gradient in Figures 6 and 7, respectively [Eqs. (9) and (10)] (30).

$$k_Q = 56 k_1 \tag{9}$$
$$k_Q = 4.9 \times 10^4 \, k_7 \tag{10}$$

The result suggests that the relative reactivities, that is, relative antioxidant activities of phenolic antioxidants in homogeneous solution, do not depend on whether singlet oxygen ($^1O_2$), peroxyl radical (LOO$^\cdot$), or substituted phenoxyl radical (PhO$^\cdot$) is the reactive species. Consequently, if we could determine the quenching rate k_Q of $^1O_2$ by phenolic antioxidants in

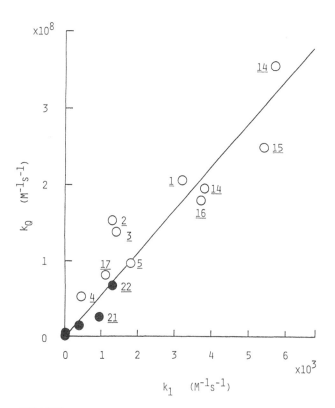

FIGURE 6 Plot of k_Q versus k_1 for tocopherols (open) and phenols (solid).

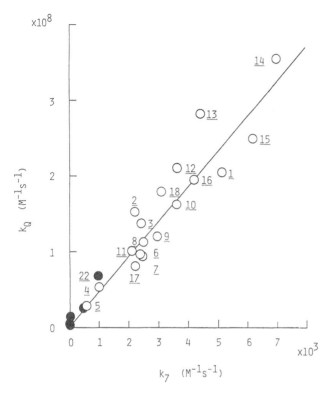

FIGURE 7   Plot of k_Q versus k_7 for tocopherols **1** through **18** (open) and phenols **21** through **25** (solid).

solution, we can presume the scavenging rates k_1 and k_7 of free radicals using Equations (9) and (10), and vice versa.

As described in a previous section, it was suggested that the transition state in the $^1O_2$ quenching reaction by tocopherols and related phenols has the property of the charge-transfer intermediate. Further, the rate of quenching of singlet oxygen by phenolic antioxidants was found to correlate with those of scavenging of peroxyl and phenoxyl radicals in solution, respectively. These facts indicate that the property of the transition states in the singlet oxygen quenching and free-radical scavenging reactions by phenolic antioxidants is similar, suggesting the charge-transfer intermediate.

MECHANISM OF THE REGENERATION REACTION OF TOCOPHEROL BY VITAMIN C

The regeneration reaction of tocopherol by vitamin C (ascorbic acid) has been studied by several investigators [reaction (3)] (3–6,34–36):

$$\text{Toc}^{\bullet} + \text{Vit C} \xrightarrow{k_3} \text{TocH} + \text{Vit C}^{\bullet} \tag{3}$$

Packer et al. (6) reported the absolute second-order rate constants k_3 for the reaction of vitamin C (sodium ascorbate) with α-tocopheroxyl radicals using pulse radiolysis methodology. They found a k_3 value of 1.55×10^6 M^{-1} s^{-1} in a system composed of an aerated, aqueous solution containing isopropanol (50%) and acetone (10%). Further, Scarpa et al. (34) measured the second-order rate constants for the reaction between α-tocopheroxyl radical in dimyristoyl PC liposomes and vitamin C in the surrounding aqueous phase (pH 7) using an EPR technique. The kinetic rate constant obtained was about $2 \times 10^5 \, M^{-1}s^{-1}$. This value is only one order of magnitude lower than that reported by Packer et al. for the same reaction in homogeneous solution.

Recently we measured the second-order rate constants for the reaction of 10 kinds of vitamin C derivatives with a vitamin E radical (5,7-diisopropyl-tocopheroxyl [8]) in organic solvent (benzene-ethanol, 2:1 vol/vol) using a stop-flow technique (4,35,36). However, detailed kinetic studies of the interaction between vitamin C derivatives and tocopheroxyl radicals, that is, the regeneration reaction of tocopherol by vitamin C, have not been performed.

Therefore, to clarify the mechanism of the reaction of vitamin C with tocopheroxyl radicals, we measured the second-order rate constants k_3 for the reaction of vitamin C with tocopheroxyl radical (7-*tert*-butyl-5-isopropyl-tocopheroxyl [11]) (see Fig. 8) in Triton X-100 micellar solution using a stop-flow technique and studied the effect of pH on the reaction rate (37). Based on the results, the structure-activity relationship for the regeneration reaction of tocopherol by vitamin C was discussed.

Generally, the tocopheroxyl radicals are unstable. However, 7-*tert*-butyl-5-isopropyl-tocopheroxyl [11] is comparatively stable in the absence of ascorbic acid and shows absorption peaks with λ_{max} = 416 and 396 nm in aqueous Triton X-100 micellar solution [10.0 weight %; 0.1 M phosphate buffer (pH 7.0); see Fig. 9]. Upon addition of excess ascorbic acid in

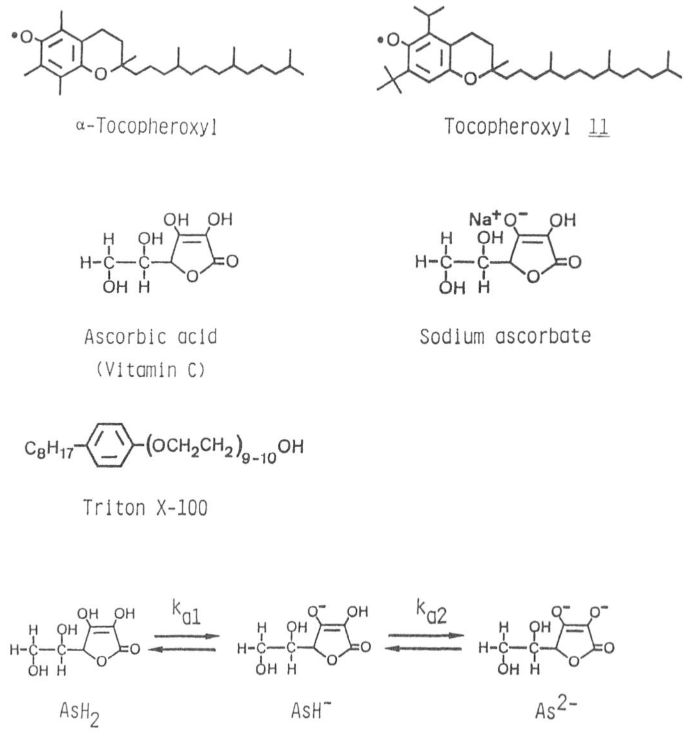

FIGURE 8   Molecular structures of ascorbic acid (AsH_2, AsH^-, and As^{2-}), sodium ascorbate (Na^+ AsH^-), α-tocopheroxyl, and tocopheroxyl **11**.

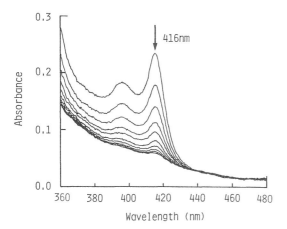

FIGURE 9 Change in electronic absorption spectrum of tocopheroxyl radical **11** during reaction of **11** with vitamin C in Triton X-100 micellar solution (0.1 M phosphate buffer, pH 7.0) at 25.0°C. [Toc $^\cdot$]$_{t=0}$ $\simeq 0.06$ mM, and [vitamin C]$_{t=0}$ = 1.15 mM. The spectra were recorded at 1000 ms intervals. The arrow indicates the decrease in absorbance maximum with time.

0.1 M phosphate buffer to Triton X-100 micellar solution containing tocopheroxyl [11] (1:1 vol/vol), the absorption spectrum of tocopheroxyl [11] disappears immediately. Figure 9 shows an example of the interaction between tocopheroxyl [11] (0.06 mM) and ascorbic acid (1.15 mM) in phosphate buffer (pH 7.0). The rate was measured by following the decrease in absorption at 416 nm of tocopheroxyl [11]. The second-order rate constant k_3 obtained for ascorbic acid at pH 7.0 is 322 M^{-1}s^{-1}.

Similar measurements were performed for the reaction of tocopheroxyl [11] with ascorbic acid at various pH values. The reaction rate for sodium ascorbate (AsH$^-$Na$^+$) with tocopheroxyl [11] in Triton X-100 micelle solution was also measured. The k_3 values obtained are summarized in Table 2 (37). The pH dependence of the second-order rate constants is represented in Figure 10. When the pH values are greater than 3, the rate constant k_3 increases rapidly from 26 M^{-1}s^{-1} at pH 3.0 to 311 M^{-1}s^{-1} at pH 6.0, remains constant ($k_3 = 318 \pm 4$ M^{-1}s^{-1}) between pH 7.0 and 9.0, and decreases to 273 M^{-1}s^{-1} at pH 10.0. This pH dependence reflects a complex mechanism that is discussed in the next section.

The second-order rate constant k_3 for the reaction of ascorbic acid with tocopheroxyl [11] was also measured in an organic solvent system. A rate constant k_3 of 49 M^{-1}s^{-1} was obtained in benzene-ethanol (2:1 vol/vol) solution. The observed k_3 value is listed in Table 2, together with those observed in micellar solution (37).

Ascorbic acid is dibasic and can exist in three different molecular forms, that is, ascorbic acid (AsH$_2$), ascorbate monoanion (AsH$^-$), and ascorbate dianion (As^{2-}), depending on the pH value (see Fig. 8). The equilibrium equations have the form

$$\text{AsH}_2 \overset{K_{a1}}{\rightleftharpoons} \text{AsH}^- \overset{K_{a2}}{\rightleftharpoons} \text{As}^{2-} \tag{11}$$

where pK_{a1} = 4.17 and pK_{a2} = 11.57.

The graph of the pH dependence of k_3 (Fig. 10) has a peak at a pH value intermediate between the two pK_a values (pK_{a1} and pK_{a2}) of the dibasic ascorbic acid. Therefore, the mole fractions f present as the AsH$_2$ molecule and the AsH$^-$ and As^{2-} ions were calculated as a function of pH. The analytical concentration C_a is given by

TABLE 2 Second-Order Rate Constants k_3 for the Reaction of Vitamin C (Ascorbic Acid) with Tocopheroxyl Radical **11** in Triton X-100 Micellar Solution and the Mole Fraction f of Ascorbate Monoanion (AsH^-)

pH	k_3	f $[AsH^-]/C_a$
3.0	25.7	0.063
3.5	56.4	0.176
4.0	123	0.403
4.5	188	0.681
5.0	248	0.871
5.5	268	0.955
6.0	311	0.985
7.0	322	0.999
8.0	319	1.000
9.0	314	0.997
10.0	273	0.974
Sodium ascorbate	300[a]	
Benzene-ethanol	49[b]	

[a]The k_3 value obtained for sodium ascorbate (Na^+AsH^-).
[b]The k_3 value obtained in benzene-ethanol (2:1 vol/vol) solution.

$$C_a = [AsH_2] + [AsH^-] + [As^{2-}] \tag{12}$$

From the ionization constant expressions for ascorbic acid (AsH_2), we obtain

$$[AsH^-] = \frac{[AsH_2]K_{a1}}{[H_3O^+]} \tag{13}$$

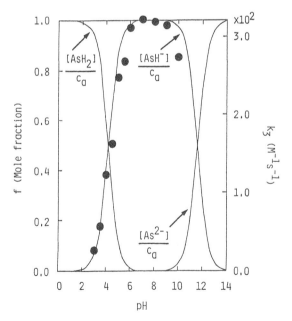

FIGURE 10 Plot of second-order rate constants k_3 versus pH and mole fraction f of three vitamin C species (AsH_2, AsH^-, and As^{2-}) versus pH.

$$[As^{2-}] = \frac{[AsH^-]K_{a2}}{[H_3O^+]} = \frac{[AsH_2]K_{a1}K_{a2}}{[H_3O^+]^2} \tag{14}$$

Substitution into the expression for the analytical concentration C_a yields Equation (15):

$$f = \frac{[AsH_2]}{C_a} = \frac{[H_3O^+]^2}{[H_3O^+]^2 + [H_3O^+]K_{a1} + K_{a1}K_{a2}} \tag{15}$$

Similarly, the expressions for the mole fractions present as AsH^- and As^{2-} can be derived from

$$f = \frac{[AsH^-]}{C_a} = \frac{[H_3O^+]K_{a1}}{[H_3O^+]^2 + [H_3O^+]K_{a1} + K_{a1}K_{a2}} \tag{16}$$

$$f = \frac{[As^{2-}]}{C_a} = \frac{K_{a1}K_{a2}}{[H_3O^+]^2 + [H_3O^+]K_{a1} + K_{a1}K_{a2}} \tag{17}$$

Fractions of total ascorbic acid present as AsH_2, AsH^-, and As^{2-} are shown as functions of pH in Figure 10.

As is clear from the results shown in Figure 10, a good correlation between the rate constants k_3 and mole fraction f of ascorbate (AsH^-) was observed. The result shows that the ascorbate monoanion (AsH^-) can regenerate tocopherol from tocopheroxyl in biological systems. In fact, as listed in Table 2, sodium ascorbate (AsH^-Na^+) and ascorbic acid at pH = 7.0 react with tocopheroxyl radical [11] at similar rates. Further, the values of k_3 have been plotted against the mole fraction ($f = [AsH^-]/C_a$) of the ascorbate monoanion. As shown in Figure 11, a good correlation between k_3 and $[AsH^-]/C_a$ was obtained. The results also indicate that ascorbic acid (AsH_2, the reduced form of ascorbic acid) does not have the ability to regenerate tocopherol from tocopheroxyl in aqueous solution.

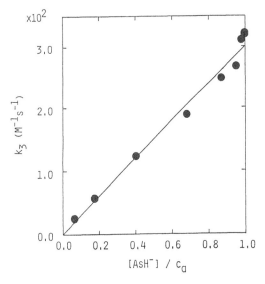

FIGURE 11 Plot of second-order rate constants k_3 versus mole fraction of ascorbate monoanion; $f = [AsH^-]/C_a$.

.In the present work we also measured the reaction rate k_3 of ascorbic acid with toco-pheroxyl radical [11] in an organic solvent system. A rate constant of 49 $M^{-1}s^{-1}$ was obtained in benzene-ethanol (2:1 vol/vol) solution. Ascorbic acid is considered to exist in the reduced form (AsH_2) in organic solvent systems. Therefore, the results indicate that the reduced form of ascorbic acid, AsH_2, has the ability to regenerate tocopherol from tocopheroxyl in organic solvent systems. However, the former reaction rate ($k_3 = 49\,M^{-1}s^{-1}$) in benzene-ethanol (2:1 vol/vol) solution is only about 15% of that ($k_3 = 322\,M^{-1}s^{-1}$) in Triton X-100 micelle solution at pH = 7.0.

KINETIC STUDY OF THE PROOXIDANT EFFECT OF VITAMIN E

As described in a previous section, several investigators showed that α-tocopherol in high concentration acts as a prooxidant during the autoxidation of polyunsaturated fatty acids (LH) (10–12). This prooxidant effect of α-tocopherol leads to an increase in hydroperoxides with a conjugated diene structure. Loury et al. (10) and Terao and Matsushita (12) proposed that tocopheroxyl radicals participate in this prooxidant effect through reactions (5) and (6), where reaction (5) is a hydrogen abstraction from the polyunsaturated fatty acids and a chain-transfer reaction. Reaction (6) is a reversal of reaction (1). These reactions are very important to understand fully the antioxidant and prooxidant properties of vitamin E.

$$Toc^{\cdot} + LH \xrightarrow{k_5} TocH + L^{\cdot} \tag{5}$$

$$Toc^{\cdot} + LOOH \xrightarrow{k_{-1}} TocH + LOO^{\cdot} \tag{6}$$

However, thus far the kinetics of these reactions have not been studied. Thus, we recently determined the rate constants k_{-1} and k_5 for these reactions and compared the values of the rate constants (38–40).

Hydrogen Abstraction by Tocopheroxyl Radical from Alkyl Hydroperoxides in Solution

We measured the second-order rate constants k_{-1} for the reaction of alkyl hydroperoxides (*n*-butyl hydroperoxide [28], *sec*-butyl hydroperoxide [29], and *tert*-butyl hydroperoxide [30]) with vitamin E radical (5,7-diisopropyl-tocopheroxyl [8]) in benzene at 25.0°C as a model for the reaction of lipid hydroperoxides with unstable vitamin E radical in biological systems. The rate constants k_{-1} for this reaction were determined by following the decrease in absorbance at 417 nm of tocopheroxyl radical [8]. The values of k_{-1} obtained are $1.34 \times 10^{-1}\,M^{-1}s^{-1}$ for [28], $2.42 \times 10^{-1}\,M^{-1}s^{-1}$ for [29], and $3.65 \times 10^{-1}\,M^{-1}s^{-1}$ for [30] and are listed in Table 3 (38). The values of k_{-1} decrease in the order t-BuOOH > sec-BuOOH > *n*-BuOOH. The result clearly indicates that the reactivity of alkyl hydroperoxides toward vitamin E radical depends on the total electron-donating capacity of alkyl substituents at the α-carbon atom of hydroperoxides rather than the steric hindrance due to alkyl substituents at the α-carbon atom. In fact, for the alkyl hydroperoxides [28], [29], and [30], log k_{-1} was found to correlate with the sum of Hammett's σ constants ($\Sigma\sigma$) and Brown's σ^+ constants ($\Sigma\sigma^+$) of alkyl sub-stituents at the α-carbon atom, but the two cases could not be distinguished (see Fig. 12).

Kinetic study for the reaction between the peroxyl radicals and hydrocarbons has been performed by many investigators (41,42). The results indicate that the reactivities of peroxyl radicals are dependent on their structure. In general, the rate constants increase along the series tertiary alkyl peroxyl < secondary alkyl peroxyl < primary alkyl peroxyl radical. This is the reverse order to that (t-BuOOH > sec-BuOOH > *n*-BuOOH) found for the value of k_{-1} for the reaction of hydroperoxides with vitamin E radical.

TABLE 3 Second-Order Rate Constants (k_{-1}) for Reaction of Tocopheroxyl Radical **8** with Alkyl Hydroperoxides **28**, **29**, and **30** in Benzene at 25.0°C and $\Sigma\sigma$ and $\Sigma\sigma^+$ Values

		k_{-1} (M^{-1} s^{-1})	$\Sigma\sigma^a$	$\Sigma\sigma^{+a}$
n-BuOOH	**28**	1.34×10^{-1}	-0.15	-0.30
s-BuOOH	**29**	2.42×10^{-1}	-0.32	-0.61
t-BuOOH	**30**	3.65×10^{-1}	-0.51	-0.93

$^a\Sigma\sigma$ and $\Sigma\sigma^+$ values were calculated by assuming that n-BuOOH, sec-BuOOH, and t-BuOOH have (1) C$_2$H$_5$-, (2) CH$_3$- and C$_2$H$_5$-, and (3) three CH$_3$ substituents at the α-carbon atom, respectively.

Recently, several investigators measured the rate constant k_1 for the reaction between peroxyl radical and vitamin E derivatives [reaction (1)]. The most reliable k_1 value was reported by Burton et al. (17) using the inhibited autoxidation of styrene method. For example, the k_1 value obtained for the reaction of α-tocopherol with poly(peroxystyryl)peroxyl radical was 3.20×10^6 M^{-1}s^{-1} in chlorobenzene at 30°C. Therefore, the k_{-1} values obtained in the present work are about seven orders of magnitude lower than this k_1 value.

Hydrogen Abstraction by Tocopheroxyl Radical from Fatty Acid Esters in Solution

We determined the second-order rate constants k_5 (39,40) for the reaction of various fatty acid esters with tocopheroxyl radicals. The rate k_5 has been compared with k_{-1} for reaction (6). The effect of substituent groups at the 5 and 7 positions of the tocopheroxyl radical was also studied. In Figure 13 we give the structures of the molecules studied in this work.

The rate constant k_5 for the preceding reaction was determined by following the decrease in absorbance at 417 nm of the tocopheroxy [8]. The rate constants k_5 for reaction (5) of the fatty acid esters with [8] are given in Table 4. Since [31], which does not contain allylic hydrogen atoms, did not react with [8], we estimated that k_5 of [31] must be $<<10^{-5}$ M^{-1}s^{-1}. As shown in Table 4, k_5 increases as the number of C$=$C double bonds in the fatty acid esters increases (**31** $<$ **32** $<$ **33** $<$ **34** $<$ **35** $<$ **36**). Accordingly, the allylic hydrogen abstraction is considered to play a major role in the high reactivity of the unsaturated fatty acid ester in reaction (5). The allylic hydrogen is activated by the π-electron system (C$=$C double bond) (39,40).

The k_5 value of [32] is about three orders of magnitude lower than that of [33]. Ethyl oleate [32] has four hydrogen atoms activated by a single π-electron system. On the other

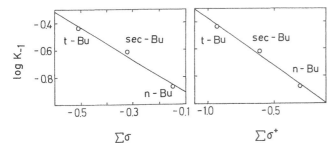

FIGURE 12 Plots of log k_{-1} versus $\Sigma\sigma$ and $\Sigma\sigma^+$ for alkyl hydroperoxides.

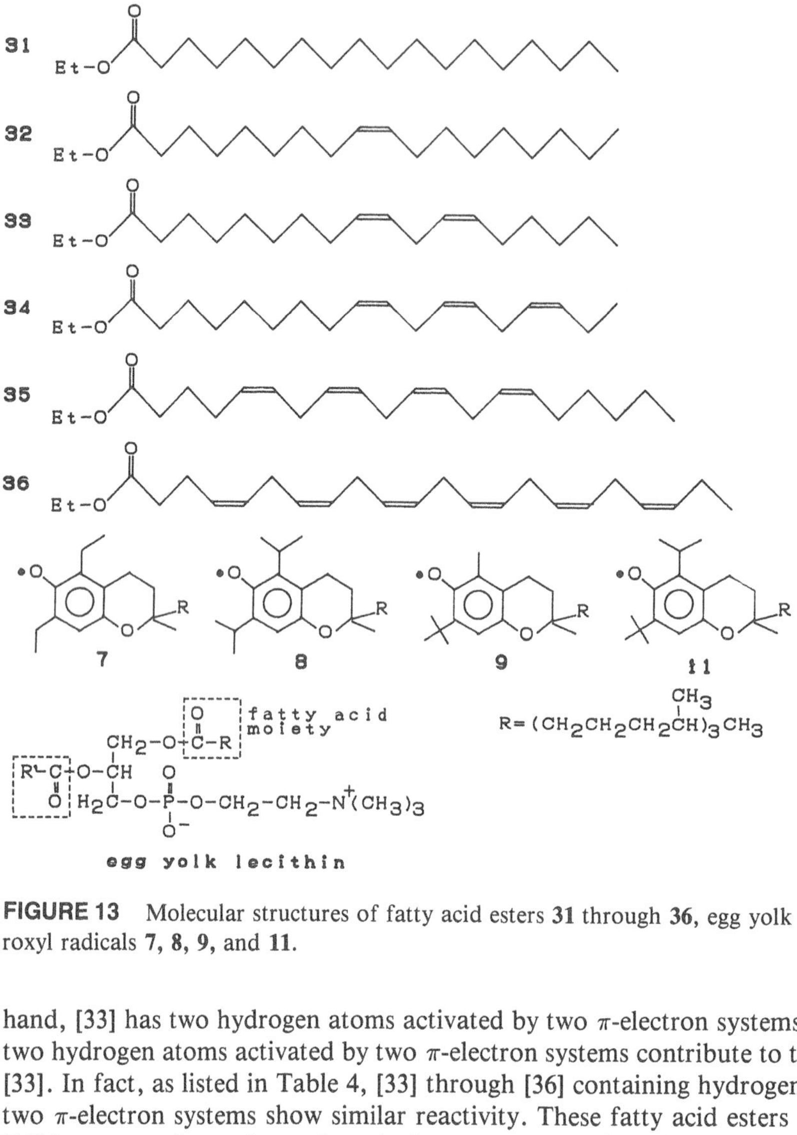

FIGURE 13 Molecular structures of fatty acid esters **31** through **36**, egg yolk lecithin, and tocopheroxyl radicals **7, 8, 9,** and **11**.

hand, [33] has two hydrogen atoms activated by two π-electron systems. Consequently, the two hydrogen atoms activated by two π-electron systems contribute to the high reactivity of [33]. In fact, as listed in Table 4, [33] through [36] containing hydrogen atoms activated by two π-electron systems show similar reactivity. These fatty acid esters [33], [34], [35], and [36] have two, four, six, and ten hydrogen atoms activated by two π-electron systems, respectively. The rate constants per an active hydrogen (k_{abstr}/H) of [33] through [36] given in Table 4 are similar to one another.

TABLE 4 Rate Constants for Reaction of Fatty Acid Ethyl Esters **31** Through **36** with Tocopheroxyl Radical **8** in Benzene at 25°C

	k_s ($M^{-1}s^{-1}$)	k_{abstr}/H ($M^{-1}s^{-1}$)
Ethyl stearate **31**	$\ll 10^{-5}$	
Ethyl oleate **32**	1.04×10^{-5}	2.60×10^{-6}
Ethyl linoleate **33**	1.82×10^{-2}	9.10×10^{-3}
Ethyl linolenate **34**	3.84×10^{-2}	9.60×10^{-3}
Ethyl arachidonate **35**	4.83×10^{-2}	8.05×10^{-3}
Cis-4,7,10,13,16,19-docosahexaenoic acid ethyl ester **36**	9.05×10^{-2}	9.05×10^{-3}

In conclusion, the second-order rate constant k_s for the hydrogen abstraction of the fatty acid esters dramatically increases as the number of π-electron systems activating the hydrogen (C = C double bonds) increases from zero to two. The rate constants per active hydrogen k_{abstr}/H for the various fatty acid esters that contain hydrogens activated by two π-electron systems are similar to one another.

To interpret the observed features of k_s and k_{abstr}/H, ab initio calculations of models for [31] through [35] were carried out (40). From the calculation results, it is shown that the observed features of k_s and k_{abstr}/H can be explained in terms of the pseudo-π conjugation between the C = C double bond and the active hydrogen-carbon bond.

The second-order rate constants k_s for the hydrogen abstraction from methyl linoleate by tocopheroxyls radical [7], [8], [9], and [11] in benzene are given in Table 5. The k_s value for the reaction of methyl linoleate with [8] is much the same as that of ethyl linoleate [33] with [8]. The bulkier the substituent groups at the 5 and 7 positions of the tocopheroxyl radical, the smaller is the k_s value. This substitution effect is likely due to steric hindrance by the substituent groups. The bulky substituent groups at the 5 and 7 positions of the tocopheroxyl radicals may inhibit hydrogen abstraction by interfering with the access of ethyl linoleate to the active site at the 6 position. The steric hindrance plays an important role in the reactions.

Reaction Between Egg Yolk Lecithin and Tocopheroxyl Radical

Egg yolk lecithin contains two fatty acid moieties. It is interesting to examine whether the rate constant of the hydrogen abstraction of egg yolk lecithin is the same as that expected from those of the fatty acid moieties contained in the lecithin. At first, we studied the reaction of mixtures of fatty acid esters to ensure that the reaction rate constant of the mixture could be estimated from those of the fatty acid esters contained in the mixture.

The pseudo–first-order rate constant k_{expd} expected for the mixture can be estimated in the following way:

$$k_{expd} = k_{LH}[LH] \qquad (18)$$

where k_{LH} refers to the second-order rate constant of a fatty acid ester LH. For example, for the mixture of 6.68×10^{-2} M of [31], 4.37×10^{-2} M of [32], 2.38×10^{-2} M of [33], 5.73×10^{-3} M of [35], and 8.86×10^{-3} M of [36], k_{expd} and k_{obsd} were estimated as 1.51×10^{-3} and 1.53×10^{-3} s^{-1}, respectively. From such results, it was shown that k_{obsd} is indeed in accord with k_{expd} based on Equation (18).

The rate constants k_{obsd} of the hydrogen abstraction from egg yolk lecithin by [8] are given in Table 6. The second-order rate constant k_s was found to be 7.02×10^{-3} M^{-1}s^{-1}. The expected value of the pseudo–first-order rate constant k_{expd} of lecithin is given by

$$k_{expd} = 2[Lec] \Sigma k_{LH} x_{LH} \qquad (19)$$

TABLE 5 Rate Constants for Reaction of Methyl Linoleate with Tocopheroxyl Radicals **7, 8, 9,** and **11** in Benzene at 25°C

	k_s (M^{-1}s^{-1})
5,7-Diethyl-tocopheroxyl **7**	4.97×10^{-2}
5,7-Diisopropyl-tocopheroxyl **8**	1.86×10^{-2}
7-*tert*-Butyl-5-methyl-tocopheroxyl **9**	1.61×10^{-2}
7-*tert*-Butyl-5-isopropyl-tocopheroxyl **11**	1.10×10^{-3}

TABLE 6 Rate Constants for Reaction of Egg Yolk Lecithin with Tocopheroxyl Radical **8** in Benzene at 25°C

[Lec](M)	k_{obsd} (s^{-1})	k_{expd} (s^{-1})
9.68×10^{-3}	8.09×10^{-5}	2.01×10^{-4}
1.94×10^{-2}	1.50×10^{-4}	4.01×10^{-4}
2.90×10^{-2}	2.19×10^{-4}	6.00×10^{-4}
3.87×10^{-2}	2.84×10^{-4}	8.02×10^{-4}

where [Lec] and x_{LH} stand for the molar concentration of the lecithin and the mole fraction of a fatty acid contained in the lecithin, respectively. Since egg yolk lecithin contains two fatty acid moieties, 2[Lec] is equal to the sum of the molar concentration of the fatty acids contained in the lecithin. We assumed that the second-order rate constant k_{LH} of a fatty acid moiety is the same as that of the corresponding ethyl ester given in Table 4. The expected values (k_{expd}) thus obtained are listed in Table 6. The observed values k_{obsd} are 35–40% of the corresponding k_{expd}. There is as yet no unambiguous explanation for this contradiction, but possible explanations are suggested as follows. First, the formation of reverse micelles in the benzene solution should be noted. Lecithins are prone to aggregate into reverse micelles in nonprotic solvents (43,44). The formation of the reverse micelles may prevent [8] from reacting with the fatty acid moiety. Another possibility of resolving this contradiction is that the steric hindrance between the two fatty acid moieties contained in the lecithin may affect the kinetics. Since the two moieties are quite close to each other in lecithin, [8] may not easily have access to the active site of these moieties.

Cause of the Prooxidant Effect of α-Tocopherol in High Concentration

As described in a previous section, the second-order rate constant for the hydrogen abstraction of the fatty acid esters (k_s) dramatically increases as the number of π-electron systems activating the hydrogen (C=C double bonds) increases from zero to two (39,40). This result is in agreement with that obtained for reactions between polyunsaturated fatty acids and peroxy radicals by Howard et al. (45) and Cosgrove et al. (46). Tocopheroxyl radicals produced in cellular membranes and edible oils are likely to react with polyunsaturated lipids or fatty acids by abstracting hydrogen atoms activated mainly by two π-electron systems.

As described earlier, we measured k_{-1} for reactions of alkyl hydroperoxides with [8], that is, the reversal of reaction (1) (38). In benzene solution at 25°C, k_{-1} was 1.34–3.65 × 10^{-1} M^{-1}s^{-1}. The values of k_{-1} are only about one order of magnitude larger than those of k_s for the reactions of [33] through [36] with [8] and of methyl linoleate with [7], [8], [9], and [11] (39,40). These results suggest that reactions (5) and (6) may relate to the prooxidant effect of α-tocopherol in high concentration. Therefore, if polyunsaturated lipids coexist with hydroperoxides in membranes or edible oils, the rate of disappearance of tocopheroxyl radical is represented by

$$\frac{-d[Toc^{\cdot}]}{dt} = k_s[LH][Toc^{\cdot}] + k_{-1}[LOOH][Toc^{\cdot}] \qquad (20)$$

where [Toc$^{\cdot}$] and [LOOH] denote the molar concentration of the tocopheroxyl radicals and the hydroperoxides, respectively.

These radical decay reactions (5) and (6) may occur competitively with reaction (2) or with radical dimer formation [reaction (21)]:

$$\text{Toc}^{\cdot} + \text{Toc}^{\cdot} \xrightarrow{k_{21}} \text{nonradical products} \qquad (21)$$

Neither reaction (2) nor (21) participates in the prooxidant effect of α-tocopherol. On the contrary, reaction (2) contributes to the antioxidant effect of α-tocopherol.

In the initial stage of lipid degradation, the concentration of hydroperoxides is much lower than that of polyunsaturated fatty acids, and thus the second term in Equation (20) is negligible. Consequently, the prooxidant effect of α-tocopherol is induced by the hydrogen abstraction [reaction (5)] between tocopheroxyl radical and polyunsaturated fatty acids. On the other hand, if the autoxidation proceeds, lipid hydroperoxides increase and reaction (6) plays an important role in the prooxidant effect of α-tocopherol. This suggests that not only reaction (5), but also reaction (6) participate in the prooxidant effect of α-tocopherol.

ACKNOWLEDGMENT

We are very grateful to all those colleagues and collaborators whose names are in the references.

REFERENCES

1. Burton GW, Ingold KU. Vitamin E: application of the principles of physical organic chemistry to the exploration of its structure and function. Acc Chem Res 1986; 19:194-201.
2. Niki E. Antioxidants in relation to lipid peroxidation. Chem Phys Lipids 1987; 44:227-53.
3. Niki E, Saito T, Kawakami A, Kamiya Y. Inhibition of oxidation of methyl linoleate in solution by vitamin E and vitamin C. J Biol Chem 1984; 259:4177-82.
4. Mukai K, Nishimura M, Nagano A, Tanaka K, Niki E. Kinetic study of the reaction of vitamin C derivatives with tocopheroxyl (vitamin E radical) and substituted phenoxyl radicals in solution. Biochim Biophys Acta 1989; 993:168-73.
5. Tappel AL. Lipid peroxidation damage to cell components. Geriatrics 1968; 23:96-105.
6. Packer JE, Slater TF, Willson RL. Direct observation of a free radical interaction between vitamin E and vitamin C. Nature 1979; 278:737-8.
7. Fahrenholtz SR, Doleiden FH, Trozzolo AM, Lamola AA. On the quenching of singlet oxygen by α-tocopherol. Photochem Photobiol 1974; 20:505-9.
8. Foote CS, Ching TY, Geller GG. Chemistry of singlet oxygen. XVIII. Rates of reaction and quenching of α-tocopherol and singlet oxygen. Photochem Photobiol 1974; 20:511-3.
9. Stevens B, Small RD Jr, Perez SR. The photoperoxidation of unsaturated organic molecules. XIII. $O_2{}^1 \Delta g$ Quenching by α-tocopherol. Photochem Photobiol 1974; 20:515-7.
10. Loury M, Bloch C, Francois R. Conditions of using tocopherol as an antioxidant in fats. Rev Fr Corps Gras 1966; 13:747-54.
11. Cillard J, Cillared P, Cormier M, Girre L. α-Tocopherol prooxidant effect in aqueous media. Increased autoxidation rate of linoleic acid. J Am Oil Chem Soc 1980; 57:252-5.
12. Terao J, Matsushita S. The peroxidizing effect of α-tocopherol on autoxidation of methyl linoleate in bulk phase. Lipids 1986; 21:255-60,.
13. Nilsson JLG, Selander H, Sievertsson H, Skanberg I. The direct effect of annelated rings in aromatic systems II. Synthesis and oxidation of 2,3-dihydro-5-benzo-furanols. Tetrahedron 1970; 26:879-86.
14. Svensson KG, Nilsson JLG. Synthesis and conformation studies of tetrahydroquinone derivatives related to tocopherol model compounds. Acta Pharm Suec 1973; 10:277-84.
15. Nakamura T, Kijima S. Studies on tocopherol derivatives. I. Conversion of β-, γ-, and δ-tocopherol to α-tocopherol. Chem Pharm Bull 1971; 19:2318-24.
16. Nakamura T, Kijima S. Studies on tocopherol derivatives. IV. Hydroxymethylation reaction of β-, γ-tocopherol and their model compounds with boric acid. Chem Pharm Bull 1972; 20:1681-6.
17. Burton GW, Doba T, Gabe EJ, et al. Autoxidation of biological molecules. 4. Maximizing the antioxidant activity of phenols. J Am Chem Soc 1985; 107:7053-65.

18. Mukai K, Yokoyama S, Fukuda K, Uemoto Y. Kinetic studies of antioxidant activity of new tocopherol model compounds in solution. Bull Chem Soc Jpn 1987; 60:2163-7.

19. Mukai K, Fukuda K, Ishizu K. Stopped-flow investigation of antioxidant activity of tocopherols. Finding of new tocopherol derivatives having higher antioxidant activity than α-tocopherol. Chem Phys Lipids 1988; 46:31-6.

20. Century B, Horwitt MK. Biological availability of various forms of vitamin E with respect to different indices of deficiency. Fed Proc Fed Am Soc Exp Biol 1965; 24:906-11.

21. Burton GW, Ingold KU. Autoxidation of biological molecules. 1. The antioxidant activity of vitamin E and related chain-breaking phenolic antioxidants in vitro. J Am Chem Soc 1981; 103:6472-7.

22. Mukai K, Watanabe Y, Uemoto Y, Ishizu K. Stopped-flow investigation of antioxidant activity of tocopherols. Bull Chem Soc Jpn 1986; 59:3113-6.

23. Mukai K, Fukuda K, Tajima K, Ishizu K. A kinetic study of reactions of tocopherols with a substituted phenoxyl radical. J Org Chem 1988; 53:430-2.

24. Mukai K, Kageyama Y, Ishida T, Fukuda K. Synthesis and kinetic study of antioxidant activity of new tocopherol (vitamin E) compounds. J Org Chem 1989; 54:552-6.

25. Mukai K, Okabe K, Hosose H. Synthesis and stopped-flow investigation of antioxidant activity of tocopherols. Finding of new tocopherol derivatives having the highest antioxidant activity among phenolic antioxidants. J Org Chem 1989; 54:557-60.

26. Howard JA, Ingold KU. The inhibited autoxidation of styrene. III. The relative inhibiting efficiencies of ortho-alkyl phenols. Can J Chem 1963; 41:2800-6.

27. Pryor WA, Strickland T, Church DF. Comparison of the efficiencies of several natural and synthetic antioxidants in aquenous sodium dodecyl sulfate micelle solutions. J Am Chem Soc 1988; 110:2224-9.

28. Saito I, Matsuura T, Inoue K. Formation of superoxide ion via one-electron transfer from electron donors of singlet oxygen. J Am Chem Soc 1983; 105:3200-6.

29. Merkel PB, Kearns DR. Radiationless decay of singlet molecular oxygen in solution. An experimental and theoretical study of electronic-to-vibrational energy transfer. J Am Chem Soc 1972; 94:7244.

30. Mukai K, Daifuku K, Okabe K, Tanigaki T, Inoue K. Structure-activity relationship in the quenching reaction of singlet oxygen by tocopherol (vitamin E) derivatives and related phenols. Finding of linear correlation between the rates of quenching of singlet oxygen and scavenging of peroxyl and phenoxyl radicals in solution. J Org Chem 1991; 56:4188-92.

31. Yamauchi R, Matsushita S. Quenching effect of tocopherols on the methyl linoleate photooxidation and their oxidation products. Agr Biol Chem 1977; 41:1425-30.

32. Kaiser S, Mascio PD, Murphy ME, Sies H. Physical and chemical scavenging of singlet molecular oxygen by tocopherols. Arch Biochem Biophys 1990; 277:101-8.

33. Thomas MJ, Foote CS. Chemistry of oxygen. XXVI. Photooxygenation of phenols. Photochem Photobiol 1978; 27:683-93.

34. Scarpa M, Rigo A, Maiorino M, Ursini F, Gregolin C. Formation of α-tocopherol radical and recycling of α-tocopherol by ascorbate during peroxidation of phosphatidylcholine liposomes. Biochim Biophys Acta 1984; 801:215-9.

35. Mukai K, Fukuda K, Ishizu K, Kitamura Y. Stopped-flow investigation of the reaction between vitamin E radical and vitamin C in solution. Biochem Biophys Res Commun 1987; 146:134-9.

36. Mukai K, Nishimura M, Ishizu K, Kitamura Y. Kinetic study of the reaction of vitamin C with vitamin E radicals (tocopheroxyls) in solution. Biochim Biophys Acta 1989; 991:276-9.

37. Mukai K, Nishimura M, Kikuchi S. Stopped-flow investigation of the reaction of vitamin C with tocopheroxyl radical in aqueous Triton X-100 micellar solutions. J Biol Chem 1991; 226:274-8.

38. Mukai K, Kohno Y, Ishizu K. Kinetic study of the reaction between vitamin E radical and alkyl hydroperoxides in solution. Biochem Biophys Res Commun 1988; 155:1046-50,.

39. Mukai K, Okauchi Y. Kinetic study of the reaction between tocopheroxyl radical and unsaturated fatty acid esters in benzene. Lipids 1989; 24:936-9.

40. Nagaoka S, Okauchi Y, Urano S, Nagashima U, Mukai K. Kinetic and ab initio study of the prooxidant effect of vitamin E. Hydrogen abstraction from fatty acid esters and egg yolk lecithin. J Am Chem Soc 1990; 112:8921-4.

41. Niki E, Kamiya Y, Ohta N. Cooxidation of hydrocarbons in lipid phase. VI. Reactivities of hydrocarbons and peroxy radicals. Bull Chem Soc Jpn 1969; 42:2312-8.

42. Howard JA. Absolute rate constants for reactions of oxyl radicals. Adv Free Radical Chem 1972; 4:49-173.

43. Barclay LRC, MacNeil JM, Van Kessel J, et al. Autoxidation and aggregation of phospholipids in organic solvents. J Am Chem Soc 1984; 106:6740-7.

44. Lehman LS, Porter NA. Product distribution of unsaturated phospholipid oxidation in organic solvent and aqueous emulsion. In: Bors W, Saran M, Tait D, eds. Berlin: Walter de Gruyter, 1984; 281.

45. Howard JA, Ingold KU. Absolute rate constants for hydrocarbon autoxidation. VI. Alkyl aromatic and olefinic hydrocarbons. Can J Chem 1967; 45:793-802.

46. Cosgrove JP, Church DF, Pryor WA. The kinetics of autoxidation of polyunsaturated fatty acids. Lipid 1987; 22:299-304.

9

Rates and Products of Reactions of Vitamin E with Oxygen Radicals

Etsuo Niki

University of Tokyo, Tokyo, Japan

Mitsuyoshi Matsuo

Tokyo Metropolitan Institute of Gerontology, Tokyo, Japan

INTRODUCTION

Vitamin E acts as an antioxidant primarily by scavenging active oxygen radicals, which initiate and/or propagate chain oxidations (1). Vitamin E scavenges the hydroxyl, alkoxyl, peroxyl, and superoxide anion radicals and it also quenches singlet oxygen. However, the major role of vitamin E in vivo must be to scavenge chain-carrying lipid peroxyl radicals to break the chain propagation of lipid peroxidation. In this chapter, the rates and products of the reaction of vitamin E with peroxyl radicals are reviewed, especially of α-tocopherol, which has the highest biological activity.

RATES OF SCAVENGING OF PEROXYL RADICALS BY VITAMIN E

The antioxidant activities of the chain-breaking antioxidants are determined primarily by how rapidly they scavenge peroxyl radicals, that is, by the rate constant k_{inh} for the reaction (1)

$$LO_2{}^{\cdot} + IH \quad \xrightarrow{k_{inh}} \quad LOOH + I^{\cdot} \tag{1}$$

where $LO_2{}^{\cdot}$ is the lipid peroxyl radical, IH the antioxidant, and $I^{\cdot}$ the radical derived from the antioxidant. This rate constant has been determined for vitamin E under both steady-state and non-steady-state conditions. Some of the literature values (2-15) for α-tocopherol are summarized in Table 1. Table 2 shows the rate constants for several antioxidants toward the poly (peroxylstyrenyl)peroxyl radical (6), which is a chain carrier in the oxidation of styrene. The second-order rate constants for the reactions of various tocopherols with stable phenoxyl radicals have been also measured in ethanol as a model for scavenging active oxygen radicals (16). The details for the measurements and the merits and demerits of each method are not discussed here. Tables 1 and 2 show that α-tocopherol is the strongest peroxyl radical scavenger, although a few phenolic compounds that have k_{inh} higher than that of α-tocopherol have been synthesized recently (6,17,18). It is also important to note that the length of a side chain and the type of a substituent at the 2-chain position of the tocopherol analog do not affect the chemical reactivity, although this side chain is essential for biological activity. Furthermore,

TABLE 1 Rate Constants k_{inh} for Scavenging of Oxygen Radicals by α-Tocopherol

Oxygen radical	Technique	Temperature (°C)	k_{inh} (M^{-1} s^{-1})	Reference
CI$_3$COO$^{\cdot}$	Pulse radiolysis		5.0×10^8	2
CF$_3$CHClOO$^{\cdot}$	Pulse radiolysis		9.2×10^7	3
Cyclohexylperoxyl	Pulse radiolysis		2.3×10^7	4
C$_6$H$_5$CH(CH$_3$)OO$^{\cdot}$	Chemiluminescence		3.3×10^6	5
X-(CH$_2$CH(C$_6$H$_5$)OO$^{\cdot}$)	Oxygen	30	3.2×10^6	6
(CH$_3$)$_3$COO$^{\cdot}$	Laser flash photolysis	23	2.6×10^6	7
C$_6$H$_5$CH(CH$_3$)OO$^{\cdot}$	Chemiluminescence	37	1.8×10^6	8
Methyl linoleoyl peroxyl	Inhibited oxidation	37	5.1×10^5	9
Methyl linoleoyl peroxyl	Inhibited oxidation		2.0×10^5	10
C$_6$H$_5$C(CH$_3$)$_2$OO$^{\cdot}$	Oxygen consumption	60	2.0×10^5	11
C$_6$H$_5$C(CH$_3$)$_2$OO$^{\cdot}$	Chemiluminescence	25	1.5×10^5	12
Methyl linoleoyl peroxyl	Pulse radiolysis		8.0×10^4	13
Linoleoyl peroxyl	Oxygen consumption	40	6.0×10^4	14
Linoleoyl peroxyl	Oxygen consumption	37	3.7×10^4	15

it should be added that all the possible eight stereoisomers of α-tocopherol have similar chemical reactivities toward peroxyl radicals, but a major component of natural vitamin E, 2R,4'R, 8'R-α-tocopherol, has greater bioactivity than the other diastereomers, due largely to a preferential uptake by tissues (19).

The high reactivity of α-tocopherol toward the peroxyl radical is ascribed to its high reactivity of OH bond at the 6-position, which stems from a large resonance stabilization energy of the α-tocopheroxyl radical. Burton and Ingold elegantly explained its high activity by stereoelectronic effects (20).

The rates of reactions of vitamin E with oxygen radicals depend on various experimental variables, such as temperature and solvent. The rates also depend markedly on the type of attacking radicals. More importantly, the rates depend on the reaction media. The rate of scavenging of lipid peroxyl radical by α-tocopherol has been found to be much smaller in membrane than in homogeneous solution (21). It has been also found that the scavenging rate becomes smaller as oxygen radicals go deeper into the interior of the bilayer membranes (22). Thus,

TABLE 2 Rate Constant k_{inh} for Scavenging Poly(peroxystyrenyl) Peroxyl Radical by Various Antioxidants at 30 °C

Antioxidant	$10^5 \times k_{inh}$ (M^{-1} s^{-1})
α-Tocopherol	32.0
β-Tocopherol	13.0
γ-Tocopherol	14.0
δ-Tocopherol	4.4
2,6-Dimethyl-4-methoxyphenol	9.4
2,6-Di-*tert*-butyl-4-methoxyphenol	1.1
2,6-Di-*tert*-butyl-4-methylphenol	0.14
2,6-Di-*tert*-butyl phenol	0.031

it should be well appreciated that the rate constants or the antioxidant activities measured in the in vitro model systems cannot be directly applied to biological systems.

PRODUCTS OF REACTION OF VITAMIN E WITH PEROXYL RADICALS

Kinetic studies have suggested that one molecule of vitamin E scavenges two molecules of peroxyl radicals (9,23), and the products consistent with this stoichiometric number have recently been identified and confirmed.

Mill and his colleagues (24) studied the products of the reaction of α-tocopherol with peroxyl radicals derived from azo compounds in the presence of oxygen and observed the formation of 8a-alkyldioxy-α-tocopherones [1]. The products of the reactions of α-tocopherol and its model compound, 2,2,5,7,8-pentamethyl-6-chromanol (PMC), with the *tert*-butylperoxyl radical have been studied in detail (25). It was found that α-tocopherol and PMC gave diastereomers of 4a,5-epoxy-4a,5-dihydro-8a-hydroperoxytocopherone [2] and 4a,5-epoxy-4a,5-dihydro-8a-*tert*-butylperoxy tocopherone (25). The structures of these products were deduced on the basis of the spectral data and, in addition, confirmed by the x-ray crystallographic analysis of the product from PMC:

1   2

More recently, Yamauchi et al. studied the products of the reaction of α-tocopherol with peroxyl radicals derived from methyl linoleate (26) as well as 2,2'-azobis-(2,4-dimethylvaleonitrile (AMVN) (27). They obtained four stereoisomers of methyl 13-(8a-peroxy-α-tocopherone)-9(Z),11(E)-octadecadienoate [3] through [6] and four stereoisomers of methyl 9-(8a-peroxy-α-tocopherone)-10(E),12(Z)-octadecadienoate [7] through [10]. When α-tocopherol was oxidized by the peroxyl radicals from AMVN in ethanol, 4a,5-epoxy-4a,5-dihydro-8a-ethoxytocopherones and 7,8-epoxy-7,8a-dihydro-8a-ethoxytocopherones were also formed. They also found spirodiene dimer [11] and two stereoisomeric trimers [12] and [13] (27). The formations of the dimers and trimers have been reported previously (25,28-30).

Furthermore, Liebler (31) more recently studied the oxidation products of α-tocopherol in the presence of AMVN in oxygenated acetonitrile. They found 8a-alkyldioxy-α-tocopherone and 4a,5-epoxy-4a,5-dihydro- and 7,8-epoxy-7,8-dihydro-8a-hydroperoxytocopherones as the principal products together with a small amount of 8a-hydroperoxytocopherone.

3   4

ACTION OF VITAMIN E AS CHAIN-BREAKING ANTIOXIDANT

The preceding experimental findings collectively suggest that vitamin E acts as a chain-breaking antioxidant, as shown in Figures 1 and 2. Vitamin E must scavenge the chain-carrying peroxyl radical before it attacks lipid and continues chain propagation. The rate constants for these two competing reactions, chain inhibition [reaction (1)] and chain propagation [reaction (2)],

$$LO_2{}^{\cdot} + LH \xoverset{k_p}{\rightarrow} LOOH + L^{\cdot} \tag{2}$$

are about 10^6 and 10^2 M^{-1} s^{-1}, respectively, at 37°C in homogeneous solution, showing that α-tocopherol scavenges the peroxyl radical about 10^4 times faster than polyunsaturated lipids react with the peroxyl radical. If the molar ratio of vitamin E to lipids is $1:10^3$, that is, 0.1% antioxidant, then the ratio of the rate of chain inhibition to that of chain propagation is 10:1,

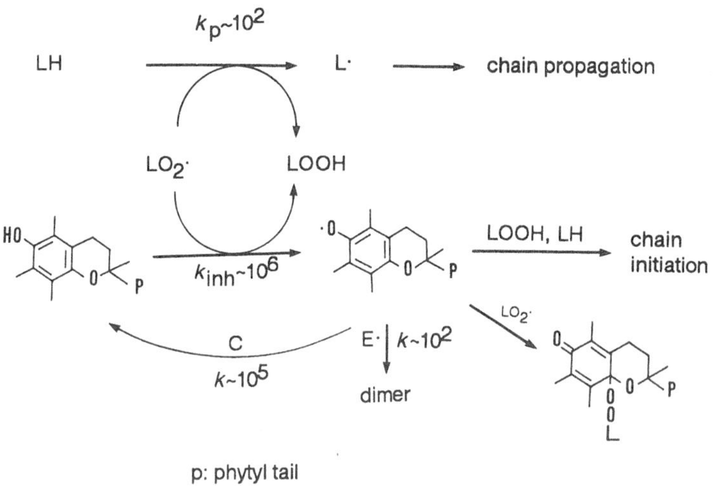

FIGURE 1 Action of vitamin E as an antioxidant. Rate constants k in $M^{-1} s^{-1}$.

suggesting that about 90% of peroxyl radicals are scavenged by vitamin E before they attack lipid.

$$\frac{\text{Rate of reaction (1)}}{\text{Rate of reaction (2)}} = \frac{k_{\text{inh}} [LO_2\text{·}][IH]}{k_p [LO_2\text{·}][LH]} = 10^4 \times \frac{1}{10^3} = 10 \tag{3}$$

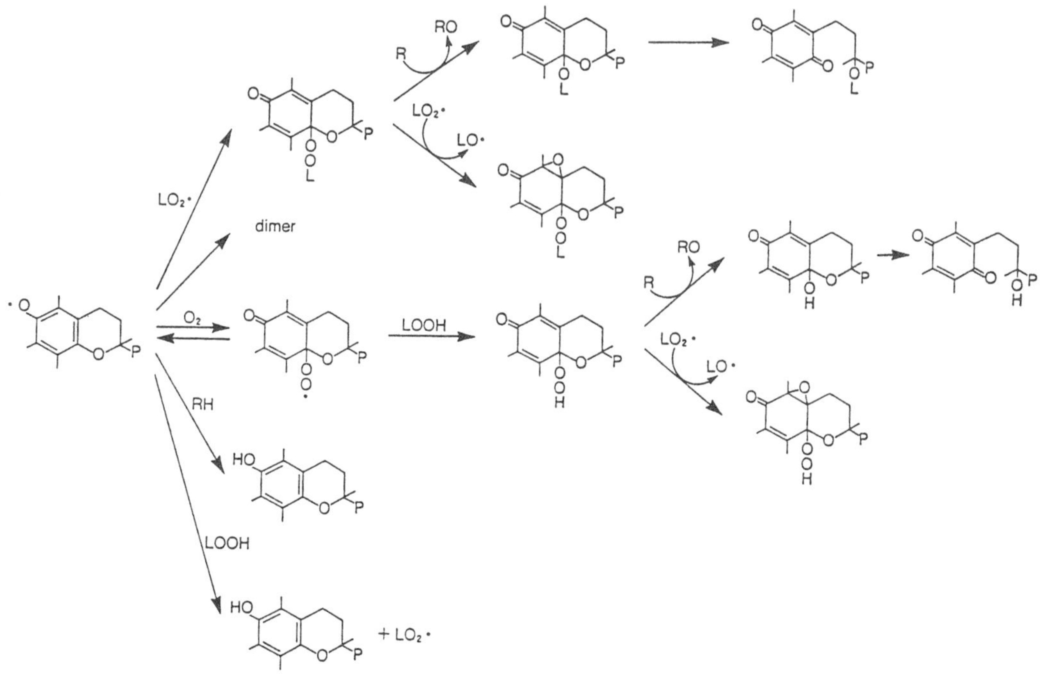

FIGURE 2 Fate of vitamin E radical.

It may be worth noting that vitamin E may also suppress the chain initiation by scavenging chain-initiating radicals. This is supported by the experimental observation that the rates of disappearance of α-tocopherol are the same when it was incorporated into soybean and dimyristoyl phosphatidylcholine liposomes in the aqueous peroxyl radical-generating systems.

The fate of the vitamin E radical formed from vitamin E depends on various experimental variables. The vitamin E radical reacts with the peroxyl radical to give 8a-alkyldioxy-α-tocopherone [1]. This reaction must proceed most predominantly in the in vitro model oxidation of lipids induced by a radical initiator in which the peroxyl radical concentration is relatively high. In such a case, the stoichiometric number of peroxyl radicals trapped by each molecule of vitamin E is two, as observed experimentally, and this reaction must be the typical chain-breaking action by vitamin E (9,23).

This adduct [1] is not stable and may be reduced to give 8a-alkoxyl-α-tocopherone [14], which cleaves easily to yield α-tocopherolquinone [15]. The adduct [1] may further react with the third peroxyl radical to give 4a,5-epoxy-4,5-dihydro-8a-alkyldioxy-α-tocopherone [16].

The bimolecular interaction of the vitamin E radicals yields dimers. The rate constant for the bimolecular reaction of the α-tocopheroxyl radicals was obtained as 3×10^3 and 1400 $M^{-1} s^{-1}$ in benzene at 23°C (32) and in ethanol at 20°C (33), respectively. They further react with the vitamin E radical to give a trimer.

The vitamin E radical reacts with molecular oxygen reversibly to form the 8a-peroxyl radical, although this reaction was suggested to proceed slowly. The persistent 2,4,6-tri-*tert*-butylphenoxyl radicals and oxygen give peroxide [17] [reaction (4)], but no such peroxide was observed for α-tocopheroxyl radical (25), although the formation of such a peroxide has been suggested in the reaction of 8a-hydroperoxy-2,2,5,7,8-pentamethyl-6(8aH)-chromanone with lead tetraacetate (34).

$$\tag{4}$$

The decay rate constant of the α-tocopheroxyl radicals was found to be the same in the presence and absence of oxygen (32). However, in the presence of excess hydroperoxide, the 8a-peroxyl radical may abstract a hydrogen atom from the hydroperoxide to give 8a-hydroperoxy-α-tocopherone [18] (25), which like 8a-alkyldioxy-α-tocopherone, reacts with such reducing agents as triphenylphosphine and dimethyl sulfide and the peroxyl radical to give the products shown in Figure 2. It may be noteworthy that this pathway must be important only in the presence of high concentration of hydroperoxide, and hence it may not be physiologically important. It has been observed that the oxidation of α-tocopherol by singlet oxygen also gives this hydroperoxide [18] (35).

Another important reaction is the reduction of the vitamin E radical with reducing agents to regenerate vitamin E (36–38). It has been found experimentally that ascorbic acid (39,40), cysteine (41), and glutathione (40) reduce the α-tocopheroxyl radical. The rate constant for the reduction of the α-tocopheroxyl radical with ascorbic acid has been reported as 1.55×10^6 M^{-1} s^{-1} in a water, 2-propanol, and acetone mixture (39). The rate constants for the interactions of related chromanoxyl radicals with ascorbic acid have also been reported (42,43). The rates of reduction of phenoxyl radical decrease in the order ascorbic acid > cysteine > glutathione (44). It was recently suggested that 8a-alkyldioxy-α-tocopherones are hydrolyzed to give an α-tocopheryl cation, which is reduced by ascorbic acid to give α-tocopherol (45). This reduction of tocopheroxyl radicals by ascorbic acid to regenerate tocopherols leads to the synergistic inhibition of oxidation by vitamin E and vitamin C, and such an effect has in fact been demonstrated in many model systems, including membrane systems (46,47) and homogeneous solutions. More recently, papers have been published that support such a synergistic action by vitamins E and C (48–52). However, it has not been unequivocally proved whether this also contributes in vivo: some support this concept (53) but others do not (54). The importance probably depends largely on the conditions.

The prooxidation effect of vitamin E has been also reported (55–57); that is, it has been observed that tocopherols accelerate the spontaneous oxidation of lipids. This effect has been attributed to the abstraction of hydrogen from lipid hydroperoxide and/or lipid itself by the tocopheroxyl radical. In fact, the prooxidant effect has been found to be enhanced with increasing hydroperoxide concentration. Interestingly, such a prooxidant effect was not observed in the presence of ascorbic acid (57), suggesting the contribution of the tocopheroxyl radical. The rate constants for the hydrogen atom abstraction by the tocopheroxyl radical from hydroperoxides and lipids have been measured (58–60). For example, the rate constants for the reactions of the 5,7-diisopropyltocopheroxyl radical with n-butyl, sec-butyl, and tert-butyl hydroperoxides have been reported as 0.134, 0.242, and 0.365 M^{-1} s^{-1}, respectively. However, there have not yet been clear data to show the importance of such reactions in vivo.

It is known that ascorbic acid initiates lipid peroxidation in the presence of iron. α-Tocopherol also reduces iron and iron(III) accelerates the oxidation, but such a reaction has not been observed for α-tocopherol in biological membranes (61).

As shown earlier, there are many competing reactions for the vitamin E radical. The antioxidant activity and efficiency depend on the relative importance of these competing reactions, which is determined by such variables as concentrations and mobilities of molecules at specific sites.

REFERENCES

1. Diplock AT, Machlin LJ, Packer L, Pryor WA, eds. Vitamin E: biochemistry and health implications. Ann N Y Acad Sci 1989; 570:1-555.
2. Packer JE, Slater TF, Willson RL. Direct observation of a free radical interaction between vitamin E and vitamin C. Nature 1979; 278:737-8.
3. Monig J, Asmus KD. Electron transfer reactions of halothane-derrived peroxyl free radicals, CF_3 $CHClO_2^-$: Measurement of absolute rate constants by pulse radiolysis. J Chem Soc Perkin Trans II 1983; 1133-7.
4. Simic MG. Kinetic and mechanistic studies of peroxyl, vitamin E and antioxidant free radicals, by pulse radiolysis. In: Simic MG, Karel M, eds. Autoxidation in food and biological systems. New York: Plenum Press, 1980; 17-26.
5. Burlakova YeB, Khrapova NG. Study of the anti-radical activity of tocopherol analogues and homologues by the method of chemiluminescence. Biophysics 1980; 24:989-93.
6. Burton GW, Doba T, Gabe EJ, et al. Autoxidation of biological molecules. 4. Maximizing the antioxidant activity of phenols. J Am Chem Soc 1985; 107:7053-65.

7. Burton GW, Cheeseman KH, Doba T, Ingold KU, Slater TF. Vitamin E as an antioxidant in vitro and in vivo. In: Biology of vitamin E. London: Pitman, 1983; 4-18.

8. Aristarkhova SA, Burlakova EB, Kharapova NG. Study of the inhibiting activity of tocopherol. Izv Akad Nauk USSR Ser Khim 1972; 2714-8.

9. Niki E, Saito T, Kawakami A, Kamiya Y. Oxidation of lipids. VI. Inhibition of oxidation of methyl linoleate in solution by vitamin E and vitamin C. J Biol Chem 1984; 259:4177-82.

10. Pryor WA. In: Armstrong G, ed. Free radicals in autoxidation and in aging. New York: Raven Press, 1984; 13-41.

11. Kharitonova AA, Kozlova ZG, Tsepalov VF. Kinetic analysis of antioxidant properties in complex mixtures using model chain reactions. Kinet Katal 1979; 20:593-9.

12. Niki E, Tanimura R, Kamiya Y. Oxidation of lipids. II. Rate of inhibition of oxidation by α-tocopherol and hindered phenols measured by chemiluminescence. Bull Chem Soc Jpn 1982; 55:1551-5.

13. Patterson LK. Studies of radiation induced peroxidation in fatty acid micelles. In: Oxygen and oxygen radicals in chemistry and biology. 1981; 89-95.

14. Castle L, Perkins MJ. Inhibition kinetics of chain-breaking phenolic antioxidants in SDS micelles. Evidence that intermicellar diffusion rates may be rate-limiting for hydrophobic inhibitors such as α-tocopherol. J Am Chem Soc 1985; 108:6381-2.

15. Pryor WA. Comparison of the efficiencies of several natural and synthetic antioxidants in aqueous sodium dodecyl sulfate micelle solutions. J Am Chem Soc 1988; 110:2224-9.

16. Mukai K, Watanabe Y, Uemoto Y, Ishizu K. Stopped-flow investigation of antioxidant activity of tocopherols. Bull Chem Soc Jpn 1986; 59:3113-6.

17. Mukai K, Fukuda K, Ishizu K. Stopped-flow investigation of antioxidant activity of tocopherols. Finding of new tocopherol derivatives having higher antioxidant activity than α-tocopherol. Chem Phys Lipids 1988; 46:31-6.

18. Mukai K, Okabe K, Hosose H. Synthesis and stopped-flow investigation of antioxidant activity of tocopherols. Finding of new tocopherol derivatives having the highest antioxidant activity among phenolic antioxidants. J Org Chem 1989; 54:557-60.

19. Ingold KU, Burton GW, Foster DO, Hughes L, Lindsay DA, Webb A. Biokinetics of and discrimination between dietary RRR- and SRR-α-tocopherols in the male rat. Lipids 1987; 11:163-72.

20. Burton GW, Ingold KU. Vitamin E: application of the principles of physical organic chemistry to the exploration of its structure and function. Acc Chem Res 1986; 19:194-201.

21. Niki E, Komuro E. Inhibition of peroxidation of membranes. In: Simic MG, Taylor KA, Ward JF, Sonntag C, eds. Oxygen radicals in biology and medicine. New York: Plenum, 1989; 561-6.

22. Takahashi M, Tsuchiya J, Niki E. Scavenging of radicals by vitamin E in the membranes as studied by spin labeling. J Am Chem Soc 1989; 111:6350-3.

23. Burton GW, Ingold KU. Autoxidation of biological molecules. 1. The antioxidant activity of vitamin E and related chain-breaking phenolic antioxidants in vitro. J Am Chem Soc 1981; 103:6472-7.

24. Winterle J, Dulin D, Mill T. Products and stoichiometry of reaction of vitamin E with alkylperoxy radicals. J Org Chem 1984; 49:491-5.

25. Matsuo M, Matsumoto S, Iitaka Y, Niki E. Radical-scavenging reactions of vitamin E and its model compound, 2,2,5,7,8-pentamethylchroman-6-ol, in a *tert*-butylperoxyl radical-generating system. J Am Chem Soc 1989; 111:7179-85.

26. Yamauchi R, Matsui T, Kato K, Ueno Y. Reaction products of α-tocopherol with methyl linoleate-peroxyl radicals. Lipids 1990; 25:152-8.

27. Yamauchi R, Matsui T, Kato K, Ueno Y. Reaction of α-tocopherol with 2,2'-azobis-(2,4-dimethyl-valeronitrile) in benzene. Agr Biol Chem 1989; 53:3257-62.

28. Skinner WA, Alaupovic P. Oxidation products of vitamin E and its model, 6-hydroxy-2,2,5,7,8-pentamethyl-chroman. V. Studies of the products of alkaline ferricyanide oxidation. J Org Chem 1963; 28:2854-8.

29. Skinner WA, Parkhurst RM. Oxidation products of vitamin E and its model, 6-hydroxy-2,2,5,7,8-pentamethylchroman. VII. Trimer formed by alkaline ferricyanide oxidation. J Org Chem 1964; 29:3601-3.

30. Ha KH, Igarashi O. The oxidation products from two kinds of tocopherols co-existing in autoxidation system of methyl linoleate. J Nutr Sci Vitaminol 1990; 36:411-21.

31. Liebler DC, Baker PF, Kaysen KL. Oxidation of vitamin E; evidence for competing autoxidation and peroxyl radical trapping reactions of the tocopheroxyl radical. J Am Chem Soc 1990; 112: 6995-7000.

32. Doba T, Burton GW, Ingold KU, Matsuo M. α-Tocopheroxyl decay: lack of effect of oxygen. J Chem Soc Chem Commun 1984; 461-2.

33. Richard CR, Richard C, Martin R. Kinetics of bimolecular decay of α-tocopheroxyl free radicals studied by ESR. FEBS Lett 1988; 233:307-10.

34. Matsumoto S, Matsuo M, Iitaka Y. Reactions of a hydroperoxide, 8a-hydroperoxy-2,2,5,7,8-penta-methylchroman-6(8aH)-one, derived from a vitamin E model compound: dimerization and x-ray crystal structure. J Chem Res 1987; miniprint 609-41, synop. 58-9.

35. Clough RL, Yee BG, Foote CS. Chemistry of singlet oxygen. 30. The unstable primary product of tocopherol photooxidation. J Am Chem Soc 1979; 101:683-6.

36. McCay PB. Vitamin E: interactions with free radicals and ascorbate. Annu Rev Nutr 1985; 5:323-40.

37. Niki E. Interaction of ascorbate and α-tocopherol. Ann N Y Acad Sci 1987; 498:186-99.

38. Chen LH. Interaction of vitamin E and ascorbate acid (review). In Vivo 1989; 3:199-210.

39. Packer J, Slater TF, Willson RL. Direct observation of free-radical interaction between vitamin E and vitamin C. Nature 1979; 278:737-8.

40. Niki E, Tsuchiya J, Tanimura R, Kamiya Y. Regeneration of vitamin E from α-chromanoxyl radical by glutathione and vitamin C. Chem Lett 1982; 789-92.

41. Motoyama T, Miki M, Mino M, Takahashi M, Niki E. Synergistic inhibition of oxidation in dispersed phosphatidylcholine liposomes by a combination of vitamin E and cysteine. Arch Biochem Biophys 1989; 270:655-61.

42. Mukai K, Fukuda K, Ishizu K, Kitamura Y. Stopped-flow investigation of the reaction between vitamin E radical and vitamin C in solution. Biochem Biophys Res Commun 1987; 146:134-9.

43. Mukai K, Nishimura M, Ishizu K, Kitamura Y. Kinetic study of the reaction of vitamin C with vitamin E radicals (tocopheroxyls) in solution. Biochim Biophys Acta 1989; 991:276-9.

44. Tsuchiya J, Yamada T, Niki E, Kamiya Y. Interaction of galvinoxyl radical with ascorbic acid, cysteine, and glutathione in homogeneous solution and in aqueous dispersions. Bull Chem Soc Jpn 1985; 58:326-30.

45. Liebler DC, Kaysen KL, Kennedy TA. Redox cycles of vitamin E: hydrolysis and ascorbic acid dependent reduction of 8a-(alkyldioxy)tocopherones. J Am Chem Soc 1989; 28:9772-7.

46. Niki E, Kawakami A, Yamamoto Y, Kamiya Y. Synergistic inhibition of oxidation of soybean phosphatidylcholine liposomes in aqueous dispersion by vitamin E and vitamin C. Bull Chem Soc Jpn 1985; 58:1971-5.

47. Doba T, Burton GW, Ingold KU. Antioxidant and co-antioxidant activity of vitamin C. The effect of vitamin C, either alone or in the presence of vitamin E or a water-soluble vitamin E analogue, upon the peroxidation of aqueous multilamellar phospholipid liposomes. Biochim Biophys Acta 1985; 835:298-303.

48. Stocker R, Hunt NH, Weidemann MJ, Clark IA. Protection of vitamin E from oxidation by increased ascorbic acid content within *Plasmodium vinckei*-infected erythrocytes. Biochim Biophys Acta 1986; 876:294-9.

49. Wefers H, Sies H. The protection by ascorbate and glutathione against microsomal lipid peroxidation is dependent on vitamin E. Eur J Biochem 1988; 174:353-7.

50. Davies MJ, Forni LG, Willson RL. Vitamin E analogue trolox C. Biochem J 1988; 255:513-22.

51. Melhorn RJ, Sumida S, Packer L. Tocopheroxyl radical persistence and tocopherol consumption in liposomes and in vitamin E-enriched rat liver mitochondria and microsomes. J Biol Chem 1989; 264:13448-52.

52. Kagan VE, Serbinova EA, Packerr L. Recycling and antioxidant activity of tocopherol homologs of differing hydrocarbon chain lengths in lever microsomes. Arch Biochem Biophys 1990; 282: 221-5.

53. Machlin LJ, Bendich A. Free radical tissue damage: protective role of antioxidant nutrients. FASEB J 1987; 1:441-5.

54. Burton GW, Wronska U, Stone L, Foster DO, Ingold KU. Biokinetics of dietary *RRR*-α-tocopherol in the male guinea pig at three dietary levels of vitamin C and two levels of vitamin E. Evidence that vitamin C does not "spare" vitamin E in vivo. Lipids 1990; 25:199-210.

55. Cillard J, Cillard P, Cormier M. α-Tocopherol prooxidant effect in aqueous media: increased autoxidation rate of linoleic acid. J Am Oil Chem Soc 1980; 57:252-5.

56. Terao J, Matsushita S. The peroxidizing effect of α-tocopherol autoxidation of methyl linoleate in bulk phase. Lipids 1986; 21:255-60.

57. Takahashi M, Yoshikawa Y, Niki E. Oxidation of lipids. XVII. Crossover effect of tocopherols in the spontaneous oxidation of methyl linoleate. Bull Chem Soc Jpn 1989; 62:1885-90.

58. Mukai K, Khono Y, Ishizu K. Kinetic study of the reaction between vitamin E radical and alkyl hydroperoxides in solution. Biochem Biophys Res Commun 1988; 155:1046-50.

59. Mukai K, Okauchi Y. Kinetic study of the reaction between tocopheroxyl radical and unsaturated fatty acid esters in benzene. Lipids 1989; 24:936-9.

60. Nagaoka S, Okauchi Y, Urano S, Nagashima U, Mukai K. Kinetic and ab initio study of the prooxidant effect of vitamin E. Hydrogen abstraction from fatty acid esters and egg yolk lecithin. J Am Chem Soc 1990; 112:8921-4.

61. Yamamoto K, Niki E. Interaction of α-tocopherol with iron; antioxidant and prooxidant effects of α-tocopherol in the oxidation of lipids in aqueous dispersions in the presence of iron. Biochem Biophys Acta 1988; 958:19-23.

10

Relative Importance of Vitamin E in Antiperoxidative Defenses in Human Blood Plasma and Low-Density Lipoprotein (LDL)

Balz B. Frei* and Bruce N. Ames

University of California, Berkeley, California

INTRODUCTION

Human blood plasma contains an array of nonenzymatic low-molecular-weight antioxidants (1,2). These antioxidants can be classified into two groups: the water-soluble antioxidants present in the aqueous phase of plasma and the lipid-soluble antioxidants associated with lipoproteins (Table 1).

The relative importance of an individual antioxidant in a biological system, such as plasma or low-density lipoprotein (LDL), depends on both the nature of the oxidant stress imposed on the biological system and the type of target molecule to be protected from oxidative damage, for example lipids, proteins, or nucleic acids. In addition, one must distinguish between the *quantitative* and the *qualitative* contribution or importance of an individual antioxidant. The *quantitative* contribution of an individual antioxidant to the total antioxidant capacity of a biological system depends on the location and concentration of the antioxidant in this system and the number of oxidant molecules scavenged by each antioxidant molecule, defined as the antioxidant's *n* value (3). The *qualitative* importance of an individual antioxidant depends on its location and concentration in the biological system and its reaction rate with the oxidant relative to the reaction rate of the oxidant with the target molecule (lipid, protein, or nucleic acid).

In this chapter we review data on the potency of vitamin E relative to the potency of other antioxidants naturally present in human plasma and LDL to prevent peroxidative damage to lipids (anti*per*oxidative potency). We do not consider antioxidative protection of proteins or nucleic acids. The studies we discuss used various types of oxidant stress, including water-soluble and lipid-soluble peroxyl radicals, activated neutrophils, transition metal ions, and the gas phase of cigarette smoke.

Lipid peroxidation in LDL recently received much attention because it is thought to initiate the atherosclerotic degeneration of arteries (for recent reviews see Refs. 4 and 5, and Chapters 45 and 46). According to this "oxidative theory of atherosclerosis," LDL present in the extracellular fluid of the subendothelial space becomes oxidatively modified through formation and breakdown of lipid hydroperoxides and subsequent covalent modification and nonenzymatic degradation of apolipoprotein B. As a consequence of such alterations, the LDL

**Present affiliation*: Harvard School of Public Health, Boston, Massachusetts.

particle is no longer recognized by the normal LDL receptor. However, macrophages avidly take up oxidized LDL via so-called acetyl-LDL, or scavenger receptors that specifically recognize the modified apolipoprotein B moiety (6). In contrast to uptake of native LDL via normal LDL receptors, uptake of oxidized LDL by macrophages via scavenger receptors is not subjected to downregulation by the intracellular cholesterol content. Therefore, in the presence of oxidized LDL, macrophages rapidly become loaded with lipids and are converted into lipid-laden so-called foam cells, the hallmark of the early atherosclerotic lesion (4,5).

An important implication of the oxidative theory of atherosclerosis is that antioxidants which can effectively prevent peroxidative damage to LDL lipids may act an antiatherogens. Therefore, it is of interest to study the antiperoxidative protection of LDL, both by lipid-soluble antioxidants associated with LDL itself and the antioxidants present in the natural environment of LDL, that is, plasma or the extracellular fluid of the subendothelial space (the composition of which is very likely to be similar to the composition of plasma).

RELATIVE IMPORTANCE OF VITAMIN E IN ANTIPEROXIDATIVE DEFENSES IN HUMAN PLASMA

Plasma Exposed to Water-Soluble Peroxyl Radicals

The most comprehensive studies on the relative importance of low-molecular-weight plasma antioxidants in protecting lipids against perioxidation used aqueous peroxyl radicals generated at a constant rate by thermal decomposition of the water-soluble radical initiator 2,2'-azobis (2-amidinopropane) hydrochloride (AAPH, also sometimes referred to as ABAP). The group of Ingold and Burton measured the peroxyl radical-trapping capability of human blood plasma and introduced the concept of the "TRAP" value (total peroxyl radical-trapping antioxidant parameter) of plasma as possible indicator of a person's antioxidant status (3,7). TRAP in 45 plasma samples of healthy individuals was found to be $820 \pm 148 \, \mu M$ (average value $\pm$ standard deviation) (3); that is, each liter of plasma on an average has the capacity to scavenge 820 μmol aqueous peroxyl radicals. TRAP correlates with the sum of the individual radical-trapping capabilities of the four major antioxidants in plasma, ascorbic acid, uric acid, protein thiols, and α-tocopherol. The individual radical-trapping capability of an antioxidant in plasma is calculated by multiplying its concentration in plasma (see Table 1) with its n value (3): TRAP ≈ 1.7 (ascorbic acid) $+ 1.3$ (uric acid) $+ 0.33$ (protein thiols) $+ 2.0$ (α-tocopherol).

On this basis, the *quantitative* contributions of the four major plasma antioxidants to TRAP of healthy individuals were calculated to be about as follows: ascorbic acid, 14%; uric acid, 56%; protein thiols, 18%; and α-tocopherol, 7% (averaged values from Refs. 3, 8, and 9). Frei et al. (10) showed that bilirubin is also an important antioxidant in human plasma and calculated that ascorbic acid, α-tocopherol, and bilirubin each contribute about 10% to TRAP, and uric acid contributes 47% and protein thiols 23% (1). All these studies (3,7-10) indicate that the *quantitative* importance of α-tocopherol in plasma under aqueous peroxyl radical stress is small (7-10%).

Although the TRAP assay gives *quantitative* information on the relative importance of the major plasma antioxidants in scavenging aqueous peroxyl radicals, no *qualitative* information can be derived from it. Wayner et al. (3) noted that the sequence of antioxidant consumption in plasma is protein thiols > uric acid > α-tocopherol. This shows that these antioxidants are differently reactive with aqueous peroxyl radicals and that the antioxidants in plasma form several lines or layers of antioxidant defenses. More detailed studies by Frei et al. (10,11) revealed that the sequence of plasma antioxidant consumption is ascorbic acid = protein thiols > bilirubin > uric acid > α-tocopherol, confirming preliminary results (3). Thus, it appears that α-tocopherol, the only lipid-soluble antioxidant measured in these experiments and generally thought to be the most effective lipid-soluble antioxidant in humans, forms the

TABLE 1 Solubility and Typical Concentrations of Protein Thiols and Antioxidants in Human Blood Plasma

Antioxidant	Plasma concentration (μM)
Water soluble	
Protein thiols	350-550
Uric acid	160-470
Ascorbic acid	30-150
Bilirubin	5-20
Lipid soluble	
α-Tocopherol	15-40
Ubiquinol-10	0.4-1.0
Lycopene	0.5-1.0
β-Carotene	0.3-0.6

Source: Modified from Reference 1.

last line of antiperoxidative defenses in plasma against aqueous peroxyl radicals. When whole blood is exposed to aqueous peroxyl radicals (12), the sequence of plasma antioxidant consumption is the same as in plasma alone (10), α-tocopherol again being of least *qualitative* importance.

As measured by a highly sensitive and selective assay that can measure lipid hydroperoxides in biological fluids at concentrations as low as 10 nM (13,14), no lipid peroxidation can be detected in AAPH-exposed plasma as long as ascorbic acid is present (10,11). Immediately after the complete consumption of ascorbic acid, however, micromolar concentrations of hydroperoxides of plasma phospholipids, cholesterol esters, and triglycerides begin to form, despite the presence of physiological or near physiological concentrations of the remaining water-soluble antioxidants and α-tocopherol. These and further data demonstrated unequivocally that ascorbic acid is the only endogenous antioxidant in plasma that can completely prevent detectable lipid peroxidation and that α-tocopherol cannot prevent substantial peroxidative damage to plasma lipids (10,11,15).

In conclusion, experiments using aqueous peroxyl radicals as oxidant revealed that uric acid *quantitatively* contributes about 50% to the antiperoxidative capacity of human blood plasma, but *qualitatively* ascorbic acid is by far most important. Ascorbic acid appears to trap the peroxyl radicals in the aqueous phase of plasma with a rate constant large enough to intercept virtually all these radicals before they can diffuse into the plasma lipids. Once ascorbic acid has been consumed completely, the remaining water-soluble antioxidants provide only a partial trap for aqueous peroxyl radicals. The peroxyl radicals that escape the antioxidants in the aqueous phase diffuse into the lipoproteins, where they initiate lipid peroxidation. Propagation of peroxidation in the lipoproteins is inhibited by lipid-soluble, chain-breaking antioxidants, including α-tocopherol. These lipid-soluble antioxidants trap chain-carrying lipid peroxyl radicals, thereby lowering the rate of detectable lipid peroxidation; unlike ascorbic acid, however, α-tocopherol cannot prevent initiation of detectable lipid peroxidation (a point further discussed in the next sections). Thus, in both *quantitative* and *qualitative* respects, α-tocopherol is relatively unimportant in protecting lipids in plasma against aqueous peroxyl radicals.

Plasma Exposed to Lipid-Soluble Peroxyl Radicals

When the lipid-soluble radical initiator 2,2'-azobis (2,4-dimethylvaleronitrile) (AMVN) rather than the water-soluble radical initiator AAPH is used as oxidant in plasma, a quite different picture emerges (15). Lipid-soluble α-tocopherol now belongs to the first line of antioxidant defenses, together with ascorbic acid. As soon as ascorbic acid has been used up, bilirubin also becomes oxidized. These findings suggest that ascorbic acid and bilirubin in the aqueous phase of plasma interact with α-tocopherol in the lipoproteins.

In marked contrast to water-soluble peroxyl radicals (10,11), lipid-soluble peroxyl radicals immediately and unsparingly induce detectable lipid peroxidation in plasma (15). Obviously, α-tocopherol is not as effective in protecting plasma lipids against lipid-soluble peroxyl radicals as is ascorbic acid against water-soluble peroxyl radicals (see earlier). α-Tocopherol appears to be unable to trap the AMVN-derived peroxyl radicals efficiently enough to prevent them from attacking plasma lipids and initiating detectable lipid peroxidation. Once initiated, propagation of lipid peroxidation in lipoproteins is inhibited by α-tocopherol, which scavenges chain-carrying lipid peroxyl radicals (LOO·). This leads to the formation of lipid hydroperoxides (LOOH) according to

$$LOO· + \alpha\text{-tocopherol} \rightarrow LOOH + \alpha\text{-tocopheroxy radical} \tag{1}$$

Therefore, consumption of α-tocopherol is associated with formation of lipid hydroperoxides.

Plasma Exposed to Pathologically Relevant Types of Oxidant Stress

The above data were obtained with aqueous and lipid-soluble peroxyl radicals chemically generated by AAPH and AMVN, respectively, and are therefore of limited physiological or pathological relevance. A pathologically very important source of oxidants in humans are activated neutrophils. These phagocytic cells are known to generate an array of oxidants once their NADPH oxidase has been activated (16). Excessive and chronic activation of neutrophils can lead to deterioration of the immune response, mutations, and tissue destruction (17,18). Neutrophils seem to play a key role in autoimmune diseases, such as rheumatoid arthritis, postischemic reoxygenation injury, and other inflammatory-immune injuries (19). In addition, oxidative modification of LDL by activated neutrophils has been implicated in atherosclerosis (20).

In plasma challenged by activated neutrophils the order of antioxidant consumption is ascorbic acid = ubiquinol-10 = protein thiols > uric acid, without significant consumption of bilirubin and α-tocopherol (10,21). Detectable lipid peroxidation is initiated immediately after completion of ascorbic acid consumption, despite the presence of high concentrations of all the remaining plasma antioxidants, including α-tocopherol. For example, a plasma sample incubated for 2 h with activated neutrophils contained 25 μM α-tocopherol and 0.68 μM lipid hydroperoxides (21), a plasma lipid hydroperoxide concentration possibly relevant to enhanced atherogenicity of LDL (see later). These findings indicate that α-tocopherol is relatively ineffective in scavenging neutrophil-derived oxidants and cannot prevent them from attacking plasma lipids, quite similar to the situation with aqueous peroxy radicals (10,11). It is noteworthy that ubiquinol-10 (the reduced form of coenzyme Q_{10}), a lipid-soluble antioxidant (22) present in low concentrations in plasma (Table 1), is consumed before α-tocopherol in neutrophil-exposed plasma (21), which suggests that ubiquinol-10 is *qualitatively* very important among the lipid-soluble antioxidants (see next section).

In other studies plasma was exposed to the gas phase of cigarette smoke (23). Cigarette smoke is known to contain a vast number of radicals and oxidants (24). One of the most abundant oxidants in the gas phase of cigarette smoke is nitric oxide, which is also generated by endothelial cells (25) and neutrophils (26) and could play a role in cigarette smoke-induced LDL oxidation and atherosclerosis (27). Oxidative damage caused by cigarette smoke has also been implicated in lung cancer and chronic obstructive pulmonary diseases, such as chronic bronchitis and emphysema (24).

In plasma exposed to the gas phase of cigarette smoke, endogenous ascorbic acid again is the first antioxidant to be totally consumed and the only one capable of preventing initiation of detectable lipid peroxidation (23). Ubiquinol-10 (Frei and Ames, unpublished observation), protein tiols, and bilirubin are also oxidized at significant rates, but oxidations of uric acid and α-tocopherol are very slow (23). LDL isolated from plasma containing as little as 0.50 μM lipid hydroperoxides but high, physiological levels of α-tocopherol (38 μM) showed increased anodic electrophoretic mobility on agarose gels (23), indicative of potentially atherogenic changes in LDL.

A third pathologically relevant type of oxidant stress imposed on plasma is the xanthine-xanthine oxidase system (Frei and Ames, unpublished results). This system is known to produce superoxide radicals, hydrogen peroxide (28), and possibly hydroxyl radicals (29); these oxidants may mediate myocardial tissue injury during postischemic reoxygenation (19). The pattern of antioxidant consumption and lipid peroxidation in plasma exposed to the xanthine-xanthine oxidase system very much resembles the pattern seen in the presence of activated neutrophils: the endogenous antioxidants are consumed in the temporal order ascorbic acid = ubiquinol-10 = protein thiols, without oxidation of bilirubin and α-tocopherol, and detectable lipid peroxidation occurs only after ascorbic acid has been used up.

In summary, under the pathologically relevant types of oxidant stress discussed above, namely activated neutrophils, the gas phase of cigarette smoke, and the xanthine-xanthine oxidase system, α-tocopherol is relatively ineffective in protecting plasma lipids and LDL against peroxidative damage; that is, α-tocopherol is *qualitatively* of minor importance.

Finally, oxidant stress has also been implicated in pathologies associated with the release of free iron, copper, heme, or hemoglobin into plasma and/or other tissues. Such conditions can occur consequent to iron or copper overload, for example in idiopathic hemochromatosis and Wilson's disease, respectively, and consequent to hemorrhage and hemolysis, for example during a stroke (19). Under these oxidative conditions, the metal binding proteins of plasma become of primary importance in preventing peroxidative damage: transferrin, lactoferrin, ceruloplasmin, albumin, hemopexin, or haptoglobin (1,2,19). The low-molecular-weight antioxidants, including α-tocopherol, are of secondary importance, with the only possible exception of uric acid, which can chelate iron (30).

RELATIVE IMPORTANCE OF VITAMIN E AMONG THE LIPID-SOLUBLE ANTIOXIDANTS IN HUMAN PLASMA AND LDL

Thus far, there are no published reports on the relative importance of the lipid-soluble, chain-breaking antioxidants in whole plasma. From Table 1 it is evident that *quantitatively* α-tocopherol contributes most to the lipid-soluble antioxidant capacity of plasma (31) because it is present in concentrations at least 15 times higher than any of the other lipid-soluble antioxidants.

Studies by Esterbauer et al. (32,33) and more recently by Stocker et al. (34) provided valuable insights into the *qualitative* importance of lipid-soluble antioxidants in isolated human LDL. Using Cu^{2+} as oxidant, Esterbauer's group showed that the antioxidants in LDL are progressively consumed in the sequence vitamin E > lycopene > phytofluene > β-carotene. Lipid peroxidation occurs continuously at a slow rate during consumption of these antioxidants. As mentioned, this can be expected because scavenging of lipid peroxyl radicals by vitamin E, and other lipid-soluble, chain-breaking antioxidants in LDL leads to formation of lipid hydroperoxides according to reaction (1) (32). When the antioxidants in LDL have been depleted, lipid peroxidation enters its uninhibited propagation phase. Added ascorbic acid fully protects the endogenous antioxidants and the lipids in LDL against Cu^{2+}-induced oxidation (32).

The study by Stocker et al. (34) revealed that ubiquinol-10 is consumed even before α-tocopherol in isolated LDL exposed to aqueous or lipid-soluble peroxyl radicals or the oxidants released from activated neutrophils. The rate of radical-mediated formation of lipid hydroperoxides in LDL is low as long as ubiquinol-10 is present but increases rapidly after its consumption, even though more than 80% of α-tocopherol, β-carotene, and lycopene are still present. These findings are in agreement with the data already discussed showing that in plasma exposed to various pathologically relevant types of oxidant stress ubiquinol-10 is consumed before α-tocopherol; they suggest that ubiquinol-10 is *qualitatively* more important than α-tocopherol in protecting LDL lipids against peroxidative damage. The latter notion is particularly interesting considering that lipid peroxidation in LDL has been implicated as causative factor in atherosclerosis (4,5) and strategies aimed at lowering the risk of heart disease by lowering plasma cholesterol levels with the drug lovastatin also lower plasma coenzyme Q levels (35).

Terao (36) investigated the effectiveness of α-tocopherol and carotenoids to trap peroxyl radicals in organic solution and found the order to be α-tocopherol > astaxanthin, canthaxanthin > β-carotene = zeaxanthin. The reactivity of lipid-soluble antioxidants in organic solvents with singlet oxygen was shown by Di Mascio et al. (37) to decrease in the order lycopene > α-carotene > β-carotene > lutein > bilirubin > α-tocopherol. Taking the concentration in plasma of these antioxidants into consideration, the authors concluded that lycopene and bilirubin are qualitatively about equally important in quenching singlet oxygen in plasma, followed by β-carotene and α-tocopherol.

In summary, although α-tocopherol is by far the most concentrated lipid-soluble antioxidant in plasma, it appears to be less effective than ubiquinol-10 in protecting LDL against lipid peroxidation. Lycopene and bilirubin appear to be qualitatively most important in quenching singlet oxygen, α-tocopherol being of secondary importance. Since chain-breaking antioxidants in the lipids scavenge lipid peroxyl radicals, consumption of lipid-soluble antioxidants is always accompanied by formation of lipid hydroperoxides.

CONCLUSIONS

The evidence discussed in this review suggests that α-tocopherol is not of primary importance in protecting lipids in human blood plasma and LDL against peroxidative damage. Quantitatively, α-tocopherol plays a minor role in the protection of plasma lipids against water-soluble oxidants (aqueous peroxyl radicals, and all or at least part of the oxidants generated by activated neutrophils, the xanthine-xanthine oxidase system, and in the gas phase of cigarette smoke). This is because relatively high concentrations of a number of natural antioxidants are present in the aqueous phase of plasma. These water-soluble antioxidants also supercede α-tocopherol in qualitative importance in protecting plasma lipids against aqueous oxidants, supposedly because of the physical proximity of the water-soluble antioxidants with these oxidants.

One might argue that α-tocopherol is not consumed in plasma exposed to pathologically relevant types of oxidant stress (10,21,23) or consumed last in plasma exposed to aqueous peroxyl radicals (10,11) because it is spared by ascorbic acid. Such a sparing effect due to regeneration of α-tocopherol by ascorbic acid at the water/lipid interface was observed in artificial and natural membranes (38,39) and has also been suggested to occur in LDL (40). However, there is no evidence for such interaction in plasma: under none of the oxidant stress conditions used was α-tocopherol consumption initiated after completion of ascorbic acid oxidation. In cases of activated neutrophils, the gas phase of cigarette smoke, and the xanthine-xanthine oxidase system, α-tocopherol was not consumed despite complete oxidation of ascorbic acid and initiation of detectable lipid peroxidation (10,21,23). In the case of aqueous peroxyl radicals, completion of ascorbic acid consumption and the start of α-tocopherol oxidation were separated by more than 30 minutes of incubation, a period during which more than 90 μM peroxyl radicals were generated (10).

Among the lipid-soluble antioxidants in plasma and LDL, α-tocopherol is quantitatively by far the most important (31). Qualitatively, however, α-tocopherol seems to be secondary to ubiquinol-10, which is consumed first in plasma and LDL (21,34) and appears to protect the lipids more effectively (34). Ubiquinol-10 may be particularly important under mild oxidant stress conditions, and α-tocopherol could become important under more drastic conditions that overwhelm the antioxidant capacity of ubiquinol-10.

It is important to reemphasize that α-tocopherol, and for that matter any other lipid-soluble antioxidant, is antiperoxidatively active during the propagation phase of lipid peroxidation, that is, the consumption of the lipid-soluble antioxidants is associated with formation of lipid hydroperoxides. In other words, the action of α-tocopherol generates lipid hydroperoxides from lipid peroxyl radicals [reaction (1)] that would otherwise attack an adjacent lipid molecule and propagate the radical chain reaction. This is not contradictory to, but actually explains why lipid peroxidation is strongly inhibited by α-tocopherol. Unfortunately, there seems to be some confusion among scientists in this respect. It is sometimes assumed that α-tocopherol can completely prevent lipid peroxidation, and measurement of lipid peroxidation by insufficiently sensitive and/or nonspecific assays appears to support this faulty view. Measurement of conjugated dienes at 234 nm (32), thiobarbituric acid reactive substances ("malondialdehyde") (41), oxygen uptake (3,7), or loss of polyunsaturated fatty acids (42,43) are all methods that may not adequately discover formation of lipid hydroperoxides during α-tocopherol consumption.

To place the in vitro findings discussed here into perspective in terms of their in vivo relevance, the following questions must be asked: What kind of oxidants are likely to play a role in human pathogenesis? Are the oxidants water soluble or lipid soluble? In the former case, ascorbic acid can obviously be expected to play a key role among the nonenzymatic extracellular antioxidants. The high efficacy of ascorbic acid against the pathologically relevant types of oxidant stress discussed earlier strongly supports the view that ascorbic acid can help in the prevention of certain human diseases caused or exacerbated by oxidant stress. In pathologies linked to lipid-soluble oxidants, α-tocopherol can also be expected to be of primary importance, together with ascorbic acid (regeneration of α-tocopherol) and ubiquinol-10.

The second important set of questions concerns the role of lipid peroxidation in human pathology. Is lipid peroxidation cause or consequence of the disease under study? Is lipid peroxidation occurring in lipoproteins or membranes? If lipid peroxidation is a causative factor, as is probably the case for atherosclerosis (4,5), what levels of lipid hydroperoxides are pathologically relevant? In other words, are pathologies triggered or exacerbated by lipid hydroperoxide levels accumulating during the inhibited phase of lipid peroxidation or only during the uninhibited phase, after all the lipid-soluble antioxidants have been consumed? In the former case, α-tocopherol would be expected to be relatively ineffective in preventing or slowing the progression of these diseases, but in the latter it should be more effective. Future research on the role of lipid peroxidation in human pathology and the preventative potency of vitamin E and other antioxidants should consider these questions.

REFERENCES

1. Stocker R, Frei B. Endogenous antioxidant defenses in human blood plasma. In: Sies H, ed. Oxidative stress: oxidants and antioxidants. Orlando, FL: Academic Press, 1991; 213-43.
2. Halliwell B, Gutteridge JMC. The antioxidants of human extracellular fluids. Arch Biochem Biophys 1990; 280:1-8.
3. Wayner DDM, Burton GW, Ingold KU, Barclay LRC, Locke SJ. The relative contributions of vitamin E, urate, ascorbate and proteins to the total peroxyl radical-trapping antioxidant activity of human blood plasma. Biochim Biophys Acta 1987; 924:408-19.
4. Steinberg D, Parthasarathy S, Carew TE, Khoo JC, Witztum JL. Beyond cholesterol. Modifications of low-density lipoprotein that increase its atherogenicity. N Engl J Med 1989; 320:915-24.
5. Steinbrecher UP, Zhang H, Loughheed M. Role of oxidatively modified LDL in atherosclerosis. Free Radic Biol Med 1990; 9:155-68.

6. Parthasarathy S, Fong LG, Otero D, Steinberg D. Recognition of solubilized apoproteins from delipidated, oxidized low density lipoprotein (LDL) by the acetyl-LDL receptor. Proc Natl Acad Sci U S A 1987; 84:537-40.

7. Wayner DDM, Burton GW, Ingold KU, Locke S. Quantitative measurement of the total, peroxyl radical-trapping antioxidant capability of human blood plasma by controlled peroxidation. The important contribution made by plasma proteins. FEBS Lett 1985; 187:33-7.

8. Thurnham DI, Situnayake RD, Koottathep S, McConkey B, Davis M. Antioxidant status measured by the TRAP assay in rheumatoid arthritis. In: Rice-Evans C, ed. Free radicals, oxidant stress and drug action. London: Richelieu Press, 1987; 169-92.

9. Lindeman JHN, Van Zoeren-Grobben D, Schrijver J, Speek AJ, Poorthuis BJHM, Berger HM. The total free radical trapping ability of cord blood plasma in preterm and term babies. Pediatr Res 1989; 26:20-4.

10. Frei B, Stocker R, Ames BN. Antioxidant defenses and lipid peroxidation in human blood plasma. Proc Natl Acad Sci U S A 1988; 85:9748-52.

11. Frei B, England L, Ames BN. Ascorbate is an outstanding antioxidant in human blood plasma. Proc Natl Acad Sci U S A 1989; 86:6377-81.

12. Niki E, Yamamoto Y, Takahashi M, et al. Free radical-mediated damage of blood and its inhibition by antioxidants. J Nutr Sci Vitaminol 1988; 34:507-12.

13. Frei B, Yamamoto Y, Niclas D, Ames BN. Evaluation of an isoluminol chemiluminescence assay for the detection of hydroperoxides in human blood plasma. Anal Biochem 1988; 175:120-30.

14. Yamamoto Y, Frei B, Ames BN. Assay of lipid hydroperoxides using high-performance liquid chromatography with isoluminol chemiluminescence detection. Methods Enzymol 1990; 186:371-80.

15. Frei B, Stocker R, England L, Ames BN. Ascorbate: the most effective antioxidant in human blood plasma. In: Emerit I, Packer L, Auclair C, eds. Antioxidants in therapy and preventative medicine. New York: Plenum Press, 1990; 155-63.

16. Hamers MN, Roos D. Oxidative stress in human neutrophilic granulocytes: host defence and self-defence. In: Sies H, ed. Oxidative stress. Orlando, FL: Academic Press, 1985; 351-81.

17. Babior BM. Oxidants from phagocytes: agents of defense and destruction. Blood 1984; 64:959-66.

18. Weiss SJ. Tissue destruction by neutrophils. N Engl J Med 1989; 320:365-76.

19. Halliwell B, Gutteridge JMC. Free radicals in biology and medicine, 2nd ed. Oxford: Clarendon Press, 1989.

20. Cathcart MK, Morel DW, Chisolm GM. Monocytes and neutrophils oxidize low density lipoproteins making it cytotoxic. J Leukoc Biol 1985; 38:341-50.

21. Cross CE, Forte T, Stocker R, et al. Oxidative stress and abnormal cholesterol metabolism in patients with adult respiratory distress syndrome. J Lab Clin Med 1990; 115:396-404.

22. Frei B, Kim MC, Ames BN. Ubiquinol-10 is an effective lipid-soluble antioxidant at physiological concentrations. Proc Natl Acad Sci U S A 1990; 87:4879-83.

23. Frei B, Forte TM, Ames BN, Cross CE. Gas phase oxidants of cigarette smoke induce lipid peroxidation and changes in lipoprotein properties in human blood plasma: protective effects of ascorbic acid. Biochem J 1991; 277:133-38.

24. Church DF, Pryor WA. Free-radical chemistry of cigarette smoke and its toxicological implications. Environ Health Perspect 1985; 64:111-26.

25. Palmer RMJ, Ashton DS, Moncada S. Vascular endothelial cells synthesize nitric oxide from L-arginine. Nature 1988; 333:664-6.

26. McCall TB, Boughton-Smith NK, Palmer RM, Whittle BJ, Moncada S. Synthesis of nitric oxide from L-arginine by neutrophils. Release and interaction with superoxide anion. Biochem J 1989; 261:293-6.

27. Yokode M, Kita T, Arai H, Kawai C, Narumiya S, Fujiwara M. Cholesteryl ester accumulation in macrophages incubated with low density lipoprotein pretreated with cigarette smoke extract. Proc Natl Acad Sci U S A 1988; 85:2344-8.

28. Link EM, Riley A. Role of hydrogen peroxide in the cytotoxicity of the xanthine/xanthine oxidase system. Biochem J 1988; 249:391-9.

29. Kuppusamy P, Zweier JL. Characterization of free radical generation by xanthine oxidase: evidence for hydroxyl radical generation. J Biol Chem 1989; 264:9880-4.

30. Davies KJA, Sevanian A, Muakkassah-Kelly SF, Hochstein P. Uric acid-iron ion complexes. A new aspect of the antioxidant functions of uric acid. Biochem J 1986; 235:747-54.

31. Burton GW, Joyce A, Ingold KU. First proof that vitamin E is major lipid-soluble, chain-breaking antioxidant in human blood plasma. Lancet 1982; 2:327.

32. Esterbauer H, Striegl G, Puhl H, et al. The role of vitamin E and carotenoids in preventing oxidation of low density lipoprotein. Ann N Y Acad Sci 1989; 570:254-67.

33. Esterbauer H, Jürgens G, Puhl H, Quehenberger O. Role of oxidatively modified LDL in atherogenesis. In: Hayaishi O, Niki E, Kondo M, Yoshikawa T, eds. Medical, biochemical and chemical aspects of free radicals. Amsterdam: Elsevier, 1989; 1203-9.

34. Stocker R, Bowry VW, Frei B. Ubiquinol-10 protects human low density lipoprotein more effectively against lipid peroxidation than does α-tocopherol. Proc Natl Acad Sci U S A 1991; 88:1646-50.

35. Folkers K, Langsjoen P, Willis R, et al. Lovastatin decreases coenzyme Q levels in humans. Proc Natl Acad Sci U S A 1990; 87:8931-4.

36. Terao J. Antioxidant activity of β-carotene-related carotenoids in solution. Lipids 1989; 24:659-61.

37. Di Mascio P, Kaiser S, Sies H. Lycopene as the most efficient biological carotenoid singlet oxygen quencher. Arch Biochem Biophys 1989; 274:532-8.

38. Doba T, Burton GW, Ingold KU. Antioxidant and co-antioxidant activity of vitamin C. The effect of vitamin C, either alone or in the presence of vitamin E or a water-soluble vitamin E analogue, upon the peroxidation of aqueous multilamellar phospholipid liposomes. Biochim Biophys Acta 1985; 835:298-303.

39. Vatassery GT, Smith WE, Quach HT. Ascorbic acid, glutathione and synthetic antioxidants prevent the oxidation of vitamin E in platelets. Lipids 1989; 24:1043-7.

40. Sato K, Niki E, Shimasaki H. Free radical-mediated chain oxidation of low density lipoprotein and its synergistic inhibition by vitamin E and vitamin C. Arch Biochem Biophys 1990; 279:402-5.

41. Steinbrecher UP, Parthasarathy S, Leake DS, Witztum JL, Steinberg D. Modification of low density lipoprotein by endothelial cells involves lipid peroxidation and degradation of low density lipoprotein phospholipids. Proc Natl Acad Sci U S A 1984; 81:3883-7.

42. Quehenberger O, Keller E, Jürgens G, Esterbauer H. Investigation of lipid peroxidation in human low density lipoprotein. Free Radic Res Commun 1987; 3:1-5.

43. Esterbauer H, Jürgens G, Quehenberger O, Koller E. Autooxidation of human low density lipoprotein: loss of polyunsaturated fatty acids and vitamin E and generation of aldehydes. J Lipid Res 1987; 28:495-509.

11

Scavenging of Singlet Molecular Oxygen by Tocopherols

Helmut Sies, Stephan Kaiser, Paolo Di Mascio*, and Michael E. Murphy[†]

University of Düsseldorf, Düsseldorf, Germany

INTRODUCTION

Tocopherols (vitamin E) are known to act as biological antioxidants (1). Attention has centered on their function as free-radical scavengers, but it has also been demonstrated that tocopherols react with singlet molecular oxygen ($^1O_2$) (2-8). The generation and possible pathological consequences of singlet oxygen in biological systems via enzymatic reactions have been described (9-12). $^1O_2$ can be produced by photoexcitation as well as by nonphotochemical processes (chemiexcitation) and has been implicated in the peroxidation of biological lipids (5,13,14). $^1O_2$ was also shown to play a role in the inactivation of cultured human cells by ultraviolet A (UV-A) and near-visible radiation (15). Furthermore, $^1O_2$ has been shown to be capable of inducing DNA damage (16-18) and to be mutagenic (19).

Therefore, tocopherols are of biological interest for their $^1O_2$-quenching capability. Scavenging of $^1O_2$ by tocopherols includes physical quenching, in which the excited state of oxygen is deactivated without light emission, and chemical quenching, which results in the formation of various oxidation products. Physical quenching by electron energy transfer almost always predominates, the rate depending on solvent polarity. This has led to the suggestion that a charge transfer intermediate may be involved in the quenching process (2). In addition, it has been pointed out that the balance between physical and chemical quenching is a sensitive function of spin-orbit coupling properties and entropy factors (20). In the quenching of free radicals and also of $^1O_2$, α-tocopherol is more effective than the other homologs in the sequence α- > β- > γ- > δ-tocopherol (21,22).

We compared the relative physical and chemical quenching ability of the tocopherol homologs toward $^1O_2$ (Table 1). This was achieved by using $NDPO_2$, the thermodissociable endoperoxide of 3,3′-(1,4-naphthylidene) dipropionate to generate the singlet oxygen (23, 24), a methodology that offers advantages in comparison to other systems of $^1O_2$ generation. Generation and quenching of $^1O_2$ were directly monitored by the monomol light emission signal arising from the transition of the excited $^1\Sigma_g$ state to the $^3\Sigma_g$ ground state at 1270 nm by the use of a liquid nitrogen-cooled germanium photodiode detector. The time course and concentration dependence of the loss of tocopherols and the formation of reaction products due to chemical reaction with $^1O_2$ were also examined.

Present affiliations:
*University of São Paulo, São Paulo, Brazil
[†]University of Vermont, Burlington, Vermont

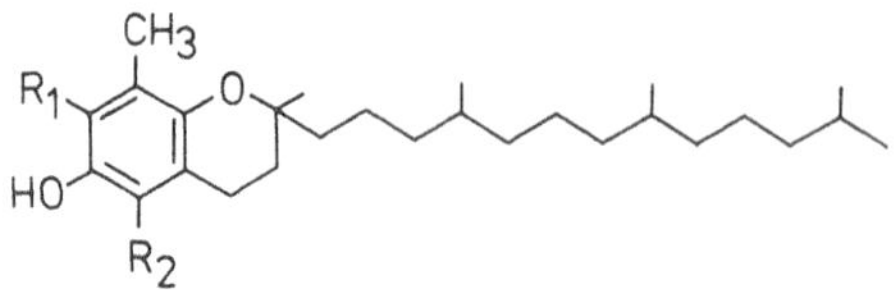

CH3

R₁ ... O ...

HO

R₂

	R₁	R₂
α-Tocopherol	CH₃	CH₃
β-Tocopherol	H	CH₃
γ-Tocopherol	CH₃	H
δ-Tocopherol	H	H

The present communication is based largely on our recent publication on this topic (25).

MATERIALS AND METHODS

Generation of singlet oxygen by $NDPO_2$, detection of singlet molecular oxygen monomol emission (1270 nm), and electrochemical detection of tocopherols and tocopherol oxidation products with high-performance liquid chromatography (HPLC) were as described in detail in References 24-26.

RESULTS

Physical Quenching of $^1O_2$ by Tocopherols

The physical quenching efficiency of the various tocopherols was assessed by their ability to decrease the monomol emission signal. The results in terms of the k_q value are shown in Table 2 and reveal that the physical quenching decreased in the sequence of α- > β- > γ- > δ-tocopherol. The corresponding half-quenching concentrations, that is, the tocopherol concentration at which 50% of the generated $^1O_2$ was quenched, were 320, 340, 420, and 580 μM, respectively (Fig. 1).

Replacing $CHCl_3$ in the solvent by D_2O to create a more polar solvent increased the quenching efficiency of α-tocopherol at concentrations of 0.3 mM from 49.6 ± 0.8 to 65.1 ± 0.3% ($n = 4$). This supports the suggestion that physical quenching of $^1O_2$ by tocopherols involves a charge transfer mechanism (2,20). Furthermore, substitution of the hydroxyl group in position 6 of the chromane ring for methyl ether or an acetyl or succinyl ester group abolished the $^1O_2$-quenching ability. The observed residual quenching with α-tocopherol methyl ether (Table 2) is explained by a 1% contamination with pure α-tocopherol as identified by electrochemical detection. Alkylation of the phenolic hydroxyl group should raise the oxidation potential and thus lower the quenching ability if a charge transfer mechanism is involved (2). Trolox, the hydrophilic tocopherol analog in which the phytyl side chain has been replaced for a carboxylic group, showed a quenching efficiency similar to that of α-tocopherol, thus excluding participation of the lipophilic side chain in the quenching process (Table 2).

The biologically occurring carotenoids lycopene and β-carotene have a $^1O_2$-quenching ability two orders of magnitude higher than the tocopherol homologs (Table 2; Refs. 3 and 27). In contrast, vitamin A derivatives did not show quenching of $^1O_2$ (Table 2), in line with

TABLE 2 Singlet Oxygen Quenching by Tocopherols and Related Compounds[a]

Compound	Loss of Ge diode signal (%) at 300 μM	k_q (10^6 M^{-1} s^{-1})	Previous values (ref.)	k_r (10^6 M^{-1} s^{-1})	Previous values (ref.)
α-Tocopherol	49.6 ± 0.8	280 / 450[b]	120–250 (2); 620 (3); 120–670 (5); 31 (8); 42 (22); 260 (30)	3.6	46 (3); 1.9 (8); 6.6 (30)
β-Tocopherol	49.6 ± 1.2	270	23 (22)	0.23	NR[c]
γ-Tocopherol	45.0 ± 1.3	230	11 (22): 180 (30)	2.8	2.6 (30)
δ-Tocopherol	37.5 ± 0.9	160	5 (22); 100 (30)	1.7	0.7 (30)
α-Tocopherol methyl ether	17.2 ± 0.3	—[d]	NR	0.94	NR
α-Tocopherol ethyl ether	ND[e]	—	NR	0.16	NR
α-Tocopherol acetyl ester	No loss	—	<1.6 (2)	—	NR
α-Tocopherol succinyl ester	No loss	—	NR	—	NR
Trolox	67.3 ± 2.4	470[b]	NR	ND	NR
DPPD	44.6 ± 1.1	270	NR	ND	NR
Lycopene	98.2 ± 0.8	ND	31,000 (27)	ND	NR
β-Carotene	96.7 ± 1.4	ND	14,000 (27)	ND	NR
Retinoic acid	9.1 ± 1.1	—	NR	ND	NR
Etretinate	7.4 ± 0.6	—	NR	ND	NR
Isotretinoin	4.0 ± 0.9	—	NR	ND	NR

[a]The loss of the Ge diode signal is expressed in percentage of the light emitted by ¹O₂ at a concentration of 300 mM of the quencher. Conditions were as described in Reference 25. Values are given as means ± standard error of the mean (n = 3 or 4).
[b]Solvent was D₂O/C₂H₅OH (1:1) for solubility reasons.
[c]Not reported previously.
[d]No reaction observed.
[e]Not determined.
Source: From Reference 25.

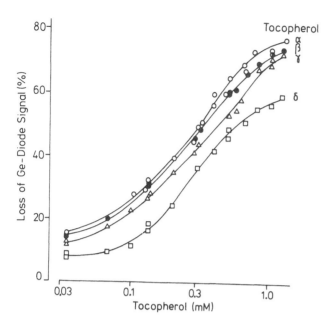

FIGURE 1 Dependence of loss of germanium diode signal on the concentration of the different tocopherol homologs. (From Ref. 25.)

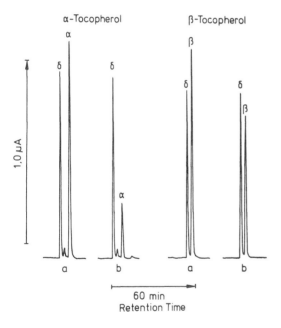

FIGURE 2 Chromatography and electrochemical detection of tocopherols after incubation with NDPO₂. Tocopherols (300 μM) were incubated in an ethanol-D₂O (1:1) solution at 37°C without (a) and with (b) NDPO₂. After 60 minutes the reaction was stopped by adding NaN₃ (2.5 mM), which quenched the remaining $^1O_2$ completely. For α- and β-tocopherols, δ-tocopherol (300 μM) was added as an internal standard. (From Ref. 25.)

the observation that structures with fewer than seven conjugated double bonds (vitamin A has four) are devoid of a $^1O_2$-quenching ability (28,29). On the other hand, the synthetic antioxidant DPPD exhibited a quenching capacity similar to that of γ-tocopherol (Table 2).

Chemical Reaction of $^1O_2$ with Tocopherols

As with physical quenching, chemical reactivity differed considerably among the various tocopherols. However, the sequence of reactivity toward $^1O_2$ was α- > γ- > δ- > β-tocopherol (Figs. 2 and 3). At low concentrations of $NDPO_2$ α-tocopherol showed the highest reactivity, but at $NDPO_2$ concentrations of more than 7 mM there was no significant difference between α-, γ-, and δ-tocopherols (Fig. 3B). Interestingly, β-tocopherol exhibited very low reactivity. However, at a reaction time of more than 2 h and at a $NDPO_2$ concentration of 20 mM all tocopherols had completely reacted (data not shown), in contrast to results reported for the microwave discharge system (22), in which the reaction of tocopherols with $^1O_2$ measured in hexadecane as solvent was incomplete, leaving about 10% of the tocopherols unreacted. The k_r value for α-tocopherol (3.6×10^6 M^{-1} s^{-1}) compares well with the value of 2.1×10^6 M^{-1} s^{-1} obtained in pyridine (2) and 6.6×10^6 M^{-1} s^{-1} observed in ethanol (30) and reveals that about 1.5% of the total quenching can be accounted for by chemical reaction. For β-tocopherol the chemical reaction accounts for only about 0.1% of the total quenching. The $NDPO_2$ concentration dependence of the relative reaction rates suggests that other components of the reaction exist.

Reaction products were determined by HPLC with electrochemical detection (Fig. 4); the fraction of each peak was collected and UV spectra and gas chromatography-mass spectrometry (GC-MS) performed. Tocopheryl quinone formation, identified by the retention times reported (26), was detected from all four homologs, whereas measurable formation of the quinone epoxide was observed with only γ-tocopherol (Fig. 4). Tocopheryl quinone formation from the tocopherols was low, accounting for 10-15% of the reacted tocopherols, but it increased consistently with time and with $NDPO_2$ concentration (Fig. 5). The relative yield of the individual tocopheryl quinones paralleled the sequence of the chemical reactivity of the tocopherol homologs, that is, α- > γ- > δ- > β-tocopheryl quinone.

The γ-tocopheryl quinone epoxide was identified by GC-MS (separation temperature 280°C, ionizing electron energy 70 eV, $m/e = 448$). The formation of the γ-tocopheryl quinone epoxide accounted for up to 80% of the reacted γ-tocopherol; the time course and concentration dependence differed markedly from those for the tocopheryl quinone (Fig. 5). The concentration of this compound rose to a maximum and then decreased with time while $^1O_2$ was still being produced. This loss at later time points was enhanced by the addition of fresh $NDPO_2$ 20 minutes after the starting the initial reaction (Fig. 5A), suggesting that the quinone epoxide may not be the stable end product of the γ-tocopherol $^1O_2$ oxidation reaction as assumed thus far (22,30-32). Rather, another product or products may be formed from a subsequent reaction of γ-tocopheryl quinone epoxide with $^1O_2$; this additional product(s) is electrochemically inactive. The formation of the γ-tocopheryl quinone epoxide and the loss of the γ-tocopherol were temperature dependent. When the reaction was carried out at 4°C for 20 h, with the $^1O_2$ released more slowly, less γ-tocopherol was lost and relatively more of the quinone epoxide was formed compared to the yield obtained at 37°C.

The γ-tocopheryl quinone epoxide with a concentration of 180 μM as quantified by electrochemical detection was incubated with rat liver microsomal epoxide hydrolase for 30 minutes at 37°C as described (33). Although active with the substrate styrene oxide, HPLC analysis of the samples showed no detectable decrease in the γ-tocopheryl quinone epoxide compared to controls, which were run with inactivated epoxide hydrolase (70°C for 30 minutes). It is concluded that the epoxide is not a substrate for the microsomal epoxide hy-

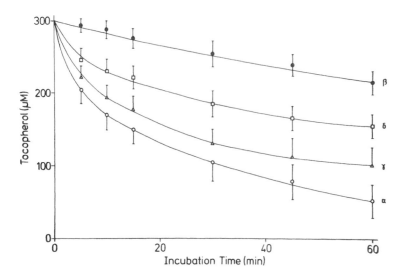

(a)

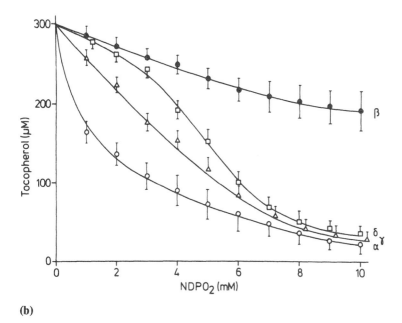

(b)

FIGURE 3   Loss of tocopherol due to chemical reaction with singlet oxygen. (a) Time course. For conditions see legend to Figure 2. $n = 4$ ($\pm$SD). (b) Dependence on $NDPO_2$ concentration. Reaction time was 30 minutes. (From Ref. 25.)

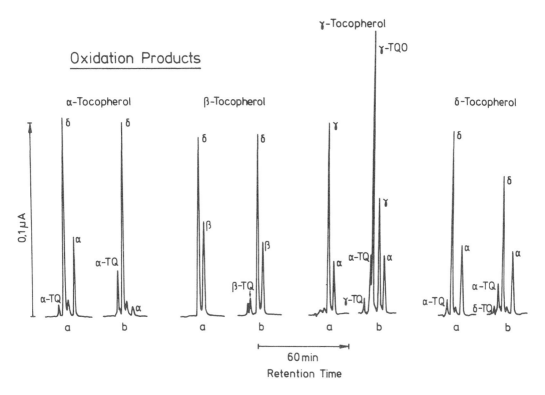

Oxidation Products

FIGURE 4 Electrochemical detection of tocopherol oxidation products after incubation with $NDPO_2$. For conditions see Figure 2. Tocopherol oxidation products were determined without (a) and with (b) $NDPO_2$. TQ, tocopheryl quinone; TQO, tocopheryl quinone epoxide. (From Ref. 25.)

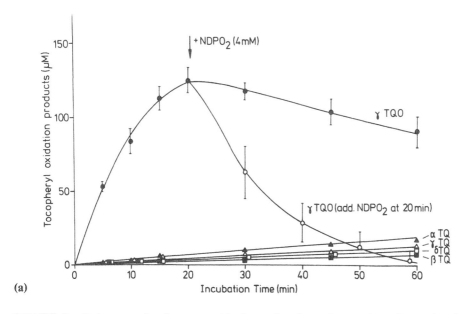

(a)

FIGURE 5 Quinone and quinone epoxide formation from the reaction of tocopherols with singlet oxygen. (a) Time course. Data were obtained as described in the legend to Figure 4. (Open circles) After formation of the γ-tocopherol quinone epoxide, a second addition of $NDPO_2$ (4 mM) was made at 20 minutes. Data are $n = 4 \pm$ standard deviation (SD). (b) Reaction of γ-tocopherol with $NDPO_2$ and corresponding products. The products not identified (right-hand scale) are expressed as percentages of the reacted γ-tocopherol. (From Ref. 25.)

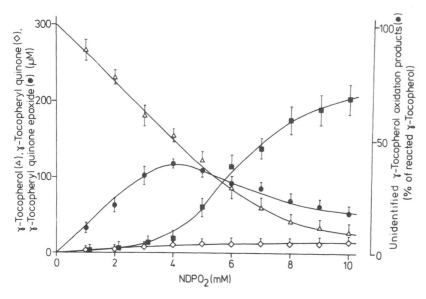

(b) FIGURE 5 (Continued)

drolase under these conditions and that the disposal of this oxidation product may occur by different means.

Apart from the quinones and the quinone epoxide, other products of the reaction of the tocopherol homologs with singlet oxygen appeared on the HPLC. Retention times under the stated HPLC conditions and corresponding UV absorbance maxima for α-tocopherol oxidation products were 15.6 h (λ = 258, 264, and 271 nm); for β-tocopherol 8.1 h (λ = 276 nm) and 17.0 h (λ = 282 nm); for γ-tocopherol 1.1 h (λ = 275 nm), 4.8 h (λ = 244 nm), 8.0 h (λ = 302 nm), and 14.8 h (λ = 295 nm) (Fig. 6); and for δ-tocopherol 1.5 h (λ = 254, 272 nm) and 12.2 h (λ = 272, 279 nm). For γ-tocopherol oxidation products the UV absorption profiles of compounds 4 and 5 (Fig. 6) were similar to those of tocopheryl quinones, whereas the profiles of compounds 6 and 7 were reminiscent of those of tocopherols. The long retention times suggest the formation of dimers or trimers, which have been shown to be formed by the alkaline ferricyanide oxidation of α-tocopherol (34). However, identification by mass spectrometry was hampered by low product volatility and perhaps low product stability. It is not known whether any product was the 8α-hydroperoxychromanone identified by Foote et al. (31).

When the γ-tocopherol was reacted with the thermolabile radical initiator AAPH (50 mM) instead of with NDPO$_2$ under otherwise identical conditions, higher amounts of the γ-tocopheryl quinone and quinone epoxide were observed, as well as compounds with the same retention time as compounds 6 and 7 of the γ-tocopherol $^1O_2$ reaction (data not shown). Thus, compounds 4 and 5 are products occurring preferentially in the γ-tocopherol $^1O_2$ reaction, not found to an appreciable extent via the radical reaction induced by AAPH.

DISCUSSION

Physical Quenching

The rate constants of physical quenching of singlet oxygen by the tocopherol homologs were found to be 280, 270, 230, and 160 $\times$ 10^6 M^{-1} s^{-1} for α-, β-, γ-, and δ-tocopherol, respectively (Table 2). This sequence of reactivity of the homologs is consistent with results obtained

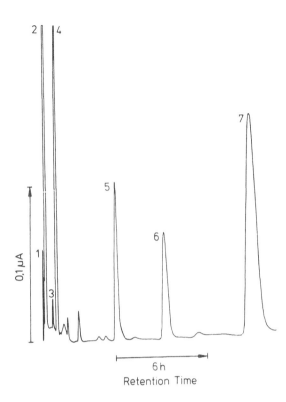

FIGURE 6 γ-Tocopherol oxidation products after incubation of γ-tocopherol with $NDPO_2$. Compound 1, γ-tocopheryl quinone epoxide; compound 2, γ-tocopherol; compound 3, γ-tocopherol quinone; compounds 4-7, unidentified products; for spectral characteristics, see text. Incubation conditions as described in the legend to Figure 2, except that the reaction was performed at 4 °C for 20 h. (From Ref. 25.)

elsewhere using other singlet oxygen-generating systems (22,30,35) and is also in line with the relative effectiveness of the homologs in preventing in vitro lipid peroxidation (21) and with their in vivo biological activity (35), as determined by prevention of respiratory decline in rat livers and erythrocyte hemolysis. The quenching ability in $C_2H_5OH/CHCl_3$ (1:1) was found to be less than in the more polar solvent of C_2H_5OH/D_2O (1:1), supporting the possibility that the quenching of $^1O_2$ by tocopherols proceeds through a charge transfer intermediate (2,20). Furthermore, esterification or ether formation at the 6-position in the chromane ring of α-tocopherol abolished the quenching ability, underscoring the requirement of a free hydroxyl group at position 6 in the quenching of $^1O_2$ by tocopherols. Trolox showed the same quenching rate as α-tocopherol in C_2H_5OH/D_2O (1:1), suggesting that the phytyl chain is not only devoid of antioxidant properties (36) but also does not possess $^1O_2$-quenching ability. Thus, physical quenching of $^1O_2$ appears to depend solely on the chromane ring with a free hydroxyl group in position 6.

The quenching ability of α-tocopherol is about 100-fold less than that of lycopene or 50-fold less than that of β-carotene, determined under similar conditions (Table 2). However, taking into account that plasma β-carotene levels in humans are only about 1/50 of the α-tocopherol levels (37-39), the overall $^1O_2$-quenching capacity of α-tocopherol may be regarded as equivalent to that of β-carotene in plasma; organs with low lycopene or β-carotene but high tocopherol levels may even depend largely on tocopherols for $^1O_2$ quenching. In addition, bilirubin has become of interest in this respect (27,40). Regarding the increasing epidemiological evidence of an inverse relation between tocopherol plasma levels and certain types of cancer (41) and the observed DNA-damaging effects of singlet oxygen (17-19), the $^1O_2$-quenching ability of tocopherols may become of importance in biology. In this respect it is of interest to note that high intracellular levels of tocopherols are not only found in microsomes and mitochondria but also in the nucles (42). Further, we recently found that trolox inhibits $^1O_2$-induced DNA damage in the plasmid pBR 322 (unpublished work).

Chemical Reactivity with Singlet Oxygen

Following the reaction of 300 μM tocopherol with 4 nm NDPO$_2$, 18% α-tocopherol, 34% γ-tocopherol, 52% δ-tocopherol, and 72% β-tocopherol remained detectable (Fig. 3A). Hence, the reactivity of the tocopherol homologs differed compared to physical quenching of $^1O_2$, with β-tocopherol showing unexpected low reactivity. This is also in contrast to the free-radical reactivity of the tocopherol homologs, shown to occur in the sequence α- > β- > γ- > δ-tocopherol (21). Characterization of the oxidation products identified quinones for all four of the tocopherol homologs and a quinone epoxide for γ-tocopherol. In previous studies (6,7,22,31,32) the formation of a quinone epoxide was also shown for α-, β-, and γ-tocopherol, but not for δ-tocopherol, via a hydroperoxide intermediate (32). It is possible that a quinone epoxide of α- and β-tocopherol actually forms as an intermediate, also under the conditions employed in the present study, but this is even more susceptible to further reaction than the γ-tocopherol quinone epoxide. An explanation for this could be the methylated 5-position of α- and β-tocopherol, which could destabilize the epoxide between positions 5 and 10. The present data indicate that the γ-tocopheryl quinone epoxide seems to react again with singlet oxygen to form a subsequent product (Fig. 5A). Such further reaction would be in contrast to previous data indicating it to be a stable end product (22,31,32). Furthermore, additional reaction products appeared on the HPLC (Fig. 6), reminiscent of tocopheryl quinones or tocopherols by their UV spectra (data not shown) but likely to be dimer and trimer structures as judged by their retention times. However, the products have not yet been isolated and their chemical structures remain to be established; as mentioned earlier, mass spectrometric data for these compounds were not obtained.

When comparing the various quenching and reaction rates of the tocopherol homologs with singlet oxygen, it is apparent that a methylated position 5 enhances physical quenching whereas chemical reactivity is favored by a methyl group in position 7. This model is based on the interesting finding that β-tocopherol has a physical quenching capacity similar to that of α-tocopherol but shows almost no chemical reactivity. Taking into account the DNA-damaging and mutagenic effects of singlet oxygen (17-19), the high concentrations of tocopherols in the cell nucleus (42), and the potentially hazardous accumulation of tocopherol oxidation products, which may not readily be disposed of enzymatically, β-tocopherol may become an import tocopherol homolog in certain biological environments. In this context it is interesting to note that higher plants often contain significant amounts of β-, γ-, and δ-tocopherol homologs (34). As they may be subject to greater exposure to light-generated $^1O_2$ production compared to animals, the accumulation of the less efficiently quenching tocopherol homologs may be an overall advantage in terms of lower chemical reaction rates.

SUMMARY

Singlet molecular oxygen ($^1O_2$) arising from the thermal decomposition of the endoperoxide of 3,3 '-(1,4-naphthylidene) dipropionate was used to assess the effectiveness of α-, β-, γ-, and δ-tocopherol in the physical quenching as well as the chemical reaction of $^1O_2$. The relative physical quenching efficiencies of the tocopherol homologs were found to decrease in the order α- > β- > γ- > δ-tocopherol. The ability of physical quenching depends on a free hydroxyl group in position 6 of the chromane ring. Chemical reactivity of the tocopherol homologs with $^1O_2$ is low, accounting for 0.1-1.5% of physical quenching, with β-tocopherol showing particularly low reactivity, resulting in the sequence α- > γ- > δ- > β-tocopherol. Tocopheryl quinones were products of all tocopherol homologs, and in addition a quinone epoxide was a major product from γ-tocopherol. It is concluded that methylation in position 5 of the chromane ring enhances physical quenching of $^1O_2$, whereas chemical reactivity is favored by a methylated position 7. Because β-tocopherol is as effective as α-tocopherol in physical quenching of $^1O_2$ but shows very low chemical reactivity, this tocopherol homolog might be particularly suitable for biological conditions in which an accumulation of oxidation products might weaken the antioxidant defense.

ACKNOWLEDGMENTS

Supported by National Foundation for Cancer Research, Bethesda, Maryland, and by Fonds der Chemischen Industrie. The generous gift of the various tocopherols by Dr. S. Wallat, Henkel Co., Düsseldorf, is gratefully acknowledged.

REFERENCES

1. Tappel AL. Vitam Horm (1962); 20:493-510.
2. Fahrenholtz SR, Doleiden FH, Trozzolo AM, Lamola AA. Photochem Photobiol (1974); 20:505-9.
3. Foote CS, Ching TV, Geller GG. Photochem Photobiol (1974); 20:511-3.
4. Stevens B, Small RD, Perez SR. Photochem Photobiol (1974); 20:515-7.
5. Carlsson DJ, Suprunchuk T, Wiles DM. J Am Oil Chem Soc (1976); 53:656-60.
6. Grams GW, Eskins K, Inglett GE. J Am Chem. Soc. (1972); 94:866-8.
7. Grams GW, Inglett GE. Lipids, (1972); 7:442-4.
8. Brabham DE, Lee J. J Phys Chem. (1976); 80:2292-6.
9. Foote CS. Science (1968); 162:963-70.
10. Krinsky NI. Science (1974); 186:363-5.
11. Khan AU. Biochem Biophys Res Commun (1984); 122:668-75.
12. Kanofsky JR. J Biol Chem (1986); 261:13546-50.
13. Wefers H. Bioelectrochem Bioenerg (1987); 18:91-104.
14. Cadenas E, Sies H. Eur J Biochem (1982); 124:349-56.
15. Tyrrell RM, Pidoux M. Photochem Photobiol (1989); 49:407-12.
16. Wefers H, Schulte-Frohlinde D, Sies H. FEBS Lett (1987); 211:49-52.
17. Epe B, Mützel P, Adam W. Chem Biol Interact (1988); 67:149-65.
18. Di Mascio P, Wefers H, Do-Thi HP, Lafleur MVM, Sies H. Biochim Biophys Acta (1989); 1007: 151-7.
19. Di Mascio P, Menck CFM, Nigro RG, Sarasin A, Sies H. Photochem Photobiol (1990); 51:17-20.
20. Gorman AA, Gould IR, Hamblett I, Standon MC. J Am Chem Soc (1984); 106:6956-9.
21. Burton GW, Ingold KU. J Am Chem Soc (1981); 103:6472-7.
22. Neely WC, Martin JM, Barker SA. Photochem Photobiol (1988); 48:423-8.
23. Saito I, Matsuura T, Inoue K. J Am Chem Soc (1983); 105:3200-6.
24. Di Mascio P, Sies H. J Am Chem Soc (1989); 111:2909-14.
25. Kaiser S, Di Mascio P, Murphy ME, Sies H. Arch Biochem Biophys (1990); 277:101-8.
26. Murphy ME, Kehrer JP. J Chromatogr Biomed Appl (1987); 421:71-82.
27. Di Mascio P, Kaiser S, Sies H. Arch Biochem Biophys (1989); 274:532-8.
28. Foote CS, Chang YC, Denny RW. J Am Chem Soc (1970); 92:5216-8.
29. Mathis P, Kleo J. Photochem Photobiol (1973); 18:343-6.
30. Yamauchi R, Matsushita S. Agr Biol Chem (1977); 41:1425-30.
31. Clough RL, Yee BG, Foote CS. J Am Chem Soc (1979); 101:683-6.

32. Yamauchi R, Matsushita S. Agr Biol Chem (1979); 43:2151-6.

33. Oesch F. Biochem J (1984); 139:77-88.

34. Schudel P, Mayer H, Isler O. In: Sebrell WH, Harris RS, eds. The vitamins, Vol. 5. New York: Academic Press, (1972); 168-217.

35. Grams GW, Eskins K. Biochemistry (1972); 11:606-8.

36. Niki E, Kawakami A, Saito M, Yamamoto Y, Tsuchiya J, Kamiya Y. J Biol Chem (1985); 260: 2191-6.

37. Cavina G, Gallinella B, Porra R, Pecora P, Saraci C. J Pharm Biomed Annal (1988); 6:259-69.

38. Stryker WS, Kaplan LA, Stein EA, Stampfer MJ, Sober A, Willett WC. Am J Epidemiol (1988); 127:283-96.

39. Comstock GW, Menkes MS, Schober SE, Vuilleumier JP, Helsing KJ. Am J Epidemiol (1988); 127:114-22.

40. Stocker R, Yamamoto Y, McDonagh AF, Glazer AW, Ames BN. Science (1987); 235:1043-6.

41. Gey KF, Brubacher GB, Stähelin HB. Am J Clin Nutr (1987); 45:1368-77.

42. Machlin LW. In: Machlin LW, ed. Handbook of vitamins. New York: Dekker, (1984); 113.

12

Relationship Between Vitamin E and a More Electronegatively Charged LDL of Human Plasma

Gabriele Bittolo-Bon, Giuseppe Cazzolato, and Pietro Avogaro

Regional General Hospital, Venice, Italy

Clinical and epidemiological studies have firmly established that the incidence of atherosclerosis and the manifestations of coronary artery disease (CAD) are related to plasma cholesterol (1); moreover, the risk is primarily related to the low-density lipoprotein (LDL) fraction (2). The pathway linking plasma LDL to the collection of cholesterol esters (CE) in the cells of parietal wall and then to formation of the atherosclerotic lesions is still poorly understood. The discovery of cellular receptors for LDL and the elucidation of the "receptorial pathway" (3,4), by which the delivery of LDL to the cell is exquisitely controlled by the intracellular CE concentration, do not help to explain all the events leading to atherosclerosis. Actually, the accumulation of monocyte-derived CE-laden macrophages (foam cells) in the parietal wall occurs even in people completely deficient of receptors, as in familial homozygous hypercholesterolemia and in Watanabe rabbits (5,6). Moreover, a familial monogenic hypercholesterolemia is present in only 5% of people complaining of a myocardial infarction (7). Furthermore, even if the risk of CAD is declining as plasma cholesterol decreases, the number of people suffering of the disease, despite being located in the bottom quintiles of plasma cholesterol levels, is still high. It is therefore necessary to think that alternative receptor mechanisms mediate the delivery of LDL to macrophages, without downregulation by intracellular CE levels.

Recently it was observed that certain postsecretive modifications of the structure of LDL could represent the key to an alternative pathway, which has been named the scavenger pathway (8). Special interest is now emerging in LDL modifications due to oxidation (9). If LDL peroxidation is a leading factor to explain LDL atherogenicity, then the antioxidant status, particularly the total plasma and LDL levels of vitamin E, the outstanding antioxidant substance, may be relevant in human disease.

Epidemiological data from 12 European populations in whom all essential antioxidants of plasma were compared show that vitamin E has the strongest inverse correlation to CAD mortality (10). In a case-control study low plasma concentrations of vitamins E and C and carotene were related to an increased risk of angina pectoris in men (11). For vitamin E the relation remained significant after adjustment for age, relative weight, total and high-density lipoprotein (HDL) cholesterol, nonfasting triglycerides, blood pressure, and smoking status.

Because of the high content of polyunsaturated fatty acids (PUFA), plasma LDL is an important target for lipid peroxidation. In vitro studies have shown that vitamin E depletion precedes the overt oxidation of LDL lipids (12). Also in our experience the incubation of LDL for 24 h at 37°C with 3 μM Cu^{2+} induces significant production of malondialdehyde (MDA) only when virtually all the vitamin E has been consumed (Fig. 1). In one sample of four, however, significant MDA production was recorded only with higher doses of copper and/or following a longer incubation. The sensitivity to oxidation by copper of the different LDL samples is therefore quite variable, some samples displaying various degrees of resistance to oxidation. A marked variation in the degree to which LDL preparations from various donors can be modified has been previously observed (13–15). In our experience, as well as in the experience of others (14,15), the endogenous vitamin E content of native LDL does not correlate with the extent of LDL oxidation. Besides α-tocopherol, other antioxidants are probably of importance (12–15) or the fatty acids composition of LDL can play a role (16).

LDL from donors treated with vitamin E supplementation becomes resistant to Cu^{2+}-induced oxidation. In a group of volunteers treated with 900 mg α-tocopherol per day for a week, the LDL vitamin E content doubled; no oxidation by copper was observed as judged by time-dependent changes in LDL electrophoretic mobility, MDA production, and variations of fluorescence intensity (Table 1). Also the somministration of probucol, another lipophilic antioxidant, induces resistance to oxidation by copper ions of human LDL (17). The antioxidant effect of probucol is apparently related to the protection afforded to vitamin E when LDL is subjected to copper-induced oxidation.

The reported data concern only in vitro experiments, thus giving only a rough indication of the physiological relationship between vitamin E and LDL oxidation.

Recently, through the use of ion-exchange chromatography, a more electronegative LDL subfraction (LDL^-) in human plasma was isolated, which displays many of the characteristics

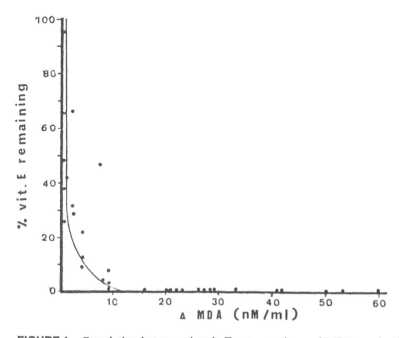

FIGURE 1   Correlation between vitamin E consumption and MDA production in Cu^{2+}-treated LDL.

TABLE 1 Mean ± SD Percentage Variation in MDA Content, Fluorescence Intensity (Excitation 360 nm, Emission 430 nm), and Electrophoretic Mobility of LDL[a]

		% Variation	
	MDA	Fluorescence intensity	Electrophoretic mobility
Before	1300 ± 700	257 ± 81	64 ± 21
After	25 ± 8	35 ± 12	0

[a]Incubated 24 h with 3 μm CuSO$_4$. LDL samples were obtained from six subjects before and after vitamin E supplementation (900 mg/day for a week).

of an oxidized LDL (18). The method of subfractionation has been improved by the use of an ion-exchange high-resolution liquid chromatographic (IE-HRLC) technique (19). Briefly, fresh plasma LDL (1.020–1.063 g/ml) is prepared and dialyzed under conditions that prevent LDL autooxidation. IE-HRLC is performed using a Bio-Rad System 700 apparatus equipped with a MA-7Q column, and LDL is eluted with a NaCl gradient from 0 to 0.3 M. Following this procedure two LDL subfractions can be obtained. The first main peak shows the peculiar chemicophysical and biological characteristics of a normal LDL (nLDL). The minor peak corresponds to LDL$^-$, whose concentration in normal subjects ranges from 0.5 to 9.8% of total plasma LDL (mean 3.9%). The LDL$^-$ to total plasma LDL ratio does not correlate with total and LDL cholesterol or with total plasma vitamin E, whereas there is a negatively correlation with the LDL phospholipids (PL) ($r = -0.59$, $P < 0.001$) and the LDL vitamin E content ($r = -0.63$, $P < 0.001$; Fig. 2). It is positively correlated with LDL proteins ($r = 0.35$, $P < 0.05$) and the content of MDA in total LDL ($r = 0.43$, $P < 0.05$). Analysis of the chemical composition of LDL$^-$ isolated through preparative IE-HPLC has shown significant differences in this subfraction compared to nLDL. Its content of CE and

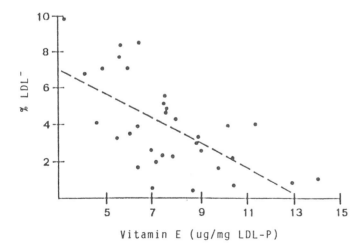

FIGURE 2 Correlation between the percentage contribution of the more electronegative LDL subfraction (LDL$^-$) to total LDL and LDL concentration of vitamin E.

PL was decreased, whereas proteins and free cholesterol (FC) were increased. The MDA content in the isolated LDL⁻ was three times higher than in nLDL (7.3 ± 2.5 versus 2.3 ± 0.6 mol/mol lipoprotein), whereas the vitamin E content is about half (4.3 ± 1.7 versus 8.2 ± 2.3 μg/mg LDL protein). Utilizing a gas chromatographic method, the presence of cholesterol oxides has been observed in LDL⁻ but not in normal LDL (Sevanian and Cazzolato, unpublished observations).

Some of the chemicophysical characteristic of LDL⁻, in particular the electrophoretic mobility, the high MDA and cholesterol oxide content, and the low vitamin E concentration, are consistent with the possibility that this subfraction represents a circulating LDL modified by a peroxidative process. The negative correlation recorded between LDL vitamin E and LDL⁻, as well as the low vitamin E content in LDL⁻, suggest that a low content of tocopherol in LDL allows a higher production of the more electronegative LDL subfraction. The peroxidative process itself could lead to complete depletion of tocopherol in LDL (12), however; thus its relatively low values in subjects with a higher concentration of plasma LDL⁻ may be an effect rather than the cause of LDL oxidation. Furthermore, because α-tocopherol is probably located in the surface part of lipoprotein and mainly transported with PL (20), the low level of vitamin E in LDL⁻ could be related to the lower content of PL that characterizes this lipoprotein subfraction.

We have previously recorded in a group of patients with angiographically proven CAD that short-term oral supplements of vitamin E significantly increase the PL content of total LDL and reduce the LDL⁻ plasma concentration by 40% (21). The low PL concentration found in LDL⁻ could be explained by the finding that oxidized PL are more susceptible to hydrolysis by phospholipases (22). The results obtained with vitamin E supplementation suggest that α-tocopherol preserves LDL PL content, preventing its oxidation.

Recently it was observed that acute smoking exerts an oxidative stress on plasma LDL, doubling its MDA content, and that supplementation of natural antioxidants, such as vitamin C and E, has a protective role (23).

Our data suggest that LDL⁻ constitutes a small subfraction of oxidized LDL circulating in vivo and that its plasma concentration is related to LDL vitamin E content.

REFERENCES

1. Lowering blood cholesterol to prevent heart disease. JAMA 1985; 253:2080-6.
2. Goldstein JL, Brown MS. The low-density pathway and its relation to atherosclerosis. Annu Rev Biochem 1977; 46:897-930.
3. Brown MS, Goldstein JL. Receptor-mediated control of cholesterol metabolism. Science 1976; 191: 150-4.
4. Brown MS, Goldstein JL. A receptor-mediated pathway for cholesterol omeostasis. Science 1986; 232:34-47.
5. Buja LM, Kita T, Goldstein JL, Brown MS. Cellular pathology of progressive atherosclerosis in the WHHL rabbit: an animal model of familial hypercholesterolemia. Arteriosclerosis 1983; 3:87-101.
6. Brown MS, Goldstein JL. Lipoprotein metabolism in the macrophage: implications for cholesterol deposition in atherosclerosis. Annu Rev Biochem 1983; 52:223-61.
7. Goldstein JL, Brown MS. Familial hypercholesterolemia. In: Stambury JB, Wyngaarden JB, Fredrickson DS, Goldstein JL, Brown MS, eds. The metabolic basis of inherited diseases, vol. V. New York: McGraw-Hill, 1983; 672-712.
8. Goldstein JL, Ho YK, Basu SK, Brown MS. Binding site on macrophage that mediates uptake and degradation of acetylated low density lipoproteins, producing massive cholesterol ester deposition. Proc Natl Acad Sci U S A 1978; 76:333-7.
9. Steinberg D, Parthasarathy S, Carew TE, Khoo JC, Witztum JL. Beyond cholesterol. Modifications of low-density lipoprotein that increase its atherogenicity. N Engl J Med 1989; 320:915-24.

10. Gey KF, Puska P. Plasma vitamin E and A inversely correlated to mortality from ischemic heart disease in cross-cultural epidemiology. Ann N Y Acad Sci 1989; 570:268-82.

11. Rirmersma RA, Wood DA, Macintyre CCA, Elton RA, Gey KF, Oliver MF. Risk of angina pectoris and plasma concentration of vitamins A, C, and E and carotene. Lancet 1991; 337:1-5.

12. Esterbauer H, Jurgens G, Quehenberger O, Koller E. Autoxidation of human low density lipoprotein: loss of polyunsaturated fatty acids and vitamin E and generation of aldehydes. J Lipid res 1987; 28:495-509.

13. van Hinsberg VHM, Scheffer M, Haveks L, Kempen HJK. Role of endothelial cells in the modification of low-density lipoproteins. Biochim Biophys Acta 1986; 878:49-64.

14. Babiy AV, Gebicki JM, Sullivan DR. Vitamin E content and low density lipoprotein oxidizability induced by free radicals. Atherosclerosis 1990; 81:1175-82.

15. Jessup W, Rankin SM, De Walley CV, Hoult JRS, Scott J, Leake DS. Alpha-tocopherol consumption during low-density lipoprotein oxidation. Biochem J 1990; 265:399-405.

16. Parthasarathy S, Khoo JC, Miller E, Barnett J, Wiztzum JL, Steinberg D. Low-density lipoprotein rich in oleic acid is protected against oxidative modification: implications for dietary prevention of atherosclerosis. Proc Natl Acad Sci U S A 1990; 87:3894-8.

17. Bittolo-Bon G, Cazzolato G, Avogaro P. Effect of probucol on a more electronegative LDL subfraction in primary hypercholesterolemia. Free Radic Biol Med 1990; 9 (Suppl 1):7.10.

18. Avogaro P, Bittolo-Bon G, Cazzolato G. Presence of a modified low density lipoprotein in humans. Arteriosclerosis 1988; 8:79-87.

19. Cazzolato G, Avogaro P, Bittolo-Bon G. Characterization of a more electronegatively charged LDL subfraction by ion exchange HPLC. Free Radic Biol Med 1991; 11:247-253.

20. Nakamura H. Distribution of free-cholesterol and alpha-tocopherol in plasma lipoprotein and tissues in humans. In: Hayaishi O, Mino M, eds. Clinical and nutritional aspects of vitamin E. New York: Elsevier, 1987; 101-8.

21. Bittolo-Bon G, Cazzolato G, Saccardi M, Avogaro P. Presence of a modified LDL in humans: effect of vitamin E. In: Hayaishi O, Mino M, eds. Clinical and nutritional aspects of vitamin E. New York: Elsevier, 1987; 109-20.

22. Sevanian A, Muakkassah-Kelly SF, Montestruque S. The influence of phospholipase A_2 and glutathione peroxidase on the elimination of membrane peroxides. Arch Biochem Biophys 1983; 223:441-52.

23. Harats D, Ben-Naim M, Dabach Y, et al. Effect of vitamin C and E supplementation on susceptibility of plasma lipoproteins to peroxidation induced by acute smoking. Atherosclerosis 1990; 85:47-54.

13

The Role of Vitamin E in Blood and Cellular Aging

Shiro Urano

Tokyo Metropolitan Institute of Gerontology, Tokyo, Japan

HEMATOLOGICAL ASPECTS OF VITAMIN E

It has been recognized that vitamin E is an essential component of blood and may function as an antioxidant and a membrane stabilizer for protecting blood components, such as erythrocytes, leukocytes, and platelets, from oxidative damage. From many observations made through in vivo and in vitro studies of animal and human blood, the role of vitamin E has become clearer. Almost all these findings have been made by taking into account phenomena of vitamin E deficiency and the effects of supplementation of the vitamin.

A variety of vitamin E deficiency syndromes are readily produced in animals, but humans have not been shown to develop symptoms related to vitamin E deprivation. Because of the widespread occurrence of tocopherol in food, nutritional vitamin E deficiency in humans is extremely uncommon. Hematological manifestations of vitamin E deficiency in humans are generally limited to prematurely delivered neonates and pathological states associated with chronic fat malabsorption (1–7). Among several disorders characterized by steatorrhea, cystic fibrosis with pancreatic achylia represents one of the most common causes of fat malabsorption syndrome (1,8). All patients with pancreatogenic steatorrhea are deficient in vitamin E, and the plasma α-tocopherol levels correlate with indices of intestinal malabsorption (9). The survival of erythrocytes in vitamin E-deficient patients is moderately but significantly decreased. The life span of red blood cells have been found to be shortened in such patients to an average half-life of 19.0 ± 1.3 days; following vitamin E supplementation, the half-life increases to 27.5 ± 0.9 days. In vivo studies of peroxide-induced hemolysis revealed that normal erythrocytes resist hemolysis during incubation, but vitamin E-deficient erythrocytes show abnormal oxidant susceptibility, evidenced by a greater than 5% hemoglobin release (9).

Another manifestation of human vitamin E deficiency has been found in patients with abetalipoproteinemia caused by fat malabsorption, attributable to the blockage of chylomicron formation and resulting in the absence of low-density lipoprotein (LDL) and very low density lipoprotein (VLDL) in blood (10–12). Since vitamin E is normally absorbed via the chylomicrons and transported mostly by LDL and VLDL, in the absence of these lipoproteins vitamin E is undetectable in the serum of patients (13). The most characteristic abnormal feature of erythrocytes in abetalipoproteinemia is the presence of acanthocytes (spiky red cells) with excess sphingomyelin (10,11,13,14). An increase in this phospholipid, which is located in the outer membrane layer, causes a decrease in erythrocyte membrane fluidity, producing acanthocytes (10,15). Although malabsorption of vitamin E has been reported to make acanthocyte lipids more sensitive than normal erythrocyte lipids to oxidative damage that can be prevented by treatment with water-soluble vitamin E (α-tocopheryl polyethylene glycol succinate), neither the extent of peroxidation in serum and erythrocytes nor the improvement of the morphological abnormality of erythrocytes by vitamin E has been investigated (16). Long-term treatment of patients with vitamin E increases serum vitamin E concentrations and reduces in vitro hemolysis (11). Since a recent report suggested that even in very severe vitamin E deficiency, the requirement for vitamin E, a chain-breaking (peroxyl radical-trapping) antioxidant, cannot be met by other exogenous or endogenous antioxidants (17), it seems likely, although not proved, that the role of vitamin E in the blood of patients with abetalipoproteinemia is not only to protect against peroxidation but also to function in membrane stabilization independently of a redox system (18,19). Morphological alterations of erythrocytes also appear in patients with sickle cell anemia (20). Although sickle cells regain their normal shape upon oxygenation, a portion of the circulating red cells retain their abnormal shape even when the blood is fully oxygenated (20,21). Sickle cells are deficient in vitamin E and spontaneously generate approximately twice the normal amount of activated oxygen; consequently, the production of hydrogen peroxide (H_2O_2) is increased (22). These cells are therefore more susceptible to oxidative stress than normal cells, resulting in the accumulation of two to three times the normal amount of malondialdehyde (MDA), an end product of lipid peroxidation (23). Treatment of sickle cell anemia patients with large amounts of vitamin E causes their plasma tocopherol levels to increase from 0.7 to 2.3 mg/g lipid and the proportion of irreversibly sickled red cells to decrease from 25 to 11% (21). It is unclear, however, whether sickle cells regain their normal shape due to the antioxidant properties of vitamin E or whether the vitamin enhances erythropagocytosis directly.

VITAMIN E IN CELLULAR AGING

One of the most neglected topics in vitamin E research concerns the role of vitamin E in aging. Although vitamin E expands the life span of a number of species, including rotifer, paramecium, turbatrix, drosophila, and rat, and actual mechanism of this life span elongation effect by vitamin E is still poorly understood (24–30). In research on cellular aging, the red blood cell provides a good model for studying the aging process because it is simple to separate aged red cells from younger cells (31). It has been found that aged red cells are more dense than younger cells and are osmotically fragile (32); also, the activities of superoxide dismutase, catalase, glutathione peroxidase, and glutathione reductase have been found to be decreased in aged red cells (33,34). In addition, membrane lipid concentrations of aged cells, especially of arachidonate in the phospholipids, decreases significantly during the aging process (35). From these observations and on the basis of the free-radical aging theory (36), it is reasonable to assume that lipid peroxidation in vivo may induce red blood cell senescence and that antioxidants, including vitamin E, may delay the aging process. Nevertheless, a recent report by Burton et al. (37) suggested that young and old human red blood cells contain about the same amount of α-tocopherol. Furthermore, young cells have a lower ratio of α-tocopherol to arachidonic acid than old cells. These results are incompatible with the preceding argument, and hence the explanation for the mechanism of red blood cell aging has become more complicated.

It has been suggested that to remove senescent red blood cells from the circulation, macrophages distinguish them from mature red cells. Kay (38) proposed that a senescent cell antigen, which develops on the surface of aged red cells, is recognized and bound by the Fab region of an IgG autoantibody in the serum; subsequently the Fc region of the IgG is recognized and bound by macrophages, resulting in phagocytosis of the red cell. This senescent cell antigen arises as a breakdown product ($M_r = 62,000$) of a membrane protein designated band 3, which is actually an anion channel in erythrocyte membranes. In addition to senescent red cells, this antigen is present in lymphocytes, platelets, and neutrophils, but not in young red cells. Erythrocytes from vitamin E-deficient rats, regardless of their age, behave like old erythrocytes from normal rats: that is, they show an increase in susceptibility to phagocytosis and IgG binding, an impairment in anion transport ability, and a decline in the activity of glyceraldehyde-3-phosphate dehydrogenase, a main segment of band 3 (39). Regrettably, these abnormalities are not reversed by the addition of exogenous vitamin E to the assay medium. Kay et al. (39) proposed that the oxidative degradation of band 3 and the generation of senescent cell antigen in vivo may provide a possible mechanism for erythrocyte aging. However, neither the effect of vitamin E supplementation on the aging process in vivo nor the content of vitamin E and lipid peroxides in the membranes of old and young red cells have been investigated. Since sickle cells generate more malondialdehyde than normal cells, these cells bind considerably more IgG and are ingested by macrophages (22). Thus, simple proteolysis of a band 3 membrane component is not enough to explain red cell aging. Consequently, it seems likely that oxidation of either erythrocyte lipids or proteins may contribute to cellular aging.

Although the mean life span of normal human platelets is thought to average 6.9–9.9 days (40), in vitamin E-deficient neonates and children platelet life span is significantly decreased to a half-life of 2.0–2.4 days. Plasma vitamin E levels during the period of E deficiency were found to be markedly decreased and MDA formation to be increased, leading to hyperaggregability. Long-term vitamin E repletion to these patients caused platelet survival times to become completely normal at 4.2–4.7 days (41). The mechanism for this effect of vitamin E, however, is not known.

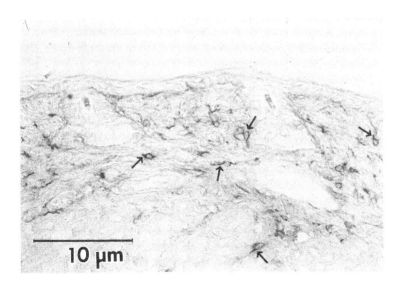

FIGURE 1   Arrows show CD4$^+$ T lymphocytes with dendrites in aged rat skin. The frozen section (8 μm thick) was stained by an avidine-biotin method with a monoclonal antibody to W3/25 (T helper cells). The antibody stained only round T lymphocytes without dendrites in young rat skin (data not shown). (Courtesy of Drs. T. Toda and J. Ohno, with permission.)

The role of vitamin E in the aging process of other cells in the blood is still unknown. Recently, however, it was shown that lymphocyte T cells are morphologically transformed to dendrite-shaped cells during aging (Fig. 1) (42). Consequently, it is interesting to consider, although firm evidence is lacking, that the mechanism for the appearance of these abnormal T cells may be related to that for the appearance of spiky red cells and sickle cells and that vitamin E may play a biologically active role in T cell aging. Since a great deal of vitamin E research is currently in progress, it can be expected that the actual link between vitamin E and cellular aging will be clarified and that the mechanism of vitamin E activity in the aging process will be made clear.

BIOCHEMICAL ASPECTS OF VITAMIN E

Vitamin E and the Susceptibility of Erythrocytes to Oxidative Stress

Erythrocyte membranes are presumed to be highly susceptible to peroxidation because of the presence of hemoglobin, which catalyzes oxidation under circumstances of high oxygen concentration, and erythrocyte membranes are rich in polyunsaturated lipids. Erythrocytes must, therefore, be protected by antioxidants and antioxidative enzymes. When these antioxidative substances are deficient in blood, the susceptibility of erythrocytes to oxidative stress is enhanced and hence the free radicals generated may attack the membranes, leading to their oxidative destruction; finally, the cells may be hemolyzed. It is well documented that even a low concentration of vitamin E can protect erythrocytes from oxidative damage and that hemolysis takes place in vitro after vitamin E depletion through oxidation (Fig. 2) (43, 44). Thus, vitamin E in erythrocyte membranes plays an important role in protection of the membrane against free radicals.

Mino et al. (44) demonstrated that when an azo compound, AAPH [2,2'-azobis-(2-amidino propane) dihydrochloride] was used as a radical initiator, the peroxidizability in vitamin E-deficient red cell membranes from neonates increased more than the peroxidizability of adult red cell membranes after depletion of vitamin E. Erythrocyte membranes from neonates are rich in polyunsaturated fatty acids, especially arachidonic acid and eicosapentatrienoic acid, compared with adult erythrocytes. To verify the peroxidizability of the membrane lipids, the

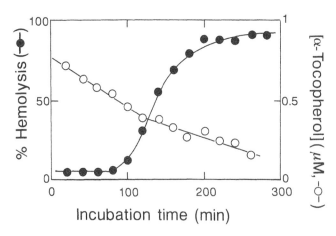

FIGURE 2   Hemolysis and consumption of α-tocopherol in rabbit erythrocytes induced by AAPH at 37°C in air. [Reproduced from E. Niki (43), with permission.]

authors calculated the amount of active hydrogen numbers (biallylic hydrogen numbers in fatty acid), which is a reliable index for assessing the extent of hydrogen abstraction by free radicals, and showed that the number is significantly higher in neonates relative to adults, resulting in high peroxidizability. Interestingly, when erythrocytes from vitamin E-deficient rats were hemolyzed using the xanthine-xanthine oxidase system, the morphological features of the damaged cells were quite different from those of the AAPH-damaged cells. As shown in Figure 3, xanthine-xanthine oxidase reaction produces spiky cells (by echinocytosis) but AAPH produces cup cells (by stomatocytosis) (45). Generally, it is considered that echinocytes are formed through the evagination of the outer leaflet of the membrane caused by changes in outer leaflet ordering; on the contrary, changes in the inner half of the bilayer induce cup formation (46). On this basis, it can be concluded that the free radicals induced by these radical initiators attack at different parts of the membrane by different mechanisms. Chain-breaking antioxidants, such as α-tocopherol, scavenge radicals arising from either initiator within the lipid core of erythrocyte membranes; consequently, the oxidative hemolysis of erythrocytes is efficiently inhibited. Similar susceptibility of erythrocytes to oxidative stress and the predominant inhibition of oxidative damage by α-tocopherol have been found in such pathological states as abetalipoproteinemia, cystic fibrosis, and aged diabetes, states in which the polyunsaturated fatty acids in the erythrocyte membranes are abnormally increased (9,16,47).

On the other hand, the attack of free radicals on erythrocytes can also lead to the oxidative destruction of membrane proteins, resulting in fragmentation, amino acid modification, cross-linking, and fluorescent pigmentation (45,48). The oxidation of rat red blood cells by either xanthine-xanthine oxidase reaction or AAPH leads not only to an increase in thiobarbituric acid-reactive substances (TBARS) and chemiluminescence in erythrocytes but also to a depletion of spectrin and the sulfhydryl groups ($--SH$) of membrane proteins (45,49–51). Although membrane tocopherol, even below a critically low level, suppresses lipid peroxidation, protein damage and the loss of SH groups are not inhibited. This difference in the eficacy of α-tocopherol can be explained by the fact that when radicals are generated outside the membranes, membrane proteins are damaged concurrently with the consumption of membrane tocopherol; on the contrary, when radicals are initiated within the lipid core of membranes, the depletion of SH groups and the formation of TBARS are predominantly suppressed

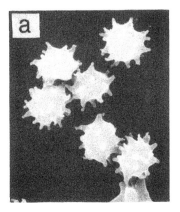 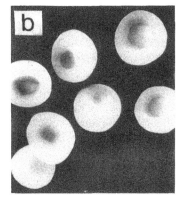

FIGURE 3 Scanning electron micrographs of erythrocytes subjected to the xanthine-xanthine oxidase reaction (a) and the AAPH reaction (b) (each ×6000). (Courtesy of Dr. M. Mino, with permission.)

by membrane tocopherol (51). The inhibition of membrane protein damage by α-tocopherol differs depending on the site of initial radical generation.

Vitamin E in Leukocyte and Platelet Function

Polymorphonuclear leukocytes (PMNs), similarly to platelets and lymphocytes, are capable of achieving a heightened state of functional and metabolic activation in response to immuno-chemical signals. This response is often altered by oxidation or vitamin E deficiency. Vitamin E-deficient rat PMNs have almost a twofold increase in peroxidized membrane lipids (MDA) due to enhanced oxygen consumption and hydrogen peroxide release; therefore, the suscepti-bility of these cells to oxidative stress is enhanced relative to that of normal PMNs. Further-more, vitamin E-deficient PMNs are unable to respond normally to chemotactic and phago-cytic stimuli. This oxidant-inflicted damage to vitamin E-deficient PMNs was suggested to occur before phagocytic events, and hence chemotactic and phagocytic functions are impaired; in contrast, autooxidative damage to vitamin E-replete PMN occurs in tempo with the pha-gocytic event, resulting in an enhancement of these functions (52). These defects are rapidly corrected by in vivo repletion of vitamin E-deficient rats with parenteral vitamin E. In the PMNs of vitamin E-replete rats, hydrogen peroxide (H_2O_2) released from the PMNs is markedly decreased, but superoxide anion (O_2^-) is slightly increased. Therefore it is concluded that H_2O_2 initiates the peroxidative damage to PMNs (53). Recent findings, however, show that a decrease of 20% or more in O_2^- occurs in vitamin E-replete rat PMNs (54). Although this discrepancy complicates the explanation for the antioxidant activity of vitamin E in PMN function, that the bactericidal functioning of vitamin E-replete PMNs is reduced (53) leads to the conclusion that vitamin E may attenuate the original oxidative activity in PMNs.

On the other hand, in infants with a congenital deficiency in glutathione synthetase ac-tivity, PMNs are unable to synthesize glutathione and hence the cellular membranes are sus-ceptible to oxidative damage, resulting in impairment of the phagocytic function. Vitamin E repletion enhances the phagocytic functioning of normal PMNs by its ability to scavenge oxy-gen by-products (55).

Platelet-activating factor (PAF) is known to be metabolized in PMNs through certain stimuli and to be a potent phospholipid mediator that causes activation of leukocytes and platelets, thus contributing to a hemostatic eent. Recent findings by Fukuzawa et al. (56) show that the synthesis of PAF is significantly enhanced in vitamin E-deficient rat PMNs (Fig. 4). Although the activity of the acetyltransferase, which transfers an acetyl moiety to 2-lyso-PAF to form PAF, is higher in vitamin E-deficient than in vitamin E-supplemented PMNs, in vitro addition of vitamin E does not inhibit the increased activity of the enzyme. From these results it can be concluded that vitamin E does not directly inhibit the activity of this enzyme but rather may inhibit the activity of phospholipase A_2, which produces 2-lyso-PAF in PMN. In any case, the mechanism of this effect of vitamin E in PMN is still unclear.

One of the most important functions of platelets is in aggregation, which under normal condition exert a hemostatic effect and may be crucial for the development of thrombosis (57). In response to various stimuli, arachidonic acid is released from platelet membranes through the activation of phospholipase A_2. Arachidonic acid is then converted enzymati-cally to thromboxane A_2 (TXA$_2$) and prostaglandins. TXA$_2$ synthesized in platelets is the most potent proaggregatory and thrombotic metabolite, whereas prostaglandin I_2 (PGI$_2$), produced in the endothelium of the vessel wall, has significant antiaggregatory and antithrom-botic effects (41,57). By a balance in the biosynthesis of these two compounds, platelet aggre-gation and its inhibition are regulated under normal conditions (Fig. 5).

In vitamin E-deficient rats, platelet aggregation induced by collagen and the contents of prostaglandin E_2 and $F_{2\alpha}$ in serum were found to be increased compared to those in normal

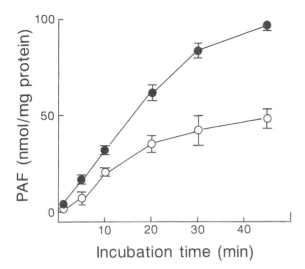

FIGURE 4 Time course for the incorporation of [³H]acetyl-CoA into lyso-PAF by PMN homogenates prepared from vitamin E-supplemented (open circles) and vitamin E-deficient (solid circles) rats. The values for the two groups were significantly different ($P < 0.001$) at all time points. [Reproduced from K. Fukuzawa (56), with permission.]

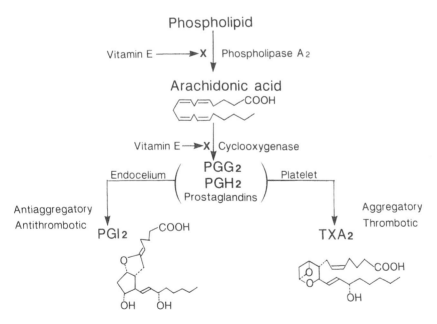

FIGURE 5 Biosynthesis of prostaglandin I_2 (PGI$_2$) and thromboxane A_2 (TXA$_2$) and its inhibition by vitamin E.

rats (58,59). Not only TXA$_2$ production is enhanced in vitamin E-deficient rabbits, but PGI$_2$ in vessel walls is significantly decreased in deficient rats (60,61). Also, in vitamin E-deficient neonates, platelet aggregability to epinephrine has been shown to be enhanced relative that in normal neonates. Platelet MDA formation in neonates induced by *N*-ethylmaleimide is increased and platelet survival is decreased to a half-life of about 2 days. Since these abnormalities are improved following therapy with vitamin E, it seems likely that by its antioxidant activity, vitamin E may inhibit the oxidative production of a "senescent antigen" in platelets (41,62,63).

Thus, the mechanism by which platelet aggregability is preferentially inhibited by vitamin E is thought to involve the inhibition of either phospholipase A$_2$ activity of arachidonic acid release from platelets or cyclooxygenase activity in the synthesis of TXA$_2$ and PGI$_2$, as shown in Figure 5.

Vitamin E in Nonoxidative Hemolysis

Retinol (vitamin A) is known to induce erythrocyte hemolysis by penetration of the molecule into membranes (64). Although it has been proposed that this hemolysis may be caused by the physical disruption of micelles in the membrane rather than by oxidative damage (18), this idea has been generally criticized because retinol, which is very unstable to oxidation because of its polyunsaturation, may undergo facile, self-initiated free-radical chain oxidation and may be capable of initiating the oxidative destruction of erythrocyte membrane lipids, resulting in hemolysis. Despite the uncertainty, there are no reports to demonstrate whether retinol-induced hemolysis arises from oxidative damage. Recently, Urano (19) proposed that erythrocyte hemolysis is actually caused by physical damage to the erythrocyte membrane. When a suspension of vitamin E-deficient rabbit erythrocytes was incubated with retinol, the erythrocytes were hemolyzed concurrently with oxygen consumption. Although the hemolysis was inhibited by vitamin E repletion and no oxygen consumption occurred within the 30 minute incubation period, the addition of ascorbate or butylated hydroxytoluene (BHT) into a suspension including vitamin E-deficient erythrocytes and retinol did not suppress hemolysis, even though oxygen consumption was inhibited (Fig. 6). After hemolysis, nonoxidized

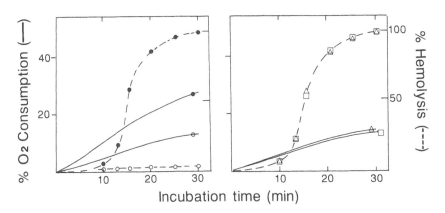

FIGURE 6   Rate of oxygen consumption and hemolysis in rabbit erythrocytes initiated by retinol and its inhibition by antioxidants. The erythrocyte suspensions were incubated with $1.7 \times 10^{-3} \mu$mol retinol (solid circles), retinol + $0.2 \times 10^{-2} \mu$mol α-tocopherol (open circles), retinol + $0.2 \times 10^{-2} \mu$mol ascorbic acid (open triangles), or retinol + $0.2 \times 10^{-2} \mu$mol BHT (open squares).

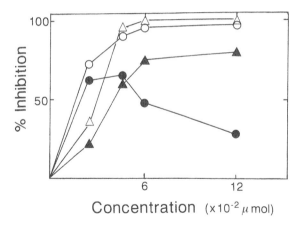

FIGURE 7 Inhibitory effect of α-tocopherol and its derivatives on retinol-induced rabbit erythrocyte hemolysis. The suspensions were incubated with α-tocopherol (solid circles), α-tocopheryl acetate (open circles), α-tocopheryl nicotinate (open triangles), and 6-deoxy-α-tocopherol (solid triangles) at 37°C for 20 minutes.

retinol and α-tocopherol were found in the broken membranes but the oxidized forms were not generally found. These results suggest that the added retinol is not oxidized in the suspension but rather penetrates into the erythrocyte membranes to cause physical damage to the erythrocytes.

Not only α-tocopherol but also its acetate and nicotinate inhibit hemolysis; in fact, the inhibitory effect of the acetate and nicotinate is higher than that of α-tocopherol by 100% (Fig. 7) (19,65). This result also shows the physical membrane-stabilizing effect of α-tocopherol. Although the mechanism of this effect is unclear as to detail, it has been reported that the chroman ring, rather than isoprenoid side chain of α-tocopherol, is important for the membrane stabilization of erythrocytes (66–69).

REFERENCES

1. Oski FA. Anemia related to nutritional deficiencies other than and folic acid. In: Williams WJ, Beutler E, Erslev AJ, Lichtman MA, eds. Hematology, 4th ed. New York: McGraw-Hill, 1990; 549-53.
2. Mino M, Kitagawa M, Nakagawa S. Red blood cell tocopherol concentrations in a normal population of Japanese children and premature infants in relation to the assessment of vitamin E status. Am J Clin Nutr 1985; 41:631-8.
3. Nitowsky HM, Hsu KS, Gordon HH. Vitamin E requirements of human infants. Vitam Horm 1962; 20:559-71.
4. Gordon H, McNamara. Fat excretion of premature infants. I. Effect on fecal fat of decreasing fat intake. Am J Dis Child 1941; 62:328-45.
5. Muller DPR, Harries JT, Lloyd LK. The relative importance of the factors involved in the absorption of vitamin E in children. Gut 1974; 15:966-71.
6. Binder HJ, Herting DC, Hurst V, Finch SC, Spiro HM. Tocopherol deficiency in man. N Engl J Med 1965; 273:1289-97.
7. Filer LJ Jr, Wright SW, Manning MP, Mason KE. Absorption of α-tocopherol and tocopheryl esters by premature and full term infants and children in health and disease. Pediatrics 1951; 8:328-39.
8. Di Sant'Agnese PA, Talamo RC. Pathogenesis and pathophysiology of cystic fibrosis of the pancreas. Fibrocystic disease of the pancreas (mucoviscidosis). N Engl J Med 1967; 277:1287-94.

9. Farrell PM, Bieri JG, Fratantoni JF, Wood RE, DiSant'Agnese PA. The occurrence and effects of human vitamin E deficiency. J Clin Invest 1977; 60:233-41.

10. Palek J. Acanthocytosis, stomathocytosis, and related disorders. In: Williams WJ, Beutler E, Erslev AJ, Lichtman MA, eds. Hematology, 4th ed. New York: McGraw-Hill, 1990; 582-90.

11. Muller DPR, Lloyd JK, Wolff OH. Vitamin E and neurological function: abetalipoproteinaemia and other disorders of fat absorption. In: Ciba foundation symposium 101, Biology of vitamin E. London: Pitman, 1983; 106-21.

12. Kayden HJ, Silber R, Kossmann CE. The role of vitamin E deficiency in the abnormal autohemolysis of acanthocytosis. Trans Assoc Am Physicians 1965; 78:334-42.

13. Ways P, Reed CF, Hanahan DJ. Red cell and plasma lipids in acanthocytosis. J Clin Invest 1963; 42:1248-60.

14. Phillips GB. Quantitative chromatographic analysis of plasma and red blood cell lipids in patient with acanthocytosis. J Lab Clin Med 1962; 59:357-63.

15. Cooper RA, Durocher JR, Leslie MH. Decrease fluidity of red cell membrane lipids in abetalipoproteinemia. J Clin Invest 1977; 60:115-21.

16. Dodge JT, Cohen G, Kayden HJ, Phillips GB. Peroxidative hemolysis of red blood cells from patients with abetalipoproteinemia (acanthocytosis). J Clin Invest 1967; 46:357-68.

17. Ingold KU, Webb AC, Witter D, Burton GW, Metcalfe TA, Muller DPR. Vitamin E remains the major lipid-soluble, chain-breaking antioxidant in human plasma even in individuals suffering severe vitamin E deficiency. Arch Biochem Biophys 1987; 259:224-5.

18. Diplock AT, Lucy JA. The biochemical modes of action of vitamin E and selenium: a hypothesis. FEBS Lett 1973; 29:205-10.

19. Urano S. Membrane-stabilizing effect of vitamin E. Vitamin (Jpn) 1989; 63:75-85.

20. Beutler E. The sickle cell diseases and related disorders. In: Williams WJ, Beutler E, Erslev AJ, Lichtman MA, eds. Hematology, 4th ed. New York: McGraw-Hill, 1990; 613-44.

21. Natta CL, Machlin LJ, Brin M. A decrease in irreversibly sickled erythrocytes in sickle cell anemia patients given vitamin E. Am J Clin Nutr 1980; 33:968-71.

22. Hebbel RP, Miller WJ. Phagocytosis of sickle erythrocytes: immunologic and oxidative determinants of hemolytic anemia. Blood 1984; 64:733-41.

23. Jain SK, Shohet SB. A novel phospholipid in irreversibly sickle cells: evidence for in vivo peroxidative membrane damage in sickle cell disease. Blood 1984; 63:362-7.

24. Sawada M, Enesco HE. Vitamin E extends lifespan in the short-lived rotifer, *Asplanchna brightwelli*. Exp Gerontol 1984; 19:179-83.

25. Enesco HE, Verdone-Smith C. α-Tocopherol increase lifespan in the rotifer *Philodina*. Exp Gerontol 1980; 15:335-8.

26. Thomas J, Nyberg D. Vitamin E supplementation and intense selection increase clonal life span in *Paramecium tetraurelia*. Exp Gerontol 1988; 23:501-12.

27. Epstein J, Himmelhoch S, Gershon D. Studies on ageing in nematodes. III. Electronmicroscopial studies on age-associated cellular damage. Mech Ageing Dev 1972; 1:245-55.

28. Zuckerman BM, Geist MA. Effects of vitamin E on the nematode *Caenorhabditis elegans*. Age 1983; 6:1-4.

29. Miquel J, Binnard R, Howard WH. Effects of dl-alpha-tocopherol on the life span of *Drosphila melanogaster*. Gerontologist 1973; 13:37.

30. Porta EA, Joun NS, Nitta RT. Effects of the type of dietary fat at two levels of vitamin E in Wister male rats during development and aging. J. Life span, serum biochemical parameters and pathological changes. Mech Ageing Dev 1980; 13:1-39.

31. Bennett GD, Kay MMB. Homeostatic removal of senescent murine erythrocytes by splenic macrophages. Exp Hematol 1981; 9:297-307.

32. Marks PA, Johnson AB. Relationship between the age of human erythrocytes and their osmotic resistance: a basis for separating young and old erythrocytes. J Clin Invest 1958; 37:1542-8.

33. Glass GA, Gershon D. Decrease enzymic protection an increased sensitivity to oxidative damage in erythrocytes as a function of cell and donor aging. Biochem J 1984; 218:531-7.

34. Glass GA, Gershon D. Enzymatic change in rat erythrocytes with increasing cell and donor age: loss of superoxide dismutase activity associated with increases in catalytically defective forms. Biochem Biophys Res Commun 1981; 103:1245-53.

35. Phillips GB, Dodge JT, Howe C. The effect of aging of human red cells in vivo on their fatty acid composition. Lipids 1969; 4:544-9.

36. Harman D. Free radical theory of aging: beneficial effect of antioxidants on the life span of male NZB mice; role of free radical reactions in the deterioration of the immune system with age and in the pathogenesis of systemic lupus erythematosus. Age 1980; 3:64-73.

37. Burton GW, Cheng SC, Webb A, Ingold KU. Vitamin E in young and old human red blood cells. Lipids 1986; 860:84-90.

38. Kay MMB. Isolation of the phagocytosis-inducing IgG-binding antigen on senescent somatic cells. Nature 1981; 289:491-4.

39. Kay MMB, Bosman GJCGM, Shapiro SS, Bendich A, Nassal PS. Oxidation as a possible mechanism of cellular aging: vitamin E deficiency causes premature aging and IgG binding to erythrocytes. Proc Natl Acad Sci U S A 1986; 83:2463-7.

40. Paulus J-M, Aster RH. Production, life-span, and fate of platelets. In: Williams WJ, Beutler E, Erslev AJ, Lichtman MA, eds. Hematology, 4th ed. New York: McGraw-Hill, 1990; 125-60.

41. Stuart MJ. Vitamin E deficiency: its effect on platelet-vascular interaction in various pathologic states. Ann N Y Acad Sci 1982; 393:277-88.

42. Toda T, Ohno J, Utumi N, Ohashi M. An immunohistochemical study on age-related changes in the cutaneous immune system. Biomed Gerontol 1990; 14:113-4.

43. Niki E, Komuro E, Takahashi M, Urano S, Ito E, Terao K. Oxidative hemolysis of erythrocytes and its inhibition by free radical scavengers. J Biol Chem 1988; 263:19809-14.

44. Mino M, Miki M, Miyake M, Ogihara T. Nutrional assessment of vitamin E in oxidative stress. Ann N Y Acad Sci 1989; 570:296-310.

45. Mino M, Miki M, Tamai H, Yasuda H, Maeda H. Membrane damage in erythrocytes induced by radical initiating reactions and the effect of tocopherol as a radical scavenger. In: Sevanian A, ed. Lipid peroxidation in biological systems. Champaign, Ill: American Oil Chemist's Society, 1988; 51-70.

46. Sceets MP, Singer SJ. Biological membrane as bilayer couples. A molecular mechanism of drug-erythrocyte interaction. Proc Natl Acad Sci U S A 1974; 71:4457-61.

47. Urano S, Hoshi-hashizume M, Tochigi N, Matsuo M, Shiraki M, Ito H. Vitamin E and the susceptibility of erythrocytes and reconstituted liposomes to oxidative stress in aged diabetics. Lipids 1991; 26:58-61.

48. Dean RT, Cheeseman KH. Vitamin E protects proteins against free radical damage in lipid environments. Biochem Biophys Res Commun 1987; 148:1277-82.

49. Yasuda H, Miki M, Takenaka Y, Tamai H, Mino M. Changes in membrane constituents and chemiluminescence in vitamin E-deficient red blood cells induced by xanthine oxidase reaction. Arch Biochem Biophys 1989; 272:81-7.

50. Miki M, Tamai H, Mino M, Yamamoto Y, Niki E. Free-radical chain oxidation of rat red blood cells by molecular oxygen and its inhibition by α-tocopherol. Arch Biochem Biophys 1987; 258:373-80.

51. Takenaka Y, Miki M, Yasuda H, Mino M. The effect of α-tocopherol as an antioxidant on the oxidation of membrane protein thiols induced by free radicals generated in different sites. Arch Biochem Biophys 1991; 285:344-50.

52. Harris RE, Boxer LA, Baehner RL. Consequences of vitamin E-deficiency on the phagocytic and oxidative functions of the rat polymorphonuclear leukocyte. Blood 1980; 55:338-43.

53. Baehner RL, Boxer LA, Ingraham LM, Butterick C, Haak RA. The influence of vitamin E on human polymorphonuclear cell metabolism and function. Ann N Y Acad Sci 1982; 237-50.

54. Mino M, Okano T, Tamai H. Leukocyte function generating superoxide and vitamine E. Ann N Y Acad Sci 1990; 587:307-8.

55. Boxer LA, Oliver JM, Spielberg SP, Allen JM, Shulman JD. Protection of granulocytes by vitamin E in glutathione synthetase deficiency. N Engl J Med 1979; 301:901-5.

56. Fukuzawa K, Kurotori Y, Tokumura A, Tsukatani H. Lipids 1989; 24:236-9.

57. Holmsen H. Metabolism of platelets. In: Williams WJ, Beutler E, Erslev AJ, Lichtman MA, eds. Hematology, 4th ed. New York: McGraw-Hill, 1990; 1200-33.

58. Machlin LJ, Filipski R, Kuhn DC. Influence of vitamin E on platelet aggregation and thrombocytopenia in rat. Proc Soc Exp Biol Med 1975; 149:275-7.

59. Hope WC, Dalton C, Machlin LJ. Influence of vitamin E on prostaglandin biosynthesis in rat blood. Prostaglandins 1975; 10:577-67.

60. Pichard KA, Karpen CW, Merola CW. Influence of dietary vitamin C on platelet thromboxane A and vascular prostacycline I_2 in rabbit. Prostagland Leuko Med 1982; 9:373-78.

61. Okuma M, Takahashi H, Uchino H. Generation of prostacycline-like substance and lipid peroxidation in vitamin E-deficient rats. Prostaglandins 1980; 19:527-36.

62. Creter D, Pavlotzky F, Savir H. Effect of vitamin E on platelet aggregation in diabetic retinopathy. Acta Haematol (Basel) 1979; 62:74-7.

63. Steiner M, Anastasi J. Vitamin E, An inhibitor of the platelet release reaction. 1976; 57:732-7.

64. Lucy JA, Dingle JT. Fat-soluble vitamins and biological membranes. Nature 1964; 204:156-204.

65. Urano S, Matsuo M. Membrane-stabilizing effect of vitamin E. Ann N Y Acad Sci 1989; 570:524-6.

66. Erin AN, Skrypin VV, Kagan VE. Formation of α-tocopherol complexes with fatty acids. Nature of complexes. Biochim Biophys Acta 1985; 815:209-14.

67. Wassall SR, Thewalt JL, Wong L, Gorrissen H, Cushley RJ. Deuterium NMR study of the interaction of α-tocopherol with a phospholipid model membrane. Biochemistry 1986; 25:319-26.

68. Urano S, Iida M, Otani I, Matsuo M. Membrane stabilization of vitamin E: interactions of α-tocopherol with phospholipids in bilayer liposomes. Biochem Biophys Res Commun 1987; 146:1413-8.

69. Urano S, Yano K, Matsuo M. Membrane stabilizing effect of vitamin E: effect of α-tocopherol and its model compounds on fluidity of lecithin liposomes. Biochem Biophys Res Commun 1988; 150:469-75.

14

Intermembrane Transfer of α-Tocopherol and Its Homologs

Valerian E. Kagan, Elena A. Serbinova, and Lester Packer

University of California, Berkeley, California

Z. Z. Zhelev, R. A. Bakalova, and S. R. Ribarov

Medical Academy, Sofia, Bulgaria

INTRODUCTION

Vitamin E is the term used for eight naturally occurring fat-soluble nutrients called tocopherols. α-Tocopherol has the highest biological activity and predominates in many species (1,2). In humans vitamin E is the most important lipid-soluble antioxidant, and its deficiency may cause neurological dysfunction, myopathies, and diminished erythrocyte life span (3-5). Most of vitamin E is located in the mitochondria and in the endoplasmic reticulum, whereas little is found in cytosol and in peroxisomes (6). In membranes vitamin E is not uniformly distributed but forms clusters in the lipid bilayer and is preferentially concentrated in domains rich in polyenoic phospholipids (7).

α-Tocopherol interacts with lipid alkoxyl and peroxyl radicals, acting as a chain-breaking antioxidant (8,9). This antioxidant function of vitamin E results in its consumption. Thus oxidative stress is usually accompanied by a loss of vitamin E (10). This may cause a localized antioxidant deficiency in the membrane: domains rich in enzymic or nonenzymic generators of active oxygen species and lipid radicals may become tocopherol-deficient microenvironments. This "local E hypovitaminosis" in the membrane may be overcome by (1) redistribution of endogenous tocopherols from regions with a high content of vitamin E to those poor in tocopherols, or (2) supplementation with exogenous vitamin E, in particular liposome-incorporated vitamin E. In both cases, the rate of inter- and intra-membrane exchange of vitamin E molecules may be crucial for the replenishment of vitamin E loss.

The efficiency of trans-bilayer tocopherol migration ("flip-flop") is very low (7,11,12). The replenishment of E-deficient membrane domains may be brought about by either lateral diffusion or intermembrane transfer and exchange of tocopherols. Since the rate of the lateral diffusion of tocopherols is sufficiently high (13), this process may not be limiting. However, opinions differ concerning the intermembrane transfer and exchange of tocopherols. It has been suggested that intermembrane exchange is only efficient in the presence of tocopherol binding proteins (14-16).

Here, the results of the studies on the intermembrane exchange of tocopherols between tocopherol-loaded (donor) and tocopherol-free (acceptor) liposomes, as well as between donor liposomes and microsomal membranes as acceptors, are reported.

EXPERIMENTS

Unilammelar donor liposomes containing α-tocopherol or its homologs (6-hydroxychromanes) and three different types of acceptor membranes (unilammelar liposomes, multilammelar liposomes, and microsomes) were utilized. Depending on the type of acceptor present, two different assays of 6-hydroxychromane transfer were used (1). In experiments with microsomes or multilammelar liposomes as acceptors, the mixture of donor and acceptor vesicles was incubated for 1, 30, 60, 90, or 120 minutes at 37 °C and then centrifuged to sediment microsomes (100,000 g × 60 minutes) or multilammelar liposomes (120,000 g × 90 minutes). The pellets were resuspended in 0.1 M phosphate buffer to give the initial concentration of protein or lipid in the acceptor suspension, and the amount of 6-hydroxychromanes transferred from donor liposomes to acceptor membranes was estimated by measuring characteristic fluorescence (λ_{excit} 298 nm, and λ_{emiss} 325 nm) (7). The fluorescence intensity readings were corrected by subtracting the fluorescence intensity of the endogenous α-tocopherol in the microsomes or microsomal lipids (2). When unilammelar liposomes were used as acceptors, the intermembrane transfer of 6-hydroxychromanes was estimated by their ability to inhibit (Fe^{2+} + ascorbate)-induced lipid peroxidation in polyunsaturated acceptor liposomes as described earlier (17,18).

To test the possibility of fusion of donor with acceptor vesicles, we used two different approaches: (1) We measured the fluorescence quenching of 6-hydroxychromanes localized in the inner monolayer of acceptor unilammelar liposomes by ferricyanide entrapped in the inner volume of donor unilammelar liposomes, as described earlier (18) (2). The fusion of donor unilammelar liposomes with acceptor microsomes or multilammelar liposomes was evaluated by changes in the amount of phospholipid phosphorus in acceptor vesicles after their separation from donor liposomes by sedimentation. Trans-bilayer migration of 6-hydroxychromanes was estimated as described earlier (7).

Intermembrane Transfer of 6-Hydroxychromanes from Unilammelar Liposomes to Unilammelar Liposomes

Inhibition of Lipid Peroxidation and Intermembrane α-Tocopherol Transfer

The accumulation of lipid peroxidation products in liposomes from rat cerebral cortex lipids in the presence or absence of tocopherol is shown in Figure 1. In the absence of α-tocopherol, a pronounced formation of lipid peroxidation products was observed. Concentration-dependent lipid peroxidation inhibition was found upon addition of α-tocopherol. When α-tocopherol was incorporated only into the donor liposomes only partial inhibition of lipid peroxidation was observed. An estimation of the amount of α-tocopherol in acceptor liposomes can be made by comparing the efficiency of lipid peroxidation inhibition in acceptor liposomes with the "calibration" curves given in Figure 1. The comparison shows that about 0.8-1.0 nmol α-tocopherol per mg lipid was present in acceptor liposomes after 60 minutes of preincubation which is 12-15% of the amount in donor liposomes (Table 1).

Similarly, when unsaturated fatty acid-containing egg yolk lecithin liposomes were used as tocopherol donors, lipid peroxidation inhibition in acceptor liposomes by α-tocopherol increased with the preincubation time of liposomes and with lipid peroxidation inducers (Table 1). After a 60 minute preincubation, lipid peroxidation corresponded to that caused by 1.0 nmol α-tocopherol per mg lipid in acceptor liposomes. This corresponds to an exchange of 15% of the α-tocopherol present in donor liposomes. This finding agrees with the data of Nakagawa et al. (15) on the rate of intermembrane transfer of [³H]α-tocopherol between liposomes from egg yolk lecithin. In experiments in which monounsaturated phosphatidylcholine (β-oleoyl-γ-palmitoyl phosphatidylcholine) donor liposomes were used, lipid peroxidation inhibition in cerebral cortex lipid acceptor liposomes was less efficient (Table 1). When lipo-

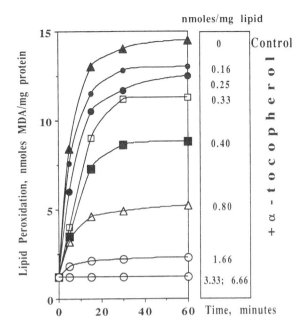

FIGURE 1 Accumulation of lipid peroxidation products in unilammelar liposomes prepared from rat cerebral cortex lipids. Incubation conditions: unilammelar liposomes (0.6 mg lipids per ml); 0.5 mM ascorbate; 40 μM Fe^{2+} in 0.1 M phosphate buffer (pH 7.4 at 25 °C). The reaction was started by adding inducers of lipid peroxidation. There were four replicates of each measurement.

TABLE 1 Intermembrane Transfer of α-Tocopherol Between Unilammelar Donor and Acceptor Liposomes [a]

Liposome composition	Preincubation time	Amount of α-tocopherol transferred (nmol/mg lipid)
DMPC	1	0.10 ± 0.02
	30	0.30 ± 0.03
	60	0.52 ± 0.03
OPPC	1	0.13 ± 0.02
	30	0.50 ± 0.03
	60	0.70 ± 0.04
Cerebral cortex lipids	1	0.33 ± 0.04
	30	0.75 ± 0.06
	60	0.98 ± 0.05
Egg yolk lecithin	1	0.62 ± 0.04
	30	0.83 ± 0.07
	60	1.08 ± 0.05
+ Cholesterol		
In donor liposomes	30	0.41 ± 0.66
In donor and in acceptor liposomes	30	0.38 ± 0.03
+ Peroxidized lipids		
In donor liposomes	30	1.10 ± 0.09
In donor and acceptor liposomes	30	1.20 ± 0.10

[a]Incubation conditions: unilammelar liposomes (0.6 mg lipid per ml), 0.5 mM ascorbate, 40 μM Fe^{2+} in 0.1 M phosphate buffer (pH 7.4 at 25 °C). The reaction was started by adding inducers of lipid peroxidation. There were four replicates of each measurement. The values are given as mean values ± standard deviation (SD). DMPC, dimiristoyl phosphatidylcholine; OPPC, oleoylpalmitoyl phosphatidylcholine. The content of cholesterol and peroxidized lipids in liposomes was 55 and 10 mol per 100 mol, respectively. The α-tocopherol concentration was 6.6 nmol/mg lipids.

somes from saturated phospholipid (dimiristoyl phosphatidylcholine, DMPC) were used as donor liposomes, the antioxidant effect in acceptor liposomes was less pronounced. After a 60 minute preincubation it did not reach the values corresponding to the presence of 0.3–0.5 nmol α-tocopherol per mg lipids in acceptor liposomes (Table 1), which corresponds to 4.5–7.5% of α-tocopherol present in donor liposomes.

Intermembrane transfer of α-tocopherol was sensitive to changes in the molecular organization of the donor and/or the acceptor liposomes. Incorporation of cholesterol into the liposomal bilayer decreased the efficiency of α-tocopherol transfer, whereas oxidized lipids stimulated the transfer of α-tocopherol from donor into acceptor liposomes (Table 1).

Test for Liposomal Fusion

In a series of experiments, the possibility that fusion of donor with acceptor liposomes had occurred was examined. Using gel filtration and ferricyanide as the oxidant, we obtained donor liposomes containing α-tocopherol only in the inner membrane monolayers and acceptor liposomes with ferricyanide trapped inside. The fluorescence intensity of α-tocopherol did not change with time during 2 h of incubation but dropped to zero values after sonication, which made the interaction of α-tocopherol with ferricyanide possible. Thus during 2 h period no fusion of donor and acceptor liposomes would occur.

Intermembrane Transfer of α-Tocopherol Homologs Between Liposomes

α-Tocopherol homologs with a shorter isoprenoid side chain (Fig. 2) possess higher intramembrane mobility and higher antioxidant activity in the lipid bilayer (11). Thus it was interesting

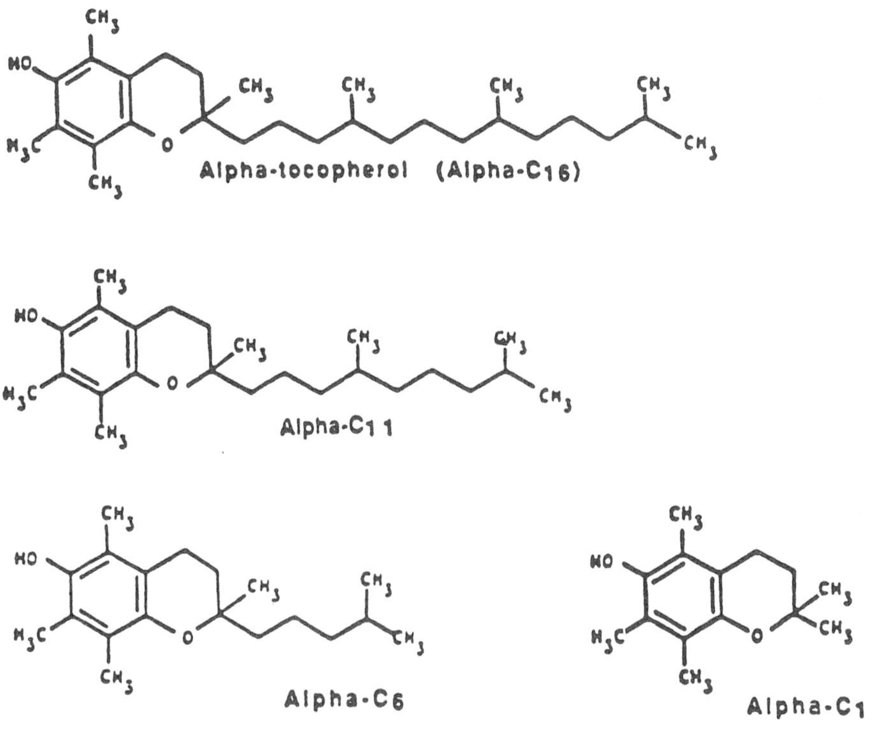

FIGURE 2   Structural formulas of α-tocopherol homologs.

to determine the efficiency of their intermembrane transfer. Fluorimetric measurements of the 6-hydroxychromane content in multilammelar acceptor liposomes from microsomal lipids incubated with unilammelar donor liposomes (prepared from the same microsomal lipids) showed that the efficiency of the transfer increased in the order C16 < C11 < C6 (Fig. 3). The amounts of 6-hydroxychromanes transferred to the acceptor liposomes increased over the course of incubation. Within the interval 0–120 minutes, the time course of the transfer showed saturation for C6 and C11 but was linear for α-tocopherol. In contrast, significant amounts of C1 (more than 30% of the total amount) were found in acceptor liposomes immediately after their addition to donor liposomes. The amounts of C1 in acceptor liposomes did not grow over time. Our earlier studies showed that C1 partitions between the aqueous and membraneous phases in such a way that only about 30% binds to the membrane, whereas 70% remains in the aqueous phase (predominantly in micelles) (7). Thus we suggest that C1 found in acceptor liposomes was not transferred from donor liposomes but rather was incorporated from its pool in the aqueous phase. In contrast, C6, C11, and C16 (α-tocopherol) are very poorly dissolved in water and partition almost entirely into the membraneous phase. Thus the presence of C6, C11, and α-tocopherol in acceptor liposomes is only possible as a result of their transfer via direct contact of the two types of liposomes. In control experiments the amount of lipid phosphorus was measured before and after incubation of the liposomal mixture (unilammelar donor liposomes and multilammelar acceptor liposomes). No increase in the content of phosphorus in acceptor liposomes was detected. This suggests that no fusion of the donor with acceptor liposomes occurred in the presence or absence of 6-hydroxychromanes. Hence the enrichment of acceptor liposomes with 6-hydroxychromanes may be due only to their intermembrane transfer.

Intermembrane Transfer of 6-Hydroxychromanes Between Unilammelar Liposomes and Microsomes

Incubation of unilammelar donor liposomes prepared from microsomal lipids or from DMPC with microsomal membranes resulted in a pronounced enrichment of microsomes with 6-hydroxychromanes (Fig. 4). Three important features were found: (1) although the amounts

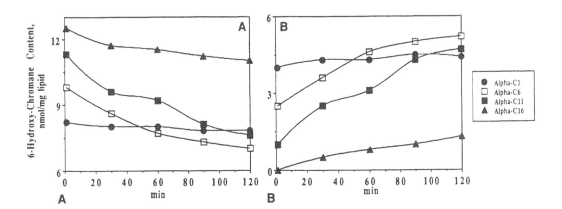

FIGURE 3 Intermembrane transfer of α-tocopherol and its homologs with a shorter side chain from unilammelar donor liposomes to multilammelar acceptor liposomes prepared from rat liver microsomal lipids. (A) Donor liposomes; (B) acceptor liposomes. Initial concentration of 6-hydrochromanes in donor liposomes was 12.5 nmol/mg lipid. There were six replicates of each measurement.

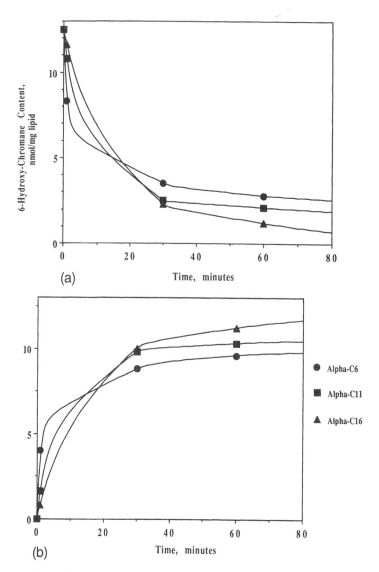

FIGURE 4   Intermembrane transfer of α-tocopherol and its homologs with a shorter side chain between unilammelar donor liposomes from microsomal lipids to acceptor microsomal membranes. (a) Donor liposomes; (b) acceptor microsomes. Initial concentration of 6-hydrochromanes in donor liposomes was 12.5 nmol/mg lipid. There were six replicates of each measurement.

of 6-hydroxychromanes in microsomes increased over the course of incubation (from 0 to 90 minutes), this increase was not monotonous and a greater part of 6-hydroxychromanes incorporation occurs after 30 minutes; (2) the amounts of 6-hydroxychromanes in microsomes at the end of 90 minute of incubation decreased in the order C6 < C11 < C16; (3) the concentrations of 6-hydroxychromanes found in microsomes after incubation with donor liposomes were much higher than the expected equilibrium concentration. When unilammelar donor liposomes from microsomal lipids were used, 78% of the initially added C6 content, 84% of C11, and 94% of C16 was detected in microsomes. Similar results were obtained with saturated DMPC donor liposomes, although the intermembrane transfer of 6-hydroxychromanes to microsomes was significantly slower. Experiments show that the transfer of 6-hydroxychromanes to microsomal membranes resulted in the inhibition of lipid peroxidation upon addition of Fe^{2+} + ascorbate.

No fusion of liposomes with microsomal membranes occurred in the absence of 6-hydroxychromanes. In the presence of 6-hydroxychromanes the amount of lipid phosphorus in microsomes increased during incubation with donor liposomes, indicating fusion occurred. α-Tocopherol was more active as a fusogenic agent than C6 or C11. However, the increment in lipid phosphorus accumulation in microsomes after 90 minutes of incubation did not exceed 45% of its initial value. This suggests that the fusion of microsomal and liposomal membranes may be only partly responsible for the transfer of 6-hydroxychromanes from liposomes to microsomes.

CONCLUDING REMARKS

Vitamin E can be transferred from donor liposomal to acceptor liposomal or natural (microsomal) membranes. The efficiency of this transfer is strongly dependent on the molecular mobility of lipids in the donor or acceptor membranes. Ordering of membrane lipids (increase in phospholipid saturation or incorporation of cholesterol) decreases the rate of α-tocopherol transfer, whereas disordering membrane lipids (increase in phospholipid unsaturation or incorporation of more polar oxidized lipids) increases the efficiency of the intermembrane transfer of α-tocopherol. Accordingly, α-tocopherol homologs with shorter side chains (C11 and C6), known to increase the molecular mobility of membrane lipids and to be more mobile in the bilayer (18), demonstrate a much higher efficiency of intermembrane transfer than α-tocopherol. Surprisingly, shortening the α-tocopherol side chain by only one isoprenoid unit dramatically increases the efficiency of the intermembrane transfer. Thus, a 16-carbon side chain in tocopherol molecule is crucial for its retention in specific membrane areas and to maintain its asymmetric distribution in the membrane (19). Interestingly, a decrease in α-tocopherol side chain length by one isoprenoid unit was shown to cause a significant loss of its vitamin activity (20).

Intermembrane exchange of α-tocopherol and its homologs results in a distribution between donor and acceptor liposomes that can only equilibrate the concentrations of 6-hydroxychromanes in the liposomes. When microsomal membranes were used as acceptors, however, the enrichment of microsomes with 6-hydroxychromanes was much greater than equilibrium concentrations. This suggests that a specific binding of 6-hydroxychromanes to microsomal proteins occurs that may be responsible for this extra binding (compared with the binding of 6-hydroxychromanes to acceptor liposomes from microsomal lipids). α-Tocopherol was much more efficient in this specific transfer and binding to microsomal proteins than its synthetic homologs. Future studies may elucidate the proteins involved in site-specific α-tocopherol binding in microsomes.

ACKNOWLEDGMENTS

This research was supported by the NIH (CA 47597) and the Bulgarian Academy of Sciences.

REFERENCES

1. Machlin LJ. Vitamin E. In: Machlin LJ, ed. Handbook of vitamins: nutritional, biochemical and clinical aspects. New York: Dekker, 1984; 99-145.
2. Horwitt MK. Status of human requirement for vitamin E. Am J Clin Nutr 1974; 27:1182-93.

3. Janero DR. Therapeutic potential of vitamin E against myocardial ischemic-reperfusion injury. Free Radic Biol Med 1991; 10:3151-324.

4. Finer NN, Grant G, Schindler RF, Hill BG, Peters KL. Effect of intramuscular vitamin E on frequency and severity of retrolental fibroplasia: a controlled trial. Lancet 1982; I:1087-91.

5. Muller DPR, Goss-Sampson MA. Role of vitamin E in neural tissue. Ann N Y Acad Sci 1989; 570: 146-55.

6. Bjorneboe A, Bjorneboe GE, Drevon CA. Absorption, transport and distribution of vitamin E. J Nutr 1990; 120:233-42.

7. Kagan VE, Bakalova RA, Serbinova EA, Stoytchev TS. Fluorescent measurements of incorporation and hydrolysis of tocopherol and its esters in biomembranes. Methods Enzymol 1989; 186: 355-67.

8. Tappel AL. Vitamin E as the biological antioxidant. Vitam Horm 1962; 20:493-510.

9. Burton GW, Ingold KU. Autoxidation of biological molecules. 1. The antioxidant activity of vitamin E and related chain-breaking phenolic antioxidants in vitro. J Am Chem Soc 1981; 103:6472-7.

10. Horwitt MK. Data supporting supplementation of humans with vitamin E. J Nutr 1990; 121:424-9.

11. Kagan VE, Serbinova EA, Bakalova RA, et al. Mechanisms of stabilization of biomembranes by alpha-tocopherol: the role of the hydrocarbon chain in the inhibition of lipid peroxidation. Biochem Pharmacol 1990; 40:2403-13.

12. Ilani A, Krakover T. Diffusion- and reaction rate-limited redox processes mediated by quinones through bilayer lipid membranes. Biophys J 1987; 51:161-7.

13. Gomes-Fernandez JC, Villalain J, Aranda FJ, et al. Localization of α-tocopherol in membranes. Ann N Y Acad Sci 1989; 570:109-20.

14. Guarnieri CP, Flamingi F, Cladarera CM. A possible role of rabbit heart cytosol tocopherol binding in the transfer of tocopherol into nuclei. Biochem J 1980; 190:469-71.

15. Murphy DJ, Mavis RD. Membrane transfer of α-tocopherol—influence of soluble α-tocopherol-binding factors from the liver, lung, heart, and brain of the rat. J Biol Chem 1981; 256:10464-8.

16. Behrens WA, Madere R. Transfer of α-tocopherol to microsomes mediated by a partially purified liver α-tocopherol binding protein. Nutr Res 1982; 2:611-8.

17. Buege J, Aust S. Microsomal lipid peroxidation. Methods Enzymol. 1978; 52:302-10.

18. Kagan VE, Bakalova RA, Zhelev ZZ, et al. Intermembrane transfer and antioxidant action of α-tocopherol in liposomes. Arch Biochem Biophys 1990; 280:147-52.

19. Kagan VE, Packer L. Servinova EA, et al. Mechanism of vitamin E control of lipid peroxidation. In: Reddy CC, Hamilton GA, Madyastha KM, eds. Biological oxidation systems. New York: Academic Press, 1990; 2:889-908.

20. Weiser H, Vecchi M. Stereoisomers of α-tocopheryl acetate. Biopotencies of all eight stereoisomers, individually or in mixtures, as determined by rat resorption-gestation tests. Intl J Vitam Nutr Res 1982; 52:351-70.

15

Vitamin E: The Antioxidant Harvesting Center of Membranes and Lipoproteins

Lester Packer and Valerian E. Kagan

University of California, Berkeley, California

INTRODUCTION

Vitamin E was discovered at the University of California at Berkeley in 1922 in the laboratory of Herbert M. Evans. From its initial discovery as an antifertility agent, it was given the name "tocopherol," meaning childbirth (*tokos*) + to carry (*pherein*). Indeed, it was the development of the rat fetal reabsorption biological assay that was used as the basis to evaluate vitamin E activity. One international unit for vitamin E activity was based upon the amount of vitamin E needed to prevent reabsorption of the fetus in a pregnant rat.

It is now known that vitamin E is a generic name for a group of naturally occurring substances found mainly in plant oils that exhibit vitamin E-like activity (1). The main substances are α-, β-, δ-, and γ-tocopherols and an almost identical family of substances found mainly in tropical plant oils known as the tocotrienols, which also have α-, β-, and δ-substituents on the chromanol nucleus but differ in that three unsaturated double bonds are present on the hydrophobic carbon tail that anchors vitamin E in membranes. The α form of vitamin E with three methyl groups on the chromanol nucleus, d-α-tocopherol, has generally been considered the most active naturally occurring form based on assays of its activity in the rat gestation model. α-Tocopherol is also the most common form of vitamin E found in the human diet except for those individuals who live in areas of the world where tropical oils are mainly utilized for cooking and as sources of food.

The designation of vitamin E status came about as a result of the discovery of natural deficiencies in vitamin E, mainly diseases of young children or young animals (2,3). Several vitamin E deficiency diseases mainly affecting newborns or young babies were discovered. Many of these diseases are based upon inadequate amounts of vitamin E in the diet, defects in the transport of vitamin E to tissues (such as abetalipoproteinemia), or poor vitamin E nutrition, such as retrolental fibroplasia and intravesicular hemmorrhage of the brain in young babies, in which mainly visual and circulation defects result in neurological disorders. Defects in bile-mediated absorption also lead to impaired vitamin E uptake and may cause muscular dystrophy in children.

With the advent and growth of the field of nutrition, it soon became recognized that it was extremely difficult if not impossible to render mature animals deficient in vitamin E even if fed diets totally devoid of vitamin E (4,5). The typical severe deficiencies of vitamin E seen in young babies and children have never been observed in adult experimental animals or humans. However, a whole host of literature ranging from epidemiological and clinical to biochemical studies provided strong circumstantial evidence that higher levels of vitamin E in the diet have beneficial health effects (6–8). Vitamin E slows the course of degenerative diseases and is useful for antioxidant protection in therapy in such acute clinical conditions as ischemia reperfusion injury or protection of tissues against the toxicity generated by redox cycling drugs (e.g., doxorubicin) used to treat infections or to target tumors in which the side effects on healthy tissues can be minimized by higher levels of vitamin E in the diet.

INHIBITION OF LIPID PEROXIDATION BY VITAMIN E

Numerous physiological and pathological changes are observed when vitamin E is deficient in experimental animals and in humans (4–8). Despite the very diverse effects of vitamin E in living systems, there is only one chemical activity with which it has been primarily identified. This is the ability of vitamin E to act as an inhibitor of the oxidation of lipids (9,10). The structure and function of membranes largely depend upon the presence of different classes of phospholipids with different types of fatty acids. It is here where vitamin E is believed to exert its fundamental and basic biological action as an antioxidant.

Many pioneers have contributed to our current knowledge of the lipid peroxidation inhibition by vitamin E, including work from Tappel's laboratory and definitive studies from the laboratories of Burton and Ingold, who put forward evidence that vitamin E is the major lipid-soluble chain-breaking antioxidant of blood plasma and of membranes (9,10). Despite this unique activity, the presence of vitamin E in natural membranes is extremely low in relation to membrane phospholipid. Frequently, the ratio of vitamin E to phospholipids is 0.05 mol per 100 mol or lower. Thus, a minute amount of vitamin E exerts important and significant effects in protecting membranes against peroxidation. This is remarkable for membranes that have electron transport reactions in which high rates of free-radical production occur that initiate lipid radicals and lipid peroxidation.

A reasonable question is why vitamin E is crucially important for the overall antioxidant protection of membranes. There are at least two answers: (1) vitamin E is considered the major if not the only chain-breaking antioxidant of membranes (Ingold and Burton) (10); (2) vitamin E functions in the membranes as a center for antioxidation that harvests (collects) the antioxidant power not only from other lipid-soluble antioxidants (like ubiquinols) but also from reductants in the cytosolic phase (like ascorbate and glutathione; see our results later). Thus vitamin E is not only the major, but is also most efficient membrane antioxidant. This is why the maintenance of a steady-state concentration of vitamin E in membranes may be crucial for their protection against antioxidant invasions.

Steady-state concentrations of vitamin E in membranes are determined by (1) the efficiency of its incorporation into membranes due to transfer from blood lipoproteins and (2) its metabolism in membranes.

The main intramembrane metabolic pathway of vitamin E is believed to be its scavenging of lipid radicals in the course of initiation and propagation of lipid peroxidation (9–11). The free radical of vitamin E is formed in the reaction

$$LOO^{\cdot} + Toc\text{-}OH \rightarrow LOOH + Toc\text{-}O^{\cdot} \qquad (1)$$

It has generally been believed that the great mobility of vitamin E in the lateral plane of the membrane and its exact positioning in the membrane are extremely important for this reaction (12,13). Vitamin E is anchored in the hydrocarbon part of the membrane bilayer by the phytol tail, which is 13 carbons long, just the right length to position the chromanol nucleus, which possesses the antioxidant activity at the membrane interface (14). There, vitamin E through its phenolic hydroxyl group quenches free radicals in the process, becoming the (phenoxyl or chromanoxyl) tocopheroxyl radical. Tocopheroxyl free radicals are less reactive than other lipid radicals (peroxyl or alkoxyl radicals) [reaction (1)] generated in membranes and thus serve to break the chain of free-radical reactions in lipid peroxidation. However, the free-radical tocopheroxyl form of vitamin E is susceptible to oxidation or to destruction by reacting with itself or by other reactions that cause it to decompose as a result of radical initiated reactions (2) and (3):

$$\text{Toc-O}^{\cdot} + \text{Toc-O}^{\cdot} \rightarrow \text{products} \tag{2}$$

$$\text{Toc-O}^{\cdot} + \text{LOO}^{\cdot} \rightarrow \text{products} \tag{3}$$

$$\text{Toc-O}^{\cdot} + \text{Red-H} \rightarrow \text{Toc-OH} + \text{Red}^{\cdot} \tag{4}$$

Unless reduced (regenerated) to its original antioxidant form [reaction (4)], it will be lost before prooxidant reactions occur. This type of vitamin E action was suspected for a long time (15–17). In fact, it was proposed almost two decades ago by Mellors and Tappel that the ubiquinone components of membranes may serve to protect vitamin E against loss by interacting with it (18). It is also known from in vitro studies that ascorbate (vitamin C) regenerates vitamin E from its free-radical form (15–17), but whether this was an important activity of membranes was not known.

Studies in Burk's laboratory revealed that reduced glutathione (GSH), the primary preventative water-soluble antioxidant in most aerobic cells, protects against lipid peroxidation of microsomal membranes in vitro (19). However, GSH does not exhibit this activity if membranes are prepared from vitamin E-deficient animals. From this finding Bast, McCay, and others have suggested the existence of a vitamin E free-radical reductase activity, that is, some enzyme or enzyme systems capable of specifically regenerating vitamin E and that glutathione may be one of the substrates for this type of activity (17,20,21). However, no direct experimental data on enzymic regeneration of vitamin E were obtained.

The only way to elucidate these types of membrane reactions of vitamin E is to dynamically follow the reactions of the vitamin E tocopheroxyl radicals in natural membranes using sensitive electron spin resonance (ESR) techniques.

ESR ASSAY OF TOCOPHEROXYL RADICAL REACTIONS AND RECYCLING ACTIVITY IN NATURAL MEMBRANES

We have developed simple and convenient methods to generate phenoxyl radicals from vitamin E and its synthetic homologs (Fig. 1) based on their (1) enzymatic oxidation by a lipoxygenase + polyunsaturated fatty acid (e.g., arachidonic or linolenic) system, or (2) nonenzymatic oxidation by the azo initiator of peroxyl radicals, 2,2′-azobis-(2,4-dimethylvaleronitrile) (AMVN), or (3) irradiation by ultraviolet B (UV-B) light, which is directly absorbed by vitamin E molecules (Fig. 2) (22,23). Lipoxygenase generates peroxyl radicals of polyunsaturated fatty acids. AMVN is thermally decomposed to produce carbon-centered radicals that are rapidly converted into peroxyl radicals in the presence of oxygen. In these two systems peroxyl radicals interact with vitamin E to generate chromanoxyl radicals. Thus both systems imitate the interaction of tocopherol with peroxyl radicals in the course of lipid peroxidation. The advantage of the UV-B induction of vitamin E radicals is the absence of other radicals (peroxyl or alkoxyl) in the system that may interact with the recycling agents.

In the first experiments (24) we used a homolog of α-tocopherol, 2,2,5,7,8-pentamethyl-6-hydroxychromane (α-C1), which is devoid of the phytol chain of vitamin E (24). This homolog is less hydrophobic than tocopherols and is more readily and uniformly distributed in

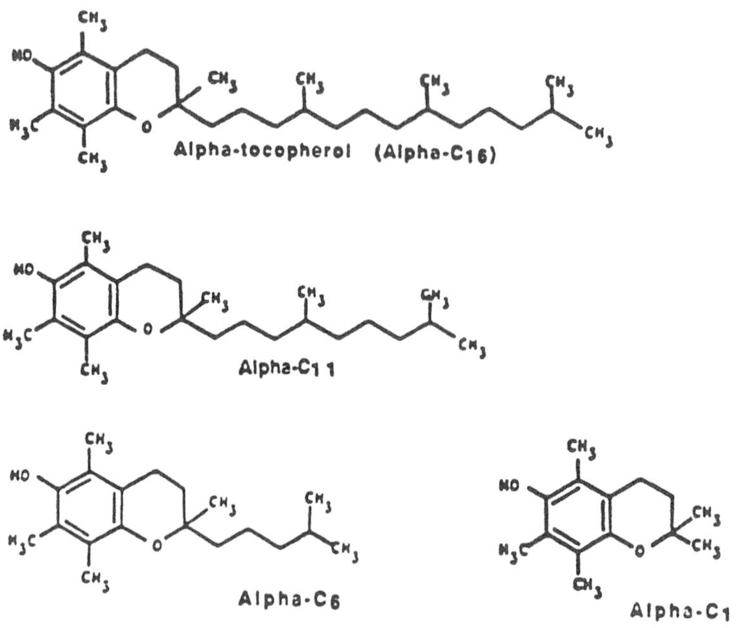

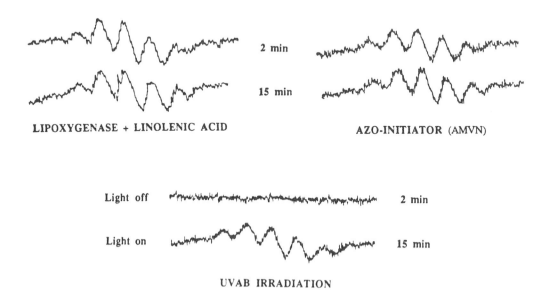

FIGURE 1 Structural formulas of α-tocopherol homologs.

FIGURE 2 ESR spectra of tocopheroxyl radicals generated from vitamin E in dioleoylphosphatidyl-choline liposomes using three different systems.

membranes. The steady-state concentrations of chromanoxyl radicals generated from α-C1 appeared to be high enough to use ESR techniques to follow the reactions of chromanoxyl radicals in cell organelles. However, normal vitamin E concentrations in natural membranes (liver microsomes and mitochondria) are too low to provide a detectable tocopheroxyl radical ESR signal. For this reason we fed animals with diets enriched in vitamin E and subsequently isolated vitamin E-enriched membranes (10 to 40-fold). At this level of enrichment the induced endogenous tocopheroxyl radical signal is detectable by ESR (Fig. 3). We also used other homologs of α-tocopherol differing in the length of their chains (Fig. 1) to enrich natural or liposomal membranes by their addition in vitro. We found that α-C6, an α-tocopherol homolog with a six-carbon side chain, gave well-resolved ESR spectra of its chromanoxyl radicals (Fig. 3).

To quantitate the efficiency of chromanoxyl radical reduction, we introduced the recycling efficiency coefficient R_e:

$$R_e = \frac{A_{-\text{red}} - A_{+\text{red}}}{A_{-\text{red}}}$$

where $A_{-\text{red}}$ and $A_{+\text{red}}$ are the magnitudes of ESR signals of chromanoxyl radicals in the absence and in the presence of a reductant, respectively. The values of recycling efficiency vary from 1.0 to 0, which corresponds to 100% reduction (complete transient disappearance of ESR signal) and to 0% reduction (no effect on ESR signal), respectively.

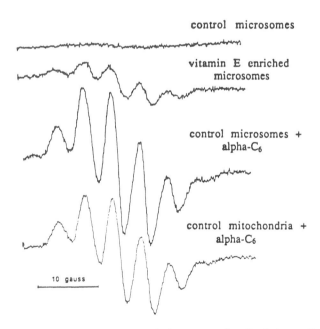

FIGURE 3   ESR spectra of chromanoxyl radicals in rat liver mitochondria and microsomes. Tocopheroxyl radicals were generated by an enzymic oxidation system (lipoxygenase + linolenic acid) in vitamin E-enriched microsomes (21.5 nmol α-tocopherol per mg protein). Chromanoxyl radicals were generated by the same enzymic oxidation system in control microsomes (0.8 nmol α-tocopherol per mg protein) or mitochondria (0.6 nmol α-tocopherol per mg protein) after addition of exogenous α-C6 (54 nmol/mg protein). Other conditions as in Figure 2.

Thus we had at our disposal simple and reliable methods of generating tocopheroxyl or homologous chromanoxyl radicals in measurable and quantifiable amounts in both natural membranes and in liposomes and of simultaneously following the time course of changes in the amounts of these radicals. This gave us an opportunity to directly study membrane reactions of chromanoxyl radicals of vitamin E and of its homologs.

SEARCH FOR A FREE-RADICAL REDUCTASE

Using these methods we were able to demonstrate that ascorbate is efficient in regeneration of chromanoxyl radicals of tocopherol and its homologs not only in liposomes but also in low-density lipoproteins and natural membranes, that is, liver microsomes and submitochondrial particles (Fig. 4). Enzyme-dependent mechanisms that prevent the accumulation of chromanoxyl radicals derived from vitamin E homologs were characterized in these membranes. NADPH or NADH in microsomes as well as NADH or succinate in mitochondria prevented the accumulation of chromanoxyl radicals until these substrates were fully consumed (Fig. 5). Thus we concluded that rat liver microsomes and mitochondria have both enzymatic electron transport-dependent and nonenzymatic mechanisms for reducing chromanoxyl radicals.

Feeding animals a vitamin E-rich diet gave us the ability to enrich membranes endogenously with vitamin E so that tocopheroxyl radical signals could be easily detected by ESR. Using these vitamin E-enriched membranes we showed that in microsomes tocopheroxyl radicals were reduced by NADH- and NADPH-dependent electron transport enzymes (Fig. 6)

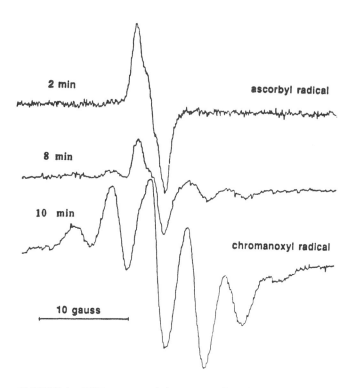

FIGURE 4 ESR spectra of chromanoxyl radicals generated from α-C6 in the presence of ascorbate. Concentration of ascorbate was 2.5 mM. Incubation medium contained microsomes, 27 mg protein per ml; linolenic acid, 14 mM; lipoxygenase, 90 U/μl; chromanols, 8 mM; NADPH, 5 mM.

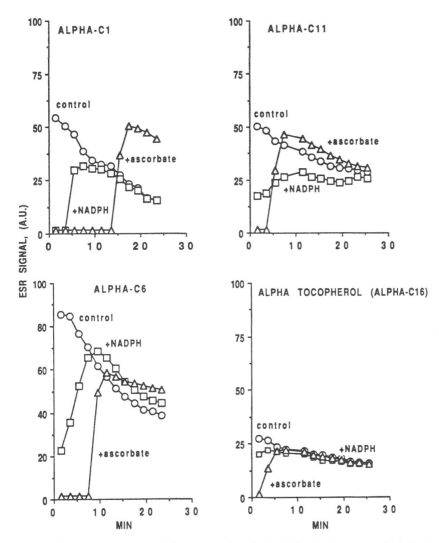

FIGURE 5 Time course of chromanoxyl radicals ESR signal generated by the lipoxygenase + linolenic acid oxidation system in the presence of rat liver microsomes and their recycling by NADPH and ascorbate.

(25). Combination of NADPH + NADH results in an additive effect, showing that these two pathways of tocopheroxyl radical reduction are independent. In submitochondrial particles NADH, NADPH, and succinate also serve as substrates for the reduction of tocopheroxyl radicals. Inhibitor studies indicate that the location of tocopheroxyl radical reduction is before cytochrome c, but reduced cytochrome c can reduce tocopheroxyl radicals. In a reconstituted proteoliposomal system containing mitochondrial complex II (succinate-ubiquinone reductase), succinate was able to regenerate vitamin E only when ubiquinone Q_{10} was incorporated together with α-tocopherol (26).

Thus the original hypothesis of a single "free-radical reductase" is simplistic. Rather, it seems that several pathways in different membranes operate to regenerate vitamin E.

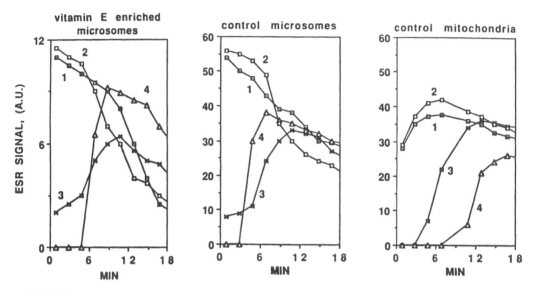

FIGURE 6 Time course of ESR signal of chromanoxyl radicals in rat liver mitochondria and microsomes: (1) control; (2) +CoQ1 (40.0 nmol/mg protein); (3) +NADPH (7.5 mM) for microsomes or NADH (7.5 mM) for mitochondria; (4) +CoQ1 + NADPH (or NADH).

REDUCED UBIQUINONES STIMULATE THE RECYCLING OF VITAMIN E

Isoprenoid quinones (CoQn) have been recognized as constituents of electron transport chains, where they act as mobile hydrogen carriers that shuttle reducing equivalents between various dehydrogenases (CoQn reductases) and the CoQH2 oxidase systems. There is now a substantial amount of experimental data that support the notion that CoQn, besides this well-recognized role as a redox component of an electron transport system, may function in its reduced or semireduced form as an antioxidant in various biological membranes (18,27). However, the specific molecular mechanisms responsible for the antioxidant activity of ubiquinones are still debated. In a direct way, reduced forms of CoQn may behave as other free-radical scavengers, donating a H atom to peroxy and alkoxy radicals. Another possibility of indirect antioxidant function of CoQn was suggested by Mellors and Tappel, who advanced a hypothesis on vitamin E recycling by ubiquinones (18).

To obtain evidence of a direct or indirect mechanism of an antioxidant action of ubiquinols (ubiquinones), we (1) compared the antioxidant efficiencies of tocopherol and ubiquinol (ubiquinone) homologs in rat liver microsomal and mitochondrial membranes under conditions of (Fe^{2+} + ascorbate)-induced lipid peroxidation, and (2) studied the recycling of tocopherol and its homologs from the chromanoxyl radicals by ubiquinone-dependent electron transport in microsomes and mitochondria. We were able to demonstrate that tocopherols are much stronger membrane antioxidants than naturally occurring ubiqinols (ubiquinones) (27). Thus direct radical-scavenging effects of ubiquinols (ubiquinones) might be negligible in the presence of comparable or higher concentrations of tocopherols. In support of this our ESR findings show that ubiquinones synergistically enhance enzymic NADH- and NADPH-dependent recycling of tocopherols by electron transport in mitochondria and microsomes (Fig. 6). Thus we conclude that the antioxidant effects of ubiquinols (ubiquinones) are not due to their direct radical-scavenging reactivity but rather result from their ability to stimulate more efficient recycling of tocopherols interacting with electron-transport enzymes (Fig. 7).

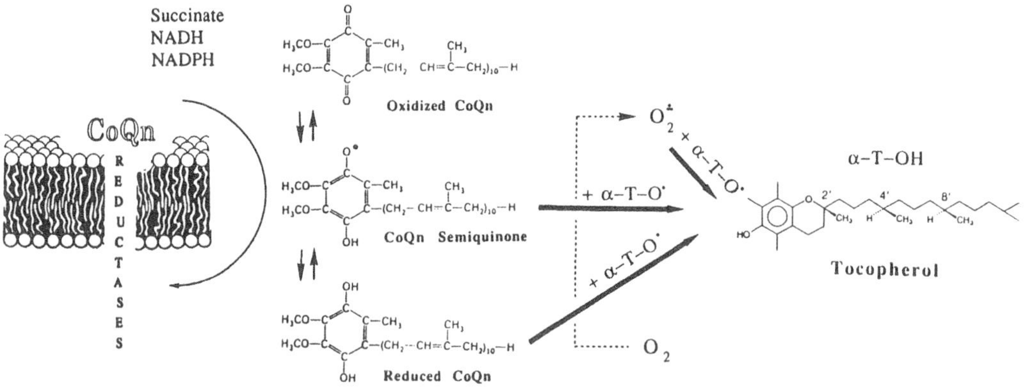

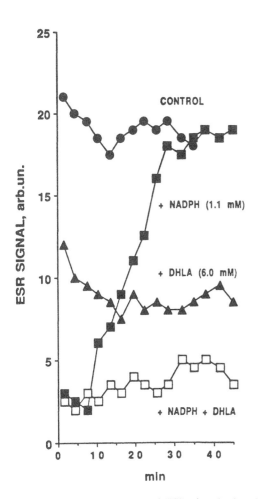

FIGURE 7   Role of ubiquinones in recycling vitamin E free radicals.

FIGURE 8   Time course of ESR signal of α-C6 chromanoxyl radical generated by the azo initiator AMVN in rat liver microsomes in the presence of NADPH and DHLA.

RECYCLING OF TOCOPHEROXYL RADICALS BY DIHYDROLIPOIC ACID (DHLA) AND REDUCED GLUTATHIONE

The antioxidant effects of DHLA in membranes might be due to its ability to regenerate vitamin E from its tocopheroxyl radicals, as suggested by Bast et al. and Scholich et al. In our ESR experiments with the chromanoxyl radicals of the α-tocopherol homolog α-C6, which were generated by the azo initiator AMVN, we found that DHLA synergistically interacted with NADH or NADPH in liver microsomes to recycle chromanoxyl radicals. In its oxidized form lipoic acid (thioctic acid) was inefficient in the reduction of chromanoxyl radicals (Fig. 8).

Similarly, reduced glutathione drastically increased the efficiency of NADPH in the recycling of chromanoxyl radicals but was without effect in the absence of NADPH. Oxidized glutathione was inefficient both in the absence and in the absence of NADPH (22).

Dihydrolipoic acid was able to maintain ascorbate in its reduced form by regenerating it from dehydroascorbate and semidehydroascorbate in both liposomes and low-density lipoproteins (Fig. 9). This recycling of ascorbate provided for a more efficient regeneration of vitamin E from its tocopheroxyl radicals. A similar stepwise ascorbate-mediated mechanism of vitamin E recycling was found to be operative with reduced glutathione. We conclude that reduced thiols can synergistically interact with ascorbic acid to enhance the antioxidant activity of vitamin E, although they are not efficient in the direct reduction of tocopherol phenoxyl radicals.

Thus, an extremely high antioxidant efficiency of tocopherols may be explained by the functioning of a vitamin E cycle in which chromanoxyl radicals can be reduced by ascorbate, NADPH-, NADH-, and succinate-dependent electron transport in microsomes and mitochondria, whereas glutathione, dihydrolipoate, and ubiquinols synergistically enhance vitamin E regeneration in membranes and lipoproteins (Fig. 10). We conclude that vitamin E molecules possess a unique ability to act as membrane free-radical–harvesting centers that collect their antioxidant power from other intramembrane and cytosolic reductants.

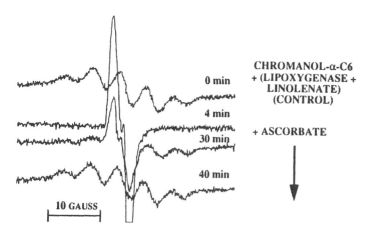

FIGURE 9 ESR spectra of radicals generated by the lipoxygenase + linolenic acid oxidation system in LDL in the presence of α-C6 and ascorbate. The incubation medium contained LDL (10 mg protein per ml) in 50 mM phosphate buffer, pH 7.4. The concentrations of α-C6 and ascorbate were 80 nmol/mg protein and 1.5 mM, respectively.

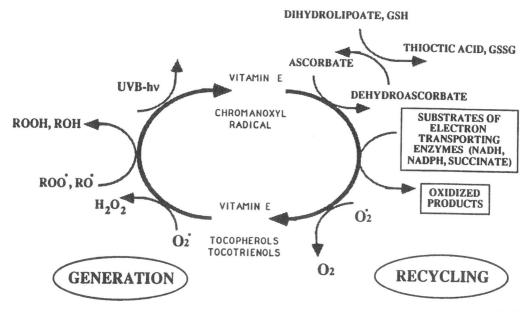

FIGURE 10   The vitamin E cycle: enzymic and nonenzymic recycling interactions of water- and lipid-soluble antioxidants with vitamin E radicals.

RECYCLING AND ANTIOXIDANT ACTIVITY OF TOCOPHEROL HOMOLOGS OF DIFFERING HYDROCARBON CHAIN LENGTHS

The antioxidant function of α-tocopherol is localized in its 6-hydroxychromane ring, and its isoprenoid chain is considered important for proper orientation of the molecule in the membrane and is not involved in its antioxidant action. However, α-tocopherol homologs with different hydrocarbon chain lengths manifest strikingly different antioxidant activity in natural membranes, α-tocopherol being about a two orders of magnitude less efficient antioxidant than 2,2,5,7,8-pentamethyl-6-hydroxychromane, a homolog devoid of the hydrocarbon chain. By contrast, in solution 6-hydroxychromanes with and without 2-alkyl substituents demonstrate the same reactivity toward peroxyl and alkoxyl radicals.

It might be suggested that the unequal recycling efficiency of chromanoxyl radicals could be at least partly responsible for the different antioxidant activity of α-tocopherol homologs with differing chain lengths against lipid peroxidation in natural membranes. To test this possibility we compared the efficiencies of ascorbate- and NADPH-supported recycling of chromanoxyl radicals with the antioxidant potency of α-tocopherol homologs in rat liver microsomes. We demonstrated that the antioxidant activity of tocopherol homologs during (Fe^{2+} + ascorbate)- or (Fe^{2+} + NADPH)-induced lipid peroxidation in rat liver microsomes increased in the order α-tocopherol (α-C16) < α-C11 < α-C6 < α-C1. Chromanoxyl radicals generated from α-tocopherol and its more polar homologs by an enzymic oxidation system (lipoxygenase + linolenic acid) can be recycled in rat liver microsomes by NADPH-dependent electron transport or by ascorbate (Figs. 5 and 6). The efficiency of recycling increased in the same order α-tocopherol (α-C16) < α-C11 < α-C6 < α-C1 (Fig. 11). This higher recycling efficiency of short-chain α-tocopherol homologs is due to their higher intramembrane mobility and more homogeneous distribution within the membrane lipid bilayer.

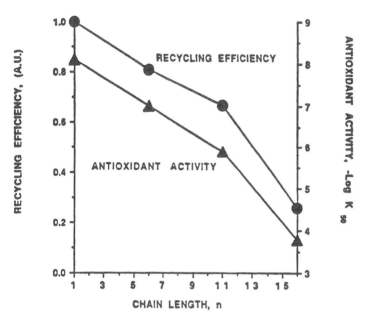

FIGURE 11 Dependence of antioxidant activity and recycling efficiency of tocopherol homologs in rat liver microsomes on the number of carbons in the hydrocarbon chain.

Thus the high efficiency of regeneration of short-chain homologs of vitamin E may account for their high antioxidant potency. Unfortunately, these short-chain α-tocopherol homologs, although highly potent, cannot be used as membrane antioxidants with prolonged action because of their pronounced membrane-perturbing effects.

α-Tocotrienol is a natural α-tocopherol homolog with the same aromatic chromanol "head" but with an unsaturated isoprenoid side chain that may increase its molecular mobility in membrane. This in turn may enhance the antioxidant potency and the recycling efficiency of tocotrienol without producing defects in the lipid bilayer. Our studies showed that α-tocotrienol possesses a markedly higher antioxidant activity in rat liver microsomal membranes and significantly higher recycling efficiency from chromanoxyl radicals than α-tocopherol (28). We suggest that α-tocotrienol should have higher physiological activity than α-tocopherol under conditions of oxidative stress. A comprehensive comparison of the antioxidant properties of tocotrienols versus tocopherols is given in Chapter 19.

CONCLUSIONS

Under physiological conditions low concentrations of vitamin E are sufficient to prevent membrane oxidative damage. We tested the hypothesis that this is due to a unique ability of vitamin E molecules (tocopherols and tocotrienols) to act as membrane free-radical–harvesting centers, their antioxidant power being derived from other intracellular reductants. Location of the chromanol nucleus of vitamin E at the membrane interface affords efficient interaction with lipid radicals and regeneration of tocopheroxyl radicals. Our ESR and HPLC data show that tocopheroxyl radicals can be reduced enzymically by NADPH-, NADH-, and succinate-dependent electron transport in microsomes and mitochondria. Ubiquinols synergistically enhance vitamin E regeneration supported by membrane electron transport enzymes.

Ascorbate is very efficient in the nonenzymic recycling of vitamin E in cell organelles and low-density lipoproteins. Reduced thiols (glutathione and dihydrolipoate) synergistically enhance the ascorbate-dependent recycling of vitamin E by regenerating ascorbate from dehydroascorbate and/or semidehydroascorbate. Thus, the well-known synergistic effects of these physiologically important antioxidants (reductants) with vitamin E are mediated via their ability to donate the electrons necessary for recycling tocopheroxyl radicals in membranes. Nontoxic reductants capable of enhancing the recycling efficiency of chromanoxyl radicals may lead to new avenues in the clinical applications of vitamin E.

ACKNOWLEDGMENTS

We acknowledge the National Institutes of Health CA47597, Palm Oil Research Institute of Malaysia (PORIM), and ASTA Pharmaceuticals (Frankfurt am Main, Germany) for research support.

REFERENCES

1. Machlin LJ. Vitamin E. In: Machlin LJ, ed. Handbook of vitamins: nutritional, biochemical and clinical aspects. New York: Dekker, 1984; 99-145.
2. Horwitt MK. Vitamin E and lipid metabolism in man. Am J Clin Nutr 1960; 8:451-61.
3. Horwitt MK. Status of human requirement for vitamin E. Am J Clin Nutr 1974; 27:1182-93.
4. Kayden HJ, Hatam LH, Traber MG. The measurement of nanograms of tocopherol from needle aspiration biopsies of adipose tissue: normal and abetalipoproteinemic subjects. J Lipid Res 1983; 24:652-6.
5. Bieri JG. Kinetics of tissue alpha-tocopherol depletion and repletion. Ann N Y Acad Sci 1972; 203: 181-91.
6. Finer NN, Grant G, Schindler RF, Hill GB, Peters KL. Effect of intramuscular vitamin E on frequency and severity of retrolental fibroplasia: a controlled trial. Lancet 1982; I:1087-91.
7. Muller DPR, Goss-Sampson MA. Role of vitamin E in neural tissue. Ann N Y Acad Sci 1989; 570: 146-55.
8. Janero DR. Therapeutic potential of vitamin E against myocardial ischemic-reperfusion injury. Free Radic Biol Med 1991; 10:3151-324.
9. Tappel AL. Vitamin E as the biological antioxidant. Vitam Horm 1962; 20:493-510.
10. Burton GW, Joyce A, Ingold KU. Is vitamin E the only lipid-soluble, chain-breaking antioxidant in human blood plasma and erythrocyte membrane? Arch Biochem Biophys 1983; 221:281-90.
11. Burton GW, Ingold KU. Autoxidation of biological molecules. 1. The antioxidant activity of vitamin E and related chain-breaking phenolic antioxidants in vitro. J Am Chem Soc 1981; 103:6472-7.
12. Niki E, Kawakami A, Saito M, Yamamoto Y, Tsuchiya J, Kamiya Y. Effect of phytyl side chain of vitamin E on its antioxidant activity. J Biol Chem 1985; 260:2191-6.
13. Kagan VE, Serbinova EA, Bakalova RA, et al. Mechanisms of stabilization of biomembranes by alpha-tocopherol. The role of the hydrocarbon chain in the inhibition of lipid peroxidation. Biochem Pharmacol 1990; 40:2403-13.
14. Kagan VE, Quinn PJ. The interaction of α-tocopherol and homologues with other hydrocarbon chains with phospholipid bilayer dispersions. Eur J Biochem 1988; 171:661-7.
15. Packer JE, Slater TF, Willson RL. Direct observation of a free radical interaction between vitamin E and vitamin C. Nature 1979; 278:737-8.
16. Niki E, Tsuchiya J, Tanimura R, Kamiya Y. The regeneration of vitamin E from alpha-chromanoxyl radical by glutathione and vitamin C. Chem Lett 1982; 6:789-92.
17. McCay PB. Vitamin E: interaction with free radicals and ascorbate. Annu Rev Nutr 1985; 5:323-40.
18. Mellors A, Tappel AL. The inhibition of mitochondrial peroxidation by ubiquinone and ubiquinol. J Biol Chem 1966; 241:4353-6.
19. Burk RF. Glutathione-dependent protection by rat liver microsomal protein against lipid peroxidation. Biochim Biophys Acta 1983; 757:21-8.

20. Haenen GRMM, Bast A. Protection against lipid peroxidation by a microsomal glutathione-dependent heat-labile factor. FEBS Lett 1983; 159:24-8.

21. Reddy CC, Scholz RW, Thomas CE, Massaro EJ. Vitamin E dependent reduced glutathione inhibition of rat microsomal lipid peroxidation. Life Sci 1982; 31:571-6.

22. Kagan VE, Serbinova EA, Packer L. Generation and recycling of radicals from phenolic antioxidants. Arch Biochem Biophys 1990; 280:33-9.

23. Kagan VE, Serbinova EA, Packer L. Recycling and antioxidant activity of tocopherol homologues of differing hydrocarbon chain length in liver microsomes. Arch Biochem Biophys 1990; 282:221-5.

24. Packer L, Maguire JJ, Melhorn R, Serbinova E, Kagan V. Mitochondria and microsomal membranes have a free radical reductase activity. Biochem Biophys Res Commun 1989; 159:229-35.

25. Maguire JJ, Wilson DS, Packer L. Mitochondrial electron transport-linked tocopheroxyl radical reduction. J Biol Chem 1989; 264:21462-5.

26. Maguire JJ, Kagan VE, Serbinova EA, Ackrell BA, Packer L. Ubiquinols quench tocopheroxyl radicals in reconstituted liposomes containing succinate-ubiquinone redictase. Arch Biochem Biophys 1991; (submitted).

27. Kagan VE, Serbinova EA, Koynova GM, et al. Antioxidant action of ubiquinol homologues with different isoprenoid chain length in biomembranes. Free Radic Biol Med 1990; 9:117-26.

28. Serbinova EA, Kagan VE, Han D, Packer L. Free radical recycling and intramembrane mobility in the antioxidant properties of alpha-tocopherol and alpha-tocotrienol. Free Radic Biol Med 1991; 10:263-75.

16

Preparation and Properties of Liposomes Containing Vitamin E

Hans-Joachim Freisleben

University Clinics, Frankfurt, Germany

Edgar Mentrup

Hoechst AG, Frankfurt, Germany

INTRODUCTION

Liposomes have been used as model membranes in biophysical and biochemical studies for many years. They are useful as a tool to investigate cell and membrane properties. Applications of small lipid vesicles have been explored in the field of controlled and targeted drug delivery in vivo. Promising results have been obtained, especially in the field of targeting to the reticuloendothelial system and in visceral leishmaniasis (1,2). In cancer chemotherapy liposomes can be used as a controlled-release system for toxic drugs. The cardiotoxicity of Adriamycin (doxorubicin) could be reduced in vitro and in vivo (3,4). Therapeutic benefits have also been reported with encapsulated macrophage activating factors for the activation of alveolar macrophages to a tumoricidal state (5,6).

The application of liposomes for drug delivery depends not only on the amelioration of therapy but also upon a reproducible and large-scale preparation process of well-defined lipid vesicles. A further limitation for pharmaceutical or medical use has been the stability of the drug carrier system: not only the physical integrity of the vesicle but also the leakage of encapsulated hydrophilic drugs and the chemical stability of all materials used must be a prerequisite.

Tocopherols play an important role in the preparation of stable liposomes:

1. They protect unsaturated fatty acid residues against oxidation. Comparing the different tocopherols, α-tocopherol exhibits the best protection against lipid oxidation (7).
2. At higher amounts tocopherol enhances the physical stability of liposomes containing free unsaturated fatty acids or phospholipids containing unsaturated fatty acid residues.
3. Tocopherol may prevent oxidation of drugs in lipid vesicles.
4. Tocopherol itself can be administered via liposomes.

Thus, in this chapter we summarize some aspects of the preparation, structure, and stability of α-tocopherol-containing liposomes.

PREPARATION OF LIPOSOMES CONTAINING α-TOCOPHEROL

Vesicle Properties and Preparation Methods

Several techniques are applied for the preparation of liposomes, from laboratory batches to the manufacturing of large-scale volumes used in cosmetics. Variations in technique and lipid composition yield liposomes of different size, charge, and physical properties. The preparation procedures influence the method of incorporating α-tocopherol into the liposomal bilayer.

Three types of liposomes are generally distinguished with regard to size and numbers of lamellae:

1. The elementary procedure for preparing inhomogeneous populations of multilamellar large vesicles (MLV) was first described more than 20 years ago (8). A lipid film is prepared by removing the solvent from a solution of lipids in organic solvents. The lipid film is dispersed by agitation with an aqueous buffer phase. MLV are formed spontaneously if the temperature is above the phase transition of the lipid. The resulting population is inhomogeneous, and hydrophilic drugs can only be encapsulated to a small extent.
2. Liposomes with one or two lamellae are more suitable for encapsulating hydrophilic drugs. Large amounts of more than 50 mol per 100 mol (drug-lipid) can be trapped inside the vesicles. The REV technique (reverse-phase evaporation method) starts by emulsifying a lipid solution in ether into an aqueous phase by sonication. The organic solvent is then removed under controlled conditions to produce the large unilamellar vesicles (LUV, > 100 nm). Although this method is very useful for laboratory trials, a scaling-up process is difficult to handle.
3. Small unilamellar vesicles (SUV, 20-100 nm) are often used for parenteral applications because of their slow clearance from the bloodstream. According to their size, the encapsulated amount of a hydrophilic drug is low. Often liposomes are prepared by sonication of a MLV dispersion either by a probe or bath-type sonicator. This procedure is not suitable for unstable drugs and may lead to oxidation of lipids. A reduction in vesicle size can also be achieved by extrusion of MLV through a French press or homogenization unit.

The detergent removal technique, which prevents the use of such organic solvents as methylene chloride or alcohols, starts with solubilization of the lipid by a surfactant. After removal of the detergent by dialysis, adsorption, or dilution, the small liposomes are formed. These vesicles are stable and homogeneous.

Techniques for Incorporation of α-Tocopherol into Liposomes

Each of the procedures just described exerts special advantages and disadvantages regarding the preparation process and the resulting vesicles. Most techniques start with dissolving the lipid in an organic solvent. Since α-tocopherol is readily soluble in ether, methylene chloride, and alcohols, this step can easily be handled. By contrast, solubilization in an appropriate mixture of lipid and detergent can only be achieved with a limited proportion of α-tocopherol to lipid. Thus using the detergent removal technique the amount of encapsulation is limited.

Small amounts of α-tocopherol (up to 5 mol per 100 mol) can easily be incorporated into liposomes with most available methods. This is sufficient for the use of α-tocopherol as an antioxidant. For therapeutic use or enhanced physical stability an increased drug content is desirable. At higher concentrations the inclusion of α-tocopherol is influenced by the preparation procedure and results in a change in the liposomal properties.

After removing an organic solvent containing lecithins and large amounts of α-tocopherol, a thick nonremovable film is formed in the flask. Above 20 mol per 100 mol of α-tocopherol a loss of lipid at the glass wall was observed; at 67 mol per 100 mol all lipids were adherent and did not enter the aqueous phase (9). A maximum of 30 mol per 100 mol of α-tocopherol could be dispersed in the aqueous phase. The total amount of liposome-encapsulated α-tocopherol is thus limited to the highest lipid concentration, which can be achieved by a defined preparation method. Encapsulated α-tocopherol is bound to the lipid bilayer in a stable manner, which was proven by gel filtration (9). Pharmeceutical application of the liposome technology requires a constant quality of the product and a homogeneous size distribution, which should be carefully monitored after inclusion of high amounts of α-tocopherol. Using the French pressure cell technique, very small (28 nm) homogeneous and unilamellar particles can be achieved; however, separation from larger multilamellar particles is necessary (9).

Using the sonication method, it was possible to obtain only a maximum content of 4.8 mol per 100 mol of α-tocopherol in liposomes of phosphatidylcholine (PC) (10). This may be due to either an insufficient preparation technique or varying lipid composition (9).

Depending on the interaction of α-tocopherol with lipids, α-tocopherol hemisuccinate (TS) can be incorporated into phospholipids in a much higher ratio than α-tocopherol. Depression of the transition temperature with increasing amounts of TS in the bilayer was observed (11). This latter derivative also forms liposomes by itself without the necessity of any other membrane-stabilizing component. These vesicles are prepared by sonication or extrusion through the French press after hydrating the material in buffer (12). After treatment with the French press procedure, homogeneous vesicles with a diameter of 25 nm or less were achieved.

One major problem in industrial scaling-up of most of the techniques discussed is the inclusion of large amounts of organic solvents or detergents and hence the necessity to remove them. For several medical applications (e.g., topical, oral, or subcutaneous) a low residual amount of ethanol or propylene glycol may be acceptable. Many other detergent or organic solvent residues cannot be accepted, especially intravenous application. In addition, the amount of residual solvent may determine the rate of drug release from the liposomes.

A suitable procedure for the preparation of large amounts of α-tocopherol-containing liposomes starts with the dissolution of high concentrations of the lipid and α-tocopherol in an adequate solvent. The solution is then mixed with the aqueous phase under stirring or with a homogenizer. The resulting liposome dispersion may be treated by French pressing, sonication, or microfluidizing techniques for further size reduction, dialyzed through membrane filters, or separated via column chromatography.

STRUCTURE AND DYNAMICS OF α-TOCOPHEROL-CONTAINING LIPOSOMES

This part and its title refer to publications on structure (13) and dynamics (14) of α-tocopherol in membranes.

Insertion of α-Tocopherol into Liposomal Membranes

α-Tocopherol itself, as well as some derivatives, is capable of forming membranes separating two aqueous phases (12,15). α-Tocopherol can be incorporated into dipalmitoylphosphatidyl-choline (DPPC) membranes up to 40 mol per 100 mol (11). The incorporation of α-tocopherol and α-tocopherolacetate up to this concentration was verified to be higher than 90% of the inset (16). Below the critical concentration of 40 mol per 100 mol tocopherolacetate appears to be completely miscible in the solid and the liquid crystalline state of DPPC membranes.

Above 40 mol per 100 mol α-tocopherol (16) and tocopherolacetate (17) were excluded from the membranes. Egg phosphatidylcholine liposomes, prepared with a French pressure cell, can contain up to 33 mol per 100 mol tocopherol (9). The penetration of tocopherols into phospholipid monolayers is facilitated by increasing the content of unsaturated fatty acid residues (18).

Hydrophobic Interaction with Liposomal Lipids

Molecular-model-building studies prompted Diplock and Lucy to propose a specific physicochemical interaction between the α-tocopherol phytyl side chain and the fatty acyl chains of polyunsaturated phospholipids, particularly those derived from arachidonic acid. It was postulated that each of the C4 and C8 methyl branches fits into a pocket formed by the cis double bonds of the "quasi-helical" arachidonic acid (19).

The fit of the methyl groups in the pockets created by the cis double bonds may permit the methylene groups in the backbones of both the phytyl and fatty acyl chains to associate closely, further promoting the stability of the complex through hydrophobic attraction forces (20). On the other hand, nuclear magnetic resonance (NMR) spin-lattice relaxation times of ^{13}C-labeled α-tocopherol in liposomes varying in their contents of arachidoyl residues proved that the segmental motion of the isoprenoid side chain of α-tocopherol tends to increase with an increasing distance from the chromanol moiety (21). The methyl groups attached on the aromatic ring were suggested to have some affinity to unsaturated fatty acid residues rather than those of the isoprenoid side chain. These results were claimed (21) not to be compatible with the hypothesis of Reference 19.

Polar Interaction

The molecular model-building studies (19) suggested that the α-tocopherol chromanol ring system interacts with phospholipid polar head groups at the membrane surface. Incorporation of α-tocopherol into liposomal membranes was compared to that of α-tocopherolacetate to learn about the role of the chromanol free phenolic hydroxyl in binding to liposomes. α-Tocopherol binds to DPPC vesicles in a mode different from α-tocopherolacetate as was demonstrated by the phase transition curves of DPPC liposomes containing one of the two compounds (22). Furthermore, ^{13}C-NMR spectroscopy revealed that α-tocopherol binds strongly to phospholipids, possibly via hydrogen bonds in which the ring-borne hydroxyl appears to be involved. α-tocopherolacetate, in which the hydroxyl is excluded from reaction by the acetate ester formation, binds to membranes only via the hydrophobic interaction between its phytyl chain and the fatty acid chains (22). Hence, α-tocopherol appears to interfere directly with the polar head groups of the phospholipids, whereas α-tocopherolacetate exerts a weaker interaction (24).

The hydroxyl group of α-tocopherol was thought to be more important for the interaction with phospholipids than the hydroxyl group in cholesterol (11). On the other hand, Gorbunov et al. (24) found the cholesterol-like influence of α-tocopherol to be similar on liposomes prepared from both phosphatidylcholine and phosphatidylethanolamine, and they suggested that the character of α-tocopherol action does not depend on the polar group structure of phospholipid molecules.

Fluidity

Order Parameters

The fluidity of the liposomal phosphatidylcholine membranes decreased with increasing concentrations of saturated phosphatidylcholine, cholesterol, and α-tocopherol. However, the physical effect of α-tocopherol was quite small at its physiological concentrations and was modified by altering the side chain at the 2-position of the chromonal ring system. The effect of tocopherols on membrane fluidity was greater in the inner part of the membranes (25). By adding tocopherol acetate the lateral diffusion coefficient for pyrene in DPPC membranes is decreased, whereas the order parameter of the fatty acid chains is slightly increased in the inner part of the membranes (17).

A higher order parameter can be interpreted as decreased fatty acid chain mobility (trans-gauche isomerization) (26). Thus, it appears that tocopherol acetate reduces the possibility for trans-gauche isomerization in the inner part of the membrane. Molecular models suggest that the side-chain methyl groups of tocopherol acetate impede trans-gauche isomerization by steric interaction, which induces increased order in the bilayer (17). The increase in order parameters by α-tocopherol acetate reported in DPPC liposomes (17) were seen in microsomes containing α-tocopherol only above the 0.2 M fraction, which is considerably above physiological concentrations (27).

Microviscosity

The term "microviscosity" was introduced (28) as a fluidity parameter of lipid regions determined by fluorescence polarization using DPH (diphenylhexatriene) and perylene. High-resolution ^{1}H-NMR spectroscopy and spin probes detected a cholesterollike influence of α-tocopherol on lipid bilayer microviscosity. α-Tocopherol increased the microviscosity of unsaturated bilayers and decreased the microviscosity of saturated bilayers (24).

α-Tocopherol increased the diphenylhexatriene (DPH) anisotropy parameter and decreased the break-point activation energy of liposomes prepared with lecithin or with lipids from frog retinal rod outer segment membranes as examined by steady-state fluorescence polarization (29).

Distribution

The fluorescence properties of α-tocopherol in bilayer membranes of dipalmitoylphosphatidylcholine or egg phosphatidylcholine indicate that α-tocopherol is not uniformly distributed in such membranes and occupies two or more sites. The fluorescence decay and depth-dependent quenching of α-tocopherol fluorescence confirm that in gel-phase lipid the chromanol group has a transverse distribution close to the head group region of the lipid. In fluid-phase lipid (in the presence of buffer) the results indicate there is more penetration of the chromanol group into the bilayer (30).

The trans-bilayer distribution of α-tocopherol and polyunsaturated fatty acid residues of phosphatidylethanolamine (PE) and phosphatidylserine (PS) in liposomes and plasma membranes of synaptosomes showed that most of the polyenoic fatty acid residues of α-tocopherol were located predominantly in the inner monolayers (31,32). No asymmetric distribution of incorporated α-tocopherol was observed in liposomes prepared from a single phospholipid, such as dioleoylphosphatidylcholine (DOPC) (31,32).

In unilamellar liposomes from saturated and unsaturated phospholipids, α-tocopherol was incorporated into both monolayers at molar ratios from 1:1000 to 1:100 (α-tocopherol to phospholipid) and was distributed in its monomeric form without forming clusters (33).

Up to 5 mol per 100 mol, the distribution of α-tocopherol in unilamellar phosphatidylcholine liposomes is uniform, whereas at higher molar ratios α-tocopherol is enriched in the inner monolayer of the vesicles. Above 7 mol per 100 mol the outer to total ratio of α-tocopherol was found to be 0.27 (10), similar to that of cholesterol.

Phase Transitions

DSC/DTA measurements with α-tocopherol and α-tocopherol acetate in DPPC multibilayer vesicles revealed that both compounds modified the thermotropic properties of the pure phospholipid, so that the pretransition disappeared even at low concentrations of either compound. The enthalpy corresponding to the main transition decreased with increasing concentrations of α-tocopherol, and the transition peak was progressively shifted to lower temperatures (11). α-Tocopherol acetate exerted the same effects as α-tocopherol, but to a lower extent (16,23). α-Tocopherol acetate and α-tocopherol also broadened the temperature range of the phase transition of PC and PE membranes (17,23).

When α-tocopherol was included in multibilayer vesicles of dimyristoylphosphatidyl-choline (DMPC), dipalmitoylphosphatidylcholine, and distearoylphosphatidylcholine (DSPC) and in equimolar mixtures of dilaurylphosphatidylethanolanine (DLPE) and DPPC, which show cocrystallization, it induced a broadening of the main transition and a shift to lower temperatures (34).

Furthermore, α-tocopherol caused a decrease in the enthalpy and cooperativity of the phase transition in DMPC liposomes as studied by DSC measurements (35). The effects were similar to those of other compounds partitioning into the hydrophobic part of the membrane (e.g., ubiquinone Q10, ionol, or vitamin K_3) and were modified by the presence of a hydrocarbon side chain.

Phase Separations

Incorporated into mixtures of PC and PE, α-tocopherol preferentially interacts with the component melting at the lower temperature, indicating a location in more fluid domains (23).

Partitioning experiments with the high-resolution electron paramagnetic resonance (EPR) probe perdeutero di-*t*-butyl nitroxide (PDDTBN) were used to investigate the effect of α-tocopherol on lecithin liposomes (36,37). The results obtained as a function both of α-tocopherol concentration and of temperature suggest the presence of two distinct hydrophobic phases in these phospholipid bilayers, one α-tocopherol poor and the other one rich in α-tocopherol (36).

Effects of α-tocopherol on saturated phosphatidylcholines and phosphatidylethanolamines (34) are summarized as follows (DPPE, dipalmitoylphosphatidylethanolamine):

Phospholipid		Effect of α-tocopherol
DPPC; DMPC; DSPC		No phase separation
DPPE; DLPE		Lateral phase separation
Equimolar mixtures	*Tectic behavior*	
DPPC/DSPC	Cocrystalline	No preference for one component
DMPC/DSPC	Monotectic	Preference for lower melting and more fluid component, DMPC
DPPC/DPPE	Cocrystalline	Lateral phase separation
DLPE/DPPC	Cocrystalline	No phase separation
DLPE/DSPC	Monotectic	Always in the lower melting and more fluid phase (DLPE or DMPC, respectively)

Versus equimolar mixtures of phosphatidylcholines α-tocopherol exhibits behavior similar to that of cholesterol and cis-unsaturated fatty acids. Regarding equimolar mixtures of PC and PE, α-tocopherol differs from cholesterol, which always prefers PC to PE, wheras α-tocopherol always prefers the lower melting and more fluid phase, independently of whether this is composed mainly of PC or PE.

Interaction with Free Fatty Acids

Using electron spin resonance (ESR) probes and ¹H-NMR spectroscopy methods, α-tocopherol was shown to remove the chaotropic action of free fatty acids on phospholipid bilayers. The stabilization effect is common and does not depend on the chemical structure of the phospholipid functional polar groups or the unsaturation degree of either fatty acyl residues or of free fatty acids entering into a phospholipid bilayer. Analogs of α-tocopherol without a phytyl chain do not show this stabilizing effect on the lipid bilayer disturbance by free fatty acids. It was thought that the interaction of α-tocopherol with free fatty acids may be one of the molecular mechanisms of lipid bilayer stabilization (38).

The effects of α-tocopherol and other tocopherols on the fluidity of liposomes composed of dipalmitoylphosphatidylcholine and fatty acids were investigated by measurement of DPH fluorescent polarization. Although all tocopherols decreased the fluidity of liposomes, being perturbed by the inclusion of unsaturated fatty acids with more than one double bond, α-tocopherol was more effective than the others. The stabilization of arachidonic acid-containing liposomes required either the chromanol moiety with methyl groups born on its aromatic ring or a side chain of appropriate length, which need not necessarily be an isoprenoid side chain of full length, and configuration of the α-tocopherol phytyl chain (39). Nevertheless, all these results indicate that in α-tocopherol both the chromanol nucleus and the phytyl chain participate in membrane and liposome bilayer stabilization.

Cholesterol Interaction

In erythrocyte membranes the ratio of arachidonic acid residues to α-tocopherol was calculated as about 500:1 (19). It was proposed that the majority of unsaturated fatty acyl residues of membranes are associated with cholesterol (40), only a few being associated with α-tocopherol. Structural alteration in the polar region of egg yolk lecithin liposomes was observed to be similar by cholesterol and α-tocopherol in pyrene eximerization studies (41).

In DPPC vesicles cholesterol (18) exhibits the same effects on the lateral diffusion coefficient for pyrene as α-tocopherol acetate (17), but to a greater extent and also in a more hydrophilic outer part of the membrane. The α-tocopherol phytyl chain appears to be necessary for the cholesterollike action (34).

It was claimed (42) that the term "antioxidant protection" for liposomes or biological membranes applies to both structural stability as well as free-radical scavenging. In this respect, α-tocopherol and cholesterol appear to play a structural antioxidant role in the membrane, whereas tocopherol alone exerts free-radical scavenging activity (40). In experiments (25) α-tocopherol inhibits lipid peroxidation at concentrations that do not affect membrane fluidity.

STABILITY OF α-TOCOPHEROL-CONTAINING LIPOSOMES

From our interpretation, α-tocopherol protects biological membranes and liposomes on three distinct levels:

1. An antioxidant by free-radical scavenging
2. A structural stabilizer of phospholipid polyunsaturated fatty acyl residues, preventing breakdown by endogenous phospholipases
3. A stabilizer of phospholipid bilayers against disturbance by unsaturated fatty acids

Level 1 needs cooperation with enzymic and nonenzymic redox systems, whereas effects (2 and 3) need cooperativity with cholesterol. The multifold stabilizing effects of tocopherol to membranes were demonstrated (43).

STABILITY OF LIPIDS

Pharmaceutical preparations of liposomes for medical application, containing unsaturated acyl chain residues, are vulnerable to lipid peroxidation during storage of raw materials and preparation and storage of end products. Depending on the lipid composition, a subsequent oxidative change in the membrane structure may impair the therapeutic value of the liposomal carrier system. To enhance the stability of the lipid vesicles, several approaches are feasible:

1. Because of its appropriate phase transition temperature, dimyristoylphosphatidylcholine is often used in liposome preparations. There was no lipid or α-tocopherol oxidation found in these preparations (44).
2. The composition of the liposomes can be selected from synthetic constituents not containing unsaturated structures (e.g., nanoparticles). The metabolism and toxicology of these substances are often the limiting factor for medical applicability.
3. Hydrogenated lecithins containing only a few unsaturated acyl chains are less susceptible to oxidation than the original lipids. The former, however, often exhibit special features that are limiting to medical application (45).
4. A new approach to the preparation of stable liposomes can be provided using the main phospholipid (MPL) of the archaebacterium *Thermoplasma acidophilum*. This phospholipid consists of two repetitively methyl-branched, saturated C_{40} chains linked to a glycerol at both ends via ether bonds, hence forming a tetraether macrocycle. Each of the hydrocarbon chains is derived from two covalently linked phytanyl residues. The third glycerol hydroxyl on each side is linked either to a sugar molecule or to a further glycerol via phosphoester linkage, respectively (Fig. 1). The length of the molecule corresponds to the width of the apolar core of common membrane or liposomal bilayers. The saturated tetraether cycle is resistant to hydrolysis and oxidation and is capable of forming stable liposomes (46). The applicability for medical use and the toxicity of these MPL liposomes are under investigation at this time (Freisleben and coworkers, unpublished).
5. Addition of hydrophilic and/or hydrophobic antioxidants ameliorates the stability of liposomes. Although the most appropriate antioxidant appears to be α-tocopherol, small amounts of up to 0.1 mol per 100 mol do not prevent lipid oxidation. During the preparation of small unilamellar vesicles by detergent dialysis, from soy lecithin stabilized with α-tocopherol under nitrogen atmosphere, an increase of the oxidation index from 0.4 to 0.75 was observed within 22 h. After storage at 2°C for 4 weeks, a further increase to over 1.2 could not be prevented (47).

This can be explained by the linear consumption of small amounts of α-tocopherol by free radicals. After the inhibition period the oxidation proceeded at a rate similar to that in the absence of tocopherol (48). For short-term storage a minimum amount of 0.1 mol per 100 mol α-tocopherol is necessary. The shelf life can be prolonged by the addition of 0.5-2 mol per 100 mol of α-tocopherol depending on the lipid composition.

During the preparation of MLV in the presence of oxygen and without the addition of an antioxidant, lipid peroxidation starts after a lag time of a few hours. The addition of α-tocopherol extends the lag period but does not prevent the further oxidation of soy bean PC (49). A higher temperature enhances the decomposition of hydroperoxides and may shorten

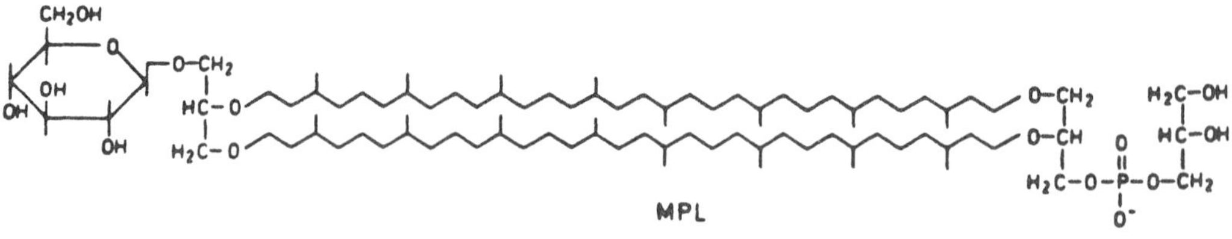

FIGURE 1 Structure of the main phospholipid from Thermoplasma acidophilum.

the induction period (48). Comparing the activity of α-tocopherol and butylated hydroxy-toluene (BHT), the latter proved to be advantageous against the oxidation of MLV from rat liver phospholipids (48). A variety of BHT homologs have also been investigated (50).

The preparation of SUV by sonication of MLV results in a doubled peroxidation rate compared to the preparation of MLV. Lipid peroxidation, however, can be completely prevented during the preparation time of both types of liposomes by adding 1 mol per 100 mol α-tocopherol or BHT (48). The efficiency of the two substances was dose dependent but comparable in these experiments. The charge of lipids controls the influence of peroxidative agents. Neutral and negatively charged liposomes are much more sensitive to peroxidation by ferrous ion and ascorbate than vesicles composed of positively charged compounds (51). Even in liposomes composed of phosphatidylcholine–dicetylphosphate and cholesterol, however, peroxidation could be drastically reduced by the insertion of 0.1-1 mol per 100 mol α-tocopherol. α-Tocopherol at 1 mol completely prevented peroxidation of about 100 mol polyunsaturated fatty acid moieties (52). The incorporation of cholesterol increased the antioxidative efficiency of α-tocopherol, although cholesterol alone had no antioxidative potential (52).

Physical Stability of Liposomes

The physical stability of liposomes is described by two parameters: (1) vesicle size and distribution and (2) leakage of hydrophilic drugs from the inner volume. Both can be influenced by incorporation of such substances as α-tocopherol into the bilayer:

1. Alteration in the size of liposomes results in different pharmacokinetic behavior: large vesicles are taken up by specific cells in tissue including the RES. Small vesicles circulate in the blood for controlled periods of time (depending on size and lipid composition) and retain their drug load (53). The physical interaction of α-tocopherol with phospholipids stabilizes the bilayer and ameliorates liposomal stability in vitro and in vivo (54). The addition of 0.1 mol per 100 mol α-tocopherol doubles the shelf life of multilamellar liposomes. A proportional extension in shelf life to increasing amounts of α-tocopherol was found. The stability in plasma was further improved by the additional use of cholesterol (54).

2. For encapsulated carboxyfluorescein an eightfold reduction in drug release from 20 to 2.5%/h was found after inclusion of 15 mol per 100 mol α-tocopherol into PC liposomes (9). The effect was comparable to the insertion of 37 mol per 100 mol cholesterol. The addition of higher amounts of α-tocopherol (29 mol per 100 mol) leads to an increase in drug release to 5%/h. The modification of membrane permeability of liposomes by α-tocopherol depends both on the type of lipid and on the type of encapsulated drug. Urea is better retained in the vesicles after the addition of α-tocopherol (2-10 mol per 100 mol), but the permeability to erythritol or glucose is unchanged (55).

The mechanism of stabilization can be explained by the insertion of the phytyl side chain of α-tocopherol into the bilayer, as already described. A further reason might be a change in the surface pressure of α-tocopherol-containing vesicles (56).

The release of glucose and chromate from liposomes prepared from lipids with varying constituents depends on the amount of arachidonic acyl chains in the lipid. Drug release from liposomes with low content is not influenced by the addition of α-tocopherol. In contrast, the addition of α-tocopherol to lipids containing a high amount of arachidoyl residues results in a reduction in drug release (57).

The permeability of liposomes composed of dipalmitoyllecithin, dimyristoyllecithin, dioleoyllecithin, egg lecithin, and soy lecithin to glucose was compared (58). Increasing amounts of α-tocopherol in liposomes composed of dipalmitoyllecithin and dimyristoyllecithin decreased the maximum rate of temperature-dependent permeability. Dioleoyllecithin and egg lecithin were influenced to a lesser extent. Using soy lecithin, which has a higher degree of unsaturation, little influence was shown.

One reason for the differences in permeability modification by tocopherol is the varying sensitivity of the lipids to peroxidation. According to the number of double bonds, arachidonic acid and linoleic acid are much more sensitive to peroxidation than oleic acid (51).

Stability of Encapsulated Drugs

The stability of an encapsulated drug depends on its chemical structure and is often a limiting factor for liposomal as for other medical applications. In many cases the adjustment of the pH in the aqueous phase or the addition of stabilizers to the liposomal composition are necessary. Thus, besides water-hydrolyzable drugs, many other labile substances can be stabilized, for example by adding an antioxidant. Because of its low toxicity and easy incorporation into bilayers, α-tocopherol is often used for this purpose.

The stability of liposomes containing hemoglobin was investigated by several groups (59-61). These "hemosomes" are able to substitute for red blood cell deficiency. Under certain conditions, hemoglobin is a potent catalyst of lipid peroxidation and increases the permeability of liposomes. Rapid oxidation occurs in hemosomes prepared from purified hemoglobin and egg phosphatidylcholine (59). The rate of hemoglobin oxidation corresponds to that of non-encapsulated hemoglobin. Using freshly prepared hemolysate, an inclusion of α-tocopherol protects against both hemoglobin oxidation and lipid peroxidation in liposomes. The addition of α-tocopherol increases the half-life of liposome-embedded heme by a factor of 8 (60). After the preparation of heme-loaded liposomes by the ether injection technique, no antioxidative influence of α-tocopherol could be found (61).

Anthraquinone drugs, such as doxorubicin, also need antioxidative protection. The oxidative damage of these drugs and liposomal lipids by free radicals is decreased by including 1.5 mol per 100 mol α-tocopherol and 50 μmol ferrioxamine (62,63).

Incorporation of 8-methoxypsoralen into sonicated egg phosphatidylcholine liposomes caused lipid peroxidation involving singlet oxygen. Addition of α-tocopherol to the liposomes inhibited the peroxidation (64).

TOCOPHEROL TRANSPORT AND INTERMEMBRANE TRANSFER PHENOMENA

In humans, tocopherol is transported in the plasma within lipoproteins. Although all lipoprotein fractions appear to participate in tocopherol transport, a large portion was found in low-density lipoproteins (LDL) (65). The high-affinity receptor for LDL was demonstrated to function as a mechanism for delivery of tocopherol to cells (66). Hence, one possibility of tocopherol uptake into cells is via lipoprotein endocytosis. A similar mechanism may exist for liposomal tocopherol delivery to cells capable of liposome endocytosis.

Intraperitoneal (IP) injection of liposomal γ-tocopherol decreased endogenous lipid peroxidation and stimulation cytochromes P450 and b5 in vitamin E-deficient rats (67). Increase of cytochromes did not occur after IP application of dispersed α- or γ-tocopherols. It was concluded that the efficiency of the liposomal γ-tocopherol could be conditioned by preventing γ-tocopherol oxidation in the lecithin vesicles during its transport and/or by an increased liposomal γ-tocopherol uptake by liver from the circulating blood. In general, liposomal administration of tocopherol may represent the most appropriate route of intravenous application.

In the adrenocortical cell membrane, specific binding sites were found for d-α-tocopherol with association constants within physiological ranges (68). Thus, in these cell membranes uptake of liberated α-tocopherol may occur without endocytosis of lipoproteins or liposomes, either via adhesion of these particles to the cell surface and a direct intermembrane exchange or via intermembrane transfer.

In rat liver cytosol a α-tocopherol binding protein with M_r of about 30,000 (69) was tested for tocopherol transfer activity (67). The protein was capable of transferring α-tocopherol from egg PC liposomes to liver microsomes. Such intermembrane transfer activities were also found in heart and in brain, but not in the lungs (70). In extracts from normal liver tissue about 60% of α-tocopherol was found in fractions also containing the 30,000 protein stimulating the transfer of α-tocopherol between membranes (71).

Although protein-induced intermembrane transfer was originally found inside the cells, it could also be demonstrated in vitro to occur between liposomes and plasma membranes (Scheme 1). The α-tocopherol transfer activity of the liver cytosolic α-tocopherol binding protein was tested in vitro from egg lecithin liposomes to human erythrocyte ghosts and to liver mitochondria. The transfer activity was inhibited by α-tocopherol acetate by only 5%, indicating the chromanol hydroxyl to be necessary for transfer (72,73). On the other hand, the phytyl side chain enhances the retainment of tocopherol in liposomes and suppresses its transfer between liposomal membranes (74).

The intermembrane transfer of α-tocopherol was also investigated from unilamellar liposomes of saturated or unsaturated phospholipids (donor liposomes) to unilamellar liposomes of rat cerebral cortex lipids (acceptor liposomes). α-Tocopherol was able to inhibit the accumulation of lipid peroxidation (LPO) products by LPO inducers in the acceptor liposomes as a result of its intermembrane transfer. The intermembrane antioxidant action of α-tocopherol was increased by preincubation of acceptor with donor liposomes for 60 minutes, and this increase was more pronounced when the donor liposomes contained unsaturated phospholipids. The intermembrane transfer of α-tocopherol did not result from the fusion of donor and acceptor liposomes during preincubation (33).

SCHEME 1 Transport and intermembrane transfer of α-tocopherol.

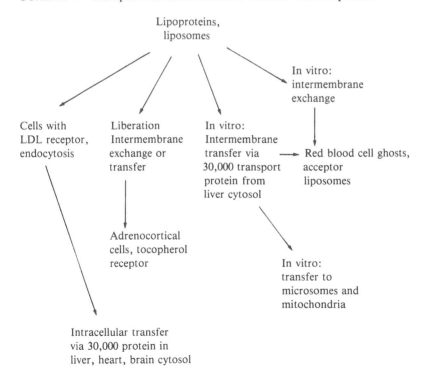

REFERENCES

1. New RRC, Chance ML, Thomas SC, Peters W. Antileishmanial activity of antimonials entrapped in liposomes. Nature 1978; 272:55.
2. Alving CA, Steck EA, Chapman WL, et al. Therapy of leishmaniasis: superior efficacies of liposome-encapsulated drugs. Proc Natl Acad Sci U S A 1978; 75:2959.
3. Forssen EA, Tokes ZA. In-vitro and in-vivo studies with adriamycin liposomes. Biochem Biophys Res Commun 1979; 91:1295-301.
4. Gabizon A, Meshorer A, Barenholz Y. Comparative long-term study of the toxicities of free and liposome-associated doxorubicin in mice after intravenous administration. J Natl Cancer Inst 1986; 77:459-69.
5. Fidler IJ, Raz A, Fogler WE, Kirsch R, Bugelski P, Poste G. Design of liposomes to improve delivery of macrophage-activating agents to alveolar macrophages. Cancer Res 1980; 40:4460.
6. Daemen T, Veninga A, Roerdink FH, Scherphof GL. In vitro activation of rat liver macrophages to tumoricidal activity by free or liposome-encapsulated muramyl dipeptide. Cancer Res 1986; 46: 4330-5.
7. Fukuzawa K, Tokumura A, Setsuhiro O, Tsukatani H. Antioxidant activities of tocopherols on Fe^{2+}-ascorbate-induced lipid peroxidation in lecithin liposomes. Lipids 1982; 17:511-3.
8. Bangham AD, Standish MM, Watkins JC. Diffusion of univalent ions across the lamellae of swollen phospholipids. J Mol Biol 165; 13:238.
9. Halks-Miller M, Guo LSS, Hamilton RL. Tocopherol-phospholipid liposomes: maximum content and stability to serum proteins. Lipids 1985; 20:195-200.
10. Bellemare F, Fragata M. Transmembrane distribution of α-tocopherol in single-lamellar mixed lipid vesicles. J Membr Biol 1981; 58:67-74.
11. Lai MZ, Düzgünes N, Szoka FC. Effects of replacement of the hydroxyl group of cholesterol and tocopherol on the thermotropic behavior of phospholipid membranes. Biochemistry 1985; 24:1646-53.
12. Janoff AS, Kurtz CL, Jablonski RL, et al. Characterization of cholesterol hemisuccinate and α-tocopherol hemisuccinate vesicles. Biochim Biophys Acta 1988; 941:165-75.
13. Burton GW, Ingold KU. Vitamin E: application and principles of physical organic chemistry to the exploration of its structure and function. Acc Chem Res 1986; 19:194-201.
14. Ekiel IH, Hughes L, Burton GW, Jovall OA, Ingold KU, Smith ICP. Structure and dynamics of α-tocopherol in model membranes and in solution: a broad-line and high-resolution NMR study. Biochemistry 1988; 27:1432-40.
15. Seufert WD, Beauchesme G, Bélanger M. Thin α-tocopherol films separating two aqueous phases. Biochim Biophys Acta 1970; 211:356-8.
16. Villalain J, Aranda FJ, Gòmez-Fernández JC. Calorimetric and infrared spectroscopic studies of the interaction of α-tocopherol and α-tocopheryl acetate with phospholipid vesicles. Eur J Biochem 1986; 158:141-7.
17. Schmidt D, Steffen H, von Planta C. Lateral diffusion, order parameter and phase transition in phospholipid bilayer membranes containing tocopherolacetate. Biochim Biophys Acta 1976; 443:1-9.
18. Maggio B, Diplock AT, Lucy JA. Interactions of tocopherols with monolayers of phospholipids. Biochem J 1977; 161:111-21.
19. Diplock AT, Lucy JA. The biochemical modes of action of vitamin E and selenium: A hypothesis. FEBS Lett 1973; 29:205-10.
20. Rosenberg A. *Euglena gracilis*: a novel lipid energy reserve and arachidonic acid enrichment during fasting. Science 1967; 157:1189-91.
21. Urano S, Iida M, Otani I, Matsuo M. Membrane stabilization of vitamin E; interactions of alpha-tocopherol with phospholipids in bilayer liposomes. Biochem Biophys Res Commun 1987; 146:1413-8.
22. Srivastava S, Phadke RS, Govil G, Rao CNR. Fluidity, permeability and antioxidant behaviour of model membranes incorporated with α-tocopherol and vitamin E acetate. Biochim Biophys Acta 1983; 734:353-62.
23. Gòmez-Fernández JC, Aranda FJ, Villalain J, Ortiz A. The interaction of coenzyme Q and vitamin E with multibilayer liposomes. Adv Exp Med Biol 1988; 238:127-39.
24. Gorbunov NV, Skrypin VI, Tyurin VA, Balevska P, Serbinova E, Kagan VE. Regulating and disturbing effects of alpha-tocopherol on lipid bilayers. Biol Nauki (Moscow) 1987; 52:27-32.

25. Takahashi M, Tsuchiya J, Niki E, Urano S. Action of vitamin E as antioxidant in phospholipid liposomal membranes as studied by spin label technique. J Nutr Sci Vitaminol (Tokyo) 1988; 34:25-34.
26. Axel F, Seelig J. Cis double bonds in liquid crystalline bilayers. J Am Chem Soc 1973; 95:7972-7.
27. Ohki K, Takamura T, Nozawa Y. Effect of α-tocopherol on lipid peroxidation and acyl chain mobility of liver microsomes from vitamin E-difficient rat. J Nutr Sci Vitaminol 1984; 30:221-34.
28. Shinitzky M, Barenholz Y. Fluidity parameters of lipid regions determined by fluorescence polarization. Biochim Biophys Acta 1978; 515:367-94.
29. Moran J, Salazar P, Pasantes-Morales H. Effect of tocopherol and taurine on membrane fluidity of retinal rod outer segments. Exp Eye Res 1987; 45:769-76.
30. Bisby RH, Ahmed S. Transverse distribution of alpha-tocopherol in bilayer membranes studied by fluorescence quenching. Free Radic Biol Med 1989; 6:231-9.
31. Tyurin VA, Kagan VE, Avrova NF, Prozorovskaia MP. Lipid asymmetry and alpha-tocopherol distribution in outer and inner monolayers of bilayer lipid membranes. Byull Eksp Biol Med (USSR) 1988; 105:667-9.
32. Tyurin VA, Korol'kov SN, Kagan VE. Transbilayer distribution of alpha-tocopherol and lipid asymmetry in nerve tissue membranes. Biokhimiia (USSR) 1989; 54:940-7.
33. Rangelova D, Zhelev Z, Bakalova RA, et al. Intermembrane transfer and antioxidant effects of α-tocopherol in liposomes. Byull Eksp Biol Med (USSR) 1990; 109:37-9.
34. Ortiz A, Aranda FJ, Gomez-Fernandez JC. A differential scanning calorimetry study of the interaction of alpha-tocopherol with mixtures of phospholipids. Biochim Biophys Acta 1987; 898:214-22.
35. Semin BK, Bautina AL, Ivanov II. The effect of membrane-tropic compounds of different structures on characteristics of phase transition of dimyristoylphosphatidylcholine. Biol Nauki (Moscow) 1989; 5:32-6.
36. Severcan F, Cannistraro S. Direct electron spin resonance evidence for alpha-tocopherol-induced phase separation in model membranes. Chem Phys Lipids 1988; 47:129-33.
37. Severcan F, Cannistraro S. Model membrane partition ESR study in the presence of alpha-tocopherol by a new spin probe. Biosci Rep 1989; 9:489-95.
38. Erin AN, Gorbunov NV, Skrypin VI, Kagan VE, Prilipko LL. Interaction of alpha-tocopherol with free fatty acids. Mechanism of stabilization of lipid bilayer microviscosity. Biol Nauki (USSR) 1987; 55:10-6.
39. Urano S, Yano K, Matsuo M. Membrane-stabilizing effect of vitamin E: effect of alpha-tocopherol and its model compounds on fluidity of lecithin liposomes. Biochem Biophys Res Commun 1988; 150:469-75.
40. Darke A, Finer EG, Flock AG, Phillips MC. Nuclear magnetic resonance study of lecithin cholesterol interactions. J Mol Biol 1972; 63:265-79.
41. Makarov TB, Korolev NP, Gangard MG, Gus'kova RA, Ivanov II. Research on high-temperature transition in liposomes made from egg yolk lecithin. Biol Nauki (USSR) 1988; 53:23-31.
42. Gutteridge JMC. The membrane effects of vitamin E, cholesterol and their acetates on peroxidative susceptibility. Res Commun Chem Pathol Pharmacol 1978; 22:563-72.
43. Kagan VE. Tocopherol stabilizes membrane against phospholipase A, free fatty acids, and lysophospholipids. Ann N Y Acad Sci 1989; 570:121-35.
44. Fukuzawa K, Takase S, Tsukatani H. The effect of concentration on the antioxidant effectiveness of α-tocopherol in lipid peroxidation induced by superoxide free radicals. Arch Biochem Biophys 1985; 240:117-20.
45. Margolis LB. Cell interactions with solid and fluid liposomes in vitro: lessons for "liposomologists" and cell biologists. In: Gregoriadis D, ed. Liposomes as drug carriers. John Wiley & Sons, Chichester, 1988; 75-92.
46. Ring K, Henkel B, Valenteijn A, Gutermann R. Studies on the permeability and stability of liposomes derived from a membrane-spanning bipolar archaebacterial tetraetherlipid. In: Schmitt KH, ed. Liposomes as drug carriers. Georg Thieme Verlag, Stuttgart, 1986; 100-23.
47. Thoma K, Schmid A. Influence of composition and manufacturing technique on size and homogeneity of small unilamellar liposomes. 5th Congr Int Technol Pharm (Assoc Pharm Galenique Ind: Chatenay Malabray) 1989; 3:15-24.
48. Konings AWT. Lipid peroxidation in liposomes. In: Gregoriadis G, ed. Liposome technology, Vol. I. Preparation of liposomes. Boca Raton, FL: CRC Press, 1984; 139-61.

49. Motoyama T, Miki M, Mino M, Takahashi M, Niki E. Synergistic inhibition of oxidation in dispersed phosphatidylcholine liposomes by a combination of vitamin E and cysteine. Arch Biochem Biophys 1989; 270:655-61.

50. Kagan VE, Serbinova EA, Packer L. Generation and recycling of radicals from phenolic antioxidants. Arch Biochem Biophys 1990; 280:33-9.

51. Kunimoto M, Inoue K, Nojima A. Effect of ferrous ion and ascorbate induced lipid peroxidation on liposomal membranes. Biochim Biophys Acta 1981; 646:169-78.

52. Fukuzawa K, Chida H, Tokumura A, Tsukatani H. Antioxidative effect of α-tocopherol incorporation into lecithin liposomes on ascorbic acid-Fe^{2+}-induced lipid peroxidation. Arch Biochem Biophys 1981; 206:173-80.

53. Senior J. Fate and behaviour of liposomes in vivo: a review of controlling factors. CRC Crit Rev Ther Drug Carrier Systems 1987; 3:123-93.

54. Hunt CA, Tsang S. α-Tocopherol retards autoxidation and prolongs the shelf life of liposomes. Int J Pharm 1981; 8:101-10.

55. Stillwell W, Bryant L. Membrane permeability changes with vitamin A/vitamin E mixed bilayers. Biochim Biophys Acta 1983; 731:483-6.

56. Fukuzawa K, Hayashi K. Effects of α-tocopherol analogs on lysosome membranes and fatty acid monolayers. Chem Phys Lipids 1977; 18:39-48.

57. Diplock AT, Lucy JA, Verrinder M, Zieleniewski A. α-Tocopherol and the permeability to glucose and chromate of unsaturated liposomes. FEBS Lett 1977; 82:341-4.

58. Fukuzawa K, Ikeno H, Tokumura A, Tsukatani H. Effect of α-tocopherol incorporation on glucose permeability and phase transition of lecithin liposomes. Chem Phys Lipids 1979; 23:13-22.

59. Szebeni J, Winterbourn CC, Carrell RW. Oxidative interactions between hemoglobin and membrane lipid. Biochem J 1984; 220:685-92.

60. Yuasa M, Tani Y, Nishide H, Tsuchida E. Stabilization effect of tocopherol and catalase on the life-time of liposome embedded heme as an oxygen carrier. Biochim Biophys Acta 1987; 900:160-2.

61. Szebeni J, Breuer JH, Szelenyi JG, Bathori G, Lelkes G, Hollan SR. Oxidation and denaturation of hemoglobin encapsulated in liposomes. Biochim Biophys Acta 1984; 798:60-7.

62. Barenholz Y, Gabizon A. Pharmaceutical liposomes containing doxorubicin and ferrioxamine and alpha-tocopherol as stabilizers. U.S. Patent 4797285, 1989.

63. Barenholz Y, Gabizon A. Liposome/anthraquinone drug composition and method. EP 274174 A1, 1988.

64. Blan QA, Grossweiner LI. Singlet oxygen generation by furocoumarins: effect of DNA and liposomes. Photochem Photobiol 1987; 45:177-83.

65. Björnson LK, Kayden HJ, Miller E, Moshell AN. The transport of alpha tocopherol and β-carotene in human blood. J Lipid Res 1976; 17:343-52.

66. Traber MG, Kayden HJ. Vitamin E is delivered to cells via the high affinity receptor for low-density lipoprotein. Am J Clin Nutr 1884; 40:747-51.

67. Lokshina EA, Abishev BK, Sagindykova SE, Aidarkhanov BB. Effect of alpha- and gamma-tocopherols on the cytochrome P-450 system in the liver of rats with E-avitaminosis. Effectiveness of liposomal forms of gamma-tocopherol. Biokhimiia (USSR) 1988; 53:1188-92.

68. Kitabchi AE, Wimalasena J, Barker JA. Specific receptor sites for α-tocopherol in purified isolated adrenocortical cell membrane. Biochem Biophys Res Commun 1980; 96:1739-46.

69. Catignani GL. An α-tocopherol binding protein in rat liver cytoplasm. Biochem Biophys Res Commun 1975; 67:66-72.

70. Murphy DJ, Mavis RD. Membrane transfer of α-tocopherol. J Biol Chem 1981; 256:10464-8.

71. Mowri H, Nojima S, Inoue K. Lack of protein-mediated alpha-tocopherol transfer between membranes in the cytoplasm of ascites hepatomas. Lipids 1988; 23:459-64.

72. Verdon CP, Blumberg JB. An assay for the alpha-tocopherol binding protein mediated transfer of vitamin E between membranes. Anal Biochem 1988; 169:109-20.

73. Verdon CP, Blumberg JB. Influence of dietary vitamin E on the intermembrane transfer of alpha-tocopherol as mediated by an alpha-tocopherol binding protein. Proc Soc Exp Biol Med 1988; 189:52-60.

74. Niki E, Kawakami A, Saito M, Yamamoto Y, Tsuchiya J, Kamiya Y. Effect of phytyl side chain of vitamin E on its antioxidant activity. J Biol Chem 1985; 260:2191-6.

17

Membrane Fluidity and Vitamin E

Guido Zimmer, Thomas Thürich, and Barbara Scheer

University of Frankfurt, Frankfurt, Germany

INTRODUCTION

Vitamin E is one of the best known antioxidants in the lipid core of cellular membranes because of its chain-breaking activity (1-3). The radical-scavenging properties of α-tocopherol add to its capacity in quenching singlet oxygen (4,5) both protecting the membrane from lipid peroxidation (6) and preventing damage of membrane-bound enzymes. Moreover, α-tocopherol has been reported to chelate transition metals, especially free iron, which is involved in the process of initiation and propagation of free-radical oxidation (7,8).

Apart from the antioxidative properties of vitamin E, an additional function may be found in the physical stabilization of lipid membranes as proposed by Diplock and Lucy (9), Erin et al. (10), and Urano and Matsuo (11).

Antioxidant protection may apply for structural stabilization as well as free-radical scavenging. In this respect both vitamin E (α-tocopherol) and cholesterol appear to have structural antioxidant roles in the membrane, whereas α-tocopherol alone has free-radical-scavenging activity (12).

The stabilizing effect of α-tocopherol is believed to be mediated by van der Waals-like interactions between the 4'a- and 8'a-methyl groups on the phytyl side chain (9) and/or the chromanol methyl groups (13,14) and the unsaturated fatty acid residues forming stable complexes within the lipid bilayer. X-ray scattering experiments suggest a distance of about 4.6 Å between the carbon chains in phosphatidylethanolamine bilayers (15). Srivastava et al. (16) assumed a hydrophilic or hydrogen-bonding interaction in the polar head group region accompanied by a reduction in the interchain distance to about 3.6 Å. Thus, the flexible phytyl tail of the tocopherol molecule is placed parallel to the hydrocarbon fatty acid chains, whereas the more hydrophilic chromanol head group is located very close to the membrane surface near the phosphate groups of the lipid molecules (17).

Although reducing the amount of free fatty acids may be an important factor in the prevention of damaging effects, especially under oxidative stress (10), there was some doubt that a specific physicochemical interaction as just described takes place in biological membranes (14). There have recently been various reports concerning the mobility of phospholipid molecules within membrane bilayers and the vitamin E effect on membrane fluidity.

The molecular mechanisms underlying the radical-scavenging process are not known in detail. However, quenching of reactive radicals by α-tocopherol leads to the production of tocopheroxyl radicals, which were restored to α-tocopherol by (membrane-bound) free-radical reductase systems using glutathione (GSH) as an electron source (18-20) or nonenzymatically by ascorbic acid (21–25). However, a diversity of biological reductants may serve to reduce the tocopheroxyl radical (26). The free-radical chain reaction is probably terminated by transferring the phenolic hydrogen atom from the 6-position of tocopherol to the oxygen radical (27).

It has been found (8) that in comparison to the C16 α-tocopherol molecule, homologs with a shorter isoprenoid chain, that is, the C1 molecule, exhibit a higher antioxidative activity in vitro due to the enhanced efficiency of regeneration of chromanoxyl radicals. This effect may be traced to the increased mobility of short-chain homologs, rendering a more efficient interaction with radicals whereas the C16 α-tocopherol mobility is restricted by strongly bonding to phospholipids. On the other hand, the C1 molecule has been determined to increase the permeability of bilayer membranes, thus causing cell damage (8). Moreover, Burton and Ingold (28) and Witting (29) pointed out that in vivo C16 α-tocopherol possesses the highest biopotency.

The distribution of tocopherol within the phospholipid bilayer is still under discussion. Murphy et al. (30) have suggested a "lateral clustering" of tocopherol in the outer and inner monolayer of rat liver microsomes. Tyurin et al. (31) determined higher levels of vitamin E in bilayer domains being rich in polyenoic phospholipids.

Concerning the vitamin E content of natural membranes, the literature offers many investigations. Buttriss and Diplock (32,33) determined the vitamin E concentration in rat liver membrane fractions using a sensitive high-performance liquid chromatography (HPLC) technique. It has been found that tocopherol is located preferentially in the mitochondrial and microsomal fractions, both containing multiple polyunsaturated fatty acid residues, especially the 20:4 arachidonic acid and the 22:6 docosahexanoic acid. Inner as well as outer mitochondrial membrane fractions contain vitamin E (1.07 and 1.38 μg tocopherol per μmol phospholipid, respectively), giving a molar ratio of about one molecule tocopherol to several thousands of fatty acids. A correlation has been confirmed between the content of membrane polyunsaturated fatty acids and the concentration of tocopherol: membrane fractions containing high levels of polyunsaturated fatty acids (PUFA) also enclose high concentrations of vitamin E.

On the other hand, Oliveira et al. (34) found tocopherol to be located in the lipoprotein complexes of the electron transport system of bovine heart mitochondria, particularly in cytochrome oxidase but apparently not in the outer membrane.

The concentration ratio of vitamin E to PUFA investigated by Buttriss and Diplock (32) is in line with the results of Poukka-Evarts and Bieri (35) (1 to about 1200, measured in different organs). Kornbrust and Mavis (36) reported ratios of 1:200 in microsomes prepared from heart and lung tissues, which are highly oxygenated and thus very susceptible to oxidative damage. As published by Fukazawa et al. (37) and Liebler et al. (38), this is in the range of the "tocopherol threshold level" (1:200 PUFA, approximately) for the prevention of iron-induced lipid peroxidation in liposomal membranes.

In normal mitochondrial membranes the tocopherol concentration has been reported to vary about 0.5–0.8 nmol/mg protein (39). It was found that the amount of α-tocopherol in rat liver microsomes is about 0.44 nmol/mg protein (30). This value is similar to the published figure of several thousand moles lipid per mole α-tocopherol (32).

WHAT IS MEMBRANE FLUIDITY?

Fluidity is the quality of ease of movement, whereas viscosity is the quality of resistance to movement in a fluid. As before, fluidity is defined as $1/\text{viscosity} = 1/n$ (40–42).

A membrane, however, differs from an ordinary isotropic fluid, such as an oil, for which the motional freedom of particles dispersed in the fluid can be described with the preceding equation. A membrane instead is an anisotropic two-dimensional fluid exhibiting lateral and rotational mobilities of proteins and lipids that cannot be adequately described by the same single parameter. Therefore, the question with regard to each method is what parameter it *mainly* measures.

It is important to note that the term "membrane fluidity" should not be reserved for the lipid site (43). For instance, one need only consider protein aggregation or disaggregation phenomena, which certainly occur in membranes. In such cases, proteins as well as lipid-protein interactions contribute to membrane fluidity (42,44). However, for lateral lipid movement in the monolayer a frequency of about $10^7–10^9$ s^{-1} has been mentioned (45), whereas the intermembrane flip-flop occurs at a much slower rate (46).

The literature on vitamin E provides us with a variety of different data about fluidity changes in the absence or presence of vitamin E. These have, however, (1) diverse cellular and membranous systems, (2) very different methods and experimental procedures, and (3) different concentration ranges of vitamin E. Fluidity decreases as well as increases have been reported. Since there is a fluidity gradient perpendicular to the plane of the membrane with relatively low fluidities at the polar interface near the phospholipid phosphates and increasing fluidity toward the center of the membrane, we must know where the method of choice registers these measurements.

The dynamics of liposomes as well as their structure are different from those of biological membranes, due to their lack of inherent protein. Liposomes therefore exhibit much higher fluidities compared to cell or intracellular membranes. Furthermore, the fluidity of liposomal bilayers depends on their respective phospholipid composition (47). The higher the concentration of saturated fatty acids and/or cholesterol incorporated into phosphatidylcholine (PC) liposomes, the more lipid fluidity is decreased. On the other hand, increased lipid mobility has been obtained using short-chain lipid molecules (48).

Intracellular membranes, like those of mitochondria, are difficult to compare with liposomes as a result of their low content of lipid (about 30%). Mitochondrial membranes are nevertheless much more fluid than plasma membranes (of red cells, for example), which are composed of about 50% lipid material and 50% protein. Thus the "milieu" in which we observe the action of vitamin E may be very important for the results of our measurements.

Membrane fluidity is thus a parameter specific for each type of membrane and important for its specific enzymatic equipment. Thus, an increase in fluidity does not necessary always imply an increase in enzymatic activities, and decreases in fluidity are not always accompanied by decreased enzymatic activities (see, for example, Kates et al. (49).

Whether the fluidity of membranes is changed during physiological activity is not yet known; in any case, an optimum fluidity exists for physiological functions. A different situation is obtained under pathological conditions. In oxidative stress (hypoxia and reoxygenation), membrane fluidity was found to be decreased (50), which could be due to the higher oxidation rate of the lipid double bonds and amino acid side chains, among other causes. This could increase the number of hydrophobic membrane interactions. Hibernating animals, microorganisms, and plants can change membrane fluidities according to environmental changes (51). If physiological functions are accompanied by small variations in pH within the physiological range (7.0–7.4), for example, such functions are invariably accompanied by membrane fluidity changes (52).

METHODS FOR INVESTIGATION OF MEMBRANE FLUIDITY

Electron Paramagnetic Resonance (EPR) Spectroscopy

Paramagnetic reporter groups are useful to elucidate the topology and dynamic state of membrane components. Probe molecules may be bound noncovalently and intercalate into the membrane (like spin-labeled fatty acids or phospholipids), or they may be bound covalently (for example, to protein-reactive groups). EPR spectroscopy of spin-labeled membranes yields information on the mobility and polarity of the environment of the reporter group. The types of molecular motion that are of interest are normally between rapid, isotropic motion and severely hindered, anisotropic motion. These are only a few basic concepts of EPR spectroscopy. The various aspects of this methodology are treated by experts in reviews (53) and books (54–56).

What information can be drawn from spin labeling? We can make use of the following:

1. Shape of the signal, indicating mobility of the label.
2. If rather mobile, h_0/h_{-1} can be taken as a semiquantitative estimation of mobility of the label within its environment; if the spectra are isotropic information about the rotational correlation time of the label can be obtained by additional measurements of W_0.
3. Measurements of (h_{-1H}/h_{-1P}) can be used if the third line partitions into more polar and more apolar portions. Different populations of spin label molecules exist under these conditions of measurement. A number of such molecules exist in the polar (mobile) portion of the system, a separate number of molecules in the hydrophobic (immobilized) portion of the system. Exchange between these spin label populations is slow compared to the EPR time scale.
4. Measurements of the coupling constant a_N. This parameter yields reliable information about polarity in the environment of the spin label molecules.
5. Measurements of the order parameter S, when $\overline{A}_{||}$ and $\overline{A}_{\perp}$ can be clearly defined. Order parameters given an indication of the degree of order within the system investigated; this may change from 0 (completely random, no order) to 1 (completely immobilized). If $\overline{A}_{||}$ cannot be clearly defined, sometimes estimation of $\overline{A}_{\perp}$ may suffice. Since order parameters essentially image differences between $\overline{A}_{||}$ and $\overline{A}_{\perp}$, which means with increasing order $\overline{A}_{||}$ becomes progressively greater and $\overline{A}_{\perp}$ progressively smaller, one can also draw some conclusions from determination of $\overline{A}_{\perp}$.
6. Measurements of the rate of signal reduction: most convenient is h_0, since maximal differences versus time should generally occur with this line. The rate of reduction yields information on differences in localization of the spin label molecule within the system. It is naturally dependent on the availability of spin label-destroying (reducing or oxidizing) molecules.

Fluorescence Spectroscopy

Certain molecules (fluorophores) emit light with a longer wavelength than that used for excitation. Incident photons can excite a fluorophore from its electronic ground state to a vibrational level of the electronic states S_1 or S_2.

Vibrational energy is lost thermally after excitation within picoseconds, and the molecule drops to the ground vibrational state of the excited electron state. The molecule then returns to one of the levels of S_0 after a short period (of the order of nanoseconds) and emits fluorescent light.

When fluorophores are intercalated into a membrane and become illuminated by polarized light, they emit polarized fluorescence. Such molecules, which wobble rapidly, emit completely depolarized fluorescence; in contrast, motionally hindered molecules emit fluorescence that is strongly polarized. Thus, the degree of polarization reflects the state of mobility of the fluorophore as a result of its more or less fluid environment. The cone angle of rotation is larger in less ordered and smaller in ordered membranes. Fluorescence intensity is measured in the probe excited by polarized light at vertical and horizontal polarization directions, $P = (I_{\parallel} - I_{\perp})/(I_{\parallel} + I_{\perp})$.

A valuable probe for detection of fluorescence polarization is diphenyl hexatriene, a rod-shaped molecule. The wobbling behavior of these molecules is actually measured since only those aligned vertically to the vertically incoming laser beam become excited. The more reorientation of the molecules prevails, the smaller is the measured value of P, the less the molecules "wobble" around their axis, and the higher is the measured value of P. Generally, P values between 0.2 and 0.4 are obtained in membranes. For reviews on this topic see Shinitzky et al. (57,58).

EFFECTS OF VITAMIN E ON THE FLUIDITY OF MEMBRANES

As a target in investigations with vitamin E, different laboratories made use of (1) phospholipid bilayers (liposomes), (2) red cell or other plasma membranes, or (3) intracellular (mitochondrial or microsomal) membranes.

The fluidity of these various types of membranes is naturally very different. In general, the mobility of the α-tocopherol molecule itself is very important for its radical-scavenging efficiency. Because the ratio of tocopherol to polyunsaturated fatty acids in natural membranes is about 1:1000, tocopherol molecules must move rapidly within the membrane to contact peroxyl radicals (17,59). Vitamin E was thought to be mobile within the lateral and perpendicular (chromanoxylphytyl "tail" interactions) directions (17). In phospholipid vesicles a lateral diffusion coefficient of 4.8×10^{-6} cm^2 s^{-1} was determined by Aranda et al. (60).

The phytyl side chain is commonly believed to reduce the intermembrane flip-flop of α-tocopherol (31,61); the lateral transfer inside the lipid monolayer is not essentially suppressed (27).

Kagan et al. (62) mentioned a specific tocopherol binding protein possibly mediating lateral movement. During lipid peroxidation, when tocopherol is used up in certain regions, a migration process from membrane parts containing much tocopherol to vitamin E-deficient regions has been assumed (8) and was also verified in rat lung tissue (63,64). In erythrocytes the natural form of vitamin E (*RRR*-α-tocopherol) is bound more tightly to membranes than *SRR*-α-tocopherol (65).

Liposomes

It has been reported that fluidity is increased or, in contrast, decreased in liposomal studies. How can we account for these apparently contradictory results?

It is well known that the fatty acid composition of lipid bilayers is one of the major parameters determining liposomal fluidity. The use of saturated fatty acids causes a more compact lipid ordering within the bilayer, thus decreasing the fluidity, whereas the double bonds of unsaturated fatty acid residues produce a "space" inside the membrane and the fluidity increases. α-Tocopherol added to dipalmitoylphosphatidylcholine (DPPC) liposomes fluidized by arachidonic acid was found to act as a "space filler," thereby producing a more compact molecular configuration (14). Accordingly, Gorbunov et al. (47) using a ^{1}H-NMR (nuclear magnetic resonance) technique, found that α-tocopherol increased the microviscosity only of unsaturated lipid bilayers.

α-Tocopherol has been observed to lower the transition temperature in PC liposomes and, furthermore, broadened the transition temperature range (66,67). This is in line with the suggestion that the addition of vitamin E to lipid layers increases order or decreases fluidity above and decreases order below the respective transition temperature, as determined by EPR studies (68,69). Similarly, diphenylhexatriene (DPH) fluorescence measurements in

DPPC liposomes revealed an increase of order parameters at concentrations of 10–40 mol per 100 mol α-tocopherol (which is 1:100–1:25 mol/mol phospholipid or 10–40 μmol/mmol phospholipid) above the phase transition (70). Such characteristics have been described for cholesterol action on fluidity in lipid bilayers (67,71).

The concentrations of α-tocopherol used in these experiments are high above the physiological level of α-tocopherol in tissue in which less than 1 nmol/mg protein was found (30, 39). This is consistent with the finding that physiological concentrations in the tissues are about several thousands moles of lipid to 1 mol α-tocopherol (32).

Porcine Intestinal Brush-Border Membranes

A fluoresence polarization study using pyrene was reported by Ohyashiki et al. (72). After induction of lipid peroxidation with ascorbic acid and iron, the mobility of the dye molecules was found to be reduced as a result of additions of α-tocopherol in increasing amounts (10–50 μM). The lowest concentration of α-tocopherol amounts to a value of 60 nmol/mg protein. Inhibitory effects on lipid peroxidation of the membranes were found, however, even at 5 nmol/mg protein.

Red Cell Membranes

It should be noted that oral administration of α-tocopherol nicotinate to healthy humans caused an increase in the red cell membrane fluidity (73). Heavy loading of red cells with α-tocopherol, in contrast, resulted in a decrease in fluidity in membranes isolated from these cells when measured with DPH fluorescence polarization. The wobbling angle was decreased from 32° to 28° (74).

Microsomes

The effect of α-tocopherol application to vitamin E-deficient lung micosomal lipids was investigated by Patel and Edwards (75). The membrane fluidity was decreased at concentrations severalfold above physiological.

During lipid peroxidation induced by the addition of NADPH to microsomes, Ohki et al. (76) found the content of arachidonic and docosahexaenoic acid decreased and the membrane order parameter, measured with 5-doxyl stearic acid, markedly increased. α-Tocopherol (0.36 mM) given before lipid oxidation almost prevented a decrease in the content of polyunsaturated fatty acids. Also, an increase in order parameters was essentially suppressed at low concentrations of α-tocopherol (0.2 M fraction). At higher molar fractions fluidity decreased.

Mitochondrial Membranes

In using mitochondria for membrane investigations one must consider the high susceptibility to oxidative damage in the inner membrane due to electron transport processes and the high concentration of unsaturated fatty acids.

In our own investigations α-tocopherol was exclusively used at very low concentrations, less than 1 nmol/mg protein, which is about the physiological range of detected vitamin E concentrations in mitochondria.

DL-α-tocopherol was used as an antioxidant against (1) hypoxia by flushing the medium with nitrogen (N_2) and subsequent reoxygenation and (2) hydrogen peroxide-induced damage of suspended mitochondria.

Fluidity measurements were carried out with EPR spectroscopy and the depolarization method, using diphenylhexatriene fluorescence. In addition, we measured the inner mito-

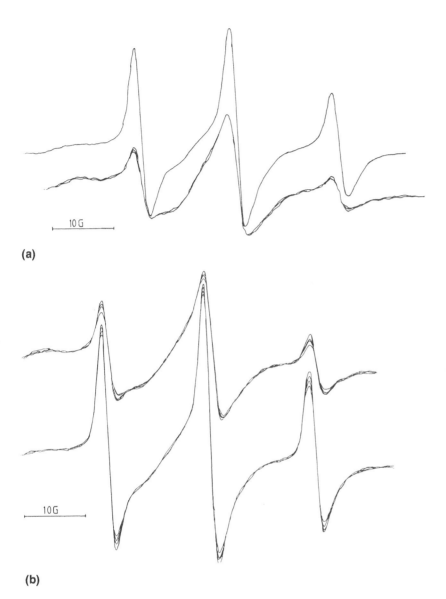

FIGURE 1 (a) Hydrogen peroxide-induced damage on mitochondrial −SH groups. Rat heart mitochondria (1 mg protein) were suspended in 1 ml reaction medium (see Table 1) containing, however, 7.5 mM sodium succinate and 60 μM ADP. Mitochondria were centrifuged for 2.5 minutes in an Eppendorf 3200 centrifuge 30 s after addition of hydrogen peroxide (1 mM). The pellets were resuspended with 50 μl reaction medium, and 1 μl 4-maleimido-TEMPO (100 μM final concentration) was added. Immediately thereafter, spectra were taken. Controls were treated similarly without addition of hydrogen peroxide (upper trace); with hydrogen peroxide (lower trace). (b) Preservation of −SH group mobility by DL-α-tocopherol. Incubation conditions and spin labeling as described in a. Upper trace, mitochondria pretreated with 0.8 nmol DL-α-tocopherol per mg protein 1 minute before addition of 1 mM hydrogen peroxide. Lower trace, DL-α-tocopherol was 1.2 nmol/mg protein.

chondrial membrane potential, with dimethylaminostyrylethylpyridinium iodide (DASPEI) giving information about the viability of the cristae membrane (77).

EPR Spin Label Studies

We were interested in investigations on the fluidity of particular regions of the membrane, which were done by EPR spectroscopy. Three types of spin labels were used:

1. 4-Maleimido-TEMPO, a conventional spin label for −SH group determination
2. 5-Proxylnonane (5-P-9), a small, not very polar (lipid) spin label (78)
3. Bis-(2,2,5,5-tetramethyl-3-imidazoline-1-oxyl-4-il)disulfide, a biradical spin label (79)

Using the 4-maleimido-TEMPO spin label in hydrogen peroxide-treated mitochondria, we found a fairly immobilized signal (Fig. 1a). After addition of α-tocopherol mitochondria behaved decisively different: The signal obtained became much more fluid (Fig. 1b) and approached the control spectra.

The increase in mobility of the labeled −SH regions observed with 4-maleimido-TEMPO after tocopherol application and nitrogen exposure (Fig. 2) corresponds to an increase in thiol reactivity detected by the biradical spin label (not shown). Increases in mitochondrial thiol reactivity are connected with an increased membrane fluidity and versus changes in pH (52).

With spin label 5-P-9 we also found an increase in the polarity of the surroundings of the label, which was apparent after adding α-tocopherol to mitochondria subsequently flushed with nitrogen (Fig. 3).

Thus, we believe that ordering effects (increase in order and decrease in membrane fluidity) that were ascribed to the physiological activities of α-tocopherol are mainly due to the

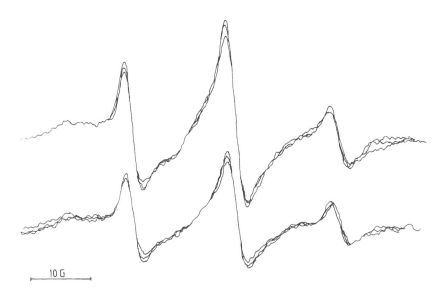

10 G

FIGURE 2   Mobility of −SH groups in mitochondria under the influence of nitrogen. Rat heart mitochondria were suspended in reaction medium as described in Figure 1a; ADP was omitted. Hypoxia was induced by incubation of mitochondria with pure nitrogen added to the medium under pH control for 3 minutes, and thereafter centrifugation, resuspension of the mitochondria, and spin labeling with 4-maleimido-TEMPO as described in Figure 1a. Upper trace, control without tocopherol. Lower trace, DL-α-tocopherol 0.625 nmol/mg protein added 1 minute before nitrogen.

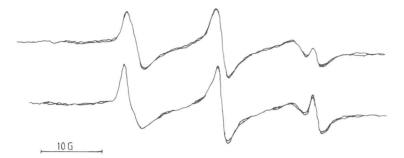

FIGURE 3 Polarity changes in mitochondrial membranes under the influence of nitrogen. For reaction medium and experimental conditions see Figure 1a. Upper trace, control (without tocopherol). Lower trace, DL-α-tocopherol, 0.625 nmol/mg protein. Mitochondria were labeled with 5-P-9 (200 μM final concentration).

nonphysiological concentrations used experimentally. Accordingly, a report of Kerimov et al. (80) suggests low concentrations (from 10 to 1 μM) to increase and high concentrations 1–10 mM) to decrease fluidity of mitochondrial lipid bilayers.

Nevertheless, a structural efficiency similar to that of cholesterol may be valid for plasma membranes. This, however, cannot hold for mitochondrial membranes, which are highly fluid in the near absence of cholesterol and a comparatively low lipid contant (about 30%). Thus, for considerations about membrane fluidity protein sites should not be neglected.

DPH Fluorescence

In hypoxically treated rat heart mitochondria the addition of α-tocopherol reveals a significant increase in membrane fluidity compared to the much less fluid damaged control (Fig. 4).

Mitochondrial Membrane Potential

Inner mitochondrial membrane potential was investigated by using the fluorochrome dimethylaminostyrylethylpyridinium iodide. After damage by nitrogen addition, the membrane potential was significant higher in α-tocopherol-treated mitochondria. Also, after addition of ADP phosphorylation in isolated mitochondria was improved (not shown). A time-dependent decline in the total membrane potential was found by exposure of mitochondria to hydrogen peroxide (Table 1). Pretreatment with α-tocopherol (≥0.8 nmol/mg protein) was found to counteract this decline, indicating a membrane-stabilizing effect that could also be mediated by radical scavenging or chelating metal ions. Scott et al. (81) observed similar effects on membrane potential with α-tocopherol supplementation in cardiac cells.

INTERACTION WITH PROTEINS

Destruction of −SH groups in amino acids as well as in proteins by peroxides of unsaturated fatty acids was found early by Lewis and Wills (82). This pioneering study also indicated that "the originally widely held view that all unsaturated fatty acids are adequately protected by vitamin E against oxidation under all conditions" is not substantiated by the work of Green et al. (83).

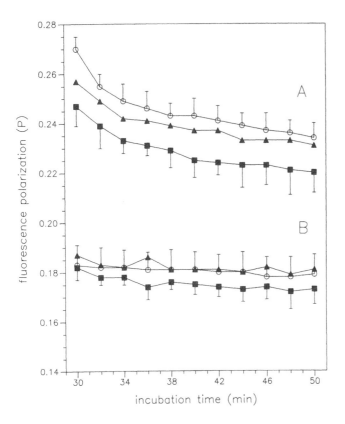

FIGURE 4   Fluorescence polarization (P) during reoxygenation of mitochondria. Rat heart mitochondria were isolated according to Mela and Seitz (108), omitting Nagarse treatment. For hypoxic incubation, 2 ml buffer containing 175 mM KCl, 25 mM Tris, and 2 nmol diphenylhexatriene, pH 7.35, was gassed with pure nitrogen for 15 minutes at 25°C (maximal oxygen concentration was 1.725 μl O_2 per ml). Mitochondrial protein (100 μg) was added to the hypoxic medium and incubated for 30 minutes in a water bath at 25°C. After 30 minutes fluorescence polarization (P) was detected at 430 nm (excitation at 360 nm). DL-α-tocopherol was dissolved in ethanol and added to mitochondria 7 minutes before incubation in the hypoxic medium. Values given are mean of five independent measurements. Significances were estimated by Student's t-test. (A) Open circles, reoxygenation phase, no tocopherol; closed triangles, reoxygenation, α-tocopherol, 0.4 nmol/mg protein; closed squares, reoxygenation, α-tocopherol, 0.8 nmol/mg protein. Values are significant: immediately at start of fluorescence measurements $p < 0.001$, decreasing to $p \leqslant 0.05$. (B) Open circles, control, no tocopherol; closed triangles, control, α-tocopherol, 0.4 nmol/mg protein; closed squares, control, α-tocopherol, 0.8 nmol/mg protein.

Oxidation of amino acid side chains to carbonyl groups and other products (84,85) results in accelerated protein degradation (86). Protein fragmentation, however, was found to become retarded by α-tocopherol (87). Repair of amino acids was also found in vitro by Trolox C, a water-soluble antioxidant (88). Similar findings were reported by Hoey and Butler (89). It should be noted that stabilization of membrane-bound enzymes, in turn, stabilizes the lipid bilayer (10).

Previously we pointed out that the loss of mitochondrial protein $-$SH groups results in deletion of overall membrane conformation and metabolic potency (90–97). Recovery of

TABLE 1 Effect of α-Tocopherol on Mitochondrial Membrane Potential (mV) During Incubation with Hydrogen Peroxide[a]

α-T (nmol/mg protein)	Incubation time (minutes)		
	2	8	14
Control, no addition	196 ± 12	106 ± 14	62 ± 15
0.4	198 ± 15	108 ± 17	64 ± 10
0.8	202 ± 11	117 ± 10	79 ± 11[b]
1.2	202 ± 14	125 ± 11[b]	88 ± 13[b]

[a]α-T, DL-α-tocopherol. Rat heart mitochondria, 0.28 mg protein (preparation as described in Fig. 4), were incubated at 25 °C in a reaction medium containing 175 mM sucrose, 90 mM mannose, 7.0 mM K_2HPO_4, 8.5 mM TES, and 30 μM ADP with further addition of DL-α-tocopherol (dissolved in dimethylsulfoxide, DMSO, concentration as indicated) and 4.5 mM sodium succinate for respiration. Nominal osmolarity was 290 mosM, pH 7.0. Fluorescence intensity was recorded after addition of 0.42 μM 2-(p-dimethylaminostyryl)-1-ethylpyridinium iodide; excitation at 471 nm, emission at 558 nm. Hydrogen peroxide (8.5 mM) was supplied 2 minutes after starting the experiment. The membrane potential was calculated according to the Nernst equation, $\Delta E = RT/nF \ln [K^+]_o/[K^+]_i$, where $[K^+]_i$ and $[K^+]_o$ are the inner and outer concentrations of potassium, R the gas constant, T, the absolute temperature, n the charge of the cation, and F the Faraday constant. For determination of ΔE, mitochondria were exposed to varying external concentrations of potassium in the presence of valinomycin; $[K^+]_i$ was assumed to be 120 mM. Data represent mV mean $\pm$ SD (standard deviation) of six measurements; values are significant as indicated[b]. $p < 0.05$ (Student's t-test).

mitochondrial protein thiols by antioxidants is a most important goal in therapeutic efforts (98-105). This is the more so because mitochondria are probably more susceptible to thiol oxidation than the rest of the cell (106). Vitamin E protects protein thiols in cells (hepatocytes) against chemical-induced toxicity (107). It may well be that the tocopherol-dependent maintenance of cellular or mitochondrial thiol mobility and reactivity is mediated via decreased lipid peroxidation (82).

CONCLUSION

Membrane fluidity is increased at low concentrations of vitamin E. This was found to be the case for most studies using plasma or intracellular membranes irrespective of the experimental procedures. This does not necessarily hold for liposomal studies, however. Liposomes are very fluid per se or were additionally fluidized by arachidonic acid.

High concentrations of vitamin E, which can be reached by in vitro loading, frequently result in decreased membrane fluidity. Physiologically important are those concentrations that can be attained by in vivo studies. In general, concentrations $\leqslant 1$ nmol/mg protein have been estimated in different membranes.

In our studies mitochondrial −SH groups damaged by oxidative stress are clearly mobilized and increased in reactivity after addition of about 1 nmol/mg protein of α-tocopherol. Under these conditions membrane fluidity measured at hydrophobic membrane sites as well as the mitochondrial membrane potential become increased.

ACKNOWLEDGMENTS

We thank Drs. Hans-Joachim Freisleben and Jürgen Fuchs for critical reading of the manuscript and for their comments.

REFERENCES

1. Tappel AL. Vitamin E as the biological lipid antioxidant. Vitam Horm 1962; 20:493-510.
2. Burton W, Cheeseman KH, Doba T, Ingold KU, Slater F. Vitamin E as an antioxidant in vitro and in vivo. In: Biology of vitamin E. Ciba foundation symposium 101. London: Pitman, 1983; 4-16.
3. Niki E. Antioxidants in relation to lipid peroxidation. Chem Phys Lipids 1987; 44:227-53.
4. Grams YW, Eskins K. Dye-sensitive photoperoxidation of tocopherols. Correlation between singlet oxygen reactivity and vitamin E activity. Biochemistry 1972; 11:606-44.
5. Fahrenholtz SR, Doleiden FH, Trozzolo AM, Lamola AA. On the quenching of singlet oxygen of alpha-tocopherol. Photochem Photobiol 1974; 20:505-9.
6. Wu GS, Stein RA, Mead JF. Autoxidation of phosphatidylcholine liposomes. Lipids 1982; 17: 403-13.
7. Stoyanovsky DA, Kagan VE, Packer L. Iron building to alpha-tocopherol-containing phospholipid liposomes. Biochem Biophys Res Commun 1989; 160:834-8.
8. Kagan V, Packer L, Serbinova E, et al. Mechanism of vitamin E control of lipid peroxidation: regeneration, synergism, asymmetry, migration and metal chelation. In: Reddy C, ed. Biological oxidation systems. New York: Academic Press, 1990; 1-20.
9. Diplock AT, Lucy JA. The biochemical modes of action of vitamin E and selenium: a hypothesis. FEBS Lett 1973; 29:205-10.
10. Erin AN, Spirin MM, Tabidze LV, Kagan VE. Formation of α-tocopherol complexes with fatty acids. A hypothetical mechanism of stabilization of biomembranes by vitamin E. Biochim Biophys Acta 1984; 774:96-102.
11. Urano S, Matsuo M. Membrane-stabilizing effect of vitamin E. Ann N Y Acad Sci 1989; 570:524-6.
12. Gutteridge JMC. The membrane effects of vitamin E, cholesterol and their acetates on peroxidative susceptibility. Res Commun Chem Pathol Pharmacol 1978; 22:563-72.
13. Urano S, Iida M, Otani I, Matsuo M. Membrane stabilization of vitamin E; interaction of α-tocopherol with phospholipids in bilayer liposomes. Biochem Biophys Res Commun 1987; 146:1413-8.
14. Urano S, Yano K, Matsuo M. Membrane-stabilizing effect of vitamin E: effect of α-tocopherol and its model compounds on fluidity of lecithin liposomes. Biochem Biophys Res Commun 1988; 150:469-75.
15. Hitchcock PB, Mason R, Thomas KM, Shipley GG. Structural chemistry of 1,2 dilauroyl-DL-phosphatidylethanolamine: molecular conformation and intermolecular packing of phospholipids. Proc Natl Acad Sci U S A 1974; 71:3036-40.
16. Srivastava S, Phadke RS, Govil G, Rao CNR. Fluidity, permeability and antioxidant behaviour of model membranes incorporated with α-topopherol and vitamine E acetate. Biochim Biophys Acta 1983; 734:353-62.
17. Burton GW, Ingold KU. Vitamin E as an in vitro and in vivo antioxidant. Ann N Y Acad Sci 1989; 570:7-22.
18. Hill KE, Burk RF. Influence of vitamin E and selenium on glutathione-dependent protection against microsomal lipid peroxidation. Biochem Pharmacol 1984; 33:1065-8.
19. McCay PB, Lai EK, Brueggemann G, Powell SR. A biological antioxidant function for vitamin E: electron shuttling for a membrane-bound "free radical reductase." Nato ASI Ser 1986; 131:145-56.
20. Scholz RW, Graham KS, Gumpricht E, Reddy CC. Mechanism of interaction of vitamin E and glutathione in the protection against membrane lipid peroxidation. Ann N Y Acad Sci 1989; 570: 514-7.
21. Tappel AL. Will antioxidant nutrients slow aging processes? Geriatrics 1968; 23:97-105.
22. Packer JE, Slater TF, Willson RL. Direct observation of a free radical interaction between vitamin E and vitamin C. Nature 1979; 278:737-8.

23. Bascetta E, Gunstone FD, Walton JC. Electron spin resonance study of the role of vitamin E and vitamin C in the inhibition of fatty acid oxidation in a model membrane. Chem Phys Lipids 1983; 33:207-10.

24. Doba T, Burton GW, Ingold KU. Antioxidant and co-antioxidant activity of vitamin C. The effect of vitamin C, either alone or in the presence of vitamin E or a water-soluble vitamin E analogue, upon the peroxidation of aqueous multilamellar phospholipid liposomes. Biochim Biophys Acta 1985; 835:298-303.

25. Packer L, Maguire JJ, Mehlhorn RJ, Serbinova E, Kagan VE. Mitochondria and microsomal membranes have a free radical reductase activity that prevents chromanoxyl radical accumulation. Biochem Biophys Res Commun 1989; 159:229-35.

26. Mehlhorn RJ, Sumida S, Packer L. Tocopheroxyl radical persistance and tocopherol consumption in liposomes and in vitamin E-enriched rat liver mitochondria and microsomes. J Biol Chem 1989; 264:13448-52.

27. Niki E, Yamamoto Y, Takahashi M, Komuro E, Miyama Y. Inhibition of oxidation of biomembranes by tocopherol. Ann N Y Acad Sci 1989; 570:23-31.

28. Burton GW, Ingold KU. Autoxidation of biological molecules. 1. The antioxidant activity of vitamin E and related chain-breaking phenolic antioxidants in vitro. J Am Chem Soc 1981; 103:6472-7.

29. Witting LA. Vitamin E and lipid antioxidants in free-radical initiated reactions. In: Pryor WA, ed. Free radicals in biology, Vol IV. New York: Academic Press, 1980; 295-319.

30. Murphy ME, Scholich H, Wefers H, Sies H. Alpha-tocopherol in microsomal lipid peroxidation. Ann N Y Sci 1989; 570:480-6.

31. Tyurin VA, Kagan VE, Serbinova EA, et al. Interaction of α-tocopherol with phospholipid liposomes: the absence of transbilayer mobility. Biull Eksp Biol Med 1986; 689-92.

32. Buttriss JL, Diplock AT. The relationship between α-tocopherol and phospholipid fatty acids in rat liver subcellular membrane functions. Biochim Biophys Acta 1988; 962:81-90.

33. Buttriss JL, Diplock AT. The α-tocopherol and phospholipid fatty acid content of rat liver subcellular membranes in vitamin E and selenium deficiency. Biochim Biophys Acta 1988; 963:61-9.

34. Oliveira MM, Weglicki WB, Nason A, Nair PP. Distribution of α-tocopherol in beef heart mitochondria. Biochim Biophys Acta 1969; 180:98-113.

35. Poukka-Evarts RP, Bieri JG. Ratios of polyunsaturated fatty acids to alpha-tocopherol in tissues of rats fed corn or soybean oils. Lipids 1974; 9:860-4.

36. Kornbrust DJ, Mavis RD. Relative susceptibility of microsomes from lung, heart, liver, kidney, brain and testes to lipid peroxidation: correlation with vitamin C content. Lipids 1980; 15:315-22.

37. Fukazawa K, Tokumura A, Ouchi S, Tsukatani H. Antioxidant activities of tocopherols on Fe^{2+}-ascorbate-induced lipid peroxidation in lecithin liposomes. Lipids 1982; 17:511-3.

38. Liebler DC, Kling DS, Reed DJ. Antioxidant protection of phospholipid bilayers by α-tocopherol. Control of α-tocopherol status and lipid peroxidation by ascorbic acid and glutathione. J Biol Chem 1986; 261:12114-9.

39. Kagan V, Serbinova E, Packer L. Antioxidant effects of ubiquinones in microsomes and mitochondria are mediated by tocopherol recycling. Biochem Biophys Res Commun 1990; 169:851-7.

40. Hildebrand JH, Lamoreaux RH. Fluidity: a general theory. Proc Natl Acad Sci U S A 1972; 69: 3428-31.

41. Lands WEM. Fluidity of membrane lipids. In: Kates M, Kuksis A, eds. Membrane fluidity. Clifton, N J: Humana Press, 1980; 69-73.

42. Zimmer G, Freisleben H-J. Membrane fluidity determinations from viscosity. In: Aloia RC, Curtain CC, Gordon LM, eds. Advances in membrane fluidity, Vol 1. New York: Liss, 1988; 297-318.

43. Cooper RA. Abnormalities of cell membrane fluidity in the pathogenesis of disease. N Engl J Med 1977; 297:371-7.

44. Zimmer G. Fluidity of cell membranes in the presence of some drugs and inhibitors. In: Kates M, Manson LA, eds. Biomembranes, Vol 12. Membrane fluidity. New York: Plenum Press, 1984; 169-203.

45. Van Ginkel G, Gerritsen HC, Korstanje L, Moormans R, van Zandvoort M, Levine YK. Spectroscopic investigation of vitamin A (retinol, β-carotine) and vitamin E (α-tocopherol). Voeding 1989; 50:179-81.

46. Marsh D. Electron spin resonance: spin labels. In: Grell E, ed. Membrane spectroscopy. New York: Springer, 1981; 51-142.

47. Gorbunov NV, Skrypin VI, Tyurin VA, Balevska P, Serbinova E. Regulating and disturbing effects of alpha-tocopherol on lipid bilayers. Biol Nauki 1987; 52:27-32.

48. Takahashi M, Tsuchiya J, Niki E, Urano S. Action of vitamin E as antioxidant in phospholipid liposomal membranes as studied by spin label technique. J Nutr Sci Vitaminol 1988; 34:25-34.

49. Kates M, Pugh EL, Ferrante G. Regulation of membrane fluidity by lipid desaturases. In: Kates M, Manson LA, eds. Biomembranes, Vol 12. Membrane fluidity. New York: Plenum Press, 1984; 379-95.

50. Victor T, van der Merwe N, Benadi AJS, LaCock C, Lochner A. Mitochondrial phospholipid composition and microviscosity in myocardial ischemia. Biochim Biophys Acta 1985; 834:215-23.

51. Aloia RC. Lipid, fluidity and functional studies of the membranes of hibernating animals. In: Alan R, ed. Advances in membrane fluidity, Vol 3. Physiological regulation of membrane fluidity. New York: Liss, 1988; 1-39.

52. Zimmer G, Freisleben H-J, Fuchs J, Influence of pH on sulfhydryl groups and fluidity of the mitochondrial membrane. Arch Biochem Biophys 1990; 282:309-17.

53. Keith AD, Sharnoff M, Cohn GE. A summary and evaluation of spin labels used as probes for biological membrane structure. Biochim Biophys Acta 1973; 300:379-419.

54. Berliner LJ, ed. Spin labeling theory and applications, Vol. I. New York: Academic Press, 1976.

55. Berliner LJ, ed. Spin labeling theory and applications, Vol. II. New York: Academic Press, 1979.

56. Berliner LJ, Reuben J, eds. Biological magnetic resonance, Vol 8. Spin labeling. Theory and applications. New York: Plenum Press, 1989.

57. Shinitzky M, Barenholz Y. Fluidity parameters of lipid regions determined by fluorescence polarization. Biochim Biophys Acta 1978; 515:367-94.

58. Shinitzky M, Yuli I. Lipid fluidity at the submacroscopic level: determination by fluorescence polarization. Chem Phys Lipids 1982; 30:261-82.

59. Miki M, Motoyama T, Takenaka Y, Mino M. Tocopherol behavior and membrane constituents in erythrocytes with oxidant stress. Basic Life Sci 1988; 49:595-604.

60. Aranda FJ, Coutinho A, Berberan-Santos MN, Prieto MJE. Gomez-Fernandez JC. Fluorescence study of the location and dynamics of α-tocopherol in phospholipid vesicles. Biochim Biophys Acta 1989; 985:26-32.

61. Niki E, Kawakami A, Saito M, Yamamoto Y, Tsuchiya J, Kamiya Y. Effect of phytyl side chain of vitamin E on its antioxidant activity. J Biol Chem 1985; 260:2191-6.

62. Kagan VE, Bakalova RA, Zhelev ZZ, et al. Intermembrane transfer and antioxidant action of α-tocopherol in liposomes. Arch Biochem Biophys 1990; 280:147-52.

63. Elsayed NM. Mobilization of vitamin E to the lung under oxidative stress. Ann N Y Acad Sci 1989; 570:439-40.

64. Elsayed NM, Mustafa MG, Mead JF. Increased vitamin E content in the lung after ozone exposure: a possible mobilization in response to oxidative stress. Arch Biochem Biophys 1990; 282:263-9.

65. Cheng SC, Burton GW, Ingold KU, Foster DO. Chiral discrimination in the exchange of α-tocopherol stereoisomers between plasma and red blood cells. Lipids 1987; 22:469-73.

66. McMurchie EJ, McIntosh GH. Thermotropic interaction of vitamin E with dimyristoyl and dipalmitoyl phosphatidylcholine liposomes. J Nutr Sci Vitaminol 1986; 32:551-8.

67. Wassall SR, Thewalt JL, Wong L, Gorrissen H, Cushley RJ. Deuterium NMR study of the interaction of α-tocopherol with a phospholipid model membrane. Biochemistry 1986; 25:319-26.

68. Severcan F, Cannistraro S. Model membrane partition ESR study in the presence of α-tocopherol by a new spin probe. Biosci Rep 1989; 9:489-95.

69. Severcan F, Cannistraro S. A spin label ESR and saturation transfer ESR study of α-tocopherol containing model membranes. Chem Phys Lipids 1990; 53:17-26.

70. Bisby RH, Birch DJS. A time-resolved fluorescence anisotropy study of bilayer membranes containing α-tocopherol. Biochem Biophys Res Commun 1989; 158:386-91.

71. Villalain J, Aranda FJ, Gomez-Fernandez JC. Calorimetric and infrared spectroscopic studies of the interaction of α-tocopherol and α-tocopherol acetate with phospholipid vesicles. Eur J Biochem 1986; 158:141-7.

72. Ohyashiki T, Ushiro H, Mohri T. Effects of α-tocopherol on the lipid peroxidation and fluidity of porcine intestinal brush-border membranes. Biochim Biophys Acta 1986; 858:294-300.

73. Koyama T, Araiso T. Effects of α-tocopherol-nicotinate administration on the microdynamics of phospholipids of erythrocyte membranes in human subjects. J Nutr Sci Vitaminol 1988; 34:449-57.

74. Araiso T, Koyama T. Dynamic properties of the phospholipid bilayer of the erythrocyte membrane in the presence of α-tocopherol. Ann N Y Acad Sci 1989; 570:409-11.

75. Patel JM, Edwards DA. Vitamin E, membrane order and antioxidant behavior in lung microsomes and reconstituted lipid vesicles. Toxicol Appl Pharmacol 1988; 96:101-14.

76. Ohki K, Takamura T, Nozawa Y. Effect of α-tocopherol on lipid peroxidation and acyl chain mobility of liver microsomes from vitamin E-deficient rat. J Nutr Sci Vitaminol 1984; 30:221-34.

77. Bereiter-Hahn J. Behavior of mitochondria in the living cell. Int Rev Cytol 1990; 122:1-63.

78. Biesert L, Adamski M, Zimmer G, et al. Anti-human immunodeficiency virus (HIV) drug HOE/BAY 946 increases membrane hydrophobicity of human lymphocytes and specifically suppresses HIV-protein synthesis. Med Microbiol Immunol 1990; 179:307-321.

79. Khramtsov VV, Yelinova VI, Weiner LM, Berezina TA, Martin VV, Volodarsky LB. Quantitative determination of SH groups in low- and high-molecular-weight compounds by an electron spin resonance method. Anal Biochem 1989; 182:58-63.

80. Kerimov RF, Goloshchapov AN, Burlakova EB, Dzhafarov AI. The influence of functional groups in tocopherol molecules on the microviscosity of mitochondrial lipids. Biull Eksp Biol Med 1987; 103:540-3.

81. Scott JA, Fischman AJ, Khaw BA, Homcy CJ, Rabito CA. Free radical-mediated membrane depolarization in renal and cardiac cells. Biochim Biophys Acta 1987; 899:76-82.

82. Lewis SE, Willis ED. The destruction of −SH groups of proteins and amino acids by peroxides of unsaturated fatty acids. Biochem Pharmacol 1962; 11:901-12.

83. Green J, Diplock AT, Bunyan J, Edwin EE, McHale D. Ubiquinone (coenzyme Q) and the function of vitamin E. Nature 1961; 190:318-25.

84. Stadtman ER. Covalent modification reactions are marking steps in protein turnover. Biochemistry 1990; 29:6323-31.

85. Berlett BS, Chock PB, Yim MB, Stadtman ER. Manganese (II) catalyses the bicarbonate-dependent oxidation of amino acids by hydrogen peroxide and the amino acid facilitated dismutation of hydrogen peroxide. Proc Natl Acad Sci U S A 1990; 97:389-93.

86. Levine RL, Oliver CN, Fulks RM, Stadtman ER. Turnover of bacterial glutamine synthetase: oxidative inactivation procedes proteolysis. Proc Natl Acad Sci U S A 1981; 78:2120-4.

87. Dean RT, Cheeseman KH. Vitamin E protects proteins against free radical damage in lipid environments. Biochem Biophys Res Commun 1987; 148:1277-82.

88. Bisby RH, Ahmed S, Cundall RB. Repair of amino acid radicals by a vitamin E analogue. Biochem Biophys Res Commun 1984; 119:245-51.

89. Hoey BM, Butler J. The repair of oxidized amino acids by antioxidants. Biochim Biophys Acta 1984; 791:212-8.

90. Zimmer G, Schneider M, Hoffmann H. ATP contents and structure of rat liver mitochondria in the presence of 2-mercaptopropionylglycine. Arzneimittelforsch 1978; 28:811-6.

91. Zimmer G, Mainka L, Ohlenschläger G. Oligomycin-sensitive ATPase from beef heart mitochondria: reaction with 2-mercaptopropionylglycine. FEBS Lett 1978; 94:223-7.

92. Zimmer G, Mainka L, Berger I. 2-mercaptopropionylglycine restores activity of oligomycin-sensitive ATPase to control values following treatment with carbonylgyanide-p-trifluoromethoxyphenyl-hydrazone. FEBS Lett 1979; 107:217-21.

93. Zimmer G, Mainka L, Ohlenschläger G. Increase of low molecular weight polypeptides of rat liver mitochondrial membranes by 2-mercaptopropionylglycine. Arzneimittelforsch 1980; 30:632-5.

94. Zimmer G, Mainka L, Heil BM. Bromobimane crosslinking studies of oligomycin-sensitive ATPase from beef heart mitochondria. FEBS Lett 1982; 150:207-10.

95. Veit P, Fuchs J, Zimmer G. Uncoupler- and hypoxia-induced damage in the working rat heart and its treatment. I. Observations with uncouplers of oxidative phosphorylation. Basic Res Cardiol 1985; 80:107-15.

96. Fuchs J, Zimmer G, Bereiter-Hahn J. A multiparameter analysis of the perfused rat heart: response to ischemia, uncouplers and drugs. Cell Biochem Funct 1987; 5:245-53.

97. Freisleben H-J, Fuchs J, Mainka L, Zimmer G. Reactivity of mitochondrial sulfhydryl groups toward dinitrobenzoic acid bromobimanes under oligomycin inhibited and uncoupling conditions. Arch Biochem Biophys 1988; 266:89-97.

98. Fuchs J, Veit P, Zimmer G. Uncoupler- and hypoxia-induced damage in the working rat heart and its treatment. II. Hypoxic reduction of aortic flow and its reversal. Basic Res Cardiol 1985; 80:231-40.

99. Fuchs J, Mainka L, Zimmer G. 2-Mercaptopropionylglycine and related compounds in treatment of mitochondrial dysfunction and postischemic myocardial damage. Arzneimittelforsch 1985; 35:1394-402.

100. Zimmer G, Beyersdorf F, Fuchs J. Decay of structure and function of heart mitochondria during hypoxia and related stress and its treatment. Mol Physiol 1985; 8:495-513.

101. Fuchs J, Zimmer G. ^{31}P NMR spectroscopic investigations and mitochondrial studies on the cardioprotective efficiency of 2-mercaptopropionylglycine. Biochem Pharmacol 1986; 35:4381-5.

102. Fuchs J, Beyersdorf F, Zimmer G. A novel cardioprotective regimen for improvement of inner mitochondrial membrane function after ischemic stress. Arzneimittelforsch 1987; 37:1030-4.

103. Zimmer G, Evers J. 2-mercaptopropionylglycine improves aortic flow after reoxygenation in working rat hearts. Basic Res Cardiol 1988; 83:445-51.

104. Fuchs J, Freisleben H-J, Mainka L, Zimmer G. Mitochondrial sulfhydryl groups under oligomycin-inhibited, aging and uncoupling conditions: beneficial influence of cardioprotective drugs. Arch Biochem Biophys 1988; 266:83-8.

105. Beyersdorf F, Fuchs J, Eberhardt B, Stauder M, Satter P, Zimmer G. Myocardial protection by 2-mercaptopropionylglycine during global ischemia in dogs. Arzneimittelforsch 1989; 39:46-9.

106. Olafsdottir K, Reed DJ. Retention of oxidized glutathione by isolated rat liver mitochondria during hydroperoxide treatment. Biochim Biophys Acta 1988; 964:377-82.

107. Pascoe GA, Olafsdottir K, Reed DJ. Vitamin E protection against chemical-induced cell injury. Arch Biochem Biophys 1987; 256:150-8.

108. Mela L, Seitz S. Isolation of mitochondria with emphasis on heart mitochondria from small amounts of tissues. Methods Enzymol 1979; 55:39-40.

18

Studies on the Interaction of Vitamin E with Phospholipid Membranes

Juan C. Gómez-Fernández, José Villalaín, and Francisco J. Aranda

Universidad de Murcia, Murcia, Spain

INTRODUCTION

α-Tocopherol has been shown to be the main component of vitamin E, and it is known to be an indispensable lipid component of biological membranes with membrane-stabilizing properties (1) and to function as an antioxidant agent, preventing the peroxidation of unsaturated fatty acids present in membrane phospholipids (2).

A role for α-tocopherol has been shown in the prevention or therapy of several conditions, including aging (3), cancer (4), circulatory conditions (5), and the effects of pollution (6). The mechanism of action of α-tocopherol that is behind such a variety of activities is not well understood. The location and dynamics of α-tocopherol in the membrane are subjects of considerable interest, not only because they may clarify its mechanism of action but also for understanding its interaction with phospholipids. We examined this question by using reconstituted systems formed by phospholipid vesicles to which α-tocopherol was incorporated and applied a number of physical techniques, including differential scanning calorimetry (DSC), nuclear magnetic resonance (^{31}P-NMR), fluorescence and Fourier-transform infrared spectroscopy (FTIR).

MODULATION OF MEMBRANE FLUIDITY BY TOCOPHEROLS

Tocopherols are able of modulating the thermotropic phase transition of phospholipids as appreciated by DSC. The effect of α-tocopherol (α-T) on the thermotropic phase transition of saturated phosphatidylcholines, like dimyristoylphosphatidylcholine (DMPC), dipalmitoyl-phosphatidylcholine (DPPC), and distearoylphosphatidylcholine (DSPC), is depicted in Figure 1. It is shown that low concentrations of α-T already introduce a significant perturbation in these phospholipids. The pretransition was abolished, and the main transition broadened and shifted to lower temperatures. The shift to lower temperatures of T_c indicates that α-T prefers to partition into the most fluid domains, which is also confirmed by other studies (see later). The main transition totally disappeared at 40 mol per 100 mol α-T in the three different phosphatidylcholines. On the other hand, the width of the transition peak at half-maximum height was increased. This may indicate that α-T disrupts the cooperative behavior of the lipid bilayer matrix. The effect of α-T on DPPC was found to be greater than that of α-tocopheryl acetate (α-TA) (7), so that the hydroxyl group seems to be very important for the interaction of tocopherols with phospholipids.

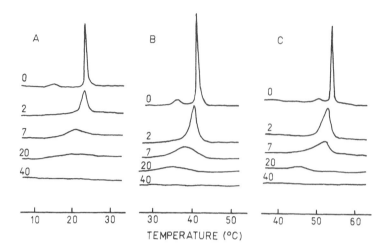

FIGURE 1   The DSC curves for systems containing pure DMPC (A), DPPC (B), and DSPC (C) plus α-tocopherol. Molar percentages of α-tocopherol in lipids are indicated on the curves.

The effect of α-T on the phase transition of DPPC has also been studied by FTIR. FTIR is a very well suited technique to characterize the degree of fluidity in membranes (8). The shift in frequency of the CH_2 stretching vibration modes of the acyl chains of the phospholipids that takes place during the main endothermic phase transition allows observation of the change from the *all-trans* to *gauche* conformers. The incorporation of increasing concentrations of α-T produces a progressive broadening of the phase transition and a shift of the onset of this transition to lower temperatures. Both below and above the phospholipid gel-to-liquid phase transition, α-T decreases the *gauche* isomers of the phospholipid, but α-TA does not appreciably affect the proportion of *gauche* conformers (7). Perhaps the very close interaction established between α-T and the phospholipid may explain why the acyl chains are further immobilized with respect to the gel phase of the pure phospholipid.

LOCATION OF α-TOCOPHEROL IN PHOSPHOLIPID VESICLES

It is very important to study the location of α-T in membranes to better understand the molecular mechanism of action of α-T as a membrane-stabilizing and antioxidant agent. We addressed this question by monitoring its interaction with phospholipids by FTIR.

It was suggested that α-T was engaged in hydrogen bonding with phospholipids through its phenolic hydroxyl group (9), and other studies supported this possibility (10,11). We addressed this question by studying the interaction of α-T and α-TA with DPPC and dimyristoylphosphatidylethanolamine (DMPE) to see if there are any differences on the effect of both tocopherols and if these differences can be adscribed to the existence of a specific interaction between tocopherols and phospholipids.

α-Tocopherol has a phenolic hydroxyl group, which can be engaged in hydrogen bonding. This type of interaction can only take place with the polar region of the phospholipids, and the perturbation of the membrane interfacial region can be followed by monitoring the frequency of the maximum of the C = O stretching band and the frequency of its components resolved by enhancement methods and also by following the frequency of the PO_2^- antisymmetric stretching band (8).

To check the possibility of hydrogen bonding, we used DPPC substituted with ^{13}C at the sn-2 C=O chain group ([sn-2-1-^{13}C]DPPC). The substitution of the sn-2 group with ^{13}C allows its independent observation from the sn-1 group since its stretching band is shifted 45 cm^{-1} to lower frequency, as a result of the vibrational isotope effect, with respect to the band corresponding to the sn-1 carbonyl group (12). Figure 2 shows the carbonyl region of [sn-2-1-^{13}C]DPPC, pure or plus tocopherols, where it can be seen that the inclusion of α-T or α-TA does not induce any appreciable change in the band contour of the sn-1 and sn-2 bands of the phospholipid. Upon band deconvolution it is possible to observe the frequencies of the component bands of both sn-1 and sn-2 carbonyl groups. The sn-1 carbonyl band of pure [sn-2-1-^{13}C]DPPC has two component bands at 1743 and 1732 cm^{-1}, whereas the two component bands of the sn-2 carbonyl band appear at 1699 and 1687 cm^{-1}. The components that appear at higher frequency of both groups can be assigned to the dehydrated carbonyl groups, whereas the components at lower frequencies come from the hydrated carbonyl groups. Upon inclusion of either α-T or α-TA, the frequencies of the different components do not change appreciably from pure [sn-2-1-^{13}C]DPPC, providing evidence that they do not establish any specific bonding with the carbonyl groups of DPPC. The PO$_2^-$ antisymmetric stretching band of DPPC that appears at 1220 cm^{-1} was also observed in the presence of α-T and α-TA, to check whether tocopherols could establish hydrogen bonding with this group. It was found that this band was not affected at all by the presence of α-T or α-TA (results not shown). It thus seems that tocopherols do not establish hydrogen bonding with DPPC.

The interaction of α-T with another type of phospholipid, DMPE, which is known to have an intermolecular hydrogen bond network, was also studied. Hydrated DMPE presents a broad C=O stretching band that after deconvolution can be resolved in two components, one at 1744 cm^{-1} and the other at 1720 cm^{-1}, which can be compared with the frequencies of 1744 and 1729 cm^{-1}, which appear in pure DPPC. The component at 1744 cm^{-1} can be assigned to a dehydrated C=O group, whereas the component at 1720 cm^{-1} can be assigned to

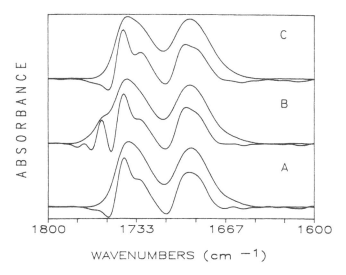

FIGURE 2 Infrared spectra of the C=O stretching region of [sn-2-1-^{13}C]DPPC. (A) Pure [sn-1-2-^{13}C]DPPC, (B) [sn-2-1-^{13}C]DPPC plus 10 mol per 100 mol α-tocopheryl acetate, and (C) [sn-2-1-^{13}C]DPPC plus 10 mol per 100 mol α-tocopherol. The bands at $\simeq$1735 and $\simeq$1690 cm^{-1} correspond at the sn-1 and sn-2 C=O groups of DPPC, respectively.

a hydrogen-bonded C=O group (13). In the presence of α-T, three main components are apparent whose frequencies are 1742, 1734, and 1720 cm^{-1}, whereas the inclusion of α-TA produces a different pattern and only two components at 1744 and 1734 cm^{-1} are seen (results not shown). The presence of a component band at 1720 cm^{-1} in the presence of α-T, which was also observed in pure DMPE, supports the contention that α-T participates in hydrogen bonding with the C=O groups of DMPE, but the absence of this component in the presence of α-TA is evidence that α-TA disrupts the hydrogen-bonding interactions occurring in pure DMPE, perhaps simply by its interposition between DMPE molecules. The effect of α-T and α-TA on DMPE has also been studied in anhydrous mixtures, and in Figure 3 the C=O stretching vibration region of anhydrous DMPE is shown. In this state, pure DMPE presents three main bands at 1745, 1735, and 1720 cm^{-1}, where the 1720 cm^{-1} band is due to an hydrogen-bonded carbonyl group. In the presence of α-T the area of this band is increased, but it is absent in the presence of α-TA. Furthermore, the PO$_2^-$ antisymmetric stretching band is split in two bands in the presence of α-T but not in the presence of α-TA (Fig. 4), which confirms that α-T disrupts the intermolecular hydrogen bonds of DMPE and probably itself forms hydrogen bonds with the phospholipid.

The situation of the hydroxyl group of α-T at the lipid/water interface is of interest in explaining its mechanism of action, because thanks to this location it could prevent the introduction to the membrane of oxidizing agents.

The effect of tocopherols on phosphatidylethanolamines has been also studied by DSC (14). α-T induced the appearance of more than one peak transition for the gel-to-liquid crystalline in DMPE but not in the case of dielaidoylphosphatidylethanolamine (DEPE) (Fig. 5), indicating that immiscibilities were taking place in the presence of DMPE, so that phases with different contents in α-T and phospholipid are present, the Tc transition temperature being lower as more α-T is present in each particular phase. α-TA did not produce this effect, however, and only one transition peak was observed for the gel-to-liquid crystalline-phase transition for both types of phospholipids, DMPE and DEPE (Fig. 5).

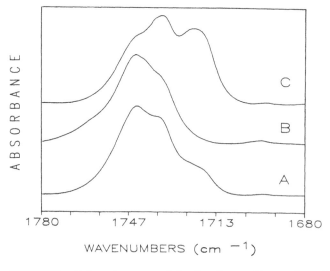

FIGURE 3 Infrared spectra of the C=O stretching region of anhydrous DMPE. (A) Pure DMPE, (B) DMPE plus 10 mol per 100 mol α-tocopherol, and (C) DMPE plus 10 mol per 100 mol α-tocopheryl acetate.

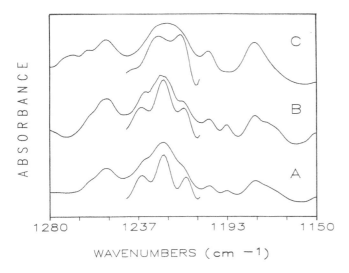

FIGURE 4 Infrared spectra of the PO_2^- antisymmetric stretching vibration band of anhydrous DMPE. (A) Pure DMPE, (B) DMPE plus 10 mol per 100 mol α-tocopherol, and (C) DMPE plus 10 mol per 100 mol α-tocopheryl acetate.

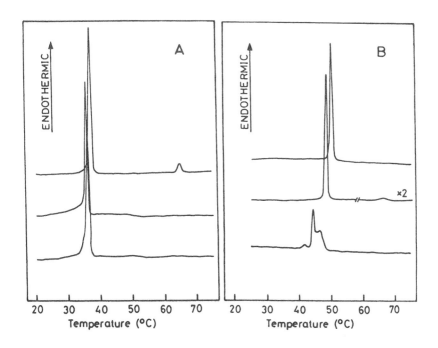

FIGURE 5 The DSC curves for systems containing (A) DEPE and (B) DMPE. The upper, middle, and lower thermograns correspond to the pure phospholipid, phospholipid plus 10 mol per 100 mol α-tocopheryl acetate, and phospholipid plus 10 mol per 100 mol α-tocopherol, respectively.

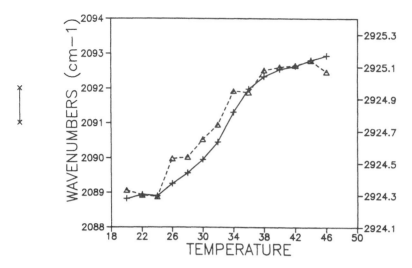

FIGURE 6   Temperature dependence of the CD_2 stretching antisymmetric band of deuterated DPPC (solid line) and the CH_2 stretching anstisymmetric band of α-tocopherol (dashed line) in the mixture of 20 mol per 100 mol α-tocopherol in deuterated DPPC.

Another important factor to be understood is whether α-T is randomly distributed in the membrane or if it shows a preference for certain regions or domains. It is pertinent to this question to remember that it has been already mentioned that α-T shows a preference for the most fluid domains in the membrane.

We monitored the phase transition of mixtures of α-T in DPPC using specifically deuterated DPPC (dDPPC), as can be seen in Figure 6. The phase transition of dDPPC was followed through the changes in frequency of the CD_2 antisymmetric stretching mode, whereas the phase transition of α-T was followed through the CH_2 antisymmetric mode. It can be observed that the transition temperatures of dDPPC and α-T observed in separate clearly agree with each other, providing evidence of good mixing between the two types of molecules.

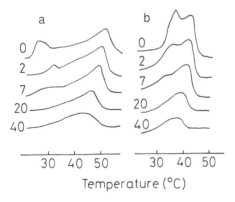

FIGURE 7   The DSC curves for mixtures of (A) DMPC/DPPE and (B) DLPE/DSPC containing different molar percentages of α-tocopherol as indicated.

The preference for fluid domains was also confirmed by other DSC studies in which α-T was included in mixtures of phospholipids showing monotectic behavior, such as DMPC/ dipalmitoylphosphatidylethanolamine (DPPE) and dilauroylphosphatidylethanolamine (DLPE)/ DSPC (Fig. 7). In all these systems α-T always most affected the component having the lowest phase transition T_c, irrespective of whether this was phosphatidylcholine (PC) or phosphatidyl-ethanolamine (PE) (15). Since it seems very likely that biological membranes are composed of dominions having different degrees of fluidity, it is very important to keep in mind that α-T is preferentially found in those that are the most fluid. Furthermore, this may have important implications with respect to the mechanism of action of α-T. That is, unsaturated fatty acyl chains are predominantly found in the most fluid domains, and at the same time they are the lipid component prone to oxidation. Since α-T prevents chain peroxidations, the regions of the membrane where its presence is meaningful are precisely these fluid domains.

DYNAMICS OF α-TOCOPHEROL IN MEMBRANES

We used the intrinsic fluorescence of α-T to determine some of its basic fluorescence parameters (16). Table 1 shows λ_{max} of the fluorescence emission spectra, quantum yield ϕ_F, and fluorescence lifetime τ_F, determined for α-T in a number of solvents and incorporated into phospholipid vesicles. It is shown that both quantum yield and fluorescence lifetime of α-T increase as solvent polarity increases. It is interesting to note that the λ_{max} and fluorescence lifetime obtained for α-T in phospholipid vesicles are similar to those obtained in protic solvents (e.g., ethanol). This observation suggests that the chromanol moiety of α-T should be located in the polar region of the model membrane.

The absorption spectra of α-T in *n*-hexane is depicted in Figure 8. It is shown that the maximum of the spectrum at a low concentration of α-T is at 283 nm (predominance of the monomeric form), but it is shifted to 295 nm at much higher concentrations (predominance of the hydrogen-bonded form). Experiments with α-T incorporated into phospholipid vesicles showed a maximum near 295 nm within a wide range of concentration (data not shown), indicating that most α-T molecules are associated when present in membranes. It could not be discerned from the data presented whether α-T molecules are associated (through hydrogen bonding) with themselves or with phospholipid or water molecules.

Figure 9 shows the fluorescence intensity of α-T at increasing concentrations of the molecule. It can be seen that a linear relationship between fluorescence intensity and concentration is obtained for α-T incorporated in phospholipid vesicles. Identical behavior was obtained in nonpolar (hexane) and polar (ethanol) solvents (data not shown). Thus we clearly show that although α-T molecules are associated, the aggregates formed are fluorescent.

The location of α-T in the bilayer was also studied through the quenching of its fluorescence by membrane probes, such as n-NS(n-(N-oxy-4,4-dimethyloxazolidin-2-yl)stearic acid). We have found that 5-NS quenches α-T fluorescence much more effectively than 16-NS, as expected if the chromanol moiety is located near the lipid/water interface.

TABLE 1 Fluorescence Parameters of α-Tocopherol in Solution and Incorporated into Phospholipid Vesicles

Medium	λ_{max} (nm)	ϕ_F	τ_F (ns)
Acetonitrile	311	ND	1.0
Ethanol	317	0.34	1.8
Cyclohexane	312	0.16	0.8
Lecithin	316	ND	1.7

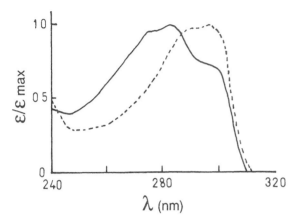

FIGURE 8   Absorption spectra of α-tocopherol on *n*-hexane at concentrations of 4.4×10^{-5} M (solid line) and 1.42×10^{-4} M (dashed line).

We also studied the location of α-T in phospholipid vesicles by using a set of *n*-(9-anthroyl-oxy)stearic acid (n-AS) probes. The explicit distance dependence (r^{-6}) of electronic energy transfer (dipolar mechanism) allowed its application as a spectroscopic ruler for determining distance in biological systems. Figure 10 shows the efficiency of energy transfer from α-T to different n-AS probes differing in the location of their 9-anthroyloxy group. It can be seen that the efficiency of energy transfer follows the order 7-AS > 2-AS > 9-AS = 12-AS. From these data, it can be concluded that the chromophore group of α-T is situated in the membrane in a region between the 9-anthroyloxy group located at carbon 7 and carbon 2, the former being the nearest. Furthermore, we found that acrylamide, which is a water-soluble fluor-escence quencher, has a very low efficiency of quenching α-T in phospholipid vesicles, but

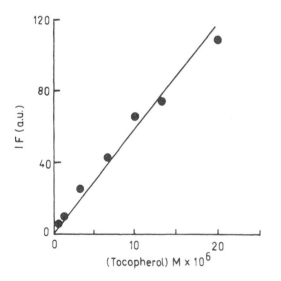

FIGURE 9   Fluorescence intensity of α-tocopherol (in arbitrary units) versus concentration of α-toco-pherol in multilamellar vesicles of egg yolk lecithin at 25°C (lipid concentration 0.13×10^{-3} M).

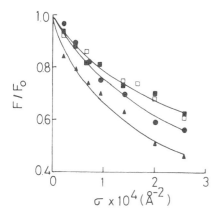

FIGURE 10 Relative yield of α-tocopherol fluorescence F/F_0 versus σ (acceptor surface concentration) of 2-AS (circles), 7-AS (triangles), 9-AS (solid squares), and 12-AS (open squares) in egg yolk PC at 25°C. Excitation and emission wavelengths were 295 and 329 nm, respectively.

acrylamide is an efficient quencher of α-T in ethanolic solution (data not shown). This indicates that although α-T may have its chromanol group relatively close to the polar part of the bilayer, it is not sufficiently exposed to allow acrylamide to reach it (acrylamide being known to have a very low capacity of penetration through phospholipid bilayers).

The conclusions of these studies is that the chromanol group is found in a position close to that occupied by 7-AS and 5-NS. The location of the chromanol moiety in the lipid/water interface could be of importance in explaining its mechanism of protection, since any oxidizing agent approaching the membrane surface should find reducing protons and its penetration should be avoided.

To understand the mechanism of action of α-T in membranes, it may be very instructive to know the lateral diffusion of this molecule when incorporated into phospholipid vesicles. This was approached through the study of quenching of the intrinsic fluorescence of α-T by 5-NS. By using the Smoluchwoski equation, we found a lateral diffusion coefficient of 4.8×10^{-6} cm²/s (16). This means that α-T may have high mobility in natural membranes and hence be quite efficient in reacting with oxidizing agents. This value is very similar to that calculated by other authors for ubiquinone-3 (5.8×10^{-6} cm²/s) (17), which is a molecule very related in structure to α-T. It is interesting to note that the lateral diffusion coefficient obtained for N-(7-nitro-2,1,3-benzoxadiazol-4-yl)-phosphatidylcholine (unpublished data) is quite similar to that of α-T.

MODULATION BY TOCOPHEROLS OF
PHOSPHATIDYLETHANOLAMINE LIPID POLYMORPHISM

Phospholipids in membranes can adopt several structures, such as the micellar phase, the bilayer and hexagonal H_{II} phases, and lipidic particles. The ability of phospholipids to adopt these different structures is known as lipid polymorphism (18). The nonbilayer structures can greatly affect the funtional behavior of the membrane (19) with a large biological potential significance. We studied the effect of α-T and α-TA on lipid polymorphism of phosphatidylethanolamine, a phospholipid that can spontaneously adopt the hexagonal H_{II} phase. We used DEPE and DMPE (14), an unsaturated and a saturated PE, DEPE, whose bilayer-to-H_{II} transition occurs at 63°C, and DMPE, which has been shown not to undergo bilayer-to-H_{II}

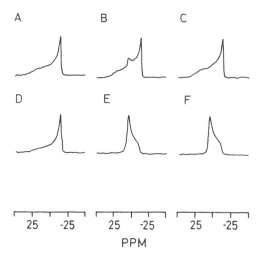

FIGURE 11   ³¹P-NMR spectra of mixtures of DMPE (A–C) and DEPE (D–F) at 53 and 55°C, respectively: (A and D) pure phospholipids, (B and E) plus 10 mol per 100 mol α-tocopherol and (C and F) plus 10 mol per 100 mol α-tocopheryl acetate.

phase transition up to 90°C. Both α-T and α-TA have almost no effect on the gel-to-liquid crystal-phase transition of DEPE, but the effect on the bilayer-to-H$_{II}$ phase transition is more significant, because increasing amounts of both molecules decrease this transition temperature (14). ³¹P-NMR confirms this effect, as seen in Figure 11, which shows that at 53°C and in the presence of α-T or α-TA, DEPE is completely in the hexagonal H$_{II}$ phase, whereas in the absence of either α-T or α-TA, DEPE is in the bilayer phase.

As seen by DSC (Fig. 5), the effect of α-T on DMPE is different from that on DEPE, whereas the effect of α-TA is similar on both phospholipids. α-TA does not significantly affect the gel-to-liquid crystal transition of DMPE, but α-T produces the presence of several transition peaks. This effect is similar to what was described previously for the interaction of α-T and other PEs (15). By ³¹P-NMR we can observe that, at 55°C, α-T induces the pres-

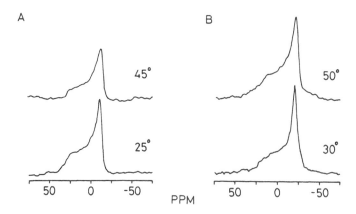

FIGURE 12   ³¹P-NMR spectra of mixtures of (A) EYL and (B) DPPC plus 20 mol per 100 mol α-tocopherol at the indicated temperatures.

ence of hexagonal H_{II} phase coexisting with bilayer structures, but the presence of α-TA at this temperature is not capable of producing the hexagonal H_{II} phase, although it produces hexagonal H_{II} phase well above the gel-to-liquid crystal transition (Fig. 11).

We also studied the effect of α-T on the polymorphic behavior of phosphatidylcholines, phospholipids that do not spontaneously adopt the hexagonal H_{II} phase, and the results are shown in Figure 12. α-T is not capable of inducing the presence of H_{II} hexagonal phase, neither in egg yolk lecithin (EYL) nor on DPPC.

These results demonstrate that tocopherols induce H_{II} phase formation specifically in PE systems, α-T being more effective than α-T acetate. The formation of H_{II} hexagonal phase would be compatible with the increasing motion on the acyl chains induced by tocopherols, as shown by FTIR (7), which would increase the hydrophobic volume of the lipid, facilitating the formation of hexagonal structures. The hydroxyl group of α-T would also facilitate its positioning in the bilayer so that the interaction between α-T and the hydrophobic part of the lipid would be maximized, explaining the differences found between α-T and α-TA, which has its hydroxyl group blocked by the acetate. It remains to be seen if the influence of tocopherols on lipid polymorphism is significant from the point of view of their biological effects.

SUMMARY

Differential scanning calorimetry showed that α-tocopherol was able to perturb the thermotropic phase transition of fully saturated phospholipids. α-Tocopheryl acetate has a less marked effect than α-tocopherol. Fourier-transform infrared spectroscopy studies remarked the importance of its phenolic group in determining the interaction of α-tocopherol with phospholipids. This hydroxyl group is not hydrogen bonded to the carbonyl group or the phosphate group of phosphatidylcholine, although it seems able to form a hydrogen bond with the carbonyl group of phosphatidylethanolamine. The intrinsic fluorescence of α-tocopherol was used to study the location and dynamics of the molecule in phospholipid vesicles. It was concluded that the chromanol moiety of the molecule is located near the water/lipid interface. The lateral diffusion coefficient of α-tocopherol in phospholipid vesicles was calculated through quenching of its fluorescence. By differential scanning calorimetry it was found that α-tocopherol is preferentially partitioned in the most fluid domains in the membrane. The effect of α-tocopherol and α-tocopheryl acetate on phosphatidylethanolamine lipid polymorphism was studied by differential scanning calorimetry and ^{31}P-nuclear magnetic resonance, and it was found that both molecules promote the formation of the hexagonal H_{II} phase in these systems, the effect of α-tocopherol being stronger than that of α-tocopheryl acetate.

ACKNOWLEDGMENTS

The work described here was supported by Grants PA86-0211 and PM90-044 from DGICYT, Spain.

REFERENCES

1. Tappel AL. Vitamin E and free radical peroxidation of lipids. Ann N Y Acad Sci 1972; 205:12-28.
2. Burton GW, Ingold KH. Vitamin E: application of the principles of physical organic chemistry to the exploration of its structure and function. Acc Chem Res 1986; 19:194-201.
3. Wartanowicz M, Panczenko-Kresowska B, Ziemlanski S, Kowalska M, Okolska G. The effect of α-tocopherol and ascorbic acid on the serum lipid peroxide level in elderly people. Am J Nutr Metab 1984; 28:186-91.
4. Watson RR, Leonard TK. Selenium and vitamins A, E and C: nutrients with cancer prevention properties. J Am Diet Assoc 1986; 86:505-10.

5. Cavarocchi NC, England NO, O'Brien JF, et al. Superoxide generation during cardiopulmonary bypass: is there a role for vitamin E? J Surg Res 1986; 40:519.

6. Sevanian A, Hacker AE, Elsayed N. Influence of vitamin E and nitrogen dioxide on lipid peroxidation in rat lung and liver microsomes. Lipids 1986; 17:269-77.

7. Villalaín J, Aranda FJ, Gómez-Fernández JC. Calorimetric and infrared spectroscopic studies of the interaction of α-tocopherol and α-tocopheryl acetate with phospholipid vesicles. Eur J Biochem 1986; 158:141-7.

8. Casal HL, Mantsch HH. Polymorphic phase behaviour of phospholipid membranes studied by infrared spectroscopy. Biochim Biophys Acta 1984; 779:381-401.

9. Srivastava S, Phadke RS, Govil G, Rao CNR. Fluidity, permeability and antioxidant behaviour of model membranes incorporated with α-tocopherol and vitamin E acetate. Biochim Biophys Acta 1983; 734:353-62.

10. Baig MMA, Laidman DL. Spectrophotometric evidence for a polar interaction between α-tocopherol and phospholipids. Biochem Soc Trans 1983; 11:600-2.

11. Perly B, Smith ICP, Hughes L, Burton GW, Ingold KW. Stimation of the location of natural α-tocopherol in lipid bilayers by carbon-13 NMR spectroscopy. Biochim Biophys Acta 1985; 819:131-5.

12. Blume A, Hübner W, Messner G. Fourier transform infrared spectroscopy of $^{13}C=O$ labeled phospholipids hydrogen bonding to carbonyl groups. Biochemistry 1988; 27:8239-49.

13. Wong PTT, Mantsch HH. High-pressure infrared spectroscopic evidence of water binding sites in 1,2-diacylphospholipids. Chem Phys Lipids 1988; 46:213-24.

14. Micol V, Aranda FJ, Villalaín J, Gómez-Fernández JC. Influence of vitamin E on phosphatidylethanolamine polymorphism. Biochim Biophys Acta 1990; 1022:194-202.

15. Ortiz A, Aranda FJ, Gómez-Fernández JC. A differential scanning calorimetry of the interaction of α-tocopherol with mixtures of phospholipids. Biochim Biophys Acta 1987; 898:214-22.

16. Aranda FJ, Coutinho A, Berberan-Santos MN, Prieto MJE, Gómez-Fernández JC. Fluorescence study of the location and dynamics of α-tocopherol in phospholipid vesicles. Biochim Biophys Acta 1989; 985:26-32.

17. Fato R, Battino M, Degli-Esposti M, Parenti-Castelli G, Lenaz G. Determination of partition and lateral diffusion coefficients of ubiquinones by fluorescence quenching of *n*-(9-anthroyloxy)steric acids in phospholipid vesicles and mitochondrial membranes. Biochemistry 1986; 25:3376-90.

18. Cullis PR, Hope MJ, De Kruijff B, Verkleij AJ, Tilcock CPS. Structural properties and functional roles of phospholipid in biological membranes. In: Kuo JF, ed. Phospholipid and cellular regulation, vol. I. Boca Ratón, FL: CRC Press, 1985; 1-60.

19. De Kruijff B, Cullis PR, Verkleij AJ, Hope MJ, Van Echteld CJA, Taraschi TF. Lipid polymorphism and membrane function. In: Martonosi AN, ed. The enzymes of biological membranes, 2nd ed., vol. I. New York: Plenum Press, 1985; 131-204.

19

Antioxidant Action of α-Tocopherol and α-Tocotrienol in Membranes

Elena A. Serbinova, Masahiko Tsuchiya, Sam Goth, Valerian E. Kagan, and Lester Packer

University of California, Berkeley, California

BIOLOGICAL ACTIVITIES OF TOCOPHEROLS AND TOCOTRIENOLS

The term "vitamin E" is now used to describe all tocopherol and tocotrienol derivatives that exhibit the general physiological activity of alleviating any symptoms related with vitamin E deficiency (1,2). At least eight compounds have been isolated from plant sources that have vitamin E activity. All have a chromanol ring and a hydrophobic side chain. The tocopherols have a phytol chain, and the tocotrienols have a similar tail but with double bonds at positions 3', 7', and 11'. Both tocopherols and tocotrienols have derivatives that differ by the number and position of methyl groups on the chromanol ring (3), designated α-, β-, γ-, and δ-isomers (4,5).

Tocopherols and tocotrienols are present in various components of human diet. Tocopherols are found in polyunsaturated vegetable oils and in the germ of cereal seeds, whereas tocotrienols are found in the aleurone and subaleurone layers of cereal seeds and in palm oil. Although the tocopherols and tocotrienols are closely related chemically, they have widely varying degrees of biological effectiveness. The potency of α-tocotrienol evaluated by gestination-resorption assays have been shown to be 32% of the potency of α-tocopherol (6).

Recently, new and unique physiological activities of tocotrienols have been discovered. In an early study, Qureshi et al. demonstrated that α-tocotrienol isolated from barley could be a potent inhibitor of cholesterol biosynthesis in vivo and in vitro, with the overall effect of reducing the levels of serum cholesterol and low-density lipoprotein (LDL) cholesterol (7). Later, significant inhibition of hepatic 3-hydroxy-3-methylglutaryl coenzyme A reductase activity, the first rate-limiting enzyme in the biosynthesis of cholesterol, and a concomitant decrease in the serum levels of cholesterol (17–28% 0 and LDL cholesterol, 13–46% reduction) was observed after addition of a tocotrienol-rich fraction of palm oil or purified α-, γ-, or δ-tocotrienol from palm oil to chicken diets (8). Dietary supplementation of tocotrienols was also reported to reduce concentrations of plasma cholesterol, apolipoprotein B, and platelet factor 4 in pigs with inherited hyperlipidemia (9). Finally, γ-tocotrienol supplementation was shown to decrease the serum cholesterol level in hypercholesterolemic subjects (10).

Tocopherol and tocotrienol were tested for chemopreventive activity in two chemically induced rat mammary tumor models (11). None of the vitamin E analogs had a major impact on mammary tumor development after tumor induction with either DMBA (7,12-dimethyl-benz[α]anthracene) or NMU (*N*-nitrosomethylurea). In an early work, Kato et al. reported that 10 successive i.p. (intraperitoneal) injections of tocotrienol in mice with IMC carcinoma gave a significant increase in the life span and the number of 60 day survivors (12). Later, the results of a comparative study of the antitumor activity of α- and γ-tocotrienol and α-tocopherol were performed. The i.p. α- and γ-tocotrienols were effective against carcinoma 180, Ehrlich carcinoma, and IMC carcinoma, and γ-tocotrienol showed a slight life-prolonging effect in mice with Meth A fibrosarcoma. Tocotrienols had no antitumor effect against P338 leukemia. In contrast, α-tocopherol had only a slight effect against carcinoma 180 and IMC carcinoma (13).

ANTIOXIDANT FUNCTION OF TOCOPHEROLS AND TOCOTRIENOLS

Even though the mechanism of physiological activity of vitamin E is not clearly understood, it is likely that at least some of the biological activities that have been demonstrated are due to its antioxidant function. The biological activity of vitamin E is generally believed to be due to its antioxidant action of inhibiting lipid peroxidation in biological membranes by scavenging chain-propagating peroxyl radical (ROO$^\cdot$) (5,14,15):

$$ROO^\cdot + Toc\text{-}OH \rightarrow ROOH + Toc\text{-}O^\cdot \tag{1}$$

The antioxidant function of vitamin E per se is localized in the chromanol nucleus, whereas the phenolic hydroxy group donates an H atom to quench lipid radicals (15–17). There appears to be widespread confusion concerning both the relative and the absolute antioxidant effectiveness in vitro of the individual tocopherols. Some researchers have reported that γ-tocopherol is more effective than α-tocopherol in preventing lipid peroxidation induction in vitro (14,15). Other reports have suggested that α-tocopherol is more effective than γ-tocopherol. It is now believed (18) that relative effectiveness depends on the experimental conditions.

In homogeneous solution the reaction rate constants of chromanols with peroxyl radicals [reaction (1)] do not depend on the length or unsaturation of the hydrocarbon chain but are mainly dependent on the number of methyl groups in the chromanol nucleus (19,20). Similarly, the reactivity of chromanoxyl radicals is mainly determined by the hindering effects of surrounding methyl groups (21,22). In microdomains of heterogeneous membranous systems, however, vitamin E owes its antioxidant potency not solely to the chemistry, as in reactions (1), but also its mobility and accessibility within the membrane (23–26).

There have been attempts to increase the antioxidant activity of α-tocopherol by modifying its chromanol nucleus or its hydrocarbon chain. It was shown that dihydrobenzofurans have a higher antioxidant activity in vitro (27). In vivo, however, these homologs did not have a higher antioxidant activity (27). α-Tocopherol homologs with shorter hydrocarbon tails manifested remarkably higher efficiency in inhibiting lipid peroxidation in different natural membranes and in liposomes compared to α-tocopherol (28–30). These short-chain α-tocopherol homologs, although highly potent in vitro, cannot be used as membrane antioxidants in vivo because of their pronounced membrane-perturbing effects, and they have no vitamin E activity (31,32). It is known that the molecular mobility of polyenoic lipids in membrane bilayers is much higher than that of saturated lipids (33). Thus we predict that tocotrienols with their unsaturated hydrocarbon side chain should be more mobile and less restricted in their interactions with lipid radicals in membranes than tocopherols. As a result, the antioxidant potency of tocotrienols in membranes is expected to be higher than that of tocopherols.

There is indirect evidence of a higher antioxidant activity of tocotrienols in comparison with tocopherols. Tatsuta reported in her studies on hemolysis that α-tocotrienol showed higher efficiency in protecting red blood cells than α-tocopherol in vitro, an effect opposite to that observed in vivo (34). Tocotrienols have been reported to possess higher protective activity against cardiotoxicity of the antitumor redox cycling drug Adriamycin (doxorubicin). The cardiotoxicity is believed to be caused by free radicals generated by Adriamycin (12). It was also found that α-tocotrienol showed a higher inhibitory effect on lipid peroxidation induced by Adriamycin in rat liver and murine microsomes than α-tocopherol (13). In liposomes prepared from microsomal phospholipids or from dipalmitoylphosphatidylcholine, however, tocopherol and tocotrienol were equally efficient in inhibiting iron-induced cholesterol 5α-hydroperoxide decomposition (35).

Radical-Scavenging Activity of α-Tocopherol and α-Tocotrienol in Hexane

The interaction of tocopherols and tocotrienols with peroxyl radicals was measured by a fluorometric assay based on the interaction of a polyunsaturated fatty acid, *cis*-parinaric acid, with peroxyl radicals generated by 2,2'-azobi-(2,4-dimethylvaleronitrile) (AMVN) in hexane (36). α-Tocopherol or α-tocotrienol added to the system prevented the oxidation of *cis*-parinaric acid (Fig. 1). The duration of the protection period was the same for α-tocopherol and α-tocotrienol. Thus the radical-scavenging activity of both antioxidants with peroxyl radicals is similar in hexane, in which random collisions of antioxidant molecules with peroxyl radicals occur.

Radical-Scavening Activity of α-Tocopherol and α-Tocotrienol in DOPC Liposomes

In membranes, lateral and transbilayer diffusion of antioxidants may be essential for their interactions with peroxyl radicals. Liposomes with incorporated α-tocopherol or α-tocotrienol inhibited AMVN-induced luminol-sensitized chemiluminescence in a concentration-dependent

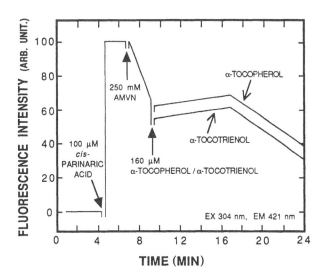

FIGURE 1 Comparison of the radical-scavenging activity of α-tocopherol and α-tocotrienol in hexane. The reaction mixture (3 ml) contained AMVN (100 mM) and *cis*-parinaric acid (30 μM) in hexane. α-Tocopherol or tocotrienol (160 μM each) dissolved in chloroform was added to the incubation medium during the course of AMVN-induced fluorescence loss of *cis*-parinaric acid.

fashion (Fig. 2). The concentrations of α-tocopherol and α-tocotrienol producing 50% inhibition of AMVN-induced chemiluminescence were 7.5 and 5 μM, respectively. Thus in liposomes α-tocotrienol was 1.5 times better at scavenging peroxyl radicals than α-tocopherol. These data are in agreement with the results reported by Yamaoka and Komiyana (37) on the antioxidant activity of α-tocopherol and α-tocotrienol in the 2,2′-azobis-(2-aminopropane)-dihydrochloride (AAPH)-initiated oxidation of dilinoleoylphosphatidylcholine (DLPC) liposomes. α-Tocotrienol added after liposome formation showed a higher antioxidant activity than α-tocopherol.

Interaction of α-Tocopherol and α-Tocotrienol with Superoxide Radicals

Initiation of lipid peroxidation often involves an oxygen activation stage during which reactive oxygen species, including superoxide, are formed (38). Although vitamin E is a lipid-soluble antioxidant, the hydroxyl group of its chromanol nucleus is exposed to the aqueous phase. Hence, interaction of vitamin E with oxygen radicals in cytosol seems feasible. To compare the efficiencies of α-tocopherol and α-tocotrienol in scavenging superoxide radicals, we measured their effects on the lucigenine-dependent chemiluminescence of superoxide radicals generated by xanthine oxidase + xanthine. α-Tocopherol and α-tocotrienol, when incorporated into dioleoylphosphatidylcholine (DOPC) liposomes, equally quenched chemiluminescence (Fig. 3). Thus α-tocopherol and α-tocotrienol when incorporated in phospholipid liposomes do not differ in the efficiency of their interaction with superoxide radical.

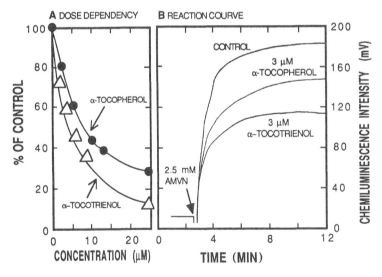

FIGURE 2 Comparison of the radical-scavenging activity of α-tocopherol and α-tocotrienol incorporated in DOPC liposomes. Dioleoyl-phosphatidylcholine (DOPC) liposomes with incorporated antioxidant were made by sonication of DOPC dispersion under nitrogen at 4°C. The incubation medium (2 ml) contained DOPC liposomes (2.5 mM), luminol (150 μM), and α-tocopherol (or tocotrienol) in concentrations as indicated in Tris-HCl buffer (pH 7.4). The reaction was started at 40°C by the addition of AMVN (2.5 mM).

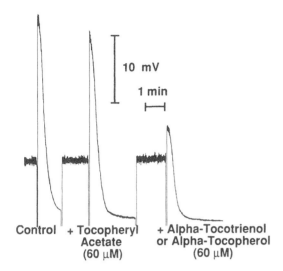

FIGURE 3 Effects of α-tocopherol or α-tocotrienol on lucigenine-dependent chemiluminescence of superoxide radicals generated by xanthine oxidase + xanthine. Incubation conditions: incubation medium (2 ml) contained DOPC liposomes (254 μM) in 0.1 M Tris-HCl, pH 7.5 at 37°C; 100 μM lucigenin (bis-*N*-methyl acridinium nitrate); 100 μM Desferal; and 50 μM xanthine. The reaction was started by adding 14 mU xanthine oxidase.

Efficiency of Inhibition of Lipid Peroxidation by α-Tocopherol and α-Tocotrienol

Comparison of the antioxidative properties of different tocopherols and tocotrienols in preventing the oxidation of lard showed that tocotrienols were more efficient than the corresponding tocopherols (39). The antioxidant activity of tocotrienol isomers measured at 110°C in the dark increased in the following order: $\alpha > \beta > \gamma > \delta$-tocotrienol.

When lipid peroxidation was induced by Fe(II) + ascorbate or Fe(II) + NADPH-induced lipid peroxidation in a more physiological system, rat liver microsomes, we showed that α-tocotrienol has higher antioxidant activity than α-tocopherol. The concentrations of α-tocopherol producing 50% inhibition (K_{50}) were 40 and 60 times higher than those for α-tocotrienol for Fe^{2+} + NADPH- and Fe^{2+} + ascorbate-dependent lipid peroxidation, respectively (40).

Uniformity of Distribution of α-Tocopherol and α-Tocotrienol in the Lipid Bilayer

In heterogeneous membrane systems the efficiency of inhibiting lipid peroxidation is dependent on the chemical reactivity of the antioxidant, its distribution between the aqueous and nonpolar phases and into the lipid bilayer (association in clusters), and the mobility of membrane lipids (32,41). Uniformity of distribution or clustering of α-tocopherol and α-tocotrienol molecules within the membrane lipid bilayer can be detected by observing changes in fluorescence intensity (42,43). The association of chromanols in clusters results in fluorescence self-quenching (decrease in fluorescence intensity); the uniform distribution of chromanol molecules (both α-tocopherol and α-tocotrienol possess characteristic fluorescence in the ultraviolet, UV, region) (33) causes an increase in fluorescence. Fluorescence measurements showed that for the molar ratios of chromanols to phospholipids from 1:1000 to 1:20, α-tocopherol demonstrated a significantly higher level of association in clusters than α-tocotrienol (40).

Comparison of the Recycling Efficiency of α-Tocopherol and α-Tocotrienol

The antioxidant potency of vitamin E is determined not only by its efficiency in reactions (1) but also by the reactivity of the resultant chromanoxyl radicals (44,45) and by the efficiency of regeneration of the active form of the antioxidant molecules from the chromanoxyl radicals due to its interaction with intracellular reductants, which do not propagate lipid peroxidation (46):

$$\text{Toc-O}^{\cdot} + \text{AH}_2 \rightarrow \text{Toc-OH} + \text{AH}$$

Chromanoxyl radicals from α-tocopherol and α-tocotrienol could be generated by an enzymic oxidation system (lipoxygenase + linolenic acid) in the presence of microsomes or liposomes (40,47). Both chromanols give characteristic pentameric chromanoxyl radical signals with g values of the components 2.0122, 2.0092, 2.0061, 2.0028, and 1.9993 both in microsomes and in liposomes (21,22).

It has been reported that α-tocotrienol radical electron spin resonance (ESR) signals were significantly higher than those of α-tocopherol in the presence of either microsomes or liposomes (40). Addition of NADPH to the microsomal suspension resulted in a decrease (but not complete disappearance) in the magnitude of the ESR signals of α-tocopherol (or α-tocotrienol). The NADPH-dependent decrease in ESR signals was much more pronounced for α-tocotrienol than for α-tocopherol. To quantitate the efficiency of NADPH-supported recycling of chromanoxyl radicals of α-tocopherol and α-tocotrienol, an index of recycling efficiency R_e was introduced (47):

$$R_e = \frac{A_{-\text{reductant}} - A_{+\text{reductant}}}{A_{-\text{reductant}}}$$

where $A_{-\text{reductant}}$ and $A_{+\text{reductant}}$ are the current magnitudes of ESR signals of chromanoxyl radicals in the absence and in the presence of exogenous reductant, respectively. The values of recycling efficiency may vary from 0 (complete transient disappearance of the ESR signal, or 100% reduction) to 1.0 (no effect on the ESR signal, or 0% reduction). Table 1 shows that in microsomes NADPH-supported recycling efficiency R_e for α-tocotrienol is higher than for α-tocopherol. Also, the delay time of chromanoxyl radical ESR signal reappearance after the addition of ascorbyl palmitate was greater for α-tocotrienol than for α-tocopherol. This shows that α-tocotrienol has a higher recycling efficiency than α-tocopherol.

TABLE 1   Recycling Efficiency and Delay Time for Reappearance of Chromanoxyl Radicals from α-Tocopherol and α-Tocotrienol in Rat Liver Microsomes[a]

	Delay time (minutes)	Recycling efficiency
d-α-Tocopherol	1.0 ± 0.2	0.23 ± 0.02
d-α-Tocotrienol	3.0 ± 0.3	0.37 ± 0.04

[a]The delay time was measured after addition of ascorbyl palmitate. Recycling efficiency was measured after addition of NADPH. The average given is for five data points.

The higher recycling efficiency of α-tocotrienol may contribute to its higher antioxidant activity compared to α-tocopherol. Even though the difference in R_e and the delay time for α-tocopherol and α-tocotrienol were about 1.6- and 2.5- to 3.0-fold, the concentrations exerting 50% inhibition of lipid peroxidation differed by 40- to 60-fold. This indicates that higher antioxidant activity of α-tocotrienol in vitro must result from the contribution of other factors in addition to its higher recycling efficiency.

In conclusion, it is recognized that differences in vivo in the antioxidant activity of tocopherols and tocotrienols may depend very much upon their pharmacokinetics. However, α-tocotrienol may have higher antioxidant activity in vivo under conditions of oxidative stress as a result of its more effective antioxidant potency in membranes.

ACKNOWLEDGMENTS

Research supported by the NIH (CA 47597) and the Palm Oil Research Institute of Malaysia (PORIM).

REFERENCES

1. Sebrell WH, Harris RS, eds. The vitamins, vol. 5. New York: Academic Press, 1972; Chap. 16.
2. International Union of Nutritional Sciences. Nutr Abstr Rev 1978; 48:831.
3. Machlin LJ. Vitamin E. In: Machlin LJ, ed. Handbook of vitamins: nutritional, biochemical and clinical aspects. New York: Marcel Dekker, 1984; 245-65.
4. Bauernfield JC. Food sources of the tocopherols. In: Machlin LJ, ed. Vitamin E. New York: Marcel Dekker, 1984; 99-135.
5. Witting LA. Vitamin E and lipid antioxidants in free-radical-initiated reactions. In: Pryor WA, ed. Free radicals in biology, vol. IV. New York: Academic Press, 1980; 295-319.
6. McHale JB, Green J, Marcinkiewicz S. Biological potency of ε- and ζ₁-tocopherol and 5-methyltocol. Br J Nutr 1961; 15:253-7.
7. Qureshi AA, Burger WC, Peterson DM, Elson CE. The structure of an inhibitor of cholesterol biosynthesis isolated from barley. J Biol Chem 1986; 261:10544-50.
8. Qureshi AA, Qureshi N, Ong A, Gapor A, deWitt GF, Chong YH. Suppression of cholesterol biosynthesis and hypocholesterolemic effects of tocotrienols from palm oil in the chicken model. In: 1987 International Oil Palm/Palm Oil Conferences, 1988; 45-7.
9. Qureshi AA, Qureshi N, Hasler-Papacz JO, et al. Dietary tocotrienols reduce concentrations of plasma cholesterol, apolipoprotein B, thromboxane B_2, and platelet factor 4 in pigs with inherited hyperlipidemias. Am J Clin Nutr 1991; 53(Suppl):1042S-6S.
10. Qureshi AA, Qureshi N, Wright JJ, et al. Lowering of serum cholesterol in hypercholesterolemic humans by tocotrienols (palmvitee). Am J Clin Nutr 1991; 53(Suppl):1021S-6S.
11. Sundram K, Khor HT, Ong AS, Pathmanathan R. Effects of dietary palm oil on mammary carcinogenesis in female rats induced by 7,12-dimethylbenz(α)anthracene. Cancer Res 49:1447-51.
12. Kato A, Yamaoka M, Tamaka A, Komyama K, Umezawa I. Physiological effects of tocotrienol. Abura Kagaku 1985; 34:375-6.
13. Komiyama K, Iizuka K, Yamaoka M, Watanabe H, Tsuchiya N, Umezawa I. Studies on the biological activity of tocotrienols. Chem Pharm Bull (Tokyo) 1989; 37:1369-71.
14. Burton GW, Ingold KU. Autooxidation of biological molecules. I. The antioxidant activity of vitamin E and related chain-breaking phenolic antioxidants in vitro. J Am Chem Soc 1981; 103:6472-7.
15. Burton GW, Joyce A, Ingold KU. Is vitamin E the only lipid-soluble chain breaking antioxidant in human blood plasma and erythrocyte membranes? Arch Biochem Biophys 1983; 221:281-90.
16. Tappel AL. Vitamin E as the biological antioxidant. Vitam Horm 1962; 20:493-510.
17. Burton GW, Ingold KU. Vitamin E: application of the principles of physical organic chemistry to the exploration of its structure and function. Acc Chem Res 1986; 19:194-201.
18. Chipault JR. In: Lundberg WO, ed. Autoxidation and antioxidants, vol. 2. New York: Interscience, 1962; 12, 477-542.

19. Burton GW, Doba T, Gabe EJ, et al. Autoxidation of biological molecules. 4. Maximizing the antioxidant activity of phenols. J Am Chem Soc 1985; 107:7053-65.

20. Burlakova YB, Kuchtina YE, Ol'khovskaya IP, Sarycheva IK, Sinkina YB, Khrapova NG. Study of the antiradical activity of tocopherol analogues and homologues by the method of chemiluminescence. Biofizika 1980; 24:989-93.

21. Mukai K, Takamatsu K, Ishizu K. Preparation and ESR-studies of new stable tocopheroxyl model radicals. Bull Chem Soc Jpn 1984; 57:3507-10.

22. Mukai K, Tsuzuki N, Ouchi S, Fukuzawa K. Electron spin resonance studies of chromanoxyl radicals derived from tocopherols. Chem Phys Lipids 1982; 30:337-45.

23. Niki E, Kawakami A, Yamamoto Y, Kamiya Y. Oxidation of lipids. VIII. Synergistic inhibition of oxidation of phosphatidylcholine liposomes in aqueous dispersion by vitamin E and vitamin C. Bull Chem Soc Jpn 1985; 58:1971-5.

24. Niki E, Kawakami A, Saito T, Yamamoto Y, Tsuchiya J, Kamiya Y. Effect of phytyl side chain of vitamin E on its antioxidant activity. J Biol Chem 1985; 260:2191-6.

25. Niki E. Antioxidants in relation to lipid peroxidation. Chem Phys Lipids 44:227-53.

26. Niki E, Komuro E. Inhibition of peroxidation of membranes. Basic Life Sci 1987; 49:561-6. 1988;

27. Ingold KU, Burton GW, Foster DO, et al. A new vitamin E analogue more active than alpha-tocopherol in rat curative myopathy bioassay. FEBS 1986; 205:117-20.

28. Kagan VE, Serbinova EA, Bakalova RA, et al. Mechanisms of stabilization of biomembranes by alpha-tocopherol: the role of the hydrocarbon chain in the inhibition of lipid peroxidation. Biochem Pharmacol 1990; 40:2403-14.

29. Kagan VE, Serbinova EA, Packer L. NADPH-supported recycling and antioxidant activity of tocopherol homologues of differing hydrocarbon chain length in liver microsomes. Arch Biochem Biophys 1990; 282:221-5.

30. Kagan VE, Serbinova EA, Novikov KN, Ritov VB, Kozlov YP, Stoytchev TS. Toxic and protective effects of antioxidants in biomembranes. Arch Toxicol (Suppl) 1986; 9:302-5.

31. Kagan VE, Packer L, Serbinova E, et al. Mechanisms of vitamin E control of lipid peroxidation: regeneration, synergism, asymmetry, migration and metal chelation. In: Reddy C, ed. Biological oxidation systems. New York: Academic Press, 1990; 889-909.

32. Kagan VE. Tocopherol stabilizes membranes against phospholipase A, free fatty acids and lysophospholipids. Ann N Y Acad Sci 1989; 570:121-35.

33. Shinitzky M. Membrane fluidity and cellular function. In: Shinitzky M, ed. Physiology of membrane fluidity, vol. 1. Boca Raton, FL: CRC Press, 1984; 1-51.

34. Tatsuta T. Relationship between chemical structure and biological activity of vitamin E. I. Free tocopherols. Vitamins (Jpn) 1971; 44:185-90.

35. Nakano M, Sugioka K, Nakamura T, Oki T. Interaction between an organic hydroperoxide and an unsaturated phospholipid and alpha-tocopherol in model membranes. Biochim Biophys Acta 1980; 619:274-86.

36. Tsuchiya M, Scita G, Freistleben H-J, Kagan V, Packer L. Antioxidant radical-scavenging activity of carotenoids and retinoids as compared to alpha-tocopherol. Methods Enzymol 1992; 190: (submitted).

37. Yamaoka M, Komiyama K. Antioxidative activities of alpha-tocotrienol and its derivative in the oxidation of dilinoleoylphosphatidylcholine liposomes. J Jpn Oil Chem Soc (Yukagaku) 1989; 38: 478-85.

38. Fridovich I. Biological effects of the superoxide radical. Arch Biochem Biophys 1986; 247:1-11.

39. Seher VA, Ivanov SA. Natural antioxidants. I. The antioxidant effects of tocotrienols. Fette Seifen Anstri 1973; 75:606-9.

40. Serbinova EA, Kagan VE, Han D, Packer L. Free radical recycling and intramembrane mobility in the antioxidant properties of alpha-tocopherol and alpha-tocotrienol. 1991; 10:263-75.

41. Kagan VE, Serbinova EA, Koynova GM, et al. Antioxidant action of ubiquinol homologues with different isoprenoid chain length in biomembranes. Free Radic Biol Med 1990; 9:117-26.

42. Kagan VE, Bakalova RA, Serbinova EA, Stoytchev TS. Fluorescence measurements of incorporation and hydrolysis of tocopherol and tocopheryl esters in biomembranes. Methods Enzymol 1990; 186:355-67.

43. Hellenius A, Simons K. Solubilization of membranes by detergents. Biochim Biophys Acta 1975; 415:29-79.

44. Emanuel NM, Lyaskovskaya YN. The inhibition of fat oxidation processes. New York: Pergamon Press, 1967.

45. Mahoney LR, Ferris FC. Evidence for chain transfer in the autooxidation of hydrocarbons retarded by phenol. J Am Chem Soc 1963; 85:2345-6.

46. McCay PB. Vitamin E interaction with free radicals and ascorbate. Annu Rev Nutr 1985; 5:323-40.

47. Kagan VE, Serbinova EA, Packer L. NADPH-supported recycling and antioxidant activity of tocopherol homologues of differing hydrocarbon chain length in liver microsomes. Arch Biochem Biophys 1990; 282:221-5.

B. Cell Biology

20

Tocotrienols: Novel Hypocholesterolemic Agents with Antioxidant Properties

Nilofer Qureshi

William S. Middleton Memorial Veterans Hospital and University of Wisconsin, Madison, Wisconsin

Asaf A. Qureshi

Advanced Medical Research, Madison, Wisconsin

INTRODUCTION

It is recognized that populations that consume large amounts of cereal grains tend to have a lower incidence of cardiovascular disease (1–3). Studies on cereal grains revealed that barley is particularly effective in lowering lipid levels in animal models (4–9). The ability of barley extracts to lower lipids in vivo prompted the purification and identification of chemical constituents responsible for the cholesterol-lowering activity. α-Tocotrienol was isolated and identified from barley extracts using state-of-the-art methods and was designated as the biologically active compound, based on subsequent in vitro and in vivo evaluation (10). Our next study was the demonstration that brewer's grain and its subfractions (bran flour and high-protein flour) caused significant lowering of serum total cholesterol and low-density lipoprotein (LDL) cholesterol in chickens (11–14).

Later, however, animal and human experiments with palm oil-enriched diets unexpectedly demonstrated that palm oil feeding does not raise serum cholesterol, but in fact it lowers it. An explanation for palm oil's hypocholesterolemic effects, despite its fatty acid composition of approximately 50% saturated fatty acids and 50% unsaturated fatty acids, was sought and explored. We discovered that the tocotrienols present in palm oil were responsible for this effect (15).

With the identification of tocotrienols as the hypocholesterolemic components of barley and palm oil, we initiated a systematic program to study the effects of the homologous series of tocotrienols from various natural sources, such as barley, other cereals, brewers' grain oil, palm oil, wheat germ oil, and rice bran oil. These oils contain variable amounts of α-, γ-, and δ-tocotrienols. These homologs were purified by high-performance liquid chromatography, and new studies into the biological actions of tocotrienols were initiated. This was soon followed by the use of chemically synthesized tocotrienols for assaying biological activities. Recent studies assessing the potency of synthetic tocotrienols in cultured cells confirmed the effect of tocotrienols as hypocholesterolemic agents.

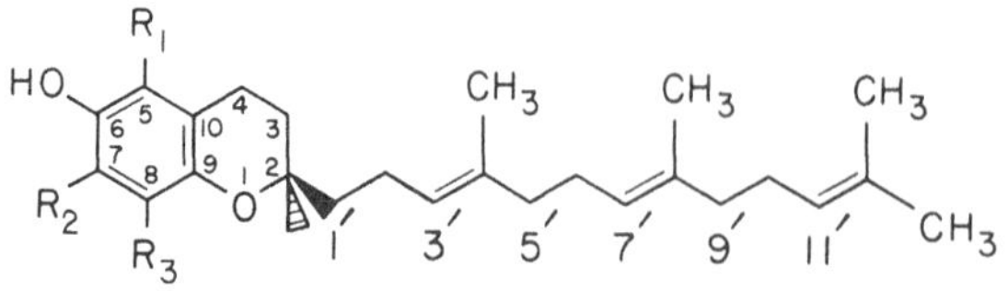

		R_1	R_2	R_3	Mw
α-T$_3$	α-tocotrienol	CH$_3$	CH$_3$	CH$_3$	424
β-T$_3$	β-tocotrienol	CH$_3$	H	CH$_3$	410
γ-T$_3$	γ-tocotrienol	H	CH$_3$	CH$_3$	410
δ-T$_3$	δ-tocotrienol	H	H	CH$_3$	396

FIGURE 1 Structures of a homologous series of tocotrienols: R_1, R_2, R_3, H, or methyl groups; T$_3$, tocotrienols; and M_w, molecular weight.

STRUCTURE AND NOMENCLATURE

Two major homologous series of tocochromanols having vitamin E activity in animals are synthesized by cyanobacteria and the eukaryotic plants. These are the tocopherols, with a saturated side chain, and the tocotrienols, with an unsaturated phytyl side chain. The natural tocotrienols are the 2R isomers, and their double bonds possess the *all-trans* configuration, as shown in Figure 1.

Vitamin E is the term used for eight naturally occurring essential fat-soluble nutrients, α-, β-, γ-, and δ-tocopherols and α-(5,7,8-trimethyl)-, β-(5,8-dimethyl)-, γ-(7,8-dimethyl)-, and δ-(8-monomethyl)tocotrienols, depending on the number and position of methyl substituents in the chroman ring. Tocopherols predominate in certain oils, such as corn oil, soybean oil, and olive oil, whereas the tocotrienol series predominates in palm oil, rice bran oil, and barley oil. Tables 1 and 2 show the composition of tocopherols and tocotrienols in commonly consumed oils and cereals, respectively. Other sources of tocotrienols have been described (16–20).

TABLE 1 Tocopherols and Tocotrienols in Different Oils and Fats

Oils and fats	Tocopherols (ppm)					Tocotrienols (ppm)					Total T + T$_3$ (ppm)
	α-T	β-T	γ-T	δ-T	%-T	α-T$_3$	β-T$_3$	γ-T$_3$	δ-T$_3$	%-T$_3$	
Corn oil	112	50	602	18	100	—	—	—	—	0	782
Soybean oil	101	—	593	264	100	—	—	—	—	0	958
Rice oil	124	40	50	—	22	184	21	570	—	78	989
Palm oil	279	—	61	—	31	274	—	398	69	69	1081
Palm oil	152	—	—	—	17	205	—	439	94	83	890
MD-RBD[a] (barley oil)	144	30	75	60	32	402	14	180	60	68	965
Olive oil	51	—	—	—	100	—	—	—	—	0	51
Coconut oil	5	—	—	6	31	5	1	19	—	69	36
Lard	12	—	7	—	73	7	—	—	—	27	26

[a]Molecular distilled-refined, bleached, and deodorized.

TABLE 2 Tocopherols and Tocotrienols in Different Cereals and Brans[a]

Cereals and their brans	Tocopherols (ppm)					Tocotrienols (ppm)					Total T + T₃ (ppm)
	α-T	β-T	γ-T	δ-T	%-T	α-T₃	β-T₃	γ-T₃	δ-T₃	%-T₃	
Wheat	14	7	0	0	36	33	0	0	0	64	59
Wheat germ	239	90	0	0	72	30	100	0	0	28	459
Wheat bran	16	10	0	0	28	13	55	0	0	72	94
Corn	6	0	45	0	86	3	0	5	0	14	59
Oat	5	1	0	0	32	11	2	0	0	68	19
Rye	16	4	0	0	47	15	8	0	0	53	43
Rice, white	1	0	1	0	42	1	0	2	0	58	4
Rice, brown	6	1	1	0	35	4	0	10	0	65	22
Rice bran	6	1	0	8	49	1	4	4	6	51	30
Rice bran[b]	3	15	4	2	24	1	14	22	29	66	90
Barley	2	4	0	1	31	11	3	2	0	69	23
Barley bran[c]	11	16	36	4	42	36	25	19	11	58	158
Brewers grain[c]	31	42	114	20	40	199	40	39	34	60	519

[a]Abbreviations used are T, tocopherols and T₃, tocotrienols.
[b]Advanced Medical Research, unpublished data.
[c]Miller Brewing Company, unpublished data.
Source: Data from references 18 and 19.

BIOSYNTHESIS

The pathway for the biosynthesis of tocopherols and tocotrienols has not been elucidated. The immediate precursors and the nature of the enzymic reactions involved are not known at present. A proposed pathway for the biosynthesis of tocotrienols and tocopherols is shown in Figure 2. The prenylation reaction with a polyprenyl phosphate takes place on homogentisic acid or its β-glucoside (21–25), which is derived from *p*-hydroxyphenylpyruvic acid. The two proposed routes for the biosynthesis of δ-tocopherol are presented. The first route involves

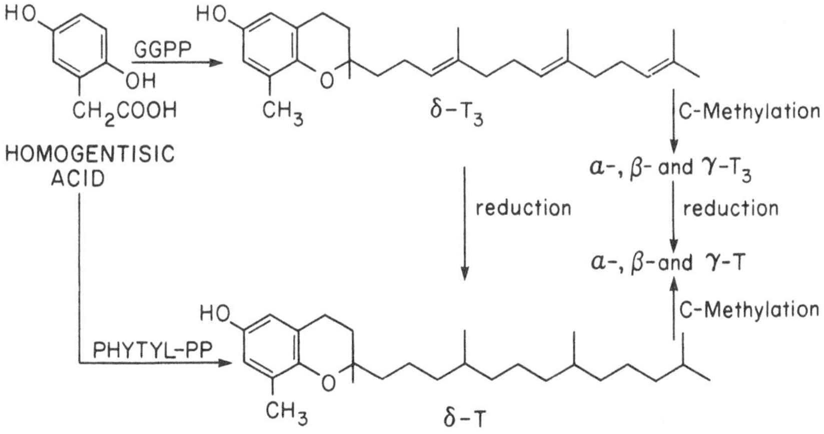

FIGURE 2 Biosynthesis of tocotrienols and tocopherols. The two proposed routes for the biosynthesis of δ-tocopherol are presented. The first route, involving geranylgeranyl pyrophosphate, is the preferred pathway for the biosynthesis of tocotrienols. GGPP, geranylgeranyl pyrophosphate; phytyl-PP, phytyl pyrophosphate; T, tocopherols; T₃, tocotrienols.

decarboxylation of homogentisic acid, followed by attachment of the geranylgeranyl group at the position *meta* to the methyl group and cyclization, which leads to the formation of δ-tocotrienol. The δ-tocotrienol then gives rise to the γ-, β-, and α-tocotrienols by successive C methylations (21–25). The δ-tocopherol is converted to the γ-, β-, and α-tocopherols.

Alternatively, the second route involves the reduced form of geranylgeranyl pyrophosphate—the phytyl pyrophosphate as the substrate giving rise to δ-tocopherol directly. The first route is the preferred pathway for the biosynthesis of tocotrienols.

EXTRACTION, PURIFICATION, AND CHARACTERIZATION

Extraction

Cereals (finely ground) and oils (1 g) are usually extracted with methanol (8.0 ml). The suspension is then centrifuged to recover the methanol layer, which is evaporated under vacuum at 40°C. The remaining dried residue is then extracted with hexane (2 ml). This hexane extract is purified to homogeneity by silica gel high-performance liquid chromatography. Tocotrienols from serum, plasma, or any other mammalian tissue (1 ml or 1 g) are extracted similarly, except instead of methanol, petroleum ether (8 ml) or hexane is used for extracting. Other methods have also been described (26,27).

Purification

Analytical instrumentation consists of a Waters pump 6000A, a Shimadzu fluorescence detector RF-535 and integrator CR3A, and a Kipp and Zonen BD41 recorder. Optimal conditions involve using a μ-Porasil (Waters column, 10 μ (microns), 4 mm × 30 cm) column and an isocratic system of hexane and isopropyl alcohol (99.7:0.3, vol/vol) at a flow rate of 1 ml/minute. Increasing the volume of isopropyl alcohol shortens the time of the run (28,29). The tocotrienols present in TRF, barley oil, and wheat germ oil are shown in Figure 3. Purified samples are then characterized by mass spectrometry.

Mass Spectrometry

Highly purified samples of tocotrienols are analyzed by either electron impact (10,30–32) or tandem mass spectrometry (33). The high-resolution mass spectral analysis of α-tocotrienol shows a molecular ion peak M^+ at m/e 424, which corresponds to the molecular formula $C_{29}H_{44}O_2$. Fragmentation peaks m/e 205 ($C_{13}H_{17}O_2$) and m/e 203 ($C_{13}H_{15}O_2$) are formed after the loss of the phytyl chain (three isoprenoid units). The fragmentation peak at m/e 165 ($C_{10}H_{13}O_5$) arises after cleavage of the chroman ring (10). The β-, γ-, and δ-tocotrienols show corresponding M^+ peaks at 410, 410, and 396, respectively. Fragmentation peaks for the tocotrienols are observed at 191 and 151 (β), 191 and 151 (γ), and 177 and 137 (δ).

HYPOCHOLESTEROLEMIC EFFECTS

The Mevalonate Pathway

In recent years, a great deal of emphasis has been laid on developing new cholesterol-lowering drugs. There is a strong causal relationship between elevated concentrations of cholesterol and atherosclerosis. β-Hydroxy-β-methylglutaryl coenzyme A (HMG-CoA) reductase catalyzes the rate-determining step in the biosynthesis of cholesterol (34). Under certain conditions, HMG-CoA synthase controls the rate of cholesterogenesis (35). Most of the early studies on HMG-CoA reductase were performed with the yeast enzyme. This enzyme was purified to

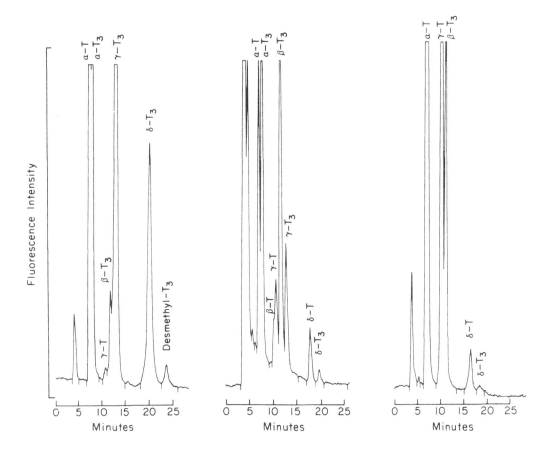

FIGURE 3 High-performance liquid chromatography of tocotrienols and tocopherols present in TRF (palm oil), barley oil, and wheat germ oil. Details of these analyses are presented in the text. T, tocopherols; T_3, tocotrienols.

homogeneity (36,37), and the mechanism of the enzyme-catalyzed reaction was determined by Qureshi et al. (38). HMG-CoA reductase catalyzes the two-step reduction of (4R,3S)-HMG-CoA (thioester) to (2R,3R)-mevalonate (alcohol) in the presence of 2 mol NADPH, as shown in Figure 4. The initial velocity patterns suggest a sequential mechanism for the addition of substrates in both the reductive steps. The HMG-CoA and NADPH bind to the reductase, and a hydride ion transfer from the A side of NADPH produces a hemithioacetal of mevaldate and CoA, which is the enzyme-bound intermediate. After replacement of NADP$^+$ with NADPH, the mevaldate hemithioacetal cleaves into mevaldate and CoA, which remain bound to the enzyme. During the second reductive step there is a hydride transfer from NADPH that reduces mevaldate to mevalonate, and the products mevalonate, CoA, and finally NADP$^+$ leave the enzyme (38). The second reductive step, particularly the release of NADP$^+$ from the enzyme, is the rate-determining step. This study predicted that groups X and Y on the enzyme act as acid-base catalysts to assist in the direct transfer of a hydride ion between the nucleotide and substrate (38). Later, Veloso et al. (39), using pH variation in kinetic parameters, showed that the X group of the yeast reductase is probably histidine and the Y group

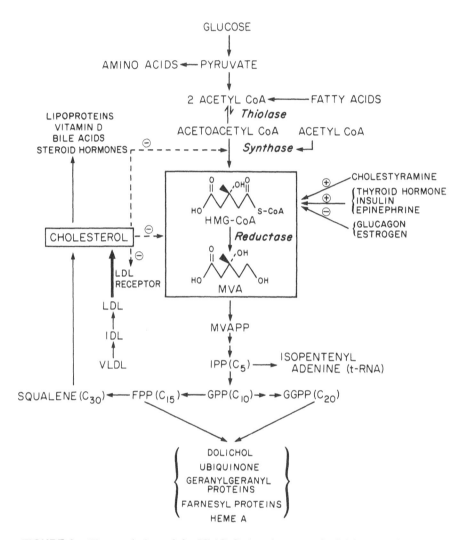

FIGURE 4   The regulation of the HMG-CoA reductase. The initial steps involved in the pathway to the biosynthesis of cholesterol, starting from acetyl CoA to isopentenyl pyrophosphate, were reviewed by Qureshi and Porter (37). The feedback regulation of the reductase was reviewed by Goldstein and Brown (43). Please see text for details. HMG-CoA, β-hydroxy-β-methylglutaryl coenzyme A; MVA, mevalonic acid; MVAPP, mevalonate pyrophosphate; IPP, isopentenyl pyrophosphate; GPP, geranyl pyrophosphate; GGPP, geranylgeranyl pyrophosphate; FPP, farnesyl pyrophosphate; LDL, low-density lipoproteins; IDL, intermediate-density lipoproteins; and VLDL, very low density lipoproteins.

has a carboxylate anion. Recently, Rodwell and associates, using mutation alteration of glutamate and aspartate residues of HMG-CoA reductase, concluded that glutamate residue 83 of the *Pseudomonas mevalonii* reductase is the principal catalytically important acid residue (*Y* group) (40).

HMG-CoA reductase is regulated by various nutritional and hormonal factors (41,42). It is also regulated by LDL receptors, as shown in Figure 4. Cholesterol is obtained from two sources, by biosynthesis starting with acetyl CoA and by receptor-mediated uptake of plasma

LDL. Both processes are controlled by the sterol-mediated feedback repression of the genes for HMG-CoA synthase, HMG-CoA reductase, and the LDL receptor (43). The posttranscriptional regulation of HMG-CoA reductase is accomplished by one of the nonsterol isoprenoids (43). The elegant study of Brown and Goldstein showed that the uptake of LDL-bound cholesterol in macrophages occurs through receptor-mediated endocytosis. This uptake then results in suppression of endogenous cholesterol biosynthesis, an enhanced rate of intracellular cholesterol esterification, and a reduction in the number of high-affinity LDL receptors expressed on the surface of the peripheral cells and liver cells (44). Subjects with homozygous familial hypercholesterolemia have a total or near total deficiency of LDL receptors and also have a tendency for premature coronary artery disease (45).

The accelerated degradation of reductase requires a nonsterol isoprenoid as well as a sterol. Sterols accelerate reductase degradation in part by diverting this mevalonate into a nonsterol regulatory product (43). The farnesylated proteins, such as *ras* protein and lamin B, may be the nonsterol regulators (43). The farnesyl group is covalently attached to the fourth cysteine (COOH end) in a thioether linkage as a result of posttranslational modification (46–50). The p^{21} *ras* proteins (M_r 20,000–30,000) are a family of guanosine triphosphate (GTP) binding proteins that regulate cell growth. Mutation in the *ras* gene that abolishes their GTPase activity leads to uncontrolled growth. The finding that the *ras* proteins can be inhibited by blocking a step in cholesterol synthesis suggests a pharmaceutical route for the control of *ras*-induced cancer (50,51). It is possible that a farnesylated protein also participates in the feedback control of HMG-CoA reductase, thus ensuring a steady mevalonate supply (43). Alternatively, it may influence the degradation of the reductase (43). Other isoprenoid-linked tRNA and proteins have also been reported. Isopentenyl adenine has been found in tRNA (52), and geranylgeranyl-modified proteins have been identified in HeLa cells (53).

Oxygenated sterols, such as 25-hydroxycholesterol and 7-ketocholesterol, exhibit a marked inhibition of HMG-CoA reductase activity and sterol synthesis from acetate in intact cells. It has been proposed that the oxysterols decrease the half-life of the reductase by increasing the rate of degradation or inactivation (54). Kandutsch et al. showed that the suppressor oxygenated sterol binds to a cytosolic protein different from the protein that binds cholesterol (55).

These oxysterols are up to 100 times more potent than cholesterol itself. Their effect on the regulation of mevalonate at the transcriptional level is very similar to that of the sterols; the enhanced potency of the oxysterols has been attributed to their greater water solubility. It was demonstrated that the sterol-resistant mutant cells retain a protein that binds to the sterol regulatory element (SRE, present in the promoters of LDL receptor and HMG-CoA synthase genes) and enhances transcription, but these cells have lost the ability to silence this protein in the presence of sterols (43).

Most of the drugs available for hypercholesterolemia are aimed at reducing cholesterol and LDL cholesterol levels. Most of these drugs have some side effects, however, and their mechanism of action at the cellular level is not known (56). The following drugs are very effective in the therapy of hypercholesterolemia and are widely used. Compactin (57) and Mevinolin (Lovastatin) (58) are competitive inhibitors of HMG-CoA reductase. These drugs cause an enhanced rate of receptor-mediated catabolism of LDL from plasma and a small reduction in the rate of very low density lipoprotein (VLDL) and LDL synthesis. Bile acid sequesterants, such as cholestyramine and colestipol, bind bile acids and increase the excretion of acid steroids in the feces. Since the bile acid pool size is decreased, this stimulates the degradation of cholesterol into bile acids, which in turn leads to an increase in cholesterol biosynthesis and an increase in the number of LDL receptors on the hepatocyte membrane (59). Clofibrate and Gemfibrosil are effective in reducing LDL cholesterol (60). Nicotinic acid reduces the synthesis of VLDL, which in turn leads to a reduction in the synthesis of LDL (61). Probucol has been shown to increase the fractional rate of catabolism of LDL and increases the biliary excretion of cholesterol (62). This chapter focuses on the hypocholesterolemic effects and antioxidant properties of the tocotrienols.

Effect of Tocotrienols in the Chicken Model

Effect of Feeding Different Oils on the
Levels of Total Serum Cholesterol of Normal Chickens

Initially, the effects of feeding different oils and fats on cholesterologenesis were investigated (63,64). Corn-based diets supplemented with 5.0% lard, coconut oil, corn oil (control), soybean oil, rice bran oil, barley oil, palm oil, and corn oil plus tocotrienol-rich fraction (TRF, isolated from palm oil, and containing α-tocopherol, 15–20%; α-tocotrienol, 12–15%; γ-tocotrienol, 35–40%; and δ-tocotrienol, 25–30%) were fed to 8-week-old female chickens for 4 weeks. The birds were fasted for 14 h before killing. The total serum cholesterol levels are shown in Figure 5.

This study showed that the rice bran oil, barley oil, and palm oil contain potent constituents that suppress cholesterologenesis, from a total serum cholesterol value of 170 mg/dl in lard-fed chickens to 106 mg/dl for corn + TRF-fed chickens. Since these oils vary in their composition of saturated versus unsaturated fatty acids and the content of tocotrienols, we

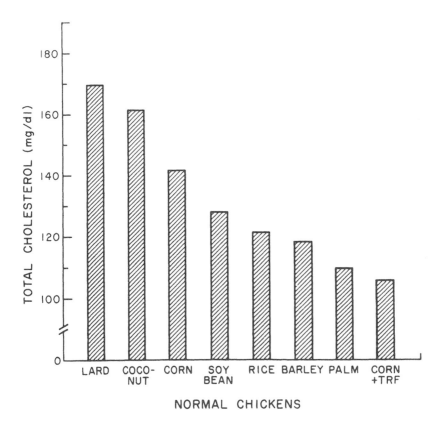

FIGURE 5 Effect of feeding different oils on the levels of total serum cholesterol of normal chickens. TRF, tocotrienol-rich fraction isolated from palm oil. [Adapted from data presented by Qureshi et al. (63,64).]

wondered if the tocotrienols were responsible for the cholesterol-lowering effect. The experiment carried out with the corn oil + TRF isolated from palm oil demonstrated that the tocotrienols are indeed responsible for the cholesterol-lowering effect.

Effect of Feeding Purified α-, γ-, and δ-Tocotrienols to Hypercholesterolemic Chickens

Once it was determined that the tocotrienols are the potent hypocholesterolemic components of the oils, we wanted to investigate which homolog of the tocotrienols is most effective. Diets supplemented with the purified α-, γ-, and δ-tocotrienols from palm oil (20 ppm) were fed to hypercholesterolemic chickens (15). The levels of total serum cholesterol decreased 29–30% upon feeding tocotrienols compared to the control corn diet (Fig. 6). LDL cholesterol and HMG-CoA reductase activity were reduced by 14–19 and 55–61%, respectively (Figs. 7 and 8). These interesting results supported the conclusion that the tocotrienols are the constituents of oils with potent hypocholesterolemic effects. The tocotrienols suppressed cholesterogenesis more effectively in hypercholesterolemic chickens than in normal chickens. Furthermore, the tocotrienols acted by reducing HMG-CoA reductase activity, and γ- and δ-tocotrienols were the most potent hypocholesterolemic agents in the avian model.

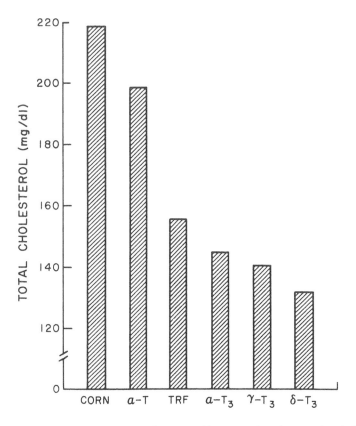

FIGURE 6 Effect of feeding corn oil, α-tocopherol, tocotrienol-rich fraction, and α-, γ-, and δ-tocotrienols on the total serum cholesterol levels of hypercholesterolemic chickens. T, tocopherols; T_3, tocotrienols; TRF, tocotrienol-rich fraction. [Adapted from data presented by Qureshi et al. (15).]

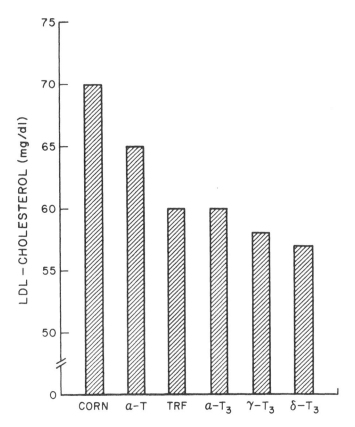

FIGURE 7   Effect of feeding corn oil, α-tocopherol, tocotrienol-rich fraction, and α-, γ-, and δ-tocotrienols on the LDL cholesterol levels of hypercholesterolemic chickens. T, tocopherols; T_3, tocotrienols; TRF, tocotrienol-rich fraction; LDL, low-density lipoprotein. (Adapted from data presented in Ref. 15.)

Effect of Tocotrienols in the Swine Model

Since pigs are very similar to humans in their cholesterol metabolism, normolipemic and genetically hypercholesterolemic pigs of defined lipoprotein genotype were fed a standard corn-soybean diet supplemented with 50 ppm molecular-distilled tocotrienol-rich fraction (TRF) isolated from palm oil (65). Hypercholesterolemic pigs fed the TRF-supplemented diet showed a 44% decrease in total serum cholesterol and a 60% decrease in LDL cholesterol. No significant changes occurred in the HDL cholesterol levels, as shown in Figure 9. However, a significant decrease of 44% in total serum cholesterol and a 60% decrease in LDL cholesterol was observed (Fig. 9).

The most interesting results were observed in swine fed the experimental diet (TRF) for 6 weeks followed by the control corn-based diet. The decrease in plasma lipid profiles obtained from the experimental diet persisted in the hypercholesterolemic swine even after 8 weeks on the control diet, whereas in the normolipemic swine the values increased to those of the control group.

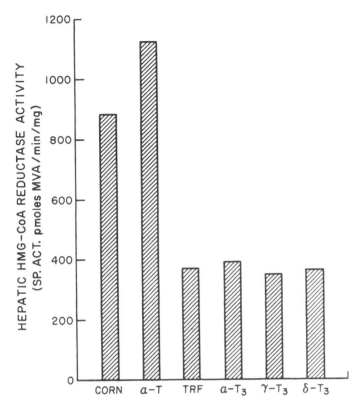

FIGURE 8 Effect of feeding corn oil, α-tocopherol, tocotrienol-rich fraction, and α-, γ-, and δ-tocotrienols on the hepatic β-hydroxy-β-methylglutaryl coenzyme A (HMG-CoA) reductase activity of hypercholesterolemic chickens. T, tocopherols; T_3, tocotrienols. [Adapted from data presented by Qureshi et al. (15).]

Effect of Tocotrienols in Hypercholesterolemic Human Subjects

Recently, the effectiveness of the tocotrienol rich fraction oil capsules from palm oil and the molecular distilled brewers' grain oil was investigated in hypercholesterolemic human subjects.

A double-blind crossover pilot study was carried out in 47 hypercholesterolemic subjects (serum cholesterol 240–310 mg/ml): 27 subjects were treated with TRF (isolated from palm oil) in corn oil; 10 subjects received 200 mg/day of molecular distilled, refined, bleached, and deodorized oil from brewers' grain (MD-RBD); and 10 received a corn oil capsule placebo (66,67). All subjects consumed their normal diets and continued their normal daily activities throughout the study. Dietary supplementation of TRF (200 mg/day, Palmvitee capsules) for 4 weeks caused significant drops of 15–22% in serum total cholesterol and 10–20% in LDL cholesterol, as seen in Figure 10. The subjects treated with MD-RBD barley oil capsules for 4 weeks experienced significant drops of 15% in serum total cholesterol and 10% in LDL cholesterol. There was no change in either of these parameters in the placebo group.

In the second leg of the study, subjects in the placebo group were treated with TRF. These subjects showed a similar decrease of 14–21% in serum total cholesterol and 13–21% in LDL

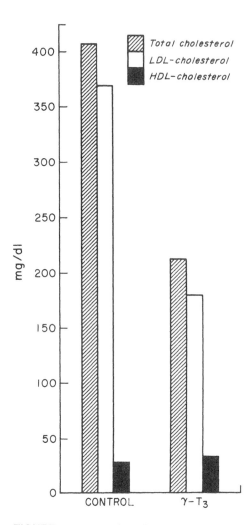

FIGURE 9 Effect of feeding tocotrienol-rich fraction (isolated from palm oil) on the total serum cholesterol, LDL cholesterol, and high-density lipoprotein (HDL) cholesterol levels in hypercholesterolemic swine. γ-T$_3$, γ-tocotrienol; LDL, low-density lipoproteins; HDL, high-density lipoproteins. (Adapted from data presented in Ref. 65.)

cholesterol. When the subjects initially given TRF capsules and MD-RBD barley oil were given placebo diets, however, their serum cholesterol levels remained low and this effect persisted even after 6 weeks. The subjects did not show any side effects. This maintained decrease in cholesterol levels was evidently due to the high levels of tocotrienols in the serum of these subjects. This was confirmed by high-performance liquid chromatography, the levels being 17 and 795 μg/dl in the sera of placebo-treated and TRF-treated human subjects, respectively, after 4 weeks of treatment.

In these studies, 40 of the 47 subjects responded well to the treatment. The 7 poor respondents, all with cholesterol > 300 mg/dl, were given additional γ- and δ-tocotrienol capsules (50 mg, 2 capsules/day), which caused a dramatic drop of 35–40% in their total serum

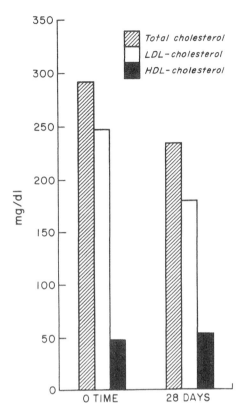

FIGURE 10 Effect of feeding tocotrienol-rich fraction (isolated from palm oil) on the total serum cholesterol, LDL cholesterol, and HDL cholesterol levels in hypercholesterolemic human subjects. 0 time refers to baseline values. LDL, low-density lipoproteins; HDL, high-density lipoproteins. (Adapted from data published in Ref. 67.)

cholesterol after 4 weeks of feeding. This indicates that γ- and δ-tocotrienols are the most potent hypocholesterolemic components in the oils.

In the pilot studies, the hypocholesterolemic effects of tocotrienols in humans were consistent with those observed in the pig and avian models. The magnitude of this effect was smaller than observed in the pig or avian model but very encouraging considering the lack of control of intake of dietary fat.

Robinson and Lupton (68) and Weber et al. (69) investigated the effect of barley flour and barley oil in 79 hypercholesterolemic men and women in a 30 day intervention trial. Each participant followed the National Cholesterol Education Program step 1 diet and was assigned to one of three treatment groups. Group 1 added 20 g cellulose fiber to the daily intake, group 2 added 3 g extracted barley oil, and group 3 added 30 g barley flour. A significant decrease in total cholesterol and LDL cholesterol was observed in both the barley flour group (-21.6 mg/dl) and the barley oil group (-18.7 mg/dl). The cellulose control group showed no change in cholesterol levels. These results showed that both barley oil and barley flour clearly contributed to and enhanced the hypocholesterolemic efficiency of the NCEP step 1 diet.

Comparison of the Hypocholesterolemic Effects of Tocotrienols and Lovastatin in Chickens

These experiments were carried out in normal chickens. Corn-based diets supplemented with 50 ppm of either TRF (from palm oil), Lovastatin, γ-tocotrienol, or Lovastatin + γ-tocotrienol were fed to normal chickens for 4 weeks. The control group showed a level of total serum cholesterol of 121 mg/dl, as shown in Figure 11. TRF and Lovastatin caused a significant decrease in total serum cholesterol to 100 and 98 mg/dl, respectively. The greatest decrease was observed with chickens fed δ-tocotrienol or γ-tocotrienol + Lovastatin, however, which showed values of total serum cholesterol as low as 82 and 76 mg/dl, respectively.

Effect of Synthetic Tocotrienols on Cholesterol Biosynthesis in Cultured Cells

Wright and associates (personal communication) carried out experiments using synthetic tocotrienols in the HepG2 cell culture model. HepG2 cells were preincubated with 10 μM γ-tocopherol or other isomers of tocotrienols for 4 h. Inhibition of cholesterol synthesis was mea-

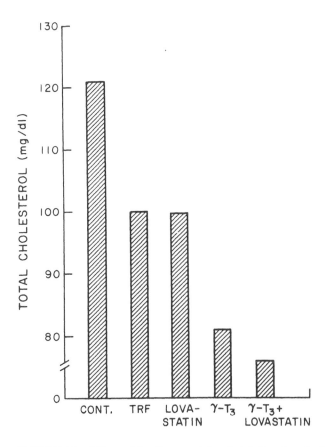

FIGURE 11   Effect of feeding tocotrienol-rich fraction (isolated from palm oil), Lovastatin, γ-tocotrienol, and a combination of a γ-tocotrienol and Lovastatin (50 ppm each) on the total serum cholesterol levels in normal chickens. CONT., control; T_3, tocotrienols; and TRF, tocotrienol-rich fraction. (From Qureshi et al., unpublished data.)

sured either by the incorporation of [^{14}C]acetate into digitonin-precipitable material or by assaying the HMG-CoA reductase. The results are shown in Table 3. Synthetic γ- and δ-tocotrienols were significantly more active than α-tocotrienols in suppressing HMG-CoA reductase. The suppression of HMG-CoA reductase protein synthesis was confirmed by immunoassay using the western blot technique. Similar experiments with rat hepatocytes gave similar results, but the effect was smaller.

Mechanism of Action for the Hypocholesterolemic Effect of the Tocotrienols

The results described here are noteworthy in many respects. The tocotrienols seem to have a profound effect on the biosynthesis of cholesterol. All studies point to the fact that the tocotrienols are very effective in lowering blood cholesterol and LDL cholesterol levels by repressing HMG-CoA reductase. It has been established that the suppression of HMG-CoA reductase requires two regulators: cholesterol delivered by receptor-mediated uptake of LDL and a nonsterol product derived from mevalonate. The former is expressed predominantly through changes in the rate of transcription of the HMG-CoA reductase gene and the latter by modulating the efficiency of the translation of HMG-CoA reductase mRNA or by degradation of the protein (43).

The exact mechanism by which the tocotrienols act at the cellular level has not been established. Some proposals have been made by Wright and associates, however, who studied the effect of tocotrienols on cholesterol synthesis in HepG2 cells (70). Inhibition of sterol synthesis correlates with the rapid (<2 h) suppression of HMG-CoA reductase. γ-Tocotrienol (10 μM) decreases the rate of HMG-CoA synthesis to 51% of control and stimulates the HMG-CoA reductase protein degradation 2.3-fold, and $t_{1/2}$ decreases from 3.7 to 1.6 h. The LDL receptor protein in HepG2 cell membranes is not cosuppressed by γ-tocotrienol, in contrast to the effects of 25-hydroxycholesterol. γ-Tocotrienol does not decrease the level of HMG-CoA reductase mRNA in HepG2 cells. These results suggest that the tocotrienols inhibit sterologenesis by suppressing HMG-CoA reductase through a novel posttranscriptional mechanism. Squalene transfer and epoxidation are also modulated by γ-tocotrienol. Squalene transfer and formation of squalene epoxide increase and the production of squalene dioxide decreases using microsomes from HepG2 cells treated with 10 μM γ-tocotrienol versus controls (71).

It is conceivable that the tocotrienols are transported in the blood in association with low- and high-density lipoproteins, as has been reported for the tocopherols (71,72,73). The tocotrienols may prevent the oxidation of lipoproteins. Tocotrienols may also bind to other proteins; such a protein has been described for α-tocopherol (74,75). The hydroxyl group on

TABLE 3 Effect of Synthetic Tocotrienols and Tocopherols on Cholesterol Synthesis and HMG-CoA Reductase in HepG2 Cells

Compound, 10 μM	% Inhibition ^{14}C-Ac	% Suppression of HMG-CoA reductase
γ-Tocopherol	3	0
d,l-α-Tocotrienol	20	19
d,l-γ-Tocotrienol	75	65
d,l-δ-Tocotrienol	69	62
d,γ-Tocotrienol	78	64
d,δ-Tocotrienol	79	65

the chroman ring may be important in the binding. With the recent discovery of the farnesylated and geranylgeranyl proteins, it is very tempting to speculate that the tocotrienols, which have a side chain derived from geranylgeranyl pyrophosphate or its derivative phytol, may bind to similar proteins, which regulate cell growth and may also serve as feedback inhibitors of HMG-CoA reductase at the posttranscriptional level. This field is becoming very interesting and exciting as mysteries surrounding the regulation of HMG-CoA reductase are unraveled.

OTHER EFFECTS

Experiments have also been carried out to investigate the effects of tocotrienols on apolipo-protein B levels in chickens, humans, and swine. Data from different laboratories have suggested that the apolipoprotein A (from HDL) to apolipoprotein B (from LDL) ratio is a better indicator than the HDL/LDL ratio for the assessment of coronary heart disease. The former ratio was reduced in tocotrienol-treated subjects (67).

The effects of tocotrienols on thromboxane B_2 levels have also been examined. Thromboxane A_2 is a potent inducer of platelet aggregation, and it also acts as a vasoconstrictor. Since its half-life is about 30 s and it is rapidly hydrolyzed to thromboxane B_2, the measurement of thromboxane B_2 is accepted as a indicator of thromboxane A_2 production. Tocotrienols cause a decrease in thromboxane B_2 levels in humans (20–26%) and a decrease in platelet aggregation against ADP, EPI, and collagen (15–30%) in platelet-rich plasma. Thus, tocotrienols may serve as antithrombotic agents by decreasing platelet aggregation significantly (67).

The α- and γ-tocotrienols were also evaluated for activity against transplantable murine tumors when inoculated intraperitoneally into mice. The α- and γ-tocotrienols are effective against sarcoma 180, Ehrlich carcinoma, and IMC carcinoma. The antitumor activity of γ-tocotrienol is higher than that of α-tocotrienol. Tocotrienols also inhibit growth of human and mouse tumor cells when the cells are incubated with tocotrienols for 72 h in vitro (76).

ANTIOXIDANT PROPERTIES

The biological activity of vitamin E is generally believed to be due to its antioxidant action, specifically, inhibition of lipid peroxidation in biological membranes. The free-radical chain mechanism of lipid peroxidation is shown in Figure 12. The peroxidation proceeds in three phases: initiation, propagation, and termination (77). In the initiation phase, a carbon-centered lipid radical $R^{\cdot}$ is produced from a polyunsaturated fatty acid RH, forming a hydroperoxide, ROOH. This $R^{\cdot}$ reacts with molecular oxygen in the propagation phase to form $ROO^{\cdot}$.

$$R^{\cdot} + O_2 \rightarrow ROO^{\cdot}$$
$$ROO^{\cdot} + RH \rightarrow ROOH + R^{\cdot}$$

This propagation process continues and consumes the valuable polyunsaturated fat. In the termination phase, the chain reaction stops when a peroxyl radical ($ROO^{\cdot}$) combines with another $ROO^{\cdot}$.

$$ROO^{\cdot} + ROO^{\cdot} \rightarrow \text{inactive products}$$

However, tocotrienols (T_3-OH) and tocopherols (T-OH) can intercept the peroxyl radical more rapidly than can polyunsaturated fatty acids by the following reactions:

$$T_3\text{-OH} + ROO^{\cdot} \rightarrow T_3\text{-O}^{\cdot} + ROOH$$
$$T_3\text{-O}^{\cdot} + ROO^{\cdot} \rightarrow \text{inactive products}$$

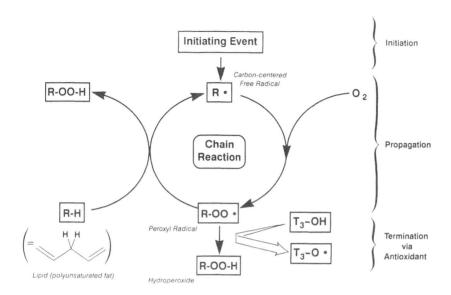

FIGURE 12 The three phases of the free-radical chain mechanism of lipid peroxidation. [Adapted from Burton and Traber (77).]

The T_3-O˙ radical is unable to continue the chain and reacts with the peroxyl to form inactive products (77–79).

α-Tocopherol has been labeled as the most efficient chain-breaking antioxidant. Recently, Packer and associates showed that compared to α-tocopherol, α-tocotrienol possesses a 40–60 times higher antioxidant activity against Fe^{2+} + ascorbate- and Fe^{2+} + NADPH-induced lipid peroxidation in rat liver microsomal membranes and 6.5 times greater protection of cytochrome P_{450} against oxidative damage (80). This higher antioxidant potency of α-tocotrienol compared to α-tocopherol is attributed to the combined effects of three properties: its higher recycling efficiency from chromanoxyl radical, its more uniform distribution in membrane bilayer, and its stronger disordering of membrane lipids, which makes interaction of chromanols with lipid radicals more efficient.

The activated neutrophils are one of the potential extracellular sources of oxygen free radicals. The membrane-bound NADPH oxidase produces superoxide that aids in the killing of any microorganism engulfed by these cells. The superoxide released amplifies the inflammatory response by activation of a latent chemotactic factor present in extracellular fluids. The chemotactic factor activates the neutrophils and causes them to attach to the endothelium. The neutrophils begin to injure the tissue and start releasing oxidative and hydrolytic enzymes. This phenomenon impairs the local circulation of blood (81). Recently we showed that γ-tocotrienol can inhibit the release of superoxide in human peripheral blood neutrophils. The neutrophils were isolated by density centrifugation on Ficoll-Hypaque gradients using conventional methods. Neutrophils were placed in a 96-well plate. γ-Tocotrienol and phorbol myristate acetate were added at the same time. The secretion of superoxide anion was measured as the superoxide dismutase-inhibitable reduction of ferricytochrome C. The superoxide was reduced from 19.7 nmol superoxide per 5×10^5 cells per h in controls to 13.5 nmol at 10^{-5} M and to 0 nmol at 10^{-3} M concentrations of γ-tocotrienols (Qureshi et al., unpublished data). This suggests that tocotrienols may also serve as excellent antiinflammatory agents.

PERSPECTIVE

The last couple of years have brought much new knowledge of the biological activities of the tocotrienols. Most of the early work was done exclusively on the tocopherols, which are the major components of vitamin E. The δ- and γ-tocotrienols seem to be very effective hypocholesterolemic agents in both in vivo and in vitro experiments. However, the exact mechanism by which the tocotrienols suppress HMG-CoA reductase needs to be clarified. The mechanism differs from that of Compactin and Lovastatin, which competitively inhibit the reductase and in cultured cells block the synthesis of mevalonate and trigger adaptive reactions that yield a 200-fold increase in reductase within a few hours. This induced protein is inactive in cells because it is blocked by the inhibitor. Tocotrienols, on the other hand, cause a 50% decrease in reductase protein in cultured cells in less than 2 h. The amount of mRNA for the reductase stays the same after the treatment. This suggests that the effect of tocotrienols is probably at the posttranscriptional level and is mediated by a nonsterol isoprenoid. Several questions need to be answered. Do the tocotrienols regulate HMG-CoA reductase at the translational level? If so, are the farnesyl or geranylgeranyl proteins involved in this posttranscriptional regulation? Whether tocotrienols have any direct inhibitory effects on the purified HMG-CoA reductase or indirect effects due to the release of another mediator remains to be established.

The other interesting impact of this research will be in the field of antioxidants. Tocotrienols may turn out to be the most potent lipid-soluble chain-breaking antioxidants, and this may have significant clinical implications. Moreover, tocotrienols may be excellent antiinflammatory agents.

Tocotrienols also decrease platelet aggregation and are good antithrombotic agents, reducing the thromboxane B_4 levels. The synthesis of leukotrienes and prostaglandins may be affected by tocotrienols. The effect of tocotrienols on the regression of tumors is also a promising line of inquiry. These are some of the questions that are currently being investigated.

ACKNOWLEDGMENTS

This review and related work were supported in part by the Miller Brewing Company, the Bristol-Myers Pharmaceutical Company, the Palm Oil Research Institute of Malaysia (PORIM), and Advanced Medical Research. We thank Carol Steinhart for editorial assistance and Debi Schaefer for typing this review. We also thank Mr. A. Gapor of PORIM for supplying pure tocotrienols and tocotrienol-rich-fraction of palm oil.

REFERENCES

1. Gould MR, Anderson JW, O'Mahony S. Biofunctional properties of oats. In: Inglett GE and Munck L, eds. Cereals for foods and beverages. New York: Academic Press, 1980; 447-60.
2. Sacks FM, Castelli WP, Donner A, Kass EH. Plasma lipids and lipoproteins in vegetarians and controls. N Engl J Med 1975; 292:1148-51.
3. Burslem J, Schonefeld G, Howard MA, Weidman SW, Miller JP. Plasma apoprotein and lipoprotein lipid levels in vegetarians. Metabolism 1978; 27:711-9.
4. Qureshi AA, Burger WC, Prentice N, Bird HR, Sunde ML. Regulation of lipid metabolism in chicken liver by dietary cereals. J Nutr 1980; 110:388-93.
5. Qureshi AA, Burger WC, Prentice N, Bird HR, Sunde ML. Suppression of cholesterogenesis and stimulation of fatty acid biosynthesis in chicken livers by dietary cereals supplemented with culture filtrate of *Trichoderma viride*. J Nutr 1980; 110:1014-22.

6. Qureshi AA, Burger WC, Prentice N, Elson CE. Regulation of lipid metabolism in chicken liver by dietary cereals supplemented with culture filtrate of *Trichoderma viride*. J Nutr 1980; 110:1473-8.

7. Qureshi AA, Burger WC, Elson CE, Benevenga NJ. Effects of cereals and culture filtrate of *Trichoderma viride* on lipid metabolism of swine. Lipids 1982; 17:924-34.

8. Prentice N, Qureshi AA, Burger WC, Elson CE. Response of hepatic cholesterol, fatty acid synthesis and activities of related enzymes to rolled barley and oats in chickens. Nutr Rep Int 1982; 26:597-604.

9. Qureshi AA, Prentice N, Din ZZ, Burger WC, Elson CE. Influence of culture filtrate of *Trichoderma viride* and barley components on lipid metabolism of feeding laying hens. Lipids 1984; 19: 250-7.

10. Qureshi AA, Burger WC, Peterson DM, Elson CE. The structure of an inhibitor of cholesterol biosynthesis isolated from barley. J Biol Chem 1986; 261:10544-50.

11. Burger WC, Qureshi AA, Prentice N, Elson CE. Effects of different fractions of barley kernel on the hepatic lipid metabolism of chickens. Lipids 1982; 17:956-63.

12. Qureshi AA, Abuirmeileh N, Fitch M, Elson CE, Burger WC, McGibbon WH. Regulation of lipid metabolism in restricted ovulator chicken by dietary supplementation with HPBF and culture filtrate. Lipids Nutr Rep Int 1983; 27:87-95.

13. Burger WC, Qureshi AA, Din ZZ, Abuirmeileh N, Elson CE. Suppression of cholesterol biosynthesis by constituents of barley kernel. Atherosclerosis 1984; 51:75-87.

14. Qureshi AA, Burger WC, Peterson DM, Elson CE. Suppression of cholesterogenesis by plant constituents: review of Wisconsin contributions to NC-167. Lipids 1985; 20:817-24.

15. Qureshi AA, Qureshi N, Wright JJK, et al. Suppression of cholesterol biosynthesis by different tocotrienols from palm oil in normolipemic and hypercholesterolemic chickens. Lipids (submitted).

16. Parrish DB. Determination of vitamin E in foods—a review. Crit Rev Food Sci Nutr 1980; 13: 161-87.

17. Syvaoja E-L, Piironen V, Varo P, Koivistoinen P, Salminen K. Tocopherols and tocotrienols in Finnish foods: human milk and infant formulas. Int J Vitam Nutr Res 1985; 55:159-66.

18. Piironen V, Syvaoja E-L, Varo P, Salminen K, Koivistoinen P. Tocopherols and tocotrienols in cereal products from Finland. Cereal Chem 1986; 63:78-81.

19. Slover HT. Tocopherols in foods and fats. Lipids 1971; 6:291-6.

20. Bourgeois CF, Pages PS, Czornomaz AM, Albrecht MJ, Cronenberger LA, George PR. Specific determination of α-tocopherol in food and feed by fluorometry. I. Manual method. J Assoc Off Anal Chem 1984; 67:627-30.

21. Threlfall DR. The biosynthesis of vitamins E and K and related compounds. Vitam Horm 1971; 29:153-200.

22. Pennock JF, Hemming FW, Kerr JD. A reassessment of tocopherol chemistry. Biochem Biophys Res Commun 1964; 17:542-8.

23. Whistance GR, Threlfall DR. Biosynthesis of phytoquinones: an outline of the biosynthetic sequences involved in terpenoid quinone and chromanol formation by higher plants. Biochem Biophys Res Commun 1967; 28:295-301.

24. Janiszowska W, Pennock JF. The biochemistry of vitamin E in plants. Vitam Horm 1976; 34:77-105.

25. Mannitto P. The isoprenoids. In: Sammes PG, ed. Biosynthesis of natural products, 1st ed. West Sussex, England: Ellis Horwood Limited, 1981; 304-6.

26. Chow CK, Draper HH, Csallany AS. Method for the assay of free and esterified tocopherols. Anal Biochem 1969; 32:81-90.

27. Ueda T, Igarashi O. Determination of vitamin E in biological specimens and foods by HPLC—pretreatment of samples and extraction of tocopherols. J Micronutrient Anal 1990; 7:79-96.

28. Piironen V, Varo P, Syvaoja E-L, Salimen K, Koivistoinen P. High performance liquid chromatographic determination of tocopherols and tocotrienols and its application to diets and plasma of Finnish men. Int J Vitam Nutr Res 1984; 53:35-40.

29. Tan B, Brzuskiewicz L. Separation of tocopherol and tocotrienol isomers using normal- and reverse-phase liquid chromatography. Anal Biochem 1989; 180:368-73.

30. Urano S, Nakano S, Madsuo M. Synthesis of dl-alpha tocopherol and dl-alpha tocotrienol. Chem Pharm Bull (Tokyo) 1983; 31:4341-5.

31. Pennock JF. The biosynthesis of chloroplastidic terpenoid quinones and chromanols. Biochem Soc Trans 1983; 11:504-10.

32. Govind Rao MK, Perkins EG. Identification and estimation of tocopherols and tocotrienols in vegetable oils using gas chromatography-mass spectrometry. J Agr Food Chem 1972; 20:240-5.

33. Walton TJ, Mullins CJ, Newton RP, Brenton AG, Beynon JH. Tanden mass spectrometry in vitamin E analysis. Biomed Environ Mass Spectrom 1988; 16:289-98.

34. Siperstein MM, Fagan VM. Feedback control of mevalonate synthesis by dietary cholesterol. J Biol Chem 1966; 241:602-9.

35. Ramachandran CK, Gray SL, Melnykovych G. Coordinate repression of cholesterol biosynthesis and cytoplasmic 3-hydroxy-3-methylglutaryl coenzyme A synthase by glucocorticoids in Hela cells. Arch Biochem Biophys 1978; 189:205-11.

36. Qureshi N, Dugan RE, Nimmannit S, Wu W-H, Porter JW. Purification of β-hydroxy-β-methyl-glutaryl-coenzyme A reductase from yeast. Biochemistry 1976; 15:4185-90.

37. Qureshi N, Porter JW. Conversion of acetyl coenzyme A to isopentenyl pyrophosphate. In: Porter JW, Spurgeon SL, eds. Biosynthesis of polyisoprenoids, Vol. 1. New York: John Wiley and Sons, 1981; 47-94.

38. Qureshi N, Dugan RE, Cleland WW, Porter JW. Kinetic analysis of the individual reductive steps catalyzed by β-hydroxy-β-methylglutaryl-coenzyme A reductase obtained from yeast. Biochemistry 1976; 15:4191-7.

39. Veloso D, Cleland WW, Porter JW. pH properties and chemical mechanism of action of 3-hydroxy-3-methylglutaryl coenzyme A reductase. Biochemistry 1981; 20:887-94.

40. Wang Y, Darnay BG, Rodwell VW. Identification of the principal catalytically important acidic residue of 3-hydroxy-3-methylglutaryl coenzyme A reductase. J Biol Chem 1990; 265:21634-41.

41. Dugan RE. Regulation of HMG-CoA reductase. In: Porter JW, Spurgeon SL, eds. Biosynthesis of polyisoprenoids, vol. 1. New York: John Wiley and Sons, 1981; 95-159.

42. Geden MJH, Gibson DM, Rodwell VW. Hydroxymethylglutaryl-CoA reductase—the rate-limiting enzyme of cholesterol biosynthesis. FEBS 1986; 201:183-6.

43. Goldstein JL, Brown MS. Regulation of the mevalonate pathway. Nature 1990; 343:425-30.

44. Brown MS, Goldstein JL. Lipoprotein metabolism in the macrophage: implications for cholesterol deposition in atherosclerosis. Annu Review Biochem 1983; 52:223-61.

45. Goldstein JL, Brown MS. Familial hypercholesterolemia. In: Stanbury et al., eds. The metabolic basis of inherited disease, 5th ed. New York: McGraw-Hill, 1983; 672-713.

46. Maltese WA. Postranslational modification of proteins by isoprenoids in mammalian cells. FASEB J 1990; 4:3319-28.

47. Casey PJ, Solski PA, Der CJ, Buss JE. p21ras is modified by a farnesyl isoprenoid. Proc Natl Acad Sci U S A 1989; 86:8323-7.

48. Schaber MD, O'Hara MB, Garsky VM, et al. Polyisoprenylation of Ras in vitro by a farnesyl-protein transferase. J Biol Chem 1990; 265:14701-4.

49. Rilling HC, Breunger E, Epstein WW, Crain PF. Prenylated proteins: the structure of the isoprenoid group. Science 1990; 247:318-20.

50. Maltese WA, Erdman RA. Characterization of isoprenoid involved in the post-translational modification of mammalian cell proteins. J Biol Chem 1989; 264:18168-72.

51. Schafer WR, Kim R, Sterne R, Thorner J, Kim SH, Rine J. Genetic and pharmacological suppression of oncogenic mutations in RAS genes of yeast and humans. Science 1989; 245:379-85.

52. Hall RH. Prog Nucleic Acid Res Molec Biol 1970; 10:57-86.

53. Farnsworth CC, Gelb MH, Glomset JA. Identification of geranylgeranyl-modified proteins in Hela cells. Science 1990; 247:320-2.

54. Bell JJ, Sargeant TE, Watson JA. Inhibition of 3-hydroxy-3-methylglutaryl coenzyme A reductase activity in hepatoma tissue culture cells by pure cholesterol and several cholesterol derivatives. Evidence supporting two distinct mechanisms. J Biol Chem 1976; 251:1745-58.

55. Kandutsch AA, Chen HW, Shown EP. Binding of 25-hydroxy-cholesterol and cholesterol to different cytoplasmic proteins. Proc Natl Acad Sci U S A 1977; 74:2500-7.

56. Illingworth DR. Lipid-lowering drugs. An overview of indications and optimum therapeutic use. Drugs 1987; 33:259-79.

57. Endo A, Tsujita Y, Kuroda M, Tanazawa K. Inhibition of cholesterol synthesis in vivo by ML236B, a competitive inhibitor of 3-hydroxy 3-methyl glutaryl coenzyme A reductase. Eur J Biochem 1977; 87:313-9.

58. Alberts AW, Chen J, Kuron G. Mevinolin: a highly potent competitive inhibitor of hydroxy methyl-glutaryl coenzyme A reductase and a cholesterol lowering agent. Proc Natl Acad Sci U S A 1980; 77:3957-61.

59. Shepherd J, Packard CJ, Bicker S, Laurie TDV, Morgan HG. Cholestyramine promotes receptor mediated low density lipoprotein catabolism. N Engl J Med 1980; 302:1219-22.

60. Kesaniemi YA, Grundy SM. Influence of gemfibrosil and clofibrate on metabolism of cholesterol and plasma triglycerides in man. JAMA 1984; 251:2241-7.

61. Grundy SM, Mok HYI, Zack L, Berman M. The influence of nicotinic acid on metabolism of cholesterol and triglycerides in man. J Lipid Res 1981; 22:24-36.

62. Nestel PJ, Billington T. Effects of probucol on low density lipoprotein removal and high density lipoprotein synthesis. Atherosclerosis 1981; 38:203-9.

63. Qureshi AA, Qureshi N, Peterson DM, et al. Effects of palm oil and other dietary fats on cholesterol regulation in chickens. Lipids 1992; (in press).

64. Qureshi AA, Qureshi N, Payne CM, et al. Suppression of plasma lipid parameters by dietary F, hybrid palm, barley and rice bran oils in chickens. Lipids 1992; (in press).

65. Qureshi AA, Qureshi N, Hasler-Rapacz JO, et al. Dietary tocotrienols reduce levels of plasma cholesterol, apolipoprotein B, thromboxane B_2 and platelet factor 4 in pigs with inherited hyperlipidemias. J Am Clin Nutr 1991; 53:1042S-6S.

66. Qureshi AA, Weber FE, Chaudhary VK, et al. Suppression of cholesterogenesis in hypercholesterolemic humans by tocotrienols of barley and palm oils (abstract). In: Proceedings of Antioxidants and Degenerative Diseases Meeting 1990;

67. Qureshi AA, Qureshi N, Wright JJK, et al. Lowering of serum cholesterol in hypercholesterolemic humans by tocotrienols. J Am Clin Nutr 1991; 53:1021S-6S.

68. Robinson MC, Lupton JR. The effects of barley flour and barley oil on hypercholesterolemic men and women (abstract). Annual meeting, October 14–16, Dallas, 1990.

69. Weber FE, Chaudhary VK, Lupton JR, Qureshi AA. Therapeutic and physiological properties of barley bran (abstract). American Association of Cereal Chemists. Annual meeting, October 14–16, Dallas, 1990.

70. Parker RA, Pearce BC, Clark RW, et al. Tocotrienols decrease cholesterol synthesis in HepG2 cells by a novel post-transcriptional suppression of HMG-CoA reductase (abstract). FASEB J 1990; 4:A1744.

71. Parker RA, Clark RW. Squalene transfer and epoxidation in HepG2 cell membranes is modulated by γ-tocotrienol (abstract). FASEB J 1991; 5:A-710.

72. Bjornson LK, Kayden HJ, Miller E, Moshell AN. The transport of alpha-tocopherol and beta carotene in human blood. J Lipid Res 1976; 17:343-51.

73. Haga P, Ek J, Kran S. Plasma tocopherol levels and vitamin E/beta lipoprotein relationships during pregnancy and in cord blood. Am J Clin Nutr 1982; 36:1200-4.

74. Catignani GL, Bieri JG. Rat liver α-tocopherol binding protein. Biochim Biophys Acta 1977; 497:349-57.

75. Behrens WA, Madere R. Transfer of α-tocopherol to microsomes mediated by a partially purified liver α-tocopherol binding protein. Nutr Res 1982; 2:611-8.

76. Komiyama K, Iizuka K, Yamaoka M, Watanabe H, Tsuchiya N, Umezawa I. Studies on the biological activity of tocotrienols. Chem Pharm Bull (Tokyo) 1989; 37(5):1369-71.

77. Burton GW, Traber MG. Vitamin E: antioxidant activity, biokinetics and bioavailability. Annu Rev Nutr 1990; 10:357-82.

78. Glover J. Free radical biology: a paradox in cancer research. J Natl Cancer Inst 1990; 82:902-3.

79. Simic MG, Jovanovic SV. Mechanisms of inactivation of oxygen radicals by dietary antioxidants and their models. Basic Life Sci 1990; 52:127-37.

80. Serbinova E, Kagan V, Han D, Packer L. Free radical recycling and intramembrane mobility in the antioxidant properties of alpha-tocopherol and alpha-tocotrienol. Free Radic Biol Med 1991; (in press).

81. Ferrari R, Ceconi C, Curello S, et al. Role of oxygen free radicals in ischemic and reperfused myocardium. Am J Clin Nutr 1991; 53:215S-22S.

21

Interaction of Vitamin E, Ascorbic Acid, and Glutathione in Protection Against Oxidative Damage

Donald J. Reed

Oregon State University, Corvallis, Oregon

INTRODUCTION

The interaction of antioxidants provides essential protection against the damaging effects of an oxygen-based metabolism (for a review see Ref. 1). Complementary antioxidant systems in cytoplasmic and membrane compartments provide an efficient cellular defense against oxidative injury (2). Vitamin E α-tocopherol (α-TH) protects cellular membranes against oxidative damage (3–5), and dietary α-TH deficiency enhances the susceptibility of biological membranes to oxidative damage in vitro (5–8) and in vivo (9–11). Nonetheless, clinical manifestations of α-TH deficiency in humans are poorly defined and are usually secondary to other disease states (3,12). Diplock (13) reviewed the status of antioxidant nutrients and disease prevention. This chapter focuses on the interaction of antioxidants for protection against oxidative damage in systems in vitro.

IN VITRO LIPOSOMES

Biochemical regulation of cellular α-TH status is only beginning to be understood. Several studies have addressed the hypothesis that soluble cellular antioxidants, particularly ascorbic acid, maintain membrane α-TH levels by regenerating α-TH from its oxidation products (Fig. 1). Indeed, the reduction of the α-TH semiquinone radical by ascorbic acid was reported to proceed in homogeneous solution with a bimolecular rate constant of 1.55×10^6 M^{-1} s^{-1} (14). Synergistic antioxidant protection by α-TH and ascorbic acid has been demonstrated in lard (15), homogeneous solution (16), and micellar (17,18) and liposome suspensions (19–22). Other reports have provided direct evidence for the α-TH-sparing action of ascorbic acid (23,24). Although these data are consonant with a hypothesis of reductive regeneration of α-TH, the reaction of ascorbic acid as well as α-TH with free radicals complicates interpretation and may account for the additive or synergistic antioxidant effects observed (25). In biological lipid peroxidation, soluble antioxidants may also exert prooxidant effects via reduction of metals (26). The role of soluble antioxidants thus appears complicated. α-TH is

a

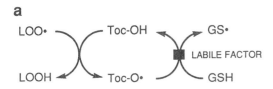

b

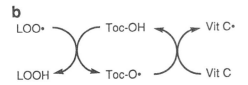

FIGURE 1 Antioxidant interactions: (a) regeneration of vitamin E (Toc-OH) via GSH-dependent labile factor; (b) ascorbic acid (Vit C) regeneration of vitamin E. LOO˙, lipid peroxyl radical; LOOH, lipid hydroperoxide.

principally consumed by lipid oxyradicals rather than by oxygen radical initiators.* Lipid oxyradical propagation is effectively inhibited by α-TH at concentrations above a threshold level. Factors affecting the balance between the pro- and antioxidant effects of ascorbic acid and glutathione in various biological preparations, such as soybean phosphatidylcholine liposomes that contain vitamin E, have been examined by challenges with such oxidant systems as Fe^{2+}/H_2O_2. Equivalent molar amounts of Fe^{2+} and H_2O_2 were added together to initiate lipid peroxidation via the celebrated Fenton reaction [Eqs. (1) through (4)] (26):

$$Fe^{2+} + H_2O_2 \rightarrow Fe^{3+} + OH^- + OH^+ \tag{1}$$

$$LH + OH^{\cdot} \rightarrow L^{\cdot} + H_2O \tag{2}$$

$$L^{\cdot} + O_2 \rightarrow LOO^{\cdot} \tag{3}$$

$$LOO^{\cdot} + LH \rightarrow LOOH + L^{\cdot} \tag{4}$$

A Fe^{2+} and lipid hydroperoxide-dependent pathway also occurs [Eqs. (5) and (6)] (26,27):

$$Fe^{2+} + LOOH \rightarrow Fe^{3+} + LO^{\cdot} + OH^- \tag{5}$$

$$LO^{\cdot} + LH \rightarrow LOH + L^{\cdot} \tag{6}$$

Both ascorbic acid and glutathione (GSH) act as prooxidants when incubated with α-TH-free liposomes and Fe^{2+}/H_2O_2 (28). The prooxidant effect of ascorbic acid but not of GSH is reversed by incorporation of α-TH into the liposomes. α-TH levels in excess of the threshold are effectively maintained by ascorbic acid, but GSH antagonizes the antioxidant effects of the α-TH–ascorbic acid combination. Effective antioxidant protection by α-tocopherol appeared to be due to efficient reaction with lipid oxyradical in the bilayer rather than to interception of initiating oxygen radicals. At concentrations above a threshold level of approximately 0.2 mol per 100 mol (based on phospholipid content), α-tocopherol completely suppressed

---

*In this chapter the term "lipid oxyradicals" refers to both lipid peroxyl radicals (LOO˙) and lipid alkoxyl radicals (LO˙). The former species are assumed to predominate during uninhibited lipid oxyradical propagation. However, relative contributions of both species to α-TH depletion may depend on α-TH concentration (see later).

lipid oxyradical propagation, which was measured by malondialdehyde (MDA) production. Both ascorbic acid and glutathione, alone or in combination, can enhance lipid oxyradical propagation (28). α-Tocopherol, incorporated into such membranes as liposomes at concentrations above its threshold protective level, can reverse the proxidant effects of 0.1–1.0 mM ascorbic acid but not those of glutathione. Ascorbic acid also prevented α-tocopherol depletion. The combination of ascorbic acid and subthreshold levels of α-tocopherol only temporarily suppresses lipid oxyradical propagation and does not maintain the α-tocopherol level. Glutathione antagonizes the antioxidant action of the α-tocopherol–ascorbic acid combination regardless of α-tocopherol concentration. These observations indicate that membrane α-tocopherol status can control the balance between the pro- and antioxidant effects of ascorbic acid. These studies also provide direct evidence that ascorbic acid interacts directly with components of the phospholipid bilayer (28).

The remarkable antioxidant efficiency of the α-TH–ascorbic acid combination may be due in part to efficient radical scavenging by these agents in hydrophobic and hydrophilic environments, respectively. Liebler et al. (28) showed that ascorbic acid also regenerates α-TH from its oxidation products. In addition, α-TH completely reversed the prooxidant effects of 0.1 or 1.0 mM ascorbic acid. The inhibition of lipid oxyradical propagation coincided with maintenance of α-TH. The ability of ascorbic acid, a relatively polar and lipid insoluble molecule, to interact directly with components of the lipid bilayer has been questioned (25). Oxyradical-bearing fatty acyl chains may partition into the hydrophilic surface of the bilayer, and extensive hydration of the liposome surface may minimize any entopic barrier to the close juxtaposition of ascorbic acid and lipid oxyradicals or α-TH (for a discussion of this point see Ref. 28). The data of Scarpa et al. (24) indicating direct reaction between ascorbic acid and liposomal oxyradicals provide further support for this view.

The limited antioxidant capacity of the α-TH–ascorbic acid combination at subthreshold α-TH concentrations is noteworthy. α-TH at 0.1 mol per 100 mol does not suppress lipid oxyradical propagation in liposomes unless 0.5–1.0 mM ascorbic acid is present (28). Even so, α-TH status was not maintained—the rate of α-TH consumption was merely decreased. Maintenance of liposome α-TH levels by reductive recycling, if operative, thus requires threshold (or higher) levels of α-TH. This is more easily understood if the nature of the threshold level is considered. α-TH, at the threshold level, lowers the efficiency of lipid oxyradical propagation to less than 1%. The amplifying effect of Fe^{2+} catalysis dramatically enhances propagation efficiency at even slightly lower α-TH concentrations (26). The threshold level thus represents a point of no return: propagation reactions cannot be effectively suppressed, nor can the α-TH level be maintained. Ascorbic acid, even at 1 mM, cannot maintain α-TH levels because the rate of reaction of the α-TH semiquinone radical with lipid oxyradicals exceeds the rate of its reduction to α-TH.

Oxidative challenge in the liposome model may mimic biological oxidative injury in which oxygen radicals are generated due to the presence of copper and iron (26,29). In the absence of α-TH, the availability of Fe^{2+} determines the extent of soybean phosphatidylcholine liposome peroxidation. Reduction of Fe^{3+} by ascorbic acid or GSH maintains an active Fe^{2+} pool and enhances both the rate and extent of lipid oxyradical propagation. The concentration-dependent prooxidant effect of GSH is thus easily explained as described by Miller et al. (29). The prooxidant action of ascorbic acid, however, was maximal at the lowest concentration studied (0.1 mM), and concentrations above 1 mM temporarily inhibited lipid peroxidation (28). Inhibition due to scavenging of hydroxyl radicals, presumably the initiating species in this model, by ascorbic acid is unlikely. GSH produces no such inhibition despite being a superior hydrogen atom donor (30). On the other hand, ascorbic acid at higher concentrations may react directly with lipid oxyradicals at the lipid/water interface. Because GSH displays no antioxidant activity in the liposome model, it does not appear to react with lipid oxyradicals to a significant extent or regenerate α-TH from its oxidation products. Both reactions may be kinetically unfavorable for GSH (30). The notable antagonism of the α-TH–ascorbic acid antioxidant combination by GSH may involve at least two mechanisms. First, GSH chelation and reduction of iron enhances and sustains the oxidative challenge (31). Second, the glutathione radicals thus produced are themselves rapidly reduced by ascorbic acid (30). The consumption of ascorbic acid would consequently exceed that expected on the basis of its antioxidant functions alone.

Ascorbic acid inhibits lipid peroxidation by regenerating α-tocopherol rather than by reacting directly with lipid oxyradicals (Fig. 1). The studies by Liebler et al. (28) represent one of the first demonstrations of a "sparing" effect of ascorbic acid upon α-tocopherol in lipid bilayers under conditions in which significant reaction between ascorbic acid and radicals can be unequivocally ruled out. Furthermore, ascorbic acid (at 0.1 mM) offers no antioxidant protection when α-tocopherol is present, but at a subthreshold level. This observation supports a regeneration hypothesis but contradicts the view that ascorbic acid reacts independently with lipid oxyradicals to spare α-tocopherol. Liebler et al. (32) provided evidence that, in vitro or in liposomes, ascorbate does not reduce α-tocopherone directly but reacts instead with the tocopherone cation. Slow rates of these reactions have led these workers to suggest that biochemical catalysis would be required to complete a two-electron tocopherol redox cycle in biological membranes.

A central conclusion is that the balance between the pro- and antioxidant effects of ascorbic acid is critically dependent upon the bilayer concentration of α-tocopherol relative to the threshold level. Ascorbic acid cannot maintain α-tocopherol at subthreshold levels, presumably because the rate of α-tocopherol consumption exceeds the rate of its reductive regeneration from a radical intermediate by ascorbic acid. The effectiveness of the α-tocopherol–ascorbic acid combination is thus intimately related to the threshold effect.

The liposome model indicates that the balance between the pro- and antioxidant effects of ascorbic acid ultimately depends on the α-TH status of the lipid bilayer. The results further imply that maintenance of membrane α-TH status may be an essential function of cellular ascorbic acid in vivo. If so, then NADH-dependent reduction of semidehydroascorbate (33) may subject α-TH recycling to a limited degree of metabolic control. Attempts to manipulate tissue α-TH levels by modifying dietary ascorbic acid intake have met with limited success (25). Ascorbic acid treatment enhanced lipid peroxidation in α-TH-deficient rats (9), however, and Litov and co-workers concluded that, in the absence of α-TH, lipid oxyradical propagation in cellular membranes proceeds unchecked by ascorbic acid. This conclusion is essentially identical to that of the liposome model; ascorbic acid at concentrations below 1 mM did not function as an antioxidant unless the α-TH concentration exceeded the threshold level.

The failure of GSH to function as an antioxidant in liposomes despite mounting evidence for such a role in biological systems (34,35) is a significant shortcoming of the liposome model. Enzymic participation in the antioxidant action of GSH is thought to be essential, and the results of these studies reinforce that view (Fig. 1). Evidence for protein- and GSH-dependent antioxidant protection of cell membranes has been reported (34,36,37), but evidence for the participation of α-TH in GSH-dependent protection is somewhat equivocal. Further study of GSH-dependent antioxidant activity in biological membranes may reveal a system whose function complements that of α-TH.

One possible mechanism by which GSH would stimulate lipid peroxidation is by reducing iron and thus promoting Fe^{2+}-dependent peroxidation. This is not a new suggestion; it is well supported by previously published work (see Refs. 38 and 39). Moreover, evidence that thiols may directly reduce iron and stimulate peroxidation has come from Aust and colleagues (31,40).

IN VITRO MICROSOMES

Lipid peroxidation, which can cause extensive damage to subcellular organelles and biomembranes, has been demonstrated to occur in isolated mitochondria, lysosomes, microsomes (41), and nuclei (42). Lipid peroxidation is an important biological consequence of oxidative cellular damage (43) that can be measured by several noninvasive techniques in humans (for a review see Ref. 44). The destruction of unsaturated fatty acids, which occurs in lipid peroxidation, has been linked with altered membrane structure (45) and enzyme inactivation (46, 47). In addition to lipid hydroperoxides and lipid radicals (48), lipid peroxidation generates activated oxygen species, such as hydroxyl radicals (49) and superoxide anions (50). The decomposition of peroxidized polyunsaturated fatty acids also generates reactive carbonyl compounds, such as MDA and hydroxyalkenals (51).

Several studies have shown that the addition of GSH protected microsomal incubations against both NADPH- and ascorbate-induced lipid peroxidation (34,36,52). The mechanism involved in this protection remains in doubt, but results suggest that a microsomal protein is required for the GSH-dependent inhibition of lipid peroxidation (52). Enzyme-mediated antioxidant action of GSH in rat liver microsomes depends at least in part on the α-TH status of the membranes (36,37). Experimental evidence for the reaction of GSH with the α-TH semiquinone radical has been reported (53). Several reports suggest that glutathione either alone (52) or in conjunction with added proteins (54,55) can protect microsomes against lipid peroxidation in a manner that is independent of glutathione transferases. A report by Burk (52) indicates that GSH inhibits microsomal lipid peroxidation. This protection did not require the addition of other proteins. However, his evidence suggested the involvement of a microsomal protein. Belouqui and Cederbaum (56) showed that the addition of GSH produced a 40% reduction in NADPH-induced peroxidation of rat liver microsomes. However, the addition of the soluble enzymes glutathione peroxidase, glutathione reductase, and glutathione transferase together with GSH did not significantly increase the protective capacity of GSH in these incubations.

Gibson et al. (57) demonstrated that the classic enzyme glutathione peroxidase is unable to inhibit lipid peroxidation of microsomal membranes. They concluded that this soluble peroxidase is unable to interact with lipid hydroperoxides while they remain associated with microsomal membranes. On the basis of these data, Gibson et al. (57) concluded that lipid hydroperoxides must be free in solution if they are to be substrates for the soluble enzymes glutathione transferase and glutathione peroxidase. In this case, the lipophilic environment of membranes apparently restricts interactions between enzymes located in the aqueous phase and lipid hydroperoxides located in the lipid bilayer.

Evidence suggests that the GSH-dependent peroxidase purified by Ursini et al. (58) is interfacial in character. Also, the protective effects of GSH described by Burk (52) in microsomes were due to a microsomal protein. Gibson et al. (55) reported that a cytosolic, GSH-dependent protein can protect microsomal membranes against peroxidation. These researchers, however, concluded that this protection was not associated with glutathione peroxidase activity but rather involved the inhibition of the initiation of peroxidation.

With microsomal lipid peroxidation that is iron induced, an antioxidant effect of vitamin E at a physiological ratio to phospholipids was observed only in the presence of phospholipid hydroperoxide glutathione peroxidase and GSH (59). The basis for this effect is the hydroperoxyl radical-scavenging capacity of vitamin E and the reduction of membrane hydroperoxides by phospholipid hydroperoxide glutathione peroxidase.

A microsomal glutathione transferase has also been purified that exhibits glutathione-dependent peroxidase activity toward cumene hydroperoxide (60). This enzyme is unique with respect to the other glutathione transferases in that it is activated by prior treatment with

N-ethylmaleimide (NEM) and has a molecular weight of 14,000. Although small amounts of this enzyme have been reported to be associated with nuclear fractions (61), the microsomal enzyme is distinct from the partially purified nuclear glutathione transferase described by Tirmenstein and Reed (62). NEM inhibited rather than activated this nuclear enzyme, and no protein was detected below a molecular weight of 21,500 when the partially purified nuclear GSH transferase was analyzed by sodium dodecyl sulfate (SDS)–polyacrylamide gel electrophoresis.

McCay et al. (35) described α-tocopherol functioning in a cyclical manner to quench free radicals in liver microsomes with the requirement for GSH and a heat-labile factor, which is assumed to be a vitamin E radical reductase. Bast and Haenen (63) implicated the involvement of liver microsomal vitamin E free-radical reductase in the regulation of lipid peroxidation by glutathione and lipoic acid. Preparations of mitochondria and microsomal membranes are prevented from chromanoxyl radical accumulation by a free-radical reductase activity that appears to be both enzymatic and nonenzymatic in nature (64). In this study, GSH prevented the accumulation of chromanoxyl radical in the presence of NADPH but was without effect in the absence of NADPH. Further, ascorbate also prevented the radical accumulation.

Mehlhorn et al. (65) used dietary means to obtain vitamin E-enriched rat liver mitochondria and microsomes that displayed a persistent tocopheroxyl radical electron spin resonance (ESR) signal and tocopherol consumption during oxidative stress-induced by continuous enzymatic oxidation with horseradish peroxidase and a hydrophilic phenol, arbutin. Ascorbic acid prevents the formation of the tocopherol radical until the ascorbyl radical ESR signal decays.

Vitamin E participates in the inhibition of lipid peroxidation of membranes by breaking of the chain propagation (66) by reactions that appear to include scavenging of lipid hydroperoxyl radicals. Murphy et al. (67) showed that the protection by ascorbate and glutathione against microsomal lipid peroxidation is dependent on vitamin E. Loss of vitamin E is related to both chemiluminescence and thiobarbituric-reactive material formation, which increased when the α-tocopherol content of the microsomes decreased to about 0.38 nmol/mg protein, which is about 86% of the initial content (67).

Dietary supplementation with vitamin E gave a 10- to 20-fold increase in vitamin E in the rat liver mitochondrial membrane (68). Treatment of submitochondrial particles with an oxidizing system composed of lipoxygenase and arachidonic acid provided evidence that reduction of the tocopheroxyl radical could occur by NADH, succinate, and reduced cytochrome c-linked oxidation. Thus, the electron transport system is proposed to have an important physiological role in recycling vitamin E by reduction of the tocopheroxyl radical to prevent its accumulation and vitamin E consumption.

Ursini et al. (58) purified an interfacial glutathione peroxidase. This enzyme has been shown to reduce linoleic acid hydroperoxides, cumene hydroperoxide, *tert*-butyl hydroperoxide, and hydrogen peroxide. However, this enzyme, which does not conjugate 1-chloro-2,4-dinitrobenzene (CDNB) with GSH (54), displays glutathione peroxidase activity toward cumene hydroperoxide, hydrogen peroxide, and lipid hydroperoxides and is distinct from the classic glutathione peroxidase (69). Evidence suggests that the enzyme is interfacial in character and can interact directly with liposomes to reduce phospholipid hydroperoxides (58). The addition of this protein to microsomal incubation mixtures inhibited lipid peroxidation (54). Substrate specificities indicate that this enzyme is distinct from the nuclear glutathione transferase (62,70).

ISOLATED NUCLEI AND THE ROLE OF GLUTATHIONE PEROXIDASE ACTIVITY

Recent studies have focused on the susceptibility of the cell nucleus to lipid peroxidation. The nuclear membrane regulates the transport of mRNA into the cytoplasm and aids in the process of nuclear division. DNA is also frequently associated with certain regions of the nuclear membrane (71), and it seems likely that nuclear membrane peroxidation may disrupt many of these critical functions. The proximity of the nuclear membrane to DNA could also contribute to the interaction of DNA with reactive compounds generated in lipid peroxidation. Several studies indicate that such lipid peroxidation products can alter the structure and function of DNA (72–75). This is of importance since hydroxyl radicals diffuse an average of only 60 Å before reacting with cellular components (76). Assays for 8-hydroxy-2'-deoxyguanosine as a biomarker of oxidative DNA damage includes in vivo studies with urine samples (77). Nuclear peroxidation may also increase interactions between more stable peroxidation products and DNA. The cytosolic enzymes aldehyde dehydrogenase (78), glutathione transferase (79), and glutathione peroxidase (80) have all been shown to metabolize various reactive lipid peroxidation products. Such cytosolic enzymes may metabolize peroxidation products generated throughout the cell before they diffuse into the nucleus and interact with DNA.

Endogenous α-tocopherol levels in isolated rat liver nuclei have been measured and found to be 0.045 mol E (mol α-tocopherol per mol phospholipid $\times$ 100) (70). This value corresponds to 970 polyunsaturated fatty acid (PUFA) moieties to 1 molecule of α-tocopherol in the nuclear membrane. These values are higher than values reported for rat liver microsomes (3313) (8) and mitochondria (2100) (81). A threshold level of 0.085 mol per 100 mol for the prevention of NADPH-induced lipid peroxidation was established for isolated nuclei. This value could be lowered to 0.040 mol per 100 mol when 1 mM GSH was added to assist in the inhibition of lipid peroxidation (70). The ability of GSH to enhance α-tocopherol-dependent protection against nuclear lipid peroxidation appears to be mainly by a "sparing" effect on the near threshold level of α-tocopherol in the nuclear membrane.

The effects of various ferric complexes on oxygen radical generation by rat liver nuclei have been examined by measurement of chemiluminescence. Vitamin E prevented NADPH-dependent but not NADH-dependent chemiluminescence (82), which supports the evidence that vitamin E is effective in preventing oxygen radical generation by processes occurring in nuclei membranes, such as lipid peroxidation, but has no effect on reactions that occur outside the membrane, including hydroxyl radical generation in solution. These observations are in agreement with the findings of Tirmenstein and Reed (42,70) on nuclei peroxidation and the sparing effect of vitamin E.

Lipid hydroperoxides located in biological membranes increase the potential for lipid peroxidation. If this potential is to be diminished, these peroxides must be removed from the membrane environment or be reduced to lipid alcohols. If glutathione peroxidase activity is associated with the phospholipid bilayer of the nuclear membrane, such an association may contribute to the ability of the peroxidase to reduce lipid hydroperoxides. Since lipid hydroperoxides can initiate lipid peroxidation, the reduction of these compounds can contribute to the inhibition of peroxidation (Fig. 2). The association of a glutathione-dependent peroxidase with membranes may encourage the reduction of lipid hydroperoxides located within lipid bilayers.

The ability of GSH to protect isolated rat liver nuclei against NADPH-induced peroxidation was examined (42). Isolated nuclei were induced to undergo lipid peroxidation by an iron-dependent system, namely, 1.7 mM ADP, 0.11 mM EDTA, 0.1 mM $FeCl_3$, and either 1 mM NADPH or 0.5 mM ascorbate. The amount of lipid peroxidation, which was determined by measuring the formation of thiobarbituric acid reactive products and the disappearance of lipid unsaturated fatty acid moieties, underwent a concentration-dependent lag period by addition of GSH. This GSH-induced lag period was abolished by pretreatment of nuclei with trypsin, thiol-modifying reagents, or disulfides or by heating nuclei at 60°C for 15 minutes. Thus GSH can protect isolated nuclei against lipid peroxidation, and this protection

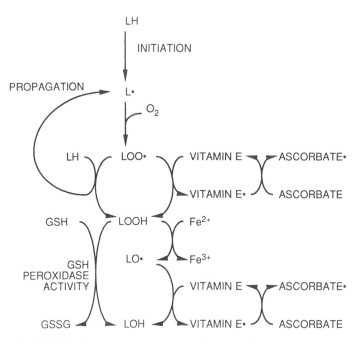

FIGURE 2   Protective interactions of vitamin E, ascorbate, and glutathione against lipid peroxidation.

involves the action of a GSH-dependent protein with peroxidase activity. The loss of unsaturated fatty acids and MDA production were both utilized to monitor NADPH-induced peroxidation and to examine the protective effects of GSH. Nuclei are susceptible to NADPH-induced lipid peroxidation, and GSH inhibits this peroxidation. Nuclei that were incubated with GSH also catalyzed the conversion of cumene hydroperoxide to cumyl alcohol. Similarly, this activity was also inhibited by thiol-modifying reagents and disulfides and by heating nuclei at 60°C for 15 minutes. The data suggested that a GSH-dependent peroxidase activity associated with rat liver nuclear membranes is capable of inhibiting lipid peroxidation.

Analysis of fatty acid composition data by Tirmenstein and Reed (42) indicated that the polyunsaturated arachidonic and docosahexaenoic acid fractions of nuclei were primarily destroyed by peroxidation to form oxidation products, including MDA, which is formed from unsaturated fatty acids containing three or more double bonds (83).

Glutathione protection of isolated rat liver nuclei against lipid peroxidation is abolished by exposing isolated nuclei to the glutathione transferase inhibitor *S*-octylglutathione (62). *S*-octylglutathione also inhibits nuclear glutathione transferase activity and glutathione peroxidase activity. A large percentage of the glutathione transferase activity associated with isolated nuclei was solubilized with 0.3% Triton X-100. Studies suggest that this treatment removes nuclear membranes while preserving the integrity of the remaining nucleus. Extraction of nuclei with 1 M NaCl accompanied by brief sonication failed to solubilize the peroxidase activity. Extraction with 0.3% Triton X-100, however, solubilized the GSH-dependent peroxidase activity (62). Electron microscopy studies conducted by Dabeva et al. (84) indicate that treatment with even higher concentrations of Triton X-100 removed the outer nuclear membrane but preserved the integrity of the remaining nucleus. Based on this information, it appears that the peroxidase activity is associated with the nuclear membrane. This activity in conjunction with GSH may contribute to the inhibition of lipid peroxidation in nuclear membranes and thereby preserves the integrity of this important membrane system. Increasing evidence suggests that this inhibition of peroxidation may in turn protect the structure and function of DNA.

Partial purification of this detergent-solubilized glutathione transferase activity was achieved with a *S*-hexylglutathione affinity column (62). Following these procedures, specific activity measurements were conducted with the partially purified fraction and a variety of substrates. Activity was detected toward CDNB, cumene hydroperoxide, lipid hydroperoxides, and *tert*-butyl hydroperoxide. However, no activity was detected toward hydrogen peroxide. This substrate specificity pattern is consistent with those exhibited by the glutathione transferases. In isolated nuclei, more glutathione peroxidase specific activity was measured toward *tert*-butyl hydroperoxide than cumene hydroperoxide. However, the partially purified nuclear glutathione transferase exhibited greater activity toward cumene hydroperoxide than *tert*-butyl hydroperoxide. The reason for this disparity is unknown but may relate to the presence of the nuclear membrane. The more lipophilic substrate cumene hydroperoxide may partition into the nuclear membrane in isolated nuclei to a greater degree than *tert*-butyl hydroperoxide and thereby limit the availability of this substrate. Such interactions may produce artificially low glutathione peroxidase activity values for the reduction of cumene hydroperoxide by isolated nuclei.

The partially purified nuclear glutathione transferase exhibited glutathione peroxidase activity toward lipid hydroperoxides in solution. Reconstitution experiments suggest that this enzyme can also inhibit the peroxidation of phosphatidylcholine liposomes. If a glutathione transferase, which is associated with the nuclear membrane, contributes to the glutathione-dependent protection of isolated nuclei against lipid peroxidation, a main question is whether it is the same transferase observed with rat liver microsomes. Molecular weight studies and the failure of *N*-ethylmaleimide to activate the partially purified nuclear glutathine transferase show that this enzyme is distinct from the microsomal glutathione transferase. Only slight glutathione-dependent peroxidase activity was detected when the nuclear glutathione transferase fraction or the selenium-containing glutathione peroxidase was incubated with peroxidized liposomes. However, this activity was increased by prior treatment of the peroxidized liposomes with phospholipase A_2. In the case of the selenium-containing glutathione peroxidase, phospholipase A_2 treatment increased the glutathione-dependent peroxidase activity sevenfold, whereas phospholipase A_2 treatment increased the activity of nuclear glutathione transferase threefold. The results with the selenium-containing glutathione peroxidase agree with those seen by Sevanian et al. (85). These researchers concluded that the selenium-containing glutathione peroxidase could not efficiently interact with lipid hydroperoxides incorporated into liposomes. Phospholipase A_2 treatment releases these lipid hydroperoxides from the liposomal membrane and allows the subsequent interaction of these compounds with the enzyme. Much remains to be understood about the status and metabolism of lipid hydroperoxides in cellular membranes.

ACKNOWLEDGMENTS

The author expresses his appreciation to Carolyn Knapp, Lynne Rogers, and Cindy Spée for their assistance in the preparation of this manuscript. Support in part for this work was by United States Public Health Service Grant ES01978 and ES00210.

REFERENCES

1. Di Mascio P, Murphy ME, Sies H. Antioxidant defense systems: the role of carotenoids, tocopherols, and thiols. Am J Clin Nutr 1991; 53:194S-200S.

2. Sies H. Introduction. In: Sies H, ed. Oxidative stress. Orlando, FL: Academic Press, 1985; 1-7.

3. Machlin L, ed. Vitamin E: a comprehensive treatise. New York: Marcel Dekker, 1980.

4. Burton GW, Joyce A, Ingold KU. Is vitamin E the only lipid-soluble, chain-breaking antioxidant in human blood plasma and erythrocyte membranes? Arch Biochem Biophys 1983; 221:281-90.

5. Tappel AL. Vitamin E and free radical peroxidation of lipids. Ann NY Acad Sci 1972; 203:12-28.

6. McCay PB, Poyer JL, Pfeifer PM, May HE, Gilliam JM. A function for α-tocopherol: stabilization of the microsomal membrane from radical attack during TPNH-dependent oxidations. Lipids 1971; 6:297-306.

7. Kornbrust DJ, Mavis RD. Relative susceptibility of microsomes from lung, heart, liver, kidney, brain and testes to lipid peroxidation: correlation with vitamin E content. Lipids 1980; 15:315-22.

8. Sevanian A, Hacker AD, Elsayed N. Influence of vitamin E and nitrogen dioxide on lipid peroxidation in rat lung and liver microsomes. Lipids 1982; 17:269-77.

9. Litov RE, Matthews LC, Tappel AL. Vitamin E protection against in vivo lipid peroxidation initiated in rats by methyl ethyl ketone peroxide as monitored by pentane. Toxicol Appl Pharmacol 1981; 59:96-106.

10. Herschberger LA, Tappel AL. Effect of vitamin E on pentane exhaled by rats treated with methyl ethyl ketone peroxide. Lipids 1982; 17:686-91.

11. Dillard CJ, Kunert KJ, Tappel AL. Effects of vitamin E, ascorbic acid and mannitol on alloxan-induced lipid peroxidation in rats. Arch Biochem Biophys 1982; 216:204-12.

12. Muller DPR, Lloyd JK, Wolff OH. In: Biology of vitamin E. Ciba Foundation Symposium 101. London: Pitman Books, 1983; 106-21.

13. Diplock AT. Antioxidant nutrients and disease prevention: an overview. Am J Clin Nutr 1991; 53:189S-93S.

14. Packer JE, Slater TF, Willson RL. Direct observation of a free radical interaction between vitamin E and vitamin C. Nature 1979; 278:737-8.

15. Golumbic C, Mattill HA. Antioxidants and the autoxidation of fats. XIII. The antioxygenic action of ascorbic acid in association with tocopherols, hydroquinones and related compounds. J Am Chem Soc 1941; 63:1279-80.

16. Niki E, Saito T, Kamiya Y. The role of vitamin C as an antioxidant. Chem Lett 1983; 631-2.

17. Tappel AL, Brown WD, Zalkin H, Maier VP. Unsaturated lipid peroxidation catalyzed by hematin compounds and its inhibition by vitamin E. J Am Oil Chem Soc 1961; 38:5-9.

18. Barclay LRC, Bailey AMH, Kong D. The antioxidant activity of α-tocopherol–bovine serum albumin complex in micellar and liposome autoxidations. J Biol Chem 1985; 260:15809-14.

19. Leung HW, Vang MJ, Mavis RD. The cooperative interaction between vitamin E and vitamin C in suppression of peroxidation of membrane phospholipids. Biochim Biophys Acta 1981; 664:266-72.

20. Fukuzawa K, Takase S, Tsukatani H. The effect of concentration on the antioxidant effectiveness of α-tocopherol in lipid peroxidation induced by superoxide free radicals. Arch Biochem Biophys 1985; 240:117-20.

21. Doba T, Burton GW, Ingold KU. Antioxidant and co-antioxidant activity of vitamin C. The effect of vitamin C, either alone or in the presence of vitamin E or a water-soluble vitamin E analogue, upon the peroxidation of aqueous multilamellar phospholipid liposomes. Biochim Biophys Acta 1985; 835:298-303.

22. Niki E, Kawakami A, Yamamoto Y, Kamiya Y. Oxidation of lipids. VIII. Synergistic inhibition of oxidation of phosphatidylcholine liposome in aqueous dispersion by vitamin E and vitamin C. Bull Chem Soc Jpn 1985; 58:1971-5.

23. Bascetta E, Gunstone FD, Walton JC. Electron spin resonance study of the role of vitamin E and vitamin C in the inhibition of fatty acid oxidation in a model membrane. Chem Phys Lipids 1983; 33:207-10.

24. Scarpa M, Rigo A, Maiorino M, Ursini F, Gregolin C. Formation of α-tocopherol radical and recycling of α-tocopherol by ascorbate during peroxidation of phosphatidylcholine liposomes. An electron paramagnetic resonance study. Biochim Biophys Acta 1984; 801:215-9.

25. McCay PB. Vitamin E: interactions with free radicals and ascorbate. Annu Rev Nutr 1985; 5:323-40.

26. Aust SD, Morehouse LA, Thomas CE. Role of metals in oxygen radical reactions. J Free Radic Biol Med 1985; 1:3-25.

27. Svingen BA, Buege JA, O'Neal FO, Aust SD. The mechanism of NADPH-dependent lipid peroxidation. The propagation of lipid peroxidation. J Biol Chem 1979; 254:5892-9.

28. Liebler DC, Kling DS, Reed DJ. Antioxidant protection of phospholipid bilayers by α-tocopherol. J Biol Chem 1986; 261:12114-9.

29. Miller DM, Buettner GR, Aust SD. Transition metals as catalysts of "autoxidation: reactions. Free Radic Biol Med 1990; 8:95-108.

30. Willson RL. In: Biology of vitamin E. Ciba Foundation Symposium 101. London: Pitman Books, 1983; 19-44.

31. Bucher JR, Tien M, Morehouse LA, Aust SD. Redox cycling and lipid peroxidation: the central role of iron chelates. Fundam Appl Toxicol 1983; 3:222-6.

32. Liebler DC, Kaysen KL, Kennedy TA. Redox cycles of vitamin E: hydrolysis and ascorbic acid dependent reduction of 8a-(alkyldioxy)tocopherones. Biochemistry 1989; 28:9772-7.

33. Ito A, Hayashi S, Yoshida T. Participation of a cytochrome b5-like hemoprotein of outer mitochondrial membrane (OM cytochrome b) in NADH-semidehydroascorbic acid reductase activity of rat liver. Biochem Biophys Res Commun 1981; 101:591-8.

34. Haenen GRMM, Bast A. Protection against lipid peroxidation by a microsomal glutathione-dependent labile factor. FEBS Lett 1983; 159:24-8.

35. McCay PB, Brueggemann G, Lai EK, Powell SR. Evidence that α-tocopherol functions cyclically to quench free radicals in hepatic microsomes. Requirement for glutathione and a heat-labile factor. Ann NY Acad Sci 1989; 570:32-45.

36. Reddy CC, Scholz RW, Thomas CE, Massaro EJ. Vitamin E dependent reduced glutathione inhibition of rat liver microsomal lipid peroxidation. Life Sci 1982; 31:571-6.

37. Hill KE, Burk RF. Influence of vitamin E and selenium on glutathione-dependent protection against microsomal lipid peroxidation. Biochem Pharmacol 1984; 33:1065-8.

38. Misra HP. Generation of superoxide free radical during the autoxidation of thiols. J Biol Chem 1974; 249:2151-5.

39. Rowley DA, Halliwell B. Superoxide-dependent formation of hydroxyl radicals in the presence of thiol compounds. FEBS Lett 1982; 138:33-6.

40. Tien M, Bucher JR, Aust SD. Thiol-dependent lipid peroxidation. Biochem Biophys Res Commun 1982; 107:279-85.

41. Tappel AL. Lipid peroxidation damage to cell components. Fed Proc 1973; 32:1870-4.

42. Tirmenstein MA, Reed DJ. Characterization of glutathione-dependent inhibition of lipid peroxidation of isolated rat liver nuclei. Arch Biochem Biophys 1988; 261:1-11.

43. Plaa GL, Witschi H. Chemicals, drugs, and lipid peroxidation. Annu Rev Pharmacol Toxicol 1976; 16:125-41.

44. Pryor WA, Godber SS. Noninvasive measures of oxidative stress status in humans. Free Radic Biol Med 1991; 10:177-84.

45. Eichenberger K, Bohni P, Winterhalter KH, Kawato S, Richter C. Microsomal lipid peroxidation causes an increase in the order of the membrane lipid domain. FEBS Lett 1982; 142:59-62.

46. Baker SP, Hemsworth BA. Effect of mitochondrial lipid peroxidation on monoamine oxidase. Biochem Pharmacol 1978; 27:805-6.

47. Thomas CE, Reed DJ. Radical-induced inactivation of kidney Na^+,K^+-ATPase: sensitivity to membrane lipid peroxidation and the protective effects of vitamin E. Arch Biochem Biophys 1990; 281:96-105.

48. Porter NA. Chemistry of lipid peroxidation. Methods Enzymol 1984; 105:273-82.

49. O'Brien PJ, Hawco FJ. Hydroxyl-radical formation during prostaglandin formation catalyzed by prostaglandin cyclo-oxygenase. Biochem Soc Trans 1978; 6:1169-71.

50. Svingen BA, O'Neal FO, Aust SD. The role of superoxide and singlet oxygen in lipid peroxidation. Photochem Photobiol 1978; 28:803-9.

51. Brambilla G, Sciaba L, Faggin P, et al. Cytotoxicity, DNA fragmentation and sister-chromatid exchange in Chinese hamster ovary cells exposed to the lipid peroxidation product 4-hydroxynonenal and homologous aldehydes. Mutat Res 1986; 171:169-76.

52. Burk RF. Glutathione-dependent protection by rat liver microsomal protein against lipid peroxidation. Biochim Biophys Acta 1983; 757:21-8.

53. Niki E, Tsuchiya J, Tanimura R, Kamiya Y. Regeneration of vitamin E from α-chromanoxyl radical by glutathione and vitamin C. Chem Lett 1982; 789-92.

54. Ursini F, Maiorino M, Valente M, Ferri L, Gregolin C. Purification from pig liver of a protein which protects liposomes and biomembranes from peroxidative degradation and exhibits glutathione peroxidase activity on phosphatidylcholine hydro-peroxides. Biochim Biophys Acta 1982; 710:197-211.

55. Gibson DD, Hawrylko J, McCay PB. GSH-dependent inhibition of lipid peroxidation: properties of a potent cytosolic system which protects cell membranes. Lipids 1985; 20:704-11.

56. Belouqui O, Cederbaum AI. Prevention of microsomal production of hydroxyl radicals, but not lipid peroxidation, by the glutathione-glutathione peroxidase system. Biochem Pharmacol 1986; 35:2663-9.

57. Gibson DD, Hornbrook KR, McKay PB. Glutathione-dependent inhibition of lipid peroxidation by a soluble, heat-labile factor in animal tissues. Biochim Biophys Acta 1980; 620:572-82.

58. Ursini F, Maiorino M, Gregolin C. The selenoenzyme phospholipid hydroperoxide glutathione peroxidase. Biochim Biophys Acta 1985; 839:62-70.

59. Maiorino M, Coassin M, Roveri A, Ursini F. Microsomal lipid peroxidation: effect of vitamin E and its functional interaction with phospholipid hydroperoxide glutathione peroxidase. Cancer Res 1989; 49:1644-8.

60. Morgenstern R, DePierre JW. Microsomal glutathione transferase. Purification in unactivated form and further characterization of the activation process, substrate specificity and amino acid composition. Eur J Biochem 1983; 134:591-7.

61. Morgenstern R, Lundqvist G, Andersson G, Balk L, DePierre JW. The distribution of microsomal glutathione transferase among different organelles, different organs, and different organisms. Biochem Pharmacol 1984; 33:3609-14.

62. Tirmenstein MA, Reed DJ. Role of a partially glutathione S-transferase from rat liver nuclei in the inhibition of nuclear lipid peroxidation. Biochim Biophys Acta 1989; 995:174-80.

63. Bast A, Haenen GR. Regulation of lipid peroxidation by glutathione and lipoic acid: involvement of liver microsomal vitamin E free radical reductase. Adv Exp Med Biol 1990; 264:111-6.

64. Packer L, Maguire JJ, Mehlhorn RJ, Serbinova E, Kagan VE. Mitochondria and microsomal membranes have a free radical reductase activity that prevents chromanoxyl radical accumulation. Biochem Biophys Res Commun 1989; 159:229-35.

65. Mehlhorn RJ, Sumida S, Packer L. Tocopheroxyl radical persistence and tocopherol consumption in liposomes and in vitamin E-enriched rat liver mitochondria and microsomes. J Biol Chem 1989; 264:13448-52.

66. Niki E, Yamamoto Y, Komuro E, Sato K. Membrane damage due to lipid oxidation. Am J Clin Nutr 1991; 53:201S-5S.

67. Murphy ME, Scholich H, Wefers H, Sies H. Alpha-tocopherol in microsomal lipid peroxidation. Ann N Y Acad Sci 1989; 570:480-6.

68. Maguire JJ, Wilson DS, Packer L. Mitochondrial electron transport-linked tocopheroxyl radical reduction. J Biol Chem 1989; 264:21462-5.

69. Nakamura W, Hosoda S, Hayashi K. Biochim Biophys Acta 1974; 358:251-61.

70. Tirmenstein MA, Reed DJ. Effects of glutathione on the α-tocopherol-dependent inhibition of nuclear lipid peroxidation. J Lipid Res 1989; 30:959-65.

71. Franke WW. Structure, biochemistry, and functions of the nuclear envelope. Int Rev Cytol Suppl 1974; 4:71-236.

72. Akasaka S. Inactivation of transforming activity of plasmid DNA by lipid peroxidation. Biochim Biophys Acta 1986; 867:201-8.

73. Ueda K, Kobayashi S, Morita J, Komano T. Site-specific DNA damage caused by lipid peroxidation products. Biochim Biophys Acta 1985; 824:341-8.

74. Brawn K, Fridovich I. DNA strand scission by enzymically generated oxygen radicals. Arch Biochem Biophys 1981; 206:414-9.

75. Mukai FH, Goldstein BD. Mutagenicity of malonaldehyde, a decomposition product of peroxidized polyunsaturated fatty acids. Science 1976; 191:868-9.

76. Roots R, Okada S. Estimation of life times and diffusion distances of radicals involved in x-ray-induced DNA strand breaks of killing of mammalian cells. Radiat Res 1975; 64:306-20.

77. Shigenaga MK, Ames BN. Assays for 8-hydroxy-2'-deoxyguanosine: a biomarker of in vivo oxidative DNA damage. Free Radic Biol Med 1991; 10:211-6.

78. Hjelle JJ, Petersen DR. Metabolism of malondialdehyde by rat liver aldehyde dehydrogenase. Toxicol Appl Pharmacol 1983; 70:57-66.

79. Alin P, Danielson UH, Mannervik B. 4-Hydroxyalk-2-enals are substrates for glutathione transferase. FEBS Lett 1985; 179:267-70.

80. Christophersen BO. Formation of monohydroxy-polyenic fatty acids from lipid peroxides by a glutathione peroxidase. Biochim Biophys Acta 1968; 164:35-46.

81. Gruger EH, Tappel AL. Reactions of biological antioxidants. III. Composition of biological membranes. Lipids 1971; 6:147-8.

82. Puntarulo S, Cederbaum AI. Vitamin E prevents NADPH-dependent but not NADH-dependent chemiluminescence by isolated rat liver nuclei. Ann N Y Acad Sci 1989; 570:503-5.

83. Pryor WA, Stanley JP. A suggested mechanism for the production of malonaldehyde during the autoxidation of polyunsaturated fatty acids. Nonenzymatic production of prostaglandin endoperoxides during autoxidation (letter). J Org Chem 1975; 40:3615-7.

84. Dabeva MD, Petrov PT, Stoykova AS, Hadjiolov AA. Contamination of detergent purified rat liver nuclei by cytoplasmic ribosomes. Exp Cell Res 1977; 108:467-71.

85. Sevanian A, Muakkassah-Kelly SF, Montestruque S. The influence of phospholipase A_2 and glutathione peroxidase on the elimination of membrane lipid peroxides. Arch Biochem Biophys 1983; 223:441-52.

22

The Role of Vitamin E in Cancer Prevention

Carmia Borek

New England Medical Center and Tufts University School of Medicine, Boston, Massachusetts

INTRODUCTION

The role of diet in the etiology of cancer has been known for some time. A number of studies have suggested that the absence of certain dietary components, notably antioxidants, contributes to a substantial portion of human malignancies (1–3). Vitamin E (α-tocopherol) is a hydrophobic, peroxyl radical trapping, chain-breaking antioxidant found in the lipid fraction of living membranes. Its main function is to protect the lipid material of the organs from oxidation (1,4). The potential of vitamin E as a cancer preventive agent has been of interest for some time (6–8).

Individuals whose intake of vitamin E was above average indicated a lower risk of lung cancer (7), whereas those with prediagnostic low serum levels of vitamin E had an increased incidence of the disease (5,6). In a follow-up study on plasma levels of antioxidants demonstrated that there was an organ-specific risk factor for lower levels of antioxidants (8). Studies in animals demonstrated that vitamin E can prevent the growth of tumors induced in animals by chemical carcinogens (9–11) and inhibits the effects of tumor promoters in enhancing carcinogenesis (12).

VITAMIN E IN CANCER PREVENTION

A protective role for vitamin E is suggested by some epidemiological reports but is not consistently supported by other studies (3). Vitamin E is particularly difficult to study epidemiologically because it is widely distributed in food and the levels vary greatly among individual items. Menkes et al. (7) found that the levels of vitamin E were low in persons who later developed squamous cell carcinoma of the lung. The results differ from a reported study by Willet et al. (13) in which no significant association was detected between cancer risk at all sites and serum levels of total carotenoids or vitamin E. The interpretation of epidemiological studies related to diet is problematic. Dietary recall information is incomplete and sometimes imprecise, and the interactive roles of a variety of dietary components are poorly understood. Furthermore, cancer is a multistep process that develops over many years; thus it is uncertain at which stage dietary intake of chemopreventive agents is the most crucial for inhibiting neoplastic development.

VITAMIN E AS AN ANTICARCINOGEN IN VITRO

Cell cultures offer powerful tools in carcinogenesis research for studying chemopreventive intervention. In these in vitro systems cells are grown under defined conditions free from complex homeostatic mechanisms that prevail in vivo. These systems afford us the opportunity to assess at a cellular and molecular level the carcinogenic potential of physical or chemical agents as well as evaluate the effect of specific genes in transforming the cells. In addition, these cell systems make it possible to identify factors that inhibit transformation and act as cancer preventive agents (14).

Work in our laboratory focused on the potential of vitamin E as an anticarcinogen agent in its ability to inhibit radiation and chemically induced transformation in cell cultures. Our studies to evaluate the role of vitamin E as an anticarcinogen were conducted with a mouse cell line C3H 10T1/2 and with hamster dermal cells.

When cells were pretreated with vitamin E (α-tocopherol succinate) and 7 μM for 1–2 days and then exposed to radiation, benzo[*a*]pyrene, or a widespread carcinogen, tryptophane pyrolsate, a carcinogen in burned food. Vitamin E dramatically inhibited transformation compared to controls that were not treated (Fig. 1). Vitamin E acted synergistically with selenium or additively with vitamin C (Refs. 15 and 16; Borek, in preparation). Selenium produced its inhibitory action by enhancing the scavenging power of the cells and increasing peroxide breakdown (15); vitamin E affected the cells in a complimentary and different manner. The effect of vitamin E (α-tocopherol succinate) as an inhibitor of transformation was correlated with its ability to reduce peroxidation initiated in the cells (16). This indicated that the mechanism of vitamin E in protecting the cells was mediated in part by inhibiting the peroxidation and suppressing damaging free-radical–mediated consequences. Vitamin E was also capable of inhibiting the effect of ultraviolet-induced transformation in hamster dermal cells (Borek et al., in preparation).

The effect of vitamin E in directly modifying processes associated with free radicals was seen in the ability of vitamin E to inhibit ozone-induced transformation (Fig. 1) as well as inhibit the synergistic action of ozone- and ultraviolet-induced transformation.

VITAMIN E INHIBITS OZONE CARCINOGENESIS

Ozone (O_3), a reactive species of oxygen, is an important natural constituent of the atmosphere and a key component of oxidant smog. We found that ozone acts as a carcinogen, can activate cellular oncogenes, and is synergistic in its cocarcinogenic action with x-irradiation (17), and additive with UV light (18). Cells pretreated with vitamin E and then exposed to ozone or to ozone along with ultraviolet light became refractory to transformation by ozone as well as by ultraviolet light. The enhanced levels of lipid peroxidation products and the free-radical intermediates found in ozone-exposed cells may be associated with its oncogenic properties. These compounds were reduced when vitamin E treatment preceded ozone exposure (16) (Borek, in preparation).

INTERACTION OF VITAMIN E AND VITAMIN C

The synergistic interaction between vitamin E and vitamin C as antioxidants is known. Vitamin C spares vitamin E by reducing the vitamin E radical to regenerate vitamin E. Vitamin E scavenges lipid peroxyl radical (LOO') and interferes with chain propagation (19,20). We conducted studies to test if vitamin E and vitamin C were synergistic in their capacity to prevent transformation. The cells were pretreated with vitamin E at 7 μM and vitamin C (0.1 mg/ml) alone or in combination. Our work indicates that vitamin E and vitamin C act in concert to inhibit transformation in the manner that appears to be synergistic in nature (16).

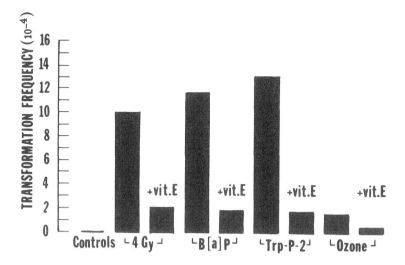

FIGURE 1 Vitamin E inhibits transformation induced by radiation and chemical carcinogens in C3H 10T1/2 cells.

CONCLUSION

As observed from experiments in vivo and in vitro and from human studies, vitamin E in its capacity as an antioxidant can protect against neoplastic development and interfere with the process of multistep carcinogenesis. Vitamin E is consumed in a linear fashion during oxidative processes. Exposure to a variety of environmental carcinogens, such as radiation, chemicals, or ozone enhances oxidant stress. It is critical to maintain a high level of vitamin E and other chemopreventive dietary factors in our tissues to increase our defense against the potential malignant process induced by a plethora of cancer-causing oxidants (14-16,21,22).

REFERENCES

1. Doll R, Peto R. The causes of cancer: quantitative estimates of avoidable risks of cancer in the United States today. J Natl Cancer Inst 1981; 66:1193-4.
2. Ames BN, Magaw R, Gold LS. Ranking possible carcinogenic hazards. Science 1987; 230:271-9.
3. Greenwald P, Nixon DW, Malone WF, Kelloff GJ, Stern HR, Witkin KM. Concepts in cancer chemoprevention research. Cancer 1990; 65:1483-90.
4. Menzel DB, Roehm NJ, Lee SD. Vitamin E: the biological and environmental antioxidant. J Agr Food Chem 1972; 20:481-6.
5. Salonen JT, Salonen R, Lappetelainen R, Maenapaa PH, Alfthan G, Puska P. Risk of cancer in relation to serum concentrations of selenium and vitamins A and E: matched case-control analysis of prospective data. Br Med J 1985; 290:917-20.
6. Haenszel W, Correa P, Lopez A, et al. Serum micronutrient levels in relation to gastric pathology. Int J Cancer 1985; 36:43-8.
7. Menkes MS, Comstock GW, Vuilleumier JP, Helsing KJ, Rider AA, Brookmeyer R. Serum beta-carotene vitamins A and E, selenium, and the risk of lung cancer. N Engl J Med 1986; 315:1250-4.
8. Stahelin HB, Gey F, Brubacher G. Preventive potential of antioxidative vitamins and carotenoids on cancer. J Vit Nutr Res 1989; 30:232-41.
9. Sadek I. Vitamin E and its effect on skin papilloma. In: Prasad KN, ed. Vitamins, nutrition, and cancer. Basel: Karger, 1984; 118-22.

10. Shklar G. Oral mucosal carcinogenesis inhibition by vitamin E. J Natl Cancer Inst 1982; 68:791-7.
11. Odukoya O, Hawach F, Shklar G. Retardation of experimental oral cancer by topical vitamin E. Nutr Cancer 1984; 6:98-104.
12. Chan JT, Black HS. The mitigating effect of dietary antioxidants on chemically-induced carcinogenesis. Experientia 1978; 34:110-1.
13. Willett W, Polk R, Underwood BA, et al. Relation of serum vitamins A and E and carotenoids to the risk of cancer. N Engl J Med 1984; 310:430-4.
14. Borek C, Ong A, Stevens VL, Wang E, Merrill AH Jr. Long-chain (sphingoid) bases inhibit multistage carcinogenesis in mouse C3H 10T1/2 cells treated with radiation and phorbol 12-myristate 13-acetate. Proc Natl Acad Sci U S A 1991; 88:1953-7.
15. Borek C, Ong A, Mason H, Donahue L, Biaglow JE. Selenium and vitamin E inhibit radiogenic and chemically induced transformation in vitro via different mechanisms of action. Proc Natl Acad Sci U S A 1986; 83:1490-4.
16. Borek C. The role of free radicals and antioxidants in cellular and molecular carcinogenesis in vitro. In: Hayaishi O, Niki E, Kondo M, Yoshikawa T, eds. Medical, biochemical and chemical aspects of free radicals. Amsterdam: Elsevier, 1989; 1461-9.
17. Borek C, Ong A, Zaider M. Ozone activates transforming genes in vitro and acts as a synergistic co-carcinogen with γ-rays only if delivered after radiation. Carcinogenesis 1989; 10:1549-1551.1.
18. Borek C, Ong A, Mason H. Ozone and ultraviolet light act as additive cocarcinogens to induce in vitro neoplastic transformation. Teratogenesis, Carcinogenesis, and Mutagenesis 1989; 9:71-74.
19. Packer JE, Slater TF, Willson RL. Direct observation of a free radical interaction between vitamin E and vitamin C. Nature 1979; 278:737-8.
20. Niki E. Antioxidants in relation to lipid peroxidation. Chem Phys Lipids 1985; 58:1971-5.
21. Borek C. The induction and regulation of radiogenic transformation in vitro: cellular and molecular mechanisms. In: Grunberger D, Goff S, eds. Mechanisms of cellular transformation by carcinogenic agents. New York: Pergamon, 1987; 151-95.

23

Vitamin E Deficiency Causes Appearance of an Aging Antigen and Accelerated Cellular Aging and Removal

Marguerite M. B. Kay

University of Arizona College of Medicine, Tucson, Arizona

INTRODUCTION

Senescent cell antigen appears on old cells and marks them for death by initiating the binding of IgG autoantibody and subsequent removal by phagocytes (1–6) in mammals and other vertebrates (4). Although the initial studies were done using erythrocytes as a model, senescent cell antigen has been found on all cells examined (4). Besides its central role in the removal of senescent and damaged cells, senescent cell antigen is involved in the removal of neural cells (7) and platelets (8) and the removal of erythrocytes in clinical hemolytic anemias (9), sickle cell anemia (10,11), and malaria (12).

Senescent cell antigen is generated by the modification of an important structural and transport membrane molecule, protein band 3 (5). Band 3 is a ubiquitous protein (13–19). It has been found in diverse cell types and tissues besides erythrocytes, including hepatocytes (13), squamous epithelial cells (13), alveolar (lung) cells (13), lymphocytes (13), kidney cells (16,17), neurons (13,14), and fibroblasts (13,19). Band 3 is also present in nuclear (13), Goldi (20), and mitochondrial membranes (21), as well as in cell membranes. Band 3-like proteins in nucleated cells participate in band 3 antibody-induced cell surface patching and capping (13). Band 3 maintains acid-base balance by mediating the exchange of anions (e.g., chloride and bicarbonate) and is the binding site for glycolytic enzymes (see Refs. 22–24). It is responsible for CO_2 exchange in all tissues and organs. Thus, it is the most heavily used anion transport system in the body. Band 3 is a major transmembrane structural protein that stabilizes the membrane and attaches the plasma membrane to the internal cell cytoskeleton by binding to band 2.1 (ankyrin) (see Ref. 22 for review). Membrane-spanning domains of band 3 are highly conserved across cells, tissues, and species, and no polymorphisms have been found (25–27).

The active antigenic sites of the aging antigen have been localized to residues 538–554 and 778–827 on band 3. Two peptides within these regions interact synergistically to generate a synthetic aging antigen that is an effective inhibitor of senescent cell IgG binding to old cells (26).

BAND 3 AGING

The demise of band 3, which is synonymous with generation of senescent cell antigen, occurs in two distinct steps. Structurally, band 3 undergoes an as yet uncharacterized initial change during cellular aging that triggeers a series of events terminating the life of the cell. We recently developed antibodies against aged band 3 that recognize this change because they bind to a distinct region of band 3 in old but not middle-aged or young cells (9). Following the change in intact band 3 with aging, band 3 undergoes degradation, presumably catalyzed by an enzyme (28). Preliminary experiments indicate that it is a calcium-dependent membrane-bound protease and suggest that the protease may be calpain. Cleavage of band 3 occurs in the transmembrane, anion transport region (5–6,26,28,29). Fragments of band 3 are detected in membranes of old but not young cells by immunoblotting with antibodies to normal band 3. Following degradation, band 3 undergoes a change in tertiary structure (28), becoming senescent cell antigen. A physiological IgG autoantibody binds to senescent cell antigen (senescent cell antigen) and initiates cellular removal.

Band 3 Structural Changes During Aging

Since senescent cell antigen and band 3 share antigenic determinants, the effect of cellular age on the accumulation of band 3 breakdown products is investigated using the immunoblotting technique (5,26,28). Freshly isolated cells are used, and membranes are prepared in the presence of protease inhibitors including diisopropylfluorophosphate (DFP), EDTA, and EGTA. Antibodies to band 3 and senescent cell antigen are used to determine the relative amount of band 3 breakdown products in the membranes of young, middle-aged, and old cells by immunoautoradiography using the gel overlay and immunoblotting procedures (5,26,28). Results revealed binding of both antibodies to band 3 and IgG eluted from senescent cells to a polypeptide migrating at M_r 62,000 in membranes of old but not young cells. The results indicated that band 3 breakdown products increase with cell age and that antigenic determinants recognized by the IgG eluted from senescent cells reside on a M_r 62,000 fragment of band 3.

Evidence supporting proteolysis of band 3 as an event initiating IgG binding, but not band 3 aging, is the finding that a senescent cell IgG column does not bind intact, nondenatured band 3. It binds only a 62,000 product of band 3. However, senescent cell IgG binds band 3 denatured by sodium dodecyl sulfate. We think that proteolysis is not the primary aging change in band 3, although data suggest that it may be required to initiate IgG binding. Data supporting this include the finding that red cells with other changes (e.g., an altered V_{max} but with all other parameters normal) and no increased breakdown products of band 3 do not have increased IgG binding. Although we have not been able to demonstrate cross-linking, we still consider it as a possibility for an aging event that may precede proteolysis and IgG binding.

Band 3 Functional Changes During Aging

Since our previous studies indicated that senescent cell antigen is derived from band 3 by cleavage in the transmembrane anion transport region (5,28,29), we suspected that anion transport is altered with cellular aging (6). If this suspicion proved to be correct, then we would have a functional assay for aging of band 3, the major anion transport protein of the erythrocyte membrane.

Transport studies on age-separated rat erythrocytes indicated that anion transport decreased with age (6). The Michaelis constant K_m increased, and the maximal velocity V_{max} decreased in old erythrocytes compared to middle-aged erythrocytes. These data provided us with another assay of cellular function to use to determine whether erythrocytes are "senescent." However, it is doubtful that the number of molecules of band 3 to which IgG is bound (100 per cell) is adequate to account for the magnitude of change in anion transport. Therefore, we suspect that another as yet unidentified change precedes events initiating IgG binding and is responsible for the observed changes in anion transport.

The following functional changes in band 3 occur as red cells age. These changes are decreased anion transport activity (increased K_m and decreased V_{max}), decreased number of high-affinity ankyrin binding sites, and binding of physiological IgG autoantibodies in situ (30). In addition, band 3 undergoes an as yet undefined change that results in binding of 980 antibodies to aged band 3 (9). These 980 antibodies recognize band 3 that has aged before its formation of senescent cell antigen. Degradation of band 3 generates senescent cell antigen.

VITAMIN E AND MECHANISMS OF CELLULAR AGING AND GENERATION OF SENESCENT CELL ANTIGEN

As an approach to the evaluation of oxidation as a possible mechanism responsible for the generation of senescent cell antigen, we studied erythrocytes from vitamin E-deficient rats. The importance of vitamin E as an antioxidant, providing protection against free-radical-induced membrane damage, has been well documented (31–33). Vitamin E is primarily localized in cellular membranes, and a major role of vitamin E is the termination of free-radical chain reactions propagated by the polyunsaturated fatty acids of membrane phospholipids. Vitamin E-deficient erythrocytes are defective in their ability to scavenge free radicals (34,35).

The erythrocyte has many potential sources for the generation of free radicals. Hemoglobin is known to catalyze lipid peroxidation as well as enhance the decomposition of lipid hydroperoxides to the corresponding free radicals (35). Autooxidation of oxyhemoglobin to methemoglobin results in the generation of a superoxide radical (36,37). The reaction of a superoxide radical with peroxides in the erythrocytes produces highly reactive intermediates, such as the hydroxyl radical ($\cdot$OH). These radicals in turn react with the lipid and protein components of the membrane, damaging its integrity and leading to eventual hemolysis of the cell (36). Lipid peroxidation in the erythrocyte membrane can result in accumulation of an aldehyde, which can cause a reduction in deformability (38,39) and formation of irreversibly sickled cells (40).

In humans, vitamin E deficiency shortens the erythrocyte life span, causing a compensated hemolytic anemia in patients with cystic fibrosis (40). In newborns, vitamin E deficiency causes a hemolytic anemia that develops by 4–6 weeks of age (40).

Specific biochemical alterations in the membrane of erythrocytes from vitamin E-deficient rhesus monkeys have been described (41,42). Furthermore, vitamin E deficiency represents a "physiological" method for rendering cells susceptible to free-radical damage and may simulate conditions encountered in situ. In contrast, methods that have been used to induce oxidative damage in vitro, such as exposing cells to malondialdehyde or peroxide, result in numerous pronounced membrane changes that are not observed in cells aged in situ (unpublished observations).

The role of free-radical damage in aging has received a great deal of attention. It is interesting that there is a correlation between life span and natural antioxidant levels in a variety of species and that the level of such antioxidants appears to correlate with metabolic activity of individual species (43). Evidence for free-radical damage associated with aging is the presence of lipofuscin and ceroid, so-called aging pigments, which represent accumulated breakdown products of polyunsaturated fatty acids and proteins. It has been suggested that free radicals may be mediators of aging and of specific pathologies, such as inflammation, arthritis, adult respiratory distress syndrome, and other conditions (44). Free radicals have also been implicated as causative agents in mutagenesis and carcinogenesis as well as causing cross-linking of macromolecules and the formation of age pigments (44).

For our experiments, weanling male Wistar-Kyoto (W/K) rats (Charles River Breeding Laboratories) were randomly separated into three groups of 12 rats each. One group received a vitamin E-deficient diet (Dyets, Bethlehem, PA). The second and third groups received the identical diet supplemented with d,l-α-tocopheryl acetate, hereafter referred to as vitamin E, at 50 and 200 mg/kg, respectively. Rats were tested at 11, 16, 20, and 30 weeks after being placed on the assigned diets. Rats fed a diet containing no vitamin E are referred to as deficient rats, those receiving vitamin E at 50 mg/kg are referred to as normal, and those receiving 200 mg/kg are referred to as high.

Clinical studies of vitamin E-deficient rats indicated accelerated destruction of erythrocytes and were consistent with the vitamin E-deficient rats having a compensated hemolytic anemia as is observed in vitamin E-deficient humans (Table 1).

The phagocytosis assay was performed on both age-separated and unseparated erythrocytes from rats fed a diet containing normal amounts of vitamin E or a diet deficient in vitamin E (Table 2). Old erythrocytes obtained from rats fed a diet containing normal amounts of vitamin E were phagocytized, whereas young and middle-aged erythrocytes were not. In contrast, young and middle-aged as well as old erythrocytes were phagocytized when obtained from vitamin E-deficient rats. There was a significant difference in phagocytosis between erythrocytes obtained from normal rats and vitamin E-deficient rats, even when unfractionated erythrocytes were used for the assay (Table 2).

To determine whether the observed defects in erythrocytes obtained from vitamin E-deficient rates were reversible and/or were the result of oxidative events occurring in vitro, vitamin E was added to aliquots of blood 24 h before the samples were tested. Addition of

TABLE 1 Hematological and Serum Indices of Vitamin E-Deficient and Control Rats[a]

Index[b]	Vitamin E diet			
	High	Normal	Deficient	Deficient + vitamin E
RBC $\times$ 10^6	10.31 $\pm$ 0.08	10.39 $\pm$ 0.08	10.35 $\pm$ 0.14	
Hgb, g/dl	16.65 $\pm$ 0.12	16.76 $\pm$ 0.12[c]	16.26 $\pm$ 0.18	
Hct, %	50.53 $\pm$ 0.38	51.26 $\pm$ 0.40[c]	49.6 $\pm$ 0.55	
MCV, μm^3	48.99 $\pm$ 0.21[d]	49.28 $\pm$ 0.14[d]	47.84 $\pm$ 0.16	
n	9	10	11	9
Haptoglobin	5.5 $\pm$ 0.2	5.9 $\pm$ 0.4		BDL[e]
Fe	177 $\pm$ 59	224 $\pm$ 122	229 $\pm$ 175	149 $\pm$ 61
TIBC, %	28.9 $\pm$ 0.4	29.7 $\pm$ 1.3	31.1 $\pm$ 0.1	30.8 $\pm$ 2.2

[a]Four experiments were performed. The serum data presented in the table are from the fourth experiment. The number of rats from which the serum data was obtained (n) precedes the serum data. Haptoglobin results on pooled samples from the third experiment were high, 5.23 mg/dl; normal, 5.13 mg/dl; deficient, <0.82 mg/dl. The reticulocytes in the same experiment were high, 0.4%; normal, 0.2%; deficient, 3.38%. Results presented are the mean $\pm$ 1 standard deviation (SD). There was no significant difference in unconjugated bilirubin or lactate dehydrogenase.
[b]Hgb, hemoglobin; Hct, hematocrit, MCV, mean cell volume; Fe, iron; TIBC, total iron binding capacity.
[c]$P \leq 0.05$ compared to deficient.
[d]$P \leq 0.001$ compared to deficient.
[e]BDL, Below the limits of detection. Detection limit is 0.83 mg/dl.

TABLE 2 Phagocytosis of Unfractionated and Age-Separated Erythrocytes from Vitamin E-Deficient and Normal Rats[a]

Vitamin E diet	Erythrocyte fraction	Phagocytosis (%)
Normal (50 mg/kg)	Young	2 ± 3
	Middle-aged	1 ± 1
	Old	72 ± 3
	Unfractionated	13 ± 3
Deficient (0 mg/kg)	Young	89 ± 1
	Middle-aged	88 ± 1
	Old	99 ± 0
	Unfractionated	87 ± 1

[a]The phagocytosis assay is performed with U927 cells. Erythrocytes are incubated with macrophages overnight at 37°C in a humidified atmosphere containing 5% CO_2. Data are presented as the mean ± 1 SD; $n = 3$. Diet is expressed as mg vitamin E per kg diet.
Source: Reproduced from Ref. 6.

vitamin E in vitro to erythrocytes from vitamin E-deficient rats did not alter the membrane defect that rendered these erythrocytes susceptible to phagocytosis (Table 3).

We suspected that anion transport may be altered with cellular aging because our previous studies indicated that senescent cell antigen is derived from band 3 by cleavage in the transmembrane anion transport region (24,35,37,38). If this suspicion proved to be correct, then we would have a functional assay for aging of band 3, the major anion transport protein of the erythrocyte membrane.

Transport studies on age-separated rat erythrocytes indicated that anion transport decreased with age (Table 4). The kinetic Michaelis constant K_m increased and the maximal velocity V_{max} decreased in old erythrocytes compared to middle-aged erythrocytes.

These data provided us with an assay of cellular function to use to determine whether erythrocytes from vitamin E-deficient rats exhibited characteristics of old erythrocytes prematurely. Results of the anion transport studies on erythrocytes from vitamin E-deficient rats revealed that their anion transport was impaired, as was transport in old erythrocytes (Table 4).

We examined the glyceraldehyde-3-phosphate dehydrogenase (G3PD) activity of red cells because it is one of the enzymes that attaches to the cytoplasmic segment of band 3 (41, 42). In addition, Shapiro's studies (41,42) showed that G3PD activity is reduced in erythrocytes from vitamin E-deficient rhesus monkeys.

TABLE 3 Phagocytosis of Unseparated Erythrocytes from Vitamin E-Deficient and Normal Rats[a]

Vitamin E diet	Vitamin E added in vitro (mg/dl)	Phagocytosis (%)
High (200 mg/kg)	0	5 ± 4
Normal (50 mg/kg)	0	25 ± 2
Deficient (0 mg/kg)	0	77 ± 4
Deficient (0 mg/kg)	1.6	64 ± 2

[a]Data are presented as the mean ± SD ($n = 3$). Macrophage source: human U937.
Source: Reproduced from Ref. 6.

TABLE 4 Effect of Cellular Aging and Vitamin E Deficiency on the Anion Transport System of Rat Erythrocytes[a]

	K_m (mM)	V_{max} (mol $\times$ 10^{-8} per 10^8 cells per minute)
Cell age		
Middle-aged	0.5 ± 0.2	41.8 ± 1.9
Old	1.6 ± 0.4[b]	17.3 ± 3.8[b]
Vitamin E diet		
Normal	0.7 ± 0.1	41.2 ± 3.7
Deficient	2.0 ± 0.3[b]	16.2 ± 2.0[b]
Deficient + vitamin E	2.0 ± 0.3[b]	12.1 ± 1.7[b]

[a]Results are presented as the mean ± 1 SD. There is no statistical difference between deficient, deficient + vitamin E, and old cells. K_m, concentration at half-maximal exchange, corresponding to an apparent Michaelis-Menten constant (mM); V_{max}, maximal flux, determined at 37°C and pH 7.2.
[b]$P \leqslant 0.01$.
Source: Reproduced from Ref. 6.

Our studies showed that G3PD did not bind to rat erythrocyte membranes. All the enzyme activity was in the cytoplasm as soluble enzyme, in contrast to humans, in whom a fraction of the enzyme is membrane bound. Studies of the soluble G3PD in rat erythrocytes showed that it is reduced in old erythrocytes and in erythrocytes from vitamin E-deficient rats (Table 5). Thus, cells from vitamin E-deficient rats behave like old cells with respect to G3PD activity.

Differences were not detected in the protein or glycoprotein composition of erythrocyte membranes from control and vitamin E-deficient rats. Although high-molecular-weight polypeptides or polymers were detected with Coomassie blue staining of 2–16% polyacrylamide gels, there were no differences in samples. Immunoblotting studies revealed increased breakdown products of band 3 in cells from vitamin E-deficient rats (Fig. 1). Thus, vitamin E deficiency leads to accelerated red cell aging, presumably through oxidation.

The results of the experiments presented here indicate that erythrocytes from vitamin E-deficient rats age prematurely. The results suggest that oxidation may accelerate cellular aging and may be a mechanism for the generation of senescent cell antigen.

As a mechanism for cellular aging and the generation of senescent cell antigen, free-radical reactions and oxidation are considered probable candidates (31,33,45). Most free-radical reactions involve the reduction of molecular oxygen, leading to the formation of highly reactive oxygen species, such as superoxide anion ($O_2\cdot$), hydroxyl radical ($\cdot OH$), hydrogen peroxide (H_2O_2), and singlet oxygen ($^1O_2$). The production of these highly reactive oxygen species as metabolic intermediates appears to be an evolutionary consequence of aerobic existence because the spin state of oxygen favors univalent pathways of reduction (45).

We used vitamin E deficiency as a model for studying oxidation because studies show that, in mammals, vitamin E functions as an antioxidant and because vitamin E deficiency simulates conditions encountered in situ more closely than does chemical treatment of cells in vitro.

The results presented here indicate that vitamin E deficiency causes premature aging of erythrocytes and IgG binding. Erythrocytes from vitamin E-deficient rats behave like old erythrocytes in the phagocytosis assay and in anion transport and glyceraldehyde-3-phosphate dehydrogenase activity. In addition, increased breakdown products of band 3 were observed in erythrocyte membranes from vitamin E-deficient rats. We have not observed high-molecular-

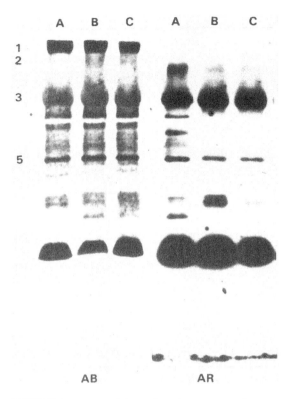

FIGURE 1 Immunoblots of erythrocyte membrane proteins from vitamin E-deficient and normal rats incubated with antibodies to band 3. AB, amido black; AR, autoradiograph. Lanes: A, vitamin E deficient; B, normal; C, high. Band designations are at left. (Reproduced from Ref. 6.)

weight complexes containing band 3 in membranes from vitamin E-deficient rats or old cells aged in situ, except under conditions that precipitate IgG (unpublished observations).

Anion transport across the membrane, which is the best-defined physiological function of band 3, is reduced by both cellular aging and vitamin E deficiency. Differences between normal unseparated rat and human erythrocytes are observed in the position of the concentration maximum [e.g., K_m = 0.5–0.8 mM for rat cells and 3–5 mM for human cells (unpublished observations)]. Also, the maximal sulfate transport activity of rat erythrocytes was found to be two to three times as high as that of human erythrocytes (unpublished observations).

The decrease in transport activity observed with vitamin E deficiency is not reversed by the addition of vitamin E in vitro. This latter observation may indicate that the observed effects are not caused by a direct effect of vitamin E on membrane fluidity.

A decrease in band 3 functioning with erythrocyte age supports the data indicating that alteration of band 3 occurs during erythrocyte aging and generation of senescent cell antigen (6,28,30). Impairment of anion transport was anticipated based on our previously published data indication that senescent cell antigen is generated by cleavage of band 3 in the anion transport region (5,9,29).

Results of the experiments on vitamin E deficiency suggest that oxidation can cause aging of band 3. We suspect that this may be one of the mechanisms of cellular aging in situ. At this time, it appears that general cellular damage, such as lysis (Kay, unpublished) and oxidation, can result in the generation of senescent cell antigen. We suspect that many different cellular insults have a final common pathway that results in the generation of senescent cell antigen.

ACKNOWLEDGMENTS

This work is supported by the Veterans Administration and by NIH Grants AG0844 and AG08574.

REFERENCES

1. Kay MMB. Mechanism of removal senescent cells by human macrophages in situ. Proc Natl Acad Sci U S A 1975; 72:3521-5.
2. Kay MMB. Role of physiologic autoantibody in the removal of senescent human red cells. J Supramol Struct 1978; 9:555-67.
3. Bennett GD, Kay MMB. Homeostatic removal of senescent murine erythrocytes by splenic macrophages. Exp Hematol 1981; 9:297-307.
4. Kay MMB. Isolation of the phagocytosis inducing IgG-binding antigen on senescent somatic cells. Nature 1981; 289:491-4.
5. Kay MMB. Localization of senescent cell antigen on band 3. Proc Natl Acad Sci U S A 1984; 81: 5753-7.
6. Kay MMB, Bosman GJCGM, Shapiro SS, Bendich A, Bassel PS. Oxidation as a possible mechanism of cellular aging: vitamin E deficiency causes premature aging and IgG binding to erythrocytes. Proc Natl Acad Sci U S A 1986; 83:2463-7.
7. Kay MMB, Hughes J, Zagon I, Lin F. Brain membrane protein band 3 performs the same functions as erythrocyte band 3. Proc Natl Acad Sci U S A 1991; 88:2778-82.
8. Khansari N, Fudenberg HH. Immune elimination of aging platelets by autologous monocytes: role of membrane-specific autoantibody. Eur J Immunol 1983; 13:990-4.
9. Kay MMB, Flowers N, Goodman J, Bosman GJCGM. Alteration in membrane protein band 3 associated with accelerated erythrocyte aging. Proc Natl Acad Sci U S A 1989; 86:5834-8.
10. Hebbel RP, Miller WJ. Phagocytosis of sickle erythrocytes. Immunologic and oxidative determinants of hemolytic anemia. Blood 1984; 64:733-41.
11. Petz LD, Yam P, Wilkinson L, Garratty G, Lubin B, Mentzer W. Increased IgG molecules bound to the surface of red blood cells of patients with sickle cell anemia. Blood 1984; 64:301-4.
12. Friedman MJ, Fukuda M, Laine RA. Evidence for a malarial parasite interaction site on the major transmembrane protein of the human erythrocyte. Science 1985; 228:75-7.
13. Kay MMB, Tracey CM, Goodman JR, Cone JC, Bassel PS. Polypeptides immunologically related to erythrocyte band 3 are present in nucleated somatic cells. Proc Natl Acad Sci U S A 1983; 80:6882-6.
14. Kay MMB, Bosman G, Notter M, Coleman P. Life and death of neurons: the role of senescent cell antigen. Ann N Y Acad Sci 1988; 521:155-69.
15. Drenckhahn D, Zinke K, Schauer U, Appell KC, Low PS. Identification of immunoreactive forms of human erythrocyte band 3 in nonerythroid cells. Eur J Cell Biol 1984; 34:144-50.
16. Demuth DR, Showe LC, Ballantine M, et al. Cloning and structural characterization of a human non-erythroid band 3-like protein. Embo J 1986; 5:1205-14.
17. Kudrycki KE, Shull GE. Primary structure of the rat kidney band 3 anion exchange protein deduced from a cDNA. J Biol Chem 1989; 264:8185-92.
18. Brosius F, Alpert S, Garcia A, Lodish H. The major kidney band 3 gene transcript predicts an amino-terminal truncated band 3 polypeptide. J Biol Chem 1989; 264:7784-7.

19. Hazen-Martin DJ, Pasternack G, Spicer SS, Sens DA. Immunolocalization of band 3 protein in normal and cystic fibrosis skin. J Histochem Cytochem 1986; 34:823-6.

20. Kellokumpu S, Neff L, Jamsa-Kellokumpu S, Kopito R, Baron R. A 115-kD polypeptide immunologically related to erythrocyte band 3 is present in Golgi membranes. Science 1988; 242:1308-11.

21. Schuster VL, Bonsib SM, Jennings ML. Two types of collecting duct mitochondria-rich (intercalated) cells: lectin and band 3 cytochemistry. Am J Physiol 1986; 251:C347-55.

22. Steck TL. The organization of proteins in human red blood cell membranes. J Cell Biol 1974; 62: 1-19.

23. Jennings M, Al-Rhaiyel S. Modification of a carboxyl group that appears to cross the permeability barrier in the red blood cell anion transporter. J Gen Physiol 1988; 92:161-78.

24. Jennings ML. Evidence for an access channel leading to the outward-facing substrate site in human red blood cell band 3. In: Hamasaki N, Jennings ML, eds. Anion transport protein of the red blood cell membrane. Proceedings of the international meeting on anion transport protein of the red blood cell membrane as well as kidney and diverse cells, Fukuoka, May 1989. New York: Elseviere, 1989; 59-72.

25. Tanner MJA, Martin PG, High S. The complete amino acid sequence of the human erythrocyte membrane anion-transport protein deduced from the cDNA sequence. Biochem J 1988; 256:703-12.

26. Kay MMB, Marchalonis JJ, Hughes J, Watababe K, Schluter SF. Definition of a physiologic aging auto-antigen using synthetic peptides of membrane protein band 3: localization of the active antigenic sites. Proc Natl Acad Sci U S A 1990; 87:5734-8.

27. Kay MMB, Lin F. Molecular mapping of the active site of an aging antigen: senescent cell antigen is located on an anion binding segment of band 3 membrane transport protein. Gerontology 1990; 36:293-305.

28. Kay MMB. Aging of cell membrane molecules leads to appearance of an aging antigen and removal of senescent cells. Gerontology 1985; 31:215-35.

29. Kay MMB. Band 3, the predominant transmembrane polypeptide, undergoes proteolytic degradation as cells age. Monogr Dev Biol 1984; 17:245-53.

30. Bosman GJCGM, Kay MMB. Erythrocyte aging: a comparison of model systems for simulating cellular aging in vitro. Blood Cells 1988; 14:19-35.

31. McCay PB, King MM. Biochemical function. I. Vitamin E: its role as a biologic free radical scavenger and its relationship to the microsomal mixed-function oxidase system. In: Machlin LJ, ed. Vitamin E, a comprehensive treatise. New York: Marcel Dekker, 1980; 289-317.

32. Walton JR, Packer L. Free radical damage and protection: relationship to cellular aging and cancer. In: Machlin LJ, ed. Vitamin E, a comprehensive treatise. New York: Marcel Dekker, 1980; 495-518.

33. Farrell PM, Bieri JG, Fratantoni JF, Wood RE, di Sant Agnese PA. The occurrence and effects of human vitamin E. deficiency. A study in patients with cystic fibrosis. J Clin Invest 1977; 60:233-41.

34. Dodge JT, Cohen G, Kayden HJ, Phillips GB. Peroxidative hemolysis of red blood cells from patients with abetalipoproteinemia. J Clin Invest 1967; 46:357-68.

35. Chiu D, Lubin D, Shohet SB. Peroxidative reactions in red cell biology. In: Pryor W, ed. Free radicals in biology. New York: Academic Press, 1982; 115-60.

36. Tapple AL. The mechanism of oxidation of unsaturated fatty acids catalyzed by hematin compounds. Arch Biochem Biophys 1953; 44:378-95.

37. Koppenol WH, Butler J. Mechanism of reactions involving singlet oxygen and the superoxide anion. FEBS Lett 1977; 83:1-6.

38. Pfafferott C, Meiselman HJ, Hochstein P. The effect of malonyldialdehyde on erythrocyte deformability. Blood 1982; 59:12-5.

39. Jain SK. The accumulation of malonyldialdehyde, a product of fatty acid peroxidation, can disturb aminophospholipid organization in the membrane bilayer of human erythrocytes. J Biol Chem 1984; 259:3391-4.

40. Oski FA. Anemia related to nutritional deficiencies other than vitamin B_{12} and folic acid. In: Williams WJ, Beutler E, Erslev AJ, Lictman MA, eds. Hematology. New York: McGraw-Hill, 1983; 532-7.

41. Shapiro SS, Mott DJ, Machlin LJ. Altered binding of glyceraldehyde-3-phosphate to its binding site in vitamin E deficient red blood cells. Nutr Rep Int 1982; 25:507-17.

42. Shapiro SS, Mott DJ, Machlin LJ. Alterations of enzymes in red blood cell membranes in vitamin E deficiency. Ann N Y Acad Sci 1982; 393:263-76.

43. Cutler RG. Cellular aging: concepts and mechanisms. Interdiscipl Top Gerontol 1976; 9:83-133.

44. Packer L. Vitamin E, physical exercise and tissue damage in animals. Med Biol 1984; 62:105-9.

45. Dean R, Cheeseman K. Vitamin E protects proteins against free radical damage in lipid environments. Biochem Biophys Res Commun 1987; 148:1277-82.

24

Adhesion of Human Platelets Inhibited by Vitamin E

Peter D. Richardson

Brown University, Providence, Rhode Island

Manfred Steiner

Brown University, Providence, and Memorial Hospital of Rhode Island, Pawtucket, Rhode Island

INTRODUCTION

The complex sequence of events that transforms the flowing blood into a clot depends upon the participation of platelets. Their adherence to exposed collagenous tissue initiates the sequence of events that begins with the activation of platelets and ends with the change of a soluble protein, fibrinogen, to an insoluble substance, fibrin. The interaction of specific receptors on the platelet surface with adhesive proteins in the environment initiates a shape change, a drastic alteration in the morphological appearance of platelets that usually turns their normally diskoid bodies into spheres from which long pseudopodia extend. This change in shape occurs with lightning speed, in fractions of a second. In the last several years great progress has been made in identifying the respective protagonists of this interaction. Glycoprotein receptors (some of which belong to the group of integrins) for many of the adhesive proteins have been identified, and knowledge of their structure and biochemical characteristics is expanding rapidly. Several reviews on this subject have been published (e.g., Refs. 1 and 2).

Compared to the wealth of information available on platelet aggregation, the literature on platelet adhesion is relatively small. Our understanding of the basic mechanisms that regulate the adhesive properties of platelets is still rather rudimentary, and little is known of how pharmacological or dietary manipulation can influence the adhesive potential of platelets. Progress in this area has been relatively slow, which is partly due to the difficulty of evaluating this platelet function in a reliable, physiologically meaningful manner.

We briefly review the methodology available to study platelet adhesion. Not only does this provide essential background information on the advantages and disadvantages of the various techniques available, but it also gives the reader the necessary information to evaluate the results that we obtained modifying platelet adhesion by dietary supplementation with vitamin E.

QUANTITATION OF PLATELET FUNCTION

Quantitative measurement of platelet aggregation and, more particularly, platelet adhesion has always been difficult because of the small size of platelets. This size makes it hard to obtain images by optical microscopy. Although the most natural site to perform experiments on platelet adhesion and aggregation would be in the circulation, experiments using the microcirculation in various species have shown difficulty with this approach. Problems include the practicalities that microcirculation tends to twitch in the viewing field of the microscope unpredictably, and this provides all sorts of difficulty in keeping track of images in a long series of frames used to record temporal processes (3). Many of the microcirculatory beds used for research purposes have fat-laden cells surrounding the arterioles and venules used for studies of interactions between platelets and vessel walls, and this fat degrades the optical images. Furthermore, the preparation of small animals for microcirculatory experiments consumes time and costs more than taking in vitro approaches. Therefore, in studies of platelet adhesion as well as aggregation, there are strong basic technical and financial incentives for using in vitro rather than in vivo studies.

The advantages of in vitro methods for study of platelet aggregation and adhesion are clear, but there has always been a consciousness that it is important to have continued motion of the fluid. Even the simple aggregometer (4), invented about 30 years ago, contains a magnetically driven stirring bar that keeps the platelet-rich plasma (PRP) in motion continuously. Only a few test methods to study platelet adhesion that do not involve continuous motion appear to have been brought forward; the one taking a short period of time is the George test (5). It does not seem to have garnered much following despite its simplicity and reproducibility. Other nonflow methods are described at the end of this section. Virtually all other methods have involved the establishment of a flow over a test surface.

Flow devices with test surfaces have mostly been applied to obtain a measure of the adhesion of platelets to the surface after exposure for a measured period of time. As such, they provide a single snapshot of the detailed transactions between platelets and the test surface. Some of these devices have been devoted to investigation of candidate "biomaterials" for possible use in the handling of blood in ex vivo circuits (such as heart-lung machines) or for implantable devices that will remain in contact with blood for extended periods, such as prosthetic heart valves (6). One of the earliest test methods of this type was the Gott ring system, in which a ring of suitable caliber was surgically implanted in the vena cava of test animals and removed after several days of implantation. The results were read somewhat simply in terms of the extent of occlusion of the lumen by thrombus.

There has always been a consciousness among hematologists that many interactions related to the activation of platelets, and the conversion of fibrinogen to fibrin, involve processes that occur in short scales of time—minutes and even seconds—following contact of whole blood or PRP with test surfaces, in view of the processes that were all too clear (although not understood in detail) in such classic tests as the Lee-White clotting test (7). This drove many investigators, eager to understand the early stages of platelet adhesion and aggregation, to develop test systems that allowed for relatively short periods of exposure to a test surface to whole blood or PRP, with subsequent examination of the test surface in considerable detail, often involving preparation for scanning electron microscopy as an approach to assessing the degree of reaction of platelets with surfaces and also measuring the number of platelets adherent per unit of surface area as another indicator of the adhesive reaction between the surface and platelets. Some experiments were conducted by prelabeling the platelets with radioisotope markers, which allowed assessment of the degree of reaction using the radioactivity counting devices already widely present in many of the laboratories, but again these experiments were based on the approach of exposing the surface for a long period of time to

PRP and then taking a snapshot at the end of this. Because platelets are excitable cells and the processing for labeling may alter the excitability and also the level of preexcitement, there have always been questions lurking about the reliability and full appropriateness of experiments conducted with such platelets compared with those in a more native state.

Extensive evidence also suggested that there were possible interactions of platelets with surfaces in the early stages that were not readily measured by taking a snapshot after a finite period of time of exposure. Experiments conducted with surfaces prepared as identically as possible but exposed for different periods of time to similar flows showed puzzling trends in the apparent relationship between platelet adhesion and time. About 15 years ago two systems for continuously observing the process of adhesion of platelets to surfaces were developed. In one of these (8), the adhesion of platelets from PRP to flat test surfaces is studied using phase-contrast microscopy with continuous video recording; in the other method (9), fluorescently labeled platelets suspended in whole blood are observed adhering to the luminal surface of a coated circular glass tube using epifluorescence microscopy, with video recording. Both methods demonstrated convincingly that the interactions of platelets with surfaces coated with collagen, fibronectin, or thrombin, for example, were indeed complicated, with platelets becoming adherent for relatively short periods of time at first to specific points on the surfaces, then departing, and sometimes being replaced on the very same sites by platelets that approach subsequently, with these also departing after some period of time. This process has been observed to repeat itself many times at some sites of adhesion.

The flow chamber for measuring the adhesion of platelets to a flat test surface involves the use of a purpose-built chamber that in plan is the size of a microscope slide (Fig. 1). As used thus far, typical chambers have an area exposed to a steady flow of PRP extending roughly 1 × 4 cm. The thickness of the blood path is about 500 μm but can be selected over a range by using gaskets of different thicknesses. It is also possible with this chamber to cause a variation in the flow rate locally along the stream lines that the laminar flow PRP follows by introducing an obstacle of a specific shape (typically a circular disk) in the center portion of

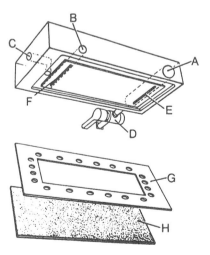

FIGURE 1 Flow chamber for continuous in vitro quantification of platelet adhesion: (A) inlet manifold; (B) outlet manifold; (C) connection to vacuum line; (D) stopcock; (E) slot that admits PRP to parallel zone; (F) exit slot; (G) gasket with holes to transmit vacuum from perimeter channel connected to C to the slide-bearing the test surface H; (H) slide-bearing test surface.

the channel and extending from one side to the other across the narrow path of flow. A convenient feature of the flow in chambers that incorporate such obstacles is that the main flow pattern is easily computed from potential flow theory (10), which has been known and understood well for many decades. As a consequence it is possible to examine effects associated with passage along stream paths such that platelets experience (for example) a steady increase in shear stress to a known maximum followed by a decrease or (for another example) flow along stream lines where the shear stress decreases at first and then increases to a maximum, decreases again, and subsequently increases to a value similar to that at which the flow began. This allows investigation of the effect of shear history upon the adhesion behavior of platelets. A further advantage of this chamber, in which the test surface for adhesion is a separate glass slide, held to the main chamber by use of a perimeter vacuum channel, is that it can be coated over the whole surface in contact with PRP or over a part of the surface in contact with a known edge location for the coating, so that effects that occur in association with change in the constitution of a coating on a surface (or the distribution of specific adhesive proteins on a surface) can be readily controlled and measured. In principle, the flat surface flow chamber can be used with the epifluorescence method, but these tests, of course, require the fluorescent labeling of all the platelets in the volume used in a typical experiment. For the flow conditions most typically found convenient with this flow chamber, the quantity of PRP required is about 15 ml, corresponding to a whole-blood volume of approximately 30 ml per experimental run, the latter lasting approximately 15 minutes. An advantage of the circular glass tube system for flow is that the aliquot of blood required for an experiment is less than that required for the flat plate chamber; however, it is not practical to achieve a circumferentially varied distribution of coating on the surface for a controlled variation in the shear rate along the stream lines. Another limitation of the circular glass tube system is that the surface to which platelets adhere is intrinsically curved, and consequently the range of the surface that can be kept in focus in a microscope operating at sufficient magnification to observe individual platelets is limited.

Some investigators have adapted a technique used more commonly to study adhesion of cells that are sessile (such as endothelial cells) or are major functional cells of organs (such as hepatocytes) to investigate platelet adhesion. This technique involves placing cells in wells that have walls specifically coated (e.g., with laminin) and leaving them there for relatively long times (many minutes) without any regular flow present at all; the cells that are introduced usually carry radioactive labels and are introduced in such numbers that a confluent layer of cells may be deposited; the extent of adhesion is quantified after removing the supernatant (carrying some of the cells) and washing (11). In some cases the adherent cells are exposed to an agent likely to remove them, such as trypsin; in other cases the adherent cells may be lysed to release their labeled contents. This technique can be used to make determinations of specific receptor sites used in adhesion between specific cell types and substrates, and in this context it is probably most useful in platelet adhesion studies (12). Otherwise it can be criticized for its lack of realism in the physical environment typical of most platelet adhesion—the lack of a regular flow over the surface. It can also be criticized for allowing too long for unchallenged adhesion at one site in comparison with what has been observed in flow chamber systems and for having a protocol that specifically excludes studies that would provide information about the phenomenon of reuse of sites.

PHENOMENOLOGY OF PLATELET ADHESION

Initial Adhesion

The pattern of initial adhesion of platelets to the surface appears to be random with respect to the location. In other words, there does not appear to be any relationship between the coordinates at which one platelet is adherent and the coordinates at which other platelets are adherent. There is simply no spatial pattern to the adhesion. This suggests that the process of deposition of coatings on the slides generates no special patterns that produce some special order in the platelet adhesion, nor do the platelets themselves appear to influence the adhesion of other platelets around them. In experiments in a flow chamber leading to platelet adhesion with distances between platelets much smaller than those typical here, there was no correlation except at distances of a few micrometers (13). It should be emphasized that this feature is characteristic dominantly of the initial adhesion phase, within the first 3 or 4 minutes. In this period of time it must be realized that the PRP entering the flow chamber displaces the priming fluid present in the chamber at the beginning of the experiment, and this displacement consists of a flow of PRP initially near the center plane of the flow channel because this is the region in which the flow moves most rapidly, so that the layers of priming fluid close to the walls of the chamber are displaced more slowly and the concentration of platelets in the immediate vicinity of the surfaces is initially zero and grows slowly until the priming solution has been flushed out thoroughly. For this reason one finds that the initial adhesion rate of platelets is much smaller than that which sets in after the first 2 or 3 minutes of flow (Fig. 2). Once the priming solution has been effectively displaced, the adhesion rate, measured in terms of the number of platelets that adhere to the surface per unit time to sites on the surface, settles to a steady value. Throughout the typical period of our experiment this process of adhesion of platelets to specific sites within the visual field, where the sites have not been previously occupied by a platelet, continues at the same uniform rate, which is itself influenced by the local flow rate.

Adhesion Site Vacation and Reuse

An important feature of platelet adhesion behavior is that almost all platelets that adhere to sites that have never been occupied before by platelets leave those sites within a finite period

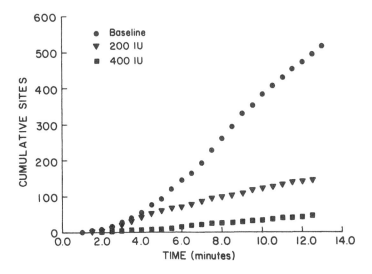

FIGURE 2 Cumulative adhesion of platelets as a function of time. Platelet adhesion to fibrinogen-coated glass slides was studied at baseline (circles), after 2 weeks of 200 IU vitamin E per day (triangles), and after an additional 2 weeks of 400 IU vitamin E per day supplementation (squares). (From Jandak et al., Ref. 14.)

of time, often less than 1 minute. Because a continuous observation system is employed and the flow chamber is not moved during the period of the experiment, it is possible to keep track of events at specific, fixed sites on the test surface. Thus in observing the test surface it is possible to know for each location whether it has never had a platelet adherent to it or whether instead it has had a platelet adherent and the platelet has subsequently vacated the site. Platelets carried by flow past the vacated sites also adhere to them at a characteristic rate. The rate at which second or subsequent occupants of a site adhere to those sites is much higher than the background rate at which platelets adhere randomly to sites not previously occupied. This rate of reuse of sites is of the order of 1000 times the background rate. The relative rate of reuse of sites is not fixed but depends upon how long the site has been vacated before another platelet adheres to it. The rate of reuse is highest for very short times of vacancy and falls with an increase in time of vacancy until, at about 7 minutes of continued vacancy, the rate of reuse has fallen to the general background level of primary adhesion rate.

Even those platelets that adhere as second occupants to specific sites also vacate them in most cases. The sites that have been vacated for a second time behave in exactly the same way as sites vacated for a first time; there is preferential platelet adhesion to them, depending upon the period for which they have been vacant. This process has been observed to repeat for up to seven times for some sites, and this number has been limited by the duration over which experiments to investigate this have been performed, so that we may suppose (lacking contradictory evidence) that the process may continue for an undetermined total number of times. The conditions typically chosen for running the experiments are such that not all the sites that have been vacated once become reoccupied by another platelet during the typical run time of an experiment, so that by the end of an experimental run one has total counts of sites that have been occupied once but not twice, twice but not three times, three times but not four times, and so on, which forms a set of numbers descending in numerical order. When this set of numbers is plotted on a semilogarithmic graph, with the number of sites occupied once, twice, three times, and so on on the logarithmic scale and the number of times occupied on a linear scale, we find that a best-fit line through the points is straight. The slope of this line we designate as the reuse of sites index. Because a logarithmic scale is used to represent numbers, this index is separate from the primary adhesion rate and indeed turns out to be a number characteristic of the adhesion process that can be influenced separately from the primary adhesion rate by pharmacological agents.

We believe that the procedural aspect by which we keep the platelet count uniform between different runs is important insofar as it affects the rate of opportunity of platelets to encounter specific sites as a function of time during the experimental procedure. If we allowed an experiment to be repeated with, say, a platelet concentration twice as high as that in a previous experiment and otherwise similar conditions (same donor and test surface), the rate at which platelets would be passing close to a vacated site per unit time would be higher and the apparent rate of reuse would be significantly different.

Reproducibility

From the beginning of experiments with the system for measuring platelet adhesion we have worked with volunteer human donors who are employees of a particular institution, and because of this it has been possible to retest with the same donors at widely separated calendar dates. It became evident on a cross plot of adhesion rate and reuse of site indexes that various specific donors show fairly reproducible values for these parameters. More recently, we conducted a study with a set of such donors that shows that the primary adhesion rate is reproducible over 6–12 month intervals within about 13% (14).

With Added ADP Thrombi Form

If adenosine-5′-diphosphate (ADP) is infused into the inlet manifold of the flow chamber during an experiment, a process that is possible by making use of a specially designed needle that has a longitudinal slot in it to permit the infusion of an ADP solution, it is found that some of the adherent platelets form the nuclei for thrombi that grow in size over time. There are two features of the thrombus growth pattern that are particularly noteworthy. One is that the thrombi tend to grow along flow stream lines in the flow chamber in the sense that there is a series of mounds, or local thrombi, along lines determined by the stream directions, and it is possible to show the flow pattern itself quite well by the lines of thrombi that form on the surface. The second feature is that the size of the thrombi varies locally along the stream lines in the flow chamber in an orderly fashion, with the thrombi found to be larger as the local stream velocity is itself larger [an effect expected in the light of the experiments of Begent and Born (15) and from gradient aggregation analysis], but with a decrease in growth rate of thrombi once some local shear rate has been exceeded, a result also in accord with the result of Begent and Born and the analysis of Richardson (16). In such experiments, when ADP has been used for a period to promote the process of thrombus formation on the test chamber wall, it has been observed that once the thrombi have grown to some typical size, the thrombi upstream and downstream of each other along stream lines begin to interact in the sense that thrombotic material begins to fill in the space between them but, after a period, embolizes, with the gap filling in over a period of time with thrombus yet again and embolizing yet again, and with this process repeated many times.

Pharmacological Influence on Adhesion

Previous studies with the laminar flow chamber system have shown that adhesion rates of human platelets to various substrates are influenced by various chemical agents with which the PRP has been incubated for a finite period before performing the experiment (17,18). The extent of influence on platelet adhesion by agents known to affect platelet aggregation does not match directly their impact on platelet aggregation. This supports the widely held view that platelet adhesion involves somewhat different processes from platelet aggregation.

Although the influence of some pharmacological agents upon platelet adhesion has been tested by incubating PRP with the agent of interest for a period of several minutes before the performance of the experiment, it has been found that the influence of some other substances is best detected when they have been consumed as part of the diet of the persons donating the blood. This is true of vitamin E and of the $n - 3$ fatty acids that have been suggested to contribute to the reduction of coronary vessel disease (19).

ROLE OF α-TOCOPHEROL

Vitamin E an Antioxidant and More

α-Tocopherol is one of the most important natural antioxidants. Because of this function it presented itself as a logical inhibitor of platelet aggregation, a process associated with consumption of oxygen and production of lipid peroxides (20,21). Whether platelet adhesion is mediated by such oxidative processes is not completely clear as yet, but the evidence obtained thus far does not favor a major role for these processes. Platelet shape change, the first morphological sign of platelet activation, is usually (under actual flow conditions) a prerequisite step for the adherence of the platelet to surfaces.

The biological function of α-tocopherol does not make it a logical candidate for attempts to prevent platelet adherence. However, the realization that oxidized forms of α-tocopherol, that is, tocopherol quinone, are as effective as α-tocopherol in the inhibition of platelet aggregation (22,23) pointed toward alternative explanations of its action than that of an antioxidant. Diplock and Lucy and their associates (24,25) proposed a hypothesis that was founded primarily on the physical-chemical interaction between the phytyl chain of vitamin E and certain polyunsaturated fatty acids. According to these authors, maximum "fit" is obtained between the natural α-tocopherol molecule containing the 15-carbon phytyl chain and the arachidonyl portion of arachidonylphosphatidylcholine. Any change in the side chain of α-tocopherol, whether an increase or a decrease in the number of carbons as well as alterations in the fatty acid chains either providing less unsaturation or shorter chain length, significantly reduced the extent of the interaction.

Although this theory has been challenged in recent years (26,27), it satisfies a basic need for an alternative mechanism by which α-tocopherol can function other than through its antioxidant activity. Questions about the general applicability of the theory formulated by Tappel (28) that the in vivo function of α-tocopherol is the prevention of destructive peroxidation of polyunsaturated lipids were raised even in the late 1960s when Green and Bunyan (29) pointed out certain inconsistencies in this hypothesis. The subsequent discovery that glutathione peroxidase, a lipid peroxide-destroying enzyme that contains selenium in its active site (30), removed some but not all the perceived inconsistencies of the antioxidant hypothesis, especially that the rate of destruction of trace quantities of α-tocopherol in tissues of vitamin E-deficient animals is not directly related to the intake of polyunsaturated fatty acids.

Effect on Platelet Adherence to Collagen

Based on these theoretical considerations and the inability to reconcile the platelet aggregation-related findings on the effect of α-tocopherol and its quinone with the antioxidant hypothesis of vitamin E, it appeared plausible to study the effect of this vitamin on platelet adhesion, a function that is not directly related to the production of the oxidative conversion products of arachidonic acid.

Spaet and Lejnieks (31) devised a method for examining platelet adhesion to collagen that utilized the turbidimetric assay system on which platelet aggregometry is based. We tried to make the method semiquantitative by winding a measured amount of fibrillar collagen around a small stirring bar agitated at the bottom of a suspension of PRP (32). The rate of platelet adherence to collagen can be analyzed quite easily from a continuous recording of changes in optical density. This system allowed us to evaluate the effect of α-tocopherol on platelet adherence. A group of normal individuals whose diet was supplemented for 6 weeks with vitamin E in doses that started at 400 IU/day and were then increased every 2 weeks by 400 IU/day gave the first indication that vitamin E may have an effect on platelet adhesiveness (33).

α-Tocopherol-enriched platelets showed a reduction in adherence to collagen that ranged from moderate to severe depending upon the administered dosage of the vitamin. A control group of individuals who received aspirin in a daily dosage of 300 mg showed no significant change in platelet adhesiveness. Although there was considerable interindividual variability, the results quite unmistakably demonstrated an inhibitory effect of α-tocopherol.

Adhesion Studies using the Laminar Flow Chamber

In subsequent studies we used the laminar flow chambers mentioned earlier to measure platelet adherence. The effect of vitamin E supplementation of the diet was tested at different intake levels ranging from 200 to 1600 IU/day (34). Only healthy, normal individuals were studied. A total of seven men and six women whose ages ranged from 23 to 55 years, all nonsmokers, participated in the studies. The volunteers were required to refrain from taking any medication, hormone, or vitamin preparation other than the one given to them for the purpose of the investigation for the duration of the study. Because of the mentioned interindividual variability of the various adhesion parameters that we measured, it was important to compare the adhesion of each individual to the baseline values obtained before vitamin E supplementations were begun. A 2 week period of administration of the dietary additive was allowed to establish optimal saturation with the vitamin at a particular level of intake. Platelet adhesion to the four adhesive surfaces—fibrinogen, collagen I, fibronectin, and glass—was consistently reduced by administration of α-tocopherol. The dose-dependent reduction in platelet adhesion appeared to reach its maximum at an intake level of approximately 400 IU/day (34,35). The magnitude of the reduction in platelet adherence was dependent on the nature of the adhesive surface studied. Collagen I exhibited the largest decrease, whereas adhesion to uncoated glass slides declined least. All reductions were statistically significant. On average, adhesion over a 2 week period of supplementation (200 IU vitamin E per day) declined by 75%. Doubling this dosage of vitamin E produced only a moderate further inhibition of the adhesion to approximately 18% of normal. Reduction in the number of cumulative sites and adhesion rates were necessarily similar. Our studies also showed that vitamin E decreased the reuse of sites index. The difference between baseline and supplemented levels was not as clear-cut as were the α-tocopherol-induced reductions in the other adhesion parameters we measured. However, there was a definite trend toward more negative indices with progressive α-tocopherol supplementation, which means that a particular adhesion site was occupied fewer times when supplemental vitamin E was given. It should be pointed out, however, that none of the changes attained statistical significance. We attribute this less to the relatively small number of individuals who were studied but rather to the great reductions in the number of times sites were reoccupied three or more times. To circumvent this problem, we measured the ratio of the cumulative number of sites occupied once to the cumulative number of sites occupied twice (O_1/O_2; Table 1). Because these ratios are calculated from greater numbers of sites, we believed that more reliance could be placed on them. Statistically significant increases in this ratio were noted when fibrinogen and collagen I were the adhesive surface, but no consistent trend was apparent when fibronectin or glass provided the adhesive coating. As yet we are reluctant to ascribe these changes to a real difference in the behavior of vitamin E-enriched platelets toward different adhesive materials. Larger numbers of individuals must be studied to clarify this point.

TABLE 1   Ratio of Adhesion Rates for Once (O_1) to Twice (O_2) Occupied Platelet Adhesion Sites at Baseline and After Vitamin E Supplementation

Adhesive surface	Baseline (O_1/O_2)	200 IU (O_1/O_2)[a]	400 IU (O_1/O_2)[a]
Fibrinogen	4.4 ± 3.1	4.7 ± 1.0	6.0 ± 2.7[b]
Collagen I	3.1 ± 0.5	5.3 ± 3.2[b]	6.2 ± 3.7[b]
Fibronectin	7.4 ± 2.7	9.5 ± 8.1	5.3 ± 2.7
Glass	8.6 ± 3.5	5.7 ± 2.7	13.4 ± 15.6

[a]Mean ± 1 SD ($n = 6$).
[b]$p < 0.05$.

Correlation of Levels of Vitamin E and Platelet Adhesion

Although the level of dietary supplementation with vitamin E constitutes an important variable that reflects upon the adhesiveness of platelets, other factors may play a role in the final expression of this effect. To search for evidence of a more direct correlation between vitamin E and inhibition of platelet adhesion, we measured the levels of α-tocopherol in circulating platelets before and after various supplementation regimens. It was interesting to note there was no direct relation between baseline levels of platelet α-tocopherol and the adhesiveness of platelets. Supplementary vitamin E, however, showed a good correlation between increased loading of platelets with α-tocopherol and a progressive decrease in platelet adhesion (Fig. 3). The slopes correlating these two parameters are quite different for each individual studied, but overall trends are clearly apparent. In general it seemed that the platelet adhesion rate fell in proportion to the inverse square of the amount of α-tocopherol incorporated into the platelets for each individual. The only exception to this relation appeared to be a postmenopausal female for whom the platelet adhesion diminished inversely with the amount of α-tocopherol incorporated (14). A larger population must be studied to determine more thoroughly the relationship and its variations for a specific cohort.

Although a 200 IU vitamin E per day supplementation represents a multiple of the normal dietary intake level, platelet tocopherol levels less than doubled. Further increases in the intake level of vitamin E to 400 IU/day raised platelet α-tocopherol levels by amounts roughly equivalent to those produced by the first 200 IU vitamin E per day. The changes in platelet α-tocopherol concentrations corresponding to the supplementary vitamin E administration far exceeded those observed in the plasma. Nevertheless, correlations of plasma levels of α-tocopherol with adhesion rates were very similar to those of platelet α-tocopherol.

Intake levels exceeding 400 IU vitamin E per day produced no further inhibition of platelet adhesiveness (34). Although it may be difficult to recognize incremental inhibitions when 400 IU vitamin E already produces an almost complete suppression of platelet adhesiveness, careful evaluation of the adhesion results reveals that platelet adherence to most adhesive surfaces, including collagen I, collagen IV, and fibronectin, remain more or less at the level at which it was when the dietary intake was supplemented by 400 IU vitamin E per day. Only adhesion to fibrinogen showed a different behavior. A distinct increase in adhesiveness was noted at 1600 IU vitamin E per day.

High Shear Rate Experiments Yield Similar Results

It should be emphasized that these results were obtained at shear rates at or about 25 s^{-1}. These calculated shear rates are not those observed in small to midsized arteries, such as those that exist in the coronary circulation. The high shear rates at such sites (200–800 s^{-1}) are difficult to reproduce with our system: large quantities of PRP are needed to perfuse the laminar flow chamber to achieve such high shear rates. For this reason, we also performed adhesion studies at a high shear rate (approximately 760 s^{-1}) under slightly different experimental conditions than those already described. Platelets prelabeled with [^{111}In]oxide were reconstituted with the red cells of the individual whose platelets were tested. Such labeled whole blood was perfused through the laminar flow chamber (see earlier) for 5 minutes, after which the flow chamber was cleaned with a washing buffer and the platelets adherent to the adhesive surface removed with a detergent solution. From the specific activity of the perfused platelets the number of adherent cells could be calculated. The results of these experiments were strikingly similar to those obtained at low shear rates (Table 2). Vitamin E again produced a marked reduction in platelet adherence. Even though we examined only one adhesive surface, that is, fibrinogen, we strongly believe the results obtained also apply to other adhesive surfaces.

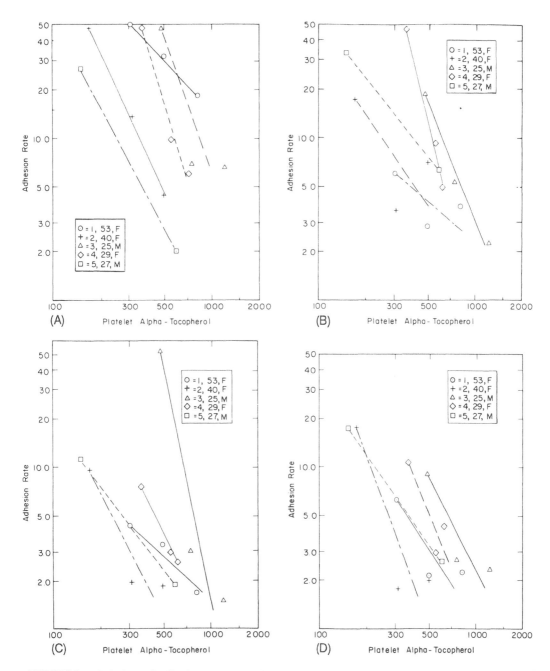

FIGURE 3   Relation of adhesion rate (cumulative number of platelets per minute) to α-tocopherol content of platelets (ng per 10⁹ platelets). Results obtained with fibrinogen as an adhesive surface are shown in A, with fibronectin in B, collagen in C, and glass in D. Data are shown for five individuals whose age and sex are given in the inset. (From Jandak et al., Ref. 14.)

TABLE 2   Effect of α-Tocopherol Supplementation on Platelet
Adhesion (High Shear Rate Study)[a]

Vitamin E supplementation (IU/day)	Adhesion (100% of control)
200	64.8 ± 8.3[b]
400	73.2 ± 9.4[b]

[a]Paired t-tests. Adhesion of ^{111}In-labeled platelets reconstituted with red
blood cells to fibrinogen-coated surfaces. Data are shown on five individuals who were tested before (control) and after 2 weeks of the respective vitamin E supplementation. Adhesion of 100% corresponds to 1.7
$\pm 0.14 \times 10^5$ platelets per 4.7×1.4 cm adhesive surface. Details of the
method have been described (36).
[b]$p < 0.005$.

Morphological Observations

Some progress has been made in trying to find the reason for the α-tocopherol-induced inhibition of platelet adhesion. Scanning electron microscopic examination of adherent platelets
under baseline and supplemented conditions revealed striking differences in their appearance
(Fig. 4). Whereas under baseline conditions platelets demonstrate long, extended, slender
pseudopodia protruding from spherical platelet bodies when they adhere to an adhesive surface, platelets enriched with supplemental vitamin E display only short, broad-based, blunt
pseudopodia. These vitamin E-induced morphological changes were found with all the adhesive surfaces tested. One can readily understand that such short pseudopodia cannot sustain
the firm adherence that is provided by the long, slender pseudopodia of platelets not enriched
by α-tocopherol. The morphological changes in the adherent platelets were dependent upon
the dosage of vitamin E administered to the individuals whose platelets were studied. A daily
dosage of 200 IU vitamin E per day resulted in less pronounced alteration than did a daily
supplementation of 400 IU/day. Although the rapidity of the platelet shape change makes it
difficult to dissect this agonist-induced event in a temporal manner, stopped flow studies (35)
have suggested that the early stages of the shape change are associated with short, blunt pseudopodia. Therefore, one may speculate that α-tocopherol could block the further evolution of
the shape change, arresting it in the initial phase. The mechanism of the action of α-tocopherol
has not yet been defined, although our preliminary experiments suggest that α-tocopherol
may in some manner alter the interaction of platelet membranes and cytoskeletons when stimulated by agonists.

CONCLUSION AND SUMMARY

Platelet adhesion, the initiating event of the hemostatic sequence, has been shown to be amenable to interdiction by supplementing the dietary intake of vitamin E. The inhibition of this
platelet function appears to be due to a fundamental alteration in the development and expression of the agonist-induced platelet shape change. A distinct dose-response relation between the concentration of α-tocopherol in platelets or plasma and the inhibition of platelet
adhesiveness could be recognized. From a clinical standpoint it is very encouraging that platelet
adhesion can be virtually inhibited with supplementary doses of vitamin E that are only one
order of magnitude above the recommended daily allowance for this vitamin. In addition,
that the efficacy of α-tocopherol as an antiadhesive agent could be shown ex vivo bodes well

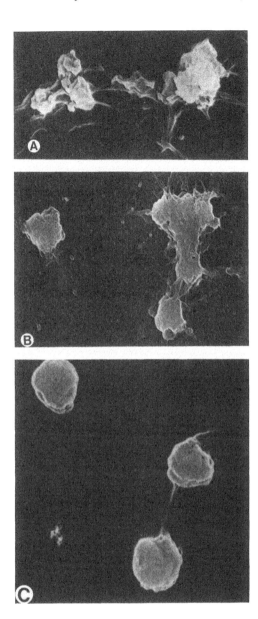

FIGURE 4 Scanning electron micrographs of adherent platelets to fibronectin-coated upper deck (H in Fig. 1) of the flow chamber. Control platelets obtained before dietary vitamin E supplementation (A), platelets after 2 weeks of 200 IU vitamin E per day (B), and platelets after 400 IU vitamin E per day for 2 weeks (C) (A, original magnification × 12,000; B, original magnification × 10,000; C, original magnification × 8000). (From Jandak et al., Ref. 14.)

for its eventual effectiveness in vivo. Although such clinical investigations have not yet been performed, we are confident that the basic mechanisms of α-tocopherol-induced inhibition of platelet adhesion are equally operative in vivo as ex vivo. The "ideal" antiplatelet regimen should provide not only a potent antiaggregating effect but also reduce platelet adhesiveness. Based on the results presented here, we suggest that a combination of acetylsalicylic acid and vitamin E may fulfill this role.

REFERENCES

1. Ginsberg MH, Loftus JC, Plow EF. Cytoadhesins, integrins and platelets. Thromb Haemost 1988; 59:1-6.
2. Bennett JS. The molecular biology of platelet membrane proteins. Semin Hematol 1990; 27:186-204.
3. Wiedeman MP, Tuma RF, Mayrovitz HN. An introduction to microcirculation. New York: Academic Press, 1981.
4. Born GVR. Aggregation of blood platelets by adenosine diphosphate and its reversal. Nature 1962; 194:927-9.
5. George JN. Direct assessment of platelet adhesion to glass: a study of the forces of interaction and the effects of plasma and serum factors, platelet function, and modification of the glass surface. Blood 1972; 40:862-74.
6. Mason RG, et al. Guidelines for blood-material interactions: report of the National Heart, Lung and Blood Institute Working Group, 1980, NIH Publication No. 80-2185.
7. Clinical methods for the measurement of disturbances of hemostatic function. In: Wintrobe MM, ed. Clinical Hematology, 6th ed. Philadelphia: Lea & Febiger, 1967; 321-44.
8. Richardson PD, Mohammad SF, Mason RG. Flow chamber studies of platelet adhesion at controlled, spatially varied flow rates. Eur Soc Artif Organs 1978; 4:175-88.
9. Adams GA, Feuerstein IA. Visual fluorescent and radioisotopic evaluation of platelet accumulation and embolization. Trans ASAIO 1980; 26:17-23.
10. Batchelor GK. An introduction to fluid dynamics. Cambridge: Cambridge University Press, 1967.
11. Ill CR, Engall E, Ruoslahti E. Adhesion of platelets to laminin in the absence of activation. J Cell Biol 1984; 99:2140-5.
12. Santoro SA. Identification of a 160,000 dalton platelet membrane protein that mediates the initial divalent cation-dependent adhesion of platelets to collagen. Cell 1986; 46:913-20.
13. Butruille YA, Leonard EF, Litwak RS. Platelet-platelet interactions and non-adhesive encounters on biomaterials. Trans ASAIO 1975; 21:609-16.
14. Jandak J, Steiner M, Richardson PD. Alha-tocopherol, an effective inhibitor of platelet adhesion. Blood 1989; 73:141-9.
15. Begent N, Born GVR. Growth rate in vivo of platelet thrombi, produced by iontophoresis of ADP, as a function of mean velocity. Nature 1970; 227:926-30.
16. Richardson PD. Effect of blood flow velocity on growth rate of platelet thrombi. Nature 1973; 245:103-5.
17. Richardson PD, Kane R, Agarwal K. Effects of drugs on adhesion of human platelets. Trans ASAIO 1981; 27:203-7.
18. Bodziak K, Richardson PD. Turnover of adherent platelets: some effects of shear rate, ASA, and reduced anticoagulation. Trans ASAIO 1982; 28:426-30.
19. Leaf A, Weber PC. Cardiovascular effects of n-3 fatty acids. N Engl J Med 1988; 318:549-57.
20. Hussain QZ, Newcomb TF. Thrombin stimulation of platelet oxygen consumption rate. J Appl Physiol 1964; 19:297-300.
21. Okuma M, Steiner M, Baldini MG. Studies on lipid peroxides in platelets. II. Effect of aggregating agents and platelet antibody. J Lab Clin Med 1971; 77:728-42.
22. Cox AC, Rao GHR, Gerrard JM, White JG. The influence of vitamin E quinone on platelet structure, function and biochemistry. Blood 1980; 55:907-14.
23. Mower R, Steiner M. Synthetic byproducts of tocopherol oxidation as inhibitors of platelet function. Prostaglandins 1982; 24:137-47.
24. Diplock AT, Lucy JA. The biochemical modes of action of vitamin E and selenium: a hypothesis. FEBS Lett 1973; 29:205-8.
25. Maggio B, Diplock AT, Lucy JA. Interactions of tocopherols and ubiquious with monolayers of phospholipids. Biochem J 1977; 161:111-21.
26. Urano S, Iida M, Otani I, Matsuo M. Membrane stabilization of vitamin E: interactions of alpha-tocopherol with phospholipids in bilayer liposomes. Biochem Biophys Res Commun 1987; 146:1414-8.

27. Urano S, Yano K, Matsuo M. Membrane stabilizing effect of vitamin E: effect of alpha-tocopherol and its model compounds on fluidity of lecithin liposomes. Biochem Biophys Res Commun 1988; 150:469-75.

28. Tappel AL. Vitamin E as a biological lipid antioxidant. Vitam Horm 1962; 20:493-510.

29. Green J, Bunyan J. Vitamin E and the antioxidant theory. Nutr Abstr Rev 1969; 39:321-45.

30. Hoekstra WG. In: Hoekstra WG, Suttie JW, Ganther HE, Mertz W, eds. Trace elements metabolism in animals. Baltimore: University Park Press, 1973; 61-77.

31. Spaet TH, Lejnieks I. A technique for estimation of platelet-collagen adhesion. Proc Soc Exp Biol Med 1969; 132:1038-41.

32. Castellan RM, Steiner M. Effect of platelet age on adhesiveness to collagen and platelet surface charge. Thromb Haemost 1976; 36:392-400.

33. Steiner M. Effect of alpha-tocopherol administration on platelet function in man. Thromb Haemost 1983; 49:636-40.

34. Jandak J, Steiner M, Richardson PD. Reduction of platelet adhesiveness by vitamin E supplementation in humans. Thromb Res 1988; 49:393-404.

35. Gear ARL. Rapid morphological changes visualized by scanning electron microscopy: kinetics derived from a quenched-flow approach. Br J Haematol 1984; 56:387-98.

36. Li X, Steiner M. Dose-response of dietary fish oil supplementations on platelet adhesion. Arteriosclerosis Thromb 1991; 11:39-46.

25

Combinations of Vitamin E and Other Antioxygenic Nutrients in Protection of Tissues

Al Tappel

University of California, Davis, California

INTRODUCTION

It is well accepted that vitamin E functions in combination with other antioxygenic nutrients in the protection of tissues of humans and animals from oxidative damage. The other antioxygenic nutrients include selenium, vitamin C, β-carotene, vitamin A, and coenzyme Q. The available information on the topic of this chapter is of such a large magnitude that only an overview can be given with reference to some reviews and papers. Publications about vitamin E used in combination with one other antioxygenic nutrient total over 1000 during the last 30 years. In the last 15 years there have been about 100 reviews on vitamin E working in combination with one other antioxygenic nutrient. This chapter focuses on research and clinical trials in which vitamin E action with one or more other antioxygenic nutrients is examined. Because of the need for quantitative integration of knowledge in this area, some of the results on the use of combinations of vitamin E and other antioxygenic nutrients are presented in simple simulation models.

PROPERTIES OF VITAMIN E AND OTHER ANTIOXYGENIC NUTRIENTS

Levels of Vitamin E and Other Antioxygenic Nutrients

The basic need for vitamin E and combinations of other antioxygenic nutrients in the human diet is expressed by the recommended dietary allowance (RDA). There have been many surveys of vitamin E and other antioxygenic nutrients in the diet of humans and animals by direct surveys of food consumed or by analyses of the blood levels of these components resulting from their consumption. The levels of some of the antioxygenic nutrients have been measured in human plasma. The approximate micromolar concentrations are vitamin E, 25; selenium, 1.3; vitamin C, 35; and β-carotene, 0.45 (1). The average dietary α-tocopherol equivalent in the United States is 9 mg/day. The approximate daily dietary intake of selenium is 0.14 mg (2). The mean dietary intake of vitamin C is approximately 120 mg (3). A recent forum focused on the RDA for the antioxygenic vitamins and selenium and the evidence for recommending increases in their intake. Bieri (4) and Draper and Bird (5) concluded that more evidence needs to be gathered before broad recommendations for increased intakes of antioxygenic vitamins and selenium are made. Diplock (6) favored supplemental vitamin E, vitamin C, and β-carotene.

Biochemical Reactions and Interactions of Vitamin E and Other Antioxygenic Nutrients

The antioxidant vitamins and selenium can have an important role in detoxifying endogenous oxygen radicals and exogenous radicals or their metabolic products. The biochemical reactions of individual antioxygenic vitamins and selenium have been elucidated (7–12). Vitamin E, the primary chain-breaking antioxidant in membranes, reduces peroxyl, hydroxyl, and superoxide radicals and singlet oxygen. β-Carotene is the most effective quencher of singlet oxygen. Vitamin C, a cytosolic chain-breaking antioxidant, scavenges hydroxyl and superoxide radicals and singlet oxygen. Selenium-glutathione peroxidase reduces hydrogen peroxide and lipid hydroperoxides in the aqueous phase. Coenzyme Q acts similarly to vitamin E. Other than the widely discussed interaction between vitamins E and C, there is meager information on biochemical interactions that may occur among the antioxidant nutrients. Most biochemical studies are done with only one antioxygenic compound. Probably at the membranous/aqueous interface, vitamin C regenerates vitamin E through reduction of the vitamin E radical formed by its antioxidant action (7–12). Reporting on the 1990 conference on antioxidants and degenerative diseases, Packer and Leibovitz (13) summarized the need for studies involving vitamin E and other antioxygenic nutrients.

Protection by Vitamin E and Other Antioxygenic Nutrients Against Lipid Peroxidation in Tissues

A tissue slice model system was employed to assess the effects of dietary antioxidant supplements on lipid peroxidation (14). In one experiment, rats were fed diets containing, either alone or in combination, vitamin E, selenium, β-carotene, or coenzyme Q_{10}, and the extent of spontaneous and induced lipid peroxidation was determined by release of thiobarbituric acid-reactive substances (TBARS) into the medium. See Table 1 for a listing of the nutrient antioxygenic components of the eight diets fed to the rats and the experimental TBARS values for liver slices. Vitamin E exhibited the greatest protection against lipid peroxidation in liver, heart, and spleen; in kidney, selenium was most protective. Coenzyme Q_{10} was active against lipid peroxidation induced by *tert*-butyl hydroperoxide. In a second experiment, rats were fed diets containing varying amounts of vitamin E, selenium, β-carotene, and coenzyme Q_{10}. See Table 2 for the experimental TBARS values for liver slices listed as a function of antioxygenic nutrients in the diet. Spontaneous lipid peroxidation in liver, kidney, and heart decreased with increasing levels of dietary antioxidants. With increasing amounts of antioxidants, there was a diminution in TBARS released by liver and kidney slices incubated with *tert*-butyl hydroperoxide; in heart, only the highest levels of antioxidants significantly decreased production of TBARS. Inverse correlations between dietary vitamin E and TBARS, tissue vitamin E and TBARS, and tissue selenium-glutathione peroxidase and TBARS were highly significant. The procedure used here can evaluate dietary supplements that may find practical applications in decreasing oxidant radical processes in vivo. The results of this research are the subject of simulation modeling as described in a later section.

DISEASE AND MULTIANTIOXYGENIC NUTRIENTS

Disease and Supplementary Multiantioxygenic Nutrients

Scientific literature contains much evidence that oxidant damage to cells and physiological mechanisms is involved in the basic cause of many diseases or in the development of disease.

TABLE 1 Components of the Model of Antioxidant Effects on Lipid Peroxidation in Liver Slices[a]

Diets	Glutathione peroxidase activity (nmol NADPH oxidized per minute/mg protein)	Vitamin E (nmol/g tissue)	Values used Q	Values used C	Experimental TBARS (nmol/g tissue)	Calculated TBARS (nmol/g tissue)
1 −Se, −E	8	5.1	0	0	53	54
2 −Se, +E	8	19	0	0	30	25
3 +Se, −E	617	5.6	0	0	40	36
4 −Se, −E, +Q	7	6.5	4.5	0	53	36
5 −Se, −E, +C	8	6.3	0	1.4	51	48
6 +Se, +E	667	21	0	0	18	15
7 +Se, +E, +Q	649	22	4.5	0	16	13
8 +Se, +E, +Q, +C	598	20	4.5	1.4	8.6	14

[a]Values that were constant for the eight diets were $A = 7$, $EE = 1$, $EC = 0.2$, $AUTOX = 1.3$; $PRF = 0.0005$, $TBARS$-$Y = 0.1$; and $LPC = 5000$.

TABLE 2 Model Components for the Effects of Increasing Dietary Antioxygenic Nutrients on Lipid Peroxidation in Liver Slices[a]

Diets	Glutathione peroxidase activity (nmol NADPH oxidized per minute/mg protein)	Vitamin E (nmol/g tissue)	Values used			Experimental TBARS (nmol/g tissue)	Calculated TBARS (nmol/g tissue)
			Q	C	EE		
1 Basal	4.7	13	0	0	1	73	32
2 Antioxygenic nutrients 1X	87	14	0.5	0.14	1	28	28
3 2X	197	14	1.4	0.42	1	15	23
4 3X	252	14	4.5	1.4	0.7	16	19
5 4X	250	37	13.5	4.2	0.7	11	10

[a]Values that were constant for the five diets were $A = 7$, $EC = 0.2$, $AUTOX = 1.3$; $PRF = 0.001$, $TBARS\text{-}Y = 0.1$; and $LPC = 5000$.

Toxic endogenous oxygen species are involved in inflammatory and degenerative diseases, hyperbaric oxygen, ionizing radiation, and ischemia (1). Toxic oxygen species of smog, some carcinogens, therapeutic drugs, and halogenated hydrocarbons are among damaging exogenous radicals or radical precursors. Physiological peroxidation also produces free radicals. The reader is referred to a current review for a delineation of oxidant interactions in various inflammatory and fibrotic diseases of the lung (7). Briefly, oxidation damage can be caused in lung by raising ambient oxygen levels, introducing oxidant toxicants, or stimulating phagocytes to release oxidant metabolites. Multiple antioxidant nutrients have been used in human studies to protect against damage by oxidants. Supplementation of American diets by antioxygenic nutrients is widespread.

Toxic oxygen species have been targeted as active in carcinogenesis (15). Epidemiological studies have shown inverse relationships between the incidence of cancer and antioxygenic vitamins and selenium. Clinical trials have been done and are ongoing. It has been reported that 10 times the nutritional requirements of vitamin E and selenium significantly decreased rat mammary carcinogenesis induced by 7,12-dimethylbenzanthracene (16).

In a group of 175 cataract patients needing surgery matched with 175 healthy controls, previous supplementation with vitamins E and C was allied with a decreased incidence of cataract (17).

Oxidants released by phagocytes in inflammatory conditions can suppress immune function. They are autotoxic to phagocytes, decrease the level of T and B lymphocytes, and depress the activity of natural killer cells. An increased incidence of cancer may result. The presence of both water- and lipid-soluble antioxidants is required to decrease damaging reactions of phagocyte formation and release of oxidants. Ascorbate plus Trolox-C, a water-soluble form of vitamin E, effectively inhibited the extracellular luminol-enhanced chemiluminescence response of activated neutrophils from smokers (18). Vitamin E also enhances immune functions by inhibiting the oxidation of vitamins C and A. Likewise, the sparing of vitamin E by vitamin C may indirectly strengthen immune responses. In an experiment to study the effects of 100% oxygen on T and B lymphocytic mitogen responses, guinea pigs fed diets high in both vitamins E and C were protected more than the vitamin E- and vitamin C-deficient animals (19). Further studies are needed to assess the role of nutritional antioxidants in maintaining immune functions.

It is evident that the diet must contain adequate levels of antioxygenic nutrients for protection of the population against oxidative damage to tissue. In addition, supplementary nutritional antioxidants have been proposed to increase protection (20). If relationships between increased levels of vitamins E and C, β-carotene, and selenium and prevention and mitigation of disease are established, then supplementation may become routine.

Epidemiological Studies of Protection by Multiantioxygenic Nutrients

A scattering of epidemiological studies has been reported that corroborate the prevailing view of the involvement of oxidants in many diseases (e.g., ischemia and inflammatory and degenerative diseases). Antioxidant nutrients would be expected to offer some protection. In a study conducted in Washington County, Maryland from 1974 to 1983, blood serum analyses of vitamin E and β-carotene were done. Levels of the antioxidant vitamins in 99 patients who developed lung cancer were compared to levels in 196 control subjects free of disease and matched for age, sex, ethnic group, month of blood collection, and smoking habits. There was an increased incidence of squamous cell carcinoma of the lung in patients with low levels of serum vitamin E and β-carotene (21).

In a 7 year study of 2707 men in Basel, Switzerland, there were 102 cancer deaths. Mean plasma levels of antioxidant vitamins showed that low levels of β-carotene, vitamin A, and a low to medium CIAVIT, an index calculated by the multiplication of micromolar quantities of β-carotene and vitamins C and E, were associated with increased risk of lung cancer. Low values of β-carotene and low to medium CIAVIT indices also indicated increased risk of lung disease other than malignancies. The Basel study also linked low levels of β-carotene, vitamins A, C, and E, and a low CIAVIT with stomach cancer and low levels of vitamins C and E with colon cancer (22).

In a Finnish 4 year epidemiological analysis of men and women, low serum selenium values and low selenium together with low vitamin E levels were indicative of increased risk of cancer, the highest risks being for men, smokers, and people with respiratory cancers (23).

A Finnish investigation involving 15,093 women in the mid-1980s showed a 10-fold incidence of breast cancer in subjects with low serum selenium in association with low vitamin E (24). In Finland a follow-up study of 150 gastrointestinal cancer patients and 276 matched control subjects showed that higher serum levels of selenium or vitamin E lowered the risk of cancers of the upper gastrointestinal tract. The major risk factors of this type of cancer are dietary (25). The correlations found in this study could reflect antioxidant vitamin and mineral protection against lipid peroxidation induced by fat in the diet.

A study of 55 lung cancer patients and their sons and daughters showed lower selenium and vitamin E blood levels in family members of adenocarcinoma patients and lower levels of these two antioxidant nutrients in the lung cancer patients. Perhaps there is a familial tendency to low serum selenium and vitamin E in close relatives of lung cancer patients (26).

In a study of middle-aged European men, low plasma levels of vitamin C correlated with increased ischemic heart disease (IHD) mortality. Low cholesterol-standardized levels of vitamin E showed an even stronger association with increased IHD mortality (1).

In these supplementary multiantioxygenic studies of palliation of disease, there appeared to be a degree of protection against oxidative damage to tissue by the antioxidant.

APPLICATION OF SIMULATION MODELING TO MULTIPLE NUTRIENT ANTIOXYGENIC COMPOUNDS

Need for Models

When multiantioxygenic components are used in research or in nutritional or therapeutic practice, there is need for conceptional models of how they quantitatively work together and for methods to assign their effectiveness in inhibiting lipid peroxidation or other oxidative reactions. As indicated in the literature (27–30), simulation modeling promises to meet some of these expressed needs. Simulation modeling has been applied to various types of nutritional problems (31–33), and empirical computations have been applied to solve some problems (34). For applications of multiple antioxygenic components, limits in our knowledge about the many reactions involved limits the complexity of simulation modeling that is appropriate at present. A study was made to test a simulation model involving vitamin E, selenium, coenzyme Q, and β-carotene against experimental data (14) that were measures of protection against tissue lipid peroxidation when these antioxygenic nutrients were fed to rats.

Conceptual Basis of Model

The simulation model applied to the effects of multiple antioxygenic nutrients is based on a simulation model of lipid peroxidation in rat tissue slices, including the effect of dietary antioxidants on the in vitro production of thiobarbituric acid-reacting substances (35). For this simulation model, the model developed previously (35) was reduced to the main parameters that apply to multiple dietary antioxidants:

$$\text{TBARS} = \left(\frac{\text{LPC}}{A + E \times EE + Q + C \times EC} - \text{SeGP} \times \text{PRF} \right)(\text{AUTOX})(\text{TBARS-}Y)$$

where TBARS = thiobarbituric acid-reacting substances produced by slices of tissues from rats fed the various antioxygenic components, and LPC = composite of the factors involved in lipid peroxidation that are not otherwise expressed in the equation. This composite can be visualized as including the amounts of polyunsaturated lipids, the peroxidizability of the polyunsaturated lipids, the effects of tissue activator enzymes, such as the cytochrome P_{450} system, and such catalysts as iron. A = basal antioxidant components not otherwise expressed in the equation; E = concentration of vitamin E in nmole per gram wet weight of tissue; and EE = an efficiency factor for vitamin E. At high levels of tissue vitamin E, the efficiency per amount of vitamin E is less than at low levels of vitamin E. Q = approximate concentration of tissue coenzyme Q in nmole per gram tissue; C = approximate amount of tissue vitamin A formed from dietary β-carotene in nmole per gram tissue; EC = factor expressing the approximate antioxidant effectiveness of vitamin A relative to vitamin E; SeGP $\times$ PRF = amount of lipid hydroperoxide reduced by selenium-glutathione peroxidase; it is the product of the activity of selenium-glutathione peroxidase (SeGP) in units of nmole NADPH oxidized per minute per mg protein times the hydroperoxide reduction factor (PRF). AUTOX = factor that provides for the appropriate amount of autoxidation; and TBARS-Y = yield of TBARS produced from hydroperoxides.

Use of the components in the model equation is justified from the literature. Likewise, values for these components were approximated from the literature. Basal antioxidant components include protein (36), vitamin E, even if an animal is fed a vitamin E-deficient diet (37), coenzyme Q (38), and other antioxidants. The value for basal antioxidant in terms of vitamin E is approximated as similar to the tissue content of vitamin E when an animal is fed a diet very low in vitamin E.

The efficiency factor for vitamin E relates to the well-known saturation effects of antioxidants. In the range of antioxidants considered in this model, the values of efficiency factor are near 1. Coenzyme Q has approximately the same lipid antioxidant strength as vitamin E (38,39). Amounts of coenzyme Q deposited in various tissues when coenzyme Q is fed have been determined (39,40). Dietary β-carotene is converted to vitamin A in the rat, and β-carotene is not deposited in the liver (41). β-Carotene and vitamin A are known to be weak antioxidants (42), and their strength was approximated as one-fifth that of vitamin E.

The action of selenium-glutathione peroxidase was approximated from the in vivo antioxygenic effects of dietary selenium when rats were fed vitamin E-deficient diets (43,44). Based on these reports, selenium-glutathione peroxidase was visualized as reducing up to one-half of the tissue hydroperoxides. In the use of the model equation, the action of glutathione peroxidase was set to reduce one-fourth of the hydroperoxides. The value for autoxidation was approximated from the literature (45) as one-fifth to one-third of the total lipid peroxidation.

Application of Model to Multiple Nutrient Antioxygenic Components

Table 1 lists some of the quantities for the components of the model equation for tissues from rats fed eight diets with four antioxygenic nutrients as variables. Glutathione peroxidase activity and vitamin E are the experimentally determined values. Values for *A, Q, C, EE, EC,* PRF, AUTOX, and TBARS-*Y* were approximated from the literature. Since all values except LPC are constrained, the value of LPC was determined by obtaining a good fit between the simulated TBARS and the experimental TBARS. Many of the approximated values were kept the same for four experimental conditions of liver slices, spontaneous and initiated, and kidney slices, spontaneous and initiated, which were described in an earlier section. As can be seen from Figure 1, good fits of simulation versus experimental values were obtained.

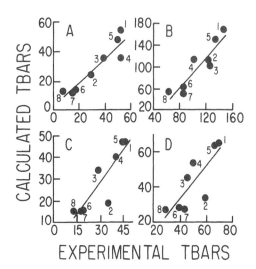

FIGURE 1 Linear regression analysis of (A) experimental TBARS from liver slices undergoing spontaneous lipid peroxidation and calculated TBARS ($r = 0.94$); (B) experimental TBARS from liver slices with induced lipid peroxidation and calculated TBARS ($r = 0.91$); (C) experimental TBARS from kidney slices undergoing spontaneous lipid peroxidation and calculated TBARS ($r = 0.89$); and (D) experimental TBARS from kidney slices with induced lipid peroxidation and calculated TBARS ($r = 0.80$). Numbers identify the corresponding diets listed in Table 1.

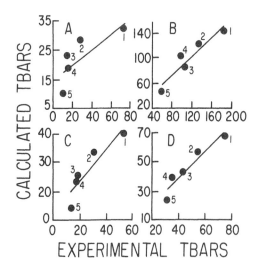

FIGURE 2 Linear regression analysis of (A) experimental TBARS from liver slices undergoing spontaneous lipid peroxidation and calculated TBARS ($r = 0.79$); (B) experimental TBARS from liver slices with induced lipid peroxidation and calculated TBARS ($r = 0.94$); (C) experimental TBARS from kidney slices undergoing spontaneous lipid peroxidation and calculated TBARS ($r = 0.91$); and (D) experimental TBARS from kidney slices with induced lipid peroxidation and calculated TBARS ($r = 0.95$). Numbers identify the corresponding diets listed in Table 2.

The model calculations were applied to a study of the effect on lipid peroxidation in slices of liver and kidney from rats when the rats were fed increasing amounts of antioxygenic nutrients. Table 2 shows values of the model as applied to liver slices. Figure 2 shows that good correlations of the simulated and experimental data were obtained.

Models that can incorporate some of the main biochemical parameters involved in testing nutrient antioxygenic components are desirable. Using only data from experiments with animals, the effects of individual antioxidants cannot be determined readily if many antioxidants are used. The experimenter cannot have control animals without any tissue antioxidants. Simulation modeling can offer a number of advantages, including the semiquantitative assignment of the effectiveness of the individual antioxygenic components.

The equation used in this study of simulation modeling seems to be appropriate for fits intended use because it expresses the relationships between lipid peroxidation and antioxygenic components that are in concert with knowledge gained experimentally and contained in the literature. Also, quantitative relationships can be approximated from the scientific literature. The main relationship, products = lipid peroxidation/antioxidants, is well established (46). At the present stage of knowledge of multiantioxidant effectiveness, the simulation model should contain only enough detail to capture the essence of the processes involved.

As shown by the plots of regression analysis if Figures 1 and 2, the results of the simulation model agree with the experimental results. These results support the information from the literature, the interpretations of the experimental results, and the concepts used in the model. The concepts indicate that vitamin E, selenium, coenzyme Q, and β-carotene can be viewed as nutrient antioxygenic components that act together in an additive manner. Vitamin E, selenium, and coenzyme Q can be viewed as strong antioxygenic agents that act to inhibit lipid peroxidation. In the rat β-carotene is not deposited, but some is converted to vitamin A. In the presence of strong antioxygenic agents, such as one or more of these three, β-carotene is much less effective than the other antioxidants.

STUDIES INVOLVING VITAMIN E AND OTHER ANTIOXYGENIC NUTRIENTS

The construction of simulation models applicable to scientific studies can increase our understanding of the results of studies. Modeling using data from the literature was applied to five studies involving vitamin E and additional antioxygenic nutrients. These five studies were chosen because vitamin E and other antioxygenic nutrients were major determinants of the results. The simple models were constructed to be in broad concert with relevant scientific knowledge and the specific literature on the topic. Otherwise, they are largely empirical. The five models are shown in Table 3 with reference to the original work. Fitting these models with the data from the original work and using the philosophy and techniques of simulation modeling gave the results shown in Figure 3. These correlations between the calculated and experimental results emphasize some broad relationships, shown in the models, which seem important for the combination of vitamin E and other antioxygenic nutrients in protecting tissues. The first general relationship is that the parameters measured that can be considered a result of the oxidant process are considered to be proportional to the reciprocal of the amount of vitamin E and other antioxygenic nutrients. The second relationship is that the effects of vitamin E and other antioxygenic nutrients can be considered additive. In a broad-based study, Gey (1) reported a linear relationship between mortality due to ischemic heart disease and an arbitrary construction of cholesterol divided by the product of vitamin C, corrected vitamin E, selenium, and β-carotene. It is proposed that the simulation model 5 of Table 3 is an advance beyond the linear relationship determined by Gey et al. because it has the analogous form of

TABLE 3 Models of Research Results in Which Vitamin E and Other Antioxygenic Nutrients are Major Determinants

Model[a]	Reference
1 TBARS of guinea pig liver	
$$\text{TBA} = \frac{\text{lipid peroxidation components}}{\text{basal antioxidants + vitamin E + vitamin C}}$$	47
2 TBA of rat liver	
$$\text{TBA} = \frac{\text{lipid peroxidation components}}{\text{basal antioxidants + vitamin E - vitamin C}}$$	47
3 Exudative diathesis	
$$\text{Exudative diathesis} = \frac{\text{exudative diathesis components}}{\text{selenium + vitamin E}}$$	48
4 Tumors	
$$\text{Tumors} = \frac{\text{basic tumor amounts at}}{\text{normal antioxidants}} + \frac{\text{tumor amounts affected by antioxidants}}{\text{log vitamin E + selenium}}$$	16
5 IHD mortality	
$$\text{Mortality} = \text{heart disease components} \times \frac{\text{cholesterol and lipid peroxidations}}{\text{vitamin E + vitamin C + } \beta\text{-carotene}} - \frac{\text{selenium}}{\text{effects}}$$	1

[a]Models 1–5 correspond to Figure 3A–E.

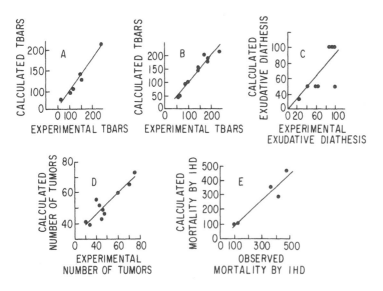

FIGURE 3 Linear regression analysis of the calculated values from the models shown in Table 3 and the corresponding experimental values given in the references. Correlation coefficients (r) for A, B, C, D, and E are 0.98, 0.97, 0.82, 0.91, and 0.94, respectively.

"chemical products equal reactants divided by antioxidants," and it follows the scientific logic, some of which is derived here, that vitamin E and other antioxygenic nutrients are additive in their protective effects.

This brief review of some of the research and clinical investigations of vitamin E and other antioxygenic nutrients shows that, compared to the practical needs, little work has been done with vitamin E in combination with several other antioxygenic nutrients, including selenium, vitamin C, β-carotene, vitamin A, and coenzyme Q. There is need for work in all areas of this complex field. The use of simple simulation models is advanced as a method of integrating information from the literature with experimental information for the purpose of understanding the relationships involving vitamin E with other antioxygenic nutrients.

REFERENCES

1. Gey KF. On the antioxidant hypothesis with regard to arteriosclerosis. Bibl Nutr Dieta 1986; 37: 53-91.
2. Olson RE, Broquist HP, Chichester CO, Darby WJ, Kolbye AC, Stalvey RM. Nutrition reviews present knowledge in nutrition, 5th ed. Washington, DC: Nutrition Foundation, 1984.
3. Burns JJ, Rivers JM, Machlin LJ. Third conference on vitamin C. Ann N Y Acad Sci 1987; Vol. 498.
4. Bieri JG. Are the recommended allowances for dietary antioxidants adequate? Free Radic Biol Med 1987; 3:193-7.
5. Draper HH, Bird RP. Micronutrients and cancer prevention: are the RDAs adequate? Free Radic Biol Med 1987; 3:203-7.
6. Diplock AT. Dietary supplementation with antioxidants. Is there a case for exceeding the recommended dietary allowance? Free Radic Biol Med 1987; 3:199-201.
7. Heffner JE, Repine JE. Pulmonary strategies of antioxidant defense. Am Rev Respir Dis 1989; 140:531-54.
8. Machlin LJ. Vitamin E. In: Machlin LJ, ed. Handbook of vitamins: nutritional, biochemical, and clinical aspects. New York: Marcel Dekker, 1984; 99-145.
9. Machlin LJ, Bendich A. Free radical tissue damage: protective role of antioxidant nutrients. FASEB J 1987; 1:441-5.
10. Sies H. Relationship between free radicals and vitamins: an overview. Int J Vitam Nutr Res Suppl (Switzerland) 1989; 30:215-23.
11. Niki E. Lipid antioxidants: how they may act in biological systems. Br J Cancer Suppl 1987; 8: 153-7.
12. Niki E. Interaction of ascorbate and alpha-tocopherol. Ann N Y Acad Sci 1987; 498:186-99.
13. Packer L, Leibovitz B. Antioxidants and degenerative diseases: highlights of conference findings. Free Radic Biol Med 1990; 8:506C-E.
14. Leibovitz B, Hu M-L, Tappel AL. Dietary supplements of vitamin E, β-carotene, coenzyme Q_{10} and selenium protect tissues against lipid peroxidation in rat tissue slices. J Nutr 1990; 120:97-104.
15. Stähelin HB, Gey F, Brubacher G. Preventive potential of antioxidative vitamins and carotenoids on cancer. Int J Vitam Nutr Res Suppl 1989; 30:232-41.
16. Ip C. Feasibility of using lower doses of chemopreventive agents in a combination regimen for cancer protection. Cancer Lett 1988; 39:239-46.
17. Gerster H. Antioxidant vitamins in cataract prevention. Z Ernährungswiss 1989; 28:56-75.
18. Anderson R, Theron AJ. Physiological potential of ascorbate, β-carotene and α-tocopherol individually and in combination in the prevention of tissue damage, carcinogenesis and immune dysfunction mediated by phagocyte-derived reactive oxidants. In: Bourne GH, ed. World review of nutrition and dietetics, Vol. 62. Basel: Karger, 1990; 27-58.
19. Bendich A. Antioxidant vitamins and their functions in immune responses. Adv Exp Med Biol 1990; 262:35-55.
20. Crary EJ, Smyrna G, McCarty MF. Potential clinical applications for high-dose nutritional antioxidants. Med Hypoth 1984; 13:77-98.

21. Menkes MS, Comstock GW, Vuilleumier JP, Helsing KJ, Rider AA, Brookmeyer R. Serum β-carotene, vitamins A and E, selenium and the risk of lung cancer. N Engl J Med 1986; 315:1250-4.

22. Gey KF, Brubacher GB, Stähelin HB. Cancer mortality inversely related to plasma levels of anti-oxidant vitamins. In: Cerutti PA, Nygaard OF, Simic MG, eds. Anticarcinogenesis and radiation protection. New York: Plenum, 1987; 259-67.

23. Salonen JT, Salonen R, Lappeteläinen R, Mäenpää PH, Alfthan G, Puska P. Risk of cancer in relation to serum concentrations of selenium and vitamins A and E: matched case-control analysis of prospective data. Br Med J 1985; 290:417-20.

24. Tolonen M. Finnish studies on antioxidants with special reference to cancer, cardiovascular diseases and ageing. Int Clin Nutr Rev 1989; 9:68-75.

25. Knekt P, Aromaa A, Maatela J, et al. Serum vitamin E, serum selenium and the risk of gastrointestinal cancer. Int J Cancer 1988; 42:846-50.

26. Miyamoto H, Araya Y, Ito M, et al. Serum selenium and vitamin E concentrations in families of lung cancer patients. Cancer 1987; 60:1159-62.

27. Sahin M, Iyengar SS, Rao RM. Computers in simulation and modeling of complex biological systems. In: Iyengar S, ed. Computer modeling of complex biological systems. Boca Raton, FL: CRC Press, 1984; 1-12.

28. Dutton JM, Starbuck WH. The plan of the book. In: Dutton JM, Starbuck WH, eds. Computer simulation of human behavior. New York: John Wiley and Sons, 1971; 3-8.

29. Law AM, Leton WD. Basic simulation modeling. In: Law AM, Kelton WD, eds. Simulation modeling and analysis. New York: McGraw-Hill, 1982; 1-58.

30. Kootsey JM. Complexity and significance in computer simulations of physiological systems. Fed Proc 1987; 46:2490-3.

31. Nathanson MH, McLaren GD. Computer simulation of iron absorption: regulation of mucosal and systemic iron kinetics in dogs. J Nutr 1987; 117:1067-75.

32. Black JL, Gill M, Thornley JHM, Beever DE, Oldham JD. Simulation of the metabolism of absorbed energy-yielding nutrients in young sheep: efficiency of utilization of lipid and amino acid. J Nutr 1987; 117:116-28.

33. Oltjen SW, Bywater AC, Baldwin RL. Simulation of normal protein accretion in rats. J Nutr 1985; 115:45-52.

34. Hegsted M, Ausman LM. Diet, alcohol and coronary heart disease in men. J Nutr 1988; 118:1184-9.

35. Tappel AL, Tappel AA, Fraga CG. Application of simulation modeling to lipid peroxidation processes. Free Radic Biol Med 1989; 7:361-8.

36. Wayner DDM, Burton GW, Ingold KU, Locke S. Quantitative measurement of the total, peroxyl radical-trapping antioxidant capability of human blood plasma by controlled peroxidation. FEBS Lett 1985; 187:33-7.

37. Herschberger LA, Tappel AL. Effect of vitamin E on pentane exhaled by rats treated with methyl ethyl ketone peroxide. Lipids 1982; 17:686-91.

38. Mellors A, Tappel AL. The inhibition of mitochondrial peroxidation by ubiquinone and ubiquinol. J Biol Chem 1966; 241:4353-6.

39. Kishi H, Kanamori N, Nishii S, Hiraoka E, Okamoto T, Kishi T. Metabolism of exogenous coenzyme Q_{10} in vivo and the bioavailability of coenzyme Q_{10} preparations in Japan. In: Folkers K, Yamamura Y, eds. Biomedical and clinical aspects of coenzyme Q. Amsterdam: Elsevier, 1984; 4:131-42.

40. Lawson DEM, Threlfall DR, Glover J, Morton RA. Biosynthesis of ubiquinone in the rat. Biochem J 1961; 79:201-8.

41. Ribaya-Mercado JD, Holmgren SC, Fox JG, Russell RM. Dietary β-carotene absorption and metabolism in ferrets and rats. J Nutr 1989; 119:665-8.

42. Cabrini L, Pasquali P, Tadolini B, Sechi AM, Landi L. Antioxidant behavior of ubiquinone and β-carotene incorporated in model membranes. Free Radic Res Commun 1986; 2:85-92.

43. Tappel AL, Dillard CJ. In vivo lipid peroxidation: measurement via exhaled pentane and protection by vitamin E. Fed Proc 1981; 40:174-8.

44. Tappel AL. Vitamin E and selenium protection from in vivo lipid peroxidation. In: Levander OA, Cheng L, eds. Micronutrient interactions: vitamins, minerals and hazardous elements. Ann N Y Acad Sci 1980; 355:18-31.

45. Tsai L-S, Smith LM. Role of the bases and phosphoryl bases of phospholipids in the autoxidation of methyl linoleate emulsions. Lipids 1971; 6:196-202.
46. Uri N. Mechanism of antioxidation. In: Lundberg WO, ed. Autoxidation and antioxidants. I. New York: Interscience, 1961; 133-69.
47. Chen LH. Interaction of vitamin E and ascorbic acid. In Vivo 1989; 3:199-210.
48. Mathias MM, Hogue DE. J Nutr 1971; 101:1399-402.

III
PHARMACOLOGY

26

Biokinetics of Vitamin E Using Deuterated Tocopherols

Graham W. Burton and Keith U. Ingold

National Research Council of Canada, Ottawa, Ontario, Canada

INTRODUCTION

The importance of vitamin E for protecting the integrity of lipid structures (especially membranes) in vivo is underscored by the finding that it is the major ($\geq 90\%$) lipid-soluble, chain-breaking antioxidant found in plasma, red cells, and tissues (1–4). This finding has been found to hold true even in the plasma of children with chronic, severe vitamin E deficiency (5).

α-Tocopherol (α-T), the most biologically active form of vitamin E, is one of the most efficient chain-breaking antioxidants known (6–10). The nearly optimal antioxidant activity of the molecule is attributable solely to the chroman group (6,8–10), whereas the dynamics of transport and retention of the molecule within membranes is determined largely by the phytyl group (see Fig. 1) (11–13).

In view of the potential of vitamin E for controlling lipid peroxidation in vivo, it has become important to know more about the dynamics (biokinetics) of its absorption, transport, and distribution within tissues. In the past, studies of the intestinal absorption, transport, and uptake into tissues of vitamin E in vivo have been done using radiolabeled tocopherol (14,15) or by using large doses of the unlabeled vitamin (16–18). The limitations of these two methods have spurred a drive to obtain biokinetic information directly and more conveniently. In this chapter a new approach is described that relies on labeling vitamin E with a stable isotope (deuterium) and mass spectrometry as a means of detection.

METHODOLOGY

Synthesis of Deuterated Tocopherols

Multigram quantities of natural and unnatural stereoisomers of α-T can be synthesized with nearly complete, selective replacement of hydrogen by deuterium in metabolically inactive, nonlabile, aromatic methyl positions [e.g., 5-CD$_3$-α-T (d$_3$-α-T), 5,7-(CD$_3$)$_2$-α-T (d$_6$-α-T), and 5,7,8-(CD$_3$)$_3$-α-T (d$_9$-α-T)] (Fig. 1) (19–21). For example, the natural stereoisomer, 2R,4'R,8'R-α-tocopherol (RRR-α-T), has been synthesized with both three and six atoms of deuterium per molecule by the deuteriomethylation of natural source γ- and δ-tocopherols, respectively (19,20). Also, a labeled γ-tocopherol has been prepared (21).

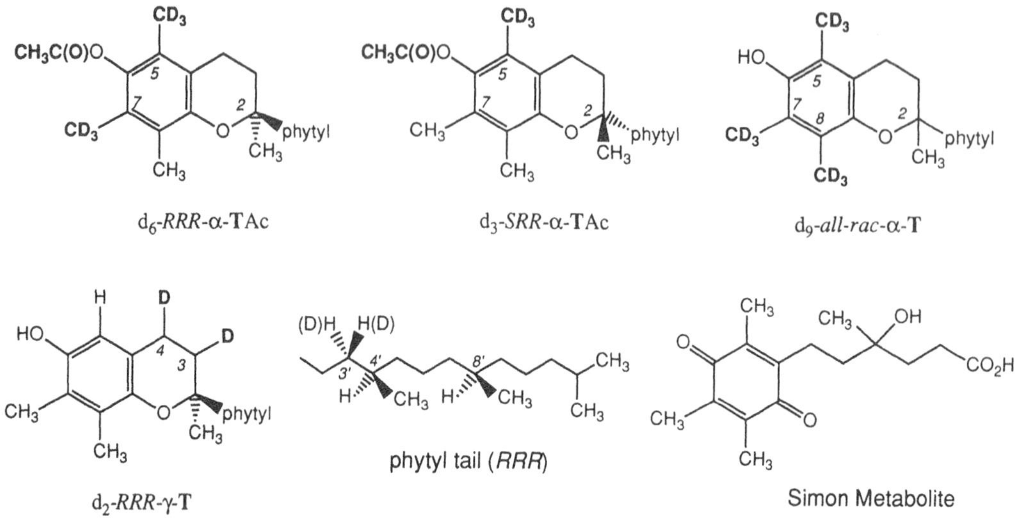

FIGURE 1 Examples illustrating the substitution of deuterium in stereoisomers or stereoisomeric mixtures of α-T or α-TAc, in γ-T, and in the phytyl tail. The sole identified catabolic product of α-tocopherol, the Simon metabolite, is also shown in its open chain—nonlactone—form.

Detection of Deuterated Tocopherols

The relative amounts of deuterated and nondeuterated α-tocopherols (d_0-α-T) present in lipid extracts of biological tissues and fluids are determined by silylating high-performance liquid chromatography (HPLC)-purified tocopherol fractions and injecting the tocopheryl silyl ether mixtures into a commercially available, bench top type, gas chromatograph–mass spectrometer (GC-MS) (19). As the α-tocopheryl silyl ether emerges from the gas chromatograph and enters the mass spectrometer, it is resolved simultaneously into its various component parent ions (Fig. 2). The absolute concentration of each tocopherol is readily determined by relating the peak area of its parent ion to that of a d_9-α-T internal standard (Fig. 1), added in known amount to the sample just before extraction. Recently, in addition to deuterated α-tocopherols, we measured unlabeled and dideuterio-γ-tocopherol (d_0- and d_2-γ-T) using an appropriately labeled γ-tocopherol that is added as an internal standard, together with d_9-α-T, immediately before lipid extraction (22). Figure 3 demonstrates that it is possible to follow the levels of three α-tocopherols and two γ-tocopherols simultaneously, in this case in plasma obtained from a cynomolgus monkey that was fed a single dose of a mixture of two deuterated α-tocopheryl acetate stereoisomers (d_3-*SRR*-α-TAc and d_6-*RRR*-α-TAc; see Fig. 1) and a deuterated γ-tocopherol (d_2-γ-T).

Advantages of the Method

A great advantage inherent in the use of deuterated tocopherols is that, unlike radiolabeled vitamin E, the compounds may be ingested without risk. Furthermore, as the deuterium appears not to undergo any measurable, metabolically mediated exchange (G. Burton, unpublished results), deuterated tocopherols can be used conveniently and readily in human and animal studies. Thus, these compounds have been used in both short-term and long-term feeding studies in laboratory animals (19,23,24) (D. Muller and G. Burton, unpublished work)

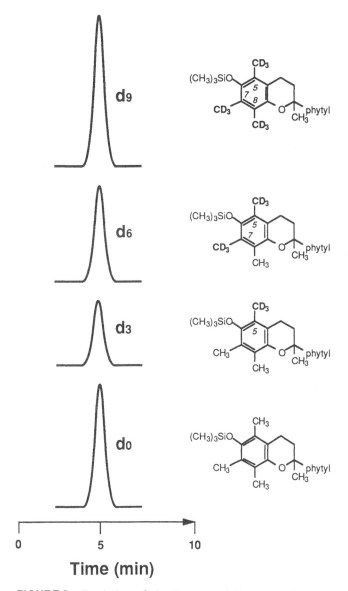

FIGURE 2 Depiction of simultaneous elution of various deuterated forms of α-T silyl ethers from the gas chromatograph of a GC-MS and the subsequent resolution into and quantitation by the mass spectrometer of the individual components using the selective ion-monitoring mode. The d_9 form is used as an internal standard.

and in single- or multiple-dose studies in humans (23,25–28), weanling piglets and calves, and large zoo animals.

Sensitivity and Limitations

In our experience, the size of a sample required to obtain reliable data requires that there be at least 100 ng ($\simeq 0.2$ nmol) of α-tocopherol present in the final, concentrated 50 μl volume from which 1 μl is drawn for GC-MS analysis. However, it is possible to detect as little as 40 pg

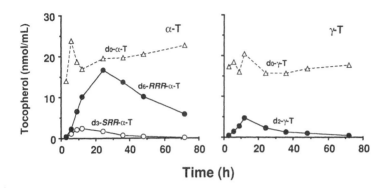

FIGURE 3 Simultaneous monitoring of three deuterium-distinguished α-T and two deuterium-distinguished γ-T in the plasma obtained from a cynomolgus monkey fed an oral dose of a 1:1:1 mixture of d_3-*SRR*-α-TAc, d_6-*RRR*-α-TAc, and d_2-γ-T. (Prepared from data in Ref. 22.)

tocopherol, and we have, for example, successfully measured uptake into human heart tissue from biopsy samples as small as 1 mg (28).

The reliability of measurements and, therefore, the range of applicability of the deuterated tocopherol technique is limited by the correction required for the contributions of naturally occurring isotopes to the peak area of the labeled tocopherol. For example, when unlabeled d_0-α-T and d_3-α-T are analyzed as their silyl ethers, the contribution from the natural abundance isotopes (^{13}C, ^{29}Si, and ^{30}Si) present in unlabeled (nominally d_0) α-TSi(CH$_3$)$_3$ to the d_3-α-TSi(CH$_3$)$_3$ parent ion peak is 2.37% of the peak area of the unlabeled parent ion. Limiting situations arise when the amount of d_3-α-TSi(CH$_3$)$_3$ is low (<5%) compared to the amount of unlabeled (d_0)α-TSi(CH$_3$)$_3$. Under these circumstances, the contribution of the d_0 silyl ether to the d_3 peak area is comparable to the true peak area of the d_3 silyl ether. In this situation it is better, if possible, to use d_6-α-tocopherol, because the d_6 parent ion peak is essentially unaffected by contributions from the d_0 ion. Another option is to analyze tocopherol directly as the free phenol. The absence of silicon substantially reduces the correction to the measured d_3-α-T peak from 2.37% to 0.63% of the unlabeled α-T peak area. The drawbacks to this option are that underivatized tocopherols tend to tail on the GC column, lowering the accuracy of the peak area integration, and, in our experience, repeated use of underivatized samples leads to a more rapid degradation of GC-MS performance.

APPLICATIONS

Turnover Studies

A unique application of deuterated α-T has been to use it as the sole source of dietary vitamin E for laboratory animals, allowing us to measure the long-term rate of its uptake into blood and tissues. Typically, young rats or guinea pigs are placed on a diet containing a known and fixed amount of unlabeled *RRR*-α-tocopheryl acetate (d_0-*RRR*-α-TAc; usually 36 mg/kg diet) for a lead-in period of 2–4 weeks and are then switched to an identical diet containing the same amount of the deuterated form (d_3- or d_6-*RRR*-α-TAc). The amount of "old" d_0-*RRR*-α-T remaining and the amount of "new" d_3- or d_6-*RRR*-α-T taken up is determined in tissues obtained from animals sacrificed at appropriate time intervals. Rates of turnover are estimated by interpolating from the data the time at which the amount of "new" d_3- or d_6-α-T becomes equal to the "old" d_0-α-T remaining in each tissue (i.e., the equalization time $t_{1:1}$).

In Figure 4, the $t_{1:1}$ values are displayed for rats and guinea pigs (19,24). The $t_{1:1}$ values allow the tissues to be classified into two broad classes. Turnover is fast in plasma, liver, lung, and kidney and is faster in the guinea pig than in the rat, but there is little difference between $t_{1:1}$ values for these four "fast" tissues within the same species of animal (i.e., rat and guinea pig). In the "slow" tissues, which include heart, testis, muscle (biceps femoris), brain, and spinal cord, turnover is substantially slower and it shows much more variability between animal species and between the various "slow" tissues within a particular species. In particular, turnover in the guinea pig brain, which has the largest $t_{1:1}$ by far, is about three times slower than in rat brain. We presume that these animal-specific differences in slow tissue turnover are associated with the underlying factor(s) responsible for the diverse range of symptoms associated with vitamin E deficiency.

Deuterated α-T has also been used in studies of patients undergoing elective heart surgery (28,29). To determine the optimal oral dose for evaluating the potential therapeutic benefits of vitamin E on heart recovery, the patients were divided into groups that took different amounts of d_3-*RRR*-α-TAc for different lengths of time immediately before surgery. Analysis of heart tissue biopsied at the time of surgery indicated that a dose of 300 mg natural α-tocopheryl acetate (approximately 400 IU), taken daily for 14 consecutive days, roughly doubled the level of vitamin E in the heart and conferred some measurable degree of protection during recovery from surgery (T. M. Yau, R. D. Weisel, D. A. G. Mickle, G. W. Burton, K. U. Ingold, J. Ivanov, M. K. Mohabeer, L. Tumiati, and S. Carson, unpublished work).

Do Vitamins C and E Interact In Vivo?

It has been clearly demonstrated in vitro that vitamin C is able to "spare" or regenerate vitamin E (30–32). Thus, if the effect is important in vivo we expect to see a decreased turnover of vitamin E. Recently we have shown, using continuous feeding of deuterated vitamin E, that the rate of turnover of α-T in plasma and tissues of guinea pigs is unaffected by dietary

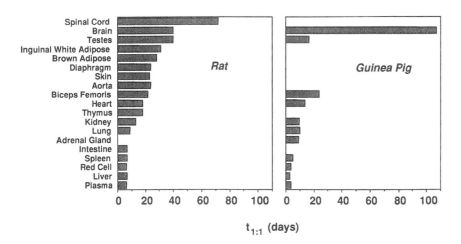

FIGURE 4 Equalization times $t_{1:1}$ reflecting the times different tissues in rats and guinea pigs take to attain equal concentrations of deuterated and unlabeled α-T after the animals were switched to a diet in which unlabeled α-TAc was replaced by an equal concentration of deuterium-labeled α-TAc. (Prepared using data obtained from studies described in Refs. 19 and 24.)

modulation of their vitamin C status, even when the level of vitamin E is declining in the animals (24). This is a surprising result in view of the in vitro findings. One possible explanation is that in oxidatively unstressed guinea pigs the flux of peroxyl radicals that enters the lipids is not large, consuming only a small fraction of the available vitamin E. This idea is supported by Tappel's finding that expired pentane levels from animals are extremely low, even for animals that are receiving inadequate or no vitamin E, relative to the levels expired by animals that are oxidatively stressed in various ways (33–36). Another possible explanation is that the tocopheroxyl radical is reduced in vivo to α-T by another, possibly enzymatic, process that is more effective than ascorbate. Indeed, there is considerable evidence for a membrane-bound, heat-labile, glutathione-dependent, free-radical reductase that appears to act in this way (37–48).

Relative Bioavailability Measurements:
The Competitive Biokinetics Method

A very useful feature of the deuterated vitamin E technique is that, because they can be analyzed simultaneously, α-tocopherols substituted with different amounts of deuterium (e.g., d_3 or d_6) can be used to measure, within the same human subject or animal, the simultaneous, competitive biokinetics of two or more forms of vitamin E. This competitive biokinetics technique eliminates most of the statistical variability between individuals that is inevitable when different compounds are compared between groups in the conventional manner. This is because the subject or animal in effect acts as its own control in the competitive experiments.

Is α-Tocopheryl Acetate a Better Source of Vitamin E
than α-Tocopherol Itself?

The relative bioavailabilities of α-T and α-TAc are of interest because the "free" form occurs naturally in food, whereas the ester, which is more air stable than the phenol, is the form most commonly used in vitamin E supplements. α-TAc itself is not an antioxidant and only becomes active biologically after it has been hydrolyzed in the gut to α-T, following which absorption of the vitamin occurs. Studies comparing these two forms of vitamin E using the traditional rat fetal gestation-resorption assay found that the phenol was only about half as potent as the acetate (49,50)!

We compared these two forms directly using our competitive biokinetics method (23). An equimolar mixture of d_6-*RRR*-α-T and d_3-*RRR*-α-TAc was given to four male rats under conditions very similar to those used in the traditional resorption bioassay (i.e., four consecutive, daily oral doses in tocopherol-stripped corn oil). The rats were sacrificed 1 day after the last dose, and the relative amounts of deuterated *RRR*-α-T taken up from the two forms of vitamin E were measured. The phenol to "acetate" (d_6/d_3) ratios were very similar for blood and tissues from all rats, irrespective of whether the rats were vitamin E deficient, and their mean value (0.49 ± 0.05) was in good agreement with the result obtained by the traditional bioassay method (Fig. 5A).

Subsequently, we have been able to show even more directly by cannulating rat lymph ducts and collecting intestinal lymph after oral administration of a single 1:1 acetate-phenol dose in corn oil that initially α-T is absorbed equally well from both forms, but eventually more α-T is derived from the acetate (Fig. 6A). After 8 h it is evident that the bulk of the tocopherol has been absorbed. Examination of the small intestine and its contents at this time reveals that there is substantially more α-T from the acetate present (Fig. 6B), a result in qualitative agreement with the relative proportions collected in the later lymph samples. The greater bioavailability of α-T from α-TAc suggests that when present in bulk the free form is subject

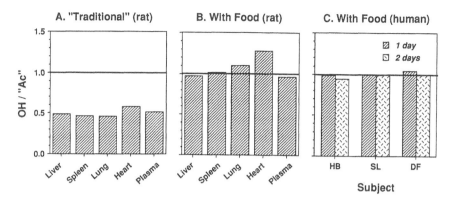

FIGURE 5   Comparison of the bioavailability of α-T and α-TAc in rats and humans using the competitive biokinetics technique. The OH/Ac ratio is the ratio of d_6-*RRR*-α-T absorbed directly to the d_3-*RRR*-α-T derived from d_3-*RRR*-α-TAc in plasma and four tissues of rats 1 day after being dosed daily for 4 days with a 1:1 mixture of the compounds dissolved in tocopherol-stripped corn oil (A) and 1 day after being dosed orally with the mixture in an aqueous bolus of laboratory diet (B). (C) The ratios are calculated from data obtained from the plasma of three human adults 24 and 48 h after ingesting 100 mg of a 1:1 mixture of the two forms of vitamin E. (Prepared from data in Ref. 23.)

to gradual destruction under the conditions existing in the rat gut. Indeed, there is evidence from earlier work that incorporation of an antioxidant into the dose reduces the loss of α-T (49).

From these findings it may be inferred that giving vitamin E to the animal in a different way could affect relative bioavailability. Indeed, when the deuterated tocopherol-corn oil mixture was instead administered in an aqueous bolus of laboratory food, the mean phenolacetate ratio in the blood and tissues 24 h later was 1.06 ± 0.11 (five rats; Fig. 5B). The relative

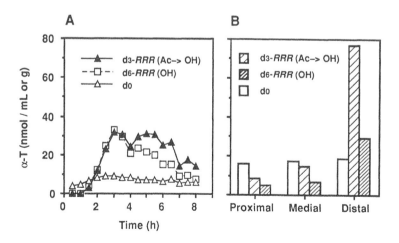

FIGURE 6   Concentrations of d_0, d_3-, and d_6-α-T after dosing a rat with a 1:1 mixture of d_3-*RRR*-α-TAc and d_6-*RRR*-α-T dissolved in tocopherol-stripped corn oil (A) in lymph and (B) in three segments of the small intestine 8 h after dosing.

bioavailability of the acetate and phenol forms therefore shows a strong dependence upon the method by which the vitamin E is presented.

In the case of humans, when five adults were each given a single oral dose of an equimolar mixture of the same deuterated phenol-acetate pair (≈ 50 mg of each) with an evening meal, the phenol-acetate ratio measured in the plasma and red cells over the ensuing 2 days was also very close to 1.0 for each individual (23) (results are presented for three of the subjects in Fig. 5C), indicating that both α-T and α-TAc are equivalent under normal dietary conditions. This finding is in good agreement with that of an earlier study in which humans were given large doses of unlabeled *RRR*-α-T and their plasma levels were compared with those of subjects given a corresponding amount of *RRR*-α-TAc (51).

Bioavailability of Tocopherol Stereoisomers Using the Competitive Biokinetics Method

Although α-tocopherol has three chiral centers (positions 2, 4′, and 8′, Fig. 1), giving rise to eight possible stereoisomers, it occurs naturally in only one form (*RRR*). Valuable insight into the mechanism of absorption, transport, uptake, and loss of vitamin E in vivo has been obtained by using the competitive biokinetics method to measure the relative bioavailabilities of natural α-T (d_6-*RRR*-α-TAc) and the 2*S*,4′*R*,8′*R* (*SRR*) stereoisomer (d_3-*SRR*-α-TAc; the configuration of the carbon at the phytyl tail-chroman ring junction has been inverted) from an orally administered 1:1 mixture.

During a long-term study of the uptake of d_3-*SRR*-α-T and d_6-*RRR*-α-T into the blood and tissues of rats continuously fed a mixture of the corresponding acetates in their diets, it was found that in blood and in all tissues except liver, there was always more of the natural stereoisomer present (i.e., d_6-*RRR*/d_3-*SRR* > 1) (19). However, in the liver the behavior of the two stereoisomers was quite different. Initially, for the first few days after the beginning of feeding, there was approximately twice as much *SRR*- as *RRR*-α-T present (i.e., d_6-*RRR*/d_3-*SRR* < 1), but over the ensuing period of weeks the *RRR/SRR* ratio changed gradually until eventually there was a slight excess of d_6-*RRR*-α-T over d_3-*SRR*-α-T.

In principle, the discrimination between the stereoisomeric tocopherols could occur first during absorption from the gut during the cholesterol esterase- and bile salt-mediated hydrolysis of the acetates. However, this possibility is ruled out by the results of experiments in which (1) rats given a single dose of a 1:1 mixture of the free tocopherols instead of the acetates showed no reduction in the extent of discrimination (H. Zahalka, G. Burton, K. Ingold, and D. Foster, unpublished work); and (2) intestinal lymph collected from a rat given a single dose of a 1:1 mixture of the acetates contained approximately equal amounts of *RRR*- and *SRR*-α-T (Fig. 7A) (52). Also, nearly equal amounts of the two (free) tocopherols were found in the small intestine 8 h after dosing (Fig. 7B).

The absence of an effect during absorption from the gut leaves the liver as the next most probable major site of discrimination. Experiments in which single doses of the 1:1 acetate mixture were given to rats show clearly that, irrespective of vitamin E status, there is an inverse relationship between the *RRR/SRR* ratios in the liver and plasma (Fig. 8) (H. Zahalka, G. Burton, K. Ingold, and D. Foster, unpublished work). Also, as the value of the ratios rises with time in the plasma, there tends to be a corresponding, albeit smaller, decrease in the liver ratio. Given the role of the liver in secreting lipids into plasma via very low density lipoproteins (VLDL), the finding of an excess of the unnatural stereoisomer in liver and an excess of the natural stereoisomer in plasma can be taken to imply the existence of a mechanism of VLDL assembly that favors selection of *RRR*-α-T. This idea receives strong support from a competitive biokinetics experiment carried out using a cynomolgus monkey fed the 1:1 tocopheryl acetate mixture. It was found that nascent VLDL, isolated from liver perfusate, were strongly enriched in *RRR*-α-T (Fig. 9) (22).

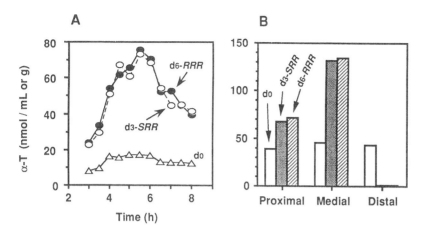

FIGURE 7 Concentrations of d_0, d_3-*SRR*-, and d_6-*RRR*-α-T after dosing a rat with a 1:1 mixture of d_3-*SRR*- and d_6-*RRR*-α-TAc dissolved in tocopherol-stripped corn oil (A) in lymph and (B) in three segments of the small intestine 8 h after dosing.

We hypothesized (19) that a cytosolic liver tocopherol binding protein, identified earlier in rats (53), is responsible in some as yet unknown way for the unique behavior of the liver. Indeed, analysis of liver fractions from rats and monkeys fed the 1:1 tocopheryl acetate mixture has shown that the fraction containing the tocopherol binding protein is the only one enriched in *RRR*-α-T (Fig. 10) (P. Dutton, G. Burton, B. Hodgkinson, K. Ingold, D. Foster, and W. Behrens, unpublished work).

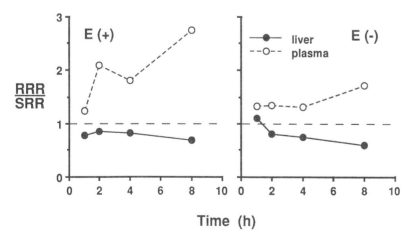

FIGURE 8 Ratios of d_6-*RRR*- to d_3-*SRR*-α-T in liver and plasma of normal (E^+) and vitamin E-deficient (E^-) rats at various times after dosing a 1:1 mixture of d_3-*RRR*- and d_6-*RRR*-α-TAc dissolved in tocopherol-stripped corn oil. Data at each time point were obtained from a single animal.

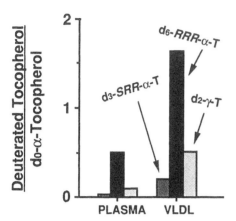

FIGURE 9 Amounts of d_3-*SRR*-α-T, d_6-*RRR*-α-T, and d_2-γ-T expressed relative to d_0-α-T in nascent VLDL isolated from the perfusate of the liver of a cynomolgus monkey fed a 1:1:1 mixture of d_3-*RRR*-α-TAc, d_6-*RRR*-α-TAc, and d_2-γ-T. The data for plasma also are presented for comparison. (Prepared from data in Ref. 22.)

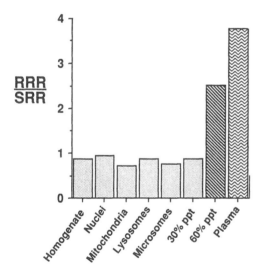

FIGURE 10 Ratios of d_6-*RRR*- to d_3-*SRR*-α-T in various liver fractions and plasma of a cynomolgus monkey 23 h after feeding of a 1:1 mixture of d_3-*SRR*- and d_6-*RRR*-α-TAc. The 30 and 60% ppt fractions are the precipitates obtained by adding 30 and 60% ammonium sulfate, respectively, to the liver cytosol. The relative enrichment of *RRR*-α-T in the precipitate obtained with 60% ammonium sulfate leads us to believe that it contains the tocopherol binding protein.

Humans also show a substantial preference for uptake of *RRR*- over *SRR*-α-tocopherols into plasma (27). Analysis of lipoprotein fractions points to the liver as the major contributor to this discrimination (27,54). This work is discussed in more detail in Chapter 3.

Bioavailability of Natural (*RRR*) and Synthetic (*all-rac*) α-Tocopherols

Synthetic, *all-rac*-α-tocopherol is a mixture of approximately equal amounts of all eight possible stereoisomers, of which the natural, *RRR* isomer is but one. Both the natural and synthetic forms of vitamin E are available commercially for use as dietary supplements, usually in the form of their acetate esters. It therefore is of some practical concern to know their relative bioavailabilities.

It is well-established that *RRR*-α-TAc is more potent, biologically, than *all-rac*-α-TAc in animal tests (55), as determined, for example, by the fetal resorption assay (50,56) or the plasma pyruvate kinase assay (57). Also, each of the seven unnatural stereoisomers has a lower activity than the *RRR* form in the resorption bioassay (56). These differences in biopotency most likely are the direct result of different bioavailabilities arising from stereochemical differences in the phytyl tail.

Animal assays provide evidence that the 2-position is the major determinant of the differences between α-tocopherol stereoisomers. Thus, 2-*ambo*- (*RRR* + *SRR*) and *all-rac*-α-TAc (*RRR* + *RRS* + *RSR* + *RSS* + *SRR* + *SRS* + *SSR* + *SSS*) have very similar biopotencies (50,56,57). That is, *all-rac*-α-TAc with four 2*R* stereoisomers and four 2*S* stereoisomers behaves as if it were a mixture of *RRR*- and *SRR*-α-TAc.

The competitive biokinetics measurements conducted using deuterated *RRR*- and *SRR*-α-TAc are particularly relevant to the question of the bioavailability of the more complex *all-rac*-α-TAc. As already indicated and provided in more detail in Chapter 3, studies carried out on humans and animals indicate that after a single dose of a 1:1 mixture of the tocopheryl acetates, the plasma level of *SRR*-α-T drops much more rapidly than that of *RRR*-α-T. Therefore, if *all-rac*-α-TAc behaves to a large extent as if it were a mixture of *RRR*- and *SRR*-α-TAc, it might be expected that in plasma, at least, the concentration of α-T derived from a dose of *all-rac*-α-TAc will be close to that resulting from a dose of *RRR*-α-TAc that is half the size.

Evidence supporting this prediction was obtained in a study we performed with five subjects who were each given a single dose of a mixture of 150 mg d$_3$-*RRR*-α-TAc and 150 mg d$_6$-*all-rac*-α-TAc (approximately 400 IU in total). The mean *RRR/all-rac* ratio in plasma rose from a value of 1.6 half a day after the dose to a value of close to 2.0 at about 3–4 days (Table 1).

TABLE 1   Mean ± SD (Standard Deviation) of *RRR/all-rac* Ratios in Plasma of Five Subjects After Taking a Single Capsule Containing 150 mg d$_3$-*RRR*-α-TAc and d$_6$-*all-rac*-α-TAc

Day	*RRR/all-rac*
0.5	1.56 ± 0.09
1.5	1.80 ± 0.07
2.5	1.86 ± 0.06
3.5	1.92 ± 0.06
4.5	1.96 ± 0.04

Source: From G. Burton and K. Ingold, unpublished work.

Bile Salt Effects Upon the Cholesterol Esterase-Catalyzed Hydrolysis of α-Tocopheryl Acetates

Pancreatic cholesterol esterase and bile salts are essential for the absorption of vitamin E (see Chap. 3). The application of the deuterated vitamin E technique to in vitro studies of the hydrolysis of 1:1 mixtures of d_0-*RRR*- and d_3-*SRR*-α-T, carried out using bovine or porcine cholesterol esterase in the presence of phospholipid and an obligatory bile salt, revealed a completely unexpected and unprecedented dependence of substrate stereoselectivity upon the nature of the bile salt used (58). In the presence of either cholate or glycocholate, *SRR*-α-TAc is hydrolyzed more rapidly than *RRR*-α-TAc, whereas when taurocholate is used the relative order of reaction is reversed. It is fascinating to note that the bile salt-modulated stereoselectivity is achieved at a site that is remarkably remote from the closest chiral center. The ester bond cleaved by the enzyme is six bonds removed from the chroman ring-phytyl junction.

The bile salt effect observed in the in vitro model contrasts sharply with the apparent lack of selectivity observed in vivo, for example, in the rat lymph study already described. However, the absence of discrimination in vivo could be explained simply if the ester-containing components that are present in individual mixed micelles, including the two tocopheryl acetate stereoisomers, undergo complete hydrolysis by cholesterol esterase before they are transferred across the intestinal wall. This explanation is not precluded by the fact that unreacted tocopheryl ester is recovered from the small intestine. The unreacted tocopheryl ester most probably is part of the triglyceride-containing corn oil emulsion that has not yet been processed by pancreatic lipase. In this two-stage scenario, lipase hydrolysis products (e.g., mono- and diglycerides) and other compounds that are esters of secondary alcohols (e.g., cholesteryl oleate) and aromatic esters (e.g., tocopheryl acetate) are incorporated into bile salt mixed micelles that subsequently are acted upon by cholesterol esterase.

Simon Metabolites of α-Tocopherol

Two products of the oxidation of vitamin E in vivo that have been unequivocally defined are 2-(3-hydroxy-3-methyl-5-carboxypentyl)-3,5,6-trimethyl-1,4-benzoquinone (Fig. 1) and its γ-lactone, the so-called Simon metabolites (59,60). These compounds have been found in the urine of rabbits (59), humans (60–62), and rats (63), largely in the form of glucuronides. For these compounds to be formed, the polyisoprenoid chain of α-T must be oxidatively cleaved at the 3' carbon atom. However, the mechanism by which this metabolic reaction occurs is unknown.

Simon et al. suggested a sequential oxidative degradation of the side chain that is preceded by hydrolytic opening of the chroman ring, presumably during oxidation of α-T to α-tocopheryl quinone (60). However, the intermediacy of the quinone has been disputed (62). Another possibility that occurred to us involves a direct enzymatic attack upon the 3' C-H bond in α-T, leading ultimately to the conversion of the 3' carbon to a carboxyl group. If scission of the C-H bond is at least partially rate limiting, replacement of the two 3' hydrogens with deuteriums slows the reaction. Therefore, 3',3'-D_2-α-T (d_2-α-T) (Fig. 1) should be lost more slowly from tissues than α-T with hydrogen bound to the 3' carbon.

We used the competitive biokinetic technique to test this possibility directly (64). Tissues and blood isolated from rats fed first a diet containing a 1:1 mixture of d_2-*RRR*-α-TAc and 5,7-$(CD_3)_2$-*RRR*-α-TAc (d_6-*RRR*-α-TAc) and then a diet containing no vitamin E were found to have d_2/d_6 ratios that varied little from the original dietary ratio during both the uptake phase and, more importantly, during the loss phase. This result, therefore, does not support a mechanism with a rate-limiting attack on the 3' C-H bond of α-T. The result, however, is consistent with Simon's mechanism involving a series of apparently irreversible oxidations.

OUTLOOK

The use of deuterated vitamin E in conjunction with mass spectrometry is a powerful, precise, sensitive, and safe technique for determining the kinetics of vitamin E in animals and humans. The technique is especially useful for evaluating simultaneously the relative bioavailabilities of two or more forms of vitamin E. In particular, the competitive biokinetics of *SRR*- and *RRR*-α-tocopherols show considerable promise as a tool for probing the details of absorption and secretion of vitamin E in vivo, especially with regard to the role of the liver cytosolic tocopherol binding protein in the intracellular transfer of vitamin E and its incorporation into newly synthesized VLDL.

At a more practical level, the tissue specificity and time dependence of the relative bioavailabilities of the *RRR*- and *SRR*-α-tocopherols observed in the long-term uptake study with rats strongly indicate that a single number, traditionally determined by the rat fetal gestation-resorption assay, cannot express adequately the relative potencies of different forms of vitamin E for nutritional purposes. Competitive biokinetics studies conducted on humans with appropriately labeled natural and synthetic vitamin E promise to give considerably more relevant information in this regard.

In a wider context, the stable isotope technique can be extended to include simultaneous evaluation of other lipids, such as cholesterol, β-carotene, triglyceride, and vitamin A, in conjunction with vitamin E. In this way, we can expect to obtain a better picture of the interactions between different lipids in vivo and achieve considerably more detailed insights into the mechanisms of their absorption and transport.

ACKNOWLEDGMENTS

The authors acknowledge the support of the Association for International Cancer Research, the National Foundation for Cancer Research, the Natural Source Vitamin E Association, Eastman Chemicals, Eisai, Ltd., Henkel, Inc., and F. Hoffmann-La Roche (Basle) and express their sincere thanks to all the individuals cited in our references whose hard work and dedication allowed us to realize our goal of quantifying the biokinetics of vitamin E.

REFERENCES

1. Burton GW, Joyce A, Ingold KU. First proof that vitamin E is major lipid-soluble, chain-breaking antioxidant in human blood plasma. Lancet 1982; 2:327.
2. Burton GW, Joyce A, Ingold KU. Is vitamin E the only lipid-soluble chain-breaking antioxidant in human blood plasma and erythrocyte membranes? Arch Biochem Biophys 1983; 221:281-90.
3. Cheeseman KH, Burton GW, Ingold KU, Slater TF. Lipid peroxidation and lipid antioxidants in normal and tumor cells. Toxicol Pathol 1984; 12:235-9.
4. Cheeseman KH, Emery S, Maddix SP, Slater TF, Burton GW, Ingold KU. Studies on lipid peroxidation in normal and tumour tissues. The Yoshida rat liver tumour. Biochem J 1988; 250:247-52.
5. Ingold KU, Webb AC, Witter D, Burton GW, Metcalfe TA, Muller DPR. Vitamin E remains the major lipid-soluble, chain-breaking antioxidant in human plasma even in individuals suffering severe vitamin E deficiency. Arch Biochem Biophys 1987; 259:224-5.
6. Burton GW, Ingold KU. Autoxidation of biological molecules. I. The antioxidant activity of vitamin E and related chain-breaking phenolic antioxidants in vitro. J Am Chem Soc 1981; 103:6472-7.
7. Burton GW, Hughes L, Ingold KU. Antioxidant activity of phenols related to vitamin E. Are there chain-breaking antioxidants better than α-tocopherol? J Am Chem Soc 1983; 105:5950-1.
8. Burton GW, Doba T, Gabe EJ, et al. Autoxidation of biological molecules. 4. Maximizing the antioxidant activity of phenols. J Am Chem Soc 1985; 107:7053-65.
9. Burton GW, Ingold KU. Vitamin E: applications of the principles of physical organic chemistry to the exploration of its structure and function. Acc Chem Res 1986; 19:194-201.

10. Burton GW, Ingold KU. Vitamin E as an in vitro and in vivo antioxidant. Ann N Y Acad Sci 1989; 570:7-22.

11. Niki E, Kawakami A, Saito M, Yamamoto Y, Tsuchiya Y, Kamiya Y. Effect of phytyl side chain of vitamin E on its antioxidant activity. J Biol Chem 1985; 260:2191-6.

12. Niki E, Komuro E, Takahashi M, Urano S, Ito E, Terao K. Oxidative hemolysis of erythrocytes and its inhibition by free radical scavengers. J Biol Chem 1988; 263:19809-14.

13. Cheng SC, Burton GW, Ingold KU, Foster DO. Chiral discrimination in the exchange of α-tocopherol stereoisomers between plasma and red blood cells. Lipids 1987; 22:469-73.

14. Gallo-Torres HE. Transport and metabolism. In: Machlin LJ, ed. Vitamin E: a comprehensive treatise. New York: Marcel Dekker, 1980; 193-267.

15. Gallo-Torres HE. Absorption. In: Machlin LJ, ed. Vitamin E: a comprehensive treatise. New York: Marcel Dekker, 1980; 170-92.

16. Bieri JG. Kinetics of tissue alpha tocopherol depletion and repletion. Ann N Y Acad Sci 1972; 203: 181-91.

17. Machlin LJ, Gabriel E. Kinetics of tissue alpha-tocopherol uptake and depletion following administration of high levels of vitamin E. Ann N Y Acad Sci 1982; 393:48-60.

18. Vatassery GT, Brin MF, Fahn S, Kayden HJ, Traber MG. The effect of high doses of dietary vitamin E upon the concentrations of vitamin E in several brain regions, plasma, liver and adipose tissue of rats. J Neurochem 1988; 51:621-3.

19. Ingold KU, Burton GW, Foster DO, Hughes L, Lindsay DA, Webb A. Biokinetics of and discrimination between dietary *RRR*- and *SRR*-α-tocopherols in the male rat. Lipids 1987; 22:163-72.

20. Ingold KU, Hughes L, Slaby M, Burton GW. Synthesis of $2R,4'R,8'R$-α-tocopherols selectively labelled with deuterium. J Labelled Comp Radiopharmacol 1987; 24:817-31.

21. Hughes L, Slaby M, Burton GW, Ingold KU. Synthesis of α- and γ-tocopherols selectively labelled with deuterium. J Labelled Comp Radiopharmacol 1990; 28:1049-57.

22. Traber MG, Rudel LL, Burton GW, Hughes L, Ingold KU, Kayden HJ. Nascent VLDL from liver perfusions of cynomolgus monkeys are preferentially enriched in *RRR*- compared with *SRR*-α tocopherol. Studies using deuterated tocopherols. J Lipid Res 1990; 31:687-94.

23. Burton GW, Ingold KU, Foster DO, et al. Comparison of free α-tocopherol and α-tocopheryl acetate as sources of vitamin E in rats and humans. Lipids 1988; 23:834-40.

24. Burton GW, Wronska U, Stone L, Foster DO, Ingold KU. Biokinetics of dietary *RRR*-α-tocopherol in the male guinea pig at three dietary levels of vitamin C and two levels of vitamin E. Evidence that vitamin C does not "spare" vitamin E in vivo. Lipids 1990; 25:199-210.

25. Traber MG, Ingold KU, Burton GW, Kayden HJ. Absorption and transport of deuterium-substituted $2R,4'R,8'R$-α-tocopherol in human lipoproteins. Lipids 1988; 23:791-7.

26. Traber MG, Sokol RJ, Burton GW, et al. Impaired ability of patients with familial isolated vitamin E deficiency to incorporate α-tocopherol into lipoproteins secreted by the liver. J Clin Invest 1990; 85:397-407.

27. Traber MG, Burton GW, Ingold KU, Kayden HJ. *RRR*- and *SRR*-α-tocopherols are secreted without discrimination in human chylomicrons, but *RRR*-α-tocopherol is preferentially secreted in very low density lipoproteins. J Lipid Res 1990; 31:675-85.

28. Weisel RD, Mickle DAG, Finkle CD, et al. Myocardial free-radical injury after cardioplegia. Circulation 1989; 80:III-14-8.

29. Weisel RD, Mickle DAG, Ingold KU, Burton GW. Prevention of free radical injury during coronary revascularization. FASEB J 1989; 3:A940.

30. Packer JE, Slater TF, Willson RL. Direct observation of a free radical interaction between vitamin E and vitamin C. Nature 1979; 278:737-8.

31. Doba T, Burton GW, Ingold KU. Antioxidant and co-antioxidant effect of vitamin C. The effect of vitamin C, either alone or in the presence of vitamin E or a water-soluble vitamin E analog, upon the peroxidation of aqueous multilamellar phospholipid liposomes. Biochim Biophys Acta 1985; 835:298-303.

32. Niki E, Kawakami A, Yamamoto Y, Kamiya Y. Oxidation of lipids. VIII. Synergistic inhibition of oxidation of phosphatidylcholine liposome in aqueous dispersion by vitamin E and vitamin C. Bull Chem Soc Jpn 1985; 58:1971-5.

33. Litov RE, Matthews LC, Tappel AL. Vitamin E protection against in vivo lipid peroxidation initiated in rats by methyl ethyl ketone peroxide as monitored by pentane. Toxicol Appl Pharmacol 1981; 59:96-106.

34. Dillard CJ, Downey JE, Tappel AL. Effect of antioxidants on lipid peroxidation in iron-loaded rats. Lipids 1984; 19:127-33.

35. Kunert K-J, Tappel AL. The effect of vitamin C on in vivo lipid peroxidation in guinea pigs as measured by pentane and ethane production. Lipids 1983; 18:271-4.

36. Dillard CJ, Tappel AL. Consequences of biological lipid peroxidation. In: Chow CK, ed. Cellular antioxidant defense mechanisms, Vol. 1. Boca Raton, FL: CRC Press, 1988; 103-15.

37. Reddy CC, Scholz RW, Thomas CE, Massaro EJ. Vitamin E dependent reduced glutathione inhibition of rat liver microsomal lipid peroxidation. Life Sci 1982; 31:571-676.

38. Haenen GRMM, Bast A. Protection against lipid peroxidation by a microsomal glutathione-dependent labile factor. FEBS Lett 1983; 159:24-8.

39. Burk RF. Glutathione-dependent protection by rat liver microsomal protein against lipid peroxidation. Biochim Biophys Acta 1983; 757:21-8.

40. Hill KE, Burk RF. Influence of vitamin E and selenium on glutathione-dependent protection against microsomal lipid peroxidation. Biochem Pharmacol 1984; 33:1065-8.

41. Franco DP, Jenkinson SG. Rat lung microsomal lipid peroxidation: effects of vitamin E and reduced glutathione. J Appl Physiol 1986; 61:785-90.

42. Haenen GRMM, Tai Tin Tsoi JNL, Vermeulen NPE, Timmerman H, Bast A. 4-Hydroxy-2,3-*trans*-nonenal stimulates microsomal lipid peroxidation by reducing the glutathione-dependent protection. Arch Biochem Biophys 1987; 259:449-56.

43. McCay PB, Lai EK, Powell SR, Brueggemann G. Vitamin E functions as an electron shuttle for glutathione-dependent "free radical reductase" activity in biological membranes. Fed Proc 1986; 45:451.

44. Bast A, Haenen GRMM. Interplay between lipoic acid and glutathione in the protection against microsomal lipid peroxidation. Biochim Biophys Acta 1988; 963:558-61.

45. Tirmenstein MA, Reed DJ. Effects of glutathione on the α-tocopherol-dependent inhibition of nuclear peroxidation. J Lipid Res 1988; 30:959-65.

46. Wefers H, Sies H. The protection by ascorbate and glutathione against microsomal lipid peroxidation is dependent on vitamin E. Eur J Biochem 1988; 174:353-7.

47. Scholich H, Murphy ME, Sies H. Antioxidant activity of dihydrolipoate against miocrosomal lipid peroxidation and its dependence on α-tocopherol. Biochim Biophys Acta 1989; 1001:256-61.

48. Packer L, Maguire JJ, Mehlhorn RJ, Serbinova E, Kagan VE. Mitochondria and microsomal membranes have a free radical reductase activity that prevents chromanoxyl radical accumulation. Biochem Biophys Res Commun 1989; 159:229-35.

49. Harris PL, Ludwig MI. Vitamin E potency of α-tocopherol and α-tocopherol esters. J Biol Chem 1949; 180:611-5.

50. Weiser H, Vecchi M, Schlachter M. Stereoisomers of α-tocopheryl acetate. IV. Units and α-tocopherol equivalents of *all-rac-*, 2-*ambo-*, and *RRR*-α-tocopherol evaluated by simultaneous determination of resorption-gestation, myopathy, and liver storage capacity in rats. Int J Vitam Nutr Res 1986; 56:45-56.

51. Horwitt MK, Elliott Wh, Kanjananggulpan P, Fitch CD. Serum concentrations of α-tocopherol after ingestion of various vitamin E preparations. Am J Clin Nutr 1984; 40:240-5.

52. Zahalka HA, Burton GW, Ingold KU. Competitive uptake of deuterium-labeled α-tocopherols in male rats. Ann N Y Acad Sci 1989; 570:533-5.

53. Catignani GL, Bieri JG. Rat liver α-tocopherol binding protein. Biochim Biophys Acta 1977; 497:349-57.

54. Burton GW, Traber MG. Vitamin E: antioxidant activity, biokinetics and bioavailability. Annu Rev Nutr 1990; 10:357-82.

55. Diplock AT. Vitamin E. In: Diplock AT, ed. Fat-soluble vitamins. Lancaster, PA: Technomic Publishing, 1985; 154-224.

56. Weiser H, Vecchi M. Stereoisomers of α-tocopheryl acetate. II. Biopotencies of all eight stereoisomers, individually or in mixtures, as determined by rat resorption-gestation tests. Int J Vitam Nutr Res 1982; 52:351-70.

57. Machlin LJ, Gabriel E, Brin M. Biopotency of α-tocopherols as determined by curative myopathy bioassay in the rat. J Nutr 1982; 112:1437-40.

58. Zahalka HA, Dutton PJ, O'Doherty B, et al. Bile salt modulated stereoselection in the cholesterol esterase catalyzed hydrolysis of α-tocopheryl acetates. J Am Chem Soc 1991; 113:2797-9.

59. Simon EJ, Gross CS, Milhorat AT. The metabolism of vitamin E. I. The absorption and excretion of d-α-tocopherol-5-methyl-C^{14}-succinate. J Biol Chem 1956; 221:797-805.

60. Simon EJ, Eisengart A, Sundheim L, Milhorat AT. The metabolism of vitamin E. II. Purification and characterization of urinary metabolites of α-tocopherol. J Biol Chem 1956; 221:807-17.

61. Schmandke H. Die Bildung von tocopheronolacton aus α-tocopherylchinon im menschlichen organismus. Int Z Vitaminforsch 1965; 35:321-7.

62. Schmandke H, Schmidt G. Zum abbaumechanismus des α-tocopherols im menslichen organismus. Int Z Vitaminforsch 1968; 38:75-8.

63. Weber F, Wiss O. Über den stoffwechsel des vitamin E in der ratte. Helv Physiol Acta 1963; 21:131-41.

64. Dutton PJ, Hughes LA, Foster DO, Burton GW, Ingold KU. Simon metabolites of α-tocopherol are not formed via a rate-controlling scission of the 3'C—H bond. Free Radic Biol Med 1990; 9:435-9.

27

Vitamin E Status of Exotic Animals Compared with Livestock and Domestics

Ellen S. Dierenfeld

New York Zoological Society, Bronx, New York

Maret G. Traber

New York University School of Medicine, New York, New York

INTRODUCTION

Vitamin E deficiency has been identified as a health problem in zoo animals for more than 50 years (1). Within the past decade there has been increased recognition of vitamin E deficiency in zoo collections because zoos have added personnel with expertise in identifying clinical or pathological signs and priority has been placed on the field of nutrition as a means of preventative medicine. Additional contributory factors have been recognized in the development of vitamin E deficiency and other disease states; these include unquantified stress components in zoo management and reduced genetic pools in zoo species.

Dietary requirements for vitamin E in most zoo species have not been established. Additionally, measurements of tissue levels of vitamin E are quite rare; therefore, blood samples provide the primary basis for quantitative evaluation of vitamin E status. However, blood samples from exotic species in zoos are often obtained only on an opportunistic basis or when animals are clinically ill or dead, with comparative normal values from healthy animals rare or nonexistent in many cases. The few values that are available from free-ranging animals provide a valuable resource, but the impact of unknown diet history remains a major variable. Pathologies of vitamin E deficiency are similar among zoo and domestic animals, yet we must be able to recognize these signs in zoo species to assess vitamin E deficiency antemortem.

Although the "production" goals of zoo management are very different from those of livestock production, three areas of focus for which comparative data from livestock and domestic models are extremely useful, discussed in detail in this chapter, include (1) pathological and clinical deficiency signs, (2) plasma and tissue concentrations, and (3) dietary evaluation.

DOMESTIC HORSES AS A MODEL FOR ZOO PERISSODACTYLS

Perissodactyls (hooved mammals with one or three toes) all possess a digestive anatomy with an enlarged hind gut (cecum and/or proximal colon) for plant fermentation (2). Domestic horses and ponies of the genus *Equus* are likely physiological models for nutritional studies of exotic equids. This genus includes approximately 17 species, including wild horses, zebras, and asses. Other perissodactyl groups include members of the tapir (4 species) and rhinoceros (5 species) families (Fig. 1). Elephants (2 species) have a similar gastrointestinal morphology as perissodactyls, and are discussed with this group for comparative purposes.

Clinical Signs of Vitamin E Deficiency

Neurological Abnormalities

Equine degenerative myoencephalopathy (EDM), a diffuse degenerative disease of the spinal cord and caudal portion of the brain stem, may be associated with vitamin E deficiency in horses (3,4). The syndrome has been reported in numerous breeds, often with an apparent familial and/or age predisposition (4,5). Severe neuronal fiber deterioration is evident upon

FIGURE 1   Horses may provide a suitable domestic model for exotic perissodactyls, such as the Indian rhinoceros (*Rhinoceros unicornis*) shown; however, plasma α-tocopherol and cholesterol levels are lower in rhinos than horses. The low α-tocopherol-cholesterol ratio (<1.0 μg/mg) observed in rhinos may indicate low vitamin E status, but a specific vitamin E deficiency has not been documented in this species. (© New York Zoological Society Photo.)

histological examination, similar to that reported in vitamin E-deficient rats (6) and humans (7).

Clinically, the disease manifests as asymmetric weakness and ataxia of the limbs, primarily affecting animals < 1 year of age. Mayhew et al. (4) reported that before dietary supplementation of affected mares and foals, circulating vitamin E levels were <1.0 μg/ml; plasma levels more than doubled following supplementation, and clinical signs were alleviated. However, in a separate study, Dill et al. (8) compared serum vitamin E levels in EDM-affected horses ($n = 40$) with unaffected controls ($n = 49$) and found no differences in serum vitamin E levels (3.8 versus 3.4 μg/ml).

Clinical and pathological findings of EDM have been reported in captive zebras (9) and Przewalskii horses (10). Plasma α-tocopherol concentrations in these zoo horses were 0.4 and 1.1 μg/ml for affected and clinically normal animals, respectively.

From the improvement observed in the neurological status in response to vitamin E, it seems likely that EDM results from vitamin E deficiency. The familial nature of EDM suggests that this disorder may be a genetic abnormality in vitamin E transport similar to that observed in humans (11). Humans with familial isolated vitamin E deficiency appear to lack or have a defective hepatic vitamin E transfer protein (11). This protein, which has been partially purified from rat liver (12,13), discriminates between forms of tocopherol (12) and is essential for the maintenance of plasma levels of *RRR*-α-tocopherol (14). Horses are likely to have a hepatic tocopherol transfer protein, as they have been shown to discriminate among a variety of dietary tocopherols and tocotrienols with a preferential enrichment of α-tocopherol in the serum, liver, and muscle (15). Further studies are required to determine whether horses have a tocopherol binding protein and whether horses with EDM lack or have a defective form of this protein. Neurological abnormalities have not been observed in elephants or rhinoceros.

Muscular Dystrophy

Vitamin E deficiency has been suggested as a cause of primary skeletal myopathy in various case studies of young foals (16,17). Decreased performance and muscle soreness in exercised horses have also been attributed to vitamin E inadequacy (18). Because oxidative tissue damage occurring during physical exercise has been linked with depleted muscle concentrations of vitamin E in rats (19), the relationship between exercise and vitamin E status was subsequently examined in horses (20). Although the exercise group had significantly lower plasma α-tocopherol levels than nonexercised controls (approximately 1.2 compared with 2.2 μg/ml, $p < 0.01$), no significant differences in muscle vitamin E concentrations (≈ 3 μg/g) were observed (20).

Skeletal muscle degeneration has been observed in vitamin E-deficient tapirs (21) and elephants (22), but muscle tissue α-tocopherol levels were not reported. Cardiomyopathy (mulberry heart disease similar to that described in swine; see subsequent discussion) associated with low circulating vitamin E (0.26 μg/ml) was also reported in a young Asian elephant (23).

Blood Vitamin E and Lipid Levels

Erythrocyte Hemolysis and Vitamin E Concentrations

Vitamin E deficiency resulting in erythrocyte hemolysis has been observed in humans (24), other primates (25), rats (26), and horses (27). Foals aged 2 months required 200 days of a vitamin E-deficient diet before erythrocytes hemolyzed, at a total serum content of < 1.2 μg/ml of α-tocopherol (27).

Low circulating levels of α-tocopherol observed in rhinoceros (28,29) were considered an underlying factor in the high incidence of hemolytic anemia reported in zoo rhinoceros

(30). However, further studies have shown no correlation between the incidence of hemolytic anemia and circulating vitamin E levels in rhinoceros (31).

Plasma α-Tocopherol and Lipid Levels

Plasma concentrations of α-tocopherol in healthy horses range from 1 to 3 μg/ml, with no significant differences due to age, gender, or time of day (32–34). The mean circulating plasma α-tocopherol concentration in clinically healthy zoo equids from five species (2.8 μg/ml; Table 1) was similar to that of domestic horses.

Horwitt et al. (39) suggested that serum vitamin E concentrations in humans be reported relative to lipid levels to standardize comparisons between subjects with high and low lipid levels. In standardbred horses, serum total lipids reported from two trials were 2.2 ± 0.4 and 2.4 ± 0.4 mg/ml (15). Wider variations in plasma lipids (range 1.7–4.0 mg/ml) have also been observed (40), and differences between ponies and horses (41) and among breeds (42) have been reported. Total vitamin E-lipid ratios in horses fed 100 mg dietary vitamin E with supplements of *all-rac-α*-tocopheryl acetate of 0, 200, 600, and 1800 IU/day ranged from 0.5 to 1.5 mg/g lipid (15).

Because cholesterol values are routinely monitored in zoo species, this lipid fraction has been suggested as a practical estimator of vitamin E status across species (1). Plasma α-tocopherol and cholesterol levels in various perissodactyls and elephants are shown in Table 1, along with some comparative values from free-ranging elephants and black rhinoceros sampled in various locations in Africa. Elephants and rhinoceros have low α-tocopherol to cholesterol ratios compared with other perissodactyls. The α-tocopherol-cholesterol ratios in free-ranging (1.2 μg/mg) and zoo (0.6 μg/mg) elephants with essentially identical cholesterol values suggests inadequate vitamin E status in zoo animals.

Vitamin E is transported in plasma lipoproteins; thus, the type of lipoprotein carrier may influence the amount of vitamin E that can be carried and distributed to the tissues (see Chap. 3). A contributing factor of the low tocopherol levels observed in elephants and rhinoceros may be an altered plasma lipoprotein distribution. High-density lipoproteins (HDL)

TABLE 1 Mean Serum or Plasma α-Tocopherol (E) and Cholesterol Levels in Perissodactyls and Elephants

Family	n	Species	α-Tocopherol (μg/ml)	Cholesterol (mg/ml)	E/cholesterol (μg/mg)
Zoo species: Proboscidea					
Elephantidae (elephant)	31	2	0.4	0.62	0.6
Free-ranging African elephant	35		0.8	0.69	1.2
Zoo species: Perissodactyla					
Equidae (horse, zebra)	93	5	2.8	1.19	2.4
Tapiridae (tapir)	5	2	3.6	1.65	2.2
Rhinocerotidae (rhinoceros)	77	3	0.4	0.79	0.5
Free-ranging black rhinos					
Namibia	4		0.8		
Zimbabwe	55		0.6		
South Africa	44		0.6		
Kenya	7		0.2		
Domestic: horse	37		2.0	0.82	2.4

Source: Data summarized from References 1, 15, and 35–38 and from U. S. Seal, unpublished.

represent more than 80% of the lipoproteins in domestic horses, Przewalskii horses, and zebras, and 60% in Malayan tapirs, but no HDL was detected in rhinoceros plasma (38,43) (Traber and Dierenfeld, unpublished observations). By way of contrast, the lipoprotein distribution (measured as cholesterol) in a captive Asian elephant was VLDL, 9%; LDL, 58%; and HDL, 32%, with a serum cholesterol of 0.36 mg/ml and α-tocopherol of 0.53 μg/ml (Traber and Dierenfeld, unpublished observations). Although the lack of HDL may be the reason for low plasma α-tocopherol levels in the rhinoceros, the lipoprotein distribution in the elephant is not different from that observed in humans (44–47).

Tissue Concentrations of Vitamin E

Defining normal tissue levels of vitamin E, and recognizing deficient levels in target tissues are essential for evaluating vitamin E status. Patterns of tissue vitamin E depletion and repletion were investigated in a single study using standardbred horses (15). Animals were fed marginal levels of vitamin E (107 IU/day) for 2½ months and then were supplemented with 0, 600, 1800, or 5400 mg *all-rac*-α-tocopheryl acetate per day for 4 months. The tocopherol contents of serum, liver, and muscle were followed throughout the study; adipose tissue concentrations were measured in a few animals sacrificed at the end of the study. Tissue responses to different levels of supplementation of vitamin E showed a linear correlation with amount of supplementation, up to an apparent saturation at approximately 600 mg/day, or about 2 mg/kg body weight. In response to supplementation of 600 mg/day, the vitamin E content of the serum was approximately 0.75 mg/g lipid, the liver and muscle each about 5 μg/g tissue, and the adipose tissue about 25 μg/g tissue.

Vitamin E levels in tissues other than blood have not been measured in most zoo species. The α-tocopherol content measured in a single tapir liver (1.3 μg/g wet weight) was considered deficient (21). A comparison of α-tocopherol concentrations in rhinoceros liver, skeletal muscle, heart, and adipose tissue (Table 2) reveals a possible vitamin E deficiency or a lower overall storage capacity for vitamin E than observed in horses. Species differences in tissue vitamin E content between the grazing white rhinoceros and the browsing black rhinoceros are also evident. The differences seen in Table 2 indicate higher overall tissue storage (thus vitamin E status) in captive black rhinoceros compared with whites, contrary to that predicted from readily obtained plasma samples, which averaged 0.20 ± 0.04 μg/ml for black rhinoceros and 1.08 ± 0.17 μg/ml for white. Current zoo husbandry practices may underlie the differences noted. In response to recognition of a potential vitamin E deficiency in black rhinoceros (28), most black rhinoceros in zoos are heavily supplemented with dietary vitamin E. A similar need for supplementation of white rhinoceros has not been specifically identified.

TABLE 2 Tissue α-Tocopherol Levels in Zoo Rhinoceros (μg/g wet weight; $\bar{x}$ ± SEM)

Tissue	Black rhino (*n*)	White rhino (*n*)
Liver	4.33 ± 1.57 (6)	1.48 ± 0.64 (3)
Skeletal muscle	1.81 ± 0.36 (6)	1.18 ± 0.30 (3)
Heart	7.72 ± 2.09 (4)	1.94 ± 0.80 (3)
Adipose	4.18 ± 1.65 (3)	4.28 ± 1.50 (3)

Source: From unpublished data by E. S. Dierenfeld.

Dietary Vitamin E Requirements

The National Research Council (NRC) suggested that horse diets contain vitamin E at a concentration of 50 mg/kg, 80–100 mg/kg for foals and reproducing females (48). Ronèus et al. (15) recommended daily supplements of 600–1800 mg *all-rac*-α-tocopheryl acetate, levels that correspond to approximately 70–200 mg/kg feed, or 1.5–4.4 mg/kg body mass.

Diets containing *all-rac*-α-tocopheryl acetate ($\approx$250 IU/kg) have been shown to maintain elephant plasma α-tocopherol levels $\approx$0.6 μg/ml (23), a level observed in free-ranging, wild elephants (0.4–1.6 μg/ml; Seal and Dierenfeld, unpublished observations). However, horses and elephants consuming similar levels of dietary supplements of vitamin E have markedly different plasma α-tocopherol concentrations (see Table 1) that appear to be independent of plasma cholesterol levels.

The influence of dietary fat upon absorption of vitamin E has not been examined in perissodactyls but appears to be an important variable in this group of animals. Stowe (27) stated that absorption of dietary α-tocopherol in horses may have been limited by the low fat content (1%) of the experimental diet. Consistent reports of higher circulating levels of α-tocopherol in horses on pasture (4,15,49,50), compared to typical commercial or dry diets, argue that horses on pasture are consuming higher dietary levels of fat, α-tocopherol, or both.

Unpublished data obtained using deuterated tocopherol fed to a captive elephant (Burton, Traber, and Dierenfeld) suggest limited absorption of this nutrient at the gut level, perhaps as a result of the low-fat diet. Furthermore, water-soluble forms of vitamin E, which bypass possible limitations of fat absorption in elephants, have been shown to effect a significant plasma response over short-term (1–6 weeks) feeding trials (51) (Stuart, Ingram, and Dierenfeld, unpublished). The effects of longer term supplementation and the impact of the forms and levels of dietary vitamin E on target tissues need to be assessed.

Nonetheless, dietary fat remains an important variable for understanding vitamin E nutrition of zoo elephants, rhinoceros, and possibly other zoo herbivores. Apparent differences between the vitamin E status of browsing versus grazing species herbivores may be a function of dietary fat levels. Essential fatty acid deficiency has been reported for free-ranging elephants as a result of habitat change and consequent reduction in dietary fat levels (52). Shrubs preferred by black rhinoceros have been shown to contain up to 25% crude fat, as well as higher α-tocopherol (>400 mg/kg) content than grasses or dried forages and pellets commonly fed zoo animals (53).

DOMESTIC RUMINANTS AS MODELS FOR ZOO ARTIODACTYLS

Exotic ruminants include all antelope (Bovidae family; Fig. 2) and deer (Cervidae) species, in addition to members of the giraffe (Giraffidae), mouse deer (Tragulidae), and camel (Camelidae) families. Although these artiodactyls possess foregut fermentation, digestive tract complexity varies considerably both within and among taxonomic groups (2). Increased ruminal microbial destruction of vitamin E (54) may result in increased dietary requirements for species eating high-grain diets; thus, natural dietary habits can considerably influence vitamin E status. Nonetheless, cattle and sheep provide the only experimental models of vitamin E metabolism for these species.

Clinical Signs of Vitamin E Deficiency

Focal or diffuse myopathies of skeletal or cardiac tissues associated with vitamin E deficiency have been reported in domestic cattle (55) and sheep (56,57). Specific muscles affected by these disorders vary, but damage is more extensive in highly active muscles, that is, heart,

FIGURE 2 Nyala antelope (*Tragelaphus angasi*), both male and female, are shown. These antelope are one of the first exotic herbivores in which vitamin E deficiency was documented (60). Pathological lesions were similar to those seen in domestic cattle and sheep; plasma α-tocopherol-cholesterol ratios (μg/mg) in clinically normal animals average 1.0–2.0. (© New York Zoological Society Photo.)

tongue (in suckling animals), or those involved with locomotion. Genetic factors have been linked with an increased incidence of vitamin E deficiency myopathy among cattle (58). Similar pathological lesions have been identified in a number of exotic ruminants, including kudu (59), nyala antelope (60), and various deer species (61).

Young animals (<6 months of age), feedlot cattle, and/or animals recently turned onto fresh pasture appear more susceptible to vitamin E deficiency than mature, free-ranging species. If indoor diets are low to marginal in vitamin E, then the additional polyunsaturated fatty acids in fresh grasses may result in lipid peroxidation (and consequent vitamin E deficiency) in animals with inadequate vitamin E stores. However, prolonged consumption of fresh pasture leads to higher tissue α-tocopherol levels (55,62,63).

Blood Vitamin E and Lipid Levels

Plasma α-Tocopherol Levels

α-Tocopherol comprises approximately 90% of total serum tocopherols in dairy cows and calves (64). Plasma α-tocopherol values > 4 μg/ml appear to indicate adequate vitamin E status and <1.5 μg/ml indicates deficiency (65). Following oral administration of different forms of vitamin E (1000 IU/day) for 1 month to cattle, circulating α-tocopherol levels increased from 1.4 μg/ml to 7.6 (*RRR-α*-tocopheryl acetate), 9.22 (*RRR-α*-tocopherol), 5.4 (*all-rac-α*-tocopherol), or 4.6 (*all-rac-α*-tocopheryl acetate), with significant differences between the *all-rac* and *RRR* forms (66).

Plasma α-tocopherol levels of 2.5 μg/ml (63) and 1.3 μg/ml (67) have been reported in sheep. In response to supplements of 400 IU vitamin E, plasma levels increased from 0.9 to $\approx$3.0 μg/ml, with no differences in the maximum plasma levels due to the vitamin E form administered (68).

White-tailed deer fed experimental diets with adequate amounts of vitamin E had serum α-tocopherol levels of 1.0–3.0 μg/ml (69,70). Plasma α-tocopherol levels recently measured in free-ranging mule deer sampled from 14 habitats in California averaged 0.8–4.2 μg/ml (71); significant differences were attributable to differences in diet, sex, age, season, and migratory behavior. α-Tocopherol values obtained from exotic ruminants displaying no clinical signs of vitamin E deficiency are shown in Table 3.

Plasma Lipid Levels

Plasma α-tocopherol levels among ruminants are related to total lipid levels, especially cholesterol. Total cholesterol in dairy cows averaged 1.4 mg/ml and was significantly correlated with serum α-tocopherol ($r = 0.44$; $p < 0.0005$) in both dams and fetuses (73). Cattle vitamin E-cholesterol ratios range from 0.95 to 1.5 μg/mg (73,74), whereas sheep vitamin E-cholesterol ratios are somewhat higher.

A strong correlation between plasma cholesterol and α-tocopherol was observed among exotic ruminants, with mean cholesterol values ranging from 0.4 to approximately 1.4 mg/ml. Ratios of α-tocopherol-cholesterol (μg/mg) of 1.0–3.0 are likely to reflect adequate vitamin E status among zoo ungulates (1). Three families for which the vitamin E-cholesterol ratio is >3 μg/mg (Camelidae, Tragulidae, and Giraffidae) include species for which digestive tract anatomy differs substantially from that of typical ruminants.

Although before supplementation camels had low plasma α-tocopherol levels, they ($n = 4$) showed a significant increase in circulating α-tocopherol levels (from 0.6 to 2.8 μg/ml) following 3 months of oral supplementation with *all-rac-α*-tocopheryl acetate at 4 IU/kg body mass (about 400 mg/kg diet) (Dierenfeld, unpublished). Although data are limited, despite the reported lack of HDL in both camels and rhinoceros (43), when supplemented

TABLE 3 Mean Serum or Plasma α-Tocopherol (E) and Cholesterol Levels in Zoo Compared with Domestic Ruminants

Family	n	Species	α-Tocopherol (μg/ml)	Cholesterol (mg/ml)	E/cholesterol (μg/mg)
Zoo species					
Camelidae (camels)	42	5	1.9	0.53	3.6
Tragulidae (mouse deer)	17	1	3.0	0.68	4.4
Cervidae (deer)	295	17	1.9	0.87	2.2
Giraffidae (giraffe)	28	2	1.4	0.40	3.5
Bovidae (antelope)	518	32	1.9	0.85	2.2
Domestic species					
Cattle	114		2.3	1.43	1.6
Sheep	22		1.9	0.50	3.8

Source: Data summarized from References 1, 37, 63, 72, and 73 and from E. S. Dierenfeld, unpublished.

with vitamin E camels have higher plasma α-tocopherol levels than rhinoceros and have α-tocopherol-cholesterol ratios similar to those of other domestic ruminants.

Tissue Concentrations of Vitamin E

Studies of vitamin E nutrition in ruminants have typically focused on plasma, but a survey of liver samples collected at slaughter revealed mean concentrations of 20 μg/g wet weight for cattle and 6 μg/g wet weight in sheep (75). Liver α-tocopherol concentrations in adult cattle fed unsupplemented rations averaged approximately 4.0 μg/g wet weight (73); those in cattle consuming approximately 1000 IU vitamin E per day for 30 days averaged 22 μg/g wet weight (66). Adrenal gland (33.3 μg/g wet weight) contained the highest tissue concentrations following the 1 month feeding trial, and skeletal and cardiac muscle concentrations were 5.6 and 18.4 μg/g wet weight, respectively (66). Liver concentrations of α-tocopherol in sheep fed rations containing 400 IU vitamin E per day for 1 month averaged 13.7 μg/g wet weight (68). Other tissue concentrations reported for sheep include skeletal muscle (3.7 μg/g wet weight), heart (9.8 μg/g wet weight), and adipose (3.0 μg/g wet weight).

Few comparative tissue data are available from exotic species. Liver α-tocopherol concentrations in a healthy compared with a vitamin E-deficient kudu were 3.7 versus 0.8 μg/g wet weight, respectively (59). These data suggest that liver α-tocopherol concentrations observed in domestic ruminants fed unsupplemented diets provide useful guidelines for exotics.

Dietary Vitamin E Requirements

Current NRC (76) requirements for beef cattle suggest feed levels of 15–60 IU/kg dry matter for young calves; no daily requirements for growing or mature beef cattle are listed. Recommended dietary levels for growing lambs range from 15 to 20 mg/kg dry matter (77).

Intakes of vitamin E ranging from 57 to 800 mg/day in 24 zoo ruminants resulted in low plasma α-tocopherol levels compared with domestic ruminants (29). In a separate study, zoo ruminants became vitamin E deficient when fed diets calculated to contain 60–70 IU vitamin E per kg dry matter (61). Thus, dietary levels of vitamin E that are adequate for domestic species appear insufficient for many exotics.

Data on α-tocopherol levels in natural forages may provide information for estimating dietary requirements of exotic species but are not currently available. Factors apart from dietary levels that may impact on the utilization of vitamin E in zoo species include environmental variables, stress, and nutrient interactions, as discussed by Dierenfeld (1). Current research suggests dietary levels of 150–200 IU vitamin E/kg dry matter as appropriate for zoo ruminants.

DOMESTIC SWINE AS A MODEL FOR NONHUMAN PRIMATES

Nine species of exotic pigs are found within the family Suidae, including bush and river pigs, wart hogs, giant forest hogs, and babirusa (Fig. 3). The stomach anatomy and feeding habits of some exotic swine (2) suggest that ruminants may provide a more appropriate comparative model of vitamin E status for zoo suids than domestic hogs, but no specific comparisons are available.

Swine have been utilized as a model for human nutrition studies because of similarities of digestive morphologies (78) and lipoprotein patterns (79). In this section, the pig is considered in relation to the vitamin E nutrition of nonhuman primates with references to other suids when possible.

FIGURE 3   Although domestic swine have been suggested as a physiological model of vitamin E nutrition for nonhuman primates, they are likely more appropriate for other suids, like the babirusa (*Babyrousa babyrussa*) shown. Circulating α-tocopherol (3.0 μg/ml) and cholesterol (0.92 mg/ml) levels in swine are generally lower than those in primates; humans appear to provide the best comparative values for assessment of status in non-human primates. (© New York Zoological Society Photo.)

Clinical Signs of Vitamin E Deficiency

Along with the more general muscular lesions described for other hoof stock, vitamin E deficiency in swine is associated with several conditions, including microangiopathy (mulberry heart disease), edema, liver necrosis, and steatitis (an inflammation of the adipose tissue) (80–82). No instances of hemolytic anemia, as seen in vitamin E-deficient humans (83) and nonhuman primates (84), have been reported in the pig.

Myocytic and myocardial fibrosis and vascular lesions induced by feeding diets low in vitamin E are similar to those described for other species (61), although histological examination is often necessary to observe these lesions in pigs (81). Myocardial hemorrhage occurs in young, fast-growing pigs and is associated with decreased tissue levels of vitamin E (82). Similarly, porcine stress syndrome has been associated with low levels of dietary antioxidants, but not necessarily vitamin E, although supplemental vitamin E can alleviate the symptomatology (85).

The only documented case of myopathy suspected to result from vitamin E deficiency in zoo swine was reported in the Vietnamese potbellied pig (86). Piglets displayed pale streaking of quadriceps and cardiac tissues; however, diets met or exceeded NRC recommendations for vitamin E and selenium. Prophylactic treatment of remaining pigs with supplemental vitamin E was implemented, and no further clinical cases were observed in this herd.

Clinical conditions associated with vitamin E deficiency in various primates include anemia, steatitis, neuronal degeneration, and muscle necrosis (84,87–90). The lesions described are essentially identical to those documented in vitamin E-deficient humans.

Plasma Vitamin E and Lipid Levels

Normal circulating α-tocopherol levels in domestic swine range from 2 to 4 μg/ml, with newborn piglets (0.3 μg/ml) displaying substantially lower concentrations than adults (91,92). Plasma α-tocopherol concentrations increase with feed tocopherol levels and have been shown to reach >5.0 μg/ml with diets containing >200 IU vitamin E per kg dry matter (93,94). Vietnamese potbellied pigs fed diets containing 15–20 IU vitamin E per kg dry matter had plasma levels averaging 2.3 μg/ml (86); babirusa ($n = 4$) fed diets containing 200 IU/kg had plasma levels of 2.4 μg/ml (Dierenfeld, unpublished), similar to those of domestic swine (Table 4).

Plasma cholesterol in pigs fed low-fat, plant-based protein diets averaged 0.89 mg/ml, compared to values in pigs fed higher fat diets (mean 1.61 mg/ml; range 1.1–2.0) (78).

Plasma α-tocopherol and cholesterol values of various zoo primates are shown in Table 4. Although similar to levels found in humans, values are in general higher compared with swine. Circulating levels of α-tocopherol < 5.0 μg/ml are considered deficient in human subjects and monkeys (90,98,99). In a survey of 74 captive lowland gorillas, 11% were below that threshold (95). Plasma α-tocopherol levels of marmosets and tamarins ranged from 5 to 10 μg/ml in healthy animals ($n = 34$) but <5 in sick ($n = 3$) monkeys (96).

Cholesterol levels measured in zoo primates range from 1.0 to >2.5 mg/ml (Table 4); these values are within ranges reported for free-ranging primates (100). Total plasma lipids in free-ranging howler monkeys (4.8 mg/ml) (100) are higher than those of squirrel monkeys (2.8 mg/ml) (101) but within the normal range for humans. Lipoprotein fractions of various nonhuman primates have been shown to be variable and are likely to be dependent upon both dietary and genetic factors (100).

TABLE 4 Mean Plasma α-Tocopherol (E) and Cholesterol Levels in Zoo Primates Compared with Swine

Family	n	Species	α-Tocopherol (μg/ml)	Cholesterol (mg/ml)	E/cholesterol (μg/mg)
Zoo species					
Lemuridae (lemurs)	9	1	6.5	1.26	5.2
Lorisidae (galagos)	15	3	10.2		
Cebidae (squirrel monkeys)	35	5	15.1	2.33	6.5
Callithricidae (marmosets, tamarins)	88	9	4.4	1.02	4.3
Cercopithidae (macaques)	136	21	7.3	1.39	5.2
Pongidae (gorillas)	152	5	9.7	2.67	3.6
Domestic species					
Swine	10		3.0	0.92	3.3

Source: Data summarized from Dierenfeld (unpublished) and from References 29, 37, 72, and 94–97.

Tissue Concentrations of Vitamin E

In swine fed diets containing 45 mg vitamin E per kg, adipose tissue levels were 10–16 μg/g wet weight, liver and heart ranged from 4 to 9 μg/g, and skeletal muscle was about 4 μg/g (94,102). A genetic predisposition for dietetic microangiopathy (DM) has been observed in pigs in Northern Ireland. The affected pigs have significantly lower liver and heart α-tocopherol concentrations than control pigs [2.0 versus 3.1 μg/g wet weight (liver) and 2.6 versus 4.9 μg/g wet weight (heart)] despite similar vitamin E and selenium intakes (82). Although Rice and Kennedy (82) discounted the likelihood of defects in absorption and transport of α-tocopherol as the causes of DM, recent studies on the importance of the tocopherol binding protein in maintaining plasma levels of α-tocopherol in humans (11,14) suggest that pigs may also have a similar genetic disorder. No genetic predisposition to vitamin E deficiency has been documented in zoo primates.

In contrast to humans (103), pigs have been shown to mobilize adipose tissue stores of vitamin E to maintain plasma levels (102). Tissue α-tocopherol levels have not been routinely quantified in nonhuman primates, making it difficult to compare the dynamics of vitamin E metabolism in these species. It is likely that human values (104) provide the most suitable guidelines of tissue norms for nonhuman primates, but few comparative values are available.

Dietary Vitamin E Requirements

Dietary concentrations of 12–15 IU/kg have been shown to prevent vitamin E deficiency symptoms in swine (82), and practical rations containing 15–60 mg/kg are recommended (105). Dietary requirements of vitamin E for nonhuman primates are currently about 70 mg/kg dry matter (106), similar to the nutrient requirements (0.7–3 mg/kg body mass) determined by Fitch and Dinning (107) for rhesus macaques fed diets of varying fat content.

From pathological observations, as well as evaluation of plasma α-tocopherol and lipid levels, it is apparent that domestic swine provide a limited model for the assessment of vitamin E status in zoo primates, and human comparisons appear much more suitable.

DOMESTIC CARNIVORES AS MODELS FOR EXOTIC CARNIVORES

Mammalian carnivores include members of the Canidae (dog), Felidae (cat; Fig. 4), Ursidae (bear), Procyonidae (raccoon), Viverridae (mongoose), Hyaenidae (hyaena), Ailuropidae (panda), and Mustelidae (weasel) families. Of these groups, experimental studies of vitamin E nutrition have only been conducted on domestic dogs or cats and economically important fur species, such as the mink. All possess similar, simple digestive tract anatomy (2). Furthermore, nutrient requirements established for domestic dogs and cats appear suitable for zoo canids and felids. Other species, which are considered in this section for comparative purposes, include pinnipeds (seals) and cetaceans (whales).

Clinical Signs of Vitamin E Deficiency

Although vitamin E deficiency has been produced experimentaly in dogs (108), the condition is more generally associated with carnivores fed diets containing high levels of polyunsaturated fatty acids. Cats and mink fed fish-based diets develop steatitis (109,110). Anemia, erythrocyte hemolysis, hind limb paralysis, and edema have also been reported in affected mink (110). Pathological skeletal and cardiac muscular lesions associated with vitamin E deficiency have been described in foxes (110) and dogs (108). Additionally, dogs with chronic vitamin E deficiency display retinal degeneration, dystrophic axons, lipofuscin pigmentation of smooth musculature, and increased erythrocyte fragility (108,111).

FIGURE 4 Exotic carnivores, such as this snow leopard (*Uncia uncia*), in general do not display signs of vitamin E deficiency unless they are fed diets containing a high degree of polyunsaturated fatty acids (i.e., fish) without adequate vitamin E supplementation. Dogs and cats appear to provide suitable nutritional models for zoo canids and felids; plasma vitamin E-cholesterol (μg/mg) ratios in these species average about 5.0. (© New York Zoological Society photo.)

Vitamin E deficiency does not appear to be a widespread observation among zoo canids and felids. Diffuse muscular degeneration and fat necrosis have been reported in seals and sea lions fed fish-based diets not supplemented with vitamin E (112,113). Seals fed diets low in vitamin E developed abnormal molt patterns and hyponatremia with subsequent incoordination leading to drowning (114).

Blood Vitamin E and Lipid Concentrations

Healthy dogs fed commercial rations (20 mg vitamin E per kg diet) had circulating α-tocopherol values from 2.7 to 12.4 μg/ml with a mean $\simeq$ 6.0 μg/ml (35,108). Adult dogs fed experimental diets supplemented with vitamin E (100 mg/kg) for over a year were found to have average levels $\simeq$ 25 μg/ml and those fed deficient diets (0.05 mg/kg) averaged ~1.5 μg/ml (111).

Values for healthy cats range from 3 to 11 μg/ml (35). Mink fed *all-rac-α-tocopheryl* acetate (7–150 mg/kg diet) had plasma levels ranging from about 11 to 22 μg/ml and were shown to discriminate between tocopherols, preferentially retaining α-tocopherol (115).

Plasma α-tocopherol values in three harp seals averaged 31 μg/ml (114); only α-tocopherol was detected in the plasma. A more detailed survey of gray seals indicated significant sex and age effects upon circulating plasma concentrations, with juveniles displaying the lowest (7.2 μg/ml) and suckling pups or adult males the highest (29.0 μg/ml) levels of α-tocopherol (116). A comparison of plasma α-tocopherol concentrations in free-ranging ($n = 14$) or aquarium beluga whales ($n = 10$) showed no significant difference between the two groups, averaging 17.9 μg/ml (117). Plasma α-tocopherol and cholesterol values for various zoo carnivores are summarized in Table 5. Carnivore α-tocopherol-cholesterol ratios (2.6–6.0 μg/mg) are similar to ratios observed in studies of individual species [i.e., beluga whales, 1.4–2.3 μg/ml (117); gray seals, 2.4–5.4 μg/mg (116)]. The α-tocopherol-cholesterol ratio of >10 μg/mg recorded for Ailuropidae represents the unique dietary habits of pandas, which are bamboo specialists. The high fiber level found in this diet may serve to lower plasma cholesterol; at the same time, bamboo is a relatively rich source of α-tocopherol (200 mg/kg dry matter) (118).

Tissue Concentrations of Vitamin E

Although plasma levels have been shown to be related to the log dietary vitamin E concentration in carnivores as well as other species (115), liver and adipose tissue concentrations provide a more comprehensive estimate of physiological status (116). Mink liver α-tocopherol content ranged from 20 to 70 μg/g wet weight; adipose tissue was threefold higher (50–150 μg/g wet weight) (115). Adipose tissue vitamin E concentrations declined from an average of 300 to <15 μg/g wet weight in dogs fed vitamin E-deficient diets for almost a year (111). Dogs remained clinically healthy, but erythrocyte fragility increased significantly.

By comparison, α-tocopherol in skeletal muscle, heart, and liver samples from three postpartum harp seals ranged from 12 to 16 μg/g wet weight, but blubber contained 90 μg/g (120). A study of α-tocopherol distribution in gray seal tissues documented liver concentrations of 21–41 μg/g wet weight in juveniles and adults but almost 200 μg/g in suckling pups.

TABLE 5 Mean Plasma α-Tocopherol (E) and Cholesterol Concentrations in Zoo Carnivores Compared with Domestic Species

Family	n	Species	α-Tocopherol (μg/ml)	Cholesterol (mg/ml)	E/cholesterol (μg/mg)
Zoo species					
Ailuropidae (pandas)	2	2	13.1	1.21	10.8
Canidae (wolves, foxes)	70	6	5.8	2.22	2.6
Felidae (lions, tigers)	187	7	8.1	1.72	4.7
Mustelidae (otters, ferrets)	21	4	11.2	2.21	5.1
Pinnipeds and Cetaceans (seals and whales)	121	11	12.3	1.16	10.6
Ursidae (bears)	44	4	16.7	2.77	6.0
Domestic species					
Dog	45		7.1	1.35	5.3
Cat	69		7.8	1.67	4.7
Mink	30		14.8		

Source: Data summarized from References 29, 35, 97, 108, 115–117, and 119 and from E. S. Dierenfeld, unpublished.

In that same study, adipose tissue concentrations ranged from 3.3 (pups) to 33.8 (nursing females) to 65.4 μg/g wet weight (adult males), as vitamin E was apparently mobilized from blubber of females to nourish pups (116).

As in other species, plasma α-tocopherol levels do not provide a particularly complete estimate of vitamin E status in carnivores. It is probable that adipose stores can be mobilized to meet vitamin E requirements to a greater extent in this group of animals compared to others that have been studied (i.e., rats, guinea pigs, and humans). In this respect, swine appear similar to other monogastrics.

Dietary Vitamin E Requirements

NRC dietary vitamin E recommendations for dogs range from 22 to 50 mg/kg (121), but cats require slightly high concentrations, 30–80 mg/kg diet (122). Mink dietary requirements of about 27 mg/kg (123) have been shown to protect against vitamin E deficiency, but practical ration concentrations of 60–80 mg/kg have been suggested (115) in diets with high polyunsaturated fatty acid contents.

Although requirements of vitamin E have not been established for exotic carnivores, NRC recommendations appear adequate for zoo canids and felids. Levels of 100 IU/kg fish (equivalent to approximately 200–300 IU/kg dry matter) have been suggested as necessary to prevent deficiencies in piscivorous species (114,124).

DOMESTIC POULTRY AS MODELS FOR ZOO AVIFAUNA

Livestock poultry species upon which vitamin E studies have been conducted include representatives of Galliformes (chickens, turkeys, quail, and pheasants) and Anseriformes (ducks and geese), only 2 of 26 existing orders of birds. Although dietary habits vary widely within the class Aves, physiological mechanisms of digestion and metabolism are similar among most birds. The use of domestic poultry as a suitable model is thus considered for examining vitamin E status in exotic species (Fig. 5).

Clinical Signs of Vitamin E Deficiency

Decreased fertility and hatchability among both domestic poultry (125,126) and exotic avians (1,127) have been linked to vitamin E deficiency. Additionally, specific disease states, including muscular dystrophy, encephalomalacia, exudative diathesis, and lipid peroxidation in the adipose tissue, have been described.

Muscular Dystrophy

Vitamin E-responsive muscle degeneration has been reported in various avian species. Affected tissues include myocardium and the gizzard in turkey poults (128). Additionally, skeletal and intestinal muscles are involved in chicks (129) and ducks (130–133).

Similar cardiac and skeletal lesions were observed in zoo birds by Liu et al. (61). These authors also reported severe degeneration of the hatching muscle (musculus complexus) in embryos that failed to pip. Clinical diagnosis of nutritional myopathy in exotic species was accompanied by elevated serum creatine phosphokinase (>40 units) (134,135) and/or abnormal electrocardiograms when cardiac muscles were involved (136). Lameness associated with skeletal muscular dystrophy was documented in ostrich chicks (137,138) and adult flamingos (135). In each case, clinical deficiency signs were alleviated following treatment with vitamin E.

Encephalomalacia

Necrosis of the cerebellum and spinal cord accompanied by neurological signs, including tremor, incoordination, opisthotonos, twisted neck, and recumbency, occurs in young

FIGURE 5 Chick survival in the Congo peacock (*Afropavo congensis*) shown and other avian species increased dramatically following dietary supplementation with vitamin E (200 IU/kg dry matter). Normal plasma α-tocopherol levels range widely among birds (~5 to >35 µg/ml); cholesterol values range from about 1.25 to 2.81 mg/ml, with higher levels in carnivorous compared to herbivorous species. (© New York Zoological Society photo.)

chicks and poults fed vitamin E-deficient diets (128,139). The condition is generally considered a primary effect of polyunsaturated lipid peroxidation; dietary fats, particularly linoleic or arachidonic acid, appear to exacerbate the disease. It is readily prevented by synthetic antioxidants or high dietary concentrations of vitamin E. Nutritional encephalomalacia has been reported in game birds (140,141) as well as zoo species (1).

Exudative Diathesis

Poultry diets deficient in both vitamin E and selenium produce exudative diathesis. Selenium is completely protective against this condition in turkey poults and chicks (128,142) and may be considered to play a primary biochemical role in the etiology of this condition. Exudative diathesis has been observed infrequently in ducklings (131).

Liu et al. (61) reported pale, edematous tissues in cardiac and skeletal muscles of zoo birds, but vitamin E, rather than selenium, was considered the primary underlying deficiency. Exudative diathesis does not appear to be a major finding in case reports of vitamin E deficiency with exotic avians.

Adipose Tissue Involvement

Lipid peroxides have been shown to increase in adipose tissue of chicks fed vitamin E-deficient diets containing at least 10% unsaturated fat for about 4 weeks. Diets containing even higher levels of unsaturated fats (20% cod liver oil) resulted in extensive deposition of pigmentation in adipose tissues. The deposition of lipofuscin was completely prevented by addition of α-tocopherol to the diets. Experimental, rather than practical diets are generally required to produce these disease conditions in domestic poultry.

Numerous cases of vitamin E deficiency have been reported among fish- and meat-eating bird species (1). In addition to cardiac and skeletal muscle degeneration, significant pathological lesions of adipose tissue, including accumulation of ceroid pigment, necrosis, and steatitis, have been found. Diets containing high levels of polyunsaturated fats, supplemental cod-liver oil, and/or low levels of vitamin E have been implicated in these examples. Because the levels and types of fat in the diet influence absorption of dietary vitamin E, as well as alter the potential for oxidation of this nutrient (143), piscivorous birds may have higher dietary requirements for vitamin E than other birds (144). This hypothesis warrants further investigation.

Blood and Tissue Vitamin E Levels

Plasma Vitamin E and Lipid Concentrations

Turkeys raised for slaughter had plasma α-tocopherol levels of 2.4 ± 1.0 μg/ml (37). In 6-week-old turkey poults fed vitamin E-deficient diets containing 2–3 IU/kg vitamin E, plasma α-tocopherol levels averaged 0.23 ± 0.03 μg/ml; birds fed diets containing no detectable vitamin E had significantly ($p < 0.005$) lower plasma concentrations (0.10 ± 0.04) and displayed encephalomalacia (139).

Ducks fed vitamin E at levels of 10 IU/kg diet had plasma concentrations of 6–7 μg/ml; those fed 22 IU/kg had levels of 13–19 μg/ml (131). Plasma tocopherol concentrations in chickens fed diets supplemented with vitamin E at levels of 0–500 IU/kg diet ranged from about 4 μg/ml (63,145,146) to >50 μg/ml (147,148), but plasma levels reached a plateau at ~ 20–30 μg/ml over a wide range of dietary vitamin E levels.

Plasma α-tocopherol levels in zoo species with no suspected vitamin E deficiency are shown in Table 6. The only values from free-ranging birds eating natural diets are from peregrine falcons and bald eagles (Falconiformes), which averaged ~ 24 μg/ml (149). By comparison, captive birds from several species with clinical or pathological signs of vitamin E deficiency had circulating α-tocopherol levels <6.0 μg/ml (1).

Total plasma cholesterol ranged from 1.8 to 3.7 μg/ml in chicks fed 100 IU vitamin E per kg diet (147) and was 1.33 μg/ml in another study (37). Mature hens ($n = 7$) had cholesterol values of 0.85 mg/ml in this latter study. In another experiment, in chicks fed diets containing 2 or 10% linoleic acid, total plasma lipids averaged 4.6 and 8.2 mg/ml, respectively (151).

Although total lipids have not been quantified in most zoo species, cholesterol values are available. Plasma α-tocopherol and cholesterol levels are higher in carnivorous birds than in herbivores, probably because of the type and amount of fat consumed. Dietary fat can range from $<3\%$ of dry matter in plant-eating species to $>60\%$ for piscivorous or carnivorous

TABLE 6   Mean Serum or Plasma α-Tocopherol and Cholesterol Levels in Domestic and Zoo Avifauna

Order	n	Species	α-Tocopherol (μg/ml)	Cholesterol (mg/ml)	E/cholesterol (μg/mg)
Zoo species					
Herbivorous birds					
Anseriformes (ducks, geese)	18	10	5.3	1.96	2.7
Columbiformes (pigeons, doves)	17	4	9.2	1.85	5.0
Cuculiformes (cuckoos)	6	2	8.8		
Galliformes (chickens, pheasants)	94	10	6.7	1.72	3.9
Gruiformes (cranes)	367	13	8.1	1.50	5.4
Passeriformes (songbirds)	5	4	5.5	1.75	3.1
Psittaciformes (parrots)	92	27	12.7	1.75	7.5
Rheiformes (rheas)	7	2	8.9	0.56	16.0
Struthioniformes (ostriches)	2	1	6.0	0.22	27.3
Carnivorous birds					
Charadriiformes (gulls, terns)	84	5	25.7	2.81	9.1
Ciconiiformes (herons, flamingos)	45	6	12.4	1.58	7.8
Coraciiformes (kingfishers)	2	2	76.7		
Falconiformes (birds of prey)	247	18	29.8	2.39	12.5
Palicaniformes (pelicans)	12	3	7.3	2.15	3.4
Sphenisciformes (penguins)	112	6	37.8	2.44	15.5
Strigiformes (owls)	5	2	23.4	2.00	11.7
Domestic species					
Chickens	15		12.5	1.63	7.7
Quail	11		1.7	1.26	1.3
Turkey	12		2.4	1.30	1.8

Source: Summarized from E. S. Dierenfeld (unpublished) and from References 1, 37, 72, 144, 147, 149, and 150.

birds. Plasma α-tocopherol-cholesterol ratios generally range from <5.0 μg/mg in herbivorous birds to >5.0–15.0 in carnivorous species (Table 6; pelicans fed unsupplemented fish comprise the only exception). Gulland et al. also reported α-tocopherol-total cholesterol ratios (μg/mg) averaging 10.4 in captive penguins (150).

Tissue Concentrations

Dietary vitamin E levels have been shown to directly influence tissue levels of poultry. In chicks fed diets supplemented with vitamin E (0–500 IU/kg diet), liver levels increased linearly from ~10 to >100 μg/g wet weight (151). Liver contained approximately 50 μg/g wet weight when chicks were fed "practical" diets (147). Heart concentrations ranged from 10 to 50 μg/g wet weight when supplemental vitamin E varied from 0 to 500 IU/kg diet (151). Muscle α-tocopherol concentrations (6–8 μg/g wet weight) were lower than organ levels in chicks fed vitamin E at up to 180 IU/kg diet (147,148). Tissue concentrations of vitamin E are not available for exotic birds, but whole quail fed diets containing approximately 200 IU/kg vitamin E averaged 31.3 ± 3.0 μg/g dry matter (Dierenfeld, unpublished).

Dietary Vitamin E Requirements

Although vitamin E requirements for domestic poultry species have been determined to be 10–25 IU/kg diet (152), levels cannot be considered independent of dietary lipids (128,143), selenium (151), or vitamin A (146). Exotic birds may require higher levels of vitamin E intake. Peregrine falcons showed signs of vitamin E deficiency when they were fed whole quail that had not been supplemented with vitamin E. These symptoms were corrected when the diets fed to the quail were adjusted to contain supplemental vitamin E at 220 IU/kg (149). Diets for exotic birds often require vitamin E levels up to 10-fold higher than NRC recommendations to protect against various deficiency diseases and/or commonly observed nutrient imbalances (1,61,144).

Vitamin E Toxicity

Excess dietary vitamin E has been implicated in a number of disease states in both domestic and exotic avifauna. Extremely high dietary levels of vitamin E (16,000–64,000 IU/kg) fed to chicks elevated plasma values to >200 μg/ml and liver concentrations to >300 μg/g wet weight (145) and resulted in decreased pigmentation and waxy feather appearance. In other studies, abnormal bone mineralization and clotting abnormalities have been reported with dietary levels > 2000 IU/kg (153,154). These accompanying clinical signs, suggestive of metabolic interactions with other fat-soluble vitamins (D and K), have resulted in recommended maximum safe dietary vitamin E levels of 1000 IU/kg dry matter for domestic poultry (155).

It has been suggested that piscivorous birds may have a higher dietary requirement for vitamin E than other species (144). Plasma α-tocopherol concentrations in healthy penguins fed approximately 1800 IU vitamin E per kg fish (dry basis) rangèd from 12.2 to 59.4 μg/ml (150) and were similar to those recorded in wild birds. Captive pelicans, however, consuming fish supplemented with an average of 5000 IU/kg dry matter, displayed vitamin K deficiency (156) and had liver α-tocopherol concentrations of 250–500 μg/g.

Another fish-eating specialist, the white-tailed sea eagle (*Haliaeetus albicilla*), displayed excessive serum α-tocopherol levels (671 μg/ml) when fed a meat- (not fish-)based diet containing 1364 IU/kg vitamin E. Poor reproduction and pathological lesions (pipping muscle edema, myodegeneration, and poor mineralization) in chicks were suggestive of a vitamin E toxicity (144). When the diet fed to the four adult birds was altered to contain lower amounts of vitamin E (48 IU/kg), serum levels dropped to 45.8 μg/ml (144) and all clinical signs disappeared.

OVERVIEW

Comparisons of vitamin E deficiency symptoms, plasma and tissue levels of α-tocopherol, and vitamin E intakes have been made between domestic and exotic animals. In many cases, detailed information from exotics is lacking entirely, and domestic examples are limited. It is clear that the diverse deficiency symptoms described for laboratory and experimental animals apply to exotic species. Variations that exist are likely dependent on physiological differences in fat metabolism and/or distribution. All groups display muscular dystrophy; unique pathologies include nerve degeneration (perissodactyls), cardiac microangiopathy (swine and elephants), and steatitis (carnivores, both mammals and birds).

Mean plasma levels of α-tocopherol in healthy animals vary widely among species [0.2 (rhinoceros) to >30.0 (penguin) μg/ml] and are highly dependent upon blood lipid levels. The distribution of various plasma lipoprotein fractions was previously shown to differ considerably among mammals; thus, plasma cholesterol is suggested as a more practical component than total lipids for comparative assessment of vitamin E status from blood samples. All species (except zoo elephants and rhinoceros) displayed plasma α-tocopherol-cholesterol ratios (μg/ml) > 1.0; ratios for herbivorous and omnivorous species ranged from about 2 to 5, and those of carnivorous species were generally between 5 and 10. Tissue levels of α-tocopherol would provide a more comprehensive evaluation of vitamin E status but are lacking

TABLE 7 Tissue α-Tocopherol Concentrations (μg/g Wet Weight) Reported for Various Domestic and Livestock Species Fed Vitamin E-Adequate Diets

Species	Tissue tocopherol (μg/g wet weight)		
	Liver	Skeletal muscle	Adipose tissue
Horse	5	2–5	25
Cattle	4–22	6	
Sheep	6–14	4	3
Swine	4–9	4	10–16
Dog			300
Mink	20–70	50–150	
Chickens	10–50	6–8	28–57
Turkeys	6	1–2	12–23

Source: Summarized from References 15, 20, 66, 68, 73, 75, 94, 111, 115, 147, 148, and 157.

for most species. A few examples of tissue levels of α-tocopherol for some common domestic species are shown in Table 7. It is clear that many exotic herbivores require higher dietary vitamin E concentrations than those recommended for domestic counterparts, but vitamin E deficiency has not been a major problem for carnivores.

Apart from feed concentration of vitamin E, fat level or type appears to be the most critical dietary variable for all species discussed and requires further study. Inadequate dietary fat may limit vitamin E absorption, thereby increasing requirements, in perissodactyls, artiodactyls, and elephants, yet high dietary levels of polyunsaturated fatty acids, while aiding absorption, also increase tissue oxidation and thus elevate dietary vitamin E requirements. High intakes of polyunsaturated fatty acids in piscivorous birds increase vitamin E requirements, but if these birds are fed nonfish diets (presumably more saturated fats) supplemented at the same level, vitamin E toxicity signs may result.

Finally, the impact of genetic diversity upon vitamin E metabolism needs to be examined. Specific horse, cattle, and swine breeds have been reported to become vitamin E deficient even though consuming apparently adequate diets, suggesting a genetic abnormality. Limited data suggest that the liver tocopherol binding protein, which has been shown to be essential for maintenance of plasma α-tocopherol levels in humans, may be lacking or defective in these animals. Such information may greatly enhance our understanding of vitamin E requirements and metabolism in exotic species with even more limited genetic variability.

REFERENCES

1. Dierenfeld ES. Vitamin E deficiency in zoo reptiles, birds and ungulates. J Zoo Wildl Med 1989; 20:3-11.
2. Stevens C. Comparative physiology of the vertebrate digestive system. Cambridge, MA: Cambridge University Press, 1988.
3. Mayhew J, Brown C, Trapp A. Equine degenerative myeloencephalopathy. Proc Fourth Annual Vet Med Forum ACVIM. Washington, DC, 1986.
4. Mayhew IG, Brown CM, Stowe HD, Trapp AL, Derksen FJ, Clement SF. Equine degenerative myeloencephalopathy: a familial disease of Equidae caused by vitamin E deficiency? J Vet Int Med 1987; 1:45-50.
5. Beech J, Haskins M. Genetic studies of neuraxonal dystrophy in the Morgan horse. J Am Vet Med Assoc 1987; 190:686.

6. Machlin LJ, Filipski R, Nelson J, Horn LR, Brin M. Effect of a prolonged vitamin E deficiency in the rat. J Nutr 1977; 107:1200-8.

7. Sokol RJ. Vitamin E deficiency and neurologic disease. Annu Rev Nutr 1988; 8:351-73.

8. Dill SG, Kallfelz FA, deLahunta A, Waldron CH. Serum vitamin E and blood glutathione peroxidase values of horses with degenerative myeloencephalopathy. Am J Vet Res 1989; 50:166-8.

9. Montali RJ, Bush M, Sauer RM, Gray CW, Xanten WA Jr. Spinal ataxia in zebras: comparison with the wobbler syndrome of horses. Vet Pathol 1974; 11:68-78.

10. Liu S, Dolensek EP, Adams CR, Tappe JP. Myelopathy and vitamin E deficiency in six Mongolian wild horses. J Am Vet Med Assoc 1983; 183:1266-8.

11. Traber MG, Sokol RJ, Burton GW, et al. Impaired ability of patients with familial isolated vitamin E deficiency to incorporate α-tocopherol into lipoproteins secreted by the liver. J Clin Invest 1990; 85:397-407.

12. Catignani GL, Bieri JG. Rat liver α-tocopherol binding protein. Biochim Biophys Acta 1977; 497:349-57.

13. Kaplowitz N, Yoshida H, Kuhlenkamp J, Slitsky B, Ren I, Stolz A. Tocopherol-binding proteins of hepatic cytosol. Ann N Y Acad Sci 1989; 570:85-94.

14. Traber MG, Burton GW, Ingold KU, Kayden HJ. *RRR-* and *SRR-*α-tocopherols are secreted without discrimination in human chylomicrons, but *RRR-*α-tocopherol is preferentially secreted in very low density lipoproteins. J Lipid Res 1990; 31:675-85.

15. Ronèus BO, Hakkarainen RVJ, Lindholm CA, Työppönen JT. Vitamin E requirements of adult standardbred horses evaluated by tissue depletion.and repletion. Equine Vet J 1986; 18:50-8.

16. Wilson TM, Morrison HA, Palmer NG, Finley GG, Drummel AA. Myodegeneration and suspected selenium/vitamin E deficiency in horses. J Am Vet Med Assoc 1976; 169:213-7.

17. Owen RR, Moore JN, Hopkins JB, Arthur D. Dystrophic myodegeneration in adult horses. J Am Vet Med Assoc 1977; 171:343-8.

18. Dewes HF. A possible vitamin E-responsive condition in adult horses (letter). N Z Vet J 1981; 29:83-4.

19. Aikawa KM, Quintanilha AT, deLumen BO, Brooks GA, Packer L. Exercise endurance-training alters vitamin E tissue levels and red-blood-cell hemolysis in rodents. Biosci Rep 1984; 4:253-7.

20. Petersson KH, Hintz HF, Schryver HF, Combs GF Jr. The effect of vitamin E on membrane integrity during submaximal exercise. Third Int Conf Equine Exercise Physiology. Uppsala, Sweden, 1990.

21. Yamini B, Schillhorn van Veen TW. Schistosomiasis and nutritional myopathy in a Brazilian tapir (*Tapirus terrestris*). J Wildl Dis 1988; 24:703-7.

22. Papas AM, Cambre RC, Citino SB. Vitamin E: considerations in practical animal feeding and case studies with elephants and rhinoceros. 8th Annual Dr. Scholl Conference on the Nutrition of Captive Wild Animals. Chicago, IL, 1989.

23. Dierenfeld ES, Dolensek EP. Circulating levels of vitamin E in captive Asian elephants (*Elephas maximus*). Zoo Biol 1988; 7:165-72.

24. Kayden HJ, Silber R. The role of vitamin E deficiency in the abnormal autohemolysis of acanthocytosis. Trans Assoc Am Phys 1965; 78:334-41.

25. Ausman LM, Hayes KC. Vitamin E deficiency anemia in old and new world monkeys. Am J Clin Nutr 1974; 27:1141-51.

26. Bieri JG, Poukka RKH. In vitro hemolysis as related to rat erythrocyte content of α-tocopherol and polyunsaturated fatty acids. J Nutr 1970; 100:557.

27. Stowe HD. Alpha-tocopherol requirements for equine erythrocyte stability. Am J Clin Nutr 1968; 21:135-42.

28. Dierenfeld ES, duToit R, Miller RE. Vitamin E in captive and wild black rhinoceros (*Diceros bicornis*). J Wildl Dis 1988; 24:547-50.

29. Ghebremeskel K, Williams G. Plasma retinol and α-tocopherol levels in captive wild animals. Comp Biochem Physiol 1988; 89B:279-83.

30. Miller RE, Boever WJ. Fatal hemolytic anemia in the black rhinoceros: case report and a survey. J Am Vet Med Assoc 1982; 181:1228-31.

31. Fairbanks V, Miller R. A beta chain hemoglobin polymorphism and hemoglobin stability in black rhinoceros (*Diceros bicornis*). Am J Vet Res 1990; 51:803-7.

32. Baker H, Handelman GJ, Short S, et al. Comparison of plasma α- and γ-tocopherol levels follow-
    ing chronic oral administration of either *all-rac*-α-tocopheryl acetate or *RRR*-α-tocopheryl acetate
    in normal adult male subjects. Am J Clin Nutr 1986; 43:382-7.

33. Butler P, Blackmore DJ. Vitamin E values in the plasma of stabled thoroughbred horses in train-
    ing. Vet Rec 1983; 112:60.

34. Matsumoto T, Ichyo S, Tatsuo K. Changes of serum tocopherol level in normal horses. Nippon
    Juishikai Zasshi 1985; 38:239-42.

35. Baker H, Schor SM, Murphy BD, DeAngelis B, Feingold S, Frank O. Blood vitamin and choline
    concentrations in healthy domestic cats, dogs, and horses. Am J Vet Res 1986; 47:1468-71.

36. Dierenfeld E. Vitamin E levels measured in rhino browse plants. Rhino Conservation Newslett
    1990; 1:1-2.

37. Schweigert F, Uehlein-Harrell S, Hegel G, Wiesner H. Vitamin A (retinol and retinyl esters) α-
    tocopherol, and lipid levels in plasma of captive wild mammals and birds. J Vet Med A 1991; 38:35-42.

38. Leat WMF, Northrop CA, Buttress N, Jones DM. Plasma lipids and lipoproteins of some mem-
    bers of the order Perissodactyla. Comp Biochem Physiol 1979; 63B:275-81.

39. Horwitt MK, Harvey CC, Dahm DH, Searcy MT. Relationship between tocopherol and serum
    lipid levels for determination of nutritional adequacy. Ann N Y Acad Sci 1972; 203:223-6.

40. Craig A, Blythe L, Lassen E, Rowe K, Barrington R, Slizeski M. Variations of serum vitamin E,
    cholesterol and total serum lipid concentrations in horses during a 72-hour period. Am J Vet Res
    1989; 50:1527-31.

41. Jeffcott LG, Field JR. Current concepts of hyperlipaemia in horses and ponies. Vet Rec 1985; 116:
    461-6.

42. Robie SM, Janson CH, Smith SC. Equine serum lipids: serum lipids and glucose in Morgan and
    thoroughbred horses and Shetland ponies. Am J Vet Res 1975; 36:1705-8.

43. Chapman M. Comparative analysis of mammalian plasma lipoproteins. In: Segrest J, Albers J, eds.
    Plasma lipoproteins. Part A. Preparation, structure and molecular biology. Orlando, FL: Academic
    Press, 1986; 70-143.

44. Behrens WA, Thompson JN, Madere R. Distribution of alpha tocopherol in human plasma lipo-
    proteins. Am J Clin Nutr 1982; 35:691-6.

45. Behrens WA, Madere R. Transport of α- and γ-tocopherol in human plasma lipoproteins. Nutr
    Res 1985; 5:167-74.

46. Ogihara T, Miki M, Kitagawa M, Mino M. Distribution of tocopherol among human plasma lipo-
    proteins. Clin Chim Acta 1988; 174:299-306.

47. McCormick EC, Cornwell DG, Brown JB. Studies on the distribution of tocopherol in human
    serum lipoproteins. J Lipid Res 1960; 1:221-8.

48. National Research Council. Nutrient requirements of horses. Washington, DC: National Academy
    Press. 1989.

49. Maylin GA, Rubin DS, Lein DH. Selenium and vitamin E in horses. Cornell Vet 1980; 70:272-89.

50. Mäenpää P, Koskinen T, Koskinen E. Serum profiles of vitamin A, E and D in mares and foals
    during different seasons. J Anim Sci 1988; 66:1418-23.

51. Papas A, Cambre R, Citino S, Baer D, Wooden G. Species differences in the utilization of various
    forms of vitamin E. In: Cambre R, ed. Proc Am Assoc Zoo Vets, South Padre Island, TX. Omni
    Press, Madison, WI, 1990; pp 186-190.

52. McCullagh K. Are African elephants deficient in essential fatty acids? Nature 1973; 242:267-8.

53. Dierenfeld E, Wareru F, duToit R, Brett R. Alpha-tocopherol levels in plants eaten by black rhi-
    noceros. FASEB J 1990; 4:abstract.

54. Alderson NE, Mitchell GE Jr, Little CO, Warner RE, Tucker RE. Preintestinal disappearance of
    vitamin E in ruminants. J Nutr 1971; 101:655-60.

55. Arthur JR. Effects of selenium and vitamin E status on plasma creatine kinase activity in calves.
    J Nutr 1988; 118:747-55.

56. Blaxter KL. Vitamin E in health and disease of cattle and sheep. Vitam Horm 1962; 20:633-43.

57. Young S, Keeler RF. Nutritional muscular dystrophy in lambs—the effect of activity on the sym-
    metrical distribution of lesions. Am J Vet Res 1962; 23:966-71.

58. Allen WM. New developments in muscle pathology:nutritional myopathies including "muscular
    dystrophy" or "white muscle disease." Vet Sci Commun 1977; 1:243-50.

59. Rüedi D, Heldstab A, Völlm J, Keller P. White muscle disease in the lesser kudu at the Basle Zoological Garden: case histories of three animals, diagnostic possibilities, and prophylactic and therapeutic measures. In: Montali RJ and G. Migaki, eds. The comparative pathology of zoo animals. Smithsonian Institute Press, Washington, DC, 1980; pp 21-26.

60. Liu SK, Dolensek EP, Herron AJ, Stover J, Doherty JG. Myopathy in the nyala. J Am Vet Med Assoc 1982; 181:1232-6.

61. Liu SK, Dolensek EP, Tappe JP. Cardiomyopathy and vitamin E deficiency in zoo animals and birds. Heart Vessels Suppl 1985; 1:228-93.

62. Stuart RL. Factors affecting the vitamin E status of beef cattle. In: The role of vitamins on animal performance and immune response. Roche Tech Symp. Daytona Beach, FL: 1987; pp 67-80.

63. Caravaggi C. Vitamin E concentrations in the serum of various experimental animals. Comp Biochem Physiol 1969; 30:585-8.

64. Lynch GP. Changes of tocopherols in blood serum of cows fed hay or silage. J Dairy Sci 1983; 66:1461-5.

65. Adams CR. Feedlot cattle need supplemental vitamin E. Feedstuffs 1982; 54:24.

66. Hidiroglou N, LaFlamme LF, McDowell LR. Blood plasma and tissue concentrations of vitamin E in beef cattle as influenced by supplementation of various tocopherol compounds. J Anim Sci 1988; 66:3227-34.

67. Storer GB. Fluorometric determination of tocopherol in sheep plasma. Biochem Med 1974; 11: 71-80.

68. Hidiroglou N, McDowell LR, Pastrana R. Bioavailability of various vitamin E compounds in sheep. Int J Vitam Nutr Res 1988; 58:189-97.

69. Brady PS, Brady LJ, Whetter PA, Ullrey DE, Fay LD. The effect of dietary selenium and vitamin E on biochemical parameters and survival of young among white-tailed deer (*Odocoileus virginianus*). J Nutr 1978; 108:1439-48.

70. Howard KA, Moore SA, Radecki SV, Shelle JE, Ullrey DE, Schmitt SM. Relative bioavailability of various sources of vitamin E for white-tailed deer (*Odocoileus virginianus*), swine, and horses. In: Cambre R, ed. Proc Am Assoc Zoo Vet. South Padre Island, TX: Omni Press, Madison, WI, 1990; pp 191-195.

71. Dierenfeld ES, Jessup DA. Variation in serum α-tocopherol, retinol, cholesterol, and selenium of free-ranging mule deer (*Odocoileus hemionus*). J Zoo Wildl Med 1990; 21:425-32.

72. ISIS. Normal physiological data. Apple Valley, MN 55124-8199: International Species Information System, 1989.

73. Van Saun RJ, Herdt TH, Stowe HD. Maternal and fetal vitamin E concentrations and selenium-vitamin E interrelationships in dairy cattle. J Nutr 1989; 119:1156-64.

74. Pehrson B, Hakkarainen J. Vitamin E status of healthy Swedish cattle. Acta Vet Scand 1986; 27: 351-60.

75. Rammell CG, Cunliffe B. Vitamin E status of cattle and sheep. 2. Survey of liver from clinically normal cattle and sheep for α-tocopherol. N Z Vet J 1983; 31:203-4.

76. National Research Council. Nutrient requirements of beef cattle. Washington, DC: National Academy Press, 1984.

77. National Research Council. Nutrient requirements of sheep. Washington, DC: National Academy Press, 1985.

78. Miller E, Ullrey D. The pig as a model for human nutrition. Annu Rev Nutr 1987; 7:361-82.

79. Black D, Davidson N. Intestinal apolipoprotein synthesis and secretion in the suckling pig. J Lipid Res 1989; 30:207-18.

80. Lannek N, Lindberg P, Nilsson G, Nordstrom G, Orstadius K. Production of vitamin E deficiency and muscular dystrophy in pigs. Res Vet Sci 1961; 2:67-72.

81. Trapp AL, Keahey KK, Whitenack DL, Whitehair CK. Vitamin E-selenium deficiency in swine: differential diagnosis and nature of field problem. J Am Vet Med Assoc 1970; 157:289-300.

82. Rice DA, Kennedy S. Vitamin E, selenium, and polyunsaturated fatty acid concentrations and glutathione peroxidase activity in tissues from pigs with dietetic microangiopathy (mulberry heart disease). Am J Vet Res 1989; 50:2101-4.

83. Horwitt MK, Harvey CC, Duncan GD, Wilson WC. Effects of limited tocopherol intake in man with relationships to erythrocyte hemolysis and lipid oxidations. Am J Clin Nutr 1956; 4:408-19.

84. Hayes KC. Pathophysiology of vitamin E deficiency in monkeys. Am J Clin Nutr 1974; 27:1130-40.

85. Duthie GG, Arthur JR, Mills CF, Morrice PC, Nicol F. Anomalous tissue vitamin E distribution in stress susceptible pigs after dietary vitamin E supplementation and effects on pyruvate kinase and creatine kinase activities. Livest Prod Sci 1987; 17:169-78.

86. Junge RE, Miller RE. Suspected nutritional myopathy in Vietnamese potbellied pigs (*Sus scrofa*). J Zoo Wildl Med 1989; 20:478-81.

87. Liu SK, Dolensek EP, Tappe JP, Stover J, Adams CR. Cardiomyopathy associated with vitamin E deficiency in seven gelada baboons. J Am Vet Med Assoc 1984; 185:1347-50.

88. Baskin GB, Wolf RH, Worth CL, Soike K, Gibson SV, Bieri JG. Anemia, steatitis, and muscle necrosis in marmosets (*Sanguinus labiatus*). Lab Anim Sci 1983; 33:74-80.

89. Nelson JS, Fitch CD, Fischer VW, et al. Progressive neuropathologic lesions in vitamin E deficient rhesus monkeys. J Neuropathol Exp Neurol 1981; 40:166-86.

90. Sehgal PK, Bronson RT, Brady PS, McIntyre KW, Elliott MW. Therapeutic efficacy of vitamin E and selenium in treating hemolytic anemia of owl monkeys (*Aotus trivirgatus*). Lab Anim Sci 1980; 30:92-8.

91. Chavez E, Patton K. Response to injectable selenium and vitamin E on reproductive performance of sows receiving a standard commercial diet. Can J Anim Sci 1986; 66:1065-74.

92. Dove CR, Ewan RC. Effect of excess dietary copper, iron or zinc on the tocopherol and selenium status of growing pigs. J Anim Sci 1990; 68:2407-13.

93. Meyer WR, Mahan DC, Moxon AL. Value of dietary selenium and vitamin E for weanling swine as measured by performance and tissue selenium and glutathione peroxidase activities. J Anim Sci 1981; 52:302-11.

94. Hoppe PP. New research findings on the function of tocopherols in swine. Technical Bulletin #24. BASF Aktiengesellschaft, Ludwigshafen, Germany, 1989.

95. McGuire JT, Dierenfeld ES, Poppenga RH, Brazelton WE. Plasma α-tocopherol, retinol, cholesterol, and mineral concentrations in captive gorillas. J Med Primatol 1989; 18:155-61.

96. Flurer CI, Schweigert FJ. Species differences in a New World monkey family in blood values of vitamins A, E, and C. J Anim Physiol Anim Nutr 1990; 63:8-11.

97. Brush PJ, Anderson PH. Levels of plasma alpha-tocopherol (vitamin E) in zoo animals. Int Zoo Yearb 1986; 24/25:316-21.

98. Bieri JG, Everts RHP. Vitamin E nutrition in the rhesus monkey. Proc Soc Exp Biol Med 1972; 140:1162-5.

99. Farrell P. Deficiency states, pharmacological effects, and nutrient requirements. In: Machlin L, ed. Vitamin E: a comprehensive treatise. New York: Marcel Dekker, 1980; 520-620.

100. Bennett Clark S, Tercyak AM, Glander KE. Plasma lipoproteins of free-ranging howling monkeys (*Alouatta palliata*). Comp Biochem Physiol 1987; 88:729-35.

101. Illingworth D. Metabolism of lipoproteins in non-human primates. Studies on the origin of low density lipoprotein apoproteins in the plasma of squirrel monkeys. Biochim Biophys Acta 1975; 388:38-51.

102. Jensen M, Hakkarainen J, Lindholm A, Jönsson L. Vitamin E requirement of growing swine. J Anim Sci 1988; 66:3101-11.

103. Schaefer EJ, Woo R, Kibata M, Bjornson L, Schreibman PH. Mobilization of triglyceride but not cholesterol or tocopherol from human adipocytes during weight reduction. Am J Clin Nutr 1983; 37:749-54.

104. Traber MG, Sokol RJ, Ringel SP, Neville HE, Thellman CA, Kayden HJ. Lack of tocopherol in peripheral nerves of vitamin E-deficient patients with peripheral neuropathy. N Engl J Med 1987; 317:262-5.

105. National Research Council. Nutrient requirements of swine. Washington, DC: National Academy Press, 1988.

106. National Research Council. Nutrient requirements of non-human primates. Washington, DC: National Academy Press, 1978.

107. Fitch CD, Dinning JS. Vitamin E deficiency in the monkey. V. Estimated requirements and the influence of fat deficiency and antioxidants on the syndrome. J Nutr 1963; 79:69-78.

108. Hayes KC. Rousseau JE Jr, Hegsted DM. Plasma tocopherol concentrations and vitamin E deficiency in dogs. J Am Vet Med Assoc 1970; 157:64-71.

109. Scott PP. The nutritional requirements of cats. Basic guide to canine nutrition. White Plains, NY: General Foods, 1977; 79-92.

110. Helgebostad A, Ender F. Vitamin E and its function in the health and disease of fur-bearing animals. Acta Agr Scand Suppl 1973; 19:79-83.

111. Pillai SR, Steiss JE, Kayden HJ, Traber MG. Vitamin E deficiency in adult dogs: clinical, electrophysiologic and erythrocyte fragility studies. FASEB J 1991; A919.

112. Wilson TM. Diffuse muscular degeneration in captive harbor seals. J Am Vet Med Assoc 1972; 161:608-10.

113. Citino SB, Montali RJ, Bush M. Nutritional myopathy in a captive California sea lion. J Am Vet Med Assoc 1985; 187:1232-3.

114. Engelhardt FR, Geraci JR. Effects of experimental vitamin E deprivation in the harp seal, *Phoca groenlandica*. Can J Zool 1978; 56:2186-93.

115. Työppönen J, Hakkarainen J, Juokslahti T, Lindberg P. Vitamin E requirement of mink with special reference to tocopherol composition in plasma, liver, and adipose tissue. Am J Vet Res 1984; 45:1790-4.

116. Schweigert FJ, Stobo WR, Zucker H. Vitamin E and fatty acids in the grey seal (*Halichoerus grypus*). J Comp Biochem Physiol 1990; 159B:649-54.

117. Cook RA, Stoskopf MK, Dierenfeld ES. Circulating levels of vitamin E, cholesterol, and selected minerals in captive and wild beluga whales (*Delphinapterus leucas*). J Zoo Wildl Med 1990; 21:65-9.

118. Dierenfeld E, Waweru F, duToit R, Brett R. α-Tocopherol levels in plants consumed by black rhinoceros (*Diceros bicornis*): native browse compared with common zoo forages. In: Cambre R, ed. Proc Am Assoc Zoo Vet. Omni Press, Madison, WI, 1990; pp 196-197.

119. ISIS. Normal physiological data. Apple Valley, MN: International Species Inventory System, 1987.

120. Engelhardt FR, Geraci JR, Walker BL. Tocopherol distribution in the harp seal, *Pagophilus groenlandicus*. J Comp Biochem Physiol 1975; 52B:561-2.

121. National Research Council. Nutrient requirements of dogs. Washington, DC: National Academy Press, 1985.

122. National Research Council. Nutrient requirements of cats. Washington, DC: National Academy Press, 1986.

123. National Research Council. Nutrient requirements of mink and foxes. Washington, DC: National Academy Press, 1982.

124. Dierenfeld E, Katz N, Pearson J, Murru F, Asper E. Retinol and α-tocopherol concentrations in whole fish commonly fed in zoos and aquariums. Zoo Biol 1991; 10:119-25.

125. Scott ML, Nesheim MC, Young RJ. Nutrition of the chicken. Ithaca, NY: M. L. Scott & Assoc., 1976.

126. Kling LJ, Soares JH Jr. Vitamin E deficiency in the Japanese quail. Poult Sci 1980; 59:2352-4.

127. Sheppard CD, Dierenfeld ES. Inbreeding depression or nutritional deficiency? Proc NE Reg Meet, Amer Assoc Zool Parks & Aquariums. Oglebay Park, WV, 1987.

128. Green J, Bunyan J. Vitamin E and the biological antioxidant theory. Nutr Abstr Rev 1969; 39: 321-45.

129. Rigdon RH. Acute myopathy in embryos and newly hatched chicks. Arch Pathol 1967; 84:633-737.

130. Rigdon RH. Spontaneous muscular dystrophy in the white Peking duck. Am J Pathol 1961; 39: 27-40.

131. Dean WF, Combs GF Jr. Influence of dietary selenium on performance, tissue selenium content, and plasma concentrations of selenium-dependent glutathione peroxidase, vitamin E and ascorbic acid in ducklings. Poult Sci 1981; 60:2655-63.

132. Dhillon AS, Winterfield RW. Selenium-vitamin E deficiency in captive wild ducks. Avian Dis 1982; 27:527-30.

133. Xu GL, Diplock AT. Glutathione peroxidase (EC 1.11.1.9), glutathione-S-transferase (EC 2.5.1.13), superoxide dismutase (EC 1.15.1.1) and catalase (EC 1.11.1.6) activities in tissues of ducklings deprived of vitamin E and selenium. Br J Nutr 1983; 50:437-44.

134. Graham DL, Halliwell WH. Malnutrition in birds of prey. In: Fowler ME, ed. Zoo and wild animal medicine. Philadelphia: W. B. Saunders, 1986; 379-85.

135. Wallach JD, Boever WJ. Diseases of exotic animals. Philadelphia: W. B. Saunders, 1983.

136. Campbell TW. Hypovitaminosis E: its effect on birds. Proc First Int Conf Zool Avian Med. Turtle Bay, HI, 1987. Omni Press, Madison, WI.

137. Van Heerden J, Hayes SC, Williams MC. Suspected vitamin E-selenium deficiency in two ostriches. J S Afr Vet Assoc 1983; 54:53-4.

138. Vorster BJ. Nutritional muscular dystrophy in a clutch of ostrich chicks. J S Afr Vet Assoc 1984; 55:39-40.

139. Jortner BS, Meldrum JB, Domermuth DH, Potter LM. Encephalomalacia associated with hypovitaminosis E in turkey poults. Avian Dis 1985; 29:488-98.

140. Mutarov L. Encephalomalacia in partridges, quails, and pheasants raised in warrens (abstract). Vet Med Nauki 1979; 16:88-92.

141. Swarbrick O, Garden NJ, Listar SA. Nutritional encephalomalacia in red legged partridges. Vet Rec 1986; 118:727-8.

142. Noguchi R, Cantor AH, Scott ML. Mode of action of selenium and vitamin E in prevention of exudative diathesis in the chick. J Nutr 1973; 103:1502-11.

143. Horwitt MK. Data supporting supplementation of humans with vitamin E. J Nutr 1991; 121:424-9.

144. Calle PP, Dierenfeld ES, Roberts ME. Serum α-tocopherol in raptors fed vitamin E-supplemented diets. J Zoo Wildl Med 1989; 20:62-7.

145. Nockels CF, Menge DL, Kienholz EW. Effect of excessive dietary vitamin E on the chick. Poult Sci 1976; 55:649-52.

146. Frigg M, Broz J. Relationships between vitamin A and vitamin E in the chick. Int J Vitam Nutr Res 1984; 54:125-34.

147. Lü J, Combs GF Jr. Excess dietary zinc decreases tissue α-tocopherol in chicks. J Nutr 1988; 118:1349-59.

148. Sheehy PJ, Morrissey PA, Flynn A. Influence of dietary alpha-tocopherol on tocopherol concentrations in chick tissues. Br Poult Sci 1991; 32:391-7.

149. Dierenfeld ES, Sandfort CE, Satterfield WC. Influence of diet on plasma vitamin E in captive peregrine falcons. J Wildl Manage 1989; 53:160-4.

150. Gulland F, Ghebremeskel K, Williams G, Olney P. Plasma vitamins A and E, total lipid and cholesterol concentrations in captive jackass penguins (*Spheniscus demersus*). Vet Rec 1988; 123:666-7.

151. Whitacre ME, Combs GF Jr, Parker RS. Influence of dietary vitamin E on nutritional pancreatic atrophy in selenium-deficient chicks. J Nutr 1987; 117:460-7.

152. National Research Council. Nutrient requirements of poultry. Washington, DC: National Academy Press, 1984.

153. March BE, Wong E, Seier L, Sim L, Biely J. Hypervitaminosis E in the chick. J Nutr 1973; 103:371-7.

154. Murphy RP, Wright RE, Pudelkiewicz WJ. An apparent rachitogenic effect of excessive vitamin E intakes in the chick. Poult Sci 1981; 60:1873-8.

155. National Research Council. Vitamin E. Vitamin tolerance of animals. Washington, DC: National Academy Press, 1987; 23-30.

156. Nichols DK. Wolff MJ, Phillips LG Jr, Montali RJ. Coagulopathy in pink-backed pelicans (*Pelecanus rufescens*) associated with hypervitaminosis E. J Zoo Wildl Med 1989; 20:57-61.

157. Marusich W. Vitamin E as an in vivo lipid stabilizer and its effect on flavor and storage properties of milk and meat. In: Machlin L, ed. Vitamin E: a comprehensive treatise. New York: Marcel Dekker, 1980, pp 445-472.

28

α-Tocopherol and Protein Kinase C Regulation of Intracellular Signaling

Angelo Manfredo Azzi, Gianna Bartoli, Daniel Boscoboinik, Carmel Hensey, and Adam Szewczyk

University of Bern, Bern, Switzerland

POSSIBLE MECHANISMS OF TOCOPHEROL ACTION AT THE MOLECULAR LEVEL

Vitamin E, the most biologically active form of tocopherol (1), behaves at a molecular level as a scavenger of alkoxy radicals formed in membranes by the action of other radicals, such as OH˙ (2,3). However, the complete molecular picture of its mechanism of action is not fully understood. In fact, a number of studies have shown some discrepancies between the anti-oxidant potency of tocopherols and their biological effects, suggesting that they may possess additional properties, such as that of a membrane stabilizer (4), a regulator of membrane fluidity (5), an inhibitor of 5-lipoxygenase activity (6), or, as we recently showed, a modulator of protein kinase C (PKC) activity (7–9). Protein kinase C belongs to a family of isoenzymes that are extremely important in cellular signal transduction. Cellular growth, differentiation, and secretion are events that are at least in part under the control of protein kinase C (reviewed in Refs. 10 and 11). Moreover, these proteins represent the major intracellular receptor for the tumor-promoting phorbol esters (12,13). The salient role of protein kinase C in cellular proliferation prompted us to investigate the effect of tocopherol on the growth of several cell lines. In particular, aortic smooth muscle cells (VSMC) were studied, whose proliferation is coupled with such vascular diseases as hypertension and atherosclerosis (14–16). If VSMC were brought to quiescence by serum deprivation and then stimulated to initiate DNA synthesis, proliferation could be selectively inhibited by physiological concentrations of α-tocopherol equivalent to those required to inhibit purified protein kinase C. This chapter describes in some detail the results indicating that α-tocopherol inhibits isolated protein kinase C and controls the proliferation of smooth muscle cells. The two phenomena appear to be correlated, indicating that α-tocopherol modulates cell proliferation by acting on protein kinase C.

TOCOPHEROL MODULATES THE ACTIVITY OF ISOLATED PROTEIN KINASE C

The activity of isolated rat brain PKC is decreased by dl α-tocopherol (Fig. 1). At around 30 μM, 50% inhibition is observed. It is reported that the physiological concentration of α-tocopherol in serum is approximately 40 μM. The effect of α-tocopherol could in principle be

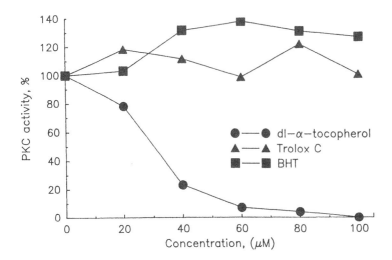

FIGURE 1 Inhibition of rat brain PKC by *d,l-α*-tocopherol, Trolox C, and butylated hydroxytoluene. Protein kinase C was purified from rat brain (17). Measurement of PKC activity was done in an assay mixture containing 20 mM Tris-HCl (pH 7.5 and 30°C), 5 mM magnesium acetate, 0.2 mg/ml of Histone III-S, 100 nM CaCl₂, 40 μg/ml of phosphatidylserine, 1 μM phorbol-12,13-dibutyrate, and 10 μM [γ-³²P]ATP (100,000 cpm/nmol). The assay mixture was added to *d,l-α*-tocopherol, Trolox, or butylated hydroxytoluene. Reactions were started by the addition of PKC and carried out for 30°C for 8 minutes. After termination of the reaction with 0.35 ml of cold 12% trichloroacetate (TCA), 2% sodium pyrophosphate, and 10 mg/ml of bovine serum albumin (BSA), the precipitated protein was quantitatively transferred to filters (Millipore HA, 0.45 μm) and counted. One unit of kinase activity is defined as the amount of enzyme catalyzing the incorporation of 1 nmol ³²P into Histone III-S per minute. The enzyme preparation had a specific activity of 500 nmol/minute/mg. Results are the mean of triplicate assays.

referred to its antioxidant properties or to some specific ligand-type interaction with the enzyme. The lack of effect on PKC activity of butylated hydroxytoluene, another lipid-soluble antioxidant, and Trolox, a water-soluble derivative of α-tocopherol, may possibly indicate a ligand interaction mechanism, although a site-specific, redox-mediated effect at the protein level cannot be excluded at the present time.

REGULATION OF PROTEIN KINASE C BY α-TOCOPHEROL IN VASCULAR SMOOTH MUSCLE CELLS

Given the central role of PKC in the transduction of signals to the cell proliferation control systems, the inhibition of protein kinase C observed with the isolated brain enzyme may lead to important consequences at the cellular level. Evidence that α-tocopherol modulates PKC also at the cellular level is presented later in this chapter. However, because of the multiple mechanism that control cellular growth, inhibition of PKC activity may not induce an obligatory inhibition of cell proliferation. Alternatively, depending, for instance, upon the importance of the PKC-dependent relative to PKC-independent pathways in a given cell, no effects or even growth activation may be expected. Growth inhibition or stimulation can thus be the result of a network of contrasting and/or cooperative events and may only be established at an empirical level. The following describes the response of VSMC to α-tocopherol. Growth-arrested cells are stimulated to proliferate by placing them in normal growth medium in the

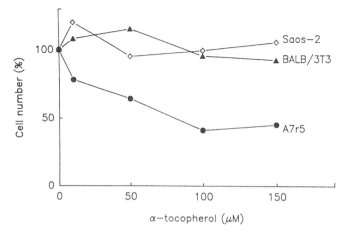

(A)

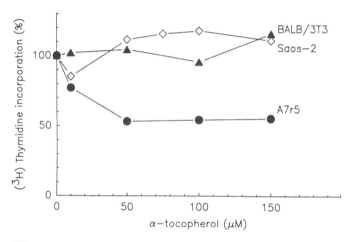

(B)

FIGURE 2 Effect of *d,l*-α-tocopherol on cell growth. Confluent and quiescent cultures of A7r5, Balb/3T3, and Saos-2 cells were stimulated to grow by placing them in 10% FCS-containing Dalbecco's modified Eagle's medium (DMEM), in the presence or absence of *d,l*-α-tocopherol at concentrations ranging from 0 to 100 μM for 48 h. (A) Triplicate cultures were tripsinized, and cell number was determined in a hemocytometer and expressed as a percentage of the control, untreated cells. The 100% represents 11.0 ± 0.9, 7.5 ± 0.6, and 13.2 ± 0.4 cells $\times 10^{-4}$ per ml for A7r5, Saos-2, and Balb/3T3 cells, respectively. (B) Cells were incubated with 0.5 μCi [³H]thymidine during the last 24 h and extracted with 5% trichloroacetic acid, and the radioactivity incorporated into acid-insoluble material was determined. The results are expressed as a percentage of the control incorporation measured in the absence of *d,l*-α-tocopherol. Similar results were obtained in at least three different experiments. The amount of [³H]thymidine incorporated in untreated A7r5, Balb/3T3, and Saos-2 cells was 71.4 ± 5.5, 51.0 ± 3.4, and 122.9 ± 7.2 cpm $\times 10^{-3}$ per well, respectively.

presence or absence of different concentrations of α-tocopherol. The presence of 100 μM α-tocopherol does not appear to be cytotoxic for this cell line, as determined by the standard Trypan blue assay, but produces approximately 50% inhibition of proliferation after 48 h as measured by cell counting (Fig. 2A) or by [³H]thymidine incorporation (Fig. 2B). At 20-30 μM α-tocopherol, which can be compared with the normal concentrations present in blood (18), 50% inhibition of proliferation is obtained. Trolox and α-tocopherol acetate are not effective in inhibiting the proliferation of VSMCs, and likewise butylated hydroxytoluene and phytol have no influence (Fig. 3).

Evidence that α-Tocopherol Inhibits Protein Kinase C in VSMC

Inhibition of cell proliferation may be the consequence of a great variety of events and, only under certain circumstances, may be solely produced by protein kinase C inhibition. However, the inhibition of the isolated enzyme and of cell proliferation occur at the same α-tocopherol concentration; moreover, the concentration available in vivo is such that a physiological modulation of PKC by α-tocopherol may be conceivable. More direct evidence is provided by the following experiments, however.

Inhibition of Phosphorylation of the 80 kD Protein

Digitonin-permeabilized VSMCs, incubated with [γ-³²P]ATP in the presence and absence of 100 nM phorbol 12,13-dibutyrate (PDBu) after pretreatment with or without 50 μM α-tocopherol, exhibit different protein phosphorylation patterns, which can be analyzed by sodium dodecyl sulfate (SDS) gel electrophoresis followed by autoradiography. Densitometer tracings of the autoradiograph in the region corresponding to the 80 kD band are particularly useful to study PKC activity in cells, since an 80 kD protein has been identified as a specific PKC substrate (19,20). Addition of PDBu results (Fig. 4) in a two- to threefold increased phosphorylation of this 80 kD protein, an effect that is completely prevented by the presence

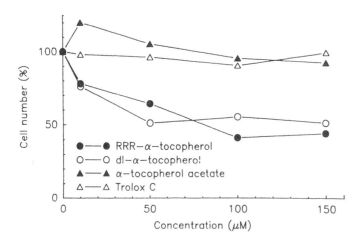

FIGURE 3   Specific inhibition of VSMC proliferation by α-tocopherol. Confluent and quiescent cultures of A7r5 smooth muscle cells were stimulated to grow by placing them in 10% FCS-DMEM. Cells were treated with *d,l*-α-tocopherol, *RRR*-α-tocopherol, Trolox C, and α-tocopherol acetate at concentrations ranging from 0 to 100 μM for 48 h. Then, triplicate cultures were tripsinized and the number of cells determined. Values are expressed as a percentage of control, untreated cells.

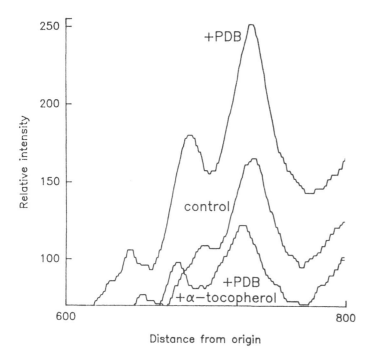

FIGURE 4 Inhibition by α-tocopherol of the phosphorylation of the 80 kD protein in aorta smooth muscle cells. Densitometer tracings of autoradiographs of dried SDS gels of the region around the 80 kD protein, which is indicated by the arrow. VSMCs were pretreated with *d,l*-α-tocopherol for 1 h at 37°C, permeabilized with digitonin, and labeled with [γ-³²P]ATP in isotonic KCl salt solution, as described in Reference 21. Both α-tocopherol-treated (lower trace) and untreated (upper trace) cells were exposed for 2 minutes to 100 nM PDBu. Control cells (middle trace) received an equivalent volume of solvent. Proteins were resolved by SDS–polyacrylamide gel electrophoresis (PAGE) and autoradiographed, and the radioactivity was determined by densitometric analysis of the labeled protein band at 80 kD.

of 50 μM α-tocopherol. The diminution of the 80 kD protein phosphorylation below the control level suggests that α-tocopherol also interferes with its nonstimulated endogenous phosphorylation.

Inhibition of Protein Kinase C Translocation

One of the molecular events leading to α-tocopherol inhibition appears to be related to the inhibition, produced by this substance, of the PKC translocation to the membrane. Such a conclusion is supported by the experiment in Table 1. Under resting conditions cells have an unequal PKC distribution, 36% being associated with the membrane and 64% being free in the cytosol. Phorbol 12-myristate 13-acetate (PMA) induces the translocation of PKC to membranes (24–26) and its downregulation (27,28), since the total amount of immunodetectable PKC after phorbol treatment is diminished and its distribution is preferentially in the membrane fraction (Table 1). In the presence of α-tocopherol, however, a two- to threefold decrease in the membrane-bound fraction is observed in favor of the free cytosolic fraction. It is apparent that α-tocopherol alters the distribution of PKC in favor of the cytosolsic form. It has been shown that the latter form is not active and that activation follows membrane

TABLE 1   Effect of α-Tocopherol on Cellular Redistribution of PKC in Smooth Muscle Cells[a]

	Cytosol		Membrane		
Treatment	Area	Relative area (%)	Area	Relative area (%)	Cytosol-membrane ratio
Control	17.9	64	9.9	36	1.8
PMA (300 nM)	0.53	4	13.6	96	0.04
Tocopherol (100 μM)	36.7	90	4.3	10	8.6

[a]Monolayer cultures of VSMCs were incubated with 300 nM PMA or 100 μM tocopherol for 1 h at 37°C. Cells were then harvested and disrupted, and cytosol as well as corresponding membrane fractions were prepared as described in Reference 22. Cytosolic and membrane fractions were subjected to immunoblot analysis (23) using anti-PKC monoclonal antibody (dilution 1:100). Quantitative data on amounts of protein kinase C present were obtained by scanning densitometry. The area under the corresponding peak and the percentage of relative area compared to control cells were determined from a representative fluorograph.

translocation. It is thus possible that α-tocopherol inhibits cell proliferation by preventing PKC translocation to the membrane and, after its activation, phosphorylation of the 80 kD protein and other membrane events.

Inhibition of EGF Receptor Transmodulation

Many factors, including phorbol esters, can affect the binding of epidermal growth factor (EGF) to certain cells by altering the receptor number and/or affinity. It is now apparent that in many situations this transmodulation of the high-affinity EGF receptor is associated with the activation of the protein kinase C. Hence, EGF binding is a useful tool to monitor the interaction of protein kinase C with its inhibitors or effectors. The effect of α-tocopherol on the EGF receptor transmodulation of different cell lines is shown in Figure 5. α-Tocopherol in smooth muscle cells is able to protect the transmodulation of EGF receptor induced by PMA, whereas α-tocopherol acetate is ineffective. In agreement with previous results, α-tocopherol is ineffective in protecting transmodulation of Saos-2 osteosarcoma cells.

Tocopherol Inhibits PKC Downregulation

Translocation of PKC to the membrane under the effect of Ca ions and phorbol esters is associated with its proteolytic attack, especially by calpain, which leads first to activation and subsequently to full digestion of the protein. This phenomenon is called downregulation. The observed lack of translocation of PKC to the membrane should have as a consequence an inhibition of downregulation of the protein. This phenomenon has been shown to occur (Table 2), indicating that α-tocopherol protects PKC from the downregulation induced by PMA.

Does α-Tocopherol Form a Complex with Protein Kinase C?

An increase in PDBu binding is surprisingly found after α-tocopherol treatment (Table 3), despite the fact that cell proliferation and PKC activity are inhibited. It has been found using cycloheximide that no de novo synthesis of PKC occurs during the course of these experiments. Thus, the increase in PKC molecules, which are titrated by [³H]PDBu in the presence of α-tocopherol, can only be the result of [³H]PDBu binding to the cytosolic form of the protein. This part of the enzyme is approximately 60% of the total (see Table 1) and normally does not bind phorbol esters (30). It may be speculated that, in the presence of α-tocopherol, the conformation of the protein is modified in such a way that phorbol esters can bind to it, as the membrane-bound form. The α-tocopherol-treated kinase remains enzymatically incompetent ("abortive complex," lack of access to the membrane resident substrates), however. We propose that α-tocopherol may form a complex with the cytosolic PKC competent

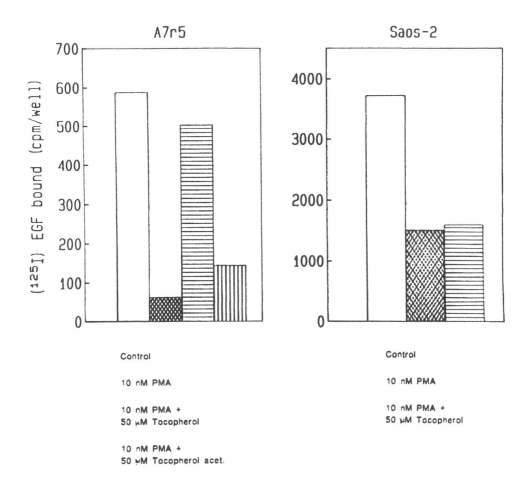

FIGURE 5 Effect of α-tocopherol on [^{125}I]EGF binding in A7r5 and Saos-2 cells. EGF binding was done essentially as described in Reference 29. Cells were preincubated at 37°C for 1 h with 50 μM of α-tocopherol, α-tocopherol acetate, or vehicle alone and then were treated with 10 nM PMA for 1 h. Cultures were then incubated for 2.5 h at 4°C with [^{125}I]EGF (0.5 ng/ml), washed, and solubilized in NaOH, and bound [^{125}I]EGF was measured using a gamma counter. Nonspecific binding was determined in the presence of a 50-fold excess of unlabeled EGF.

TABLE 2 α-Tocopherol Prevention of PKC Downregulation[a]

Treatment	[^{3}H]PDBu binding (fmol per 10^6 cells)
Control	122 ± 26
PMA (18 h)	14 ± 3
PMA (18 h) + α-tocopherol (18 h[b])	110 ± 31

[a]A7r5 cells were treated with PMA (0.4 μM) and d,l-α-tocopherol (50 μM) for the indicated time. Cells were harvested and washed thoroughly to remove PMA, and the specific [^{3}H]PDBu binding was measured. Values in parentheses represent incubation times.
[b]Added at the same time as PMA.

TABLE 3 Effect of *d,l*-α-Tocopherol, Tocopherol Derivatives, and BHT on Binding of [³H]PDBu in A7r5 and Saos-2 Cells[a]

	A7r5		Saos-2	
Compounds	50 μM	100 μM	50 μM	100 μM
Control	100	100	100	100
d,l-α-Tocopherol	208 ± 24	220 ± 7	127 ± 15	120 ± 10
RRR-α-Tocopherol	200 ± 30	351 ± 20	ND	135 ± 20
α-Tocopherol Acetate	93 ± 5	125 ± 15	117 ± 7	128 ± 13
α-Tocopherol Succinate	95 ± 5	100 ± 10	145 ± 12	145 ± 3
Trolox	120 ± 15	110 ± 16	116 ± 6	107 ± 6
Phytol	125 ± 10	115 ± 12	ND	ND
Butylated hydroxytoluene	81 ± 3	Nonviable	ND	ND

[a]After 24-well plates of confluent smooth muscle (A7r5) and osteosarcoma (Saos-2) cells were incubated with the indicated compounds for 60 minutes at 37°C, [³H]phorbol dibutyrate binding was measured. Cells were incubated for 30 minutes in serum-free medium containing 10 nM [³H]PDBu in the presence or absence of 10 μM unlabeled PDBu for determination of nonspecific binding. Assays were terminated by aspirating the binding mixture, rinsing the plates with phosphate-buffered saline (PBS) and solubiling the cells with 0.4 ml of 0.1 M NaOH and 2% Na_2CO_3. Values are expressed as percentage of control cells measured in the absence of compounds and are the mean ± SD (standard deviation) of three separate experiments. The amount of [³H]PDBu bound in control cultures was 91 ± 15 and 468 ± 44 (fmol per 10^6 cells) for A7r5 and Saos-2 cells, respectively. Values are expressed as percentage of control. ND = not determined.

in phorbol ester binding but not susceptible to translocation to the membrane and unable in this location to phosphorylate membrane-bound substrates. Preliminary experiments have indicated that a monoclonal antibody against PKC is able to coprecipitate [³H]α-tocopherol together with PKC in VSMC.

Inhibition of Cell Proliferation Depends on the Mitogen Employed for Activation

Fetal calf serum growth stimulation is produced by the presence of a number of factors activating proliferation through multiple pathways. Stimulation of cell proliferation by defined mitogens, in a serum-free medium, has begun to clarify the issue of which mitogen pathways are neutralized and which are still active in the presence of α-tocopherol. The effect of α-tocopherol on the proliferation of VSMC induced by different growth-promoting agents is reported in Figure 6. α-Tocopherol (100 μM) fully inhibits growth in response to PDGF and endothelin stimulation but is not effective in blocking the mitogenic response to lysophosphatidic acid or bombesin.

Chemical and Cellular Specificity of α-Tocopherol Effects

The effect of α-tocopherol on inhibiting cell proliferation exhibits three orders of specificity: chemical, cellular, and biochemical. A series of compounds (Trolox and BHT) that have radical-scavenging properties have already been discussed, but they do not appear to be effective in cell growth inhibition. Similarly, α-tocopherol esters (succinate and acetate) were also without effect. Of the cell lines we tested, the most sensitive to α-tocopherol was A7r5 smooth muscle. Balb/3T3 fibroblasts did not respond to a single dose of added α-tocopherol, although inhibition of growth is obtained by several subsequent additions of α-tocopherol. Saos-2 osteosarcoma was never found to be inhibited by α-tocopherol, but NB2A neuroblastoma cell proliferation was affected by higher concentrations of α-tocopherol (approximately

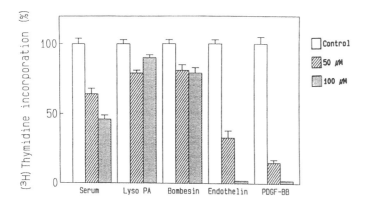

FIGURE 6 α-Tocopherol inhibits a PKC-dependent proliferation pathway in VSMCs. Growth-arrested VSMCs were incubated in serum-free DMEM containing the following concentrations of mitogens: 2% serum, 50 μM lysophosphatidic acid, 20 nM bombesin, 80 nM endothelin, and 20 ng/ml of PDGF-BB in the absence (control) or presence of 50 or 100 μM d,l-α-tocopherol. [³H]thymidine (1 μCi/ml) was present during the last 24 h of incubation. Cells were then processed for scintillation counting as described in Figure 2. The results are expressed as a percentage of the control incorporation for each mitogen measured in the absence of d,l-α-tocopherol and are the mean of triplicate determinations from a representative experiment. Similar results were obtained in three additional experiments. The amount of [³H]thymidine incorporated in unstimulated and FCS-, lysophosphatidic acid- (LPA), bombesin-, endothelin-, or PDGF-BB-treated cells was 12.9 ± 1.1, 43.7 ± 2.5, 21.7 ± 0.6, 21.9 ± 0.9, 19.3 ± 2.3, and 21.4 ± 1.8 cpm × 10⁻³ per well, respectively.

100 μM). In conclusion, the interpretation of the α-tocopherol effects in terms of a nonspecific hydrophobic interaction or a general radical-scavenging phenomenon is not consistent with the present experimental data that suggest rather that α-tocopherol-induced events have cell specificity and chemical specificity and that they are only realized when cell proliferation is activated by certain mitogens.

The Point of Regulation of α-Tocopherol Relative to the Cell Cycle and the Reversibility of α-Tocopherol Effects

As mentioned, α-tocopherol inhibits the proliferation responses to fetal calf serum (FCS), platelet-derived growth factor (PDGF), or endothelin. The location of the α-tocopherol-sensitive points in the cell cycle and the effect of α-tocopherol on the kinetics of the G_0-S transition can be established by a cell cycle analysis of VSMCs. VSMCs are synchronized by a combination of serum deprivation and hydroxyurea treatment. Upon removal of hyudroxyurea, the cells quickly enter the S phase, becoming synchronized at the G_1/S boundary (31). To study the kinetics of the G_0-S transition, growth-arrested cells are released from the G_0 block by placing them in normal growth medium. [³H]thymidine uptake is employed to establish that the cells enter into the S phase 12 h after the release of the G_0 block (not shown). α-Tocopherol (100 μM) does not alter the kinetics of the G_0-S transition but, in synchronized cultures, inhibits approximately 50% proliferation. Such an inhibition occurs with both FCS and PDGF (Fig. 7). The effect is reversible since treated cells regain their proliferation capacity after incubation in the absence of α-tocopherol for 30 h (Table 4). The α-tocopherol-sensitive point in the cell cycle is established by a "delayed addition" experiment. Quiescent VSMCs are released from the G_0 block by placing them in normal growth medium, and at

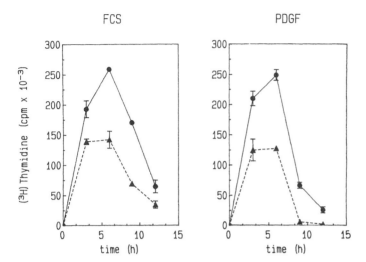

FIGURE 7   Effect of 100 μM α-tocopherol on VSMC synchronized by a combination of serum deprivation and hydroxyurea treatment. VSMC were plated on 35 mm dishes for 2 days. Cells were then cultured in medium containing 0.2% serum for 72 h. Fresh medium containing 10% FCS in the presence or the absence of 100 μM α-tocopherol was then given, and hydroxyurea (1.5 mM) was added 6 h later. Hydroxyurea was removed after 14 h treatment, and fresh medium containing 10% FCS (A) or 25 ng/ml of PDGF (B), in the presence (dashed line, treated) or the absence (solid line, control) of 100 μM α-tocopherol was added to the cells (zero time on the abscissa).

the indicated times (Fig. 8), 100 μM α-tocopherol is added. Cells are then synchronized by hydroxyurea treatment, and their ability to synthesize DNA is measured. Inhibition by α-tocopherol can be observed upon its addition up to 6 h before the *S* phase. α-Tocopherol added afterward does not inhibit DNA synthesis. This experiment indicates that the point of regulation of the cell cycle by α-tocopherol is located in the mid- to late G_1 phase.

CONCLUSIONS

The possibility that certain cells, in particular vascular smooth muscle cells, are especially sensitive to α-tocopherol permits some speculation about the physiological role of α-tocopherol.

TABLE 4   Reversibility of the α-Tocopherol Effect on VSMC[a]

	[³H]thymidine uptake (cpm × 10³ per dish)			
Time (h)	Control cells	Treated cells	Washed cells	Recovery (%)
3	121.46 ± 20.45	61.23 ± 0.82	80.60 ± 7.88	35
6	191.36 ± 6.26	110.91 ± 1.74	136.17 ± 10.67	35
9	33.01 ± 0.99	1.19 ± 0.29	24.64 ± 6.61	75
30	60.69 ± 5.48	24.48 ± 11.25	57.23 ± 2.55	92

[a]Low-serum arrested VSMCs were restimulated to growth by the addition of 10% FCS in the presence or absence of 100 μM α-tocopherol. After 6 h, cells were treated for 14 h with hydroxyurea, and subsequently cells pretreated with α-tocopherol were split between two sets of dishes: one received an additional dose of 100 μM α-tocopherol (α-tocopherol-treated cells), and the other received fresh medium containing serum (washed cells). [³H]thymidine incorporation (3 h pulse) was determined at the indicated times after the hydroxyurea treatment. Recovery was calculated as the difference between the counts in washed and treated cells divided by the difference between the counts in control and treated cells.

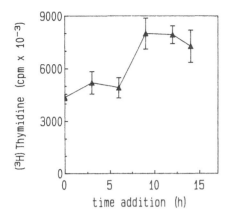

FIGURE 8 Delayed addition of α-tocopherol indicates a α-tocopherol-sensitive point in G_1. Growth-arrested VSMC were placed in normal growth medium containing 10% FCS. At the indicated times after the release from G_0, 100 μM α-tocopherol was added to the cultures. Hydroxyurea was added 9 h after the release from G_0, and cells were incubated for an additional 14 h. Then, hydroxyurea was removed and [³H]thymidine uptake (3 h pulse) was determined.

Although it is fully documented that α-tocopherol works in vivo as a radical scavenger, an additional and complementary reaction mechanism may be inferred from the data reported here. Whether the effect can be solely attributed to a ligand or, in addition, to a site-directed antioxidant function of α-tocopherol cannot be estimated at the present time. A systematic study of α-tocopherol analogs is being carried out in our laboratory to clarify this point. The proliferation of certain cells may be partially inhibited at physiological concentrations of blood α-tocopherol. Consequently, a decrease in α-tocopherol in blood (an outcome of pro-oxidant states and correlated with a higher risk of atherosclerosis) may result in an increase in vascular smooth muscle cell proliferation, an event constantly observed as part of atherosclerotic lesions. The observation that the growth of some tumor cells is diminished by the presence of α-tocopherol is also in line with epidemiological studies indicating a negative correlation between tumor risk and blood α-tocopherol level (32). Finally, that α-tocopherol inhibits cellular growth indicates a possible way of developing compounds capable of controlling certain pathways leading to pathological cell proliferation.

ACKNOWLEDGMENTS

Bernische and Schweizerische Krebsliga and F. Hoffmann-La Roche AG are gratefully acknowledged for generous financial support. The supply of R,R,R-α-tocopherol by Henkel, Ltd. (USA) and the generous gift of pPDGF by Dr. M. Pech are thankfully acknowledged. A FEBS long-term fellowship in part supported the stay of Adam Szewcyk.

REFERENCES

1. Burton GW, Ingold KH. Vitamin E: application of the principles of physical organic chemistry to the exploration of its structure and function. Acc Chem Res 1986; 19:194-201.
2. Tappel AL. Vitamin E and free radical peroxidation of lipids. Ann N Y Acad Sci 1972; 203:12-27.
3. Packer L, Landvik S. Vitamin E: introduction to biochemistry and health benefits. Ann N Y Acad Sci 1989; 570:1-6.

4. Urano S, Yano K, Matsuo M. Membrane-stabilizing effect of vitamin E: effect of α-tocopherol and its model compounds on fluidity of lecithin liposomes. Biochem Biophys Res Commun 1988; 150:469-75.

5. Niki E. In: Hayaishi O, Mino M, eds. Clinical and nutritional aspects of vitamin E. Amsterdam: Elsevier, 1986; 3-14.

6. Redanna P, Whelan J, Burgess JR, et al. The role of vitamin E and selenium on arachidonic acid oxidation by way of the 5-lipoxygenase pathway. Ann N Y Acad Sci 1989; 570:136-45.

7. Mahoney CW, Azzi A. Vitamin E inhibits protein kinase C activity. Biochem Biophys Res Commun 1988; 154:694-7.

8. Boscoboinik D, Szewczyk A, Azzi A. α-Tocopherol (vitamin E) regulates vascular smooth muscle cell proliferation and protein kinase C activity. Arch Biochem Biophys 1991; 286:264-69.

9. Boscoboinik D, Szewczyk A, Hensey C, Azzi A. Inhibition of cell proliferation by α-tocopherol: role of protein kinase C. J Biol Chem 1991; 266:6188-94.

10. Nishizuka Y. The molecular heterogeneity of protein kinase C and its implications for cellular regulation. Nature 1988; 334:661-5.

11. Kikkawa U, Kishimoto A, Nishizuka Y. The protein kinase C family: heterogeneity and its implications. Annu Rev Biochem 1989; 58:31-44.

12. Niedel JE, Kuhn LJ, Vandenbark GR. Phorbol diester receptor copurifies with protein kinase C. Proc Natl Acad Sci U S A 1983; 80:36-40.

13. Nishizuka Y. The role of protein kinase C in cell surface signal transduction and tumor promotion. Nature 1984; 308:693-8.

14. Ross R, Glomset JA. The pathogenesis of atherosclerosis. N Engl J Med 1976; 295:369-77.

15. Owens GK, Reidy MA. Hyperplastic growth response of vascular smooth muscle cells following induction of acute hypertension in rats by aortic coaretation. Circ Res 1985; 57:695-705.

16. Schwartz SM, Campbell GR, Campbell JH. Replication of smooth muscle cells in vascular disease. Circ Res 1986; 58:427-44.

17. Huang EL, Joshida Y, Nakabayashi H, Huang KP. Differential distribution of protein kinase C isozymes in the various regions of brain. J Biol Chem 1987; 262:15714-20.

18. Gey FK, Puska P. Plasma vitamins E and A inversely correlated to mortality from ischemic heart disease in cross-cultural epidemiology. Ann N Y Acad Sci 1989; 570:268-82.

19. Rozengurt E, Rodriguez-Pena A, Smith KA. Phorbol esters, phospholipase C, and growth factors rapidly stimulate the phosphorylation of a M_r:80,000 protein in intact quiescent 3T3 cells. Proc Natl Acad Sci U S A 1983; 80:7244-8.

20. Blackshear PJ, Witters LA, Girard PR, Kuo JF, Kuamo SN. Growth factor-stimulated protein phosphorylation in 3T3-L1 cells. Evidence for protein kinase C-dependent and -independent pathways. J Biol Chem 1985; 260:13304-15.

21. Erusalimsky JD, Friedberg I, Rozengurt E. Bombesin, diacylglycerols, and phorbol esters rapidly stimulate the phosphorylation of an $M_r = 80,000$ protein kinase C substrate in permeabilized 3T3 cells. Effect of guanine nucleotides. J Biol Chem 1988; 263:19188-94.

22. Regazzi R, Fabbro D, Costa SD, Borner C, Eppenberger U. Effects of tumor promoters on growth and on cellular redistribution of phospholipid/Ca^{2+}-dependent protein kinase in human breast cancer cells. Int J Cancer 1986; 37:731-7.

23. Towbin H, Staehelin T, Gordon J. Electrophoretic transfer of proteins from polyacrylamide gels to nitrocellulose sheets: procedure and some applications. Proc Natl Acad Sci U S A 1979; 76: 4350-4.

24. Kraft AS, Anderson WB. Phorbol esters increase the amount of Ca^{2+}, phospholipid-dependent protein kinase associated with plasma membrane. Nature 1983; 301:621-3.

25. Tapley PM, Murray AW. Modulation of Ca^{2+}-activated, phospholipid-dependent protein kinase in platelets treated with a tumor-promoting phorbol ester. Biochem Biophys Res Commun 1984; 122:158-64.

26. Hirota K, Hirota T, Aguilera G, Catt KJ. Hormone-induced redistribution of calcium-activated phospholipid-dependent protein kinase in pituitary gonadotrophs. J Biol Chem 1985; 260:3243-6.

27. Solanki V, Slaga TJ, Callaham M, Huberman E. Downregulation of specific binding of [20-³H]phorbol 12,13-dibutyrate and phorbol ester-induced differentiation of human promyelocytic leukemia cells. Proc Natl Acad Sci U S A 1981; 78:1722-5.

28. Rodriguez-Pena A, Rozengurt E. Disappearance of Ca^{2+}-sensitive, phospholipid-dependent protein kinase activity in phorbol ester-treted 3T3 cells. Biochem Biophys Res Commun 1984; 120:1053-9.

29. Olson JE, Pledger WJ. Transmodulation of epidermal growth factor binding by platelet-derived growth factor and phorbol 12-13-myristate acetate is not sodium dependent in Balb/c/3T3 cells. J Biol Chem 1990; 265:1847-51.

30. Trilivas I, Brown JH. Increases in intracellular Ca^{2+} regulate the binding of [³H]phorbol 12,13-dibutyrate to intact 132N1 astrocytoma cells. J Biol Chem 1989; 264:3102-7.

31. Ashihara T, Baserga R. Cell synchronization. Methods Enzymol 1979; 58:248-62.

32. Stahelin HB, Gey KF, Eichholzer M, Ludin E, Brubacher G. Cancer mortality and vitamin E status. Ann N Y Acad Sci 1989; 570:391-9.

29

Vitamin E Action in Modulating the Arachidonic Acid Cascade

David G. Cornwell and Rao V. Panganamala

Ohio State University, Columbus, Ohio

INTRODUCTION

A review of the role of vitamin E in the arachidonic acid cascade is important for two reasons. In the first place it demonstrates some of the unique properties of vitamin E itself, and in the second place it helps to identify subtleties in the process, usually called the arachidonic acid cascade, that controls the metabolism of arachidonic acid leading to its many eicosanoid derivatives. We accept the proposition that vitamin E functions primarily as a lipid antioxidant, but we propose that vitamin E has properties that distinguish it from many other lipid antioxidants. We accept the proposition that the arachidonic acid cascade is modified in several ways by oxygen radicals, but we propose that oxygen-centered radicals have widely different properties depending on their structures, site of origin, and the nature of the carrier. Our studies have generally focused on vitamin E and arachidonic acid metabolism in smooth muscle cells from the aorta, but other studies from our laboratory and elsewhere indicate that it is possible to generalize from this cell type to other tissues.

LOCALIZATION OF VITAMIN E IN MEMBRANES AND LIPOPROTEINS AT THE LIPID/WATER INTERFACE

The physical properties of vitamin E restrict its distribution within tissues and cells. Vitamin E and its quinone oxidation product are amphipathic molecules containing both a hydroxyl group and an aliphatic side chain. These molecules form stable monolayers at the air/water interface even in the compressed state (1,2). Force-area isotherms for vitamin E and its quinone that are reproduced in Figure 1 show that vitamin E is oxidized slowly to the quinone and that both molecules are highly insoluble since the area per molecule in the monolayer does not decrease with time. Monolayers of even slightly soluble amphipathic molecules are unstable, and these monolayers demonstrate unusually small surface areas over time that are the result of desorption or solution in the aqueous phase (3). The hydroxyl group is particularly important in establishing the surface properties of vitamin E and its quinone. Alkaline potassium ferricyanide oxidizes vitamin E to a spirodienone ether that does not contain a hydroxyl group, and this compound collapses immediately to an insoluble oil that forms a droplet at the air/water interface (2). Since there is a formal similarity between an oil/water interface or membrane and the air/water interface, any spirodienone ether formed in a cell is extruded from a membrane into a bulk lipid phase.

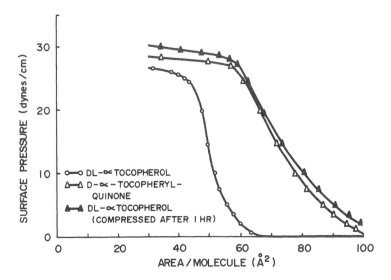

FIGURE 1 Force-area isotherms at 22–24 °C of DL-α-tocopherol and DL-α-tocopherylquinone spread on a subphase containing 0.1 M sodium chloride and 0.01 N hydrochloric acid. Compounds were spread as gaseous films and compressed immediately or after 1 h. Films were compressed at 11 Å² per molecule per minute (α-tocopherol) and 21 Å² per molecule per minute (α-tocopherylquinone). [Reproduced with permission from the *Journal of Lipid Research* (2).]

Free cholesterol and vitamin E share many surface properties. Both molecules contain hydroxyl groups, form stable (highly insoluble) monolayers, expand on oxidation, and collapse into a bulk lipid phase when the hydroxyl group is removed or blocked. Many of these properties are demonstrated by studies on the surface properties of long-chain cholesterol esters (4). Cholesterol, like other amphipathic molecules, is localized in organized structures represented by membranes and plasma lipoproteins, and it is important to recognize that the cholesterol content varies in specific membranes and lipoproteins. Free cholesterol is concentrated in the plasma membrane, and small amounts of free cholesterol are found in widely different concentrations in other cellular membranes (5–7). Cytosolic free cholesterol is evidently bound to nonspecific transfer proteins (8–10) and is not in a true solution in the aqueous phase. Plasma lipoproteins contain different amounts of total cholesterol (11), which is divided into free and esterified fractions, with free cholesterol concentrated at the lipoprotein surface and cholesterol esters, the larger fraction, concentrated in a bulk lipid phase or neutral lipid core (12–14). The free cholesterol concentration is low in adipose tissue, but the size of this pool is so large that a significant part of total-body free cholesterol is stored in adipose tissue (5).

The distribution of vitamin E is in most respect similar to the distribution of free cholesterol, but there is one important difference, the accumulation of vitamin E in the bulk lipid phase of the plasma lipoproteins, which is shown by the direct correlation between total plasma vitamin E and total plasma lipid (15). Vitamin E and free cholesterol are only slightly soluble in the bulk lipid phase, but the small amount of vitamin E relative to free cholesterol results in a disproportionate partition of vitamin E into the lipoprotein bulk lipid phase. Vitamin E-cholesterol ratios are actually greater in bulk lipid phases than in membranes (16,17). Since mobilization from a bulk phase is slow (18), it is possible to have a localized vitamin E deficiency at the membrane level even though tissues have a high overall vitamin E content.

Vitamin E is, like free cholesterol, concentrated in plasma membrane and subcellular membrane compartments rather than the cytosol (16,19,20). Vitamin E is probably localized in distinct regions of the anisotropic membrane lipid phase since membranes discriminate between different sterioisomers of vitamin E (21) and this would be an unusual property for a random distribution. Cytosolic vitamin E is, again like free cholesterol, protein bound, but in this instance there are specific carrier proteins (16,22,23).

Most of the vitamin E in blood is carried in the plasma membrane of the red blood cell (24). Vitamin E is also found in other formed elements, and indeed platelets are a sensitive indicator of vitamin E status (16,25,26). Dietary vitamin E is transported by the plasma lipoproteins (27,28), and although small amounts of vitamin E (27,29), like free cholesterol (10), exchange between lipoproteins, most vitamin E is transported to other tissues by low-density lipoprotein uptake through a high-affinity receptor (30).

Natural and synthetic antioxidants vary widely both in lipid solubility and the capacity to function as amphipathic molecules in anisotropic lipid phases. As a consequence, it should not be surprising that antioxidants have highly specific and very different effects on biomembranes and model systems (31,32). The natural antioxidant vitamin E, a local and focal antioxidant, furnishes a number of examples of antioxidant specificity that are discussed in subsequent sections of this review.

OXIDANT SPECIFICITY IN BIOLOGICAL SYSTEMS

It is usually assumed that oxygen-centered radicals derived from dioxygen, and its reduction products provide common pathways for the oxidation of polyunsaturated acyl chains in membrane lipids, and sulfydryl groups in membrane proteins and cytosolic enzymes. However, the nature of the prooxidant and the site of the redox reaction result in highly selective oxidation reactions (33). This selectivity is readily demonstrated in synthetic and natural membranes.

Alloxan and the reducing agent ascorbic acid may be used as a system for the controlled generation of small amounts of H_2O_2 through the alloxan-dialuric acid cycle (34). Monolayers containing unsaturated acyl chains are expanded, and the resistance of synthetic bilayers is reduced by oxidation with H_2O_2 generated by the alloxan-dialuric acid cycle (35). Vitamin E, as expected, stabilizes synthetic bilayers (Fig. 2), and it is reasonable to assume that this general process explains the protective effect of vitamin E on natural membranes in, for example, oxidative hemolysis.

A number of agents are available for the generation of H_2O_2. For example, the oxidation of reduced glutathione at pH 8 results in the formation of H_2O_2 (36). Red blood cell hemolysis with this system varies directly with lipid peroxidation, and it is prevented either by catalase or dimethyl sulfoxide, which acts as an antioxidant by trapping radicals with the formation of methane (36). However, hydrophilic dimethyl sulfoxide prevents hemolysis without having any effect on membrane lipid peroxidation, showing that a selective oxidation reaction leading to hemolysis is not coupled specifically to membrane lipid peroxidation.

The separation between generalized lipid peroxidation and hemolysis is even more apparent in studies with menadione, another agent that generates H_2O_2 (37). Menadione initiates the oxidation of intracellular reduced glutathione and the formation of methemoglobin in red blood cell (38). These effects occur without lipid peroxidation and well before hemolysis. Indeed, studies with thiodione, the 3-glutathionyl ether of menadione, show that it is possible to oxidize reduced glutathione and hemoglobin without inducing the hemolysis that is apparently caused by a specific effect on membrane sulfhydryl groups (38). Furthermore, menadione may function both as an oxidizing agent, perhaps through the menadione-menadiol cycle, and a quinone antioxidant that prevents lipid peroxidation (38,39). H_2O_2 is a final common intermediate in oxidation reactions involving alloxan, reduced glutathione, and menadione, and yet these agents have very different effects on membrane lipids and membrane

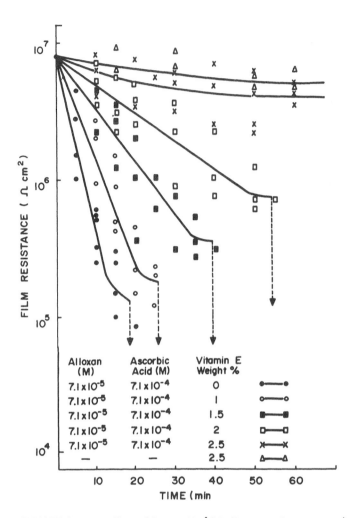

FIGURE 2 The effect of 7.1×10^{-5} M alloxan and 7.1×10^{-4} M ascorbic acid on the dc resistance and stability of lipid bilayers formed from solutions containing egg yolk lecithin and different weight percentages of vitamin E. The membranes were generated in 0.1 N NaCl, and the reagents were added to one layer of the bilayer cell. Temperature was 24–26°C. The weight percentages of vitamin E were calculated on the amount of lecithin. The dotted line indicates the point of film rupture. [Reproduced with permission from the *Journal of Membrane Biology* (35).]

hemolysis. Thus the oxidizing agent itself confers specificity on what are apparently reactions involving dioxygen and its reduction products.

OXIDATION AND THE ARACHIDONIC ACID CASCADE

The arachidonic acid cascade is a series of reactions that begins with the release of free fatty acid from membrane phospholipids and ends with the formation of prostanoids and other eicosanoid metabolites (Fig. 3). Early investigators suggested that fatty acid release through phospholipase activity was the rate-limiting step in the cascade (40–42). Release is indeed required for product formation, but once release has occurred it is also possible to affect the

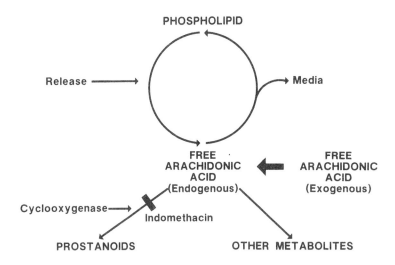

FIGURE 3 The arachidonic acid cascade showing the fatty acid release/cyclooxygenase sequence. Type 1 agents act at the release site, and Type 2 agents act at the cyclooxygenase site in the cascade.

amount of product formed through the action of agents that stimulate or inhibit the cyclooxygenase and lipoxygenase steps in the cascade.

Agents that act at the phospholipase step (type 1) and agents that act at the cyclooxygenase step (type 2) have very different properties (43,44). Type 1 antagonists block fatty acid release and all subsequent steps in the cascade. Type 2 antagonists block only the formation of cyclooxygenase metabolites. Type 1 and 2 agonists both enhance prostanoid synthesis. The effect of the type 1 agonist is eliminated by exogenous free arachidonic acid since any effect from endogenous release is overwhelmed by exogenous fatty acid. Type 2 agonists and exogenous free arachidonic acid interact and have a synergistic effect on prostanoid synthesis. Type 1 agonists increase and type 2 agonists have no effect on the level of labeled free arachidonic acid in cells that are prelabeled by the incorporation of [^{14}C]arachidonic acid into cellular phospholipids. This effect with type 1 agonists is even more pronounced in the presence of indomethacin. Metabolite profiles are very different with type 1 and type 2 agonists. Type 1 agonists increase the formation of both prostanoids and other metabolites that are not cyclooxygenase products (Fig. 4), and type 2 agonists affect only cyclooxygenase products (Fig. 5). Compounds I and II (Fig. 4) may be lipoxygenase products formed when cells are activated by A23187 (45). Interestingly enough, cyclosporine A, another type 1 agonist, stimulates the formation of compounds I and II (43).

Provocative studies suggesting a role for oxidants in enhancing fatty acid release have appeared, but the evidence for a significant effect on the cascade is not strong. Oxidizing agents promote the action of phospholipase A$_2$ on fatty acid release in mitochondria (46) and platelets (47). There may be phospholipase-susceptible domains in membranes (48), and these domains could have sulfhydryl groups that are oxidized (cross-linked) by such agents as diazenedicarboxylic acid(bis)dimethylamide (47) and perhaps lipid peroxides in ways that alter interfacial binding to improve the efficiency of phospholipase A$_2$ catalysis (49).

The enzymatic hydrolysis of phospholipids in low-density lipoproteins is stimulated by oxidation (50,51), but this may not affect the cascade. If the mechanism for the lipid peroxide effect involves the oxidative cleavage of arachidonic acid and other polyunsaturated fatty

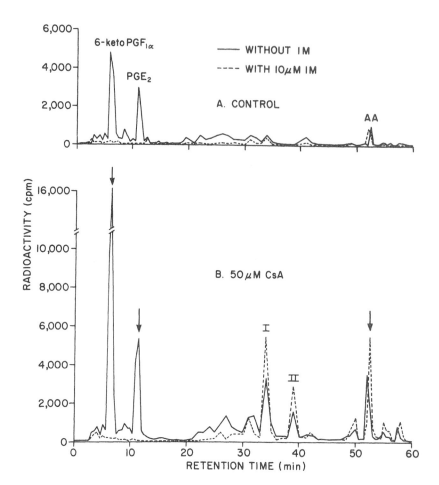

FIGURE 4 Increased amounts of labeled 6-keto-PGF$_{1\alpha}$, PGE$_2$, and free arachidonic acid (AA) are formed and two new metabolites, I and II, are identified when cyclosporine A (CsA) is added to the incubation media. Note the increase in free AA relative to PGE$_2$. Indomethacin (IM) increases free AA and metabolites I and II. An identical metabolite profile was obtained with A 23187 (43). Smooth muscle cells were prelabeled with [^{14}C]AA and then incubated with CsA. Metabolites were separated by HPLC. [Reproduced with permission from *Transplantation* (43).]

acids in the *sn*-2 position forming a phospholipid derivative susceptible to platelet-activating factor acetyhydrolase (51–53), then arachidonic acid would be destroyed before release as a substrate for the arachidonic acid cascade.

Oxidative fragmentation at the *sn*-2 position forms phospholipid derivatives that are not only susceptible to platelet-activating factor acetylhydrolase (see earlier) but also have physiological effects similar to those of platelet-activating factor (53). Since platelet-activating factor stimulates arachidonic acid release (54), oxidants may stimulate the arachidonic acid cascade by forming a "pseudo–platelet-activating factor" that acts on cells with the appropriate receptor.

There is compelling evidence for the catalytic effect of oxygen-centered radicals on the cyclooxygenase reaction. Studies with antioxidants (radical trapping agents) show that oxygen-

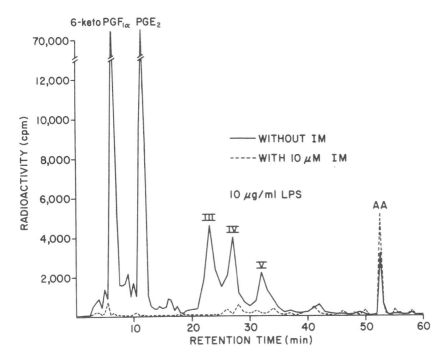

FIGURE 5 Increased amounts of 6-keto-PGF$_{1\alpha}$ and PGE$_2$ are formed, and three new metabolites, III, IV, and V, are identified when bacterial lipopolysaccharide (LPS) is added to the incubation media. Note the absence of an increase in free AA relative to PGE$_2$ (see Fig. 4 for comparisons). IM blocks the formation of metabolites III, IV, and V. An identical metabolite profile was obtained with bradykinin (44). Smooth muscle cells were prelabeled with [^{14}C]AA and then incubated with LPS. Metabolites were separated by HPLC. [Reproduced with permission from *Transplantation* (43).]

centered radicals promote prostanoid synthesis (55–58). Oxidants are complex (58), with both stimulatory and inhibitory effects that depend on variables that include (59) antioxidant concentration, incubation time, and the physical state of the enzyme complex (detergent effect). These data suggest that antioxidants under certain conditions inhibit synthesis by competing for oxygen-centered radicals that promote the cyclooxygenase reaction (55–59) and under other conditions promote synthesis by protecting the enzyme (59–63) from its known susceptibility to oxidative inactivation (64). This dual effect is shown in a time study with dipyridamole (Fig. 6), an antioxidant (61–63) that functions as a hydroxyl radical scavenger (65). This antioxidant initially diminishes prostanoid synthesis, probably by competing for oxygen-centered radicals (early time effect in Fig. 6), and then enhances prostanoid synthesis by prolonging enzyme activity (later time effect in Fig. 6). Other investigators have recently shown that 6-hydroxy-2,5,7,8-tetramethylchromane-2-carboxylic acid and two isoflavanone antioxidants, but not vitamin E, also protected prostaglandin H synthase from autodeactivation (66).

A number of studies have identified a key role for lipid hydroperoxides in the activation of the cyclooxygenase enzyme complex (64,67,68), and indeed prostanoid synthesis has been used as a highly sensitive assay for the generation of lipid hydroperoxides (69). It is worth noting that synergy exists between hydroperoxy fatty acids and free arachidonic acid (33,70), showing that hydroperoxy fatty acids satisfy one important criterion for a type 2 agonist.

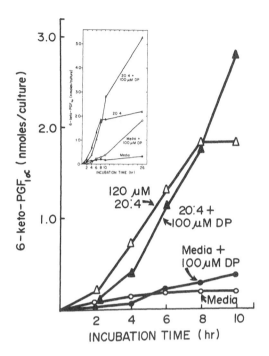

FIGURE 6 Time study of PGI₂ biosynthesis in confluent cultures of smooth muscle cells treated with media alone or 120 μM arachidonic acid (20:4). Data were obtained with and without 100 μM dipyridamole (DP). Short-time incubation intervals are expanded and all incubation times included in insert. [Adapted from and reproduced with permission from *Artery* (61).]

TABLE 1 Hydroperoxy Fatty Acids and Free AA have a Synergistic Effect on PGI₂ Synthesis in SMC[a]

	6-Keto-PGF$_{1\alpha}$ per plate (nmol)	
Treatment	No AA	120 μM AA
15-HPEPE		
Media	0.36 ± 0.01 (12)[b]	1.46 ± 0.10 (3)[b]
5 μM	0.45 ± 0.02 (11)	2.47, 2.14
15-HPETE		
Media	0.26 ± 0.03 (8)[c]	0.74 ± 0.04 (4)[c]
5 μM	0.37 ± 0.05 (4)	0.98, 0.86
25 μM	0.43 ± 0.02 (4)	1.16, 1.19
50 μM	0.42 ± 0.03 (4)	1.47, 1.47

[a]SMC were incubated for 24 h in media alone and media containing [5Z,8Z,11Z,13E,15(s)]-15-hydroperoxyeicosatetraenoic acid (15-HPETE) or [5Z,8Z,11Z,13E,15(S),17Z]-15-hydroperoxyeicosapentaenoic acid (15-HPEPE) in the presence and absence of 120 μM AA.
[b]A two-way analysis of variance showed that both 15-HPEPE (F 122.02, $P < 0.0001$) and AA (F 1215.95, $P < 0.0001$) enhanced the 6-keto-PGF$_{1\alpha}$ level. 15-HPEPE and AA interacted (F 77.99, $P < 0.0001$) to enhance the 6-keto-PGF$_{1\alpha}$ level.
[c]A two-way analysis of variance showed that both 15-HPETE (F 43.27, $P < 0.0001$) and AA (F 479.45, $P < 0.0001$) enhanced the 6-keto-PGF$_{1\alpha}$ level. 15-HPETE and AA interacted (F 16.62, $P < 0.0001$) to enhance the 6-keto-PGF$_{1\alpha}$ level.
Source: Reproduced with permission from the *Journal of Lipid Research* (70).

Data on the interaction between hydroperoxy fatty acids and free arachidonic acid are analyzed in Table 1.

PARADOXICAL EFFECTS OF VITAMIN E ON PROSTANOID SYNTHESIS IN CELLS

The numerous interactions between oxidants and prostanoid synthesis led us to postulate in 1982 that vitamin E would have multiple direct and indirect effects that would modulate the arachidonic acid cascade (71). Surprisingly, positive results with vitamin E are limited and interpretations are difficult. Early studies involving the addition of antioxidants to isolated microsomes and tissue homogenates have been summarized previously (71). These studies showed that vitamin E, unlike other antioxidants, such as α-naphthol, propyl gallate, butylated hydroxyanisole, butylated hydroxytoluene, guaiacol, and nordihydroguaiaretic acid, had little effect on prostanoid synthesis. Studies with microsomes isolated from animals maintained on vitamin E-deficient and vitamin E-supplemented diets yielded conflicting results. Cyclooxygenase activity was reduced significantly in microsomes from the muscle of E-deficient rabbits (72) and testes of E-deficient rats (73). Prostanoid synthesis was also decreased in aorta from vitamin E-deficient rabbits and rats (74–76). Cyclooxygenase activity was unchanged in microsomes from the lung and homogenates of liver kidney and lung of E-deficient rats (71,77).

When vitamin E and other antioxidants were added to isolated platelets, arachidonic acid-induced aggregation was inhibited, but vitamin E, unlike other antioxidants, did not discriminate between arachidonic acid- and ADP-induced platelet aggregation (78), and it is reasonable to assume that vitamin E in this instance acted nonspecifically to stabilize the platelet membrane (78–80). Vitamin E has been reported to diminish platelet prostanoid release (81–84) and inhibit platelet phospholipase A_2 activity (85). The effect on phospholipase may be related to the previously described effect of lipid peroxidation on platelet-activating factor acetylhydrolase (51,52) and therefore unrelated to the arachidonic acid cascade. However, a recent study with polymorphonuclear leukocytes shows that vitamin E deficiency has no effect on platelet-activating factor acetylhydrolase activity (86), so this explanation appears unlikely.

Several studies report that A23187-induced fatty acid release is potentiated by vitamin E (87) and inhibited by vitamin E quinone (88). A23187 is a cytotoxic agent that probably acts by disrupting membranes, as do other agonists for fatty acid release, such as cyclosporine A and 12-O-tetradecanoylphorbol-13-acetate (89,90). Since vitamin E and vitamin E quinone both function effectively as antioxidants (91,92) but differ in such properties as surface area (2) and, consequently, their interactions with membranes (92), it is reasonable to assume that their effects on A23187 release are related to surface properties that may stabilize or destabilize the platelet membrane.

Vitamin E has profound effects on cells in culture. Early studies identified vitamin E as a cell proliferation factor (93,94), and recent studies have amply confirmed this observation both for vitamin E (91,92,95–97) and its quinone, which also functions as a radical-trapping agent in a quinone-hydroquinone cycle (91,92). Recent studies explain the cell proliferation effect by showing that antioxidants decrease the population doubling time and do not merely prevent cell death (97). Increased cell number is a factor that should be considered in any long incubation experiments with vitamin E in cell cultures.

Studies from several laboratories have shown that vitamin E stimulates prostacyclin synthesis in endothelial cells (98–101). Prostanoid synthesis and lysophosphatidylcholine formation were both enhanced, and as a consequence it was proposed that vitamin E acted through fatty acid release (101). This interpretation may be questioned since vitamin E potentianted prostanoid synthesis in the presence of exogenous arachidonic acid (98,101), an effect that is not consistent with the properties of type 1 releasing agents (43,44). Lysophosphatidylcholine formation could be enhanced by a type 2 agent that acted through cyclooxygenase to channel arachidonic acid to prostanoids, thereby decreasing the availability of arachidonic acid for the resynthesis of phospholipid (Fig. 3). The concept of vitamin E as a releasing agent for endothelial cells is difficult to support because studies with *tert*-butylhydroperoxide show that oxidants promote the release of [^{14}C]arachidonic acid in prelabeled endothelial cells and this action is blocked by vitamin E (102).

Interactions of lipid peroxides and antioxidants with the arachidonic acid cascade have been studied extensively in smooth muscle cells from guinea pig aorta (103,104). Smooth muscle cells in confluent cultures underwent phenotypic modulation. These cells had large nuclei, two to six prominent nucleoli, and a granular perinuclear area that accumulated refractile granules that stained for acid phosphatase when cells were treated with fatty acid (96). Treated cells also contained large numbers of discrete lipid bodies that stained with oil red O and neutral red (96). The fatty acids in these cells were stored as triglyceride, and indeed cells treated with arachidonic acid contained large amounts of triarachidonin (96). Vitamin E had no effect on lipid accumulation or cell morphology.

Fatty acids in treated cells underwent chain elongation but were not further desaturated (105). Cells accumulated large amounts of lipid peroxides measured as thiobarbituric acid reactants (TBAR) that were bound to cellular lipids and did not distribute into the tissue culture media (91,106). A small amount of material that eluted with 15-hydroperoxyeicosatetraenoic acid on high-performance liquid chromatography was found in the media, and since this compound disappeared when prostanoid synthesis was blocked it is reasonable to conclude that it was a product of cooxidation during enhanced prostanoid synthesis (106).

Smooth muscle cells synthesized prostacyclin (PGI$_2$) and prostaglandin E$_2$ (PGE$_2$), and their formation, like the formation of bound lipid peroxides, varied directly with the concentration of exogenous fatty acid added to the tissue culture (107). This correlation is described in Figure 7. Some polyunsaturated fatty acids, for example, 6,9,12-octadecatrienoic acid, 11,14,17-eicosatrienoic acid, and 5,8,11,14,17-eicosatetraenoic acid, formed large amounts of lipid peroxides that had no effect on prostanoid synthesis. Other polyunsaturated fatty acids, 7,10,13,16-docosatetraenoic acid and 4,7,10,13,16,19-docosahexaenoic acid, formed large amounts of lipid peroxides and inhibited prostanoid synthesis, but this effect was caused by the fact that fatty acids with the 7,10,13,16 double-bond configuration were competitive inhibitors for prostanoid substrates with the 5,8,11,14 double-bond configuration (107), an observation later confirmed for the docosahexaenoic acid (108).

Vitamin E and vitamin E quinone were highly effective inhibitors of lipid peroxidation in confluent cultures of smooth muscle cells treated with exogenous arachidonic acid, but the inhibition of lipid peroxidation with these agents had little effect on prostanoid synthesis (Table 2). Antioxidants that inhibited microsomal prostanoid synthesis, α-naphthol and propyl gallate, also inhibited prostanoid synthesis in smooth muscle cells, and their effect on prostanoids was much greater than their effect on lipid peroxidation (Table 2).

The studies already summarized show that vitamin E inhibits prostanoid synthesis in platelets, stimulates prostanoid synthesis in endothelial cells, and has no effect on prostanoid synthesis in isolated microsomes and smooth muscle cells. These observations are difficult to resolve with a direct vitamin E effect on either fatty acid release (type 1 agent) or cyclooxygenase (type 2 agent). Vitamin E may stabilize membranes and therefore affect fatty acid release from these membranes, but the evidence for an effect of this nature is not strong. It is difficult to reconcile a direct effect for vitamin E with stimulation, inhibition, and lack of any effect on prostanoid synthesis in specific cell types. However, these differences may be

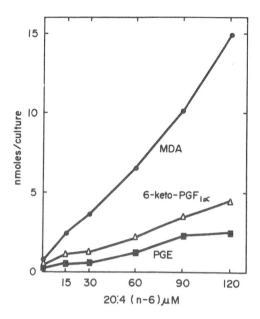

FIGURE 7 Formation of lipid peroxides (MDA), immunoreactive PGE, and immunoreactive 6-keto-PGF$_{1\alpha}$ when confluent smooth muscle cells were incubated for 24 h with different concentrations of arachidonic acid (20:4, *n*-6). [Reproduced with permission from *Lipids* (107).]

TABLE 2 Inhibitory Effect of Antioxidants on the Relative Concentrations of Lipid Peroxides and Prostanoids[a]

Antioxidant	Relative concentration (%)	
	Lipid peroxides	Prostanoid
Vitamin E		
10 μM	42	110
50 μM	4	86
Vitamin E quinone		
10 μM	13	89
α-Naphthol		
50 μM	65	3
100 μM	43	1
200 μM	23	1
Propyl gallate		
50 μM	88	35
100 μM	68	14
200 μM	48	2

[a]Smooth muscle cells were treated with 120 μM arachidonic acid in the presence and absence of the specified antioxidant. Lipid peroxides were estimated as thiobarbituric acid reactants, and prostanoids were measured by the radioimmunoassay for 6-keto-PGF$_{1\alpha}$. Relative concentration refers to the concentration relative to cultures that did not contain the antioxidant.
Source: Data are calculated from a previous study (106).

explained if vitamin E acts indirectly through the inhibition of lipid peroxidation. Endothelial cells are highly susceptible to oxidative stress (109), and vitamin E may potentiate prostanoid synthesis in these cells by protecting them against oxidative reactions that inhibit cyclooxygenase activity. Smooth muscle cells form bound lipid peroxides from endogenous synthesis initiated by exogenous arachidonic acid, and these lipid peroxides, which are not found in tissue culture media, differ from hydroperoxy fatty acids in having had no effect on prostanoid synthesis (106). As a consequence, any effect of vitamin E on bound lipid peroxides is not translated to an effect on prostanoid synthesis. The bound lipid peroxides found with exogenous arachidonic acid (106) are quite different from the lipid peroxides found with strong free-radical–generating systems, which yield insoluble membrane phospholipid-protein adducts and lipid-soluble TBAR (110).

PARADOXICAL EFFECTS OF VITAMIN E ON PROSTANOID SYNTHESIS IN THE WHOLE ANIMAL

The role of dietary vitamin E in modulating the arachidonic acid cascade is the subject of many studies that have generated, as have studies in vitro with cells and tissues, conflicting results. An early study on weanling rats (111) provided convincing evidence for the inhibitory effect of dietary vitamin E on prostanoid synthesis. This observation was confirmed in studies with insulin-dependent diabetes mellitus that first showed that subjects with this disease have increased prostanoid release from platelets (112) and later showed a relative decrease in platelet vitamin E (113–115).

The interactions of vitamin E with prostanoid metabolism in diabetics are analyzed in a recent masterful review (116), and it is unnecessary to repeat here a thorough description of the primary literature. Two paradoxical observations emerge from these studies. The synthesis of a platelet prostanoid thromboxane A_2 is enhanced and the synthesis of a tissue prostanoid, PGI_2, is inhibited by the relative vitamin E deficiency found in diabetes mellitus. Conversely, dietary vitamin E lowers the formation of platelet thromboxane A_2 and increases the formation of tissue PGI_2 in both vitamin E-deficient and diabetic animals (74,75,81,113–119). Thromboxane A_2 synthesis is clearly the result of fatty acid release since the effect of a dietary or diabetes-induced vitamin E deficiency on thromboxane A_2 synthesis is eliminated by the addition of free arachidonic acid (117–119). It is reasonable to assume that this effect is the indirect result of increased peroxides since these agents promote fatty acid release in endothelial cells (102). However, the positive effect of peroxides on fatty acid release in endothelial cells may be overwhelmed by the negative effect of lipid peroxides on cyclooxygenase or PGI_2 synthetase in cells highly susceptible to lipid peroxidation (109), giving as an overall result decreased PGI_2 synthesis. Cyclooxygenase in platelets is evidently less susceptible to inhibition by lipid peroxides so thromboxane A_2 synthesis in platelets is not decreased along with PGI_2 synthesis in other tissues.

The data on thromboxane A_2 and PGI_2 levels in vitamin E-deficient and diabetic animals do not define completely the interaction between vitamin E and the arachidonic acid cascade. Other studies, particularly in normal humans and experimental animals, fail to find any significant effect of dietary vitamin E on prostanoid metabolism (120–123). These data may be explained by the very interesting observation that low concentrations of either hydrogen peroxide or homocysteine, an agent that generates oxygen-centered radicals, stimulate PGI_2 synthesis but high concentrations of hydrogen peroxide and homocysteine inhibit PGI_2 synthesis (124). It is possible that other oxidants susceptible to vitamin E act in a similar manner, and as a consequence vitamin E does not always function exclusively to enhance prostanoid synthesis by protecting the cyclooxygenase from the destructive effects of lipid peroxides or to inhibit prostanoid synthesis by eliminating the lipid peroxides necessary to catalyze the cyclooxygenase reaction.

SEARCH FOR LIPID PEROXIDE MEDIATORS OF THE ARACHIDONIC ACID CASCADE

We believe that experimental results summarized earlier prove that some but not all lipid peroxides are among the oxidant species that downregulate and upregulate the arachidonic acid cascade and that vitamin E acts indirectly through its effects on specific lipid peroxides that are susceptible to its antioxidant properties. Vitamin E is highly localized in membrane and lipid phases and as a consequence vitamin E has little influence on cytosolic oxidants. Some lipid peroxides are also highly localized in membrane and lipid phases, and these peroxides, which are sensitive to vitamin E, have little effect on the cyclooxygenase complex, which is sensitive to cytosolic oxidants. We propose that the arachidonic acid cascade is modulated by circulating lipid peroxides that are regulated by vitamin E and that the plasma lipoproteins appear to be molecular complexes that satisfy these criteria.

Plasma lipoproteins are highly unstable molecular complexes that are readily susceptible to chemical modification by such processes as oxidation (125). Oxidized lipoproteins undergo chemical and physical changes, including decreased flotation rate, increased boundary spreading, and increased electrophoretic mobility (11,126). Lipoproteins are vehicles for the transport of carotenoids (127,128) and vitamin E (27–30). Lipid peroxides are found in serum and isolated lipoproteins (129–131) and recent studies (132,133) have identified oxidized lipoproteins under carefully controlled isolation conditions that should prevent the formation of oxidation artifacts (134–139).

Many recent studies have been undertaken with lipoproteins that were oxidized through either metal ion catalysis or incubation with endothelial cells (52). Oxidation was extensive, particularly with metal ions, and there should be some question about the same level of lipoprotein oxidation in nature (33,70,125,135–139). However, it is possible to control peroxidation in lipoproteins and generate reproducible low levels of lipid peroxides by incubating purified lipoproteins in 0.15 M NaCl at 37°C under 96% air and 4% CO_2 for different time intervals (33,70,135–138). The degree of lipid peroxidation was approximated by the disappearance of carotenoid color and quantitated by the thiobarbituric acid reaction (136). The rate of lipid peroxidation varied dramatically for lipoproteins isolated from the serum of different individuals (136,137). The vitamin E content of the lipoprotein decreased during lipid peroxidation (138,139). Mild oxidation, which did not alter physical properties, generated lipid peroxides in neutral lipids and lysophosphatidylcholine (136). Complex oxidation products were formed from free fatty acids and included a number of short-chain aldehydes (138) and soluble materials that cross-reacted with antibodies to prostaglandin E_2 but not 6-keto-PGF$_{1\alpha}$ (136). Cross-reacting materials may represent the noncyclooxygenase-derived prostanoids that have been recently described (140), and these compounds are probably formed from the intramolecular rearrangement of peroxyl radicals (141).

Mildly oxidized lipoproteins were taken up by cells in culture, which then showed refractile lipid droplets in phase-contrast microscopy and yellow-gold fluorescence after Nile red staining (70). These lipoproteins had dose-associated effects on the viability and proliferation of smooth muscle cells in culture. Increasing concentrations of oxidized lipoproteins led to increased numbers of dead cells (135) measured by a fluorescein diacetate and propidium iodide staining procedure (142). Dead cells retained their outline, but the cytoplasm was shrunken and almost all nuclei were pyknotic (70,125,135). Dead cells tended to detach from the tissue culture plate. The number of mitotic figures decreased and thymidine uptake was diminished in viable cells before the onset of cell death, showing that proliferation was diminished in viable cells (33,70,125,135). Antioxidants prevented the effects of mildly oxidized lipoproteins on both viability and proliferative potential (33,70,125,135).

Our early studies with smooth muscle cell cultures established that bound lipid peroxides from endogenous synthesis, unlike soluble hydroperoxy fatty acids, had little effect on prostanoid synthesis. Furthermore, vitamin E in smooth muscle cells, unlike vitamin E in the whole animal, had little effect on the arachidonic acid cascade. Cell cultures and the whole animal differ in many ways, and one major difference is the absence of circulating lipoproteins that have undergone mild oxidation in the cell cultures since the culture media has an antioxidant capacity (136) that evidently diminishes the peroxidation of the small amount of lipid already present in fetal bovine serum. Mildly oxidized lipoproteins had a number of effects when they were added to cultures, and we proposed that lipoproteins were the vehicle that presented lipid peroxides to cells in a way that allowed the lipid peroxides to modulate the arachidonic acid cascade (33,70,125,135). This hypothesis explained the different effects of vitamin E in the whole animal and cell culture. Vitamin E, which had little effect on the arachidonic acid cascade in cultures, would have a profound effect in the animal by regulating lipoprotein oxidation. The hypothesis, which also explains the difference between relatively soluble antioxidants that act directly on cyclooxygenase and vitamin E, which acts indirectly through lipoproteins (Fig. 8), was tested in studies with mildly oxidized lipoproteins in smooth muscle cells (33,70,125,135).

Mildly oxidized low-density lipoproteins had a number of concentration (oxidant or TBAR level) effects on prostanoid synthesis in smooth muscle cells in culture (33,70,125, 135), and these effects are summarized in Figure 9. Unoxidized lipoproteins had no effect on prostanoid synthesis. As with other type 2 agents that affect the cyclooxygenase complex, small amounts of mildly oxidized lipoprotein (low-TBAR-LDL$_{OXID}$) stimulated prostanoid synthesis as measured by radioimmunoassay and large amounts (high-TBAR-LDL$_{OXID}$) inhibited

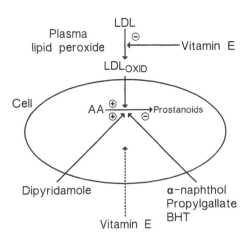

FIGURE 8   Hypothesis explaining the paradoxical effect of vitamin E on prostanoid synthesis in animals and in tissue culture. Vitamin E in plasma blocks extracellular lipid peroxidation and indirectly diminishes prostanoid synthesis by abolishing the stimulatory effects of extracellular lipid peroxides. Vitamin E also blocks intracellular lipid peroxidation (dashed line), but vitamin E, unlike other antioxidants in tissue culture, has no direct effect on prostanoid synthesis. [Reproduced with permission from Plenum Press (125).]

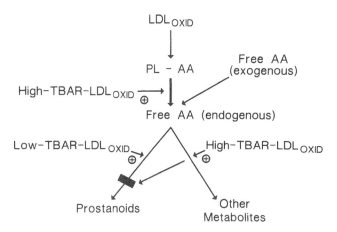

FIGURE 9 Effect of low-TBAR-LDL$_{OXID}$ and high-TBAR-LDL$_{OXID}$ on arachidonic acid metabolism in smooth muscle cells. [Reproduced with permission from the *Journal of Lipid Research* (70).]

prostanoid synthesis (Fig. 10). Low-TBAR-LDL$_{OXID}$ did not enhance the release of labeled prostanoids from prelabeled cells (Fig. 10), and since unlabeled exogenous arachidonic acid had a very similar effect it was reasonable to conclude that lipoproteins not only affected the cyclooxygenase complex but also supplied unlabeled arachidonic acid as a substrate for the arachidonic acid cascade, thereby diluting the pool of labeled endogenous arachidonic acid (70).

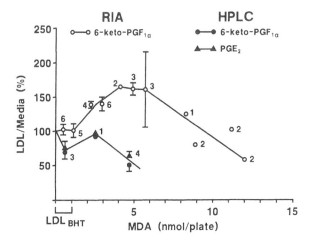

FIGURE 10 The relative prostanoid content of smooth muscle cell cultures is a function of the amount of lipid peroxide supplied by LDL$_{OXID}$. In the radioimmunoassay (RIA) experiments, confluent cultures were incubated with media alone or with LDL$_{OXID}$ preparations, which supplied different amounts of lipid peroxides. In the HPLC experiments, cells prelabeled with [^{14}C]arachidonic acid were incubated with media lone or LDL$_{OXID}$ preparations. The number of cultures for each data point is listed. Data are reported as the relative prostanoid level treatment/media alone (%). [Reproduced with permission from the *Journal of Lipid Research* (70).]

Further support for the role of oxidized lipoproteins acting as a type 2 agent at the cyclo-oxygenase step in the arachidonic acid cascade was provided by the synergism between low-TBAR-LDL$_{OXID}$ and free arachidonic acid in promoting prostanoid synthesis (70). High-TBAR-LDL$_{OXID}$ blocked prostanoid synthesis and enhanced cell death. Since time and pulse recovery experiments showed that these effects were unrelated, it was apparent that high-TBAR-LDL$_{OXID}$ had a distinct inhibitory effect on cyclooxygenase that was not secondary to cell death. Similar dose- and time-dependent effects on prostanoid synthesis were also found when hyperlipidemic serum containing lipid peroxides was added to endothelial cell cultures (143).

Low-TBAR-LDL$_{OXID}$ did not appear to act as a type 1 agent in stimulating arachidonic acid release and the formation of other oxidative metabolites (Fig. 11), but these effects may have been masked by increased utilization of free arachidonic acid through enhanced cyclo-oxygenase activity. Other agents, however, such as 12-*O*-tetradecanoylphorbol-13-acetate, clearly enhanced both cyclooxygenase and lipoxygenase activity in other cell lines (reviewed in Ref. 90), and it is unlikely that any type 1 effect of low-TBAR-LDL$_{OXID}$ would be masked

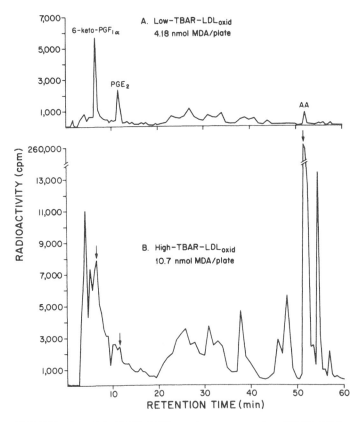

FIGURE 11   High-TBAR-LDL$_{OXID}$ in contrast to low-TBAR-LDL$_{OXID}$ diminishes radioactivity released in prostanoid fractions and greatly enhances radioactivity released as free arachidonic acid (AA) and other labeled AA derivatives (see Figs. 4 and 5 for comparisons). Smooth muscle cells were prelabeled with [^{14}C]AA and then incubated with LDL$_{OXID}$. Metabolites were separated by HPLC. [Reproduced with permission from the *Journal of Lipid Research* (70).]

in smooth muscle cells. High-TBAR-LDL$_{OXID}$ dramatically enhanced both free arachidonic acid release and the formation of many as yet unidentified oxidative metabolites of arachidonic acid, but high-TBAR-LDL$_{OXID}$, unlike true type 1 agents (Fig. 5), blocked cyclooxygenase activity (Fig. 11). Lipoxygenase seems to be absent in unstimulated smooth muscle cells (106), but these cells are capable of forming many oxidative metabolites through stimulation by both type 1 agents (Fig. 5) and high-TBAR-LDL$_{OXID}$. Controversy surrounds the purported effects of vitamin E on various lipoxygenases (122,144–151), but there is no doubt that vitamin E would diminish the formation of many indomethacin-insensitive derivatives of arachidonic acid by preventing the formation of high-TBAR-LDL$_{OXID}$.

CONSEQUENCES OF VITAMIN E STATUS THROUGH ITS EFFECT ON LIPOPROTEIN OXIDATION

The stimulatory and inhibitory effects of mildly oxidized lipoproteins on the arachidonic acid cascade help to resolve controversies surrounding the role of oxidized lipoproteins in a number of cellular processes. Some studies showed that lipoproteins inhibited prostanoid synthesis, and other studies showed that lipoproteins enhanced prostanoid synthesis (130,131, 143,152–155). Our studies showed that either effect was possible depending on the degree of lipoprotein oxidation (33,70,125,135), which is undoubtedly a function of vitamin E status (137–139) but not vitamin E status alone (137). Similarly, a number of studies showed that lipoproteins were either cytotoxic or mitogenic (156–164). Our studies again showed that either effect was possible (33,70,125,135) depending on the degree of lipoprotein oxidation, which is an expression of antioxidant status. Cytotoxicity is a direct effect of lipid peroxides, and positive and negative proliferative effects are the indirect result of lipid peroxides on prostanoid synthesis since prostanoids both promote (low-level effect) and inhibits (high-level effect) cell proliferation. Various properties of lipid peroxides and prostanoids related to the biology of the smooth muscle cell are described in a number of experimental studies (33,61, 62,70,89,90,92,95–97,106,107,165) and reviews (71,103,104,125,135) from our laboratory. Experimental data suggest that cells are most vulnerable to lipid peroxidation during mitosis (96), and there is a mathematical model that describes different lipid peroxide levels during the stages of the cell cycle (166,167) that may regulate mitosis through the recently described cell cycle-dependent inhibition of DNA synthesis by PGI$_2$, a process that is highly selective for different cell types (168,169).

Evidence is accumulating that vitamin E affects prostanoid metabolism by controlling lipoprotein oxidation and ultimately the arachidonic acid cascade. Lipid peroxides themselves and prostanoids have a number of regulatory functions, and it is not surprising that vitamin E has many and sometimes paradoxical effects in animals and cells (103,104,125,135). Indeed, vitamin E and antioxidants in general are disappointing in most pathological conditions (170), even though there is ample evidence to propose a therapeutic role (116).

Atherosclerosis provides a good model for the vitamin E-antioxidant dilemma (103, 104,125,135). There are many reasons to implicate lipid peroxides in general and oxidized lipoproteins in particular in the pathogenesis of atherosclerosis (52,171,172). Relatively high levels of lipid peroxides cause endothelial cell injury, diminish PGI$_2$ synthesis, promote platelet aggregation, and thereby elevate the level of platelet-derived growth factor. Since these events initiate and promote the disease, it is expected that antioxidants function as antiatherogenic agents, and indeed the antioxidant probucol slows progression of atherosclerosis in the Watanabe heritable hyperlipidemic rabbit (173,174). It would be interesting to study the effect of probucol on cellular injury, PGI$_2$ synthesis, and cell proliferation.

There are data to suggest that too much antioxidant has an adverse effect on atherosclerosis. Dipyridamole is a particularly attractive therapeutic agent since it functions as both

TABLE 3 Hypothesis for the Paradoxical Effects of Vitamin E on the Pathogenesis of Atherosclerosis

Vitamin E	TBAR	PGI_2	Proliferation	Atherosclerosis
Low	High	Absent	Up (injury)	Initiation
Intermediate	Low	High	Down	Controlled
High	Absent	Low	Up	Progression

Source: Adapted from and reproduced with permission from Plenum Press (125).

an antioxidant and an agent that elevates PGI_2 levels by protecting cyclooxygenase from oxidative inactivation (61–63). Interestingly enough, dipyridamole enhanced plaque formation in animals fed atherogenic diets (175–178). In a recent study, a massive dose of vitamin E was found to potentiate spontaneous atherosclerotic lesions in rabbits subjected to balloon catheterization and a high-cholesterol diet (179). Since the atherosclerotic lesion involves uncontrolled cell proliferation in response to injury and cell proliferation is regulated in part by lipid peroxides and prostanoids, we predicted that excess vitamin E would enhance atherosclerosis particularly in states in which injury had already occurred or been initiated by a process other than lipid peroxidation (103,104,125,135). These are conditions that obtain with atherogenic diets and mechanical injury.

Other data suggest that an antioxidant deficiency has an adverse effect on atherosclerosis. A thorough review of the early literature on vitamin E in ischemic heart disease was published in 1980 (170). Initial reports of the beneficial effects of vitamin E-supplemented diets were followed by a number of uncontrolled studies and fully controlled double-blind studies with supplemented diets that did not support the claim for the beneficial effect of vitamin E. However, recent data implicate a vitamin E deficiency in the development of atherosclerosis. A number of studies have shown that ischemic heart disease is inversely correlated with plasma vitamin E and that correlations for plasma cholesterol, triglyceride, vitamins A and C, and smoking, all factors that could be involved in oxidant stress, improve the correlation between a vitamin E deficiency and ischemic heart disease (180–183).

Oxidant stress and atherosclerosis do not have a simple cause-and-effect relationship. Low concentrations of linoleic acid and its desaturation product 6,9,12-octadecatrienoic acid in plasma and adipose tissue are correlated with increased heart disease even though decreased levels of these fatty acids are expected to diminish oxidant stress (180,184). Interestingly enough, desaturation-chain elongation metabolites of linoleic acid are prostanoid precursors.

We propose that there are specific levels of vitamin E necessary for optimal prostanoid synthesis through the arachidonic acid cascade (Table 3). When the vitamin E level is low or insufficient (unavailable in bulk lipid), oxidized lipoproteins injure the endothelium, block PGI_2 synthesis through the arachidonic acid cascade, and initiate atherosclerosis (Table 3). When the vitamin E level is high, oxidized lipoproteins are not available to stimulate the arachidonic acid cascade and modulate mitosis through this and other means, and as a consequence atherosclerosis is enhanced through uncontrolled cell proliferation (Table 3). There is an intermediate or optimal vitamin E level in which both the initiation of injury to the endothelium and uncontrolled cell proliferation are suppressed (Table 3). It will be important to identify the parameters that define the optimal vitamin E status for different dietary regimens and disease states, such as diabetes and hyperlipemia, before the therapeutic potential of vitamin E is realized.

REFERENCES

1. Weitzel G, Fretzdorff AM, Heller S. Grenzflächenuntersuchungen an Tokopherol-Verbindungen und am vitamin K_1. Hoppe-Seylers Z Physiol Chem 1956; 303:14-26.

2. Patil GS, Cornwell DG. Interfacial oxidation of α-tocopherol and the surface properties of its oxidation products. J Lipid Res 1978; 19:416-22.

3. Patil GS, Matthews RH, Cornwell DG. Kinetics of the processes of desorption from fatty acid monolayers. J Lipid Res 1973; 14:26-31.

4. Kwong CN, Heikkila RE, Cornwell DG. Properties of cholesteryl esters in pure and mixed monolayers. J Lipid Res 1971; 12:31-5.

5. Gibbons GF, Mitropoulos KA, Myant NB. Biochemistry of cholesterol. Amsterdam: Elsevier Biomedical Press, 1982; 109-30.

6. Lange Y, Ramos BV. Analysis of the distribution of cholesterol in the intact cell. J Biol Chem 1983; 258:15130-4.

7. Lange Y. Membrane cholesterol. In: Scanu AM, Spector AA, eds. Biochemistry and biology of plasma lipoproteins. New York: Marcel Dekker, 1986; 179-99.

8. Bloj B, Zilversmit DB. Rat liver proteins capable of transferring phosphatidylethanolamine. J Biol Chem 1977; 252:1613-9.

9. Crain RC, Zilversmit DB. Two nonspecific phospholipid exchange proteins from beef liver. 1. Purification and characterization. Biochemistry 1980; 19:1433-9.

10. Norum KR, Berg T, Helgerud P, Drevon CA. Transport of cholesterol. Physiol Rev 1983; 63:1343-419.

11. Oncley JL, Walton KW, Cornwell DG. A rapid method for the bulk isolation of β-lipoproteins from human plasma. J Am Chem Soc 1957; 79:4666-71.

12. Morrisett JD, Jackson RL, Gotto AM. Lipoproteins structure and function. Annu Rev Biochem 1975; 44:183-207.

13. Shen BW, Scanu AM, Kézdy FJ. Structure of human serum lipoproteins inferred from compositional analysis. Proc Natl Acad Sci U S A 1977; 74:837-41.

14. Kézdy FJ. Structural models and overview. In: Scanu AM, Wissler RW, Getz GS, eds. The biochemistry of atherosclerosis. New York: Marcel Dekker, 1979; 145-8.

15. Gallo-Torres HE. Transport and metabolism. In: Machlin LJ, ed. Vitamin E: a comprehensive treatise. New York: Marcel Dekker, 1980; 193-267.

16. Bjorneboe GE-A, Dreven CA. Absorption, transport and distribution of vitamin E. J Nutr 1990; 120:233-42.

17. Traber MG, Kayden HJ. Tocopherol distribution and localization in human adipose tissue. Am J Clin Nutr 1987; 46:488-95.

18. Bjornson LK, Gniewkowski C, Kayden HJ. Comparison of exchange of α-tocopherol and free cholesterol between rat plasma lipoproteins and erythrocytes. J Lipid Res 1975; 16:39-53.

19. Murphy DJ, Mavis RD. A comparison of the in vitro binding of α-tocopherol to microsomes of lung, liver, heart and brain of the rat. Biochim Biophys Acta 1981; 663:390-400.

20. Buttriss JL, Diplock AT. The relationship between α-tocopherol and phospholipid fatty acids in rat liver subcellular membrane fractions. Biochim Biophys Acta 1988; 962:81-190.

21. Cheng SC, Burton GW, Ingold KU, Foster DO. Chiral discrimination in the exchange of α-tocopherol stereoisomers between plasma and red blood cells. Lipids 1987; 22:469-73.

22. Catignani GL, Bieri JG. Rat liver α-tocopherol binding protein. Biochim Biophys Acta 1977; 497:349-57.

23. Verdon CP, Blumberg JB. An assay for the α-tocopherol binding protein mediated transfer of vitamin E between membranes. Anal Biochem 1988; 169:109-20.

24. Chow CK. Distribution of tocopherols in human plasma and red blood cells. Am J Clin Nutr 1975; 28:756-60.

25. Omaye ST, Chow FI. Distribution of vitamins A and E in blood and liver of rats depleted of vitamin A or vitamin E. Lipids 1986; 21:465-9.

26. Lehmann J, Rao DD, Canary JJ, Judd JT. Vitamin E and relationships among tocopherols in human plasma, platelets, lymphocytes, and red blood cells. Am J Clin Nutr 1988; 47:470-4.

27. McCormick EC, Cornwell DG, Brown JB. Studies on the distribution of tocopherol in human serum lipoproteins. J Lipid Res 1960; 1:221-8.

28. Bjornson LK, Kayden HJ, Miller E, Moshell AN. The transport of α-tocopherol and β-carotene in human blood. J Lipid Res 1976; 17:343-52.

29. Massay JB. Kinetics of transfer of α-tocopherol between model and native plasma lipoproteins. Biochim Biophys Acta 1984; 793:387-92.

30. Traber MG, Kayden HJ. Vitamin E is delivered to cells via the high affinity receptor for low-density lipoprotein. Am J Clin Nutr 1984; 40:747-51.

31. Niki E. Inhibition of oxidation of liposomal- and bio-membranes by vitamin E. In: Hayaishi O, Mino M, eds. Clinical and nutritional aspects of vitamin E. Amsterdam: Elsevier, 1987; 3-13.

32. Fukuzawa K. Site-specific induction of lipid peroxidation by iron in charged micelles and anti-oxidant effect of α-tocopherol. In: Hayaishi O, Mino M, eds. Clinical and nutritional aspects of vitamin E. Amsterdam: Elsevier, 1987; 25-36.

33. Zhang H, Jones KH, Davis WB, Whisler RL, Panganamala RV, Cornwell DG. Heterogeneity in lipid peroxides: cellular arachidonic acid metabolism and DNA synthesis. In: Kabara JJ, ed. The pharmacological effects of lipids, Vol. III. Champaign, IL: American Oil Chemists' Society, 1990; 255-65.

34. Deamer DW, Heikkila RE, Panganamala RV, Cohen G, Cornwell DG. The alloxan-dialuric acid cycle and the generation of hydrogen peroxide. Physiol Chem Phys 1971; 3:426-30.

35. Van Zutphen H, Cornwell DG. Some studies on lipid peroxidation in monomolecular and bimolecular lipid films. J Membr Biol 1973; 13:79-88.

36. Brownlee NR, Huttner JJ, Panganamala RV, Cornwell DG. Role of vitamin E in glutathione-induced oxidant stress: methemoglobin, lipid peroxidation, and hemolysis. J Lipid Res 1977; 18: 635-44.

37. Cohen G, Hochstein P. Generation of hydrogen peroxide in erythrocytes by hemolytic agents. Biochemistry 1964; 3:895-900.

38. Mezick JA, Settlemire CT, Brierley GP, Barefield KP, Jensen WN, Cornwell DG. Erythrocyte membrane interactions with menadione and the mechanism of menadione-induced hemolysis. Biochim Biophys Acta 1970; 219:361-71.

39. Smith MT, Evans CG, Thor H, Orrenius S. Quinone-induced oxidative injury to cells and tissues. In: Sies H, ed. Oxidative stress. London: Academic Press, 1985; 91-113.

40. Kunze H, Vogt W. Significance of phospholipase A for prostaglandin formation. Ann N Y Acad Sci 1971; 180:123-5.

41. Flower RJ, Blackwell GJ. The importance of phospholipase-A_2 in prostaglandin biosynthesis. Biochem Pharmacol 1976; 25:285-91.

42. Irvine RF. How is the level of free arachidonic acid controlled in mammalian cells? Biochem J 1982; 204:3-16.

43. Zhang H, Kaseki H, Davis WB, Whisler RL, Cornwell DG. Mechanisms for the stimulation of prostanoid synthesis by cyclosporine A and bacterial lipopolysaccharide. Transplantation 1989; 47:864-71.

44. Zhang H, Gaginella TS, Chen X, Cornwell DG. Action of bradykinin at the cyclooxygenase step in prostanoid synthesis through the arachidonic acid cascade. Agents Actions 1991; 34:397-404.

45. Lewis RA, Austen KF, Soberman RJ. Leukotrienes and other products of the 5-lipoxygenase pathway. N Engl J Med 1990; 323:645-55.

46. Yasuda M, Fujita T. Effect of lipid peroxidation on phospholipase A_2 activity of rat liver mitochondria. Jpn J Pharmacol 1977; 27:429-35.

47. Silk ST, Wong KTH, Marcus AJ. Arachidonic acid releasing activity in platelet membranes: effects of sulfydryl-modifying reagents. Biochemistry 1981; 20:391-7.

48. Kannagi R, Koizumi K, Masuda T. Limited hydrolysis of platelet membrane phospholipids. On the proposed phospholipase susceptible domain in platelet membranes. J Biol Chem 1981; 256: 1177-84.

49. Scott DL, White SP, Otwinowski Z, Yuan W, Gelb MH, Sigler PB. Interfacial catalysis: the mechanism of phospholipase A_2. Science 1990; 250:1541-6.

50. Steinbrecher UP, Parthasarathy S, Leake DS, Witztum JL, Steinberg D. Modification of low density lipoprotein by endothelial cells involves lipid peroxidation and degradation of low density lipoproteins phospholipids. Proc Natl Acad Sci U S A 1984; 81:3883-7.

51. Steinbrecher UP, Pritchard PH. Hydrolysis of phosphatidylcholine during LDL oxidation is mediated by platelet-activating factor acetylhydrolase. J Lipid Res 1989; 30:305-15.

52. Steinbrecher UP, Zhang H, Lougheed M. Role of oxidatively modified LDL in atherosclerosis. Free Radic Biol Med 1990; 9:155-68.

53. Prescott SM, Zimmerman GA, McIntyre TM. Platelet-activating factor. J Biol Chem 1990; 265: 17381-4.

54. Glaser KB, Asmis R, Dennis EA. Bacterial lipopolysaccharide priming of P338D, macrophage-like cells for enhanced arachidonic acid metabolism. J Biol Chem 1990; 265:8658-64.

55. Panganamala RV, Brownlee NR, Sprecher H, Cornwell DG. Evaluation of superoxide anion and singlet oxygen in the biosynthesis of prostaglandin from eicosa-8,11,14-trienoic acid. Prostaglandins 1974; 7:21-8.

56. Panganamala RV, Sharma HM, Sprecher H, Geer JC, Cornwell DG. A suggested role for hydrogen peroxide in the biosynthesis of prostaglandins. Prostaglandins 1974; 8:3-11.

57. Panganamala RV, Sharma HM, Heikkila RE, Geer JC, Cornwell DG. Role of hydroxyl radical scavengers dimethyl sulfoxide, alcohols and methional in the inhibition of prostaglandin biosynthesis. Prostaglandins 1976; 11:599-607.

58. Deby C, Deby-Dupont G. Oxygen species in prostaglandin biosynthesis in vitro and in vivo. In: Bannister WH, Bannister JV, eds. Biological and clinical aspects of superoxide and superoxide dismutase, Vol. 11B. New York: Elsevier/North Holland, 1980; 84-97.

59. Panganamala RV, Gavino VC, Cornwell DG. Effect of low and high methional concentration on prostaglandin biosynthesis in microsomes from bovine and sheep vesicular glands. Prostaglandins 1979; 17:155-62.

60. Weiss SJ, Turk J, Needleman P. A mechanism for the hydroperoxide-mediated inactivation of prostacyclin synthetase. Blood 1979; 53:1191-6.

61. Morisaki N, Stitts JM, Bartels-Tomei L, Milo GE, Panganamala RV, Cornwell DG. Dipyridamole: an antioxidant that promotes the proliferation of aorta smooth muscle cells. Artery 1982; 11:88-107.

62. Cornwell DG, Lindsey JA, Zhang H, Morisaki N. Fatty acid metabolism and cell proliferation. VI. Properties of antithrombotic agents that influence metastasis. In: Thaler-Dao H, Crastes de Paulet A, Paoletti R, eds. Icosanoids and cancer. New York: Raven Press, 1984; 205-22.

63. Deckmyn H, Gresele P, Arnout J, Todisco A, Vermylen J. Prolonging prostacyclin production by nafazatrom or dipyridamole. Lancet 1984; 2:410-1.

64. Gale PH, Egan RW. Prostaglandin endoperoxide synthase-catalyzed oxidation reactions. In: Pryor WA, ed. Free radicals in biology, Vol. VI. Orlando, FL: Academic Press, 1984; 1-38.

65. Iuliano L, Violi F, Ghiselli A, Alessandri C, Balsano F. Dipyridamole inhibits lipid peroxidation and scavenges oxygen radicals. Lipids 1989; 24:430-3.

66. Seeger W, Moser U, Roka L. Effects of alpha-tocopherol, its carboxylic acid chromane compound and two novel antioxidant isoflavanones on prostaglandin H synthase activity and autodeactivation. Naunyn Schmiedebergs Arch Pharmacol 1988; 338:74-81.

67. Hemler ME, Cook HW, Lands WEM. Prostaglandin biosynthesis can be triggered by lipid peroxides. Arch Biochem Biophys 1979; 193:340-5.

68. Lands WEM, Kulmacz RJ, Marshall PJ. Lipid peroxide actions in the regulation of prostaglandin biosynthesis. In: Pryor WA, ed. Free radicals in biology, Vol. VI. Orlando, FL: Academic Press, 1984; 39-61.

69. Marshall PJ, Warso WA, Lands WEM. Selective microdetermination of lipid hydroperoxides. Anal Biochem 1985; 145:192-9.

70. Zhang H, Davis WB, Chen X, Jones KH, Whisler RL, Cornwell DG. Effects of oxidized low density lipoproteins on arachidonic acid metabolism in smooth muscle cells. J Lipid Res 1990; 31:551-65.

71. Panganamala RV, Cornwell DG. The effects of vitamin E on arachidonic acid metabolism. Ann N Y Acad Sci 1982; 393:376-90.

72. Chang AC, Allen CE, Hegarty PVJ. The effects of vitamin E depletion and repletion on prostaglandin synthesis in semitendinosus muscle of young rabbits. J Nutr 1980; 110:66-73.

73. Carpenter MP. Antioxidant effects on the prostaglandin endoperoxide synthetase product profile. Fed Proc 1981; 40:189-94.

74. Okuma M, Takayama H, Uchino H. Generation of prostacyclin like substance and lipid peroxidation in vitamin E-deficient rats. Prostaglandins 1980; 19:527-36.

75. Chan AC, Leith MK. Decreased prostacyclin synthesis in vitamin E-deficient rabbit aorta. Am J Clin Nutr 1981; 34:2341-7.

76. Valentovic MA, Gairola C, Labawy WC. Lung, aorta and platelet metabolism of ^{14}C-arachidonic acid in vitamin E deficient rats. Prostaglandins 1982; 24:215-24.

77. Wu-Wang C-Y, Craig-Schmidt MC, Faircloth SA. Conversion of arachidonate to prostanoids by lung microsomes from rats fed varying amounts of vitamin E. Prostaglandins Leukot Med 1987; 26:291-8.

78. Panganamala RV, Miller JS, Gwebu ET, Sharma HM, Cornwell DG. Differential inhibitory effects of vitamin E and other antioxidants on prostaglandin synthetase, platelet aggregation and lipoxidase. Prostaglandins 1977; 14:261-71.

79. Agradi E, Petroni A, Socini A, Galli C. In vitro effects of synthetic antioxidants and vitamin E on arachidonic acid metabolism and thromboxane formation in human platelets and platelet aggregation. Prostaglandins 1981; 22:255-66.

80. Srivastava KC. Vitamin E exerts antiaggregatory effects without inhibiting the enzymes of the arachidonic acid cascade in platelets. Prostaglandins Leukot Med 1986; 21:177-85.

81. Hamelin SS, Chan AC. Modulation of platelet thromboxane and malonaldehyde by dietary vitamin E and linoleate. Lipids 1983; 18:267-9.

82. Toivanen JL. Effects of selenium, vitamin E and vitamin C on human prostacyclin and thromboxane synthesis in vitro. Prostaglandins Leukot Med 1987; 26:265-80.

83. Ali M, Gudbranson CG, McDonald JWD. Inhibition of human platelet cyclooxygenase by alpha-tocopherol. Prostaglandins Med 1980; 4:79-85.

84. Mower R, Steiner M. Biochemical interaction of arachidonic acid and vitamin E in human platelets. Prostaglandins Leukot Med 1983; 10:389-403.

85. Douglas CE, Chan AC, Choy PC. Vitamin E inhibits platelet phospholipase A_2. Biochim Biophys Acta 1986; 876:639-45.

86. Fukuzawa K, Kurotori Y, Tokumura A, Tsukatani H. Vitamin E deficiency increases the synthesis of platelet-activating factor (PAF) in rat polymorphonuclear leukocytes. Lipids 1989; 24:236-9.

87. Butler AM, Gerrard JM, Peller J, Stoddard SF, Rao GHR, White JG. Vitamin E inhibits the release of calcium from a platelet membrane fraction in vitro. Prostaglandins Med 1979; 2:203-16.

88. Cox AC, Rao GHR, Gerrard JM, White JG. The influence of vitamin E quinone on platelet structure, function and biochemistry. Blood 1980; 55:907-14.

89. Lindsey JA, Morisaki N, Stitts JM, Zager RA, Cornwell DG. Fatty acid metabolism and cell proliferation. IV. Effect of prostaglandin biosynthesis from endogenous fatty acid release with cyclosporin-A. Lipids 1983; 18:566-9.

90. Morisaki N, Tomei LD, Milo GE, Cornwell DG. Role of prostanoids and lipid peroxides as mediators of the 12-0-tetradecanoylphorbol-13-acetate effect on cell growth. Lipids 1985; 20:602-10.

91. Gavino VC, Miller JS, Ikharebha SO, Milo GE, Cornwell DG. Effect of polyunsaturated fatty acids and antioxidants on lipid peroxidation in tissue cultures. J Lipid Res 1981; 22:763-9.

92. Lindsey JA, Zhang H, Morisaki N, Sato T, Kaseki H, Cornwell DG. Fatty acid metabolism and cell proliferation. VII. Antioxidant effects of tocopherols and their quinones. Lipids 1985; 20: 151-7.

93. Juhász-Schäffer A. Arbeiten über das E-vitamin. II. Wirkung des E-vitamins auf explante in vitro. Virchows Arch Pathol Anat Physiol 1931; 281:35-45.

94. Mason KE. Differences in testes injury and repair after vitamin A-deficiency, vitamin E-deficiency, and inanition. Am J Anat 1933; 52:153-239.

95. Cornwell DG, Huttner JJ, Milo GE, Panganamala RV, Sharma HM, Geer JC. Polyunsaturated fatty acids, vitamin E, and the proliferation of aortic smooth muscle cells. Lipids 1979; 14:194-207.

96. Miller JS, Gavino VC, Ackerman GA, et al. Triglycerides, lipid droplets and lysosomes in aorta smooth muscle cells during the control of cell proliferation with polyunsaturated fatty acids and vitamin E. Lab Invest 1980; 42:495-506.

97. Gavino VC, Milo GE, Cornwell DG. Image analysis for the automated estimation of clonal growth and its application to the growth of smooth muscle cells. Cell Tissue Kinet 1982; 15:225-31.

98. Boogaerts MA, Van de Broeck J, Deckmyn H, Roelant C, Vermylen J, Verwilghen RL. Protective effect of vitamin E on immune triggered, granulocyte mediated endothelial injury. Thromb Haemost 1984; 51:89-92.

99. Kunisaki M, Umeda F, Inoguchi T, Ono H, Sako Y. Effect of vitamin E on prostacyclin production from cultured aortic endothelial cells. In: Chien S, ed. Vascular endothelium in health and disease. New York: Plenum Press, 1988; 113-7.

100. Huang N, Lineberger B, Steiner M. Alpha-tocopherol, a potent modulator of endothelial cell function. Thromb Res 1988; 50:547-57.

101. Tran K, Chan AC. *R,R,R,*-α-tocopherol potentiates prostacyclin release in human endothelial cells. Evidence for structural specificity of the tocopherol molecule. Biochim Biophys Acta 1990; 1043:189-97.

102. Chakraborti S, Gurtner GH, Michael JR. Oxidant-mediated activation of phospholipase A_2 in pulmonary endothelium. Am J Physiol (Lung Cellular Mol Physiol) 1989; 257:L430-7.

103. Cornwell DG, Morisaki N. Fatty acid paradoxes in the control of cell proliferation: prostaglandins, lipid peroxides and co-oxidation reactions. In: Pryor WA, ed. Free radicals in biology, Vol. VI. Orlando, FL: Academic Press, 1984; 95-148.

104. Cornwell DG, Zhang H. Fatty acid metabolism and cell proliferation. In: Sevanian A, ed. Lipid peroxidation in biological systems. Champaign, IL: American Oil Chemists' Society, 1988; 163-95.

105. Gavino VC, Miller JS, Dillman JM, Milo GE, Cornwell DG. Polyunsaturated fatty acid accumulation in the lipids of cultured fibroblasts and smooth muscle cells. J Lipid Res 1981; 22:57-62.

106. Morisaki N, Lindsey JA, Stitts JM, Zhang H, Cornwell DG. Fatty acid metabolism and cell proliferation. V. Evaluation of pathways for the generation of lipid peroxides. Lipids 1984; 19:381-94.

107. Morisaki N, Sprecher H, Milo GE, Cornwell DG. Fatty acid specificity in the inhibition of cell proliferation and its relationship to lipid peroxidation and prostaglandin biosynthesis. Lipids 1982; 17:893-9.

108. Corey EJ, Shih C, Cashman JR. Docosahexaenoic acid is a strong inhibitor of prostaglandin but not leukotriene biosynthesis. Proc Natl Acad Sci U S A 1983; 80:3581-4.

109. Hennig B, Chow CK. Lipid peroxidation and endothelial cell injury: implications in atherosclerosis. Free Radic Biol Med 1988; 4:99-106.

110. Parinandi NL, Weis BK, Natarajan V, Schmid HHO. Peroxidative modification of phospholipids in myocardial membranes. Arch Biochem Biophys 1990; 280:45-52.

111. Hope WC, Dalton C, Machlin LJ, Filipski RF, Vane FM. Influence of dietary vitamin E on prostaglandin biosynthesis in rat blood. Prostaglandins 1975; 10:557-71.

112. Halushka PY, Lurie D, Colwell J. Increased synthesis of prostaglandin-E-like material by platelets from patients with diabetes mellitus. N Engl J Med 1977; 297:1306-10.

113. Karpen CW, Pritchard KA Jr, Arnold JH, Cornwell DG, Panganamala RV. Restoration of prostacyclin/thromboxane A_2 balance in the diabetic rats. Influence of dietary vitamin E. Diabetes 1982; 31:947-51.

114. Karpen CW, Cataland S, O'Dorisio TM, Panganamala RV. Interrelation of platelet vitamin E and thromboxane synthesis in type I diabetes mellitus. Diabetes 1984; 33:239-43.

115. Watanabe J, Umeda F, Wakasugi H, Ibayashi H. Effect of vitamin E on platelet aggregation in diabetes mellitus. Thromb Haemost 1984; 51:313-6.

116. Gisinger C, Watanabe J, Colwell JA. Vitamin E and platelet eicosanoids in diabetes mellitus. Prostaglandins Leukot Essential Fatty Acids 1990; 40:169-76.

117. Karpen CW, Merola AJ, Trewyn RW, Cornwell DG, Panganamala RV. Modulation of platelet thromboxane A_2 and arterial prostacyclin by dietary vitamin E. Prostaglandins 1981; 22:651-61.

118. Karpen CW, Pritchard KA, Merola AJ, Panganamala RV. Alterations of the prostacyclin-thromboxane ratio in streptozotocin induced diabetic rats. Prostaglandins Leukot Med 1982; 8:93-103.

119. Pritchard KA, Karpen CW, Merola AJ, Panganamala RV. Influence of dietary vitamin E on platelet thromboxane A_2 and vascular prostacyclin I_2 in rabbit. Prostaglandins Leukot Med 1982; 9:373-8.

120. Gomes JAC, Venkatachalapathy D, Haft JI. The effect of vitamin E on platelet aggregation. Am Heart J 1976; 91:425-9.

121. Hwang DH, Donovan J. In vitro and in vivo effects of vitamin E on arachidonic acid metabolism in rat platelets. J Nutr 1982; 112:1233-7.

122. Stampfer MJ, Jakubowski JA, Faigel D, Vailancourt R, Deykin D. Vitamin E supplementation effect on human platelet function, arachidonic acid metabolism, and plasma prostacyclin levels. Am J Clin Nutr 1988; 47:700-6.

123. Kockmann V, Vericel E, Croset M, Lagarde M. Vitamin E fails to alter the aggregation and the oxygenated metabolism of arachidonic acid in normal human platelets. Prostaglandins 1988; 36:607-20.

124. Panganamala RV, Karpen CW, Merola AJ. Peroxide mediated effects of homocysteine on arterial prostacyclin synthesis. Prostaglandins Leukot Med 1986; 22:349-56.

125. Cornwell DG, Zhang H, Davis WB, Whisler RL, Panganamala RV. Paradoxical effects of vitamin E: oxidized lipoproteins, prostanoids and the pathogenesis of atherosclerosis. In: Crastes de Paulet A, Douste-Blazy L, Paoletti R, eds. Free radicals, lipoproteins, and membrane lipids. New York: Plenum Press, 1990; 215-37.

126. Ray BR, Davisson EO, Crespi HL. Experiments on the degradation of lipoproteins from serum. J Phys Chem 1954; 58:841-6.

127. Krinsky NI, Cornwell DG, Oncley JL. The transport of vitamin A and carotenoids in human plasma. Arch Biochem Biophys 1958; 73:233-46.

128. Cornwell DG, Kruger FA, Robinson HB. Studies on the absorption of beta-carotene and the distribution of total carotenoid in human serum lipoproteins after oral administration. J Lipid Res 1962; 3:65-70.

129. Yagi K. Assay for serum lipid peroxide level and its clinical significance. In: Yagi K, ed. Lipid peroxides in biology and medicine. New York: Academic Press, 1982; 223-42.

130. Szczeklik A, Gryglewski RJ. Low density lipoproteins (LD) are carriers for lipid peroxides and inhibit prostacyclin (PGI$_2$) biosynthesis in arteries. Artery 1980; 7:488-95.

131. Beitz J, Panse M, Fischer S, Hora C, Förster W. Inhibition of prostaglandin I$_2$ (PGI$_2$) formation by LDL-cholesterol or LDL-peroxides? Prostaglandins 1983; 26:885-92.

132. Bittolo-Bon G, Cazzolato G, Saccardi M, Avogaro P. Presence of a modified LDL in humans: effect of vitamin E. In: Hayaishi O, Mino M, eds. Clinical and nutritional aspects of vitamin E. Amsterdam: Elsevier Science Publishers, 1987; 109-20.

133. Avogaro P, Bon GB, Cazzolato G. Presence of a modified low density lipoprotein in humans. Arteriosclerosis 1988; 8:79-87.

134. Lee DM. Malondialdehyde formation in stored plasma. Biochem Biophys Res Commun 1980; 95:1662-72.

135. Zhang H, Jones KH, Davis WB, Whisler RL, Panganamala RV, Cornwell DG. Oxidized low density lipoproteins in smooth muscle cell cultures: differential effects on prostanoid synthesis and viability. In: Hayaishi O, Mino M, eds. Clinical and nutritional aspects of vitamin E. Amsterdam: Elsevier Science Publishers, 1987; 89-100.

136. Zhang H, Davis WB, Chen X, Whisler RL, Cornwell DG. Studies on oxidized low density lipoproteins. Controlled oxidation and a prostaglandin artifact. J Lipid Res 1989; 30:141-8.

137. Esterbauer H, Dieber-Rotheneder M, Striegl G, Waeg G. Role of vitamin E in preventing the oxidation of low-density lipoprotein. Am J Clin Nutr 1991; 53:314S-21S.

138. Esterbauer H, Jürgens G, Quehenberger O, Koller E. Autoxidation of human low density lipoprotein: loss of polyunsaturated fatty acids and vitamin E and generation of aldehydes. J Lipid Res 1987; 28:495-509.

139. Jessup W, Rankin SM, deWhalley CV, Hoult JRS, Scott J, Leake DS. α-tocopherol consumption during low density lipoprotein oxidation. Biochem J 1990; 265:399-405.

140. Morrow JD, Roberts LJ. Quantification of noncyclooxygenase derived prostanoids as a marker of oxidative stress. Free Radic Biol Med 1991; 10:195-200.

141. Gardner HW. Oxygen radical chemistry of polyunsaturated fatty acids. Free Radic Biol Med 1989; 7:65-86.

142. Jones KH, Senft JA. An improved method to determine cell viability by simultaneous staining with fluorescein diacetate-propidium iodide. J Histochem Cytochem 1985; 33:77-9.

143. Wang J, Zhen E, Guo Z, Lu Y. Effect of hyperlipidemic serum on lipid peroxidation, synthesis of prostacyclin and thromboxane by cultured endothelial cells: protective effect of antioxidants. Free Radic Biol Med 1989; 7:243-9.

144. Gwebu ET, Trewyn RW, Cornwell DG, Panganamala RV. Vitamin E and the inhibition of platelet lipoxygenase. Res Commun Chem Pathol Pharmacol 1980; 28:361-76.

145. Goetzl EJ. Vitamin E modulates the lipoxygenation of arachidonic acid in leukocytes. Nature 1980; 288:183-5.

146. Reddanna P, Rao MK, Reddy CC. Inhibition of 5-lipoxygenase by vitamin E. FEBS Lett 1985; 193:39-43.

147. Karpen CW, Cataland S, O'Dorisio TM, Panganamala RV. Production of 12-hydroxyeicosate-traenoic acid and vitamin E status in platelets from type I human diabetic subjects. Diabetes 1985; 34:526-31.

148. Pritchard KA, Greco NJ, Panganamala RV. Effect of dietary vitamin E on the production of platelet 12-hydroxyeicosatetraenoic acid (12-HETE). Thromb Haemost 1986; 55:6-7.

149. Chan AC, Tran K, Pyke DD, Powell WS. Effect of dietary vitamin E on the biosynthesis of 5-lipoxygenase products by rat polymorphonuclear leukocytes (PMNL). Biochim Biophys Acta 1989; 1005:265-9.

150. Chan AC, St Maurice ST, Douglas CE. Effect of dietary vitamin E on platelet tocopherol values and 12-lipoxygenase activity. Proc Soc Exp Biol Med 1988; 187:197-201.

151. Colette C, Pares-Herbute N, Monnier LH, Cartry E. Platelet function in type I diabetes: effects of supplementation with large doses of vitamin E. Am J Clin Nutr 1988; 47:256-61.

152. Pomerantz KB, Tall AR, Feinmark SJ. Cannon PJ. Stimulation of vascular smooth muscle cell prostacyclin and prostaglandin E$_2$ synthesis by plasma high and low density lipoproteins. Circ Res 1984; 54:554-65.

153. Giessler C, Beitz J, Mentz P, Förster W. The influence of lipoproteins (LDL and HDL) on PGI$_2$ formation by isolated aortic preparations of rabbits. Prostaglandins Leukot Med 1986; 22:221-34.

154. Triau JE, Meydani SN, Schaefer EJ. Oxidized low density lipoproteins stimulate prostacyclin (PGI$_2$) production by adult human vascular endothelial cells. Arteriosclerosis 1988; 8:810-8.

155. Yokode M, Kita T, Kikawa Y, Ogorochi T, Narumiya S, Kawai C. Stimulated arachidonate metabolism during formal cell transformation of mouse peritoneal macrophages with oxidized low density lipoprotein. J Clin Invest 1988; 81:720-9.

156. Hessler JR, Robertson AL, Chisolm GM. LDL-induced cytotoxicity and its inhibition by HDL in human vascular smooth muscle and endothelial cells in culture. Atherosclerosis 1979; 32:213-29.

157. Henriksen T, Evensen SA, Carlander B. Injury to human endothelial cells in culture induced by low density lipoproteins. Scand J Clin Lab Invest 1979; 39:361-8.

158. Hessler JR, Morel DW, Lewis LJ, Chisolm GM. Lipoproteins oxidation and lipoprotein-induced cytotoxicity. Arteriosclerosis 1983; 3:215-22.

159. Morel DW, Hessler JR, Chisolm GM. Low density lipoprotein cytotoxicity induced by free radical peroxidation in lipids. J Lipid Res 1983; 24:1070-6.

160. Morel DW, DiCorleto PE, Chisolm GM. Endothelial and smooth muscle cells alter low density lipoprotein in vitro by free radical oxidation. Arteriosclerosis 1984; 4:357-64.

161. Brown BG, Mahley R, Assmann G. Swine aortic smooth muscle in tissue culture (some effect of purified swine lipoprotein on cell growth and morphology). Circ Res 1976; 39:415-24.

162. Fisher-Dzoga K, Fraser RA, Wissler RW. Stimulation of proliferation in stationary primary cultures of monkey and rabbit aortic smooth muscle cells. I. Effects of lipoprotein fractions of hyperlipemic serum and lymph. Exp Mol Pathol 1976; 24:346-59.

163. Layman DL, Jelen BJ, Illingworth DR. Inability of serum from abetalipoproteinemic subjects to stimulate proliferation of human smooth muscle cells and dermal fibroblasts in vitro. Proc Natl Acad Sci U S A 1980; 77:1511-5.

164. Tauber JP, Cheng J, Gospodarowicz D. Effect of high and low density lipoproteins on proliferation of cultured bovine vascular endothelial cells. J Clin Invest 1980; 66:696-708.

165. Huttner JJ, Gwebu ET, Panganamala RV, et al. Fatty acids and their prostaglandin derivatives: inhibitors of proliferation in aortic smooth muscle cells. Science 1977; 197:289-91.

166. Chernavskii DS, Polezhaev AA, Volkov EI. Cell surface and cell division. Cell Biophys 1982; 4:143-61.

167. Mustafin AT, Volkov EI. The role of lipid and antioxidant exchange in cell division synchronization (mathematical model). Biol Cybern 1984; 49:149-54.

168. Morisaki N, Kanzaki T, Motoyama N, Saito Y, Yoshida S. Cell cycle-dependent inhibition of DNA synthesis by prostaglandin I$_3$ in cultured rabbit aortic smooth muscle cells. Atherosclerosis 1988; 71:165-71.

169. Morisaki N, Kanzaki T, Sato Y, Yoshida S. Lack of inhibition of DNA synthesis by prostaglandin I_2 in cultured intimal smooth muscle cells from rabbits. Atherosclerosis 1988; 73:67-9.

170. Farrell PM. Deficiency states, pharmacological effects, and nutrient requirements. In: Machlin LJ, ed. Vitamin E, a comprehensive treatise. New York: Marcel Dekker, 1980; 520-620.

171. Steinberg D. Metabolism of lipoproteins and their role in the pathogenesis of atherosclerosis. Atherosclerosis Rev 1988; 18:1-23.

172. Steinberg D. Arterial metabolism of lipoproteins in relation to atherogenesis. Ann N Y Acad Sci 1990; 598:125-35.

173. Carew TE, Schwanke DC, Steinberg D. Antiatherogenic effect of probucol unrelated to its hypocholesterolemic effect: evidence that antioxidants in vivo can selectively inhibit low density lipoprotein degradation in macrophage-rich fatty streaks slow the progression of atherosclerosis in the Watanabe heritable hyperlipidemic rabbit. Proc Natl Acad Sci U S A 1987; 84:7725-9.

174. Kita T, Nagano Y, Yokode M, et al. Probucol prevents the progression of atherosclerosis in Watanabe heritable hyperlipidemic rabbit, an animal model for familial hypercholesterolemia. Proc Natl Acad Sci U S A 1987; 84:5928-31.

175. Dembinska-Kièc A, Rucker W, Schönhöfer PS. Effects of dipyridamole in experimental atherosclerosis. Action on PGI_2, platelet aggregation and atherosclerotic plaque formation. Atherosclerosis 1979; 33:315-27.

176. Dembinska-Kièc A, Rucker W, Schönhöfer PS. Effects of dipyridamole in vivo on ATP and CAMP content in platelets and arterial walls and on atherosclerotic plaque formation. Naunyn Schmiedebergs Arch Pharmacol 1979; 309:59-64.

177. Koster JK Jr, Tryka AF, H'Doubler P, Collins JJ Jr. The effect of low-dose aspirin and dipyridamole upon atherosclerosis in the rabbit. Artery 1981; 9:405-13.

178. Hollander W, Kirkpatrick B, Paddock J, Colombo M, Nagraj S, Prusty S. Studies on the progression and regression of coronary and peripheral atherosclerosis in the cynomolgus monkey. I. Effects of dipyridamole and aspirin. Exp Mol Pathol 1979; 30:55-73.

179. Godfried SL, Combs GF Jr, Saroka JM, Dillingham LA. Potentiation of atherosclerotic lesions in rabbits by a high dietary level of vitamin E. Br J Nutr 1989; 61:607-17.

180. Riemersma RA, Wood DA, Macintyre CCA, Elton R, Gey KF, Oliver MF. Low plasma vitamin E and C increased the risk of angina in scottish men. Ann N Y Acad Sci 1989; 570:291-5.

181. Duthie GG, Arthur JR, James WPF, Vint HM. Antioxidant status of smokers and nonsmokers: effect of vitamin E supplements. Ann N Y Acad Sci 1989; 570:435-8.

182. Gey KF, Puska P, Jordan P, Moser UK. Inverse correlation between plasma vitamin E and mortality from ischemic heart disease in cross-cultural epidemiology. Am J Clin Nutr 1991; 53:326S-45S.

183. Thurnham DI, Davies JA, Crump BJ, Situnayake RD, Davis M. The use of different lipids to express serum tocopherol: lipid ratios for the measurement of vitamin E status. Ann Clin Biochem 1986; 23:514-20.

184. Riemersma RA, Wood DA, Butler S, et al. Linoleic acid content of adipose tissue and coronary heart disease. Br Med J 1986; 292:1423-7.

30

The Safety of Oral Intake of Vitamin E: Data from Clinical Studies from 1986 to 1991

Adrianne Bendich and Lawrence J. Machlin

Hoffmann-La Roche, Inc., Nutley, New Jersey

INTRODUCTION

In 1988, a comprehensive review of the safety of vitamin E was published (1) that included references until 1986. This chapter concentrates on 29 peer-reviewed articles published since 1986 dealing with the human use of vitamin E supplements. These include 14 studies in which vitamin E supplements were administered in either unblinded or non–placebo-controlled protocols and 15 involving placebo-controlled, single- or double-blind investigations. None of the intervention studies reported any adverse effects of vitamin E, confirming the conclusions reached in the 1988 review that vitamin E supplementation at high doses is not associated with any clinically relevant adverse effects.

PLACEBO-CONTROLLED, SINGLE- OR DOUBLE-BLIND STUDIES

Table 1 outlines the 15 study groups that were given dosages of 250–2400 IU vitamin E for durations ranging from 28 to 1643 days (4.5 years). None of the studies reported any adverse effects of vitamin E supplementation. The subjects' ages ranged from children to the elderly and included healthy individuals and those with chronic diseases. In several studies, vitamin E was administered along with another nutrient or drug. Regardless of these differences in population groups, none experienced side effects attributable to vitamin E.

Only two of the studies (2,3) specifically evaluated vitamin E safety issues, such as thyroid hormone levels, which were previously reported to decrease with supplementation (4); serum creatinine levels, which were previously reported to increase with supplementation (5); or serum lipid levels, with which the response to supplementation has been variable (1). The two studies (2,3) are reviewed in detail because the participants were from dissimilar cultural backgrounds with different dietary intakes, involved adults with ages spanning over 40 years, included only healthy individuals, and were at dosage levels commonly taken by vitamin E supplement users.

Meydani et al. (2), in a double-blind study, gave healthy adults, mainly of European descent (whites) from the Boston area over 60 years of age, either 800 IU vitamin E or placebo for 30 days. All subjects were housed in a metabolic ward throughout the study period so that

TABLE 1 Lack of Adverse Effects of Vitamin E Supplementation in Placebo-Controlled Studies

Dose (IU)	Duration (days)	Total study population	Group	Reference
250	90	12 adult, 6 children epileptics	Sullivan et al., 1990	12
400	28	13 adult mountaineers	Simon-Schnass and Korniszewski, 1990	13
400 + 400 mg vitamin C	720	185 adults with colo-rectal polyps	McKeown-Eyssen et al., 1988	14
400	14 followed by	15 neuroleptic	Lohr et al., 1988	15
800	14 followed by	inpatients		
1200	14			
600 + 600 μg selenium	90	60 adults with atopic dermatitis	Fairris et al., 1989	16
600	60	30 insulin-dependent diabetics	Ceriello et al., 1991	17
800	30	32 adults $\geqslant$ 60 years	Meydani et al., 1990	2
900	42	60 adult smokers	Richards et al., 1990	18
900	84	14 college students	Kitagawa and Mino, 1989	3
1000	35	18 insulin-dependent diabetics	Colette et al., 1988	19
1200	60	30 insulin-dependent diabetics	Ceriello et al., 1991	17
1200 + 325 mg aspirin	120	100 coronary care patients	DeMaio et al., 1991	20
1200 + penicillamine	540	38 patients with Duchenne muscular dystrophy	Fenichel et al., 1988	21
2000 + Deprenyl	1,643	800 patients with Parkinson's disease	Parkinson's Study Group, 1989	6
2400 + penicillamine	540	68 patients with Duchenne muscular dystrophy	Fenichel et al., 1988	21

diets were matched for nutrient intakes. Hepatic function, as determined by serum albumin, liver enzymes (SGPT, SGOT, LDH, and alkaline phosphatase), and total bilirubin (including conjugated and unconjugated) were unaffected by vitamin E supplementation, and values in the supplemented group did not differ from those seen in the placebo group. Hematological analyses included total and individual white blood cell counts, red blood cell and platelet counts, hemoglobin, and hematocrit, which were all unaffected by vitamin E. Renal function, as measured by urinary and serum creatinine, creatinine clearance, and blood urea nitrogen (BUN) levels, were unchanged by vitamin E. Thyroid function, as reflected by T_3 and T_4 uptakes and ratios, as well as free thyroxine, were also unchanged. Two important changes were seen: a significant decrease in serum lipid peroxides and a significant increase in serum zinc values in the vitamin E-supplemented group. There were also improvements in clinically relevant immunological parameters in the supplemented group.

In the second study, Kitagawa and Mino (3) gave 14 Japanese college students 900 IU vitamin E for 12 weeks and gave 5 students matched placebos in a single-blinded study. There were no changes in serum liver enzymes (SGOT and SGPT), white blood cell, red blood cell, and platelet counts, hemoglobin level, creatinine phosphokinase level, or BUN in either group during the study. Serum thyroid hormone levels (T_3, T_4, and TSH) were also unchanged. Bleeding times, as measured by percentage prothrombin time, partial thrombin time, hepatoplastin test, and PIVKA-II (protein induced by vitamin K absence or antagonist), were unaffected by vitamin E supplementation. Serum indices of cholesterol, phospholipids, triglycerides, and total lipids were similar in the placebo and supplemented groups and were unchanged following vitamin E supplementation. Using a questionnaire, it was shown that there were no differences between the placebo and vitamin E-supplemented group in reports of muscle weakness, gastrointestinal discomfort, headache, hypertension, or visual changes.

In 1987 the DATATOP (deprenyl and tocopherol) study was initiated (6). Adults ($n = 800$) with early signs of Parkinson's disease were entered into a placebo double-blind protocol

TABLE 2 Lack of Adverse Effects of Vitamin E Supplementation in Unblinded Studies: Single-Dose Studies

Dose (IU)	Duration (days)	Study population	Comments	Group	Reference
200	60	30 female oral contraceptive users age 29–42 years	Normalizes hyperactive platelet aggregation and clotting activity	Renaud et al., 1987	22
200	182	26 adults	—	Harman and	23
400	185	25 adults	—	Miller, 1986	
800	14	13 smokers	Breath pentane decreased	Hoshino et al., 1990	24
800	42	8 adults	Significant increase in HDL in 3 individuals	Muckle and Nazir, 1989	25
800	60	36 male children (3–12 years) with G6PD deficiency	Increased red cell half-life, and hemoglobin level and decreased reticulocytosis; "no side effects."	Hafez et al., 1986	26
800	56-189	23 males (0.8–22 years) with G6PD deficiency	Increased hemoglobin level and decreased reticulocytosis	Eldamhougy et al., 1988	27
800	182	45 myeloplastic patients treated with 13-*cis*-retinoic acid	"There were no reports of intolerance . . ."; improved tolerance to 13-*cis*-retinoic acid	Besa et al., 1990	28
1000	10	10 adults	Decreased pentane exhalation	Van Gossum et 1988	29
1200	72	8 adults	—	Handelman et al., 1985	30
1600	120	20 cancer patients	"No toxicity attributable to alpha tocopherol was observed"	Perez et al., 1986	31
2000	10	6 males	No influence on bleeding time or platelet aggregation	Huijgens et al., 1981	11

in which the vitamin E supplement was 2000 IU/day. The code for the vitamin E portion of the study has not been broken as yet; however, approximately 400 adults have been using 2000 IU vitamin E daily for 4½ years and no clinically relevant adverse effects have been reported by the study's safety committee.

NONBLINDED OR UNCONTROLLED STUDIES

In 14 studies outlined in Tables 2 and 3, no deleterious effects were reported with doses of vitamin E ranging from 200 to 1200 IU per day for 14–189 days. The effect of vitamin E in decreasing platelet adhesion was not considered an adverse effect but a potential health benefit by reducing risk of thrombosis (7,8). However, Churukian et al. (9) reported nose bleeding following rhinoplasty in two subjects taking 800–1200 IU/day. No bleeding was found in 98 patients taking 400 IU or less per day, suggesting a possible association between high intake of vitamin E and postoperative nose bleeding. No bleeding disorders or abnormalities in clotting factors have been previously reported unless subjects were simultaneously on anticoagulant (warfarin) therapy (1). A retrospective epidemiological study of over 2000 elderly individuals showed no differences in bleeding disorders between vitamin E users and nonusers (10). Huijgens et al. (11) found no effect on bleeding time in 6 healthy volunteers supplemented with 2000 IU for 10 days, and Kitagawa and Mino (3) in a placebo-controlled study found no effect on bleeding time in 14 subjects given 900 IU/day for 12 weeks. Nevertheless, in view of the known effect of high doses of vitamin E on platelet adhesion and the report of postoperative bleeding by Churukian et al. (9), it would be advisable to discourage ingestion of high doses of vitamin E for 2 weeks before and following surgery.

In a retrospective epidemiological study of ambulatory elderly including 369 vitamin E users and 1861 nonusers, no differences in vaginal bleeding, hypertension, headache, fatigue,

TABLE 3 Lack of Adverse Effects of Vitamin E Supplementation in Unblinded Studies: Multiple-Dose Studies

Dose (IU)	Duration (days)	Study population	Comments	Group	Reference
200 and 400	14 (at each level)	6 adults	200 IU dose followed by 400 IU dose in same subjects; decreased platelet adhesion at both levels	Jandak et al., 1989	7
200, 800, and 1200	14 (at each level)	6 males and 6 females	Doses given sequentially; no effect on platelet aggregation, but decreased platelet adhesion; same regime with aspirin also with no adverse effects reported	Steiner, 1983	8
400, 800, and 1200	28 28 28	8 adults 23 adults 18 adults	—	Dimitrov et al., 1991	32

dizziness, or hypo- or hyperthyroidism were observed (10). Men using vitamin E complained more often of shortness of breath and angina, but the men using vitamin E were more likely to be taking digoxin, diuretics, and propanalol than men who did not use vitamin E. Therefore, it is likely that many of these men were already experiencing symptoms of angina and shortness of breath, and they may have sought to prevent the symptoms with vitamin E.

CONCLUSIONS

Vitamin E has been shown to decrease platelet adhesion, a potential health benefit due to reducing risk of thrombosis. However, in view of the two reported cases of postsurgery bleeding associated with intakes of 800–1200 IU of vitamin E per day, it is advisable to discourage vitamin E supplementation 2 weeks before and following surgery. In addition, based on earlier studies indicating synergistic activities between vitamin K antagonists and vitamin E, high levels of vitamin E are contraindicated in individuals receiving anticoagulant therapy.

Review of clinical studies in which doses of vitamin E of 200 mg/day or more for periods up to 4½ years have consistently shown no adverse effects associated with supplementation. The current recommended daily allowance (RDA) for an adult male is equivalent to 15 IU or 15 mg of *d,l-α*-tocopheryl acetate, the form usually found in vitamin supplements. Clearly, the data in this review and the earlier publication (1) show that doses of vitamin E of more than 50 times the RDA are safe.

REFERENCES

1. Bendich AB, Machlin LJ. Safety of oral intake of vitamin E. Am J Clin Nutr 1988; 48:612-9.
2. Meydani SN, Barklund MP, Liu S, et al. Vitamin E supplementation enhances cell-mediated immunity in healthy elderly subjects. Am J Clin Nutr 1990; 52:557-63.
3. Kitagawa M, Mino M. Effects of elevated d-alpha(*RRR*)-tocopherol dosage in man. J Nutr Sci Vitaminol 1989; 35:133-42.
4. Tsai AC, Kelley JJ, Peng B, Cook N. Study on the effect of megavitamin E supplementation in man. Am J Clin Nutr 1978; 31:831-7.
5. Briggs M. Vitamin E supplements and fatigue. N Engl J Med 1974; 290:579-80.
6. Parkinson study group. Effect of deprenyl on the progression of disability in early Parkinson's Disease. N Engl J Med 1989; 321:1364-71.
7. Jandak J, Steiner M, Richardson PD. Alpha-tocopherol, an effective inhibitor of platelet adhesion. Blood 1989; 73:141-9.
8. Steiner M. Effect of alpha-tocopherol administration on platelet function in man. Thromb Haemost 1983; 49:73-7.
9. Churukian MM, Zemplenyi J, Steiner M, Kramer FM, Cohen A. Postrhinoplasty epistaxis. Arch Otolaryngol Head Neck Surg 1988; 114:748-50.
10. Hale WE, Perkins LL, May FE, Marks RG, Stewart RB. Vitamin E effect on symptoms and laboratory values in the elderly. J Am Diet Assoc 1986; 86:625-9.
11. Huijgens PC, van de Berg CAM, Imandt LMFM, Langenhuijsen MMAC. Vitamin E and platelet aggregation. Acta Haematol (Basel) 1981; 65:217-8.
12. Sullivan C, Capaldi N, Mack G, Buchanan N. Seizures and natural vitamin E. Med J Aust 1990; 152:613-4.
13. Simon-Schnass I, Korniszewski L. The influence of vitamin E on rheological parameters in high altitude mountaineers. Int J Vitam Nutr Res 1990; 60:26-34.
14. McKeown-Eyssen G, Holloway C, Jazmaji V, Bright-See E, Dion P, Bruce WR. A randomized trial of vitamins C and E in the prevention of recurrence of colorectal polyps. Cancer Res 1988; 48:4701-5.
15. Lohr JB, Cadet JL, Lohr MA, et al. Vitamin E in the treatment of tardive dyskinesia: the possible involvement of free radical mechanisms. Schizophr Bull 1988; 14:291-6.

16. Fairris GM, Perkins PJ, Lloyd B, Hinks L, Clayton BE. The effect of atopic dermatitis supplementation with selenium and vitamin E. Acta Derm Venereol (Stockh) 1989; 69:359-62.

17. Ceriello A, Giugliano D, Quatraro A, Donzella C, Dipalo G, Lefebvre PJ. Vitamin E reduction of protein glycosylation in diabetes. Diabetes Care 1991; 14:68-72.

18. Richards GA, Theron AJ, van Rensburg CEJ, et al. Investigation of the effects of oral administration of vitamin E and beta-carotene on the chemiluminescence responses and the frequency of sister chromatid exchanges in circulating leukocytes from cigarette smokers. Am Rev Respir Dis 1990; 142:648-54.

19. Colette C, Pares-Herbute N, Monnier LH, Cartry E. Platelet function in type I diabetes: effects of supplementation with large doses of vitamin E. Am J Clin Nutr 1988; 47:256-61.

20. DeMaio SJ, King SB, Lembo NJ, et al. Vitamin E supplementation, plasma lipids and incidence of restenosis after percutaneous transluminal coronary angioplasty (PTCA). (in press).

21. Fenichel GM, Brooke MH, Griggs RC, et al. Clinical investigation in Duchenne muscular dystrophy: penicillamine and vitamin E. Muscle Nerve 1988; 11:1164-8.

22. Renaud S, Ciavatti M, Perrot L, Berthezene F, Dargent D, Condamin P. Influence of vitamin E administration on platelet functions in hormonal contraceptive users. Contraception 1987; 36: 347-58.

23. Harman D, Miller RW. Effect of vitamin E on the immune response to influenza virus vaccine and the incidence of infectious disease in man. Age 1986; 9:21-3.

24. Hoshino E, Shariff R, van Gossum A, et al. Vitamin E suppresses increased lipid peroxidation in cigarette smokers. J Parenteral Enteral Nutr 1990; 14:300-5.

25. Muckle TJ, Nazir DJ. Variation in human blood high-density lipoprotein response to oral vitamin E megadosage. Am J Clin Pathol 1989; 91:165-71.

26. Hafez M, Amar E-S, Zedan M, et al. Improved erythrocyte survival with combined vitamin E and selenium therapy in children with glucose-6-phosphate dehydrogenase deficiency and mild chronic hemolysis. J Pediatr 1986; 108:558-61.

27. Eldamhougy S, Elhelw Z, Yamamah G, Hussein L, Fayyad I, Fawzy D. The vitamin E status among glucose-6-phosphate dehydrogenase deficient patients and effectiveness or oral vitamin E. Int J Vitam Nutr Res 1988; 58:184-8.

28. Besa EC, Abrahm JL, Bartholomew MJ, Hyzinski M, Nowell PC. Treatment with 13-*cis*-retinoic acid in transfusion-dependent patients with myelodysplastic syndrome and decreased toxicity with addition of alpha-tocopherol. Am J Med 1990; 89:739-47.

29. Van Gossum A, Kurian R, Whitwell J, Jeejeebhoy KN. Decrease in lipid peroxidation measured by breath pentane output in normals after oral supplementation with vitamin E. Clin Nutr 1988; 7:53-7.

30. Handelman GJ, Machlin LJ, Fitch K, Weiter JJ, Dratz EA. Oral alpha tocopherol supplements decrease plasma gamma tocopherol levels in humans. J Nutr 1985; 115:807-13.

31. Perez JE, Macchiavelli M, Leone BA, et al. High-dose alpha tocopherol as a preventive of doxorubicin-induced alopecia. Cancer Treat Rep 1986; 70:1213-4.

32. Dimitrov NV, Meyer C, Gilliand D. Ruppenthal M, Chenoweth W, Malone W. Plasma tocopherol concentrations in response to supplemental vitamin E. Am J Clin Nutr 1991; 53:723-9.

31

Reduction in the Toxicity of Doxorubicin (Adriamycin) in the Heart by Vitamin E

José Milei

Juan A. Fernández Hospital, Buenos Aires, Argentina

Susana Llesuy

University of Buenos Aires, Buenos Aires, Argentina

INTRODUCTION

Role of Anthracyclines in Oncology

Adriamycin is an antibiotic belonging to the anthracyclinic structures, obtained by cultivation of *Streptomyces peucetius* var. *caesius* (1). Anthracyclines play a major role in cancer chemotherapy (2,3). Adriamycin (ADM) is effective as an antineoplastic drug. Unfortunately, the use of these agents has been hampered by conventional toxicities (nausea, vomiting, alopecia, and hematopoietic suppression), as well as an unique cardiotoxicity manifested by congestive cardiomyopathy. Despite these side effects, ADM is widely employed, the anthracyclinic most sold all over the world.

From the oncological point of view ADM is one of the most effective drugs active against a wide variety of human neoplasms, in particular against solid tumors (4).

Because of the importance of ADM in the treatment of human tumors, several analogs of this agent have been used in attempts to obtain drugs with similar antitumoral activity but devoid of its toxicity. Currently, nine anthracycline analogs are in various stages of clinical trial (5). Among these compounds 4'-epiadriamycin (4'-ADM), with only a change in the chiral position of the hydroxyl group in the sugar moiety (6), appears to have less myocardial toxicity and similar antitumor effectiveness (7). Animal (8–10) and human studies (11–15) indicate that 4'-ADM is less cardiotoxic than ADM but is also an effective antineoplastic agent (16).

The antitumoral action of the quinone-type antitumor agents ADM (doxorubicin), 4'-ADM, daunomycin, rubidazone, mitomycin C, and streptonigrin can be explained by the fact that the drugs become intercalated in the duplex structure of the DNA molecule and act directly without a free radical event (17).

Chemical Structure of Adriamycin (Doxorubicin)

Chemical name: (8S-*cis*)-10-[(3-amino-2,3,6-trideoxy-α-L-lyxohexopyranosyl)oxy]-7,8, 9,10-tetrahydro-6,8,11-tryhydroxy-8-(hydroxyacetil)-1-methoxy-5,12-naphtracenedione; 14-hydroxydaunomycin, hydrochloride (see Fig. 1)

Chemical formula: $C_{27}H_{30}ClNO_{11}$

Molecular weight: 579.98

Solubility: sparingly soluble in water and dimethylsulfoxide (DMSO); slightly soluble in absolute ethanol; practically insoluble in nonpolar solvents

Adriamycin-Induced Biochemical Changes

The biochemical basis as well as the exact relationship between doses and ADM-induced myocardial lesions have been established. Mitochondrial damage (18), decreased activity of Na,K-adenosine triphosphatase (19), ADM binding to DNA (20), inhibition of coenzyme Q_{10} (21), and ADM binding to bivalent cations (Ca^{2+}, Mg^{2+}, Cu^{2+}, *and* Zn^{2+}) (22) have been suggested as ADM effects involved in the pathogenesis of the chronic type of cardiomyopathy. Besides all these mechanisms, ADM-generated oxy radicals would lead to lipid peroxidation and membrane damage (23–26). This appears to be the most important mechanism.

Adriamycin-Induced Myocardial Toxicity

As already mentioned, ADM is effective in the treatment of a wide variety of malignant tumors (27). Unfortunately, its clinical use has been compromised by an unusual and potentially lethal myocardial toxicity. Two main possible mechanisms of cardiac damage have been proposed: (1) an increase calcium concentration in the interior of myocardial fibers (28,29) and (2) damage to cell and organelle membranes by ADM-generated oxy radicals that produce an increase in the rate of endogenous lipid peroxidation (30,31). Both mechanisms can obviously be sequentially ordered; the ADM-generated oxy radicals would lead, through lipid peroxidation and membrane damage, to a loss of membrane-selective permeability and to increased calcium levels in the myocardial fibers (26).

The final consequence is that the severe alterations in cardiac function limit the safe dose of drug that can be administered. The incidence and probably the severity of congestive cardiac insufficiency are directly related to the cumulative dose of the drug (32). A relatively significant number of patients (7%) given 550 mg/m² developed dilated (congestive) cardio-

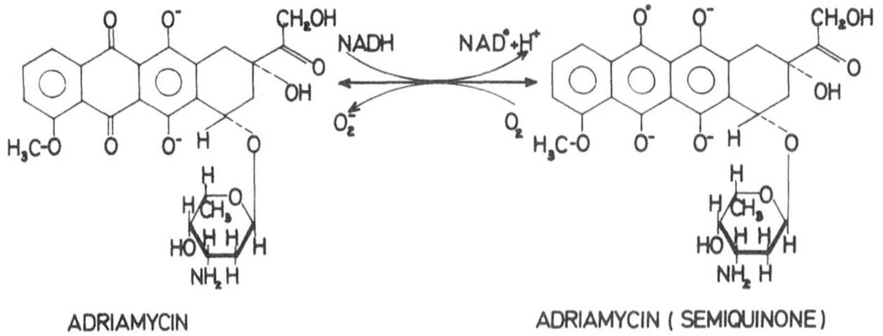

ADRIAMYCIN   ADRIAMYCIN (SEMIQUINONE)

FIGURE 1 Structural formula of Adriamycin.

myopathy (33); the percentage increases up to 20%, when the doses reaches more than 550 mg/m² and even heart failure has been reported to occur in about one-third of patients exceeding this dose (34–36).

Unfortunately, discontinuation of ADM administration at this arbitrary cumulative dose may deprive most patients of further worthwhile treatment. Congestive myocardiopathy, however, may occur below 550 mg/m², especially in patients with preexisting heart disease (35,37), in patients receiving mediastinal irradiation or synergistic cardiotoxic chemotherapy, or in elderly patients (5,35,36,38–41).

ATTEMPTS TO REDUCE ADRIAMYCIN-INDUCED MYOCARDIAL TOXICITY

Previous Attempts to Protect or Ameliorate ADM-Induced Myocardiopathy

Several approaches have been used to reduce the incidence of the cardiac toxicity of the anthracycline drugs (16). These include dose limitation, close cardiac monitoring, particularly in patients with risk factors and those receiving more than the empirical dose (42), alteration of dosage schedules (43,44), the development of new anthracycline analogs that retain chemotherapeutic efficacy but have reduced cardiac toxicity (45), and the administration of protective agents. Among them, a variety of drugs have been used to prevent the occurrence of ADM-induced myocardiopathy. The list includes prednisone, digitalis, vitamin E (46), ubiquinone (47), selenium-vitamin E (48,49), verapamil, and propranolol (50). Prevention was not obtained with digitoxin, ouabain, or strophanthin (51). In mice, partial protection was obtained with prenylamine (25) and in the dog with diphenydramine, cimetidine, phentolamine, and propranolol (52).

On the other hand, Rahman et al. (53,54) entrapped ADM within liposomes and found that the cardiac uptake of liposome-bound ADM in mice was reduced compared to the cardiac uptake of free drug. The decrease in cardiac drug concentration resulted in a decrease in histological cardiotoxicity after chronic ADM administration. Similar effects in humans are not known at present. Initial trials of free scavengers, such as N-acetylcysteine (55) and ICRF 159 and 187 (56,57), provided promising experimental studies. Methylene blue amelioration of ADM-induced myocardiopathy was reported by Hrushesky et al. (58) in mice. They hypothesized that oxidation of NADPH by the dye may interfere with the in vivo reduction of ADM, thereby slowing the chain of oxidative events responsible for the toxicity of the drug.

Previous Attempts to Protect or Ameliorate ADM-Induced Myocardiopathy with Vitamin E (See Table 1)

In mice, partial protection was obtained with vitamin E (31–46). In fact, vitamin E in excessive doses was found to be effective in ameliorating ADM-induced cardiotoxicity in mice without affecting the antitumor activity (51). Sonneveld (49), in the rat, demonstrated that pretreatment with D-α-tocopherol 24 h before a high ADM dose (7.5 mg/kg on day 1 and 10 mg/kg on day 15) diminished cardiotoxic effects as judged by electrocardiographic (ECG) changes, decreases in heart weight, and histological lesions.

Breed et al. (59) reported the failure of vitamin E (300 mg/kg body weight per day) to protect against ADM-induced cardiotoxicity in rabbits (ADM dose, 1 mg/kg body weight twice a week). Conversely, Herman and Ferrans (60) found in miniature swine that vitamin E decreased the severity of myocardial lesions during the chronic administration of ADM (total dose 9.6 mg/kg).

Also, Wang et al. (61) demonstrated in the rabbit the protective effects of vitamin E against ADM-induced toxicity. The trials were based upon the idea that the dose-dependent

TABLE 1 Attempts to Ameliorate or Inhibit ADM-Induced Cardiomyopathy in Animals

Authors and reference	Animal	Type of experiment	Doses ADM	Doses Vitamin E	Methods of evaluation	Results
Myers et al., 1976 (46)	Mice	Acute	Single IP dose of 15 mg/kg	85 IU, 24 h before ADM IP injection	Mortality	Success
Myers et al., 1977 (31)	Mice	Acute	Single IP dose of 15 mg/kg	85 IU, 24 h before ADM IP injection	Malonaldehyde formation	Success
Sonneveld, 1978 (49)	Rat	Chronic	7.5 and 10 mg/kg 10 days after	3200 IU/kg 24 h before ADM	ECG; heart weight; histological studies	Success
Van Vleet et al., 1978 (48)	Rabbit	Chronic	1.2 mg/kg twice a week for 9 weeks (total ≈ 200 mg/m²)	8.5 mg/kg 24 h before ADM injection	Body weight; light and electron microscopy	Success
Breed et al., 1980 (59)	Rabbit	Chronic	1 mg/kg twice a week (total dose 400 mg/m²)	300 mg/kg day five times in week before ADM injection	Histological studies	Failure
Wang et al., 1980 (61)	Rabbit	Acute	Single IV dose of 7 mg/kg body weight	200 mg (4–14 days) before ADM injection	Glutathione levels; light and electron microscopy	Success
Herman and Ferrans, 1983 (60)	Miniature swine	Chronic	9.6 mg/kg (total dose)	5000 IU/day 14 days); 1000 IU (17 days)	Electron microscopy	Success
Llesuy et al., 1985 (74)	Mice	Acute	Single IP dose of 15 mg/kg body weight	85 IU, 24 h before ADM injection	Malonaldehyde formation; chemiluminescence; light microscopy	Success
Milei et al., 1986 (26)	Rabbit	Chronic	0.6 mg/kg twice a week for 9–11 weeks	40 mg every day during the experiment	ECG; chemiluminescence; light and electron microscopy	Success

oxidation of reduced glutathione by ADM could be prevented by the addition of adequate amounts of vitamin E (200 mg IM, intramuscular, for 4–14 days before ADM injection). These animals showed no decrease in reduced glutathione levels in erythrocytes or cardiac tissue. The final conclusion was that the oxidation of free sulfhydryl groups represents a peroxidative insult induced by ADM cardiotoxicity. Light and electron microscopy studies supported this conclusion.

Legha et al. (62) found in humans that α-tocopherol used at an oral dose of 2 g/m^2 daily resulted in a six- to eightfold increase in vitamin E levels in serum. Congestive heart failure was observed in 14% and significant pathological changes in endomyocardial biopsies in approximately 50% of patients treated with a cumulative dose of 550 mg/m^2 of ADM. The authors concluded that α-tocopherol did not offer sustantial protection against ADM-induced cardiac toxicity. However, it was recently shown that vitamin E deficiency accentuates ADM-induced cardiomyopathy and cell surface changes in the rat (63). Animals were treated with a cumulative dose of 15 mg/kg of ADM. Mortality and myocardial malondialdehyde content were significantly higher in animals on a vitamin E-deficient diet plus ADM treatment. Accordingly, the total sialic acid content of the sarcolemma was diminished in these animals, showing a characteristic sarcolemmal injury, probably mediated by increased free-radical activity and lipid peroxidation.

With regard to these last studies, Ferrans and Van Vleet (64) unequivocally demonstrated the importance of selenium-vitamin E deficiency in cardiac lesions in animals; Guarnieri et al. also showed an increased formation of oxygen radicals in mitochondria from α-tocopherol-deficient rabbit hearts (65) and a decreased superoxide dismutase activity (66).

AMELIORATION OF ADRIAMYCIN-INDUCED CARDIOTOXICITY BY VITAMIN E

In this section, a summary of our experience on this subject is given.

Acute Murine Experiments

Weanling male and female Swiss mice weighing 20–23 g were used. ADM (Montedison Farmacéutica Argentina), dissolved in distilled water, was given at a dose of 15 mg/kg of body weight in a single intraperitoneal (IP) administration 4 days before killing the mice. Vitamin E dissolved in corn oil was administered at the dose of 85 IU/mice IP 24 h before giving the ADM. In some animals, 5000 IU vitamin A was added to vitamin E. Control animals were injected with corn oil. Mice were sacrificed by cervical dislocation, and a complete autopsy was performed.

Mice hearts were immediately placed in saline solution at 0–2°C, homogenized in 140 mM KCl and 20 mM potassium phosphate buffer, pH 7.3, at a ratio of 1 g organ to 9 ml solution, and centrifuged at 600 $\times$ g for 10 minutes at 0–2°C. The supernatant, which was constituted of a suspension of preserved organelles, was termed the homogenate (67). Mouse liver homogenates were similarly prepared.

The homogenate was assayed for indirect detection of oxydative stress using (1) the test of hydroperoxide-initiated chemiluminescence, and (2) malonaldehyde formation after aerobic incubation. For the assay of hydroperoxide-initiated chemiluminescence the homogenates were diluted to 1 mg protein per ml in 140 mM KCl and 20 mM potassium phosphate buffer, pH 7.3, added with 3 mM *tert*-butyl hydroperoxide, and assayed for chemiluminescence in a Packard Tri-Carb model 3320 scintillation counter in the out-of-coincidence mode (68–71). The homogenate suspension (4 ml) was placed into 19 mm diameter, 45 mm long glass tubes placed into 25 mm diameter, 60 minute long scintillation vials. Tubes and vials filled with the

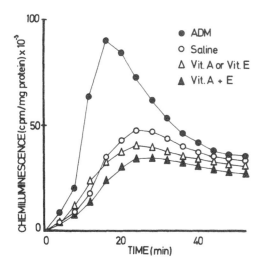

FIGURE 2 Effect of vitamins A and E on the hydroperoxide-initiated chemiluminescence of mouse heart homogenates.

suspension (without hydroperoxide) were checked for the background. For the assay of malonaldehyde formation (TBA method), the homogenate was diluted to 1 mg protein per ml in 140 mM KCl and 20 mM potassium phosphate buffer, pH 7.3, and incubated for 2 h at 37°C under oxygen bubbling. Samples of 0.3 ml were diluted to 1.5 ml with distilled water and added with 1.5 ml of 10% trichloroacetic acid. The precipitate was removed by centrifugation and the supernatant incubated with 0.67% thiobarbituric acid for 10 minutes at 100°C. The amount of malonaldehyde formed was expressed in nmol/g organ by using $E_{535} = 156\ \text{mM}^{-1}\ \text{cm}^{-1}$. Proteins were assayed by the method of Lowry et al. (72).

Slices of heart and other organs were processed for light microscopy. The areas of the damaged myocardium were assessed on one transverse section of both ventricles by stereological differential point count methods (73). The frequency and severity of ADM-induced myocardial lesions were based on the number of myofibers showing myofibrillar loss and cytoplasmic vacuolization (60). Administration of ADM produced an increase in the hydroperoxide-initiated chemiluminescence of heart homogenates of 38% (Fig. 2 and Table 2). The percentage was based on the relative maximal emission observed 20 minutes after the addition of hydroperoxide. The production of malonaldehyde upon aerobic incubation of heart homogenates was increased by 370% in the animals treated with ADM. These values correspond to the malonaldehyde formation detected after 2 h of aerobic incubation of the homogenates (Fig. 3 and Table 2).

TABLE 2 Hydroperoxide-Initiated Chemiluminescence and Malonaldehyde Formation in Mice Treated with ADM[a]

Animal treatment	Addition to this in vitro assay (μg/g organ)	Hydroperoxide-initiated chemiluminescence (cpm/mg protein) $\times 10^{-3}$	Malonaldehyde formation (nmol/g organ)
ADM	None	$72.0 \pm 6.0\ (8)^b$	$74.0 \pm 7\ (8)$
Saline	None	$52.0 \pm 2.0\ (9)$	$20.0 \pm 4\ (9)$
Saline-ADM	2.2	$51.0 \pm 1.0\ (6)$	$18.0 \pm 3\ (6)$

[a]Values are means ± SEM, with the number of animals indicated in parentheses.
[b]$p < 0.01$.

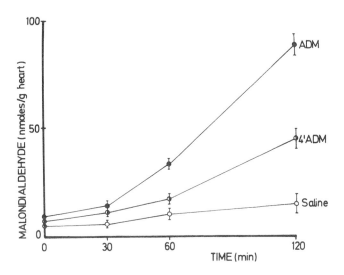

FIGURE 3 Effect of ADM and 4'-ADM on the malonaldehyde formation of mouse heart homogenates.

Because of their chemical structure, ADM can interfere with the values of hydroperoxide-initiated chemiluminescence and TBA-reactive material measured in homogenate. The amount of doxorubicin present in the heart of mice after 4 days of treatment with ADM was calculated by use of the data of Casazza et al. (8) and was 42 pg/g organ. These amounts of ADM were added to the in vitro assays for lipoperoxidation and were found to be without effect.

Mice pretreated with vitamin E or A plus E showed a decrease in the extent of the ADM effect in increasing the hydroperoxide-stimulated chemiluminescence. This effect amounted to an inhibition of 26 and 44% in the ADM-dependent stimulation of chemiluminescence for vitamins E and A plus E (Fig. 2 and Table 3). Microscopically, the hearts of ADM-treated mice showed scarce isolated microvacuolated subendocardial fibers, as well as a mild loss of succinate dehydrogenase activity.

Chronic Rabbit Experiments

A combination of vitamins A and E was selected for this study. The choice was based on the nonuniform results obtained with plain vitamin E (48,59,60) and the satisfactory results obtained in mice (see earlier) (74). A total of 24 New Zealand white rabbits were used. Animals

TABLE 3 Hydroperoxide-Initiated Chemiluminescence in Mice Treated with ADM and Vitamin E

Groups	Chemiluminescence (cpm/mg protein) $\times 10^{-3}$
Control	52.0 ± 2 (8)
Corn oil	50.4 ± 4 (8)
Vitamin E	48.4 ± 1 (7)
Vitamins A and E	44.2 ± 2 (7)
ADM	72.0 ± 6[a] (7)
ADM + vitamin E	53.4 ± 1 (7)
ADM + vitamins A + E	40.3 ± 5 (7)

[a]$p < 0.001$ compared with control values.

weighed between 1.9 and 2.1 kg at the beginning of the trial. ADM was injected intravenously through the dorsal ear veins twice a week for 9–11 weeks at a dose of 0.6 mg/kg body weight. ADM was suspended in saline solution at 1 mg/ml immediately before injection. The total dose given (10.8 mg/kg body weight) was known to produce cardiac lesions of significant to moderate severity (47,48). Vitamins A and E were jointly given orally every day during the experiment at doses of 250 IU (vitamin A) and 40 mg (vitamin E; d-α-tocopherol acetate; Raymos Laboratory, Buenos Aires, Argentina). The ECGs were recorded at the beginning and at the end of the experiment. Rabbits were killed under light ether anesthesia between 9 and 11 weeks after the beginning of the trial. Cubes of heart (1 cm^3) were removed from the rabbits and immediately processed for chemiluminescence as for the mice (see earlier). The ventricles and atria were cut into 2 mm thick transverse slices. These slices were processed for optical and electron microscopy. The relative areas of the damaged myocardium were assessed on one transverse section of both ventricles by stereological differential point count method (73).

From the third week until the end of the trial, ADM-treated rabbits developed crustiness over the skin of the ears used for injections. These changes were taken as an indication to discontinue injections for 1 or 2 weeks. Some of these animals simultaneously developed alopecia, crustiness of the skin in the thighs and feet, bilateral mucopurulent ocular discharge, and ulcers in the muzzles and feet. The abnormalities are listed as clinical alterations in Table 4. In ADM-treated rabbits, 8 of 10 showed post-treatment ECG changes. Sinus tachycardia (1 of 10), marked right axis deviation (1 of 10), conduction disturbances with enlargement of QRS (1 of 10), increase in voltage of the P wave (1 of 10), and increases in duration of P (4 of 10) and T waves (2 of 10) were observed. Rabbits treated with ADM that also received vitamins A and E showed only one of six ECG changes (sinus arrhythmia or intraventricular conduction disturbances). Control animals, as well as those that received vitamins A and E, did not show any significant ECG changes. Statistical comparison showed significant differences between ADM-treated rabbits and those that also received vitamins A and E ($p < 0.05$). Heart homogenates from ADM-treated rabbits compared with control animals showed an enhanced hydroperoxide-initiated chemiluminescence (Fig. 4 and Table 5) compared with the control animals. Administration of vitamins A and E produced a decrease in the hydroperoxide-initiated chemiluminescence of heart homogenates in ADM-treated and control rabbits.

Microscopically, the ADM-treated rabbits showed damaged myocardial fibers and mild to severe hydropic vacuolization of sarcoplasm, which led to progressive myocytolysis (Fig. 5). The total incidence of myocardial damage in ADM-treated rabbits (103 ± 23) was lowered by vitamins A and E (28 ± 8; $p < 0.01$). The topographic distribution of myocardial

TABLE 4 Clinical Alterations and ECG Changes in ADM-Treated Rabbits

Treatment	Weight gain (% of initial weight)	Clinical alterations[a] (fraction of total animals)	Number of animals with ECG changes	Total number of ECG changes
None	53 ± 5	0/6	0/6	0
ADM	39 ± 8	6/10	8/10[b]	9
ADM + vitamins A + E	47 ± 8	0/6	1/6[c]	2
Vitamins A + E	50 ± 7	0/6	0/6	0

[a]For definition see text.
[b]$p < 0.01$ compared with control rabbits.
[c]$p < 0.05$ compared with ADM-treated rabbits.

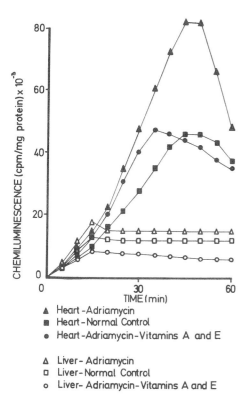

FIGURE 4 Effects of the administration of ADM and of vitamins A and E on the hydroperoxide-initiated chemiluminescence of rabbit heart homogenates.

damage is detailed in Table 6. Electron microscopy showed three types of myocardial alterations: sarcoplasmic vacuolization, mitochondrial swelling or mitochondrial disruption, and lysis of myofibrils. Vacuoles containing faint granular material was the main finding in injured myocytes. Perinuclear space edema and hydropic distention of sarcoplasmic reticulum and T tubules were frequently seen and diffusely scattered in affected fibers. Mitochondrial alterations resulted from swelling, disruption of cristae, the appearance of dense spherical

TABLE 5 Hydroperoxide-Initiated Chemiluminescence of Heart Homogenates of ADM-Treated Rabbits

Treatment	Hydroperoxide-initiated chemiluminescence (cpm/mg protein) $\times 10^{-3}$
None	52.8 ± 0.5
ADM	77.2 ± 3.9[a]
ADM + vitamins A + E	42.1 ± 2.0[b]
Vitamins A + E	40.4 ± 1.0[a]

[a] $p < 0.01$ compared with control animals.
[b] $p < 0.01$ compared with ADM-treated animals.

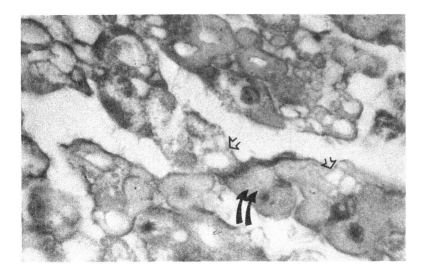

FIGURE 5 Multiple vacuolization of some fibers (arrow) and hyalinic degenerated changes (double arrows). Barbeito-Lopez trichome stain, ×400.

bodies or myelin figures in the matrix space, and membrane disruption in the most advanced stages of degeneration (Fig. 6). Accumulation of granular lysate with fragments of filaments was present in the sarcoplasm, resulting in atrophy of fibers. Rabbits treated with ADM and vitamins A and E showed similar alterations but to a markedly lesser degree than those treated only with ADM.

TABLE 6 Topographic Distribution of ADM-Induced Myocardial Damage[a]

	Treatment	
	ADM (n = 10)	ADM + vitamins A and E (n = 6)
Left ventricle		
Subendocardial	16.3 ± 3.6	6.5 ± 1.1[b]
Intramural	9.1 ± 2.4	1.7 ± 0.3[c]
Septum intraventricular		
Left subendocardial	23.1 ± 4.6	7.0 ± 1.5[c]
Intramural	5.3 ± 1.7	0.5 ± 0.2[c]
Right subendocardial	23.0 ± 4.9	5.3 ± 1.2[c]
Right ventricle		
Subendocardial	18.0 ± 3.6	4.7 ± 1.4[c]
Intramural	8.1 ± 2.2	2.0 ± 1.2[b]
Total myocardium	103 ± 23	28 ± 8[c]

[a]The values in the table indicate number of damaged fibers per 100 fibers as mean values ± SEM.
[b]$p < 0.05$.
[c]$p < 0.01$.

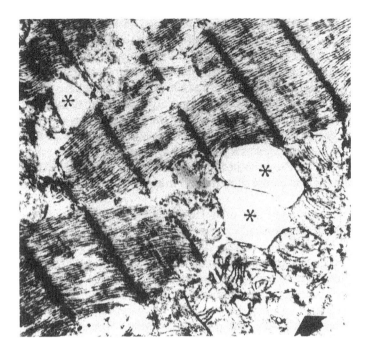

FIGURE 6 Early stage of degeneration in cardiac muscle cell. Myofibrils are preserved, but hydropic distention of elements of sarcoplasmic reticulum and T tubules are seen (asterisk), as well as swollen and disrupted mitochondria (arrow) ($\times 19,500$).

ADM-Induced Lipid Peroxidation and the Effects of Vitamin E

Our experience and that of others (46–48,60,61,63) show that vitamin E alone or in combination with vitamin A (26–74) has a protective effect against ADM-induced cardiotoxicity. Both an acute murine and a chronic rabbit model of ADM-induced myocardial damage were used. This information is of interest from both theoretical and practical points of view considering the limitation in the use of ADM as an antineoplastic drug because of its cardiotoxicity. The myocardial morphological alterations observed in ADM-treated rabbits were similar to those previously reported (47,48,75).

Hydroperoxide-initiated chemiluminescence has been used to detect decreased levels of endogenous antioxidants. The assay seems adequate for organs in which increased lipid peroxidation occurs. The increased hydroperoxide-dependent chemiluminescence in the hearts of ADM-treated rabbits indicates a decrease in the antioxidant defense of the tissue that could be interpreted as a consequence of the oxidative stress caused by ADM treatment. ADM is a lipophilic anthraquinone that generates O_2^- after being reduced to the semiquinone form by NADH at the mitochondrial membranes (76). An increased steady state of oxy radicals certainly exerts oxidative pressure in the tissue, leading to lipid peroxidation and exhaustion of antioxidant defenses. Some studies have suggested that the cytotoxic effects of anthracycline compounds may be related to the formation of semiquinone free-radical intermediates in vivo (74). The semiquinone reacts with O_2 and produces O_2^-, which seems responsible for starting the lipid peroxidation process in the unsaturated fatty acids of the cell membranes, a process that seems mediated by the hydroxyl radical (77).

Doroshow (78) examined the effects of ADM, daunorubicin, and various anthracycline analogs in heart homogenates of sarcoplasmic reticulum, mitochondria, and cytosol. These drugs significantly increased both O_2^- and H_2O_2 production in the aforementioned subcellular fractions, suggesting that ADM-induced disruption of heart mitochondrial and sarcoplasmic membranes may be explained by free-radical formation. The cells possess a protective system to maintain low levels of lipid peroxidation. This is achieved mainly through three enzymes: catalase, glutathione peroxidase, and superoxide dismutase. Myocardial cells have a limited capacity to detoxify oxygen radicals enzymatically; therefore, the heart is particularly susceptible to damage from reactive oxygen species generated as a result of ADM treatment. Conversely, enhancement of endogenous defenses of the myocardium against these species with free-radical scavengers (such as vitamins A and E and glutathione) may reduce ADM-generated membrane damage.

In a previous paper (26) we demonstrated that prenylamine, a well-known antianginal drug (79), protected against chronic ADM-induced cardiotoxicity in the rabbit but did not alter the ADM prooxidant effect, that is, the increased hydroperoxide-initiated chemiluminescence of heart homogenates. The general interpretation of the effects of prenylamine and those of vitamin E alone or associated with vitamin A is that the latter prevents lipid peroxidation and subsequent membrane damage, but prenylamine only antagonizes Ca^{2+}-increased permeability. This explanation is in agreement with the idea that Ca^{2+} excess is a consequence rather than a cause of ADM-induced lesions. Accordingly, it hardly seems necessary to point out that ADM-induced lesions do not appear ultrastructurally similar to those related to calcium overloading.

In summary, vitamin E is a relatively innocuous intracellular free-radical scavenger that has proven to inhibit or ameliorate ADM-induced cardiotoxicity.

As shown in this chapter, these properties have been demonstrated through mortality ratio, electrocardiographic studies, indirect measurements of lipid peroxidation (hydroperoxide-initiated chemiluminescence and malonaldehyde formation), diminution of the oxidation of reduced glutathione, and light and electron microscopic changes.

REFERENCES

1. Di Marco A, Gaetani M, Scarpinato B. Adriamycin (NSC-123127): a new antibiotic with antitumor activity. Cancer Chemother Rep 1969; 53:33-7.
2. Young RC, Ozols RF, Myers CE. The anthracycline antineoplastic drugs. N Engl J Med 1981; 305: 139-53.
3. Casazza AM. Antitumor activity of anthracyclines: experimental studies. In: Muggia FM, Young CW, eds. Anthracycline antibodies in cancer therapy. Boston: Martinus Nijhoff, 1982; 13-29.
4. Blum RH. An overview of studies with adriamycin (NSC-123127) in the United States. Cancer Chemother Rep 1975; 6:247-51.
5. Von Hoff DD, Rozencweig M, Piccart M. The cardiotoxicity of anticancer agents. Semin Oncol 1982; 9:23-33.
6. Bonfante V, Bonadonna G, Villani F, Di Fronzo G, Martini A, Casazza AM. Preliminary phase I study of 4'-epiadriamycin. Cancer Treat Rep 1979; 63:915-8.
7. Arcomone F, Penco S, Vigevani A, et al. Synthesis and antitumor properties of new glycosides of daumonycine and adriamycine. J Med Chem 1975; 18:703-7.
8. Casazza AM, Di Marco A, Bertazzoli C. Antitumor activity, toxicity and pharmacological properties of 4'-epi-adriamycin. In: Siegenthaler A, Luthy R, eds. Current chemotherapy. Washington, DC: American Society for Microbiology, 1988; 1257-60.
9. Casazza AM, Giuliani FC. Preclinical properties of epirubicin. In: Bonadonna G, ed. Advances in anthracycline chemotherapy: epirubicin. Milan: Masson, 1984; 31-40.
10. Ganzina F. 4'-Epi-Doxorubicin, a new analogue of doxorubicin: a preliminary overview of preclinical and clinical data. Cancer Treat Rev 1983; 10:1-22.

11. Young CW. Epirubicin, a therapeutically active doxorubicin analogue with reduced cardiotoxicity. In: Bonadonna G, ed. Advances in anthracycline chemotherapy: epirubicin. Milan: Masson, 1984; 183-8.
12. Jain KK, Casper ES, Geller NL. A prospective randomized comparison of epirubicin and doxorubicin in patients with advances breast cancer. J Clin Oncol 1985; 3:818-26.
13. Young CW, Casper ES, Geller NL. Clinical and cineangiographic comparison of the cardiotoxic effects of epirubicin and doxorubicin. International Symposium on Advances in Anthracycline Chemotherapy. Milan, Italy, 1984; 18.
14. Cersosimo RJ, Hong WK. Epirubicin: a review of the pharmacology, clinical activity and adverse effects of an Adriamycin analogue. J Clin Oncol 1986; 4:425-39.
15. Bonfante V, Villani F, Bonadonna G. Toxic and therapeutic activity of 4'-epi-doxorubicin. Tumori 1982; 68:105-11.
16. Dardir M, Ferrans V, Mikhall Y, et al. Cardiac morphologic and functional changes induced by epirubicin chemotherapy. J Clin Oncol 1989; 7:947-58.
17. Fialboff H, Goodman M, Seraydarian MW. Differential effect of adriamycin on DNA replicative and repair synthesis in cultured neonatal rat cardiac cell. Cancer Res 1979; 39:1321-7.
18. Ferrero ME, Ferrero E, Gaja G. Adriamycin: energy metabolism and mitochondrial oxidations in the heart of treated rabbits. Biochem Pharmacol 1976; 25:125-30.
19. Gozalvez M, Blanco M. Inhibition of NA-K ATPase by the antitumor antibiotic adriamycin. 5th International Biophysics Congress, Copenhagen, 1975.
20. Di Marco A. Adriamycin (NSC-123127): mode and mechanism of action. Cancer Chemother Rep 1975; 6:91-106.
21. Kishi T, Folkers K. Prevention by coenzyme Q10 of the inhibition of adriamycin of coenzyme Q10 enzymes. Cancer Treat Rep 1976; 60:223-8.
22. Lenaz L, Page JA. Cardiotoxicity of adriamycin and related anthracyclines. Cancer Treat Rev 1976; 3:111-20.
23. Tomasz M. Hydrogen peroxide generation during the redox cycle of mitomycin C and DNA bound mitomycin C. Chem Biol Interact 1976; 13:89-92.
24. Kwon TW, Olcott HS. Reduction and modification of ferricytochrome c by malonaldehyde. Biochim Biophys Acta 1966; 130:528-34.
25. Milei J, Bolomo NJ, Marantz A. Prenylamine inhibition of Adriamycin-induced cardiomyopathy in mice. Medicina (B Aires) 1982; 42:409-14.
26. Milei J, Boveris A, Llesuy S, et al. Amelioration of adriamycin-induced cardiotoxicity in rabbits by prenylamine and vitamins A and E. Am Heart J 1986; 111:95-102.
27. Blum A, Carter S. A new anticancer drug with significant clinical activity. Ann Intern Med 1974; 80:249-59.
28. Henderson IC, Frei E. Adriamycin and the heart. N Engl J Med 1979; 300:310-1.
29. Olson HM, Young DM, Prieur DJ. Electrolyte and morphologic alterations of myocardium in Adriamycin-treated rabbits. Am J Pathol 1974; 77:439-54.
30. Ferrans V. Overview of cardiac pathology in relation to anthracycline cardiotoxicity. Cancer Treat Rep 1978; 62:955-61.
31. Myers CE, McGuire WP, Liss RH. Adriamycin: the role of lipid peroxidation in cardiac toxicity and tumor response. Science 1977; 197:165-6.
32. Henderson IC, Frei E. Adriamycin cardiotoxicity. Am Heart J 1980; 99:671.
33. Appelbaum FR, Strauchen JA, Graw RG Jr, et al. Acute lethal carditis caused by high-dose combination chemotherapy. A unique clinical and pathological entity. Lancet 1976; 1:58.
34. Lefrak EA, Pitha J, Rosenheim S. A clinico-pathologic analysis of Adriamycin cardiotoxicity. Cancer 1973; 32:302-14.
35. Minow RA, Benjamin RS, Gottlieb JA. Adriamycin (NSC-123127) cardiomyopathy: an overview with determination of risk factors. Cancer Chemother Rep 1975; 6:195-201.
36. Minow RA, Benjamin RS, Lee ET. Adriamycin cardiomyopathy: risk factors. Cancer 1977; 39:1397-402.
37. Cortes EP, Lutman G, Wanka J. Adriamycin (NSC-123127) cardiotoxicity: a clinicopathological correlation. Cancer Chemother Rep 1975; 6:215-25.

38. Von Hoff DD, Layard MW, Basa P. Risk factors for doxorubicin-induced congestive heart failure. Ann Intern Med 1979; 91:710-7.

39. Bristow MR, Mason JW, Billingham ME. Doxorubicin cardiomyopathy: evaluation by phonocardiography, endomyocardial biopsy, and cardiac catherization. Ann Intern Med 1978; 88:168-75.

40. Gilladoga AC, Manuel C, Tan CC. Cardiotoxicity of Adriamycin (NSC-123127) in children. Cancer Chemother Rep 1975; 6:209-14.

41. Prout NM, Richards MJS, Chung JK. Adriamycin cardiotoxicity in children: case reports. Cancer 1977; 39:62-5.

42. Bristow MR, Lopez MB, Mason JW. Efficacy and cost of cardiac monitoring in patients receiving doxorubicin. Cancer 1982; 50:32-41.

43. Torti FM, Bristow MR, Howes AE. Reduced cardiotoxicity of doxorubicin delivered on a weekly schedule: assessment by endocardial biopsy. Ann Intern Med 1983; 99:745-9.

44. Legha SS, Benjamin RS, Mackay B. Reduction of doxorubicin cardiotoxicity by prolonged continuous intravenous infusion. Ann Intern Med 1982; 96:133-9.

45. Weiss RB, Clagett-Carr K, Russo M. Anthracycline analogs: the past, present, and future. Cancer Chemother Pharmacol 1985; 18:186-97.

46. Myers CE, McGuire WP, Young R. Adriamycin: amelioration of toxicity by tocopherol. Cancer Treat Rep 1976; 60:961-2.

47. Bertazzoli C, Sala L, Solcia E. Experimental Adriamycin cardiotoxicity prevented by ubiquinone "in vivo" rabbits. Int Res Commun Syst Med Sci 1975; 3:468.

48. Van Vleet JF, Greenwood L, Ferrans VJ, Rebar AH. Effect of selenium and vitamin E on adriamycin-induced cardiomyopathy in rabbits. Am J Vet Res 1978; 39:997-1010.

49. Sonneveld P. Effect of α-tocopherol on the cardiotoxicity of Adriamycin in the rat. Cancer Treat Rep 1978; 672:1033-6.

50. Daniels JR, Bilingham ME, Gelbart A, Bristow M. Effect of verapamil and propranolol on Adriamycin-induced cardiomyopathy in rabbits (abstract). Circulation 1976; 53(Suppl II):20.

51. Philips FS, Gilladoga A, Marquardt H. Some observations on the toxicity of Adriamycin (NSC-123127). Cancer Chemother Rep 1975; 6:177-81.

52. Bristow MR, Billingham ME, Minobe WA, Masek MA, Daniels JR. Demonstration that Adriamycin cardiotoxicity is mediated by vasoactive amines. J Mol Cell Cardiol 1979; 2(Suppl I):10.

53. Rahman A, Kessler A, More N, et al. Liposomal protection of Adriamycin-induced cardiotoxicity in mice. Cancer Res 1980; 40:1532-7.

54. Herman EH, Rahman A, Ferrans VJ, Vick JA, Schein PS. Prevention of chronic doxorubicin cardiotoxicity in beagles by liposomal encapsulation. Cancer Res 1983; 43:5427-32.

55. Doroshow J, Locker G, Myers C. Prevention of doxorubicin cardiotoxicity in the mouse by *N*-acetylcysteine. J Clin Invest 1981; 68:1053-64.

56. Giuliani F, Casazza AM, Di Marco A, Savi G. Studies in mice treated with ICRF-159 combined daunorubicin or doxorubicin. Cancer Treat 1981; 65:267-73.

57. Herman EH, Ferrans VJ. Reduction of chronic doxorubicin cardiotoxicity in dogs pretreatment with ($\pm$)-1,2-bis-(3,5-dioxopiperazinyl-1 (ICRF-187). Cancer Res 1981; 41:3436-40.

58. Hrushesky WJM, Olshefski RS, Wood PA, Mechnik S, Eaton JW. Methylene blue ameliorate anthracycline toxicity. Clin Res 1983; 31:778A-85A.

59. Breed JGS, Zimmerman ANB, Dormans JAMA, Pinedo HM. Failure of the antioxidant vitamin E to protect against Adriamycin-induced cardiotoxicity in the rabbit. Cancer Res 1980; 40:2033-8.

60. Herman EH, Ferrans VJ. Influence of vitamin E and ICRF-187 on chronic doxorubicin cardiotoxicity in miniature swine. Lab Invest 1983; 49:69-77.

61. Wang YM, Madanat FF, Kimball JC, et al. Effect of vitamin E against adriamycin induced toxicity in rabbits. Cancer Res 1980; 40:1022-6.

62. Legha S, Wang Y, Mackay B, et al. Clinical and pharmacologic investigation of the effects of α-tocopherol on adriamycin cardiotoxicity. In: Lubin B, Machlin L, eds. Vitamin E: biochemical, hematological and clinical aspects. Ann N Y Acad Sci 1982; 393:411-7.

63. Singal PK, Tong JG. Vitamin E deficiency accentuates adriamycin-induced cardiomyopathy and cell surface changes. Mol Cell Biochem 1988; 84:163-71.

64. Ferrans VJ, Van Vleet JF. Cardiac lesions of selenium-vitamin E deficiency in animals. Heart Vessels 1985; (Suppl 1):294-7.

65. Guarnieri C, Flamigni F, Rossoni-Caldarera C. Myocardial mitochondrial functions in α-tocopherol-deficient and refed rabbits. In: Chazov E, Smirnov V, Dhalla NS, eds. Advances in myocardiology, Vol. 3. New York: Plenum Publishing, 1982; 621-7.

66. Flamigni F, Guarnieri C, Toni C, Caldarera CM. Effect of oxygen radicals on heart mitochondrial function in α-tocopherol deficient rabbits. Int J Vitam Nutr Res 1982; 52:402-6.

67. Boveris A, Cadenas E, Reiter R, Filipkowsky M, Nakase Y, Chance B. Organ chemiluminescence: noninvasive assay for oxidative radical reactions. Proc Natl Acad Sci U S A 1980; 77:347-51.

68. Boveris A, Fraga CG, Varsavsky AT, Koch OR. Increased chemiluminescence and superoxide production in the liver of chronically ethanol-treated rats. Arch Biochem Biophys 1983; 227:534-41.

69. Cadenas E, Sies H. Low level chemiluminescence of liver microsomal function initiated by *tert*-butylhydroperoxide. J Biochem 1982; 124:349-56.

70. Cadenas E, Boveris A, Chance B. Low level chemiluminescence of hydroperoxide, supplemented cytochrome c. Biochem J 1980; 187:131-40.

71. Cadenas E, Varsavsky AI, Boveris A, Chance B. Oxygen or organic hydroperoxide-induced chemiluminescence of brain and liver homogenates. Biochem J 1981; 198:645-52.

72. Lowry OH, Rosebrough AL, Farr AL, Randall RJ. Protein measurement with the folin phenol reagent. J Biol Chem 1951; 193:265-75.

73. Weibel ER. Principles and methods for the morphometric study of the lung and other organs. Lab Invest 1963; 12:131-55.

74. Llesuy S, Milei J, Molina H, Boveris A, Milei S. Comparison of lipid peroxidation and myocardial damage induced by adriamycin and 4'epiadriamycin in mice. Tumori 1985; 71:241-9.

75. Jaenke R. An anthracycline antibiotic-induced cardiomyopathy in rabbits. Lab Invest 1974; 30:292-394.

76. Sato S, Iwaizmi M, Hauda K, Tamura J. Electron spin resonance study on the mode of generation of free radicals of daunomycin, Adriamycin and carboquone in NAD(P) H-microsome system. Gann 1977; 68:603-8.

77. Chance B, Sies H, Boveris A. Hydroperoxide metabolism in mammalian organs. Phys Rev 1979; 59:527-605.

78. Doroshow JH. Effect of anthracycline antibiotics on oxygen radical formation in rat heart. Cancer Res 1983; 43:460-72.

79. Murphy JE. Drug profile: Synadrin (prenylamine). J Int Med Res 1973; 1:204-8.

IV
PATHOPHYSIOLOGICAL CONDITIONS AND VITAMIN E

A. Metabolism and Aging

32

Vitamin E and the Aging Process

Abraham Z. Reznick and Bruce Rappaport

The Bruce Rappaport Faculty of Medicine and Rappaport Institute for Medical Research, Technion–Israel Institute of Technology, Haifa, Israel

Sharon V. Landvik

Vitamin E Research and Information Service, Edina, Minnesota

Irene Simon-Schnass

Hermes Arzneimittel GmbH, Grosshesselohe, Munich, Germany

Lester Packer

University of California, Berkeley, California

INTRODUCTION

Nature of the Aging Process

Aging can be defined as the overall reduction in biological functions of an organism with time, which may lead to increased vulnerability to environmental insults and ultimately to aging and death.

The biological world can be divided in general terms into single-cell organisms, the Protozoa, and multicellular organisms, the Metazoa. Usually, under optimal living conditions (including nutrients, density, and temperature), Protozoa theoretically do not age and can divide almost indefinitely. Metazoa, on the other hand, are different, because they age. Metazoa, through the process of evolution, underwent differentiation that subsequently resulted in acquiring cells and tissues that do not divide: postmitotic cells. In Metazoa, however, both mitotic cells and postmitotic cells show signs of aging. Interestingly enough, dividing cells can replace injured or old cells, but what about nondividing cells? They still have protective and repair mechanisms to encounter aberrant processes. For example, on the molecular level, defective proteins are removed by protein degradation systems. DNA aberrations can be replaced by DNA repair mechanisms. On the cellular level, all internal structures, such as organelles and membranes, including the plasma membrane, are constantly replaced by dynamic turnover processes. Nevertheless, the cells and tissues are aging, probably as a result of the decreased efficiency of the molecular and cellular protective and repair mechanisms over time.

In the not too distant past, few people lived long enough to suffer from the so-called infirmities of old age. Death usually occurred earlier as a result of infection, poor nutrition, or accidents. Today, however, the so-called old-age disorders constitute a considerable proportion of all illnesses in developed countries, and this fraction continues to grow at a steady rate. It is thus inevitable that aging phenomena assume an ever-growing importance in research.

At the present time, the mean life expectancy (MLE) in the developed countries is relatively high. It has grown rapidly over the last 100 years from 47 years in 1900 to 67 years in the middle 1950s and, somewhat more gradually, to its current level of 75 years. This is because of the reduction in infant mortality, improved nutrition, and the control of infectious diseases. Under the most favorable conditions it would be possible to further increase mean life expectancy by 10 years by completely eliminating all major diseases. By estimation, triumph over cancer would add about 2 years to the MLE and victory over cardiovascular disorders another 7 years. Thus, current opinions consider the highest mean life expectancy to be attained in the future by humans is about 85 years of age.

In the absence of any concomitant disorders, the aging process itself is the cardinal factor that impedes any further extension of mean life expectancy. This process is responsible for the progressive accumulation of changes that occur with the passing of time and that culminate in "old age" diseases and, ultimately, in death.

Further prolongation of a healthy life span can probably only be achieved by slowing the aging process. Unfortunately, our understanding of the aging process is currently very limited. As yet very little has been undertaken toward drawing up scientifically substantiated intervention programs for preventing and treating the phenomena associated with old age. This is partly a result of the complexity of the aging process itself. Nevertheless, there is now a range of theoretical approaches that claim to explain the aging phenomenon and that, at the present time, are under active discussion and research interest.

Theories of Aging

In essence, about 10 different theories of aging have been advanced in the last several decades, and these can be divided into two main categories: (1) molecular and cellular theories, and (2) systemic and physiological theories. The first group can be further divided into two additional categories: genetic and nongenetic theories. The genetic theories are associated with changes in the genetic material and with the mistransfer of genetic information; the nongenetic theories are associated with random events that are posttranscriptional and posttranslational in nature. The physiological theories, on the other hand, are associated with the biology of entire systems and the interactions of these systems. Examples are the immunological theory, the central nervous system (CNS)-hormone theory, and the extracellular cross-linking theory. This division of theories is illustrated in Figure 1.

It is not within the scope of this review to describe and elaborate these various theories. For a more comprehensive review on the subject, the reader should refer to other sources (1,2). The free-radical theory of aging was originally formulated by Harman and colleagues (3–6). This theory can be grouped under the nongenetic category, although some elements of the theory—the involvement of antioxidant enzymes (superoxide dismutase, catalase, and peroxidases)—are controlled genetically. Nonetheless, the free-radical theory is one of the most attractive theories because some of its aspects can be tested experimentally.

Free-Radical Theory of Aging

As the organism ages, progressive defects in protection against free-radical reactions allow an increase in tissue damage and eventually lead to various aging processes and phenomena.

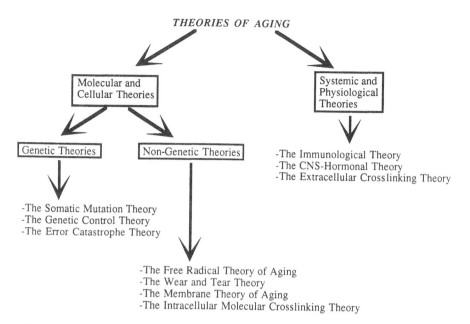

FIGURE 1   Theories of aging.

Antidefense mechanisms (enzymatic and nonenzymatic) are most probably sufficient for protection against normal rates of free-radical generation. However, these mechanisms may not be sufficient to cope with increased rates of free-radical generation under pathological and aging processes.

Theoretical Framework

An important starting point in research into the aging process is to find out why the species *Homo sapiens* has a life expectancy that is virtually twice that of other highly developed mammals. It is assumed that the processes that determine aging and longevity are the same for all species, as are the mechanisms that counteract these processes. According to this theory, whatever leads to the so-called maximum life expectancy (life span potential, LSP) of a species is the result of the intensity of the processes that trigger aging and the effectiveness of the protective mechanisms against these processes.

As seen in Figure 2, regardless of the elongation of mean life span expectancy (MLE = 50% survival), maximum life span has a fixed value. Animals that live in the wild usually die before aging strongly affects their biological capacity. This was also true of humans until recent history. Since aging is probably an expression of normal metabolic processes, a wide range of papers have been published concentrating on the role of oxygen metabolism in the cause of the aging process. The starting point for this was the empirical understanding of the relationship between the maximum life span potential of a species and its specific metabolic rate (SMR). SMR is defined as the quantity of oxygen per unit weight that a species consumes at rest per unit of time (7).

As shown in Figure 3, taken from Reference (7), the product of LSP and SMR is remarkably similar for almost all mammals. This product is known as the life span energy potential (LEP). It usually lies at about 220 cal/g. There are, however, a few exceptions, such as cats, which achieve a value of approximately 460 kcal/g, and a group that includes the capuchin monkey, a few lemurs, and the human, with a far higher figure of approximately 780 kcal/g.

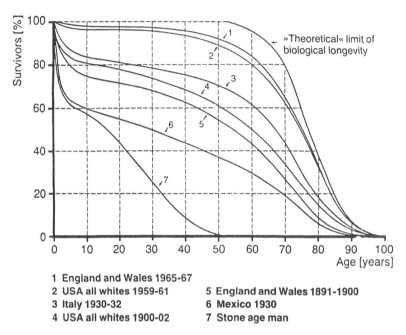

FIGURE 2 Life expectancy profile in humans.

1 England and Wales 1965-67
2 USA all whites 1959-61   5 England and Wales 1891-1900
3 Italy 1930-32   6 Mexico 1930
4 USA all whites 1900-02   7 Stone age man

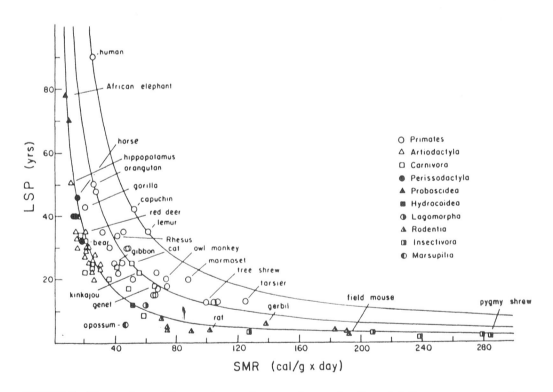

FIGURE 3 The maximum life expectancy of mammals as a function of daily energy consumption. Reprinted with permission from *Free Radicals in Biology*, Vol. VI, Pryor, WA, ed. 1984; 381.

Figure 3 shows that there is a close correlation between oxygen consumption per kilogram body weight and life expectancy and thus that oxygen metabolism must somehow be connected with the aging process. If this assumption is valid, species with a high LEP must be more resistant to aging processes caused by oxygen metabolism.

Another line of evidence that supports the free-radical theory of aging is the correlation between plasma levels of certain antioxidants and the life span of species. Thus species with longer life spans were found to have higher plasma concentrations of these antioxidants. Correlations were demonstrated for the enzyme superoxide dismutase (SOD), uric acid, and vitamins C and E, which are discussed later. On the other hand, other important antioxidants, such as glutathione and glutathione peroxidase, showed a negative correlation with species age, and it was concluded that these antioxidants are not longevity determinants (7).

Another interesting phenomenon is the accumulation of lipofuscin pigments in aging tissues. These pigments, also called age pigments, are produced largely by oxidative polymerization of mitochondrial lipids and proteins. Lipofuscin accumulation was shown to increase in a linear fashion with age in humans as well as in laboratory animals (8–10). Investigations in rats showed that there is a positive correlation between lipofuscin production and the intake of highly unsaturated fatty acids in their food. Such antioxidants as butylated hydroxytoluene (BHT) and vitamin E can decrease lipofuscin accumulation (11), and vitamin E deficiency accelerates the production of these age pigments (12).

BIOLOGY OF FREE RADICALS AND ANTIOXIDANT MECHANISMS

A radical is a substance with an unpaired electron in an outer orbit. Several substances possess such a configuration naturally, for example, NO_2. Others can be induced with varying degrees of simplicity by ultraviolet (UV) or x-ray irradiation or certain chemicals. Because of the presence of their free electrons, radicals are extremely unstable and, as a consequence, constantly endeavor to gain an electron or to shed the unpaired electron to restore their stable configurations. During this process, however, other substances that are deprived of an electron become radicals in turn. In this manner, chain reactions may be triggered that can spread over cell structures like a two-dimensional fuse (13).

Free-Radical Formation and Effects

Table 1 illustrates a number of substances that are either themselves radicals that can easily be transformed into radical forms or that may trigger the formation of free radicals.

Metabolic reactions also continually produce free radicals that are highly reactive and thus react with numerous substances in cells, such as nucleic acids, proteins, and lipids. They may cause cumulative and possibly irreversible changes in the structure and function of such molecules, including so-called cross-linking, bridgelike bonds between two or more macromolecules (13).

Reactions involving free radicals are important for a wide range of biological metabolic reactions, including electron transport in mitochondrial respiration (14), prostaglandin synthesis (15), phagocytosis (16), the cytochrome P_{450} system (17), and glucogenesis by chlorophyll (18). As long as free radicals occur and react under controlled conditions, they are of pivotal importance for certain metabolic processes; however, they may lead to degenerative processes if they are not controlled (19–22).

Antioxidant Defense Mechanisms

In the course of the evolution of aerobic life forms, a broad spectrum of antioxidants was also developed as protection against free radicals. The outstanding characteristic of an antioxidant

TABLE 1 Examples of Substances Involved in Free-Radical Generation and Their Toxic Effects

Factors and substances inducing free radicals	Sources	Toxic effects
NO_2, NO, O_3, CO, SO_2, lead	Exhaust	Changes in membranes
	Tobacco smoke	Lung damage
UV radiation	Sun	Skin irritation
	X-ray equipment	Cytoplasmic radiolysis
	Radiotherapy	
Ethanol	Alcoholic beverages	Liver damage
CCl_4, benzyprenes	Environmental pollution	Liver damage
Adriamycin, paracetamol, nitrofuran derivatives	Drugs	Heart, liver, and lung damage
Oxygen, physiological concentration and hyperbaric	Reperfusion	Hemolytic anemia
	Oxygen therapy	Lung damage
		Retrolental fibroplasia
		Tissue damage
Nitrosamines	Smoked and pickled foods	Gastrointestinal cancer

is that it can release an electron without becoming a reactive radical itself, or it can be recycled efficiently back to its reduced form. In this manner, free-radical reactions can be halted and/or be brought under control. In general, the nonenzymatic defense agents are divided into two groups: water-soluble antioxidants, including vitamin C and glutathione, and lipid-soluble antioxidants, including vitamin E and β-carotene. The enzymatic mechanisms include several enzyme systems, including superoxide dismutase, catalase, and peroxidases. Demopoulos from the New York Medical Center summed it up succinctly: "Antioxidants that evolution added to aerobic life are what make the difference between aerobic life and free radical pathology."

Oxygen Metabolism and Free Radicals

Nature has found a way of using the high energy potential of hydrogen in the mitochondrial respiration chain by using its links with oxygen. Nevertheless, nature was not able to prevent the formation of other reduced forms of oxygen with toxic properties, some of which are radicals. Approximately 3–10% of molecular oxygen is not reduced to water but instead yields various forms of reactive oxygen species. These include O_2^-, H_2O_2, and the $OH^{\pm}$ radical, which occur in this order as each loses an electron. They are interchangeable as a result of different reactions (20).

In terms of chemical reactivity, the oxygen free radicals O_2^- and $OH^{\pm}$ differ widely in their potential to damage tissue. O_2^- and $OH^{\pm}$ can react with many more targets in tissue than H_2O_2, which can only directly participate chemically in redox reactions. However, all the different forms can be transformed into one another (23). Not all the structures in a cell are equally susceptible to the reactions of free radicals. Probably the most sensitive structures susceptible to free-radical attacks are the polyunsaturated fatty acids in the cell membrane.

Effects of Free Radicals on Lipid Peroxidation

As seen in Figure 4, an aggressive substance like hydroxyl radical $OH^{\pm}$ ($t_{1/2} \sim 10^{-9}$ m/s) extracts hydrogen from a fatty acid, giving rise to a fatty acid radical. In turn, this may appear either as another aggressive substance or, by accumulating oxygen, as a fatty acid peroxyl

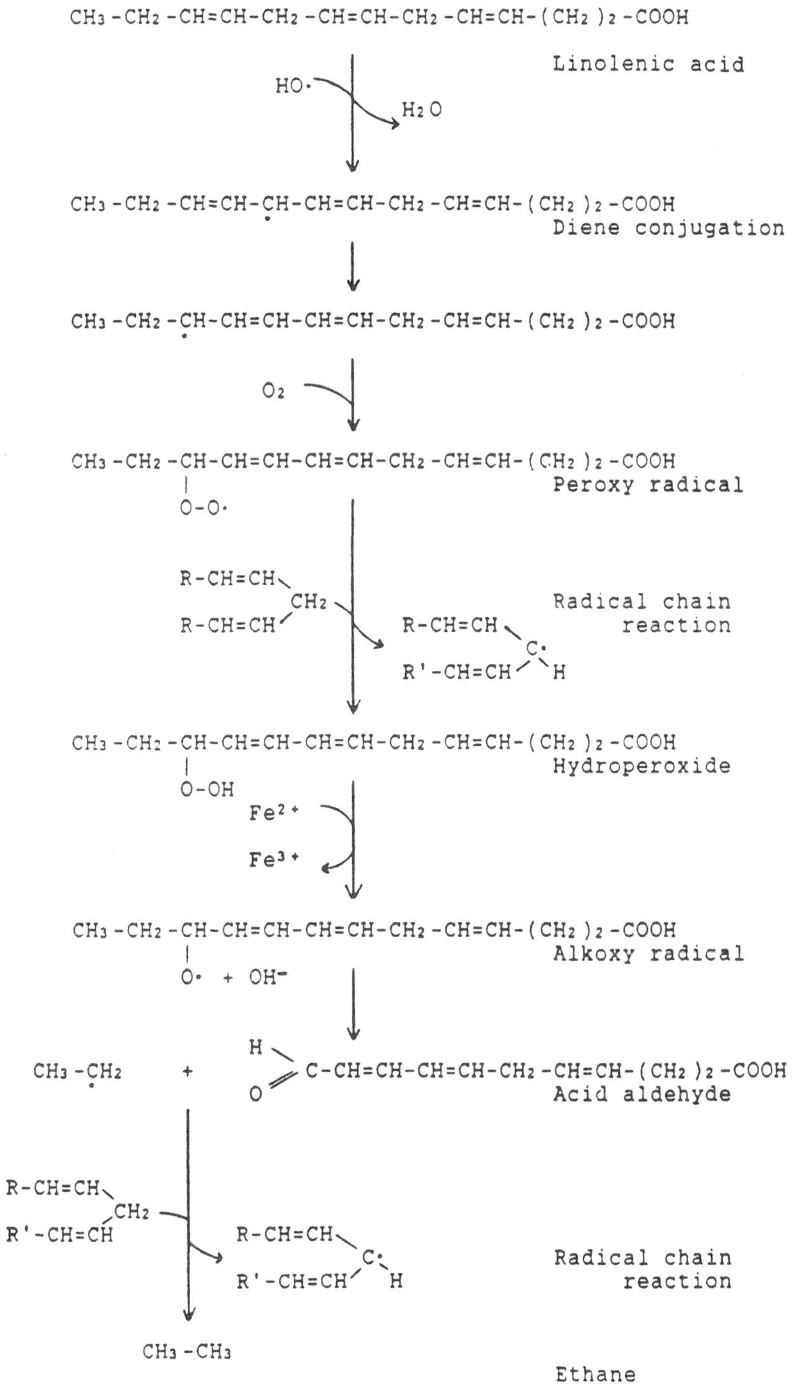

$$CH_3-CH_2-CH=CH-CH_2-CH=CH-CH_2-CH=CH-(CH_2)_2-COOH$$

Linolenic acid

HO· H_2O

$$CH_3-CH_2-CH=CH-\overset{\bullet}{C}H-CH=CH-CH_2-CH=CH-(CH_2)_2-COOH$$

Diene conjugation

$$CH_3-CH_2-\overset{\bullet}{C}H-CH=CH-CH=CH-CH_2-CH=CH-(CH_2)_2-COOH$$

O_2

$$CH_3-CH_2-CH-CH=CH-CH=CH-CH_2-CH=CH-(CH_2)_2-COOH$$
$$|$$
$$O-O\cdot$$

Peroxy radical

R-CH=CH
       CH₂
R-CH=CH

R-CH=CH
       C·
R'-CH=CH H

Radical chain reaction

$$CH_3-CH_2-CH-CH=CH-CH=CH-CH_2-CH=CH-(CH_2)_2-COOH$$
$$|$$
$$O-OH$$

Hydroperoxide

Fe^{2+}

Fe^{3+}

$$CH_3-CH_2-CH-CH=CH-CH=CH-CH_2-CH=CH-(CH_2)_2-COOH$$
$$|$$
$$O\cdot + OH^-$$

Alkoxy radical

$CH_3-\overset{\bullet}{C}H_2$ +

H
   C-CH=CH-CH=CH-CH₂-CH=CH-(CH₂)₂-COOH
O

Acid aldehyde

R-CH=CH
       CH₂
R'-CH=CH

R-CH=CH
       C·
R'-CH=CH H

Radical chain reaction

CH_3-CH_3

Ethane

FIGURE 4   Lipid peroxidation.

radical, which is also able to trigger chain reactions. The addition of hydrogen results in a hydroperoxyl radical. In accordance with the principle of β oxidation, this may result in two fragments: a fatty acid aldehyde and an alkane. Depending on the type of fatty acid involved, this ends up as either methane or pentane (24). It has become possible to assay both substances, either separately or jointly, by gas chromatography, providing the opportunity to determine the degree of lipid peroxidation in vivo (24–26).

In addition to unsaturated fatty acids, other unstable high-molecular-weight substances can also be attacked by radicals. These include certain proteins: possibly enzymes, collagen, or elastin and, under the worst conditions, the nucleic acids in the cell nucleus. These reactions are shown diagrammatically in Figure 5.

Free-Radical Damage to Biological Systems

It has been demonstrated that free radicals lead to a whole range of biochemically relevant changes. The following is only a partial list of biological systems in which free radicals have been implicated in causing significant changes and damage.

1. Loss of the cell membrane as a site of exchange for cell metabolism. Lipid peroxidation leads to reduced fluidity of the cell membrane. This may impair receptor function and disrupt the buildup of membrane potentials which regulate ionic transport, especially of calcium and magnesium (27–29).
2. Inactivation of enzymes may have a particularly adverse effect on energy metabolism (30).
3. Changes in the nuclear membrane are basically also related to the mechanism of lipid peroxidation. However, since the genetic material becomes more accessible as a result, there is an enhanced risk of mutation (31,32).
4. Vascular changes. Free radicals may contribute to damage to the intima, which is regarded as one of the most significant factors in the etiology of atherosclerosis. Furthermore, lipid peroxides inhibit the synthesis of prostacyclins, resulting in a tendency

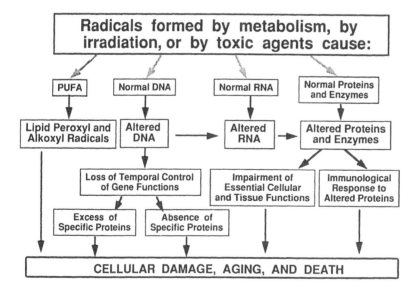

FIGURE 5 Effect of radicals on biological macromolecules.

to vasospasm and thrombosis. Free radicals lead to oxidation products of cholesterol, which are transported together with the normal cholesterol esters in the low-density lipoprotein (LDL) fraction but are not recognized as such by the appropriate receptors in the vessel cells. The elimination of the feedback mechanism may result in the accumulation of modified cholesterol in cells. This causes them to become so-called foam cells, which also play a significant part in the genesis of atherosclerosis (22,33–35).

5. Cross-linking between macromolecules. Cross-linkage between DNA and protein can overload the cell's repair mechanisms. These defects can result in cancer or other genetically induced disorders. The formation of lipofuscin or ceroid is less profound but can nevertheless have serious consequences for health. These are dark-colored links between macromolecular fragments that can no longer be broken down and are deposited as "metabolic waste products," thus disturbing cell metabolism (31,36,37).

6. Stimulation of guanyl cyclase. As a rule, changes in the cell membrane are associated with the stimulation of guanyl cyclase, which leads to enhanced cell division and possibly promotes carcinogenic activity (38).

There is growing evidence that essentially everyone in our society is exposed to free radicals, more now than ever before. It may also be assumed that the ubiquitous reactions of free radicals have an increasingly injurious influence with the passage of time. The changes caused by these reactions, which affect more or less all of us, could trigger some of the normal aging processes.

The "free-radical disorders" have been associated with the major causes of death in the industrialized world—cancer and atherosclerosis. Other degenerative disorders, such as inflammatory rheumatic diseases, osteoarthritis, essential hypertension, amyloidosis, Parkinson's disease, certain autoimmune disorders, and the Alzheimer form of senile dementia, have also been discussed in connection with free radicals (5,6).

VITAMIN E AND THE AGING PHENOMENON

Vitamin E and Life Span

Antioxidants that are normally present in cells and tissues are possibly conected with the maximum life expectancy and the LSP of a species. According to this working hypothesis, all mammals may have the same protective antioxidants, and their maximum life expectancy should show a positive correlation with the concentration of the latter. This was indeed found for plasma vitamin E levels and the maximum life expectancy in various species (Fig. 6) and for the quotient of the plasma vitamin E concentrations and specific metabolic rates with the maximum life expectancy (Fig. 7).

Human plasma contains considerably more vitamin E than plasma of any other mammals, and there is a positive correlation between both the maximum life expectancy and the LSP, showing that vitamin E is an important antioxidant that may play a part in influencing life expectancy. Unfortunately, very little is known about how tissue vitamin E concentrations are regulated genetically or biochemically. Thus, the question of whether the high plasma vitamin E level in humans is an expression of a special regulatory gene remains unanswered.

Up to this point, antioxidants have largely been referred to in connection with the body's protective mechanisms against free radicals. The following illustrates why vitamin E plays such an important role.

Vitamin E is a double-ring system with a side chain consisting of a number of isoprene units. Depending on the number of methyl groups, we distinguish between α-, β-, γ-, and δ-tocopherol. The OH^- group is of central importance for its action as an antioxidant. Vitamin

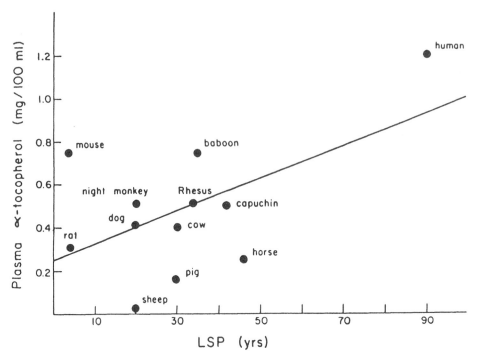

FIGURE 6 Plasma levels of vitamin E as a function of LSP in mammalian species. (Reprinted with permission from *Free Radicals in Biology*, Vol. VI, WA Pryor, ed. 1984; 397.)

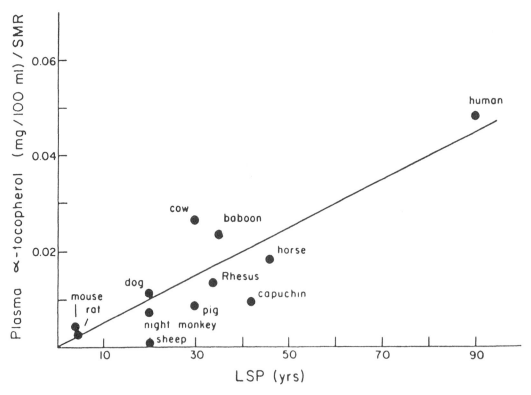

FIGURE 7 Plasma levels of vitamin E per SMR as a function of LSP in mammalian species. (Reprinted with permission from *Free Radicals in Biology*, Vol. VI. WA Pryor, ed. 1984; 397.)

E can release an electron to the free radical to become a tocopherol radical. By the action of the glutathione reductase or ascorbic acid, the tocopherol radical is transformed back to tocopherol. On the other hand, in another reaction it may release a further electron to become tocopherol quinone. This reaction is no longer reversible. Tocopherol quinone is one of the eliminatory metabolites of vitamin E. In summary, one molecule of vitamin E can capture two radicals and may, in part, be regenerated (39).

Since vitamin E is fat soluble, it accumulates in structures containing fat. For instance, in plasma, vitamin E is found in the various lipoprotein fractions and above all in the low-density lipoprotein fraction. In cells, the highest concentration is found in the cell membrane, where the side chain pierces deep into the membrane and the ring system remains on the surface. The ring system is exposed to the highest risk of peroxidation (40).

Early studies on invertebrate systems like the nematode *Caennorhabditis briggsae* showed that vitamin E supplementation significantly extended the maximum life span of these animals (41). This effect was pronounced when feeding started at an early age but was much less pronounced when supplementation was initiated at later stages of the animal's life. Similar observations of extension of maximum life span with vitamin E supplementation were reported for fruit flies (42).

In this connection, it is interesting that feeding studies on mice showed that both the vitamin E and the polyenic acid contents of the feed are important. An adequate intake of polyunsaturated fatty acids results in an increased average survival time under the proviso that the vitamin E intake is correspondingly high. Unless vitamin E intake is adequate, a high polyenic acid intake is more likely to have negative effects. The prolongation of the mean survival time in those animals with a vitamin E-supplemented diet was primarily attributed to a lower mortality rate due to malignant neoplasms. The maximum life span was not affected (43). Several other studies on vitamin E feeding for long periods in mammals showed only a slight increase in MLE and no effect on maximum life span (44).

Another very important area that is growing in significance is radiation exposure. Ionizing radiation causes so-called radiolysis of the intracellular fluid. In this process free radicals are produced. Particularly important is the incorporation of isotopes into the normal cell substance since this is not only one single external contamination as, for instance, in x-ray radiation, but the radiation source is transferred into the cell and the body itself becomes a source of radiation. The effects of radiation exposure on survival time are shown in Figure 8 (45).

A number of study designs have demonstrated the protective effects of vitamin E on radiation-induced changes. Irradiated mice that received vitamin E supplements lived significantly

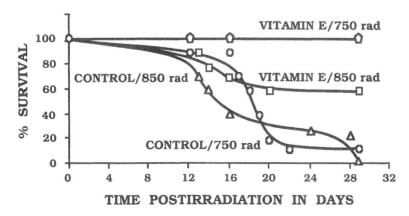

FIGURE 8 Shortening of the survival time of mice in relation to the radiation dose.

longer than those in the corresponding reference group (45–47). Elevated levels of vitamin E significantly increased the mitogenic response of splenic lymphocytes following irradiation of the animals and caused a large increase in the proliferative response (48). The erythrocytes of volunteers who received vitamin E over a period of 16 days were shown to have a marked increase in resistance to radiation-induced hemolysis (49).

Vitamin E and Aging: Animal Studies

Studies on the Immune System

Studies initiated in the 1970s and early 1980s showed positive effects of vitamin E supplementation on the immune system. Yasunaga et al. (50) studied the effects of various dosages of vitamin E on the lymphoproliferative capacity of mice using concanavalin A (ConA), phytohemagglutinin (PHA), and lipopolysaccharide (LPS). The optimal dosage of vitamin E needed to achieve optimal immunopotentiation was between 5 and 20 IU/kg body weight per day, and dosages over 80 IU/kg per day were toxic to mice. Immunopotentiation effects were achieved when serum tocopherol levels were about twice control levels. Similar results were also obtained by Corwin and Gordon (51), although quantities of 50 IU/kg diet fed to mice were not sufficient for optimal immune response. In other studies, addition of vitamin E to normal laboratory chow enhanced the antibody response to several antigens (52). Indeed, many other aspects of the immune system, including resistance to infection, were shown to be enhanced by vitamin E supplementation (52–57).

In a study of the effects of α-tocopherol in the immune system of aging animals, Meydani et al. (58) showed that vitamin E supplementation to 24-month-old mice for a period of 6 weeks significantly increased spleen cell proliferation by ConA and LPS but not by PHA in comparison to mice on a low vitamin E intake. Concomitantly, with increasing splenocyte proliferation, there was an increase in interleukin-2 production and a decrease in prostaglandin E_2 (PGE_2) production (58). In another study, natural killer (NK) cells activities were shown to be reduced in old mice and PGE_2 levels were higher. When old mice were supplemented with tocopherol and immunized with sheep red blood cell, however, their NK-mediated cytotoxicity was significantly increased (53). On the other hand, old lambs subjected to vitamin E deficiency showed immunological malfunctions that were associated with low selenium and vitamin E uptake (55).

In vitro studies of mitogen-induced splenocyte proliferation initiated by the addition of vitamin E showed a significant increase in cells of both young (3 months) and old (24 months) mice. However, the percentage increase was always higher in "young" cells compared to "old" cells (59).

Studies on the Nervous System

Several reports of the effects of either vitamin E-deficient or E-supplemented diets on the parameters of the nervous system have recently been published.

Nakashima and Esashi (60) studied the effect of a vitamin E-deficient diet on levels of urinary and other organ catecholamines as measured by sympathetic nervous system activity. Overall, the urinary excretion of catecholamines (norepinephrine, epinephrine, and dopamine) increased with the age of rats and was even further enhanced in vitamin E-deficient rats. However, these results could be a manifestation of a general stress reaction rather than a direct response to the vitamin E insufficiency.

Meydani et al. (61) studied the effect of vitamin E-deficient or E-supplemented diets on the content of tocopherol in several brain regions of young and old rats. In general, cerebrum and midbrain regions of all ages of rats showed the highest concentration of vitamin E and the cerebellum showed the lowest concentration. However, the cerebellum at the brain stem showed a selective decrease in tocopherol content with age. In old rats, vitamin E deficiency resulted in greater depletion of α-tocopherol, especially in the cerebellum of the brain stem. The conclusion was that the cerebellum and brain stem appear to have higher turnover of vitamin E than the cerebrum and midbrain and older rats need higher levels of vitamin E intake to maintain a steady state of α-tocopherol in these brain regions. A similar study examined the effect of vitamin E supplementation or deficiency on lipid peroxidation in the brain of young and old rats. After 20 weeks of vitamin E supplementation to old rats, there was an eightfold increase in plasma vitamin E levels and a twofold increase in cerebrum vitamin E levels compared to vitamin E-deficient old rats. These differences in vitamin E content between vitamin E-supplemented and vitamin E-deficient animals were much more pronounced in the brains of young rats. Determination of "endogenous" lipid peroxidation by the thiobarbituric acid (TBA) method was not well correlated with tissue vitamin E or the age of the animals. However, stimulation of lipid peroxidation by oxidant stressors in incubated cerebrum homogenates revealed marked effects of both diet and age (62). Koistinaho et al. (63) described the effects of vitamin E on the content of age pigment in ganglions of aging peripheral neurons. The results suggested that vitamin E deficiency induced a threefold increase in age pigment of dorsal root ganglion but not in autonomic ganglia like superior cervical and vagal ganglia. Vitamin E supplementation was associated with lesser amounts of age pigments in old rats. The author concluded that exogenous vitamin E provision may play a more crucial role in lipid peroxidation of age pigment accumulation in dorsal root ganglion than in autonomic ganglia.

Studies of Blood Components and Other Systems

Other systems, including serum parameters, blood vessels, red blood cells, endothelial cells, and muscle and liver tissues, were also reported to be influenced by vitamin E supplementation in aged animals.

A long-term study was conducted by Porta et al. (64) on Wistar rats using dietary fat combinations with two levels of vitamin E supplementation of either 2 or 200 mg per 100 mg. Even at high levels of vitamin E intake, serum biochemical parameters, such as triglycerides, phospholipids, and cholesterol levels, did not change beyond some age-related notable changes. Although there was no effect on maximum life span in the various dietary groups, animals fed a diet high in safflower oil supplemented with vitamin E had a significantly longer 50% survival time than all other groups, including the control group (64). Another study by Kay et al. (see Ref. 89) on red blood cell aging showed that red blood cells from vitamin E-deficient rats displayed much earlier what was termed "senescent cell antigen," a membrane protein that signals the immune system to destroy the red cell. Red blood cells from all ages of vitamin E-deficient rats behaved like old cells from control animals. Thus, the conclusion drawn by the researchers was that vitamin E deficiency caused premature aging of red blood cells, probably due to free-radical damage.

A study of spontaneous injuries in the aortic endothelium of inherited cataract rats by morphological techniques showed that the age-related injuries can be reduced by administration of α-tocopherol to these rats (66). Similar studies of endothelial cell cultures subjected to linoleic acid hydroperoxide oxidative stress were conducted by Hennig et al. (67). In their study, the endothelial barrier function (as measured by an increased albumin transfer) was affected by oxidative stress. However, addition of vitamin E significantly decreased the albumin transfer at all cell passages tested (up to 50) independently of cell age.

A study to assess the general α-tocopherol distribution in rat tissues and systems as a function of age was conducted by Hollander and Dadufalza (65). Three age groups (4-, 14-, and 24-month-old male Sprague-Dawley rats) were choosen for their study. With aging, there was a significant increase in lymphatic transport of vitamin E compared to bile transport with a concomitant increase in vitamin E levels in such organs as the small intestine and the liver. The authors concluded that their study demonstrated an age-related increase in the total amount of vitamin E absorbed and a shift from portal to lymphatic transport.

A study was conducted by Starness et al. (68) on lipid peroxidation in exercised- and food-restricted rats during aging. The extent of lipid peroxidation determined in muscle homogenates was not affected by either age or exercise. Using oxidant stress lipid oxidation to assess antioxidant capacity and total peroxidizable lipid, however, it was found that the antioxidant capacity was significantly decreased with aging but tocopheroxyl quinone, the oxidized form of α-tocopherol, increased with age. α-Tocopherol levels were not affected by exercise but were reduced by about 20% in gastrocnemius muscles of old animals. Total peroxidizing capacity was not affected by age but was increased after training.

Vitamin E and Aging: Human Studies

Studies on the Immune System

Early studies on the effect of "megadose" supplementation of several vitamins, including vitamin E, revealed no effect on the immune cells of healthy elderly subjects. Thus, no change was found in the levels of polymorphonuclear leukocytes or lymphocyte count at various levels of vitamin intake (69). Similar observations of minimal effects of vitamin E on the immune system, as measured by the antibody immune response to influenza virus vaccine, were also reported (70).

Nevertheless, more recent studies by Meydani et al. (71,72) indicated that vitamin E supplementation may have a beneficial effect on the immune response of healthy elderly subjects consuming 800 mg vitamin E per day for 30 days. There was a significant improvement in delayed-type hypersensitivity skin test response in the vitamin E-supplemented group. Vitamin E supplementation also resulted in an enhanced response of isolated lymphocytes to the T cell mitogen concanavalin A. There was also a significant decrease in phytohemagglutinin-stimulated prostaglandin E_2 production and in plasma lipid peroxide levels in vitamin E-supplemented subjects. This study showed a positive effect of vitamin E on the immune system of elderly people. However, the authors were careful to conclude that (72)

> Although our study suggests that many elderly individuals might benefit from supplementary intake of vitamin E, such public health recommendations can only be considered after longer-term studies with lower amounts of tocopherol are completed. Nevertheless, it is encouraging to note that a single nutrient supplement can enhance immune responsiveness in healthy elderly subjects consuming the recommended amount of all nutrients. This is especially significant because dietary intervention represents the most practical approach for delaying or reversing the rate of decline of immune function with age.

In another recent study of 82 healthy, free-living elderly individuals, nutrition factors were assessed in relation to cellular and regulatory immune variables. Cytotoxic activity of NK cells could not be correlated with nutrition factors, but dietary intake of vitamin E and D negatively influenced the activity of interleukin-2 (73).

From these studies on the influence of vitamin E on the immune systems of both animals and humans, it is quite obvious that more work is needed to establish these interesting effects of vitamin E on the immune system of elderly individuals.

Studies on the Circulatory System and Blood Components

Research in the early 1980s on the effect of vitamin E on blood peroxides was reported by Wartanowiz et al. (73). In this study, elderly subjects 60–100 years old were supplemented with 200 IU vitamin E, 400 mg vitamin C, or both vitamins for 12 months. The findings showed that blood peroxide level were reduced by 26% in the vitamin E-supplemented group, 13% in the vitamin C-supplemented group, and 25% in the group receiving both vitamins. The conclusion drawn by the investigators was that small doses of vitamin E and C administered over a long period of time could decrease the accumulation of blood peroxides in elderly people.

In a study of elderly nursing home residents by Tolonen et al. (75), there was a significant decrease in serum levels of lipid peroxides (TBA reactants) after supplementation with antioxidants (vitamins C, E, and B_6, β-carotene, zinc, and selenium) for 3 months. Serum lipid peroxide levels were initially higher in the elderly subjects than in younger control subjects but decreased to levels of the younger adults when the elderly were supplemented with antioxidants. Serum lipid peroxide levels remained higher in the elderly subjects on placebos. The group of antioxidant-supplemented elderly subjects also improved slightly in several psychological tests.

A study by Clausen et al. of elderly nursing home residents evaluated biochemical and clinical effects of daily antioxidant supplementation (vitamins A, C, E, and B_6, selenium, and zinc) or a placebo for 1 year (76). During antioxidant therapy, erythrocyte levels of the age pigment lipofuscin decreased significantly and serum vitamin E and whole blood selenium levels increased significantly. There was a nearly significant improvement in psychological scores in the antioxidant-supplemented group compared to control subjects on placebos. Blood flow measurements showed slight improvements in all brain areas tested in the subjects on antioxidant treatment and deterioration in blood flow in the placebo group.

In another study, general aspects of selenium (Se) and vitamin E status were compared in elderly and younger subjects by Campbell et al. (77). Healthy elderly subjects had significantly lower levels of Se in plasma and whole blood than younger subjects, but these concentrations in erythrocytes and platelets and erythrocyte glutathione peroxidase activity did not change with age. Plasma levels of vitamin E increased with age up to 60 years and decreased above 80 years. The vitamin E to LDL cholesterol ratio was not changed with age. Interestingly, institutionalized elderly people had a lower vitamin E status than healthy free-living elderly subjects.

An investigation of the level of lipid peroxidation, as measured by the thiobarbituric acid assay, in plasma of young and old women was conducted by Shafer and Thorling (78). Although no difference was found in vitamin E levels between young and old women, there was a significant difference in TBA values and plasma lipid concentrations in old women compared to young women. The authors concluded that the TBA values in plasma primarily were determined by the lipid composition rather than by the levels of the antioxidants.

In a double-blind study by Meydani et al. (79), the effect of vitamin E supplementation (800 IU/day) for 48 days on exercise-induced oxidative stress was examined in young adult (in their twenties) and in old adults (over 55). Plasma lipid peroxidation levels were measured by the TBA assay, which showed that the older individuals tended to have higher baseline levels of TBA-reactive substances than young adults, and in most of the subjects (18 of 21), extensive exercise caused an additional increase in TBA-reactive substances in plasma as well as in urine. The urinary increase was lower in the vitamin E-supplemented subjects. The conclusion was that vitamin E may provide some protection from exercise-induced oxidative stress.

The increase in plasma creatine kinase and neutrophil levels was investigated by the same authors in the same group of exercising young and older subjects (80). Plasma creatine kinase levels were increased more pronouncedly in the younger subjects following exercise than in the older subjects. The researchers attributed this to increased muscle protein turnover in young people compared to older people. Peak increases in circulatory neutrophils were observed about 6 h after initiation of exercise, and the increase was much smaller in the older subjects on placebo than those on vitamin E supplementation. The conclusion was that vitamin E may increase the response of white blood cells in aging subjects due to exercise and tends to eliminate the differences between the two age groups.

Blood platelets of elderly people on low intakes of eicosapentaenoic acid (EPA) showed a higher level of α-tocopherol compared to a placebo group. This increase of platelet vitamin E was associated with a decrease in platelet aggregation in elderly subjects. Therefore, low intake of EPA could increase platelet vitamin E and inhibit platelet aggregation, thus providing some protection against platelet hypersensitivity, which is associated with vascular cerebral damage in elderly people (81).

Studies of the Nervous System and Cataract Formation

Several reports have discussed the important role of vitamin E for neural function in humans (81) as well as in experimental animals (83). Thus, a long-term deficiency in vitamin E results in an array of neurological disorders, from ataxia of limbs and general propioceptive loss to ophthalmoplegia and general muscle weakness (81). Since vitamin E is the most important lipid-soluble antioxidant, it is probably very crucial for the proper function and neurophysiology of nerve cells (83).

Metcalfe et al. (84) measured the levels of α-tocopherol in a sample of cerebral cortex from patients with Alzheimer's disease, fetuses with Down's syndrome, and a group of centenarians. There were no marked differences in the vitamin E levels of these various groups compared to control subjects. The conclusion drawn by the investigators was that "Neither the normal aging processes, Alzheimer's disease, nor the increase in in vitro lipid peroxidation reported in fetuses with Down's syndrome resulted from gross lack of α-tocopherol."

A different approach to the understanding of the role of vitamins in the aged nervous system was taken by Fahn (85). In this study, patients with Parkinson's disease (PD) were supplemented with vitamins E and C, 3200 IU/day and 3000 mg/day, respectively, for a period from several months to several years. Supplementation with vitamins E and C in high dosages delayed the time when L-DOPA was needed to control symptoms of Parkinson's disease. However, the study had a large drawback because a control placebo group of PD patients was not included.

The formation of senile cataract has been suggested to involve light-associated oxidative stress. Thus, antioxidants like vitamin E could potentially protect against cataract formation. However, Sharma et al. (86) did not observe improvement or slowing of cataract formation in vitamin E-treated eyes compared to a control group. In a study by Jacques et al. (87) of the relationship between antioxidant status and senile cataract in adults 40–70 years of age, subjects with high plasma levels of at least two of the three antioxidant vitamins (vitamin C, vitamin E, and carotenoids) had a decreased cataract risk relative to subjects with low levels of one or more of these vitamins. The estimated odds ratio for cataract occurrence was 0.6 for subjects with moderate plasma levels of the antioxidant vitamins and 0.2 for subjects with high levels. As noted by the researchers, their results appear to support the hypothesis that the lens antioxidant defense may be a factor in cataract development.

Robertson et al. (88) studied the effect of vitamin C and vitamin E supplementation on cataract development in humans. Their results showed that these vitamins could reduce senile cataracts by at least 50% compared to control groups who were not supplemented with these vitamins. This study suggested that oxidative stress is involved in the pathogenesis of cataract and that the process can be slowed by antioxidants.

CONCLUSIONS

Vitamin E was shown to correlate with species maximum life span. However, only studies of vitamin E supplementation in invertebrates, which are cold-blooded animals (poikilotherms), showed a positive effect on increase in maximum life span. Studies in mammals failed to show a positive effect of vitamin E supplementation. Since poikilotherms have relatively higher basal metabolic rates than mammals, oxygen consumption and free-radical reactions are probably more crucial for their survival and aging. Therefore, in this case, vitamin E may be limited in concentration and vitamin E supplementation may enhance the antioxidant defense mechanism against free-radical reactions in aging animals.

This discussion has covered only selected animals and human aging systems and organs in which vitamin E status and supplementation were studied. Since the free-radical theory implicates free radicals in impairment of the immune system, it seems logical that vitamin E has an important role in controlling free-radical reactions, which play an important part in our metabolism, as well as in our immune system. The concentrations of components of the antioxidant defense system that are produced by the body, such as superoxide dismutase and the catalases, fall continually with aging. Thus, the deficit in oxidation protection that may occur under these circumstances, can only be compensated for by the intake of other antioxidant components. Therefore, with increasing age, and also under conditions of heightened oxidative stress, attention should be paid to the intake of antioxidants, in particular vitamin E.

Finally, the question of how much vitamin E should be recommended to elderly people is of great interest. Studies of supplementation of 1200 IU/day of vitamin E for several months to people of various ages showed no adverse side effects. On the other hand, the RDA (recommended dietary allowance) for adults is only 15 IU/day of vitamin E, which is very low and probably should be reconsidered in view of the latest research findings. It has been suggested that vitamin E supplementation of up to 400 IU/day may provide a host of beneficial effects to active healthy older individuals.

REFERENCES

1. Hayflick L. Theories of biological aging. Exp Gerontol 1985; 20:145-59.
2. Warner HR, Butler RN, Sprott RL, Schneider EL, eds. Modern Biological Theories of Aging. Raven Press, NY (1987).
3. Harman D, Eddy DE, Seibold J. Free radical theory of aging: effect of dietary fat on central nervous system function. J Am Geriatr Soc 1977; 24:301-7.
4. Harman D, Heidrick ML, Eddy DE. Free radical theory of aging: effect of free-radical-reaction inhibitors on the immune response. J Am Geriatr Soc 1977; 25:400-7.
5. Harman D. The free-radical theory of aging. In: Pryor WA, ed. Free radicals in biology, Vol. V. Orlando, FL: Academic Press, 1982; 255-75.
6. Harman D. Free radical theory of aging: role of free radicals in the origination and evolution of life, aging and disease processes. In: Johnson JE, ed. Free radicals, aging, and degenerative diseases. New York: Alan R. Liss, 1986; 3-49.
7. Cutler RG. Antioxidants, aging, and longevity. In: Pryor WA, ed. Free radicals in biology, Vol. VI. New York: Academic Press, 1984; 371-428.
8. Mann DMA, Yates PO. Lipoprotein pigments—their relationship to aging in the human nervous system. I. The lipofuscin content of nerve cells. Brain 1974; 97:481-8.
9. Brizzee KR, Johnson FA. Depth distribution of lipofuscin pigment in cerebral cortex of albino rats. Acta Neuropathol (Berl) 1970; 16:205-19.
10. Brizzee KR, Ordy JM. Age pigments, cell loss and hippocampal function. Mech Ageing Dev 1979; 9:143-62.
11. Tappel AL, Fletcher B, Deamer D. J Gerontol 1973; 28:415-24.
12. Hayes KC. Pathophysiology of vitamin E deficiency in monkeys. Am J Clin Nutr 1974; 27:1130-40.
13. Slater TF. Disturbances of free radical reactions: a cause or consequence of cell injury? In: Hayaishi O, Niki E, Kondo M, Yoshikawa T, eds. Medical, biochemical and chemical aspects of free radicals. Amsterdam: Elsevier, 1989; 1-9.
14. Harman D. The biological block: the mitochondria? J Am Geriatr Soc 1972; 20:145-7.
15. Hemler ME, Lands WEM. Evidence for a peroxide-initiated free radical mechanism of prostaglandin biosynthesis. J Biol Chem 1980; 225:6253-61.

16. Klebanoff SJ. Oxygen metabolism and the toxic properties of phagocytes. Ann Intern Med 1980; 93:480-9.

17. Yamazaki I. Free radical mechanisms in enzyme reactions. Free Radic Biol Med 1987; 3(6):397-404.

18. Loach PA, Hales B. Free radicals in photosynthesis. In: Pryor WA, ed. Free radicals in biology, Vol. I. New York: Academic Press, 1976; 199-237.

19. Demopoulos HB, Pietronigro DD, Seligman ML. The development of secondary pathology with free radical reactions as a threshhold mechanism. J Am Coll Toxicol 1983; 2(3):173-84.

20. Demopoulos HB, Santomier JP, Seligman ML. Pietronigro DD. Free radical pathology; rationale and toxicology of antioxidants and other supplements in sports medicine and exercise science. In: Katch FI, ed. Sport, health, and nutrition; the 1984 Olympic scientific congress proceedings, Vol. 2. Champaign, IL: Human Kinetics, 1986; 139-90.

21. Hammond B, Hess ML. The oxygen free radical system: potential mediator of myocardial injury. J Am Coll Cardiol 1985; 6:215-20.

22. Harman D. Vitamin E—effect on serum cholesterol. Circulation 1960; 22(1):151-3.

23. Stier A. Freie Radikale als Ursache von Erkrankungen. In: Bohlau V, Reimann J, eds. Freie Radikale: Vitamin E—Therapeutische Bedeutung. Munchen: Otto Hoffmanns Verlag, 1986; 35-45.

24. Cohen G. Production of ethane and pentane during lipid peroxidation: biphasic effect of oxygen. In: Yagi K, ed. Lipid peroxides in biology and medicine. New York: Academic Press, 1982; 199-211.

25. Dillard CJ, Litov RE, Savin WM, Dumelin EE, Tappel AL. Effects of exercise, vitamin E, and ozone on pulmonary function and lipid peroxidation. J Appl Physiol 1978; 45(6):927-32.

26. Tappel AL. Measurement of in vivo lipid peroxidation via exhalation of pentane and protection by vitamin E. In: Yagi, K, ed. Lipid peroxides in biology and medicine. New York: Academic Press, 1982; 213-22.

27. Hegner D. Age dependence of molecular and functional changes in biological membrane properties. Mech Ageing Dev 1980; 14:101-18.

28. Hendricks LC, Heidrick ML. Susceptibility to lipid peroxidation and accumulation of fluorescent products with age is greater in T-cells than in B-cells. Free Radic Biol Med 1988; 5:145-54.

29. Patel JM, Block ER. The effect of oxidant gases on membrane fluidity and function in pulmonary endothelial cells. Free Radic Biol Med 1988; 4:121-34.

30. Elmadfa I, Bosse W. In: Vitamin E: Eigenschaften, Wirkungsweise und therapeutische Bedeutung, Stuttgart: Wissenschaftliche Verlagsgesellschaft mbH, 1985.

31. Pietronigro DD, Jones WBG, Kalty K, Demopoulos HB. Interaction of DNA and liposomes as a model for membrane-mediated DNA damage. Nature 1978; 267:78-9.

32. Sies H. Biochemie des oxidativen Stress. Angew Chem 1986; 98:1061-75.

33. Esterbauer H, Dieber-Rothendeder M, Striegl G, Waeg G. Role of vitamin E in preventing oxidation of LDL. Am J Clin Nutr 1991; 53(suppl. 1):314S-21S.

34. Parker JC, Martin DJ, Rutili G, McCord J, Taylor AE. Prevention of free radical mediated vascular permeability increases in lung using superoxide dismutase. Chest 1983; 83S:52S-3S.

35. Simon-Schnass I, Korniszewski L. The influence of vitamin E on rheological parameters in high altitude mountaineers. Int J Vitam Nutr Res 1990; 60(1):26-34.

36. Clausen J. Demential syndromes and the lipid metabolism. Acta Neural Scand 1984; 70(5):345-55.

37. Miquel J, Oro J, Bensch KG, Johnson JE Jr. Lipofuscin: fine-structural and biochemical studies. In: Pryor WA, ed. Free radicals in biology, Vol. III. New York: Academic Press, 1977; 133-82.

38. Murad F, Arnold WP, Mittal C, Braughler JM. Properties and regulation of guanylate cyclase and some proposed functions for cyclic GMP. Adv Cyclic Nucleotide Res 1979; 11:175-204.

39. Simon-Schnass I. Zur Bedeutung von Vitamin E in der Menschlichen Ernshrung. Ernahrungs-Umschau 1984; 31:395-9.

40. Gallo-Torres HE. Transport and metabolism. In: Machlin LJ, ed. Vitamin E—a comprehensive treatise. New York: Marcel Dekker, 1980; 193-267.

41. Epstein J, Gershon D. Studies on ageing in nematodes, IV, the effect of antioxidants on cellular damage and life span. Mech Age Dev 1972; 1:257-264.

42. Miquel J, Binnard R, Witt WH. Effects of dl-α-tocoherol on lifespan of *Drosophilia melanogaster* exposed to high O_2 tension. Gerontologist 1973; 3:37-43.

43. Blackett AD, Hall DA. Vitamin E—its significance in mouse ageing. Age Ageing 1981; 10:191-5.
44. Blumberg TB, Meydani SN. Role of dietary antioxidants in aging. In: Munro H, Hutchinson, M, eds. Nutrition and aging, Vol. 5. New York: Academic Press, 1986; 85-97.
45. Srinivasan V, Jacobs AJ, Simpson SA, Weiss JF. Radioprotection by vitamin E: effects on hepatic enzymes, delayed type hypersensitivity, and postirradiation survival of mice. In: Meyskens FL, Prasad KN, eds. Modulation and mediation of cancer by vitamins. Basel: Karger, 1983; 119-31.
46. Malick MA, Roy RM. Effect of vitamin E on post-irradiation death in mice. Experientia 1978; 34:1216-7.
47. Pellegri-Formentini U, Poy C. Experiments on mice treated with alpha-tocopherol and subjected to whole-body irradiation. Chem Abstr 1973; 78:67079m.
48. Roy RM, Petrella M, Shateri H. Effects of administering tocopherol after irradiation on survival and proliferation of murine lymphocytes. Pharmacol Thera 1988; 39:393-5.
49. Brown MA. Resistance of human erythrocytes containing elevated levels of vitamin E to radiation-induced hemolysis. Radiat Res 1983; 95:303-16.
50. Yasunaga T, Kato H, Ohgaki K, Inamoto T, Hikasa Y. Effect of vitamin E as an immunopotentiation agent for mice at optimal dosage and its toxicity at high dosage. J Nutr 1982; 1120:1075-84.
51. Corwin LM, Gordon RK. Vitamin E and immune regulation. Ann N Y Acad Sci 1982; 393:437-51.
52. Walford RL. The 120 years diet. New York: Simon and Schuster, 1986; 395.
53. Meydani SN, Yogeeswaram G, Lin S, Baskar S, Meydani M. Fish oil and tocopherol-induced changes in natural killer cell mediated cytotoxicity and PGE 2 synthesis in young and old mice. J Nutr 1988; 118:1245-52.
54. Meydani SN, Meydani M, Blumberg JB. Antioxidants and the aging immune response. Adv Exp Med Biol 1990; 262:59-62.
55. Turner RJ, Finch JM. Immunological malfunctions associated with low selenium-vitamin E diets in lambs. J Comp Pathol 1990; 102:99-109.
56. Bendich A, Gabriel E, Machlin LJ. Dietary vitamin E requirement for optimum immune response in the rat. J Nutr 1986; 116:675-81.
57. Campbell PA, Cooper HR, Veinzerling RA, Tengerdy RP. Vitamin E enhances in vitro immune response by normal and non-adherent spleen cells. Proc Soc Exp Biol Med 1974; 146(2):465-9.
58. Meydani SN, Meydani M, Verdon CP, Shapiro AC, Blumberg JB, Hayes KC. Vitamin E supplementation suppresses prostaglandin E_2 synthesis and enhances the immune response of aged mice. Mech Ageing Dev 1986; 34:191-201.
59. Meydani SN, Blumberg JB. Vitamin E and immune function in the elderly. In: Rundles S, ed. Nutritional modulation of immune responses. New York: Marcel Dekker, In press.
60. Nakashima Y, Esashi T. Age-related changes in sympathetic nervous activity of rats receiving vitamin E-deficient diet. J Nutr Sci Vitaminol 1986; 32:569-79.
61. Meydani M, Macauley JB, Blumberg JB. Influence of dietary vitamin E, selenium, and age on regional distribution of α-tocopherol in the rat brain. Lipids 1986; 21:786-91.
62. Meydani M, Verdon CP, Blumberg JB. Effect of vitamin E, selenium and age on lipid peroxidation events in rat cerebrum. Nutrition Res 1985; 5:1227-36.
63. Koistinaho J, Alho H, Hervonen A. Effect of vitamin E and selenium supplement on aging peripheral neurons of the male Sprague-Dawley rat. Mech Ageing Dev 1990; 51(1):63-72.
64. Porta E, et al. Effects of the type of dietary fat at two levels of vitamin E in Wistar male rats during development and aging. I. Life span, serum biochemical parameters and pathological changes. Mech Ageing Dev 1980; 13:1-39.
65. Hollander D, Dadufalza V. Lymphatic and portal absorption of vitamin E in aging rats. Dig Dis Sci 1980; 34(5):768-72.
66. Masuda M, Thara N, Kuriki H, et al. Spontaneous injuries in the aortic endothelium of the inherited cataract rats and their prevention by tocopherol. A study by scanning electron microscopy. Atherosclerosis 1989; 75(1):23-30.
67. Hennig B, Boissonneault GA, Wang Y. Protective effects of vitamin E in age-related endothelial cell injury. Int J Vitam Nutr Res 1989; 59(3):273-9.
68. Starnes JW, Cantu G, Farrar RP, Kehrer JP. Skeletal muscle lipid peroxidation in exercised and food-restricted rats during aging. J Appl Physiol 1989; 67(1):69-75.

69. Goodwin JS, Garry TJ. Relationship between megadose vitamin supplementation and immunological function in a healthy elderly population. Clin Exp Immunol 1983; 51:647-53.

70. Harman D, Miller RW. Effect of vitamin E on the immune response to influenza virus vaccine and incidence of infectious disease in man. Age 1986; 9:21.

71. Meydani SN, Barklund PM, Liu S, et al. Effect of vitamin E supplementation on immune responsiveness of the aged. Ann N Y Acad Sci 1989; 570:283-90.

72. Meydani SN, Barklund MP, Liu S, et al. Vitamin E supplementation enhances cell-mediated immunity in healthy elderly subjects. Am J Clin Nutr 1990; 52(3):557-63.

73. Payette H, Rola-Pleszczynski M, Ghadirian P. Nutrition factors in relation to cellular and regulatory immune variables in a free-living elderly population. Am J Clin Nutr 1990; 52(5):927-32.

74. Wartanowicz M, et al. The effect of alpha-tocopherol and ascorbic acid on the serum lipid peroxide level in elderly people. Ann Nutr Metab 1984; 28:186-91.

75. Tolonen M, Sarna S, Halme M, et al. Antioxidant supplementation decreases TBA reactants in serum of elderly. Biol Trace Elem Res 1988; 17:221-8.

76. Clausen J, Nielsen SA, Kristensen M. Biochemical and clinical effects of an antioxidative supplementation of geriatric patients. Biol Trace Elem Res 1989; 20:135-51.

77. Campbell D, Bunker VW, Thomas AJ, Clayton BE. Selenium and vitamin E status of healthy and institutionalized elderly subjects: analysis of plasma, erythrocytes and platelets. Br J Nutr 1989; 62(1):221-7.

78. Schafer L, Thorling EB. Lipid peroxidation and antioxidant supplementation in old age. Scand J Clin Lab Invest 1990; 50(1):69-75.

79. Meydani M, Cannon JG, Meydani SN, et al. Protective effect of vitamin E on exercise-induced oxidative damage in young and elderly subjects. Tufts University, Boston, MA: USDA Human Nutrition Research Center on Aging, 1990.

80. Cannon JG, Orencole SF, Fielding RA, et al. Acute phase response in exercise: interaction of age and vitamin E on neutrophils and muscle enzyme release. Am J Physiol 1990; 259:R1214-9.

81. Croset M, Vericel E, Rigaud M, et al. Functions and tocopherol content of blood platelets from elderly people after low intake of purified eicosapentaenoic acid. Thromb Res 1990; 57(1):1-12.

82. Muller DPR, Lloyd JK, Wolff OH. Vitamin E and neurological function. Lancet 1983; 1:225-8.

83. Muller DPR, Goss-Sampson MA. Role of vitamin E in neural tissue. Ann N Y Acad Sci 1985; 570: 147-55.

84. Metcalfe T, Bowen DM, Muller DP. Vitamin E concentrations in human brain of patients with Alzheimer's disease, fetuses with Down's syndrome, centenarians, and controls. Neurochem Res 1989; 14(12):1209-12.

85. Fahn S. The endogenous toxin hypothesis of the etiology of Parkinson's disease and a pilot trial of high-dosage antioxidants in an attempt to slow the progression of the illness. Ann N Y Acad Sci 1989; 570:186-96.

86. Sharma YR, Vajpayee RB, Bhatnagar R, et al. Systemic aspirin and systemic vitamin E in senile cataracts: cataract V. Ind J Ophthalmol 1989; 37(3):134-41.

87. Jacques PF, Chylack LT, McGaudy RB, Hartz SC. Antioxidant status in persons with and without senile cataract. Arch Ophthalmol 1988; 106:337-40.

88. Robertson JM, Donner AP, Trevithick JR. Vitamin E intake and risk of cataracts in humans. Ann N Y Acad Sci 1989; 570:372-81.

89. Kay M, et al. Oxidation as a possible mechanism of cellular aging: vitamin E deficiency causes premature aging and IgG binding to erythrocytes. Proc Natl Acad Sci U S A 1986; 83:2463-7.

33

Vitamin E and High-Altitude Exercise

Irene Simon-Schnass

Hermes Arzneimittel GmbH, Grosshesselohe, Munich, Germany

Significant physiological problems are associated with high-altitude mountain climbing. They amplify the demands placed upon physiological function and nutritional status. Therefore, the high-altitude mountain climber, who may be in a state of nutrition deprivation and with significant needs beyond the normal levels of nutritional support, provides a unique subject to evaluate the effect of specific nutrients on physiological functions.

One of the main problems at high altitude is the low O_2 partial pressure. As an aerobic energy supply is necessary for all higher life forms, an inadequate oxygen supply leads at least to unnoticed metabolic impairment and may even pose a direct threat to life. Prolonged exposure to hypoxia is known to affect both aerobic and anaerobic metabolism. The altered oxidative process suppresses ATP synthesis and impairs anaerobic glycolysis. Furthermore, the low pCO_2 reduces the body's buffer capacity (1). During hypoxia oxidative injury may be enhanced as a consequence of alterations in the oxidation-reduction potential (2). This may lead to increased oxidative stress.

The evolution of aerobic reactions brought significant advantages for energy production, but at the expense of considerable risks. A correlation has recently been identified between increased metabolic rate as a consequence of exercise and increased tissue damage due to free-radical pathology (3–8). Vitamin E not only economizes energy production but also reduces free-radical reactions.

By its stabilizing effect on various components of the respiratory chain, vitamin E contributes to aerobic energy production (9–11). A local vitamin E deficiency leads to disturbances in electron transport and this to reduced cell respiration (12–14). This is especially apparent when the available oxygen is also limited, as can occur as a result of high demand, poor local supply, or low partial pressure of oxygen. It can be expected that the impairment of metabolism is especially pronounced under conditions of increased physical load at high altitude. Recent investigations have shown that a prolonged stay at extreme altitudes leads to a loss of activity of succinate and lactate dehydrogenase (1). The activity of both enzymes in skeletal muscle is also decreased by vitamin E deficiency (15–17). This can be explained by their labile SH groups.

Special investigations on the anaerobic threshold confirmed the previous observation that a prolonged stay at high altitude leads to reduced physical performance, apparent as a decreased anaerobic threshold (see Fig. 1). Supplementation with 200 mg vitamin E twice daily prevented the falling of the anaerobic threshold (18).

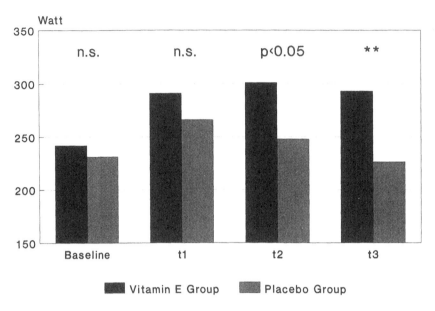

FIGURE 1 Anaerobic threshold during a prolonged stay at high altitude (5100 m) with and without supplementation with vitamin E compared to preexpedition value. t_1, t_2, and t_3 were measured 2, 4, and 6 weeks after baseline.

In the same experiment it could be shown that insufficient supply with antioxidants leads to an increased lipid peroxidation. The exhalation of pentane has been demonstrated in experimental animals and in humans to be related to the vitamin E status (5,10). In animals, exericse-induced lipid peroxidation could be prevented by vitamin E administration (4,8). In this experiment this could be confirmed in the human. As shown in Figure 2, the pentane expiration was maintained constant in the vitamin E group, indicating that no appreciable additional lipid peroxidation had occurred. In the control group, the mean expired pentane rose by more than 100% (18). It can be concluded from this study that high-altitude climbing incurs a considerable risk of metabolically induced cell damage.

The lack of available oxygen at high altitude moreover causes some important changes in blood composition and rheological properties. One of the main tasks of the blood, substance exchange (including oxygen exchange) between the individual tissues, is performed in the terminal vessels of the capillary system. The local capillary blood supply is governed mainly by cardiac output, vascular resistance (determined by vascular tone, vasomotion, and vascular wall elasticity), and the rheological properties of the blood or its constituents.

The special rheological properties of the blood arise from its two-phase composition of plasma and blood cells. The viscosity of the blood depends largely on the packed cell volume, plasma viscosity, the deformability of the erythrocytes, and their tendency to aggregate (19, 20). This is why blood is a liquid without Newtonian properties; instead, its viscosity varies as a function of flow conditions. Within the same individual blood can at one time be highly fluid and flow rapidly, at another highly viscous with sluggish flow (21). Erythrocytes are able to change their shape because, among other things, of their membrane fluidity. The loss of fluidity can be influenced by such factors as acidosis, hyperthermia, immobilization (stasis) as a result of aggregation, for example, membrane defects, and aging of the cell (22). One of

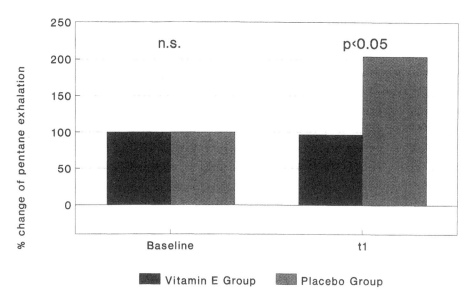

FIGURE 2 Pentane exhalation during a prolonged stay at high altitude (5100 m) with and without supplementation with vitamin E compared to preexpedition value. t_1 was measured 4 weeks after baseline.

the important underlying phenomena is seen to be an oxidative change in membrane lipids, which may be triggered by free radicals (23). In addition to blood viscosity, an important role is played in the capillary blood supply by the elasticity and integrity of the vascular wall. Here, too, oxidative changes are discussed as a pathogenetic factor (23).

The release of tissue hormones, such as histamine, kinins, and prostaglandins, also plays an important part in damage to the vascular wall (24). Endothelial lesions may lead to disturbances in microcirculation. These may be compounded by activation of the coagulation system, with resultant consumption coagulopathy. This can lead to the formation of microthrombi and, consequently, increased reactive fibrinolysis. This in turn results in an increased tendency to hemorrhage (25).

These disturbances are known to occur in the capillary bed in a number of arteriosclerotic diseases (25). Phenomena of this kind can also be found at high altitudes. Here, the pathological changes are mainly in the area of the pulmonary and cerebral capillaries, but also in those of the retina and the mucosa (26). Hypothermia may be an additional source of damage to the blood vessels and may induce a deterioration in the rheological properties of the blood; the degree of deterioration varies but invariably is clearly temperature dependent (24). Both the radical binding properties of vitamin E and its involvement in the metabolism of eicosanoids indicate that this vitamin has an effect on these phenomena (27–29).

At high altitudes the body attempts to boost the blood's oxygen transport capacity by increasing erythropoiesis. Because of the special two-phase composition of blood, this does not, under otherwise normal conditions, produce any changes in flow characteristics in the terminal vessels, as the local hematocrit in the capillaries is much lower (with a high flow rate) than in the larger blood vessels. There are limits to even this mechanism, however. In measurements in vitro an elevated hematocrit level is discernible immediately, whereas in vivo a critical level may be reached at a hematocrit of 55% (30). Thus, it is nowadays also assumed that from hematocrit levels of over 50–55%, the oxygen transport capacity falls again, even in pulmonary diseases (31). In the event of a general or localized reduction in flow velocity, however, the capillary hematocrit level approaches the venous hematocrit, and aggregation occurs. In this case the hematocrit becomes an important influencing factor.

The viscosity of the blood is determined not only by the hematocrit level, however, but also by plasma viscosity and the flexibility of erythrocytes. In two recent experiments a decreased filterability of the erythrocytes could be detected (see Fig. 3) (32,33). This may be due to an increased lipid peroxidation of membrane lipids. This is supported by the finding that there was also a negative correlation between erythrocyte filterability and the amounts of thiobarbituric reactive substances (TBARs) (32).

Besides erythrocytes, leukocytes likewise have an important influence on the rheological properties of blood. Because of their rigidity and spherical shape, they cannot pass through the terminal vessels as easily as erythrocytes. Even under physiological conditions there may be a reduction in flow velocity or even temporary stasis in the passage of leukocytes through the capillaries (34). If the perfusion pressure falls, pronounced disturbances of microcirculation occur, mainly due to the leukocytes. Thus, there may also be occlusions of the arterioles and venules, for example, due to adhesions to the vascular walls (35,36). After stimulation, for example by activation of complement, endotoxins, immune complexes, or leukotriene B_4, there is a particularly large rise in the tendency of the leukocytes, in particular the granulocytes, to aggregate. This in turn leads not only to a further increased risk of occlusion, but also to an increased release of intracellular proteases (37).

Against this background, the increase in leukocytes (Fig. 4) that was found in the control group during an experiment in the Himalayas can be seen as a risk factor. The rise in the vitamin E group was appreciably smaller but the differences were not significant. The positive correlation also found in this study between the leukocyte count and the viscosity of the

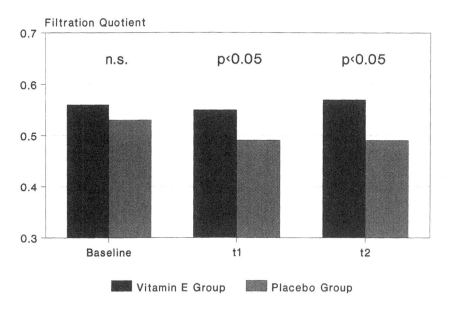

FIGURE 3 Filterability of red blood cells during a prolonged stay at high altitude (4300 m) with and without supplementation with vitamin E compared to preexpedition value. t_1 and t_2 were measured 2 and 4 weeks after baseline.

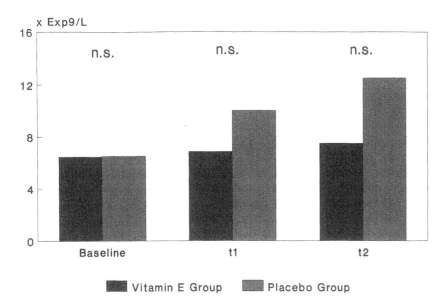

x Exp9/L

FIGURE 4 Change in leukocytes during a prolonged stay at high altitude (4300 m) with and without supplementation with vitamin E compared to preexpedition value. t_1 and t_2 were measured 2 and 4 weeks after baseline.

whole blood, standardized on a hematocrit of 45% (compare Fig. 5), implies that both the red and the white blood cells have an important effect on the flow characteristics of the blood (33). The question of whether there is a causal connection between the smaller increase in leukocytes in the active medication group and vitamin E substitution cannot be answered without further investigations.

Moreover, there was a disturbance in the coagulation system. As shown in Figure 6, the activity of protein C was significantly reduced, which may lead to an increased susceptibility to coagulation (33). On the basis of data in the literature it can be assumed that increased granulocyte stimulation occurred (38). The proteases then released can split not only the endothelial cells and the proteins bound to the endothelial cells but also proteins free in the plasma (38,39). This could explain the drop in protein C observed in the control group. Another possible cause seems to be modulation of endothelial cell function in hemostasis. On the one hand, endotoxins (the presence of which is indicated by the rise in leukocytes) can reduce the concentration of available thrombomodulin, so that there is only reduced protein C activation (40). Blockade of the binding sites could then result in increased protein C clearance. On the other hand, increased formation of the inflammation mediator interleukin-1 may cause similar reactions (41). As no drop in protein C activity could be found in the vitamin E group, it can be concluded that supplementation with vitamin E seems to lead to stabilization of both the leukocytes and the endothelial cells and to protect against splitting of proteins.

The data as a whole show that a relatively rapid rise in altitude and continued presence at high altitude produce changes in a wide range of parameters (32,33). Particular note should be taken of the fact that all the changes in the different parameters point clearly to impaired blood rheology (33). In this connection it should be borne in mind that mountain sickness is often attributed to general microcirculation disturbances. This explains the tendency to

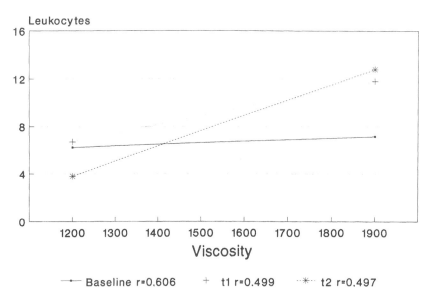

FIGURE 5 Correlation between leukocytes and viscosity of whole blood standardized on RBC 45% before expedition (baseline) and at high altitude (4300 m). t_1 and t_2 were measured 2 and 4 weeks after baseline.

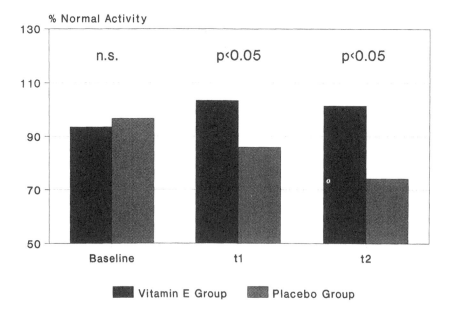

FIGURE 6 Change in protein C activity during a prolonged stay at high altitude (4300 m) with and without supplementation with vitamin E compared to preexpedition value. t_1 and t_2 were measured 2 and 4 weeks after baseline.

frostbite but also to retinal hemorrhage and cerebral and pulmonary edema. There seems to be a combination of increased precapillary vascular resistance and microrheological and hemostatic disturbances (42). Left ventricular failure is not generally found.

To my knowledge these were the first studies that investigated whether vitamins have any effect on the phenomena described. This seems to be the case. Taking two 200 mg vitamin E doses daily (43) prevented the deterioration in parameters observed in the control group. The body was obviously better able to tolerate the strain most probably caused by oxidative stress (32,33). This supposition is supported by the results of a similar expedition that pointed in the same direction; there, too, it was shown that under medication with vitamin E there was no deterioration from baseline values of the kind seen in the control group (18). It seems to us desirable to check whether this positive effect can also be found in other hypoxic conditions, such as chronic obstructive pulmonary diseases, cardiac and cerebral ischemia, and peripheral circulation disturbance.

Evidence is accumulating that oxidative injury mediated by free radicals is an important factor in various pathologies. The protection against these damaging free-radical processes is provided by cellular antioxidant defense systems (2,23,28,44,45), vitamin E, among other substances. Given the relationship of free-radical damage to both hypoxia (causative) and cellular antioxidant defense systems (preventive), antioxidants, especially vitamin E, may play a significant role in the prevention of high-altitude deterioration.

REFERENCES

1. Cerretelli P, Di Prampero PE. Aerobic and anaerobic metabolism during exercise at altitude. In: Jokl E, Hebbelinck M, eds. High altitude deterioration; medicine and sport science, Vol. 19. Basel: Karger, 1985; 1-19.
2. Jones DP. The role of oxygen concentration in oxidative stress: hypoxic and hyperoxic models. In: Sies H, ed. Oxidative stress. London: Academic Press, 1985; 151-95.
3. Demopoulos HB, Santomier JP, Seligman ML, Pietronigro DD. Free radical pathology: rationale and toxicology of antioxidants and other supplements in sports medicine and exercise science. In: Katch FI, ed. Sport, health, and nutrition; the 1984 Olympic scientific congress proceedings, Vol. 2. Champaign, IL: Human Kinetics, 1986; 139-90.
4. Dillard CJ, Dumelin EE, Tappel AL. Effect of dietary vitamin E on expiration of pentane and ethane by the rat. Lipids 1977; 12(1):109-14.
5. Dillard CJ, Litov RE, Savin WM, Dumelin EE, Tappel AL. Effects of exercise, vitamin E, and ozone on pulmonary function and lipid peroxidation. J Appl Physiol 1978; 45(6):927-32.
6. Dillard CJ, Litov RE, Tappel AL. Effects of dietary vitamin E, selenium, and polyunsaturated fats on in vivo lipid peroxidation in the rat as measured by pentane production. Lipids 1978; 13: 396-402.
7. Dillard CJ, Kunert KJ, Tappel AL. Lipid peroxidation during chronic inflammation induced in rats by Freund's adjuvant: effect of vitamin E as measured by expired pentane. Res Commun Chem Pathol Pharmacol 1982; 37:143-6.
8. Packer L. Vitamin E, physical exercise and tissue damage in animals. Med Biol 1984; 62:105-9.
9. Cormier M. Regulatory mechanisms of energy needs: vitamins in energy utilization. Prog Food Nutr Sci 1977; 2:347-56.
10. Schwarz K. Vitamin E, trace elements, and sulfhydryl groups in respiratory decline. Vitam Horm 1962; 20:463-84.
11. Schwarz K. The cellular mechanisms of vitamin E action: direct and indirect effects of alpha-tocopherol on mitochondrial respiration. Ann N Y Acad Sci 1972; 203:42-52.
12. Carabello F, Liu F, Eames O, Bird J. Effect of vitamin E deficiency on mitochondrial energy transfer. Fed Proc 1971; 30:639.
13. Carabello FB. Role of tocopherol in the reduction of mitochondrial NAD. Can J Biochem 1974; 52(8):679-88.

14. Fedelesova MP, Sulakhe PV, Yates JC, Dhalla NS. Biochemical basis of heart function. IV. Energy metabolism and calcium transport in hearts of vitamin E deficient rats. Can J Physiol Pharmacol 1971; 49:909-18.

15. Bertolotti E, Loidodice G, Quazza GF, Vento R. Richerche sperimentali istologica ed istochimiche nel ratto d dieta carenziata di vitamina E. Minerva Pediatr 1965; 17:873-7.

16. Chen LH, Lin CT. Some enzymatic changes associated with pathological changes in rats with long-term vitamin E deficiency. Nutr Rep Int 1980; 21:387-95.

17. Tureen L, Simons R. Enzyme changes within muscle fibers in genetic and nutritional muscular dystrophy. Proc Soc Exp Biol Med 1968; 129:384-90.

18. Simon-Schnass I, Pabst H. Influence of vitamin E on physical performance. Int J Vitam Nutr Res 1988; 58:49-54.

19. Ernst E, Matrai A, Aschenbrenner E. Blood rheology in athletes. J Sports Med Phys Fitness 1985; 25:207-10.

20. Ernst E. In: Ernst E, ed. Hämorheologie für den Praktiker. München: W. Zuckschwerdt Verlag, 1986.

21. Schmid-Schönbein H. Blood rheology in hemoconcentration. In: Brendel W, Zink RA, eds. High altitude physiology and medicine. New York: Springer-Verlag, 1982; 109-16.

22. Thews G, Mutschler E, Vaupel P. In: Anatomie, Physiologie, Pathophysiologie des Menschen. Stuttgart: Wissenschaftliche Verlagsgesellschaft mbH, 1982.

23. Kappus H. Lipid peroxidation: mechanisms, analysis, enzymology and biological relevance. In: Sies H, ed. Oxidative Stress. London: Academic Press, 1985; 273-310.

24. Schmid-Schönbein H. Pathophysiology of cutaneous frost injury: disturbed microcirculation as a consequence of abnormal flow behaviour of the blood. In: Jokl E, Hebbelinck M, eds. High altitude deterioration; medicine sport science, Vol. 19. Basel: Karger, 1985; 20-38.

25. Hiller E, Riess H. Erworbene plasmatische Gerinnungsstörungen. In: Hiller E, Riess H, eds. Hämorrhagische Diathese und Thrombose. Stuttgart: Wissenschaftliche Verlagsgesellschaft mbH, 1988; 86-109.

26. Volger E. Einflüsse der Hypoxie auf die Erythropoese, Sauerstoffbindung des Hämoglobins, Gerinnung und Rheologie des Blutes. In: Daum S, ed. Hypoxia—Pathophysiologie, Klinik und Therapie. München: Dustri-Verlag, 1984; 225-40.

27. Chow CK. Nutritional influence on cellular antioxidant defense systems. Am J Clin Nutr 1979; 32:1066-81.

28. Leibovitz BE, Siegel BV. Aspects of free radical reactions in biological systems: aging. J Gerontol 1980; 35:45-56.

29. Simon-Schnass I, Koeppe HW. Vitamin E und Arteriosklerose. Z Allgemeinmed 1983; 59(27): 1474-6.

30. Oelz O. Höhenhypoxie—Physiologische und medizinische Aspekte. In: Daum S, ed. Hypoxie—Pathophysiologie, Klinik und Therapie. München: Dustri-Verlag, 1984; 205-20.

31. Winslow RM. Hypoxia. Man at altitude. Stuttgart: Thieme Verlag, 1982.

32. Simon-Schnass I. Nutrition and exercise at high altitude. In: AIN symposium on nutrition and exercise, FASEB meeting. April 21–25, 1991.

33. Simon-Schnass I, Korniszewski L. The influence of vitamin E on rheological parameters in high altitude mountaineers. Int J Vitam Nutr Res 1990; 60(1):26-34.

34. Asano M, Branemark PI, Castenholz A. A comparative study of continuous qualitative and quantitative analysis of microcirculation in man. Adv Microcirc 1973; 5:1.

35. Bagge U, Blixt A, Braide M. Macromodel experiments on the effect of wall-adhering white cells on flow resistance. Clin Hemorrheol 1986; 6:365.

36. Lipowsky H, Usami S, Chien S. In vivo measurements of ''apparent viscosity'' and microvessel hematocrit in the mesentery of the rat. Microvasc Res 1980; 19:297.

37. Harlan JM, Killen PD, Harker LA. Neutrophil-mediated endothelial injury in vitro: Mechanisms of cell detachment. J Clin Invest 1981; 68:1394.

38. Benjamini E, Keskowitz S. Überempfindlichkeitsreaktionen. In: Benjamini E, Leskowitz S, ed. Immunologie. Stuttgart: Schwer Verlag, 1988; 197-2125.

39. Weiss SJ, Regiani S. Neutrophils degrade subendothelial matrices in the presence of alpha-1-proteinase inhibitor. Cooperative use of lysosomal proteinase and oxygen metabolites. J Clin Invest 1984; 73:1297.

40. Moore KL, Andreoli SP, Esmon NL, Esmon CT, Bang NU. Endotoxin enhances tissue factor and suppresses thrombomodulin expression of human vascular endothelium in vitro. J Clin Invest 1987; 79:124-30.

41. Nawroth PP, Handley DA, Esmon CT, Stern DM. Interleukin 1 induces endothelial cell procoagulant while suppressing cell-surface antocoagulant activity. Proc Natl Acad Sci U S A 1986; 83: 3460-4.

42. Hultgren HN. High altitude pulmonary edema. Adv Cardiol 1970; 5:42.

43. Simon-Schnass I, Reimann J, Böhlau V. Vitamin-E-Therapie: Zur Frage der Resorption von Vitamin E. Notabene Med 1984; 14:793-4.

44. Julicher RHM, Sterrenberg L, Loomen JM, Bast A, Noordhoek J. Evidence for lipid peroxidation during the calcium paradox in vitamin E deficient rat heart. Naunyn Schmiedebergs Arch Pharmacol 1984; 326:87-9.

45. Tappel AL. Vitamin E and free radical peroxidation of lipids. Ann N Y Acad Sci 1972; 203:12-28.

34

Significance of Vitamin E for the Athlete

Lester Packer

University of California, Berkeley, California

Abraham Z. Reznick

The Bruce Rappaport Faculty of Medicine and Rappaport Institute for Medical Research, Technion–Israel Institute of Technology, Haifa, Israel

Irene Simon-Schnass

Hermes Arzneimittel GmbH, Munich, Germany

Sharon V. Landvik

Vitamin E Research and Information Service, Edina, Minnesota

INTRODUCTION

Optimal physical performance capacity is inconceivable without optimum cellular function. Physical performance always creates situations that subject the cellular metabolism to the limits of stress. The supply, turnover, and regeneration of substrates for producing energy are only one part of the whole. The effects on the structural components of the cell, however, are especially important and in part not fully recognized or taken into account. Although it is up to the individual, even with optimal training and nutrition, whether to stress his or her body beyond its given limits and to accept possible damage or sacrifice in performance, there is no doubt that targeted nutritional planning in conjunction with optimized training can clearly extend these limits (1,2).

PHYSIOLOGICAL EFFECTS OF EXERCISE

During physical exercise, oxygen consumption may rise severalfold and is usually associated with a higher rate of lipid peroxidation (3). Analysis of hind limb skeletal muscle of endurance-trained rats demonstrated a twofold increase in the mitochondria and associated enzyme activities. Thus, it could be anticipated that trained animals produce higher levels of free radicals than sedentary animals since mitochondria are increased in amount and activity per unit of muscle weight (3,4). Indeed, in animals exercised to exhaustion, free-radical concentrations are increased two- to threefold in muscle and liver following exercise to exhaustion (5).

During exhaustive exercise, muscle damage occurs even in the highly trained athlete. In marathon runners, skeletal muscle damage was estimated by analysis of blood serum creatine kinase activity. Following a 42 km marathon race, skeletal muscle injury was significantly greater in men than in women, with or without correction for body surface area. Although the release of creatine kinase from skeletal muscle results from muscle injury, it is not known whether it is related to reversible and/or irreversible injury (6).

MECHANISM OF VITAMIN E ACTION

Polyunsaturated fatty acids (PUFA) may undergo oxidative modification by oxidatively reactive species (ROS), which results in various lipid radicals (R˙, RO˙, and ROO˙). Unless destroyed, these radicals could propagate the process of lipid peroxidation, which has a serious destabilization effect on the physical and conformational properties of membrane lipids. Vitamin E was shown to be an inhibitor of free-radical oxidation, including lipid hydroxy and peroxy radicals. As seen in Figure 1, a peroxy radical is reduced by α-tocopherols or tocotrienols to an organic peroxide.

At the same time, tocopherol or tocotrienol ends as tocopheroxyl and tocotrienoxyl radicals, which can undergo decomposition, or by enzymatic and nonenzymatic regeneration pathways revert to their reduced form of tocopherol and tocotrienols (Fig. 1). By stopping lipid peroxidation and free-radical propagations, vitamin E acts to regulate lipid peroxidation, thus stabilizing the lipid bilayer membranes.

EFFECTS OF EXERCISE ON VITAMIN E STATUS

In animal studies, there was more rapid depletion of vitamin E from liver and muscle of rats undergoing endurance training than in unexercised control animals (7,8). Vitamin E concentrations in muscle tissue were significantly lower in endurance-trained animals than in controls, regardless of vitamin E intake (Table 1). It was concluded by the researchers that these results in rats demonstrated evidence of vitamin E depletion in skeletal muscle during endurance training (8). In other animal studies, there was no corresponding rise in vitamin E content in association with the increased mitochondrial content in skeletal muscles of endurance-trained animals. Therefore, it appears that endurance training results in a substantial

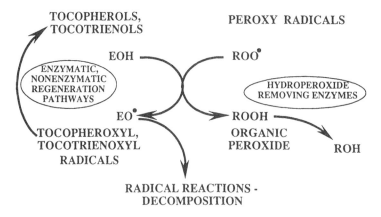

FIGURE 1 Vitamin E destroys free radicals.

TABLE 1 Vitamin E Levels in Liver and Hind Leg Muscle of Trained and Sedentary Rats

	Vitamin E (μg/g wet tissue)	
	Muscle	Liver
Endurance trained		
Vitamin E deficient	0.0 $\pm$ 0.0	0.2 $\pm$ 0.2
Control diet	4.9 $\pm$ 1.1	37.8 $\pm$ 7.4
Sedentary		
Vitamin E deficient	1.2 $\pm$ 0.6	3.0 $\pm$ 1.6
Control diet	7.2 $\pm$ 1.2	34.6 $\pm$ 7.6

decrease in the vitamin E content of proliferating muscle mitochondrial membranes and depletes the muscle mitochondria of their major lipid-soluble antioxidant (9,10). Since increased physical exercise results in consumption of vitamin E by body tissues, these research findings suggest a higher vitamin E requirement during endurance training (3,11).

VITAMIN E AND ENERGY SUPPLY

In the mitochondria of vitamin E-malnourished laboratory animals, reduced cellular respiration is a regular occurrence that does not result from a blockage of the cytochrome c redox series (12–14). Various investigations have shown that blockage of the electron flow in vitamin E deficiency occurs at the beginning of the respiratory chain during the transfer of hydrogen from $NADH_2$ to an Fe-S protein (complex I) or during the transfer of succinate to an Fe-S protein (complex II). Furthermore, the channeling of hydrogen into the respiratory chain is inhibited by a reduction in the activity of lipoic acid amide dehydrogenase (LADH). LADH catalyzes the transfer of hydrogen of lipoic acid amide to NAD and thus produces an important connective pathway between the final breakdown of the C skeleton (pyruvate) and the respiratory chain (15).

Interestingly, all vitamin E-sensitive enzymes are proteins with unstable SH groups. The transfer of these sulfhydryl groups into the oxidized state leads to the loss of biological activity. It has been demonstrated that vitamin E is localized in the mitochondrial membrane in the immediate vicinity of the enzymes containing SH groups and has a protective influence on these SH groups. In this way, vitamin E plays a major if indirect role in electron transport to and within the respiratory chain and thus in the production of energy. A significant influence of vitamin E on energy metabolism can correspondingly only be expected when performance is accomplished with the help of aerobic energy supply (16,17).

EFFECTS OF VITAMIN E DEFICIENCY IN EXERCISE

In an investigation of the effects of vitamin E deficiency in exercise, muscle and liver tissue of vitamin E-deficient, endurance-trained animals had significantly lower vitamin E levels and higher free-radical concentrations than vitamin E-deficient, sedentary rats (6,11). In vitamin E-deficient rats, strenuous exercise also resulted in increased membrane fragility and a decreased respiratory control ratio of muscle mitochondria. Vitamin E deficiency resulted in a 40% decrease in running time to exhaustion in untrained animals. As noted by the researchers, their results imply that the reduced endurance capacity in vitamin E-deficient rats was associated with peroxidative damage to mitochondria (5). Endurance capacity was decreased 38% in vitamin E-deficient animals and 33% in vitamin C-supplemented, vitamin E-deficient animals, demonstrating that vitamin C did not prevent the reduction in endurance capacity associated with vitamin E deficiency (7).

PROTECTIVE ROLE OF VITAMIN E IN EXERCISE

In recent studies, rats were supplemented for 4 weeks with a high vitamin E diet (10,000 IU/ kg diet) or with palm oil-rich vitamin E and tocotrienol concentrate (7000 mg tocotrienol/ kg diet). Other animals were supplemented with a control diet of low levels of vitamin E (30 IU/kg body weight). After 4 weeks, some of the animals were exercised to exhaustion on a treadmill, sacrificed, and their organs were removed for various biochemical studies. Table 2 shows the level of protein carbonyls as a measure of protein oxidation. In rats supplemented with a normal diet, the protein carbonyl content of gastrocnemius muscles increased by 17.23% as a result of exercise, but in vitamin E-supplemented animals the increase due to exercise was only between 5.8 and 6.3%. Overall, tocopherol- and tocotrienol-supplemented animals had a significantly lower carbonyl content compared to animals fed the control diet (− 32.3 and − 19.8%, respectively).

In a number of human studies evaluating the effects of strenuous exercise, breath pentane excretion as a measure of free-radical reactions can be significantly reduced with vitamin E supplementation. In a study of human volunteers, strenuous physical exercise resulted in increased pentane production. Vitamin E supplementation (1200 IU/day for 2 weeks) resulted in significantly reduced breath pentane excretion during exercise and at rest (18).

In male college students, exhaustive exercise was associated with a slight but significant rise in serum lipid peroxide concentrations immediately after exercise. There was a significant increase in leakage of enzymes from tissues to the blood during exercise, which was considered indicative of exercise-induced oxidative membrane damage. When subjects were supplemented with vitamin E (300 mg/day for 4 weeks), serum lipid peroxide concentrations significantly decreased immediately after exhaustive exercise and the increase in serum enzyme activities was also lower. Based on these results, it was concluded that lipid peroxidation associated with strenuous exercise can be inhibited by vitamin E supplementation (19).

In a study of male volunteers in two age ranges (22–29 and 55–74 years), subjects were monitored for 12 days after running downhill on an inclined treadmill. Before exercise, subjects

TABLE 2 Carbonyl Content in Gastrocnemius Muscle of Control Rats and Rats Subjected to a Single Bout of Endurance Exercise to Exhaustion: Effect of Various Diets (nmol/mg Protein)

		Vitamin E-supplemented diet	
Animals	Normal diet	Tocopherol	Tocopherol + tocotrienol[a]
Control	2.14 ± 0.23	1.45 ± 0.06	1.73 ± 0.13
	$n = 3$	$n = 5$ (− 32.3%)	$n = 4$ (− 19.8%)
Exercised	2.51 ± 0.11	1.57 ± 0.13	1.83 ± 0.15
	$n = 3$	$n = 6$ (− 37.5%)	$n = 4$ (− 27.1%)
% Change in exercised versus control	+ 17.28%	+ 8.27%	+ 5.78%

[a]Palm oil vitamin E concentrate.

TABLE 3 Effect of Vitamin E on Pentane Exhalation of High-Altitude Mountain Climbers

	Pentane exhalation change (%)	
	Vitamin E (400 mg/day)	Placebo
Median	− 3.0	104.0
Lower quartile	− 7.4	25.5
Upper quartile	2.9	121.5

were randomized to receive vitamin E (800 IU/day for 48 days) or placebo. Mean plasma creatine kinase levels were significantly increased the morning after damaging eccentric exercise. The peak increase in plasma creatine kinase levels was similar in both groups under 30 years old, but levels returned to baseline more rapidly in the vitamin E-supplemented group. Peak plasma creatine kinase levels after exercise were significantly lower in the older subjects on placebo than in the younger subjects. The older vitamin E-suplemented subjects exhibited plasma creatine kinase levels similar to those of younger subjects following exercise. According to the researchers, the finding that creatine kinase release was more pronounced in young people, who are usually considered more resilient to physical stress, challenges the assumption that efflux of myocellular enzymes represents undesirable damage to muscle membranes. An alternative theory is that creatine kinase is a manifestation of increased muscle protein turnover, which is needed to clear partially damaged proteins. The researchers noted that dietary vitamin E supplementation tended to eliminate the differences between the two groups, primarily by increasing the responses of the older subjects (20).

The effects of vitamin E on athletic performance were evaluated in swimmers. In two studies of trained swimmers, swimming speed was not significantly different in vitamin E-supplemented swimmers compared to swimmers on placebo (21,22). In a study in mice, swimming endurance was significantly prolonged by pretreatment with vitamin E or other compounds that trap free radicals (23). The effects of vitamin E on tissue damage and physical performance were also studied in a group of high-altitude mountain climbers. In unsupplemented mountain climbers, prolonged exposure to physical exertion at high altitudes while climbing resulted in significantly increased breath pentane output and decreased physical performance as demonstrated by a significant decrease in anaerobic threshold. In contrast, vitamin E supplementation (400 IU/day) prevented the increase in breath pentane exhalation and the decrease in anaerobic threshold, and it was concluded that vitamin E has a beneficial effect on cell protection and physical performance, at least at high altitude (Table 3) (24). In another study of mountain climbers, the viscosity of whole blood increased markedly during the ascent. The filterability of red blood cells dropped significantly in control subjects but remained unchanged in mountain climbers supplemented with 400 mg vitamin E daily, leading to the conclusion that prolonged physical exertion at high altitude can impair blood flow characteristics and that deterioration in blood parameters can be prevented by vitamin E (25).

RECOMMENDATION OF VITAMIN E PROVISION FOR ACTIVE ATHLETES

In view of the preceding discussion, the question arises as to how much vitamin E should be recommended to people engaged in moderate to heavy exercise.

The U.S. recommended daily allowance (RDA) published by the Food and Drug Administration (FDA) in 1968 was 30 International Units (IU) (26). This was later reduced by the FDA in 1980 to 15 IU (26). In the same year, however, it was estimated that in the United States, the amount of vitamin E supplied by a "normal" diet is only about 7.4 mg (11 IU) (27). Obviously these numbers are quite low and are not sufficient for very active athletes. However, levels of vitamin E supplementation in exercise studies ranged from 300 to 1200 IU, and in a well-controlled study no side effects of vitamin E supplementation were observed (28). Thus to be on the safe side, provisions of up to 400 IU/day of vitamin E may be a reasonable recommendation for active athletes engaged in moderate to heavy exercise.

CONCLUSION

Vitamin E is directly or indirectly involved in electron transport and protection of cellular structures and thus shows a broad relation to cellular functions that could be connected with adequate performance or susceptibility to injury. Research continues on the protective role of vitamin E in exercise, but the results to date demonstrate an increased vitamin E requirement to prevent the free-radical–related tissue damage associated with strenuous exercise.

REFERENCES

1. Keul J, Ber A, Lehmann M, Dickhuth HH, Schmid P, Jakob E. Edrschopfung und regeneration des muskels in training und Weltkamp. Leistungssport 1984; 5:13-8.
2. Keul J, Jakob E, Berg A, Dickhuth HH, Lehmann M. Zur wirking von vitaminen auf die leistungs- und erholengsfahigkeit des menschen phrmazeut. Rundschau 1987; 9:94-8.
3. Packer L. Vitamin E, physical exercise and tissue damage in animals. Med Biol 1984; 62:105-9.
4. Davies KJA, Packer L, Brooks GA. Biochemical adaptation of mitochondria, muscle and whole animal respiration to endurance training. Arch Biochem Biophys 1981; 209:539-54.
5. Davies KJA, Quintanilha AT, Brooks GA, Packer L. Free radicals and tissue damage produced by exercise. Biochem Biophys Res Commun 1982; 107:1198-205.
6. Apple FS, Rhodes M. Enzymatic estimation of skeletal muscle damage by analysis of changes in serum creatine kinase. J Appl Physiol 1988; 65:2598-600.
7. Gohil K, Packer L, de Lumen B, Brooks GA, Terblanche SE. Vitamin E deficiency and vitamin C supplements: exercise and mitochondrial oxidation. J Appl Physiol 1986; 60:1986-91.
8. Aikawa KM, Quintanilha AT, de Lumen BO, Brooks GA, Packer L. Exercise endurance training alters vitamin E tissue levels and red blood cell hemolysis in rodents. Biosci Rep 1984; 4:253-7.
9. Gohil K, Rothfuss L, Lang J, Packer L. Effect of exercise training on tissue, vitamin E and ubiquinone content. J Appl Physiol 1987; 63:1638-41.
10. Lang J, Gohil K, Rothfuss L, Packer L. Exercise training effects on mitochondrial enzyme activity, ubiquinones and vitamin E. In: Anticarcinogenesis and radiation protection. New York: Plenum Press, 1987; 253-7.
11. Quintanilha AT. Effects of physical exercise and/or vitamin E on tissue oxidative metabolism. Biochem Soc Trans 1984; 12:403-4.
12. Carabello F, Liu F, Eames O, Bird J. Effect of vitamin E deficiency on mitochondrial energy transfer. Fed Proc 1971; 30:639.
13. Carabello FB. Role of tocopherol in the reduction of mitochondrial NAD. Can J Biochem 1974; 52:679-88.
14. Fedelesova MP, Sulakhe PV, Yates JC, Dhalla NS. Biochemical basis of heart function. IV. Energy metabolism and calcium transport in hearts of vitamin E deficient rats. Can J Physiol Pharmacol 1971; 49:909-18.
15. Cormier M. Regulatory mechanisms of energy needs: vitamins in energy utilization. Prog Food Nutr Sci 1977; 2:347-56.

16. Schwarz K. The cellular mechanisms of vitamin E action: direct and indirect effects of alpha tocopherol on mitochondrial respiration. Ann N Y Acad Sci 1972; 203:42-52.
17. Schwarz K. Vitamin E, trace elements and sulfhydryl groups in respiratory decline. Vitam Horm 1962; 20:463-84.
18. Dillard CJ, Litov RE, Savin WM, Dumelin EE, Tappel AL. Effects of exercise, vitamin E and ozone on pulmonary function and lipid peroxidation. J Appl Physiol 1978; 45:927-32.
19. Sumida S, Tanaka K, Kitao H, Nakadomo F. Exercise-induced lipid peroxidation and leakage of enzymes before and after vitamin E supplementation. Int J Biochem 1989; 21:835-8.
20. Cannon JG, Orencole SF, Fielding RA, et al. Acute phase response in exercise: interaction of age and vitamin E on neutrophils and muscle enzyme release. Am J Physiol 1990; 259:R1214-9.
21. Sharman IM, Down MG, Norgan NG. The effects of vitamin E on physiological function and athletic performance of trained swimmers. J Sports Med 1976; 16:215-25.
22. Lawrence JD, Bower RC, Riehl WP, Smith JL. Effects of alpha tocopherol acetate on the swimming endurance of trained swimmers. Am J Clin Nutr 1975; 28:205-8.
23. Novelli GP, Bracciotti G, Falsini S. Spin-trappers and vitamin E prolong endurance to muscle fatigue in mice. Free Radic Biol Med 1990; 8:9-13.
24. Simon-Schnass I, Pabst H. Influence of vitamin E on physical performance. Int J Vitam Nutr Res 1988; 58:49-54.
25. Simon-Schnass I, Korniszewski L. The influence of vitamin E on rheological parameters in high altitude mountaineers. Int J Vitam Nutr Res 1990; 60:26-34.
26. Vitamin E Research and Information Service (VERIS). The vitamin E fact book. LaGrange, IL, 1989.
27. Bauerfeind JB. Tocopherols in foods. In: Vitamin E: a comprehensive treatise. 1980; 125-6.
28. Bendich A, Machlin LJ. Safety of oral intake of vitamin E. Am J Clin Nutr 1988; 48:612-9.

35

Plasma α-Tocopherol and Ubiquinone and Their Relations to Muscle Function in Healthy Humans and in Cardiac Diseases

Jan Karlsson

Karolinska Hospital, Stockholm, Sweden

Bertil Diamant

Copenhagen University, Copenhagen, Denmark

Henning Theorell

General Practitioner, Stockholm, Sweden

Kurt Johansen

Kabi-Pharmacia, Stockholm, Sweden

Karl Folkers

University of Texas at Austin, Austin, Texas

INTRODUCTION

In recent reports, we have documented a linear relationship between plasma α-tocopherol (α-T), ubiquinone (UQ) (coenzyme Q_{10} or CoQ_{10}), and free cholesterol (FC) levels in healthy humans and in many patient groups (1,2). In patients with severe ischemic heart disease and candidates for coronary bypass surgery, a linear relationship is also present between UQ and α-T but not in endurance-trained athletes (1,2).

The three compounds α-T, UQ, and FC are all extremely lipophilic and are consequently exclusively confined to lipoidal systems, predominantly in blood lipoproteins. The presence in plasma can be related to (1) the bloodstream as a means of transport and (2) an intrinsic activity in the bloodstream and/or its constituents.

α-Tocopherol, in contrast to both UQ and FC, is an entirely exogenous compound and is provided by nutritional intake. The main amounts of UQ and FC, on the other hand, are synthesized in all cells and organ systems (3,4). The body receives at the most 30% of its cholesterol from food. The corresponding figure for UQ is most probably small. Experiments on rats have revealed that approximately 3% originates from food intake (4). Corresponding figures for humans are not yet available, but treatment data of healthy volunteers and patients reveal that substantial plasma, heart, and muscle levels can be obtained by an oral intake of UQ (5–8).

It was suggested that α-T and UQ in plasma could have a synergistic activity in lipoproteins of the blood as protection from peroxidation (1,2,9). Lipoprotein protection was proposed for α-T (10). Experimental evidence was found showing that at exhaustion of lipoprotein α-T stores, peroxidation takes place, which activates the scavenger receptors in macrophages (11). This is a significant step in foam cell formation and atherogenesis (12). α-T is a potent antioxidant in most cellular systems (13) and is frequently used as a standard to evaluate other endogenous or exogenous compounds and their scavenging potentials (14,15).

The possible role of UQ as an antioxidant in biological systems was originally suggested for mitochondria function by the investigators Ernster and coworkers (16–19). The antioxidation has also been studied under more physiological conditions in whole animals by Beyer et al. (20) and Pedersen et al. (21). In a recent review, Beyer and Ernster presented in detail the background for UQ as an antioxidant, which includes both biochemical aspects and clinical potentials (22).

This review summarizes our data concerning plasma α-T and UQ interrelations and how they are affected in different diseases. This is done in the context of skeletal muscle metabolic profiles, which are essential both in respect to molecular oxygen metabolism and regulation of the central and peripheral circulation (23,24).

BIOCHEMICAL AND BIOPHYSICAL PROPERTIES OF α-T AND UQ OF SIGNIFICANCE FOR THEIR BIOLOGICAL ACTIVITIES

α-T, UQ, and FC are lipophilic and consequently retained in lipids, lipoid areas, that is, mostly cellular membranes, and plasma-borne lipoproteins. Of the different plasma lipoprotein fractions the very low and low-density lipoproteins (VLDL + LDL) contain 83% of the total cholesterol (TC) and 84% of the triglycerides (TG) (25). FC is constant over a wide range of TC and amounts to 10% (10).

TC and/or FC are consequently representative markers for the magnitude of levels of lipophilic UQ and α-T. This was the rationale to apply measurements of UQ and α-T and to normalize them as expressed by FC, for example (26).

PLASMA α-TOCOPHEROL AND UBIQUINONE LEVELS

Plasma Values, Normalization, and Interpretations

The lipophilic properties of α-T and UQ and how to relate their plasma or tissue levels has been a concern for both α-T (27,28) and UQ investigators (26,29,30). Edlund (26) suggested the computation of the ratios UQ and α-T over FC, respectively, to normalize for the lipids available to store α-T and UQ.

The rationale for this suggestion was the linear relationship between α-T and UQ, on one hand, and FC on the other in healthy humans and many patient groups (1,2,26). In the remaining part of this review, the ratios α-T and UQ over FC are referred to as α-T and UQ normalized for FC or N-α-T and N-UQ, respectively.

In Table 1, mean values $\pm$ SEM (standard error of the mean) for the actual plasma levels (α-T, UQ, and FC) and the corresponding normalized values (N-α-T, UQ, and FC) in various groups of healthy subjects and in different patient groups are presented. When plotted versus FC, mean α-T for the groups appears to be linearly distributed (Fig. 1A and B). UQ showed

TABLE 1   Mean values ± 1 SEM for ubiquinone (UQ or coenzyme Q_{10}, CoQ_{10}), alpha-tocopherol (AT or vitamin E) and free cholesterol (FC) in plasma from healthy sedentary (SM) and physically active males (PAM), endurance trained males (ETM), dilated cardiomyopathy (CMP), male patients with ischemic heart disease (IHD), patients treated with radical producing drugs as Adriamycin[R], Cordalorone[R], etc (RPD), heart transplanted patients (HTX), patients undergoing cardiac artery by-pass grafting (CABG) and CMP undergoing UQ treatment (QTP). The actual reference or whether data are unpublished (UP) are also included.

Patient/subject group	UQ	AT	FC	N-UQ	N-AT	Reference(s)
	μg × ml^{-1}		mg × ml^{-1}		Units	
Sedentary males (SM)	0.78 ± .04	11.5 ± .7	0.57 ± .02	1.4 ± .1	20.5 ± 1.0	2
Physically active males						
(PAM) I	0.71 ± .06	10.3 ± 1.8	0.73 ± .03	1.0 ± .1	13.9 ± .8	1
II	0.69 ± .02	6.7 ± .4	0.59 ± .02	1.2 ± .0	11.1 ± .5	UP
Endurance trained males (ETM, mixed)	0.56 ± .06	10.2 ± .1	0.57 ± .03	1.0 ± .1	18.5 ± 2.4	1
Endurance trained males (ETM): Competition season	0.86 ± .10	9.6 ± .6	0.64 ± .06	1.3 ± .1	15.6 ± 1.0	UP
Heavy training season	0.56 ± .05	10.0 ± 1.2	0.55 ± .03	1.0 ± .1	18.4 ± 2.5	UP
Patients with cardio- myopathy (CMP)	0.63 ± .08	7.4 ± .9	0.48 ± .05	1.3 ± .1	15.1 ± 1.0	2
Patients with ischemic heart disease (IHD)	0.79 ± .09	11.7 ± 1.2	0.72 ± .08	1.2 ± .1	16.2 ± .7	1
Patients treated with rad prod drugs (RPD)	0.69 ± .06	8.2 ± 1.0	0.60 ± .06	1.2 ± .1	14.2 ± 1.5	1
Heart transplanted patients (HTX)						
min	0.50 ± .09	6.0 ± 1.0	0.46 ± .08	1.0 ± .1	14.2 ± 2.0	UP
max	0.85 ± .09	10.8 ± 2.5	0.63 ± .13	1.4 ± .3	19.1 ± 5.8	UP
Cardiac surgery:						
Onset of art circ	0.70 ± .08	8.4 ± 1.1	0.60 ± .05	1.2 ± .1	14.2 ± 1.2	UP
Term of art circ	0.67 ± .08	9.4 ± 1.2	0.64 ± .06	1.1 ± .1	14.9 ± .9	UP
At cath withdrawal	0.64 ± .06	—	0.59 ± .06	1.1 ± .1	—	UP
CoQ_{10} treatment						
Placebo	0.23 ± .06	4.1 ± .7	0.31 ± .10	0.8 ± .0	16.2 ± 4.7	55
100–400 mg × day^{-1} (QTP)	0.94 ± .57	6.2 ± .6	0.33 ± .07	2.6 ± 1.1	21.3 ± 4.7	55

the same pattern as α-T: linearity versus FC, with the exception of a patient group treated with 100–400 mg UQ a day for more than 4 weeks.

It is evident that α-T is linearly related to UQ (Fig. 1C). Notably, the UQ-treated group shows a significant α-T increase, although not to the magnitude expected from the relationships found in the other groups (Table 1). Even if it can be criticized on theoretical grounds, the regression lines in Figure 1A and B were computed for the sake of comparison. Thus, the α-T equation is

$$y = 15.0x + 0.42 \qquad r = 0.85 \qquad p < 0.001$$

where the intercept = 0.42; slope = 15.1; and the 95% CI range is 9.1–21.0. The corresponding UQ equation is

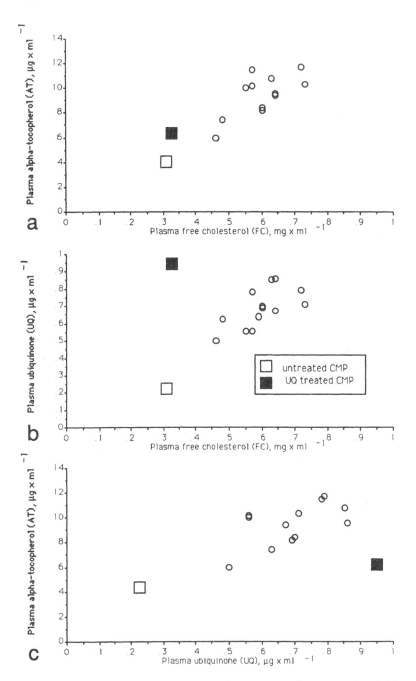

FIGURE 1   The relationships between plasma means for α-tocopherol (a) and ubiquinone (b) versus plasma free cholesterol and (c) α-T versus UQ for the different groups of apparently healthy humans and some patient groups, presented in Table 1. One group is particularly emphasized: patients with dilated cardiomyopathy (CMP) and treated orally with UQ (open squares before and closed squares after more than 6 weeks treatment with 100–400 mg/day).

$$y = 1.25x - 0.07 \qquad r = 0.83 \qquad p < 0.001$$

where the intercept $= -0.07$; the slope $= 1.26$; and the 95% CI range is 0.74–1.78.

The corresponding α-T values for sedentary but healthy, young males (SM, Table 1) are

$$y = 19.9x + 0.22 \qquad r = 0.66 \qquad p < 0.001$$

where the intercept $= 0.22$; slope $= 19.9$; and the 95% CI range is 11.3–28.6.

Finally, for UQ,

$$y = 1.20x + 0.10 \qquad r = 0.72 \qquad p < 0.001$$

with intercept $= 0.10$; slope $= 1.20$; and the 95% CI range 0.76–1.65.

The equations for the regression lines calculated from the means in Table 1 are almost identical to the equations for healthy sedentary males based in individual data. This agreement is in congruence with Edlund's suggestion to normalize the gross plasma α-T and UQ values for the corresponding FC values (26).

N-α-T and N-UQ values can then be used to evaluate the degree of saturation of this particular compartment to store α-T and UQ. This approach does not exclude the possibility that even other compartments can be used under certain conditions. This situation is exemplified with the N-UQ value with UQ treatment (Table 1 and Fig. 1A and B). This value is 2.6 compared to the other values, which have the range 0.8–1.4 units (Table 1). It seems reasonable that this unphysiologically high value discloses such compartmentalization, although its nature is not known.

Variations in α-T and N-α-T in Healthy Subjects and Different Patient Groups

The highest plasma α-T levels were in sedentary males and patients with ischemic heart disease (IHD, 11.5 ± 0.7 and 11.7 ± 1.2 μg/ml). The lowest levels were in heart-transplanted patients (HTX) in distress as a result of symptoms of rejection (6.0 ± 1.0 μg/ml or 50% of peak values; Table 1).

Normalized α-T peaks were 21.3 ± 4.7 in the UQ-treated patients and 20.5 ± 1.0 units in the sedentary males, whereas the lowest peaks appeared among the physically active males (PAM, 13.9 ± 0.8). It is obvious from Table 1 that those who had low N-α-T values are also low in α-T. Thus, there seems to be a relationship between the absolute value and the level of saturation. The only exception to this pattern is the UQ-treated group, which had one of the lowest α-T levels but the highest with respect to N-α-T.

This general pattern between N-α-T and α-T remains on an individual level, which is exemplified in Figure 2A for patients with severe ischemic heart disease. Those who have low α-T levels have also low N-α-T values. In Figure 2B another two groups are added: physically active and sedentary males, and polynomial regression analysis reveals that a curvilinear relationship is present. This means that those who have high α-T levels disclose a leveling-off phenomenon (Fig. 2C). The existence of a plateau may be indicative that physiological saturation is present in those subjects with high α-T levels.

Variations in UQ and N-UQ in Healthy Subjects and Different Patient Groups

UQ and N-UQ follow approximately the same patterns as α-T and N-α-T. Thus, those with low UQ levels are also low in N-UQ (Fig. 3A). In contrast to N-α-T, N-UQ does not level off (Fig. 3B). This difference could indicate that UQ in comparison with α-T never reaches a level of physiological saturation, not even in supposedly well-nourished healthy humans. This

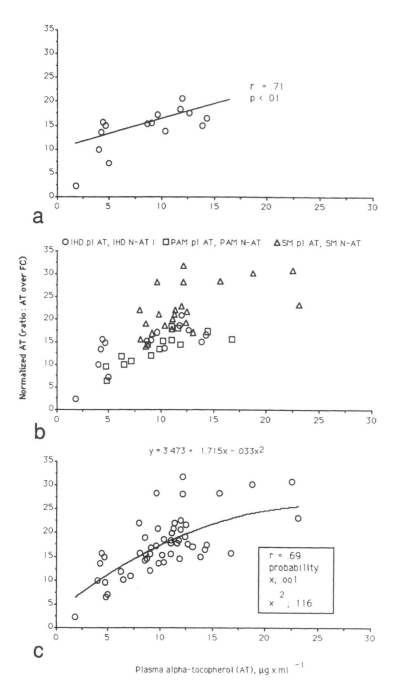

FIGURE 2 Individual relationships between plasma α-T normalized for its suggested site of deposition (FC) by computation of the ratio of plasma α-T to FC (N-α-T) versus plasma α-T level. This was done (a) for patients with severe ischemic heart disease (IHD) and (b) for IDH plus physically active males (PAM) and healthy but sedentary males (SM). (c) All individual data are pooled and the polynomial regression line tested for the relationship N-α-T versus plasma α-T.

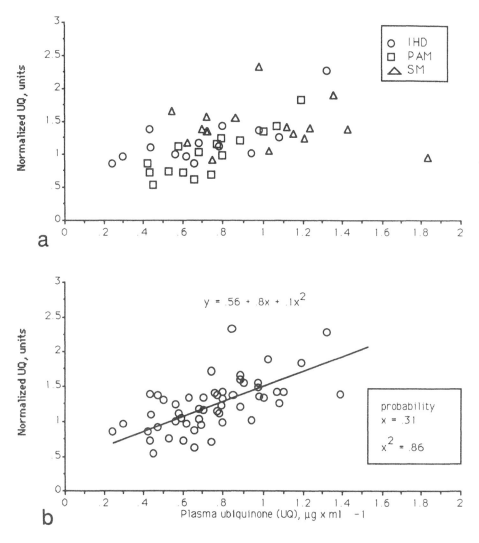

FIGURE 3 (a) Normalized plasma UQ (N-UQ; see Fig. 2) versus plasma UQ for the groups IHD, PAM, and SM. (b) The polynomial regression line is tested as in Figure 2c.

interpretation is further supported by the relationship between normalized α-T and UQ in Figures 2 and 3 (N-α-T and N-UQ) when plotted versus each other as in Figure 4. Patients with severe dilated cardiomyopathy (CMP) were treated with oral UQ (100–400 mg/day). As a group they deviate from others in Table 1 in many respects. Table 1 and Figure 1A and B show that they have an elevated α-T level in both absolute and relative terms ($p < 0.001$). This observation is further emphasized by the changes in N-α-T and N-UQ as depicted in Figure 4. With UQ treatment the mean N-α-T changed in parallel with the pooled IHD, PAM, and SM groups.

It thus seems reasonable to conclude that the dynamics of plasma α-T regulation takes into account the level of other plasma antioxidants, at least with respect to UQ. This implies that the relationship as advocated by Ernster and Lee is also valid for the extracellular space

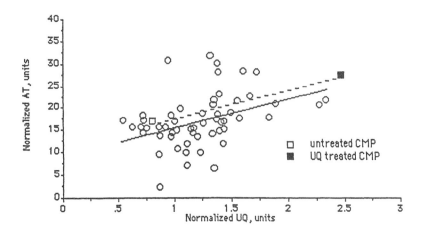

FIGURE 4   N-α-T versus N-UQ in the groups IHD, PAM, and SM. The UQ-treated group is denoted by squares (see Fig. 1).

(31). The hypothesis and the corresponding relationship couples α-T to UQ in a functional fashion and drains the lipids of reactive radicals from the midst of the lipid layers (UQ location site) to the outer section of the layer. There, α-T transports radicals by means of its hydrophilic entities to exclusively hydrophilic antioxidants, such as ascorbate, for further metabolism and neutralization by different enzymatic processes. This metabolism reduces the possibility of peroxidation, which could lead to cascadelike reactions and to a threat to maintenance of the integrity, ultimately to the very existence, of the cell or lipoprotein particle.

α-T AND UQ ALLOCATION TO DIFFERENT PLASMA LIPOPROTEINS

Lipoprotein Fractions and Their Atherogenicity

With the exception of free fatty acids, plasma lipids exist in the form of lipoprotein particles. These lipids constitute an outer membrane with apoproteins, phospholipids, and FC encompassing a core of TG and cholesterol esters (32). By means of preparative ultracentrifugation, six lipoprotein classes have been identified (33). In most epidemiological studies only four of them are cited: chylomicrons, VLDL, LDL, and HDL (34). Frequently VLDL and LDL are pooled in one class (35). The rationale for this classification is that VLDL and LDL are interrelated and represent a different metabolic pathway than HDL (36). The VLDL + LDL pool contains the majority of TC and TG (34). The FC percentage is constant over a wide range of TC, which is the rationale to standardize plasma α-T and UQ levels with the ratios N-α-T and N-UQ and make valid interindividual comparisons (10,26).

HDL and VLDL + LDL have different atherogenic profiles, and TG peroxidation was suggested by Goldstein and Brown to be crucial to activate the scavenger receptors (37). This is a significant sequence in the process of atherosclerosis. Antioxidant treatment can evidently reduce and/or abolish this state (38). The data of Esterbauer and coworkers (10) indicated the significance of α-T as a protective, antioxidant means in that respect. Analyses of both UQ and α-T in the HDL and LDL lipoproteins were therefore performed.

α-T and UQ in HDL and LDL

Edlund (10) reported after selective plasma VLDL and LDL precipitation that 79% of the total human UQ content was present in VLDL + LDL lipids, but there were no data concerning the corresponding α-T content. Elmberger et al. (39) showed after ultracentrifugation that human plasma HDL corresponded to 20% of TC and 63% of UQ. They consequently concluded that a relatively larger portion of plasma UQ is stored and carried by HDL than by VLDL + LDL. These two reports are thus not in agreement. No indication was given about the selection of subjects or patients, which could differentially affect allocation and exploration of the plasma antioxidants α-T and UQ.

Karlsson (29) showed that the blood UQ level varies considerably in health depending on training status. Nikkila et al. (40) showed that although training increases HDL both in absolute and relative terms, TC and correspondingly total plasma UQ decrease.

Plasma free cholesterol and the HDL fraction levels in healthy physically active individuals amount to 0.59 ± 0.02 and 0.17 ± 0.00 mg/ml, respectively. Plasma α-T and UQ values

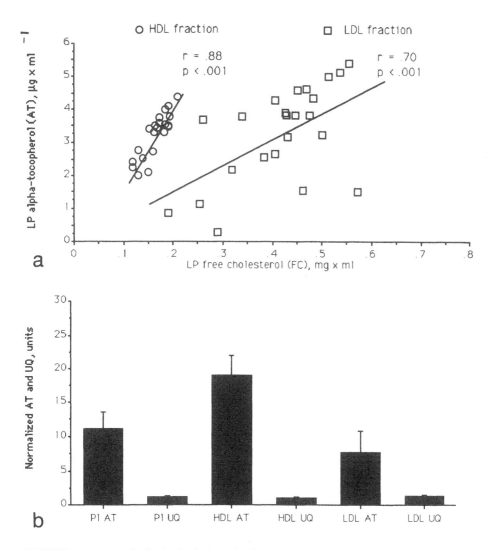

FIGURE 5 (a) The individual relationships between α-T levels in the plasma HDL and LDL lipoprotein fractions versus corresponding FC levels in healthy sedentary males. (b) Mean values + SEM for N-α-T and N-UQ in plasma and the HDL and LDL lipoprotein fractions, respectively.

average 6.7 ± 0.4 and 0.69 ± 0.02 and the corresponding HDL values 3.2 ± 0.15 and 0.17 ± 0.01 μg/ml. In relative terms, those levels correspond the HDL lipoprotein fraction to 29, 48, and 25% of the plasma FC, α-T, and UQ levels. The qualitative differences between α-T and UQ levels become even more obvious when α-T in HDL and LDL is plotted versus the corresponding FC levels (Fig. 5A). Similarly, correlations are seen for UQ in HDL and LDL ($r = 0.83$ and 0.78; $p < 0.001$). In contrast to α-T, the equation lines for UQ are practically identical.

Another example of differences between α-T and UQ levels is in respect to the N-α-T and N-UQ means in the different pools (Fig. 5B). N-UQ is approximately the same in all three pools, but N-α-T is approximately twice in the HDL compared to the LDL lipoproteins.

FC in LDL, expressed as the fraction of plasma FC, can be used to express an "atherogenic factor." A negative relationship is present between N-UQ in LDL and this atherogenic factor ($r = -0.52, p < 0.01$), but no relationship could be seen with respect to N-α-T. Thus, there seems to be a selective depletion of UQ in the LDL lipoprotein with an increased atherogenic potential. This observation is compared with the reduced N-α-T value in LDL (Fig. 5B).

Individual data revealed that α-T in both the HDL and LDL pools show leveling-off patterns with increased α-T levels, but UQ increases linearly. This pattern is identical to that for plasma α-T and UQ levels (Figs. 2C and 3B) and is indicative of those subjects low in α-T and UQ who are not saturated with respect to storage capacity. One may question whether supplementation programs could alter this pattern. Studies on patients on oral UQ treatment (100–400 mg/day) show not only that UQ increases but α-T is also affected. This is most probably due to an enhanced scavenging capacity with elevated UQ levels and with a sparing effect on α-T.

SKELETAL MUSCLE QUALITIES OF SIGNIFICANCE FOR PLASMA α-T AND UQ LEVELS

Healthy Volunteers

Needle muscle biopsies were obtained in the lateral portion of the thigh muscles (vastus lateralis) and used for histological and biochemical analysis. Muscle quality was evaluated based on muscle fiber typing (41,42), and the percentage distribution of the type I, slow-twitch (ST) or "red" muscle fibers was determined. As frequently demonstrated during the last two decades, muscle quality, expressed as percentage distribution of ST fibers (% ST), is a marker of the oxidative potential as represented by the activity of the mitochondrial enzyme citrate synthase (CS, Fig. 6A) (29,41,42). An increased mitochondrial function is, in the majority of cases, related to an increased mitochondrial volume (43). As a consequence, the deposition space for UQ is elevated and a positive relationship is present between % ST and muscle UQ level (Fig. 6B) (8,29). UQ in mitochondria is cited under the name coenzyme Q_{10} (CoQ_{10}) (44). The reason for this name (frequently misused) is that this substance was originally found in mitochondria and its function in electron transport was explored first (22,45).

Healthy males and their % ST and muscle UQ levels range 22–89 % ST and 0.08–0.35 mg UQ/mg dry weight (means 46 ± 3% and 0.19 ± 0.02 μg/mg (SEM), respectively). These data are in agreement with earlier observations in the healthy human (29). The regression line equation corresponds to an intercept of 0.04 μg/mg dry weight. It seems reasonable to suggest that this figure represents the minimum amount of UQ stored in the muscle in compartments other than mitochondria (3).

Increased % ST and/or muscle UQ is a matter of both endowment and local adaptation to increased muscle activity ("training") (8,23,24,29). Increased % ST results in a higher muscle lipoprotein lipase (LPL) activity and lowered plasma TG and TC levels (46). These

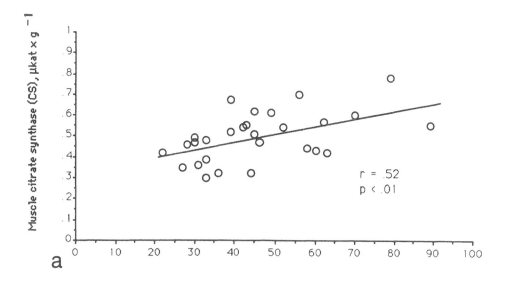

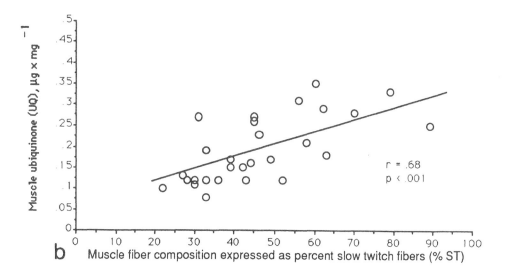

FIGURE 6 The individual relationships in PAM (Table 1) between the mitochondrial activity marker citrate synthase (a) and the muscle UQ content (b) versus the corresponding histochemical quality of the muscle expressed as percentage distribution of slow-twitch muscle fibers in the lateral portion of the thigh (vastus lateralis). In heart and skeletal muscle tissue UQ has an intrinsic position in mitochondrial electron transport. This is the reason for its alternative name, coenzyme Q_{10}.

changes result in a negative relationship between plasma FC and % ST (Fig. 7A). In contrast to muscle UQ, plasma UQ decreases with % ST muscle fibers (Fig. 7B). In addition, it was found that UQ normalized for its deposition sites (plasma FC, N-UQ) decreases (Fig. 7C). If % ST is exchanged with muscle UQ, the relationships to the same variables are $r = -0.46$ ($p < 0.01$), $r = -0.76$ ($p < 0.001$), and $r = -0.63$ ($p < 0.001$).

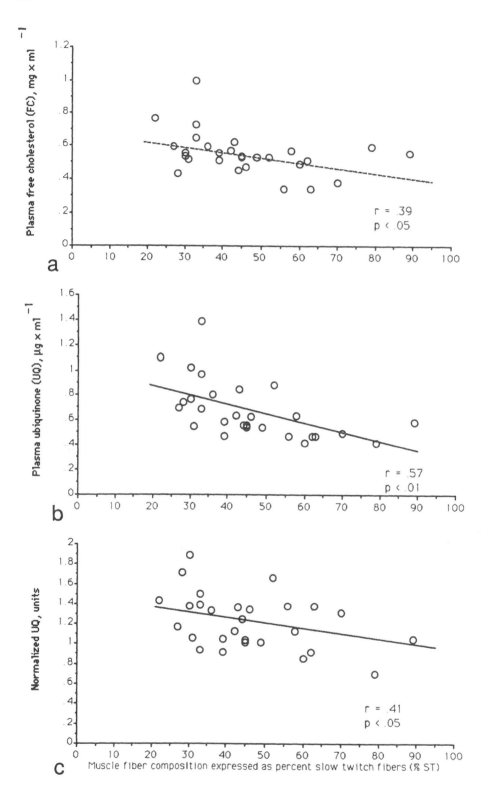

FIGURE 7   Individual relations for PAM as in Figure 6 but for (a) plasma FC, (b) plasma UQ, and (c) N-UQ versus muscle fiber composition.

Because of the lack of biopsy tissue, the muscle α-T level was not obtained. Plasma α-T decreased with % ST ($r = -0.38$, $p < 0.05$), but when α-T was normalized for FC (N-α-T) no relationship was present.

The results indicate that in the healthy human, plasma α-T is not related to % ST, but plasma UQ decreases. A high % ST means, by definition, a large combustive capacity in the muscle, a dense capillary network, and a high muscle LPL activity (46–48). All these features are prerequisites for lowering the atherogenic factor as expressed by the LDL/HDL ratio (49). Because α-T and UQ are differently allocated to HDL and LDL, respectively, the ratio α-T over UQ (α-T/UQ) was calculated and compared to % ST (Fig. 8). This ratio increased with % ST. It seems reasonable to suggest that the relatively higher relationship can be explained by both the relatively larger allocation of α-T to the HDL lipoprotein (Fig. 5B) but also the decreased N-UQ in the healthy human (Fig. 7C).

As is obvious from Table 1, plasma α-T and UQ are both higher in sedentary males (SM) compared to endurance-trained man (ETM). This is in line with the lower % ST fibers in SM, their lower muscle LPL activity, and higher plasma FC levels to retain α-T and UQ. However, variations were observed in UQ but not in α-T in the ETM group related to the type of physical activity undertaken at the time of plasma sampling. At the height of heavy endurance or OBLA training (23,48), UQ was lower ($p < 0.001$) compared to the competition season. It is well established that energy turnover is much higher during the former period. It is also evident that "rugged muscle fibers" are more frequent in endurance-trained than in strength-trained subjects and that the peak frequency of traumatized fibers occurs during this period (1). It is suggested that the appearance of rugged fibers reflects excessive radical formation and signs of a pathological cascade reaction of peroxidation in different cell compartments or membranes (1,24,29). Some ETM had equally low values as seen in HTX or RPD patients. This further strengthens the preceding suggestion that in health UQ is relatively more limited than α-T.

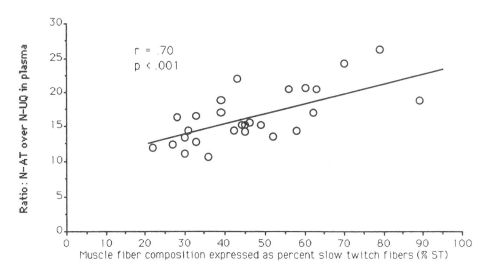

FIGURE 8 The ratio of plasma α-T to UQ versus individual muscle fiber composition as in Figures 6 and 7. The rationale for this ratio computation is that α-T allocation is HDL dependent (see Fig. 5). % ST promotes higher HDL levels.

Patients with Ischemic Heart Disease

Patients with heart failure have been shown by means of nuclear magnetic resonance techniques and "noninvasive" muscle fiber typing, as well as by conventional muscle biopsy techniques, that their muscles are relatively richer in type II, fast-twitch (FT) or "white" fibers than those of age-matched controls (50–52). The background of this observation was not

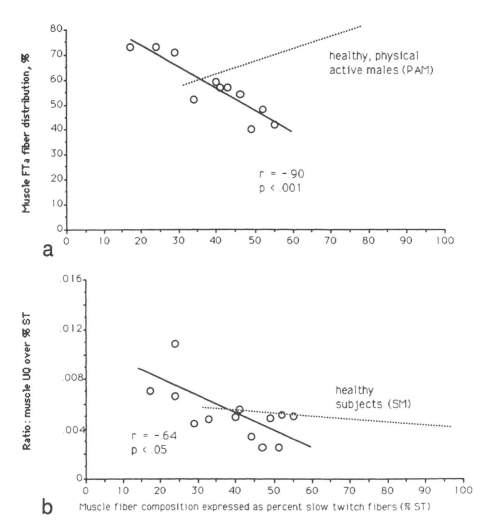

FIGURE 9   (a) Patients with severe IHD (Table 1) were tested for their muscle histological profile in respect to distribution of the intermediate fiber type FTa and % ST (see Fig. 6). This subgroup of the glycogenolytic, fast-twitch fibers (FT) has by means of adaptation obtained oxidative properties as seen in ST. In healthy humans the % FTa increases with the % ST. (b) Because the quality of the muscle expressed as % ST is of significance for muscle UQ deposition, the ratio of muscle UQ to % ST was computed and plotted versus % ST (first derivative computation). This ratio decreases in the healthy human (from Fig. 6B) and significantly faster in IHD as a sign of a ST fiber-selective UQ shortage, which may represent a sign of a muscle fiber metabolic lesion. UQ also represents the antioxidant machinery, and thus the oxidative ST fiber in both health and IHD, at least with respect to UQ, has an hampered scavenging function.

elucidated, although indications of local histochemical trauma confined to the "oxidative muscle fiber population" (ST fibers) was reported (8,51–53).

Muscle fiber composition expressed as % ST in patients with severe ischemic heart disease and candidates for coronary artery bypass grafting was 35–40% compared to 60% for an age-matched control group of healthy volunteers (48,51,52). Another indication of a different histochemical organization or regulation in these patients concerns the relative distribution of the subgroup FTa (% FTa; Fig. 9A). FTa fibers are referred to as an "intermediate fiber population," with oxidative properties like ST and glycogenolytic properties like FT fibers (41,42). In health, there is a positive relationship between percentage figures for ST and FTa, but not in IHD patients, in whom a negative relationship is seen. Moreover, their muscle UQ level decreased versus % ST when expressed as the first derivative (Fig. 9B). A decrease in the muscle UQ/% ST ratio represents a relative depletion of UQ the higher the % ST. This result confirms earlier findings concerning muscle UQ and IHD (51,52) and add support to the view that metabolic lesions could be the causal background to the changed muscle fiber composition in heart failure and IHD in particular.

The depressed muscle UQ/% ST ratio in IHD can be the result of both a deteriorated local UQ production (4) and/or transportation or uptake from the blood. As pointed out, UQ seems never to be close to physiological saturation either in plasma or in the HDL and LDL lipoproteins (Fig. 2C). These data could be the basis for speculation that in healthy individuals and cardiac patients, UQ levels are relatively more depleted than α-T levels. This relative lack of UQ compared to α-T is further stressed by relating the plasma α-T/UQ ratio to the muscle UQ/% ST ratio as in Figure 10. In IHD patients, the plasma α-T/UQ ratio decreases versus the relative muscle UQ levels. In health, the ratios are relatively constant over the whole range of muscle fiber compositions. In health (Fig. 8) the plasma α-T/UQ ratio increases with % ST because HDL contains more α-T. IHD patients show the opposite pattern: a decrease in the plasma α-T/UQ ratio. This decrease could be explained by their higher LDL level, which is lower in α-T than HDL, but also a different dependence on α-T than in health. Although there are reasons to explain a different reliance on α-T versus UQ in IHD,

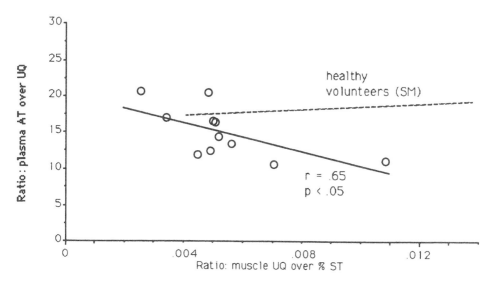

FIGURE 10 In patients with IHD but not in health, the relative muscle UQ content (see Fig. 9b) decreases the plasma α-T/UQ ratio (see Fig. 8), indicative of a different role of α-T in IHD than in health.

the reasons do not affect the qualitative relationship between plasma α-T and UQ (Fig. 11A). Ernster et al. (1991) suggested a functional coupling between UQ, α-T, and ascorbate (vitamin C) with UQ positioned in the deep lipid layers and α-T, in accordance with a section of the molecule being hydrophilic, in the outer layers. Because ascorbate is entirely hydrophilic, it is able to translocate radicals, originating in lipids or lipid-rich membranes, to the different cytosolic enzyme systems for further scavenging. A functional coupling thus exists between UQ, α-T, and ascorbate to drain membranes of reactive radicals, which otherwise could develop cascade reactions, membrane ruptures, cell trauma loss of cell integrity ("protein leakage" as indicated by e.g. elevated plasma creatine kinase (CK) activity), and possibly cell death. Previous data showed that IHD patients, who are extremely low in plasma UQ, have

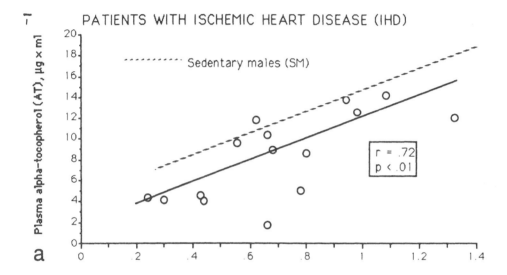

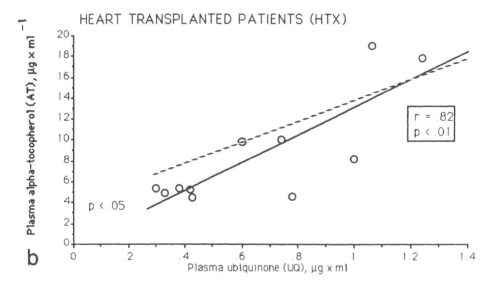

FIGURE 11   Plasma α-T versus plasma UQ in severe IHD (a), and HTX without clinical signs of rejection (b). The dashed line denotes the corresponding equation line in healthy but sedentary males (SM). Only HTX differed from SM and with respect to the intercepts for $x = 0$ ($p < 0.05$).

a higher percentage of the FTc muscle fiber subgroup (8,51,51). This fiber subgroup has been linked to pathological changes at the cell level, synonymous with muscle fiber trauma (41, 42). Blood and muscle UQ were both relatively low. It was suggested that the low muscle UQ/% ST ratio represented a sign of a metabolic lesion and could be involved in elevating the trauma-specific FTc presence.

In conclusion, patients with IHD have a different metabolic and histochemical profile in skeletal muscle compared to healthy subjects. This difference seems to affect UQ relatively more than α-T and could be a cause of the local trauma to skeletal muscle fibers.

Heart-Transplanted Patients

It has been shown that well-controlled HTX patients have normal heart muscle UQ but depressed plasma α-T and UQ levels (1,2,55,56). This depression is maintained when plasma values are normalized for FC levels. However the plasma values are expressed, they are further depressed with signs of graft rejection. Rejection leads to neutrophil invasion and subsequent elevated local radical formation (8,31,44,55–57). The lowest heart muscle UQ and plasma α-T and UQ levels were seen in those who expired after irreversible rejection trauma (2,55).

In contrast to IHD patients, HTX patients seem to have a different relationship between plasma α-T and UQ with extremely low UQ levels (Fig. 11B). Plasma α-T levels were relatively more reduced in HTX than SM and IHD. This difference may indicate a relatively larger dependence on α-T in HTX compared to IHD under conditions similar to metabolic stress.

α-T was found in a rat model to prevent autoxidation in heart muscle (58). Patients with cardiomyopathy respond positively to oral UQ treatment, most probably due to protection against autooxidation and peroxidation (5–7,59). The majority of HTX patients are recruited from this patient group. To our knowledge no objective pharmacological studies have been undertaken on HTX patients to elucidate whether α-T and/or UQ treatments has a prophylactic significance to rejection or other threats to the transplant. Most information in that respect is anecdotal in nature.

CONCLUSIONS

α-Tocopherol and ubiquinone are both biologically important intra- and extracellular antioxidants. UQ is in addition, as coenzyme Q_{10}, known as an intrinsic component of the mitochondrial electron transport.

The lipophilic properties of α-T and UQ confine them to lipoid areas in different cell membranes, lipid droplets, lipoprotein particles, and so on.

To make intraindividual comparisons possible and to correctly interpret plasma values of α-T and UQ, free cholesterol was suggested as a marker for the plasma lipids available for storage. α-T and UQ normalized for FC (N-α-T and N-UQ) were therefore computed for such comparisons.

Studies in healthy subjects and many patient groups reveal that N-α-T levels off with increased plasma α-T and UQ content. This relationship is interpreted as a sign of plasma α-T saturation related to the FC compartment. Such leveling does not exist with respect to plasma UQ. On the contrary, a UQ treatment program showed that UQ can be stored or trapped elsewhere or by unknown mechanisms retained in existing compartments.

α-T and UQ allocation is different in the plasma HDL lipoprotein fraction compared to the LDL fraction. N-UQ is deposited in the same magnitude in HDL and LDL; the former fraction is twice as rich in N-α-T as in LDL and displays different storage kinetics.

HDL in absolute and relative amounts increases with physical training. Trained individuals have more oxidative muscle fibers (ST) than untrained. The ST fiber contains three to four times more UQ than the other main fiber type, the glycogenolytic FT fiber. Similar data are not available for fiber types and α-T.

Healthy volunteers display a negative relationship between N-UQ and % ST but not between N-α-T and % ST. This could mirror the nonsaturated properties of plasma UQ, but it must also be taken into account that the ST fiber, richer in UQ, may also have a higher turnover of UQ with increased physical activity.

The lower FC values and/or the higher HDL levels in these UQ-rich ST fibers mean a less pronounced atherogenic potency compared to tissues rich in FT fibers. As a consequence of the elevated α-T-rich HDL fraction, the computed ratio plasma α-T over UQ increases with % ST.

It is also possible that the oxidative profile of the ST fiber in health selectively promotes a higher muscle and plasma UQ turnover than in FT fiber-rich untrained individuals (see above).

The ratio of muscle UQ to % ST also decreases the higher the % ST. This adds further support to the concept that ST-rich healthy individuals have signs of a relative shortage of plasma and muscle UQ.

Patients with IHD reveal a more pronounced negative relationship between relative muscle UQ content (muscle UQ over % ST) versus % ST than healthy volunteers as an indication of selective ST muscle fiber trauma. Histochemical evidence is also present in skeletal muscle of IHD patients showing signs of a different muscle fiber composition than in healthy subjects and the presence of traumatized fibers.

In contrast to healthy volunteers, IHD patients display a negative relationship between the ratio plasma α-T over UQ and relative muscle UQ levels. IHD patients are recognized for their higher plasma LDL values, which in itself grants a low ratio (see above), but the ratio decreases with relative muscle UQ and indicates a different relationship to plasma α-T in IHD compared to health.

Heart-transplanted patients who are well controlled have normal heart muscle UQ but low plasma α-T and UQ levels even when normalized for FC. Further depletion of normalized α-T and UQ is seen in relation to signs of rejection of the transplant. The relationship between plasma α-T and UQ reveals a relatively larger depression of and possibly reliance on α-T with low UQ levels in HTX than IHD patients. This could be related to the double function of UQ in contrast to α-T: to serve as an electron carrier in the mitochondria in addition to its activity as an unspecific intra- and extracellular antioxidant in parallel with and/or coupled to α-T.

REFERENCES

1. Karlsson J, Diamant B, Folkers K. Skeletal muscle characteristics of significance for metabolism in health and disease. Adv Myochem 1989; 2:283-91.
2. Karlsson J, Diamant B, Folkers K, Edlund P-O, Lund B, Theorell H. Plasma ubiquinone and cholesterol contents with and without ubiquinone treatment. In: Lenaz G, Barnabe O, Rabbi A, Battino M, eds. Highlights in ubiquinone research. London: Taylor and Francis, 1990; 296-302.
3. Elmberger PG, Kale'n A, Appelkvist E-L, Dallner G. In vitro and in vivo synthesis of dolichol and other main mevalonate products in various organs of the rat. Eur J Biochem 1987; 168:1-11.
4. Appelkvist E-L, Kale'n A. Dallner biosynthesis and regulation of coenzyme Q. In: Folkers K, Yamamura Y, eds. Biomedical and clinical aspects of coenzyme Q, Vol. 6. Amsterdam: Elsevier, 1991; xx-yy.
5. Mortensen SA, Vadhanavikit S, Baandrup U, Folkers K. Long-term coenzyme Q_{10} therapy: a major advance in the management of resistant myocardial failure. Drugs Exp Clin Res 1985; 11:581-93.
6. Mortensen SA, Bouchelouche P, Muratso K, Folkers K. Clinical decline and relapse of cardiac patients on coenzyme Q withdrawal. In: Folkers K, Yamamyra Y, eds. Biomedical aspects of coenzyme Q, Vol. 5. Amsterdam: Elsevier, 1986; 281-90.

7. Mortensen SA. Endomyocardial biopsy. Köbenhavn: Laegeforeningens Forlag, 1989; 1-36.
8. Karlson J, Diamant B, Folkers K, Edlund P-O, Lund B, Theorell H. Skeletal muscle and blood CoQ$_{10}$ in health and disease. In: Lenaz G, Barnabei, Rabbi A, Battino M, eds. Highlights in ubiquinone research. London: Taylor and Franzis, 1990; 288-95.
9. Packer L. Oxygen radicals and antioxidants in endurance exercise. In: Benzi G, Packer L, Siliprandi N, eds. Biochemical aspects of physical exercise. Amsterdam: Elsevier, 1986; 73-92.
10. Esterbauer H, Dieber-Rothender M, Striegl G, Waeg G. Role of vitamin E in preventing the oxidation of low-density lipoprotein. Am J Clin Nutr 1991; 53:314-321.
11. Kita T, Yokode M, Watanabe Y, Narumiya S, Kawai C. Stimulation of cholesteryl ester synthesis in mouse macrophages by cholesterol-rich very low density lipoproteins from the Watanabe heritable hyperlipidemic rabbit, an animal model of familial hypercholesterolemia. J Clin Invest 1986; 77:1460-5.
12. Quinn MT, Parthasarathy S, Fong LG, Steinberg D. Oxidatively modified low density lipoproteins: a potential role in recruitment and retention of monocyte/macrophages during atherogenesis. Proc Natl Acad Sci U S A 1987; 84:2995-8.
13. Lewis DH, Del Maestro R. Free radicals in medicine and biology. Acta Physiol Scand Suppl 1980; 492:1-168.
14. Lea CH, Kwietny A. The antioxidant properties of ubiquinone and related compounds. Chem Ind 1962; 24:1245-6.
15. Fitch CD, Folkers K. Coenzyme Q and the stability of biological membranes. Biochem Biophys Res Commun 1967; 26:128-31.
16. Hochstein P, Nordenbrand K, Ernster L. Evidence for involvement of iron in the ADP-activated peroxidation of lipids in microsomes and mitochondria. Biochem Biophys Res Commun 1964; 14: 323-8.
17. Ernster L, Nordenbrand K. Microsomal lipid peroxidation. Methods Enzymol 1967; 10:574-80.
18. Ernster L, Lee I-Y, Norling B. Studies with ubiquinone-depleted submitochondrial particles. Essentiality of ubiquinone for the interaction of succinate dehydrogenase, NADH dehydrogenase and cytochrome b. Eur J Biochem 1969; 9:299-310.
19. Ernster L, Glaser E, Norling B. Extraction and reincorporation of ubiquinone in submitochondrial particles. Methods Enzymol 1978; 53:573-9.
20. Beyer RE, Noble WM, Hirschfield TJ. Coenzyme Q (ubiquinone) levels of tissues of rats during acclimation to cold. Can J Biochem Physiol 1962; 40:511-8.
21. Pedersen S, Tata JR, Ernster L. Ubiquinone (coenzyme Q) and the regulation of basal metabolic rate by thyroid hormones. Biochim Biophys Acta 1963; 69:407-9.
22. Beyer RE, Ernster L. The antioxidant role of coenzyme Q. In: Lenaz G, Barnabe O, Rabbi A, Battino M, eds. Highlights in ubiquinone research. London: Taylor and Francis, 1990; 296-302.
23. Karlsson J. Muscle exercise, energy metabolism and blood lactate. In: Tavazzi L, di Prampero PE, eds. The anaerobic threshold: physiological and clinical significance. Adv Cardiol 1986; 7: 36-46.
24. Karlsson J. Muscle fiber composition, metabolic potentials, oxygen transport and exercise performance in man. In: Benzi G, Packer L, Siliprandi N, eds. Biochemical aspects of physical exercise. Amsterdam: Elsevier, 1986; 3-13.
25. Nilsson-Ehle P, Garfinkel AS, Schotz MC. Lipolytic enzymes and plasma lipoprotein metabolism. Annu Rev Biochem 1980; 49:667-93.
26. Edlund PO. Determination of coenzyme Q$_{10}$, alpha-tocopherol and cholesterol in biological samples by coupled-column liquid chromatography with colometric and ultraviolet detection. J Chromatogr 1988; 425:87-97.
27. Horwitt MK, Harvey CC, Dahm CH, Searcy MT. Relationship between tocopherol and serum lipid levels for determination of nutritional adequacy. Ann N Y Acad Sci 1972; 203:233-36.
28. Laryea MD, Biggemann B, Cieslicki P, Wendel U. Plasma tocopherol to lipids ratios in a normal population of infants and children. Int J Vitam Nutr Res 1989; 59:269-272.
29. Karlsson J. Heart and muscle ubiquinone or CoQ$_{10}$ as a protective agent against radical formation in man. Adv Myochem 1987; 1:305-18.
30. Okamoto T, Matruya T, Fukunaga Y, Kishio T, Yamagami T. Humans serum ubiquinol-10 levels and relationship to serum lipids. Int J Vitam Nutr Res 1989; 59:288-92.

31. Ernster L, Lee C-P. Thirty years of mitochondrial pathophysiology: from Luft's disease to oxygen toxicity. In: Kim CH, Ozawa T, eds. Bioenergetics: molecular biology, biochemistry and pathology. New York: Plenum Press, 1990; 451-465.

32. Thompson GR. Plasma lipids and lipoproteins, and the hyperlipoproteinaemias. In: Elkeles RS, Tavill AS, eds. Biochemical aspects of human disease. Oxford: Blackwell, 1983; 85-123.

33. Alaupovic P. Apolipoproteins and lipoproteins. Atherosclerosis 1971; 13:141.

34. Jackson RL, Morrisett JD, Gotto AM Jr. Lipoprotein structure and metabolism. Physiol Rev 1976; 56:259-316.

35. Thompson GR. A handbook of hyperlipidaemia. London: Curr Sci 1989; 43-58.

36. Eisenberg S, Chajek T, Deckelbaum RJ. The plasma origin of low density and high density lipoproteins. In: Carlsson LA, Pernow B, eds. Metabolic risk factors in ischemic cardiovascular disease. New York: Raven Press, 1982; 59-67.

37. Goldstein JL, Brown MS. Regulation of low density lipoprotein receptors: implications for pathogenesis and therapy of hypercholesterolemia and atherosclerosis. Circulation 1987; 76:504-7.

38. Kita T, Nagano Y, Yokode M, et al. Prevention of atherosclerosis progression in Watanabe rabbits by Probucol. Am J Cardiol 1998; 62:13B-9B.

39. Elmberger PG, Kale'n A, Brunk U, Dallner G. Discharge of newly-synthesized dolichol and ubiquinone with lipoproteins to rat liver perfusate and to the bile. Lipids 1989; 24:919-30.

40. Nikkila EA, Kuusi T, Taskinen M-R. Role of lipoprotein lipase and hepatic endothelial lipase in the metabolism of high density lipoproteins, a novel concept on cholesterol transport in HDL cycle. In: Carlsson LA, Pernow B, eds. Metabolic risk factors in ischemic cardiovascular disease. New York: Raven Press, 1982; 205-15.

41. Karlsson J. Localized muscular fatigue: role of muscle metabolism and substrate depletion. In: Hutton RS, Miller DI, eds. Exerc Sports Sci Rev 1979; 7:1-42.

42. Karlsson J, Smith HJ. Muscle fibers in human skeletal muscle and their metabolic and circulatory significance. In: Hunyor S, Ludbrook J, Shaw J, McGrath H, eds. The peripheral circulation. Amsterdam: Excerpta Medica, 1984; 67-77.

43. Kiessling K-H, Pilstrom L, Karlsson J, Piehl K. Mitochondrial volume in skeletal muscle from young and old physically untrained and trained healthy men and from alcoholics. Clin Sci 1973; 44:547-54.

44. Beyer RE. Recent contributions to the role of coenzyme Q in the clinical etiology and treatment of myopathies and remarks on its nomeclature. In: Lenaz G, Barnabe O, Rabbi A, Battino M, eds. Highlights in ubiquinone research. London: Taylor and Francis, 1990; 258-61.

45. Crane FL. Physiological coenzyme Q function and pharmacological reactions. In: Folkers K, Yamamura Y, eds. Biomedical and clinical aspects of coenzyme Q, Vol. 5. Amsterdam: Elsevier, 1986; 3-14.

46. Lithell H, Karlsson J. Some factors determining the lipoprotein-lipase activity of skeletal muscle during heavy exercise. In: Komi PV, ed. Exercise and sport biology. Champaign, IL: Human Kinetics, 1982; 26-33.

47. Lithell H, Cedermark M, Froberg J, Tesch P, Karlsson J. Increase in lipoprotein lipase activity in skeletal muscle during heavy exercise. Relation to epinephrine excretion. Metabolism 1981; 30: 1130-4.

48. Karlsson J. Metabolic adaptations to exercise: a review of potential beta-adrenoceptor antagonist effects. Am J Cardiol 1984; 55:48D-58D.

49. Stubbe I, Gustafson A, Nilsson-Ehle P. Alterations in plasma proteins and lipoproteins in acute myocardial infarction: effects on activation of lipoprotein lipase. Scand Clin Lab Invest 1982; 42: 437-44.

50. Just H, Drexler H, Zelis R. A symposium: regional blood flow in congestive heart failure. Am J Cardiol 1988; 62:1E-114E.

51. Karlsson J. Exercise capacity and muscle fibre types in effort angina. Eur Heart J 1987; 8:51G-7G.

52. Karlsson J. Onset of blood lactate accumulation, exercise capacity, skeletal muscle fibers and metabolism before and after coronary artery bypass grafting. Am J Cardiol 1988; 62:108E-14E.

53. Karlsson J, Diamant B, Folkers K, Lund B. Muscle fiber types, ubiquinone content and exercise capacity in hypertension and effort angina. Ann Med 1991; 23:339-344.

54. Larsson L. Morphological and functional characteristics of the ageing skeletal muscle in man. Acta Physiol Scand Suppl 1978; 457:1-36.
55. Karlsson J, Diamant B, Theorell H, Folkers K. Skeletal muscle coenzyme Q_{10} in healthy man and selected patient groups. In: Folkers K, Yamagami T, Littaru GP, eds. Biomedical and clinical aspects of coenzyme Q, Vol. 6. Amsterdam: Elsevier, 1991; in press.
56. Mortensen SA, Heidt P, Sehested J. Clinical perspectives in the treatment of cardiovascular diseases with coenzyme Q_{10}. In: Lenaz G, Barnabe O, Rabbi A, Battino M, eds. Highlights in ubiquinone research. London: Taylor and Francis, 1990; 226-31.
57. Ernster L, et al. Title to be reported In: Folkers K, Yamagami T, Littaru GP, eds. Biomedical and clinical aspects of coenzyme Q, Vol. 6. Amsterdam: Elsevier, 1991; in press.
58. Sylve'n C, Glavind J. Peroxide formation, vitamin E and myocardial damage in rat. Int J Vitam Nutr Res 1977; 47:9-16.
59. Langsjoen PH, Langsjoen PH Jr, Folkers K. Long-term efficacy and safety of coenzyme Q_{10} therapy for idiopathic dilated cardiomyopathy. Am J Cardioliol 1990; 65:521-3.

B. Cancer

36

Effects of Vitamin E on Oral Carcinogenesis and Oral Cancer

Gerald Shklar and Joel L. Schwartz

Harvard School of Dental Medicine, Harvard University, Boston, Massachusetts

INTRODUCTION

α-Tocopherol (vitamin E) is a potent anticancer agent in terms of both cancer prevention and cancer regression. This has been established in numerous animal experiments and in a variety of cell culture studies with cell lines of human cancer. Furthermore, vitamin E has a synergistic effect together with other anticancer agents and can play a significant future role in cancer prevention and cancer therapy in humans. The mechanism of vitamin E anticancer activity is currently being explored and gradually clarified. Vitamin E acts directly on cancer cells and also acts as an immunoenhancer. The immune response appears to play a major role in cancer prevention, and direct cytotoxicity may be the major mechanism in cancer regression.

Although there were reports of the anticancer activity of vitamin E as far back as 1934, these studies were inconclusive and contradictory (1,2). Davidson (1) found a delay in skin carcinogenesis in mice on a diet high in vitamin E; Telford (2) found that vitamin E supplements increased the percentage of lung tumors in a tumor strain of mice. More recently, Harman (3) reported that dietary vitamin E inhibited dibenzanthracene-induced tumors. Slaga and Bracken (4) found that vitamin E and other antioxidants were effective inhibitors of tumor initiation induced by DMBA (7,12-dimethylbenz[*a*]anthracene) and attributed their activity to their ability to prevent the in vivo activation of polycyclic aromatic hydrocarbons to carcinogen epoxides or other intermediates. Cook and McNamara (5) found that mice receiving a diet with a high vitamin E content developed fewer adenomas and fewer invasive carcinomas of the colon than mice receiving dimethylhydrazine and a normal diet. Our research using the hamster buccal pouch carcinoma model definitively established the anticancer activity of vitamin E. This experimental cancer model has proven to be a superior animal model for the study of solid malignant tumors, probably the best animal model currently available. It is the one animal model with an almost total similarity to the human counterpart, both in development and in biological behavior. The epidermoid carcinomas are induced by topical application of DMBA three times per week in strengths ranging from 0.1 to 0.5% solution in heavy mineral oil. The technique of painting the carcinogen onto the pouches is rapid and simple, not requiring anesthesia (6,7). With a 0.5% solution gross tumors arise in 12–14 weeks and, histologically, are indistinguishable from human oral epidermoid carcinomas of the well to moderately differentiated variety (8).

The carcinomas are preceded at 8–12 weeks by dysplastic and hyperkeratotic lesions entirely similar to precancerous human oral leukoplakia, both grossly and microscopically (9). The dysplastic lesions gradually become carcinoma in situ lesions and develop into frank invasive and proliferative lesions by 12 weeks and grossly visible lesions by 14 weeks. This tumor model allows continuous gross observation of the developing lesions. The individual lesions can be counted and measured and an overall figure obtained for tumor burden, expressed in cubic millimeters of tumor for each experimental animal or for each experimental group (10). There are no spontaneously arising tumors of the hamster buccal pouch that could complicate the interpretation of tumor development stimulated by the chemical carcinogen. The hamster pouch epidermoid carcinomas have been shown to have an altered metabolic pattern similar to human oral carcinomas, with increased lactic dehydrogenase activity and decreased succinic dehydrogenase activity (11). Solt and associates (12–14) showed that both hamster pouch carcinomas and human oral carcinomas express GGT (gammaglutamyltranspeptidase), a well-established marker for malignant cellular activity (15).

Furthermore, Wong and Biswas clearly established the expression of a specific oncogene (c-erb B_1) during hamster pouch carcinogenesis (16,17), and this gene is also expressed in human oral carcinomas (18). Other oncogenes (H-ras and K-ras) also appear to play a role and show an increased expression in hamster pouch carcinogenesis (19,20). Both animal and human oral cancer show an increased expression for the cytokine transforming growth factor α (TGF-α) (21–23).

Hamster pouch carcinogenesis has been shown to be sensitive to systemic factors, in particular to the animal's immune response. Agents acting as immunosuppressives, such as cortisone, methotrexate, and antilymphocyte serum, enhance carcinogenesis (24–26); agents acting as immunoenhancers, such as levamisole and bacillus Calmette-Guérin (BCG) (27–29) depress carcinogenesis. These studies effectively repudiate a previously held fallacious concept of the hamster pouch as an immunologically privileged site (30).

More recently, Shin et al. (31) found that the hamster buccal pouch cancer model could be used to assess the expression of epidermal growth factor receptor, transglutaminase type 1, polyamine (putrescine, spermidine, and spermine) levels, ornithine decarboxylase activity, and the occurrence of micronuclei. They suggest that these biological markers could be excellent intermediate end points in assessing the effects of various chemopreventive agents to be tested in both hamster and in human clinical trials (31). Gimenez-Conte et al. found that alterations in the pattern of keratin expression were a common feature of hamster buccal pouch carcinogenesis and could be used to study chemoprevention (32).

EXPERIMENTAL EVIDENCE FOR VITAMIN E ANTICANCER ACTIVITY ANIMAL STUDIES

Following the reports of the anticancer activity of 13-cis-retinoic acid on skin carcinogenesis by Bollag (33) and on bladder carcinogenesis by Sporn and associates (34), we demonstrated that 13-cis-retinoic acid could inhibit carcinogenesis in the hamster buccal pouch model (35) and in a hamster tongue model (36). At any given period during the development of carcinogenesis the animals receiving 10 mg retinoid twice weekly demonstrated fewer and smaller tumors. The process of carcinogenesis was inhibited, but tumor development was not prevented in any of the animals. Higher doses of 13-cis-retinoic acid could not be used because of substantial toxicity. It was believed that relatively nontoxic agents would be more successful, and our attention focused on vitamin E (37) and on β-carotene (38).

Vitamin E was shown to inhibit carcinogenesis in the hamster buccal pouch model. A total of 80 young adult male and female hamsters were divided into four equal groups. Groups 1 and 2 had the left buccal pouches painted three times weekly with a 0.5% solution of DMBA

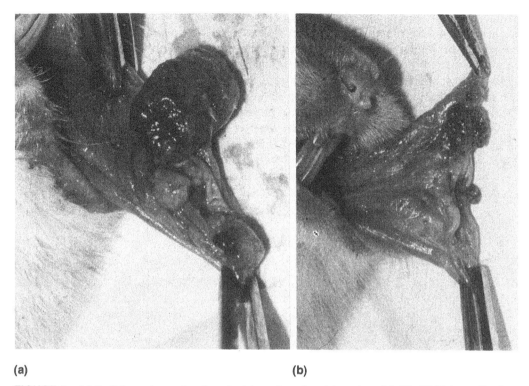

(a)   **(b)**

FIGURE 1 (a) Left buccal pouch of control hamster after 14 weeks of 0.5% DMBA applications. The pouch contains numerous large tumors. (b) Left buccal pouch in animal after 14 weeks of 0.5% DMBA applications and systemically administered vitamin E. Tumors are fewer in number and smaller, indicating an inhibition of carcinogenesis.

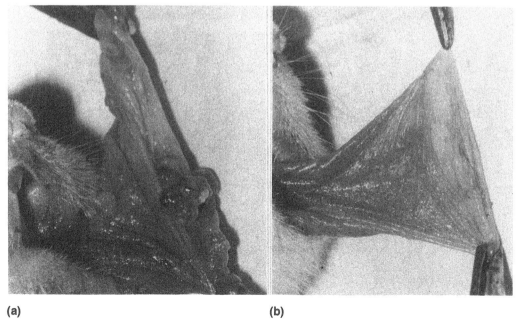

(a)   **(b)**

FIGURE 2 (a) Left buccal pouch of control hamster after 28 weeks of 0.1% DMBA applications. Several gross tumors are observed. (b) Left buccal pouch of animal after 28 weeks of 0.1% DMBA applications and systemically administered vitamin E. The pouch is free of tumors.

in mineral oil. Group 2 animals were given 10 mg *d,l*-α-tocopherol in peanut oil twice weekly. Groups 3 and 4 were vitamin E and untreated controls. There was a significant delay in carcinogenesis, but tumors developed in all the vitamin E animals, although fewer and smaller than in the animals not receiving vitamin E (Fig. 1) (39). Vitamin E was also found to delay hamster buccal pouch carcinogenesis when applied topically to the pouch three times weekly, with each application representing 47.5 μg *d,l*-tocopherol (40). It was postulated that 0.5% DMBA was so potent a carcinogen that actual prevention of tumors could not be accomplished.

A new model was developed using the hamster buccal pouch and the application of a 0.1% solution of DMBA in oil rather than a 0.5% solution (41). In this tumor model the entire process of carcinogenesis was extended from 12 to 28 weeks. Leukoplakia developed at 16–20 weeks rather than 6–8 weeks, carcinoma in situ at 20–24 weeks rather than 8–10 weeks, frank carcinoma at 24–26 weeks rather than 10–12 weeks, and large gross tumors at 26–28

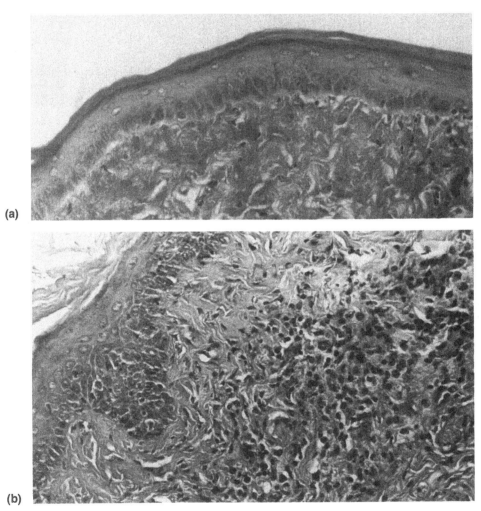

(a)

(b)

FIGURE 3 (a) Microscopic appearance of normal buccal pouch epithelium. (b) Microscopic appearance of epithelium in DMBA-vitamin E animal in Figure 2. An area of dysplasia is seen with some cellular degeneration. These is a dense infiltrate of lymphocytes and macrophages adjacent to the area of epithelial dysplasia.

weeks rather than 12–14 weeks. This model probably is closer to human carcinogenesis, in which the carcinogenic influences are relatively mild but act over a long period of time. In this model vitamin E was able to completely prevent the development of carcinomas (42) at 28 weeks when the animals were sacrificed. All the control animals had grossly visible epidermoid carcinomas of buccal pouch, and none of the animals receiving vitamin E had grossly visible carcinomas (Fig. 2). The animals receiving vitamin E had foci of dysplasia and carcinoma in situ histologically, but these foci demonstrated degenerative changes and accumulations of lymphocytes and macrophages were adjacent to the foci (Fig. 3). With immunohistochemical techniques it was shown that many of the macrophages were positive for tumor necrosis factor and many of the lymphocytes were positive for tumor necrosis factor B (Figs. 4 and 5). These findings suggested that vitamin E was an immunoenhancer, stimulating the mobilization of cytotoxic macrophages and cytotoxic lymphocytes to kill cancer cells in developing foci. This could be interpreted as enhanced immunosurveillance by vitamin E to prevent the development of foci of cancer cells (43).

ACTION ON ESTABLISHED CANCERS

It was found that the injection of β-carotene in minimum essential medium close to epidermoid carcinomas of hamster buccal pouch could result in regression of the cancers (44). Vitamin E was also found to have the capability of regressing epidermoid carcinomas of hamster buccal pouch when injected into the pouch in minimum essential medium (45). The carcinomas gradually could be completely destroyed and replaced by normal tissue (Fig. 6). If the vitamin E injections (twice weekly) were stopped during the course of tumor regression, the tumors again increased in size. However, if the tumors were completely obliterated, the cessation of vitamin E injections did not result in tumor growth at that site. Thus, vitamin E is capable of total destruction of established carcinomas. In experimental cancer regression there is also an enhanced presence of macrophages with tumor necrosis factor α (46).

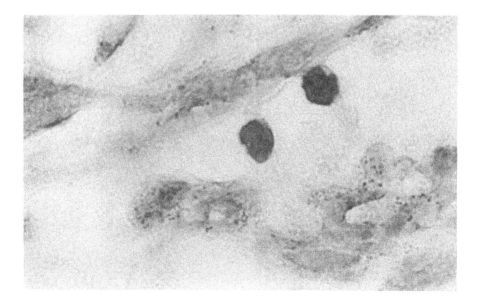

FIGURE 4 Immunohistochemical staining for lymphocytes positive for tumor necrosis factor β in infiltrate from Figure 3.

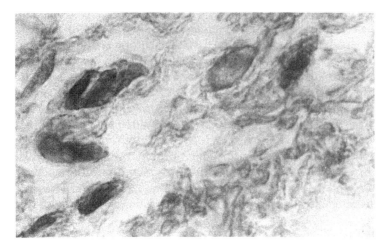

FIGURE 5   Immunohistochemical staining for macrophages positive for tumor necrosis factor α in infiltrate from Figure 3.

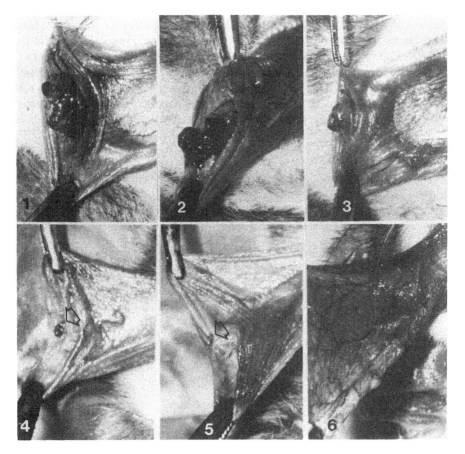

FIGURE 6   Established epidermoid carcinomas of left buccal pouch (1) regressing after 1 week (2), 2 weeks (3), 3 weeks (4), 4 weeks (5), and 6 weeks (6) of twice weekly injections of vitamin E into the pouch. Tumors are completely gone after 6 weeks, and pouch musoca is normal.

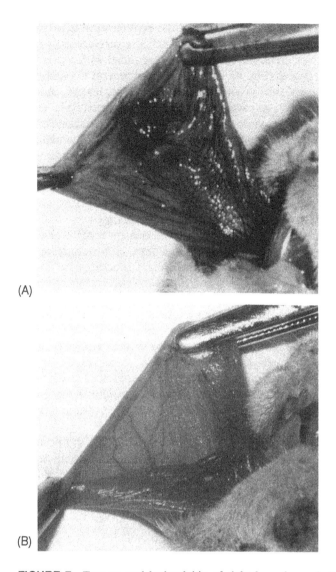

FIGURE 7 Tumors and leukoplakia of right buccal pouch (A) regressing after 6 weeks of oral administration of combined vitamin E and β-carotene (B).

COMBINED SYSTEMIC TREATMENT WITH β-CAROTENE AND VITAMIN E

Neither β-carotene nor vitamin E given systemically can prevent cancer development in the standard hamster pouch model using 0.5% DMBA, but the combination was found to be capable of doing this (47). Even as potent a carcinogen as 0.5% DMBA can be overcome by a combination of β-carotene and vitamin E (Fig. 7). They appear to act synergistically, and this may be because they act at different oxygen partial pressures.

STUDIES WITH CELL CULTURES OF HUMAN CARCINOMAS

In addition to our various animal studies with vitamin E, we have studied α-tocopherol's in vitro cytolytic activity upon a number of cell lines derived from human cancer. We found that α-tocopherol (70 and 300 μM) decreased the viability and the proliferation after incubation for 6 h with the squamous carcinoma tumor cell line SK-MES, a carcinoma of lung origin. SCC-25, an oral carcinoma derived from the tongue, also responded with a decrease in tumor cell growth but to a different degree (Fig. 8). All cells were grown on the same 24-well plate in supplemented media for normal cells (hydrocortisone, thyroxine, insulin, cholera toxin, pituitary extract, and MCDF media). The viability of the cells was determined using trypan blue (0.25%) and dye exclusion. Microscopic examination revealed an accumulation of lipid material (70 μM α-tocopherol) following 6 h of incubation with the tumor cell line SCC-25 (Figs. 9 and 10). There was a decrease in cell density and MTT levels consistent with the values observed in the SK-MES cell line.

The MTT assay level is based on the activity of the mitochondrial enzyme succinic dehydrogenase. Another assay, the [³H]thymidine incorporation assay, demonstrated an increased proliferative capacity for these tumor cells. An explanation for this apparent discrepancy may be related to the types of assays performed. Both the cell density analysis and the MTT assay were performed by incubating the cells for 5–6 h and then, either counting for viable cells per millimeter or analyzing directly for a tetrazolium salt precipitation. In the thymidine incorporation study the cells were treated for 5–6 h and the percentage of viable cells incorporating [³H]thymidine 20 h later was determined. This assay would then allow tumor cells resistant to α-tocopherol to proliferate. Another explanation may be that in certain cell lines α-tocopherol may actually enhance the growth of tumor cells. In a previous study we showed that at the concentrations utilized here α-tocopherol acid succinate could increase the colony-forming capacity of the oral cancer line HCPC-1, a carcinoma derived from the buccal pouch of the Syrian hamster. Higher concentrations were found to decrease tumor cell activity (48). A chemotherapy-resistant variant of an oral cancer cell line was found to exhibit greater viability and proliferation than the standard SCC-25 line. It demonstrated a greater response to α-tocopherol.

The cell line SSC-25, parent (P) or cisplatinum-resistant variant (CDDP), were plated (× 10⁴ cells per well) onto a 96-well plate and treated with α-tocopherol acid succinate (70 μM) for 2, 5, 12, and 24 h, in tumor medium (Dulbecco's modified Eagle's medium, DMEM,

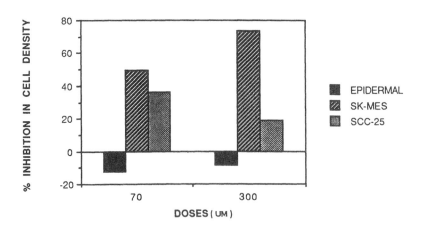

FIGURE 8   Effect of vitamin E on cell growth in cultures of normal epidermoid cells. and two lines of cancer cells, a lung carcinoma (SK-MES) and a tongue carcinoma (SCC-25). A significant inhibition occurred in the tumor cell lines but not in the normal epidermoid cell culture. Two dosages of vitamin E were used (70 and 300 μM).

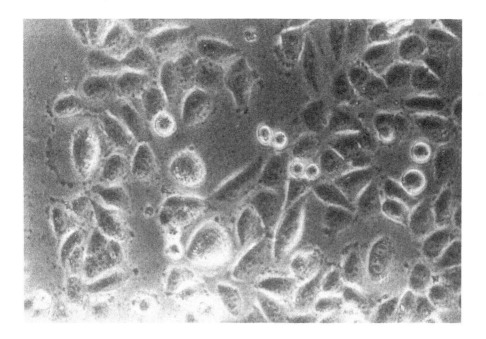

FIGURE 9 An inverse-phase photomicrograph of SCC-25, an oral squamous cell carcinoma ($\times$ 400).

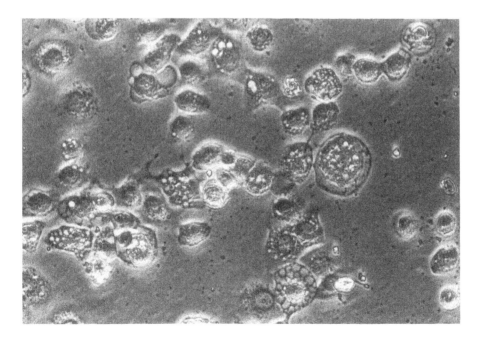

FIGURE 10 Upon treatment with vitamin E acid succinate (70 μM, 6 h) there was an accumulation of α-tocopherol in the tumor cells noted by the lipid vacuoles ($\times$ 400).

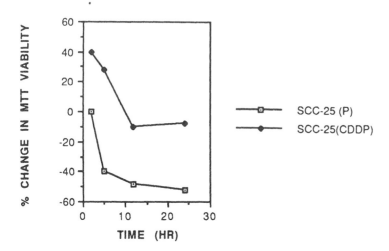

FIGURE 11   Greater decrease in viability of cisplatinum-resistant (CDDP) tumor cell line than the regular tumor cell line (P).

+ 10% fetal calf serum, FCS, + 10 M hydrocortisone). The determination of viability of the tumor cells following incubation was determined with tetrazolium salt precipitation as a result of the activity of succinic dehydrogenase in the mitochondria defined as the MTT assay. The results confirm the previous determination that α-tocopherol reduced SCC-25 (P) viability. In addition, the significant difference between the viability level of the resistant cell line and the parent was observed. After extended incubation with α-tocopherol there was a significant reduction in the viability of the varient compared to that of untreated controls ($p < 0.01$; Fig. 11).

We also investigated the effect of α-tocopherol on a variety of cancer cell lines to see whether the antitumor effect could be related to the origin of the tumor cell line. We found

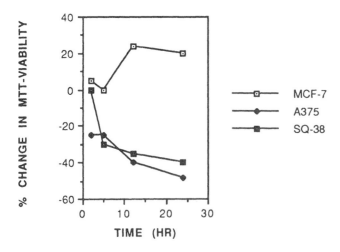

FIGURE 12   Greater antitumor effect on melanoma (A375) and oral carcinoma (SQ 38) than on breast carcinoma.

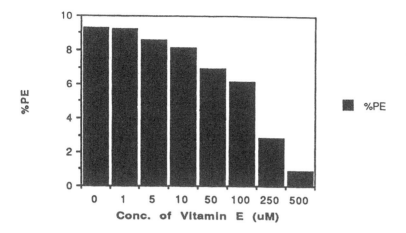

FIGURE 13 Tumor cells treated with α-tocopherol (0–500 μM) for 6 h, washed, formed into a single-cell suspension, replated, and colonies allowed to form for 5 days. The percentage plating efficiency was determined as the number of cells present per number cells originally plated $\times$ 100.

that a breast tumor cell line (MCF-7) exhibited only a slight antitumor effect, but a malignant melanoma cell line (A375) and another oral carcinoma line (SQ-38) demonstrated a greater antitumor effect of α-tocopherol, after incubation for 2, 5, 12, and 24 h. MTT values indicated that the breast cancer cells were not affected, but the melanoma and epidermoid carcinoma cells were strongly affected (Fig. 12).

In an additional assay, α-tocopherol was found to decrease the number of colonies of SCC-25 tumor cells after a 6 h incubation in a dose-dependent relationship, using concentrations of 0–500 μM. The tumor cells in this assay were washed free of α-tocopherol after the 6 h incubation period and the viable cells allowed to proliferate for 5 days, at which time the colonies were counted. The results indicated that the antitumor effect of α-tocopherol was persistent for a considerable number of cell cycles (Fig. 13).

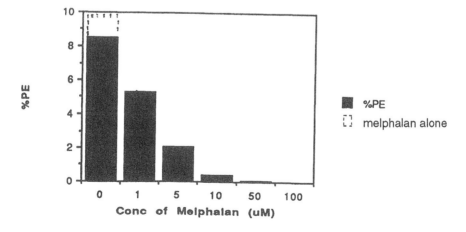

FIGURE 14 Effect of vitamin E and melphalan on the plating efficiency of a cancer cell line (SCC-25).

SYNERGISM OF α-TOCOPHEROL AND CHEMOTHERAPEUTIC DRUGS

α-Tocopherol was found to have a synergistic effect with the alkylating agent melphalan. There was a significant reduction in viable colony formation even when a concentration of 1 μM melphalan was used upon SCC-25 cells (Fig. 14). At 100 μM no viable cells remained. Melphalan was added to the cultures following 6 h of α-tocopherol treatment.

Having observed this synergistic effect with a potent alkylating agent, the question arose whether α-tocopherol, could protect normal human epidermal cells from the cytotoxic damage of melphalan. Human epidermal cells derived from foreskin or obtained during mammoplasties were grown under various conditions, that is, with a feeder layer of 3T3 NIH fibroblasts or in supplemented serum-free medium without the feed layer. In addition, the tumor cells and the normal cells were grown either in the normal supplemented media or the tumor-supplemented media to eliminate the possibility that any selective effect of the α-tocopherol on the cells was a result of the in vitro conditions. The in vitro studies with human epidermal keratinocytes following α-tocopherol treatment demonstrated a proliferation and increase in MTT activity. In addition, when these keratinocytes were incubated with α-tocopherol before treatment with melphalan, there was protection of the chemotherapy's cytotoxic effects.

From the studies of colony growth inhibition, it was postulated that the tumor cells may have been altered in their growth through a direct change in their cell cycle. An analysis of the propidium iodine-stained nuclei from SSC-25 using flow cytometry indicated that α-tocopherol treatment of this tumor cell resulted in accumulation of the tumor cells in G_1 phase (Fig. 15). This becomes important when we consider that one of the critical steps in the cell cycle is the transition from G_1 to S phase, an event supposedly characterized by changes in c-*myb* expression. Using a c-*myb* probe hybridized to the mRNA from the carcinoma cell line SCC-25, a relative decrease in c-*myb* expression was found. Further studies are required to confirm this relationship in a dose-dependent manner and in other cell lines.

MECHANISM OF ANTICANCER ACTIVITY OF VITAMIN E

Now that the anticancer activity of vitamin E has been well established, the mechanism or mechanisms of action will gradually become clarified. At this time two major roles of vitamin E in carcinogenesis and tumor regression are understood: (1) enhancement of the immune response to developing cancer cells and (2) direct action of vitamin E upon the cancer cells. The immunoenhancement of vitamin E was well documented by Tengerdy (50). Vitamin E supplementation was found to enhance both humoral and cell-mediated immunity and augment the efficiency of phagocytosis in laboratory animals, farm animals, and humans. Schwartz et al. (51) found that α-tocopherol could prevent the dramatic decrease in Langerhans cells during chemical carcinogenesis of hamster buccal pouch. Langerhans cells are essentially macrophage-equivalent cells found in skin and oral mucosa. Immunoenhancement may play the major role in cancer prevention, with developing cancer cells destroyed by cytotoxic lymphocytes and macrophages producing tumor necrosis factors β and α. A direct action of vitamin on cancer cells is an obvious mechanism, since immune factors do not apply to the cell culture studies. Vitamin E enters cancer cells and destroys them by altering the cellular metabolism and perhaps by inactivating the oncogenes or oncoproteins that maintain the viability of the abnormal cells.

Vitamin E has nearly optimal activity as a chain-breaking antioxidant (52) and is a well-known trapper of free oxygen radicals. It is found in the lipid fraction of living organisms, and its main function may be to protect the lipid material of an organism from oxidation. Borek found that vitamin E protected cells from carcinogenic chemicals by inhibiting lipid peroxidation and its damaging free-radical–mediated consequences (53).

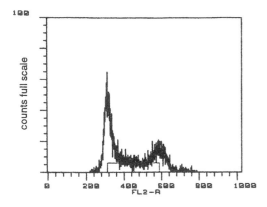

CONTROL UNTREATED CARCINOMA CELLS

<u>Cell Cycle Statistics</u>

Phase		Events	Percent
G1	:	2247	46.1
S	:	1664	34.2
G2+M	:	958	19.7
Total:		4869	100.0

Total Events in Histogram: 5000

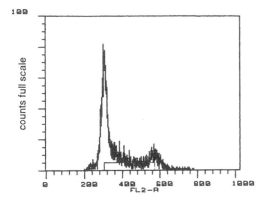

VITAMIN E (70 UM)

<u>Cell Cycle Statistics</u>

Phase		Events	Percent
G1	:	2897	60.0
S	:	1419	29.4
G2+M	:	512	10.6
Total:		4828	100.0

Total Events in Histogram: 5000

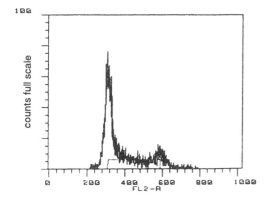

VITAMIN E (20 UM)

<u>Cell Cycle Statistics</u>

Phase		Events	Percent
G1	:	2813	57.7
S	:	1631	33.5
G2+M	:	431	8.8
Total:		4875	100.0

Total Events in Histogram: 5000

FIGURE 15   The effect of α-tocopherol on the cell cycle of SCC-25 carcinoma cells.

REFERENCES

1. Davidson JR. An attempt to inhibit the development of tar-carcinomata in mice. Can Med Assoc J 1934; 31:486-7.
2. Telford IR. The effects of hypo-and-hyper-vitaminosis E in lung tumor growth in mice. Ann N Y Acad Sci 1949; 52:132-4.
3. Harman D. Dibenzanthracene-induced cancer. Inhibition effect of dietary vitamin E. Clin Res 1969; 17:125-9.
4. Slaga TJ, Bracken WM. The effects of antioxidants on skin tumor initiation and aryl hydrocarbon hydroxylase. Cancer Res 1977; 37:1631-5.
5. Cook MG, McNamara P. Effect of dietary vitamin E on dimethylhydrazine-induced colonic tumors in mice. Cancer Res 1980; 40:1329-32.
6. Salley JJ. Experimental carcinogenesis in the cheek pouch of the Syrian hamster. J Dent Res 1954; 33:253-8.
7. Silberman S, Shklar G. The effect of a carginogen (DMBA) applied to the hamster's buccal pouch in combination with croton oil. Oral Surg 1963; 16:1344-55.
8. Shklar G. Experimental oral pathology in the Syrian hamster. Prog Exp Tumor Res 1972; 16:518-38.
9. Santis H, Shklar G, Chauncey H. The histochemistry of experimentally induced leukoplakia and carcinoma of the hamster buccal pouch. Oral Surg 1964; 17:207-18.
10. Malament DS, Shklar G. Inhibition of DMBA carcinogenesis of hamster buccal pouch by phenanthrene and dimethylnapthalene. Carcinogenesis 1981; 2:723-9.
11. Shklar G. Metabolic characteristics of experimental hamster carcinomas. Oral Surg 1965; 20:336-9.
12. Solt DB. Localization of gammaglutamyltranspeptidase in hamster buccal pouch epithelium treated with 7,12-dimethylbenz[*a*]anthracene. J Natl Cancer Inst 1981; 67:193-200.
13. Solt DB, Shklar G. Rapid induction of γ glutamyl transpeptidase-rich intraepithelial clones in 7,12-dimethylbenz[*a*]anthracene treated hamster buccal pouch. Can Res 1982; 42:285-91.
14. Calderon-Solt L, Solt DB. Gamma-glutamyl transpeptidase in precancerous lesions and carcinomas of oral, pharyngeal mucosa. Cancer 1985; 56:138-43.
15. Solt DB, Medline A, Farber E. Rapid emergence of carcinogen-induced hyperplastic lesions in a new model for the sequential analysis of liver carcinogenesis. Am J Pathol 1977; 88:595-618.
16. Wong DTW, Biswas DK. Activation of c-*erb* B oncogene in the hamster cheek pouch during DMBA-induced carcinogenesis. Oncogene 1987; 2:67-72.
17. Wong DTW. Amplification of c-*erb* B$_1$ oncogene in chemically-induced oral carcinofmas. Carcinogenesis 1987; 8:1963-5.
18. Merlino GT, Xu HY, Ishii S, Clark AJL, et al. Elevated epidermoid growth factor receptor gene copy number and expression in a squamous carcinoma cell line. Science 1974; 224:417-9.
19. Husain Z, Fei Y, Roy S, Solt DB, Polverini PJ, Biswas DK. Sequential expression and cooperative interaction of c-Ha-*ras* and c-*erb* B genes in in vivo chemical carcinogenesis. Proc Natl Acad Sci U S A 1989; 86:1264-8.
20. Wong DTW, Gertz R, Chow P, et al. Detection of Ki-*ras* mRNA in normal and chemically transformed hamster oral keratinocytes. Cancer Res 1989; 49:4562-7.
21. Todd R, Donoff B, Gertz R, et al. TGF-α and EGF-receptor mRNSa in human oral cancer. Carcinogenesis 10:1553-6.
22. Chang LC, Chou MY, Chow P, et al. Detection of TGF-mRNA in normal and chemically-transformed hamster oral epithelium. Cancer Res 1989; 49:6700-7.
23. Wong DTW, Gallagher GT, Gertz R, Chang ALC, Shklar G. Transforming growth factor-alpha in chemically-transformed hamster oral keratinocytes. Cancer Res 1988; 48:3130-4.
24. Shklar G. Cortisone and hamster buccal pouch carcinogenesis. Cancer Res 1966; 26:246-63.
25. Shklar G, Cataldo E, Fitzgerald A. The effect of methotrexate on chemical carcinogenesis of hamster buccal pouch. Cancer Res 1966; 2218-24.
26. Giunta j, Shklar G. The effect of antilymphocyte serum on experimental hamster buccal pouch carcinogenesis. Oral Surg 1971; 31:344-55.
27. Eisenberg E, Shklar G. Levamisole and hamster buccal pouch carcinogenesis. Oral Surg 1977; 43:562-74.

28. Shklar G, Eisenberg E, Flynn E. Immunoenhancing agents and experimental leukoplakia and carcinoma by the hamster buccal pouch. Prog Exp Tumor Res 1979; 24:269-82.

29. Giunta J. Reif AE, Shklar G. Bacillus Calmette-Guerin and antilymphocyte serum in carcinogenesis. Arch Pathol 1974; 98:237-40.

30. Billingham RL, Silvers WK. Syrian hamster and transplantation immunity. Plastic Reconstruct Surg 1964; 34:329-53.

31. Shin DM, Gimenez IB, Lee JS, et al. Expression of epidermal growth factor receptor, polyamine levels, ornithine decarboxylase activity, micronuclei, and transglutaminase I in a 7,12-dimethylbenz[a]anthracene-induced hamster buccal pouch carcinogenesis model. Cancer Res 1990; 50: 2505-10.

32. Gimenez-Conti IB, Shin DM, Bianchi AB, et al. Changes in keratin expression during 7,12-dimethylbenz[a]arthracene-induced hamster cheek pouch carcinogenesis. Cancer Res 1990; 50:4441-5.

33. Bollag W. Therapeutic effects of an aromatic retinoic acid analog on chemically induced skin papillomes and carcinomas of mice. Eur J Cancer 1974; 10:731-7.

34. Sporn MB, Dunlop NM, Newton DL, Smith JM. Prevention of chemical carcinogenesis by vitamin A and its synthetic analogs (retinoids). Fed Proc 1976; 35:1332-8.

35. Shklar G, Schwartz J, Grau D, Trickler DP, Wallace DK. Inhibition of hamster buccal pouch carcinogenesis by 13-cis retinoic acid. Oral Surg 1980; 50:45-53.

36. Shklar G, Flynn E, Szabo G, Marefat P. Retinoid inhibition of experimental lingual carcinogenesis: ultrastructural observations. J Natl Cancer Inst 1980; 65:1307-16.

37. Weerapradist W, Shklar G. Vitamin E inhibition of hamster buccal pouch carcinogenesis. A gross, histologic and ultrastructural study. Oral Surg 1982; 54:304-12.

38. Suda D, Schwartz H, Shklar G. Inhibition of experimental oral carcinogenesis by topical beta carotene. Carcinogenesis 1986; 7:711-5.

39. Shklar G. Inhibition of oral mucosal carcinogenesis by vitamin E. J Natl Cancer Inst 1982; 68: 791-7.

40. Odukoya O, Hawach F, Shklar G. Retardation of experimental oral cancer by topical vitamin E. Nutr Cancer 1984; 6:98-104.

41. Odukoya O, Shklar G. Initiation and promotion in experimental oral carcinogenesis. Oral Surg 1984; 58:315-23.

42. Trickler D, Shklar G. Prevention by vitamin E of oral carcinogenesis. J Natl Cancer Inst 1987; 78:165-9.

43. Shklar G, Schwarz JL, Trickler DP, Reid S. Prevention of experimental cancer and immunostimulation by vitamin E (Immunosurveillance). J Oral Pathol Med 1990; 19:60-4.

44. Schwartz J, Shklar G. Regression of experimental oral cancinomas by local injection of beta carotene and canthaxanthin. Nutr Cancer 1988; 11:35-40.

45. Shklar G, Schwartz J, Trickler DP, Niukian K. Regression by vitamin E of experimental oral cancer. J Natl Cancer Inst. 1987; 78:987-92.

46. Shklar G, Schwartz J. Tumor necrosis factor in experimental cancer regression with alpha tocopherol, beta carotene and algae extract. Eur J Cancer Clin Oncol 1988; 24:839-50.

47. Shklar G, Schwartz J, Trickler D, Reid S. Regression of experimental cancer by oral administration of combined alpha-tocopherol and beta carotene. Nutr Cancer 1989; 12:321-5.

48. Odukoya O, Schwartz J, Weichselbaum R, Shklar G. An epidermoid carcinoma cell line derived from hamster 7,12-dimethylbenz[a]anthracene-induced buccal pouch tumors. J Natl Cancer Inst 1983; 71:1253-64.

49. Odukoya O, Schwartz J, Shklar G. The effect of vitamin E on the growth of an epidermoid carcinoma cell line in culture. Nutr Cancer 1986; 8:101-6.

50. Tengerdy RP. The role of vitamin E in immune response and disease resistance. Ann N Y Acad Sci 1990; 587:24-33.

51. Schwartz J, Odukoya O, Stoufi E, Shklar G. Alpha tocopherol alters the distribution of langerhans cells in DMBA treated hamster cheek pouch epithelium. J Dent Res 1985; 64:117-21.

52. Burton GW, Ingold KU. Vitamin E as an in vitro and in vivo antioxidant. Ann N Y Acad Sci 1990; 570:7-22.

53. Borek C. Vitamin E as an anticarcinogen. Ann N Y Acad Sci 1990; 570:417-20.

37

Epidemiology of Vitamin E: Evidence for Anticancer Effects in Humans

Paul Knekt

Social Insurance Institution, Helsinki, Finland

INTRODUCTION

The hypothesis has been advanced that vitamin E (α-tocopherol) protects against cancer because of its function as a lipid antioxidant and free-radical scavenger and because it is a potential blocker of nitrosation (1–4). Animal experiments have given somewhat contradictory results, showing vitamin E to inhibit buccal pouch and ear duct carcinogenesis, to inhibit or have no effect on mammary gland, liver, or stomach carcinogenesis, to inhibit or enhance skin carcinogenesis, and to inhibit, have no effect on, or enhance colon carcinogenesis (5–9). The effect may depend on the amount of vitamin E, type and dosage of the carcinogen used, and the level of modifying factors, such as selenium and fat (5).

Despite the membrane-protective action of vitamin E and the promising findings that emerged from some animal studies, the relation between vitamin E and cancer in humans needs to be clarified. In fact, during the past decade a growing body of epidemiological studies has focused on the protective effect of vitamin E against cancer. The studies have generally been observational (ecological studies, case-control studies, and cohort studies), but intervention trials have also been started and are partly completed. In the observational studies, exposure to vitamin E has been determined on the basis of dietary recall methods or by serum or plasma vitamin E measurements. In the trials the intervention groups have been exposed to vitamin E by supplementation. The present study is an update of the findings of these human studies (9).

ECOLOGICAL STUDIES

Ecological studies compare mean exposure in a number of geographical areas to the incidence of cancer in the same areas. Each area contributes one unit of analysis in the design, whereas the other designs are based on the individuals themselves. Several ecological studies have presented strong correlations between dietary factors and the occurrence of cancer (10), suggesting that these factors may be involved in the development of cancer. Hardly any studies specifically investigating the association between vitamin E status and cancer risk have been carried out thus far, however.

One correlation study based on 65 population groups investigated the association between the vitamin E level in pooled plasma samples and the mortality rate from esophageal cancer and found only a weak association (11). Studies comparing intake or serum level of vitamin E between a few populations with a high or low risk of esophageal (12,13) or gastric cancer (14) reported lower vitamin E status in the high-risk populations. For reasons of methodology, however, such as a weak control for potential determinants of cancer, ecological studies should only be used in aggregating hypotheses, and the findings need to be corroborated by results from studies with other designs.

CASE-CONTROL STUDIES

Dietary Studies

In a case-control study on vitamin E status and cancer occurrence, a group of cancer cases is compared to a group of controls with respect to prior exposure to vitamin E. Case-control studies on diet and cancer have been carried out in which the intake of foods rich in vitamin E (vegetable oils, margarine, fruits and vegetables, cereals, and so on) was associated with the occurrence of cancer (10). For example, studies on the intake of foods high in polyunsaturated fats, the main source of vitamin E in the diet, present a modest inverse association or no association with colorectal cancer and no association with breast cancer (15). However, since these foods also include other substances (e.g., carotenoids, vitamin C, fiber) that may modify cancer risk and since different health behaviors may be associated with the intake of these foods, no firm conclusions about the effect of vitamin E can be drawn from these studies. Correspondingly, the finding that the use of vitamin supplements, which included several vitamins (16), provided a highly protective effect against colorectal cancer is no strong evidence of any effect of vitamin E.

In some of the dietary studies, the association between the calculated total intake of vitamin E and cancer risk was also reported (Table 1). A nonsignificant inverse gradient between vitamin E intake and lung cancer risk was observed among men, but not among women, in a population-based case-control study (17). Also, a study on lung cancer in a small sample using hospitalized persons as controls showed an inverse association (18). One study carried out in a region in France that has a high incidence of esophageal cancer reported a strong inverse

TABLE 1 Summary of Case-Control Studies on the Relationship Between Vitamin E Intake and Cancer Risk

Population	No. of cancer cases	Site of cancer	Mean case-control difference (%)	Relative risk (low versus high level)	Statistical significance	Reference
United States	9	Lung	− 57		NS	Lopez-S et al. (18)
New York	450	Lung		1.3, 1.1	NS	Byers et al. (17)
France	743	Esophageal		3.1	<0.001	Tuyns et al. (19)
Canada	246	Stomach		1.0	NS	Risch et al. (21)
Italy	1016	Gastric		1.7	<0.05	Buiatti et al. (20)
New York	428	Colon			NS	Graham et al. (22)
New York	422	Rectum		1.0, 1.3	NS	Freudenheim et al. (23)
Massachusetts	204	Melanoma	− 5	2.0	<0.01	Stryker et al. (24)
Italy, France	290	Breast	3		NS	Gerber et al. (25)
France	120	Breast	14		NS	Gerber et al. (26)
Italy	250	Breast	− 2	1.0	NS	Toniolo et al. (27)
United States	271	Cervix		1.4	NS	Ziegler et al. (28)

association between vitamin E intake and the occurrence of the disease (19), even after adjustment for smoking and alcohol consumption. Of two studies on gastric cancer patients compared with randomly selected population controls, one (20) reported a significantly higher daily intake of vitamin E among controls than among cancer cases, but the other found no corresponding differences (21). Also, two studies on colon (22) and rectum cancer (23), carried out in New York counties, found no association with vitamin E intake.

One study (24) showed that increased dietary intake of vitamin E was associated with a decreased risk of melanoma. No similar association with respect to the supplemental use of vitamin E was found in that study, however. None of three studies on breast cancer (25–27) showed any inverse associations with vitamin E intake, whereas one study (28) presented a weak downward trend in the risk of cervical cancer with duration of use of vitamin E supplements.

Although case-control studies provide information rapidly, there is an inherent potential for methodological bias in these studies. The choice of an appropriate control group is problematical, especially in studies using hospital controls. Several potential sources of bias are also connected with the use of dietary recall data. One such source is bias caused by misclassification, since the dietary recall does not give dietary exposure before the onset of cancer because of too short recall, changes in eating patterns, and other reasons. The presence of cancer may also alter a subject's diet or his or her description of past diet. The vitamin E content in the same foods may vary in different circumstances, and thus accurate assessment of vitamin E intake is hindered by the lack of adequate food tables. It is difficult to distinguish the effect of vitamin E because of its strong association with several foods and nutrients, such as fiber and carotenoids, which may have a protective effect. Furthermore, dietary recall is not necessarily a reliable measure of available vitamin E because of individual differences in bioavailability due to metabolic factors, disease, or dietary interactions, for example.

Serum Studies

Serum α-tocopherol level is responsive to both dietary and supplemental vitamin E intake (29), and a relatively high long-term repeatability of serum α-tocopherol has been reported (30). Thus, a single measurement of serum vitamin E can represent long-term vitamin E or α-tocopherol intake to a reasonable degree (31). Some of the weaknesses in the use of the dietary recall method can thus be eliminated by using serum or plasma level of vitamin E as a measure of vitamin E exposure.

Case-control studies on the association between serum and plasma vitamin E or α-tocopherol levels and cancer risk have given somewhat contradictory results, showing significantly inverse association in six, nonsignificant associations in ten, and a significant positive association in two studies (Table 2). Four studies suggested lower serum levels among lung cancer cases than among matched controls (18,32–34), whereas one observed no association (35). One study presented a nonsignificant inverse association between serum vitamin E and oral cancer (36), and another presented no association with respect to laryngeal cancer (37). One study presented significantly lower serum vitamin E levels among digestive cancer cases than among controls (38), the differences being similar for esophageal, stomach, pancreas, and colon or rectum cancer. In another study there was a nonsignificant inverse association between serum vitamin E level and gastric cancer observed (33).

Single studies on melanoma (24) and prostate cancer (39) presented no associations. One study investigating several cancer sites revealed inverse associations with respect to cancers of the lung, gastrointestinal tract, and nervous system but not for hormonally related cancers (cancers of the prostate, breast, uterus, and ovary) combined (40). Generally, no inverse associations have emerged in studies on gynecological cancers, such as cancers of the breast

TABLE 2 Summary of Case-Control Studies of the Relationship Between Serum and Plasma Vitamin E Level and Cancer Risk

Population	No. of cancer cases	Site of cancer	Mean case-control difference (%)	Statistical significance	Reference
United States	26	Lung	−3	NS	Atukorala et al. (35)
United States	29	Lung	−21	<0.05	Lopez-S et al. (18)
Japan	37	Lung	−22	<0.001	Miyamoto et al. (32)
New Orleans	59	Lung	−33	<0.01	LeGardeur et al. (34)
Korea	20	Lung	−33	NS	Yeum et al. (33)
	15	Gastric	−31	NS	
France	70	Digestive	−17	<0.001	Charpiot et al. (38)
India	30	Oral	−30	NS	Krishnamurthy et al. (36)
Poland	22	Larynx	+7	NS	Drozdz et al. (37)
France	190	Gastrointestinal, lower	−42	<0.001	Rougereau et al. (40)
	570	Gastrointestinal, other	−29		
	197	Hormonal	+46		
	71	Lung	−41		
	40	Brain	−81		
Massachusetts	204	Melanoma	0	NS	Stryker et al. (24)
The Netherlands	134	Prostate	−7	NS	Hayes et al. (39)
Italy, France	314	Breast	+9	<0.001	Gerber et al. (25)
France	120	Breast	+16	<0.01	Gerber et al. (26)
Bethesda	30	Breast	−11	NS	Basu et al. (41)
Finland	88	Gynecological	?	NS	Heinonen et al. (44)
Finland	11	Ovary	+12	NS	Heinonen et al. (42)
Finland	6	Vulva	0	NS	Romppanen et al. (43)
London	70	Cervix	−13	<0.05	Cuzick et al. (45)

(25,26,41), vulva, cervix, endometrium, or ovary (42–44). The exception was a study (45) that found an inverse relationship that was stronger with high grades of cervical intraepithelial neoplasia.

Although several of the studies have suggested an inverse association between serum vitamin E levels, especially cancers of the lung and the gastrointestinal tract, the serum studies have the same weaknesses as the dietary studies, possibly biasing the results. A notable weakness in such studies is that prevalent disease may affect the measurement because blood from cancer cases is obtained after the disease has occurred, and it is thus impossible to know whether a low level of serum vitamin E preceded cancer or vice versa. Furthermore, the majority of the studies have been based on hospitalized controls not presenting the general population, and control for confounding has been unsatisfactory in some studies.

COHORT STUDIES

In a cohort study, a group of persons whose vitamin E exposure has been ascertained is followed over a period of time with respect to the occurrence of cancer. When a sufficient number of cancer cases has occurred, cancer incidence among persons with high vitamin E exposure is compared with that among persons with low exposure. The major advantage of cohort studies in comparison with case-control studies is that the information on vitamin E exposure has been collected before diagnosis of cancer, and thus the possibility that the disease has affected the serum vitamin E level or the answers to the dietary recall is smaller.

Thus far cohort studies on vitamin E exposure and cancer occurrence have been reported only scantily. The Social Insurance Institution study on vitamin E intake and incidence of lung cancer among 4500 men in Finland presented an inverse association among nonsmokers but not among smokers (46). A systematic prospective screening of commonly used prescription drugs for a possible carcinogenic effect, based on 143,574 persons in the San Francisco Bay area, reported an inverse association between prescribed vitamin E and all sites of cancer combined (47). One 7 year cancer mortality follow-up among men of the Basel study observed a significantly inverse association between the plasma α-tocopherol level and the risk of stomach cancer and of all cancers combined. Also, a tendency for a higher risk of lung and colorectal cancer associated with a low vitamin E level was observed (48). In a 12 year follow-up of the same cohort, however, lower vitamin E concentrations among stomach cancer cases or among all cancer cases combined was no longer detected (49). The discrepant results may be due to differences with respect to lipid adjustment, to the statistical methods used, or to the loss of power of the plasma vitamin E results to predict the occurrence of cancer during the last years of follow-up.

NESTED CASE-CONTROL STUDIES

In a case-control study nested within a cohort study, the cancer cases arising in the cohort during the follow-up period are compared to controls randomly selected from the entire cohort. The major advantages of the nested case-control design in comparison with the usual case-control design are that the vitamin E intake and the serum vitamin E levels are measured before the occurrence of cancer and the controls are selected from the entire cohort at risk. In the nested case-control studies carried out thus far, the controls have generally been selected by pair matching, mainly using the date of serum collection, sex, age, and survival of the individuals up to the date of onset of those who developed cancer as matching criteria.

Dietary Studies

Only a few nested case-control studies on vitamin E intake and cancer incidence have been carried out. The Multiple Risk Factor Intervention Trial (50) reported on 156 cancer cases of all sites and 66 lung cancer cases, the Norwegian JANUS study (51) on 29 thyroid cancer cases, and the Honolulu Heart Program on 162 colorectal cancer cases (52). None of these studies revealed any significant associations between vitamin E intake and the incidence of respective cancers. That in two of these studies the dietary data were collected by the 24 h recall method, thus having low repeatability, may have contributed to the negative findings.

Serum Studies

Since it has become possible to store serum samples of large cohorts for several years and to determine vitamin E concentration in large number of blood specimens, it has also been possible to carry out studies on serum vitamin E level and cancer risk using a nested case-control design.

This design is particularly efficient when serum samples have been collected from the entire cohort, but it would be very expensive to analyze all specimens. Results on the association between serum and plasma α-tocopherol or vitamin E levels and the subsequent risk of cancer have been reported from 11 cohorts (Table 3). The sizes of the cohorts varied from 4000 to 36,000 and the number of cancer cases during a follow-up of 3–14 years from 25 to 766. A total of 16 different sites of cancer were studied. In 10 studies, the vitamin E determinations were made from stored frozen serum or plasma samples collected at baseline and thawed for analysis at the end of follow-up; only one of the studies was performed using fresh plasma samples (53).

All Sites of Cancer

Most of the eight studies reporting on all sites of cancer combined found lower serum vitamin E levels in cancer cases than in controls (Table 4). The differences were, however, statistically significant in the total cohort in only two of the studies (64,67,68). The relatively weak associations generally observed may be because different factors, such as intake of some nutrients, may modify the absorption, metabolism, or requirement of vitamin E and the association may thus vary in different subpopulations and among people with different living conditions. Associations between serum vitamin E and cancer risk in different subgroups of populations have rarely been studied, probably because of the small number of cancer cases available. In one study no interaction between serum vitamin E and serum vitamin A was found (70). On the other hand, a higher risk was particularly observed in the presence of low cholesterol and among young subjects in that study. In studies on the association between serum vitamin E level and cancer risk, conducted separately for smokers and nonsmokers, an inverse association was primarily observed among nonsmokers (58,65,68).

In accordance with the suggested synergy between vitamin E and selenium in protecting against peroxidative cell damage (1), subjects with both low serum selenium and low serum α-tocopherol levels were reported to have an elevated risk of cancer of all sites combined (58, 72) in two studies and an elevated risk of cancer unrelated to smoking in one study (70). In site-specific studies, with few exceptions (70), no such interactions have been observed (59–61).

The weak association between serum vitamin E level and the occurrence of all sites of cancer may also be due to vitamin E having a protective effect against certain specific exposures, and thus the strength of the association may differ for cancers at distinct anatomic sites. Unfortunately, however, with few exceptions, the number of site-specific cancer cases in such studies has been small.

Lung Cancer

Lung cancer has been examined in nine studies (Table 4) (50,53,56–59,64,65,68). Generally, the lung cancer cases presented lower mean vitamin E levels than the controls, but only one study demonstrated a significant protective effect of vitamin E (59). In another study there was an inverse association between serum vitamin E and lung cancer combined with other cancers regarded as smoking related (68) among nonsmokers but not smokers. This may be because smoking is the dominant risk factor for lung cancer; some substance in tobacco smoke, or health behavior associated with smoking, overrides the possible effect of vitamin E in some circumstances.

Colorectal Cancer

The association between serum vitamin E level and the occurrence of colorectal cancers has been studied in five cohorts (Table 4) (53,57,60,65,69). All the studies reported lower mean serum vitamin E levels among cancer cases than among controls. The only exception

TABLE 3 Summary of the Nested Case-Control Studies of the Relationship Between Plasma and Serum Vitamin E or α-Tocopherol Levels and Cancer Risk[a]

Study (ref.)	Cohort size	No. of cancer cases	Sex	Age at entry (years)	No. of years of follow-up	Site of cancer	Matching variables	Storage temperature (°C)	Control mean vitamin E (mg/L)
Basel study, Switzerland (53)	4,224	115	Male	26–71	7–9	All, 3 sites	Age, sex	Fresh	16.2
Guernsey study, England (54,55)	10,090	69	Female	26–88	7–14	Breast	Age, gynecological variables, breast disease, time of blood collection	−20	6.1
Hypertension Detection and Follow-up Program (56)	4,480	111	Both	30–69	5	All, 5 sites	Age, sex, ethnic group, time of blood collection, smoking	−70	12.6
Honolulu Heart Program (57)	6,860	284	Male	52–71	10	5 sites	Randomly selected controls	−75	12.3
Eastern Finland Heart Survey (58)	12,155	51	Both	30–64	4	All, 2 sites	Age, sex, smoking	−20	5.0
Washington County study (59–63)	25,802	331	Both	11–98	8–13	5 sites	Age, sex, ethnic group, time of blood collection, hours since last meal, smoking	−73	12.0
Zoetermeer study, The Netherlands (64)	10,532	69	Both	5–99	6–9	All, lung	Age, sex, smoking	−20	8.5
British United Provident Association (65)	22,000	271	Male	35–64	2–9	All, 6 sites	Age, smoking, duration of storage of the serum sample	−40	10.3
Malmö study, Sweden (66)	10,000	25	Male	46–48	3–8	All	Restricted sex, age, population group	−20	3.0
Social Insurance Institution study, Finland (67–71)	36,265	766	Both	15–99	7–10	All, 16 sites	Age, sex, time of blood collection	−20	9.1
The Multiple Risk Factor Intervention Trial (50)	12,866	156	Male	35–57	10	All, lung	Age, smoking, treatment group, date of randomization into the trial	−50	13.0

[a]Update of Table 2 in Knekt (70).

TABLE 4 Summary of Case-Control Differences of Serum Vitamin E Levels by Site of Cancer in Nested Case-Control Studies[a]

Study	% Case-control difference/number of cancer cases								
	All	Lung	Colorectal	Stomach	Breast	Prostate	Bladder	Pancreas	Nervous system
Basel study, Switzerland (53)	−6/115	−12/35	−13/14	0/19					
Guernsey study, England (54,55)					−22[b]/39, +5/30				
Hypertension Detection and Follow-up Program (56)	−8/111	+10/17		−12/11[c]	−13/14	−7/11			
Honolulu Heart Program (57)		+4/74	−1/81,[d] −6/32[e]	−1/70			+3/37		
Eastern Finland Heart Survey (58)	−2/51	?/15		?/18[c]					
Washington County study (59–63)		−12[f]/99	−8/72[d]			−9/103	−10/35	+9/22	
Zoetermeer study, The Netherlands (64)	−15[b]/69	−9/18							
British United Provident Association (65)	−2/271	−4/50	−4/30	+5/13			+2/15		0/17
Malmö study, Sweden (66)	−11/25								
Social Insurance Institution study, Finland (67–71)	−3[b]/766	−4/144	−3/58	−5/76	0/67	−3/37	+4/15	−8/28	
The Multiple Risk Factor Invervention Trial (50)	+2/156	−4/166							−6/18

[a]Test for difference from zero is indicated.
[b]p < 0.01.
[c]Gastrointestinal.
[d]Colon.
[e]Rectum.
[f]p < 0.01.

was observed with respect to males in the Finnish Social Insurance Institution study (69). All the differences were nonsignificant, however. To reach greater statistical power in the test, the original data from the five studies were pooled (73). The pooled results presented a significant inverse association that decreased slightly, however, after adjustment for serum cholesterol.

Stomach Cancer

The association with stomach cancer has been studied in relatively small samples, and the results are consequently inconsistent, presenting both nonsignificant inverse (57,69) and positive (65) associations (Table 4). Although the Basel study reported significantly lower mean α-tocopherol levels among stomach cancer cases than among controls, when the study was carried out as a cohort study, also including all noncancerous individuals (48), no association was found when using the nested case-control design (53). Studies in the United States and Finland based on small numbers of gastrointestinal cancers combined have not reported any notable associations (56,58). The inconsistent results may be due mainly to the small numbers of stomach cancer cases or possibly the result of confounding by risk factors, such as saturated fats and alcohol intake, that were not adjusted for in these studies.

Gynecological Cancers

A significant inverse association was reported between serum vitamin E and breast cancer in one study based on an English population (54), but another on the same population failed to confirm this finding (Table 4) (55). The Finnish Social Insurance Institution study found no association with hormone-related female cancers (e.g., cancers of the breast, endometrium, or ovary) (67). In contrast, however, the women in this study who had a low level of both vitamin E and selenium showed an elevated risk of both breast cancer and all hormone-related cancers combined. Also, cervix cancer cases had nonsignificantly lower vitamin E levels than corresponding controls in this study.

Other Sites of Cancer

Individual studies involving small numbers of cases with other sites of cancer either yield contradictory results or reveal only weak associations (Table 4), and no conclusion can be based on them. Three studies reporting on the association between serum vitamin E level and occurrence of prostate cancer demonstrated a nonsignificant inverse association (56,63,68). The results for bladder cancer, which was included in four studies (57,62,65,68), were contradictory. The results with respect to pancreatic cancer were also contradictory. A low level of vitamin E was associated with a significantly elevated risk among men but not among women in one study (67,68) and with a protective effect in another study (61).

No significant associations have been reported with respect to leukemias and lymphomas (56,68) or cancer of the nervous system (65,71). Melanoma was inversely associated in a small sample (71), whereas basal cell carcinoma (68) was not. In one study the mean α-tocopherol levels were nonsignificantly smaller among cases than among controls with respect to laryngeal and esophageal cancer but not with respect to oral, liver, or kidney cancer (71).

Methodological Aspects of Serum Studies

Although some of the weaknesses in case-control studies are not present in nested case-control studies, there are still factors making interpretation of the results difficult. In principle, the inverse association observed between serum vitamin E level and cancer risk may be due to preclinical cancer, confounding, or biases in the vitamin E determinations.

Undetected cancer may depress the serum levels of vitamin E at the baseline examination, thus causing an artifactual inverse association. This possibility was investigated in some studies by excluding the cancer cases occurring during the first years of follow-up from the sample and by comparing vitamin E levels among cancer cases with those of matched controls according to years between blood drawn and case diagnosis. Inverse associations were generally found even after exclusion of the cancer cases occurring during the first years of follow-up (54,64,68), and no trends in the strength of the association were detected in favor of a stronger association among cancers occurring in the early part of the follow-up (59,60, 73,74). Although one nested case-control study reported the inverse association to be concentrated only in the first 2 years of follow-up (65) and a cohort study over a 7 year but not over a 12 year follow-up (48,49), the inverse association between vitamin E level and the occurrence of cancer cannot fully be explained by preclinical cancer.

Although sex, age, smoking, and time of blood collection have been controlled for in the majority of studies, by either matching or modeling, it is still possible that the association observed between serum vitamin E level and cancer risk is not causal but due to some other dietary or nondietary factors, which are also correlated with serum vitamin E level or vitamin E intake. Vitamin E is contained in several foods, and the observed inverse association could thus be due to other protective substances in these foods or in other foods taken simultaneously. It may also be due to behavior associated with the consumption of such foods. Generally, confounding caused by such factors has scarcely been taken into account. Several longitudinal studies have shown evidence of an association between other micronutrients, such as serum β-carotene and selenium, and the occurrence of cancer (70). Adjustments for β-carotene, retinol, and selenium have not materially altered the association between α-tocopherol and cancer (58,64,67,68), leaving the possibility of an independent association between the serum α-tocopherol level and cancer risk.

Because vitamin E is transported in lipoproteins, thus being highly correlated with serum cholesterol (75), and because serum cholesterol level and cancer risk are inversely associated (76), it is possible that the serum vitamin E-cancer association is secondary to the serum cholesterol-cancer association. Generally, the serum cholesterol level has been adjusted for in different studies. Some of these studies (56,65,73) reported a slight reduction in the vitamin E-cancer association, but some reported no effect (57,60,68) of adjustment. Also, a significant independent association for both the serum cholesterol and the serum α-tocopherol level was reported in one study (77). These results suggest that the reported serum vitamin E-cancer association is probably not fully secondary to the serum cholesterol-cancer association.

The true quantitative relationship between vitamin E level and risk of cancer is not necessarily that observed due to factors affecting the serum vitamin E level. Although the serum α-tocopherol level has been observed to be relatively stable over a period of 4 years (30), serum vitamin E may be only a weak measure of the body's vitamin E status over a lengthy period of time (78). In principle, it is also possible that systematic variation in the laboratory measurements may have affected case-control comparisons. Because serum samples of each cancer case and its matched controls were generally handled similarly and analyzed in random order in the different studies, this source of bias is diminished, however.

The mean serum α-tocopherol levels varied between the studies from 10 to 16 mg/L when the serum samples were fresh or stored at least at $-45°C$, and within 3–10 mg/L when stored at $-20°C$ (Table 3). Thus some degradation may have occurred among the samples stored at $-20°C$. In this case a different loss among cases and controls may bias the results. Two of the studies based on serum samples stored at $-20°C$ reported significant associations between serum vitamin E level and the occurrence of cancer (54,67,68). Of these, the Social Insurance Institution study in Finland used a study design that minimized possible bias in the case of equal degradation among cases and controls, so it seems reasonable to conclude that loss of

α-tocopherol during storage did not bias the findings (70). In the Guernsey study, however, it is possible that the association reported may be due to different losses among cases and controls (79). Also, that the findings in the prime study were not confirmed in a later study based on samples from the same population (55) supports this conclusion.

Thus, a weak inverse association between serum α-tocopherol level and the occurrence of several cancers has been observed in the majority of nested case-control studies carried out. Although several sources of bias have been adjusted for in these studies, no definite conclusions about causality can be drawn on their basis.

INTERVENTION TRIALS

Existing evidence from experimental and observational epidemiological studies on the possible preventive effect of vitamin E against cancer has encouraged the initiation of clinical intervention trials on different sites of cancer. A clinical trial is the most reliable design to test the hypothesis that vitamin E reduces the risk of cancer. A clinical trial is a prospective study in which individuals who have no history of cancer are randomly assigned in a double-blind fashion to either an intervention group receiving vitamin E supplements or to a control group receiving a placebo. The occurrence of cancer within the groups is then followed for a given time period. The risk of confounding is minimized in clinical trials, as the distributions of potential confounding factors also are randomized between the groups.

On the basis on several trials using a limited number of patients with mammary dysplasia, it was suggested that the risk of breast cancer may be reduced by vitamin E supplementation (7,80). Later, larger double-blind, randomized, placebo-controlled trials on benign breast disease, however, failed to confirm these results (81–83). Similar trials on the recurrence rate of colorectal polyps reported a small nonsignificant inhibitory effect in an intervention group receiving vitamins E and C (84,85). The weak association observed may be true but may also be due to several methodological issues, for example, the small samples, short period of follow-up, and that higher doses of vitamin E would be required. Thus, a larger trial is required to ensure that the reduction in the rate of polyp regression was not a chance finding.

To obtain reliable evidence on whether vitamin E intake can reduce the future risk of different cancers, a number of large-scale, randomized placebo-controlled trials using vitamin E and other micronutrient supplements (i.e., β-carotene, vitamin A, vitamin C, and selenium) have been started (Table 5) (86–88). The trials are mainly concentrated in groups at high risk of colon, lung, skin, or esophageal cancer, thus also possibly giving some information

TABLE 5 Ongoing Clinical Intervention Trials Concerning Vitamin E and Cancer

Target site	Target/risk group	Inhibitory agents in addition to vitamin E	Investigator
Colon	Adenomatous polyps	Vitamin C, β-carotene	Greenberg
Colon	Normal subjects	Vitamin C	Colacchio
Lung	Smoking men	β-Carotene	Albanes
Lung	Tin miners	β-Carotene, retinol, selenium	Schatzkin
Skin	Basal cell carcinoma	Vitamin C, β-carotene	Safai
All sites	Dentists, nurses	Vitamin B_6, retinol, selenium	Hennekens
Esophagus	Dysplasia	Multiple vitamins and minerals	Taylor
Esophagus	High-risk population	Multiple vitamins and minerals	Taylor

Source: Modified from DeWys et al. (86) and Cullen (88).

on other sites. Two of the trials are focusing on the prevention of colon cancer, one in a target group of individuals with adenomatous polyps and the other among normal subjects. In two trials of lung cancer prevention, the study populations include male smokers and tin miners. There is also one study on skin cancer prevention based on persons with prior basal cell carcinomas. In a trial focusing on all sites of cancer, the study population includes nurses and dentists. Results from several of the chemopreventive trials will be available in a few years.

SUMMARY AND CONCLUSIONS

In accordance with finding from animal experiments, some epidemiological studies on humans have also suggested that vitamin E may protect against cancer. The most consistent associations have been reported for cancers of the lung, esophagus, and colorectum. In contrast, possibly with the exception of cervical cancer, gynecological cancers have generally not been found to be associated with vitamin E status.

The associations observed do not convincingly satisfy criteria for causality, since they are generally relatively weak and only occasionally present a dose-response gradient. Nor are the results very consistent in various populations. The inconclusiveness of the results may be partly due to methodological issues but also because the preventive effect of vitamin E may interact with other causes of cancer and thus be concealed in some populations. It is possible, for example, that a high intake of polyunsaturated fatty acids or a selenium deficiency may increase the vitamin requirement and that the strength of the vitamin E-cancer association in a population thus depends on the distribution of these variables.

To define risk groups in which vitamin E may have a protective effect, results are needed from large-scale observational studies carried out under varying circumstances. Also, meta-analyses should be carried out, pooling data from existing nested case-control studies to obtain results with a higher power. To establish the possible anticarcinogenic effect of vitamin E in the human, several intervention trials have been started. Until the results of these studies are available, no definite conclusions about the possible protective effect of vitamin E against cancer can be drawn, however.

REFERENCES

1. Tappel AL. Vitamin E and selenium protection from in vivo lipid peroxidation. Ann N Y Acad Sci 1980; 355:18-29.
2. Ames BN. Dietary carcinogens and anticarcinogens. Science 1983; 221:1256-64.
3. Mirvish SS. Effects of vitamins C and E on N-nitroso compound formation, carcinogenesis, and cancer. Cancer 1986; 58(suppl):1842-50.
4. Mergens WJ. Efficacy of vitamin E to prevent nitrosamine formation. Ann N Y Acad Sci 1982; 393:61-9.
5. Birt DF. Update on the effects of vitamins A, C, and E and selenium on carcinogenesis. Proc Soc Exp Biol Med 1986; 183:311-20.
6. Shamberger RJ. Chemoprevention of cancer. In: Reddy BS, Cohen LA, eds. Diet, nutrition and cancer: a critical evaluation, Vol. II. Boca Raton, FL: CRC Press, 1986; 43-62.
7. Chen LH, Boissonneault GA, Glauert HP. Vitamin C, vitamin E and cancer (review). Anticancer Res 1988; 8:739-48.
8. Mergens WJ, Bhagavan HN. Alpha-tocopherols (vitamin E). In: Moon TE, Micozzi MS, eds. Nutrition and cancer prevention. New York: Marcel Dekker, 1989; 305-40.
9. Knekt P. Role of vitamin E in the prophylaxis of cancer. Ann Med 1991; 23:3-12.
10. Committee on Diet, Nutrition and Cancer. National Research Council. Diet, nutrition and cancer. Washington, DC: National Academy Press, 1982.

11. Guo W, Li J-Y, Blot WJ, Hsing AW, Chen J, Fraumeni JF Jr. Correlations of dietary intake and blood nutrient levels with esophageal cancer mortality in China. Nutr Cancer 1990; 13:121-7.

12. Yang CS, Sun Y, Yang Q, et al. Vitamin A and other deficiencies in Linxian, a high esophageal cancer incidence area in northern China. J Natl Cancer Inst 1984; 73:1449-53.

13. van Helden PD, Beyers AD, Bester AJ, Jaskiewicz K. Esophageal cancer: vitamin and lipotrope deficiencies in an at-risk South African population. Nutr Cancer 1987; 10:247-55.

14. Correa P, Cuello C, Fajardo LF, Haenszel W, Bolanos O, Ramirez de B. Diet and gastric cancer: nutrition survey in a high-risk area. J Natl Cancer Inst 1983; 70:673-8.

15. Rogers AE, Longnecker MP. Dietary and nutritional influences on cancer: a review of epidemiologic and experimental data. Lab Invest 1988; 59:729-59.

16. Kune S, Kune GA, Watson LF. Case-control study of dietary etiological factors: the Melborne colorectal cancer study. Nutr Cancer 1987; 9:21-42.

17. Byers TE, Graham S, Haughey BP, Marshall JR, Swanson MK. Diet and lung cancer risk: findings from the western New York diet study. Am J Epidemiol 1987; 125:351-63.

18. Lopez-S A, LeGardeur BY. Vitamins A, C, and E in relation to lung cancer incidence. (abstract). Am J Clin Nutr 1982; 35:851.

19. Tuyns AJ, Riboli E, Doornbos G, Péquignot G. Diet and esophageal cancer in Calvados (France). Nutr Cancer 1987; 9:81-92.

20. Buiatti E, Palli D, Decarli A, et al. A case-control study of gastric cancer and diet in Italy. II. Association with nutrients. Int J Cancer 1990; 45:896-901.

21. Risch HA, Jain M, Choi NW, et al. Dietary factors and the incidence of cancer of the stomach. Am J Epidemiol 1985; 122:947-59.

22. Graham S, Marshall J, Haughey B, et al. Dietary epidemiology of cancer of the colon in western New York. Am J Epidemiol 1988; 128:490-503.

23. Freudenheim JL, Graham S, Marshall JR, Haughey BP, Wilkinson G. A case-control study of diet and rectal cancer in western New York. Am J Epidemiol 1990; 131:612-24.

24. Stryker WS, Stampfer MJ, Stein EA, et al. Diet, plasma levels of beta-carotene and alpha-tocopherol, and risk of malignant melanoma. Am J Epidemiol 1990; 131:597-611.

25. Gerber M, Cavallo F, Marubini E, et al. Liposoluble vitamins and lipid parameters in breast cancer. A joint study in northern Italy and southern France. Int J Cancer 1988; 42:489-94.

26. Gerber M, Richardson S, Crastes de Paulet P, Pujol H, Crastes de Paulet A. Relationship between vitamin E and polyunsaturated fatty acids in breast cancer. Nutritional and metabolic aspects. Cancer 1989; 64:2347-53.

27. Foniolo P, Riboli E, Protta F, Charrel M, Cappa APM. Calorie-providing nutrients and risk of breast cancer. J Natl Cancer Inst 1989; 81:278-86.

28. Ziegler RG, Brinton LA, Hamman RF, et al. Diet and the risk of invasive cervical cancer among white women in the United States. Am J Epidemiol 1990; 132:432-45.

29. Willett WC, Stampfer MJ, Underwood BA, Taylor JO, Hennekens CH. Vitamins A, E, and carotene: effects of supplementation on their plasma levels. Am J Clin Nutr 1983; 38:559-66.

30. Knekt P, Seppänen R, Aaran RK. Determinants of serum alpha-tocopherol in Finnish adults. Prev Med 1988; 17:725-35.

31. Willett W. Nutritional epidemiology. New York: Oxford University Press, 1990.

32. Miyamoto H, Araya Y, Ito M, et al. Serum selenium and vitamin E concentrations in families of lung cancer patients. Cancer 1987; 60:1159-62.

33. Yeum KJ, Lee-Kim YC, Lee KY, Kim BS, Russell RM. The serum levels of retinol, beta-carotene and alpha-tocopherol of cancer patients in Korea (abstract). 14th International Congress of Nutrition 1989; 665.

34. LeGardeur BY, Lopez-S A, Johnson WD. A case-control study of serum vitamins A, E, and C in lung cancer patients. Nutr Cancer 1990; 14:133-40.

35. Atukorala S, Basu TK, Dickerson JWT, Donaldson D, Sakula A. Vitamin A, zinc and lung cancer. Br J Cancer 1979; 40:927-31.

36. Krishnamurthy S, Jaya S. Serum alpha-tocopherol, lipo-peroxides, and ceruloplasmin and red cell glutathione and antioxidant enzymes in patients of oral cancer. Indian J Cancer 1986; 23:36-42.

37. Drozdz M, Gierek T, Jendryczko A, Piekarska J, Pilch J, Polanska D. Zinc, vitamins A and E, and retinol-binding protein in sera of patients with cancer of the larynx. Neoplasma 1989; 36:357-62.

· 38. Charpiot P, Calaf R, Di-Costanzo J, et al. Vitamin A, vitamin E, retinol binding protein (RBP) and prealbumin in digestive cancers. Int J Vitam Nutr Res 1989; 59:323-8.

39. Hayes RB, Bogdanovicz JFAT, Schroeder FH, et al. Serum retinol and prostate cancer. Cancer 1988; 62:2021-6.

40. Rougereau A, Person O, Rougereau G. Fat soluble vitamins and cancer localization associated to an abnormal ketone derivative of D_3 vitamin: carcinomedin. Int J Vitam Nutr Res 1987; 57:367-73.

41. Basu TK, Hill GB, Ng D, Abdi E, Temple N. Serum vitamins A and E, beta-carotene, and selenium in patients with breast cancer. J Am Coll Nutr 1989; 8:524-8.

42. Heinonen PK, Koskinen T, Tuimala R. Serum levels of vitamins A and E in women with ovarian cancer. Arch Gynecol 1985; 237:37-40.

43. Romppanen U, Tuimala R, Punnonen R, Koskinen T. Serum vitamin A and E levels in patients with lichen sclerosus and carcinoma of the vulva—effect of oral etretinate treatment. Ann Chir Gynaecol 1985; 74(suppl 197):27-9.

44. Heinonen PK, Kuoppala T, Koshinen T, Punnonen R. Serum vitamins A and E and carotene in patients with gynecologic cancer. Arch Gynecol Obstet 1987; 241:151-6.

45. Cuzick J, De Stavola BL, Russell MJ, Thomas BS. Vitamin A, vitamin E and the risk of cervical intraepithelial neoplasia. Br J Cancer 1990; 62:651-2.

46. Knekt P, Järvinen R, Seppänen R, et al. Dietary antioxidants and the risk of lung cancer. Am J Epidemiol 1991; 134:471-9.

47. Friedman GD, Selby JV. Epidemiological screening for potentially carcinogenic drugs. Agents Actions (Suppl) 1990; 29:83-96.

48. Gey KF, Brubacher GB, Stähelin HB. Plasma levels of antioxidant vitamins in relation to ischemic heart disease and cancer. Am J Clin Nutr 1987; 45:1368-77.

49. Stähelin HB, Gey KF, Eichholzer M, Lüdin E, Brubacher G. Cancer mortality and vitamin E status. Ann N Y Acad Sci 1989; 570:391-9.

50. Connett JE, Kuller LH, Kjelsberg MO, et al. Relationship between carotenoids and cancer. The multiple risk factor intervention trial (MRFIT) study. Cancer 1989; 64:126-34.

51. Glattre E, Thomassen Y, Thoresen SO, et al. Prediagnostic serum selenium in a case-control study of thyroid cancer. Int J Epidemiol 1989; 18:45-9.

52. Heilbrun LK, Nomura A, Hankin JH, Stemmermann GN. Diet and colorectal cancer with special reference to fiber intake. Int J Cancer 1989; 44:1-6.

53. Stähelin HB, Rösel F, Buess E, Brubacher G. Cancer, vitamins, and plasma lipids: prospective Basel study. J Natl Cancer Inst 1984; 73:1463-8.

54. Wald NJ, Boreham J, Hayward JL, Bulbrook RD. Plasma retinol, beta-carotene and vitamin E levels in relation to the future risk of breast cancer. Br J Cancer 1984; 49:321-4.

55. Russell MJ, Thomas BS, Bulbrook RD. A prospective study of the relationship between serum vitamins A and E and risk of breast cancer. Br J Cancer 1988; 57:213-5.

56. Willett WC, Polk BF, Underwood BA, et al. Relation of serum vitamins A and E and carotenoids to the risk of cancer. N Engl J Med 1984; 310:430-4.

57. Nomura AMY, Stemmermann GN, Heilbrun LK, Salkeld RM, Vuilleumier JP. Serum vitamin levels and the risk of cancer of specific sites in men of Japanese ancestry in Hawaii. Cancer Res 1985; 45:2369-72.

58. Salonen JT, Salonen R, Lappeteläinen R, Mäenpää PH, Alfthan G, Puska P. Risk of cancer in relation to serum concentrations of selenium and vitamins A and E: matched case-control analysis of prospective data. Br Med J 1985; 290:417-20.

59. Menkes MS, Comstock GW, Vuilleumier JP, Helsing KJ, Rider AA, Brookmeyer R. Serum beta-carotene, vitamins A and E, selenium, and the risk of lung cancer. N Engl J Med 1986; 315:1250-4.

60. Schober SE, Comstock GW, Helsing KJ, et al. Serologic precursors of cancer. I. Prediagnostic serum nutrients and colon cancer risk. Am J Epidemiol 1987; 126:1033-41.

61. Burney PGJ, Comstock GW, Morris JS. Serologic precursors of cancer: serum micronutrients and the subsequent risk of pancreatic cancer. Am J Clin Nutr 1989; 49:895-900.

62. Helzlsouer KJ, Comstock GW, Morris JS. Selenium, lycopene, alpha-tocopherol, beta-carotene, retinol, and subsequent bladder cancer. Cancer Res 1989; 49:6144-8.

63. Hsing AW, Comstock GW, Abbey H, Polk BF. Serologic precursors of cancer. Retinol, carotenoids, and tocopherol and risk of prostate cancer. J Natl Cancer Inst 1990; 82:941-6.

64. Kok FJ, van Duijn CM, Hofman A, Vermeeren R, de Bruijn AM, Valkenburg HA. Micronutrients and the risk of lung cancer. N Engl J Med 1987; 316:1416.

65. Wald NJ, Thompson SG, Densem JW, Boreham J, Bailey A. Serum vitamin E and subsequent risk of cancer. Br J Cancer 1987; 56:69-72.

66. Fex G, Pettersson B, Åkesson B. Low plasma selenium as a risk factor for cancer death in middle-aged men. Nutr Cancer 1987; 10:221-9.

67. Knekt P. Serum vitamin E level and risk of female cancers. Int J Epidemiol 1988; 17:281-6.

68. Knekt P, Aromaa A, Maatela J, et al. Serum vitamin E and risk of cancer among Finnish men during a 10-year follow-up. Am J Epidemiol 1988; 127:28-41.

69. Knekt P, Aromaa A, Maatela J, et al. Serum vitamin E, serum selenium and the risk of gastrointestinal cancer. Int J Cancer 1988; 42:846-50.

70. Knekt P. Serum alpha-tocopherol and the risk of cancer. Helsinki: Publications of the Social Insurance Institution, Finland, ML:83, 1988; 1-148.

71. Knekt P, Aromaa A, Maatela J, et al. Serum alpha-tocopherol, beta-carotene, retinol, retinol-binding protein and selenium and risk of cancers of low incidence in Finland. Am J Epidemiol 1991; 134:356-61.

72. Willett WC, Polk BF, Morris JS, et al. Prediagnostic serum selenium and risk of cancer. Lancet 1983; 2:130-4.

73. Longnecker MP, Martin-Moreno J-M, Knekt P, et al. Serum alpha-tocopherol concentration in relation to risk of colorectal cancer in pooled data from 5 prospective cohort studies. (Submitted).

74. Knekt P, Aromaa A, Maatela J, et al. Vitamin E and cancer prevention. Am J Clin Nutr 1991; 53: 283s-6s.

75. Takahashi Y, Uruno K, Kimura S. Vitamin E binding proteins in human serum. J Nutr Sci Vitaminol 1977; 23:201-9.

76. McMichael AJ, Jensen OM, Parkin DM, Zaridze DG. Dietary and endogenous cholesterol and human cancer. Epidemiol Rev 1984; 6:192-216.

77. Knekt P, Reunanen A, Aromaa A, et al. Serum cholesterol and risk of cancer in a cohort of 39,000 men and women. J Clin Epidemiol 1988; 41:519-30.

78. Tangney CC, Shekelle RB, Raynor W, Gale M, Betz EP. Intra- and interindividual variation in measurements of beta-carotene, retinol, and tocopherols in diet and plasma. Am J Clin Nutr 1987; 45:764-9.

79. Wald NJ, Nicolaides-Bouman A, Hudson GA. Plasma retinol, beta-carotene and vitamin E levels in relation to the future risk of breast cancer. Br J Cancer 1988; 57:235.

80. London RS, Murphy L, Kitlowski KE. Hypothesis: breast cancer prevention by supplemental vitamin E. J Amer Coll Nutr 1985; 4:559-64.

81. Ernster VL, Goodson WH, Hunt TK, Petrakis NL, Sickles EA, Miike R. Vitamin E and benign breast "disease": a double-blind, randomized clinical trial. Surgery 1985; 97:490-4.

82. London RS, Sundaram GS, Murphy L, Manimekalai S, Reynolds M, Goldstein PJ. The effect of vitamin E on mammary dysplasia: a double-blind study. Obstet Gynecol 1985; 65:104-6.

83. Meyer EC, Sommers DK, Reitz CJ, Mentis H. Vitamin E and benign breast disease. Surgery 1990; 107:549-51.

84. McKeown-Eyssen G, Holloway C, Jazmaji V, Bright-See E, Dion P, Bruce WR. A randomized trial of vitamins C and E in the prevention of recurrence of colorectal polyps. Cancer Res 1988; 48:4701-5.

85. DeCosse JJ, Miller HH, Lesser ML. Effect of wheat fiber and vitamins C and E on rectal polyps in patients with familial adenomatous polyposis. J Natl Cancer Inst 1989; 81:1290-7.

86. DeWys WD, Malone WF, Butrum RR, Sestili MA. Clinical trials in cancer prevention. Cancer 1986; 58:1954-62.

87. Bertram JS, Kolonel LN, Meyskens FL Jr. Rationale and strategies for chemoprevention of cancer in humans. Cancer Res 1987; 47:3012-31.

88. Cullen JW. The National Cancer Institute's intervention trials. Cancer 1988; 62:1851-64.

38

Antitumor Activity of Tocotrienols

Kanki Komiyama

The Kitasato Institute, Shirokane, Minato-ku, Tokyo, Japan

Masakazu Yamaoka

National Chemical Laboratory for Industry, Tsukuba, Ibaraki, Japan

INTRODUCTION

The biological activity of α-tocopherol has been widely studied, but the physiological activity of tocotrienol has not been studied very much although it is present in many vegetable oils. During a search for new physiological activities of tocotrienols isolated from palm oil, it was found that tocotrienols showed antitumor activity against mouse transplantable tumors.

ANTITUMOR ACTIVITY

Tumor cells were inoculated intraperitoneally (IP) into mice, and tocotrienols were given IP successively on days 1-5 and 7-11. Thereafter, the number of days of survival of the mice was checked every day, and the activity was evaluated by the increase in life span (ILS). As shown in Table 1, γ-tocotrienols showed life prolongation effects in tumor-bearing animals with four different cell lines at doses of 20 and 40 mg/kg, and α-tocotrienols showed the activity on three cell lines. The percentage of ILS of γ-tocotrienol was higher than that of α-tocotrienol. However, α-tocopherol did not show any antitumor activity.

The effects of tocotrienols on the growth of sarcoma 180 solid tumor were examined. Tumor cells were inoculated subcutaneously (SC) and samples were given IP on days 1-5 and 7-11. α-Tocopherol showed very slight growth inhibition of the tumor.

The antitumor activity of a local injection of tocotrienol was examined using sarcoma 180 inoculated SC into ICR mice. Tocotrienol inhibited the tumor growth of solid form of S180, as shown in Table 2.

TOXICITY

The dose showing antitumor activity resulted in suppression of body weight gain. To examine the cause of this suppression, mice received 20 or 40 mg/kg of α- and γ-tocotrienol for 14 consecutive days and were sacrificed for histopathological examination. Adhesive peritonitis and atrophy of the thymus were observed in mice treated with higher doses of γ-tocotrienols. However, no remarkable histopathological changes were observed in the main organs, such

Table 1 Antitumor Activity of Tocotrienols on Murine Tumors

Group	Drug dose (mg/kg/day)	Increase in life span[a] (%)				
		Ehrlich	S180	IMC	Meth A	P388
Control[b]	—	(17.9)	(11.7)	(13.4)	(18.7)	(11.3)
α-T	40	10	40	35	0	0
	20	4	0	39	2	0
α-T3	40	65	82	46	0	0
	20	0	56	34	0	0
γ-T3	40	92	126	97	37	9
	20	78	117	63	35	0
MMC	0.5	134	174	>348	65	>256

[a]Tumor/mouse: Ehrlich carcinoma/ICR, Sarcoma 180/ICR, IMC carcinoma/CDF$_1$, Meth A fibrosarcoma/Balb/c, P388 leukemia/CDF$_1$.
[b]Numbers in parentheses for the control group represent mean survival days.
Source: Adapted from Komiyama et al. (1).

as the liver, kidneys, lungs, and heart. Therefore, it appeared that at least adhesive peritonitis was involved in the loss of body weight in the mice.

CYTOCIDAL AND ANTIMICROBIAL ACTIVITY

The direct cytotoxicity of tocotrienols was examined using cultured tumor cells derived from human or mouse tumors. Tumor cells were placed on a tissue culture plate and were exposed to tocotrienols for 3 days. The cells (HeLa human cervix carcinoma, H69 human lung carcinoma, and P388 mouse leukemia) were then treated with trypan blue and the cells were counted with a hemocytometer. The IC$_{50}$ values (50% growth inhibition) of tocotrienols were 23-47 μg/ml.

Since known cytotoxic antitumor agents usually show both antibacterial activity and cytotoxic activity, the antimicrobial activity of tocotrienols was examined using the paper disk method on agar plates. No antimicrobial activity was observed in many kinds of bacteria, including a Gram-positive and Gram-negative bacteria, fungi, and yeast, even at a concentration of 1000 μg/ml. Therefore, it is considered that the mode of action of tocotrienols differs from that of conventional nucleic acid synthesis inhibitors.

Table 2 Antitumor Activity of Intratumor Injection of Tocotrienols on Sarcoma 180[a]

Drug	Tumor size (mm²)		
	Day 7	Day 12	Day 17
Control	63 ± 16	165 ± 34	185 ± 50
α-Tocophrol	45 ± 22	121 ± 66	165 ± 84
α-Tocotrienol	43 ± 16	85 ± 23	141 ± 61
γ-Tocotrienol	49 ± 14	85 ± 15	129 ± 26

[a]Mice were subcutaneously inoculated with sarcoma 180 on day 0 and received intratumor injection of 0.5 mg/mouse of tocotrienol on days 5-14. Tumor sizes were measured by calipers.

EFFECT ON CARCINOGENESIS

Sundram et al. (3) reported the effects of dietary palm oil on mammary carcinogenesis induced in female rats by dimethylbenzanthracene (DMBA). Female rats were treated with carcinogen (DMBA) and fed semisynthetic diets containing different fats and oils during a 5-month experimental period. The results showed that high palm oil diets did not promote chemically induced mammary tumorigenesis in female rats compared with high corn oil or soybean oil diets. Palm oil differs greatly from corn oil and soybean oil in its content of tocopherols, tocotrienols, and carotenes. Palm oil has high levels of tocotrienols and carotenes.

We also examined the effects on skin carcinogenesis induced in mice by DMBA and tetra-decanoylphorbol acetate (TPA). The skin of the mice was coated with DMBA (200 μg/mouse) as an initiator, and 1 week later twice weekly applications of 10 μg/mouse of TPA were used as a promoter. After application of TPA, tocopherol or α-tocotrienol was also applied to the same sites. Of seven control mice three showed papillomas on the skin 7 weeks later, whereas the skin of mice administered α-tocotrienol or tocopherol did not have any papillomas. In the next week, however, five of seven mice in each group showed papillomas. Thus, tocopherol and α-tocotrienol slightly inhibited the development of skin papillomas.

ANTIOXIDANT ACTIVITY

Since the antitumor antibiotic doxorubicin induces severe cardiac toxicity associated with peroxidation of cardiac lipids in mice and humans, we attempted to determine if tocotrienols inhibit the formation of lipid peroxidate in mice treated with doxorubicin. Mice were given a single dose of doxorubicin and three successive injections of tocotrienols or tocopherol. The level of malonaldehyde in the cardiac tissue was measured 3 days later by the thiobarbituric acid (TBA) method. As shown in Table 3, α- and γ-tocotrienols and tocopherol inhibited the formation of lipid peroxidate by about 45, 31, and 69%, respectively.

CONCLUSION

Antitumor activity of tocotrienols was observed when tocotrienols were injected IP into mice bearing ascites tumors or intratumorly. Since the direct cytotoxic activity is weaker then that of known cytotoxic antitumor agents, the mechanism of action of tocotrienols may differ from that of antitumor agents. The antioxidant activity of tocotrienols in vivo was weaker than that of tocopherol, but it would be interesting to demonstrate that the reduction in toxicity of known antitumor agents producing radicals and combination therapy consisting of tocotrienols and antitumor agents would improve their efficacy.

Table 3 Inhibitory Effect on Malonaldehyde Generation in Heart Tissue of Mice Treated with Doxorubicin[a]

Drug	Dose (mg/kg/day)	Malonaldehyde (nmol/g wet tissue)	Inhibition (%)
Saline		8.2 ± 0.97	0
α-Tocophrol	40	2.5 ± 0.83*	69
α-Tocotrienol	40	4.5 ± 0.86*	45
γ-Tocotrienol	40	5.7 ± 0.98*	31

[a]Mice received IP 15 mg/kg of doxorubicin on day 0 and tocopherol or tocotrienols on days 0, 1, and 2. They were sacrificed on day 3 to examine the amount of malondialdehyde by the TBA method.
Source: Adapted from Komiyama and Yamaoka (2).

REFERENCES

1. Komiyama K, Iizuka K, Yamaoka M, Watanabe H, Tsuchiya N, Umezawa I. Studies on the biological activity of tocotrienols. Chem Pharm Bull 1989; 37:1369-71.
2. Komiyama K, Yamaoka M. Application of tocotrienols to cancer chemotherapy. *In* Lipid-Soluble Antioxidants: Biochemistry and Clinical Applications. *Ed.,* Ong ASH and Packer L. Birkhauser Verlag, Basel/Switzerland.
3. Sundram K. Khor HT, Ong ASH, Pathmanathan R. Effect of dietary palm oils on mammary carcinogenesis in female rats induced by 7, 12-dimethylbenz[a]anthracene. Cancer Res 1989; 49:1447-51.

39

Impact of Palm Oil
on Experimental Carcinogenesis

Charles E. Elson

University of Wisconsin, Madison, Wisconsin

INTRODUCTION

Increasing the total fat intake of experimental animals increases the incidence of chemically induced and spontaneous cancer at certain sites, predominantly the mammary gland, pancreas, and colon, and increases the growth of transplanted tumors. When total fat intake is low but adequate in essential fatty acids, an increase in linoleic acid intake is more effective than an increase in saturated fatty acid intake in enhancing tumorigenesis (1-6). The quantity of linoleic acid required for full support of tumorigenesis, in studies of induced and transplanted mammary tumors, and of induced and spontaneous pancreatic and colon tumors exceeds that required to support normal growth and reproduction of tumor-free animals (7).

Semipurified diets containing 5% fat, either a blend of coconut oil and a vegetable oil high in linoleic acid content or palm oil as the sole fat, provide sufficient linoleic acid to meet the essential fatty acid requirement (>1 en%) of experimental animals. These diets provide a base against which the impact on experimental carcinogenesis of other fats and oils can be measured. Reviewed here are studies in which dietary protocols employed palm oil alone or in blends with other oils. Studies of the influence of dietary fat on chemical carcinogenesis utilize both direct-acting carcinogens, for example methylnitrosourea (MNU), and others that require metabolic activation, such as 7,12-dimethylbenz[a]anthracene (DMBA). Use of the latter type of carcinogen permits the study of the impact of type and quantity of dietary fat during the initiation and the promotion-progression stages of tumorigenesis.

EXPERIMENTAL CARCINOGENESIS

Palm Oil Fatty Acids

Initiation Stage

Sylvester et al. (12) investigated the influence of dietary fat level and composition on the initiation stage of DMBA-initiated mammary carcinogenesis. Five diets, one with 11 en% corn oil and the remainder with palm oil, corn oil, beef tallow, or lard providing 45 en%, were fed to groups of Sprague-Dawley rats for 4 weeks before and for 1 week after DMBA administration. All groups were then fed the 11 en% corn oil diet during the promotion-progression stage. At 19 weeks following DMBA administration, the number of tumors in the group of 32 rats fed 11 en% corn oil during initiation was 39, and in the groups fed 45 en% palm oil, corn oil, beef tallow, and tard, the numbers were 40, 51, 67, and 83, respectively. A second study revealed that rats fed the 11 en% corn oil diet had delayed sexual maturation; estrous cycle regularity was not affected by the diets.

Promotion—Progression

Ip et al. (13) studied the role of linoleic acid in the promotion-progression phase of DMBA-induced mammary tumorigenesis. Following administration of this host-activated carcinogen, Sprague-Dawley rats were fed diets containing 20% (45 en%) fat provided by corn oil blended with palm oil to provide 4.5, 9.2, 14.6, and 27.2 en% dietary linoleic acid. Mammary tumorigenesis, reflecting both incidence of tumors and number of tumors per dietary group, was substantially lower in rats receiving 4.5 en% linoleic acid. The experiment was repeated using mixtures of corn oil and coconut oil to provide linoleic acid intake over the range 1.1-25.9 en%. Mammary carcinogenesis during the promotion-progression stage was very sensitive to linoleic acid intake and increased proportionally in the range of 1.12-9.90 en% dietary linoleic acid (13,14). Sundram et al. (15) fed a semipurified diet with corn oil, soybean oil, RBD palm oil, metabisulfite-treated palm oil, or crude palm oil providing 45% of energy to DMBA-induced Sprague-Dawley rats for 5 months. In order, the tumors per group of 20 rats at the conclusion of the study were 71, 57, 30, 28, and 25. The significantly lower number of tumors in the palm oil groups presumably traces to the lower linoleic acid content (4.5% of total energy) of the diet. Thompson et al. (16) investigated the effects of moderate treadmill exercise and type and amount of dietary fat on the promotion and progression stage of DMBA-initiated rat mammary carcinogenesis. Moderate intensity exercise enhanced tumorigenesis in Sprague-Dawley rats fed a diet with corn oil providing either 11% (7.2 en% linoleic acid; leg 1) or 46% (28 en% linoleic acid; leg 2) of total energy. A third leg of the study tested the effect of exercise on carcinogenesis when the diet was modified with a mixture of corn oil and palm oil to provide fat at the level of the second leg (46% of calories) and linoleic acid at the level of the first leg (7.2% of calories). Contrary to the responses noted in legs 1 and 2, exercise failed to enhance carcinogenesis. In moderately exercised rats, increasing the level of fat in the diet from 11 en% (leg 1) to 46 en% (leg 2) with the addition of corn oil enhanced carcinogenesis; increasing the level of dietary fat from 11 to 46 en% with the addition of the 89:11 palm oil-corn oil mixture (leg 3) did not enhance carcinogenesis.

The enhancement of tumorigenesis by an increase in dietary linoleic acid intake is disputed by Beth et al. (17). They used two blends of palm oil, lard, and sunflower seed oil in formulating fats moderate (75:14:11) and high (40:20:40) in linoleic acid. The fat blends provided 35% of the calories in two semisynthetic diets differing in linoleic acid content (5.5 and 11% of calories). Tumorigenesis was initiated in Sprague-Dawley rats with a direct-acting carcinogen, MNU and then the diets were fed both ad libitum and energy restricted (30%) for 6 months. Consistent with other studies, energy restriction resulted in extended latency and fewer tumors. Differing from the consensus reported earlier (1-6), doubling the linoleic acid content of the diet failed to change tumor latency and the numbers of tumors per tumor-bearing rat.

Transplanted Tumors

Latency is defined as the interval between the administration of the carcinogen and the appearance of a palpable tumor. A dietary linoleic acid-mediated decrease in tumor latency therefore reflects an increased tumor growth rate. This modulation of tumor growth rate by high levels of dietary linoleic acid is demonstrated in a report by Buckman et al. (18). Five groups of mice were fed semipurified diets containing at least the minimum of essential fatty acids. The basal low-fat diet contained 0.5% (1.2% of total calories) corn oil. Four of the diets were modified by the addition of safflower oil or palm oil to raise the total fat content to 5% (12% of total calories) or 20% (48% of total calories). The diets provided 0.8% (basal diet), 1.9% (5% palm oil), 4.6% (20% palm oil), 8.9% (5% safflower oil), and 31.1% (20% safflower oil) of energy as linoleic acid. The diets were fed for 4 weeks. Tumor cells derived from a spontaneously arising mammary tumor were then injected into the thigh area of the mice and the diets were continued. After 21 days, the mean volume of tumors in mice fed the 20% safflower oil diet was double that of tumors in mice fed the 5% safflower oil and 20% palm oil diets and six-fold that of mice fed the basal low-fat and 5% palm oil diets. Small tumors in mice fed the palm oil diets had significantly greater infiltrations of mast cells, which may be involved in tumor cell cytotoxicity (19). No differences in natural killer cell activity were noted (20).

Targets

Membranes. The fatty acid composition of tissue lipids is modulated by the composition of the dietary fat. Studies comparing the impact of dietary palm oil with that of more highly polyunsaturated fatty acids show that the major change in the composition of cellular lipids due to the latter is reflected in an elevated linoleic acid-oleic acid ratio (21-25). Diet-mediated changes in the fluidity of plasma membranes may alter carrier-mediated transport, the binding of ligands to cellular receptors, and signal transduction. Numerous reports comparing differences in the degree of membrane fluidity effected by lipids extremely high in saturated or polyunsaturated fatty acids demonstrate influences on tumor-host interactions of both humoral and cellular immunological types, the immunogenicity of tumor cells, and their sensitivity toward certain chemotherapeutic drugs or hypothermia. These reports led Damen et al. (21) to compare the impact of diets providing 30% of energy as corn oil and as palm oil:safflower oil (11:1) on the composition and fluidity of plasma membranes of lymphoid cells from normal and leukemic mice. The linoleic acid-oleic acid ratios of the resulting membrane phospholipids were 1.83 and 0.66; neither membrane fluidity measured by fluorescence polarization nor the expression of cell surface antigens by ascites tumor cells differed between groups. Van Amelsvoort et al. (22-24) examined insulin function in epididymal fat cells from rats fed diets providing 30% of energy as sunflower seed oil or palm oil. The respective linoleic acid-oleic acid ratios of the epididymal cell phospholipids were 2.7 and 0.4; changes in membrane fluidity induced by the dietary fats were small. Nevertheless, both the number of insulin cell receptors and the post-receptor cell action of insulin were elevated in cells from rats fed the sunflower seed oil diet. These responses appear to be independent of differences in epididymal cell size; subsequent studies demonstrated the differential responses independently of changes in the carbohydrate component of the diet and of the energy contributed by the fat component of the diet. These dietary fat-mediated changes in insulin function presumably influence plasma triglyceride clearance and other anabolic activities, some of which are likely to modulate tumor growth.

Prostanoids. Inhibitors of eicosanoid synthesis block the linoleic acid-mediated enhancement of chemically induced mammary carcinogenesis (26) and the growth of transplanted mammary tumors (27). Prostaglandin E_2 has an immunosuppressive effect, inhibiting lymphokine production, lymphocyte proliferation, antibody production, and natural killer cell cytotoxicity (28). The lipoxygenase product of arachidonic acid metabolism, leukotriene B_4, is a potent chemotactic and chemokinetic agent (29). In rodents, linoleic acid is readily converted to arachidonic acid, the eicosanoid precursor (30). Because of these metabolic actions, diets high in linoleic acid content have a profound effect on the development of normal immune responses as well as on the pathogenesis of inflammatory and neoplastic diseases in experimental animals (31).

The influence of palm oil-supplemented diets on prostanoid production has been compared with that of diets richer in polyunsaturated fatty acid content. Activated blood platelets produce prothrombotic thromboxane A_2 and blood vessels, antithrombotic prostaglandin I_2 (prostacyclin). Hornstra and his colleagues (32–36) report that rats fed diets richer in polyunsaturated fatty acids tend to have less tendency to develop arterial thrombosis, which indirectly points to enhanced prostacyclin production. Anomalously, animals fed palm oil, a relatively saturated fat, and those fed rapeseed, linseed, and sunflower seed oils had similar thrombotic tendencies. Rand et al. (35) measured prostanoid production by collagen-activated platelet in whole blood collected from rats fed high (50 en%) of palm oil and sunflower seed oil diets. Prostacyclin production was not influenced by the dietary fat; thromboxane production and the thromboxane-prostacyclin ratio were significantly lower in blood of the palm oil group. In a test of dietary palm oil and evening primrose oil (74% linoleic acid and 9% γ-linolenic acid) the more unsaturated oil increased thromboxane production and concomitantly, the thromboxane (pg/ml plasma) to prostacyclin (pg/mg aorta) ratio (37). Mold oil (10% linoleic acid and 6% γ-linolenic acid) also elevated both thromboxane and prostacyclin production, the latter to the greater degree; the thromboxane-prostacyclin ratio was marginally lower than that calculated for rats fed palm olein (38). In general, these observations are consistent with observations of the impact of the saturated dietary fats on prostanoid production (39-41). Consistent with their lower P/S ratios, dietary palm oil and lard tended to elicit lower thromboxane (pg/ml plasma) to prostacyclin (pg/mg aorta) ratios than olive oil (25); a follow-up with palm oil, palm olein, and safflower oil (P/S ratios 0.16, 0.41, and 10.02) failed to reveal a dietary effect.

Summary

At the midpoint of this review, the reports attribute the differences between the impact of palm oil and that of more polyunsaturated oils on tumorigenesis to differences in the linoleic acid content of the oils. Increasing the linoleic acid content of diets held constant in fat-derived energy increases tumorigenicity (1-6,13,14,16). It is generally accepted that increasing fat-derived energy in the diet increases tumorigenicity (1-6,16). A study already reviewed revealed that the addition of palm oil to increase the fat-derived energy content of diets held constant in linoleic acid content did not enhance tumorigenesis in moderately exercised rats (16). Linoleic acid may modulate tumorigenesis through its influence on prostanoid metabolism, immune response, or cell membrane structure and function. These considerations cannot explain the difference in the tumorigenic actions of lard and palm oil, two fats equal in linoleic acid content (11). When compared with fats higher in linoleic acid content, these fats elicit similar changes in the thromboxane-prostacyclin ratio (25). Data presented by Sundram et al. (14) further suggest that crude palm oil is more effective than the refined, bleached, and deodorized (RBD) palm oil in increasing the latency of DMBA-initiated tumorigenesis.

Minor Constituents of Palm Oil

Carotenoids

Crude palm oil is an extremely rich source of carotenoids and tocols (Table 1); 50-80% of the tocols are retained during processing (8). The concentration of tocotrienols in RBD palm oil and palm olein, 200–700 ppm, far exceeds that present in other commercial oils (11, 42-47). Red palm oil, the unbleached palm oil product preferred in Asian and African markets, has the highest concentration of agriculturally derived carotenoids (10), all of which are destroyed by bleaching in preparing the product for European and American markets (8). Epidemiological studies relating nutrient intake to cancer risk, predominantly lung cancer, reveal a protective effect of a high intake of carotene; anomalously, a high level of retinol

Table 1 Composition of Crude Malaysian Palm Oil and the Single-Stage Fractionation Products Palm Olein and Stearin

	Palm oil	Palm olein		Palm stearin	
		Range	Neutralized	Range	Neutralized
C12:0	0.1	0.1-1.1		0.1-0.6	
C14:0	1.0	0.9-1.4		1.1-1.9	
C16:0	43.7	37.9-41.7	39.0	47.2-73.8	55.2
C16:1	0.1	0.1-0.4		0-0.2	
C18:0	4.4	4.0-4.8	4.4	4.4-5.6	4.9
C18:1	39.9	40.7-43.6	41.7	15.6-37.0	29.9
C18:2	10.3	10.4-13.4	11.6	3.2-9.8	7.8
C18:3	0.1	0.1-0.6		0.1-0.6	
C20	0.2	0.2-0.5		0.1-0.6	
Minor constituents, ppm					
Carotenoids	450-820	—	—	—	—
Tocols	730 ± 70	864 ± 77		372 ± 95	
Physical properties					
Iodine value		56-58	57.2	25-49	40.1
Melting point, °C			21.5		49.7
Yield, %			72		28

Source: From References 8-11.

intake has no effect (48–50). De Vet (51) attributes this anomaly to the metabolic fate of dietary retinol, storage in the liver, which precludes the attainment of chemopreventive levels of retinol in peripheral tissues.

The more powerful relationship suggested for total rather than β-carotene intake by epidemiological and case-control studies (48-52) is supported by studies of Nishino et al. (53). α-Carotene isolated from palm oil proved to be more effective than β-carotene in suppressing the proliferation of human malignant tumor cells; both α- and β-carotene suppressed promotion in a two-stage mouse skin model. Although β-carotene is eightfold more effective than α-carotene in supporting, in vitro, the microsomal detoxification of benzo[a]pyrene, the latter may be the more powerful in vivo anticarcinogenic agent (54). Tan and Chu (54) examined the in vivo impact of 0.5 ml neutralized palm oil (0.35 mg carotenoids) and 0.5 ml RBD palm oil with and without 0.5 mg β-carotene on the metabolic activation of benzo[a]pyrene to the ultimate carcinogen, benzo[a]pyrene-7,8-diol-9,19-epoxide, by microsomes harvested from rats 6 days following treatment. Activation was suppressed to the greater extent by the neutralized palm oil and to the least by the RBD palm oil. The vitamin E of the palm oil had little effect on the metabolism of the carcinogen.

Strong negative associations between lung cancer risk and consumption of foods rich in lycopene and lutein (48,55,56) and the results of studies of experimental carcinogenesis (57-59) suggest that carotenoids without vitamin A activity also play protective roles. Epidemiological studies cannot identify associations between cancer risk and the intake of individual carotenoids as their distribution in foods is not incorporated into food composition data banks (50). In addition to α- and β-carotene (53) and phytoene (57,58), one or more of the minor carotenoid constituents of unbleached palm oil, γ, δ- ζ, and ϵ-carotene, ζ-carotene dione, aurochrome, ζ-carotene 1,2-epoxide, α-carotene 5,8-epoxide, α-carotene 5,6-epoxide, β-carotene 5,6-epoxide, *cis*-γ-carotene, citroxanthin, phytofluene, 3-dehydroretinal, neurosporene,

α- and β-zeacarotene, and lycopene (10,60), may modulate carcinogenesis through antioxidant activities (49). β-Carotene acts as a singlet oxygen scavenger and inhibits free-radical-initiated lipid peroxidation (61). These carotenoids and the tocols contribute to the oxidative stability of unbleached palm oil.

Tocols

Antioxidant Activity. The unsaturated homologs of vitamin E, the tocotrienols, account for 85% of the tocols in palm oil. Synthesis of the tocotrienols is distinguished from that of the tocophrerols by the addition of a prenyl-PP chain (geranylgeranyl-PP) rather than phytyl-PP to homogentisic acid. The tocotrienols are generally considered to have only a fraction of the biological vitamin E activity of the saturated homologs, the tocopherols (62-64). In biological systems, the chromanol nucleus of tocopherol or tocotrienol lies at the polar surface of cellular and subcellular membranes. Within the lipoid core the side chain of the tocol interacts hydrophobically with the acyl chains of phospholipids. Based on their studies of artificial membranes, Serbinova et al. (65) predict that tocopherols have a cholesterol-like effect, increasing the order of biological membranes, whereas the isoprenoid group of the tocotrienols decreases the ordering. The increased mobility of the tocotrienols within the membrane provides for a more uniform distribution of the antioxidant activity and greater radical-scavenging activity (65,66). There is a considerable discrepancy between the relative in vitro antioxidant activities of tocopherol and tocotrienol (46,64,67) and conventional estimates of their vitamin activities.

Chemical Carcinogenesis. We postulated that differences in the subcellular distribution of the tocopherols and tocotrienols would modulate their anticarcinogenic activities (68). In one study we compared the effectiveness of the AIN-76A diet (0.116 μmol α-tocopherol/g diet), the AIN-76A diet with added α-tocopherol (3.349 μmol α-tocopherol/g diet), and the AIN-76A diet supplemented with a carotenoid-free vitamin E mixture isolated from palm oil (3.381 μmol tocols/g diet). The tocol mixture in the finished diet consisted of 19.90% α-tocotrienol, 30.66% γ-tocotrienol, 11.51% δ-tocotrienol, 32.07% α-tocopherol, and <1% γ-tocopherol. Sprague-Dawley female rats were fed the diets for 2 weeks before and for 18 weeks following treatment with the indirectly acting carcinogen, DMBA. The medium latency of tumor development in the control group of rats was 73 days. Rats fed the tocopherol-enriched diet had a latency of 83 days ($P < 0.80$ versus controls); those fed the tocotrienol-enriched diet had a median tumor latency of 93 days ($P < 0.03$ versus controls). There was no significant difference between the latency of the two experimental groups. The dietary aspect of the study was repeated with the direct-acting carcinogen, N-nitrosomethylurea. Tumor latency was increased significantly, but only by the tocopherol-supplemented diet.

The tocotrienols reduced the severity of hepatocarcinogenesis induced by treatment with 2-acetylaminofluorene (69). Plasma γ-glutamyltranspeptidase activity, a marker of early neoplastic activity, was significantly lower in rats fed a diet providing 0.7 mg of a tocotrienol-enriched isolate of palm oil. Ngah et al. (69) also examined gammaglutamyltranspeptidase and uridyldiphosphate glucuronyltransferase activities in microsomes isolated from livers of carcinogen-treated rats. These activities, which are elevated in preneoplastic nodules, were modestly lower, at 1 and 20 weeks in livers of rats fed the tocotrienols. An earlier report (70) indicated that tocopherol was effective only during the very early events of 2-acetylamino-fluorene-induced hepatic carcinogenesis.

Transplanted Tumors. More promising responses to tocotrienols were reported by Kato et al. (71). They treated mice with 10 successive injections of tocol mixtures following the intraperitoneal (IP) transfer of IMC carcinoma cells. The survival of mice receiving injections (100 mg/kg/day) of a mixture consisting of 20% α-tocopherol, 25% α-tocotrienol, and 55%

γ-tocotrienol (40% cures) or of α-tocotrienol (80% cures) was significantly increased. In a comprehensive study, these investigators (67) evaluated α-tocotrienol, γ-tocotrienol, and α-tocopherol for activities against transplanted murine tumors inoculated IP into mice. The α- and γ- and tocotrienols injected IP produced significant, dose-responsive increases in the mean survival times of mice following the transfer of sarcoma 180 and IMC carcinoma cells; γ-tocotrienol also increased the mean survival time of mice injected with Erlich carcinoma cells and Meth A fibrosarcoma cells. The antitumor activity of γ-tocotrienol was higher than that of α-tocotrienol. α-Tocopherol produced a modest increase in the survival time of mice following the IP injection of sarcoma 180 and IMC carcinoma cells. Komiyama et al. (67) suggest that the antitumor action of the tocotrienols is mediated either through a direct cytotoxic activity or through the stimulation of the host immune system. The investigators suggest that the lower body weights of mice receiving the most effective treatments are consistent with a cytotoxic action. More likely, the elevated body weights associated with the less effective treatments reflect the accumulation of ascites fluid. The results of in vitro experiments point to a tocotrienol-mediated action other than that of stimulation of the host immune system. Although α-tocopherol tested at concentrations to 2.3 mM had no marked effect, γ- (0.05 mM) and α- (0.10 mM) tocotrienol inhibited by 50% the growth of cultured human H69 and HeLa cells and mouse P388 cells.

Suppression of Mevalonate Synthesis. These considerations lead to the suggestion of a molecular action fostered by the tocotrienols. α-Tocotrienol exerts a dose-dependent inhibition of 3-hydroxy-3-methylglutaryl coenzyme A reductase (HMGR) activity (72), whereas supplemental α-tocopherol elicited an increase in HMGR activity (73). Using a molecular approach, Parker et al. (74) found that α- and, more potently, γ- and δ-tocotrienol, act posttranscriptionally to lower the mass of HMGR in cultured cells. This posttranscriptional action may occur secondarily to an altered squalene metabolism. Preliminary studies of squalene metabolism by microsomes from α-tocotrienol-treated HepG2 cells reveal that the ratio of squalene dioxide to squalene epoxide is decreased (75). Nonsterol, mevalonate-derived products are required for the growth of normal and neoplastic tissues (76-78). In sterologenic tissues, mevalonate synthesis is modulated primarily through the sterol feedback regulation of the transcription of HMGR mRNA (79). An elevated HMGR activity, resistant to sterol feedback regulation, is characteristic of many neoplastic tissues (80); posttranscriptional regulation plays the major role in the modulation of HMGR activity in proliferating cells (76).

Two studies suggest that dietary palm oil provides a level of tocotrienol sufficient to suppress hepatic HMGR activity. We found that HMGR activity (203 ± 49 pmol mevalonate/mg microsomal protein/minute) in livers of single-cross white Leghorn pullets fed a diet containing 5% RBD palm olein for 1 month was 80% that of the activity (255 ± 53 pmol mevalonate/mg microsomal protein/minute) in livers of birds fed a 5% corn oil diet (81). Using a different approach, Haave et al. (82) fed diets containing 20 wt% palm oil, olive oil, or safflower oil to pregnant rats. The "active" HMGR activities in the livers of fetuses of dams fed olive oil and safflower oil were 13.4- and 4.3-fold higher, respectively, than that in fetuses of dams fed palm oil. Palm oil contains a number of volatile monoterpenoid constituents (83, 84), some of which are structurally related to the monoterpenes that suppress (85), by a posttranscriptional action (86), HMGR activity. A representative monoterpene, geraniol, has recently been shown to exert a dose-dependent, mevalonate-reversed increase in the population doubling time of murine tumor cells in vitro and, as previously demonstrated with the tocotrienols (71,72), to prolong the survival of mice following the IP transfer of tumor cells (87).

Sterologenesis. Human and animal studies demonstrate that dietary fats with iodine values below 90 elevate blood cholesterol. Regression equations predicting the cholesterol-elevating impact of the substitution of saturated fatty acids for unsaturated fatty acids (88-

91) have generally stood the test of time. A puzzling aspect of the equations has been their failure to predict the impact of palm oil, a relatively saturated oil with a low iodine value (Table 1), on serum cholesterol levels of humans (92-99) and animals (15,100-105). The Keys equation (90), for example, predicts that the replacement of half the dietary fat in current Western diets with palm oil would result in an increase in 10-20 mg/dl in the average plasma cholesterol level. Human studies demonstrate that palm oil feeding results in serum cholesterol concentrations that are higher than those found after administration of highly unsaturated oils; when reported, cholesterol levels of subjects during the palm oil leg of the studies are lower than their entry levels (106). Pre- and postprandial serum cholesterol levels of animals are elevated only when palm oil diets provide an excess of cholesterol (25,107,108). The dietary contribution to the body pool of cholesterol is a fraction of that provided by the sterologenic pathway, in which mevalonic acid is the first committed substrate (76). The suppression of HMGR activity effected by minor constituents of palm oil appears to neutralize the predicted cholesterol-elevating action of its saturated fatty acids.

SUMMARY

For foods in which a fat or oil is not the predominant ingredient and for which a manufacturer may wish to substitute one oil for another, the manufacturer is permitted to list the alternative oils that might be present. This use of and/or labeling in the ingredient lists of a variety of foods has apparently given the impression that palm oil is widely used in the food industry. A scholarly estimate of the contribution of palm oil to the daily energy intake of Americans, 0.5% of calories, points to its very limited presence in processed foods (109).

Few studies have examined the contributions of palm oil to the risk of chronic disease. The major share of the data reviewed in the preceding sections was drawn from studies in which palm oil served as the saturated oil for comparisons with more highly unsaturated oils. These "single-fat" studies suggest that an experimental diet high in linoleic acid content (7-28% of calories), independent of total fat content (5-20%), enhances chemically induced and spontaneous tumorigenesis to a greater degree than that noted when a 20% palm oil diet providing 5% of calories as linoleic acid is fed. Other high-fat diets low in linoleic acid content, such as contaminant-free beef tallow and lard, fail to provide a similar protection. An intake of 3-5% of energy as linoleic acid results in the saturation of tissues with arachidonic acid and ensures normalization of icosanoid synthesis. A lower intake places the animal at risk for an essential fatty acid deficiency; a higher intake tends to lower plasma cholesterol and to enhance tumorigenesis. This carefully delineated impact of linoleic acid intake on experimental carcinogenesis is not documented by data drawn from epidemiological surveys (7). This lack of consistency may be explained by the attenuating presence of α-linolenic acid in oils consumed by humans (31). Another rationale recognizes the role of food-borne anticarcinogens that are missing in the semipurified diets used in studies of experimental carcinogenesis. RBD palm oil introduces into the experimental diet two classes of isoprenoids, the tocotrienols and monoterpenes, both of which demonstrate anticarcinogenic activity. Crude (and red) palm oil is the richest agriculturally derived source of carotenoids, some of which exert more potent anticarcinogenic actions than those reported for β-carotene and retinol. The availability of red palm oil in the American food supply is extremely limited. RBD palm oil is the richest source of the tocotrienols, the unsaturated tocols generally, and likely erroneously, perceived to have limited vitamin E activity. The tocotrienols and certain of the monoterpenes, products of the plant mevalonate pathway, suppress mevalonate synthesis and, concomitantly, the synthesis of intermediates in the sterologenic pathway. The recent recognition that these mevalonate-derived intermediates of the sterologenic pathway play a fundamental role in neoplastic growth may explain the negative association between chronic disease risk and habitual consumption of the foods emphasized in recent dietary guidelines (110-112). Fruits, vegetables, and whole grains are the primary dietary sources of the isoprenoid products of plant metabolism that suppress mevalonate synthesis. Palm oil, the subject of this and another recent review (113), is an underutilized resource.

REFERENCES

1. Committee on Diet, Nutrition and Cancer. Diet, Nutrition and Cancer. National Research Council, National Academy of Sciences. Washington, DC: National Academy Press, 1982; 73-93.
2. Carroll KK. Summation: which fat/how much fat—animals. Prev Med 1987; 16:510-5.
3. Birt DF. Fat and calorie effects on carcinogenesis at sites other than the mammary gland. Am J Clin Nutr 1987; 45:203-9.
4. Birt DF. The influence of dietary fat on carcinogenesis: lessons from experimental animals. Nutr Rev 1990; 48:1-5.
5. Welsch CW. Enhancement of mammary tumorigenesis by dietary fat: review of potential mechanisms. Am J Clin Nutr 1987; 45:192-202.
6. Erickson KL, Hubbard NE. Dietary fat and tumor metastasis. Nutr Rev 1990; 48:6-14.
7. Dupont J, White PJ, Carpenter MP, et al. Food uses and health effects of corn oil. J Am Coll Nutr 1990; 9:438-70.
8. Berger KG. Palm oil. In: Chan HT Jr, ed. Handbook of tropical oils. New York: Marcel Dekker, 1983; 448.
9. Deffense E. Fractionation of palm oil. JAOCS 1985; 62:376-85.
10. Tan B, Grady CM, Gawienowski AM. Hydrocarbon carotenoid profiles of palm oil processed fractions. JAOCS 1986; 63:1175-9.
11. Tan B. Palm carotenoids, tocopherols and tocotrienols. JAOCS 1989; 66:770-6.
12. Sylvester PW, Russell M, Ip MM, Ip C. Comparative effects of different animal and vegetable fats fed before and during carcinogen administration on mammary tumorigenesis, sexual maturation, and endocrine function in rats. Cancer Res 1986; 46:757-62.
13. Ip C, Carter CA, Ip MM. Requirement of essential fatty acid for mammary tumorigenesis in the rat. Cancer Res 1985; 45:1997-2001.
14. Ip C. Fat and essential fatty acids in mammary carcinogenesis. Am J Clin Nutr 1987; 45:218-24.
15. Sundram K, Khor HT, Ong AS, Pathmanathan R. Effect of dietary palm oils on mammary carcinogenesis in female rats induced by 7,12-dimethylbenz[a]anthracene. Cancer Res 1989; 49:1447-51.
16. Thompson HJ, Ronan AM, Ritacco KA, Tagliaferro AR. Effect of type and amount of dietary fat on the enhancement of rat mammary tumorigenesis by exercise. Cancer Res 1989; 49:1904-8.
17. Beth M, Berger MR, Aksoy M, Schmahl D. Comparison between the effects of dietary fat level and of calorie intake on methylnitrosourea-induced mammary carcinogenesis in female SD rats. Int J Cancer 1987; 39:737-44.
18. Buckman DK, Erickson KL, Ross BD. Dietary fat modulation of murine mammary tumor metabolism studied by in vivo 31-P-nuclear magnetic resonance spectroscopy. Cancer Res 1987; 47:5631-6.
19. Hubbard NE, Erickson KL. Influence of dietary fats on cell populations of line 168 mouse mammary tumors: a morphometric and ultrastructural study. Cancer Lett 1987; 35:281-94.
20. Erickson KL, Schumacher LA. Lack of an influence of dietary fat on murine natural killer cell activity. J Nutr 1989; 119:1311-7.
21. Damen J, De Widt J, Hilkmann H, Van Blitterswijk WJ. Effect of dietary lipids on plasma lipoproteins and fluidity of lymphoid cell membranes in normal and leukemic mice. Biochim Biophys Acta 1988; 943:166-74.
22. van Amelsvoort JM, van der Beek A, Stam JJ, Houtsmuller UM. Dietary influence on the insulin function in the epididymal fat cell of the Wistar rat. I. Effect of type of fat. Ann Nutr Metab 1988; 32:138-48.
23. van Amelsvoort JM, van der Beek A, Stam JJ. Dietary influence on the insulin function in the epididymal fat cell of the Wistar rat. II. Effect of type of carbohydrate. Ann Nutr Metab 1988; 32:149-58.

24. van Amelsvoort JM, van der Beek A, Stam JJ. Dietary influence on the insulin function in the epididymal fat cell of the Wistar rat. III. Effect of the ratio of carbohydrate to fat. Ann Nutr Metab 1988; 32:160-9.

25. Imaizumi K, Nagata JI, Sugano M, Maeda H, Hashimoto Y. Effects of dietary palm oil and tocotrienol concentrate on plasma lipids, eicosanoid productions and tissue fatty acid compositions in rats. Agr Biol Chem 1990; 54:965-72.

26. Carter CA, Milholland RJ, Shea W, Ip MM. Effect of the prostaglandin synthetase inhibitor indomethacin on 7,12-dimethylbenz[a]anthracene-induced mammary tumorigenesis in rats fed different levels of fat. Cancer Res 1983; 43:3559-62.

27. Kollmorgen GM, King MM, Kosanke SK, Cuong D. Influence of dietary fat and indomethacin on the growth of transplantable mammary tumors in rats. Cancer Res 1983; 43:4714-9.

28. Hwang D. Essential fatty acids and immune responses. FASEB J 1989; 3:2052-61.

29. Rola-Pleszczynski M. Immunoregulation by leukotrienes and other lipoxygenase metabolites. Immunol Today 1985; 6:302-7.

30. Smith DL, Willis AL. Eicosanoids, their dietary precursors and drugs that modify their production or actions: implications in cancer. In: Abraham, S. ed. Carcinogenesis and dietary fat. Boston: Kluwer Academic Press, 1989; 53-81.

31. Meydani SN, Lichtenstein AH, White PJ, et al. Food uses and health effects of soybean and sunflower oils. J Am Coll Nutr 1991; 10:406-28.

32. Hornstra G, Vendelmans-Starrenburg A. Induction of experimental arterial thrombi in rats. Atherosclerosis 1973; 17:369-82.

33. Hornstra G, Lussenburg RN. Relationship between the type of dietary fat and arterial thrombosis tendency in rats. Atherosclerosis 1975; 22:499-516.

34. Hornstra G, Hennissen AAHM, Tan DTS, Kalafusz R. Unexpected effects of dietary palm oil on arterial thrombosis (rat) and atherosclerosis (rabbit): comparison with other vegetable oils and fish oil. In: Galli C, Fedeli E, eds. Fat production and consumption. Technologies and nutritional implications. NATO Adv Ser, Ser A Life Sci. New York: Plenum Press, 1987; 131:69-82.

35. Rand ML, Hennissen AAHM, Hornstra G. Effects of dietary palm oil on arterial thrombosis, platelet responses and platelet membrane fluidity. Lipids 1988; 23:1019-23.

36. Hornstra G. Dietary lipids and cardiovascular disease: effects of palm oil. Oleagineux 1988; 43:75-81.

37. Lee JH, Sugano M, Ide T. Effects of various combinations of omega 3 and omega 6 polyunsaturated fats with saturated fat on serum lipid level and eicosanoid production in rats. J Nutr Aci Vitaminol 1988; 34:117-29.

38. Sugano M, Ishida T, Koba K. Protein-fat interaction on serum cholesterol level, fatty acid desaturation and eicosanoid production in rats. J Nutr 1988; 118:548-54.

39. De Deckere EAM, Nugteren DH, Hoor F. Influence of type of dietary fat on the prostaglandin release from isolated rabbit and rat hearts and from rat aorta. Prostaglandins 1979; 17:947-55.

40. Galli C, Agradi E, Petroni A, Tremoli E. Differential effects of dietary fatty acids on the accumulation of arachidonic acid and its metabolic conversion through the cyclooxygenase and lipoxygenase in platelets and vascular walls. Lipids 1981; 16:165-72.

41. Abeywardena MY, McLennan PL, Charnock JS. Long-term saturated fat supplementation in the rat causes an increase in PGI_2/TBX_2 ratio of platelet and vessel wall compared to $n - 3$ and $n - 6$ dietary fatty acids. Atherosclerosis 1987; 66:181-9.

42. Slover HT. Tocopherols in foods and fats. Lipids 1971; 6:291-6.

43. Jacobsberg B, Deldime P, Gapor AB. Tocopherols and tocotrineols of palm oil. Oleagineux 1978; 33:239-47.

44. Goh SH, Choo YM, Ong ASH. Minor constituents of palm oil. JAOCS 1985; 62:237-40.

45. Syvaoja E-L, Piironen V, Varo P, Koivistoinen P, Salminen K. Tocopherols and tocotrienols in Finnish foods: Oils and fats. JAOCS 1986; 63:328-9.

46. Gabor A, Ong ASH, Kato A, Watanabe H, Kawada T. Antioxidant activities of palm vitamin E with special reference to tocotrienols. Elaeis 1989; 1:63-7.

47. Goh SH, Hew NF, Ong ASH, Choo YM, Brumby S. Tocotrienols from palm oil: Electron spin resonance spectra of tocotrienoxyl radicals. JAOCS 1990; 67:250-4.

48. Le Marchand L, Yoshizawa CN, Kolonel LN, Hankin JH, Goodman MT. Vegetable consumption and lung cancer risk: a population-based case-control study in Hawaii. J Natl Cancer Inst 1989; 81:1158-64.

49. Ziegler R. A review of epidemiologic evidence that carotenoids reduce the risk of cancer. J Nutr 1989; 119:116-22.

50. Fontham ETH. Protective dietary factors and lung cancer. Int J Epidemiol 1990; 19:S32-42.

51. De Vet HCW. The puzzling role of vitamin A in cancer prevention (review). Anticancer Res 1989; 9:145-52.

52. Connett JE, Kuller LH, Kjelsberg MO, Polk BF, Collins G, Rider A, Hulley SB. The relationship between carotenoids and cancer. The multiple risk factor intervention trial (MRFIT) study. Cancer 1989; 64:126-34.

53. Nishino H, Takayasu J, Murakoshi M, Imanishina J. Anticarcinogenesis activity of natural carotenes. CR Soc Biol (Paris) 1989; 183:85-9.

54. Tan B, Chu FL. Effects of palm carotenoids in rat hepatic cytochrome P_{450}-mediated benzo[a]pyrene metabolism. Am J Clin Nutr 1991; 53:1071S-15S.

55. Kvale G, Bjelke E, Gart JJ. Dietary habits and lung cancer risk. Int J Cancer 1983; 31:397-405.

56. Ziegler RG, Mason TJ, Stemhagen A, et al. Carotenoid intake, vegetables, and the risk of lung cancer among white men in New Jersey. Am J Epidemiol 1986; 23:1080-93.

57. Mathews-Roth MM. Antitumor activity of beta-carotene, canthaxanthin and phytoene. Oncology 1982; 39:33-7.

58. Mathews-Roth MM. Carotenoid pigment administration and delay in development of UV-B-induced tumors. Photochem Photobiol 1983; 37:509-11.

59. Schwartz J, Shklar G. Regression of experimental hamster cancer by beta carotene and algae extracts. J Oral Maxillofac Surg 1987; 45:510-5.

60. Ng JH, Tan B. Analysis of palm oil carotenoids by HPLC with diode-array detector. J Chromatogr Sci 1988; 26:463-9.

61. Burton GW, Ingold KU. Beta-carotene: an unusual type of lipid antioxidant. Science 1984; 224: 569-73.

62. Bunyan J, McHale D, Green G, Marcinkiewicz S. Biological potencies of alpha- and gamma-tocopherol and 5-methyltocol. Br J Nutr 1961; 15:253-7.

63. Leth T, Sondergaard H. Biological activity of vitamin E compounds and natural materials by the resorption gestation test and chemical determination of the vitamin E activity in foods and feeds. J Nutr 1977; 107:2236-43.

64. Hakkarainen RVJ, Jonsson SRL, Lindberg PO. Biopotency of vitamin E in barley. Br J Nutr 1984; 52:335-49.

65. Serbinova E, Kagan V, Han D, Packer L. Free radical recycling and intramembrane mobility in the antioxidant properties of alpha-tocopherol and alpha-tocotrienol. Free Radic Biol Med 1991; 10:263-75.

66. Yamaoka M, Carrillo MJH. Effect of tocopherols and tocotrienols on the physicochemical property of the liposomal membrane in relation to their antioxidant activity. Chem Phys Lipids 1990; 55:295-300.

67. Komiyama K, Lizuka K, Yamaoka M, Watanabe H, Tsuchiya N, Umezawa I. Studies on the biological activity of tocotrienols. Chem Pharm Bull 1989; 37:1369-71.

68. Gould MN, Haag JD, Kennan WS, Tanner MA, Elson CE. A comparison of tocopherol and tocotrienol for the chemoprevention of chemically-induced mammary tumors. Am J Clin Nutr 1991; 53:1068S-70S.

69. Ngah WZW, Jarien Z, San MM, et al. Effect of tocotrienols on hepatocarcinogenesis induced by 2-acetylaminofluorene in rats. Am J Clin Nutr 1991; 53:1076S-81S.

70. Ura H, Denca A, Yokose Y, Tsutsumi M, Konishi Y. Efficacy of vitamin E on the induction and evolution of enzyme altered foci in the liver of rats treated with diethylnitrosamine. Carcinogenesis 1987; 8:1595-600.

71. Kato A, Yamaoka M, Tanaka A, Komiyama K, Umezawa I. Physiological effect of tocotrienol. J Jpn Oil Chem Soc 1985; 34:375-6.

72. Qureshi AA, Burger WC, Peterson DA, Elson CE. The structure of an inhibitor of cholesterol biosynthesis isolated from barley. J Biol Chem 1986; 261:10544-50.

73. Qureshi AA, Peterson DM, Elson CE, Mangels AR, Din ZZ. Stimulation of avian cholesterol metabolism by alpha tocopherol. Nutr Rep Int 1989; 40:993-1001.

74. Parker RA, Pearce BC, Clark RW et al. Tocotrienols decrease cholesterol synthesis in HepG2 cells by a novel post-transcriptional suppression of HMGCoA reductase. FASEB J 1990; 4:A1744.

75. Parker RA, Clark RW. Squalene transfer and epoxidation in HepG2 cell membranes is modulated by gamma-tocotrienol. FASEB J 1991; A710.

76. Goldstein JS, Brown MS. Regulation of the mevalonate pathway. Nature 1990; 343:425-30.

77. Rine J, Kim S-H. A role for isoprenoid lipids in the localization and function of an oncoprotein. New Biol 1990; 2:219-26.

78. Maltese WA. Posttranslational modification of proteins by isoprenoids in mammalian cells. FASEB J 1990; 4:3319-28.

79. Panini SR, Schnitzer-Polokoff R, Spencer TA, Sinensky M. Sterol-independent regulation of 3-hydroxy-3-methylglutaryl-CoA reductase activity by mevalonate in Chinese hamster ovary cells. J Biol Chem 1989; 264:11044-52.

80. Siperstein MD, Fagan VW. Feedback control of mevalonate synthesis by dietary cholesterol. Cancer Res 1964; 24:1108-15.

81. Elson CE. Potential technologies for value-added products fro mprotein and co-products. Am Oil Chemists Symposium, Cincinnati: May 3-6, 1989.

82. Haave NC, Nicol LJ, Innis SM. Effect of dietary fat content and composition during pregnancy on fetal HMG CoA reductase activities and lipids in rats. J Nutr 1990; 120:539-43.

83. Dirinck P, Schreyan L, De Schoenmacker L, Wychuyse F, Schamp N. Volatile components of crude palm oil. J Food Sci 1977; 42:645-8.

84. Kuntom AHJ, Dirinck PJ, Schamp NM. Identification of volatile compounds that contribute to the aroma of fresh palm oil and oxidized oil. Elaeis 1989; 1:53-67.

85. Fitch ME, Mangels AR, Altmann WA, El Hawary M, Qureshi AA, Elson CE. Microbiological screening of mevalonate-suppressive minor plant constituents. J Agr Food Chem 1989; 37:686-91.

86. Clegg RJ, Middleton B, Bell GD, White DA. The mechanism of cyclic monoterpene inhibition of 3-hydroxy-3-methylglutaryl coenzyme A reductase in vivo in the rat. J Biol Chem 1982; 257:2294-9.

87. Shoff SM, Grummer M, Yatvin MB, Elson CE. Concentration-dependent increase of murine P388 and B16 population doubling time by the acyclic monoterpene geraniol. Cancer Res 1991; 51:37-42.

88. Keys A, Anderson JT, Grande F. Prediction of serum cholesterol responses of man to changes in fats in the diet. Lancet 1957; 2:959-66.

89. Keys A, Anderson JT, Grande F. Serum cholesterol response to changes in the diet. I. Iodine value of fat versus 2S-P. Metabol Clin Exp 1965; 14:747-58.

90. Keys A, Anderson JT, Grande F. Serum cholesterol response to changes in the diet. IV. Particular saturated fatty acids. Metabol Clin Exp 1965; 14:776-87.

91. Hegsted DM, McGandy RB, Myers ML, Stare FJ. Quantitative effects of dietary fat on serum cholesterol in man. Am J Clin Nutr 1965; 17:281-95.

92. Ahrens EH, Hirsch J, Insull W Jr, Tsaltas TT, Blomstrand R, Peterson ML. The influence of dietary fats on serum-lipid levels in man. Lancet 1957; 1:943-53.

93. Laine DC, Snodgrass CM, Dawson EA, Ener MA, Kuba K, Frantz ID Jr. Lightly hydrogenated soy oils versus other vegetable oils as a lipid-lowering dietary constituent. Am J Clin Nutr 1982; 35:683-90.

94. Baudet MF, Dachet C, Lasserre M, Esteva O, Jacotot B. Modification in the composition and metabolic properties of human low density and high density lipoproteins by different dietary fats. J Lipid Res 1984; 25:456-68.

95. Mattson FH, Grundy SM. Comparison of effects of dietary saturated, monounsaturated and polyunsaturated fatty acids on plasma lipids and lipoproteins in man. J Lipid Res 1985; 26:194-202.

96. Grundy SM. Comparison of monounsaturated fatty acids and carbohydrates for lowering plasma cholesterol. N Engl J Med 1986; 314:745-8.

97. Baudet MF, Esteva O, Lasserre M, Jacotot B. Dietary modifications of low-density lipoprotein fatty acids in humans: their effect on low-density lipoprotein-fibroblast interactions. Clin Physiol Biochem 1986; 4:173-86.

98. Bonanome A, Grundy SM. Effect of dietary stearic acid on plasma cholesterol and lipoprotein levels. N Engl J Med 1988; 318:1244-8.

99. Kesteloot H, Oviasu VO, Obasohan AO, Olomu A, Cobbaert C, Lissens W. Serum lipid and apo-protein levels in a Nigerian population sample. Atherosclerosis 1989; 78:33-9.
100. Kris-Etherton PM, Ho CY, Fosmire MA. The effect of dietary fat saturation on plasma and hepatic lipoproteins in the rat. J Nutr 1984; 114:1675-82.
101. Sugano M, Ishida T, Koba K. Protein-fat interaction on serum cholesterol level, fatty acid desaturation and eicosanoid production in rats. J Nutr 1988; 118:548-54.
102. Damen J, De Widt J, Hilkmann H, Van Blitterswijk WJ. Effect of dietary lipids on plasma lipoproteins and fluidity of lymphoid cell membranes in normal and leukemic mice. Biochem Biophys Acta 1988; 943:166-74.
103. Lee JH, Taguchi S, Ikeda I, Sugano M. The P/S ratio of dietary fats and lipid metabolism in rats: Gamma linolenic acid as a source of polyunsaturated fatty acid. Agr Biol Chem 1988; 52:3137-42.
104. Schouten JA, van der Veen EA, Spaaij CJK, van Gent CM, Popp-Snijders C, Beynen AC. Influence of dietary fat type on serum lipoids in rhesus monkeys. Nutr Rep Int 1989; 39:487-92.
105. Sundram K, Khor HT, Ong AS. Effect of palm oil and its fractions on rat plasma and high density lipoprotein lipids. Lipids 1990; 25:187-93.
106. Nutr Rev. New findings on palm oil. 1987; 45:205-7.
107. Lee JH, Fukumoto M, Nishida H, Ikeda I, Sugano M. The interrelated effects of n - 6/n - 3 and polyunsaturated/saturated ratios of dietary fats on the regulation of lipid metabolism in rats. J Nutr 1989; 119:1893-9.
108. Zhang X, Meijer GW, Beynen AC. Liver cholesterol concentrations in rats fed diets containing various fats of plant origin. Internat J Vitam Res 1990; 60:275-8.
109. Park YK, Yetley E. Trend in changes in use and current intakes of tropical oils in the United States. Am J Clin Nutr 1990; 51:738-48.
110. U.S. Dept. of Agriculture/Dept. of Health and Human Services. Nutrition and your health: Dietary guidelines for Americans. Pueblo, CO: Consumer Information Center, Dept. 622N, 1985.
111. U.S. Dept Health and Human Services. Surgeon General's report on Nutrition and Health. Summary and recommendations. DHHS (PHS) Publication No. 88-50211. Washington, DC. US Govt Printing Office, 1988.
112. National Research Council. Diet and health: Implications for reducing chronic disease risk. Washington DC: National Academy Press, 1989.
113. Elson CE. Tropical oils: Nutritional and Scientific issues. CRC Rev Food Sci Nutr 1992; 79-102.

C. Immunity and Disease Resistance

40

Vitamin E and Immune Response

Simin Nikbin Meydani

United States Department of Agriculture–Human Nutrition Research Center on Aging at Tufts University, Boston, Massachusetts

Robert P. Tengerdy

Colorado State University, Fort Collins, Colorado

INTRODUCTION

Substantial evidence from tissue culture and animal studies indicate that vitamin E is essential for normal function of the immune system. In experimental models different aspects of the immune system, such as resistance to infections, antibody production in response to antigens, delayed cutaneous hypersensitivity skin test (DCH), in vitro mitogenic response of lymphocytes, phagocytic index, and reticuloendothelial system clearance, have been shown to be altered by vitamin E deficiency. Furthermore, intakes of vitamin E beyond recommended levels appear to be needed to achieve optimal response in several of these indices of the immune function (1-5).

The effect of vitamin E on the immune system depends on interaction with other antioxidant and prooxidant nutrients, especially with that of polyunsaturated fatty acids (PUFA), and on other factors modulating the immune response, such as age and stress. The extent to which these vitamin E-induced immunological changes are associated with meaningful clinical outcomes has not been well studied. However, it is encouraging to note that epidemiological studies show high plasma vitamin E levels to be associated with a low incidence of infectious disease (6) and colon cancer (7-9). Further support is obtained from animal studies and clinical trials in which a beneficial effect of vitamin E supplementation was observed in reducing the pathogenesis of autoimmune disorders (10-13). Previous publications have reviewed the role of vitamin E in animal models (5,14-17). This review emphasizes recent human studies.

VITAMIN E DEFICIENCY

Animal Studies

It has been demonstrated that a balanced nutrient status is necessary for the normal maintenance of the immune response (18). A deficiency in any essential nutrient, including vitamin E, would predictably lead to impaired immune response (19).

Studies in different species of experimental animals using deficient as well as higher than recommended levels of vitamin E indicate that tocopherol is involved in the maintenance of immune function. Tengerdy et al. (20) showed that vitamin E-deficient mice had lower plaque-forming cells (PFC) and hemagglutination (HA) titer in response to sheep red blood cell (SRBC) injection than mice fed a sufficient amount of vitamin E. Supplementation with additional amounts of vitamin E further increased both PFC and HA. The adverse effect of vitamin E deficiency on humoral immunity was restored to normal by vitamin E supplementation but not by the synthetic antioxidant *N,N*-diphenyl-*p*-phenylenediamine (DPPD). In rats vitamin E deficiency impaired T cell function (16). Moderate impairment of B cell function was also reported (16). Furthermore, macrophage membranes of vitamin E-deficient rats expressed fewer Ia antigens (21) than those fed the control diet. Phagocytic function was also diminished in vitamin E-deficient animals (19,22).

Eskew et al. (23) found that either vitamin E or selenium deficiency suppressed lymphocyte proliferation and antibody-dependent chicken red blood cell lysis in rats. A double deficiency had the most suppressive effect and also reduced cell viability during the culture period. It was suggested that the nutrient deficiencies increased the formation of oxygen intermediates. Eskew et al. (24) also looked at the combined effect of vitamin E deficiency and ozone exposure on different cell-mediated immune responses and found that although vitamin E deficiency or ozone exposure alone did not have a significant effect on splenic antibody-dependent cell-mediated cytotoxicity (ADCC), a combined treatment significantly depressed spleen ADCC. No such interaction was observed on the mitogenic responses of splenocytes. It therefore appears that the vitamin E requirement for optimal immune function is even higher under conditions in which lipid peroxidation and oxidative stress are increased.

Vitamin E deficiency in dogs decreased the blastogenic response of lymphocytes to concanavalin A (attributable to a serum factor that could be washed from the cell surface of depressed lymphocytes) (25).

Vitamin E-deficient lambs (26) and pigs (4,27) showed depressed mitogenic response to T cell mitogens, a defect that was reversable following vitamin E repletion. Vitamin E-deficient pigs (27) had higher mortality and more pronounced clinical pathology from experimentally induced swine dysentery than those receiving adequate levels of vitamin E.

Human Studies

In contrast to the animal models investigated, the effect of vitamin E deficiency on the immune response of humans has not been directly studied. Vitamin E deficiency has been suggested to contribute to alterations in neonatal polymorphonuclear leukocyte (PMN) function via peroxidative damage to the cell membrane (28). Newborn infants, especially those who are premature and have a low vitamin E and neonatal PMN level, have been shown to be deficient in phagocytosis, bactericidal activity, and chemotaxis (29). These defects may play a role in determining the high incidence of newborn susceptibility to infections by impaired host defense. Chirico et al. (30) administered 120 mg/kg *d,l*-α-tocopherol (divided into single doses of 20 mg/kg intramuscularly on days 2, 3, 4, 7, 10, and 13 after birth) to 10 of 20 healthy premature infants and assessed neutrophil phagocytosis at 2, 5, 14, and 30 days of age. No differences were observed in PMN function in any of the children before vitamin E treatment. However, phagocytosis, bactericidal activity, and chemotaxis were lower compared to that of the 30 adults used as control group. At 5 days of age, the untreated infants maintained a low index and frequency of phagocytosis but these parameters were significantly increased in the group receiving vitamin E. At days 14 and 30, however, phagocytosis was normal in both groups of infants, with no differences noted in bactericidal activity, NBT reduction, random movement, chemotaxis, or metabolic activity. Thus, vitamin E appeared to accelerate the normalization of phagocytic function during the first week of life in the newborn.

Kowdley et al. (31) evaluated the in vitro and in vivo indices of T cell-mediated function in a patient who developed severe vitamin E deficiency as a secondary presentation of an intestinal malabsorption disorder. At the time of evaluation, the patient had an undetectable plasma vitamin E level and abnormal RBC peroxide hemolysis and was anergic to a battery of seven recall antigens administered in the form of multitest CMI. Her mitogenic responses to phytohemagglutinin (PHA) and concanavalin A (ConA, 846 ± 59 and 338 ± 100 cpm for optimal doses of PHA and ConA, respectively) and interleukin-2 (IL-2) production (2.6 units/ml) were very low. Vitamin E supplementation induced a rapid (after 1 week) improvement in both mitogenic responses and IL-2 production. The subject exhibited positive responses to three of the antigens 4 months after supplementation, with a mean induration score of 11.5 mm. Furthermore, mitogenic responses to PHA and ConA and IL-2 production showed a dramatic improvement (22,490 ± 2171 and 29,818 ± 908 cpm for optimal doses of PHA and ConA and 115 units/ml for IL-2 production, respectively). This study indicates that vitamin E deficiency causes a reversible impairment of T cell function in humans.

Sokol et al. (32) reported that vitamin E deficiency in children with chronic cholestasis impairs neutrophil chemotaxis, which is correctable by vitamin E repletion.

VITAMIN E SUPPLEMENTATION

Animal Studies

A dietary suplementation of 2–10 times higher than currently recommended levels of vitamin E significantly increased humoral and cell-mediated immune responses and phagocytic functions in laboratory and farm animals and increased their resistance to infectious diseases (14, 15, and 19, 33). Vitamin E supplementation may counteract the immunosuppressive effect of stress in animals (34–36).

Other antioxidants nutrients may interact with vitamin E in these processes. Vitamin E and C interact in enhancing immune responses and protection from infectious diseases and cancer (27,37,38). Recently we showed that vitamin E in combination with β-carotene (but not with vitamin A) significantly increased disease protection and reduced hepatomegaly caused by *Escherichia coli* infection (Table 1) (39). The decreased mortality was accompanied by elevated antibody titers but without a significant correlation between titer and mortality.

These studies indicate that the dietary tocopherol requirement for maintenance of optimal immune responsivness in animals may be higher than the levels recommended for normal growth and reproduction. The effective level of dietary supplemention is difficult to define because it depends on several factors, such as the composition of the diet, food and feed consumption, rate of growth or animal production (meat, egg, and milk in farm animals), and living conditions (stress, crowding, and environment). In our experience, 150–300 mg/kg diet of vitamin E consistently provided immunoenhancement and disease protection in chickens and mice (15). Morrill and Reddy (40) reported immune protection in cattle with a vitamin E supplement of 125 mg/head per day over a 24 week feeding period. The supplement was effective when fed for at least 3–4 weeks. It is important to note that to determine the efficacy of vitamin E, feeding of vitamin E should always be accompanied by an adequate test of plasma and tissue levels of vitamin E. Furthermore, vitamin E supplementation should be done always in the presence of an adequate Se level.

The main role of vitamin E in enhancing immune response and phagocytosis is the prevention of lipid peroxidation of cell membranes. The rapidly proliferating cells of the stimulated immune and phagocytic systems are particularly prone to peroxidative damage by free radicals, peroxides, and superoxides. The antioxidant effect also modulates the biosynthesis and activity of important cell regulators, prostaglandins (PG), thromboxane, and leukotrienes (41–43). In this regard, Meydani et al. (42) showed that vitamin E supplementation (500 ppm for 30 days) significantly decreased PGE_2 production while significantly increasing mitogenic response, IL-2 production, and delayed-type hypersensitivity (DTH) in old mice.

Adjuvant Action of Vitamin E

D,L-α-tocopheryl acetate is a viscous oil that can be emulsified with aqueous antigens to form water in oil, oil in water, or double emulsions, microspheres, or even artificial liposomes. Such adjuvants can be delivered directly to target organs. A β-carotene-liposome preparation was injected directly into hamster cheek pouch tumors and caused the regression of such tumors (44). In vaccination, such adjuvants may boost the immune response to weakly immunogenic antigens, as is the case with many protein antigens (45).

Water-in-oil adjuvants remain at the site of a subcutaneous injection as a stable depot from which the antigen is slowly and gradually released over a long period of time. The ordered orientation of antigen molecules in the oil/water boundary helps the optimal presentation of the antigen to antigen-processing cells, macrophages and lymphocytes. When vitamin E is the oil phase in such an adjuvant, it is in close contact with those antigen-processing cells and accessory cells that are attracted to the sites by chemotaxis (PMN, lymphocytes, and macrophages). This contact and the high concentration of vitamin E at the site of action assures the most favorable immunoenhancing effect of the vitamin, far more favorable than it is possible by dietary administration.

In sheep, a vitamin E adjuvant vaccine of *Brucella ovis* significantly reduced overall infectivity and enhanced the protective humoral immunity, as seen in Table 2 (46,47). Vitamin E adjuvants were also successfully tested in chickens, as shown in Table 2. Vitamin E adjuvant formulations were tested in mice with a view to potential human applications (48). Such adjuvants may be used increasingly in modern vaccination practice with very specific but weakly immunogenic subunit vaccines.

Human Studies

Baehner et al. (49) showed that in vivo supplementation with vitamin E could depress the bactericidal activity of leukocytes without affecting other blood chemistry and hematological parameters. They hypothesized that directed movement and phagocytosis by PMN are attenuated by the autooxidative damage to the cell membrane by endogenously derived H_2O_2 and that the administration of vitamin E may prevent this damage via scavenging of reduced oxygen radicals. These investigators administered 1600 IU daily for 7 days to three volunteers and found their isolated PMN were hyperphagocytic but killed *Staphylococcus aureus* 502A less effectively than controls, suggesting that less H_2O_2 was available to damage PMN or kill bacteria. H_2O_2-dependent stimulation of the hexose monphosphate shunt, H_2O_2 release from phagocytizing PMN, and fluoresceinated ConA cap formation promoted by H_2O_2 dam-

TABLE 1 Effect of Dietary Antioxidants on Mortality of *E. coli*-Infected Chickens[a]

Supplement per kg diet	% Mortality
Control	50 (12/24)
Vitamin E, 300 mg	25 (6/24)
Vitamin A, 60,000 IU	21 (5/24)
Beta carotene, 0.5 mg	33 (8/24)
Vitamin E + A	38 (9/24)
Vitamin E + β-carotene	17(4/24)

[a]Numbers in parentheses are the number of dead chicks/total.
Source: From Reference 39.

TABLE 2 Correlation of Humoral Immunity and Infection in Rams Vaccinated with *Brucella ovis* Against Epididymitis

Groups	No. in group	No. infected	Overall infectivity	Peak ELISA[a] titers ± SD
Bacterin[b]	9	4	44.4	0.27 ± 0.15
B. ovis-Vitamin E	9	2	22.2	0.42 ± 0.25
B. ovis-FIA[c]	9	4	44.4	0.35 ± 0.16
Vitamin E placebo	8	3	37.5	0.14 ± 0.08
FIA placebo	8	5	62.5	0.15 ± 0.09
Aluminum	9	6	66.7	0.26 ± 0.18

[a]Enzyme-linked immunosorbent assay: titer is expressed in OD (optical density) units at 485 nm and 1:200 serum dilution.
[b]Killed *B. ovis* cells in $Al(OH)_3$.
[c]Freund's incomplete adjuvant.
Source: From Reference 46.

age to microtubules were all diminished, but the release of superoxide from phagocytizing PMN was not reduced in the vitamin E group. This finding was confirmed by the study of Engle et al. (50), who showed that in vitro addition of tocopherol (5–10 mg/dl) decreased phorbal myristate acetate (PMA)-stimulated superoxide anion production by human PMN. It is difficult to assess the dichotomy between the depressant effect on PMN killing in vitro and the reduced PMN autotoxicity leading to improved phagocytosis. Boxer et al. (51) suggest that these results indicate the occurrence of a reduction in H_2O_2 that may be responsible for modulating PMN motile functions.

Yamada et al. (52) noted that immune-triggered neutrophils adhere to and damage endothelial cells. Boogaert et al. (53) found that human umbilical vein endothelial cells pretreated with α-tocopherol were protected against the cytotoxic effect of complement-activated neutrophils. The cytotoxic effect of neutrophils, presumably due to reactive oxygen radicals produced during their activation, is enhanced in the presence of platelets. Therefore, vitamin E not only protected endothelial cells from the damage caused by neutrophils, it also abolished the exaggerating effect of platelets. This characteristic of vitamin E appears due to an increased production of PGI_2 via endothelial cells. When endothelial cells are exposed to oxidative stress (53,54), they synthesize more PGI_2, perhaps as a defensive or protective response against injury. Meydani et al. (43) showed that vitamin E supplementation in mice increases lung PGI_2 synthesis.

Leb et al. (55) demonstrated that preincubation of human mononuclear phagocytes with α-tocopherol decreased PMA-induced cytotoxicity. Furthermore, target erythrocytes incubated with α-tocopherol were also more resistant to PMA-induced monocyte cytotoxicity. This effect of tocopherol appears due to a reduction in H_2O_2 formation. However, α-tocopherol was not effective in reducing monocyte ADCC, presumably because nonoxidative injury is more important in ADCC.

Prasad (56) studied the effect of a daily, 3 week treatment of 300 mg d,l-α-tocopheryl acetate on DCH in five boys (13–18 years) and on the bactericidal activity of peripheral leukocytes and cell-mediated immunity in 13 men (25–30 years). The vitamin E supplementation decreased leukocyte bactericidal activity, the release of acid phosphatase activity, and PHA-stimulated lymphocyte proliferation but did not alter the DCH of the skin to an intradermal injection of PHA. Interestingly, two subjects showed clinical improvement in their symptoms of asthmatic attacks and nasal allergy, respectively.

Ziemlanski et al. (57) supplemented 20 institutionalized elderly women (63–93 years) with 100 mg *d,l*-α-tocopheryl acetate twice daily and assessed serum proteins and immunoglobulin concentrations after 4 and 12 months. Vitamin E increased total serum protein with the principal effect on α_2- and β_2-globulin fractions occuring at 4 months. No significant effects were noted in the levels of either the immunoglobulins or complement C3, although another group to whom vitamin C (400 mg daily) was administered with vitamin E displayed significant increases in IgG and complement C3 levels.

Harman and Miller (58) supplemented 103 elderly patients from a chronic care facility with 200 or 400 mg daily α-tocopheryl acetate but did not observe any beneficial effect on antibody development against influenza virus vaccine. Unfortunately, data on the health status, medication use, antibody levels, and other relevant parameters were not reported.

Meydani et al. (59) showed via a placebo-controlled, double-blind study that vitamin E supplementation (800 mg/day) of healthy elderly individuals for a 30 day period significantly improves mitogenic responses to T cell mitogen ConA, IL-2 production, and the DCH (Table 3). This finding is particularly important since immune responsiveness declines with age, a phenomenon that contributes to the increased incidence of infectious disease and tumors in the elderly. In fact, DCH has been shown to be a good predictor of mortality in healthy elderly subjects (60). The immunostimulatory effect of vitamin E was associated with decreased PGE_2 production and plasma lipid peroxide levels (Table 3). Vitamin E supplementation did not have an effect on the basal production of IL-1. However, in a recent study examining the interaction of age, vitamin E, and exercise on cytokine production and exercise-induced muscle damage, Cannon et al. (61) noted that although vitamin E supplementation (800 IU *d,l*-α-tocopheryl acetate for 48 days before exercise) did not affect the basal production of IL-1 or TNF, it prevented the eccentric exercise-induced rise in IL-1. Vitamin E supplementation, however, inhibited the production of another cytokine, IL-6. Since IL-1 and IL-6 have been implicated in exercise-induced muscle proteolysis and damage, their inhibition by vitamin E during damaging exercise could have practical implications. In the same study, the young subjects fed placebo capsules had a significantly greater neutrophilia and plasma creatine kinase in response to eccentric exercise compared to old subjects receiving placebos (62). However, vitamin E supplementation tended to eliminate the difference in neutrophilia and creatine kinase between the two age groups. These studies indicate that in addition to its enhancement of cell-mediated immunity, vitamin E, by its modulation of cytokine production, can affect the catabolic consequences of inflammatory process and the acute-phase response.

TABLE 3   Immunostimulatory Effect of Vitamin E in Healthy Elderly Subjects[a]

	Plasma α-tocopherol	PBMC α-tocopherol	PBMC PGE_2[b]	Plasma lipid peroxide	Mitogenic response to ConA[c]	IL-2 production[d]	DCH[e]
	(µmol/L)	(nmol)	(lnpmol/L)	(µmol/L)	(ccpm)	(U/ml)	(mm)
(n)	(17)	(5)	(18)	(18)	(18)	(18)	(16)
Before	25.6 ± 1.4	0.12 ± 0.02	9.1 ± 1.6	2.76 ± 0.67	20551 ± 1927	35.6 ± 9.1	14.2 ± 2.9
After	70.9 ± 6.3*	0.39 ± 0.02*	8.5 ± 1.5*	1.2 ± 0.60*	23770 ± 2991*	49.6 ± 12.6*	18.9 ± 3.5*

[a]Mean ± SEM, healthy elderly subjects supplemented their diet with 800 IU/day of dl-α-tocopherol acetate for 30 days in a double blind, placebo controlled residential trial (Meydani et al., 1990).
[b]1×10^6 to 5×10^6 cells/ml were stimulated with 2.5 to 10 µg/ml PHA for 48 hrs. PGE_2 was measured by radioimmunoassay.
[c]1×10^6 cells/ml were cultured in presence or absence of optimal concentration of ConA for 68 hr. Cell proliferation was measured by incorporation of ^{3}H thymidine into newly synthesized DNA following a 4 hr pulse. Values represent corrected counts per minute which is the cpm of stimulated cultures minus the cpm of unstimulated cultures.
[d]1×10^6 cells/ml were cultured in presence of ConA for 48 hrs. IL-2 production was measured in supernatant using a bioassay as described in Meydani et al. (1990).
[e]DCH was measured using Multi-Test CMI which administers 7 recall antigens simultaneously. Values represent cumulative index which is the sum of all the positive responses.
*Significantly different from before values by paired Student's t-test or paired Wilcoxon's signed-rank test.

The effect of vitamin E on the immune response has also been studied in association with diseases. Evidence has been accumulating that oxidant damage to tissues underlies the pathology of several human diseases, including many associated with altered immune function, such as amyloidosis, rheumatoid arthritis, and cancer. Vitamin E deficiency in premature infants is now recognized as a key factor in hemolytic anemia, retrolental fibroplasia syndrome, and bronchopulmonary dysplasia. The precise role played by free radicals and H_2O_2 in such disorders or immune injury to the kidney and lung is not resolved, although they are formed and they interact with PG, leukotrienes, interleukins, and other modulators of immune function. Vitamin E may be beneficial to host defense mechanisms by preventing the infection-induced increases in tissue PG production from arachidonic acid (42,63,64). Swartz et al. (65) administered 800 IU vitamin E or placebo daily to 30 healthy adults for 8 weeks and observed a dramatic decline in plasma 6-keto-PGF$_{1\alpha}$, the principal stable degradation product of prostacyclin, in the treatment group. A similar inverse relationship was noted between serum vitamin E and other PGs, such as PGE$_2$ and PGE$_{2\alpha}$, in animal studies (15,66, 67). However, Lauritsen et al. (68) found oral supplements of 1920 IU α-tocopherol daily for 2 weeks in eight patients with active ulcerative colitis did not affect the disease-induced elevations of PGE$_2$ or leukotriene B$_4$ in the lumen. Goetzl (69) reported that vitamin E bidirectionally modulates the activity of the lipoxygenase pathway of human neutrophils in vitro. He further noted that normal plasma concentrations of vitamin E enhanced the lipoxygenation of arachidonic acid, whereas higher concentrations exerted a suppressive effect, consistent with α-tocopherol's role as a hydroperoxide scavenger.

Mowat and Baun (70), Hill et al. (71), and others have demonstrated abnormal PMN function in diabetic subjects, including defective phagocytic uptake and chemotactic responses. Kitahara et al. (72) found that monocytes from diabetic patients produced higher chemiluminescence peaks and generated more superoxide anion upon phagocytic challenge than controls. Hill et al. (71) administered 25 IU α-tocopherol/kg/day orally for 2–3 weeks to seven diabetic patients with consistently depressed monocyte chemotactic responses. The vitamin E treatment doubled monocyte random motility and chemotactic responsiveness toward zymosan-activated serum to levels comparable to normal controls, suggesting that this defective function may partially be a result of autooxidative membrane damage.

Taccone-Gallucci et al. (73) reported that PBMC from 10 patients on maintenance hemodialysis produced significantly more malonylaldehyde (MDA) and had a lower tocopherol content. After 15 days of parenteral vitamin E supplementation the MDA level decreased and the vitamin E level remained low. Vitamin E supplementation did not have an effect on natural killer (NK) activity or the PHA-stimulated mitogenic response. However, a reduction in the percentage of suppressor-cytotoxic (OKT[8+]) T cells was observed. Kayden et al. (74) also reported a lower level of tocopherol in the B cells of patients with chronic lymphocytic leukemia compared to normal subjects.

Spielberg et al. (75) described an infant boy with congenital glutathione synthetase deficiency who had episodic profound neutropenia accompanying infections. Boxer et al. (76) treated this patient with 400 IU (approximately 30 IU/kg) α-tocopherol acetate as an oral, single daily dose for 3 months. Before the treatment, the abnormal leukocytes exposed to phagocytic challenge released more H_2O_2, fixed less opsonized zymosan particles, killed bacteria less effectively, and demonstrated impaired microtubule assembly. These functional abnormalities disappeared after administration of vitamin E and no neutropenia occurred during subsequent infections. The patient was continued on the vitamin E therapy and was noted to have no further episodes of bacterial infections for 18 months (compared to six episodes the year before treatment). The investigators suggested this evidence supports the idea that in glutathione-deficient leukocytes vitamin E protects against nonspecific oxidant damage to membranes and nonmembranous structures (e.g., microtubules) by hastening the destruction of excess peroxide during phagocytosis. This suggestion was supported by Oliver (77), who further studied the cytoskeletal defects associated with glutathione synthetase deficiency in this patient and found normal microtubule assembly during the resting phase but damage to the structure from excessive peroxide production during stimulated phagocytic activity. Vitamin E protected against cytoskeletal damage incurred during phagocytosis but did not normalize other cellular functions dependent upon glutathione.

Epidemiological Studies

Goodwin and Garry (78) studied a population of healthy older adults (54–94 years) consuming megadoses of vitamin supplements and did not see any correlation between vitamin E intake and DCH, mitogen stimulation, serum antibodies, or circulating immune complexes. Persons taking megadoses of vitamin E (>5 times the RDA) had lower absolute circulating lymphocyte counts than the rest of the population. However, the study was complicated by the fact that several vitamin supplements at megadose levels were used by each subject and the interaction between different nutrients presented confounding variables. It was suggested that some megadose nutrients may act as nonspecific immunoadjuvants, the effects of which diminish with time.

Chavance et al. (6,79) conducted a community-based survey on the relationship between nutritional and immunological status in 100 healthy subjects over 60 years of age. They reported that plasma vitamin E levels were positively correlated with positive DCH responses to Diphtheria toxoid, *Candida*, and *Trichophyton*. In men only, positive correlations were also observed between vitamin E levels and the number of positive DCH responses. Subjects with tocopherol levels greater than 135 mg/L were found to have higher helper-inducer–cytotoxic-suppressor ratios. Blood vitamin E concentrations were also negatively correlated with the number of infectious disease episodes in the preceding 3 years.

Payatte et al. (80) reported a negative correlation between dietary vitamin E and ex vivo IL-2 production in free-living elderly Canadians. This study suffers from a major problem in that 70% of their presumably "healthy" elderly had undetectable IL-2 levels. This is rather unusual, as in the experience of Meydani et al. (59) and Nagel et al. (81) the level of IL-2 production in healthy elderly is about one-half to two-thirds that of young subjects. The study is further complicated by using dietary vitamin E level rather than plasma vitamin E as indicator of tocopherol status. Nutrient data bases for vitamin E are incomplete and do not necessarily represent the tocopherol status.

Free-radical damage is thought to be involved in the initiation and promotion of many cancers. The increased incidence of cancer among older adults has been postulated to be due, in part, to the increasing level of free-radical reactions with age and the diminishing ability of the immune system to eliminate the altered cells. Although controlled human studies on vitamin E and cancer are very limited, epidemiological data suggest that higher intakes of vitamin E (and other dietary antioxidants) may decrease the risk for certain cancers, particularly cancers of the breast, colon, lung, and stomach (8,9,82–85). Recently, Knekt et al. (7) assessed blood vitamin E levels and subsequent cancer incidence in a longitudinal study of 21,172 men in Finland. Vitamin E was measured from stored blood samples of 453 subjects who developed cancer during the 6–10 year study period and 841 matched controls. Adjusted relative risks in the two highest quintiles of blood vitamin E concentrations compared to all other quintiles were 0.7 for all cancers and 0.6 for cancers unrelated to smoking.

SUMMARY AND CONCLUSIONS

Evidence from animal and human studies indicate that vitamin E plays an important role in the maintenance of the immune system. Vitamin E deficiency impairs immune response; supplementation with higher than recommended dietary levels enhances humoral and cell-mediated immunity and augments the efficiency of phagocytosis in laboratory and farm animals as well as in humans. This immunostimulatory effect of vitamin E in animals is associated with increased resistance to some bacterial and viral infections and tumors. Indirect evidence in humans also indicate a disease protection effect for vitamin E. This evidence has led to questions about the adequacy of the current recommended dietary guidelines for vitamin E, especially under conditions in which environmental stress is increased. The level of vitamin E required for optimal host defense depends on many factors, including age of the host, interaction with other nutrients, environmental pollutants, physical activity, and stress, and thus has to be established accordingly. In addition to its beneficial effect as a dietary supplement, vitamin E, as an adjuvant to vaccines, provides better protection against bacterial infections than conventional vaccines.

The mechanism of the immunostimulatory effect of vitamin E needs to be determined. White blood cells are particularly rich in polyunsaturated phospholipids prone to oxidative destruction. The mounting of an immune response requires membrane-bound receptor-mediated communication between cells as well as between protein and lipid mediators, which can be directly or indirectly affected by tocopherol status. Tocopherol could effect the immune system through its antioxidant function either by decreasing reactive oxygen metabolites, such as H_2O_2, and/or by altering the formation of arachidonic acid metabolites, such as PG, both of which have been shown to suppress immune responsiveness. The immunostimulatory effect of tocopherol cannot be fully explained by its antioxidant function because other antioxidants do not produce similar actions. The mechanisms of the vitamin E effect on immune responsiveness need to be further explored. More research is needed to characterize the effect of vitamin E on the immune response of humans and related diseases. A substantial amount of work has been completed in animal models, but very few human trials have been conducted. Data are especially lacking in elderly subjects and in individuals exposed to environmental stress.

REFERENCES

1. Bendich A, Gabriel E, Machlin LJ. Dietary vitamin E requirement for optimum immune response in the rat. J Nutr 1986; 116:675-81.
2. Reddy PG, Morrill JL, Frey RA. Vitamin E requirements of dairy calves. J Dairy Sci 1987; 69:164-71.
3. Reddy PG, Morrill JL, Minocha HC, Stevenson JS. Vitamin E is immunostimulatory in calves. J Dairy Sci 1987; 70:993-9.
4. Jensen M, Fossum C, Ederoth M, Hakkarainen RV. The effect of vitamin E on the cell-mediated immune response in pigs. J Vet Med Br 1988; 35:549-55.
5. Tengerdy RP. The role of vitamin E in immune response and disease resistance. In: Bendich A, Chandra RK, eds. Micronutrients and immune functions. Ann N Y Acad Sci 1990; 587:24-33.
6. Chavance M, Brubacher G, Herbeth B, et al. Immunological and nutritional status among the elderly. In: De Wick AL, ed. Lymphoid cell functions in aging. Paris: Eurage, 1984; 231-7.
7. Knekt P, Aromaa A, Maatela J, et al. Serum vitamin E and risk of cancer among Finnish men during a ten-year follow-up. Am J Epidemiol 1988; 127:28-41.
8. Menkes MS, Comstock GW, Vuillemier JP, Helsing KJ, Rider AA, Brookmezer R. Serum beta-carotene, vitamins A and E, selenium, and the risk of lung cancer. N Engl J Med 1986; 315:1250-4.
9. Wald NJ, Boreham J, Hayward JL, Bulbrook RD. Plasma retinol, beta-carotene and vitamin E levels in relation to the future risk of breast cancer. Br J Cancer 1984; 49:321-4.
10. Blankenhorn G. Efficacy of Spondyvit® (vitamin E) in activated arthroses. A multicenter, placebo-controlled, double-blind study. Z Orthop 1986; 24:340.
11. Machtey I, Quaknine L. Tocopherol in osteoarthritis: a controlled pilot study. J Am Geriatr Soc 1978; 26:328.

12. Pletsityi K, Nikushkin E, Askerov M, Ponomareva L. Inhibition of development of adjuvant arthritis in rats by vitamin E. Bull Exp Med 1987; 103:49.

13. Meydani SN, Cathcart ES, Hopkins RE, Meydani M, Hayes KC, Blumberg JB. Antioxidants in experimental amyloidosis of young and old mice. In: Glenner GG, Asserman EP, Benditt E, Calkins E, Cohen AS, Zucker-Franklin D, eds. Fourth international symposium of amyloidosis. New York: Plenum Press, 1986; 683-92.

14. Tengerdy RP. Vitamin E, immune response and disease resistance. In: Diplock AT, Machlin JL, Packer L, Pryor WA, eds. Vitamin E biochemistry and health implications. Ann N Y Acad Sci 1989; 570:335-44.

15. Tengerdy RP. Immunity and disease resistance in farm animals fed vitamin E supplement. In: Bendich A, Phillips M, Tengerdy R, eds. Antioxidant nutrients and the immune response. New York: Plenum Press, 1989; 103-10.

16. Bendich A. Antioxidant vitamins and immune responses. In: Chandra RK, ed. Nutrition and immunology. New York: Alan R. Liss, 1988; 125-48.

17. Carpenter MP. Effects of vitamin E on the immune system. In: Meyskens FL, Prasad KN, eds. Vitamins and cancer. Basel: Karger, 1986; 199-211.

18. Axelrod AE. Nutrition in relation to immunity. In: Goodhart RS, Shils ME, eds. Modern nutrition in health and disease. Philadelphia: Lea and Febiger, 1980; 171-96.

19. Bendich A. Antioxidant micronutrients and immune responses. In: Bendich A, Chandra RK, eds. Micronutrients and immune functions. Ann N Y Acad Sci 1990; 587:168-80.

20. Tengerdy RP, Heinzerling RH, Brown GL, Mathias MM. Enhancement of the humoral immune response by vitamin E. Intern Arch Allergy 1973; 44:221-32.

21. Gebremichael A, Levy EM, Corwin LM. Adherent cell requirement for the effect of vitamin E on in vitro antibody synthesis. J Nutr 1984; 114:1297-305.

22. Dillard CJ, Kunert KJ, Tappel AL. Lipid peroxidation during chronic inflammation induced in rats by Freund's adjuvant: effect of vitamin E as measured by expired pentane. Res Commun Chem Pharmacol 1982; 37:143-6.

23. Eskew ML, Scholz RW, Reddy CC, Todhunter DA, Zarkower A. Effects of vitamin E and selenium deficiencies on rat immune function. Immunology 1985; 54:173-80.

24. Eskew ML, Scheuchenzuker WJ, Scholz RW, Reddy CC, Zarkower A. The effects of ozone inhalation on the immunological response of selenium and vitamin E-deprived rats. Environ Res 1986; 40:274-84.

25. Langweiler M, Schultz RD, Sheffy BE. Effect of vitamin E deficiency on the proliferative response of canine lymphocytes. Am J Vet Res 1981; 42:1681.

26. Finch JM, Turner RJ. Enhancement of bovine lymphocyte responses: a comparison of selenium and vitamin E supplementation. Vet Immunol Immunopathol 1989; 23:245-56.

27. Teige J, Saxegaard F, Froslie A. Influence of diet on experimental swine dysentery. Acta Vet Scand 1978; 19:133-46.

28. Oski FA. Anemia in infancy: iron deficiency and vitamin E deficiency. Pediatr Rev 1980; 1:247-53.

29. Miller ME. Phagocytic function in the neonate: selected aspects. Pediatrics 1979; 64:5709-12.

30. Chirico G, Marconi M, Colombo A, Chiara A, Rondini G, Ugazio AG. Deficiency of neutrophil phagocytosis in premature infants: effect of vitamin E supplementation. Acta Paedtr Scand 1983; 72:521-4.

31. Kowdley KV, Meydani SN, Cornwall S, Grand RJ, Mason JB. Reversal of depressed T-lymphocyte function with repletion of vitamin E deficiency. Gastroenterology (in press).

32. Sokol FJ, Harris RE, Heubi JE. Effect of vitamin E on neutrophil chemotaxis in chronic childhood cholestasis (abstract). Hepatology 1984; 4:1048.

33. Moriguchi S, Kobayashi N, Kishino Y. High dietary intakes of vitamin E and cellular immune functions in rats. J Nutr 1990; 120:1096-102.

34. Julseth DR. Evaluation of vitamin E and disease stress on turkey performance. M.S. Thesis, Colorado State University, Fort Collins, 1974.

35. Guptill DR. Stress, vitamin E and immune response. M.S. Thesis, Colorado State University, Fort Collins, 1979.

36. Watson RR, Petro TM. Cellular immune response, corticosteroid levels and resistance to *Listeria monocytogenes* and murine leukemia in mice fed a high vitamin E diet. Ann N Y Acad Sci 1982; 393:205-10.

37. Bendich A, D'Apolito P, Gabriel E, Machlin LJ. Interaction of dietary vitamin C and vitamin E on guinea pig immune responses to mitogens. J Nutr 1984; 114:1588-93.

38. Watson RR, Leonard TK. Selenium and vitamins A, E and C, nutrients with cancer prevention properties. J Am Diabet Assoc 1986; 85:505-10.

39. Tengerdy RP, Lacetera NG, Nockels CF. Effect of beta carotene on disease protection and humoral immunity in chickens. Avian Dis 1990; 34:848-54.

40. Morrill JL, Reddy PG. Effect of vitamin E on immune responses and performance of diary calves. Roche Tech Symp, March 11, Daytona, FL, 1987; 34-46.

41. Tengerdy RP, Mathias MM, Nockels CF. Effect of vitamin E on immunity and disease resistance. In: Prasad KN, ed. Vitamins, nutrition and cancer. Basel: Karger, 1984; 118-22.

42. Meydani SN, Meydani M, Verdon CP, Shapiro AC, Blumber JB, Hayes KC. Vitamin E supplementation suppresses prostaglandin E_2 synthesis and enhances the immune response in aged mice. Mech Ageing Dev 1986; 34:191-201.

43. Meydani SN, Shapiro AC, Meydani M, Blumberg JB. Effect of fat type and tocopherol supplementation on tocopherol status and eicosanoid synthesis in lung. In: Lands WEM, ed. Proc Am Oil Chem Soc on polyunsaturated fatty acids and eicosanoids. Champaign, IL: AOCS, 1987; 438-41.

44. Schwartz JL, Shklar G, Flynn E, Trickler D. Beta carotene to prevent and regress oral carcinoma. In: Bendich A, Phillips M, Tengerdy R, eds. Antioxidant nutrients and the immune response. New York: Plenum Press, 1990; 77-94.

45. Nervig RM, Gough PM, Kaeberle ML, Whetstone CA, eds. Advances in carriers and adjuvants for veterinary biologics. Ames, IA: Iowa State University, 1986.

46. Afzal M, Tengerdy RP, Ellis RP, Kimberling CV, Morris CJ. Protection of rams against epididymitis by a *B. ovis*-vitamin E adjuvant vaccine. Vet Immunol Immunopathol 1984; 7:293-304.

47. Tengerdy RP, Ameghino E, Riemann H. Serological responses of rams to a Brucella ovis-vitamin E adjuvant vaccine. Vaccine 1991; 9:273-276.

48. Lacetera NG, Tengerdy RP. Preliminary study on vitamin E adjuvant vaccine against *E. coli* infection in chicken. Chim Oggi 1989; 7:11-2.

49. Baehner RL, Boxer LA, Allen JM, Davis J. Autooxidation as a basis for altered function by polymorphonuclear leukocytes. Blood 1977; 50:327-35.

50. Engle WA, Yoder MC, Baurley JL, Yu P. Vitamin E decreases superoxide anion production by polymorphonuclear leukocytes. Pediatr Res 1988; 23:245-8.

51. Boxer LA, Oliver JM, Spielberg SP, Allen JM, Schulman JD. Protection of granulocytes by vitamin E in glutathione synthetase deficiency. N Engl J Med 1979; 301:901-5.

52. Yamada O, Moldow CF, Sachs T, Craddock PR, Boogaerts MA, Jacob HS. Deleterious effects of entoxin on cultured endothelial cells. Inflammation 1981; 5:115-9.

53. Boogaerts MA, Van De Broeck J, Deckmyn H, Roellant C, Vermylen J, Verwilghen RL. Protective effect of vitamin E on immune triggered granulocyte mediated endothelial injury. Thromb Haemost 1984; 51:89-92.

54. Triau JE, Meydani SN, Schaefer EJ. Oxidized low density lipoproteins stimulate prostacyclin production by adult human vascular endothelial cells. Arteriosclerosis 1988; 8:810-818.

55. Leb L, Beatson P, Fortier N, Newburger PE, Snyder LM. Modulation of mononuclear phagocyte cytotoxicity by alpha-tocopherol (vitamin E). J Leuk Biol 1985; 37:449-59.

56. Prasad JS. Effect of vitamin E supplementation on leukocyte function. Am J Clin Nutr 1980; 33:606-8.

57. Ziemlanski S, Wartanowicz M, Klos A, Raczka A, Klos M. The effects of ascorbic acid and alpha-tocopherol supplementation on serum proteins and immunoblobulin concentrations in the elderly. Nutr Int 1986; 2:1-5.

58. Harman D, Miller RW. Effect of vitamin E on the immune response to influenza virus vaccine and incidence of infectious disease in man. Age 1986; 9:21-3.

59. Meydani SN, Barklund MP, Liu S, et al. Vitamin E supplementation enhances cell-mediated immunity in healthy elderly subjects. Am J Clin Nutr 1990; 52:557-63.

60. Wayne SJ, Rhyne FL, Garry PJ, Goodwin JS. Cell-mediated immunity as a predictor of morbidity and mortality in the aged. J Gerontol Med Sci 1990; 45:M45-8.

61. Cannon JG, Meydani SN, Fielding RA, et al. The acute phase response in exercise. II. Associations between vitamin E, cytokines and muscle proteolysis. Am J Physiol 1991; 24:R1235-R1240.

62. Cannon JG, Orencole SF, Fielding RA, et al. Acute phase response in exercise: interaction of age and vitamin E on neutrophils and muscle enzyme release. Am J Physiol 1990; 259:R1214-9.

63. Likoff RO, Mathias MM, Neckles CP. Vitamin E enhancement of immunity mediated by the prostaglandins. Fed Proc 1978; 37:829.

64. Tengerdy RP, heinzerling RH, Mathias MM. Effect of vitamin E on disease resistance and immune responses. In: de Duve C, Hayaishi O, eds. Tocopherol, oxygen, and biomembranes. Amsterdam: Elsevier/North-Holland, 1978; 191-200.

65. Swartz SL, Willett WC, Hennekens CH. A randomized trial of the effect of vitamin E on plasma prostacyclin (6-keto-PGF$_1$.) levels in healthy adults. Prostaglandins Leuk Med 1985; 18:105-11.

66. Hope WC, Dalton C, Machlin LJ, Filipski RJ, Vane FM. Influence of dietary vitamin E on prostaglandin synthesis in rat blood. Prostaglandins 1975; 10:557-61.

67. Machlin L. Vitamin E and prostaglandins. In: de Duve C, Hayaishi O, eds. Tocopherol, oxygen, and biomembranes. Amsterdam: Elsevier/North-Holland, 1978; 179-89.

68. Lauritsen K, Laursen LS, Bukhave K, Rask-Madsen J. Does vitamin E supplementation modulate in vivo arachidonate metabolism in human inflammation? Pharmacol Toxicol 1987; 61:246-9.

69. Goetzel EJ. Vitamin E modulates lipoxygenation of arachidonic acid in leukocytes. Nature 1980; 288:183-7.

70. Mowat AG, Baum J. Chemotaxis of polymorphonuclear leukocytes from patients with diabetes mellitus. N Engl J Med 1971; 284:621-7.

71. Hill HR, Augustine NH, Rallison ML, Santos JI. Defective monocyte chemotactic responses in diabetes mellitus. J Clin Immunol 1974; 3:70-7.

72. Kitahara M, Eyre HJ, Lynch RE, Rallison ML, Hill HR. Metabolic activity of diabetic monocytes. Diabetes 1980; 29:251-256.

73. Taccone-Gallucci M, Giardini O, Ausiello C, et al. Vitamin E supplementation in hemodialysis patients: effects on peripheral blood mononuclear cells lipid peroxidation and immune response. Clin Nephol 1986; 25:81-6.

74. Kayden HJ, Hatam L, Traber MG, Conklyn M, Liebes LF, Silber R. Reduced tocopherol content of B cells from patients with chronic lymphocytic leukemia. Blood 1984; 63:213-5.

75. Spielberg SP, Boxer LA, Oliver JM, Allen JH, Schulman JD. Oxidative damage to neutrophils in glutathione synthetase deficiency. Br J Haematol 1979; 42:215-23.

76. Boxer LA, Harris RE, Baehner RL. Regulation of membrane peroxidation in health and disease. Pediatrics 1979; 61(1):S713-8.

77. Oliver J. Reduced glutathine in specific cellular processes. Ann Intern Med 1980; 93:337-40.

78. Goodwin JS, Garry TJ. Relationship between megadose vitamin supplementation and immunological function in a healthy elderly population. Clin Exp Immunol 1983; 51:627-53.

79. Chavance M, Brubacher G, Herberth B, et al. Immunological and nutritional status among the elderly. In: Chandra RK, ed. Nutrition, immunity, and illness in the elderly. New York: Pergamon Press, 1985; 137-42.

80. Payette H, Rola-Pleszczynski M, Ghadirian P. Nutrition factors in relation to cellular and regulatory immune variables in a free-living elderly population. Am J Clin Nutr 1990; 52:927-32.

81. Nagel JE, Chopra RK, Chrest FJ, et al. Decreased proliferation, interleukin 2 synthesis, and interleukin 2 receptor expression are accompanied by decreased mRNA expression in phytohemagglutinin-stimulated cells from elderly donors. J Clin Invest 1988; 81:1096-102.

82. Salonen JT, Salonen R, Lappetelainen R, Maenpaa PH, Alfthan G, Puska P. Risk of cancer in relation to serum concentrations of selenium and vitamin A and E: Matched case-control analysis of prospective data. Br Med J [Clin Res] 1985; 290(6466):417-20.

83. Stahelin HB, Rosel F, Buess E, Brubacher G. Cancer, vitamins and plasma lipids: prospective Basel study. J Natl Cancer Inst 1984; 73:1463-8.

84. Kok FJ, Van Duijn C, Hofman A, Vermeeren R, de Bruijn AM, Valkenburg HA. Micronutrients and the risk of lung cancer. N Engl J Med 1987; 316:1416.

85. Lopez-S A, Le Gardeur BY, Johnson WD. Vitamins and lung cancer. Am J Clin Nutr 1985; 41: 854.

41

Human Studies of Vitamin E and Rheumatic Inflammatory Disease

Gunter Blankenhorn

R. P. Scherer GmbH, Eberbach, Germany

Sabine Clewing

Brenner-Efeka Pharma GmbH, Münster, Germany

INTRODUCTION

Scientific research has established vitamin E, α-tocopherol, as the major lipid-soluble, chain-breaking, natural antioxidant in tissues protecting cells against free-radical damage (1).

When given orally vitamin E is absorbed through gastrointestinal and lymphatic pathways as a part of a lipoprotein complex (2). Vitamin E is well tolerated and safe, being neither mutagenic nor carcinogenic nor teratogenic (3).

Tests of biological availability and biopotency demonstrated α-tocopherol to be the most potent isomer of the tocopherol family (4–6). In human blood $2R,4R,8R$-α-tocopherol, the natural stereoisomer of eight possible α-tocopherols, was determined as the major factor with radical-trapping ability (7,8).

New findings indicate that free radicals may be responsible for or are at least one of the promoting factors of chronic multiple inflammation, such as rheumatoid arthritis (9–14), suggesting that vitamin E may be a promising new medication for the treatment or cotreatment of rheumatic disease.

Because of those findings, clinical studies aimed at demonstrating the antiphlogistic efficacy of vitamin E must be based on RRR-α-tocopherol derivatives. The free tocopherol is quite vulnerable to autoxidation, but its esters, in particular α-tocopherol acetate, are stable toward oxygen (15–17). In the gut, α-tocopherol acetate is hydrolized, yielding native tocopherol before absorption. Hence, mostly RRR-α-tocopherol acetate has been used in the clinical studies described here.

VITAMIN E STATUS IN PATIENTS WITH RHEUMATIC DISEASE

Reactive oxygen species that are released either by triggering the arachidonic acid cascade or by phagocytosis are a characteristic part of the inflammatory process (18).

During phagocytosis the activation of a specific antigen-antibody complex leads to an increase in oxygen consumption that is well known as "respiratory burst" (19). As a consequence, reactive oxygen species, such as superoxide radicals, peroxide, and hydroxyl radicals, are generated and are potentially the cause of damage to responsible cells and tissues.

It is an important question whether chronic inflammatory disease is associated with decreased levels of vitamin E in the serum of patients.

Honkanen et al. demonstrated that patients suffering from rheumatoid arthritis showed significantly lower vitamin E serum levels than healthy subjects (20). A correlation between the vitamin E levels and the status of the disease could not be established.

A further study compared the serum levels of patients with juvenile chronic arthritis to those of healthy probands of about the same age (21). A significant decrease in vitamin E serum levels ($p < 0.001$) was shown in the group of the arthritic patients.

Hence, there is good evidence suggesting that vitamin E supplementation may be useful to regenerate the antioxidant protection in cells by means of increasing reduced vitamin E serum levels.

PILOT STUDIES

Among a number of diseases that might be treated with vitamin E, we focused on those related to inflammation. To our knowledge, the first clinical studies with vitamin E in the therapy of inflammatory diseases were published by Jensen and Spuler in 1962 (22). In this study patients suffering from different spinal syndromes were treated with 400 mg vitamin E per day over a period of 6–8 weeks. The clinical symptoms treated were upper cervical syndromes, lower cervical, lumbar syndromes, thoracal syndromes, and ischialgia. A total of 242 patients took part in the study; 62 were suffering from a specific syndrome, and the other 180 patients were suffering from mixed syndromes. In 58% of the cases a beneficial effect could be achieved in those patients suffering from only one of the syndromes, whereas 93% of the cases with combinated syndromes responded positively. For the statistical analysis the cases were divided into four subgroups matched to the predominant syndrome. Figure 1 summarizes the results.

The upward bars represent the responder group to the treatment with vitamin E. The downward bars symbolize those patients who did not show any obvious improvement.

These positive results were supported by a further study by Oloffs in 1977 with patients suffering from recurrent spinal syndromes (23). In this study vitamin E was given in combination with the neurotropic vitamins B_1, B_6, and B_{12} for more than 8 weeks. In 13 of 20 cases there was a considerable improvement. The patients showed prolonged intervals free from pain and without recurrence.

In a multicentric study including 60 patients suffering from degenerative spondylopathies and arthropathies as well as extraarticular rheumatic diseases, Fischer and Seuss (24) studied the effect of vitamin E medication. Of the doctors involved 62% reported the therapeutic result as "good" and 22% as "satisfactory" after a period of 5 weeks. Such criteria as "pain," "mobility," and "morning stiffness" were compared. The concomitant medication with analgesics and nonsteroidal antirheumatics could be significantly reduced during treatment. In the beginning 73% of the patients had to use additional medication; at the end of the study this was decreased to 35%. Patients needing corticosteroids showed even better improvement: at the beginning of the study 10% had to be given corticosteroid drugs; after 6 weeks only 3% still needed corticosteroids in addition to vitamin E (Fig. 2).

A long-term study over a period of 3 years was published by Klein and Kaufhold (25). Patients suffering from morbus Bechterew were treated with a daily dose of 400 mg d-α-tocopherol acetate. The parameters studied were "pain," "function," "number and degree of severity of acute attacks," and "additional need of analgesics." After 52 weeks the average duration of morning stiffness could be reduced by 50%. The clinical results could be

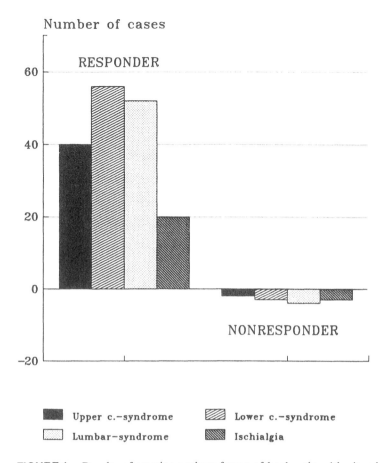

FIGURE 1   Results of treating various forms of back pain with vitamin E.

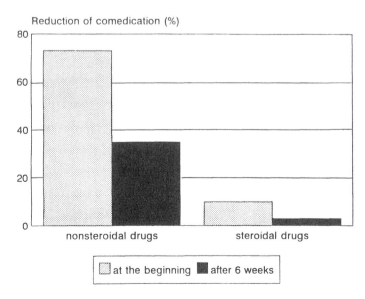

FIGURE 2   Need of additional treatment with nonsteroidal or steroidal drugs during therapy with vitamin E.

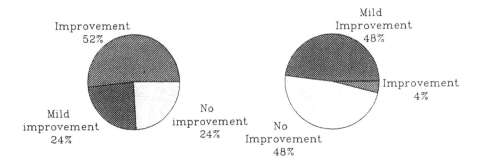

Vitamin E   Placebo

FIGURE 3   Comparative presentation for treating osteoarthritis with vitamin E or placebo.

correlated with the radiological results. This study also indicates the efficacy of vitamin E in the therapy of inflammatory diseases.

PLACEBO-CONTROLLED STUDIES

The successful use of vitamin E in pilot studies suggested that placebo-controlled studies should be conducted to verify efficacy. This was particularly important because medical treatment of arthritic pain may be associated with a very significant placebo effect of up to 50% (26).

In a study on the efficacy of mefenamic acid in rheumatic diseases, vitamin E was used as placebo (27). Surprisingly, the results indicated that vitamin E may also be effective. A

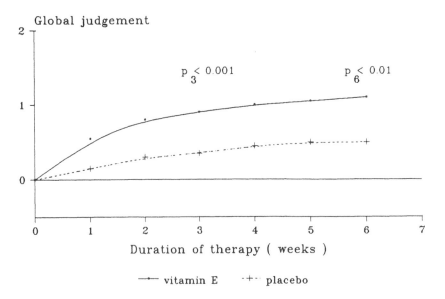

FIGURE 4   General judgment by the doctor (status of disease: 0 = unchanged, 1 = improved, 2 = free of pain).

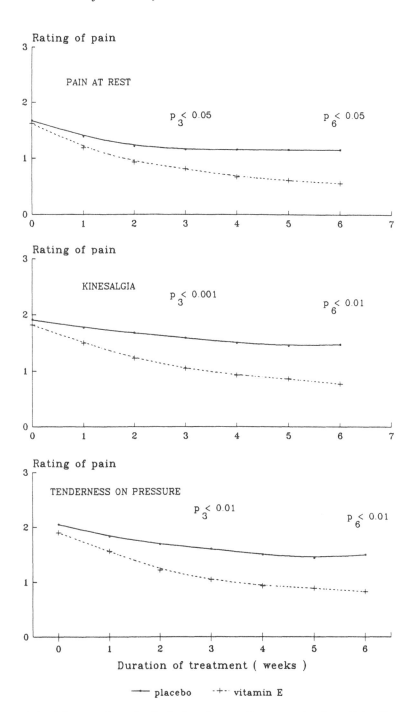

FIGURE 5   Rating of pain during therapy (0 = no pain, 1 = slight, 2 = severe, 3 = very severe).

further single-blind crossover study with 32 patients suffering from osteoarthritis followed. The results of treatment with α-tocopherol versus placebo are summarized in Figure 3.

A significant improvement in symptoms ($p < 0.01$) was observed in 51.7% of the cases and a minor improvement in 24.2%. Thus, it can be concluded that the osteoarthritis improved in 75.9% of the cases ($p = 0.05$) with treatment with vitamin E (28).

In 1986 Blankenhorn published the first placebo-controlled double-blind study on the efficacy of vitamin E in patients suffering from acute arthritic pain. This multicentric randomized study with 50 patients showed that vitamin E medication was effective in osteoarthritis, coxarthritis, and omarthritis, as well as arthritis of wrist, ankle, and toe joints. These results are probably the first clinical proof for the antiphlogistic efficacy of vitamin E in rheumatic inflammatory processes (29).

Increased steady-state blood levels of vitamin E were achieved by a dose of 1200 IU *d-α*-tocopherol acetate per day, which was reduced to 800 IU/day after a period of 5 days. The efficacy of vitamin E medication is shown in Figure 4.

Pain intensity was assessed by means of a four-stage scale. Moreover, the pain was differentiated between "pain at rest," "pain during movement," and "tenderness on pressure." Mobility was improved in the vitamin E group compared to the placebo group. Additionally, the onset of relief of pain was faster with vitamin E than with placebo.

Figure 5 shows the changes in pain ratings during treatment. Statistical significance is included in each graph. The relief of pain is also demonstrated by a reduction in additional analgetic drug therapy (Fig. 6).

That 25% of the analgetics could also be reduced in the placebo group underscores the placebo effect already mentioned. However, the reduction in additional analgesic therapy in the verum group is 50%, a statistically significant difference from the placebo group.

In the medical treatment of chronic inflammatory diseases, in particular when using steroidal or nonsteroidal antiinflammatory drugs, consideration of adverse side effects may limit possible choices. Hence, this question was addressed in the present study.

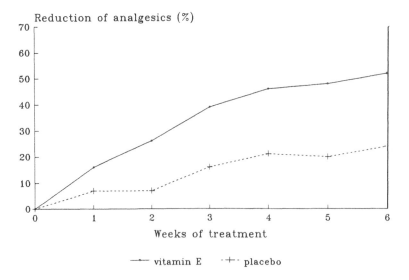

FIGURE 6   Reduced need for analgesics during treatment with vitamin E.

As it turns out, one of the remarkable properties of vitamin E is its low incidence of side effects. Adverse reactions, such as nausea, emesia, stomach ache, hemorrhagic stool, dizziness, and sleepiness, occurred in 7.1% of the cases under treatment with vitamin E compared to 10.5% when placebo was applied. No significant gastrointestinal side effects were observed in either group. Because both steroidal and nonsteroidal antiinflammatory drugs used in the treatment of rheumatic disease show a high incidence of gastrointestinal side effects, this is a very important result.

It appears that vitamin E side effects at the doses applied do not significantly differ from those of placebo. Hence vitamin E should be particularly useful in chronic treatment but also for patients who do not tolerate steroidal or nonsteroidal antiinflammatory drugs.

STUDIES VERSUS ANTIINFLAMMATORY DRUGS

The evidence derived from pilot studies and placebo-controlled studies suggested that the efficacy of vitamin E should also be tested versus standard antiinflammatory drugs. Since diclofenac sodium is most widely used in Europe, it was selected as standard in double-blind controlled studies versus vitamin E.

Gonarthritis

In a double-blind prospective clinical trial the antiphlogistic and analgesic effects of vitamin E (300 IU/per day), diclofenac sodium (75 mg daily), and a combination of both drugs were compared (30). To confirm the effectiveness of the therapy, several parameters, such as "articular swelling," "pain," and "Keitel's function test," were determined. All parameters were significantly improved during therapy with both drugs. There were no significant differences. Treatment with only one of the drugs compared to the combination showed a beneficial effect of the combination. However, statistical significance could not be demonstrated. Figure 7 shows the change during therapy as a function improvement of the sum of scores.

Morbus Bechterew

A further clinical study investigated the efficacy of vitamin E in the treatment of the stages I and II of spondylitis ankylosans (31). Vitamin E therapy in the inflammatory phases of morbus Bechterew was compared to antiphlogistic therapy with diclofenac sodium.

A group of 24 patients were treated with either 544 IU vitamin E per day or with 50 mg diclofenac sodium per day for 12 weeks. Test criteria were extent of bending measured by the distance of hands from the floor, morning stiffness, pain, general condition, tolerance, and side effects, as well as various laboratory parameters. Statistical analysis showed that there was no significant difference between both medical treatments with regard to the parameters pain, morning stiffness, and general condition. In addition, the fingers to floor distance improved significantly after 6 weeks under treatment with both drugs. After a period of 12 weeks there was a further improvement by diclofenac sodium. This could not be shown for vitamin E. The patients in the diclofenac group complained about gastrointestinal side effects, and 1 patient had to drop out for this reason.

Interestingly, a significant improvement in the α_2-globulin values could be shown for the vitamin E group. Because an increase in this serum protein fraction is regarded as the marker for inflammatory reactions, a reduction in the pathologically increased values can be considered further proof of the antiphlogistic activity of vitamin E.

Because no satisfactory medical treatment has been found for morbus Bechterew, vitamin E in association with manual and physical treatment seems to offer a safer regimen than

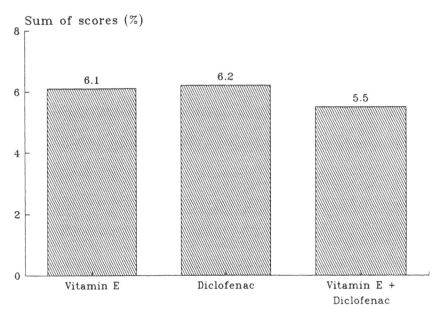

FIGURE 7   Average improvement in Keitel's function test during treatment for osteoarthritis of the knees.

diclofenac sodium. Unfortunately, treatment offers only help in pain relief; it does not impede progression of the disease.

Rheumatoid Arthritis

Finally we report the results of three double-blind controlled three-center clinical studies with patients suffering from rheumatoid arthritis and osteoarthritis, respectively (32–34). Patients were carefully selected according to a number of criteria. Only those patients entered the study who complained of morning stiffness lasting more than 45 minutes and whose blood sedimentation rate was about 28 mm/hr. Additionally, they had to be sensitive to pressure at a minimum of six joints and had to have swelling at a minimum of three joints. Patients with gastrointestinal disorder or serious heart, kidney, or liver damage were not included or patients with allergies toward antirheumatic drugs. Furthermore, pregnant women and patients treated with vitamin K antagonists were excluded. Ongoing basic therapy would be maintained nor would basic therapy be allowed within 4 months before the start of the study.

A period of 5–7 days was chosen as the washout phase during which additional medication with paracetamol was allowed. In the study by Kolarz et al., the patients received double-blind three times per day (32) 400 mg *d*-α-tocopherol acetate or 50 mg diclofenac sodium. Laboratory data as well as clinical data, such as morning stiffness and Ritchie index, were monitored carefully. Analysis of the clinical parameters resulted in a significant improvement in both groups. A significant difference between the diclofenac and the vitamin E group could not be observed. As an example of the many parameters analyzed, Figure 8 shows a comparison of morning stiffness, swelling, Ritchie index, and pain for post-washout, after 1 week, and at the end of therapy. The data for the postwashout phase were set to 100%. The relative changes in the parameters are expressed as percentages of these values.

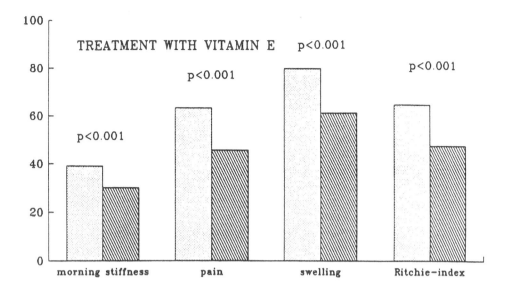

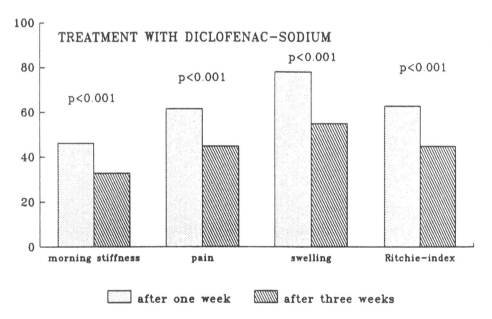

FIGURE 8 Follow-up of clinical parameters during treatment (initial value = 100%).

This study had to be limited to 3 weeks of treatment because patients left the hospital after this period. Although further studies over extended time periods will have to be carried out, these initial results are very encouraging. In a short time treatment vitamin E showed efficacy comparable relative to diclofenac sodium in patients with rheumatoid arthritis. Since vitamin E has fewer side effects, it appears to be a genuine relative to standard antiinflammatory drugs.

Osteoarthritis

A number of studies have been conducted assessing the benefit of vitamin E therapy in osteo-arthritis.

A double-blind study was performed by Scherak et al. with 53 patients suffering from osteoarthritis of hip and knee joints (33). The patients were treated with 400 mg *d*-*α*-tocopherol acetate or 50 mg diclofenac sodium three times per day. As described for the studies already mentioned, a number of clinical and laboratory data were gathered. During the test any additional local therapy was prohibited and standard physical therapy was performed. One patient in the vitamin E group dropped out because the therapy was ineffective. However, the results with the other patients were very successful. Upon evaluation of the side effects vitamin E showed a clear superiority. The efficacy of both drugs was comparable. The onset of action was quicker for diclofenac, as expected. Some of the clinical data summarizing the results of this study are presented in Figure 9.

Link compared the effect of vitamin E versus diclofenac sodium in patients suffering from gonarthritis and coxarthritis (34). Both drugs were compared with regard to such clinical parameters as pain, movability and circumference of a joint, joint swelling, and Keitel's function test. Laboratory parameters were also monitored but did not provide any unexpected results.

Both groups showed a highly significant improvement in the clinical data. Some parameters favored vitamin E over diclofenac, but the difference was not statistically significant. These were the hip joint movability and the examination of the knee joint. However, with regard to Keitel's function test, the difference was significant ($p < 0.001$). Because of the relatively small number of patients ($n = 30$), further long-term studies must be carried out to evaluate vitamin E in the therapy of osteoarthritis.

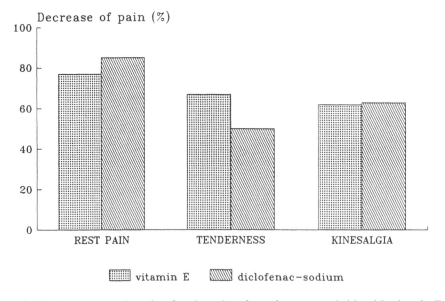

FIGURE 9   Decrease in pain after 3 weeks of treating osteoarthritis with vitamin E and diclofenac sodium.

Side Effects and Benefit-Risk Evaluation

Summarizing our own studies, we found that daily doses of up to 1200 IU were well tolerated. Specific vitamin E side effects were not observed and typical side effects observed with placebo were also seen with vitamin E (for a more detailed discussion see Ref. 35). Although standard antiinflammatory drugs, such as diclofenac sodium, appear to be marginally more effective than vitamin E, they suffer from the disadvantage of having a higher incidence of serious, specific side effects. Hence, medical therapy of inflammatory rheumatic disease should include serious consideration of using vitamin E in daily doses varying between 400 and 1200 IU. Our results indicate that a number of patients do not respond to vitamin E therapy. In these instances alternative antiinflammatory medication is mandatory.

Finally, vitamin E does not solve the principal problem that causal medical therapy is not available for rheumatic disease at the present time. Successful treatment depends on a number of important factors, which complement each other: most importantly a well-established doctor-patient relationship guarantees a maximal response by the patient (a specific form of placebo effect). In addition, physical therapy (specific exercises) and radiation (infrared and laser) therapy may assist the success achieved with proper medication.

In view of the reported results, vitamin E should not be accepted as promising medication because its established efficacy coupled with its low incidence of side effects makes it a very promising antiinflammatory drug.

Efficacy of *d*-α-Tocopherol Versus *d,l*-α-Tocopherol

In view of the commercial availability of stereoisomers, namely *d*-α- and *l*-α-tocopherol, and their derivatives, it is important to discuss their differing efficacy if in fact there is any difference. In our view the antioxidant activity of the stereoisomers does not differ in homogeneous solution. When bound to biological cell membranes, a completely different situation might evolve.

The interaction of the phytyl side chain of tocopherol with the lipid phase of cell membranes depends on steric factors. Using molecular models it appears that *d*-α-tocopherol may interact in a way that leads to tighter binding to the cell membrane than *l*-α-tocopherol. In other words, *d*-α-tocopherol may partition to a higher extent into the lipid membrane relative to blood plasma or intercellular fluid.

When considering the effectiveness of protecting the membrane against oxygen-induced damage, a major factor must be orientation and flexibility of the phenolic vitamin E part relative to the membrane/water interface of biological cells. A high degree of flexibility ensures optimal orientation toward oxygen radicals released by inflammatory processes into the intercellular space.

We propose that *d*-α-tocopherol may be bound more tightly to the cell membrane than its *l*-α-counterpart. In addition, better flexibility exposure and orientation in the membrane/water interface may be the cause for improved reactivity. It would be very interesting to test these proposals in model systems using liposomes or vesicles in connection with an aqueous superoxide generating system in the presence of varying concentrations of *d*- and *d,l*-α-tocopherol.

Schmidt and Hemmerich (36,37) have been able to demonstrate that the reactivity of amphiphilic flavins bound to artificial membrane vesicles strongly depends on their accessibility in the lipid/water interface. Similar results allowing differentiation of tocopherol stereoisomers and enantiomers, respectively, can be expected from experiments measuring the reactivity of membrane-bound tocopherols with oxygen radicals generated in the aqueous phase.

SUMMARY

Vitamin E has been tested successfully in a series of studies using rats and rabbits suffering from inflammatory diseases (38,39). Because of the antioxidative and lipid-binding properties of vitamin E, two mechanisms for the action of tocopherol can be discussed:

1. As a radical scavenger vitamin E inhibits lipid peroxidation and represses the release of inflammatory mediators. In addition, the oxygen and hydroxyl radicals with their direct cytotoxic effects are inactivated.
2. The biosynthesis of prostaglandines is inhibited by interactions between vitamin E and membrane-bound phospholipids. Thus, tocopherol may act as a weak inhibitor of lipooxygenase. The release of membrane-bound arachidonic acid would be prevented, interfering with the metabolism of further eicosanoids.

The studies discussed in this review provide evidence for the efficacy of vitamin E in the treatment of inflammatory diseases. In view of the excellent side effect profile, in which vitamin E is similar to placebo, further clinical studies, in particular long-term studies over periods of up to 2 years, are now necessary. If the clinical results presented in this chapter can be substantiated, vitamin E will become standard medical treatment in rheumatic inflammatory disease.

ACKNOWLEDGMENTS

Thanks are due to Dr. G. Scriba for critically reading the manuscript and for his valuable comments. Appreciation is also expressed to Mrs. N. Mührmann for her skillfull scretarial work.

REFERENCES

1. Burton G, Joyce A, Ingold KU. First proof that vitamin E is the major lipid-soluble, chain breaking antioxidant in human blood plasma. Lancet 1982; 2:327.
2. Gallo-Torres HE. Absorption. In: Machlin LJ, ed. Vitamin E—a comprehensive treatise, 1st ed. New York: Basel, 1980; 170-92.
3. Bendich A, Machlin LJ. Safety of oral intake of vitamin E. Am J Clin Nutr 1988; 48:612-9.
4. Century B, Horwitt MK. Biological availability of tocopherols. Fed Proc 1965; 24:906-11.
5. Horwitt MK, Ellioh WH, Kanjananggulpan PH, Fitch CD. Serum concentrations of α-tocopherol after investigation of various vitamin E preparations. Am J Clin Nutr 1984; 40(2):240-5.
6. Weiser H, Vecchi M. Stereoisomers of α-tocopheryl acetate. Int J Vitam Nutr Res 1982; 52:351-70.
7. Ingold KU, Burton GW, Foster DO, Hughes L, Lindsay DA, Webb A. Biokinetics of and discrimination between dietary *RRR*- and *SRR*-α-tocopherols in the male rat. Lipids 1987; 22(3):163-72.
8. Horwitt MK. Relative biological values of *d*-α-tocopheryl acetate and *all-rac*-α-tocopheryl acetate in man. Am J Clin Nutr 1980; 33:1856-60.
9. Flohe L, Beckmann R, Giertz H, Loschen G. Oxygen-centered free radicals as mediators of inflammation. In: Sies H, ed. Oxidative stress, 1st ed. London: 1985; 403-28.
10. Draht DB, Karnowsky ML. Superoxide production by phagocytic leukocytes. J Exp Med 1975; 141:257-63.
11. Fridovich I. The biology of oxygen radicals. Science 1978; 201:875-80.
12. McCord JM, Wong K. Phagocyte-producted free radicals: roles in cytotoxicity and inflammation. In: Oxygen free radicals and tissue damage. Ciba Foundation Symposium 65. Amsterdam: Excerpta Medica, 1979; 343-60.
13. Kappus H. Lipid peroxidation: mechanismus, analysis, enzymology and biological relevance. In: Sies H, ed. Oxidative stress, 1st ed. London: 1985; 273-303.

14. Blankenhorn G, Fischer I, Seuss J. Möglichkeiten und Grenzen der Antioxidans-Therapie entzündlicher rheumatischer Erkrankungen mit hochdosiertem Vitamin E. Akt Rheumatol 1985; 10:125-8.

15. Burton GW, Traber MG. Vitamin E: antioxidant activity, biokinetics and bioavailability. Annu Rev Nutr 1990; 10:357-82.

16. Harris PL, Ludwig MI. Relative vitamin E potency of natural and of synthetic α-tocopherol. Biol Chem 1949; 179:1111.

17. Harris PL, Ludwig MI. Vitamin E potency of α-tocopherol and α-tocopherol esters. Biol Chem 1949; 180:611.

18. Meyer-Franzen P, Kaufhold M. Wirkungsweise von Vitamin E in der Behandlung rheumatischer Erkrankungen. Jahrbuch der Orthopädie Biermann Verlag FRG, 1989:185-90.

19. Hoffstein ST. Intra- and extracellular secretion from polymorphonuclear leukocytes. In: Weissmann G, ed. The cell biology of inflammation, 2nd ed. Amsterdam: Elsevier/North-Holland, Oxford, 1980; 387-430.

20. Honkanen V, Konttinen YT, Mussalo-Rauhamaa H. Vitamins A and E retinol binding protein and zinc in rheumatoid arthritis. Clin Exp Rheumatol 1989; 7:465-9.

21. Honkanen V, Pelkonen P, Konttinen YT, Mussalo-Rauhamaa H, Lehto J, Coestermarck T. Serum cholesterol and vitamins A and E in juvenile chronic arthritis. Clin Exp Rheumatol 1990; 8:187-91.

22. Jensen HP, Spuler H. Neuralgien und Vitamin E. Kapsel 1962; 2.

23. Oloffs J. Zur Therapie rezidivierender Wirbelsäulensyndrome mit hochdosiertem Vitamin E. Therapiewoche 1977; 27:1941-3.

24. Fischer J, Seuss J. Antioxidans-therapierheumatische Erkrankungen. Heilkunst 1985; 3:145-8.

25. Klein KG, Kaufhold M. Wirksamkeit und Verträglichkeit von Vitamin E. Therapiewoche 1990; 40:1414-21.

26. Fricke U. Placebo—ein Aspekt der Pharmakotherapie. Med Mol Pharmacol 1983; 6:356.

27. Machtey I. Mefenamic acid in osteoarthritis. Harefuah 1969; 76:280.

28. Machtey I, Ouaknine L. Tocoperol in osteoarthritis. J Am Geriatr Soc 1978; 26:328.

29. Blankenhorn G. Klinische Wirksamkeit von Spondyvit (Vitamin E) bei aktivierten Arthrosen Z Orthop 1986; 124:340.

30. Bartsch M, Bartsch H, Toloczyki Ch. Behandlung von entzündlichen Gonarthrosen. Therapiewoche 1989; 39:1839-45.

31. Klein KG, Blankenhorn G. Vergleich der Klinischen Wirksamkeit von Vitamin E und Diclofenac-Natrium bei Spondylitis ankylosans (Morbus Bechterew). VitaMinSpur 1987; 3:137-42.

32. Kolarz G, Scherak O, Shohoumi MEL, Blankenhorn G. Hochdosiertes Vitamin E bei chronischer Polyarthritis. Akt Rheumatol 1990; 15:233-7.

33. Scherak O, Kolarz G, Schödl C, Blankenhorn G. Hochdosierte Vitamin E Therapie bei Patienten mit aktivierter Arthrose. Z Rheumatol 1990; 49:369-73.

34. Link P, Dreher R. D-α-Tocopherolacetat (Vitamin E) versus Diclofenac-Na in der Therapie der aktivierten Arthrose. Dtsch Ärztemagazin 1990; 22:48-52.

35. Blankenhorn G, Fischer I. Vitamin E (α-Tocopherol) als Arzneimittel—eine Bestandsaufnahme. Fette, Seifen, Anstrichmittel 1985; 87:577-9.

36. Schmidt W, Hemmerich P. On the redox reactions and accessibility of amphiphilic flavins in artificial membrane vesicles. J Membr Biol 1981; 60:129-40.

37. Schmidt W, Hemmerich P. Further photophysical and photochemical characterization of flavins associated with single-shelled vesicles. J Membr Biol 1983; 76:73-82.

38. Yoshikawa T, Tanaka H, Kondo M. Effect of vitamin E on adjuvant arthritis in rats. Biochem Med 1983; 29:227-9.

39. Kamimura M. Antiinflammatory activity of vitamin E. J Vitaminol 1972; 18:204-9.

42

Free Radicals and Antioxidants in Multiple-Organ Failure

Toshikazu Yoshikawa, Hirohisa Takano, and Motoharu Kondo

Kyoto Prefectural University of Medicine, Kyoto, Japan

INTRODUCTION

Significant attention is paid to multiple organ failure (MOF), because it is a serious condition that often causes death of patients in intensive care units and because it can be controlled by appropriate treatment. MOF is defined as a functional condition in which multiple organs or systems rapidly fall into dysfunction. MOF sometimes results in the pathological damage of multiple organs and finally in death of the patients. Typical symptoms of MOF involve jaundice, oliguria, respiratory disturbance, such as dyspnea, gastrointestinal bleeding, consciousness disturbance, bleeding tendency, hypotension, and tachycardia, and its target organs or systems often include the liver, kidney, lung, gastrointestinal tract, central nervous system, coagulation system, and cardiovascular system. MOF is induced by various types of shock, adult respiratory distress syndrome (ARDS), disseminated intravascular coagulation (DIC), serious inflammation or infections, including peritonitis and sepsis, pancreatitis, severe trauma, and surgical operations. Among these pathophysiological causes of MOF, attention is focused in this chapter on free-radical involvement and the role of antioxidants in shock, ARDS, DIC, and peritoneal inflammation with special reference to endotoxin-induced models.

LIPID PEROXIDATION AND ANTIOXIDANTS IN SHOCK

Attention has been focused on the role of free radicals, phagocytes as their source, and lipid peroxidation they induce in the pathogenesis of various types of shock (1–3). The participation of free radicals in shock has been investigated using an experimental endotoxin shock model in rats (4,5). In endotoxin-treated rats, increased lipid peroxidation is observed not only in the serum but also in multiple organs, such as aortic wall, gray matter of the brain, liver, stomach, and small intestine. An increase in lipid peroxidation in the serum precedes that of serum lysosomal enzymes, including acid phosphatase, β-glucuronidase, and cathepsin B, which are some of the indices suggesting the severity of shock state. Lipid peroxidation may play a role in lysosomal enzyme release. The vitamin E/cholesterol ratio in the serum of our experimental model demonstrates the gradual decrease, and the reverse is true of the serum lipid peroxide. Judging from these results, lipid peroxidation should be involved in the genesis and aggravation of endotoxin shock, and vitamin E is suggested to be exhausted to attenuate the accelerated lipid peroxidation.

FREE-RADICAL INVOLVEMENT AND ROLE OF ANTIOXIDANTS IN ENDOTOXIN-INDUCED DAMAGE

In our experimental endotoxin shock model, superoxide and the other oxygen-derived free radicals are suggested to participate in its pathogenesis, aggravation, and lipid peroxidation. Superoxide dismutase (SOD) and catalase attenuate the shock state, including hypotension, tachycardia, and increased serum lysosomal enzymes, as well as increased serum lipid peroxide. SOD or catalase can also inhibit the progress of lipid peroxidation in organs, which suggests the involvement of oxygen-derived free radicals in tissue lipid peroxidation. According to the literature, the vitamin E level significantly decreases not only in the serum but also in the liver of endotoxin-treated mice, but vitamin E-deficient mice exhibit a higher formation of lipid peroxide after endotoxin treatment than E-sufficient mice (6). Vitamin E is thought to be exhausted in the process of protection against free-radical–induced damage as one of the important antioxidants in endotoxin-treated animals. It is also reported that vitamin E and CoQ10 improve the survival rates of endotoxin-treated mice, suppress the increased lipid peroxide level in the liver, and preserve the hepatic ATP level, which is the best index of the energy state in the normal range (7). One of the protective mechanisms of vitamin E against free-radical injury in endotoxicosis may be stabilization of lysosomal membrane against lipid peroxidation. It is reported that lysosomal membrane is destroyed by free-radical attack and its damage can be prevented by vitamin E (8). The results of these reports correspond with our experimental data that SOD or catalase can inhibit the increased lysosomal enzyme in the serum of endotoxin-treated rats.

ROLE OF PMNs AND ANTIOXIDANTS IN ENDOTOXIN-INDUCED SHOCK AND ARDS

In our experimental endotoxin shock model, polymorphonuclear leukocytes (PMNs) may be the main source of oxygen-derived free radicals because PMN depletion by anti-PMN antibody also attenuates the shock state. The number of PMNs in the blood decreases after endotoxin treatment. PMNs accumulate into the pulmonary vasculature, which is confirmed by measurement of tissue myeloperoxidase activity (9). Migrated PMNs injure the pulmonary microcirculation by oxygen-derived free radicals and the other phagocyte products, such as platelet-activating factor (PAF) and tumor necrosis factor (TNF). Both PAF and TNF can augment PMN-derived superoxide production (10,11). Our recent studies by vascular labeling using Monastral blue B show deposition of the pigment in the pulmonary microvasculature, suggesting disturbed or leaky microvessels. Endotoxin treatment can also provoke ARDS in addition to shock. SOD and catalase inhibit the deposition of Monastral blue B in the lung, exhibiting the protective effects of these antioxidative enzymes against microcirculatory injury in the lung and the participation of oxygen-derived free radicals in the pulmonary microcirculatory damage. The effects of vitamin E on ARDS have been previously reported. Vitamin E-treated patients with ARDS had a higher survival rate than control patients in a clinical trial (12). It was also reported that vitamin E-deficient rats suffer from more serious injury in the alveolar or interalveolar space than control rats (13).

FREE RADICALS AND ANTIOXIDANTS IN DIC

Our previous investigations demonstrate that sustained infusion of endotoxin induces an experimental DIC model in rats (14,15). Lipid peroxidation progresses in the serum and several organs, such as the abdominal aorta and ileum. Administration of vitamin E prevents aggravation of the DIC model, and vitamin E deficiency results in augmented injury. SOD and catalase also attenuate severity of the experimental DIC, suggesting the involvement of oxygen-derived free radicals in its pathogenesis.

The role of platelets cannot be disregarded in the genesis of intravascular coagulation. It was reported by the other investigators that vitamin E inhibits the platelet aggregation in vitro induced by collagen or hydrogen peroxide (16,17) and that vitamin E deficiency increases the activity of platelet aggregation (18). Decreased prostacyclin synthesis in the aorta and increased thromboxane A_2 synthesis in the aorta and platelet were observed in vitamin E-deficient rats by several investigators (19,20). Free radicals may play some role in intravascular coagulation, partly by modulating platelet aggregation and the prostaglandin pathway.

FREE RADICALS AND ANTIOXIDANTS IN MOF

Shock, ARDS, and DIC with a variety of causes often develop into MOF and finally to death. For instance, our experimental endotoxin shock model demonstrates hemorrhagic lesions in the lung and small intestine, and increased serum amylase and blood urea nitrogen (BUN) in its early stage, but it shows elevated GOT and creatinine levels in the late period, indicating the development from a shock model to an MOF model. The roles and effects of antioxidants in endotoxin-induced damage, including shock, ARDS, and DIC, which were described here, may also be applied to our MOF model induced by endotoxin.

Another well-known MOF model with less relation to endotoxin is created by the intraperitoneal administration of sterile zymosan in rats or mice (21,22). Also in this model, an ARDS-like change precedes the development of MOF and lipid peroxidation progress in the serum, liver, and lung. Autodestructive and self-sustaining activation of PMNs and macrophages, especially in oxygen-derived free-radical production, was suggested as responsible for its pathogenesis. The role of antioxidants in this model is now under further investigation.

REFERENCES

1. Lefer AM, Araki H, Okamatsu S. Beneficial actions of a free radical scavenger in traumatic shock and myocardial ischemia. Circ Shock 1981; 8:273-82.
2. Cunningham SK, Keaveny TV. Effect of a xanthine oxidase inhibitor on adenine nucleotide degradation in hemorrhagic shock. Eur Surg Res 1978; 10:305-13.
3. Crowell JW, Jones CE, Smith EE. Effect of allopurinol on hemorrhagic shock. Am J Physiol 1969; 216:744-8.
4. Seto O. Oxygen derived free radicals and lipid peroxidation in the pathogenesis of endotoxin shock. J Kyoto Pref Univ Med 1988; 97:1141-54.
5. Yoshikawa T, Murakami M, Seto O, et al. Effect of superoxide dismutase and catalase on endotoxin shock in rats. J Clin Biochem Nutr 1986; 1:165-70.
6. Sakaguchi O, Kanada N, Sakaguchi S, Hsu Cheng-Chin Abe H. Effect of α-tocopherol on endotoxicosis. Microbiol Immunol 1981; 25:787-99.
7. Sugino K, Dohi K, Yamada K, Kawasaki T. The role of lipid peroxidation in endotoxin-induced hepatic damage and the protective effect of antioxidants. Surgery 1987; 101:746-52.
8. Moore T, Sharman IM, Stanton MG, Dingle JT. Nutrition and lysosomal activity. Biochem J 1967; 103:923-8.
9. Yoshikawa T. Antioxidant therapy in shock and multiple organ failure. Free Radic Biol Med in press.
10. Yoshida N. Role of oxygen radicals and polymorphonuclear leukocytes in platelet activating factor-induced gastric mucosal injury in rats. J Kyoto Pref Univ Med 1989; 98:1141-51.
11. Takano H, Yoshikawa T, Ichikawa H, et al. Augmentative effects of immunomodulators and cytokines on superoxide production from polymorphonuclear leukocytes. Advances in mucosal immunology. Tsuchiya M, ed. in press.

12. Wolf HRD, Seeger HW. Experimental and clinical results in shock lung treatment with vitamin E. In: Lubin B, Machlin LJ, eds. Vitamin E: biochemical, hematological, and clinical aspects. New York: Academy of Science, 1982; 392-409.

13. Weibel ER. Oxygen effect on lung cell. Arch Intern Med 1971; 128:54-6.

14. Yoshikawa T, Murakami M, Furukawa Y, Kato H, Takemura S, Kondo M. Lipid peroxidation and experimental disseminated intravascular coagulation in rats induced by endotoxin. Thromb Haemost 1983; 49:214-6.

15. Yoshikawa T, Furukawa Y, Murakami M, Watanabe K, Kondo M. Effect of vitamin E on endotoxin-induced disseminated intravascular coagulation in rats. Thromb Haemost 1982; 48:235-7.

16. Higashi O, Kikuchi Y. Effect of vitamin E on the aggregation and the lipid peroxidation of platelets exposed to hydrogen peroxide. Tohoku J Exp Med 1974; 112:271-8.

17. Fong JSC. Alpha tocopherol. Its inhibition on human platelet aggregation. Experimentia 1976; 32:639-41.

18. Machlin L. In: deDuve C, Hayaishi O, eds. Vitamin E and prostaglandins. Amsterdam: Elsevier/North Holland, 1978; 179-89.

19. Okuma M, Takayama H, Uchino H. Generation of prostacyclin-like substance and lipid peroxidation in vitamin E-deficient rats. Prostaglandins 1980; 19:527-36.

20. Karpen CW, Merola AJ, Trewyn RW. Modulation of platelet thromboxane A_2 and arterial prostacyclin by dietary vitamin E. Prostaglandins 1981; 22:651-9.

21. Bebber IPT, Boekholz MD, Goris RJA, et al. Neutrophil function and lipid peroxidation in a rat model of multiple organ failure. J Surg Res 1989; 47:471-5.

22. Goris RJA. Multiple organ failure: whole body inflammation? Schweiz Med Wochenschr 1989; 119:347-53.

43

Vitamin E and Arthritis

Toshikazu Yoshikawa and Motoharu Kondo

Kyoto Prefectural University of Medicine, Kyoto, Japan

INTRODUCTION

Rheumatoid arthritis is a chronic disease of joints. Beginning with inflammation in the synovium, it may go into remission, run an intermittent course, or progress to a proliferative synovitis capable of erosion of bone and destruction of cartilage and tendons. However, the cause of rheumatoid arthritis remains unidentified. Adjuvant-induced arthritis is widely used in the assessment of antiinflammatory drugs used in the treatment of rheumatoid arthritis. The arthritis is only one of a variety of lesions that appear in rats after the injection of Freund's complete adjuvant into the hind foot pad. The polyarthritis can be assessed by scoring the joints daily. In this chapter, the role of vitamin E on the model of arthritis is described.

INFLAMMATION AND OXYGEN-DERIVED FREE RADICALS

Active oxygen species, such as superoxide anion, hydroxyl radical, hydrogen peroxide, and singlet oxygen, are thought to be involved in inflammatory reactions since they are produced from phagocytosing cells by the stimulation of bacteria, chemical mediators, and cytokines. These chemical mediators and cytokines are produced by polymorphonuclear leukocytes (PMNs) in inflammatory joints. In addition to killing pathogenic organisms, these reactive molecules, when released into the surrounding area, can cause inflammation of surrounding tissues (Fig. 1).

RHEUMATOID ARTHRITIS AND LIPID PEROXIDATION

It is now generally conceded that lysosomal instability resulting in the release of enzymes in tissues of the rheumatoid joints plays some role in the rheumatoid process. Recent investigations have shown that free-radical reactions and lipid peroxidation reactions play a part in the mechanisms of cellular damage. Biomembranes and subcellular organelles are major sites of lipid peroxidation damage. Of importance to cellular damage is the lability of lysosomal membranes to rupture with concurrent release of an array of hydrolytic enzymes and with the capacity to initiate intracellular digestion and catabolism. The release of lysosomal enzymes initiates further damage to the structural and functional parts of the cell. We found that thiobarbituric acid (TBA)-reactive substances, an index of lipid peroxidation in synovial fluid, were elevated in patients with rheumatoid arthritis and showed a significant correlation with the levels of lysosomal enzymes (Figs. 2 through 4) (1).

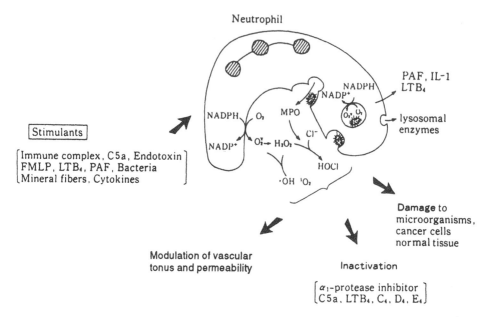

FIGURE 1 Active oxygen generation by polymorphonuclear leukocytes and their action.

ADJUVANT ARTHRITIS AND OXYGEN-DERIVED FREE RADICALS

McCord showed that superoxide radicals degrade various high-molecular-weight polymers, such as polysaccharides, in the synovial fluid of joints and may be an important factor in rheumatoid arthritis (2). Since phagocytizing PMNs present in the synovial fluid produce superoxide radicals with attendant generation of hydrogen proxide and hydroxyl radicals,

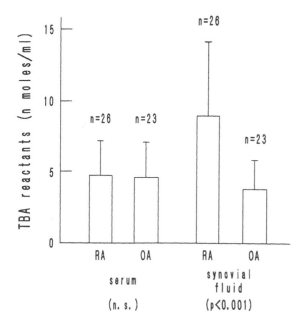

FIGURE 2 TBA reactants in serum and synovial fluid. Mean ± SD. (RA, rheumatoid arthritis; OA, osteo-arthritis.)

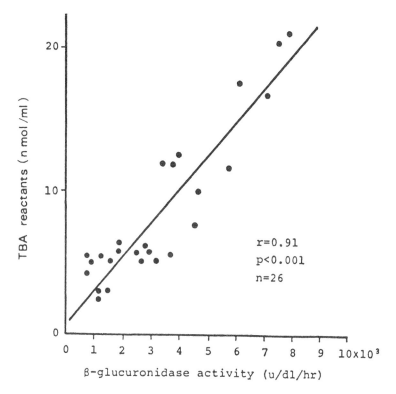

FIGURE 3 Correlation between TBA reactants and β-glucuronidase. r = correlation coefficient.

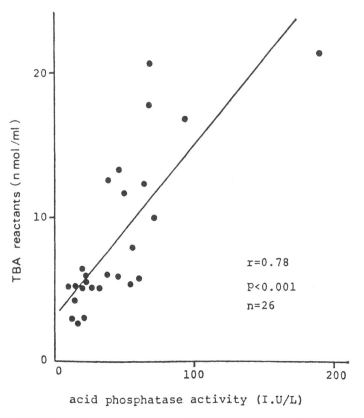

FIGURE 4 Correlation between TBA reactants and acid phosphatase. r = correlation coefficient.

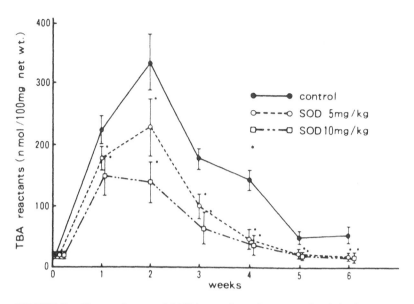

FIGURE 5 Changes in synovial TBA-reactive substances after injection of adjuvant. Each value represents the mean ± SD of nine rats. *P < 0.001, difference from control rats.

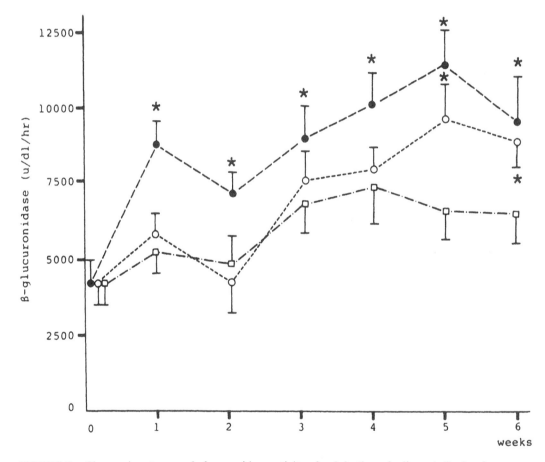

FIGURE 6 Changes in rat serum β-glucuronidase activity after injection of adjuvant. Each value represents the mean ± SD of 10 rats: (solid circles) vitamin E-deficient diet group; (open circles) control diet group; (open squares) vitamin E-supplemented diet group. *P < 0.001 for difference from rats fed on a diet supplemented with vitamin E.

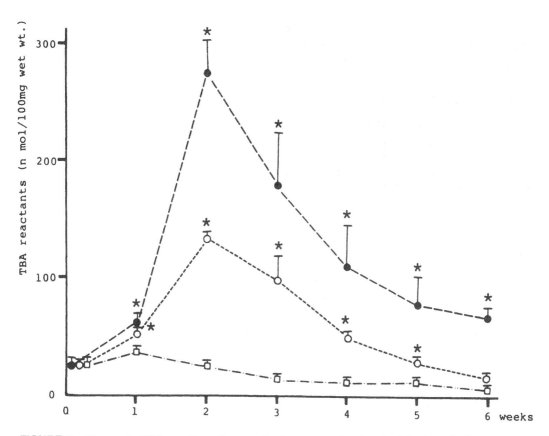

FIGURE 7 Changes in TBA-reactive substances in the rat synovia after the injection of adjuvant. Each value represents the mean ± SD of 10 rats: (solid circles) vitamin E-deficient diet group; (open circles) control diet group; (open squares) vitamin E-supplemented diet group. *$P < 0.001$ for difference from rats fed on a diet supplemented with vitamin E.

the reaction was suggested as the in vivo mechanism of synovial fluid degradation in inflamed joints (3). Fridovich suggested that superoxide dismutase (SOD), injected into an inflamed area, might minimize the damage caused by superoxide anion secreted by phagocytes (4).

Adjuvant arhtiritis was induced in rats by the injection of *Mycobacterium tuberculosis*, and its severity was scored according to the macroscopic findings of the legs, tail, and ears. The average score so obtained was lower in SOD-injected rats than in the control group (5). The depression of albumin-globulin (A/G) ratio, an index of the severity of inflammation, was significantly inhibited in rats treated with SOD. The levels of lysosomal enzymes, such as acid phosphatase and β-glucuronidase, were elevated after administration of an adjuvant, but the elevation was inhibited by treatment with SOD. The levels of TBA-reactive substances in the sera and synovia were elevated at 2 weeks after the injection of adjuvant and decreased thereafter. The increase in both serum and synovial levels of TBA reactants was significantly inhibited by treatment with SOD (Fig. 5). These observations suggest that the aggrevation of adjuvant arthritis may be associated with lipid peroxidation due to oxygen-derived free radicals and that SOD may be beneficial for the treatment of arthritis.

ADJUVANT ARTHRITIS AND VITAMIN E

To obtain more information on the role of lipid peroxidation in relation to arthritis, an investigation was undertaken using vitamin E as an antioxidant. Vitamin E has been known to possess a strong antioxidant action and to inhibit peroxidation of lipids. Rats were maintained on a diet deficient in or supplemented with vitamin E, and its effect on experimental adjuvant arthritis was determined (6). The average score, an index of the severity of adjuvant arthritis, was higher in the vitamin E-deficient diet group than in the group of rats supplemented with vitamin E. Whereas the A/G ratio remained depressed in vitamin E-deficient rats, rats on a vitamin E-supplemented diet showed a fast recovery from the depression of the ratio. The serum levels of lysosomal enzymes showed a remarkable increase in rats fed a vitamin E-deficient diet, but the elevation in rats fed a vitamin E-supplemented diet was inhibited (Fig. 6). The levels of TBA-reactive substances in the synovia were elevated in rats fed a vitamin E-deficient or control diet; on the other hand, the increase was inhibited in rats maintained on a diet supplemented with vitamin E (Fig. 7). From these results, it was suggested that antioxidants, such as vitamin E, may be beneficial for arthritis.

REFERENCES

1. Yoshikawa T, Yokoe N, Takemura S, Kato H, Hosokawa KP, Kondo M. Studies on the pathogenesis of rheumatoid arthritis: lipid peroxide and lysosomal enzymes in rheumatoid joints. Jpn J Med 1979; 18:199-204.
2. McCord JM. Free radicals and inflammation protection of synovial fluid by SOD. Science 1974; 185:529-531.
3. Richardson T. Salicylates, copper complexes, free radicals and arthritis. J Pharm Pharmacol 1976; 28:666.
4. Fridovich I. A free radical pathology: superoxide radical and superoxide dismutases. Annu Rep Med Chem 1975; 10:257-264.
5. Yoshikawa T, Tanaka H, Kondo M. The increase of lipid peroxidation in rats adjuvant arthritis and its inhibition by superoxide dismutase. Biochemical med 1985; 33:320-326.
6. Yoshikawa T, Tanaka H, Kondo M. Effect of vitamin E on adjuvant arthritis in rats. Biochem Med 1983; 29:227-235.

D. Ischemic Heart Disease and Atherosclerosis

44

Vitamin E and Other Essential Antioxidants Regarding Coronary Heart Disease: Risk Assessment Studies

Epidemiological Basis of the Antioxidant Hypothesis of Cardiovascular Disease

K. Fred Gey

University of Bern, Bern, Switzerland

INTRODUCTION

Limitations of Classical Risk Factors of Arteriosclerosis

Coronary heart disease (CHD) with underlying arteriosclerosis is a major cause of death in adults of industrialized countries. The disease is multifactorial and has several stages during several decades of life (1–6); that is, it proceeds relatively slowly except the final event of myocardial infarction. The expression of arteriosclerosis varies enormously within a single artery (presumably because of differences of hemodynamic factors), between individuals, and between populations. The risk to the majority of individuals of dying from CHD can only in part be ascribed to the well-established but previously preferentially evaluated classical risk factors, that is, an elevated plasma level of total cholesterol, of LDL (low-density lipoprotein) cholesterol, or of the LDL/HDL (high-density lipoprotein) ratio, hypertension, smoking, and so on, even if combined in multivariate analysis. According to a recent expert review (6), "few would claim that hypercholesterolemia is the exclusive causative factor of atherosclerosis. It certainly interacts with other variables that importantly influence the disease process." On the one hand CHD can develop in individuals with unsuspicious levels of classical risk factors, and on the other hand prevalence of the latter cannot reliably predict the actual risk of any individual even in familial hypercholesterolemia within the same kindred (6). The approximately sixfold difference in CHD mortality in industrialized nations could neither be fully ascribed to classical risk factors in Ancel Key's seven-country study (6a) or in the current core study of the WHO/MONICA Project (MONICA was designed for monitoring of determinants and trends of cardiovascular disease; 7). The MONICA Project is the first highly standardized and by far greatest cross-cultural comparison of CHD (7a). Thereby, the combination of total cholesterol, blood pressure, and smoking has explained the differences in CHD risk of 33 study populations only to roughly 20% ($r^2 = 0.203$) (A. Stewart and K. Kuulasmaa, evaluations of the MONICA Data Centre Helsinki, oral presentation at Congress of

European Society of Cardiology, Nice, 1989). In consequence, the clinically established CHD clearly also depends on other factors, among which environmental, particularly dietary factors may deserve special interest, not least because of their potential preventive aspects (8,8a).

Nutritonal Hints of Essential Antioxidants Regarding Arteriosclerosis

There is fair epidemiological evidence that various diets are associated with a lower morbidity and mortality from CHD, for example, the vegetarian diet (characterized by a preponderance of vegetables of all kinds and low levels of mammalian protein and fat) or the Mediterranean diet (rich in olive oil, legumes, seafood, and so on) (6a,9–12). Potential benefits may therefore be expected on the one hand from bulky nonessential although typical components of plants, such as fibers, potassium and other minerals, and many types of phenols (13,14). On the other hand, a greater impact is expected from compounds that might be very vital for plants, that is, polyunsaturated fatty acids (PUFAs), and monoenic fatty acids, such as oleic acid (6a, 15–20), micronutrients, such as trace elements, for example, selenium, and particularly plant-protecting antioxidants, such as vitamin A [retinol, scavenging peroxy and particularly thyil radicals (21)], vitamin C [ascorbic acid, the only essential water-soluble oxygen radical scavenger (22,23)], carotenoids [e.g., β-carotene and lycopene as typical singlet oxygen quenchers (24)], and last but not least vitamin E [RRR-α-tocopherol, the principal chain-breaking liposoluble oxygen radical scavenger (25)], the requirement for which is multiplied by the number of double bonds of PUFAs in any lipid compartment (26). The peroxy radical-trapping ability of plasma is not only best predicted by vitamin C and E (27,28), but both single hydrogen donors are also unique regarding redox cycling (29) and their synergistic interaction (30,31). PUFAs can provide eicosanoids and affect membrane properties and may thus in different ways modulate physiological regulations. In mammals the dietary replacement of PUFA-poor animal fat by PUFA-rich vegetable oils, which under certain experimental conditions reduces the mortality by CHD (32,33), or particularly olive oil may automatically increase the dietary supply of vitamin E, provided that the latter was not previously been lost by industrial procedures. Unrefined vegetable oils are generally known to be the major source of vitamin E in common diets (34a), and in vegetarians, increased plasma levels of lipid-standardized vitamin E have been demonstrated (34b,c). The absolute dietary vitamin E intake can also be increased by the frequent consumption of fresh sea fish, which was reported to be associated with fewer deaths from CHD (35). Fresh fish has relatively higher absolute vitamin E concentrations than common meat (34). PUFAs, similar to proteins, proteoglycans, and DNA, for example, require protection against aggressive oxygen species (such as superoxide anion, hydroxyl radicals, singlet oxygen, and nitric acid), which are regularly formed in the body (by many enzymatic reactions) or occur as environmental hazards (e.g., ozone and nitrogen dioxide) (36,37). The body's multilevel defense system against radical injury consists of defense enzymes (superoxide dismutase, catalase, and glutathione peroxidase, among others), as well as of endogenous radical scavengers (e.g., glutathione, uric acid, bilirubin, ubiquinol-10, and proteins) and of exogenous, essential radical scavengers. The state of the latter, in contrast to that of other defense lines, depends on the one hand on the individual requirement and, on the other hand, on their dietary intake and can thus relatively easily be controlled. As far as essential antioxidants of the diet can be calculated, they correlate inversely with CHD mortality in several countries. This seems to be the case for vitamin C in England, Wales, Scotland, Norway, and Israel (38–43). Therefore, the negative correlation between vitamin C-containing fresh fruit and vascular death rates was not only the highest (r about 0.7–0.9) of all calculated major food items but also higher than the correlation to the classical risk factor. This inverse correlation was suggested to explain, at least in part, well-known geographical differences (e.g., the north-south gradient) as well as the great socioeconomic differences of CHD. The recently increased consumption of fruits and ascorbic acid has even been considered to be linked to recent local trends of the decrease in CHD (38,44). Regarding vitamin E in the United States, in preliminary analysis of a large population of healthy professionals of both sexes, a much lower risk of CHD and stroke was seen among persons in the highest quintile of vitamin E intake (which then depended frequently on vitamin supplements) (W. C. Willett, personal communication, 1991), but the risk of CHD of females was also significantly lower at a high dietary intake of β-carotene and/or of vitamin A respectively (44a). Regression analysis of trends of the mortality in Israel also revealed an inverse relationship between both overall mortality and death by ischemic heart disease and the calculated sum of dietary β-carotene and vitamin A (41). A very preliminary report on the U.S. physicians' study (enrolling 333 coronary cases) states a substantial reduction in coronary events by supplements of β-carotene (50 mg every second day) and of acetylsalicylic acid (100 mg every alternate day) for more than 5 years (45; C. H. Hennekens, personal communication, 1991). This finding is not only interesting regarding β-carotene but also with respect to salicyclic acid, which is actually a very simple plant phenol. Since already the common diet in the United States and Europe provides daily at least 1 g of (although only in part easily absorbable) phenols and polyphenols (46) of different chemical classes (47), their conceivable role in vegetarian diets remains to be further elucidated. Actually, many plant phenols or polyphenols are highly potent oxygen radical scavengers as well as metal chelators, and salicyclic acid can even be used as an analytical hydroxy radical trap (36,48). Bioflavonoids inhibit in vitro the oxidative modification of LDL as well as essential antioxidants and preserve vitamin E (49–51). Conceivable synergistic antioxidant functions between common plant phenols, such as salicyclic acid, and specific phenols, such as the (mostly poorly absorbable) bioflavonoids, as well as the nonphenolic vitamin C and the phenolic vitamin E, await further exploration in plants and mammals.

Further Suggestive Evidence for the Antioxidant Hypothesis of CHD

Based on the fundamentals of LDL uptake by macrophages (52), the oxidative modification hypothesis of CHD (3,4,6,53–63) (see Chapter 46) postulates initiation of foam cells or fatty streaks as the first cellular stage of arteriosclerotic plaques by oxidatively modified LDL. The complementary antioxidant hypothesis (8,8a,64,65) stipulates that cardiovascular health requires protection of plasma lipoproteins (6,53–63) and/or of arterial cells by essential antioxidants (66–68) against free-radical injury, that is, that cardiovascular risk is due to an imbalance between antioxidant potential and radicals and thus is merely a special case of radical injury. The latter may be due to the onset of either primary endogenous oxygen radicals and secondarily propagated radicals from the chain reaction of PUFA peroxidation or of exogenous radicals. The principal aggressive endogenous oxygen species are superoxide (O_2^-) and hydroxyl radicals (mainly $HO^·$), which can, however, also derive from H_2O_2, whereas singlet oxygen is an aggressive oxygen species without the chemical characteristics of a radical; that is, it lacks the typical unpaired electron (36). Oxygen radicals are produced in considerable quantity in the normal metabolism (e.g., by flavoenzymes, such as xanthine oxidase, and by P_{450} cytochromes) and can be increased by ischemia, inflammatory and degenerative processes, hyperbaric oxygen, ionizing radiation, and other trauma (36).

The evaluation of the antioxidant hypothesis by exploration of essential antioxidants regarding CHD may be justified by a series of arguments aside from the nutritional hints already mentioned.

Chronic marginal deficiency of vitamins C or E was reported to cause arteriosclerosis-like lesions in experimental animals and humans. Vitamin E deficiency (which can generally be characterized by an enhanced susceptibility toward lipid peroxidation to corresponding lipofuscinosis or ceroid accumulation (69,70) results in rats, first, in weakening of the basement membranes of muscle capillaries and in a breakdown of endothelial cells (71) and, later, in subendothelial fibrosis combined with fibrous and calcified lesions to necroses in the media of the aorta (72). Arteries of other mammals, including primates, are modified as well (73, 74). In the (vitamin C-dependent) guinea pig with acute deprivation or with chronic marginal deficiency of ascorbic acid, various changes in arterial biochemistry and morphology and atherosclerosis-like alterations have been found with and without simultaneous fat feeding in many previous (8) and very recent studies (75). The vitamin C deficiency aggregates the intimal thickening and lipid infiltration of cholesterol-fed rabbits (8). In piglets, vitamin E deficiency leads to a combination of myocardial necrosis with widespread thrombosis of the myocardial microcirculation (76). In the human, vitamin E deficiency causes a proliferative vasculopathy (retrolental fibroplasia) in prematurely born infants, as well as severe neuropathological disturbances, cardiomyopathy, and hematological disorders in children and adults (77,78) similarly to numerous animal species (69). In volunteers with prescorbutic plasma levels of vitamin C due to dietary deprivation of vitamin C, sudden cardiomegaly, various electrocardiographic abnormalities, and acute cardiac emergency have been observed, which were normalized by ascorbic acid therapy (79–81). Sudden death by scurvy may be related to heart failure (82).

Vitamin E reduces in cholesterol-fed primates (*Macaca fascicularis*) atheromatosis of the aorta as well as stenosis of the carotid artery (83). Correspondingly vitamin E diminishes in triacylglycerol-fed rabbits atheromatosis of the aorta (84,85), perhaps in part by a depression of plasma lipids. Unfortunately, a lipid-lowering effect of larger doses of vitamin E in rabbits (86,87) as well as in rats (88) but not in the human (89–92) limits the use of these experimental models considerably. In the WHHL (Watanabe heritable hyperlipemic) rabbit, moderate doses of vitamin E reduced the plaque surface of the aorta significantly in a pilot experiment (K. F. Gey, U. Moser, and E. Bühler, unpublished morphometric evaluations of aortic surface after moderate dosage of vitamin E, 1990). Both rabbit models differ further from the human with normal plasma lipids or with common familial hypercholesterolemia by an almost exclusive increase in β-VLDL (93,94), which can yield foam cells without previous modification, in contrast to LDL (94a). Supplements of vitamins A and/or E counteract the spontaneous arteriosclerosis of the middle-aged hen in the egg-laying period (95). Vitamin E supplements diminish elevated plasma TBARs (later) and increased intimal thickness of restricted (nonlaying) ovulatory chicken to the level of normal laying hens despite that vitamin E does not affect the hyperlipidemia of the restricted ovulator chicken (96).

Hypoxia with subsequent reoxygenation (as conceivable after a transitory impairment of coronary flow, as by vasoconstriction, thrombus formation, heart failure, or cardiac fibrillation) produces a sudden "respiratory burst" of oxygen radicals that can overload protective enzymes (36,97,98). Oxygen radicals seriously damage arteries in animal models as soon as the physiological radical scavengers, such as vitamins E and C and glutathione, are exhausted (8,98–103). Pretreatment with vitamins C and/or E and/or concurrent infusion of antioxidants diminish the reperfusion injury in situ and in vivo, as demonstrated by numerous laboratories (see last section).

Peroxidized diets are toxic in rodents (104,105) and cause sudden death with degenerative changes in heart muscle in the pig (106). Since oxidized lipids can, at least in part, be absorbed from the intestine, peroxides of arachidonic acid or other PUFAs have a high toxicity by the parenteral or oral route (107,108). Exogenous peroxidized PUFAs (and probably endogenous ones derived from a "respiratory burst") damage the endothelium and heart muscle cells (36,37,109) and provoke the proliferation of smooth muscle cells (110,111). All this cytotoxicity can be prevented by vitamins C and E (110,112,113). The reaction of cholesterol with singlet oxygen yields primarily 5-α-hydroperoxide, whereas free-radical oxydation gives 7α- and 7β-hydroperoxides aside from epoxides, such as cholesterol-5α,6α-epoxide (36). Autoxidized cholesterol (8,114) and oxidatively modified LDL are as cytotoxic to endothelial cells and fibroblast as peroxidized PUFAs (115–118). Autoxidized cholesterol has also been thought to be causally related to atherosclerosis in the cholesterol-fed rabbit (8,119,120) and perhaps even in the human (120). Although common food items can contain at least trace amounts of oxidized cholesterol, their atherogenic importance is obscure.

LDL, the plasma lipoprotein with the strongest correlation with arteriosclerosis, is also most susceptible to autoxidation, as well as to oxidation by endothelium, macrophages, other arterial cells, and blood platelets (4,6,63,121–130), as well as by transition metals, such as copper (53) (see Chapter 46), and to modification by hydroperoxides (117). Lipid peroxidation modifies at least PUFAs and ApoB (apoprotein B) of LDL. Oxidatively modified LDL (OM-LDL), in contrast to native LDL, is cytotoxic, is recognized by macrophages, is taken up by several receptors or phagocytosis, and causes subsequent lipid accumulation and ceroid deposition in cultured macrophages (4,6,63,66–68,116,129). Subendothelial clusters of lipid-laden, in part ceroid-accumulating foam cells, called fatty streaks, represent the initial stage of arteriosclerotic plaques. The occurrence of OM-LDL was unequivocally demonstrated in arterial lesions of animals and humans (4,6,130–138) and may be related to the fact that the accumulation of lipid peroxides and hydroperoxides in arteriosclerotic plaques is positively correlated to the degree of arteriosclerosis in postmortem material (139–142) and in arteries freshly obtained by surgery (143). Interestingly, OM-LDL in macrophage-rich arteriosclerotic lesions is colocalized with mRNA for 15-lipoxygenase as well as for the m-LDL-scavenger receptor (144,145). Oxygen radicals and lipoxygenase are able to transform LDL into toxic OM-LDL, which is taken up by macrophages (127,128,146a,b,147). Taken together, LDL can oxidatively be modified in arteries, although the site and nature of the lesion are still unknown (5,66,148). LDL oxidation by endothelial cells can be reduced by vitamin E (124), and LDL of vitamin E-supplemented humans is more resistant to oxidation in vitro (49,149) (see Chap. 46). Vitamins E and C counteract the cytotoxicity of OM-LDL on endothelial cells (115,117, 129,150–155), similar to that of PUFA hydroperoxides (155a). Vitamin E attenuates the intracellular accumulation of cholesteryl esters as well as of ceroid in cultured macrophages (66–68,156–161) and decreases the oxidative modification of cholesterol by macrophages (66–68) and the secretion of catabolites of LDL-ApoB (162). Vitamins E and C have other beneficial effects on cultured macrophages, as well as on their migrational and phagocytic potentials (163–169), on the respiratory burst of macrophages (170,171), and on the oxidative modification of LDL (172), as well as on the secretion of proteolytic lysosomal enzymes (173) and cell-cell interactions, for example by inhibition of the expression and secretion of IL-1 (interleukin-1) by macrophages (174–176).

Feeding synthetic antioxidants to rabbits yielded equivocal results. Several laboratories showed that probucol, a cholesterol-lowering drug with a radical-scavenging bisphenol structure, reduces (in part presumably independent of its hypocholesterolemic potential) LDL modification and lipid storage in macrophages, as well as the progression of arteriosclerosis in the WHHL rabbit (93,126,151,177,177a,178–180). It has been debated whether these anti-atherosclerotic properties of probucol in the WHHL rabbit are due only to the antioxidative prevention of LDL oxidation (181) or also to the antioxidative help of macrophages against OM-LDL (180,182). Interestingly, probucol can preserve plasma vitamin E (B. Finckh, M. Rath, A. Niendorf, and U. Beisiegel, unpublished work, 1989). Probucol seems, however to lack corresponding antiatherosclerotic properties in the cholesterol-fed rabbit (183). Butylated hydroxytoluene (BHT), the structural unit of probucol and a common food antioxidant preservative) counteracts the atheromatosis of cholesterol-fed rabbits as well as the accumulation of cholesterol $5\alpha,6\alpha$-epoxide in plasma (184), but these effects could, at least in part, be due to antioxidant protection of cholesterol in the feed.

HUMAN PLASMA LEVELS OF
ANTIOXIDANT VITAMINS REGARDING CHD

Comparison of Populations

Cross-cultural comparisons provide statistical medians of any parameter in populations who differ regarding the prevelance of CHD. When such a statistically phantom parameter differs strongly between populations, it can reveal important correlations to the disease.

In the current Optional Study on Antioxidant Vitamins and PUFAs of the WHO/MONICA Project (initiated as the International Collaborative Study on the Fatty Acid-Antioxidant Hypothesis of Arteriosclerosis), the well-known up to about sixfold differences in CHD mortality in 16 European populations were compared with the medians of plasma levels of all essential antioxidants in random samples of middle-aged males. The study samples consisted, in principle, of about 100 randomized apparently healthy males, 40–49 years of age, who represented (with a response rate > 60%) regions with different incidences of CHD mortality (mean values of at least 3 preceding years):

1. Highest incidence, yearly > 260 deaths from CHD (ICD 410–414) per 100,000 males 40–59 years of age: North Karelia, Finland (481 in 1983 in rural North Karelia and 469 in 1987 in rural and semiurban North Karelia) > Glasgow, Scotland (381) > Southwest Finland (359) > Edinburgh, Scotland (298) > Aberdeen, Scotland (270).
2. Medium incidence, that is, approximately 150–259 per 100,000: peripheral Belfast, Northern Ireland (254) > Glostrup or Copenhagen, Denmark (208) = Schleiz, Germany (208) > Cottbus, Germany (182) = Schwedt, Germany (186) > Tel Aviv, Israel (154).
3. Low incidence, approximately 65–115 per 100,000: Thun, foothills of the Alps in Switzerland (112) > Sapri, southern Italy (107) > Haute Garonne-Toulouse, France (72) and Catalonia, Spain (66).

The available data indicate (8,8a,64,65,185–188) that in uni- and multivariate analysis vitamin E is the strongest inverse predictive factor of cross-cultural CHD mortality, but vitamins A, C, and carotene at least to some extent seem to behave synergistically with vitamin E. In some populations (with common levels of cholesterol and blood pressure) the relative risk due to a low vitamin E status may have even greater quantitative importance than that of these classical risk factors as obvious from their comparison in uni- and multivariate analysis.

Populations with Common and Similar Levels
of Plasma Cholesterol and Blood Pressure

In the majority—in 12 of 16 European study populations—the total plasma cholesterol ranged in a narrow band of usual levels, 5.7–6.2 mmol/L (220–240 mg/dl). In this subgroup, total cholesterol lacked any statistically significant direct correlation with CHD mortality ($r^2 = 0.04$, Table 1) as expected, but this was also true for the blood pressure (diastolic $r^2 = 0.08$, systolic $r^2 = 0.01$; Table 1) and for the combination in the multivariate analysis of total cholesterol and diastolic blood pressure ($r^2 = 0.10$, $P = 0.62$). Since the effect of these classical risk factors is almost identical in this subgroup, its sixfold differences in CHD mortality must mainly be due to other factors.

Among the essential antioxidants tested, vitamin E (either absolute or lipid-standardized levels) showed a very strong inverse correlation with CHD. In study populations with a very low CHD mortality (<115/100,000), the absolute vitamin E level varied between 27 and 30 μmol/L (Fig. 1) and the α-tocopherol to cholesterol ratio (μM/mM) was almost 5 (8,185). In contrast, populations with a high CHD risk (>300/100,000) revealed vitamin E levels of around 20 μmol/L, i.e. about 1.35-fold lower values, and the α-tocopherol-cholesterol ratio

TABLE 1 Pearson's Correlation Coefficients in Univariate Analysis of Age-Specific CHD Mortality of Study Populations and Various Parameters of the Vitamin Substudy of the WHO/MONICA Project

Parameter	Populations with unusual cholesterol ($n = 12$)		All populations ($n = 16$)	
	r^2	P	r^2	P
Cholesterol	0.04	0.53	0.29	0.03
Blood pressure, systolic	0.01	0.80	0.19	0.09
Blood pressure, diastolic	0.08	0.36	0.25	0.05
% Smokers	0.002	0.90	0.01	0.65
% Smokers (folded log)	0.002	0.89	0.01	0.67
Cigarettes/study subject	0.004	0.85	0.02	0.64
Vitamin A, absolute	0.22	0.13	—	—
Vitamin A, lipid-standard	0.16	0.19	0.24	0.05
Vitamin E, absolute	0.63	0.002	—	—
Vitamin E, lipid-standard	0.73	0.0004	0.92	0.0003
Vitamin C	0.41	0.03	0.11	0.22
Carotene, absolute	0.21	0.14	0.04	0.48
Selenium ($n = 8/12$)	0.05	0.59	0.02	0.69

was around 3.5 (8). In univariate analysis vitamin E was clearly in the leading rank ($r^2 = 0.63$ and 0.73, respectively; $P = 0.002$; Fig. 1) before the moderately strong correlated vitamin C ($r^2 = 0.41$, $P = 0.03$; Fig. 2) and the marginally correlated vitamin A ($r^2 = 0.22$, $P = 0.13$; Fig. 2) and carotene, respectively ($r^2 = 0.21$, $P = 0.14$), whereas selenium lacked any correlation (Table 1). In partial regression analysis of combined vitamins with fixation of the three classical risk factors (65), vitamin C improved the correlation to CHD, at least in part, with statistical significance. Thus, for instance, the combination of vitamin C and lipid-standardized vitamin E yielded a stronger correlation ($r^2 = 0.87$, $P = 0.039$ for difference) in comparison with vitamin E alone, and correspondingly the combination of vitamins A, C, and E had a stronger correlation ($r^2 = 0.90$, $P = 0.019$ for difference) compared to combined lipid-standardized vitamins A and E ($r^2 = 0.79$).

In summary, in the major cluster of European populations whose three classical risk factors of CHD were very similar, the severalfold differences in CHD mortality could in uni- and multivariate analysis be explained at least to about 60% by differences in the plasma levels of vitamin E and up to 90% by the combination of vitamins E, A, and C.

All Populations Disregarding the Levels of Classical Risk Factors

Of the currently available 16 study populations of the Vitamin Substudy of the MONICA Project, 4 had either markedly higher levels of total cholesterol (6.4–6.9 mmol/L in North Karelia, Finland) or lower levels (Israel and Italy), and the median of diastolic blood pressure was in two study populations lower and in three higher than that of the major cluster (65). In consequence, on consideration of all available populations, two classical risk factors of CHD, that is, total plasma cholesterol ($r^2 = 0.29$, $P = 0.03$; Table 1) and blood pressure (diastolic $r^2 = 0.25$, $P = 0.05$; systolic $r^2 = 0.19$; Table 1), revealed in univariate analysis the expected direct association with CHD, whereas the relatively small differences (24–55% smokers) in cigarette smoking (Table 1) lacked any significant correlation. When total cholesterol and diastolic blood pressure were combined in multivariate regression analysis, their

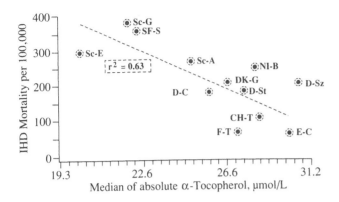

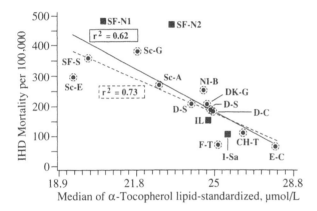

FIGURE 1 Relation between CHD mortality and plasma α-tocopherol in European study popula-tions of the Vitamin Substudy of the WHO/MONICA Project. Inverse correlations between age-specific CHD mortality and the medians of logarithms of absolute α-tocopherol (above; encircled solid center points), as well as lipid-standardized α-tocopherol (below) in 12 study samples with usual cholesterol (5.7–6.2 mmol/L = 220–240 mg/dl) and with lipid-standardized α-tocopherol, respectively (round solid center points, dotted regression lines). The solid regression line represents all 16 study samples, that is, with inclusion of hypo- and hypercholesterolaemic study samples. Abbreviations: CH-T, Switzer-land, Thun; D-C, D-St, and D-Sz, Germany, Cottbus, Schwedt, and Schleiz; DK-G, Denmark, Glostrup/Copenhagen; E, Spain, Catalunia; F, France, Toulouse/Haute Garonne; IL, Israel, Tel Aviv; I-Sa, Italy, Sapri; NI-B, Northern Ireland, semiurban Belfast; SC-A, SC-E and SC-G, Scotland, Aberdeen, Edinburgh, and Glasgow; SF-N1, SF-N2, SF-S, Finland, rural North Karelia 1 (1983), rural/semiurban North Karelia 2 (1987), and rural Southwest Finland. (From Ref. 65 with permission.)

correlation with CHD mortality remained moderate (r^2 = 0.44, P = 0.02). The same was true for the combination of cholesterol, blood pressure, and smoking (r^2 = 0.46, P = 0.051). Thus, in the present 16 European study populations (65), the relative importance of the classical risk factors was even higher than in the 33 study populations of the WHO/MONICA Core Study from all over the world (A. Stewart and K. Kuulasmaa, unpublished evaluations of the MONICA Data Centre Helsinki, 1989).

It is widely known that the absolute plasma level of vitamin E depends in part on the coexistent concentration of its carriers, that is, among the plasma lipoproteins mainly on LDL: in the present study, including almost 2000 subjects, a Spearman's rank correlation coefficient of r_S = 0.54 was obtained for α-tocopherol and total cholesterol, and r_S = 0.47 for triglycerides. Therefore, and because of compelling biological reasons, vitamin E requires in different lipid levels lipid standardization for a specific anraveled interpretation of the vitamin E status (8,65,185–189). Although HDL also contains α-tocopherol, the variability in plasma vitamin E is mainly due to differences in α-tocopherol in LDL (189,190), and the transfer of α-tocopherol from LDL to human muscle cells is at least one order more effective than that from HDL (191). For statistical reasons vitamin A requires the same lipid standardization as viamin E (65,188), although the biological justification is still uncertain. In the current vitamin study, plasma vitamin E (α-tocopherol) as well as vitamin A (retinol) were standardized in two steps (based on a two-dimensional regression plane from 1950 plasma samples of the pooled data from all study populations) to a plasma level of 5.7 mmol/L of cholesterol per L (220 mg/dl) and independently to a plasma level of 1.25 mmol/L of triacylglycerols per L (110 mg/dl), that is, to the most common "normal" lipid levels of European populations (65). The standardization of vitamin E for both cholesterol and triglycerides reflects its plasma status as powerfully as the previously suggested α-tocopherol-total lipid ratio and clearly better than the α-tocopherol-cholesterol ratio (8,185,190,192). The latter may only be recommended as a compromise for the fasting state when triglycerides levels were not available. After stepwise adjustment, the regression coefficient of vitamins E and A, respectively, varied practically independently of cholesterol and triglycerides ($r_s = 0.0075$; $P = 0.74$ and -0.02; $P = 0.08$, for lipid-standardized α-tocopherol, and -0.021; $P = 0.40$ and -0.01; $P = 0.57$ for lipid-standardized retinol). Therefore both vitamins also became practically independent of each other (from $r_s = 0.33$, $P = 0.0001$ to $r_s = 0.10$, $P = 0.0001$). As previously described (193) vitamin E varied practically independently from the three classical risk factors of CHD (65,194–196), although some studies reported inverse correlations between blood pressure and seasonal vitamin C intake (197,198).

In all populations ($n = 16$), the lipid-standardized antioxidant vitamins E and A showed by univariate analysis a similar inverse correlation with CHD mortality, as absolute levels in the subgroup with common cholesterol and blood pressure ($n = 12$). Therefore vitamin E again took the uncontested top position ($r^2 = 0.62$; $P = 0.0003$; Fig. 1) with lipid-standardized vitamin E levels above 25 μmol/L in all low-CHD populations (mortality <115/100,000). Vitamin E was followed at a great distance by total cholesterol ($r^2 = 0.29$; $P = 0.03$), diastolic blood pressure ($r^2 = 0.25$, $P = 0.05$), and vitamin A ($r^2 = 0.24$, $P = 0.05$; Table 1 and Fig. 2). This order was very similar to that of the stepwise regression analysis by which the combination of lipid-standardized vitamin E and cholesterol made a statistically significant improvement in the correlation to CHD mortality ($r^2 = 0.79$, $P = 0.008$ for "enlarging" the model). The combination of lipid-standardized vitamin E and cholesterol was further but only slightly complemented by lipid-standardized vitamin A ($r^2 = 0.83$) and finally by diastolic blood pressure ($r^2 = 0.87$). Thus, these four variables predicted the actual CHD mortality to 87% (Fig. 3). Partial regression analysis (with fixation of the classical risk factors, cholesterol, diastolic, blood pressure, and smoking (65) confirmed the top rank of vitamin E ($r^2 = 0.65$, $P = 0.0002$ for the F test), which tended to be complemented by lipid-standardized vitamin A ($r^2 = 0.72$) and/or vitamins A and C ($r^2 = 0.74$). Considering all populations, no consistent striking correlation with CHD mortality was found for plasma carotene, selenium, uric acid, bilirubin, iron, or transferrin. This of course by no means excludes any moderate role of the latter in atherogenesis particularly in individuals. If however special genetic risk factors existed in Finland, as conceivable, for instance, for a higher frequency of detrimental isoforms of apolipoproteins B and/or E, the relative rank order of essential antioxidants could become vitamin E >> carotene = vitamin C > vitamin A, independently of classical IHD-risk factors (198a).

In conclusion, all currently available data reconfirm the predominant role of vitamin E with a consistent inverse correlation in univariate analysis of both absolute and lipid-standardized values in stepwise regression analysis as well as partial regression analysis with fixation of classical risk factors) and again reveal additional weak inverse correlations of lipid-standardized vitamin A and of vitamin C (8,8a,65,186,188). The CHD mortality in the currently available European study populations is by far more strongly correlated to vitamin E

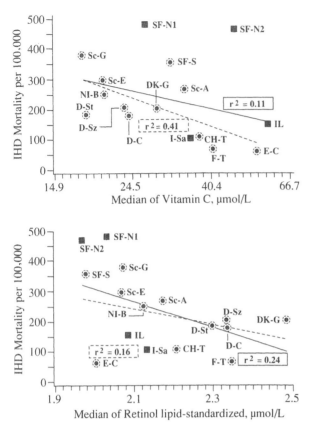

FIGURE 2 Relation between CHD mortality and plasma vitamin C and lipid-standardized vitamin A (retinol) in European study populations of the Vitamin Substudy of the WHO/MONICA Project. Major group of 12 populations with common cholesterol and blood pressure: encircled points, dotted regression line. All 16 study populations, that is, with inclusion of hyper- and hypocholesterolemic populations. Further details as in Figure 1. (From Ref. 65 with permission.)

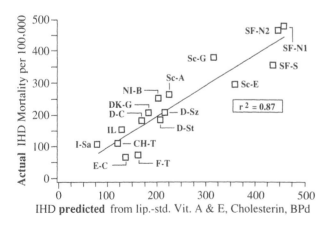

FIGURE 3 Relation between actual and predicted CHD mortality in European study populations of the Vitamin Substudy of the WHO/MONICA Project. Correlation between actual age-specific CHD mortality in middle-aged males of 16 study populations (ordinate) and the CHD mortality as predicted by multiple-regression analysis with four variables, that is, by lipid-standardized vitamins E and A, total cholesterol, and diastolic blood pressure (abscissa). Abbreviations as in Figure 1. (From Ref. 65 with permission.)

than to the classical risk factors total plasma cholesterol and blood pressure, which are thus substantially complemented by the plasma level of vitamin E and other essential antioxidants.

Synergistic Links Between Vitamins E and C

An interaction between vitamin E and C may be assumed because of several arguments for synergistic interactions.

Sparing Effects. Under long-term steady-state conditions vitamin C is able to spare vitamin E and the reverse is true in vitro (199) as well as in vivo in animals and humans (200–202; K. F. Gey and E. Bühler, unpublished data on guinea pigs kept at marginal vitamin C deficiency for 2 years, 1987), whereas short-term experiments of only a few months fail to detect this sparing effect (203). The administration of moderate doses of ascorbic acid also decreases the lipid peroxidation induced by vitamin E deficiency (199,204–210), and vice versa (206,208,212). In vitro the mixture of vitamin C and E has been found more effective than the sum of both antioxidants alone (206,213,214). Supplementation by 3 g ascorbic acid daily in the human was even reported significantly to increase initially normal levels of absolute vitamin E as well as the vitamin E-cholesterol ratio (215).

Complementary Action in Different Compartments. Vitamin C, unique because of its high reactivity with all aggressive oxygen radicals (and even with molecular oxygen), is a major antioxidant and the only essential one in the aqueous compartment (8,36), whereas vitamin E is the major chain-breaking radical scavenger in the liposoluble compartment, which thus represents the major specific defense line against lipid peroxidation (25,216). In the aqueous phase vitamin C (despite other concurrent nonessential antioxidants) is clearly the first line of antioxidative defense (59,217–220). As long as vitamin C [and the minor quantities of ubiquinol-10 (221)] are available, lipid peroxides produced by radicals (from cells, tobacco smoke, and other substances) are reduced to the relatively stable hydroperoxides, and as long as the chain reaction of lipid peroxidation cannot proceed vitamin E is saved (see Chapter 46 of this book). This synergism may be crucially important for the antioxidant protection of LDL (31).

Regeneration of Lipid-Soluble Vitamin E. Regeneration of the lipid-soluble vitamin E from its radical (222) by vitamin C (related to the sparing action already mentioned) may occur at the interphase. It is very well documented (30,31,37,213,214,223–233) and may be of major physiological importance, although vitamin E regeneration by other (e.g., NADH- or glutathione-dependent) reactions are also conceivable (222,234–236).

Inertness of Vitamin C Radical (Ascorbyl Free Radical). When vitamin C reacts with oxygen radicals or vitamin E radical, it becomes a radical itself. Since the ascorbate free radical has a very high, almost unique stability and very low chemical reactivity (22), it can be observed by electron spin resonance in tissues (237,238) and in coronary sinus blood after coronary infarction (239). Its occasional accumulation in tissues may suggest that ascorbate free radical is some transient aqueous radical sink or radical "buffer" from which vitamin C can be regenerated at convenience by alternative physiological mechanisms (8).

Potential Risks from "Prooxidant" Properties of Critically Low Levels of Any Anti-oxidant. All antioxidants can, as well known and generally accepted, reveal in vitro prooxidative properties at very low levels that are completely suppressed by an appropriate increase in the antioxidant concentration (8,36,240,241). Demonstration of the prooxidative behavior of very low levels of antioxidants in vitro requires, however, the presence of such metals as ferrous ions, hematin compounds, or copper ions. Therefore the low level of the antioxidant promotes peroxidation by keeping the transition metal in the reduced state, and in the case of vitamin C, independently of the formation of ascorbyl free radical (8,36,242).

An imbalance of antioxidants might be undesirable, too. For instance, in vitamin E-deficient microsomes, even normal levels of vitamin C (243) or glutathione (244) can have prooxidative effects, whereas at regular vitamin E levels both water-soluble antioxidants reveal only antioxidative properties. Synchronous and balanced optimization of all essential antioxidants can also ameliorate the status of endogenous antioxidants. Thus larger supplements of vitamin E in humans improve the glutathione status by an increase in erythrocyte glutathione and by a decrease in the plasma ratio of oxidized-reduced glutathione (245). Vitamin C spares and protects glutathione, too (246).

Epidemiological Correlations. In individuals of the cross-cultural MONICA study, the plasma vitamins C and (lipid-standardized) vitamin E were moderately correlated (r_S = 0.33, P = 0.0001; n = 1796) whereas the population medians clearly tended to correlate stronger (r_P = 0.43; although only with P = 0.10 at n as low as 16). In detail and more informative, at the lowest end of CHD mortality all four populations with an "optimal" vitamin E status had simultaneously high vitamin C levels, whereas at the upper end of CHD mortality vitamin C was only in three of six populations comparably as poor as vitamin E. In the intermittent range of CHD mortality four of six populations had a rather low vitamin C status (65). Thus the CHD risk may become minimal only after optimization of both vitamin E and C, whereas medium or fair levels of vitamin C may help preserve vitamin E when the latter is not consumed in excess. Even an optimum level of vitamin C may be unable to compensate for a genuinely poor vitamin E status, however.

Interindividual Variation in Lipid-Standardized Vitamin E in Populations

In the cross-cultural vitamin comparisons of the WHO/MONICA Project, the interindividual vitamin E level varied in a relatively narrow band without substantial overlap between populations at high and low risk of CHD (Fig. 4). Therefore, a great risk difference between populations can clearly be attributed to a great difference in their median antioxidant status, but within a given population the interindividual variation in antioxidant status can be too small to be detectable as CHD factor. This demands caution in the interpretation of prospective studies within a singly study population (see later) as well as for the design of forthcoming intervention studies (which should aim for study subjects with initially poor antioxidant levels to be substantially improved by the intervention).

Age Dependence of Vitamin E Status

The myocardial infarct with or without subsequent death as a common hard end point of CHD-related studies is mostly due to more or less acute arterial occlusion by a thrombus or vasospasm usually superimposed on top of an arteriosclerotic plaque. The latter, however, results from a multifactorial multistage process starting silently in early life. Therefore all risk factors of CHD remain be explored at least for a substantial period before the coronary infarct, ideally of course for the whole life span. Thus far the variation in plasma vitamin E with age has been studied in Swiss and French (whose plasma antioxidant status is rather high and associated with low CHD mortality) and all over Britain (where vitamin E levels may vary from fair in England to poor in Scotland, with a corresponding inverse gradient of CHD mortality). About 1000 clinically healthy Swiss males 30–65 years of age (the 40–49 year stratum of which formed the subalpine study population of the MONICA vitamin study) did not reveal any statistically significant age dependence of vitamin E (Fig. 5). The same was true for 3000 urban males who volunteered in the prospective study in Basel, Switzerland and who had an even slightly higher vitamin E status (247; K. F. Gey, P. Jordan, and H. B. Stähelin, unpublished data, 1990). Similarly, in 200 apparently healthy French males aged 18–45, the plasma α-tocopherol at levels of about 28 μmol/L did not vary substantially (r = 0.04) in multiregression analysis after adjustment for cholesterol and triglycerides (248). Correspondingly,

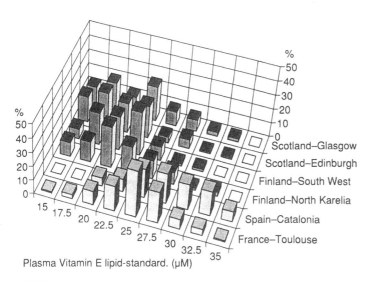

FIGURE 4 Variation of lipid-standardized plasma vitamin E in population with high and low CHD risk from the Vitamin Substudy of the WHO/MONICA Project. The altitude of the columns represents the %-age of subjects within any study population with different grades of plasma vitamin E.

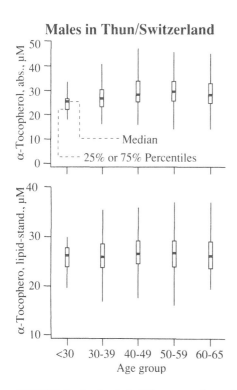

FIGURE 5 Lack of significant age dependence of lipid-standardized plasma vitamin E in Swiss with fairly high vitamin E status and low CHD mortality. The box plot shows means (solid bar), medians as well as 25 and 75% percentiles of 97 clinically healthy males of the subalpine cohort of the Vitamin Substudy of the WHO/MONICA Project.

in British males 18–64 years old the medians of α-tocopherol levels rose from 21 to 27 μmol/L, but again without any changes in the α-tocopherol-cholesterol ratio (249). This suggests that in these three European populations the status of vitamin E (as well of vitamin C) may have been similar in all decades of life, that is, including those in whom initiation of fatty streaks and their progression to proliferating, complicated plaques is most frequent and critical. In consequence, if vitamin E were, presumably in conjunction with other antioxidants, causally related to CHD, it could be involved in any stage and mechanism of CHD. If a high status of essential antioxidants, particularly of vitamin E, were indeed a CHD-preventive factor, it should prudently be optimized during all decades of life, being critical for CHD, that is, since initiation of fatty streaks in childhood as previously suggested for cholesterol-lowering diets.

Summary of Cross-Cultural Comparisons of European Study Populations

The up to sixfold interregional differences in the age-specific CHD mortality of males (40–59 years of age) revealed by uni- and multivariate analysis a very strong inverse correlation to plasma vitamin E and moderate correlations to synergistic antioxidants, such as vitamins C and A. Since vitamin E alone may explain about 60% of the differences in CHD mortality, it seems to be a markedly stronger predictor of CHD than the three classical risk factors total cholesterol, blood pressure, and smoking. If all the variables mentioned are combined in multivariate analysis, they can for the first time predict as much as about 90% of the differences in CHD mortality. Therefore, the influence of essential antioxidants, particularly of vitamin E, may hitherto have been underrated and deserves further exploration with regard to preventive potentials.

Multiple risk factor intervention (with simultaneous minimization of all currently known prominent risk factors) has been the most logical approach for primary and secondary prevention of CHD since the combination of classical CHD risk factors results in an overadditive increase in the CHD risk. Accordingly, the CHD-preventive potential of an optimized antioxidant status may best be explored by simultaneous optimization of all synergistic essential antioxidants. Although vitamin E has clearly the first rank it might require support from its synergists.

Plasma Antioxidants in Individuals

The interpretation of cross-cultural comparisons meets two major obstacles. First, since populations can differ in very many, in part eventually unknown respects, confounders are difficult to exclude. Migration studies indicate that at least genetic differences of populations may be of marginal importance. Second, the cross-cultural design assumes that risk factors occur in the apparently healthy randomized representatives of a study population to the same degree as in the actual cases of the disease. In consequence, population characteristics require complimentary data in individuals and vice versa.

Case-Control Studies

Middle-aged (35–54 years old) males of Edinburgh, Scotland, where CHD mortality is at its upper end among all countries, were screened for relatively early but hitherto undiagnosed angina pectoris (250,251); since these subjects had previously disregarded the symptoms of this first manifestation of CHD they had neither changed their life style nor received medical treatment. The measurement of antioxidant vitamins in the plasma of 110 cases (with mean total cholesterol as common as 6.2 mmol/L and an unsuspicious blood pressure) revealed in comparison to 394 matched controls significantly ($P > 0.01$) lower mean values of plasma vitamin C, carotene, and vitamin E-cholesterol ratio. The relative risk (odds ratio) of the quintile with lowest vitamin E values (< 19 μmol α-tocopherol per L) was after adjustment for other accepted risk factors 2.68-fold higher (95% confidence interval 1.07–6.70) than in the quintile with the highest vitamin E content, the latter having actually at least 1.5-fold higher vitamin E values (<28 μmol/L) than the lowest quintile (Table 2). The inverse correlation of vitamin E to angina pectoris was linear ($P = 0.02$). Therefore the high risk related to low vitamin E was independent of the generally accepted major classical risk factors as well as of age, relative body weight, and iron status (251; K. F. Gey, unpublished data, 1990). The last could be a confounder since a relative lack of this transition metal, as by iron deficiency (253) or after administration of iron chelators (36), reduces the oxygen radical damage. The relative risk of angina pectoris in the lowest quintile of plasma vitamin C (<13.1 μmol/L), that is, a level at the margin of classical vitamin C deficiency (8,187) but quite frequent in Edinburgh as well as of carotene [<0.26 μmol/L, very low according to European standards (8) and frequently associated with an increased risk of cancer of the lung and of other sites (64,187,247)] were as high as the risk of low vitamin E levels when vitamin C and carotene were not adjusted for cigarette smoking, but the relative risk dropped to 1.63 and 1.41 after adjustment (Table 2). Cigarette smoking is known to decrease plasma levels of vitamin C and of carotene, actually by both an increased requirement and a lower dietary intake (23,195,254–257). The relative risk of low vitamin A [<1.93 μmol/L, clearly below the European average (8,258)] increased after these adjustments to 2.73 but lacked a linear, statistically significant trend (Table 2).

In the United States, the arteriographic findings in patients with advanced coronary lesions were also inversely related to leukocyte vitamin C, which was a stronger determinant of coronary abnormalities than smoking and/or age (259). This study is of course less conclusive

TABLE 2 Relative Risk (Odds Ratios) of Middle-aged Scottish Men for Previously Undiagnosed Angina Pectoris[a]

Quintile	Vitamin A μmol/L	Odds ratio	Carotene μmol/L	Odds ratio	Vitamin C μmol/L	Odds ratio	Vitamin E μmol/L	Odds ratio
1	<1.93		<0.26		<13.1		<18.9	
Unadjusted		1.67		2.64		2.35		2.51
Adjusted		2.73		1.41		1.63		2.68
2	1.93–2.16		0.26–0.37		13.1–23.8		19.0–21.8	
Unadjusted		1.11		1.42		1.66		1.04
Adjusted		1.34		1.00		1.32		1.69
3	2.17–2.37		0.38–0.49		23.9–41.4		21.9–24.2	
Unadjusted		1.11		1.30		1.75		1.00
Adjusted		1.58		0.98		1.56		1.18
4	2.38–2.68		0.50–0.67		41.5–57.3		24.3–28.1	
Unadjusted		1.39		1.16		0.81		1.63
Adjusted		1.84		0.95		0.87		1.64
5	≥2.69		≥0.68		≥57.4		≥28.2	
Unadjusted		1.00		1.00		1.00		1.00
Adjusted		1.00		1.00		1.00		1.00

[a]Quintiles of essential plasma concentrations of antioxidants in controls, with and without adjustment for classical risk factors of CHD, body weight, and season, by logistic regression for age, systolic and diastolic blood pressure, total cholesterol, HDL cholesterol, nonfasting triglycerides, relative weight, smoking habit, and season for 105 cases and 382 controls with complete data. Linear trend after adjustment in logistic regression not statistically significant, except for vitamin E ($p = 0.02$).
Source: From Reference 251.

because of two serious problems of all case-control studies on rather advanced and thus usually medically treated disease. Thus, any biochemical alteration could be the result of the disease (rather than its cause) and/or of the therapy, which may have had many components and unknown effects on the antioxidant balance. For instance, calcium channel blockers as well as β-blockers and ACE (angiotensin converting enzyme) inhibitors have antioxidant properties (260). Calcium channel blockers counteract lipid accumulation in macrophages (261) and can improve at least experimental arteriosclerosis (262). The β-blocker propanolol enhances LDL metabolism and alters cholesterol metabolism in fibroblasts (263). Cholesterol-lowering agents of the fibrate type can raise vitamin E and the oxidation resistance of LDL (264). Hypocholesterolemic treatments have been reported to improve the regression of plaques (265,266). It is also possible that patients with fully advanced CHD recently replaced their former diet, which may have been involved in the initiation or promotion of CHD.

Outside Britain two case-control studies have failed to reveal an inverse correlation of plasma antioxidant vitamins to existing CHD. First, in eastern Finnish men (54 years of age) who are known for hypercholesterolemia or hypertriglyceridemia and high CHD mortality, neither plasma vitamin C (in the fair range around 40 μmol/L) nor the vitamin E-cholesterol ratio (of an order as poor as 3.4–3.8) differed between men with and without CHD (194,195). The Finnish patients might have been more heterogeneous than the Scottish men with early angina pectoris (251), since in Finland CHD was defined on the basis of either history (thus apparently including advanced and treated CHD), present symptoms (as in Scotland), or ischemia indicated only by an abnormal electrocardiogram on a bicycle ergometer test. Interestingly, Finnish subjects with abnormal electrocardiograms were related to different aspects of oxygen radicals; that is, they revealed lower levels of blood glutathione peroxidase (-5%), and selenium (-7%) (194). It cannot be excluded that the vitamin E-related discrepancy between the Finnish and Scottish data is due to hitherto unknown genetic factors and/or some special importance of hypercholesterolemia for angina pectoris in Finland, or that vitamin E is related only to special aspects of angina pectoris, such as painful vasospasms. More likely, in the Finnish study, in contrast to the Scottish, the interindividual variation in the α-tocopherol-cholesterol ratio (with a standard error of the mean below 10% in the major group with angina pectoris; $n = 445$) may have been too small to detect an interindividual relationship in the Finnish subjects, that is, the CHD risk of all Finns may have been similarly high in comparison to populations with a substantially higher vitamin E status. In the second study in the Netherlands, a country with medium to low CHD mortality, middle-aged patients (53–55 years, with hypercholesterolemia of 7.0–7.4 mmol/L) were grouped into those with the severest degree of coronary stenosis (>85% obstruction in at least one vessel) and with the second strongest stenosis (<50% in all three vessels). Both groups differed, similarly to the Finnish individuals, significantly regarding plasma selenium (-12%) but again not regarding the plasma α-tocopherol-cholesterol ratio (-6%) (267). Since the latter was, however, fairly to "optimally" high (with 4.84 and 5.13, respectively), presumably above its critical level in all subjects, as suggested by cross-cultural comparisons (185,187), any correlation regarding vitamin E was unlikely as suggested by prospective studies (later).

In summary, in individuals with early and advanced CHD an inverse correlation can be found to vitamins C and/or E, too, at least in regions with both common plasma cholesterol and with a considerable (at least 1.5-fold) variation in the antioxidant status, such as the United Kingdom. Such an inverse relation, however, may not necessarily apply to other communities, as characterized by hypercholesterolemia and/or by a lack of interindividual variation in plasma antioxidants or by a fairly high, possibly optimal level of vitamin E.

Prospective Studies

Findings in case-control studies can be erroneous inherent weaknesses of the protocol, as discussed. Therefore, further complementary evidence is desirable from the prospective approach. Blood samples are collected in a clinically healthy study cohort at baseline, and after an adequate observation period of several years the test variables are measured in cases with meanwhile obvious disease in comparison with still healthy matched controls. There have been three prospective studies in males with measurement of plasma or serum levels of essential antioxidants, all of which lacked any significant correlation of the vitamin E level at baseline with subsequent CHD mortality, but all three studies may be inconclusive, although for different reasons.

In a Finnish cohort with very high CHD mortality (92 male CHD cases with a mean age of 54 years compared to 55 matched surviving controls) and initially probably rather low levels of serum vitamin E, the latter was analyzed after 7 years storage at $-20°C$, which is known to cause a considerable loss of antioxidants [with interindividual differences (189; K. F. Gey, unpublished storage tests, 1985), and unfortunately the data were not individually adjusted for the concurrent cholesterol- and triglyceride-rich vitamin E carriers (268). If in this study cohort the interindividual vitamin E variation had been as small as in a subsequent case-control study in the same Finnish region [< 10% SEM, standard error of the mean, in angina pectoris (194,195)] it may have again been too small to reveal vitamin E as a conceivable individual risk factor.

A Dutch cohort with medium to low CHD mortality (56 CHD cases with a mean age of 68 years compared to 168 controls) had similar methodological drawbacks but in addition there was a wide age range (37–87 years) at entry and a mean age of cases (68 years) with fast-growing multimorbidity (269). Furthermore, if the vitamin E status at baseline was as high as in the subsequent Dutch case-control study [α-tocopherol-cholesterol ratio at about 5, that is, presumably beyond an "optimal" status (185)] any CHD risk attributable to vitamin E might have been as negligible as in the prospective Basel study.

The prospective Basel study described a cohort with very low CHD mortality (3000 males with 67 CHD cases with 62 years mean age compared to 2700 survivors with an initial mean age of 59 years). In this study population the plasma vitamin C at baseline tended to be slightly correlated to subsequent CHD mortality [relative adjusted risk 1.6, but $P > 0.05$ (186)]. In contrast, plasma vitamin E, either lipid standardized or expressed as α-tocopherol-cholesterol ratio, was certainly not inversely correlated to subsequent CHD death (186,187; E. Lüdin, H. B. Stähelin, and K. F. Gey, unpublished evaluations with the Cox model of proportional hazards after adjustment for age and classical risk factors, 1990), although the plasma vitamins were almost immediately assayed at baseline and properly lipid-standardized. This lack of CHD correlation to vitamin E may at least in part be due to the fact that the vitamin E status in Basel (median 34.6 μmol/L in men 40–59 years of age, lipid adjusted as in the MONICA vitamin substudy, or with an α-tocopherol-cholesterol ratio of 6.07) is in the highest range of European countries and is even higher than that in the Swiss population of the MONICA vitamin substudy (median 27 μmol/L or α-tocopherol-cholesterol ratio 4.99). Furthermore, in the Basel study the interindividual variation in vitamin E was minimal (practically no subject < 30 μmol/L lipid-standardized vitamin E). Since both Swiss study populations have a very low CHD mortality, the Basel data conceivably support the working hypothesis of CHD-protective potentials of vitamin E. It may be speculated that CHD risk factors other than a poor vitamin E status prevail in Switzerland. This would correspond to the behavior of the Basel population regarding vitamin E-related cancer. Thus, in Basel the notoriously high vitamin E levels also failed to correlate with subsequent cancers, in contrast to other study populations with poorer vitamin E levels in the United States and Finland (247,270). With regard to CHD, it seems worthwhile to mention that in the Basel cohort raised cholesterol levels (>6.5 mmol/L) were able still to increase the relative risk of subsequent CHD mortality, actually parallel to that in the vast MRFIT Study (K. F. Gey, H. B. Stähelin, and E. Lüdin, unpublished data, 1991), although in the Basel cohort the age-specific mortality by CHD is markedly smaller. This could suggest that in Basel the relatively high plasma status of vitamin E either counteracts the risk of pathological cholesterol levels independently, or if not that, the risk of excessive hypercholesterolemia (e.g., >7 mmol/L) can be significantly counteracted only by a still higher vitamin E status. Low levels of carotene or vitamin C increased cardiovascular mortality even at fair vitamin E levels (198a).

In summary, because of different drawbacks of the currently available prospective studies, more conclusive data on essential antioxidants are warranted. They can presumably best be obtained by the simultaneous comparison of different study populations with either fairly high or very low CHD risk but similar levels of classical risk factors and with a broad variation in new parameters, that is, with differences of essential antioxidants by a factor of at least 1.5. Such a study (the PRIME study of various MONICA centers comparing Irishmen with a CHD-proportioned greater number of Frenchmen) is going to embark in 1991–1992. The ideal protocol would of course include all currently conceivable risk factors for a final multivariate analysis.

Intervention Studies

Epidemiological data neither from the comparison of populations nor from case-control studies, not even from follow-ups of individuals in prospective studies, permit us fully to exclude confounding factors. Therefore, the protective properties of essential antioxidants, primarily of vitamin E, can conclusively be tested only by randomized double-blind intervention trials, probably best in study populations at high CHD risk but with poor antioxidant status and with common levels of classical risk factors. Therefore the attempted improvement in the antioxidant status must be ascertained to correlate any beneficial effects with steady plasma levels.

A CHD-related intervention trial in volunteers of the medical profession with very low CHD risk (U.S. physicians' study) revealed in a preliminary evaluation a 44% reduction in all major coronary events (of myocardial infarction, revascularization, or cardiac death) by the self-reported consumption of 50 mg β-carotene every second day alternately with acetyl-salicyclic acid (actually an esterified plant phenol) over 5 years, after adjustment for age and acetylsalicylic acid (45), but this effect was only clear with the combination (C. H. Hennekens, personal communication, 1991), and the results of this very special study cohort may be difficult to extrapolate to common populations.

The potential prevention of cancer by essential antioxidants is at present already being tested in more than 20 intervention studies (mostly initiated by the U.S. National Cancer Institute (270a). At least one, the Finnish smoker study, might provide some accidental information on the conceivable reduction of CHD in male Finnish smokers who receive daily 50 mg vitamin E and/or 15 mg β-carotene. Unfortunately there is no indication that the 15–20% rise in plasma vitamin E as it has actually occurred in these Finnish smokers (J. Virtamo, personal communication, 1991) will be sufficient to reduce the very high rate of CHD in this population with very high cholesterol levels. A substantial rise in plasma vitamin E, for example, by a factor 1.5–2.0 as conceivably needed in Finland, was virtually found (196) after doses of 300 mg vitamin E (220 IU, in combination with vitamin C and β-carotene). Another study project in postmenopausal women (probably sponsored by the National Institutes of Health) is going to test the prevention of breast cancer as well as cardiovascular events by β-carotene, vitamin E, and/or acetylsalicyclic acid (C. H. Hennekens, personal communication, 1991).

Since at present no primarily CHD-related intervention studies test the preventive potentials of natural essential antioxidants in appropriate doses, an expert meeting of the National Heart, Lung and Blood Institute, Bethesda, considered options in late 1991 (270b). It was felt that the epidemiological evidence is (although still incomplete) sufficiently congruent to plan appropriately size randomized double-blind primary intervention studies in high risk subjects. This is of course also justified by the fact that at present there is no real treatment of CHD and cancer, but only prevention (Dr. Mahler, former Director General of WHO, statement, 1987). The previous population-based multifactorial intervention campaign (focusing on the reduction of classical risk factors, such as atherogenic cholesterol-rich lipoproteins, hypertension, and advice to abstain from cigarettes) have lacked strong benefits (271), and thus the previously attempted reduction in classical risk factors may hardly be expected fully to eliminate CHD. Unfortunately, the unequivocally desirable "normalization" of LDL cholesterol has been limited. Thus, the present dietary recommendations (by international and many national nutrition boards) to reduce total calories, particularly those from animal fat (19), are by many individuals not easy to follow continuously, and even if applied they are insufficient in many patients to decrease cholesterol levels to that of small risk, 5.1–5.7 mmol/L. All previously tested cholesterol-lowering drug classes had other problems. For instance, clofibrate reduced coronary events at the expense of cancer mortality, which could in part be related to the fact that in westernized societies very low cholesterol levels (actually below 4.7 mmol/L) can be associated with an increased cancer risk. Bile sequestrants are not easy to consume and potentially reduce the intestinal absorption of essential nutrients, such as PUFAs and liposoluble vitamins. The very potent inhibitors of the first key enzyme of cholesterol biosynthesis (HMG-CoA reductase) also block the synthesis of ubiquinol-10 (272), which participates not only in the respiratory chain but is also an endogenous antioxidant protector of LDL (221) and can endanger heart functions. Nicotinic acid reduced both coronary and total mortality without serious side effects, but compliance is limited because of flush, and thus improved preparations are needed (272a). In contrast, the exploration of CHD-preventive potentials of essential antioxidants may be favored by their lack of safety problems (23,273) as regular food ingredients as well as by the possibility to optimize the antioxidant status easily and substantially by dietary supplements, either by capsules (e.g., for initial experimental purposes) or by food supplementation (for population-based provision as common nowadays in many countries, e.g., for vitamin A, D, or B_1).

If the multifactorial multistage process of arteriosclerosis demands a multirisk factor intervention, it seems most logical simultaneously to optimize all pertinent essential antioxidants, in case of need, together with the reduction in classical risk factors and an improvement in the life style in general (274). The evaluation of the potentials of an optimal antioxidant status (instead of a single although prominent component, such as vitamin E) may be based on two facts:

1. Epidemiological studies on CHD (as well as cancer) have revealed correlations bebetween vitamins E and C as well as A and carotene.
2. Experimental data have indicated corresponding synergistic links in the body's multilevel defense system against radical injury.

Thus, the synchronous and balanced optimization of all antioxidants may have the best chance to minimize the CHD hazards of radicals.

Pilot Studies on the Treatment of CHD with Antioxidants

Early claims that vitamin E was effective in the treatment of angina pectoris or heart failure have not been confirmed (70). If the design was double-blind and randomized, the number of patients was inconclusively small (77,275), and unfortunately clinical tests have never included measurement of the antioxidant plasma status before and/or during treatment.

Recently CHD patients subjected to percutaneous transluminal coronary angioplasty (dilatation of coronaries by a balloon catheter) were subsequently given vitamin E supplements (in addition to calcium blockers and aspirin). The initially poor plasma vitamin E level of 21 μmol α-tocopherol per L doubled and reduced the remaining angiographically documented coronary stenosis from 47.5% in controls to 35.5%, and an corresponding improvement was found for an abnormal angiogram or thallium test (scintographic measurement of ischemic regions) or exercise stress test, but the difference did not reach statistical significance ($P = 0.06$) since the sample size of 100 randomized patients was too small with regard to the great qualitative and quantitative differences in advanced coronary lesions between individuals (92). The latter would also be a major handicap for testing the preventive properties of antioxidants in an unavoidable surgical coronary "reperfusion injury," that is, in the common bypass operation, as recently suggested (98).

Prudent Antioxidant Intake

In Britain, a country with a very high CHD mortality, the national intake of vitamin E before 1971 was assumed to be at most 5 IU daily (276), resulting in a critically low vitamin E-cholesterol ratio of about 3 (277), and recently a vitamin E intake of 8.4 and 9.3 mg/person/day was calculated with a downhill gradient from England to Scotland (249,278), that is, an intake below the present U.S. RDA for vitamin E of 15 IU (10 mg RRR-α-tocopherol) for male adults. Although a recent official survey (249) reports a calculated median intake of 59 mg vitamin C per male per day for the United Kingdom, the actual supply of vitamin C may be lower, at least in Scotland (43,279). The particularly low antioxidant status in Scottish cities (65) is not unexpected since in Glasgow every third male adult never consumes fresh fruits or vegetables (W. C. S. Cairns Smith, personal communication, 1990). Also in the United States, however, a nation with medium to high CHD mortality, only a very few individuals approach the recommended level of antioxidant intake: actually 41% of the U.S. population has no fruit on a survey day (280), and 50% of the male population consume only 7.3 mg vitamin E and 23% do not reach the desirable vitamin E/PUFA ratio > 0.4 (281). The adequacy of the present RDA for vitamin E has, however, again and again been questioned (282–285). Thus, in 1968 twice the present RDA was recommended by the U.S. National Academy of Sciences on behalf of balance studies in men (286), but in 1974 the RDA was reduced to the present level since it was thought to correspond to the actual intake in the United States. When the current dietary recommendations were followed to reduce PUFA and monoenic fatty acids (from vegetable oils, the main natural vitamin E carriers) together with 20% of total calories (13–15) it would be very difficult to approach the present RDA of 15 IU vitamin E daily from a common diet (287,288). On the other hand, in case of a continued increase in the PUFA consumption, a reevaluation of the vitamin E/PUFA ratio is warranted as well (26,70,284). Regarding the plasma level of the European study populations with very low CHD mortality (Fig. 1) and of Scottish subjects with the lowest risk of angina pectoris (Table 2), as well as the diminished exhalation of pentane (an in vivo marker of PUFA peroxidation) in vitamin E-supplemented normal adults (289,290), plasma levels of at least 25–30 μmol α-tocopherol per L may be desirable for CHD-related "optimum health" as defined by WHO. Considering the inverse correlation between plasma vitamin E and subsequent cancer risk, as established in several prospective studies (247,270), at least 30–35 μmol lipid-standardized α-tocopherol per L may be recommended (258).

A poor status of any antioxidant vitamin may still prevent overt deficiency syndromes but might not yet provide optimum body functions and/or health as defined by WHO, including protection against CHD. This may not only be postulated regarding CHD and cancer (as discussed earlier) but also in regard to other vital functions. Thus the vitamin E requirement for the optimal response of the immune system seems to be several times higher than that needed to prevent frank, classical vitamin E deficiency in the animal (69,70,291,292). Similarly, as little as 10–15 mg ascorbic acid prevents human scurvy (79), but an optimum vitamin C status requires about 75–100 mg ascorbic acid daily, that is, the most commonly recommended daily allowance for healthy adults (8,23,258,293,294). In consequence, the proper balance between common health hazards, dietary PUFAs, and optimal levels of natural antioxidants remains to be reevaluated with respect to CHD.

In the epidemiological studies the vitamin E status resulted almost exclusively from a regular diet. The actual vitamin E consumption of individuals is very difficult to estimate because of the well-known high variation in the vitamin E content in industrially processed vegetable oils (34,295) and the unknown level in popular prefabricated deep-freezer menus and in otherwise industrially modified food items. Unfortunately, legislation is yet missing for a food label indicating the content of absolute vitamin E, of the vitamin E/PUFA ratio, and/or of "net" vitamin E [extra over minimal requirement for PUFA protection (34a)]. In relatively short-term experiments vitamin E levels of 25–30 μmol/L require in nonsmokers daily supplements of at least 30–60 mg vitamin E or, more safely, at least 100 mg, depending on the concomitant status of PUFAs and of adequate quantities of vitamin C (187,189,282–285,287,288,296–299). Such levels have experimentally been achieved by daily doses of about 60–100 IU. It is conceivable, however, that maximum antioxidant protection requires higher vitamin E doses, such as 400 IU daily (284,289,290), which roughly double the "common" plasma level of α-tocopherol and the α-tocopherol-cholesterol ratio (284,285,300). A corresponding increase in the vitamin E status may also be achieved by 220 IU daily in combination with vitamin C, β-carotene, and small amounts of selenium (196). Studies on breath pentane (a marker of lipid peroxidation in vivo) in vitamin E-deficient children suggests that supplements become effective only after simultaneous normalization of the selenium status (300a). Clearly, after oral supplmentation of vitamin E, the resulting plasma status does not depend only on synergistic antioxidants but on the dietary fat content as well since the intestinal absorption of the liposoluble vitamin E is to some extent controlled by dietary lipids (287,288). Of course, concurrent optimization of all essential antioxidants may be particularly desirable where CHD mortality is high and the overall antioxidant status tends to be poor, for example, geographically in the north of the British Isles, or at increased radical stress, such as by cigarette smoking for vitamin C and β-carotene (254,256,257,301) or vitamin E (173,302,303,309).

BENEFICIAL ACTIONS OF ANTIOXIDANT VITAMINS ON ATHEROGENESIS

It is beyond the scope of this review to discuss all conceivably beneficial effects of essential antioxidants in the multifactorial multistage development of arteriosclerosis. Nevertheless, some intriguing findings should be listed to show that essential antioxidants can potentially modulate almost all processes of crucial importance in atherogenesis.

I. Circulating plasma lipoproteins
  A. "Electronegative" LDL attenuated by vitamin E in vivo (304).
  B. TBARs [Thiobarbituric acid-reactive material (304a,304b) indicating lower antioxidative resistance of LDL] decreased by vitamins C and/or E in vivo (8,89,109,196,305–309,311,312).
  C. Lipoprotein(a) diminished in vivo by vitamin E (313) and vitamin C (75).
  D. Glycosylated proteins attenuated by vitamin E in vivo (315).
  E. HDL$_3$ increased by vitamin E in vivo (90,91,316,317,317a), but not in all subjects.
II. Fatty streaks
  A. Endothelial modulation or dysfunction
    1. Uptake of OM-LDL reduced by vitamin E in vitro and in vivo (318).
    2. Injury due to oxygen radicals, OM-LDL, or MO diminished by vitamin E in cultures and in vivo (112,117,129,153,154) and by vitamins C and E in vivo (155).
    3. Endothelial barrier function preserved by vitamin E in vitro (154, 319).
    4. Calcium release attenuated from storage sites by vitamin E in vitro (320).
    5. Phospholipase A$_2$ inhibited by vitamin E in vitro and in vivo (321).

6. Prostacyclin production increased by vitamin E in normal endothelium and restored in atherosclerotic endothelium (89,322,323,325–327,327a,328–333) at simultaneously unaltered or depressed thromboxane formation in vivo (196,334). Vitamin C acts correspondingly in vitro and in vivo (335–337).

7. Lipoxygenases (in part of other sources) inhibited by vitamin E in vitro and in vivo (338–345), by vitamin C directly and/or indirectly (233,346–348), and possibly by β-carotene and vitamin A in vitro (344,345).

8. Endothelium-induced LDL modification reduced by vitamin E in vitro (124).

B. Proteoglycan structure: Control of biosynthesis in vitro and in vivo by vitamin C (23,293,349,349a), vitamin E (350,405), and vitamin A (351).

C. Oxidative susceptibility of isolated LDL

1. Oxidation resistance prolonged in vitro and ex vivo by substantial vitamin E increase, that is, to levels comparable to populations of low CHD risk (see Chapter 46) and thus less uptake of OM-LDL by MO (49,57–62,149,277). Corresponding effects of vitamin C in vitro (49,59,217,219,220,220a,b).

2. Transition metals (copper and iron) kept in reduced state and chelated by optimal levels of vitamin C in vitro (8,23,36,240). Iron- and copper-induced lipid peroxidation and toxicity are counteracted in vivo by vitamins E and C (242,354–357). Ceruloplasmin activity is diminished by vitamin C (358).

D. Immunological and inflammatory responses

1. Improvements, for example, of mitogenic response of T cells ex vivo by vitamin E (273,291,292,311,358a,359–369), vitamin C (23,370), β-carotene, and vitamin A (291,292,371–373).

2. Prostaglandin production and lipoxygenase inhibited in immune cells by vitamin E ex vivo (311,325,374,375) and by vitamin C (346).

3. IL-2 production by PMN increased by vitamin E ex vivo (311) or inversely related to low vitamin E intake (371), depending on challenging mitogen.

4. Cytotoxicity of OM-LDL on lymphoblasts reduced by vitamin E in vitro (375a).

E. Monocytes-Macrophages

1. Oxygen radical-induced damage reduced by vitamin E in vitro (376).

2. Chemotactic response and migration improved in vitro by vitamin E (163,165,378), vitamin C (166–168,377), and vitamin A (164).

3. Phagocytic functions improved by vitamin E ex vivo (165,169,366).

4. Respiratory burst reduced by vitamin E ex vivo (170,171).

5. MO-challenged LDL modification reduced in vitro by vitamin E (172).

6. Secretion of proteolytic enzymes reduced by vitamin E (173).

7. Intracellular accumulation of cholesteryl esters as well as of ceroid and chemoluminescent lipid peroxides counteracted by vitamin E in vitro (66–68,156,158–161) and in vivo (161), and with vitamin E-induced activation of cholesteryl ester hydrolysis and ACAT inhibition ex vivo (156) as well as attenuation of intracellular lipid peroxidation in vitro (67,68,159) with corresponding deactivation of lysosomal enzymes by vitamin E ex vivo (171) and diminution of OM-ApoB catabolites (162).

8. Lipoxygenase inhibited by vitamin E (379).

9. IL-1 expression prevented by vitamin E in vitro and ex vivo (174–176).

10. Protein kinase C activation prevented by vitamin E ex vivo (170).

III. Fibrous plaques and smooth muscle cell proliferation

A. Smooth muscle cells

1. Proliferation inhibited by vitamin E in vitro (113,380–382) parallel to reduced activation of protein kinase C (382; see Chapter 28).

2. Calcium efflux attenuated by vitamin E in vitro (383).
3. Lipoxygenase inhibited by vitamin E in vitro (383).
4. Cholesteryl ester accumulation counteracted by vitamin E in vitro (113).

B. Fibroblasts: prostaglandin synthesis stimulated by vitamin E in vitro (384).

IV. Myocardial infarction

A. Arrythmia and myopathy
  1. Catacholamine-provoked effects attenuated by vitamin E in vivo (324,385–387).
  2. Occlusion-induced fibrillations decreased by vitamin E in vivo (388).
  3. Left ventricular function abnormalities improved by vitamin E in vivo (389).
  4. Injury of myocytes by oxidized fat reduced by vitamins E and C in vitro (110).

B. Platelets
  1. Adhesion onto collagen, fibrinogen, and fibronectin prevented by vitamin E ex vivo (390,391).
  2. Aggregation inhibited by vitamin E probably only in poor vitamin E status (196, 323,392–394) or in contraceptive-using women (395) but in synergism with aspirin (395a).
  3. Prostacyclin production increased by vitamin E in vivo (396).
  4. Thromboxane formation inhibited by vitamin E in poor states of vitamin E (323, 392,397).
  5. Lipoxygenase inhibited by vitamin C in vitro and in vivo (393,398,399).
  6. PAF (platelet-activating factor) production in PMN reduced by vitamin E in vivo (400).

C. Coagulation factors
  1. Fibrinogen level decreased by vitamin E in vivo (317a,401). Fibrinolytic activity increased by vitamin C (403,404).
  2. Antithrombotic proteoglycans maintained by vitamin E in vivo (405).
  3. Aggregatory factor synthesis inhibited by vitamin E in vitro in whole blood (406).

D. "Reperfusion injury": infarct area, tissue damage, and peroxides decreased by vitamins E and/or C in vitro, in situ and in vivo (92,98,99,100,101,242,332,407–418).

Taken together essential antioxidants can potentially prevent even the initial pathophysiological transformations of all hallmark cells in the multistage atherogenesis of endothelial cells, of monocytes-macrophages (and T lymphocytes), and of smooth muscle cells (and fibrocytes), as well as of platelets, and may later sustain their vital functions.

From the present data three protective mechanisms of essential antioxidants may be inferred that differ by site.

1. Common radical scavenging in all aqueous and liposoluble compartments where vitamins E and C may act as mobile "commodity" antioxidants. This is obvious for vitamin C but is also true for vitamin E. Thus, α-tocopherol, being intercalated in the lipid layers of LDL or of membranes (25,30,70,229,419), can scavenge any lipid peroxides easily and in time by fast lateral diffusion into the neighboring lipid area of about 1000–3000 double bonds.

2. Site-specific modulations of redox functions by antioxidants vitamins, for example, inhibition of lipoxygenases or phospholipase A_2 by vitamins C and E via a dose-dependent regulation of the "peroxide tone" but possibly also by keeping the active center in the inactive ferric state (338,340,341,342,343,420,429) or indirectly by preventing the oxygen radical-induced release of calcium from the storage vesicles (320). The control of the lipid peroxide tone may be easy for radical scavengers since lipid peroxides are regular intermediates in PUFA peroxidation by lipoxygenases (421).

3. Site-specific modulation of "second messengers" or signal transducers as well as of messengers for cell-cell interaction by compounds with antioxidant structure but without

information yet on the functional importance of their radical-scavenging properties. This applies, for instance, to the inhibition of IL-1 expression in cultured macrophages (174), which could have greatest consequences for the progression of the fatty streak into the fibrous lesion with muscle cell proliferation (2,422–424). Furthermore, although protein kinase C can be activated by oxidative modification independently of calcium and phospholipids (425,426), only structure specificity has thus far been demonstrated for the prevention of protein kinase C activation by α-tocopherol in macrophages ex vivo (170) and in cultured smooth muscle cells (113,380,382; see Chapter 28). Since protein kinase C was assumed to play a pivotal role in modulating relationships between the signal transduction pathways of many intracellular processes, vitamin E could as the site-specific ligand of protein kinase C potentially control many crucial intracellular phenomena of atherogenesis, such as dysfunctions and lipid accumulation in endothelial cells (427), the "respiratory burst" of macrophages (170,428), smooth muscle cell proliferation (380–382; see Chapter 28), reverse cholesterol transport (429), and the synthesis of PAF (430).

Of course, the relative importance of such beneficial actions of essential antioxidants requires further elucidation and in most instances reconfirmation in the human with and without CHD. Therefore previous experience on the resistance of isolated LDL (against copper-induced oxidation) recommends not only rigorous analytical quality control but also caution in design and interpretation. Thus, all laboratories found consistently (see Chapter 46;49,61,221, 431) that the oxidation resistance of isolated LDL from clinically healthy volunteers (without vitamin E substitution) shows great interindividual differences, which have also been found for macrophage-induced LDL modifications (146b), as well as for cholesterol peroxidation by monocytes from different blood donors (67,68). When initially in these relatively small study groups the level of α-tocopherol in LDL did not clearly correlate with the oxidation resistance of LDL, components other than vitamin E were considered relatively more important, such as ubiquinol-10 (221) and/or preformed hydroperoxides, PUFAs, and/or the antioxidant/PUFA ratio, the mobility of α-tocopherol, the structure of ApoB, unknown antioxidants (see Chapter 46; 49,60–62,431), possibly the relative proportion of the dense, more easily oxidizable subfraction of LDL (432,433,433a), and others. An obvious correlation between vitamin E and the oxidation resistance of LDL, however, can be missed for several reasons. First, in subjects with a poor vitamin E status, vitamin E below a critical level does not prevent LDL oxidation effectively (431). Second, in study groups with a fair vitamin E status (without vitamin E substitution) the interindividual differences can (at least at a locally similar life style and diet) be too small to detect any strong correlation. Actually even in larger epidemiological study populations (Fig. 4) the interindividual variation in lipid-standardized vitamin E is mostly only moderate. When, however, the α-tocopherol levels of LDL were substantially increased in vitro, the oxidation resistance of LDL was significantly prolonged as expected (see Chapter 46; 62,431). Correspondingly, in vivo a dose-dependent, up to 2.5-fold increase in vitamin E in isolated LDL (by oral supplements of vitamin E of 125 up to 1200 mg daily) doubled the oxidation resistance of isolated LDL in comparison to baseline levels (see Chapter 46; 49,60,61,149). When in the greatest study group all data before and after vitamin E substitution were statistically evaluated, at least half of the variation (r^2 = 0.51 in univariate analysis) in the oxidation resistance of isolated LDL depended on its vitamin E content, despite that the improvement in the oxidation resistance of LDL by vitamin E enrichment still showed considerable individual differences (see Chapter 46; 60–62, 149). Since total antioxidants predicted only about 60% of the oxidative susceptibility of LDL (149), vitamin E may be valued as the potentially strongest improver of the oxidation resistance of LDL. Interestingly, the greatest improvement by vitamin E substitution occurred in subjects with a negligible protection from vitamin E-independent components (see Chapter 46; 149). The oxidative resistance may be an indicator of pathophysiological importance since uptake of OM-LDL by macrophages occurs only after complete consumption of vitamin E (49). Of course, the inverse and highly concentration-dependent ex vivo correlation between α-tocopherol and the oxidation resistance of LDL may at least in part be related to the predominant predictive power of lipid-standardized plasma vitamin E with regard to the great differences of CHD mortality in European populations as well as in early angina pectoris in Scotland (above) in which the lipid-adjusted vitamin E status differed at least by a factor of 1.35–1.5, too. Of course, the intriguing correlation between the plasma vitamin E status and oxidation resistance of LDL does not exclude the possibility that other biological actions of vitamin E (as listed earlier) simultaneously play a crucial role in atherogenesis as well.

Clearly, the considerable remainder of at least 40% of the oxidative susceptibility of LDL (see Chapter 46; 60,61,149) depended (again with considerable interindividual differences) on vitamin E-independent variables, as mentioned. Their differential measurement will be needed for a multivariate analysis of the oxidation resistance of LDL, that is, for a conclusive assessment of vitamin E, too. Anyway the PUFA pattern of LDL, or more precisely the vitamin E/PUFA ratio, likely plays some role. Thus, LDL from animals fed a special monoene(oleic acid)-rich sunflower oil was found less susceptible than LDL from animals fed regular PUFA-rich sunflower oil (434). It remains to be considered that monoene-rich oils, such as olive oil, have a higher α-tocopherol-PUFA ratio (olive oils of about 1–2) than common polyen-rich vegetable oils, such as regular sunflower oil (ratio about 0.2–0.7): the net vitamin E supply from cold-pressed olive oils can be about 3- to 5-fold higher (435,436, K. F. Gey and E. Bühler, unpublished comparisons of olive oils, 1989). Correspondingly, the lower cytotoxicity of OM-LDL from animals fed an oleic acid-rich diet seems primarily to be due to an increase of vitamin E in LDL (437).

Since the actual CHD-preventive potential of an optimized antioxidant status remains to be tested by a conclusive intervention study, it is highly desirable simultaneously to incorporate mechanistic substudies ex vivo to test some very intriguing actions of essential antioxidants. Thus it would be relatively easy in subjects of intervention trails to compare ex vivo the oxidation resistance of isolated LDL and the catabolism of OM-LDL, as well as the production of cytokines by monocyte-derived macrophages, the adhesion of thrombocytes, and so on.

SUMMARY

The risk of coronary heart disease can only in part be ascribed to classical risk factors, such as hypercholesterolemia, hypertension, and smoking. A protective role of essential antioxidants is suggested by experimental data in animals and humans from deficiency and supplements of antioxidant vitamins, as well as from dietary surveys and more specifically from recent epidemiological comparison of major essential antioxidants in plasma of men:

In 16 European populations (of the Optional Study on Antioxidant Vitamins and PUFAs, WHO/MONICA Project) with up to sixfold differences in the age-specific mortality from CHD. In the major subgroup of 12 study populations having common total plasma cholesterol in the medium range (5.7–6.2 mmol/L), usual blood pressure, and comparable smoking habits, these classical risk factors lacked a statistically significant correlation to CHD mortality by univariate analysis, but the absolute level of plasma vitamin E (α-tocopherol) showed a very strong inverse correlation ($r^2 = 0.63$, $P = 0.002$), plasma vitamin C a moderate inverse correlation ($r^2 = 0.41$, $P = 0.03$), and carotene and vitamin A weak correlations ($r^2 = 0.21$ and 0.16). After inclusion of the 4 hyper- and hypocholesterolemic study populations, total cholesterol and diastolic blood pressure had a moderately direct association with CHD ($r^2 = 0.29$, $P = 0.03$, and $r^2 = 0.25$, $P = 0.05$, respectively), but their relative importance remained inferior to that of vitamin E. In stepwise regression and multiple-regression analysis, the CHD mortality of all 16 study populations was predictable to 62% by lipid-standardized vitamin E, to 79% by vitamin E and total cholesterol, to 83% after inclusion of lipid-standardized vitamin A (retinol), to 87% by all these parameters plus diastolic blood pressure, and up to 90% after further inclusion of vitamin C (ascorbic acid).

In individuals with previously undiagnosed angina pectoris in Edinburgh, Scotland who had common levels of plasma cholesterol and blood pressure, the lowest quintile of plasma vitamin E predicted a 2.68-fold higher risk of angina pectoris ($p = 0.02$) independently of generally accepted risk factors, in comparison to the quintile with highest vitamin E levels. Low plasma concentrations of vitamin C and carotene revealed relative higher risks, too (2.35 and 2.35, respectively, before adjustment for classical risk factors), but were confounded, presumably in part due to cigarette smoking (relative risk after adjustment 1.63 and 1.41, respectively), whereas the adjusted relative risk of low vitamin A (2.73) was clearly increased although without statistical significance.

In conclusion, in these epidemiological data in European study populations and Scottish individuals essential antioxidants in plasma were inverse, that is, presumably protective factors of CHD that seemed to be relatively more important than the major direct risk factors. The gross rank order of essential antioxidants was vitamin E > vitamin C > vitamin A = carotene. Potentially protective vitamin E levels may be above 25–30 μmol/L of lipid-standardized α-tocopherol, presumably depending on the concurrent status of synergistic antioxidants. Potentially protective vitamin C levels may be above 35 μmol/L.

Since antioxidant vitamins seem to be substantially complemented by the well-established classical risk factors, the latter should be conclusively tested as potentially protective factors of CHD. This requires properly designed prospective studies, particularly double-blind intervention studies in randomized CHD-prone subjects of poor antioxidants status who await optimization by supplements of all essential antioxidants but particularly of high amounts of vitamin E.

The oxidative modification hypothesis of CHD, as recently emerging from data in experimental animals, in cell cultures, and in the human, postulates subintimal extracellular oxidative modification of LDL as a major prerequisite for its rapid and uncontrolled uptake by macrophages and thus for the initiating of foam cells and fatty streaks (56). The complementary antioxidant hypothesis (8,64) focused on antiatherosclerotic properties of natural essential antioxidants, such as vitamin C and E regarding antioxidant protection of LDL, as well as demonstrated in vitro and ex vivo (see Chapter 46), but considers also the present experimental evidence for beneficial modulation by antioxidant vitamins of almost all steps of the multifactorial multistage process of atherogenesis. Thus, the antioxidant vitamins E, A, and C may potentially already hinder the initial endothelial transitions, and the adhesion of monocytes to the endothelium, as well as their subsequent chemotactic migration into the subendothelial space. Vitamins C and E may furthermore diminish the hydroperoxide load of LDL by inhibition of lipoxygenases, modify the effect of transition metals involved in oxidative modifications of LDL, and improve the general response of macrophages but especially their LDL catabolism, resulting in a reduction of intracellular cholesteryl esters and of ceroid. Antioxidant vitamins may also downregulate cell-cell interactions of growth- and/or immunoresponse-related cytokines (e.g., by inhibition of interleukin-1 expression by macrophages), increase the endothelial prostacyclin production, inhibit smooth muscle proliferation (e.g., by preventing the activation of protein kinase C), modulate arterial proteoglycans and thus crucial physicochemical properties of arteries, and inhibit platelet adhesion and in poor antioxidant states also platelet aggregation, and antioxidant vitamins may finally reduce the "reperfusion injury" related to myocardial infarction. Since the actual CHD-preventive potential of an optimized antioxidant status remains to be tested by a conclusive intervention study, it is highly desirable simultaneously to include mechanistic substudies ex vivo to test or differentiate the most intriguing actions of essential antioxidants in arteriosclerosis.

ACKNOWLEDGMENTS

This work is related to a research grant from the Swiss Foundation of Cardiology. The competent and perseverant secretarial help of Mrs. Kathrin Gfeller and Mrs. Barbara Bangerter is gratefully acknowledged.

REFERENCES

1. Ross R. The pathogenesis of atherosclerosis—an update. N Engl J Med 1986; 314:488-500.
2. Steinberg D. Lipoproteins and atherogenesis: current concepts. In: Hallgren B, et al. eds. Diet and prevention of coronary heart disease and cancer. New York: Raven Press, 1986; 95-112.
3. Steinberg D. Metabolism of lipoproteins and their role in the pathogenesis of atherosclerosis. Atherosclerosis Rev 1988; 18:1-23.
4. Steinberg D, Parthasarathy S, Carew TE, Khoo JC, Witztum JL. Beyond cholesterol: modifications of low-density lipoprotein that increases its atherogenicity. N Engl J Med 1989; 320:915-24.
5. Ross R, Masuda J, Raines EW. Cellular interactions, growth factors, and smooth muscle proliferation in atherogenesis. Ann N Y Acad Sci 1990; 598:102-12.
6. Steinberg D, Witztum JL. Lipoproteins and atherogenesis. Current concepts. JAMA 1990; 264:3074-52.
6a. Keys A. Seven countries, a multivariate analysis of death and coronary heart disease. Cambridge, MA: Harvard University Press, 1980.
7. WHO (World Health Organization)/MONICA Project Principal Investigators. MONICA sites and personnel. Wld hlth statist annu 1988:138-40.
7a. WHO (World Health Organization)/MONICA Project Principal Investigators. The WHO MONICA project: a worldwide monitoring system for cardiovascular diseases. Wld hlth statist annu 1989: 27-149.
8. Gey KF. On the antioxidant hypothesis with regard to arteriosclerosis. Bibl Nutr Dieta 1986; 37:53-91.
8a. Gey KF. The antioxidant hypothesis of cardiovascular disease: epidemiology and mechanisms. Biochem Soc Transact 1990; 18:1041-5.
9. Phillips RL, Lemon FR, Beeson WL, et al. Coronary heart disease mortality among Seventh-Day Adventists with differing dietary habits: a preliminary report. Am J Clin Nutr 1978; 31:191-8.
10. Kushi LH, Lew LA, Stare FJ, et al. Diet and 20-year mortality from coronary heart disease. The Ireland-Boston diet-heart study. N Engl J Med 1985; 312:811-8.
11. Arntzenius AC, Kromhout D, Barth JD, et al. Diet, lipoproteins, and the progression of coronary atherosclerosis. The Leiden intervention trial. N Engl J Med 1985; 312:805-11.
12. Reference deleted
13. National Research Council. Diet and health: implications for reducing chronic disease risk. Washington, DC: National Academy Press, 1989.
14. US Department of Agriculture, US Department of Health and Human Services. Nutrition and your health: dietary guidelines for Americans. Washington, DC: Government Printing Office, 1980 (Home and Garden Bulletin 232).
15. Miettinen TA, Naukkarienen V, Huttunen JK, Mattila S, Kumlin T. Fatty acid composition of serum lipids predicts myocardial infarction. Br Med J 1982; 285:933-6.
16. Valek J, Hammer J, Kohout M, et al. Serum linoleic acid and cardiovascular death in postinfarct middle-aged men. Atherosclerosis 1985; 54:11-8.
17. Riemersma RA, Wood DA, Butler S, et al. Linoleic acid content in adipose tissue and coronary heart disease. Br Med J 1986; 292:1423-7.
18. Wood DA, Riemersma RA, Butler S, et al. Linoleic acid and eicosapentanoic acids in adipose tissue and platelets and risk of coronary heart disease. Lancet 1987; 1:17-80.
19. Brown WV. Dietary recommendations to prevent coronary heart disease. Ann NY Acad Sci 1990; 598:376-88.
20. Katsouyanni K, Skalkidis Y, Petridou E, Polychronopoulou-Trichopoulou A, Willett W, Trichopoulos D. Diet and peripheral arterial occlusive disease: the role of poly-, mono-, and saturated fatty acids. Am J Epidemiol 1991; 133:24-31.

21. D'Aquino M, Dunster C, Wilson R. Vitamin A and glutathione-mediated free radical damage: competing reactions with polyunsaturated fatty acids and vitamin C. Biochem Biophys Res Commun 1989; 161:1199-203.

22. Bielski BHJ. Chemistry of ascorbic acid radicals. In: Seib PA, Tobert BM, eds. Ascorbic acid: chemistry, metabolism and uses. Adv Chem Ser, Am Chem Soc 1982; 200:81-100.

23. Moser U, Bendich A. Vitamin C. In: Machlin LJ, ed. Handbook of vitamins, 2nd ed. New York: Marcel Dekker, 1990; 195-232.

24. Di Mascio P, Kaiser S, Sies H. Lycopene as the most efficient singlet oxygen quencher. Arch Biochem Biophys 1989; 274:532-8.

25. Burton GW, Ingold KU. Vitamin E as in vitro and in vivo antioxidant. Ann N Y Acad Sci 1989; 570:7-22.

26. Muggli R. Dietary fish oil increases the requirement for vitamin E in humans. In: Chandra RK, ed. Health effects of fish oils. St. John's, Newfoundland: ARTS Biomed. Publ., 1989; 203-310.

27. Mulholland CW, Strain JJ. Total peroxyl radical trapping ability of serum: relationship to secondary antioxidant concentrations. Biochem Soc Trans 1990; 18:1169-70.

28. Wayner DDM, Burton GW, Ingold KW, Barclay LRC, Locke SJ. The relative contributions of vitamin E, urate, ascorbate and proteins to the total peroxyl radical-trapping anti-oxidant activity of human blood plasma. Biochim Biophys Acta 1987; 924:408-19.

29. Njus D, Kelley PM. Vitamins C and E donate single hydrogen atoms in vivo. FEBS Lett 1991; 284:147-51.

30. Niki E, Yamamoto Y, Kumuro E, Sato K. Membrane damage due to lipid oxidation. Am J Clin Nutr 1991; 53:201S-5S.

31. Sato K, Niki E, Shimasaki H. Free radical-mediated chain oxidation of low density lipoprotein and its synergistic inhibition by vitamin E and vitamin C. Arch Biochem Biophys 1990; 279:402-5.

32. Hjermann I, Velve Byre K, Holme I, Leren P. Effect of diet and smoking intervention on the incidence of coronary heart disease. Report from the Oslo study group of a randomized trial in healthy men. Lancet 1981; 2:1303-10.

33. Miettinen M, Turpeinen O, Karvonen MJ, Elosuo R, Paavilainen E. Effect of cholesterol-lowering diet on mortality from coronary heart disease and other causes. Lancet 1972; 2:835-8.

34. McLaughlin PJ, Weihrauch JL. Vitamin E content of foods. J Am Diet Assoc 1979; 75:649-65.

34a. Bässler KH. On the problematic nature of vitamin E requirements: net vitamin E. Z Ernährungswiss 1991; 30:174-80.

34b. Kumpusalo E, Karinpaa A, Jauhiainen M, Laitinen M, Lappetelainen R, Maenpaa PH. Multivitamin supplementation of adult omnivores and lactovegetarians: circulating levels of vitamin A, D and E, lipids, apoproteins and selenium. Int J Vitam Nutr Res 1990; 60:58-66.

34c. Pronzcuk A, Kipervag Y, Hayes KC. Vegetarians have higher plasma alpha-tocopherol relative to cholesterol than do have nonvegetarians. J Am Coll Nutr 1992; 11:55-5.

35. Kromhout D, Bosschieter EB, de Leszenne Coulander C. The inverse relation between fish consumption and 20-year mortality from coronary heart disease. N Engl J Med 1985; 312:1205-9.

36. Halliwell B, Gutteridge JMC. Free radicals in biology and medicine, 2nd ed. Oxford: Clarendon Press, 1989.

37. Sies H, ed. Oxidative stress, 1st and 2nd ed. London: Academic Press, 1987, 1991.

38. Acheson RM, Williams DRR. Does consumption of fruit and vegetables protect against stroke? Lancet 1983; 1:1191-3.

39. Armstrong BK, Mann JI, Adelstein AM, Eskin F. Commodity consumption and ischemic heart disease mortality, with special reference to dietary practices. J Chron Dis 1975; 28:455-69.

40. Knox EG. Ischemic-heart-disease mortality and dietary intake of calcium. Lancet 1973; 1:1465-8.

41. Palgi A. Association between dietary changes and mortality rates: Israel 1949 to 1977; a trend-free regression model. Am J Clin Nutr 1981; 34:1569-83.

42. Vollset SE, Bjelke E. Does consumption of fruit and vegetables protect against stroke? Lancet 1983; 2:742.

43. Crombie IK, Smith WCS, Tavendale R, Tunstall-Pedoe H. Geographical clustering of risk factors and lifestyle for coronary heart disease in the Scottish Heart Health Study. Br Heart J 1990; 64: 199-203.

44. Ginter E. Decline of coronary mortality in the United States and vitamin C. Am J Clin Nutr 1979; 32:511-2.

44a. Manson JA, Stampfer MJ, Willett WC, et al. A prospective study of antioxidant vitamins and incidence of coronary heart disease in women. Circulation 1991; 84(Suppl. II-546):Abstract 2168.

45. Gaziano JM, Manson JE, Ridker PM, Buring JE, Hennekens CH. Beta carotene therapy for chronic stable angina. Circulation 1990; 82(Suppl. III-201):Abstract 796.

46. Pierpoint WS. Flavonoids in the human diet. Prog Clin Biol Res 1986; 213:125-40.

47. Harborne JB, ed. The flavonoids: advances in research since 1980. London: Chapman and Hall, 1988.

48. Udassin R, Ariel I, Haskel Y, Kitrossky N, Chevion M. Salicylate as an in vivo free radical trap: studies on ischemic insult to the rat intestine. Free Radic Biol Med 1991; 10:1-6.

49. Jessup W, Rankin SM, DeWhalley CV, Hoult JRS, Scott J, Leake DS. α-Tocopherol consumption during low-density-lipoprotein oxidation. Biochem J 1990; 265:399-405.

50. DeWhalley CV, Rankin SM, Hoult JBS, et al. Modification of low-density lipoproteins by flavonoids. Biochem Soc Trans 1990; 18:1172-3.

51. DeWhalley CV, Rankin SM, Hoult JRS, Jessup W, Leake DS. Flavonoids inhibit the oxidative modification of low density lipoproteins by macrophages. Biochem Pharmacol 1990; 39:1734-50.

52. Brown MS, Goldstein JL. Lipoprotein metabolism in the macrophage: implications for cholesterol deposition in atherosclerosis. Annu Rev Biochem 1983; 52:223-61.

53. Heinecke JW, Rosen H, Chait A. Iron and copper promotes modification of low density lipoprotein by human arterial smooth muscle cells in culture. J Clin Invest 1984; 74:1890-4.

54. Jürgens G, Hoff HF, Chisolm GM, Esterbauer H. Modification of human serum low density lipoprotein by oxidation-characterization and pathophysiological implications. Chem Phys Lipids 1987; 45:315-36.

55. Steinberg D. Arterial metabolism of lipoproteins in relation to atherogenesis. Ann N Y Acad Sci 1990; 598:125-35.

56. Steinberg D. Antioxidants and atherosclerosis: a current assessment. Circulation 1991; 84:1420-5.

57. Esterbauer H, Jürgens G, Quehenberger O, Koller E. Autoxidation of human low density lipoprotein: loss of polyunsaturated fatty acids and vitamin E and generation of aldehydes. J Lipid Res 1987; 28:495-509.

58. Esterbauer H, Rotheneder M, STriegl G, et al. Vitamin E and other lipophilic antioxidants protect LDL against oxidation. Fat Sci Technol 1989; 91:316-24.

59. Esterbauer H, STriegl G, Puhl H, et al. The role of vitamin E and carotenoids in preventing oxidation of low density lipoprotein. Ann N Y Acad Sci 1989; 570:254-61.

60. Esterbauer H, Dieber-Rotheneder M, Waeg G, Puhl H, Tatzber F. Endogenous antioxidants and lipoprotein oxidation. Biochem Soc Trans 1990; 18:1059-61.

61. Esterbauer H, Dieber-Rotheneder M, Waeg G, Striegl G, Jürgens G. Biochemical, structural, and functional properties of oxidized low-density lipoprotein. Chem Res Tox 1990; 3:77-92.

62. Esterbauer H, Dieber-Rotheneder M, Striegl G, Waeg G. Role of vitamin E in preventing the oxidation of low-density lipoprotein. Am J Clin Nutr 1991; 53:314S-21S.

63. Steinbrecher UP, Zhang H, Lougheed M. Role of oxidatively modified LDL in atherosclerosis. Free Radic Biol Med 1990; 9:155-68.

64. Gey KF. Epidemiological correlations between poor plasma levels of essential antioxidants and the risk of coronary heart disease and cancer. In: Ong AHS, Packer L, eds. Lipid-soluble antioxidants: biochemistry and clinical applications. Basel: Birkhäuser, 1992; in press.

65. Gey KF, Puska P, Jordan P, Moser UK. Inverse correlation between plasma vitamin E and mortality from ischemic heart disease in cross-cultural epidemiology. Am J Clin Nutr 1991; 53:326S-34S.

66. Mitchinson MJ, Ball RY, Carpenter KLH, Enright JH, Brabbs CE. Ceroid, macrophages and atherosclerosis. Biochem SocTrans 1990; 18:1066-8.

67. Carpenter KHL, Ballentine JA, Fussell B, Enright JH, Mitchinson MJ. Oxidation of cholesteryl linoleate by human monocyte-macrophages in vitro. Atherosclerosis 1990; 83:217-29.

68. Carpenter KLH, Ball RY, Carter NP, et al. Modulation of ceroid accumulation in macrophages in vitro. In: Porta EA, ed. Lipofuscin and ceroid pigments. New York: Plenum, 1990; 333-43.

69. Machlin LJ. Vitamin E. A comprehensive treatise. New York: Marcel Dekker, 1980.

70. Machlin LJ. Vitamin E. In: Machlin LJ, ed. Handbook of vitamins, 2nd ed. New York: Marcel Dekker, 1990; 99-144.

71. Amemiya T. Effects of vitamin E and selenium deficiencies on rat capillaries. Int J Vitam Nutr

Res 1989; 59:122-6.

72. Loewi G. A lesion in the aorta of vitamin E-deficient animals. J Pathol Bacteriol 1955; 70:246-9.

73. Nelson JS. Pathology of vitamin E deficiency. In: Machlin LJ, ed. Vitamin E, a comprehensive treatise. New York: Marcel Dekker, 1980; 397-428.

74. Liu SK, Dolensek EP, Tappe JP. Cardiomyopathy associated with vitamin E deficiency in seven gelada baboons. J Am Vet Med Assoc 1984; 185:1347-50.

75. Rath M, Pauling L. Immunological evidence for the accumulation of lipoprotein (a) in the atherosclerotic lesion of the hypoascorbic guinea pig. Proc Natl Acad Sci U S A 1990; 87:9388-90.

76. Nafstad I. Endothelial damage and platelet thrombosis associated with PUFA-rich, vitamin E deficient diet fed to pig. Thromb Res 1974; 5:25-6.

77. Bieri JG, Corash L, Hubbard VS. Medical uses of vitamin E. N Engl J Med 1983; 308:1063-71.

78. Farrell PM. Deficency states, pharmacological effects, and nutrient sequirements. In: Machlin LJ, ed. Vitamin E. A comprehensive treatise. New York: Marcel Dekker, 1980; 520-620.

79. Bartley W, Krebs HA, O'Brien JRP. Vitamin C requirement of human adults. Med Res Council, Spec Rep Ser 1953; 280:1-179.

80. Shafar J. Rapid reversion of electrocardiographic abnormalities after treatment in two cases of scurvy. Lancet 1967; 2:176-8.

81. Singh D, Chan W. Cardiomegaly and generalized oedema due to vitamin C deficiency. Singapore Med J 1974; 15:60-3.

82. Follis RH. Sudden death in infants with scurvy. J Pediatr 1942; 20:347-57.

83. Verlangieri AJ, Bush M. Prevention and regression of primate atherosclerosis by d-α-tocopherol. Free Radic Biol Med 1990; 9(suppl. 1):73.

84. Wilson RB, Middleton CC, Sun GY. Vitamin E antioxidants and lipid peroxidation in experimental atherosclerosis of rabbits. J Nutr 1978; 108:1858-67.

85. Wojcicki J, Rozewicka B, Barcew-Wiszniewska L, et al. Effect of selenium and vitamin E on the development of experimental atherosclerosis in rabbits. Atherosclerosis 1991; 87:9-16.

86. Viswanathan M, Bhakthan NMG, Rockerbie RA. Effect of dietary supplementation of vitamin E on serum lipids and lipoproteins in rabbits fed a cholesterolemic diet. Int J Vitam Nutr Res 1979; 49:370-5.

87. Phonpanichrasamee C, Komaratat P, Wilairat P. Hypocholesterolemic effect of vitamin E on cholesterol-fed rabbit. Int J Vitam Nutr Res 1990; 60:240-4.

88. Paul J, Bai NJ, Devi GL. Effect of vitamin E on lipid components of atherogenic rats. Int J Vitam Nutr Res 1989; 59:35-9.

89. Szczeklik A, Grylewski B, Domagala B, Dworski R, Batista M. Dietary supplementation with vitamin E in hyperlipoproteinemias: effects on plasma lipid peroxides, antioxidant activity, prostacyclin generation and platelet aggregability. Thromb Hemost 1985; 54:425-30.

90. Howard DR, Rundell CA, Batsakis JG. Vitamin E does not modify HLD-cholesterol. Am J Clin Pathol 1982; 77:86-9.

91. Serfontein WJ, Ubbink JB, de Villiers LS. Further evidence in the effect of vitamin E on the cholesterol distribution of lipoproteins with special reference to HDL subfractions. Am J Clin Pathol 1983; 79:604-6.

92. De Maio SJ, King SB, Lembo NJ, et al. Vitamin E supplementation, plasma lipids and incidence of restenosis after percutaneous transluminal coronary angioplasty (PTCA). J Am Coll Nutr 1992; 11:68-73.

93. Kita T, Yokode M, Ishii K, Arai H, Nagano Y. The role of atherogenic low density lipoproteins (LDL) in the pathogenesis of atherosclerosis. Ann N Y Acad Sci 1990; 598:188-93.

94. Ishii K, Kita T, Yokode M, et al. Characterization of very low density lipoprotein from Watanabe heritable hyperlipidemic rabbits. J Lipid Res 1989; 30:1-7.

94a. Tanimura N, Asada Y, Hayashi T, Kisanuki A, Sumiyoshi A. Aortic endothelial cell damage induced by β-VLDL and macrophages in vitro. Atherosclerosis 1990; 85:161-7.

95. Weitzel G, Schön H, Gey KF, Buddecke E. Lipid-soluble vitamins and atherosclerosis. Hoppe-Seylers Z Physiol Chem 1956; 304:247-72.

96. Smith TL, Kummerow FA. Effect of dietary vitamin E on plasma lipids and atherogenesis in restricted ovulator chicken. Atherosclerosis 1989; 75:105-9.

97. McCord JM. Oxygen-derived free radicals in postischemic tissue injury. N Engl J Med 1985; 312: 159-63.

98. Janero DR. Therapeutic potential of vitamin E against myocardial ischemic reperfusion injury. Free Radic Biol Med 1991; 10:315-24.

99. Ferrari R, Ceconi C, Curello S, Cargnoni A, Condorelli E, Raddino R. Role of oxygen in myocardial ischemic and reperfusion damage: effect of tocopherol. Acta Vitaminol Enzymol 1987; 7(suppl.):61-70.

100. Ferrari R, Curello S, Boffa GM, et al. Oxygen free radical-mediated heart injury in animal models and during bypass surgery in humans. Effects of α-tocopherol. Ann N Y Acad Sci 1989; 570: 237-53.

101. Gauduel Y, Duvelleroy MA. Role of oxygen radicals in cardiac injury due to reoxygenation. J Mol Cell Cardiol 1984; 16:459-70.

102. Marabayashi S, Dohi K, Sugino K, Kawasaki T. The protective effect of administered α-tocopherol against hepatic damage caused by ischemia-peperfusion or endotoxemia. Ann N Y Acad Sci 1989; 570:208-18.

103. Yoshida S. Brain injury after ischemia and trauma. The role of vitamin E. Ann N Y Acad Sci 1989; 570:219-36.

104. Andrews JS, Griffith WH, Mead JF, Stein RA. Toxicity of air-oxidized soybean oil. J Nutr 1960; 70:199-210.

105. Gabriel HG, Alexander JC, Valli VE. Effects of intubating rats with fractions from thermally oxidized corn oil and olive oil. Nutr Rep Int 1979; 19:515-26.

106. Thafvelin B. Role of cereal fat in the production of nutritional disease in pigs. Nature 1960; 188: 1169-72.

107. Cortesi RA, Privett OS. Toxicity of fatty ozonides and peroxides. Lipids 1972; 7:715-21.

108. Horgan VJ, Philpot JSL, Porter BW, Roodyn DB. Toxicity of autoxidized squalene and linoleic acid, and of simpler peroxides, in relation to toxicity of radiation. Biochem J 1957; 67:551-8.

109. Yagi K. Lipid peroxides and human diseases. Chem Phys Lipids 1987; 45:337-51.

110. Bird RP, Alexander JC. Effects of vitamin E and ascorbyl palmitate on cultured myocardial cells exposed to oxidized fats. J Toxicol Environ Health 1981; 7:59-68.

111. Cutler RG, Schneider R. Linoleate oxidation products and cardiovascular lesions. Atherosclerosis 1974; 20:383-94.

112. Hennig B, Enoch C, Chow CK. Protection by vitamin E against endothial cell injury by linoleic acid hydroperoxides. Nutr Res 1987; 7:1253-9.

113. Orekhov AN, Tertov SA, Kudryashov SA, Khashimov KA, Smirnov VN. Primary culture of human aortic intima cells as a model for testing antiatherosclerotic drugs. Effects of cyclic AMP, prostaglandins, calcium antagonists, antioxidants, and lipid-lowering agents. Atherosclerosis 1986; 60:101-10.

114. Sevanian A, Berliner J, Peterson H. Uptake, metabolism, and cytotoxicity of isomeric cholesterol-5,6-epoxides in rabbit aortic endothelial cells. J Lipid Res 1991; 32:147-55.

115. Evensen SA, Galdal KS, Nilsen E. LDL-induced cytotoxity and its inhibition by antioxidant treatment in cultured human endothelial cells and fibroblasts. Atherosclerosis 1983; 49:23-30.

116. Van Hinsbergh VWM. LDL toxicity. The state of the art. Atherosclerosis 1984; 53:113-8.

117. Morel DW, Hessler JR, Chisolm GM. Low density lipoprotein cytotoxicity induced by free radical peroxidation of lipid. J Lipid Res 1983; 24:1070-6.

118. Nordoy A, Svensson B, Wiebe D, Hoack JC. Lipoprotein and the inhibitory effect of human endothelial cells on platelet function. Circ Res 1978; 43:527-34.

119. Morin RJ, Peng SK. The role of cholesterol oxidation products in the pathogenesis of atherosclerosis. Ann Clin Lab Sci 1989; 19:225-37.

120. Gray MF, Lavrie TDV, Brooks CJW. Isolation and identification of cholesterol-α-epoxide and other minor sterols in human serums. Lipids 1971; 6:836-43.

121. Henriksen T, Mahoney EM, Steinberg D. Enhanced macrophage degradation of low density lipoprotein previously incubated with cultured endothelial cells: recognition by receptors for acetylated low density lipoproteins. Proc Natl Acad Sci U S A 1981; 78:6499-503.

122. Henriksen T, Mahoney EM, Steinberg D. Enhanced macrophage degradation of biologically modified low density lipoprotein. Arteriosclerosis 1983; 3:149-59.

123. Morel DW, DiCorleto PE, Chisolm GM. Endothelial and smooth muscle cells alter low-density lipoprotein in vitro by free radical oxidation. Arteriosclerosis 1984; 4:357-64.

124. Van Hinsbergh VWM, Scheffer M, Havekes L, Kempen HJM. Role of endothelial cells and their

products on the modification of low-density lipoproteins. Biochim Biophys Acta 1986; 878:49-64.

125. Hiramatsu K, Rosen H, Heinecke JW, Wolfbauer G, Chait A. Superoxide initiates oxidation of low density lipoprotein by human monocytes. Arteriosclerosis 1987; 7:55-60.

126. Parthasarathy S, Young SC, Witztum JL, Pitman RC, Steinberg D. Probucol inhibits oxidative modification of low density lipoprotein. J Clin Invest 1986; 77:641-4.

127. Parthasarathy S, Quinn MT, Schwenke DC, Carew TE, Steinberg D. Oxidative modification of beta-very low density lipoprotein. Potential role in monocyte recruitment and foam cell formation. Arteriosclerosis 1989; 9:398-404.

128. Parthasarathy S, Wieland E, Steinberg D. A role for endothelial cell lipoxygenase in the oxidative modification of low density lipoprotein. Proc Natl Acad Sci U S A 1989; 86:1046-50.

129. Hessler JR, Morel DW, Lewis LJ, Chisolm GM. Lipoprotein oxidation and lipoprotein-induced cytotoxicity. Arteriosclerosis 1983; 3:215-22.

130. Haberland ME, Fogelman AM. The role of altered lipoproteins in the pathogenesis of atherosclerosis. Am Heart J 1987; 113:573-7.

131. Haberland ME, Fogelman AM, Edwards PA. Specificity of receptor-mediated recognition of malondialdehyde modified low density lipoprotein. Proc Natl Acad Sci U S A 1982; 79: 1712-6.

132. Haberland ME, Fong D, Cheng L. Malondialdehyde-altered protein occurs in atheroma of Watanabe heritable hyperlipemic rabbits. Science 1988; 241:215-8.

133. Palinski W, Rosenfeld ME, Ylä-Herttuala S,, et al. Low density lipoprotein undergoes oxidative modification in vivo. Proc Natl Acad Sci U S A 1989; 86:1372-6.

134. Palinski W, Ylä-Herttuala S, Rosenfeld ME, et al. Antisera and monoclonal antibodies specific for epitopes generated during oxidative modification of low density lipoprotein. Arteriosclerosis 1990; 10:325-35.

135. Ylä-Herttuala S, Palinski W, Rosenfeld ME, et al. Evidence for the presence of oxidatively modified low density lipoprotein in atherosclerotic lesions of rabbit and man. J Clin Invest 1989; 84:1086-95.

136. Shaikh M, Martini S, Quiney JR, et al. Modified plasma-derived lipoproteins in human atherosclerotic plaques. Atherosclerosis 1988; 69:165-72.

137. Hoff HF, Lie JT, Titus JL, et al. Lipoproteins in atherosclerotic lesions. Localization by immunofluorescence and apo-low density lipoproteins in atherosclerotic arteries from normal and hyperlipoproteinemics. Arch Pathol 1975; 99:253-8.

138. Hoff H, O'Neil J, Osborne A, Pepin J. Lesion-derived LDL and oxidized LDL share an enhanced aggregability that leads to phagocytosis but unefficient processing by macrophages. Arteriosclerosis 1990; 10:783a.

139. Glavind J, Hartmann S, Clemmensen J, Jessen KE, Dam H. Studies on the role of lipoperoxides in human pathology. Acta Pathol Microbiol Scand 1952; 30:1-6.

140. Goto Y. Lipid peroxides as a cause of vascular disease. In: Yagi K, ed. Lipid peroxides in biology and medicine. Orlando, FL: Academic Press, 1982; 295-303.

141. Harland WA, Gilbert JD, Brooks CJW. Lipids of human atheroma. VII. Oxidised derivatives of cholesteryl linoleate. Biochim Biophys Acta 1973; 316:378-85.

142. Wilson RB. Lipid peroxidation and atherosclerosis. Crit Rev Food Sci Nutr 1976; 7:325-37.

143. Piotrowski JJ, Hunter GC, Eskelson CD, Dubick MA, Bernhard VM. Evidence for lipid peroxidation in atherosclerosis. Life Sciences 1990; 46:715-21.

144. Ylä-Herttuala S, Rosenfeld ME, Parthasarathy S, et al. Colocalization of 15-lipoxygenase mRNA with epitopes of oxidized low-density lipoprotein in macrophage-rich areas of atherosclerotic lesions. Proc Natl Acad Sci U S A 1990; 87:6959-63.

145. Ylä-Herttuala S, Rosenfeld ME, Parthasarathy S, et al. Gene expression in macrophage-rich human atherosclerotic lesions: 15-lipoxygenase and acetyl-LDL receptor mRNA colocalizes with oxidation specific lipid-protein adducts. J Clin Invest 1991; 87:1146-52.

146. Cathcart MK, McNally AK, Morel DW, Chisolm GM III. Lipoxygenase-mediated transformation of human low density lipoprotein to an oxidized and cytotoxic complex. J Lipid Res 1991; 32:63-70.

146a. Cathcart MK, McNally AK, Morel DW, Chisolm GM III. Superoxide anion participation in human monocyte-mediated oxidation of low-density lipoprotein and conversion of low-density lipoprotein to a cytotoxin. J Immunol 1989; 142: 262:707-12.

146b.   Cathcart MK, Chisolm GM III, McNally AK, Morel DW. Oxidative modification of low density (LDL) by activated human monocytes and cell lines U937 and HL60. In Vitro Cell Dev Biol 1988; 24:1001-8.

147.   Sparrow CP, Parthasarathy S, Steinberg D. Enzymatic modification of low density lipoprotein by purified lipoxygenase plus phospholipase A_2 mimics cell-mediated oxidative modification. J Lipid Res 1988; 29:745-53.

148.   Jürgens G. Modified serum lipoproteins and atherosclerosis. Annu Rep Med Chem (London: Academic Press) 1989; 25 (Johns ed.):169-76.

149.   Dieber-Rotheneder M, Puhl H, Waeg G, Striegel G, Esterbauer H. Effect of oral supplementation with D-α-tocopherol on the vitamin E content of human low density lipoproteins and resistance to oxidation. J Lipid Res 1991; 32:1325-32.

150.   Cathcart MK, Morel DW, Chisolm GM III. Monocytes and neutrophils oxidize low density lipoproteins making it cytotoxic. J Leukocyte BIOL 1985; 38:341-50.

151.   Kuzuya M, Naito M, Funaki C, Hayashi T, Asai K, Kuzuya F. Probucol prevents oxidative injury to endothelial cells. J Lipid Res 1991; 32:197-204.

153.   Masuda M, Ihara N, Kuriki H, et al. Spontaneous injuries in the aortic endothelium of the inherited cataract rats and their prevention by tocopherol. A study by scanning electron microscopy. Atherosclerosis 1989; 75:23-30.

154.   Hennig B, Boissonneault GA, Wang Y. Protective effects of vitamin E in age-related endothelial cell injury. Int J Vitam Nutr Res 1989; 59:273-9.

155.   Hladovec J. Protective effect of oxygen-derived free radical scavengers on the endothelium in vivo. Physiol Bohemoskov 1986; 35:97-103.

155a.   Kaneko T, Nakano SI, Matsuo M. Protective effect of vitamin E on linoleic acid hydroperoxide-induced injury to human endothelial cells. Lipids 1991; 26:345-8.

156.   Saito Y, Shinomiya M, Kumagai A. Effect of tocopherol on low-density lipoprotein cholesterol ester metabolism in the arterial wall. Ann N Y Acad Sci 1982; 393:183-5.

157.   Bowyer DE, Mitchinson MJ. The role of macrophages in atherosclerosis. In: Zembala M, Asherson GL, eds. Human monocytes. London: Academic Press, 1989; 439-58.

158.   Shimasaki H, Maeba R, Ueta N. Lipid peroxidation and storage of fluorescent products by macrophages in vitro as a model of ceroid-like pigment formation. Adv Exp Med Biol 1989; 266:323-30.

159.   Shimakasi H, Maeba R, Ueta N. Lipid peroxidation and storage of fluorescent products by macrophages in vitro as a model of ceroid-like pigment formation. In: Porta EA, ed. Lipofuscin and ceroid pigments. New York: Plenum, 1990; 323-31.

160.   Sharmanov AT, Aidarkhanov BB, Burmangaliev SM. Effect of vitamin E on the oxidative metabolism of macrophages. Biull Eksp Biol Med 1986; 101:723-5.

161.   Shirai K, Matsuoka N, Morisaki N, et al. Effect of tocopherol deficiency on lipid metabolism in the arterial walls of rats on normal and high cholesterol diets. Artery 1980; 6:484-506.

162.   Fruchart JC, Sauzieres J, Clavey V, Plancke MO. Antioxidant therapy and uptake of human oxidized LDL by macrophages. Ann N Y Acad Sci 1989; 570:447-8.

163.   Baehner RL, Boxer LA. Role of membrane vitamin E and cytoplasmatic glutathione in the regulation of phagocytic functions of neutrophils and monocytes. Am J Pediatr Hematol Oncol 1979; 1:71-6.

164.   Hatchigan EA, Santos JI, Broitman SA, Vitale JJ. Vitamin A supplements improve macrophage function and bacterial clearance during experimental salmonella infection. Proc Soc Exp Biol Med 1989; 191:47-54.

165.   Rocha NP. Vitamin E stimulates endotoxin-inhibited monocyte migration and phagocytosis in vivo. Braz J Med Biol Res 1989; 22:1401-3.

166.   Ganguly R, Waldman RH. Macrophage functions in aging: effects of vitamin C deficiency. Allerg Immunol (Leipz) 1985; 31:37-43.

167.   Lopez AJ, Palomo DG, Castrillon JL, Arellano JL. "In vitro" improvement of defective chemotaxis in intravenous drug abusers after incubation with ascorbic acid. Klin Wochenschr 1987; 65:625-6.

168.   Scheinberg MA. The effect of vitamin C on certain monocyte cell functions. An in vitro and in vivo approach. Int J Vitam Nutr Res 1983; 23:199-206.

169. Moriguchi S, Kobayashi N, Kishino Y. Effects of vitamin E deficiency on the functions of splenic lymphocytes and alveolar macrophages. J Nutr Sci Vitaminol 1989; 35:419-30.

170. Sakamoto W, Fujie K, Handa H, Ogihara T, Mino M. In vivo inhibition of superoxide production and protein kinase C activity in macrophages from vitamin E-treated rats. Int J Vitam Nutr Res 1990; 60:338-42.

171. Sakamoto W, Yoshikawa K, Shindo M, et al. In vivo effect of vitamin E on peritoneal macrophages and T-kininogen level in rats. Int J Vitam Nutr Res 1989; 59:131-9.

171a. Cathcart RF. A unique function for ascorbate. Med Hypotheses 1991; 35:32-7.

172. Leake DS, Rankin SM. The oxidative modification of LDL by macrophages. Biochem J 1990; 270:741-8.

173. Pacht ER, Kaseki H, Mohammed JR, Cornwell DG, Davis WB. Deficiency of vitamin E in the alveolar fluid of cigarette smokers. J Clin Invest 1986; 77:789-96.

174. Akeson AL, Woods CW, Mosher LB, Thomas CE, Jackson RL. Inhibition of IL-1β expression in THP-1 cells by probucol and tocopherol. Atherosclerosis 1991; 86:261-70.

175. Ku G, Doherty NS, Schmidt LF, Jackson RL, Dinerstein RJ. Ex vivo lipopolysaccharide-induced interleukin-1 secretion from murine peritoneal macrophages inhibited by probucol, a hypocholesterolemic agent with antioxidant properties, inhibits interleukin (IL-1) secretion. FASEB J 1990; 4:1645-53.

176. Ku G, Doherty NS, Wolos JA, Jackson RL. Inhibition by probucol of interleukin-1 secretion and its implication in atherosclerosis. Am J Cardiol 1988; 62:77B.

177. Yamamoto A, Takaichi S, Hara H, et al. Probucol prevents lipid storage in macrophages. Atherosclerosis 1986; 62:209-17.

177a. Yamamoto A, Hara H, Takaichi S, Wakasugi JI, Tomikawa M. Effect of probucol on macrophages, leading to regression of xanthomas and atheromatous vascular lesions. Am J Cardiol 1988; 62:31B-6B.

178. Kita T, Nagano Y, Yokode M, et al. Prevention of atherosclerotic progression in Watanabe rabbits by probucol. Am J Cardiol 1988; 62:13B-9B.

179. Kita T, Nagano Y, Yokode M, et al. Probucol prevents the progression of atherosclerosis in Watanabe heritable hyperlipemic rabbit, an animal model for familial hypercholesterolemia. Proc Natl Acad Sci U S A 1987; 84:5928-31.

180. Carew TE, Schwenke DC, Steinberg D. Antiatherogenic effect of probucol unrelated to its hypocholesterolemic effect: evidence that antioxidants in vivo inhibit low density lipoprotein degradation in macrophage-rich fatty streaks and slow the progression of atherosclerosis in the Watanabe heritable hyperlipemic rabbit. Proc Natl Acad Sci U S A 1987; 84:7725-9.

181. Nagano Y, Kume N, Otani K, et al. Probucol does not act on lipoprotein metabolism of WHHL rabbit macrophages. Ann N Y Acad Sci 1990; 598:530-1.

182. Goldberg RB, Mendez A. Probucol enhances cholesterol efflux from cultured human fibroblasts. Am J Cardiol 1988; 62:109-17.

183. Stein Y, Stein O, Delplanque B, Fesmire JD, Lee DM, Alaupovic P. Lack of effect of probucol on atheroma formation in cholesterol-fed rabbits kept at a comparable plasma cholesterol levels. Atherosclerosis 1989; 75:145-66.

184. Björkhem I, Henriksson-Freyschuss A, Breuer O, Diczfalusy U, Berglund L, Henriksson P. The antioxidant butylated hydroxytoluene protects against atherosclerosis. Arterioscl Thromb 1991; 11:15-22.

185. Gey KF. Inverse correlation of vitamin E and ischemic heart disease. Int J Vitam Nutr Res Suppl 1989; 30:224-31.

186. Gey KF, Stähelin HB, Puska P, Evans A. Relationship of plasma vitamin C to mortality from ischemic heart disease. Ann N Y Acad Sci 1987; 498:110-23.

187. Gey KF, Brubacher GB, Stähelin HB. Plasma levels of antioxidant vitamins in relation to ischemic heart disease and cancer. Am J Clin Nutr 1987; 45:1368-77.

188. Gey KF, Puska P. Plasma vitamins E and A inversely related to mortality from ischemic heart disease in cross-cultural epidemiology. Ann N Y Acad Sci 1989; 570:268-82.

189. Willett WW. Nutritional epidemiology. New York: Oxford University Press, 1990.

190. Rubba P, Mancini M, Fidanza F, Leccia G, Riemersma RA, Gey KF. Plasma vitamin E, apoprotein B and HDL-cholesterol in middle-aged men from Southern Italy. Atherosclerosis 1989; 77:25-9.

191. Gurusinghe A, DeNiese M, Renaud JF, Austin L. The binding of lipoproteins to human muscle cells: binding and uptake of LDL, HDL, and α-tocopherol. Muscle Nerve 1988; 11:1231-9.

192. Thurnham DI, Davies JA, Crump BJ, Situnayake RD, Davis M. The use of different lipids to express serum tocopherol: lipid ratios for the measurement of vitamin E status. Ann Clin Biochem 1986; 23:514-20.

193. Ellis N, Lloyd B, Lloyd RS, Clayton BE. Selenium and vitamin E in relation to risk factors for coronary heart disease. J Clin Pathol 1984; 37:200-6.

194. Salonen JT, Salonen R, Ihanainen M, et al. Blood pressure, dietary fats, and antioxidants. Am J Clin Nutr 1988; 48:1226-32.

195. Salonen JT, Salonen R, Seppänen K, et al. Relationship of serum selenium and antioxidants to plasma lipoproteins, platelet aggregability and prevalent ischemic heart disease in eastern finnish men. Atherosclerosis 1988; 70:155-60.

196. Salonen JT, Salonen R, Seppänen K, et al. Effects of antioxidant supplementation on platelet function: a randomized pair-matched, placebo-controlled, double-blind trial in men with low antioxidant status. Am J Clin Nutr 1991; 53:1222-9.

197. Bulpitt CJ. Vitamin C and blood pressure. J Hypertension 1990; 8:1071-5.

198. Trout DL. Vitamin C and cardiovascular risk factors. Am J Clin Nutr 1991; 53:322-5.

198a. Gey KF, Moser UK, Jordan P, Stähelin HB, Eichholzer M, Lüdin E. Increased risk of cardiovascular disease at suboptimal plasma levels of essential antioxidants: an epidemiological update with special attention to carotene and vitamin C. Am J Clin Nutr 1992; in press.

199. Vatassery GT, Smith WE, Quach HT. Ascorbic acid, glutathione and synthetic antioxidants prevent the oxidation of vitamin E in platelets. Lipids 1989; 24:1043-7.

200. Hruba F, Novakowa V, Ginter E. The effect of chronic marginal vitamin C deficiency on the α-tocopherol content of the organs and plasma of guinea pigs. Experientia 1982; 38:1454-5.

201. Kanazawa K, Takeuchi S, Hasegawa R, et al. Influence of ascorbic acid deficiency on the level of non-protein SH compounds and vitamin E in the blood and tissues of guinea pigs. Nihon Univ J Med 1981; 23:257-65.

202. De AK, Darad R. Physiological antioxidants and antioxidative enzymes in vitamin E-deficient rats. Toxicol Lett 1988; 44:47-54.

203. Burton GW, Wronska U, Stone L, Foster DO, Ingold KU. Biokinetics of dietary *RRR*-α-tocopherol in the male guinea pig at three dietary levels of vitamin C and two levels of vitamin E. Evidence that vitamin C does not "spare" vitamin E in vivo. Lipids 1990; 25:199-210.

204. Bai NJ, Kumar PS, George T, Krishunamurthy S. Effect of dietary protein and hypervitaminosis A or C on tissue peroxidation and erythrocyte lysis of vitamin E deficiency. Int J Vitam Nutr Res 1982; 52:386-92.

205. Chen LH, Barnes KJ. Nutritional relationship of vitamin E and vitamin C in guinea pigs. Nutr Rep Int 1976; 14:89-96.

206. Chen LH, Chang ML. Effect of dietary vitamin E and vitamin C on respiration and swelling of guinea pig liver mitochondria. J Nutr 1978; 108:1616-20.

207. Chen LH, Chang ML. Effects of high level of vitamin C on tissue antioxidant status of guinea pigs. Int J Vitam Nutr Res 1979; 49:87-91.

208. Chen LH, Chow CK. Effect of cigarette smoking and dietary vitamin E on plasma level of vitamin C in rats. Nutr Rep Int 1980; 22:301-9.

209. Ginter E, Kostinova A, Hudecova A, Majaric A. Synergism between vitamins C and E: effect of microsomal hydroxylation in guinea pig liver. Int J Vitam Nutr Res 1982; 52:54-7.

210. Tappel AL. Will antioxidant nutrients slow aging processes? Geriatrics 1968; 23:97-105.

211. Reference deleted.

212. Kunert KJ, Tappel AL. The effect of vitamin C on in vivo lipid peroxidation in guinea pigs as measured by pentane and ethane production. Lipids 1983; 18:271-4.

213. Leung HW, Vang MJ, Mavis RD. The cooperative interaction between vitamin E and vitamin C in suppression of peroxidation of membrane phospholipids. Biochim Biophys Acta 1981; 664:266-72.

214. Tappel AL, Brown WD, Zalkin H, Maier VP. Unsaturated lipid peroxidation catalyzed by hematin compounds and its inhibition by vitamin E. J Am Oil Chem Soc 1961; 38:5-9.

215. Fidanza A, Audisio M, Mastroacovo P. Vitamin C and cholesterol. Int J Vitam Nutr Res Suppl 1982; 23:153-71.

216. Ingold KU, Webb AC, Witter D, Burton GW, Metcalfe TA, Muller DPR. Vitamin E remains the major lipid-soluble, chain-breaking antioxidant in human plasma even in individuals suffering severe vitamins E deficiency. Arch Biochem Biophys 1987; 259:224-5.

217. Esterbauer H, Striegl G, Puhl H, Rotheneder M. Continuous monitoring of in vitro oxidation of human low density lipoprotein. Free Radic Res Commun 1989; 6:67-75.

218. Jessup W, Bedwell S, Kwok K, Dean RT. Oxidative modification of low-density lipoprotein: initiation by free radicals and protection by antioxidants. Agents Actions Suppl 1988; 26:241-6.

219. Frei B, Stocker R, Ames AN. Antioxidant defenses and lipid peroxidation in human plasma. Proc Natl Acad Sci U S A 1988; 85:9748-52.

220. Frei B, England L, Ames BN. Ascorbate is an outstanding antioxidant in human blood plasma. Proc Natl Acad Sci U S A 1989; 86:6377-81.

220a. Jialal I, Grundy SM. Preservation of the endogenous antioxidants in low density lipoprotein by ascorbate but not by probucol during oxidative modification. J Clin Invest 1991; 87:597-601.

220b. Jialal I, Vega GL, Grundy SM. Physiologic levels of ascorbate inhibit the oxidative modification of low density lipoprotein. Atherosclerosis 1990; 82:185-91.

221. Stocker R, Bowry VW, Frei B. Ubiquinol-10 protects human low density lipoprotein more efficiently against lipid peroxidation than does α-tocopherol. Proc Natl Acad Sci U S A 1991; 88:1646-50.

222. Maguire JJ, Wilson DS, Packer L. Mitochondrial electron transport-linked tocopheroxyl radical reduction. J Biol Chem 1989; 264:21462-5.

223. Barclay LRC, Locke SJ, MacNeil JM. The antioxidation of unsaturated lipids in micelles. Synergism of inhibitors vitamins C and E. Can J Chem 1983; 61:1288-90.

224. Golumbic C, Mattill HA. Antioxidants and the antioxidation of fats. XIII. The antioxygenemic action of ascorbic acid in association with tocopherols, hydrochinones and related compounds. J Am Chem Soc 1941; 63:1279-80.

225. Lambelet P, Loeliger J. The fate of antioxidant radicals during lipid autoxidation. I. The tocopheroxyl radicals. Chem Phys Lipids 1984; 35:185-98.

226. Niki E, Saito T, Kawakami A, Kamiya Y. Inhibition of oxidation of methyl linoleate in solution by vitamin E and vitamin C. J Biol Chem 1984; 259:4177-82.

227. Niki E, Yamamoto Y, Kamiya Y. Oxidation of phosphatidylcholine and its inhibition by vitamin E and vitamin C. In: Bors W, Saran M, Tait D, eds. Oxygen radicals in chemistry and biology. Berlin: Walter de Gruyter, 1984; 274-8.

228. Niki E. Interaction of ascorbate and α-tocopherol. Ann N Y Acad Sci 1987; 498:186-99.

229. Niki E, Yamamoto Y, Takahashi M, Komura E, Miyama Y. Inhibition of oxidation by biomembranes by tocopherol. Ann N Y Acad Sci 1989; 570:23-31.

230. Packer JE, Slater TF, Willson RL. Direct observation of a free radical interaction between vitamin E and vitamin C. Nature 1979; 278:737-8.

231. Scarpa M, Rigo A, Maiorino M, Ursini F, Gregolin C. Formation of α-tocopherol radical and recycling of α-tocopherol by ascorbate during peroxidation of phosphatidylcholine liposomes. Biochim Biophys Acta 1984; 801:215-9.

232. Liebler DC, Kaysen KL, Kennedy TA. Redox cycles of vitamin E: hydrolysis and ascorbic acid dependent reduction of 8A-(alkyldioxy)-tocopherones. Biochemistry 1989; 28:9772-7.

233. Kagan VE, Serbinova EA, Packer L. Recycling and antioxidant activity of tocopherol homologs of different hydrocarbon chain length in liver microsomes. Arch Biochem Biophys 1990; 282:221-5.

234. McCay PB, Lai EK, Brueggemann G, Powell SR. A biological antioxidant function for vitamin E: electron shuttling for a membrane-bound "free radical reductase." NATO ASI Ser, Ser A 1987; 131:145-56.

235. McCay PB, Brueggemann G, Lai EK, Powell SR. Evidence that α-tocopherol functions cyclically to quench free radicals in hepatic microsomes: requirement for glutathine and a heat-labile factor. Ann N Y Acad Sci 1989; 570:32-45.

236. Packer L, Maguire J, Melhorn R, Serbinova E, Kagan V. Mitochondria and microsomal membranes have a free radical reductase activity that prevents chromanoxyl radical accumulation. Biochem Biophys Res Commun 1989; 159:229-35.

237. Demopoulos HB, Flamm ES, Pietronigro DD, Seligmann ML. The free radical pathology and the microcirculation in the major central nervous system disorders. Acta Physiol Scand Suppl 1980; 492:91-119.

238. Swartz HM, Dodd NJF. The role of ascorbic acid on radical reactions in vivo. In: Rodgers MAJ, Powers EL, eds. Oxygen and oxy radicals in chemistry and biology. New York: Academic Press, 1981; 161-8.

239. Rao PS, Mueller HS. Adv Esp Med Biol 1983; 161:347-67.

240. Borg DC, Schaich KM. Pro-oxidant action of antioxidants. In: Miquel J, Quintanilha AT, Weber H, eds. CRC Handbook of free radicals and antioxidants in biomedicine, Vol. I. Boca Raton, FL: RCR Press, 1989; 63-80.

241. Wayner DDM, Burton GW, Ingold KW. The antioxidant efficiency of vitamin C is concentration dependent. Biochim Biophys Acta 1986; 884:119-23.

242. Link G, Pinson A, Kahane I, Hershko C. Iron loading modifies the fatty acid composition of cultured rat myocardial cells and liposomal vesicles: effect of ascorbate and α-tocopherol on myocardial lipid peroxidation. J Lab Clin Med 1989; 144:243-9.

243. Wefers H, Sies H. Antioxidant effects of ascorbate and glutathione in microsomal lipid peroxidation are dependent on vitamin E. Adv Biosci 1988; 76:309-16.

244. Graham KS, Reddy CC, Scholz RW. Reduced glutathione effects on α-tocopherol concentration of rat liver microsomes undergoing NADPH-dependent lipid peroxidation. Lipids 1989; 24:909-14.

245. Costagliola C, Menzione M. Effect of vitamin E on the oxidative state of glutathione in plasma. Clin Physiol Biochem 1990; 140:140-3.

246. Mrtensson J, Meister A. Glutathione deficiency decreases tissues ascorbate levels in newborn rats: ascorbate spares glutathione and protects. Proc Natl Acad Sci U S A 1991; 88:4656-60.

247. Stähelin HB, Gey KF, Eichholzer M, et al. Plasma antioxidant vitamins and subsequent cancer mortality in the 12-year follow-up of the prospective Basel study. Am J Epidemiol 1991; 133:766-75.

248. Herberth B, Didelot Barthelemy L, Le Devehat C, Lemoine A. Plasma retinol, carotenoids and tocopherols: biological variation factors between 18-45 years. Ann Nutr Metab 1988; 32:297-304.

249. Gregory J, Foster K, Tyler H, Wiseman M. The dietary and nutritional survey of British adults. A survey of the dietary behaviour, nutritional status and blood pressure of adults aged 16 to 64 living on Great Britain, carried out by the Social Division of OPCS and the Ministry of Agriculture, Fisheries and Food and the Department of Health. London: HMSO, 1990.

250. Riemersma RA, Wood DA, Macintyre CCA, Elton R, Gey KF, Oliver MF. Low plasma vitamins E and C: increased risk of angina pectoris in Scottish men. Ann N Y Acad Sci 1989; 570:291-5.

251. Riemersma RA, Wood DA, Macintyre CCA, Elton RA, Gey KF, Oliver MF. Risk of angina pectoris and plasma concentrations of vitamins A, C and E and carotene. Lancet 1991; 337:1-5.

252. Reference deleted

253. Sullivan JL. Antioxidants and coronary heart disease. Lancet 1991; 337:432-3.

254. Subar AF, Harlan LC, Mattson ME. Food and nutrient intake differences between smokers and non-smokers in the US. Am J Public Health 1990; 80:1323-9.

255. Hornig DH, Moser U, Glatthaar BE. Ascorbic acid. In: Shils ME, Young VR, eds. Modern nutrition in health and disease, 7th ed. Philadelphia: Lea & Febriger, 1988; 417-35.

256. Kallner AB, Hartmann D, Hornig DH. On the requirements of ascorbic acid in man: steady-state turnover and body pool in smokers. Am J Clin Nutr 1981; 34:1347-55.

257. Anderson R. Assessment of the roles of vitamin C, vitamin E, and beta-carotene in the modulation of oxidant stress mediated by cigarette smoke-activated phagocytes. Am J Clin Nutr 1991; 53:358-61.

258. Gey KF. Vitamin E: the functional significance of suboptimal plasma levels. In: Pietrzik K, ed. Modern lifestyles, lower energy intake and micronutrient status. London: Springer, 1991; 126-33.

259. Ramirez J, Flowers NC. Leucocyte ascorbic acid and its relationship to coronary heart disease in man. Am J Clin Nutr 1980; 33:2079-87.

260. Weglicki WB, Mak IT, Simic MG. Mechanisms of cardiovascular drugs as antioxidants. J Mol Cell Cardiol 1990; 22:1199-208.

261. Islam S, Houtia NE, Mazière C, Polonovski J. Comparative study of the effect of beta-blockers with different pharmacological properties on cholesteryl ester formation in mouse peritoneal macrophages. Biochem Pharmacol 1987; 36:847-9.

262. Atkinson JB, Swift LL. Nifedipine reduces atherogenesis in cholesterol-fed heterozygous WHHL rabbits. Atherosclerosis 1990; 84:195-201.

263. Mazière JC, Salmon S, Mora L, Auclair M. The antihypertensive drug propranolol enhances LDL catabolism and alters cholesterol metabolism in human cultured fibroblasts. Atherosclerosis 1990; 81:151-60.

264. Naruszewicz M, Mirkiewicz E, Klosiewiczf-Latoszek L. Modification of low-density lipoproteins from hypertriglyceridemic patients by macrophages in vitro and the effect of benzafibrate treatment. Atherosclerosis 1989; 79:261-5.

265. Kane JP, Malloy MJ, Ports TA, Phillips NR, Diehl JC, Havel RJ. Regression of coronary atherosclerosis during treatment of familial hypercholesterolemia with combined drug regimens. JAMA 1990; 264:3007-12.

266. Brown G, Albers JJ, Fisher LD, et al. Regression of coronary artery disease as a result of intensive lipid-lowering therapy in men with high levels of apolipoprotein B. N Engl J Med 1990; 323:1289-98.

267. Kok FJ, van Poppel G, Melse J, et al. Do antioxidants and polyunsaturated fatty acids have a combined association with coronary atherosclerosis? Atherosclerosis 1991; 31:85-90.

268. Salonen JT, Salonen R, Penttilae L, et al. Serum fatty acids, apolipoproteins, selenium and vitamin antioxidants and the risk of death from coronary artery disease. Am J Cardiol 1985; 56:226-31.

269. Kok F, de Bruijn AM, Vermeeren R, et al. Serum selenium, vitamin antioxidants and cardiovascular mortality: a 9-year follow-up study in the Netherlands. Am J Clin Nutr 1987; 45:462-8.

270. Stähelin HB, Gey KF, Eichholzer M, Lüdin E, Brubacher G. Cancer mortality and vitamin E status. Ann N Y Acad Sci 1989; 570:391-9.

270a. Malone W. Studies evaluating antioxidants and β-carotene as chemopreventives. Am J Clin Nutr 1991; 53:305S-13S.

270b. Steinberg D. Summary of the proceedings of a NHLBI workshop: "Antioxidants in the prevention of human atherosclerosis." September 5-6, 1991—Bethesda MD. Circulation 1992. In press.

271. McCormick J, Skrabanek P. Coronary heart disease is not preventable by population intervention. Lancet 1988; 2:839-41.

272. Folkers K, Langsjoen P, Willis R, et al. Lovastation decreases coenzyme Q levels in humans. Proc Natl Acad Sci U S A 1990; 87:8931-4.

272a. Keenan JM, Fontaine PL, Wenz JB, Myers S, Huang Z, Ripsin CM. Niacin revisisted. Arch Intern Med 1991; 151:1424-32.

273. Bendich A, Machlin LJ. Safety of oral intake of vitamin E. Am J Clin Nutr 1988; 48:612-9.

274. Ornish D, Brown SE, Scherwith LW, et al. Can lifestyle changes reverse coronary heart disease? Lancet 1990; 336:129-33.

275. Kleijnen J, Knipschild P, Riet G. Vitamin E and cardiovascular disease. Eur J Clin Pharmacol 1989; 37:541-4.

276. Smith CL, Kelleher J, Losowsky MS, Morrish N. The content of vitamin E in British diets. Br J Nutr 1971; 26:89-96.

277. Davies T, Kelleher J, Kosowski MS. Interrelation of serum lipoprotein and tocopherol levels. Clin Chim Acta 1969; 24:431-6.

278. Lewis J, Buss DH. Trace nutrients. 5. Minerals and vitamins in the British household food supply. Br J Nutr 1988; 60:413-24.

279. Schorah CH. In: Counsell JN, Hornig DH, eds. Vitamin C. London: Applied Science Publishers, 1981; 23-47.

280. Block G. Dietary guidelines and the results of food consumption surveys. Am J Clin Nutr 1991; 53:356-7.

281. Murphy SP, Subar AF, Block G. Vitamin E intakes and sources in the United States. Am J Clin Nutr 1990; 52:361-7.

282. Horwitt MK. Status of human requirements for vitamin E. Am J Clin Nutr 1974; 27:1182-93.

283. Horwitt MK. Supplementation with vitamin E. Am J Clin Nutr 1988; 47:1088-9.

284. Horwitt MK. Data supporting supplementation of humans with vitamin E. J Nutr 1991; 121:424-9.

285. Horwitt MK, Elliott WH, Kanjananggulpan P, Fitch CD. Serum concentrations of α-tocopherol after ingestion of various vitamin E preparations. Am J Clin Nutr 1988; 40:240-5.

286. Horwitt MK, Century B, Zeman AA. Erythrocyte survival and reticulocyte levels after toco-pherol depletion in man. Am J Clin Nutr 1963; 12:99-106.

287. Lehmann J, Marshall MW, Slover HT, Jacono JM. Influence of dietary fat level and dietary tocopherols on plasma tocopherols in human subjects. J Nutr 1977; 107:1006-15.

288. Lehmann J, Martin HL, Lashley EL, Marshall MW, Judo JF. Vitamin E in foods from high and low linoleic acid diets. J Am Diet Assoc 1986; 86:1208-16.

289. Lemoyne M, van Gossom A, Kurian R, Ostro M, Axler J, Jeejeebhoy KN. Breath pentane analysis as index of lipid peroxidation: a functional test of vitamin E status. Am J Clin Nutr 1987; 46:267-72.

290. Van Gossum A, Kurian R, Whitwell J, Jeejeebhoy KN. Decrease in lipid peroxidation mea-sured by breath pentane output in normals after oral supplementation with vitamin E. Clin Nutr 1988; 7:53-7.

291. Bendich A. Antioxidant micronutrient and immune response. Ann N Y Acad Sci 1990; 587: 168-80.

292. Bendich A. Antioxidant vitamins and their functions in immune responses. Adv Exp Med Biol 1990; 262:35-55.

293. Buzina R, Aurer-Kozelj J, Srdak-Jorgic K, Bühler E, Gey KF. Increase of gingival hydroxy-proline and proline by improvement of ascorbic acid status in man. Int J Vitam Nutr Res 1986; 56:367-72.

294. Suboticanec-Buzina K, Buzina R, Brubacher G, Sapunar J, Christeller S. Vitamin C status and physical work capacity in adolescents. Int J Vitam Nutr Res 1984; 54:55-60.

295. Frankel EN. The antioxidant and nutritional effects of tocopherols, ascorbic acid and beta-carotene in relation to processing of edible oils. Bibl Nutr Dieta 1989; 43:297-312.

296. Diplock AT. Dietary supplementation with antioxidants. Is there a case for exceeding the recom-mended dietary allowance? Free Radic Biol Med 1989; 3:199-201.

297. Horwitt MK. Relative biological values of *d-α*-tocopheryl acetate and *all-rac-α*-tocopheryl ace-tate in man. Am J Clin Nutr 1980; 33:1856-60.

298. Willett WC, Stampfer MJ, Underwood BA, Taylor JO, Hennekens CH. Vitamins A, E and beta-carotene: effects of supplementation on their plasma levels. Am J Clin Nutr 1983; 38:559-66.

299. Lehmann J, Rao DD, Canary JJ, Judd JT. Vitamin E and relationship among tocopherols in human plasma, platelets, lymphocytes and red blood cells. Am J Clin Nutr 1988; 47:470-4.

300. Dimitrov NV, Meyer C, Gilliland D, Ruppenthal M, Chenoweth W, Malone W. Am J Clin Nutr 1991; 53:723-9.

300a. Refeat M, Moore TJ, Kazui M, Risby TH, Perman JA, Schwarz KB. Utility of breath pentane ethane as a noninvasive biomarker of vitamin E status in children. Pediatr Res 1991; 30:396-403.

301. Schectman G, Byrd JC, Gruchow HW. The influence of smoking on vitamin C status in adults. Am J Public Health 1989; 79:158-62.

302. Hoshino E, Shariff R, Van Gossum A, et al. Vitamin E suppresses increased lipid peroxida-tion in cigarette smokers. J Parenter Enteral Nutr 1990; 14:300-5.

303. Duthie GG, Arthur JR, James WPF, Vint HM. Antioxidant status of smokers and nonsmok-ers: effect of vitamin E supplements. Ann N Y Acad Sci 1989; 570:435-8.

304. Bittolo-Bon G, Cazzolato G, Saccarde M, Avogaro P. Presence of a modified LDL in humans: effect of vitamin E. In: Mayaishi EO, Mino M, eds. Clinical and nutritional aspects of vitamin E. Amsterdam: Elsevier, 1987; 109-20.

304a. Yagi K. Assay for serum lipid peroxide level and its clinical significance. In: Yagi K, ed. Lipid peroxides in biology and medicine. New York: Academic Press, 1982; 223-42.

304b. Janero DR. Malondialdehyde and thiobarbituric acid-reactivity as diagnostic indices of lipid peroxidation and peroxidative tissue injury. Free Radic Biol Med 1990; 9:515-40.

305. Wartanowicz M, Panczenkof Kresowska B, Ziemlanski S, Kowalska M, Okolska G. The effect of α-tocopherol and ascorbic acid on the serum lipid peroxide level in elderly people. Ann Nutr Metab 1984; 28:186-91.

306. Cordova C, Musca A, Violi F, et al. Influence of vitamin E on plasma malondialdehyde-like material in man. Thromb Hemost 1984; 51:347-8.

307. Tolonen M, Sarna S, Halme M, et al. Antioxidant supplementation decreases TBA reactants in serum of elderly. Biol Trace Elem 1988; 17:221-8.

308. Tolonen M, Sarna S, Westermarck T, et al. Reduktion von Lipidperoxiden im Serum bei älteren Menschen durch Supplementierung mit Antioxidantien. VitaMinSpur 1987; 2:181-6.

309. Harats D, Ben-Naim M, Dabach Y, et al. Effect of vitamin C and E supplementation on susceptibility of plasma lipoproteins to peroxidation induced by acute smoking. Atherosclerosis 1990; 85:47-54.

310. Reference deleted

311. Meydani SN, Barklund MP, Liu S, et al. Vitamin E supplementation enhances cell-mediated immunity in healthy elderly subjects. Am J Clin Nutr 1990; 52:557-63.

312. Sumida S, Tanaka K, Kitao H, Nakadomo F. Exercise-induced lipid peroxidation and leakage of enzymes before and after vitamin E supplementation. Int J Biochem 1989; 21:835-8.

313. Noma A, Maeda S, Okuno M, Abe A, Muto Y. Reduction of serum lipoprotein (a) levels in hyperlipidaemic patients with α-tocopheryl nicotinate. Atherosclerosis 1990; 84:213-7.

314. Reference deleted

315. Ceriello A, Giugliano D, Quatraro A, Donzella C, Dipalo G, Lefebvre PJ. Vitamin E reduction of protein glycosylation in diabetes. New prospect for prevention of diebatic complications? Diabetes Care 1991; 14:68-72.

316. Muckle J, Nazir DJ. Variation in human blood high-density lipoprotein response to oral vitamin E megadosage. Am J Clin Pathol 1989; 91:165-71.

317. Cloarec MJ, Perdriset GM, Lamberdiere FA, et al. α-Tocopherol: effect on plasma lipoproteins in hypercholesterolemic patients. Isr J Med Sci 1987; 23:669-72.

317a. Haglund O, Luostarinen R, Wallin R, Wibell L, Saldeen T. The effect of fish oil on triglycerides, cholesterol, fibrinogen and malondialdehyde in humans supplemented with vitamin E. J Nutr 1991; 121:165-9.

318. Görög P, Kakkar VV. Increased uptake of monocyte-treated low-density lipoproteins by aortic endothelium in vivo. Atherosclerosis 1987; 65:99-107.

319. Boissonneault GA, Hennig B, Wang Y, Wood CL. Aging and endothelial barrier function in culture: effects of chronic exposure to fatty acid hydroperoxides and vitamin E. Mech Ageing Dev 1990; 56:1-9.

320. Butler AM, Gerrard JM, Peller J, Stoodard PF, Rao GHR, White JG. Vitamin E inhibits the release of calcium from a platelet membrane fraction in vitro. Prostaglandins Leukot Med 1979; 2:203-16.

321. Douglas CE, Chan AC, Choy PC. Vitamin E inhibits platelet phospholipase A_2. Biochim Biophys Acta 1986; 876:639-45.

322. Chan AC, Leith MK. Decreased prostacyclin synthesis in vitamin E-deficient rabbit aorta. Am J Clin Nutr 1981; 34:2341-7.

323. Karpen CW, Merola AJ, Trewyn RW, Cornwell DG, Panganamala RV. Modulation of platelet thromboxane A_2 and arterial prostacyclin by dietary vitamin E. Prostaglandins 1981; 22:651-61.

324. Aliev MA, Bakbolotova AK, Kostiuchenko LS, Lemeshenko VA. Changes in lipid peroxidation and the antioxidative system of the myocardium in adrenaline-induced damage of the heart. Kardiologia 1989; 29:77-81.

325. Meydani SN, Stocking LM, Shapiro AC, Meydani M, Blumberg JB. Fish oil and tocopherol induced changes in ex-vivo synthesis of spleen and lung leukotriene B_4 in mice. Ann N Y Acad Sci 1988; 524:395-8.

326. Okuma M, Takayama H, Uchino H. Generation of prostacyclin-like substance and lipid peroxidation in vitamin E-deficient rats. Prostaglandins 1980; 19:527-36.

327. Stuart MJ. Deficiency of plasma PGI_2-like regenerating activity in neonatal plasma. Reversal by vitamin E in vitro. Pediatr Res 1981; 15:971-3.

327a. Stuart MJ. Vitamin E deficiency: its effect on platelet-vascular interaction in various pathologic states. Ann N Y Acad Sci 1982; 393:277-88.

328. Valentovic MA, Gairola C, Lubawy CC. Lung, aorta and platelet metabolism of 14C-arachidonic acid in vitamin E-deficient rats. Prostaglandins 1982; 24:215-24.

329. Weimann BJ, Steffen H, Weiser H. Effects of α- and γ-tocopherol (α-T, γ-T) and α-tocotrienol (α-TT) on the spontaneous and induced prostacyclin (PGI_2) synthesis from cultured human endothelial cells (HEC) and rat aorta segments ex vivo. Ann N Y Acad Sci 1989; 570:530-2.

330. Weimann BJ, Gey KF. Beziehungen von Vitamin E zur Arteriosklerose. Fat Sci Technol 1990; 92:29-37.

331. Umeda F, Kunisaki M, Inoguchi T, Nawata H. Vitamin E enhances prostacyclin production by cultured aortic endothelial cells. J Clin Biochem Nutr 1990; 8:175-83.

332. Pyke DD, Chan AC. Effects of vitamin E on prostacyclin release and lipid composition of the ischemic rat heart. Arch Biochem Biophys 1990; 277:429-33.

333. Wang J, Zhen E, Guo Z, Lu Y. Effect of hyperlipidemic serum on lipid peroxidation, synthesis of prostacyclin and thromboxane by cultured endothelial cells: protective effect of antioxidants. Free Radic Biol Med 1989; 7:243-9.

334. Fitzgerald GA, Brash AR. Endogenous prostacyclin and thromboxane biosynthesis during chronic vitamin E therapy in man. Ann N Y Acad Sci 1982; 393:209-11.

335. Beetens JR, Coene MC, Verheynen A, Zonnekeyn L, Hermann AG. Influence of vitamin C on the metabolism of arachidonic acid and the development of aortic lesions during experimental atherosclerosis in rabbits. Biomed Biochem Acta 1984; 43:273-6.

336. Sharma SC. Relationship of total ascorbic acid to prostaglandins $F_{2\alpha}$ and E_2 levels in the blood of women during the 3rd trimester of normal pregnancy. Int J Vitam Nutr Res 1982; 23:239-56.

337. Lee JW, Lee TY, Mo S, et al. Effects of L-ascorbic acid on the plasma 2-thiobarbituric acid value prostaglandin biosynthesis, photohemolysis, superoxide dismutase and catalase activities in guinea-pigs. Korean Biochem J 1987; 20:378-88.

338. Reddanna P, Whelan J, Burgess JR, et al. The role of vitamin E and selenium on arachidonic acid oxidation by way of 5-lipoxygenase pathway. Ann N Y Acad Sci 1989; 570:136-45.

339. Grossman S, Waksman EG. New aspects of the inhibition of soybean lipoxygenase by α-tocopherol. Evidence for the existence of a specific complex. Eur J Biochem 1984; 16:281-9.

340. Cucurou C, Battioni JP, Daniel R, Mansuy D. Peroxidase like activity of lipoxygenase: different substrate specificity of potato 5-lipoxygenase and soybean 15-lipoxygenase and particular affinity of vitamin E derivatives for the 5-lipoxygenase. Biochim Biophys Acta 1991; 1081:99-105.

341. Bakalova RA, Nekrasov AS, Lankin VZ, et al. Mechanism of the inhibitory action of α-tocopherol and its synthetic derivatives on linoleic acid oxidation catalyzed by lipoxygenase from reticulocytes. Dokl Akad Nauk SSSR 1988; 299:1008-11.

342. Egan RW, Tischler AN, Baptista EM, Ham EA, Soderman DD, Gale PH. Specific inhibition and oxydative regulation of 5-lipoxygenase. Prostgland Thromboxane Leukot Res 1983; 1: 151-7.

343. Chamulitrat W, Mason RP. Lipid peroxyl radical intermediates in the peroxidation of polyunsaturated fatty acids by lipoxygenase. J Biol Chem 1989; 264:20968-73.

344. Gordon MH, Barimalaa IS. Co-oxidation of fat soluble vitamins by soybean lipoxygenase. Food Chem 1989; 32:31-7.

345. Barimalaa IS, Gordon MH. Cooxidation of beta-carotene by soybean lipoxygenase. J Agr Food Chem 1988; 36:685-7.

346. Schmidt KH, Steinhilber D, Moser U, Roth HJ. L-Ascorbic acid modulates 5-lipoxygenase activity in human polymorphonuclear leucocytes. Int Arch Allergy Appl Immunol 1988; 85:441-5.

347. Takahama U. Inhibition of lipoxygenase-dependent lipid peroxydation by quercetin: mechanism of antioxydative function. Phytochemistry 1985; 24:1443-6.

348. Kagan VE, Serbinova EA, Packer L. Generation and recycling of radicals from phenolic antioxidants. Arch Biochim Biophys 1990; 280:33-9.

349. Kao J, Huey G, Kao R, Stern R. Ascorbic acid stimulates production of glycosaminoglycans in cultured fibroblasts. Exp Mol Pathol 1990; 53:1-10.

349a. Fisher E, McLennan SV, Tada H, Heffernan, Yue DN, Turtle J. Interaction of ascorbic acid and glucose on production of collagen and proteoglycan by fibroblasts. Diabetes 1991; 40:371-6.

350. Mamutov ZI. Effect of thymosin and vitamins A and E on the content of sialic acids in the blood serum and liver of thymectomized rats. Uzb Bioil Zh 1990; 1:4-8.

351. Olson JA. Vitamin A. In: Machlin LJ, ed. Handbook of vitamins. New York: Marcel Dekker, 1990; 1-57.

354.   Dougherty JJ, Croft WA, Hoekstra WG. Effects of ferrous chloride and iron-dextran on lipid peroxidation in vivo in vitamin E and selenium adequate and deficient rats. J Nutr 1981; 111: 1784-96.

355.   Dougherty JJ, Hoekstra WG. Effects of vitamin E and selenium on copper-induced lipid peroxidation in vivo and on acute copper toxicity. Proc Soc Exp Biol Med 1982; 169:201-8.

356.   Dillard CJ, Downey JE, Tappel AL. Effect of antioxidants on lipid peroxidation in iron-loaded rats. Lipids 1984; 19:127-33.

357.   Messripour M, Haddady H. Effect of ascorbic acid administration on copper-induced changes of rat brain hypothalamic catecholamine contents. Acta Neurol Scand 1988; 77:481-5.

358.   Jacob RA, Skala JH, Omaye ST, Turnland JR. Effect of varying ascorbic acid intakes on copper absorption and ceruloplasmin levels of young men. J Nutr 1987; 117:2109-15.

358a. Weimann BJ, Weiser H. Functions of vitamin E in reproduction and in prostacyclin and immunoglobulin synthesis in rats. Am J Clin Nutr 1991; 53:1056-60.

359.   Tengerdy RP. Vitamin E immune response, and disease resistance. Ann N Y Acad Sci 1989; 570: 335-44.

360.   Tengerdy RP. The role of vitamin E in immune response and disease resistance. Ann N Y Acad Sci 1990; 587:24-33.

361.   Tengerdy RP. Immunity and disease resistance in farm animals fed vitamin E supplement. Adv Exp Med Biol 1990; 262:103-10.

362.   Tengerdy RP, Mathias MM, Nockels CF. Effect of vitamin E on immunity and disease resistance. In: Prasad A, ed. Vitamins, nutrition and cancer. Basel: Karger, 1986; 123-33.

363.   Penn ND, Purkins L, Kelleher J, et al. The effect of dietary supplementation with vitamins A, C and E on cell-mediated immune function in elderly long-stay patients: a randomized trial. Age Ageing 1991; 20:169-74.

364.   Meydani SN. Micronutrients and immune function in the elderly. Ann N Y Acad Sci 1990; 587: 197-207.

365.   Meydani SN, Meydani M, Barklund PM, et al. Effect of vitamin E supplementation on immune responsiveness of the aged. Ann N Y Acad Sci 1989; 570:283-90.

366.   Moriguchi S, Kobayashi N, Kishino Y. High dietary intakes of vitamin E and cellular immune function in rats. J Nutr 1990; 120:1096-102.

367.   Afonina GB, Bordonos VG. Role of free-radical oxidation of lymphocyte membrane lipids in the development of immunodeficiency and its correction by α-tocopherol. Immunologiya 1990; 5:33-5.

368.   Roy RM, Petrella M, Shateri H. Effects of administering tocopherol after irradiation to survival and proliferation of murine lymphocytes. Pharmacol Ther 1988; 39:393-5.

369.   Jensen M, Fossum C, Ederoth M, Hakkarainen RVJ. The effect of vitamin E on the cell-mediated immune response in pigs. J Vet Med (B) 1988; 35:549-55.

370.   Siegel BV, Morton JI. Vitamin C and immunity: influence of ascorbate on prostaglandin E_2 synthesis and implications for natural killer cell activity. J Vitam Nutr Res 1984; 54:39-42.

371.   Payette H, Rola-Pleszczynski M, Ghardirian P. Nutrition factors in relation to cellular and regulatory immune variables in a free-living elderly population. Am J Clin Nutr 1990; 52:927-32.

372.   Bowman TA, Goonewardene IM, Pasatiempo AMG, Ross AC, Taylor CE. Vitamin A deficiency decreases natural killer cell activity and interferon production in rats. J Nutr 1990; 120:1264-73.

373.   Watson RR, Prabhala RH, Plezia PM, Alberts DS. Effect of beta-carotene on lymphocyte subpopulations in elderly humans: evidence for a dose-response relationship. Am J Clin Nutr 1991; 53:90-4.

374.   Goetzel EJ. Vitamin E modulates the lipoxygenation of arachidonic acid in leukocytes. Nature 1981; 288:193-5.

375.   Chan AC, Tran K, Pyke DD, Powell WS. Effects of dietary vitamin E on the biosynthesis of 5-lipoxygenase products in rat polmorphonuclear leukocytes (PMNL). Biochim Biophys Acta 1989; 1005:265-9.

375a. Negre-Salvayre A, Alomar Y, Troly M, Salvayre R. Ultraviolet-treated lipoproteins as a model system of the biological effects of lipid peroxides on cultured cells. III. The protective effect of antioxidants (probucol, catechin, vitamin E) against cytotoxicity. Biochim Biophys Acta 1991; 1096:291-300.

376. Alink GM, Rietjens IMCM. Mechanisms of ozone and nitrogen dioxide toxicity in lung cells in vitro. In: Seemayer NH, Radnay W, eds. Enviro Hyg. Berlin: Springer, 1988; 1:7-14.

377. Boura P, Tsapas G, Papadopoulou A, Magoula I, Kountouras G. Monocyte locomotion in anergic chronic brucellosis patients: in vivo effect of ascorbic acid. Immunopharmacol Immunotoxicol 1989; 11:119-29.

378. Dietert R, Combs GF Jr, Lin HK, Puzzi JV, Golemboski KA, Marsh JA. Impact of combined vitamin E and selenium defiency on chicken macrophage function. Ann N Y Acad Sci 1990; 587:281-2.

379. Eskew ML, Zarkower A, Scheuchenzuber WJ, et al. Effects of inadequate vitamin E and/or selenium nutrition on the release of arachidonic acid metabolites in rat alveolar macrophages. Prostaglandins 1989; 38:79-89.

380. Boscoboinik D, Szewczyk A, Azzi A. α-Tocopherol (Vitamin E) regulates vascular smooth muscle cell proliferation and protein kinase C activity. Arch Biochem Biophys 1991; 286:264-9.

381. Boscoboinik D, Swewczyk A, Hensey C, Azzi A. Inhibition of cell proliferation by α-tocopherol. Role of protein kinase C. J Biol Chem 1991; 266:6188-94.

382. Azzi A. Boscoboinik D, Chatelain E. Modulation of cell proliferation by tocopherols and tocotrienols: role in arteriosclerosis. In: Ong AHS, Packer L, eds. Lipid-soluble antioxidants: biochemistry and clinical applications. Basel: Birkhäuser, 1992; in press.

383. Phoenix J, Edwards RH, Jackson MJ. Inhibition of Ca^{2-}-induced cytosolic enzyme efflux from skeletal muscle by vitamin E and related compounds. Biochem J 1989; 257:207-13.

384. Diplock A, Xu GL, Yeow CL, Okikola M. Relationship of tocopherol structure to biological activity, tissue uptake, and prostaglandin biosynthesis. Ann N Y Acad Sci 1989; 570:72-84.

385. Singal PK, Kapur N, Beamish RE. Antioxidant protection against epinephrine-induced arrhythmia. Dev Cardiovasc Med 1985; 60:190-201.

386. Bekbolotova AK, Varvashtyan VM, Aliev MA. α-Tocopherol influence on the conditioned reflex activity in adrenalin-induced myocardial dystrophy. Byull Eksp Biol Med 1990; 109:211-3.

387. Kirshenbaum LA, Gupta M, Thomas TP, Singal PK. Antioxidant protection against adrenaline-induced arrhythmias in rats with chronic heart hyperthrophy. Can J Cardiol 1990; 6:71-4.

388. Fuenmayor AJ, Fuenmayor AM, Lopez T, Winterdaal DM. Vitamin E and ventricular fibrillation threshold in myocardial ischemia. Jpn Circ J 1989; 53:1229-32.

389. Dzizinskii AA, Ana'nev AA, Fuks AR. Effect of membrane protectors on the diastolic function of the heart in patients with acute myocardial infarction. Kardiologiia 1989; 29:52-5.

390. Jandak J. Steiner M, Richardson PD. α-Tocopherol, an effective inhibitor of platelet adhesion. Blood 1989; 73:141-9.

391. Jandak J, Steiner M, Richardson PD. Reduction of platelet adhesiveness by vitamin E supplementation in humans. Thromb Res 1988; 49:393-404.

392. Reference deleted.

393. Karpen CW, Cataland S, O'Dorosio TM, Panganamala RV. Production of 12-hydroxyeicosatetraenoic acid and vitamin E status in platelets from type I human diabetic subjects. Diabetes 1985; 34:526-31.

394. Watanabe J, Umeda F, Wakasugi H, Ibayashi H. Effect of vitamin E on platelet aggregation in diabetes mellitus. Thromb Haemost 1984; 51:313-6.

395. Renaud S, Ciavatti M, Perrot L, et al. Influence of vitamin E administration on platelet functions in hormonal contraceptive users. Contraception 1987; 36:347-58.

395a. Viola F, Pratico D, Ghiselli A, et al. Inhibition of cyclooxygenase-independent platelet aggregation by low vitamin E concentration. Atherosclerosis 1990; 82:247-52.

396. Salonen JT. Antioxidants and platelets. Ann Med 1989; 21:59-62.

397. Karpen CW, Cataland S, O'Dorisio TM, Panganamala RV. Interrelation of platelet vitamin E and thromboxane synthesis in type I diabetes mellitus. Diabetes 1984; 33:239-43.

398. Cordova C, Musca A, Violi F, et al. Vitamin C inhibits platelet lipoxygenase and cyclooxygenase pathways. Prostaglandins 1984; 27(suppl):103.

399. Gwebu ET, Trewyn RW, Cornwel DG, Panganamala RV. Vitamin E and inhibition of platelet lipoxygenase. Res Commun Chem Pathol Pharmacol 1980; 28:361-76.

400. Fukuzawa K, Kurotori Y, Tokumura A, Tsukatani H. Vitamin E deficiency increases the synthesis of platelet-activating factor (PAF) in rat polymorphonuclear leukocytes. Lipids 1989; 24:236-9.

401. Skjaerlund JM. Exocrine pancreatic degeneration during vitamin E and selenium deficiency in rats. Nutr Rep Int 1989; 40:151-9.

402. Reference deleted.

403. Bordia AK. The effect of vitamin C on blood lipids, fibrinolytic activity and platelet adhesiveness in patients with coronary artery diseae. Atherosclerosis 1980; 35:181-7.

404. Kudryashov BA, Pastorova VE, Lyapina LA, Londashevskaya MV, Kobozeva LP. Fibrin depolymerization and nonenzymic fibrinolysis in blood plasma of rabbits kept on atherogenic rations with antioxidants. Biull Eksp Biol Med 1990; 110:419-21.

405. Iwama M, Honda A, Ohohashi Y, Sakai T, Mori Y. Alterations in glycosaminoglycans of the aorta of vitamin E-deficient rats. Atherosclerosis 1985; 55:115-23.

406. Kakishita E, Suehiro A, Oura Y, Nagai K. Inhibitory effect of vitamin E (α-tocopherol) on spontaneous platelet aggregation in whole blood. Thromb Res 1990; 60:489-99.

407. Klein HH, Pich S, Lindert S, Nebendahl K, Niedmann P, Kreuzer H. Combined treatment with vitamins E and C in experimental myocardial infarction in pigs. Am Heart J 1989; 118:667-73.

408. Massey KD, Burton KP. α-Tocopherol attenuates myocardial membrane-related alterations resulting from ischemia and reperfusion. Am J Physiol 1989; 256:H1192-9.

409. Massey KD, Burton KP. Free radical damage in neonatal rat cardiac myocyte cultures: effects of α-tocopherol, trolox, and phytol. Free Radic Biol Med 1990; 8:449-58.

410. Mickle DAG, Li RK, Weisel RD, et al. Myocardial salvage with trolox and ascorbic acid for an acute evolving infarction. Ann Thorac Surg 1989; 47:546-57.

411. Janero DR, Burghardt B. Oxidative injury to myocardial membrane direct modulation by endogenous α-tocopherol. J Mol Cell Cardiol 1989; 21:1111-24.

412. Barsacchi R, Coassin M, Maiorino M, Pelosi G, Siminelli C, Ursini F. Increased ultra weak chemiluminescence emission from rat heart at postischemic reoxygenation: protective role of vitamin E. Free Radic Biol Med 1989; 6:573-9.

413. Axford-Gatley RA, Wilson GJ. Dietary loading with vitamin E reduces experimental myocardial infarct size. Cardiovasc Res 1991; 25:89-92.

414. Axford-Gatley RA, Wilson GJ. Myocardial infarct size reduction by vitamin E hypersupplementation. J Mol Cell Cardiol 1990; 22(Suppl III):S.28, Abstract PT3.

415. Cavarocchi NC, England MD, O'Brien JF, et al. Superoxide generation during cardiopulmonary bypass: is there a role for vitamin E? J Surg Res 1986; 40:519-27.

416. Buchwald A, Klein HH, Lindert S, et al. Effect of α-tocopherol (vitamin E) in a porcine model of stunned myocardium. J Cardiovasc Pharmacol 1989; 14:46-52.

417. Matevosyan RS, Amatuni VG. The effect of pretreatment with α-tocopherol and intal on the course of experimental myocardial necrosis. Kardiologiya 1989; 29:94-6.

418. Weisel RD, Mickle DA, Finkle CO, et al. Myocardial free-radical injury after cardioplegia. Circulation 1989; 80(III):14-8.

419. Stoyanovski D, Kagan V, Packer L. Iron binding to α-tocopherol-containing phospholipid liposomes. Biochem Biophys Res Commun 1989; 160:834-8.

420. Hatzelmann A, Schatz M, Ullrich V. Involvement of glutathione peroxidase activity in the stimulation of 5-lipoxygenase activity by glutathione-depleting agents in human polymorphonuclear leukocytes. Eur J Biochem 1989; 180:527-33.

421. Camulitrat W, Mason RP. Lipid peroxyl radical intermediates in the peroxidation of polyunsaturated fatty acids by lipoxygenase direct ESR investigations. J Biol Chem 1989; 264:20968-73.

422. Cole EH, Levy GA. Interaction of monocytes with vascular endothelium. In: Zembala M, Asherson GL, eds. Human monocytes. London: Academic Press, 1989; 353-60.

423. Dinarello CA, Savage N. Interleukin-1 and its receptor. Crit Rev Immunol 1989; 9:1-20.

424. Nagai Y, Yamane T, Watanabe H, Yoshida Y. Diffuse intimal thickening and other mesenchymal changes. Ann N Y Acad Sci 1990; 598:71-6.

425. Gopalakrishna R, Anderson WB. Ca^{2+}- and phospholipid-independent activation of protein kinase C by selective oxidative modification of the regulatory domain. Proc Natl Acad Sci U S A 1989; 86:6758-62.

426. O'Brian CA, Ward NE, Weinstein IB, Bull AW, Marnett LJ. Activation of rat brain protein kinase C by lipid oxidation products. Biochem Biophys Res Commun 1988; 155:1374-80.

427. Smirnov VN, Voyno-Yasenetskaya TA, Antonov AS, et al. Vascular signal transduction and atherosclerosis. Ann N Y Acad Sci 1990; 598:167-81.

428. Watson F, Robinson J, Edwards SW. Protein kinase C-dependent and -independent activation of the NADPH oxidase of human neutrophils. J Biol Chem 1991; 266:7432-9.

429. Mendez AJ, Oram JF, Bierman EL. Protein kinase C as a mediator of high density lipoprotein receptor-dependent efflux of intracellular cholesterol. J Biol Chem 1991; 266:10104-11.

430. Elstad MR, Mcintyre TM, Prescott SM, Zimmermann GA. Protein kinase C regulates their synthesis of platelet-activating factor by human monocytes. Am J Respir Cell Mol Biol 1991; 4:148-55.

431. Babiy AV, Gebicki FM, Sullivan DR. Vitamin E content and low density lipoprotein oxidizability induced by free radicals. Atherosclerosis 1990; 81:175-82.

432. De Graaf J, Hak Lemmers HLM, Hectors MPC, Demacker PNM, Hendiks JCM, Stanlenhoef AFH. Enhanced susceptibility to in vitro oxidation of the dense low density lipoprotein subfraction in healthy subjects. Arteriosclerosis 1991; 11:298-306.

433. Shimano H, Yamada N, Ishibashi S, et al. Oxidation-labile subfraction of human plasma low density lipoprotein isolated by ion-exchange chromatography. J Lipid Res 1991; 32:763-73.

433a. Nigon F, Lesnik kP, Rouis M, Chapman MJ. Discrete subspecies of human low density lipoproteins are heterogeneous in their interaction with the cellular LDL receptor. J Lipid Res 1991; 32:1741-53.

433b. Knipping G, Rotheneder M, Striegl G, Esterbauer H. Antioxidants and resistance against oxidation of porcine LDL fractions. J Lipid Res 1990; 31:1965-72.

434. Parthasarathy S, Khoo JH, Miller E, Barnett J, Witztum JL, Steinberg D. Low-density lipoprotein enriched in oleic acid is protected against oxidative modification: implication for dietary prevention of atherosclerosis. Proc Natl Acad Sci U S A 1990; 87:3894-8.

435. Andrikopoulos NK, Hassapidou MN, Manoukas AG. The tocopherol content of Greek olive oils. J Sci Food Agr 1989; 46:503-9.

436. Kiritsakis A, Markakis P. Olive oil: a review. Adv Food Res 1987; 31:453-82.

437. Hayes C. In: Ong AHS, Packer L, eds. Lipid-soluble antioxidants: biochemistry and clinical applications. Basel: Birkhäuser, 1992; in press.

45

Lipoprotein Oxidation and Atherosclerosis

Gérald Luc and Jean-Charles Fruchart

Pasteur Institute, Lille, France

INTRODUCTION

Atherosclerotic cardiovascular disease is the leading cause of mortality and morbidity in industrialized countries as indicated by the international statistics on causes of death (1). Atherosclerosis is a metabolic disease that leads to reductions in the luminal diameter of the main arteries. Several theories—endothelial injury, thrombogenesis, monoclonal antibodies, permeability alteration, virus, hemodynamics, and lipids—were advanced to explain the appearance of atherosclerotic lesions, but there is no unanimous agreement about its pathogenesis. The lipid theory was sustained by basic clinical, epidemiological, and experimental studies and pharmacological trials. Particularly, prospective epidemiological studies have identified a number of risk factors for atherosclerosis. Among these is a high concentration of low-density lipoprotein (LDL) cholesterol (2) and a low concentration of high-density lipoprotein cholesterol (3).

The complexity of atherosclerotic lesions increases as a function of their age (4,5). The first step of atherosclerosis is the appearance of fatty streak lesions, characterized by recruitment of macrophages in the intima of arteries just beneath the endothelium (4–6). These macrophages play a significant role in atherogenesis (7) because they are capable of taking up and degrading considerable quantities of lipoprotein. They then appear as foam cells, this aspect being the consequence of the accumulation of cholesteryl esters in the cytoplasm of cells. Macrophages present in the intima of arteries are derived from blood monocytes and cholesteryl esters essentially from plasma low-density lipoproteins or from abnormal lipoproteins, such as β-VLDL (very low density lipoproteins). The progression of fatty streaks toward more advanced lesions is characterized by the migration and proliferation of smooth muscle cells (8) and the accumulation of lipids, carbohydrates, and constituents of blood. The presence of lymphocytic cells and peripheral fibrosis may involve different processes. Thus an understanding of the initiation and development of fatty streaks, the first step in atherosclerosis, is required. Because the earliest step in atherosclerosis appears to be the penetration of monocytes between endothelial cells and the internal elastic lamina and therefore the differentiation of these cells into macrophages, it is of major importance to elucidate the biological mechanisms of these phenomena.

For similar levels of cholesterol, the extension of coronary atherosclerosis and the incidence of coronary heart disease are very variable (9). This variation is probably the consequence of different biological properties of arterial cells and lipoproteins among individuals. Some of these differences reflect a variation in modifications of LDL during their metabolism.

STRUCTURE AND METABOLISM OF LDL

Structure of LDL

LDL, which constitutes the major vehicle for cholesterol transport in normal human plasma (10), are quasi-spherical pseudomicellar particles with a hydrophobic core of apolar constituents, primarily cholesteryl esters and triglycerides, surrounded by a polar content of phospholipids, some free cholesterol, and proteins (11). The protein moiety consists essentially of apolipoprotein (apo) B-100, a high-molecular-weight protein, which plays a determining role both in the molecular structure of LDL particles and in their in vivo metabolism (12,13).

Heterogeneity in physical, chemical, hydrodynamic, and immunological properties is an inherent characteristic of the LDL particles that comprise the density (usually 1.019–1.063 g/ml) profile, both in normolipidemic (14–16) and hyperlipidemic (17,18) individuals. The different properties of LDL subclasses isolated by various procedures with regard to particle size, hydrated density, and molecular weight have been attributed to discrete changes in the lipid and protein content of lipoprotein molecules. The triglyceride content in particular decreases when the density increases. In contrast, the relative content of protein increases in LDL of higher density. The inverse relation between density and the size of LDL has also been well demonstrated. The size of the predominant subclass of LDL is variable among normo- and hyperlipidemic subjects, however, the variation in size being a result of modifications in chemical structure (17–19). The LDL structure depends on genetic (20) and environmental (21) factors.

Metabolism of LDL

The study of patients with homozygous familial hypercholesterolemia has shown the essential function played by the LDL receptor in LDL metabolism (13,22). Apo B-100 is the ligand of LDL for the LDL receptor, and basic amino acids of this protein, particularly arginine and lysine residues, play an essential role in binding of LDL to the receptor (23). Indeed, methylation, which modifies lysine, or cyclohexanedione, which reacts with arginine, blocks LDL binding to its receptor. About two-thirds of LDL is removed from plasma by the LDL receptor pathway, this removal appearing essentially in the liver. However, if the majority of cholesterol found in the artery wall comes from LDL, the uptake of LDL-producing foam cells in the artery wall must be independent of the LDL receptor pathway. Indeed, patients with homozygous familial hypercholesterolemia have no LDL receptor activity but clearly have an accelerated atherosclerosis, with cholesterol deposits in macrophages of the artery wall (22). Similarly, rabbits deficient in LDL receptor (Watanabe heritable hyperlipidemic rabbits, WHHL) demonstrate early atherosclerosis with lesions rich in macrophage-derived foam cells (24). Therefore, the accumulation of cholesterol in macrophages could not occur by the LDL receptor pathway but by another pathway called the scavenger pathway (25,26). The capacity for removal of LDL by the scavenger pathway appears quantitatively very important, because if the catabolism of LDL is very slow in receptor-negative patients or animals, the LDL production and removal are very high, about threefold that of normolipidemic subjects. Second, incubation of native LDL with monocyte-macrophages in culture does not lead to the formation of foam cells. This observation could be explained by a low number of LDL receptors on the surface of macrophages. Moreover, these LDL receptors, as LDL receptors present in other tissues, can be downregulated in the presence of LDL, and no accumulation of cholesterol appeared during this experiment.

Thus, apparently contradictory observations of the absence of native LDL uptake by monocyte-macrophages, on the one hand, and the accumulation of cholesterol in monocyte-macrophages occurring essentially from LDL, on the another hand, have led to the hypothesis that the LDL may be undergoing one or more modifications and that this modified LDL could be taken up by these cells.

MODIFICATIONS OF LDL

Goldstein et al. and Mahley et al. demonstrated that cultured mouse peritoneal macrophages accumulate massive amounts of cholesterol when exposed to chemically modified LDL, such as acetylated or acetoacetylated modified LDL (27,28). These studies have been extended to other modifications of LDL, such as malondialdehydation (29) or by incubation with some cells. The capacity for cells to modify LDL varies with the type of cells and the species. Thus, modification of LDL has been reported for endothelial cells from rabbits (30), human umbilical veins and adult arteries and veins (31), bovine and human smooth muscle cells (32,33), and human monocyte-macrophages (34,35), but it was not found with bovine aorta endothelial cells or human fibroblasts (33,36). Some of these in vitro modifications (malondialdehydation and modification by cells) could appear in vivo. Modifications of LDL induced by incubation with cells have been extensively studied since Henricksen et al. demonstrated structural changes in LDL under these conditions (30). The structural and biological properties of cell-modified LDL can be mimicked by incubating LDL in a serum-free medium in the presence of copper or iron (32,37,38). These modifications appear to be through an oxidative phenomenon, the polyunsaturated fatty acids (PUFAs) of LDL being the substrates.

Structural Changes in Modified LDL

Incubation of LDL with different types of cells (32–35) or in the presence of copper induces a large number of structural modifications. The most evident changes in the endothelial cell-modified LDL are their physical properties. Indeed, the hydrated density increased with the degree of modification. The buoyant density of native LDL was 1.030–1.040 g/ml and that of endothelial cell LDL increased to 1.060–1.078 g/ml after 24 h of incubation (36). Simultaneously, the electrophoretic migration of modified LDL was faster as a result of the modification of the apparent electrical charge at the surface. These two parameters showed a time-dependent increase and plateaued by 24 h (30,36,37).

During LDL oxidation, the chemical composition of the lipoprotein was greatly modified, with the disappearance of some components and the synthesis of others. Studies of these modifications as a function of time have shown that the earliest change was the fast decrease in the content of vitamin E (39,40). LDL preparations contained variable quantities of vitamin E, from 3.15 to 9.90 mol/mol LDL, with a mean of 6.4. The vitamin E was completely consumed during LDL oxidation in 1–6 h, this time being a function of the vitamin E content in LDL and of the experimental conditions. Other antioxidants in lesser quantities in LDL (γ-tocopherol, β-carotene, lycopine, and retinyl stearate) also disappeared during LDL oxidation. The antioxidant consumption phase can be defined as a lag phase, because no oxidation of PUFAs occurs. The second phase is the propagation phase. When LDL were completely depleted of its antioxidants, PUFAs are rapidly oxidized. Thus, PUFAs at 18:2 and 20:4 decreased dramatically during incubation of LDL with copper 10 μM. Indeed, the content of these two fatty acids in LDL dropped to 10% of the value present in native LDL during 24 h of incubation (41). The importance of PUFAs in the susceptibility of LDL to peroxidation was suggested by the resistance of LDL to oxidation when these particles were enriched by monounsaturated fatty acids, such as oleic acid (42). Furthermore, the modification of LDL by endothelial cells appeared to be accompanied by extensive hydrolysis of the phosphatidylcholine in LDL to lysophosphatidylcholine (37). About 40% of the phosphatidylcholine disappeared during the 24 h of incubation.

Lysophosphatidylcholine remains associated with LDL, and the phospholipid to protein ratio remains constant in cell-modified LDL (43). Studies using radiolabeled phosphatidylcholine showed that the 2-position fatty acid was cleaved during modification by incubation with endothelial cells or Cu^{2+}. The specificity of hydrolysis suggested the presence of phospholipase A_2 activity.

Phospholipase activity and the peroxidation of fatty acids could favor one another. Indeed, phospholipase could release peroxidized fatty acids that would propagate peroxidation reactions of other fatty acids. Further, the oxidation of fatty acyl chains of phosphatidylcholine could make a better substrate for phospholipase activity. Thus, the irreversible inhibition of phospholipase A_2 by *p*-bromophenacyl bromide blocked phosphatidylcholine hydrolysis but only partially inhibited peroxidation (44). Lipid peroxidation therefore does not require hydrolysis of phosphatidylcholine.

The lipid peroxidation of fatty acids of LDL, essentially linoleic and arachidonic acids, generated a number of new products, the main mechanism for their formation being the so-called β-cleavage reaction of the lipid alkoxyl radicals. When PUFAs are more or less completely oxidized during the decomposition phase, the lipids are converted to a great variety of other products, including complex hydroperoxy and hydroxy acid derivatives, such as 13- or 9-hydroperoxyoctodecadienoate derivatives produced by the oxidation of linoleic acid and hydroxyeicosatetraenoate derivatives produced by that of arachidonic acid (45) and lower molecular weight derivatives, such as malonaldehyde, hexanal, propanal, 4-hydroxynonenal, butanal, hytadienal, pentanal, 4-hydoxyhexenal, and 4-hydroxyoctenal.

The kinetics of the concentration of the various aldehydes were different. Although the levels of propanal, hexanal, and 4-hydroxyoctenal increased during the first 3 h of incubation of LDL in oxygenated buffer, the concentrations of 4-hydroxynonenal, 4-hydroxyhexenal, 2,4-heptadienal, pentanal, and butanal only increased after this time. The linoleate- and arachidonate-derived hydroxy acids increased with the incubation time of LDL with Cu^{2+} (45).

Some of these products were essentially hydrophilic, such as malonaldehyde, and were found in the aqueous phase of the incubation mixture; others were lipophilic and remained mostly in the lipid core of oxidized LDL (40).

The oxidation of LDL is accompanied by an increase in thiobarbituric acid-reactive substances (32,37), which is a routine method to evaluate LDL oxidation, but this procedure is not specific and detects less than 10% of the hydroperoxides derived from linoleic and arachidonic acids (45), the main PUFAs in LDL.

The protein moiety of LDL, the apolipoprotein B-100, was degraded during oxidation. The band corresponding to apo B-100 with a molecular weight of 550,000 daltons on sodium dodecyl sulfate (SDS)–polyacrylamide gel electrophoresis disappeared almost completely in oxidized LDL, and numerous bands of lower molecular weight ranging from 14,000 to 200,000 were detected after incubation of LDL with endothelial cells (37) or Cu^{2+} (46), but no discrete band was seen. The disappearance of apo B-100 was rapid: indeed, no intact apo B-100 could be detected by 4 h. This breakdown of apo B-100, whether cell induced or Cu^{2+} catalyzed, is not mediated by proteolytic enzymes and could be the result of the oxidative cleavage of the polypeptide chain, because it only appeared in media supporting oxidation and was inhibited by such antioxidants as butylated hydroxytoluene (BHT) or EDTA but not by proteolytic enzymes (46). This modification is accompanied simultaneously by a decrease in LDL amino group reactivity, as shown by the TNBS (trinitrobenzene sulfonic acid) reactivity that determined the free amino groups (47). Amino acids already shown to be susceptible to oxidation—lysine, proline, and histidine—decrease more than 10%.

The chemical modification (acetylation) of LDL has shown that this modification involves lysine residues that present ϵ-amino groups (27). The oxidation of LDL is similarly accompanied by derivatization of lysine ϵ-amino groups that neutralized the positive charge in LDL, explaining the increase in electrophoretic migration of oxidized LDL. The fragmentation of apo B-100 and modification of amino acids is the consequence of the reactivity of lipid peroxidation product with lysine. The decomposition products of PUFAs could be candidates for this reaction (48). Indeed, the incubation of one of the most reactive lipid peroxidation products, 4-hydroxynonenal, with LDL revealed that the majority of this aldehyde is found bound to the protein. 4-hydroxynonenal reacts mainly with lysine but also with serine, tyrosine, cysteine, and histidine (49). However, the incubation of LDL with 4-hydroxynonenal leads to higher molecular weight forms than apo B-100, these forms not being detected in endothelial cell-modified LDL.

The structural modification of apo B-100 during oxidation was reflected by a modified immunoreactivity. Indeed, the immunoreactivity of three different epitopes of apo B-100 decreases during oxidation of LDL with Cu^{2+}, but the immunoreactivity of another epitope located in the C-terminal 20 amino acids of apo B-100 increased during the first 6 h of oxidation and thereafter diminished (50). Using monoclonal antibodies developed against malondialdehyde-modified LDL, 4-hydroxynonenal LDL, it was demonstrated that malondialdehyde lysine and 4-hydroxynonenal lysine were generated during LDL oxidation in the presence of Cu^{2+} (51).

It appeared that the extent of modification was extremely variable for each LDL. As mentioned, LDL isolated from rabbits fed an oleic-rich diet were remarkably resistant to oxidation compared to LDL isolated from animals fed a diet rich in PUFAs. The composition of the diet was partly reflected in LDL because LDL contained 49.5 and 24% oleate and 16.5 and 39% linoleate, respectively (42). The density range in which LDL were isolated also influences the susceptibility to oxidation, dense LDL being less well protected against oxidation than lighter (52). Some LDL preparations could not be modified at all (53), perhaps related to the presence of antioxidants in the LDL. A high concentration of antioxidants in lipoproteins could be dependent on vitamin E or other antioxidants in the diet.

Mechanisms of LDL Oxidation

LDL oxidation occurred in the presence of oxygen and in the absence of EDTA. Thus, the lipid peroxidation and fragmentation of apo B-100 are consistent with the role of such metals ions as Cu^{2+} or Fe^{2+} in this process. This process was thus inhibited in the presence of EDTA, which chelates metals ions, or by the free-radical scavenger BHT (37). The occurrence of lipid peroxidation and the inhibition of LDL modification by free-radical scavengers strongly suggest that cells modify LDL by a free-radical process. A first possibility could be that cells secrete oxidants into the medium. Several oxidants generated by the cells—hydrogen peroxide, superoxide, and hydroxy radical—could be candidates for the oxidative effect. Catalase, an enzyme that disrupts hydrogen peroxide, and mannitol, a hydroxy radical scavenger, had no inhibitory effect on the oxidative modification of LDL, suggesting that these two oxidant agents are not responsible for the oxidative process of LDL (54). On the other hand, superoxide anions in the presence of Cu^{2+} could be implicated in the oxidation of LDL (54,55). The secretion of superoxide anions by arterial smooth cells and monocytes was demonstrated, and superoxide dismutase inhibited modification of LDL by smooth cells or monocytes (54,55). However, the involvement of superoxide anion secretion in LDL modification is currently under discussion since superoxide dismutase has been reported to inhibit endothelial cell modification of LDL (54-56), but other workers have failed to obtain this inhibition (53). An argument against the role of superoxide anion in LDL oxidation is the absence of LDL modification when these lipoproteins are incubated in the presence of xanthine oxidase and xanthine (57). A second possibility could be the oxidation of LDL during LDL-cell contact. This LDL-cell contact seems essential to oxidation because LDL in dialysis tubing incubated in the presence of endothelial cells does not undergo oxidative modification (57). Furthermore, of the two known means by which endothelial cells oxidize lipids, lipoxygenase and cyclooxygenase, only the first enzyme could directly induce the peroxidation of LDL lipids. Indeed, cyclooxygenase seems not to be involved in the oxidative process, because its inhibition by aspirin or indomethacin had no effect on the modification of LDL induced by endothelial cell (53, 57). In contrast, inhibitors of lipoxygenase reduced LDL oxidation by as much as 70–85% (57).

However, cellular lipoxygenase is essentially cytosolic and is not situated at the cell surface. Lipoxygenase could therefore generate peroxy lipids from cell lipids, and these oxidized lipids could then be transferred to LDL during LDL-cell contact. It was noticed that the oxidation of LDL induced by endothelial cells was accompanied by the association of endothelial cell proteins (53), indicating the contact of LDL with cells. It is also possible that the enzyme comes partially to the cell surface under these experimental conditions to directly oxidize LDL lipids. The possible role of lipoxygenase in the oxidation of LDL was suggested by the demonstration of colocalization of this enzyme with oxidized LDL in the macrophage-rich fatty streak lesions of WHHL rabbits (58).

The hydroperoxides in LDL could prime a chain reaction, leading to the oxidation of the fatty acids of LDL. The propagation phase could be dependent upon the presence of metal ions or sulfydryl compounds (32,59), these last residues (L-cysteine and glutathione) that are released from cells and capable of generating a superoxide anion from oxygen in the presence of metal ions.

Biological Properties of Oxidized LDL

Oxidized LDL present several properties that could explain their role in atherosclerosis.

Cytoxicity of Oxidized LDL

Several tests for the appreciation of cytoxicity, such as enumeration of attached cells, cell loss of lactate dehydrogenae into the culture medium, and trypan blue uptake, have shown that oxidized LDL were cytotoxic for fibroblasts, smooth muscle cells, and endothelial cells (60–62). This toxicity was related to the oxidized LDL itself, not to the free radicals generated in cell culture by the oxidized LDL. Indeed, the presence of antioxidants in the cell culture medium did not inhibit the oxidized LDL-induced cytotoxicity (61). The toxicity was associated with the formation of thiobarbituric acid-reacting substances. However, the malondialdehyde itself was not cytotoxic. The toxic substances were in the lipid extract of the oxidized LDL (62). The suppression of toxicity of oxidized LDL by pretreatment of lipoprotein with a lysolecithinase suggested that the toxicity was related to lysolecithin (63).

Effects of Oxidized LDL on Monocytes-Macrophages

Chemotactic Activity for Monocytes and Inhibition of Migration of Resident Macrophages. Endothelial cell-modified LDL and Cu^{2+}-oxidized LDL show a chemotactic activity for human monocytes, but native LDL do not have this activity (64). The acetyl LDL receptor present at the surface of monocyte-macrophages was not involved in the chemotactic activity because monocytes express a very small number of these receptors. Second, substances (acetyl LDL, polyinosinic acid, and fucoidin) interacting with the acetyl LDL have no chemotactic activity for circulating human monocytes (64). The chemotactic activity resides in the polar fractions of the lipid extracts of the oxidized LDL, and lysophosphatidylcholine, one of the products synthesized during oxidation, appeared to be a potent chemotactic factor (65). Synthetic lysophosphatidylcholine showed chemotactic activity in a dose-dependent manner. The chemotactic activity of lysophosphatidylcholine is dependent on the acyl chain length: 1-caproyl lysophosphatidylcholine had no chemotactic activity, and 1-myristoyl lysophosphatidylcholine had only 70% of the activity of 1-palmitoyl lysophosphatidylcholine. The platelet-activating factor (PAF), which is a known chemotactic factor for eosinophils and neutrophils and has a structure similar to that of phosphatidylcholine, lacked chemotactic activity but lyso-PAF presented a dose-dependent chemotactic activity (65).

The chemotactic activity of mouse peritoneal macrophages was strongly inhibited by oxidized LDL, whether induced by incubation with endothelial cells or Cu^{2+}, but endothelial cell-conditioned medium or acetyl LDL had no effect (63).

Uptake and Degradation of Oxidized LDL by Macrophages.   The primary biological properties of endothelial cell-modified LDL was their quantitatively important uptake and degradation by macrophages and, simultaneously, a decrease in their affinity for the LDL receptor (30,36). The uptake and degradation of oxidized LDL delivers a sufficient amount of cholesterol into macrophages to produce foam cells. This uptake and degradation is receptor mediated, including a saturable component and a high affinity of oxidized LDL for this receptor and esterification of cholesterol in macrophages induced by the uptake of lipoproteins. Competitive studies showed that acetyl LDL partially inhibits the uptake of oxidized LDL, but native LDL did not effectively compete (37,38), revealing that oxidized LDL receptor and acetylated LDL are two different receptors (66). Indeed, the structures of two types of scavenger receptor have been determined but the binding of oxidized LDL to these receptors is unknown (67,68). The ligand to the scavenger receptor is included in fragments of the degraded apo B-100. Indeed, solubilized apo B-100 fragments obtained by oxidation of LDL bound to the scavenger receptor and the degradation of these fragments were competitively inhibited by acetyl LDL. The binding domain of oxidized LDL included C-terminal and N-terminal parts of the apo B-100. Indeed, monoclonal antibodies against C- and N-terminal parts of apo B-100 inhibited the uptake and degradation of oxidized LDL by macrophages, but a monoclonal antibody against the middle of apo B-100 had no effect (69). The macrophage degradation of acetyl LDL was not affected by these monoclonal antibodies (69). Furthermore, the lipids did not seem to play a role in this binding (70).

HYPOTHESIS (Fig. 1)

Prospective studies and intervention assays have shown that a high level of plasma LDL is a major risk factor for atherosclerosis (71,72). The metabolism of LDL in the artery wall could partly explain the relationship between LDL and atherosclerosis. LDL must pass through the vascular endothelium. For the most part, LDL is transported across endothelial cells by transcytosis, and a relatively small amount of LDL is taken up by endothelial cells by receptor-dependent and independent endocytosis (73). The LDL could be oxidized during passage through the endothelial cell barrier or in the immediate subendothelial space. Oxidized LDL could contribute to the recruitment of circulating monocytes and stimulate endothelial cells to synthesize a monocyte binding protein (74). The monocytes thus bind to endothelial cells and thereafter pass between the endothelial cells toward the subendothelial space. Oxidized LDL inhibit the migration of macrophages present in the intima, and moreover, activated macrophages could secrete a chemotactic factor for monocytes (75), increasing the number of macrophages in the intima. Macrophages then rapidly take up oxidized LDL, and as the scavenger receptor is not downregulated, macrophages accumulate a considerable amount of cholesterol. The transformation of macrophages into foam cells modifies the metabolism of arachidonate when they are incubated with oxidized LDL, the synthesis of prostaglandin E_2

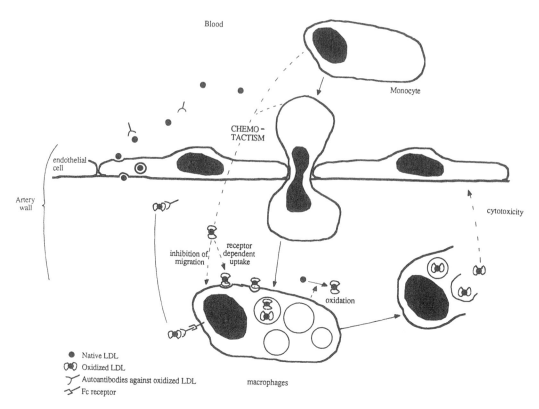

FIGURE 1 Mechanisms by which the oxidation of LDL may contribute to atherogenesis. The mono-cyte binds to an endothelial cell and thereafter passes toward the subendothelium space between two endothelial cells and differentiate into a macrophage. LDL pass through the vascular endothelium and are then oxidized and taken up by macrophages. The accumulation of oxidized LDL in macrophages transforms these cells into foam cells.

and leukotriene being multiplied 10-fold (76). These products increase vasopermeability and thereby the flux of LDL into the artery wall. When the foam cells die, the intracellular contents, including oxidized materials, are discharged into the extracellular space.

Cytotoxic factors are then released and could then attack endothelial cells. Thus, fatty streaks that are covered by endothelial cells advanced toward fibrous lesions denuded of endo-thelial cells. Therefore, platelets can adhere to the subendothelium. Thereafter, platelets re-lease several constituents, such as platelet-derived growth factor (PDGF), that stimulate the migration and proliferation of smooth muscle cells.

OXIDATION OF LDL IN VIVO

Although workers have detected in plasma a LDL subfraction resembling oxidized LDL (77), it now seems unlikely that oxidized LDL exist in plasma. First, these modified lipoproteins disappeared from the circulation within minutes after their injection into the bloodstream (49,78,79). The rate of removal of endothelial cell-modified LDL was in relation to the degree of modification (79). A rapid clearance of oxidized LDL by the liver has been demonstrated. A few minutes after the injection of radiolabeled oxidized LDL, the radioactivity was essen-tially in Küpffer's cells and endothelial cells (79,80).

Second, oxidation of LDL was inhibited by HDL, an HDL subfraction called LpAI being particularly efficient for this protection (80). Moreover, the cytoxocity of LDL for smooth muscle and endothelial cells was inhibited by HDL (82,83). The LDL used in these last experiments were not oxidized before, but were probably partly oxidized during their preparation. The absence of oxidized LDL in plasma does not mean that LDL do not occur in vivo. The presence of oxidized LDL was first suggested by the detection of peroxidized lipids in the atherosclerotic aorta (84,85). Derivatives of linoleic acid, the main PUFA of LDL, such as 13-hydroxyoctodecadienoic acid, were in particular raised in atherosclerotic aorta (84).

To be oxidized, LDL must be in a favorable environment: in the proximity of the cells, during the transcytosis of LDL through the endothelial cells, or in the subintimal space.

Evidence has been increasingly accumulating that oxidative modification of LDL occurs in vivo. This phenomenon has been essentially demonstrated by immunocytochemical techniques using specific monoclonal antibodies against modified LDL, such as malondialdehyde LDL, 4-hydroxynonenal-conjugated LDL, and fragments of apo B-100 generated by Cu^{2+} oxidation. Using these different monoclonal antibodies, several workers have detected malondialdehyde LDL and 4-hydroxynonenal LDL in atherosclerotic lesions of WHHL rabbits, an animal model for human familial hypercholesterolemia (86–89). Furthermore, the localization of epitopes specific for oxidized LDL was different from that of intact apo B-100. In the fatty streaks, oxidized LDL were strongly detected in macrophages and apo B-100 was present in the extracellular space. In fibrous lesions, oxidized LDL and apo B-100 colocalized in the extracellular environment (89). The results of these studies could be interpreted as follows. LDL is oxidized by the macrophages present in the fatty streak lesion, the interaction of LDL with arterial proteoglycans increasing the susceptibility of LDL to oxidative modification (90) and, therefore, uptake by these cells. When cells die, cellular oxidized LDL are released and trapped in the extracellular matrix. The small number of cells in the fibrous advanced atherosclerotic lesion does not permit the oxidation of LDL. This interpretation is consistent with findings that relatively intact LDL and lipoprotein particles resembling oxidized LDL can be gently extracted from the normal or atherosclerotic artery wall (87,91,92). Finally, human and rabbit serum contain autoantibodies against malondialdehyde LDL, suggesting in these subjects and animals the presence of such modified LDL. The immune complexes made by autoantibodies and oxidized LDL could be rapidly taken up by macrophages by way of the Fc receptor, another pathway for macrophages to take up oxidized LDL.

The importance of oxidation of LDL in the occurrence and development of atherosclerosis is indirectly demonstrated by the antiatherogenic effect of probucol, a hypolipidemic drug that is also an antioxidant. Indeed, probucol blocks all the changes of LDL induced by cells or Cu^{2+} in vitro (93). Probucol reduces to about one-half the degradation of LDL by the macrophage-rich fatty streaks in the probucol-treated WHHL rabbit compared with WHHL rabbits with similar cholesterol levels (94). As it was demonstrated that probucol does not affect lipoprotein metabolism in macrophages (95), it seems likely that the effect of probucol on atherosclerosis would be related to its antioxidative property.

REFERENCES

1. Uemura K, Piza Z. Tendances récentes de la mortalité par maladies cardiovasculaires dans 27 pays industrialisés. Rapport trimestriel de Statistiques Sanitaires Mondiales, Vol. 38, No. 2. Genèva: OMS, 1985.
2. Gordon T, Kannel WB, Castelli WP, Dawber TR. Lipoproteins, cardiovascular disease and death; the Framingham Study. Arch Intern Med 1981; 141:1128-31.
3. Miller GJ, Miller NE. Plasma high density lipoprotein concentration and development of ischemic heart disease. Lancet 1975; 1:16-9.

4. Faggiotto A, Ross K, Harker L. Studies of hypercholesterolemia in the non human primate. I. Changes that lead to fatty streak formation. Arteriosclerosis 1984; 4:323-40.

5. Faggiotto A, Ross K. Studies of hypercholesterolemia in the non human primate. II. Fatty streak conversion to fibrous plaque. Arteriosclerosis 1984; 4:341-56.

6. Gerrity RG. The role of the monocyte in atherogenesis. I. Transition of blood-borne monocytes into foam cell in fatty lesions. Am J Pathol 1981; 103:181-90.

7. Schaffner R, Taylor K, Bartucci EJ, et al. Arterial foam cells with distinctive immunomorphologic and histochemical features of macrophages. Am J Pathol 1980; 100:57-80.

8. Tsukada T, Rosenfeld M, Ross R, Gown AM. Immunocytochemical analysis of cellular components in atherosclerotic lesions. Use of monoclonal antibodies with the Watanabe and fat-fed rabbit. Arteriosclerosis 1986; 6:601-13.

9. Holme I, Enger SC, Helgeland A. Risk factors and raised atherosclerosis lesions in coronary and cerebral arteries. Statistical analysis from the Oslo study. Atherosclerosis 1981; 1:250-6.

10. Nichols AV. Human serum lipoprotein ans their interrelationships. Adv Biol Med Phys 1967; 11: 109-58.

11. Deckelbaum RJ, Shipley GG, Small DM. Structure and interactions of lipids in human plasma low density lipoproteins. J Biol Chem 1977; 252:744-54.

12. Chen GC, Chapman MJ, Kane JP. Secondary structure and thermal behaviour of trypsin-treated low density lipoproteins from human serum, studied by circular dichroism. Biochim Biophys Acta 1983; 754:51-6.

13. Brown MS, Kovanen PT, Goldstein JL. Regulation of plasma cholesterol by lipoprotein receptors. Science 1981; 212:628-35.

14. Lindgren FT, Jensen LC, Wills RD, Freeman NK. Flotation rates, molecular weight and hydrated densities of the low density lipoproteins. Lipids 1969; 4:337-44.

15. Shen MS, Krauss RM, Lindgren FT, Forte TM. Heterogeneity of serum low-density lipoproteins in normal human subjects. J Lipid Res 1981; 22:236-44.

16. Chapman MJ, Laplaud PM, Luc G, et al. Further resolution of the low-density lipoprotein spectrum in normal plasma: physicochemical characteristics of discrete subspecies separated by density gradient ultracentrifugation. J Lipid Res 1988; 29:442-58.

17. Luc G, Chapman MJ, De Gennes JL, Turpin G. A study of the structural heterogeneity of low density lipoproteins in two patients homozygous for familial hypercholesterolemia, one of phenotype E2/2. Eur J Invest 1986; 16:329-37.

18. Luc G, De Gennes JL, Chapman MJ. Further resolution and comparison of the heterogeneity of plasma low density lipoproteins in hyperlipoproteinemias: type III hyperlipoproteinemia, hypertriglyceridemia and familial hypercholesterolemia. Atherosclerosis 1988; 71:143-56.

19. La Belle M, Krauss RM. Differences in carbohydrate content of low density lipoproteins associated with low density lipoprotein subclass patterns. J Lipid Res 1990; 31:1577-88.

20. Jiao S, Cole TG, Kitchens RT, Pfleger B, Schonfeld G. Genetic heterogeneity of plasma lipoproteins in the mouse: control of low density lipoprotein particle sizes by genetic factors. J Lipid Res 1990; 31:467-77.

21. Pownall HJ, Shepherd J, Mantulin WW, Sklar LA, Gotto AM. Effect of saturated and polyunsaturated fat diets on the composition and structure of human low density lipoproteins. Atherosclerosis 1980; 36:299-314.

22. Goldstein JL, Brown MS. Familial hypercholesterolemia. In: Scriver CR, Beaudet AL, Sly WS, Valle D, eds. The metabolic basis of inherited disease, 6th ed. 1215-50.

23. Mahley RW, Weisgraber KU, Melchior GW, Innerarity TL, Holcombe KS. Inhibition of receptor mediated clearance of lysine and arginine modified lipoproteins from the plasma of rats and monkeys. Proc Natl Acad Sci U S A 1988; 77:225-9.

24. Tanzawa K, Shim Ada Y, Kuruda M, Tsujita Y, Arai M, Watanabe Y. WHHL-rabbit: a low density lipoprotein receptor-deficient animal model for familial hypercholesterolemia. FEBS Lett 1980; 118:81-4.

25. Brown MS, Basu SK, Falck JR, Hu YK, Goldstein JL. The scavenger cell pathway for lipoprotein degradation: specificity of the binding site that mediates the uptake of negatively-charged LDL by macrophages. J Supramol Struct 1980; 13:67-81.

26. Brown MS, Goldstein JL. Lipoprotein metabolism in the macrophage: implicatins for cholesterol deposition in atherosclerosis. Annu Rev Biochem 1983; 52:223-61.

27. Goldstein JL, Ho YK, Basu SK, Brown MS. Binding site on macrophages that mediates uptake and degradation of acetylated low density lipoprotein, producing massive cholesterol deposition. Proc Natl Acad Sci U S A 1979; 76:333-7.

28. Mahley RW, Innerarity TL, Weisgraber KV, Oh SY. Altered metabolism (in vivo and in vitro) of plasma lipoproteins after selective chemical modification of lysine residues of the lipoproteins. J Clin Invest 1979; 64:743-50.

29. Fogelman AM, Schechter J, Seager J, Hokom M, Child JJ, Edwards PA. Malondialdehyde alteration of low density lipoproteins leads to cholesteryl ester accumulation in human monocyte-macrophages. Proc Natl Acad Sci U S A 1980; 77:2214-8.

30. Henriksen T, Mahoney EM, Steinberg D. Enhanced macrophage degradation of low density lipoprotein previously incubated with cultured endothelial cells: recognition by receptor for acetylated low density lipoproteins. Proc Natl Acad Sci U S A 1981; 78:6499-503.

31. Van Hinsbergh VWM, Scheffer M, Havekes L, Kempen HJM. Role of endothelial cells and their products in the modification of low-density lipoproteins. Biochim Biophys Acta 1986; 878:49-64.

32. Heinecke JW, Rosen H, Suzuki LA, Chait A. The role of sulfur-containing amino acids in superoxide production and modification of low density lipoprotein by arterial smooth muscle cells. J Biol Chem 1987; 262:10098-103.

33. Morel DW, Dicorleto PE, Chisolm GM. Endothelial and smooth muscle cells alter low density lipoprotein in vitro by free radical oxidation. Arteriosclerosis 1984; 4:357-64.

34. Hiramatsu K, Rosen H, Heinecke JW, Wolfbauer G, Chait A. Superoxide initiates oxidation of low density lipoprotein by human monocytes. Arteriosclerosis 1987; 7:55-60.

35. Cathcart MK, Morel DW, Chisolm GM. Monocytes and neutrophils oxidize low density lipoprotein making it cytotoxic. J Leukoc Biol 1985; 38:341-50.

36. Henriksen T, Mahoney EM, Steinberg D. Enhanced macrophage degradation of biologically modified low density lipoprotein. Arteriosclerosis 1983; 3:149-59.

37. Steinbrecher VP, Parthasarathy S, Leake DS, Witztum JL, Steinberg D. Modification of low density lipoprotein by endothelial cells involves lipid peroxidation and degradation of low density lipoprotein phospholipids. Proc Natl Acad Sci U S A 1984; 81:3883-7.

38. Steinbrecher VP, Witztum JL, Parthasarathy S, Steinberg D. Decrease in reactive amino groups during oxidation or endothelial cell modification of LDL: correlation with changes in receptor-mediated catabolism. Arteriosclerosis 1987; 1:135-43.

39. Esterbauer H, Rotheneder M, Striegl G, et al. Vitamine E and other lipophylic antioxidants protect LDL against oxidation. Fat Sci Technol 1989; 8:316-24.

40. Esterbauer H, Jurgens G, Quehenberger O, Koller E. Autoxidation of human low density lipoprotein: loss of polyunsaturated fatty acids and vitamin E and generation of aldehydes. J Lipid Res 1987; 28:495-509.

41. Esterbauer H, Jurgens G, Quehenberger O. Modification of human low density lipoprotein by lipid peroxidation. In: Sinue MB, Taylor KA, Ward JF, Von Sonntal C, eds. Oxygen radical biology and medicine. 1989; 369-73.

42. Parthasarathy S, Khoo JC, Miller E, Barnett J, Witztum JL, Steinberg D. LDL from rabbits on oleic acid-rich diets strongly resists oxidative modification. Circulation 1990; 82:4 (Suppl. III): abstract 2219.

43. Henriksen T, Mahoney EM, Steinberg D. Interactions of plasma lipoproteins with endothelial cells. Ann N Y Acad Sci 1982; 401:102-16.

44. Parthasarathy S, Steinbrecher UP, Barnett J, Witztum JL, Steinberg D. Essential role of phospholipase A_2 activity in endothelial cell-induced modification of low density lipoprotein. Proc Natl Acad Sci U S A 1985; 82:3000-4.

45. Lenz ML, Hughes H, Mitchell JR, et al. Lipid hydroperoxy and hydroxy derivatives in copper catalyzed oxidation of low density lipoprotein. J Lipid Res 1990; 31:1043-50.

46. Fong LG, Parthasarathy S, Witztum JL, Steinberg D. Nonenzymatic oxidative cleavage of peptide bonds in apolipoprotein B100. J Lipid Res 1987; 28:1466-77.

47. Steinbrecher UP, Witztum JL, Parthasarathy S, Steinberg D. Decrease in reactive amino groups during oxidation or endothelial cell modification of LDL. Correlation with changes in receptor-mediated catabolism. Arteriosclerosis 1987; 7:135-43.

48. Steinbrecher UP. Oxidation of human low density lipoprotein results in derivatization of lysine residues of apolipoprotein B by lipid peroxide decomposition products. J Biol Chem 1987; 262: 3603-8.

49. Jurgens G, Lang J, Esterbauer H. Modification of human low density lipoprotein by the lipid peroxidation product 4-hydroxynonenal. Biochim Biophys Acta 1986; 875:103-14.

50. Zawadzki Z, Milne KW, Marcel YL. An immunochemical marker of low density lipoprotein oxidation. J Lipid Res 1989; 30:885-91.

51. Palinski W, Yla-Herttuala S, Rosenfeld ME, et al. Antisera and monoclonal antibodies specific for epitopes generated during oxidative modification of low density lipoprotein. Arteriosclerosis 1990; 10:325-35.

52. De Graaf J, Demacker PNM, Stalenhoef AFH. Enhanced susceptibility to oxidation of the dense low density lipoprotein subfraction in vitro. Circulation 1990; 82(Suppl. III):abstract 2217.

53. Van Hinsbergh VWM, Scheffer M, Kavekes L, Kempen HJM. Role of endothelial cells and their products in the modification of low-density lipoproteins.

54. Heinecke JW, Baker L, Rosen H, Chait A. Superoxide-mediated modification of low density lipoprotein by arterial smooth muscle cells. J Clin Invest 1986; 77:757-61.

55. Hiramatsu K, Rosen H, Heinecke JW, Wolfbauer G, Chait A. Superoxide initiates oxidation of low density lipoprotein by human monocytes. Arteriosclerosis 1987; 7:55-60.

56. Steinbrecher UP. Role of superoxide in endothelial-cell modification of low-density lipoproteins. Biochim Biophys Acta 1988; 959:20-30.

57. Parthasarathy S, Wieland E, Steinberg D. A role for endothelial cell lipoxygenase in the oxidative modification of low density lipoprotein. Proc Natl Acad Sci U S A 1989; 86:1046-50.

58. Yla-Herttula S, Rosenfeld ME, Parthasarathy S, Sigal E, Witzum JL, Steinberg D. Colocalization of 15-lipoxygenase with epitopes characteristic of oxidized LDL in macrophage-risk areas of atherosclerotic lesions. Circulation 1990; 82(Suppl. III):abstract 2218.

59. Parthasarathy S. Oxidation of low density lipoprotein by thiol compounds leads to its recognition by the acetyl LDL receptor. Biochim Biophys Acta 1987; 917:337-40.

60. Evensen SA, Galdal KS, Nilsen E. LDL-induced cytotoxicity and its inhibition by anti-oxydant treatment in cultured human endothelial cells and fibroblasts. Atherosclerosis 1983; 49:23-30.

61. Morel DW, Hessler JR, Chisolm GM. Low density lipoprotein cytotoxicity induced by free radical peroxidation of lipid. J Lipid Res 1983; 24:1070-6.

62. Hessler JR, Morel DW, Lewis LJ, Chisolm GM. Lipoprotein oxidation and lipoprotein-induced cytotoxicity. Arteriosclerosis 1983; 3:215-22.

63. Quinn MT, Parthasarathy S, Fong LG, Steinberg D. Oxidatively modified low density lipoproteins: a potential role in recruitement and retention of monocytes/macrophages during atherogenesis. Proc Natl Acad Sci U S A 1987; 84:2995-8.

64. Mangin EL, Kerns S, Nguy J, Eisenberg SB, Li Z, Henry PD. Endothelial toxicity of oxidized LDL is suppressed by treatment of the lipoprotein with a lysolecithinase (phospholipase B). Circulation 1990; 82(Suppl. III):abstract 675.

65. Quinn MT, Parthasarathy S, Steinberg D. Lysophosphatidylcholine: a chemotactic factor for human monocytes and its potential role in atherogenesis. Proc Natl Acad Sci U S A 1988; 85:2805-9.

66. Sparrow CP, Parthasarathy S, Steinberg D. A macrophage receptor that recognizes oxidized low density lipoprotein but not acetylated low density lipoprotein. J Biol Chem 1989; 264:2599-604.

67. Kodama T, Freeman M, Rohrer L, Zabrecky J, Matsudaira P, Krieger M. Type I macrophage scavenger receptor contains a-helical and collagen-like coiled coils. Nature 1990; 343:531-5.

68. Rohrer L, Freeman M, Kodama T, Penman M, Krieger M. Coiled-coil fibrous domains mediate ligand binding by macrophage scavenger receptor type II. Nature 1990; 343:570-2.

69. Keidar S, Brouck GJ, Aviram M. Macrophage binding of oxidized-LDL is mediated via epitopes on both the amino and the carboxy terminus domain of apo B100. Circulation 1990; 82(Suppl. III):abstract 2220.

70. Parthasarathy S, Fong LG, Otero D, Steinberg D. Recognition of solubilized apoproteins from delipidated, oxidized low density lipoprotein (LDL) by the acetyl-LDL receptor. Proc Natl Acad Sci U S A 1987; 84:537-40.

71. Kannel WB, Castelli WP, Gordon T. Cholesterol in the prediction of atherosclerotic disease. New prospectives based on the Framingham study. Ann Intern Med 1979; 90:85-.

72. Lipid Research Clinics Program Trial. II. The relationship of reduction in incidence of coronary heart disease to cholesterol lowering. JAMA 1984; 251:365-74.

73. Vasile E, Simionescu M, Simionescu N. Visualization of the binding, endocytosis and transcytosis of low density lipoprotein in the arterial endothelium in situ. J Cell Biol 1983; 96:1677-89.

74. Berliner JA, Territo MC, Sevanian A, et al. Minimally modified low density lipoprotein stimulates monocyte endothelial interactions. J Clin Invest 1990; 85:1260-6.

75. Mazzone T, Jensen M, Chait A. Human arterial wall cells secrete factors that are chemotactic for monocytes. J Biol Chem 1983; 80:5094-7.

76. Yokode M, Kita T, Kikawa Y, Ogorochi T, Narumiya S, Kawai C. Stimulated arachidonate metabolism during foam cell transformation of mouse peritoneal macrophages with oxidized low density lipoprotein. J Clin Invest 1988; 81:720-9.

77. Avogaro P, Bittolo Bon G, Cazzolato G. Presence of a modified low density lipoprotein in humans. Arteriosclerosis 1988; 8:79-87.

78. Nagelkerke JF, Havekes L, Van Hinsbergh VWM, Van Berkel TJC. In vivo and in vitro catabolism of native and biologically modified LDL. FEBS Lett 1984; 171:149-53.

79. Nagelkerke JF, Havekes L, Van Hinsbergh VWM, Van Berkel TJC. In vivo catabolism of biologically modified LDL. Ateriosclerosis 1984; 4:256-64.

80. Van Berkel TJC, De Rijke YB, Bijsterbosch MK, Kar Kruijt J. Different fate in vivo of oxidatively modified-LDL and acetylated-LDL. Circulation 1990; 82(Suppl. III):abstract 2225.

81. Ohta T, Takata K, Horiuchi S, Morino Y, Matsuda I. Protective effect of lipoproteins containing apoprotein AI on Cu^{2+} catalyzed oxidation of human low density lipoprotein. FEBS Lett 1989; 257:435-8.

82. Henriksen T, Evensen A, Carlander B. Injury to cultured endothelial cells induced by low density lipoprotein: protection by high density lipoproteins. Scand J Clin Lab Invest 1979; 39:369-75.

83. Robertson AL,, Chisolm GM. LDL-induced cytotoxicity and its inhibition by HDL in human vascular smooth muscle and endothelial cells in culture. Atherosclerosis 1979; 32:213-29.

84. Mowri H, Chinen K, Ohkuma S, Takano T. Perodized lipids isolated by HPLC from atherosclerotic aorta. Biochem Int 1986; 12:347-52.

85. Piotrowski JJ, Hunter GC, Eskelson CD, Dubick MA, Bernhard VM. Evidence for lipid peroxidation in atherosclerosis. Life Sci 1990; 46:715-21.

86. Haberland ME, Fong D, Cheng L. Malondialdehyde-altered protein occurs in atheroma of Watanabe heritable hyperlipidemic rabbits. Science 1988; 241:215-8.

87. Palinski W, Rosenfeld ME, Yla-Herttuala S, et al. Low density lipoprotein undergoes oxidative modification in vivo. Proc Natl Acad Sci U S A 1989; 86:1372-6.

88. Boyd HC, Gown AM, Wolfbauer G, Chait A. Direct evidence for a protein recognized by a monoclonal antibody against oxidatively modified LDL in atherosclerotic lesions from Watanake heritable hyperlipidemic rabbit. Am J Pathol 1989; 135:815-26.

89. Rosenfeld ME, Palinski W. Yla-Herttula S, Butler S, Witztum JL. Distribution of oxidation specific lipid-protein adducts and apolipoprotein B in atherosclerotic lesions of varying severity from WHHL rabbits. Arteriosclerosis 1990; 10:336-49.

90. Hurt-Camejo E, Camejo G, Rosengren B, Wirklund O, Bonjers G. Arterial proteoglycans increase the rate of oxidation of low density lipoprotein and its uptake by macrophages. Circulation 1990; 82(Suppl. III):abstract 2221.

91. Hoff HF, Bradley WA, Heideman CL, Gaubatz JW, Karagas MD, Gotto AM. Characterization of low density lipoprotein-like particle in the human aorta from grossly normal and atherosclerotic regions. Biochim Biophys Acta 1979; 573:361-74.

92. Morton RE, West GA, Hoff HF. A low density lipoprotein-sized particle isolated from human atherosclerotic lesions is internalized by macrophages via a non-scavenger-receptor mechanism. J Lipid Res 1986; 27:1124-34.

93. Parthasarathy S, Young SG, Witztum JL, Pittman RC, Steinberg D. Probucol inhibits oxidative modification of low density lipoprotein. J Clin Invest 1986; 77:641-4.

94. Carew TE, Schwenke DC, Steinberg D. Antiatherogenic effect of probucol unrelated to its hypo-cholesterolemic effect: evidence that antioxidants in vivo can selectively inhibit low density lipoprotein degradation in macrophage-rich fatty streaks and slow the progression of atherosclerosis in the Watanabe heritable hyperlipidemic rabbit. Proc Natl Acad Sci U S A 1987; 84:7725-9.

95. Kita T, Yokode M, Ishii K, et al. Probucol does not affect lipoprotein metabolism in macrophages of Watanabe heritable hyperlipidemic rabbits. Arteriosclerosis 1989; 9:453-61.

46

The Role of Vitamin E in Lipoprotein Oxidation

Hermann Esterbauer, Herbert Puhl, George Waeg, Angelica Krebs, and Martina Dieber-Rotheneder

University of Graz, Graz, Austria

INTRODUCTION

Oxidatively modified low-density lipoprotein (LDL) shows in vitro properties that could explain several phenomena of the chain of events leading to the development of atherosclerosis (for review see Refs. 1–5). The hypothesis that oxidation of LDL is an event triggering the localized deposition of cholesterol in the arteries is based on three major discoveries. First, oxidized LDL (oLDL) is recognized and taken up by the scavenger receptor of macrophages; this uptake is not downregulated by the internalized cholesterol and leads to conversion of macrophages to lipid-laden foam cells (6,7). The accumulation of foam cells in the arterial intima is a characteristic early event in atherosclerosis (8). Immunohistochemical methods indicate that most of these cells are derived from monocyte-macrophages, although some may develop from smooth muscle cells. As atherosclerosis progresses, the massive accumulation of foam cells can lead to the development of fatty streaks and finally to plaques (Fig. 1). Second, oLDL contains highly cytotoxic components (9,10), and its presence in the arterial intima would therefore lead to damage of the endothelial cell layer, which in turn provokes a number of deleterious effects, such as platelet aggregation, release of growth factors, increased infiltration of LDL, disturbance of eicosanoid homeostasis, and accumulation of inflammatory cells (11,12). Finally, oLDL possesses chemotactic properties toward blood monocytes (13). Its presence in the arterial wall could therefore stimulate the immigration of more monocyte-macrophages in the intima. Since macrophages also have a high capacity to oxidize LDL (14,15), this chain of events exhibits all components of a self-sustaining and accelerating process.

A number of other findings support the hypothesis that oLDL plays a role in atherogenesis. Increased thiobarbituric acid reactants (TBARs) were found in serum of atherosclerotic patients (16); the Folch extract of aortic tissue from atherosclerotic patients contains lipid peroxides and fluorescent compounds characteristic for lipid peroxidation (16,17) (for review see also Ref. 3). Particularly impressive are investigations in some animal models. for instance, the LDL from hyperlipidemic nonlaying hens has a 15-fold increased content of TBARs compared to control laying hens (18). The LDL isolated from the vasculature of Watanabe heritable hyperlipidemic rabbits has 8 times more TBARs than the corresponding plasma LDL (19). The lipoprotein fraction containing very low density lipoprotein (VLDL)

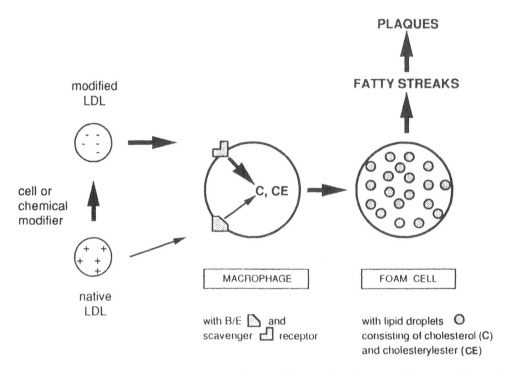

FIGURE 1 Conversion of macrophages to foam cells. LDL is endocytosed by macrophages via the B/E receptor, which recognizes positive charges on the apo B. This uptake is controlled by intracellular cholesterol (C) and does not lead to foam cells. Cell-modified LDL has an increased negative surface charge and is endocytosed in an uncontrolled fashion. This uptake leads in vitro to cells with accumulated cholesterol (C) and cholesterol esters (CE) in lipid droplets. The foam cell can develop into fatty streaks and plaques. (From Ref. 4, with permission.)

and LDL obtained from rats with streptozotocin diabetes contains 10 times more TBARs than normal VLDL + LDL, that is, 25 versus 2.5 nmol/mg cholesterol (20). The VLDL + LDL from diabetic rats is cytotoxic, and antioxidant treatment of diabetic rats inhibits lipoprotein oxidation and cytotoxicity. It has been repeatedly reported that the antiatherogenic effect of the drug probucol is at least in part due to its antioxidant activity and its ability to inhibit LDL oxidation in vivo (21,22).

Some remarkable contributions also come from immunological studies with polyclonal and monoclonal antibodies directed against various epitopes of oxidatively modified LDL (23) or native LDL modified by the lipid peroxidation products malonaldehyde (24) or 4-hydroxynonenal (23,25,26). These immunological studies show that oLDL is indeed present in the arterial wall of hyperlipidemic WHHL rabbits; furthermore, autoantibodies against oLDL appear to be present in plasma of rabbits and humans.

The processes that could initiate oxidation of LDL in vivo are largely unknown. In vitro, oxidation of LDL can be mediated by incubation with certain endothelial cells, smooth muscle cells, and monocyte-macrophages or in cell-free systems by exposing LDL to copper ions (for review see Refs. 3 and 27), lipoxygenase (28), ultraviolet (UV) irradiation (29), γ-irradiation (30), azo compounds (31), or defined oxygen radicals produced by pulse radiolysis (32). Independent of how oxidation of LDL is initiated, it always leads to the degradation of the polyunsaturated fatty acids bound in LDL lipids by a lipid peroxidation process, and all studies agree that lipid peroxidation is fundamental for most if not all altered biological properties exhibited by oLDL. If lipid peroxidation and the concomitant or subsequent decomposition of lipid hydroperoxides is the key event, then antioxidants should play a critical role in protecting LDL against oxidative modification in vitro as well as in vivo.

FATTY ACIDS AND ANTIOXIDANTS IN LDL

LDL is a large spherical particle with an average molecular weight of 2.5 million. It consists of a central core of about 1600 molecules cholesteryl ester and 170 molecules triglycerides. The core is surrounded by a monolayer comprised of about 700 molecules phospholipids and 600 molecules free cholesterol (3). Embedded in the monolayer is a peripheral protein termed apolipoprotein B (apo B), which mediates recognition of the LDL by the classic LDL receptor (B/E receptor) present on the surface of most cells. Binding to the LDL receptor is mediated by a certain pattern of positive surface charges on the apo B, for example, ϵ-amino groups of lysine residues (33,34). If a certain fraction of these charges are neutralized, for instance by reaction with aldehydes (35,36) or other reagents, such as acetic acid anhydride, binding to the LDL receptor is lost whereas recognition by the scavenger receptors strongly increases (for review see Ref. 37). The current hypothesis assumes that lipid peroxidation products (e.g., malonaldehyde, 4-hydroxynonenal, or others) generated within the LDL particle itself cause the conversion of LDL into a form rapidly endocytosed in an unregulated manner by the macrophage scavenger receptor (Fig. 1).

The total number of fatty acids bound in the different lipid classes of a LDL molecule is 2700 on average, of which about 1300 are polyunsaturated fatty acids (PUFAs), mainly linoleic acid (86%), with minor amounts of arachidonic acid (12%) and docosehexaenoic acid (2%) (Table 1). By far the major endogenous antioxidant in LDL is α-tocopherol; on average about six molecules of α-tocopherol are present in an LDL particle. All other compounds with potential antioxidant activity, that is, carotenoids, oxycarotenoids, and ubiquinol-10, are present in much lower amounts (Table 1). The value of 0.29 mol β-carotene per mol LDL means indeed that on average only about 30% of the LDL molecules contain β-carotene whereas the others do not. It has been argued (38) that this low content rather suggests that α-tocopherol is the only significant antioxidant in LDL and that carotenoids play no or only a minor role in protecting LDL against oxidation. According to Stocker et al. (39), ubiquinol-10 is the most powerful antioxidant in LDL. It appears to us, however, that its contribution to the total antioxidant capacity of LDL is negligible because of its low concentration. The ubiquinol-10 values we determined (Table 1) agree with those reported by this group.

The kinetics of transfer of α-tocopherol and β-carotene between native plasma lipoproteins has been investigated with tritium-labeled α-tocopherol (40). It has been found that α-tocopherol transfers spontaneously with a rate about 3 times slower than that of cholesterol transfer but at least 20 times faster than that of β-carotene. For HDL, VLDL, and LDL the intermolecular exchange results in an equilibrium distribution of α-tocopherol within about 1 h (40). For isolated LDL the α-tocopherol can therefore be considered a common antioxidant pool, shared by all LDL particles through rapid intermolecular exchange, whereas on the other hand, an intermolecular exchange of carotenoids is unlikely. Since plasma contains a great variety of water-soluble and lipid-soluble antioxidants [more than 20 different carotenoids have been reported, for example (41)], it may well be that LDL contains in addition to the antioxidants listed in Table 1, other not yet identified antioxidants. An important factor for in vitro studies likely is the method of LDL isolation. Stocker et al. (39), for example, found that an LDL isolated by a rapid ultracentrifugation method contains significant amounts of ascorbic acid and urate, which would of course affect the in vitro oxidation of LDL. On

TABLE 1 Basal Values of Cholesterol, PUFAs, Antioxidants, and Lag Phases Determined for Human LDL Samples[a]

| | n | Mol/mol LDL | | |
		Mean	SD	Range
Total cholesterol	3	2100	±40	2000–2100
Linoleic acid	31	1101	±298	680–1832
Arachidonic acid	31	153	±55	48–250
Docosahexaenoic acid	15	29	±17	15–62
α-Tocopherol	87	6.37	±1.84	2.9–14.9
γ-Tocopherol	88	0.51	±0.20	0.18–1.26
β-Carotene	122	0.29	±0.26	0.03–1.87
α-Carotene	28	0.12	±0.14	0.02–0.52
Lycopene	136	0.16	±0.11	0.03–0.70
Cryptoxanthine	114	0.14	±0.13	0.03–0.70
Cantaxanthine	53	0.02	±0.04	0.01–0.24
Lutein + zeaxanthine	113	0.04	±0.03	0.01–0.16
Phytofluene	10	0.05	±0.03	0.02–0.11
Ubiquinol-10	7	0.10	±0.10	0.03–0.35
Total antioxidants		7.8		
Total PUFAs		1283		
Antioxidants-PUFAs		1:165		
Lag Phase, min	72	68	±15	34–114

[a]The LDL was prepared from the blood of clinically healthy male and female donors in aged 20–30 years. n is the number of LDL samples investigated. This is an updated version of a previous presented report (3) with a larger number n.

the other hand, these compounds are undetectable in LDL isolated by conventional density gradient ultracentrifugation followed by dialysis or size-exclusion chromatography to remove EDTA, which is used to complex transition metal ions throughout all preparation steps (Refs. 42 and 43 and unpublished observations by the authors). Finally, the possibility should be considered that components of apolipoprotein B (amino acid residues and/or carbohydrates) may also have antioxidant properties. In case of the lipoprotein Lp(a) it was found, for example, that N-acetylneuraminic acid, which is a part of the glycosylated protein, protects to some extent against Cu^{2+}-mediated oxidation (44).

KINETICS OF OXIDATION OF LDL BY CELLS OR COPPER IONS

In the experiments leading to the discovery that cells can oxidatively modify LDL, LDL was incubated with cultured rabbit aortic endothelial cells in Ham's F10 medium for 24 h; after that time the medium contained TBARs and the cell-conditioned LDL was more rapidly degraded by macrophages than the native LDL (7). Principally the same procedure with a 24–48 h end point of modification was later used in many studies with endothelial cells, smooth muscle cells, and macrophages and also for oxidation in cell-free systems. The time course of the formation of TBARs during incubation of LDL with cells has with few exceptions only occasionally been measured (for review see Ref. 4). Examples for which TBARs are given for at least a few time points are human aortic smooth muscle cells (45), human umbilical vein endothelial cells (46), rabbit thoratic aortic endothelial cells (47), and stimulated human monocytes (48). From these experiments one can conclude that TBARs must reach a threshold level of at least about 10 nmol TBARs per mg total LDL to obtain a form of oLDL that is cytotoxic, chemotactic, and degraded by macrophages (3,9,10,13). Such a degree of oxidation is, depending on the cell type, obtained after about 12–24 h of incubation.

The exact time course of modification of LDL by macrophages into a form endocytosed by macrophages was first investigated by Leake and Rankin (Fig. 2) (14). These experiments showed clearly a lag phase of about 4–6 h before the onset of modification; thereafter modification commenced and continued for various periods of time (12–24 h), depending on cell preparation, before a plateau was reached. The kinetics of modification of LDL into a high-uptake form by Cu^{2+} ions (5 μM) in the absence of cells resembled very closely cell modification (14), which is in agreement with the general assumption that cell-oxidized LDL has properties very similar if not identical to those of Cu^{2+}-oxidized LDL (3). In distinction to copper, ferrous ions appear not to be able to convert LDL into a form endocytosed by macrophages. A very significant study on the temporal relationship of oxidation of LDL (i.e., increase of peroxides and loss of vitamin E) by macrophages and Cu^{2+} or Fe^{2+} ions and the uptake rate of the modified LDL by macrophages was made by Jessup et al. (15). These studies fully confirmed the findings made by Esterbauer et al. (3,42), that the process of LDL oxidation by cells or in cell-free systems can be divided into three consecutive time phases: a lag phase, during which the LDL becomes depleted from its α-tocopherol content; a propagation phase, during which the lipid hydroperoxide content of LDL rapidly increases; and a decomposition phase, during which the lipid hydroperoxides decrease again because of decomposition reactions. The significance of the study by Jessup et al. (15) lies in the demonstration that modification of LDL into a form recognized and taken up by the macrophage scavenger receptor is temporarily linked with the decomposition of the lipid hydroperoxides (Fig. 3). No measurable modification of LDL into high-uptake forms occurred during the lag phase, and modification was also minimal during the propagation period, when the lipid hydroperoxides increased to a maximum value. LDL conditioned by macrophages had a maximum lipid hydroperoxide level of about 450 mol/mol LDL at 12 h of incubation; thereafter the lipid hydroperoxides decreased as a result of decomposition reactions and the decrease was

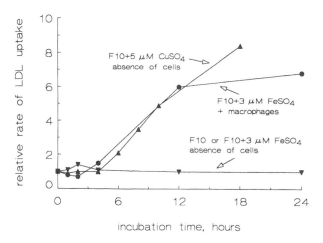

FIGURE 2 Time course of conversion of LDL by macrophages or copper ions into a form taken up by the scavenger receptor. LDL was incubated with (circles) or without macrophages (triangles) for various times up to 24 h; thereafter the rate of its uptake by other macrophages was determined. The uptake rate relative to native LDL is given. (Adapted from Ref. 14, with permission.)

macrophage uptake

lipid hydroperoxides

α–Tocopherol

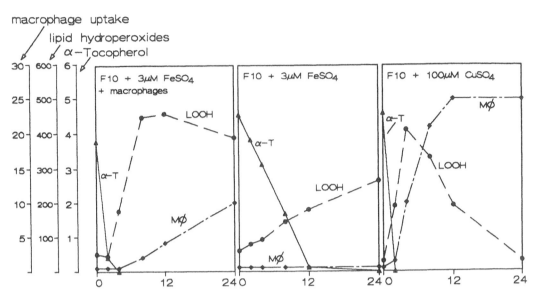

incubation time, hours

FIGURE 3 Temporal relationship between degree of oxidation LDL and its uptake by macrophages. LDL was incubated with macrophages in Ham's F10 medium containing 3 μM FeSO$_4$ (left) or in cell-free F10 medium containing 3 μM FeSO$_4$ (middle) or 100 μM CuSO$_4$ (right). At the indicated time points the LDL was separated from the medium and its α-tocopherol (α-T) and lipid hydroperoxide (LOOH) content and the uptake by macrophages (Mϕ) was determined. α-Tocopherol and lipid hydroperoxides are given in mol/mol LDL; macrophage uptake is in μg LDL protein degradation per mg cell protein in 20 h. (Adapted from Ref. 15, with permission.)

associated with a strong increase in the macrophage uptake rate. In the absence of cells, high concentrations of copper ions (100 μM) lead to about the same maximum of lipid hydroperoxides (420 mol/mol LDL) already within 4 h and caused a very rapid subsequent decomposition of the lipid hydroperoxides. Concomitant with the decomposition was the development of a high-uptake LDL. The copper-oxidized LDL was degraded by macrophages about 2.5 times more rapidly than the macrophage-modified LDL and 25 times more rapidly than native LDL. Concentrations of copper ions in the range of 1.6–10 μM oxidize LDL with a rate comparable to macrophages (3,14). On the other hand, incubation of LDL with FeSO$_4$ in the absence of cells generates only lipid hydroperoxides yet not a high-uptake LDL, probably because FeSO$_4$ does not mediate decomposition of the lipid hydroperoxide to the necessary extent.

That the presence of lipid hydroperoxides in LDL is per se not sufficient to affect its rapid uptake by the scavenger receptor is also supported by studies using selected radicals generated by γ-irradiation. Hydroxyl (OH˙) and hydroperoxyl (HO$_2$˙) free radicals led to heavy oxidation of the PUFAs in LDL, yet the radical-oxidized LDL was not taken up by the mouse peritoneal macrophage receptor (32). LDL heavily oxidized by γ-irradiation was also not a good substrate for the scavenger receptor, but when additionally conditioned for a short time with Cu^{2+}, it was converted into a high-uptake form (30). All these results suggest that decomposition of the lipid hydroperoxides is a necessary prerequisite to generate the characteristic epitopes on apo B recognized by this scavenger receptor.

The mechanism by which macrophages and other cells initiate oxidation of LDL are largely unknown, and not surprisingly this subject is currently receiving much attention. From the available data one can conclude that oxidation requires viable cells, oxygen, redox-active transition metal ions, and a reducing agent (5). Whichever mechanism ultimately proves to be involved in the initiation of oxidative modification of LDL, it already seems clear that the subsequent processes following initiation are principally always the same: lipid peroxidation and decomposition of the lipid hydroperoxides to aldehydes and other reactive products. An LDL that has reached the end point of decomposition always has more or less the same biological, functional, and chemical properties, independently if oxidation was initiated by cells or in the absence of cells by prooxidative agents.

The time course of the changes occurring in LDL during the lag, propagation, and decomposition phase was studied in detail by Esterbauer et al. (3,27,42,49) and Jürgens et al. (50). Immediately after addition of Cu^{2+} to an LDL solution, the endogenous antioxidants of LDL start to disappear, vitamin E first, followed by the carotenoids (Fig. 4). During this period (lag phase) no or only minimal lipid peroxidation in LDL occurs as evidenced by measurements of fatty acids, TBARs, lipid hydroperoxides, conjugated dienes, fluorescence, or aldehydes. The duration of the lag phase varied considerably from person to person from about 35 to 115 min (Table 1), which indicates a high individual variability in the oxidation resistance of LDL. If the LDL is depleted from its antioxidants, it is, as expressed by Brown and Goldstein (51), left to the mercy of oxygen, and the rate of lipid peroxidation rapidly accelerates to a maximum rate of about 3 mol lipid hydroperoxides formed per mole LDL in 1 min.

This propagating lipid peroxidation phase lasts for about 1–2 h, and during this time the abundant fraction of the PUFAs in LDL are oxidized in the sequence docosahexaenoic acid, arachidonic acid, linoleic acid. As the PUFAs decrease, the lipid hydroperoxides and conjugated dienes increase to a maximum level. Shortly after the onset of accelerated lipid peroxidation, numerous other changes occur in LDL, for example an increase in aldehydic lipid peroxidation products, formation of lysophosphatides and oxodienes, fragmentation of apo B, generation of fluorescent chromophores (emission maximum 430 nm, excitation 360 nm) in the apo B and in the lipids, and increase in the negative surface charge, that is, increase in the relative electrophoretic mobility (Figs. 5 through 7). The lipid hydroperoxides generated during the propagation phase must be seen as labile intermediates, which reach a transient maximum level (see Fig. 5) when the rate of their decomposition exceeds the rate of their formation. In copper-stimulated oxidation this maximum is reached when about 70–80% of the PUFAs are oxidized. Thereafter, decomposition reactions become prevalent and consequently the concentration of lipid hydroperoxides or conjugated dienes start to decrease again.

The period following the peroxide peak is termed the decomposition phase (phase 3, Fig. 5), but one should keep in mind that in most cases this phase temporarily overlaps with the propagation phase. This is clearly evident from the fact that in case of Cu^{2+} stimulation aldehydes (malonaldehyde, 4-hydroxynonenal, and alkanals) are also formed to some extent during the propagation phase. Most of the aldehydes, however, are generated during the decomposition phase, when the concentration of 4-hydroxynonenal, for example, increased from about 14 to 63 mol/mol LDL and hexanal increased from 29 to 160 mol/mol LDL (Fig. 6). The second increase in the 234 nm absorption during the decomposition phase is not due to newly formed dienes but to other compounds exhibiting UV absorption in this region, such as α,β-unsaturated aldehydes.

Various lines of research suggest that the changes occurring in LDL during decomposition result mainly from the aldehydes and their reaction with amino acid residues in apo B. The strong increase in the 430 nm fluorescence of apo B is, for example, likely due to chromophores

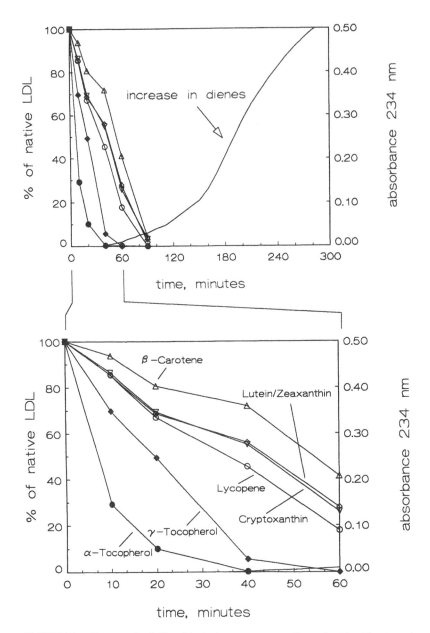

FIGURE 4   Temporal relationship between consumption of antioxidants and onset of lipid peroxidation in copper-stimulated oxidation of LDL. To an LDL solution (1 mg LDL per ml) in oxygen-saturated PBS, pH 7.4, CuCl₂ was added (6.7 μM final concentration). At the indicated time points the antioxidants were determined by HPLC and the degree of oxidation was determined by the conjugated diene absorption at 234 nm. The lower panel is an expansion showing the sequence during the first 60 min.

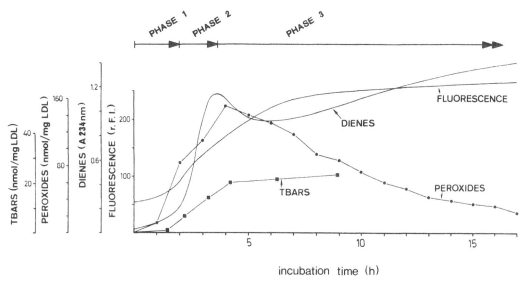

FIGURE 5 Kinetics of Cu^{2+}-stimulated oxidation of LDL. The oxidation was followed by measuring the change in the 430 nm fluorescence (excitation 360 nm), dienes (UV absorbance 234 nm), lipid hydroperoxides, and TBARs. LDL (0.25 mg/ml) was incubated in PBS. Phases 1–3 represent the lag, propagation, and decomposition phases, respectively. (Adapted from Ref. 27.)

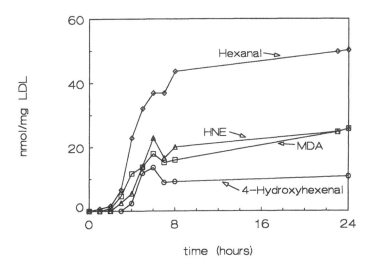

FIGURE 6 Kinetics of the formation of aldehydes during Cu^{2+}-stimulated oxidation of LDL. LDL (1 mg/ml) in PBS was incubated with 6.7 μM $CuCl_2$. MDA = malonaldehyde, HNE = 4-hydroxynonenal. The aldehyde concentrations are given as nmol/mg total LDL.

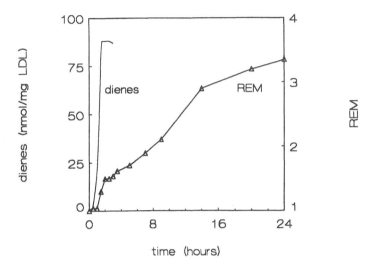

FIGURE 7   Kinetics of the increase in the relative electrophoretic mobility (REM) of LDL during oxidation. LDL (1.5 mg/ml) in PBS was incubated with 10 μM CuCl$_2$. The dienes (given as nmol/mg total LDL) were determined from the 234 nm absorbance. The electrophoretic mobility was determined by lipid electrophoresis. The change relative to native LDL is given ($= 1.0$).

newly formed by reaction of aldehydes with lysine residues of the apo B (50). Similarly, the strong increase in the negative surface charge of LDL (Fig. 7) may be produced by the reaction of aldehydes with positively charged amino groups (R – CHO + protein – NH$_3^+$ → R – CH = N – protein + H$_2$O + H$^+$). The reaction of aldehydes with the apo B is also likely, at least in part, responsible for the generation of the characteristic epitopes that are recognized by the macrophage scavenger receptor (1,3,35,52).

From the temporal change in the chemical composition of LDL during oxidation (Figs. 3 through 7), it is clear that the chemical and functional properties of oLDL continuously change during the lag, propagation, and decomposition phase until the LDL is so heavily oxidized that no further significant modifications are possible. The rate at which LDL enters the propagation phase and approaches the end point depends, among others, on the length of the lag period in which the LDL resists oxidation.

ROLE OF VITAMIN E AND OTHER ANTIOXIDANTS IN RETARDING THE OXIDATION OF LDL

In Vitro Investigations

A number of researchers observed that inclusion of high contents of vitamin E, butylated hydroxytoluene (BHT), probucol, or other chain-breaking antioxidants into the culture medium prevents oxidative modification of LDL by cells and its conversion into a form taken up by macrophages or that is cytotoxic or chemotactic (7,15,22,53–55). To test and quantify the contribution of the endogenous antioxidants to the oxidation resistance of LDL, a standardized and highly reproducible in vitro procedure was developed, which allows the exact determination of the length of the lag phase, preceding the onset of lipid peroxidation (42). This procedure is based on measurement of the conjugated dienes by recording the change in 234 nm UV absorption. The 234 diene absorption develops in the LDL when the PUFAs with isolated double bonds are converted to lipid hydroperoxides with conjugated double bonds. It has been proven that the diene absorption exhibited by LDL during the lag and propagation phase is linearly correlated with its content of lipid hydroperoxides and TBARs. LDL is fully soluble in phosphate-buffered saline and remains in solution when it becomes oxidized. The diene absorption can therefore be measured directly in the LDL solution without prior extraction of the lipids. The whole procedure is very simple and is briefly as follows (42,56). To an EDTA-free solution of LDL (0.25 mg total LDL per ml = 0.055 mg LDL protein per ml) in oxygen-saturated phosphate buffer at pH 7.4, CuCl$_2$ or CuSO$_4$ is added as prooxidant to give a final concentration of 1.66 μM Cu^{2+}. The rate of LDL oxidation is then followed by measuring in an UV spectrometer the change in the 234 nm absorption. If more than one LDL sample is measured at the same time, an automatic cuvette changer is used and the absorption in each cuvette is measured at 30 s intervals. Typical experiments for different LDL samples are shown in Figure 8. From the diene versus time profile the lag phase is obtained by the intercept of the tangent to the time axis, as shown in Figure 8. In experiments in which loss of antioxidants was followed in parallel assays, it was always found that α- and γ-tocopherol disappear first; thereafter the carotenoids decrease to zero, with lycopene first and β-carotene as last (Fig. 4). Shortly after LDL was depleted from its antioxidants, the formation of diene rapidly accelerated, indicative of the onset of a propagating lipid peroxidation chain reaction. This sequence of events suggests that the oxidation resistance of an LDL sample as measured by the lag phase is determined by its content of vitamin E and carotenoids. If this is so, loading of LDL with α-tocopherol should lead to an increase in the lag phase. The best procedure to prepare in vitro LDL with an increased vitamin E content proved to be incubation of plasma with vitamin E (0.125–1 μmol vitamin E per ml plasma, 3 h) before isolation of LDL (56). In this way the vitamin E content of LDL could be increased severalfold

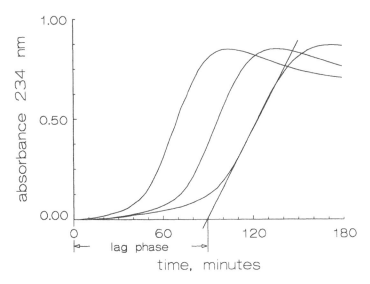

FIGURE 8   Examples for continuous measurement of LDL oxidation by recording the increase in diene absorption. To LDL (0.25 mg/ml) in PBS 1.67 μM CuCl$_2$ was added and the change in the 234 nm absorption was recorded in a spectrophotometer in 1 cm cuvettes. The samples were measured in parallel, with automatic cuvette changing at 30 s intervals. The curves represent LDL samples from three different donors.

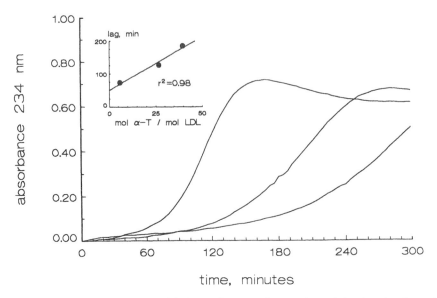

FIGURE 9 Increase in the oxidation resistance of LDL through in vitro loading with vitamin E. The plasma for a single donor was divided into three aliquots and supplemented with 0, 500, and 1000 μM *RRR*-α-tocopherol. After 3 h incubation at 37°C under N_2, the LDL samples were isolated and their α-tocopherol content determined. The oxidation kinetics of the three LDL samples were measured under the conditions as given in Figure 8. Insert: shows the dependence of the lag phase in minutes on the α-tocopherol content in LDL.

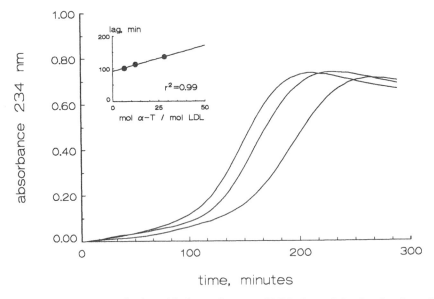

FIGURE 10 Increase in the oxidation resistance of LDL through in vitro loading with vitamin E. The conditions are the same as in Figure 9, except that plasma from another donor was used.

to about 30 mol/mol LDL. Typical experiments with LDL isolated from vitamin E-supplemented plasma of two donors are shown in Figures 9 and 10. For each of these two LDL samples loaded with vitamin E in this way, the lag phase increased proportionally with the vitamin E content, as shown in the insert of Figures 9 and 10. In 90% of the cases investigated thus far ($n = 24$), the correlation coefficient r^2 obtained from a linear regression analysis was better than 0.83. The dependence of the lag phase (y) from the vitamin E content of a particular LDL sample (x) can therefore be described by the equation $y = kx + a$. Note that in LDL the vitamin E content resembles very closely the α-tocopherol content and y only marginally changes if x is taken as α-tocopherol only or as α- + γ-tocopherol (Table 1). Interestingly, however, the values for k (range 0.7–34) and a (range -49 to 97) varied considerable from donor to donor (Table 2). This suggests that the efficiency (k) of vitamin E in increasing the oxidation resistance of LDL, as well as the vitamin E-independent component (a) of the lag phase, is an LDL-specific constant and varies from person to person. The individual values for k and a obtained in this in vitro study are given in Table 2 (donors 1–24).

In addition to these experiments with vitamin E, we also loaded LDL with other lipophilic antioxidants (to be published by Puhl et al.). Figure 11 shows, for example, the results obtained with the vitamin E analog α-tocotrienol. This antioxidant proved to be very efficient in protecting LDL against oxidation, with an increase in the lag time of 6 min/mol α-tocotrienol per mol LDL.

Ex Vivo Investigations

Besides the in vitro experiments already described, we made an ex vivo study with vitamin E supplementation to show how an increased vitamin E intake would affect the oxidation resistance of LDL in different persons (43,57). A group of 12 clinically healthy male and female volunteers aged 20–30 years participated in this single-blind study and took daily for 3 weeks either placebo capsules (4 persons) or capsules with 150 IU (2 persons), 225 IU (2 persons), 800 IU (2 persons), or 1200 IU *RRR*-α-tocopherol (2 persons). The lipid status of all subjects, as well as the antioxidants (α-tocopherol, γ-tocopherol, β-carotene, lycopene, cryptoxanthine, and lutein/zeaxanthine) in the plasma and LDL were determined 3 days before supplementation. These analysis were repeated in five intervals (days 3, 5, 10, 12, and 18) during supplementation and 1 week after termination of the supplementation (day 26). For the two participants (codes MK and DK) taking 1200 IU *RRR*-α-tocopherol per day, the protocol for the change in α- and γ-tocopherol and carotenoids in LDL and the lag phases is given in Table 3.

All persons receiving *RRR*-α-tocopherol responded with an increase in α-tocopherol in plasma and LDL and a decrease in γ-tocopherol. From the time dependence of the increase (Fig. 12), it appears that in plasma plateau levels of vitamin E were reached after about 10 days of supplementation. The average (mean of days 3, 5, 10, 12, and 18) increases in plasma α-tocoherol based on the initial value at day -3 ($=100\%$) were 146, 165, 183, and 248% for the doses of 150, 225, 800, and 1200 IU (Table 4). The LDL α-tocopherol did not fully follow the change in the plasma (Fig. 13). First, the 150 and 800 IU led to a saturation of LDL with α-tocopherol even after 3 days of supplementation. Second, with the large dose of 1200 IU, the LDL α-tocopherol continued to increase during the whole period of supplementation, and third, an end point was obviously also not reached with the dose of 225 IU. Both subjects receiving this dose responded with a second increase in LDL α-tocopherol between days 12 and 18. The mean increase in α-tocopherol during the 3 weeks was 138, 158, 144, and 205% with the doses of 150, 225, 800, and 1200 IU (Table 4). The plasma and LDL α-tocopherol

TABLE 2   Effect of α-Tocopherol on the Oxidation Resistance of LDL from 36 Single Subjects[a]

Subject	n	k	a	r^2
1	3	17.14	−49.3	0.999
2	3	9.99	29.4	0.875
3	3	4.46	60.3	0.988
4	3	3.02	35.0	0.967
5	3	0.70	75.4	0.574
6	3	3.65	6.0	0.998
7	3	4.42	11.6	0.998
8	3	6.13	53.5	0.981
9	3	4.16	44.6	0.836
10	3	1.50	97.3	0.687
11	3	3.23	39.5	0.960
12	3	3.24	50.1	0.977
13	3	1.57	92.3	0.993
14	3	4.49	77.1	0.954
15	3	2.43	44.9	0.999
16	3	3.79	1.0	0.999
17	3	3.12	91.6	0.848
18	3	6.34	40.9	0.947
19	3	5.13	44.2	0.932
20	3	3.82	59.8	0.899
21	3	2.97	43.4	0.878
22	3	4.60	32.4	0.922
23	5	4.16	42.2	0.749
24	5	2.59	31.3	0.967
25	7	4.30	22.7	0.735
26	7	9.93	−31.8	0.848
27	7	5.77	11.3	0.885
28	7	5.05	28.9	0.574
29	7	5.32	30.5	0.277
30	7	1.38	57.5	0.065
31	7	8.92	29.4	0.595
32	7	3.46	64.4	0.577
33	7	1.87	58.6	0.057
34	7	2.88	49.8	0.500
35	7	4.41	54.1	0.901
36	7	2.62	55.6	0.116
Mean 1–24		4.44 ± 3.21	43.9 ± 31.5	
Mean 25–36		4.66 ± 2.50	35.9 ± 26.1	

[a]Given is the effectiveness of α-tocopherol (k) as well as the vitamin E-independent component (a) of the lag phase (see formula in Fig. 17). n is the number of LDL samples with different α-tocopherol contents investigated for each subject; r^2 is the correlation coefficient obtained by linear regression analysis. For subjects 1–24, the plasma was loaded with α-tocopherol (see Figs. 9 and 10 and text); subjects 25–36 are those participating in the 3 week vitamin E supplementation study (see Figs. 13–15), including those receiving placebo. Subjects 1–20 were aged 50–80 years; all others were aged 20–30 years.

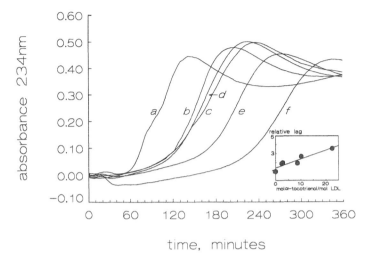

FIGURE 11 Increase in the oxidation resistance of LDL through in vitro loading with α-tocotrienol. The conditions were as in Figure 9, except that instead of α-tocopherol its analog α-tocotrienol was used. (a) No supplementation, (b) 2.34, (c) 2.8, (d) 8.6, (e) 10.15, and (f) 22.48 mol α-tocotrienol per mol LDL respectively.

levels were decreased again 7 days after termination of the supplementation, very close to the initial level before supplementation. In all α-tocopherol-supplemented persons the γ-tocopherol content in plasma and LDL decreased to about 0–15% of the initial values. The carotenoids also slightly decreased in the supplemented persons. In the placebo group no significant changes in the plasma and LDL α-tocopherol levels occurred (Table 4). To facilitate

TABLE 3 Typical Protocol of the Ex Vivo Study in Which Two Volunteers (MW and DK) Received 1200 IU *RRR*-α-Tocopherol Daily for 3 Weeks

| Day of study | Antioxidants (mol/mol LDL) | | | | | | Lag phase (min) | |
| | α-Tocopherol | | γ-Tocopherol | | Carotenoids | | | |
	MW	DK	MW	DK	MW	DK	MW	DK
−3	8.4	6.9	0.7	0.6	0.7	1.1	74	69
3	12.0	13.6	0.1	0.1	0.2	0.9	118	128
5	16.9	14.6	0.1	0.1	0.4	0.6	132	119
10	15.9	12.9	0.1	0.1	0.3	0.7	120	110
12	20.6	14.2	0.1	0.1	0.4	0.9	141	112
18	26.0	18.8	0.1	0.1	0.4	0.8	170	118
26	8.7	7.3	0.5	0.4	0.7	1.0	103	101
Rel. mean ± SD[a]	2.15 ± 0.47		0.14 ± 0		0.59 ± 0.17		1.75 ± 0.21	

[a]Mean change for days 3–18 relative to day −3 (= 1.0).

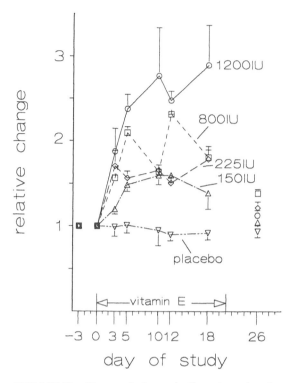

FIGURE 12   Temporal change in the α-tocopherol content in plasma through daily dietary intake of *RRR*-α-tocopherol over 3 weeks. The relative change $\pm$ SD compared to day -3 is given ($= 1.0$).

TABLE 4   Effect of Dietary α-Tocopherol on the α-Tocopherol Content of Plasma and LDL and Its Oxidation Resistance[a]

Dose (IU)	n	Vitamin E		Lag phase
		Plasma	LDL	
0	20	0.95 $\pm$ 0.11	0.98 $\pm$ 0.17	97 $\pm$ 30
150	10	1.46 $\pm$ 0.18	1.38 $\pm$ 0.12	118 $\pm$ 17
225	10	1.65 $\pm$ 0.15	1.58 $\pm$ 0.32	156 $\pm$ 22
800	10	1.83 $\pm$ 0.23	1.44 $\pm$ 0.12	135 $\pm$ 23
1200	10	2.48 $\pm$ 0.53	2.15 $\pm$ 0.47	175 $\pm$ 21

[a]Volunteers took daily over 3 weeks capsules with the indicated doses of *RRR*-α-tocopherol. Values are expressed as the relative change $\pm$ standard deviation compared to the basal level at day -3 ($= 1.0$).

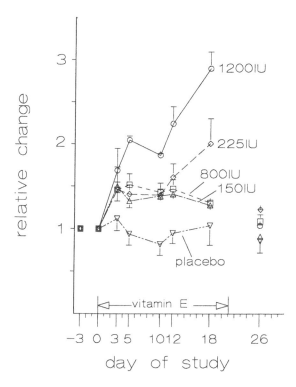

FIGURE 13 Temporal change in the α-tocopherol content in LDL through daily dietary intake of *RRR*-α-tocopherol over 3 weeks. The relative change $\pm$ SD compared to day -3 is given ($= 1.0$); study group as in Figure 12.

for the reader the comparison of the efficiency of dietary vitamin E as determined in this study, the average relative change together with the standard deviation for each individual dose and the placebo group is given in Table 4 and graphically shown in Figure 14. This comparison shows that doses of 150–800 IU are about equally effective and increase the LDL α-tocopherol on average by about 50%. To double the LDL α-tocopherol content megadoses in the range of 1200 IU are required.

The major objective of this supplementation study was to find out whether dietary vitamin E would have a positive effect on the oxidation resistance of LDL. Figure 15 shows an "ideal" responder, receiving 1200 IU/day. The temporal change in the oxidation resistance of LDL as determined by the duration of the lag phase followed exactly the time profile of the α-tocopherol content of LDL; furthermore, the lag phase increased strictly linearly ($r^2 = 0.90$) with the LDL α-tocopherol. A similar good correlation was found for the two persons receiving 800 IU ($r^2 = 0.85, 0.74$) and for one person receiving 150 IU ($r^2 = 0.89$). For four persons (one at 1200 IU, two at 225 IU, and one receiving placebo) the correlation between lag phase and α-tocopherol levels was less ($r^2 = 0.50$–0.60), and for the remaining four persons (one at 150 IU and three at placebo), no statistical correlation existed ($r^2 = 0.06$–0.28). The first conclusion that can be drawn from this ex vivo study is that the majority of the subjects (seven of eight) could increase the oxidation resistance of the LDL by oral supplementation

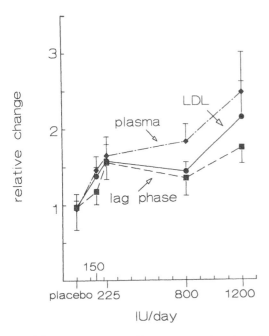

FIGURE 14 Effect of dietary α-tocopherol on the α-tocopherol content of plasma and LDL and its oxidation resistance. Volunteers took daily over 3 weeks capsules with the indicated doses of *RRR*-α-tocopherol. The relative change compared to day −3 is given (= 1.0); study group as in Figures 12 and 13.

with *RRR*-α-tocopherol. The second important point is the individual variation in the effectiveness of α-tocopherol. As mentioned previously, a direct measure for the effectiveness is the slope k of the regression line (see Fig. 15 and inserts in Figs. 9 and 10), which gives the incremental increase in the lag phase in minutes produced by 1 mol α-tocopherol per mol LDL. In seven persons the effectiveness was nearly identical with 4.5 ± 1.0 min (mean ± SD), in

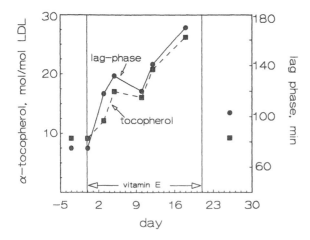

FIGURE 15 Temporal change in the oxidation resistance (lag phase) and the α-tocopherol content of a single subject, receiving 1200 IU *RRR*-α-tocopherol over 3 weeks.

two persons the effectiveness was clearly better (k = 8.9 and 9.9 min, respectively), and in the remaining three persons α-tocopherol was less effective (k = 2.0 ± 0.6 min). It is of interest that the two persons in whom vitamin E showed the highest effectiveness had a very low value for the vitamin E-independent parameter of the lag phase, that is, the intercept a for a theoretical zero α-tocopherol concentration. This fully agrees with the in vitro supplementation, in which it was also found that individuals showing the strongest dependence of the lag phase from the LDL α-tocopherol had no or only negligible protection from the vitamin E-independent component (Table 2).

CONCLUSIONS

In our ongoing study on the importance of antioxidants for the oxidation resistance of LDL, we have thus far determined the length of the lag phase preceding the onset of lipid peroxidation and the antioxidants in 182 LDL samples. We are aware that exposing LDL in vitro to Cu^{2+} ions as prooxidant, as done in these in vitro assays, is an unphysiological condition of oxidative stress. Nevertheless, we think that the oxidation resistance determined in this manner is a useful and reliable index for comparing the relative susceptibility of LDL toward oxidative stress in vivo. When all data estimated for the LDL α-tocopherol and the associated lag phases (basal values supplemented in vitro and in vivo) are evaluated by linear regression analysis, a highly significant (p < 0.001, r^2 = 0.57) positive correlation between the duration of the lag phase in minutes (y) and the α-tocopherol content (mol/mol LDL, x) is obtained with y = 3.43x + 49.3 (Fig. 14). Statistically, one α-tocopherol molecule per LDL particle prolongs the lag phase by 3.43 min, and the vitamin E-independent residual lag phase is 49.3 min. With these values and basal LDL α-tocopherol content (6.37 mol/mol LDL, Table 1), the LDL samples in our study group (nonsupplemented subjects, Table 1) should have an average lag phase of 71 min; indeed, this agrees fully with the experimentally determined mean value ± SD (standard deviation of 68 ± 15 min (Table 1).

Our statistical results on the relationship between lag phase and α-tocopherol are based on a rather small sample size and on subjects living in the same area and having more or less the same living habits. It would be worth investigating whether the same relationship is applicable to other study groups. An interesting aspect in this regard comes from a cross-cultural epidemiological study of the European population, which revealed a highly significant (r^2 = 0.55) inverse correlation between mortality from ischemic heart disease and plasma vitamin E 58). This is in agreement with the hypothesis that oxidized LDL is atherogenic and that better protection of LDL by elevated vitamin E could reduce the risk of oxidative modification in vivo.

Of course, the statistical medians (Fig. 16) allow prediction only for a large sample size, not for individuals. Our investigations of individuals (Figs. 9, 10, and 15 and Table 2) clearly indicate that the effectiveness of α-tocopherol in increasing the oxidation resistance (=lag phase), as well as the vitamin E-independent component of the lag phase, varies significantly from subject to subject. The vitamin E content of the LDL is therefore per se not sufficient to predict the lag phase of an individual LDL. For that the subject-specific constants k and a must be known. This is in full agreement with reports by others (15,30) that the basal level of vitamin E in LDL is not predictive of its oxidizability in vitro. The large individual variation in k and a also explains why the 72 cases for whom we determined the basal α-tocopherol level (i.e., without supplementation) and lag phases gave in the linear regression analysis a much weaker correlation (r^2 = 0.07, p < 0.01) than found in the analysis in which the supplementation experiments were included. The importance of individual analysis is clearly evident from the three examples shown in Figure 17. If for one reason or another the vitamin E level falls to 2 mol α-tocopherol per mol LDL in all three persons, the response of the LDL

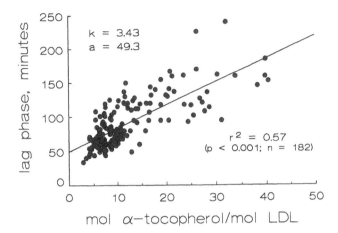

FIGURE 16 Statistical relationship between oxidation resistance of LDL as measured by the lag phase and the α-tocopherol content. The plot contains all data, that is, not vitamin E supplemented and supplemented in vitro or in vivo.

to an oxidative stress would be very different. One ($k = 17.1$, $a = -49.3$) would be virtually unprotected and quickly oxidized (lag phase -15 min); the other ($k = 4.4$, $a = 11.6$) would have a residual oxidation resistance of 20 min, and the third one ($k = 1.57$ and $a = 92.3$) would still be well protected with a lag phase of 95 min.

At present it is not known which factors determine the effectiveness of α-tocopherol (k) and the vitamin E-independent protection (α). Important contributions might come from the amount of polyunsaturated fatty acids, the ratio of antioxidants to polyunsaturated fatty acids, preformed peroxides, the mobility of vitamin E, the structure of apo B, and the amount of other antioxidants. The comparably low concentration of carotenoids suggests that their contribution to the total antioxidant capacity of LDL is rather low. If a linear regression analysis is made for the dependence of the lag phase from the total LDL antioxidants (vitamin E + carotenoids), more or less the same relationship is obtained as for α-tocopherol alone

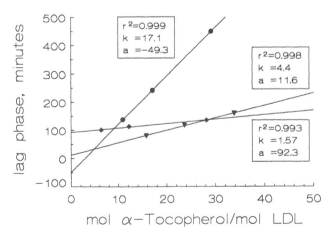

FIGURE 17 Comparison of the effectiveness of vitamin E (k) and the vitamin E-independent parameter (a) of three single donors. Before LDL isolation the plasma was supplemented with *RRR*-α-tocopherol as described in Figure 9.

(Fig. 16). Although the underlying mechanism of the variations in the effectiveness of vitamin E and the vitamin E-independent protection are not yet understood, measurement of these parameters is valuable additional information that could be of medical interest in vitamin E supplementation.

ACKNOWLEDGMENT

The authors have been supported by the Association for International Cancer Research (AICR), UK.

REFERENCES

1. Steinberg D, Parthasarathy S, Carew TE, Khoo JC, Witztum JL. Beyond cholesterol. Modifications of low-density lipoprotein that increases its atherogenicity. N Engl J Med 1989; 320:915-24.
2. Steinbrecher UP, Zhang H, Lougheed M. Role of oxidatively modified LDL in atherosclerosis. Free Rad Biol Med 1990; 9:155-68.
3. Esterbauer H, Dieber-Rotheneder M, Waeg G, Striegl G, Jürgens G. Biochemical, structural, and functional properties of oxidized low-density lipoprotein. Chem Res Toxicol 1990; 3:77-92.
4. Lang J, Esterbauer H. Oxidized LDL and atherosclerosis. In: Vigo-Pelfrey C, ed. Membrane lipid oxidation, Vol. III. Boca Raton, FL: CRC Press, 1990; 266-82.
5. Gebicki JM, Jürgens G, Esterbauer H. Oxidation of low-density lipoprotein. In: Sies H, ed. Oxidative stress: oxidants and antioxidants. London: Academic Press, 1991; 371-97.
6. Henriksen T, Mahoney EM, Steinberg D. Enhanced macrophage degradation of low density lipoprotein previously incubated with cultured endothelial cells: recognition by receptors for acetylated low density lipoproteins. Proc Natl Acad Sci U S A 1981; 78:6499-503.
7. Steinbrecher UP, Parthasarathy S, Leake DS, Witztum JL, Steinberg D. Modification of low density lipoprotein by endothelial cells involves lipid peroxidation; degradation of low density lipoprotein phospholipids. Proc Natl Acad Sci U S A 1984; 81:3883-7.
8. Ross R. The pathogenesis of atherosclerosis—an update. N Engl J Med 1986; 314:488-500.
9. Hessler JR, Morel DW, Lewis LJ, Chisholm GM. Lipoprotein oxidation and lipoprotein-induced cytotoxicity. Arteriosclerosis 1983; 3:215-22.
10. Morel DW, Hessler JR, Chisolm GM. Low density lipoprotein cytotoxicity induced by free radical peroxidation of lipid. J Lipid Res 1983; 24:1070-6.
11. Stam H, Hülsmann JF, Jongkind JF, Van der Kraaij AMM, Koster JF. Endothelial lesions, dietary composition and lipid peroxidation. Eicosanoids 1989; 2:1-14.
12. Rosenfeld ME, Palinski W, Ylä-Herttuala S, Carew TE. Macrophages, endothelial cells, and lipoprotein oxidation in the pathogenesis of atherosclerosis. Toxicol Pathol 1990; 18:560-71.
13. Quinn MT, Parthasarathy S, Fong LG, Steinberg D. Oxidatively modified low density lipoproteins: a potential role in recruitment and retention of monocytes/macrophages during atherogenesis. Proc Natl Acad Sci U S A 1987; 84:2995-8.
14. Leake DS, Rankin SM. The oxidative modification of LDL by macrophages. Biochem J 1990;
15. Jessup W, Rankin SM, De Whalley CV, Hoult JRS, Scott J, Leake DS. Alpha-tocopherol consumption during low-density-lipoprotein oxidation. Biochem J 1990; 265:399-405.
16. Ledwozyw A, Michalak J, Stepien A, Kadziolka A. The relationship between plasma triglycerides, cholesterol, total lipid and lipid peroxidation products during human atherosclerosis. Clin Chim Acta 1986; 155:275-84.
17. Piotrovski JJ, Hunter GC, Eskelson CD, Dubick MA, Bernhard VM. Evidence for lipid peroxidation in artherosclerosis. Life Sci 1990; 46:715-21.
18. Smith TL, Kumerow FA. Plasma lipid peroxidation and atherosclerosis in restricted ovulator chickens. Basic Life Sci 1988; 49:941-4.
19. Daugherty A, Zweifel BS, Burton SE, Schonfeld G. Isolation of low density lipoprotein from atherosclerotic vascular tissue of Watanabe heritable hyperlipidemic rabbits. Arteriosclerosis 1988; 8:768-77.
20. Morel DW, Chisolm GM. Antioxidant treatment of diabetic rats inhibits lipoprotein oxidation and cytotoxicity. J Lipid Res 1989; 30:1827-.

21. Kita T, Nagano Y, Yokode M, et al. Prevention of atherosclerotic progression in watanabe rabbits by probucol. Am J Cardiol 1988; 62:13B-9B.

22. Parthasarathy S, Young SG, Witztum JL, Pittman RC, Steinberg D. Probucol inhibits oxidative modification of low density lipoprotein. J Clin Invest 1986; 7:641-4.

23. Palinski W, Rosenfeld ME, Ylä-Herttuala, S, et al. Low density lipoprotein undergoes oxidative modification in vivo. Proc Natl Acad Sci U S A 1989; 86:1372-6.

24. Haberland ME, Fong D, Cheng L. Malondialdehyde-altered protein occurs in atheroma of Watanabe heritable hyperlipidemic rabbits. Science 1988; 241:215-8.

25. Rosenfeld ME, Palinski W, Ylä-Herttuala S, Butler S, Witztum JL. Distribution of oxidation specific lipid-protein adducts and apolipoprotein B in atherosclerotic lesions of varying severity from WHHL rabbits. Arteriosclerosis 1990; 10:336-49.

26. Palinski W, Ylä-Herttula S, Rosenfeld ME, et al. Antisera and monoclonal antibodies specific for epitopes generated during oxidative modification of low density lipoprotein. Arteriosclerosis 1990; 10:325-35.

27. Esterbauer H, Rotheneder M, Striegl G, et al. Vitamin E and other lipophilic antioxidants protect LDL against oxidation. Fat Sci Technol 1989; 91:316-24.

28. Sparrow CP, Parthasarathy S, Steinberg D. Enzymatic modification of low density lipoprotein by purified lipoxygenase plus phospholipase A_2 mimics cell-mediated oxidative modification. J Lipid Res 1988; 29:745-53.

29. Salmon S, Maziere JC, Santus R, Morliere P, Bouchemal N. UVB-induced photoperoxidation of lipids of human low and high density lipoproteins—a possible role of tryptophan residues. Photochem Photobiol 1990; 52:541-5.

30. Babiy AV, Gebicki JM, Sullivan DR. Vitamin E content and LDL oxidizability induced by free radicals. Atherosclerosis 1990; 81:175-82.

31. Mino M, Miki M, Miyake M, Ogihara T. Nutritional assessment of vitamin E. In: Diplock AT, Machlin LJ, Packer L, Pryor WA, eds. Vitamin E: biochemistry and health implications. Ann N Y Acad Sci 1989; 570:296-310.

32. Bedwell S, Dean RT, Jessup W. The action of defined oxygen-centred free radicals on human low-density lipoprotein. Biochem J 1989; 262:707-12.

33. Brown MS, Goldstein JL. A receptor-mediated pathway for cholesterol homeostasis. Science 1986; 232:34-47.

34. Weisgraber KH, Innerarity TL, Mahley RW. Role of lysine residues of plasma lipoproteins in high affinity binding to cell surface receptors on human fibroblasts. J Biol Chem 1978; 253:9053-62.

35. Haberland ME, Fogelman AM, Edwards PA. Specificity of receptor-mediated recognition of malondialdehyde modified low density lipoprotein. Proc Natl Acad Sci U S A 1982; 79:1712-6.

36. Hoff HF, O'Neill J, Chisolm GM III, et al. Modification of LDL with 4-hydroxynonenal, a propagation product of lipid peroxidation, induces uptake of LDL by macrophages. Arteriosclerosis 1989; 9:538-49.

37. Haberland ME, Olch CL, Fogelman AM. Role of lysines in mediating interaction of modified low density lipoproteins with the scavenger receptor of human monocyte macrophages. J Biol Chem 1984; 259:11305-11.

38. Halliwell B. How to characterize a biological antioxidant. Free Rad Res Comms 1990; 9:1-32.

39. Stocker R, Bowry VW, Frei B. Ubiquinol-10 protects human low density lipoprotein more efficiently against lipid peroxidation than does alpha-tocopherol. Proc Natl Acad Sci U S A 1991; 88:1646-50.

40. Massay JB. Kinetics of transfer of alpha-tocopherol between model and native plasma lipoproteins. Biochim Biophys Acta 1984; 793:387-92.

41. DiMascio P, Kaiser S, Sies H. Lycopene as the most efficient singlet oxygen quencher. Arch Biochem Biophys 1989; 274:

42. Esterbauer H, Striegl G, Puhl H, Rotheneder M. Continuous monitoring of in vitro oxidation of human low density lipoprotein. Free Rad Res Comms 1989; 6:67-75.

43. Dieber-Rotheneder M, Puhl H, Waeg G, Striegl G, Esterbauer H. Effect of oral supplementation with *d*-alpha-tocopherol on the vitamin E content of human low density lipoproteins and its oxidation resistance. J Lipid Res 1991; 32:1325-32.

44. Sattler W, Kostner GM, Waeg G, Esterbauer H. Oxidation of lipoprotein Lp(a). A comparison with low-density lipoprotein. Biochim Biophys Acta 1991; 1081:65-74.

45. Heinecke JW, Rosen H, Chait A. Iron and copper promote modification of low density lipoprotein by human arterial smooth muscle cells in culture. J Clin Invest 1984; 4:1890-4.

46. Morel DW, DiCorleto PE, Chisolm GM. Endothelial and smooth muscle cells alter low density lipoprotein in vitro by free radical oxidation. Arteriosclerosis 1984; 4:357-64.

47. Fong LG, Partharasarathy S, Witztum JL, Steinberg D. Nonenzymatic oxidative cleavage of peptide bonds in apoprotein B-100. J Lipid Res 1987; 28:1466-77.

48. Hiramatsu K, Rosen H, Heinecke JW, Wolfbauer G, Chait A. Superoxide initiates oxidation of low density lipoprotein by human monocytes. Arteriosclerosis 1987; 7:55-60.

49. Esterbauer H, Jürgens G, Quehenberger O, Koller E. Autoxidation of human low density lipoprotein: loss of polyunsaturated fatty acids and vitamin E and generation of aldehydes. J Lipid Res 1987; 28:495-509.

50. Jürgens G, Hoff HF, Chisolm GM, Esterbauer H. Modification of human serum low density lipoprotein by oxidation-characterization and pathophysiological implications. Chem Phys Lipids 1987; 45:315.

51. Brown MS, Goldstein JL. Scavenging for receptors. Nature 1990; 343:506-7.

52. Hoff HF, Cole TB. Macrophage uptake of low-density lipoprotein modified by 4-hydroxynonenal. An ultrastructural study. Lab Invest 1991; 674:254-64.

53. Morel DW, DiCorleto PE, Chisolm GM. Endothelial and smooth muscle cells alter low density lipoprotein in vitro by free radical oxidation. Arteriosclerosis 1984; 4:357-64.

54. Cathcart MK, McNally AK, Morel DW, Chisholm GM III. Superoxide anion participation in human monocyte-mediated oxidation of low-density lipoprotein and conversion of low-density lipoprotein to a cytotoxin. J Immunol 1989; 142:1963-9.

55. Van Hinsbergh VWM, Van Scheffer M, Havekes L, Kempen HJM. Role of endothelial cells and their products in the modification of low-density lipoproteins. Biochim Biophys Acta 1986; 878:49-64.

56. Esterbauer H, Dieber-Rotheneder M, Striegl G, Waeg G. Role of vitamin E in preventing the oxidation of low-density lipoprotein. Am J Clin Nutr 1991; 53:314S-21S.

57. Esterbauer H, Dieber-Rotheneder M, Waeg G, Puhl H, Tatzber F. Endogenous antioxidants and lipoprotein oxidation. Biochem Soc Trans 1990; 18:1059-61.

58. Gey KF, Puska P, Jordan P, Moser UK. Inverse correlation between plasma vitamin E and mortality from ischemic heart disease in cross-cultural epidemiology. Am J Clin Nutr 1991; 53:326S-34S.

47

Antioxidant Therapy in Cardiac Surgery

D. A. G. Mickle

Toronto General Hospital, Toronto, Ontario, Canada

INTRODUCTION

A number of different biochemical mechanisms act to damage the reperfused ischemic myocardium during cardiac surgery despite the use of cold blood cardioplegia. Oxyradical injury has been implicated as an important component of this injury (1–5). High oxygen tensions in cardiopulmonary bypass, reperfusion of the myocardium after cross-clamp removal, and activated polymorphonuclear neutrophils all contribute to this injury.

The primary initiator of myocardial free-radical injury is the superoxide ion ($O_2^{\dot-}$), which is generated from single-electron transfer to molecular O_2 by a variety of mechanisms, which include xanthine oxidase, cytochromes, ubisemiquinone, and heme. Although eukaryotic cells do not produce a high level of $O_2^{\dot-}$ under normal conditions, a small fraction (perhaps 1–2% of the O_2 consumed by respiring cells) appears to be converted to $O_2^{\dot-}$, that is, $O_2^{\dot-}$ that escapes from its site of formation and is not immediately reduced to H_2O_2 by a second electron transfer. The Fe^{2+} form of heme can react with H_2O_2 to generate $OH^{\cdot}$ via a Fenton-like reaction. The Fe^{3+} form of the myoglobin heme (ferrimyoglobin) can be further oxidized by H_2O_2 to form what is thought to be a porphyrin cation radical. All oxygen-consuming cells, particularly mitochondria, contain the enzymes necessary to scavenge $O_2^{\dot-}$. Although $O_2^{\dot-}$ is not in itself the most reactive oxygen species, its destructive potential lies in its ability to generate $OH^{\cdot}$ radicals. These are the most reactive radicals that can be formed in vivo, so reactive that they react with essentially any organic molecule immediately. These reactions occur at rates that are limited only by the diffusion together of the $OH^{\cdot}$ radical and the organic molecule.

Superoxide dismutase (SOD) provides the first line of intracellular defense against radical-induced tissue damage, but SOD may be "overwhelmed," particularly in a localized subcellular area. In such a case $OH^{\cdot}$ radical formation can become possible. Oxygen is generally believed to be concentrated in the hydrophobic milieu of membranes. Since membranes contain high concentrations of compounds that are susceptible to attack by radicals, such as polyunsaturated fatty acids, a large portion of the eventual destructive power of the oxyradicals is unleashed upon membrane lipids and proteins. This occurs irrespective of whether the initial generation of the $O_2^{\dot-}$ is in the cytosol or in the lipid material itself. If the effects of lipid peroxidation on the myocyte membranes are sufficient to affect membrane structure and function, a "stunned" myocardium (reversible injury) with a loss of contractility or myocardial necrosis (irreversible injury) can result.

Stimulated neutrophils have an important role in determining the amount of ischemic myocardial necrosis in cardiac surgery (6,7). In a canine model of LAD regional ischemia and reperfusion designed to simulate urgent revascularization for acute myocardial ischemia, myocardial leukocyte deposition peaked in the LAD region at about 30 min of reperfusion and was almost back to baseline levels by 60 min or reperfusion (8).

Cardiac arrhythmias produce disturbances in cardiac output that can terminate in cardiac arrest with sudden death. Reperfusion of the ischemic myocardium after cardioplegic ischemic arrest can precipitate arrhythmias. Biochemical factors, which are believed to be important in ischemia and reperfusion-induced arrhythmias, are (a) increased lipolysis and phospholipolysis producing fatty acids and lysophosphatides (9), and (b) lipid peroxidation from free-radical formation during reperfusion or reoxygenation of ischemic or hypoxic myocardium (10,11).

Lipid peroxidation disrupts cellular and subcellular membranes and yields some highly toxic molecular products that can diffuse appreciable distances from their sites of formation. It is not surprising to find that biological defenses against peroxidation are well developed in the myocardium. These defensive mechanisms rely on various types of antioxidants, which include (1) preventive antioxidants that convert molecules that can yield radicals (H_2O_2) to inert products (H_2O); and (2) chain-breaking antioxidants that trap the free radicals that actually cause lipid peroxidation. In biological systems, classic glutathione peroxidase, phospholipid hydroperoxide glutathione peroxidase, and catalase are preventive antioxidants. Superoxide dismutase, vitamin C, uric acid, possibly bilirubin, reduced glutathione, and protein sulfhydryl groups are water-soluble chain-breaking antioxidants. Vitamin E is the major and possibly the only lipid-soluble chain-breaking antioxidant in the myocardium.

CYANOTIC HEART PATIENTS

In an important study of intracardiac repair of congenital cardiac anomalies from the Hospital for Sick Children, Burrows et al. (12) showed that a significantly ($p < 0.05$) greater percentage of children with tetralogy of Fallot (TOF) have decreased postoperative function than children with a ventricular septal defect or transposition. The TOF children had a downslope in both cardiac index and left ventricular stroke work index of 31 and 23%, respectively, compared to 8 and 14% for ventricular septal defect children and 8 and 15% for transposition.

We have shown that cold blood cardioplegia does not prevent marked ATP loss, lactate accumulation, or focal necrosis in the myocardium of cyanotic TOF patients undergoing elective surgical repair (13). The myocardial accumulation of lactate during the first minutes of reperfusion is consistent with significant mitochondrial dysfunction, which could explain the postoperative contractile failure that sometimes occurs after technically successful corrective surgery of TOF patients. Based on this research, we hypothesized that chronic cyanosis in TOF patients predisposes the myocardium to oxyradical damage during cardiopulmonary bypass and during reperfusion. In our next study (2), we showed that myocardial phospholipid hydroxyconjugated dienes, the chemical signature of oxygen-mediated free-radical injury, were detectable before ischemia on bypass, at end ischemia, and during reperfusion in cyanotic TOF patients. Since control canine hearts on cardiopulmonary bypass had undetectable levels of conjugated dienes, we hypothesized that the cyanotic TOF myocardium was susceptible to free-radical injury both during cardiopulmonary bypass with its high oxygen tension and during reperfusion.

To better understand the antioxidant defenses of the cyanotic myocardium, we did in vitro studies using cultured tetralogy of Fallot myocytes (14) and fibroblasts and vascular endothelial cells. The myocyte activities of the antioxidant enzymes, superoxide dismutase (SOD), catalase, and glutathione peroxidase (GPx), were measured and found to be regulated by oxygen tension. The cells exposed to a pO_2 of 40 mmHg had lower activities of antioxidant enzymes than those cells cultured at a pO_2 of 150 mmHg. Similar oxygen tension regulation of antioxidant enzymes was found in vascular endothelial and myocardial fibroblast (papers in preparation). We have evidence that this cellular regulation of antioxidant levels with oxygen tension is at a transcriptional level. The cells cultured at a pO_2 of 40 mmHg were more susceptible to free-radical injury than those cells cultured at a pO_2 of 150 mmHg. In addition, the endothelial cells cultured at a pO_2 of 40 mmHg were significantly more susceptible to damage. Specifically, approximately 30% of the endothelial cells necrosed from a sudden increase in oxygen tension to 150 mmHg. To determine whether these in vitro antioxidant enzyme results were clinically relevant, a study of TOF ($n = 14$) and coronary artery bypass (CABG; $n = 11$) patients (15) was done to determine the in vivo relationship between the antioxidant enzymes SOD, catalase, and GPx in the TOF myocardium and the patient's resting preoperative arterial blood oxygen tension. The TOF patients scheduled for elective surgery had baseline arterial blood gas measurements done before surgery. During surgery right ventricular biopsies were taken for enzyme analysis immediately before cold blood cardioplegic arrest and 20 min after cross-clamp removal. There were no changes in myocardial antioxidant enzyme activities during surgery. The myocardial SOD, catalase, and GPx activities correlated (0.82, 0.68, and 0.89, respecitvely) significantly ($p < 0.01$, $p < 0.05$, and $p < 0.01$, respectively) with the preoperatuive arterial oxygen tensions. The myocardial GPx activities in the TOF patient were fivefold less ($P < 0.01$) than those in adult patients having elective CABG surgery. No dramatic differences in SOD and catalase activities were found. This research provides putative evidence that the chronically cyanotic myocardium is at greater risk of oxygen-derived free-radical injury than the noncyanotic myocardium and may be at higher risk for cardiovascular surgery.

Antioxidant Therapy

Unfortunately, intravenous antioxidant therapy given during cardiac surgery is unlikely to protect the myocardial intracellular organelles, such as mitochondria, sarcoplasmic reticulum, and sarcolemma. SOD, catalase, and GPx molecules are too large to pass through an intact nonpermeabilized sarcolemma. Oral administration of lipid-soluble antioxidants will take too long to diffuse through the sarcolemma to reach the sarcoplasmic reticulum and mitochondria to provide antioxidant protection for the urgent cardiac surgical patient. Therefore, intravenous antioxidant therapy is unlikely to affect postoperative cardiac contractility. The cyanotic TOF myocardium, with its decreased antioxidant enzyme activities, is susceptible to intracellular oxyradical damage. A therapeutic approach suitable for the surgical TOF patient is (1) to maintain arterial oxygen tension during cardiopulmonary bypass close to 100 mmHg (the high oxygen tensions of 150 mmHg or greater that commonly occur when the patient is on cardiopulmonary bypass must be avoided); (2) to administer α-tocopherol for 14 consecutive days before elective surgery (see CABG section); and (3) to increase the intracellular myocardial SOD and GPx levels by passive diffusion of the enzymes from cold blood cardioplegia before reperfusion of the ischemic myocardium.

A technique that can increase the intracellular myocardial SOD and GPx levels should protect intracellularly the postoperative TOF myocardium from oxyradical reperfusion damage. In a preliminary study of uncomplicated aortocoronary bypass surgery, cold cardioplegic arrest appeared to permeabilize the ischemic myocardium to proteins. Human ventricular myosin light-chain 1 (HVLC1) and creatine kinase MB isozyme (CK-MB) were measured in early reperfusion after elective CABG surgery. The release of either protein in routine CABG surgery is specific for the myocardium. Since none of the elective CABG patients studied ($N = 20$) had any complications before, during, or after surgery, it is unlikely that any or sufficient myocardial necrosis would occur to increase the blood levels of these proteins during cold cardioplegic ischemic arrest or reperfusion. However, both HVLC1 and CK-MB blood levels increased almost immediately with reperfusion. HVLC1 was released into the circulation in all the CABG patients at levels similar to those found in patients having a large myocardial infarction. The CK-MB release was less marked in the same patients. The early increase in blood CK-MB and HVLC1 levels with the onset of reperfusion is consistent with these proteins being washed out of a reversibly permeabilized myocardium. In an earlier CABG study that serially monitored CK-MB release (16), we showed that the blood CK-MB level was increased with the first measurement, which was taken 20 min after cross-clamp removal, peaked (four times the upper limit of normal) at 60 min of reperfusion, and returned to normal at 8 h. This time course of CK-MB release is quite different from that of a myocardial infarction, with the blood CK-MB level increasing at 3.5 h and peaking between 8 and 12 h after the onset of chest pain. HVLC1 (MW 22 kD) and CK-MB (MW 82 kD) were chosen because both proteins are found inside myocytes in soluble pools that can readily be released into the circulation with increased myocyte permeability. Low-molecular-weight proteins (HVLC1) are released more readily than high-molecular-weight proteins (CK-MB). For example, 50% patients with unstable angina release HVLC1 but not CK-MB (17,18), which is consistent with reversible ischemia permeabilizing the sarcolemma to permit leakage of the soluble small molecular weight proteins into the circulation. Berger and Johnson (19) showed that cold shock with buffers can permeabilize eukaryotic cells, permitting exogenous compounds to enter viable cells for normal metabolism when the cell's temperature has returned to normal. If the cold cardioplegia increased the permeability of the ischemic myocardium to release proteins of molecular weight greater than 22 kD, but usually less than 82 kD, it should be possible to increase the myocardial content of SOD (molecular weight 32 kD, two identical subunits of molecular weight 16,000 each), and GPx (molecular weight 84,000 kD, four identical subunits of molecular weight 21,000 each) by the addition of high concentrations of the enzymes or their subunits to the blood cardioplegia. Catalase, with its molecular weight of 240 kD, is probably too large to enter the ischemic myocardium. Beckman et al. (20), in an excellent study, showed that polyethylene glycol (PEG) conjugation to SOD and catalase decreased the susceptibility of these enzymes to proteolysis and enhanced their uptake by cultured vascular endothelial cells, which became more resistant to oxyradical-mediated injury. The mechanism of normothermic endothelial cell augmentation of SOD-PEG (molecular weight approximately 100 kD) and catalase-PEG (molecular weight > 240 kD) is unknown. Beckman et al. proposed that the endothelial cell normothermic nonischemic uptake of the polyethylene glycolated enzymes was both by membrane binding and by endocytosis. An advantage to the cardiac surgeon of cold blood cardioplegia permeabilization is that cold ischemic arrest should increase the sarcolemmal permeability to large-molecular-weight proteins by a nonendocytosis mechanism. The addition of native enzyme and/or their subunits to cold blood cardioplegia should increase their myocardial intracellular content during the ischemic arrest period. Little breakdown of the native enzymes should occur in the cardioplegic solution because of its cold temperature. An increase in the myocyte intracellular content of native SOD and GPx by a nonendocytosis mechanism results in antioxidant enzymes in the cytosol that are not trapped within endosomes or lyososomes. The intracellular enzymes may have half-lives between 3 and 7 days. If the half-lives are too short because of intracellular proteolysis, I will study the polyethylene glycolated forms of the native enzymes and their subunits. The myocardium will be able to physiologically adjust its antioxidant enzymes levels within 7 days to physiological arterial oxygen tensions, with minimal postoperative intracellular free-radical damage.

CORONARY ARTERY BYPASS GRAFT PATIENTS

Although cold blood cardioplegia provides excellent myocardial protection for elective coronary bypass patients, myocardial metabolic recovery is delayed postoperatively. To determine whether oxyradical injury occurs during aortocoronary bypass surgery of stable angina patients, we measured blood phospholipid-conjugated dienes in the arterial and coronary sinus blood (21). We found peak releases of conjugated dienes at 3 and 60 min after reperfusion. These times of release are consistent with ischemia-reperfusion injury (22) and myocardial neutrophil accumulation (8). For optimal antioxidant therapy, the myocardial subcellular organelle most affected by ischemia-reperfusion injury must be determined. We quantitated at 10 min of reperfusion organelle phospholipid-hydroxyconjugated diene formation after 45 min of normothermic reversible global ischemia in the canine myocardium (23). Our results showed that the sarcolemma was the organelle most injured by oxyradicals. Both the sarcoplasmic reticulum and the mitochondrial membranes were less peroxidized. It is important to note that these studies were done with noncyanotic animals that had normal myocardial levels of antioxidant enzymes. Since the sarcolemma is a lipid membrane, a lipid-soluble chain-breaking antioxidant, such as α-tocopherol, should provide the myocardium with optimal protection against oxyradical injury.

Antioxidants

To test the efficacy of orally administered α-tocopherol in myocardial ischemia-reperfusion injury, we (24,25) did a clinical study in 24 stable angina patients having aortocoronary bypass surgery. Our objectives were to determine (1) whether orally administered α-tocopheryl acetate was effective in increasing myocardial α-tocopherol levels, (2) the effect of cardioplegic arrest followed by reperfusion on the myocardial α-tocopherol levels, and (3) the effect of doubling the α-tocopherol levels on early postoperative myocardial metabolism and function. We showed that at least 300 mg α-tocopherol must be taken orally for 14 consecutive days to double the myocardial α-tocopherol levels. Since the myocardial α-tocopherol levels did not further increase despite preoperative treatment of 900 mg for 14 consecutive days, oral administration to further increase the myocardial levels may be impractical. There was a decrease in myocardial α-tocopherol levels with the onset of reperfusion (cross-clamp removal). The α-tocopherol levels had returned to control levels when the next myocardial biopsy was taken at 20 min of reperfusion. When the myocardium was on bypass and paced 25 min after cross-clamp removal, myocardial lactate extraction decreased in the control patients compared to the α-tocopherol-treated patients. When bypass was discontinued and the heart paced again, myocardial lactate extraction was the same for the control and treated patients. Postoperative myocardial performance was the same for both groups, except for diastolic compliance, which was increased at 4 h postoperatively in the α-tocopherol patients. The disadvantage of studying elective CABG patients with stable angina is that they are surgically low-risk patients. More than 24 patients needed to be studied to determine whether administration of oral α-tocopherol provided additional metabolic and functional benefits.

The clinical disadvantage of oral administration of α-tocopherol is the time required to increase myocardial levels. Patients who need emergency cardiac surgery cannot be treated with oral α-tocopherol. This major limitation is avoided through the use of water-soluble analogs of α-tocopherol, which can be given safely intravenously.

For effective water-soluble antioxidant therapy in TOF and CABG cardiac surgery, the protective effect of water-soluble antioxidants against oxyradical injury must be understood in vitro before being studied in vivo. The antioxidant effectiveness of SOD, catalase, ascorbic acid, and Trolox (6-hydroxy-2,5,7,8-tetramethylchroman-2-carboxylic acid), a water-soluble analog of α-tocopherol, was assessed in protecting cultured human ventricular myocytes and fibroblasts and saphenous vein endothelial cells from oxyradical injury (26). The cells were cultured at oxygen tension of 150 mmHg. Passage P1–4 cells were injured by a hypoxanthine/xanthine oxidase free-radical generation system. Fibroblasts were more resistant to free-radical injury than myocytes, which were more resistant than endothelial cells. Trolox and ascorbic acid were effective antioxidants for myocytes; SOD and catalase were ineffective. SOD and catalase were more effective than ascorbic acid as antioxidants for endothelial cells and fibroblasts, but Trolox was ineffective. In summary, each cultured cell type has a different susceptibility to free-radical damage, and extracellular water-soluble antioxidants were not effective for all cell types. Therefore, if the in vitro cell culture results are true for the in vivo situation with CABG and TOF patients having cardiac surgery, intravenous therapeutic regimes will need to be "tailormade" to prevent oxyradical damage during surgery and for the first 7 postoperative days. Based on the antioxidants studied, optimal antioxidant intravenous therapy to protect the reperfused ischemic myocardium against extracellularly generated oxyradicals would be a combination of Trolox, SOD, and catalase. Trolox will protect the myocytes and SOD with catalase will protect the endothelial cells against oxyradical damage to the plasma membranes. My preliminary cell culture data indicate that the Trolox blood levels should be between 1.5 and 2.0 mM, and SOD and catalase need to be about 100,000 IU/L each to provide optimal therapeutic effectiveness.

We tested the efficacy of intravenous therapy with Trolox and ascorbic acid in an animal ischemia-reperfusion model intended to simulate a patient with an acute myocardial infarction (27). Our pharmacological dose-response cell culture studies showed that Trolox at concentrations of 1.5 mM should be effective. In preincubation cell culture studies, we also showed that the antioxidant efficacy of Trolox was enhanced when it was incorporated into the myocyte. In a canine model of 2 h of occlusion of the left anterior descending coronary artery followed by 4 h of reperfusion, an intraarterial injection of Trolox and ascorbic acid reduced the area of infarction to the area of risk from $30.4 \pm 5.1\%$ in the controls to $8.7 \pm 4.0\%$ in the treated animals. There were no statistical differences in collateral blood flow in either group. Our in vivo research was independently confirmed by Lee et al. (28), who showed that Trolox, not ascorbic acid, was the effective antioxidant in reducing myocardial necrosis and that SOD and catalase were ineffective antioxidants. In a global ischemia-reperfusion liver model, we showed that intravenous Trolox prevented the formation of hepatic phospholipid-conjugated dienes with reperfusion and decreased liver necrosis 88% (29). The clinical disadvantages of Trolox as antioxidant in cardiac surgery are that it appears to protect only the myocytes, not the vascular endothelial cells, from oxyradical injury and that millimolar levels must be obtained for in vivo efficacy.

FUTURE ANTIOXIDANTS

Research on many fronts is actively proceeding to develop water-soluble injectable analogs of α-tocopherol that are potent in the micromolar range and can protect all myocardial cell types from oxyradical damage. Encouraging results are now appearing in the literature (30).

REFERENCES

1. Menasche P, Pasquier C, Bellucci S, Lorente P, Jaillon P, Piwnica A. Deferoxamine reduces neutrophil-mediated free radical production during cardiopulmonary bypass in man. J Thorac Cardiovasc Surg 1988; 96:582-9.

2. Del Nido PJ, Mickle DAG, Wilson GJ, et al. Evidence of myocardial free radical injury during elective repair of tetralogy of Fallot. Circulation 1987; 76(Suppl. V):174-9.

3. Royston D, Fleming JS, Desai JB, Westaby S, Taylor KM. Increased production of peroxidation products associated with cardiac operations. J Thorac Cardiovasc Surg 1986; 91:759-66.

4. Cavarocchi NC, England MD, Schaff HV, et al. Oxygen free radical generation during cardiopulmonary bypass: correlation with complement activation. Circulation 1986; 74(Suppl. III):130-3.

5. England MD, Cavarocchi NC, O'Brien JF, et al. Influence of antioxidants (mannitol and allopurinol) on oxygen free radical generation during and after cardiopulmonary bypass. Circulation 1986; 74(Suppl. III):134-7.

6. Dinerman JL, Mehta JL. Endothelial, platelet and leukocyte interactions in ischemic heart disease—insights into potential mechanisms and their clinical relevance. J Am Coll Cardiol 1990; 16:207-22.

7. Ratliff NB, Young GW, Jackel DB, Mikat E, Wilson JW. Pulmonary injury secondary to extracorporeal circulation. An ultrastructural study. J Thorac Cardiovasc Surg 1973; 65:425-32.

8. Teoh KH, Christakis GT, Weisel RD, et al. Dipyridamole reduced myocardial platelet and leukocyte deposition following ischemia and cardioplegia. J Surg Res 1987; 42(6):642-52.

9. Man RY, Choi PC. Lysophosphatidylcholine causes cardiac arrhythmias. J Mol Cell Cardiol 1982; 14(3):173-5.

10. Meerson FZ, Belkina LM, Sazontova TG, Salykova VA, Archipenko YV. The role of lipid peroxidation in the pathogenesis of arrhythmias and prevention of cardiac fibrillation with antioxidants. Basic Res Cardiol 1987; 82:123-7.

11. Hearse DJ, Tosaki A. Reperfusion-induced arrhythmias and free radicals: studies in the rat heart with DMPO. J Cardiovasc Pharmacol 1987; 9:641-50.

12. Burrows FA, Williams WG, Teoh KH, et al. Myocardial performance after repair of congenital defects in infants and children. J Thorac Cardiovasc Surg 1988; 96:548-56.

13. Del Nido PJ, Mickle DAG, Wilson GJ, et al. Inadequate myocardial protection with cold cardioplegic arrest during repair of tetralogy of Fallot. J Thorac Cardiovasc Surg 1988; 95(2):223-9.

14. Li R-K, Mickle DAG, Weisel RD, et al. Effect of oxygen tension on the anti-oxidant enzyme activities of tetralogy of Fallot ventricular myocytes. J Mol Cell Cardiol 1989; 21:567-75.

15. Teoh KH, Mickle DAG, Weisel RD, et al. Effect of oxygen tension on the myocardial antioxidant enzyme activities in tetralogy of Fallot and adult aortocoronary bypass patients. J Thorac Cardiovasc Surg, 1992.

16. Weisel RD, Goldman BS, Lipton IH, Teasdale S, Mickle DAG, Baird RJ. Optimal myocardial protection. Surgery 1978; 84(6):812-21.

17. Katus HA, Yasuda T, Gold H, et al. Diagnosis of acute myocardial infarction by detection of circulating cardiac myosin light chains. Am J Cardiol 1984; 54:964-70.

18. Katus HA, Diederich KW, Hoberg E, Kubler W. Circulating cardiac myosin light chains in patients with angina at rest: identification of a high risk subgroup. J Am Coll Cardiol 1988; 11:487-93.

19. Berger NA, Johnson ES. DNA synthesis in permeabilized mouse L cells. Biochim Biophys Acta 1976; 425:1-17.

20. Beckman JS, Minor RL Jr, White CW, Repine JE, Rosen GM, Freeman BA. Superoxide dismutase and catalase conjugated to polyethylene glycol increases endothelial cell activity and antioxidant resistance. J Biol Chem 1988; 263:6884-92.

21. Weisel RD, Mickle DAG, Finkle DC, et al. Myocardial free radical injury following cardioplegia. Circulation 1989; 80(Suppl. III):14-8.

22. Romaschin AD, Rebyka I, Wilson GJ, Mickle DAG. Conjugated dienes in ischemia and reperfused myocardium: an in vivo chemical signature of oxygen free radical mediated injury. J Mol Cell Cardiol 1987; 19:289-302.

23. Romaschin AD, Wilson GJ, Thomas U, Feitler DA, Tumiati LC, Mickle DAG. Subcellular distribution of peroxidized lipids in myocardial ischemia-reperfusion injury. Am J Physiol 1990; 28:H116-23.

24. Mickle DAG, Weisel RD, Burton GW, Ingold KU. Effect of orally administered alpha-tocopheryl acetate on human myocardial alpha-tocopherol levels. Cardiovasc Drugs Ther 1991; 5:309-12.

25. Yau TM, Weisel RD, Mickle DAG, et al. Vitamin E reduced perioperative ischemic injury. Circulation, in press 1991.

26. Mickle DAG, Li R-K, Weisel RD, Tumiati LC, Wu T-W. Water-soluble antioxidant specificity against free radical injury using cultured human ventricular myocytes and fibroblasts and saphenous vein endothelial cells. J Mol Cell Cardiol 1990; 22:1297-304.

27. Mickle DAG, Li R-K, Weisel RD, et al. Myocardial salvage with Trolox and ascorbic acid for an evolving infarction. Ann Thorac Surg 1989; 47:553-7.

28. Lee K, Canniff P, Hamel D, Silver P, Ezrin A. Trolox but not superoxide dismutase or ascorbic acid limits myocardial infarction in dogs. Clin Pharmacol Ther 1990; 47(2):174.

29. Wu T-W, Hashimoto N, Au J-W, Mickle DAG, Carey D. Trolox protects rat hepatocytes against oxyradical damage and the ischemic rat liver from reperfusion injury. Hepatology 1991; 13(3):575-80.

30. Petty MA, Grisar JM, Dow J, De Jong W. Effects of an alpha-tocopherol analogue on myocardial ischemia and reperfusion injury in rats. Eur J Pharmacol 1990; 179:241-2.

E. Pulmonary and Respiratory Diseases

48

Vitamin E and Cigarette Smoking-Induced Oxidative Damage

Ching K. Chow

University of Kentucky, Lexington, Kentucky

INTRODUCTION

Epidemiological and experimental evidence has implicated cigarette smoking as a significant contributing factor in the etiology of many disorders (1–7). Lung cancer and chronic emphysema are the most serious respiratory diseases attributed to smoking. Smokers have a much higher mortality rate attributable to lung cancer than nonsmokers (2). The death rate from bronchitis and emphysema in men aged 45–65 is five times higher in smokers than nonsmokers (3). Coronary thrombosis and myocardial infarction are more prevalent in smokers than in nonsmokers (4). Also, the death rate from coronary heart disease for both men and women is much higher in smokers than in nonsmokers (3). In addition, smoking is associated with higher secretion of adrenal hormones, decreased birth weight of children, and higher death rate from peptic ulcer (3,5,8).

Since nutrients are essential for all fundamental cellular processes, it is conceivable that their status may modulate the metabolism as well as the actions of chemical and physical agents in the environment, including cigarette smoke. On the other hand, the nutritional status and requirement for a nutrient may be altered following exposure to environmental agents. During the past decade, many studies dealing with the cellular effects of cigarette smoking have been reported. This chapter deals primarily with the possible role of oxidative damage and vitamin E in the biological effects of cigarette smoking, with an emphasis on information obtained from animal studies.

CIGARETTE SMOKING AND OXIDATIVE DAMAGE

Although cigarette smoking has been implicated in the causation of a variety of disorders, the underlying mechanisms involved remain poorly understood. This is partly due to the complex nature of cigarette smoke, which is a composite of numerous pollutants inhaled in high concentrations. For example, the concentration of carbon monoxide in cigarette smoke may be as high as 42,000 ppm (9). Over 1000 constituents of smoke, including nicotine, phenols, acetaldehyde, and cadmium, have been identified (10,11).

As cigarette smoke contains a large variety of compounds, it is not surprising that many of them are oxidants or prooxidants. Nitrogen dioxide, one of the major oxidant air pollutants present in photochemical smog, is found in cigarette smoke at levels of up to 250 ppm (7). Animal experiments indicate that nitrogen dioxide can cause irreversible damage to the respiratory system at concentrations as low as 1 ppm (12). Pryor and associates (13,14) have identified two different population of free radicals, one in the tar and one in the gas phase, of cigarette smoke. The principal radical in the tar phase, a quinone-hydroquinone complex, is capable of reducing molecular oxygen to superoxide radicals, which may eventually lead to the generation of hydrogen peroxide and hydroxyl radicals. The gas phase of cigarette smoke contains small oxygen- and carbon-centered radicals that are much more reactive than the tar-phase radicals. Nitrogen dioxide formed in the flame may also interact with other smoke components in smoke and generate more reactive species. Thus, cigarette smoke contains various oxidants, free radicals, and metastable products derived from radical reactions that may react with or inactivate essential cellular constituents.

In smokers, the cumulative smoking history is highly correlated with both leukocytosis and elevation of acute-phase reactants (15). This appears to reflect a smoking-induced inflammatory response of increasing accumulation of alveolar macrophages and neutrophils in the lungs (16). In addition, smoking causes an increase in oxidative metabolism of macrophages and neutrophils (17–20). The increased oxidative metabolism of phagocytes is accompanied by increased generation of reactive oxygen species, such as hydrogen peroxide, hydroxyl radicals, and superoxide radicals. Also, smokers have a higher neutrophil myeloperoxidase activity than nonsmokers (21). Wayner et al. (22) suggested that hydrogen peroxide and a myeloperoxidase product, hypochlorous acid, may be generated in biologically relevant quantities. These agents, if released extracellularly, can damage host tissue cells. Thus, smokers are subjected to oxidative stress resulting from oxidants and free radicals present in smoke, as well as reactive oxygen species generated by increased, activated phagocytes (Fig. 1).

Results obtained from in vitro studies support the view that components of cigarette smoke can cause oxidative damage. For example, incubation of sonicated rabbit alveolar

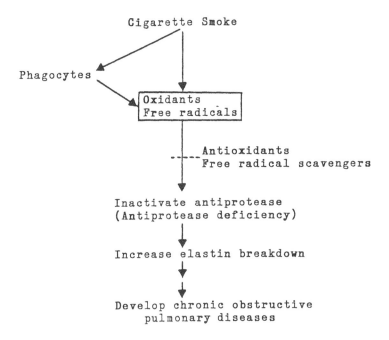

FIGURE 1 Possible role of cigarette smoking-induced oxidative damage in the development of chronic obstructive pulmonary diseases. The dashed line denotes interruption of the process or event.

macrophages and pulmonary protective factor with an aqueous extract of cigarette smoke results in increased formation of lipid peroxidation products (23). An unidentified factor in cigarette smoke has been shown to oxidize thiols (24). Also, Schwalb and Anderson (25) showed that following activation with the synthetic chemotactic tripeptide, *N*-formylmethionine-leucyl-phenylalanine, potentiated by cytochalasin B, blood neutrophils from smokers inflicted increased damage to the DNA of cocultured mononuclear leukocytes. They also found that the damage to DNA was preventable by the inclusion of superoxide dismutase and catalase, individually or in combination. These observations suggest that cigarette smoking may prime phagocytes to generate increased amounts of potentially carcinogenic reactive oxidants. Data obtained from in vivo studies is not conclusive, but oxidative damage mechanisms have been linked to the development of lung cancer, emphysema, and other smoking-related disorders. The possible association of oxidative inactivation of α_1-antiprotease (antielastase or antitrypsin) and the incidence of emphysema in smokers (Fig. 1) has received the most attention.

VITAMIN E AND CIGARETTE SMOKING

Biological Functions of Vitamin E

Vitamin E deficiency is associated with many pathological and biochemical abnormalities in several species of animals (26–28). The precise mechanism by which vitamin E prevents various deficiency symptoms remains unclear, but it is generally accepted that the primary role of vitamin E is preventing free-radical–initiated peroxidative tissue damage (29). The suggestion that vitamin E may exert its biological function in relation to its membrane localization property (30) does not conflict with the free-radical–scavenging mechanism of the antioxidant function of vitamin E.

The function of vitamin E is closely related to the status of several micronutrients, notably selenium and vitamin C (31). Several symptoms of vitamin E deficiency, such as muscular degeneration, eosinophilic enteritis, and liver necrosis in rats, are preventable by selenium (32). Also, the antioxidant function and requirement for vitamin E may be related to the status of vitamin C (31,33,34). Recent studies suggest that the antioxidant function of vitamin E in vivo may be fortified by the presence of vitamin C or glutathione, and/or an unidentified system may be involved in the restoration or regeneration of vitamin E at the radical stage before its conversion to tocopheryl quinone (35–38). However, since vitamin E is lipid soluble, and glutathione and vitamin C are water soluble, whether vitamin E is indeed regenerated or spared in vivo by vitamin C or glutathione has been questioned (39,40).

Dietary Vitamin E and Cellular Effects of Cigarette Smoking

The nutritional status of vitamin E in experimental subjects may play a role in modulating the action and metabolism of chemicals, drugs, and environmental agents. Administration of vitamin E has been shown to lessen the toxicity of a large variety of environmental agents, including ozone (41,42), oxygen (43), lead (44), paraquat (45), adriamycin (46), nitrosamine (47), nitrofurantoin (48), methyl mercury (49), and cadmium (50). Although the mechanism of such an effect is not yet clear, vitamin E may protect cellular components from the adverse effects of those compounds either via a free-radical–scavenging mechanism or as a component of cell membranes (29,30). As discussed previously, cigarette smoke contains many oxidants and free radicals that are capable of initiating oxidative damage. Also, oxidative damage may be caused by reactive oxygen species generated by activated phagocytes following smoking exposure. Thus, vitamin E may protect against the adverse effects of cigarette smoking.

The underlying mechanism involved in the biological effects of cigarette smoking is very complex. The task of understanding the underlying mechanism can be simplified somewhat by using animals as experimental subjects. However, the factors involved remain so numerous that no simple or clear conclusion can be made from a single experiment. In addition to a large number and variety of components present in cigarette smoke, species, age, sex, and nutritional status of experimental animals, as well as types of cigarette and exposure conditions employed, have made the task difficult to accomplish. For example, rats, mice, hamsters, and guinea pigs have been used as experimental animals for examining the biological effects of cigarette smoking. Animals of different age and sex have either been maintained on a commercial chow diet or fed a synthetic diet for various lengths of time. Experimental animals have been exposed to whole smoke, the gaseous phase of the smoke, or sham smoke or serve as unhandled room controls. Besides, animals have been exposed to different strengths (degree of dilution) of cigarette smoke generated from different types of smoke machines or exposed to mainstream (active) or sidestream (passive) smoke (51,52). Furthermore, animals have been exposed to a dose ranging from 1 to over 100 puffs (approximately 10 puffs/cigarette; 1 puff/min) of smoke in one or more sessions daily for up to 1 year or longer. Different types of cigarettes have very different levels of nicotine and other chemical constituents.

In studies relating to vitamin E, experimental animals have either been maintained on a commercial chow diet or fed a synthetic diet that contains a defined amount of vitamin E. A diet that contains 10 mg vitamin E per kg diet is considered marginally adequate, and over 100 mg/kg is a therapeutic or protective dose. Because of the functional interrelation between vitamin E and several other nutrients (31), the status of selenium, vitamin C, and the sulfur-containing amino acids needs to be considered as well. Partly as a result of the difficulty in deriving easily interpretable data, information concerning the influence of dietary vitamin E on cigarette smoking is rather limited. The results obtained from animal studies are summarized briefly and discussed here along with certain data obtained from human studies.

Animal Mortality, Body Weight, and Tissue Weight

Sudden exposure to a large dose of cigarette smoke can be fatal. In an acute study, 5 of 16 vitamin E-deficient rats died within 24 h following exposure to smoke generated from reference cigarette 2R1 (high nicotine) for 120 puffs. This compares to 1 of 13 in the vitamin-supplemented group (53). However, the cause of the higher mortality in the vitamin E-deficient group was not established. No animal mortality occurred in either dietary group when animals were exposed to 60 puffs or less daily for 7 days (53,54). If a break-in or adjustment period (e.g., less than 5 puffs of diluted smoke daily for 3–7 days) is allowed, animal mortality due to exposure to higher doses of cigarette smoke and/or for a prolonged period of time can be avoided or minimized. Graziano et al. (55), for example, observed no mortality in male C57BL mice exposed to 20 puffs of smoke generated from reference cigarette 2A1 (low nicotine) daily for 8 weeks while maintained on a diet containing 0, 5, or 100 ppm (mg/kg) vitamin E. Also, all animals maintained on a chow diet survived when they were exposed to mainstream or sidestream smoke daily for up to 32 weeks (56,57).

In relation to the sham group, the body weight of cigarette-smoked animals changed very little during the first few weeks but was significantly lower after 8 weeks (56,57). Similarly, the weight of lung, liver, heart, kidney, spleen, and testes was not altered during initial smoke exposure. The tissue weight was significantly lower in animals exposed to cigarette smoke continuously for 8 weeks or longer, however, reflecting a decrease in body weight (56,57). Dietary vitamin E did not have a significant effect on body or tissue weight during short-term studies (1 month or less) (53).

Blood Parameters

The lung is vulnerable to injury from a variety of agents, including cigarette smoke, because its airways and alveolar surface are directly exposed to the external environment. Animals exposed to either whole smoke or the gaseous phase of smoke exhibited a marked increase in blood levels of carboxyhemoglobin measured immediately after each exposure. The levels of carboxyhemoglobin are directly related to the dose of smoke inhaled and are not affected by the nutritional status of vitamin E (53). Also, smoked animals have higher red blood cell counts, mean cell volume, hematocrit, and hemoglobin levels. The magnitude of the alteration was more profound in the vitamin E-deficient animals than in the supplemented group.

Compared to the supplemented group, the activities of pyruvate kinase, lactate dehydrogenase, glutamate oxaloacetate transaminase, alkaline phosphatase, and creatine phosphatase are significantly higher in the plasma of vitamin E-deficient rats (58). The activities of these enzymes in plasma are not affected significantly by cigarette smoking (53,54). Thus, despite the potential increase in oxidative stress, short-term cigarette smoking does not enhance or aggravate vitamin E deficiency.

Vitamin E and Vitamin C

Since the oxidative damage mechanism may be involved in the toxicity of cigarette smoking, the nutritional status of vitamin E may play a role in mediating the development of smoking-related disorders. Also, increased utilization of vitamin E may be associated with cigarette smoking. Pacht et al. (59) showed that the levels of vitamin E were significantly lower in alveolar fluid of smokers than in nonsmokers. They suggested that smokers had a faster rate of vitamin E utilization and that smoking may predispose them to enhanced oxidant attack on their lung parenchymal cells. However, it is not clear why nonsmokers had a higher content of tocopheryl quinone in bronchoalveolar lavage fluid than smokers and why smokers had higher levels of vitamin E in alveolar macrophages than nonsmokers (59).

Efforts to correlate the status of vitamin E with the incidence of smoking-related respiratory diseases have resulted in largely negative findings. There is no significant correlation between plasma vitamin E levels and indices of smoking effects, such as pulmonary function abnormalities and cytogenetic changes (60,61). Also, there is no significant difference in plasma vitamin E levels between smokers and nonsmokers (59,62). In addition, cigarette smoking did not significantly alter the plasma levels of vitamin E in either vitamin E-deficient or supplemented rats (53). Besides, vitamin E and selenium deficiencies do not enhance lung inflammation from cigarette smoke in the hamster (63).

Since the respiratory system is the primary target of cigarette smoke, the vitamin E content of pulmonary tissue is likely to be adversely affected. Contrary to this expectation, however, the levels of vitamin E in the lungs of chronically smoked animals are higher, rather than lower, than in the controls. For example, relative to the sham group, the vitamin E concentration was increased over threefold in the lungs of guinea pigs exposed to mainstream smoke and over twofold in the group exposed to sidestream smoke for 17 or 20 weeks (57). The levels of vitamin E in the plasma, liver, and kidney were not significantly different between the treatment groups. Similarly, the levels of vitamin E were significantly increased in the lungs, but not in the plasma, liver, or kidney, of rats exposed to mainstream smoke for 8, 16, 24, or 32 weeks (56). The increased levels of vitamin E observed in the lungs of chronically smoked guinea pigs and rats suggest an adaptive response that may protect pulmonary tissues against oxidative damage. Since those experimental animals were fed a diet containing an adequate amount of vitamin E, it is possible that higher levels of vitamin E found in the lungs of smoked animals may result from mobilization of body stores. The ability of pulmonary tissue to increase vitamin E during cigarette smoking may explain some of the conflicting results obtained from animal and human studies. Similarly, increased levels of vitamin E have been found in the blood of human subjects following exercise (64) and in the lungs of ozone-exposed animals (65).

It has long been established that human smokers have lower blood vitamin C levels and decreased urinary excretion of vitamin C (66,67). Both increased utilization and decreased intake and bioavailability of vitamin C in smokers may be responsible. Similar to the human studies, the plasma levels of ascorbic acid were lower in cigarette-smoked rats than in the sham group, and the degree of decline was relatively greater in animals fed a vitamin E-deficient diet than in the supplemented group (53,68). Vitamin E supplement appears to aid in maintaining higher levels of plasma ascorbic acid in smoked animals. On the other hand, the levels of ascorbic acid were significantly increased in the lungs of animals on vitamin E-deficient diet, but not in supplemented animals, following smoking exposure. It appears that increased amounts of ascorbic acid may be synthesized by smoked rats to meet an increased need. A stimulation of hepatic ascorbic acid biosynthesis in rats and mice has been reported following exposure to various xenobiotics (69,70). Guinea pigs, like humans, cannot synthesize ascorbic acid. In a study using guinea pigs as experimental animals, the levels of vitamin C in plasma, livers, kidneys, and lungs were not different significantly among cigarette-smoked, sham-smoked, or room control groups (57). Since the guinea pigs employed were fed a nutritionally adequate diet, sufficient ascorbic acid was probably available to replenish the vitamin consumed. It would be of interest to investigate the consequence of smoking exposure for guinea pigs or primates maintained on a diet deficient or inadequate in vitamin C.

Lipid Peroxidation and the Glutathione Peroxidase System

As discussed earlier, smokers are being subjected to oxidative stress directly from oxidants and free radicals present in cigarette smoke (12–14) and indirectly from reactive oxygen species generated by activated phagocytes (17–21). However, experimental evidence for increased oxidative damage, such as formation of oxidation products, in smoked subjects is not conclusive. On the one hand, Hoshino et al. (71) showed that smokers had significantly higher outputs of breath pentane than nonsmokers and that supplement with vitamin E (800 mg/day for 2 weeks) decreased exhaled pentane in smokers. Richards et al. (20) measured the levels of chemoluminescence as an indicator for the release of reactive oxidants and found that cigarette smoking was associated with elevated intracellular, and especially extracellular, chemiluminescence responses. Also, Duthie et al. (72) observed that the levels of conjugated dienes in plasma, but not malonaldehyde, were significantly higher in the plasma of smokers than nonsmokers. Furthermore, Kiyosawa et al. (100) found increased formation of 8-hydroxydeoxyguanosine, one of the oxidation products of DNA damage, in human peripheral leukocytes. On the other hand, smokers have lower levels of lipid hydroperoxides in bronchoalveolar lavage fluid than nonsmokers (73) and the production of oxygen free radicals by neutrophils is decreased by smoke components nicotine and cotinine (74). Also, the levels of lipid peroxidation products, thiobarbituric acid reactants (TBARs), in freshly isolated low-density lipoproteins (LDL) are no different in smokers and nonsmokers (101). However, the levels of TBARs are higher in the oxidized LDL of smokers than in nonsmokers; possibly reflecting lower levels of vitamin E in LDL of smokers (101).

Whether cigarette smoking causes enhanced lipid peroxidation in pulmonary tissue has been examined in experimental animals. The levels of TBARs were either decreased or unchanged, rather than increased, in the lungs of cigarette-smoked rats compared with those of the sham smoke-exposed animals (53). Similar results were observed whether rats were fed a vitamin E-deficient or supplemented diet (53). Thus, despite the possible attack by many free radicals and oxidants, increased formation of lipid peroxidation products did not seem to occur in the lungs of smoked animals. The apparent lack of increased formation of lipid peroxidation products in the lungs of smoked rats can be attributed partly to the increased levels of vitamin E in the lungs of these animals (56,57). Similarly, Duthie et al. (72) found that plasma levels of malonaldehyde were not different between smokers and nonsmokers, although the levels of conjugated dienes were higher in smokers. More studies, such as those employing experimental animals that are maintained on a vitamin E-deficient or marginally adequate diet and using specific and sensitive methods to measure products of lipid peroxidation, are needed to have a better understanding of the smoking-induced mobilization of vitamin E.

The enzyme glutathione peroxidase, which utilizes the reducing equivalent of reduced glutathione (GSH), is an important system responsible for the reduction of lipid hydroperoxides that may be formed (Fig. 2). The levels of GSH and activities of GSH peroxidase and metabolically related enzymes are either higher or unchanged in lungs of cigarette-smoked rats, depending upon the exposure conditions (53,75). The magnitude of the changes is in

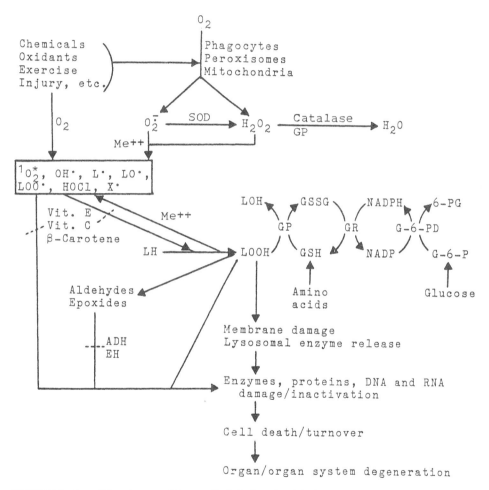

FIGURE 2   Possible scheme of free-radical–induced oxidative damage and antioxidant defense. LH, membrane or polyunsaturated lipids; LOOH, lipid hydroperoxides; LOH, hydroxy acid; LOO$^\cdot$, peroxyl radical; L$^\cdot$, alkyl radical; LO$^\cdot$, alkoxy radical; OH$^\cdot$, hydroxyl radical; O$_2^{\bar{\cdot}}$, superoxide radical; X$^\cdot$, other radicals; ^{1}O$_2^*$, singlet oxygen; H$_2$O$_2$, hydrogen peroxide; SOD, superoxide dismutase; GSH, reduced glutathione; GSSG, oxidized glutathione; GP, glutathione peroxidase or phospholipid hydroperoxide GSH peroxidase, GR, GSSG reductase; G-6-PD, glucose-6-phosphate dehydrogenase; 6-PG, 6-phosphogluconate; ADH, aldehyde dehydrogenases; EH, epoxide hydralase; vit., vitamin; NADPH or NADP, reduced or oxidized nicotinamide adenine dinucleotide phosphate; Me^{++}, metal catalyst. The dashed line denotes interruption of the process or event.

general greater in the lungs of animals fed a vitamin E-deficient diet than in those of the supplemented group (53). The activity of glucose-6-phosphate dehydrogenase is increased in the lungs of cigarette-smoked rats fed a vitamin E-deficient diet but not in those fed the supplemented diet. The activity of GSH reductase is not significantly altered by cigarette smoking in either dietary group.

In addition to being the source of reducing equivalents for GSH peroxidase, GSH plays an important role in detoxification of xenobiotics (102 and 103) and may be involved in the regeneration of vitamin E (37,38). The levels of GSH were increased in the lungs of rats exposed to cigarette smoking for 7 days, and the differences were more profound in vitamin E-deficient rats than the supplemented group (53,54). On the other hand, Cotgreave et al. (104) showed that acute cigarette smoke inhalation for 1 h caused significant depletion of GSH in the lungs, lavaged cells, and lavage fluid of rats. They also showed that acute cigarette smoke inhalation had no effect on the blood or hepatic GSH redox balance and that the levels of total cysteine were unaffected in the lungs, increased in liver, and decreased in plasma. These results suggest that acute cigarette smoke exposure causes a transient depletion of GSH and that the lungs may respond to the stress by increasing cysteine uptake.

McCuster et al. (76) showed that the activities of superoxide dismutase and catalase from alveolar macrophages of smokers and smoke-exposed hamsters were twice that found in control subjects, but there was no change in the activity of GSH peroxidase. They also found that smoked hamsters had prolonged survival in normobaric hyperoxia ($>95\%$ O_2). Thus it appears that the adaptive response of augmented antioxidant activity may serve as a mechanism to limit oxidant-mediated damage to alveolar structure.

Microsomal Enzymes

Aryl hydrocarbon hydroxylase is the most thoroughly studied microsomal monooxygenase and is considered a good indicator of cytochrome P_{450}-mediated metabolism. Induction of this mixed-function oxidase is a biochemical change that occurs in the lung and other tissues of animals and humans as a direct response to cigarette smoking (77,78). The activity of aryl hydrocarbon hydroxylase was increased significantly in the lungs of cigarette-smoking rats fed vitamin E-deficient or supplemented diet, although the increase was significantly less in animals maintained on the deficient diet (55). Also, the activity of this enzyme increased significantly in the kidney and liver of smoked rats.

GSH-S-transferases are important enzymes that catalyze the metabolism of a diversified group of foreign compounds, including carcinogens (79). They are also known to bind certain drugs nonenzymatically, both reversibly and covalently, and thereby decrease the drugs' potential toxicity (80). The activity of GSH-S-transferase increased significantly in the lungs of rats exposed to cigarette smoke for 7 days and maintained on a vitamin E-deficient diet but not in the supplemented group (53,54). The enzyme activity in the kidney was not significantly altered by smoking in either dietary group of animals. Exposure of mice to cigarette smoke for 8 weeks did not significantly alter hepatic activities of GSH-S-transferase, uridine 5'-diphosphoglucuronyltransferase, parathione desulfurase, parathione hydralase, or aryl hydrocarbon hydroxylase in smoked animals fed either a vitamin E-deficient or supplemented diet (55).

α_1-Antielastase

Emphysema is a disease characterized by specific alterations in physiological lung function related to the loss of elastic recoil. The findings that emphysema patients have reduced levels of α_1-antiprotease (81) and that emphysema can be induced by proteases from neutrophil granulocytes in animal models (82) led to the hypothesis that emphysema results from degradation of lung elastin (Fig. 1). α_1-Antiprotease, the major plasma protease inhibitor of elastase, is important in protecting the lung from proteolytic damage, particularly from the elastase of neutrophils. Rodriguez et al. (83) showed that alveolar macrophages in broncho-alveolar lavages obtained from smokers produced substantial amounts of elastase, whereas those from nonsmokers did not. Janoff et al. (84) showed that lung lavage collected from cigarette-smoked rats shows a 39% reduction in elastase inhibitory capacity. They also showed that the loss of elastase inhibitory activity could be reversed by exposure of the lavage fluid to a reducing agent, thereby supporting the oxidant mechanism of cigarette smoke in vivo. Thus, decreased α_1-antiprotease activity caused by oxidants or free radicals derived from cigarette smoke or activated phagocytes may play a role in the pathogenesis of obstructive lung disease in smokers (85,86). Cox and Billingsley (87), however, did not find a difference in the plasma inhibitory activity of α_1-antiprotease between smokers and nonsmokers. They suggest that sufficient antioxidant in the plasma may prevent the detection of oxidized in-activated α_1-antiprotease. Also, cigarette smoking did not significantly alter the elastase in-hibitory capacity in the plasma and lungs of rats with or without vitamin E in the diet (88). Thus, the role of smoking-induced oxidative inactivation of α_1-antiprotease requires further investigation.

Inflammation

The development of pulmonary disease may be related to cell populations in the broncho-alveolar lavage of the patients (89,90). Cigarette smoking is capable of inducing an inflam-matory response, as evidenced by the pulmonary accumulation of macrophages and neutro-phils (16). Smokers have a greater number of lavagable cells than nonsmokers and exhibit significant higher proportions of polymorphonuclear neutrophils and lymphocytes in their bronchoalveolar lavage cells (88,91). Alveolar macrophages, which account for most of the increase in lavagable cells from smokers, have the ability to release increased amounts of superoxide radicals (17) and have higher levels of lysosomal enzymes (92,93). An increase in free cell population in bronchoalveolar lavage has also been found in cigarette-smoked animals. However, there is a species-dependent response to cigarette smoke. For example, an onset of inflammatory cell response in smoked mice occurs within 6 weeks, compared to none in rats, suggesting a greater susceptibility of mice to cigarette smoke (94). A more pro-nounced murine macrophage response to cigarette smoke was also demonstrated by morpho-metric analysis of lung tissue (95). On the other hand, increased numbers of phagocytes were recovered in the lavage fluid of smoked Syrian hamsters (63).

Oxidative damage from cigarette smoke may be mediated indirectly through inflam-matory reactions. Chronic inflammation is accompanied by the migration of successive waves of neutrophils into host tissues (22). Localized generation and extracellular release of super-oxide, hydrogen peroxide, and hypochlorous acid may overwhelm antioxidant defense and lead to the potential oxidative injury to tissue cells. Hypochlorous acid may cause tissue dam-age indirectly by potentiating the extracellular proteolytic activity of neutrophil-derived pro-tease, elastin, collagenase, and gelatinase, which attack key components of the extracellular matrix (22). Although activated neutrophils generate reactive oxygen species, oxidative damage is usually prevented under normal conditions. Also, inflammatory responses are usually self-limiting, and various biological antioxidant mechanisms (Fig. 2) can neutralize the actions of oxidants and free radicals (31).

If cigarette smoke-induced inflammation and damage is oxidant mediated, then cellular antioxidant defense systems may play a protective role. Decreased plasma and leukocyte con-centrations of vitamin E and vitamin C are associated with increased numbers and activity of neutrophils in smokers (60,96,97). This suggests an increased consumption of these micro-nutrients during neutralization of phagocyte-derived extracellular oxidants (98). Richards et al. (20) observed that pulmonary dystrophy in smokers was highly correlated with the extent of generation of oxygen species by activated neutrophils. Plasma levels of vitamin E, vitamin C, and β-carotene, however, are not correlated with the release of reactive oxidants from circulating phagocyte or spirometric abnormalities in cigarette smokers (60). After exposing vitamin E- and selenium-deficient Syrian hamsters to cigarette smoke for 8 weeks, Niewoehner et al. (63) did not observe enhanced inflammation. Compared to supplemented animals, no alterations were found in the histological appearance of smoke-induced inflammatory lesions, in the number of phagocytes recruited, or in the oxidative metabolism of these phagocytes. They suggested that vitamin E and selenium are unimportant in protecting against cigarette smoke-induced lung injury.

DOES CIGARETTE SMOKING ENHANCE OR CAUSE OXIDATIVE DAMAGE?

Cigarette smoke contains many oxidants and free radicals (13,14). Also, increased reactive oxygen species may be generated by activated phagocytes following cigarette smoking (15–20). The oxidative damage mechanism has been implicated to play a role in the pathogenesis of smoking-associated disorders. Results obtained from in vitro studies (23,76) generally support the view that increased oxidative damage or lipid peroxidation is a consequence of cigarette smoking. However, data obtained from in vivo studies are not conclusive. Cigarette smoking did not cause increased formation of lipid peroxidation products in the lungs of vitamin E-deficient or supplemented rats (53). Nor did cigarette smoking alter the activities of pyruvate kinase, lactate dehydrogenase, glutamate oxaloacetate transaminase, or glutamate pyruvate transaminase in the plasma of vitamin E-deficient or supplemented rats (53). These findings suggest that the smoking did not potentiate the severity of vitamin E deficiency or enhnace oxidative stress under the experimental conditions. Similarly, Niewoehner et al. (63) showed that dietary vitamin E and selenium deficiency did not significantly alter the inflammatory response in smoked hamsters.

However, the failure to detect increased formation of lipid peroxidation products in smoked subjects does not indicate lipid peroxidation or oxidative damage did not occur. An adaptive response may explain this seemingly paradoxical view. Higher levels of vitamin E in the lungs of chronically cigarette-smoked animals (56,57) suggest that vitamin E may be mobilized to the pulmonary tissue to counteract the effect of increased oxidative stress. Also, the activities of glutathione peroxidase and related enzymes are higher in the lungs of smoked rats (53,54,75). Similarly, the activities of superoxide dismutase and catalase in the alveolar macrophages of smokers or smoked hamsters are higher than in control subjects (76). Furthermore, smokers have a higher pulmonary activity of dithiodiaphorase, which catalyzes the two-electron reduction of quinones to less harmful hydroquinones, than nonsmokers (99). Thus, the enhanced antioxidant defense potential observed following smoke exposure may enable experimental subjects to resist further damage or limit oxidant-mediated damage to alveolar structures.

SUMMARY AND CONCLUSIONS

Cigarette smoke contains a large number of compounds, including many oxidants and free radicals, that are capable of initiating oxidative damage. Also, oxidative damage may result from reactive oxygen species generated by activated phagocytes following cigarette smoking. In vitro studies are generally supportive of the theory that cigarette smoke can cause or promote oxidative damage. However, information obtained from in vivo studies is not conclusive. The adaptive responses observed, such as accumulation of vitamin E and increased activities of GSH peroxidase and related enzymes in the lung and increased activities of superoxide dismutase and catalase in alveolar macrophages of smoked animals, may be responsible for the lack of evidence of lipid peroxidation or oxidative damage in studies in vivo. Because of the complexity of smoking exposure and lack of sensitive and specific methods for measuring lipid peroxidation products, the role of vitamin E and oxidative damage in the etiology of smoking-related disorders remains to be elucidated.

REFERENCES

1. U.S. Department of Public Health Service. Smoking and health: The report of the Surgeon General. Washington, DC: Government Printing Office, 1979; U.S. DHEW Publication (PHS) 79-50066.
2. Wynder EL. The epidemiology of cancer of the bronchus: facts and suppositions. Arch Otolaryngol Rhinol Laryngol 1967; 76:228-36.
3. Hammond EC. Smoking in relation to the death rates of 1 million men and women. Monograph 19, National Cancer Institute, January, 1966; 127-204.
4. Doyle JT, Dawler TR, Kannel WB, Kinch SH, Kahn HA. The relationship of cigarette smoking to coronary heart disease. JAMA 1964; 190:886-90.
5. Diehl HS. Tobacco and your health: the smoking controversy. New York: McGraw-Hill, 1969.
6. U.S. Department of Health and Human Service. Nicotine action: a report of the Surgeon General. Washington, DC: Government Printing Office, 1988; U.S. DHEW Publication (CDC) 88-8406.
7. U.S. Department of Health and Human Service. The health consequences of smoking. Chronic obstructive lung disease. A report of the Surgeon General. Washington, DC: Government Printing Office, 1984; DHHS Publication 84-50205.
8. MacMahon B, Alpert M, Salber EJ. Infant weight and parental smoking habits. 1965; Am J Epidemeol 82:247-61.
9. Abelson PH. A damaging source of air pollution. Science 1967; 158:1527.
10. Schumacher JM, Green CR, Best FW, Newell MP. Smoke composition: an extensive investigation of the water-soluble portion of cigarette smoke. J Agr Food Chem 1977; 25:310-20.
11. Sakuma H, Ohsumi T, Sugawara S. Particular phase of cellulose cigarette smoke. Agr Biol Chem 1980; 44:555-61.
12. Parkinson DR, Stephens RJ. Morphological surface changes in the terminal bronchial regions of NO_2-exposed rat lung. Environ Res 1973; 6:37-51.
13. Church DF, Pryor WA. Free radical chemistry of cigarette smoke and its toxicological implications. Environ Health Perspect 1985; 64:111-26.
14. Pryor WA, Hales BJ, Premovic PI, Church DF. The radicals in cigarette tar: their nature and suggested physiological implications. Science 1983; 220:425-7.
15. Bridges RB, Chow CK, Rehm SR. Micronutrients and immune function in smokers. Ann N Y Acad Sci 1990; 587:218-31.
16. Hunninghake CW, Crystal RG. Cigarette smoking and lung destruction: accumulation of neutrophils in the lungs of cigarette smokers. Am Rev Respir Dis 1983; 128:833-8.
17. Hoidal JR, Fox RB, LeMarbre PA, Perri R, Repine JE. Altered oxidative metabolic responses in vitro of alveolar macrophages from asymptomatic cigarette smokers. Am Rev Respir Dis 1981; 123:85-9.
18. Hoidal JR, Niewoehner DE. Cigarette-smoke-induced phagocyte recruitment and metabolic alterations in human and hamsters. Am Rev Respir Dis 1982; 126:548-52.
19. Ludwig PW, Hoidal JR. Alterations in leukocyte oxidative metabolism in cigarette smokers. Am Rev Respir Dis 1982; 126:977-80.
20. Richards GA, Theron AJ, Van der Merwe CA, Anderson R. Spirometric abnormalities in young smokers correlate with increased chemiluminescence responses of activated blood phagocytes. Am Rev Respir Dis 1989; 139:181-7.
21. Bridges RB, Fu MC, Rehm SR. Increased neutrophil myeloperoxidase activity associated with cigarette smoking. Eur J Respir Dis 1985; 67:84-93.
22. Wayner DD, Burton GW, Ingold KW, Barclay LRC, Locke SJ. The relative contributions of vitamin E, urate, ascorbate and protein to the total peroxy radical-trapping antioxidant activity of human blood plasma. Biochim Biophys Acta 1987; 924:408-19.

23. Lentz PE, Di Luzio NRD. Peroxidation of lipids in alveolar macrophages: production by aqueous extracts of cigarette smoke. Arch Environ Health 1974; 28:279-82.

24. Fenner ML, Braven J. The mechanism of carcinogenesis by tobacco smoke: further experimental evidence and a prediction from the thio-defense hypothesis. Br J Cancer 1968; 22:474-9.

25. Schwalb G, Anderson R. Increased frequency of oxidant-mediated DNA strand breaks in mononuclear leukocytes exposed to activated neutrophils from cigarette smokers. Nutr Res 1989; 225:95-9.

26. Scott ML. Studies on vitamin E and related factors in nutrition and metabolism. In: DeLuca HF, Suttie JW, eds. The fat-soluble vitamins. Madison, WI: University of Wisconsin Press, 1969; 355-68.

27. Machlin L. Vitamin E. In: Machlin L, ed. Handbook of vitamins. New York: Marcel Dekker, 1984; 99-145.

28. Chow CK. Vitamin E and blood. World Rev Nutr Dietet 1985; 45:133-66.

29. Tappel AL. Vitamin E and free radical peroxidation of lipids. Ann N Y Acad Sci 1972; 203:12-28.

30. Lucy JA. Functional and structural aspects of biological membranes: a suggested structural role of vitamin E in the control of membrane permeability and stability. Ann N Y Acad Sci 1972; 203:4-11.

31. Chow CK. Interrelationship of cellular antioxidant defense systems. In: Chow CK, ed. Cellular antioxidant defense mechanisms, Vol. 2. Boca Raton, FL: CRC Press, 1988; 217-37.

32. Hong CB, Chow CK. Induction of eosinophilic enteritis and eosinophils in rats by vitamin E and selenium deficiency. Exp Mol Pathol 1988; 48:182-92.

33. Bendich A, Machlin LJ, Scandurra O, Burton GW, Wayner DDM. The antioxidant role of vitamin C. Adv Free Radic Biol Med 1986; 2:419-44.

34. Chen LH, Chang HM. Effects of high level of vitamin C on tissue antioxidant status of guinea pigs. J Int Vitam Nutr Res 1979; 49:87-91.

35. Niki E, Tsuchiya J, Tanimura R, Kamiya Y. Regeneration of vitamin E from α-chromanoxy radical by glutathione and vitamin C. Chem Lett 1982; 789-92.

36. Packer JE, Slater TF, Willson RL. Direct observation of a free radical interaction between vitamin E and vitamin C. Nature 1979; 278:737-8.

37. Reddy RC, Scholz RW, Thomas CE, Massaro EJ. Vitamin E dependent reduced glutathione inhibition of rat liver microsomal lipid peroxidation. Life Sci 1982; 31:571-6.

38. Wefers H, Sies H. The protection by ascorbate and glutathione against microsomal lipid peroxidation is dependent on vitamin E. Eur J Biochem 1988; 174:353-7.

39. Burton GW, Ingold KU. Vitamin E: an application of the principles of physical organic chemistry to the exploration of its structure and function. Acc Chem Res 1986; 19:194-201.

40. Burton GW, Wronska U, Stone L, Foster DO, Ingold KU. Biokinetics of dietary *RRR*-α-tocopherol in the male guinea pig at three dietary levels of vitamin C and two levels of vitamin E. Evidence that vitamin C does not "spare" vitamin E in vivo. Lipids 1990; 25:199-210.

41. Chow CK. Influence of dietary vitamin E on susceptibility to ozone exposure. In: Lee SD, Mustafa MG, Mehlman MA, eds. The biochemical effects of ozone and photochemical oxidants. Princeton, NJ: Princeton Sci. Publ., 1983; 75-91.

42. Chow CK, Plopper CG, Chiu M, Dungworth DL. Dietary vitamin E and pulmonary biochemical and morphological alterations of rats exposed to 0.1 ppm ozone. Environ Res 1981; 24:315-24.

43. Mino M. Oxygen poisoning and vitamin E deficiency. J Nutr Sci Vitaminol 1973; 19:95-104.

44. Levander OA, Morris VC, Ferretti RJ. Comparative effects of selenium and vitamin E in lead-poisoned rats. J Nutr 1977; 107:378-82.

45. Block ER. Potentiation of acute paraquat toxicity by vitamin E deficiency. Lung 1979; 156:195-203.

46. Doroshow JH, Locker GY, Myers CE. Experimental animal models of adriamycin cardiotoxicity. Cancer Treat Rep 1979; 63:855-60.

47. Dashman T, Kamm JJ. Effects of high doses of vitamin E on dimethylnitrosamine hepatotoxicity and drug metabolism in the rat. J Nutr 1979; 109:1485-90.

48. Boyd MR, Catignani GL, Sasame HA, Mitchell JR, Stiko AW. Acute pulmonary injury in rats by nitrofurantoin and modification by vitamin E, dietary fat and oxygen. Am Rev Respir Dis 1979; 120:93-9.

49. Welsh SO. The protective effect of vitamin E and *N,N'*-diphenylenediamine (DPPD) against methyl mercury toxicity in the rat. J Nutr 1979; 109:1673-81.

50. Korkeala H. The effect of vitamin E on the toxicity of cadmium in cadmium-sensitive *Staphylococcus aureus*. Acta Vet Scand 1980; 21:224-8.

51. Griffith RB, Hancock R. Simultaneous mainstream-sidestream smoke exposure systems. 1. Equipment and procedure. Toxicology 1985; 34:123-38.

52. Griffith RB, Standafer S. Simultaneous mainstream-sidestream smoke exposure systems. II. The rat exposure system. Toxicology 1985; 35:13-24.

53. Chow CK. Dietary vitamin E and cellular susceptibility to cigarette smoking. Ann N Y Acad Sci 1982; 393:426-36.

54. Chow CK, Chen LH, Thacker RR, Griffith RB. Dietary vitamin E and pulmonary biochemical responses of rats to cigarette smoking. Environ Res 1984; 34:8-17.

55. Graziano MJ, Gairola C, Dorough HW. Effects of cigarette smoke and dietary vitamin E levels on selected lung and hepatic biotransformation enzymes in mice. Drug-Nutrient Interact 1985; 3:213-22.

56. Chow CK, Airriess GR, Changchit C. Increased vitamin E content in the lungs of chronic cigarette-smoked rats. Ann N Y Acad Sci 1989; 570:425-7.

57. Airriess G, Changchit C, Chen LC, Chow CK. Increased levels of vitamin E in the lungs of guinea pigs exposed to mainstream or sidestream smoke. Nutr Res 1988; 8:653-61.

58. Chow CK, Hong CB, Reese M, Gairola C. Effect of dietary vitamin E on nitrite-treated rats. Toxicol Lett 1984; 23:109-17.

59. Pacht ER, Kaseki H, Mohammed JR, Cornwell DG, Davis WR. Deficiency of vitamin E in the alveolar fluid of cigarette smokers. Influence on alveolar macrophage cytotoxicity. J Clin Invest 1988; 77:789-96.

60. Theron AJ, Richards GA, Van Rensburg AJ, Van Der Merwe CA, Anderson R. Investigation of the role of phagocytes and antioxidant nutrients in oxidant stress mediated by cigarette smoke. Int J Vitam Nutr Res 1990; 60:261-6.

61. Van Resnburg CEJ, Theron AJ, Richards GA, Van Der Merwe CA, Anderson R. Investigation of the relationships between plasma levels of ascorbate, vitamin E and β-carotene and the frequency of sister-chromatid exchanges and release of reactive oxidants by blood leukocytes from cigarette smokers. Mutat Res 1989; 215:167-72.

62. Chow CK, Thacker R, Bridges RB, Rehm SR, Humble J, Turbek J. Lower levels of vitamin C and carotenes in plasma of cigarette smokers. J Am Coll Nutr 1986; 5:305-12.

63. Niewoehner DE, Peterson FJ, Hoidal JR. Selenium and vitamin E deficiencies do not enhance lung inflammation from cigarette smoke in the hamster. Am Rev Respir Dis 1983; 127:227-30.

64. Pincemail J, Deby C, Camus G, et al. Tocopherol mobilization during intensive exercise. Eur J Appl Physiol 1988; 57:189-91.

65. Elsayed NM, Mustafa MG, Mead JF. Increased vitamin E content in the lung after ozone exposure: a possible mobilization in response to oxidative stress. Arch Biochem Biophys 1990; 282: 263-9.

66. Pelletier O. Vitamin C status of cigarette smokers and nonsmokers. Am J Clin Nutr 1970; 23:520-28.

67. Pelletier O. Smoking and vitamin C levels in humans. Am J Clin Nutr 1969; 21:1259-67.

68. Chen LH, Chow CK. Effect of cigarette smoking and dietary vitamin E on plasma levels of vitamin C in rats. Nutr Rep Int 1980; 22:301-9.

69. Boyland E, Grove PL. Stimulation of ascorbic acid synthesis and excretion by carcinogenic and other foreign compounds. Biochem J 1961; 81:163-8.

70. Conney AH, Burns JJ. Stimulatory effects of foreign compounds on ascorbic acid biosynthesis and on drug-metabolizing enzymes. Nature 1959; 184:363-4.

71. Hoshino E, Shariff R, Van Gossum A, et al. Vitamin E suppresses increased lipid peroxidation in cigarette smokers. J Parenter Enter Nutr 1990; 14:300-5.

72. Duthie GG, Arthur JR, James WP, Vint HM. Antioxidant status of smokers and nonsmokers. Effects of vitamin E supplementation. Ann N Y Acad Sci 1989; 570:435-8.

73. Kawakami M, Kameyama S, Takizawa T. Lipid peroxidation in bronchoalveolar lavage fluid in interstitial lung diseases in relation to other components and smoking. Nippon Kyobu Shikkan Gakkai Zasshi 1989; 27:422-7.

74. Srivastava ED, Hallett MB, Rhodes J. Effect of nicotine and cotinine on the production of oxygen free radicals by neutrophils in smokers and non-smokers. Hum Toxicol 1989; 8:461-3.

75. York GK, Pierce TH, Schwartz LS, Cross CE. Stimulation by cigarette smoke of glutathione peroxidase system enzyme activities in rat lung. Arch Environ Health 1976; 31:286-90.

76. McCuster K, Hodil J. Selective increase of antioxidant enzyme activity in the alveolar macrophages from cigarette smokers and smoke-exposed hamsters. Am Rev Respir Dis 1990; 141:678-82.

77. Akin FJ, Benner JF. Induction of aryl hydrocarbon hydroxylase in rodent lung by cigarette smoke: a potential short-term bioassay. Toxicol Appl Pharmacol 1976; 36:331-7.

78. Contrell ET, Warr GA, Busbee DL, Martin RR. Induction of aryl hydrocarbon hydroxylase in human alveolar macrophage by cigarette smoking. J Clin Invest 1973; 52:1881-4.

79. Smith GJ, Ohl VS, Litwack G. Ligandin, the glutathione S-transferases, and chemically induced hepatocarcinogenesis: a review. Cancer Res 1977; 37:8-14.

80. Chasseaul LF. Conjugation with glutathione and mercapturic acid excretion. In: Arias IM, Jacob BB, eds. Glutathione: metabolism and function. New York: Raven Press, 1976; 77-114.

81. Laurell CB, Eriksson S. The electrophoretic alpha-globulin pattern of serum in alpha-1-antitrypsin deficiency. Scand J Clin Lab Invest 1963; 15:132-40.

82. Mass B, Ikeda T, Meranze DR, Weinbaugh G, Kimbel P. Induction of experimental emphysema: cellular and species specificity. Am Rev Respir Dis 1972; 106:384-91.

83. Rodriguez RJ, White RR, Senior RM, Levine EA. Elastase release from human alveolar macrophages: comparison between smokers and nonsmokers. Science 1977; 198:313-4.

84. Janoff A, Carp H, Lee DK. Cigarette smoking induces functional antiprotease deficiency in the lower respiratory tract of humans. Science 1979; 206:1313-4.

85. Carp H, Janoff A. Inactivation of bronchial mucous protease inhibitor by cigarette smoke and phagocyte-derived oxidants. Exp Lung Res 1980; 1:225-37.

86. Carp H, Janoff A. Possible mechanisms of emphysema in smokers: in vitro suppression of serum elastase inhibitory capacity by fresh cigarette smoke and its prevention by antioxidants. Am Rev Respir Dis 1978; 118:617-21.

87. Cox AW, Billingsley GD. Oxidation of plasma alpha-1-antitrypsin in smokers and nonsmokers and by an oxidizing agents. Am Rev Respir Dis 1984; 130:594-9.

88. Janoff A, Chow CK. Unpublished results.

89. Hunninghake CW, Gadek JE, Kawanami O, Ferrans VJ, Crystal RG. Inflammatory and immune processes in the human lung in health and disease: evaluation by bronchoalveolar lavage. Am J Pathol 1979; 97:149-206.

90. Haslam PL, Turton CWG, Heard B, et al. Bronchoalveolar lavage in pulmonary fibrosis: comparison of cells obtained with lung biopsy and clinical features. Thorax 1980; 35:9-18.

91. Plowman PN. The pulmonary macrophage population of human smokers. Ann Occup Hyg 1982; 25:393-405.

92. Harris JO, Olssen GN, Castle JR, Maloney AS. Comparison of proteolytic enzyme activity in pulmonary alveolar macrophages and blood leukocytes in smokers and nonsmokers. Am Rev Respir Dis 1975; 111:579-80.

93. Martin RR. Altered morphology and increased acid hydrolase content of pulmonary alveolar macrophages from cigarette smokers. Am Rev Respir Dis 1973; 107:596-601.

94. Gairola CG. Free lung cell response of mice and rats to mainstream cigarette smoke exposure. Toxicol Appl Pharmacol 1986; 84:567-75.

95. Matulionis DH. Effects of cigarette smoke generated by different smoking machines on pulmonary macrophages in mice and rats. J Anal Toxicol 1984; 8:187-91.

96. Barton GM, Roath OS. Leukocyte ascorbic acid in abnormal leukocyte states. Int J Vitam Nutr Res 1976; 46:271-4.

97. Sakamoto W, Yoshikawa K, Shindoh M, et al. Int J Vitam Nutr Res 1989; 59:131-9.

98. Hemila H, Roberts P, Wikstrom M. Activated polymorphonuclear leukocytes consume vitamin C. FEBS Lett 1984; 178:25-30.

99. Schlager JJ, Powis G. Cytisolic NAD(P)H:(quinone-acceptor)oxidoreductase in human normal and tumor tissue: effects of cigarette smoking and alcohol. Int J Cancer 1990; 45:403-9.

100. Kiyosawa H, Suka M, Okudaira H, et al. Cigarette smoking induces formation of 8-hydroxy-deoxyguanosine, one of the oxidative products of DNA damage in human peripheral leukocytes. Free Radic Res Commun 1990; 11:1-3.

101. Harats D, Naim M, Dabach Y, Hollander G, Stein O, Stein Y. Cigarette smoking renders LDL susceptible to peroxidative modification and enhanced metabolism by macrophages. Athero-sclerosis 1989; 79:245-52.

102. Orrenius S, Moldeus P. The multiple roles of glutathione in drug metabolism. Trends Pharmacol Sci 1984; 5:432-5.

103. Meister A, Andersson M. Glutathione. Annu Rev Biochem 1983; 52:711-60.

104. Cotgreave IA, Johansson U, Moldeus P, Brattsand R. The effect of acute cigarette smoke in-halation on pulmonary and systemic cystieine and glutathione redox states in the rat. Toxicology 1987; 45:203-12.

49

Modulation of Pulmonary Vitamin E by Environmental Oxidants

Nabil M. Elsayed

Walter Reed Army Institute of Research, Washington, D.C., and University of California, Los Angeles, California

INTRODUCTION

In recent years the health effects of environmental pollution have been of increasing concern. Ozone (O_3) and nitrogen dioxide (NO_2) are two natural constituents of the earth's atmosphere, but at higher concentrations they become major oxidant air pollutants with potentially adverse health effects [1–3]. Increased levels of these pollutants can be attributed to the rapid growth in the human population and the consequent increase in the demand and production of fossil fuel-dependent energy and means of transportation, particularly the automobile. Air pollution was initially a problem associated with urban areas, but it is increasingly encountered in rural areas, possibly as polluted air drifts from the cities [4].

The lung is the target for attack by environmental oxidants, because it has the largest surface area in contact with the atmosphere. Experimental studies have shown that exposure to O_3 and NO_2 can produce a variety of biochemical, morphological, and physiological alterations in the lungs of animals and humans, depending upon the concentration, duration, and mode of exposure [5–10]. Extrapulmonary effects of exposure to oxidants have also been reported [11–16]. In addition, studies of chronic, low-level exposures to oxidants in experimental animals suggest that prolonged exposures or repeated exposures over a long period of time may be carcinogenic [17–19].

The mechanisms of O_3 and NO_2 toxicity are based on their ability either to initiate free-radical–mediated reactions (O_3), reactions (1a) and (b) or to behave as free radicals (NO_2), reactions (2a) through (2c), with O_3 displayhing 10–15 times the toxicity of NO_2 [20–23].

$$O_3 + RCH = CHR' + H_2O \rightarrow \rightarrow \rightarrow RCHO + {}'RCHO + H_2O_2 \tag{1a}$$

$$H_2O_2 + Fe^{2+} \rightarrow HO^- + HO^{\cdot} + Fe^{3+} \tag{1b}$$

$$NO_2 + RHC = CHR' \rightarrow RHC - {}^{\cdot}CHR' \xrightarrow{O_2} RHC - CHR' \tag{2a}$$
$$\qquad\qquad\qquad\qquad\quad | \qquad\qquad\qquad | \quad |$$
$$\qquad\qquad\qquad\quad NO_2 \qquad\qquad NO_2\ O-O^{\cdot}$$

$$NO_2 + RHC = CHR' \rightarrow HNO_2 + \overline{RC - CHR'} \tag{2b}$$

$$2\,HNO_2 \rightarrow NO + NO_2 + H_2O \tag{2c}$$

The lung, however, normally exists in an oxygen-rich environment and thus has developed several intracellular and extracellular antioxidant defense systems, one of which is vitamin E (24).

Vitamin E is a lipophilic vitamin with antioxidant properties and is an excellent free-radical quencher. Since its discovery more than 60 years ago, few vitamins have received such widespread interest. The different roles and functions of vitamin E have been extensively studied (25–29). Despite a large body of research that has accumulated over the years, there is still controversy as to its mechanism of action and its requirement in the human (28,29). The difficulty in resolving these issues is due on the one hand to the lack of an apparent human disease associated with vitamin E deficiency and, on the other, to the diversity of effects this vitamin displays in experimental animals.

The biological antioxidant function of vitamin E was first proposed by Mattill (30) a few years after its discovery but was disputed for a long time, particularly by Green (31). For many years, the antioxidant role of vitamin E was articulated and defended by Tappel until it received general acceptance (32–34). Vitamin E is also thought to play another role contributing to membrane stability and fluidity (35,36). This function is based on the structural characteristics of the vitamin and its ability to be inserted within the fatty acid layers of the cell membranes.

In a series of studies investigating the pulmonary response to inhaled oxidants and the influence of vitamin E on that response, we observed that the vitamin E concentration tended to increase after exposure. We suggested that this increase may reflect a mobilization of the vitamin to the lung under oxidative stress. Increased vitamin E concentration in the lung was observed later by other investigators. Moreover, other antioxidants were also reported to increase after oxidant exposure. This chapter discusses the lung response to oxidant exposure and modulation of vitamin E concentration in the lung resulting from such exposures.

METHODS

This section deals with some of the peculiarities and precautions incorporated in the experimental design and the biochemical analyses associated with vitamin E studies as they were applied in our laboratory.

Animals and Dietary Conditioning

In the course of our studies, two strains of rats and two methods of dietary conditioning were used. In general, the rats were purchased pregnant, 10 days from term, from local commercial suppliers. We used Sprague-Dawley rats for the NO_2 studies but used Long-Evans rats for the O_3 studies. Long-Evans rats were selected because rigorous depletion of their vitamin E content can be achieved in a relatively shorter period of time and to a far greater extent than in other rats. The rats were fed test diets with a general composition as shown in Table 1. The literature does not uniformly list the dietary composition, particularly the fat source and whether it is mostly saturated or unsaturated. This inconsistency does not permit a proper evaluation accounting for the role of lipids. We usually purchased the diets in meal form (coarsely ground) without fat or vitamin E and then added vitamin E-stripped corn oil and defined amounts of vitamin E (when required). The diets were mixed weekly and kept refrigerated at 4°C until used. The diets were pelleted manually to avoid the possibility of both lipid peroxidation, particularly in the vitamin E-deficient diet, and vitamin E oxidation in the vitamin E-supplemented diets. These changes could result from the heat and pressure generated during automatic pelleting. All rats received fresh diet every day to further minimize the ingestiion of peroxidized lipids. After birth, the dams along with their pups were maintained on either of two dietary regimens, vitamin E-deficient and vitamin E-supplemented

TABLE 1 General Dietary Composition

Ingredient	Composition (%)
Sucrose	50–60
Casein (vitamin free)	20–30
Corn oil (tocopherol stripped)	5
Fiber	5
Mineral mixture	4
Vitamin mixture (vitamin E omitted)	1
Vitamin E	0, 50, or 1000 IU/kg diet

diets, or only a vitamin E-deficient diet. About 3 weeks following birth (when the pups were weaned), the dams were discarded. In the NO_2 studies, the pups were fed according to the first dietary regimen just described for 5–8 weeks, then half the rats were exposed to NO_2 and the other half received filtered room air and served as matched controls. In the O_3 studies, all rats were fed vitamin E-deficient diet only and then divided into two groups, and for 2 additional weeks (total of 10 weeks of dietary conditioning) one group continued to receive the same vitamin E-deficient diet, but the other group was fed a diet supplemented with *dl*-α-tocopheryl acetate (1000 IU/kg diet). Half the rats from each dietary group were exposed to O_3, and the other half served as matched controls. The vitamin E deficiency status of the animals during the dietary conditioning period was monitored by sacrificing randomly selected rats and evaluating vitamin E content in plasma and a few other tissues.

Throughout these studies, the rats were housed in stainless steel cages (one rat per cage). The daily food and water intakes were carefully monitored. The cages were kept in a laminar airflow isolation unit fitted with a 4 inch thick charcoal filter to eliminate atmospheric oxidants and a 0.3 μm pore size high-efficiency particulate air (HEPA) filter to remove airborne particulates and/or bacterial infectious agents. The room housing the animals was automatically controlled to provide a 12 h photoperiod and a 23 $\pm$ 2°C environmental temperature.

Oxidant Exposures

Following the dietary conditioning period, we exposed rats from both vitamin E-deficient and vitamin E-supplemented groups to NO_2 (3.0 $\pm$ 0.1 ppm, 5640 $\pm$ 188 mg/m³), O_3 (0.5 $\pm$ 0.05 ppm, 980 $\pm$ 98 mg/m³), or filtered room air (control) continuously for 5–7 days. NO_2 was usually purchased as a mixture of 15% NO_2 in nitrogen then diluted to the desired concentration, and O_3 was generated by passing 100% oxygen at a rate of 0.5 L/min through a silent electric arc ozonizer (Sander, Am Osterberg, Germany). The exposures took place in 300 L stainless steel inhalation chambers (Young and Burtke, Cincinnati, OH) with the airflow adjusted to permit 20 chamber changes per hour. Control rats were placed in identical chambers supplied with filtered room air. While in the inhalation chambers, all rats were housed in open-mesh stainless steel cages (one rat per cage) and allowed free access to water and their respective diets. The food was replaced daily to minimize the possibility of the rats ingesting peroxidized lipids in food exposed to NO_2 or O_3. The level of NO_2 in the chamber was continuously monitored using an oxides of nitrogen analyzer (Columbia Scientific Instruments, Austin, TX), and the level of O_3 was monitored using a Daisibi ozone monitor (model 1003-PC; Environmental Corp., Glendale, CA).

[14]C-Labeled Vitamin E Studies

To examine the distribution of vitamin E in vitamin E-deficient and vitamin E-supplemented tissues and to determine whether oxidant exposure would significantly alter this distribution, we injected rats from both dietary regimens (vitamin E-deficient and vitamin E-supplemented), 1 h before exposure, intraperitoneally, with a single dose of [^{14}C]*dl*-α-tocopheryl acetate (10 μCi per rat) dissolved in 0.1 ml absolute ethanol. The *dl*-α-[3,4-^{14}C$_2$]-tocopheryl acetate, specific activity 75.8 μCi/mg vitamin E, was a generous gift from Dr. Lawrence J. Machlin (Hoffman-La Roche, Nutley, NJ). Injection of an equivalent volume of absolute ethanol in a separate experiment was found to have no effect on vitamin E content.

Tissue Preparation and Vitamin E Assay

After exposures, we weighed both exposed and control rats, then anesthetized them with intraperitoneal injections of sodium pentobarbital (60 mg/kg body weight), opened the chest, and excised the lungs free of major airways and connective tissues. The lungs were rinsed with cold saline, blotted gently on dry gauze and weighed, and then stored at −70°C until analyzed. All samples were analyzed within a week after exposure, and the analysis was performed in the dark at 4°C to prevent the oxidation of vitamin E during extraction. We determined vitamin E fluorometrically by HPLC, essentially as described by Vatassery and Hagan (37) with minor modifications. Briefly, portions of lung tissue (0.2–0.3 g) were mixed with a 3 ml absolute ethanol, 3 ml of 10% aqueous ascorbic acid, and 6 ml hexane containing 26 mg per 100 mg of butylated hydroxytoluene (BHT) as an antioxidant to protect vitamin E during extraction (38). The samples were then homogenized at 4°C in two 15 s bursts using a Polytron homogenizer (Brinkman Instruments, Westbury, NY). We transferred the homogenized samples to screw-cap tubes and centrifuged them at 30,000 × *g* for 10 min at 4°C. The pellet was discarded, and the upper layer was transferred to a conical test tube and evaporated to dryness with a stream of nitrogen while on ice. The dried extract was redissolved in 150 ml of a chloroform and methanol (1:1, vol/vol) mixture containing 5 mg per 100 mg BHT. The tubes were capped tightly and mixed in a vortex mixer for 30 s. An aliquot was taken to count ^{14}C-labeled vitamin E using a liquid scintillation counter; another aliquot of the same extract was analyzed for vitamin E content. We used a Varian 5000 liquid chromatograph and a Fluorichrom fluorescent detector fitted with 220 nm excitation and 360 nm emission filters (Varian Instruments, Palo Alto, CA). The column used was a 25 cm × 4.6 mm, μBondapak, reversed-phase, C18, 10 mm particle size (Waters Associates, Milford, MA). The eluting solvents were methanol and water (98:2, vol/vol) at a flow rate of 1.3 ml/min. Vitamin E content was calculated by extrapolation from standard curves constructed each day using stock solutions of *dl*-α-tocopherol in ethanol.

DISCUSSION

Pulmonary Response to Environmental Oxidants

Inhalation of high levels of oxidants (above 1 ppm O$_3$, or 10 ppm NO$_2$) can cause severe damage and may even lead to death in some laboratory animals. Ambient or near ambient levels of the same oxidants, however, are not lethal and cause only lung injury rather than death. Manifestations of lung injury from oxidants can be grouped broadly into two phases. First, an initial injury phase is characterized morphologically by damaged bronchiolar ciliated cells and alveolar epithelial type I cells (7–9) and is characterized biochemically by oxidation of several cellular components, such as unsaturated fatty acids and sulfhydryl-containing amino acids and proteins (10,20,21,39,40). Second, a repair or adaptive phase is characterized by increased cell replication and metabolic activities. The magnitude of such increases is proportional to the oxidant dose. The second phase is characterized morphologically by proliferation of Clara cells replacing injured ciliated cells in the airways and epithelial type II cells replacing damaged type I cells (7–9). Biochemically, the repair phase is associated with increased nucleic acid synthesis and induction of antioxidant enzyme activities such as those of the gluta-thione redox cycle and the pentose phosphate cycle (6,20,21,39,41). Recently, a number of animal studies reported that the concentrations of several antioxidants were significantly elevated in the lung after oxidant exposure (42–46). Other studies in humans showed that a few hours of exposure to 0.1–0.5 ppm O_3 can produce detectable changes in respiratory function (47,48).

General Model for the Pulmonary Response to Oxidants

A quantitative relationship between oxidant exposure and the pulmonary response was shown biochemically by Mustafa et al. (39) and corroborated by the morphological observations of Evans (8). Based on these observations, a general model for a typical pulmonary response to oxidant exposure is presented in Figure 1. In this model, lungs from animals exposed to an oxidant at a specified dose are analyzed for biochemical or morphological alterations, such as changes in metabolic activities or incorporation of tritiated thymidine into newly synthe-sized DNA of proliferating cells (e.g., labeling indexes). The resulting alterations measured over time typically decrease below control levels in the first day following exposure and then progressively increase to reach a peak about the third or fourth day. Thereafter, the rate plateaus and remains constant as long as the exposure continues at the same concentration. This ob-servation suggests that the magnitude of increase above control, in this model, is propor-tional to the exposure dose; that is, oxidant exposure results in a pulmonary response, but continued exposure generally does not result in a continued unlimited increase in that response. If, on the other hand, the exposure is stopped, the rate decreases gradually to control levels. Upon reexposure for a second time, however, a comparable increase similar to that resulting from the first exposure would occur provided the concentration of oxidant is also comparable. This model can be generalized to assess the pulmonary response to different oxidants as well as the influence of antioxidants on that response by comparing the magnitude of attenuation of the rate of increase in the model.

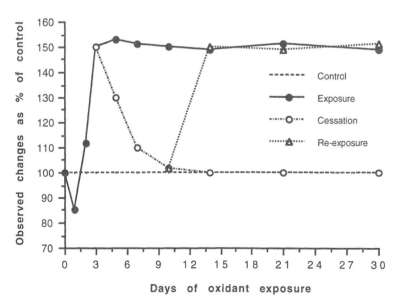

FIGURE 1 A general model for the lung response to oxidant exposure. The abscissa shows hypothetical biochemical or morphological quantitative changes observed after exposure to an oxidant, and the ordinate days after exposure. Full discussion of the model is found in the text.

Effects of Vitamin E on the Pulmonary Response to Oxidants

As the exposure to environmental O_3 and NO_2 was found to cause oxidative stress, with lipid peroxidation as a major manifestation of that stress, it was logical to investigate the potential for antioxidant protection. Vitamin E, a lipid-soluble antioxidant with very low toxicity even at high doses (49), has been extensively studied in animals as an effective prophylactic agent that potentially can protect humans. Studies using in vitro systems have demonstrated in model membranes that vitamin E can protect polyunsaturated fatty acids (PUFAs) from peroxidation by oxidants (50–53). In an earlier in vivo study using both O_3 and NO_2, Fletcher and Tappel (54) reported that dietary supplementation with vitamin E protected rats from both low-level mediated lipid peroxidation and high-level mediated mortality. Most in vivo reports published to date strongly support a protective role for vitamin E (55–63). These findings have led the U.S. Environmental Protection Agency to conclude that in vivo protection from O_3 by vitamin E was clearly demonstrated in rats and mice (64). Some studies, however, found that under specific experimental conditions vitamin E offers no such protection in animals (65–67) or humans (68,69).

Despite the large number of studies that have examined vitamin E, several issues remain unanswered. The first issue concerns the relevancy of animal data to humans, because vitamin E deficiency seldom occurs in humans, particularly in adults. One of the few adverse health effects in humans associated with vitamin E deficiency is retrolental fibroplasia (RLF), which afflicts some infants and premature babies receiving oxygen therapy. This condition was found to be ameliorated by vitamin E supplementation. The second issue is that most studies compare the response of deficient and supplemented animals. The question then remains as whether vitamin E supplementation above the recommended daily allowance (RDA) would offer increasing protection or whether it is needed only in situations associated with increased oxidative stress. The third issue centers around the process of making animals vitamin E deficient. Do the animals become artificially more sensitive than if they were otherwise adequately receiving vitamin E? Can this potentially heightened sensitivity confound the effect of oxidant exposure? These questions about the benefit of vitamin E supplementation indicate the need for more studies comparing the response to oxidants by animals fed adequate amounts of the vitamin to those that are supplemented with therapeutic doses. Such studies will invariably narrow the gap in knowledge between animal studies and human relevancy.

Effect of Oxidants on Pulmonary Vitamin E Content

In a previous study we observed (42) that after exposure to NO_2, vitamin E content increases significantly in the lungs of vitamin E-supplemented but not vitamin E-deficient rats (Fig. 2). In that study, we fed pregnant rats (10 days from birth) diets either deficient or supplemented with vitamin E until they gave birth, and then we fed the offspring similar diets until maturity (about 8 weeks) before exposure to NO_2. Similar increases in lung vitamin E after NO_2 exposure were observed by Sagai et al. (44) in rats fed only a commercial diet adequate in vitamin E (Fig. 3). As discussed in Methods, animals raised on a vitamin E-deficient diet may acquire structural or biochemical impairments, rendering them artificially more susceptible to oxidant treatment than those raised on a supplemented diet. To avoid this possibility, we modified the experimental design in the subsequent experiments. In the modified design, we first depleted all animals of vitamin E, then supplemented half of them with the vitamin while keeping the other half deficient before exposing both groups to the oxidant. Thus, any alterations acquired by dietary manipulation would be shared by both groups, and the effect of exposure would not then be confounded. Using this modified dietary regimen, we exposed Long-Evans rats to O_3 and observed significant increases in lung vitamin E content of supplemented animals similar to the responses of Sprague-Dawley rats with NO_2 (61,70). We suggested that the increase

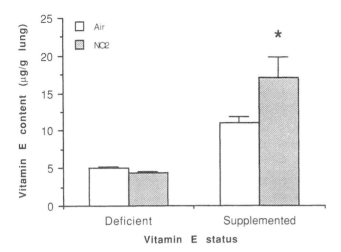

FIGURE 2 Vitamin E content in rat lungs after continuous exposure to 3 ppm NO_2 for 7 days. Pregnant Sprague-Dawley rats (10 days from term) were fed a test diet (Teklad Test Diets, Madison, WI) containing either 0 or 50 IU vitamin E per kg diet. After birth, the offspring were fed the same diets for 8 weeks and then exposed to NO_2. (*) Significantly different from control, $P < 0.05$. (Data presented as mean ± SD are calculated and redrawn from Elsayed and Mustafa. Toxicol Appl Pharmacol 1982; 66:319-28.)

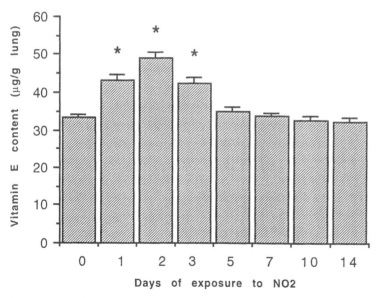

FIGURE 3 Vitamin E content in rat lungs after exposure to 10 ppm NO_2. Wistar male rats, 8 weeks old, were fed a commercial stock diet (CD-2, Japan Clea, Ltd., Tokyo) containing vitamin E. (*) Significantly different from 0 days, $P < 0.05$. (Data presented as mean ± SEM are calculated and redrawn from Sagai et al. J Toxicol Environ Health 1982; 9:153-64.)

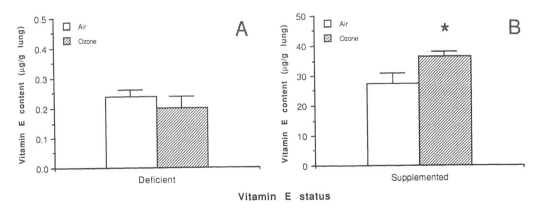

FIGURE 4 Vitamin E content in rat lungs after continuous exposure to 0.5 ppm O_3 for 5 days. (A) Vitamin E-deficient and (B) vitamin E-supplemented rats. Note the extent of depletion achieved in Long-Evans rats compared to Sprague-Dawley rats shown in Figure 2. Pregnant Long-Evans rats (10 days from term) were fed AIN-76 test diet (ICN, Nutritional Biochemicals, Cleveland, OH) containing no vitamin. After birth, the offspring were fed the same diets for 8 weeks and then divided into two groups, and for 2 additional weeks (total of 10 weeks of dietary conditioning) one group received the same diet and the other was supplemented with *dl*-α-tocopheryl acetate (1 g/kg diet) and then exposed to O_3. (*) Significantly different from control, $P < 0.05$. (Data presented as mean ± SD are calculated and redrawn from Elsayed et al. Arch Biochem Biophys 1990; 282:263-9.)

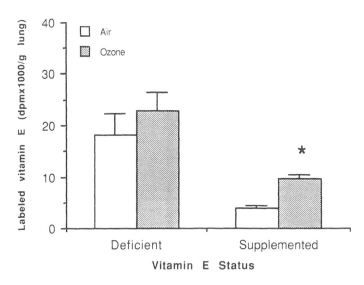

FIGURE 5 Effect of exposure to 0.5 ppm O_3 continuously for 5 days on lung tissue uptake and retention of ^{14}C-labeled vitamin E. [*dl*-α-^{14}C]Tocopheryl acetate (10 μCi per rat) was dissolved in 0.1 ml absolute ethanol and then injected intraperitoneally 1 h before the start of O_3 exposure. Other conditions and symbols are the same as in Figure 4.

may represent a *mobilization* of the vitamin (when the vitamin is sufficiently present) from other body sites to the lung under oxidative stress. To test this hypothesis further, we repeated the experiment, but both deficient and supplemented rats were injected with an equal dose of ^{14}C-labeled vitamin E shortly before exposure to O_3 (43). The results again showed a mobilization of vitamin E content to the lungs of supplemented rats (Fig. 4), concomitant with a similar accumulation of ^{14}C-labeled vitamin E (Fig. 5) after O_3 exposure. Another observation was made in the course of this experiment, that of an inverse relationship between the uptake of ^{14}C-labeled vitamin E and the overall nutritional status. Thus, the supplemented rats retained less ^{14}C-labeled vitamin than the deficient rats, suggesting the possible presence of saturable binding sites for vitamin E. This increase in pulmonary vitamin E was not observed in any one of seven other organs tested during the study.

These observations thus suggest that the proposed mobilization of vitamin E under environmental oxidant stress may apply to other antioxidants and types of oxidative stress. For example, Arriess et al. (71) found an increase in lung vitamin E content after exposure to cigarette smoke. Studies from our laboratory showed increased levels of selenium in mice lungs after O_3 exposure (72,73). Bankes et al. (46) also reported increased taurine levels in isolated rat alveolar macrophages. Taurine has been proposed to act directly as a primary antioxidant by scavenging free radicals and indirectly as a secondary antioxidant capable of preventing increases in membrane permeability resulting from oxidant damage (46). Pincemail et al. (74) and Camus et al. (75) reported that vitamin E levels in the blood of human volunteers increase transiently for several hours after intensive exercise (which is considered to cause oxidative stress) before returning to control levels. In addition to vitamin E, preliminary studies at the U.S. Environmental Protection Agency reported that significant increases in vitamins E and C and uric acid (a proposed antioxidant) concentrations were observed in lung lavage fluids and/or tissues from humans, rats, and guinea pigs after exposure to O_3 (76,77). All these observations support the hypothesis that vitamin E and possibly other antioxidants as well are mobilized to the lung under oxidative stress such as that caused by exposure to environmental oxidants. However, the mechanism of such mobilization and the signal that triggers it remain to be elucidated.

CONCLUSIONS

In conclusion, vitamin E can be mobilized to the lung from other body sites in response to oxidative stress. This occurs whenever the lung is a target organ for that stress and when the vitamin is sufficiently available in the body. The implication of these observations to human health is that moderate increases in vitamin E intake to supplement the average daily diet above the RDA may be beneficial in highly polluted areas where oxidant stress occurs frequently, such as in southern California. Levels of vitamin E beyond those required for normal body functions make antioxidants available to organs likely to be targets of environmental oxidants. The results obtained from most animal studies, however, remain inconclusive and should be approached with caution before extrapolation to humans, because they compare the responses of deficient and supplemented animals. To establish relevance to human health, more studies should be conducted, but these should compare vitamin E-adequate to vitamin E-supplemented animals or humans.

DISCLAIMER

The opinions and assertions contained here are the private views of the author and are not to be construed as official, nor do they reflect the views of the Department of the Army or the Department of Defense (AR 360-5). In conducting the research described in this report, the author adhered to the Guide for the Care and Use of Laboratory Animals, DHEW Publication (NIH) 85-23.

REFERENCES

1. National Research Council. Ozone and other photochemical oxidants. Vol 19. Committee on Medical and Biological Effects of Environmental Pollutants. Subcommittee on Ozone and Other Photochemical Oxidants. Washington, DC: National Academy of Sciences, 1977.
2. U.S. Environmental Protection Agency. Air quality criteria for ozone and other photochemical oxidants. Report No. EPA-600/8-79-004. Research Triangle Park, NC: U.S. EPA, 1978.
3. World Health Organization. Photochemical oxidants: environmental health criteria 7. Geneva: WHO, 1979.
4. Bruntz SM, Cleveland WS, Graedel TE, Kleiner B, Warner JL. Science 1974; 186:257-9.
5. Stokinger HE, Coffin DL. In: Stern AC, ed. Air pollution, 2nd ed. New York: Academic Press, 1968; 445-546.
6. Mustafa MG, Tierney DF. Am Rev Respir Dis 1978; 118:1061-1090.
7. Plopper CG, Dungworth DL, Tyler WS. Am J Pathol 1973; 71:375-94.
8. Evans MJ, In: Witschi HP, Nettesheim P, eds. Mechanisms in respiratory toxicology, Vol. I. Boca Raton, FL: CRC Press, 1982; 189-218.
9. Crapo JD, Barry BE, Chang L-Y, Mercer RR. J Toxicol Environ Health 1984; 13:301-21.
10. Mudd JB, Freeman BA. In: Lee SD, ed. Biochemical effects of environmental pollutants. Ann Arbor, MI: Ann Arbor Science, 1977; 97-133.
11. Goldstein DB. In: Oxygen free radical tissue damage, Vol. 65, CIBA Foundation Symposium. New York: Elsevier, 1979; 295-319.
12. Clemons GK, Garcia JF. J Environ Pathol Toxicol 1980; 4:359-69.
13. Kuraitis KV, Richters A, Sherwin RP. J Toxicol Environ Health 1981; 7:851-9.
14. Hassett C, Mustafa MG, Coulson WF, Elashoff RM. Toxicol Lett 1985; 26:139-44.
15. Mehlman MA, Borek C. Environ Res 1987; 42:36-53.
16. Farahani H, Hasan M. Pharmacol Toxicol 1991; 69:56-9.
17. Hassett C, Mustafa MG, Coulson WF, Elashoff RM. J Natl Cancer Inst 1985; 75:771-7.
18. Mustafa MG, Hassett CM, Newell GW, Schrauzer GN. Ann N Y Acad Sci 1988; 534:714-23.
19. Witschi HP. Toxicology 1988; 48:1-20.
20. Menzel DB. In: Pryor WA, ed. Free radicals in biology, Vol. II. New York: Academic Press, 1976; 181-2302.
21. Pryor WA, Dooley MM, Church DF. In: Lee SD, Mustafa MG, Mehlman MA, eds. The biochemical effects of ozone and related photochemical oxidants. Princeton, NJ: Princeton Scientific Publishers, 1983; 7-19.
22. Mustafa MG. Free Radic Biol Med 1990; 9:245-65.
23. Pryor WA, Das B, Church DF. Chem Res Toxicol 1991; 4:341-8.
24. Heffner JE, Repine JE. Am Rev Respir Dis 1989; 140:531-54.
25. de Duve C, Hayaishi O, eds. Tocopherol, oxygen, and biomembranes. New York: Elsevier, 1978.
26. Witting LA. In: Pryor WA, ed. Free radicals in biology, Vol. IV. New York: Academic Press, 1980; 295-319.
27. Machlin LJ, ed. Vitamin E—a comprehensive treatise. New York: Marcel Dekker, 1980.
28. Bieri JG, Corash L, Hubbard VS. N Engl J Med 1983; 308:1063-71.
29. Diplock AT. In: Biology of vitamin E. CIBA Foundation Symposium, Vol. 101, London: Pitman, 1983; 45-55.
30. Mattill HA. J Biol Chem 1931; 90:141-51.
31. Green J. Ann N Y Acad Sci 1972; 203:29-44.
32. Tappel AL. Vitam Horm 1962; 20:493-510.
33. Tappel AL. Fed Proc 1965; 24:73-8.
34. Tappel AL. Ann N Y Acad Sci 1972; 203:12-28.
35. Lucy JA. Ann N Y Acad Sci 1972; 203:4-11.
36. Patel JM, Edwards DA. Toxicol Appl Pharmacol 1988; 96:101-14.

37. Vatassery GT, Hagen DF. Anal Biochem 1977; 79:129-34.
38. Chow FI, Omaye ST. Lipids 1983; 18:837-41.
39. Mustafa MG, Hacker AD, Ospital JJ, Hussain MZ, Lee SD. In: Lee SD, ed. Biochemical effects of environmental pollutants. Ann Arbor, MI: Ann Arbor Science, 1977; 59-96.
40. Last JA, Greenberg DB, Castleman WL. Toxicol Appl Pharmacol 1979; 51:247-58.
41. Chow CK, Tappel AL. Lipids 1972; 7:518-24.
42. Elsayed NM, Mustafa MG. Toxicol Appl Pharmacol 1982; 66:319-28.
43. Elsayed NM, Mustafa MG, Mead JF. Arch Biochem Biophys 1990; 282:263-9.
44. Sagai M, Ichinose T, Oda H, Kubota K. J Toxicol Environ Health 1982; 9:153-64.
45. Hatch GE, Slade R, Selgrade MK, Stead AG. Toxicol Appl Pharmacol 1986; 82:351-9.
46. Banks MA, Porter DW, Martin WG, Castranova V. J Nutr Biochem 1991; 2:308-13.
47. Folinsbee LJ, Bedi JF, Horvath SM. Am Rev Respir Dis 1980; 121:431-9.
48. Lippman M. Annu Rev Public Health 1989; 10:49-67.
49. Bendich A, Machlin LJ. Am J Clin Nutr 1988; 48:612-9.
50. Thomas HV, Muller PK, Lyman RL. Science 1967; 159:532-4.
51. Goldstein BD, Buckley RD, Cardenas R, Balchum OJ. Science 1970; 169:605-6.
52. Roehm JN, Hadley JG, Menzel DB. Arch Intern Med 1971; 24:237-42.
53. Menzel DB. Arch Environ Health 1971; 23:149-53.
54. Fletcher BL, Tappel AL. Environ Res 1973; 6:165-75.
55. Mustafa MG, Nutr Rep Intl 1975; 11:473-6.
56. Sato S, Kawakama M, Maeda S, Takishima T. Am Rev Respir Dis 1976; 113:809-21.
57. Chow CK, Plopper CG, Chiu M, Dungworth DL. Environ Res 1981; 24:315-24.
58. Evans MJ, Cabral-Anderson L, Dekker NP, Freeman G. Chest 1981; 80 (Suppl.):5S-8S.
59. Sevanian A, Elsayed N, Hacker AD. J Toxicol Environ Health 1982; 10:743-56.
60. Sevanian A, Hacker AD, Elsayed N. Lipids 1982; 17:269-77.
61. Elsayed NM. Arch Biochem Biophys 1987; 255:392-9.
62. Elsayed NM, Kass R, Mustafa MG, et al. Drug Nutr Interact 1988; 5:373-86.
63. Pryor WA. Am J Clin Nutr 1991; 53:702-22.
64. U.S. Environmental Protection Agency. Air quality criteria for ozone and other photochemical oxidants, Vol. I. Research Triangle Park, NC: EPA-600/8-84-020 aF, 1986.
65. Ramazzotto LJ, Engstrom R. Environ Physiol Biochem 1975; 5:226-34.
66. Stephens RJ, Buntman DJ, Negi DS, Parkhurst RM, Thomas DW. Chest 1983; 83:37S-40S.
67. Warren DL, Hyde DM, Last JA. Toxicology 1988; 53:113-33.
68. Posin CI, Clark KW, Jones MP, Buckley RD, Hackney JD. J Toxicol Environ Health 1979; 5:1049-58.
69. Hackney JD, Linn WS, Buckley RD, et al. J Toxicol Environ Health 1981; 7:383-90.
70. Elsayed NM. Ann N Y Acad Sci 1990; 570:439-40.
71. Airriess GR, Chagchitt C, Chen L-C, Chow CK. Nutr Res 1988; 8:653-61.
72. Elsayed NM, Hacker AD, Kuehn K, Mustafa MG, Schrauzer GN. Toxicol Appl Pharmacol 1983; 71:398-406.
73. Elsayed NM, Mustafa MG, Hacker AD, Kuehn K, Schrauzer GN. Biol Trace Elem 19??; 6:249-61.
74. Pincemail J, Deby C, Camus G, et al. Eur J Appl Physiol 1988; 57:189-91.
75. Camus G, Pincemail J, Roesgen A, Dreezen E, Sluse FE, Deby C. Arch Int Physiol Biochim 1990; 98:121-6.
76. Norwood J, Grissman K, Slade R, Hatch G. Toxicologist 1989; 9:45 (abstract).
77. Grissman K, Norwood J, Slade R, Highfill J, Koreen H, Hatch G. Toxicologist 1989; 9:46 (abstract).

50

Antioxidant Status in Smokers

Garry Graeme Duthie

Rowett Research Institute, Aberdeen, Scotland

INTRODUCTION

Free radicals are implicated in the pathogenesis of many diseases, including heart disease and cancer (1). Smokers, a group at high risk of developing such diseases, inhale large amounts of potentially injurious free radicals derived from tobacco (2). This chapter considers whether such a sustained free-radical load affects the antioxidant defense system of smokers and contributes to the development of smoking-related diseases.

FREE RADICALS IN CIGARETTE SMOKE

The free-radical chemistry of cigarette smoke was recently reviewed (2–5). In brief, each puff of a cigarette contains approximately 10^{14} free radicals in the tar phase and 10^{15} in the gas phase (2). Long-lived quinone-semiquinone (Q⋅) radicals associated with the particulate phase are generated by the oxidation of polycyclic hydrocarbons. In aqueous medium, Q⋅ can reduce oxygen to superoxide (O_2^-) and hydrogen peroxide (H_2O_2) and catalyze the conversion of H_2O_2 to the highly reactive hydroxyl (OH⋅) radical. In the smoke phase reactive carbon- and oxygen-centered radicals (ROO⋅) are continuously generated by a reaction between nitrogen dioxide (NO_2) and aldehydes and olefins. Moreover, NO_2 may react with H_2O_2 to produce OH⋅. Pulmonary macrophages activated by nicotine may provide a source of H_2O_2. Copper or iron catalysis could then also promote the formation of OH⋅ by a Fenton reaction (5).

EFFECTS OF SMOKING ON LUNG ANTIOXIDANTS

The abnormally low concentrations of vitamin E in lung lavages from smokers (6) presumably result from a sustained exposure to the free radicals associated with tobacco. Paradoxically, exposure of rats and guinea pigs to mainstream tobacco smoke increases the vitamin E content of lung tissue, possibly reflecting an adaptive response to oxidative damage (7–9). Moreover, alveolar macrophages of smokers have a higher total ascorbate content and an enhanced ability to accumulate ascorbate in vitro (10), again suggesting adaptation to oxidative stress. Animal studies indicate similar adaptive responses in lung tissue, exposure to cigarette smoke increasing the activities of glucose-6-phosphate dehydrogenase, reduced glutathione, glutathione reductase, and glutathione peroxidase (8,11). Nutritional vitamin E status influences this cellular response to cigarette smoke as compensatory increases in lung antioxidant enzymes are more pronounced in vitamin E-deficient animals (8).

EFFECTS OF SMOKING ON BLOOD ANTIOXIDANTS

Plasma vitamin E concentrations of smokers and nonsmokers are similar (12,13). However, there is a well-recognized decrease in plasma and serum vitamin C in long-term smokers (Fig. 1) (12–16), that may reflect a lower intake of foods rich in vitamin C (14). In addition, increased plasma concentrations of dehydroascorbate (17) suggest that smokers have an enhanced turnover of vitamin C (16), possibly in response to a sustained oxidant load. Plasma carotenoid concentrations are also depressed in smokers (13), whereas plasma ceruloplasmin activity is increased (17–20). This latter change could reflect increased copper intake from the tar components of cigarettes or an acute stress response (17). Moreover, raised ceruloplasmin may provide antioxidant protection by preventing the oxidative inactivation of α_1-proteinase inhibitor (19,20).

In the erythrocyte, catalase and superoxide dismutase activities are unaffected by smoking (12,17,18). However, a decrease in the activity of glucose-6-phosphate dehydrogenase in the erythrocytes of smokers may reflect inhibition of the enzyme by increased concentrations of extracellular or intracellular lipid hydroperoxides (12). A decrease in glutathione peroxidase in smokers' erythrocytes has been inconsistently observed (12,17,18,21) and may therefore reflect differences in selenium status rather than a direct response to a sustained oxidative stress. However, as the increased free-radical activity induced by exercise or vitamin E deficiency causes higher tissue concentrations of the antioxidant peptide, glutathione (22,23), the increase in erythrocyte glutathione in smokers (12,17,24) could reflect a similar adaptive response.

LIPID PEROXIDATION AND VITAMIN E

Plasma concentrations of indicators of lipid peroxidation, such as conjugated dienes, are increased in smokers (12), suggesting enhanced peroxidation of endogenous fatty acids. Increased hydrocarbon expiration in smokers compared with nonsmokers also suggests that lipid peroxidation is stimulated in vivo. For example, smokers have enhanced ethane expiration, indicative of peroxidation of omega-3 fatty acids, which falls to levels detectable in

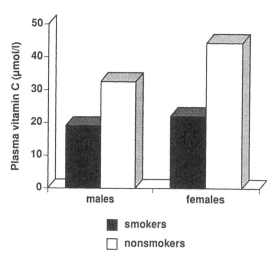

FIGURE 1 Plasma vitamin C concentrations of smokers and nonsmokers (Duthie, unpublished data). Smoking significantly ($P < 0.01$) reduces plasma vitamin C levels in both sexes.

TABLE 1 Effects of Smoking on Antioxidants and Related Enzymes

Tissue or cell	Antioxidant	References
Lung	↑ Vitamin E	7–9
	↑ Glucose-6-phosphate dehydrogenase	8, 11
	↑ Glutathione peroxidase	8, 11
	↑ Glutathione reductase	8, 11
	↑ Reduced glutathione	8, 11
Plasma	↓ Vitamin C	12–16
	↓ β-Carotene	13
	↑ Ceruloplasmin	17–20
Erythrocyte	↓ Glucose-6-phosphate dehydrogenase	12
	↓ Glutathione peroxidase	17, 21
	↑ Reduced glutathione	17, 24

nonsmokers only after 6 months of abstinence (25). Increased peroxidation of omega-6 fatty acids in smokers, as suggested by increased breath pentane output, can be reduced by supplementation with vitamin E (800 mg/day for 2 weeks) (26,27). Similarly, the increased susceptibility of washed erythrocytes of smokers to hydrogen peroxide-induced peroxidation can be abolished by supplementing with 1000 mg vitamin E per day for 2 weeks (12,17). Because the plasma and erythrocyte concentrations of unsupplemented smokers and nonsmokers (12) are similar, such observations suggest that smokers have a greater requirement for vitamin E.

CONCLUSION

Smoking causes several changes in antioxidant status (summarized in Table 1). Although some of these appear to be compensatory adaptations to a sustained oxidant load, lipid peroxidation is still enhanced in long-term smokers. Moreover, vitamin E supplements suppress such enhanced peroxidation. Because of the desirability of keeping the degree of lipid peroxidation in cells to a minimum (28), it has been suggested that the recommended daily allowance (RDA) for vitamin E should be increased (28,29). Such advice to the general population is contentious (30), but smokers who are unable to break their addiction may nevertheless benefit from an increased vitamin E intake. Thus there is a strong correlation between the number of cigarettes smoked and the incidence of heart disease in northern Europe, but the relationship is much weaker in southern Europe and Japan (31). This may reflect differences in consumption of food containing antioxidants like vitamin E (32).

REFERENCES

1. Halliwell B. Oxidants and human disease: some new concepts. FASEB J 1987; 1:358-64.
2. Church DF, Pryor WA. Free-radical chemistry of cigarette smoke. Environ Health Perspect 1985; 64:111-26.
3. Pryor WA. The free radical chemistry of cigarette smoke and the inactivation of alpha-1-proteinase inhibitor. In: Taylor J, ed. Pulmonary emphysema and proteolysis. New York: Academic Press, 1986; 369-92.
4. Pryor WA. Cigarette smoke and the involvement of free radical reactions in chemical carcinogenesis. Br J Cancer 1987; 55(Suppl VIII):19-23.
5. Duthie GG, Wahle KJ. Smoking, antioxidants, essential fatty acids and coronary heart disease. Biochem Soc Trans 1990; 18:1051-4.

6. Pacht ER, Kaseki H, Mohammed JR, Cornwell DG, Davis WB. Deficiency of vitamin E in the alveolar fluid of cigarette smokers. J Clin Invest 1986; 77:789-96.

7. Chow CK, Airriess GR, Changchit C. Increased vitamin E content in the lungs of chronic cigarette-smoked rats. Ann N Y Acad Sci 1989; 570:425-7.

8. Chow CK, Chen LH, Thacker RR, Griffith RB. Dietary vitamin E and pulmonary biochemical responses of rats to cigarette smoking. Environ Res 1984; 34:8-17.

9. Chow CK. Dietary vitamin E and cellular susceptibility to cigarette smoking. Ann N Y Acad Sci 1984; 560:426-35.

10. McGowan E, Parenti CM, Hoidal JR, Niewoehner DE. Ascorbic acid content and accumulation by alveolar macrophages from cigarette smokers and non-smokers. J Lab Clin Med 1984; 104: 127-34.

11. York GK, Pierce TH, Schwartz LW, Cross CE. Stimulation by cigarette smoke of glutathione peroxidase system enzyme activities in rat lung. Arch Environ Health 1976; 29:286-90.

12. Duthie GG, Arthur JR, James WPT, Vint HM. Antioxidant status of smokers and non-smokers. Effects of vitamin E supplementation. Ann N Y Acad Sci 1989; 570:435-8.

13. Chow CK, Thacker RR, Changchit C, et al. Lower levels of vitamin C and carotenes in plasma of cigarette smokers. J Am Coll Nutr 1986; 5:305-12.

14. Smith JL, Hodges RE. Serum levels of vitamin C in relation to dietary and supplemental intake of vitamin C in smokers and non-smokers. Ann N Y Acad Sci 1987; 498:144-52.

15. Pelletier O. Smoking and vitamin C levels in humans. Am J Clin Nutr 1968; 21:1259-67.

16. Kallner AB, Hartman D, Hornig DH. On the requirements of ascorbic acid in man: steady state turnover and body pool in smokers. Am J Clin Nutr 1981; 34:1347-55.

17. Duthie GG, Arthur JR, James WPT. Effects of smoking and vitamin E on blood antioxidant status. Am J Clin Nutr 1990; 42:273-287.

18. Strain JJ, Carville DGM, Barker ME, et al. Smoking and blood antioxidant enzyme activities. Biochem Soc Trans 1989; 17:497-8.

19. Galdstone M, Feldman JG, Levytska V, Magnusson B. Antioxidant activity of serum ceruloplasmin and transferrin available iron-binding capacity in smokers and non-smokers. Am Rev Respir Dis 1987; 783-7.

20. Bridges RB, Rehm SR. Serum antioxidant activity as a determinant of pulmonary dysfunction in cigarette smokers. Basic Life Sci 1988; 49:631-4.

21. Ellis N, Lloyd B, Lloyd RS, Clayton BE. Selenium and vitamin E in relation to risk factors for coronary heart disease. J Clin Pathol 1984; 37:213-7.

22. Duthie GG, Robertson JD, Maughan RJ, Morrice PC. Blood antioxidant status and erythrocyte lipid peroxidation following distance running. Arch Biochem Biophys 1990; 282:1-6.

23. Meister A. Selective modification of glutathione metabolism. Science 1983; 220:472-7.

24. Duthie GG, Arthur JR. Vitamin E supplementation of smokers and non-smokers. Fat Sci Technol 1990; 92:456-8.

25. Sakamoto M. Ethane expiration among smokers and non-smokers. Jpn J Hyg 1985; 40:835-40.

26. Shariff R, Hoshino E, Allard J, Pichard C, Kurian R, Jeejebhoy KN. Vitamin E supplementation in smokers (abstract). Am J Clin Nutr 1988; 4:758.

27. Hoshino E, Shariff R, Van Gossum A, et al. Vitamin E suppresses increased lipid peroxidation in cigarette smokers. J Parenter Enter Nutr 1990; 14:300-5.

28. Horwitt MK. Supplementation with vitamin E. Am J Clin Nutr 1988; 47:1088-9.

29. Diplock AT. Dietary supplementation with antioxidants. Is there a case for exceeding the recommended dietary allowance? Free Radic Biol Med 1987; 3:199-201.

30. Jacobson HN. Dietary standards and future developments. Free Radic Biol Med 1987; 3:209-13.

31. Waterlow JC. General discussion. Diet and coronary heart disease. Proc Nutr Soc 1988; 47:3-8.

32. Duthie GG, Wahle KWJ, James WPT. Oxidants antioxidants and cardiovascular disease. Nutr Res Rev 1989; 2:51-62.

51

The Role of Vitamin E in the Protection of In Vitro Systems and Animals Against the Effects of Ozone

William A. Pryor

Louisiana State University, Baton Rouge, Louisiana

INTRODUCTION

Ozone, the most powerful oxidant in photochemical smog, places an oxidative burden on the entire organism as well as specifically on the lung. Half the population of the United States lives in areas that fail to meet the standards for ozone mandated by the Clean Air Act, and it is unlikely that some of these cities will be brought into compliance by the year 2005, when new, more restrictive controls will be fully implemented (1,2).

Antioxidant defense mechanisms can protect us against many types of oxidative stress (3–9), and it is possible that dietary supplementation with antioxidant vitamins may provide significant protection to humans against ozone. The literature on the protective effects of vitamin E against ozone is far more extensive than that on any of the other antioxidant nutrients, even though very limited evidence suggests that vitamin C may be even more potent than vitamin E. This chapter reviews the effects of vitamin E on ozone-induced damage in chemical systems, cells in culture, animals, and humans.

NONRADICAL MECHANISMS INVOLVED IN OZONE-INDUCED PATHOLOGY

Ozone causes damage by both radical and nonradical mechanisms. We recently suggested that ozone reacts with olefins, such as unsaturated fatty acids (UFA), to yield aldehydes and hydrogen peroxide, as shown in Equation (1), and that these species diffuse through the air-tissue boundary and cause damage to both pulmonary and nonpulmonary cells (such as the red blood cell, RBC) (10–12).

$$R-CH=CH-R' + O_3 + H_2O \rightarrow R-CH=O + R'-CH=O + H_2O_2 \tag{1}$$

For example, ozonized lipids cause lysis of human RBC, and the extent of lysis can be duplicated by a 2:1 mixture of aldehydes and hydrogen peroxide; since neither catalase nor iron chelators influence the extent of hemolysis, radicals do not appear to be involved (11). In addition, several authors have shown that lipid peroxidation may not necessarily accompany the damage caused by ozone. Even in these cases, in which ozone damage does not appear to involve radicals, vitamin E may provide some protection, although the mechanisms by which it might do so are largely unknown. It is true, however, that oxidative stress can be pervasive and that increases in the pool of reducing equivalents can be protective, even if radicals are not clearly involved in the changes that can be measured.

RADICAL MECHANISMS OF OZONE TOXICITY

Vitamin E protects polyunsaturated fatty acids (PUFA) in vitro from ozone-induced autoxidation. In vivo, ozone reacts with and damages biological target molecules, including PUFA, and vitamin E often prevents these reactions. These facts have led most investigators to infer that a substantial portion of the damage caused by ozone is mediated by free radicals. For this reason it is important to understand the mechanisms by which ozone (which is not a free radical) can react with biological target molecules to produce radicals.

Direct Reactions of Ozone with Olefins to Produce Radicals

Ozone reacts particularly rapidly with olefins, such as the monounsaturated and polyunsaturated fatty acids in lung lipids, and these unsaturated fatty acids are thought to be a primary target for the reactions of inhaled ozone. In agreement with this hypothesis, PUFA have been shown to undergo reaction with ozone in rats in vivo (13) and to be primary targets for attack by ozone in mouse lung fibroblasts (14). Ozone reacts with unsaturated compoundslike PUFA in a process called Criegee ozonation.

That vitamin E protects PUFA against ozone-induced damage in vitro and animals against virtually all types of ozone-induced morbidity and pathology in vivo has been interpreted to mean that ozone induces the production of free radicals in biological systems. Radical production in ozone-olefin reactions occurs both in organic, lipophilic solvents and in aqueous solutions. Radical formation in organic solvents can be demonstrated by bubbling ozone-containing air through olefins or PUFA and detecting the radicals formed using the electron spin resonance (ESR) spin trap method (15–18). More surprisingly, radical formation can also be demonstrated by the spin trap method for 3-hexenoic acid in homogeneous aqueous solutions and for solutions of unsaturated fatty acids in aqueous emulsions, micelles, and liposomes (19).

The primary role of vitamin E is the scavenging of free radicals within lipophilic regions of the cell, in particular within cell membranes. Thus, vitamin E should protect membrane lipids against the autoxidation that is initiated by these ozone-induced radicals. This protection against PUFA autoxidation by vitamin E can be seen in simple model system experiments; mixtures of PUFA and vitamin E that are exposed to ozone-containing air undergo less oxidation than samples that are not protected by vitamin E (20–23).

Ozone Does Not React Directly with Vitamin E

The protection that vitamin E provides to simple PUFA model systems could arise by two different mechanisms. In the first, vitamin E may react directly with ozone in a sacrificial way to destroy the ozone. In the second mechanism, ozone may react with PUFA to form radicals, these radicals could initiate autoxidation of the PUFA, and vitamin E could act in its usual radical-trapping mode, scavenging these PUFA-derived radicals and stopping the autoxidation. To distinguish these two possibilities, the rate constants for the reaction of ozone with typical PUFA and with vitamin E were measured both in homogeneous solution and in aqueous micellar systems (24,25). As expected, the rates of reaction of ozone with PUFA are extremely fast, but the rates of reaction with vitamin E were discovered to be similarly fast. However, because the number of PUFA molecules in a membrane exceeds that of vitamin E by a factor of 100–1000, ozone is expected to react exclusively with PUFA. Thus, these kinetic studies imply that vitamin E protects PUFA because it acts as a free-radical scavenger rather than as a sacrificial target for attack by ozone (24,25). This distinction is quite significant: if vitamin E were to destroy ozone directly in a sacrificial reaction, it would stop both ozone-initiated PUFA autoxidation and the direct ozonation of PUFA to form the Criegee ozonide. However, if vitamin E stops ozone-initiated PUFA autoxidation, then the rate of the Criegee ozonation of PUFA should not be affected by vitamin E.

Criegee Ozonation and the Carbonyl Oxide

The most rapid reaction between ozone and an unsaturated compound is a process called Criegee ozonation that involves an insertion of ozone into a double bond; Figure 1 illustrates this reaction for a 1,2-disubstituted unsaturated compound (such as an unsaturated fatty acid) [1], and shows the first product, the 1,2,3-troxolane [2] (26). The preceding section suggests that vitamin E cannot prevent this ozonation process since it does not react with ozone sufficiently rapidly to compete with the ozone-olefin reaction. [In contrast, vitamin C may be able to block even Criegee ozonation (25).] The 1,2,3-trioxolane, which has a trioxide bond, undergoes rapid O−O bond homolysis to form a diradical [3], which can rapidly decompose to give a species called the carbonyl oxide [4] and a mole of aldehyde [5]. Alternatively, the 1,2,3-trioxolane [2] may decompose directly to the carbonyl oxide and another aldehyde [8], as shown in Figure 1 (Ref. 27, p. 260). The carbonyl oxide is unstable and can undergo several reactions, two of which are shown. It can recombine with an aldehyde to form the Criegee ozonide [6], and if water is present, it can react to form a hydroxyhydroperoxide [7]. The

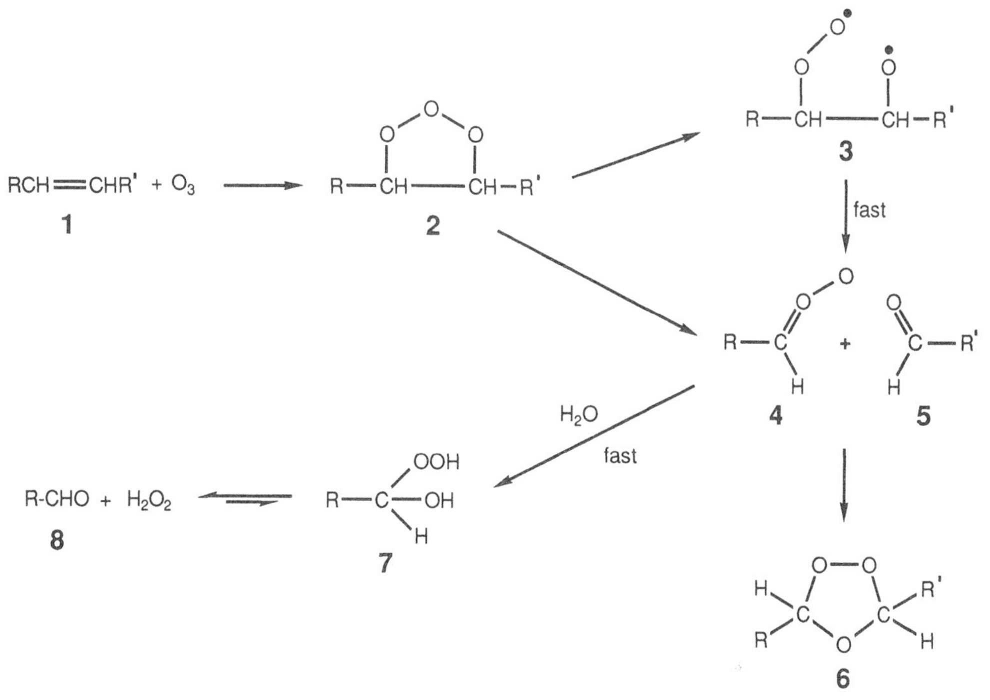

FIGURE 1 Mechanism of ozonation of a 1,2-disubstituted olefin (such as an unsaturated fatty acid) in the presence and absence of water. Compound 2 is the 1,2,3-trioxolene; 4 is the carbonyl oxide; 6 is the Criegee ozonide; and 7 is the hydroxyhydroperoxide.

Criegee ozonide [6] is quite stable in the absence of solvents that can hydrolyze it or reductants that can reduce its peroxidic bond (26), and although it undergoes homolysis to form free radicals, it does so very slowly (28). In contrast, the hydroxyhydroperoxide [7] is rapidly hydrolyzed to give a second mole of aldehyde and a mole of hydrogen peroxide.

Consequences of the Fact That PUFA Autoxidation Is a Chain Reaction

Even a small yield of radicals that can initiate the autoxidation of PUFA may be significant. Humans breathe about 1 L air per minute or 10^{-2} L/s. If air containing 0.1 ppm ozone is breathed and all the ozone reacts with PUFA, then unsaturated fatty acids are destroyed by Criegee ozonation at a rate of about 10^{-9} mol/L air inhaled, or 10^{-11} mol/s. If only 1% of the ozone initiates autoxidation of PUFA, then lipid peroxidation would occur in a lung of 1 L volume at a rate of 10^{-9} mol/L/s (29). This is a sufficiently rapid rate to produce the rapid destruction of PUFA and unsaturated lipids (29). Thus, because of the chain nature of the autoxidation of PUFA, even an extremely low yield of free radicals from ozone-PUFA reactions could result in appreciable amounts of PUFA being destroyed by autoxidation in addition to that destroyed by direct Criegee ozonation. This undoubtedly explains why vitamin E, which stops chain autoxidation reactions, can be so effective in preventing ozone-induced pathology both in vitro and in animals.

Electron Transfer Reactions to Ozone

Ozone adds to the double bond in molecules, such as unsaturated fatty acids, with very large rate constants; however, ozone also reacts very rapidly with several other types of molecules. In particular, thiols, such as glutathione (GSH), and thiol-containing proteins are particularly reactive toward ozone (30–32).

Amines react faster with ozone if they are unprotonated, whereas phenols, such as vitamin E, and thiols, such as GSH, react faster when they are in their ionized form (24,25,30, 33); thus, all these species react with ozone more rapidly when they are converted to a form that is a better electron donor. This suggests that these molecules might react with ozone by a one-electron transfer reaction, as shown in reaction (2) for a generalized electron donor.

The transfer of an electron to ozone, reaction (2), produces the ozone radical anion. This species is known to lead to the production of the extremely reactive hydroxyl radical, as shown in reaction (3) (34,35).

$$O_3 + X^{\cdot \cdot} \rightarrow O_3^{\dot{-}} + X^{\dot{+}} \tag{2}$$

$$O_3^{\dot{-}} + H^+ \rightarrow HO^{\cdot} + O_2 \tag{3}$$

Thus, ozone may react with electron donors, such as phenols, thiols, and amines, to produce the hydroxyl radical. (Pryor et al., unpublished observations, 1990).

Role of Water-Soluble Antioxidants

It is important to recognize that water-soluble compounds, such as ascorbate and glutathione, can act as scavengers of ozone and/or of radicals produced from ozone-PUFA or ozone-protein reactions; in fact, in vitro data suggest that vitamin C, for which there are few data, is even more important than vitamin E as a scavenger of ozone in the lung (25).

Indirect Mechanisms for Radical Production

Ozone causes the recruitment of phagocytic cells to the lung, and these cells release superoxide and hydrogen peroxide, which could lead to higher concentrations of radicals in the lungs of ozone-exposed animals. This mechanism for ozone-induced radical production is quite different from those already discussed in that the radicals produced are formed endogenously in enzyme-controlled processes that are activated by the effects of ozone, rather than from direct, non-enzymatic reactions of ozone.

Increased numbers of phagocytes are found in the lungs of ozone-exposed rats (36). Furthermore, Esterline et al. (37) found that rat lung inflammatory cells isolated by lavage and exposed to 2 ppm ozone show a strong induction of myeloperoxidase-derived oxidants, suggesting that neutrophils recruited to the lung in ozone-exposed animals generate increased amounts of activated oxygen species. Thus, increased numbers of phagocytic cells could be recruited to the lung in ozone-exposed animals and release increased amounts of superoxide, leading to increased levels of hydrogen peroxide and hypochlorite. Since antioxidants, such as vitamin E, can modulate the responses of neutrophils (28), the protective effects of vitamin E against ozone exposure may be partly explained by an antiinflammatory effect by vitamin E on an ozone-induced phagocyte response. If this were true, that vitamin E protects animals against ozone would be explained not because of direct radical-producing reactions of ozone with biotarget molecules, but instead because of complex biological cascades of reactions triggered by ozone and modulated by vitamin E. This phagocyte-mediated theory of ozone toxicity cannot explain the direct radical production observed in the reactions of ozone with PUFA in studies in which phagocytic cells are not present (17–23,39,40). It is difficult to estimate whether radicals produced directly from ozone-olefin reactions or from phagocytic cells are the more important.

Hydrogen peroxide is produced in animals exposed to ozone (41–44). Hydrogen peroxide is also produced in model systems or animals exposed to elevated oxygen concentrations, presumably because of more rapid production of superoxide by mitochondria (45,46). Thus, the hydrogen peroxide produced as a result of inhalation of ozone could arise from reaction (1) or could arise as a generalized response to the oxidative stress caused in cells exposed to ozone-containing air.

Calculation of the numbers of radicals produced in the lung from reaction of a toxin with biological target molecules versus oxyradicals formed endogenously from the activation of phagocytic cells is a general problem. For example, cigarette smoke (47), as well as smoke from plastics and other sources (48–50), contains oxyradicals and also causes recruitment and activation of phagocytes to the lung, and the relative numbers of radicals from the smoke itself and from activated macrophages are unknown (51).

ACUTE TOXICITY OF THE CRIEGEE OZONIDE

Rapid nonpulmonary changes are observed in animals and humans exposed very briefly to low levels of ozone. For example, short ozone exposures cause changes in blood chemistry at sites distal to the lung (52,53). Ozone itself is far too reactive to be able to diffuse through the air/blood interface without reacting (11). Thus, either a reaction product of ozone with a biomolecule has sufficient stability to circulate in the bloodstream, or a cascade of biochemical changes transmits the oxidant stress on the lung to distal sites (11).

If an ozone reaction product were formed from ozone-PUFA reactions and were able to diffuse into the bloodstream, the Criegee ozonide might be thought to be a likely possibility, since in a nonprotonic milieu the Criegee ozonide is the major product of ozone-PUFA reactions and these ozonides are quite stable in the absence of hydrolysis or peroxide-destroying compounds (54). However, there are two arguments that suggest that the Criegee ozonide is not an important mediator of ozone toxicity. First, as discussed earlier, the carbonyl oxide is probably trapped by water, leading to the formation of hydrogen peroxide and aldehydes, rather than the Criegee ozonide. Second, even if formed, Criegee ozonides appear to have quite a low toxicity, at least in the acute studies that have been done (55–57).

ANIMAL DATA ON VITAMIN E AND OZONE

All the data that have been published strongly support the hypothesis that vitamin E protects animals against the harmful effects of ozone. For example, the review by the U.S. Environmental Protection Agency (EPA) concludes (Ref. 58, p. 161) that the "antioxidant properties of vitamin E in preventing ozone-initiated peroxidation . . . in vivo are clearly demonstrated in rats and mice."

However, a point that appears to have been overlooked is that most of the studies that have been published in this field compare vitamin E-deficient to vitamin E-sufficient animals (59). Furthermore, most of the published experiments use rats, and standard rat chow provides more vitamin E than the absolute minimum necessary to prevent at least some deficiency diseases. (The National Research Council, NRC, recommended level of vitamin E for rats is 30 mg/kg feed.) Thus, some of the studies using rats compare vitamin E-deficient animals to relatively highly supplemented animals, leaving unanswered the question of whether a lesser amount of vitamin E would have been equally effective in protecting animals against the effects of ozone.

Thus, the published studies show that the absence of vitamin E potentiates the harmful effects of ozone. However, few of the studies ask whether an increased dietary intake of vitamin E protects animals better than the minimal amount needed to prevent vitamin E deficiency diseases. This question is vitally important, since it is generally agreed that overt vitamin E deficiency is extremely rare in developed countries. Thus, for humans, we need to know whether supplementation with vitamin E above that in the typical U.S. diet can provide increased protection against the deleterious effects of ozone. [The typical U.S. diet is thought to provide an amount of tocopherols near the recommended dietary allowance, RDA, which is 10 tocopherol equivalents (TE), where 10 TE equals 10 mg d-α-tocopherol, or 15 IU (60).]

Increases in Measures of Oxidative Stress and Changes in the Levels of Reducing Equivalents

In one of the early publications in this area, Goldstein et al. (52) exposed both vitamin E-sufficient and E-deficient rats to 10 ppm ozone; eight of eight of the vitamin E-deficient rats died of pulmonary edema after 6–7 h of exposure whereas all of the control animals survived the experiment. Rats exposed to 3.5 ppm ozone for 4 h/day for 10 days had reduced serum vitamin E levels, suggesting an ozone-induced depletion of vitamin E, a measure of oxidative stress. This reduction in serum vitamin E levels was observed regardless of whether the animals were partially vitamin E deficient, but it must be noted that this early experiment used 10 ppm ozone, a high level relative to ambient exposures.

Moore et al. (61) studied blood chemistry parameters for C57L/J mice exposed to 0.3 ppm ozone for 6 h. They found that serum vitamin E, GSH, and glucose-6-phosphate dehydrogenase (G6PD) levels all increase; in these experiments the oxidative stress of ozone induces an increase in blood serum concentrations of chemical species that can contribute reducing equivalents.

A preliminary report by Elsayed (62) compares the mobilization of vitamin E in vitamin E-deficient rats and rats supplemented with 1000 IU/kg for 2 weeks. The rats were exposed to 0.5 ppm ozone continuously for 5 days, and both the vitamin E content of lung tissue and the uptake of vitamin E using ^{14}C-labeled d,l-α-tocopherol acetate were determined. Both the vitamin E-deficient and E-sufficient rats take up vitamin E more rapidly after exposure to ozone, by 28 and 147%, respectively. The vitamin E content of the lung tissue in the vitamin E-deficient animals decreased by 38% after exposure to ozone, whereas that in the vitamin E-sufficient animals increased by 35%. These differences in response may explain the discrepancies in the older work discussed earlier. It should also be mentioned that Chow and his coworkers report that vitamin E content increases in the lungs of rats (63) and guinea pigs (64) exposed to cigarette smoke. Thus, there appears to be a mobilization of vitamin E to the lungs of animals exposed to some types of oxidative stress.

Chow and Tappel (65) reported the exposure of rats to 0.8 ppm ozone for 7 days and found the activities of the pentose shunt enzymes in the lungs are increased, enzymes that are used to provide reducing equivalents. They reported that 45 mg/kg of *d,l*-α-tocopherol partially retarded the ozone-induced elevation of glucose-6-phosphate dehydrogenase and 6-phosphogluconate dehydrogenase. The activities of GSH peroxidase and GSH reductase decreased initially and then increased continuously over a 29 day exposure to ozone. These findings suggest, as one might expect, that the time point at which enzyme activities are measured might be critical; many workers have measured enzyme activities at only one time point and have not studied the time dependence of the changes. Chow and Tappel suggest their results can be explained by the increased need for the organism to produce reducing equivalents, which are necessary to detoxify peroxidic compounds produced in ozone-exposed rats.

Morgan et al. (66) suggest that changes in RBC deformability may be an early marker of oxidative stress. They exposed mice to 0.3, 0.7, or 1.0 ppm ozone for 4 h and showed that RBC deformability, as measured by the rate of RBC filterability, decreases in ozone-exposed mice. Supplementation with 105 mg α-tocopheryl acetate per kg diet partially protected the mice against this ozone-induced decrease in deformability. An increase in hematocrit was also found in ozone-exposed mice, and again vitamin E protected against this increase. No Heinz body formation, a measure of hemoglobin oxidation that could cause a decrease in RBC deformability, was found. Decreases in GSH or increases in thiobarbituric acid-reactive substances (TBARS) were not found.

Changes in Thiol Levels

Thiol-containing compounds are rapidly oxidized by ozone and may therefore be expected to be depleted in the lung of ozone-exposed animals (30,53,58,67,68). For example, Menzel (69) reported that thiols, such as GSH and cysteine, are rapidly oxidized by ozone in aqueous buffers and that the GSH oxidation product is not reducible by GSH reductase, suggesting that the thiol group is oxidized above the disulfide stage (possibly to a sulfonic acid). DeLucia et al. (70) also suggested that the loss of thiol groups may be a critical consequence of ozone exposure, since thiols and thiol-dependent enzyme levels were significantly decreased in rat lung homogenates exposed to ozone.

Freeman, Mudd, and coworkers (31,71) studied the oxidation of GSH in the human RBC. About 5% of the ozone caused GSH oxidation; the products were mainly GSSG, but appreciable amounts of higher oxidation products were also formed. If ozone can penetrate the cellular membrane without reacting, which seems unlikely, this intracellular oxidation of GSH could be due to ozone itself; if ozone cannot penetrate a membrane, then the oxidation of GSH results from an ozone-derived oxidant (perhaps hydrogen peroxide) produced by the reaction of ozone with a biotarget molecule. Interestingly, the Criegee ozonide reacts with GSH and GSH S-transferases (72,73).

As discussed, Moore et al. (61) found that GSH levels in serum slightly increase in C57 mice exposed to 0.3 ppm ozone for 6 h. In contrast, Chow and Kaneko (74) reported a decrease in RBC GSH in vitamin E-deficient Sprague-Dawley rats but not in vitamin E-sufficient rats. They also find that G6PD, catalase, superoxide dismutase (SOD), and TBARS are not significantly altered by ozone exposure or by the nutritional status of vitamin E.

In 1981, Chow et al. (75) reported the exposure of rats on either 0, 11, or 110 ppm vitamin E to 0.1 ppm ozone continuously for 7 days. They studied changes in the levels of proteins in the lung, rather than in serum or RBC, which is probably a more direct measure of the oxidative effects of ozone. They found that GSH, G6PD, and GSH peroxidase were significantly increased in the lungs of rats on the vitamin E-deficient diet and the 11 ppm vitamin E diet, but that 110 ppm vitamin E protected the rats from these changes. They also reported by scanning electron microscopy (SEM) examination that vitamin E protects against pathological changes in a dose-dependent fashion. Two 1979 publications by Chow and his coworkers are consistent with these findings on lung GSH changes (76,77).

Some of the data just described suggest differences in the response of different species and/or strains of animals to ozone. Joshi et al. (78) reported the effects of cigarette smoke on GSH and GSSG levels in perfused rat and rabbit lungs, and there are profound species differences.

Increased Exhalation of Ethane and/or Pentane

Ethane is produced during the autoxidation of $n-3$ PUFA and pentane from the autoxidation of $n-6$ PUFAA, and the increased exhalation of either of these volatile hydrocarbons is thought to be a marker for increased rates of lipid autoxidation that may occur in animals exposed to oxidative stress (79–81). Several workers reported that ethane or pentane is exhaled in increased amounts by vitamin E-deficient animals exposed to ozone (82,83).

Dumelin et al. (83) exposed rats to 1 ppm ozone for 1 h and measured the amounts of ethane and pentane in exhaled breath. The rats were kept on diets that included 0, 11, or 40 IU vitamin E acetate for 8 weeks before ozone exposure. Pentane was found to be increased in the vitamin E-deficient rats exposed to ozone relative to room air controls. The implication from this study is that exposure to even relatively low levels of ozone increases the rate of lipid autoxidation in ozone-exposed rats and that vitamin E reduces the rate of this process.

Dumelin et al. (84) exposed bonnet monkeys to 0, 0.5, or 0.8 ppm ozone for 7, 28, or 90 days and measured the effects of ozone exposure on exhaled ethane and pentane. No differences were found in the levels of expired gases for any of the ozone exposures. The authors point out that the commercial monkey chow used to feed these animals provided 85 mg vitamin E per kg diet. They calculate that this is about eightfold more vitamin E than required, based on the need for vitamin E to protect dietary linoleic acid from oxidation in humans, as calculated by Horwitt (85), and assuming a similar vitamin E requirement in monkeys and humans. They then hypothesized that the absence of elevated ethane and pentane levels following ozone exposure resulted from the fact that the monkeys were well protected by vitamin E. The authors point out that only vitamin E-deficient rats in their earlier study (83) exhaled increased amounts of alkane gases on exposure to ozone, as described earlier.

Dutta et al. (86) studied the evolution of ethane and pentane from isolated, perfused rat lungs exposed to about 25 ppm ozone. The efflux of GSH in the lung perfusate was increased by ozone, but there was not a statistically significant increase in either ethane or pentane evolution on exposure to ozone.

These findings suggest that ethane and/or pentane evolution is not sufficiently sensitive to detect low-level, early changes caused by ozone. Since oxidation of lung lipids is presumably accelerated by ozone, this seems surprising. However, regardless of its sensitivity and usefulness to measuring oxidative stress in animals exposed to ozone alone, ethane and pentane evolution is not sufficiently sensitive to probe the protection that vitamin E may afford against the effects of ozone in humans, as reviewed subsequently.

TBARS

Chow and Tappel (87) exposed rats to 0.7 or 0.8 ppm ozone continuously for 5 or 7 days measured TBARS in lung lipids. The rats were fed diets containing 0, 10, 45, 150, or 1500 mg/kg of α-tocopherol. Increases in TBARS, as well as increases in the activities of GSH peroxidase and G6PD, were observed and were partially inhibited by vitamin E; however, the activity of GSH reductase was not affected by α-tocopherol. Chow and Kaneko (88) reported that vitamin E-deficient rats exposed to 0.8 ppm ozone continuously for 7 days undergo an increase in several enzyme activities, including glutathione peroxidase; however, catalase and SOD activities and amounts of TBARS were not increased.

Morgan et al. (66) exposed vitamin E-deficient and E-sufficient male CD-1 mice to 1.0, 0.7, or 0.3 ppm ozone for 4 h. They found no detectable TBARS in the ozone-exposed animals. However, they reported increases in RBC filtration times for ozone-exposed mice and suggested that measurement of RBC filterability may be useful "as a clinical test for short-term injury from exposure to oxidant gases."

Eicosanoids

The data just summarized suggest that TBARS may not be sufficiently sensitive to serve as a marker of early or low-level damage by ozone. However, there is another aspect of the oxidation of PUFA to yield malonaldehyde (MDA) that is worth notice. In 1976, we suggested that ozone-initiated autoxidation of PUFA produces prostaglandin (PG)-like endoperoxides that are the source of the MDA detected in the TBA test (89). The report of Rabinowitz and Bassett (13) suggested that arachidonic acid undergoes ozonation to a far greater extent than any other lung PUFA; this could be interpreted as suggesting that ozonation activates lipases that liberate arachidonic acid from membrane lipids so that it is more accessible to ozone than other PUFA. Both Roycroft et al. (90) and Leikauf et al. (91) suggest that ozone increases the activities of both lipoxygenase and cyclooxygenase products in the lung. Recently, Morrow et al. (92) showed that some autoxidation-derived products of arachidonic acid, in particular an epimere of $PGF_{2\alpha}$, have extremely potent biological properties, consistent with our 1976 suggestion.

Leikauff et al. reported that ozone exposure increases both cyclooxygenase and lipoxygenase products in bovine tracheal cells exposed to ozone concentrations as low as 0.1 ppm for 2 h (91). These authors found that arachidonic acid in the absence of cells was not oxidized in their system. Schelegle et al. (93) report that plasma $PGF_{2\alpha}$ is significantly increased in ozone-sensitive humans exposed to 0.35 ppm ozone for 80 minutes.

These reports suggest that further studies of the oxidation of arachidonic acid, the non-enzymatic production of MDA, and the possible role of arachidonic acid oxidation products in ozone toxicity may be rewarding. It is possible that changes in eicosanoids could be used as an early marker of ozone-induced damage.

Changes in Enzyme Levels

Enzymes, particularly thiol-dependent enzymes, may be expected to be inactivated by direct reaction with ozone (32,65,94,95). Chow et al. (76) studied the exposure of rats to 0.7 or 0.8 ppm ozone continuously for 7 days and found changes in both vitamin E-deficient and E-sufficient rats. Enzyme activities and/or levels of proteins or polypeptides changed in the order GSH peroxidase > lactate dehydrogenase > G6PD and pyruvate kinase > GSH > GSH reductase > malic dehydrogenase. Lesions in the lungs from the vitamin E-deficient animals were ranked as more severe by two independent investigators who examined the tissues without being aware of whether they came from ozone-exposed or control animals (76).

Changes in Morphology

Sato et al. (96) reported that rats exposed to 0.3 ppm ozone for 3 h/day for up to 16 days show pathological changes in lung morphology. Chow reported similar findings for rats exposed to 0.7 ppm ozone for 7 days (97). In both studies, vitamin E-sufficient animals were significantly protected relative to E-deficient animals. The report by Chow et al. (75) described earlier in which SEM data were reported for rats exposed to 0.1 ppm ozone and supplemented with 0, 11, or 110 ppm vitamin E is particularly interesting. This group reported lesions in five of six of the rats on the vitamin E-deficient diet, four of six of the rats supplemented with 11 ppm, and only one of six rats on the 110 ppm vitamin E diet. Since the NRC recommendation for rat chow is 30 ppm vitamin E, as discussed earlier, this study suggests that amounts of vitamin E significantly less than this level lead to greater sensitivity to ozone than in rats supplemented with amounts considerably above this amount.

Studies That Report Several Levels of Vitamin E

Most studies compare the effects of ozone exposure on animals that either are vitamin E deficient or are fed one level of dietary vitamin E. However, few humans are vitamin E deficient. Therefore, we wish to know if persons eating a "typical" diet can benefit from taking supplemental vitamin E. (The vitamin E content of a typical diet depends on the subject group. Americans are thought to eat a diet that supplies an amount of vitamin E approximately equal to the RDA; for other countries, particularly northern European countries (98) and undeveloped nations, the diet supplies less than the U.S. RDA.) Thus, studies that report the effects of more than one dietary level of vitamin E on ozone-induced changes in animals are of particular interest.

In 1972, Chow and Tappel (87) reported what appears to be the first study of this type; they exposed rats fed 0, 10, 45, 150, or 1500 mg/kg vitamin E to 0.7 or 0.8 ppm ozone continuously for 7 days. Some measures of oxidative stress, such as TBARS and the activity of GSH peroxidase, were correlated with the vitamin E status of the rats in a dose-dependent fashion.

In 1973, Fletcher and Tappel (99) reported a study in which they exposed rats to increasing ozone levels, starting with 2.5 ppm, then 5, 8, 12, and finally 16 ppm, with an 8 h exposure followed by a 16 h recovery period between each ozone exposure. Rats were divided into five groups supplemented with 0, 10, 45, 150, or 1500 mg vitamin E per kg feed. Fletcher and Tappel studied the survival of the animals, and in general, the more vitamin E in the feed, the longer the animals survived. For example, all the vitamin E-deficient animals died by the second day (during or following the 5 ppm exposure), whereas four of eight animals were still alive even after day 5 (and the 16 ppm exposure) for the 1500 mg/kg E group. (The feed of the 1500 mg/kg vitamin E group also contained 1500 mg/kg methionine and ascorbate and 45 mg/kg BHT, butylated hydroxytoluene.)

It would be of interest to replicate studies similar to those described by Chow and Tappel (87), by Fletcher and Tappel (99), and by Chow et al. (75), for several reasons. (1) It is desirable to confirm that more vitamin E protects rats against ozone better than lesser amounts of vitamin E. (2) In general, these studies use small numbers of animals. (3) In two of these studies (87,99), high levels of ozone were used; levels near or below 1 ppm should be studied.

Thus, a need exists for studies in which a large number of animals on several vitamin E diets are repeatedly exposed to levels of ozone near ambient levels for a period of weeks or months. A number of biochemical and morphological parameters should be measured, particularly those parameters that may be sensitive to early, low levels of ozone-induced damage.

Elsayed et al. (100) reported the exposure of rats to 0.8 ppm ozone for 8 h/day for 7 days. Rats were fed 0 or 50 IU vitamin E per kg diet. Significant increases in the activities of several pulmonary enzymes were found in both dietary groups upon exposure to ozone, but the increases were greater in the 0 than in the 50 IU vitamin E group. Vitamin E protection against the ozone-induced increases in enzyme activities was observed for succinate oxidase, succinate-cytochrome c reductase, GSH peroxidase, GSH reductase, SOD, and G6PD. In a second experiment by these workers (100), rats were given 10, 50, or 500 IU vitamin E per kg diet for 2 months and then exposed to 0.8 ppm ozone for 4 days. Again, all the ozone-exposed groups showed increases in enzyme activities relative to air-exposed controls. The group fed 10 IU was protected less against these ozone-induced increases in enzyme activities than the groups fed 50 or 500 IU vitamin E, but the differences between the 50 and 500 IU groups were small. The authors conclude that vitamin E "protects [against lung injury, but] the magnitude of the protective effect does not increase proportionately with increased dietary vitamin E supplementation beyond a certain level." This publication is important both because it reports several vitamin E levels and because of the plateau effect in enzyme activation that appears to have been observed with increasing vitamin E intake.

ENVIRONMENTALLY REALISTIC LEVELS OF OZONE

Many of the animal studies already described used ozone concentrations far above those actually found in smog. Many of the ozone and vitamin E studies were done at an early stage in the study of ozone toxicity, and at that time it was not uncommon to use levels of ozone as high as 30 ppm. These acute exposures allow shorter studies than would be required if lower levels of ozone were used; for example, the Fletcher and Tappel study described earlier could be performed in 4 or 5 days to probe the protective role of vitamin E against ozone-induced death (99). However, the question that must now be addressed is whether long-term, chronic exposure to levels of ozone near 1 ppm cause clinically relevant changes that affect human health. It is not clear that high concentrations of ozone, which very rapidly kill animals, cause the same biochemical transformations as low concentrations. High levels of ozone may overwhelm the antioxidant defense systems, whereas lower levels may allow continual regeneration of reducing equivalents and repair of low-level damage. For these reasons, the EPA now recommends (Ref. 58, pp. 1–126): "If animal models are to be used to reflect the toxicological response in humans [to ozone], then the endpoint for . . . such studies should be morbidity rather than mortality." Very few studies (59,75,77,101) probe the effects of vitamin E on chronic exposure to environmentally realistic concentrations of ozone. This clearly is an area in which data are badly needed, since it is possible that ozone levels even below 0.12 ppm may cause permanent changes (102). For example, Barry et al. (36) exposed rats for 6 weeks to either 0.25 or 0.12 ppm ozone for 12 h/day and then performed a morphological study of the area of the lung thought to be most sensitive to ozone-induced damage. Significant changes were found by ultrastructural morphometric analysis, suggestive of increased cellular turnover in the ozone-exposed group. Both type 1 and type 2 cells increased in a dose-dependent manner, with greater changes in the 0.25 ppm ozone-exposed group. The numbers of macrophages also increased, doubling in the 0.25 ppm group. The authors (36) conclude that "low concentrations of ozone cause a chronic epithelial injury in the proximal alveolar region . . . in a concentration-dependent manner, even at concentrations as low as 0.12 ppm."

CONTROLLED HUMAN EXPOSURES TO OZONE

The effects of the exposure of human volunteers to low levels of ozone for short times in an ozone exposure chamber have been reviewed by several authors (103,104). Human exposure to ozone and the effects of many variables are covered in detail in the EPA document (see Ref. 58, pp. 142ff); the EPA concludes that "no evidence indicates . . . that man would benefit from increased vitamin E intake relative to ambient ozone exposures."

Chaney et al. (105) reported a study of 44 humans exposed to 0.4 ppm ozone and 30 controls exposed to air for 4 h on 5 consecutive days. Nine biochemical blood parameters were measured; only vitamin E levels, RBC G6PD, and complement C_3 showed significant changes with exposure to ozone. Groups that underwent exercise during ozone exposure had lower serum vitamin E levels than control groups.

The effect of ozone exposure on pentane exhalation was studied in a group of 10 humans exposed to 0.3 ppm ozone while exercising (106). As described earlier, these alkane gases are thought to be exhaled in greater amounts by animals and humans undergoing oxidative stress (79–81,83,107–111). This study found that exercise increased the amount of exhaled pentane; however, ozone exposure was not found to further increase pentane. This is consistent with the reports of these same investigators that only vitamin E-deficient animals exhale increased amounts of pentane or ethane (83,84). Since the humans enrolled in this study were not vitamin E deficient, it might be expected that ozone would not influence the yields of exhaled hydrocarbon gases. It is noteworthy, however, that exercise itself produced an increase in exhaled pentane in these humans.

Hackney and his collaborators have published several articles that report the results of ozone exposures of humans who were given vitamin E supplements. All the studies reported by this group used a placebo-controlled, double-blind protocol. In 1979, experiments were reported using 26 females and 9 males (112). After 9–11 weeks of supplementation with 800 IU d,l-α-tocopherol or a placebo, both the vitamin E and placebo groups were exposed to 0.5 ppm ozone for 2 h (112). No significant differences were found in the blood chemistry of the vitamin E versus the placebo group before ozone exposure; parameters studied include RBC fragility, hematocrit, RBC GSH concentration, and the activities of typical RBC enzymes (e.g., G6PD). These parameters also did not change in either the vitamin E or the placebo group upon ozone exposure; for example, neither RBC fragility nor RBC GSH levels were changed by ozone exposure. In neither the control nor the vitamin E-supplemented humans did ozone exposure lessen the serum concentration of vitamin E, but a 2 h exposure may not be sufficient to lower serum vitamin E. Cigarette smokers have vitamin E concentrations that are normal in serum but are lower in fluid obtained by lung lavage (80), suggesting that stresses that place an oxidative burden on the lung may produce changes that are only detectable in lung lavage fluids, not in blood.

This publication (112) also reported a study of a group of 22 male volunteers supplemented with 0, 800, or 1600 IU d,l-α-tocopherol for 9 weeks and exposed to 0.5 ppm ozone for 2 h. No significant differences were found between the supplemented and placebo groups in RBC fragility, hematocrit or hemoglobin values, RBC GSH, G6PD, or lactic acid dehydrogenase (112).

The results reported in this 1979 study (112) are in partial disagreement with a 1975 study by the same group (113) that found some blood chemistry changes in subjects exposed to the same (0.5 ppm) concentration of ozone for a slightly longer time period (2.75 h).

In contrast to these results in humans, and as discussed previously, Morgan et al. (66) found that RBC deformability decreases in mice exposed to ozone in a dose-dependent fashion and that vitamin E partially prevents this change. The lack of agreement between the rodent and human experiments is probably due in part to the fact that the human controls in the experiments referred to earlier were not vitamin E deficient. It is known, for example, that hemolysis is not observed in blood from normal, vitamin E-sufficient humans. It is also true that the animal and human exposures differ in that the animal experiments use higher ozone exposures for much longer times. The short ozone exposures used in these human studies may not produce sufficient oxidative stress to alter the types of noninvasive parameters that can be studied using human subjects.

A second report using the same 22 males used in the previous study (112) was published in 1981 (114). The group was divided into 11 controls and 11 subjects who were given 800 IU d,l-α-tocopherol for 9 or 10 weeks; both groups were then exposed to 0.5 ppm ozone for 2 h. No significant differences in pulmonary function tests (such as forced expiratory volume in one second, FEV_1, and forced vital capacity, FVC) were found for the group taking vitamin E versus the placebo group.

In another study using the same 22 males (114), the subject group was given 1600 IU *d,l*-α-tocopherol for 11 or 12 weeks. During this time the mean serum tocopherol concentrations increased 140%; however, there was also an unexpected and unexplained 30% increase in serum vitamin E levels in the placebo group. The experimental group was then exposed to a purified air control followed by a 2 h exposure to 0.5 ppm ozone on 2 successive days. It was anticipated that the second ozone exposure and the higher vitamin E supplementation level would lead to a clear-cut difference between the vitamin E and placebo groups. However, all the pulmonary function tests again showed no significant differences between the supplemented and placebo groups.

Hackney and his collaborators conclude (114) that their studies "offer little support for the hypothesis that supplemental vitamin E has protective effects against short-term responses in people exposed to ozone." They go on to state that "Since controlled clinical studies tend to be difficult and are limited by ethical considerations, additional work [using] laboratory animals is indicated." To this might be added the following caveats. Most of the significant results in these two publications (112,114) are based on studies of the same 22 male subjects. As stressed earlier, short-term exposures to ozone were used: 0.5 ppm for 2 or 4 h. Longer or more frequent exposures to ozone may result in changes in the lung that might be resisted by vitamin E-supplemented subjects. Additionally, it must be remarked that there is no consensus on the ideal early marker for low levels of ozone-induced pathology, and it may be difficult to probe the effects of ozone and vitamin E on humans, given the low ozone levels that must be used and the short exposure times, without this tool.

CHANGES IN THE IMMUNE RESPONSE

Effects of Ozone

A large body of evidence shows that ozone exposure inhibits bactericidal activity, decreases macrophage function, alters immune function, and increases susceptibility to infection (Ref. 58, Table 1-15). For example, Bergers et al. reported increased susceptibility to infection of mice exposed to 0.2–0.7 ppm ozone and virulent *Klebsiella pneumonia* or nonvirulent *Staphylococcus aureus* bacteria (115). A review by Ehrlich (116) stated that "decreased resistance to respiratory infections as a result of inhalation of photochemical oxidant pollutants [is] observed in various species of animals, including mice, hamsters and monkeys." Ehrlich also reported that ozone increases the susceptibility of humans to bacterial infections (116). These effects may only be seen in long exposures to ozone. As Ehrlich pointed out (116), "at low concentrations [of ozone-nitrogen dioxide mixtures, effects on the immune status were not seen] until 6- to 9-month exposures [of mice] to the pollutants."

Vitamin E may protect humans only against the long-term changes in the lung that occur only with chronic ozone exposures. This could explain the lack of protective effect of vitamin E that is observed in the human studies reviewed earlier, in which vitamin E supplementation was only given for several weeks and ozone exposures were limited to no more than 5 h. It may also suggest that immune competence is a sensitive measure of changes that occur with chronic ozone exposures in humans.

Effects of Vitamin E

In many animal species vitamin E has been found to improve the immune response (117–119). Strikingly, and as discussed earlier, the immune response in rats was found to correlate with the serum level of vitamin E even up to levels of vitamin E far greater than those necessary for the prevention of RBC hemolysis, correlating with serum levels of vitamin E from 0.04 to 18 μg/ml (119).

This improvement in immune response for animals given large amounts of vitamin E has a parallel in an ex vivo study of humans. Meydani and her collaborators (120) reported the immune response in 32 human volunteers 60 years old or older in a double-blind study using 800 IU *d,l*-α-tocopherol. Delayed hypersensitivity responses to skin tests were significantly improved in the vitamin E-supplemented group. Peripheral blood T cells were tested against a variety of antigens and mitogens, and most tests showed significant improvement for those subjects on vitamin E supplementation.

Chavance et al. also reported that French healthy subjects over 60 years old show correlations between nutritional and immunological status (121). In particular, vitamin E serum concentration was negatively correlated with the number of infections in the 3 years preceding the study ($p < 0.01$).

Summary

Ozone decreases and vitamin E improves the immune competency of animals. Thus, there is the possibility that vitamin E provides protection against the subtle, long-term effects of exposure to ozone.

CANCER, SMOG, AND VITAMIN E

Effects of Ozone

Oxidative stress, which often produces a higher concentration of oxyradicals in tissue, appears to contribute to tumor initiation and/or promotion, and antioxidants often protect against these effects (4,122–126). Thus, smog could lead to higher cancer rates, and this effect could be mitigated by vitamin E (104,127). Furthermore, ozone exposure causes changes in enzymes associated with xenobiotic metabolism, suggesting that carcinogenic substances are metabolized differently by animals exposed to smog (128,129).

Cells in culture have been used to study the carcinogenicity of ozone and the protective effects of vitamin E. These studies show that ozone may interact with other environmental hazards to increase tumor yields (130,131), and vitamin E may protect against some of these chemically induced cell transformations (132).

Hassett et al. (133) reported that mice exposed to either 0.3 or 0.5 ppm ozone intermittently for 6 months had an increase in lung tumor number relative to controls. Some of the methodology used by Hassett et al. has been criticized (134). However, Last et al. (135) took these criticisms into account and repeated a similar study, and they reached similar conclusions.

Epidemiological studies in 31 California counties suggest that both overall male mortality and female cancer mortality show positive correlations with air pollution indices, although a poorer correlation was found with oxidant levels than with some other indices of pollution (136).

A critical review by Witschi (127) of all the evidence for ozone carcinogenicity, inclduing the reports of Hassett et al. and Last et al., concluded that ''[although] there is little evidence to implicate ozone or nitrogen dioxide directly as pulmonary carcinogens . . . they might modify and influence . . . carcinogenic processes in the lung [perhaps by having tumor promotional activity].'' A more recent review of ozone toxicity by Lippmann (104) reached a similar conclusion.

Effects of Vitamin E

A review by Birt (137) of animals exposed to known carcinogens and supplemented with vitamin E supported the hypothesis that vitamin E can reduce chemically induced tumor formation.

The review covered tumors in a number of animal species in different organs and induced by several types of carcinogens (but not ozone). A similar review by Mirvish came to the same conclusion (138).

The potential health effects of a number of nutrients and dietary regimens are currently under study in 32 National Cancer Institute (NCI)-funded prospective intervention trials, a few of which involve vitamin E. The results of these studies are not yet known. However, several retrospective studies have been published in which serum was collected and stored and then serum levels of nutrients compared for subjects who later develop cancer and matched controls who do not. For example, Menkes et al. found that both β-carotene and vitamin E levels were lower in patients who subsequently developed lung cancer than in controls who did not in a large study using serum collected in 1974 (139). An ongoing study in Basel, Switzerland found a correlation between low vitamin E levels and colon cancer (140). Knekt et al. have published extensive data of a similar nature (141).

Very recently, Wald et al. reviewed many of the retrospective studies (including two by his group) that have correlated the cancer rates for humans versus their serum vitamin E levels. They found an overall difference for serum vitamin E levels between subjects and matched controls of 0.4 mg/L (9×10^{-4} mM; $p = 0.003$), with the controls (who did not develop cancer) having the higher serum vitamin E levels and the patients who develop cancer having the lower serum vitamin E levels (142).

There are no epidemiological data on the incidence of cancer or other diseases for humans who live in areas with differing smog levels and are supplemented to different levels with vitamin E. Data that would bear on this question would be extremely valuable.

SUMMARY

The studies discussed here suggest that supplementation with vitamin E may partially mitigate the harmful effects of long-term exposure to ozone in smog. However, none of the studies conducted to date, either in animals or in human volunteers, adequately tests the extent to which supplemental vitamin E above the recommended level can protect against chronic exposure to ozone. The animal data are inconclusive since often vitamin E-deficient and E-sufficient animals were compared, rather than animals supplemented to differing levels with vitamin E. Most of the animal studies clearly demonstrate that the lack of vitamin E potentiates damage from ozone. Some of the animal studies suggest that supplementation with vitamin E above the recommended levels may provide additional protection. The human studies use small numbers of subjects and for ethical reasons only very short ozone exposures (of the order of 2 h). Under these conditions, which do not model the chronic exposures that would be experienced by many urban populations, supplemental amounts of vitamin E do not appear to provide increased protection.

Studies of the immune system indicate that ozone lowers immune competence and that vitamin E improves it, suggesting that longer term studies with vitamin E might show protective effects against ozone-induced susceptibility to infectious diseases or even cancer.

RECOMMENDED FURTHER STUDIES

At least four types of studies should be conducted to probe the potential beneficial effects of vitamin E supplementation in protecting humans against ozone (1). Chemical model system studies are required to elucidate the mechanisms involved in ozone-induced damage and to clarify how vitamin E can serve a protective role. In particular, biological targets for ozone must be identified and the products identified from ozone reactions with biological target molecules and their toxicities determined. Marker molecules that measure ozone damage must be identified. It is also important to probe the role of free radicals in ozone-induced changes. (2) Animal studies should be done comparing the protective effects of multiple dietary levels of vitamin E that are below, slightly above, and considerably above the amount of vitamin E in standard laboratory animal feeds. These studies should be done using chronic exposures to ozone for weeks or months and at levels of ozone near 0.1 ppm. A large number of measures of ozone-induced morbidity should be followed. (3) These animal experiments may require the development of new, sensitive markers of ozone-induced morbidity. It would be particularly useful to identify one or several early markers of ozone-induced injury, particularly changes that can be detected noninvasively and used in humans. There is some evidence that changes in immune competency may be a useful early marker of ozone damage in animals and in humans. (4) Epidemiological studies of the effects of ozone coupled with both a dietary questionnaire and monitoring of the vitamin E status of the subjects might test the applicability of the conclusions from the animal and chemical studies to humans. However, the difficulty and expense of these experiments, which would combine dietary studies and epidemiology, would be considerable.

A FINAL NOTE

Our inability to give a conclusive and satisfying answer to the question of whether vitamin E can protect us against ozone in smog is not unique. Many questions with regard to the toxic effects of ozone are similarly unanswered. The recent summary of the effects of smog by the Office of Technology Assessment (1) concludes that "an intelligent approach to ozone requires a broader understanding of its effects. . . . We cannot evaluate ozone's true risks . . . without knowing much more about the chronic effects of long-term exposure."

Finally, it should be noted that the potential protective effects of *all* the antioxidant vitamins and micronutrients should be tested. This review deals with vitamin E because it is the antioxidant that has been studied in the greatest detail. A recent three-volume treatise on vitamin C cites only three publications on the effects of ascorbate on lung damage caused by ozone, two from 1957 and one from 1980 (143). A 1985 publication (which is not cited in that treatise) reports that ozone reacts with vitamin C almost 100-fold faster than with PUFA (at pH near 7), whereas ozone reacts with vitamin E at about the same rate as with PUFA (25). This publication (25) concludes that "ascorbate in the lung may undergo direct ozonation [destroying ozone in a sacrificial way, whereas] . . . α-tocopherol [does not react directly with ozone, but only can] scavenge radicals produced from ozone-PUFA reactions." This suggests that vitamin C may be a more effective antioxidant against ozone than vitamin E, scavenging ozone before it can react with more critical biological target molecules (24,25).

Ozone is the most potent oxidant to which humans are routinely exposed and ozone-containing smog is ubiquitous. The potential deleterious health consequences of the polluted air to which we are now exposed and the delay that will occur before our air is cleaner emphasize the need for biochemical, animal, and human experiments that provide a conclusive answer to the question of whether dietary supplements can protect us against this type of oxidative stress.

ACKNOWLEDGMENTS

Work in our laboratories on ozone and vitamin E over the past two decades was supported by grants from the National Institutes of Health and the National Science Foundation and the Health Effects Institute.

REFERENCES

1. Office of Technology Assessment. Catching our breath: next steps for reducing urban ozone. Washington, DC: U.S. Government Printing Office, 1980; 1-24.
2. U.S. Environmental Protection Agency. Air quality criteria for ozone and other photochemical oxidants, Vol. III (Report No. EPA/600/8-84/020cF). Springfield, VA: National Technical Information Service, 1986.
3. Lubin B, Machlin LJ, eds. Vitamin E: biochemical, hematological, and clinical aspects, Vol. 393. New York: New York Academy of Sciences, 1982.
4. Pryor WA. Free radical biology: xenobiotics, cancer, and aging. In: Lubin B, Machlin LJ, eds. Vitamin E: biochemical, hematological, and clinical aspects. Ann N Y Acad Sci 1982; 393:1-22.
5. Machlin LJ. Vitamin E. New York: Marcel Dekker, 1980.
6. Pryor WA. Views on the wisdom of using antioxidant vitamin supplements. Free Radic Biol Med 1987; 3:189-91.
7. Ames BN. Endogenous DNA damage as related to cancer and aging. Mutat Res 1989; 214:41-6.
8. Boutwell RK. An overview of the role of diet and nutrition in carcinogenesis. In: Tryfiates GP, Prasad KN, eds. Nutrition, growth, and cancer. New York: Alan R. Liss, 1988; 81-104.
9. Sies H, ed. Oxidative stress. New York: Academic Press (Harcourt Brace Jovanovich), 1985.
10. Pryor WA. The role of free radicals in smog-induced pathology and protection by vitamin E (abstract). Free Radic Biol Med 1990; 9(suppl. 1):113.
11. Pryor WA. How far does ozone penetrate into the pulmonary air/tissue boundary before it reacts? Free Radic Biol Med 1992; 12:83-88.
12. Pryor WA, Church DF. Aldehydes, hydrogen peroxide, and organic radicals as mediators of ozone toxicity. Free Radic Biol Med 1991; 11:41-46.
13. Rabinowitz JL, Bassett DJP. Effect of 2 ppm ozone exposure on rat lung lipid fatty acids. Exp Lung Res 1988; 14:477-89.
14. Konings AWT. Mechanisms of ozone toxicity in cultured cells. I. Reduced clonogenic ability of polyunsaturated fatty acid-supplemented fibroblasts. Effect of vitamin E. J Toxicol Environ Health 1986; 18:491-7.
15. Pryor WA, Dooley MM, Church DF. Mechanisms for the reaction of ozone with biological molecules: the source of the toxic effects of ozone. In: Lee SD, Mustafa MG, Mehlman MA, eds. Advances in modern environmental toxicology. International symposium on the biomedical effects of ozone and related photochemical oxidants, Vol. V. Princeton, NJ: Princeton Scientific Publishers, 1982; 7-19.
16. Pryor WA, Lightsey JW, Prier DG. The production of free radicals in vivo from the action of xenobiotics: the initiation of autoxidation of polyunsaturated fatty acids by nitrogen dioxide and ozone. In: Yagi K, ed. Lipid peroxides in biology and medicine. New York: Academic Press, 1982; 1-22.
17. Pryor WA, Prier DG, Church DF. Radical production from the interaction of ozone and PUFA as demonstrated by electron spin resonance spin-trapping techniques. Environ Res 1981; 24:42-52.
18. Pryor WA, Prier DG, Church DF. Detection of free radicals from low-temperature ozone-olefin reactions by ESR spin trapping: evidence that the radical precursor is a trioxide. J Am Chem Soc 1983; 105:2883-8.
19. Church DF, McAdams M, Pryor WA. Radical production from the ozonation of simple alkenes, fatty acid emulsions and phosphatidylcholine liposomes. In: Davies KJA, ed. Oxidative damage and repair. New York: Pergamon Press, 1991; 517-522.
20. Roehm JN, Hadley JC, Menzel DB. Oxidation of unsaturated fatty acids by ozone and nitrogen dioxide: a common mechanism of action. Arch Environ Health 1971; 23:142.
21. Pryor WA, Stanley JP, Blair E, Cullen GB. Autoxidation of polyunsaturated fatty acids. Part I. Effect of ozone on the autoxidation of neat methyl linoleate and methyl linoleate. Arch Environ Health 1976; 31:201-10.
22. Menzel DB. The role of free radicals in the toxicity of air pollutants (nitrogen oxides and ozone). In: Pryor WA, ed. Free radicals in biology, Vol. II. New York: Academic Press, 1976; 181-200.
23. Goldstein BD, Lodi C, Collinson C, Balchum OJ. Ozone and lipid peroxidation. Arch Environ Health 1969; 18:631-5.

24. Giamalva DH, Church DF, Pryor WA. Kinetics of ozonation. 4. Reactions of ozone with alpha-tocopherol and oleate and linoleate esters in carbon tetrachloride and in aqueous micellar solvents. J Am Chem Soc 1986; 108:6646-51.

25. Giamalva DH, Church DF, Pryor WA. A comparison of the rates of ozonation of biological antioxidants and oleate and linoleate esters. Biochem Biophys Res Commun 1985; 133:773-9.

26. Bailey PS. Ozonation in organic chemistry, Vol. I. Olefinic compounds. New York: Academic Press, 1978; 220-37.

27. Benson SW. Thermochemical kinetics. New York: John Wiley & Sons, 1976; 1-320.

28. Ewing JW, Cosgrove JP, Giamalva DH, Church DF, Pryor WA. Autoxidation of methyl linoleate initiated by the ozonide of allylbenzene. Lipids 1989; 24:609-15.

29. Pryor WA. Free radicals in autoxidation and in aging. Part I. Kinetics of the autoxidation of linoleic acid in SDS micelles: calculations of radical concentrations, kinetic chain lengths, and the effects of vitamin E. Part II. The role of radicals in chronic human diseases and in aging. In: Armstrong D, Sohal RS, Cutler RG, Slater TF, eds. Free radicals in molecular biology, aging and disease. New York: Raven Press, 1984; 13-41.

30. Pryor WA, Giamalva DH, Church DF. Kinetics of ozonation. 2. Amino acids and model compounds in water and comparisons to rates in nonpolar solvents. J Am Chem Soc 1984; 106:7094-100.

31. Knight KL, Mudd JB. The reaction of ozone with glyceraldehyde-3-phosphate dehydrogenase. Arch Biochem Biophys 1984; 229:259-69.

32. Peters RE, Mudd JB. Inhibition by ozone of the acylation of glycerol 3-phosphate in mitochondria and microsomes from rat lung. Arch Biochem Biophys 1982; 216:34-41.

33. Bailey PS. Ozonation in organic chemistry, Vol. II. Nonolefinic compounds. New York: Academic Press, 1982; 238-497.

34. Koppenol WH. Generation and thermodynamic properties of oxyradicals. In: Vigo-Pelfrey C, ed. Focus on membrane lipid oxidation. Boca Raton, FL: CRC Press, 1989; 1-13.

35. Koppenol WH. The reduction potential of the couple ozone and the ozonide radical anion. FEBS Lett 1982; 140:169-72.

36. Barry BE, Miller FJ, Crapo JD. Effects of inhalation of 0.12 and 0.25 parts per million ozone on the proximal alveolar region of juvenile and adult rats. Lab Invest 1985; 6:692-704.

37. Esterline RL, Bassett DJP, Trush MA. Characterization of the oxidant generation by inflammatory cells lavaged from rat lungs following acute exposure to ozone. Toxicol Appl Pharmacol 1989; 99:229-39.

38. Boxer LA. The role of antioxidants in modulating neutrophil functional responses: In: Bendich A, Phillips M, Tengerdy RP, eds. Antioxidant nutrients and immune functions. New York: Plenum Press, 1990; 19-33.

39. Pryor WA, Ohto N, Church DF. Reaction of cumene with ozone to form cumyl hydrotrioxide and the kinetics of decomposition of cumyl hydrotrioxide. J Am Chem Soc 1983; 105:3614-22.

40. Greenwood FL, Rubinstein H. Ozonolysis. IX. The alkene ozonation oligomer. J Org Chem 1967; 32:3369-74.

41. Goldstein BD. Hydrogen peroxide in erythrocytes. Detection in rats and mice inhaling ozone. Arch Environ Health 1973; 26:279-80.

42. Yusa T, Beckman JS, Crapo JD, Freeman BA. Hyperoxia increases hydrogen peroxide production by brain in vivo. J Appl Physiol 1987; 63:353-8.

43. Pryor WA, Das B, Church DF. The ozonation of unsaturated fatty acids: aldehydes and hydrogen peroxide as products and possible mediators of ozone toxicity. Chem Res Toxicol 1991; 4:341-348.

44. Pryor WA, Miki M, Das B, Church DF. The mixture of aldehydes and hydrogen peroxide produced in the ozonation of dioleoyl phosphatidylcholine causes hemolysis of human red blood cells. Chem Biol Interact 1991; 79:41-52.

45. Teige B, McManus TT, Mudd JB. Reaction of ozone with phosphatidylcholine liposomes and the lytic effect of products on red blood cells. Chem Phys Lipids 1974; 12:153-71.

46. Turrens JF, Freeman BA, Crapo JD. Hyperoxia increases hydrogen peroxide release by lung mitochondria and microsomes. Arch Biochem Biophys 1982; 217:411-21.

47. Pryor WA, Prier DG, Church DF. Electron-spin resonance study of mainstream and sidestream cigarette smoke: nature of the free radicals in gas-phase smoke and in cigarette tar. Environ Health Perspect 1983; 47:345-55.

48. Lachocki TM, Church DF, Pryor WA. Persistent free radicals in the smoke of common household materials: biological and clinical implications. Environ Res 1988; 45:127-39.

49. Lachocki TM, Nuggehalli SK, Scherer KV, Church DF, Pryor WA. The smoke produced from the oxidative pyrolysis of perfluoro polymers: an ESR spin-trappin study. Chem Res Toxicol 1989; 2: 174-80.

50. Pryor WA, Nuggehalli SK, Scherer KV, Church DF. An electron spin resonance study of the particles produced in the pyrolysis of perfluropolymers. Chem Res Toxicol 1990; 3:2-7.

51. Pryor WA, Dooley MM, Church DF. The mechanisms of the inactivation of human alpha-1-proteinase inhibitor by gas-phase cigarette smoke. Free Radic Biol Med 1986; 2:161-88.

52. Goldstein BD, Buckley RD, Cardenas R, Balchum OJ. Ozone and vitamin E. Science 1970; 169: 605-6.

53. Goldstein BD, Pearson B, Lodi C, Buckley RD, Balchum OJ. The effect of ozone on mouse blood in vivo. Arch Environ Health 1968; 16:648-50.

54. Weenen H, Porter NA. Autoxidation of model membrane systems: the cooxidation of polyunsaturated lecithins with steroids, fatty acids, and α-tocopherol. J Am Chem Soc 1982; 104:5216-21.

55. Menzel DB, Slaughter RJ, Byrant AM, Jauregui HO. Heinz bodies formed in erythrocytes by fatty acid ozonides and ozone. Arch Environ Health 1975; 30:296-301.

56. Calabrese EJ, Moore GS, Grinberg-Funes R. Ozone induced hematological changes in mouse strains with differential levels of erythrocyte G-6-PD activity and vitamin E status. J Environ Pathol Toxicol Oncol 1985; 6:283-92.

57. Cortesi R, Privett OS. Toxicity of fatty ozonides and peroxides. Lipids 1972; 7:715-21.

58. U.S. Environmental Protection Agency. Air quality criteria for ozone and other photochemical oxidants, Vol. I. EPA-600/8-84-020aF. Research Triangle Park, NC: U.S. Environmental Protection Agency, 1986; 1-237.

59. Pryor WA. Can vitamin E protect us against the pathological effects of ozone in smog? Am J Clin Nutr 1991; 53:702-722.

60. U.S. Department of Health and Human Services. Surgeon General's Report on Nutrition and Health. Washington, DC: U.S. Government Printing Office, 1988; 1-727.

61. Moore GS, Calabrese EJ, Grinberg-Funes RA. The C57L/J mouse strain as a model for extrapulmonary effects of ozone exposure. Bull Environ Contam Toxicol 1980; 25:578-85.

62. Elsayed NM. Mobilization of vitamin E to the lung under oxidative stress. In: Diplock AT, Machlin LJ, Packer L, Pryor WA, eds. Vitamin E: biochemistry and health implications. New York: New York Academy of Sciences, 1989; 439-40.

63. Chow CK, Airriess GR, Changchit C. Increased vitamin E content in the lungs of chronic cigarette-smoked rats. In: Diplock AT, Machlin LJ, Packer L, Pryor WA, eds. Vitamin E: biochemistry and health implications. New York: New York Academy of Sciences, 1989; 425-7.

64. Airriess G, Changchit C, Chen LC, Chow CK. Increased levels of vitamin E in the lungs of guinea pigs exposed to cigarette smoke. Nutr Res 1988; 8:653-61.

65. Chow CK, Tappel AL. Activities of pentose shunt and glycolytic enzymes in lungs of ozone-exposed rats. Arch Environ Health 1973; 26:205-8.

66. Morgan DL, Dorsey AF, Menzel DB. Erythrocytes from ozone-exposed mice exhibit decreased deformability. Fundam Appl Toxicol 1985; 5:137-43.

67. Pace DM, Landolt PA. Effects of ozone on cells in vitro. Arch Environ Health 1969; 18:165-70.

68. Stokinger HE. Ozone toxicology. Arch Environ Health 1965; 10:719-31.

69. Menzel DB. Oxidation of biologically active reducing substances byozone. Arch Environ Health 1971; 23:149-53.

70. DeLucia AJ, Hoque PM, Mustafa MG, Cross CE. Ozone interaction with rodent lung: effect on sulfhydryls and sulfhydryl-containing enzyme activities. J Lab Clin Med 1972; 80:559-66.

71. Freeman BA, Mudd JB. Reaction of ozone with sulfhydryls of human erythrocytes. Arch Biochem Biophys 1981; 208:212-20.

72. Vos RME, Rietjens IMCM, Stevens LH, Van Bladeren PJ. Methyl linoleate ozonide as a substrate for rat glutathione S-transferases: reaction pathway and isoenzyme selectivity. Chem Biol Interact 1980; 69:269-78.

73. Rietjens IMCM, Lemmink HH, Alink GM, Van Bladeren PJ. The role of glutathione and glutathione S-transferases in fatty acid ozonide detoxification. Chem Biol Interact 1987; 62:3-14.

74. Abraham WM, Delehunt JC, Yerger L, Marchette B, Oliver W Jr. Changes in airway permeability and responsiveness after exposure to ozone. Environ Res 1984; 34:110-9.

75. Chow CK, Plopper CG, Chiu M, Dungworth DL. Dietary vitamin E and pulmonary biochemical and morphological alterations of rats exposed to 0.1 ppm ozone. Environ Res 1981; 24:315-24.

76. Chow CK, Plopper CG, Dungworth DL. Influence of dietary vitamin E on the lungs of ozone-exposed rats. Environ Res 1979; 20:309-17.

77. Plopper CG, Chow CK, Dungworth DL, Tyler WS. Pulmonary alterations in rats exposed to 0.2 and 0.1 ppm ozone: a correlated morphological and biochemical study. Arch Environ Health 1979; 34:390-5.

78. Joshi UM, Dodavanti PRS, Mehendale HM. Glutathione metabolism and utilization of external thiols by cigarette smoke-challenged, isolated rat and rabbit lungs. Toxicol Appl Pharmacol 1988; 96:324-35.

79. Lemoyne M, Van Gossum A, Kurian R, Ostro M, Axler J, Jeejeebhoy KN. Breath pentane analysis as an index of lipid peroxidation: a functional test of vitamin E status. Am J Clin Nutr 1987; 46:267-72.

80. Shariff R, Hoshino E, Allard J, Pichard C, Kurian R, Jeejeebhoy KN. Vitamin E supplementation in smokers (abstract). Am J Clin Nutr 1988; 47:758.

81. Tappel AL, Dillard CJ. In vivo lipid peroxidation: measurement via exhaled pentane and protection by vitamin E. Fed Proc 1981; 40:174-8.

82. Lawrence G, Cohen G, Machlin LJ. Ethane exhalation by vitamin E-deficient rats. In: Lubin B, Machlin LJ, eds. Ann N Y Acad Sci 1982; 393:227-8.

83. Dumelin EE, Dillard CJ, Tappel AL. Effect of vitamin E and ozone on pentane and ethane expired by rats. Arch Environ Health 1978; 33:129-34.

84. Dumelin EE, Dillard CJ, Tappel AL. Breath ethane and pentane as measures of vitamin E protection of *Macaca radiata* against 90 days of exposure to ozone. Environ Res 1978; 15:38-43.

85. Horwitt MK. Interrelations between vitamin E and polyunsaturated fatty acids in adult men. Vitam Horm 1962; 20:541-58.

86. Dutta S, Zimmer M, Sies H, Carlson RW, Cullen WJ, Thayer JW. Comparative toxicology of ozone and *t*-butyl hydroperoxide on isolated rat lung. Free Radic Res Commun 1989; 8:27-35.

87. Chow CK, Tappel AL. An enzymatic protective mechanism against lipid peroxidation damage to lungs of ozone-exposed rats. Lipids 1972; 7:518-24.

88. Chow CK, Kaneko JJ. Influence of dietary vitamin E on the red cells of ozone-exposed rats. Environ Res 1979; 19:49-55.

89. Pryor WA, Stanley JP, Blair E. Autoxidation of polyunsaturated fatty acids. Part II. A suggested mechanism for the formation of TBA-reactive materials from prostaglandin-like endoperoxides. Lipids 1976; 11:370-9.

90. Roycroft JH, Gunter WB, Menzel DB. Ozone toxicity: hormone-like oxidation products from arachidonic acid by ozone-catalyzed autoxidation. Toxicol Lett 1977; 1:75-82.

91. Leikauf GD, Driscoll KE, Wey HE. Ozone-induced augmentation of eicosanoid metabolism in epithelial cells from bovine trachea. Am Rev Respir Dis 1988; 1;37:435-42.

92. Morrow JD, Harris TM, Roberts LJ II. Noncyclooxygenase oxidative formation of a series of novel prostaglandins: analytical ramifications for measurement of eicosanoids. Anal Biochem 1990; 184:1-10.

93. Schelegle ES, Adams WC, Giri SN, Siefkin AD. Acute ozone exposure increases plasma prostaglandin $F_{2\alpha}$ in ozone-sensitive human subjects. Am Rev Respir Dis 1989; 140:211-6.

94. Dooley MM, Mudd JB. Reaction of ozone with lysozyme under different exposure conditions. Arch Biochem Biophys 1982; 218:459-71.

95. Whiteside C, Hassan HM. Role of oxyradicals in the inactivation of catalase by ozone. Free Radic Biol Med 1988; 5:305-12.

96. Sato S, Kawakami M, Maeda S, Takishima T. Scanning electron microscopy of the lungs of vitamin E-deficient rats exposed to a low concentration of ozone. Am Rev Respir Dis 1976; 113:809-21.

97. Chow CK. Influence of dietary vitamin E on susceptibilit to ozone exposure. In: Lee SD, Mustafa MG, Mehlman MA, eds. The biomedical effects of ozone and related photochemical oxidants. Princeton, NJ: Princeton Scientific Publishers, 1982; 75-93.

98. Gey KF, Puska P. Plasma vitamins E and A inversely correlated to mortality from ischemic heart disease in cross-cultural epidemiology. Ann N Y Acad Sci 1989; 570:268-82.

99. Fletcher BL, Tappel AL. Protective effects of dietary α-tocopherol in rats exposed to toxic levels of ozone and nitrogen dioxide. Environ Res 1973; 6:165-75.

100. Elsayed NM, Kass R, Mustafa MG, et al. Effect of dietary vitamin E level on the biochemical response of rat lung to ozone inhalation. Drug Nutr Interact 1988; 5:373-86.

101. Mustafa MG, Lee SD. Pulmonary biochemical alterations resulting from ozone exposure. Ann Occup Hyg 1976; 19:17-26.

102. Harkema JR, Hotchkiss JA, Henderson RF. Effects of 0.12 and 0.80 ppm ozone on rat nasal and nasopharyngeal epithelial mucosubstances: quantitative histochemistry. Toxicol Pathol 1989; 17:525-35.

103. Melton CE. Effects of long-term exposure to low levels of ozone: a review. Aviat Space Environ Med 1982; 53:105-11.

104. Lippmann M. Health effects of ozone: a critical review. J Air Pollut Control Assoc 1989; 39: 672-95.

105. Chaney S, DeWitt P, Blomquist W, Muller K, Bruce R, Goldstein G. Biochemical changes in humans upon exposure to ozone and exercise. Chapel Hill, NC 27514: Health Effects Research Laboratory, U.S. Environmental Protection Agency, 1979; EPA-600/1-79-026:

106. Dillard CJ, Litov RE, Savin WM, Dumelin EE, Tappel AL. Effects of exercise, vitamin E, and ozone on pulmonary function and lipid peroxidation. J Appl Physiol 1978; 45:927-32.

107. Van Gossum A, Shariff R, Lemoyne Mf, Kurian R, Jeejeebhoy K. Increased lipid peroxidation after lipid infusion as measured by breath pentane output. Am J Clin Nutr 1988; 48:1394-9.

108. Morita S, Snider MT, Inada Y. Increased *n*-pentane excretion in humans: a consequence of pulmonary oxygen exposure. Anesthesiology 1986; 64:730-3.

109. Pincemail J, Deby C, Dethier A, Bertrand Y, Lismonde M, Lamy M. Pentane measurement in man as an index of lipoperoxidation. Bioelectrochem Bioenerg 1987; 18:117-25.

110. Sagai M, Ichinose T. Age-related changes in lipid peroxidation as measured by ethane, ethylene, butane and pentane in respired gases of rats. Life Sci 1980; 27:731-8.

111. Snider MT, Morita S, Verma TK, Kibelian G. Kinetics of *n*-pentane excretion in the expired gas of man (abstract). Fed Proc 1981; 40:894.

112. Posin CI, Clark KW, Jones MP, Buckley RD, Hackney JD. Human biochemical response to ozone and vitamin E. J Toxicol Environ Health 1979; 5:1049-58.

113. Buckley RD, Hackney JD, Clark K, Posin C. Ozone and human blood. Arch Environ Health 1975; 30:40-3.

114. Hackney JD, Linn WS, Buckley RD, et al. Vitamin E supplementation and respiratory effects of ozone in humans. J Toxicol Environ Health 1981; 7:383-90.

115. Bergers WAA, Gerbrandy JLF, Stap JGMM, Dura EA. Influence of air polluting components viz ozone and the open air factor on host-resistance towards respiratory infection. In: Lee SD, Mustafa MG, Mehlman MA, eds. Advances in modern environmental toxicology. International symposium on the biomedical effects of ozone and related photochemical oxidants, Vol. V. Princeton, NJ: Princeton Scientific Publishers, 1982; 459-67.

116. Ehrlich R. Changes in susceptibility to respiratory infection caused by exposures to photochemical oxidant pollutants. In: Lee SD, Mustafa MG, Mehlman MA, eds. Advances in modern environmental toxicology. International symposium on the biomedical effects of ozone and related photochemical oxidants, Vol. V. Princeton, NJ: Princeton Scientific Publishers, 1982; 273-85.

117. Ball SS, Weindruch R, Walford RL. Antioxidants and the immune response. In: Johnson JE Jr, Walford R, Harman D, Miquel J, eds. Free radicals, aging, and degenerative diseases. New York Alan R. Liss, 1986; 427-56.

118. Bendich A. Antioxidant vitamins and immune responses. In: Chandra RJ, ed. Nutrition and immunology. New York: Alan R. Liss, 1988; 125-47.

119. Bendich A, Gabriel E, Machlin LJ. Dietary vitamin E requirement for optimum immune responses in the rat. J Nutr 1986; 116:675-81.

120. Meydani SN, Meydani M, Barklund PM, et al. Effect of vitamin E supplementation on immune responsiveness of the aged. In: Diplock AT, Machlin LJ, Packer L, Pryor WA, eds. Vitamin E: biochemistry and health implications. New York: New York Academy of Sciences, 1989; 283-90.

121. Chavance M, Brubacher G, Herberth B, et al. Immunological and nutritional status among the elderly. In: Chandra RK, ed. Nutrition, immunity, and illness in the elderly. New York: Pergamon Press, 1985; 137-42.

122. Meyskens FL Jr, Prasad KN, eds. Vitamins and cancer. Clifton, NJ: Humana Press, 1986; 1-481.

123. Gairola CC, Chow CK. Dietary vitamin E, hepatic aryl hydrocarbon hydroxylase and the metabolic activation of procarcinogens to mutagens. Nutr Cancer 1980; 2:125-8.

124. Pryor WA. The free-radical theory of aging revisited: a critique and a suggested disease-specific theory. In: Warner HR, Butler RN, Sprott RL, eds. Modern biological theories of aging. New York: Raven Press, 1987; 89-112.

125. Pryor WA. Cigarette smoke and the involvement of free radical reactions in chemical carcinogenesis. Br J Cancer 1987; 55:19-23.

126. Pryor WA. Cancer and free radicals. In: Shankel D, Hartman P, Kada T, Hollaender A, eds. Antimutagenesis and anticarcinogenesis mechanisms. New York: Plenum Press, 1986; 45-59.

127. Witschi H. Ozone, nitrogen dioxide and lung cancer: a review of some recent issues and problems. Toxicology 1988; 48:1-20.

128. Reddy CC, Thomas CE, Scholz RW, Labosh TJ, Massaro EJ. Effects of ozone exposure under inadequate vitamin E and/or selenium nutrition on enzymes associated with xenobiotic metabolism. In: Lee SD, Mustafa MG, Hehlman MA, eds. Advances in modern environmental toxicology. International symposium on the biomedical effects of ozone and related photochemical oxidants, Vol. V. Princeton, NJ: Princeton Scientific Publishers, 1983; 395-410.

129. Takahashi Y, Miura T. Effects of nitrogen dioxide and ozone in combination on xenobiotic metabolizing activities of rat lungs. Toxicology 1989; 56:253-62.

130. Borek C, Ong A, Mason H. Ozone and ultraviolet light act as additive cocarcinogens to induce in vitro neoplastic transformation. Teratogenesis Carcinog Mutagen 1989; 9:71-4.

131. Borek C, Ong A, Zaider M. Ozone activates transforming genes in vitro and acts as a synergistic co-carcinogen with gamma-rays only if delivered after radiation. Carcinogenesis 1989; 10:1549-51.

132. Borek C, Ong A, Mason H, Donahue L, Biaglow JE. Selenium and vitamin E inhibit radiogenic and chemically induced transformation in vitro via different mechanisms. Proc Natl Acad Sci U S A 1986; 83:1490-4.

133. Hassett C, Mustafa MG, Coulson WF, Elashoff RM. Murine lung carcinogenesis following exposure to ambient ozone concentrations. J Natl Cancer Inst 1985; 75:771-7.

134. Hassett C, Mustafa MG, Coulson WF, Elashoff RM. Murine lung carcinogenesis following exposure to ambient ozone concentrations (correction). J Natl Cancer Inst 1986; 77:991.

135. Last JA, Warren DL, Pecquet-Goad E, Witschi H. Modification by ozone of lung tumor development in mice. J Natl Cancer Inst 1987; 78:149-54.

136. Jacobson BS. The role of air pollution and other factors in local variations in general mortality and cancer mortality. Arch Environ Health 1984; 39:306-13.

137. Birt DF. Update on the effects of vitamins A, C, and E and selenium on carcinogenesis. Proc Soc Exp Biol 1986; 183:311-20.

138. Mirvish SS. Effects of vitamins C and E on N-nitroso compound formation, carcinogenesis, and cancer. Cancer 1986; 58:1842-50.

139. Menkes MS, Comstock GW, Vuilleumier JP, Helsing KJ, Rider AA, Brookmeyer R. Serum beta-carotene, vitamins A and E, selenium, and the risk of lung cancer. N Engl J Med 1986; 315:1250-4.

140. Stähelin HB, Rösel F, Buess E, Brubacher G. Cancer, vitamins, and plasma lipids: prospective Basel study. J Natl Cancer Inst 1984; 73:1463-8.

141. Knekt P, Aromaa A, Maatela J, et al. Serum vitamin E, serum selenium and the risk of gastrointestinal cancer. Int J Cancer 1988; 42:846-50.

142. Wald NJ, Thompson SG, Densem JW, Boreham J. Serum vitamin E and human cancer. In: Hayaishi O, Mino M, eds. Clinical and nutritional aspects of vitamin E. Amsterdam: Elsevier Science Publishers, 1987; 73-82.

143. Clemetson CAB. Vitamin C, Vol. 1. Boca Raton, FL: CRC Press, 1989; 1-318.

V
MEDICAL APPLICATIONS
A. Dermatology

52

Vitamin E in Dermatological Therapy

Jürgen Fuchs

Johann Wolfgang Goethe University, Frankfurt, Germany

Lester Packer

University of California, Berkeley, California

INTRODUCTION

Vitamin E is an essential nutrient that is receiving growing attention for its antioxidant function in human biological systems. In 1922 an unknown substance was discovered in lettuce by Evans and Bishop that prevented fetal death in animals fed a rancid diet. Subsequently, Sure designated this unknown compound a fertility vitamin in 1924. The antioxidant properties of the new vitamin were first studied by Olcott and Emerson. In 1936, Evans isolated the fertility vitamin from wheat germ oil, and in the same year Reffy (1) and Leranth and Laszlo (2) published reports on clinical application of the substance for treatment of leg ulcers and other skin diseases. Fernholtz revealed the chemical structure of vitamin E in 1938, and shortly thereafter it was synthesized by Karrer. The availability of large amounts of the vitamin led to clinical trials, including dermatological studies. In 1948, Burgess reported the successful treatment of patients diagnosed with lupus erythematosus by tocopherol (3,4). Subsequently, wider application of tocopherol in the treatment of various skin diseases ensued. The therapeutic euphoria of the 1950s and 1960s has passed and is succeeded by a more critical and restricted use of the vitamin. Only recently have we understood the pharmacodynamic and pharmacokinetic properties of tocopherol, in particular its antioxidant mechanism of action in skin. Today, studies on the biological significance of oxidative injury in skin and the potential use of antioxidants, such as tocopherol, in dermatological therapy are a topic of high priority in many research laboratories and clinics. In view of all this, it is useful to present a review on the current status of the clinical use of tocopherol in dermatology.

BIOCHEMICAL AND PHARMACODYNAMIC PROPERTIES

Vitamin E is a generic term that includes all entities that exhibit the biological activity of natural vitamin E, d-α-tocopherol. Biological activity is assessed by studying the inhibition of reactive oxidant-mediated erythrocyte hemolysis in vitro or by analyzing the pregnancy-protecting (prevention of fetal resorption in the rat) effect in animals deficient of vitamin E. Eight natural substances have vitamin E activity: d-α-, d-β-, d-γ-, and d-δ-tocopherol and d-α-, d-β-, d-γ-, and d-δ-tocotrienol. Synthetic tocopherols also have vitamin E activity, as do the succinate and acetate derivatives of tocopherols. In the rat resorption gestation test, the natural d-α-tocopherol (RRR-α-tocopherol) has the highest biopotency (1.49 IU/mg = 100%), the l-enantiomer of α-tocopherol (SRR-α-tocopherol) has only 31% bioactivity (0.46 IU/mg) (5). Synthetic vitamin E (d,l-α-tocopherol), a mixture of eight stereoisomers, has a biopotency of 74%.

Most of the biological effects of vitamin E can be explained by its antioxidant properties, presumably including interactions with prostanoid metabolism. Membrane-modulating effects may also contribute to its biological activity to an unknown extent. Antioxidant and membrane-modulating effects are the rationale used to explain antiinflammatory, photoprotective, and mesenchymal metabolism-modulating properties. Other pharmacodynamic properties are anti-carcinogenic, antiarteriosclerotic, and immunomodulating effects.

Antioxidant Effects

Vitamin E is among the oldest recognized biological antioxidants, and its redox and free-radical chemistry are well documented (6,7). Direct and indirect evidence is convincing that α-tocopherol is a powerful antioxidant in subcellular structures and in isolated cells and or-gans in animals and in humans. Tocopherol is a primary and secondary antioxidant. It reacts directly with reactive oxygen species in biological membranes and inhibits already initiated lipid peroxidation (radical chain reaction) by capturing the propagating species (peroxyl radi-cal). Highly toxic free-radical species of xenobiotics, such as trichloromethyl and trichloro-peroxymethyl radical, are also scavenged and detoxified rapidly by tocopherol. Tocopherols also possess iron ion binding activity, thereby limiting the amount of biologically active transi-tion metal ions in biomembranes (8).

The initial oxidation product of tocopherol is the metastable tocopheroxyl radical, which can be further oxidized to tocopherol quinone. Other oxidation products are epoxides as well as oligomers. The tocopheroxyl radical possesses a central role for antioxidant potency. It can be reduced under physiological conditions to tocopherol, hence regenerating the anti-oxidant. Since the physiological mole ratio of tocopherol to polyunsaturated phospholipid is about $1: >1000$ in most biological membranes, tocopherol regeneration is essential for the high antioxidant efficiency of tocopherol in vivo. Physiological low-molecular-weight com-pounds, such as ascorbate, glutathione, dihydrolipoate, and ubiquinoles, contribute to toco-pherol regeneration. Because of the long half-life of the tocopheroxyl radical (9), enzymatic regenerative processes seem to be plausible. Enzymatic regeneration may occur via micro-somal and mitochondrial factors (10–14). The exact understanding of tocopherol recycling is of high interest. Dysfunction of the regeneration process may cause tocopherol deficiency in cellular microcompartments and result in a prooxidative environment.

Under certain in vitro conditions tocopherol may also exert prooxidative effects, which have also been described for other antioxidants, such as ascorbate and dihydrolipoate. High concentrations of α-tocopherol ($>>0.005$ mol tocopherol) per mole unsaturated fatty acid accelerate lipid autooxidation in vitro (15,16). Other authors also reported the prooxidant in vitro effects of α-tocopherol (17,18). Although it seems reasonable to speculate that high tocopherol concentrations may also cause prooxidative effects in vivo, no animal experiments or clinical reports support this assumption.

Membrane-Modulating Effects

The effect of α-tocopherol on membrane fluidity and permeability of phospholipids is quite complex and depends upon the composition of the phospholipid and the tocopherol-phospho-lipid mole ratio (19). α-Tocopherol decreases the membrane fluidity of phosphatidylcholine liposomes (20) and reduces the membrane permeability for uncharged molecules, such as glucose (21). Tocopherol in high concentrations, however, increases the membrane fluidity of human erythrocytes and thrombocytes (22). In contrast to these reports, small concentrations of tocopherol increase the fluidity of mitochondrial membranes, and high concentrations produce a decrease in lipid movement (23). Similar results were obtained by Zimmer (Chap. 17). It has been suggested that the antioxidant properties of tocopherol in biomembranes depend on its free-radical–capturing activity as well as on membrane-stabilizing effects (24). However, the inhibition of microsomal lipid peroxidation by tocopherol has been suggested to be explained by the antioxidant activity of the compound rather than by an indirect effect via membrane stabilization (25). Tocopherol-induced membrane alterations in model membranes are usually observed only in very high unphysiological tocopherol concentrations. Physiological tocopherol concentrations are suggested to influence membrane structures probably only marginally, so that the biological relevance of the membrane-modulating effects are rather questionable (26) (Ingold, personal communication). The recent studies of Zimmer, however, suggest that physiological concentrations of tocopherol significantly influence biophysical membrane properties (Chap. 17). However, the mechanism by which 1 molecule tocopherol per 1000 molecules phospholipid modulates membrane structure is not yet understood.

Photoprotective Effects

Several in vitro animal and human studies support the concept that tocopherol has photoprotective properties. The antioxidant properties provide a rationale for the use of tocopherol as a photoprotective agent. Ultraviolet (UV) irradiation of skin induces the formation of reactive oxygen species and lipid peroxidation products, which participate in mediating the inflammatory reaction. It has been documented that reactive oxygen species (27,28) and lipid peroxidation products can cause severe skin damage (29–32). In tocopherol-deficient rats there is a marked increase in skin lipid peroxide content after ultraviolet irradiation in comparison to control rats (33). In addition to its antioxidant properties tocopherol, also possesses a weak ultraviolet filter effect, but this is probably of minor importance. Inhibition of ornithine decarboxylase activity and modulation of prostanoid metabolism may further contribute to its photoprotective effects.

In Vitro Studies

Squalene, a sterol compound with six unsaturated double bonds originating from the sebaceous glands, is readily photoperoxidized. Large amounts of squalene accumulate in human skin; squalene represents about 10–15% of total skin surface lipids. Peroxidized squalene causes strong irritation of stripped human skin and growth inhibition of human keratinocytes in culture (34). It has been suggested that oxidation of squalene may induce comedogenesis (35), and may participate in ultraviolet-induced erythema. α-Tocopherol inhibits the photoperoxidation of squalene in vitro (32).

Human keratinocytes are significantly protected from simulated solar light when treated with liposomal α-tocopherol (36). Cultivated fibroblasts of patients with actinic reticuloid are photoprotected by a water-soluble α-tocopherol derivative (37). α-Tocopherol inhibits the photohemolysis of erythrocytes derived from patients with erythropoietic protoporphyria (photosensitization of erythrocytes by a defect in erythrocytic porphyrin synthesis) (38).

Animal Studies

In hairless mice the photoprotective potency of the *d*-tocopherol homologs is comparable to their antioxidant activity in vitro (39). The topical treatment of hairless mouse skin with α-tocopherol acetate suppresses the UVB-induced increase in epidermal ornithine decarboxylase activity (40) and causes a reduction in polyamine biosynthesis (41). Ultraviolet (280–365 nm) erythema of the rabbit is prevented by topical application of tocopherol (application before irradiation or 2 minutes after irradiation). Local therapy 5 h after irradiation does not reduce the erythema. It was concluded that tocopherol influences dark-phase reactions of erythema formation (42). Similar results were obtained by Ohzawa (43). PUVA (psoralen UVA)-induced erythema of the rabbit is inhibited by topical pretreatment with α-tocopherol. Topical application of tocopherol immediately after irradiation has only a minor antiinflammatory effect (44). It was suggested that the local tocopherol concentration plays an important role in its antiinflammatory efficiency. Low tocopherol concentrations have stronger antiinflammatory effects in PUVA-induced erythema than higher concentrations (45). A prooxidant in vivo effect was suggested to be responsible for a significantly lower skin antiinflammatory activity of topical preparations with a high tocopherol content (45). This effect can also be explained by the nonspecific, irritating potential of the oily substance, however.

Human Studies

Parenteral administration of α-tocopherol in humans reduces the symptom complex caused by ultraviolet irradiation (46). Although oral tocopherol did not cause any clinical improvement in skin lesions in patients with lupus erythematosus, significant photoprotection was observed in most cases (47). Prophylactic topical α-tocopherol application to human skin inhibits PUVA-induced erythema at a threefold minimal erythemal dose (MED), but the local protective effect is very small if higher UV doses (10-fold MED) are used (44). In contrast to cyclooxygenase inhibitors, such as indomethacin (48,49), which affect only the inflammatory response, tocopherol also inhibits UVB-induced sunburn cell formation in human epidermis (50).

Antiinflammatory Effects

The antiinflammatory effect of α-tocopherol has been demonstrated in several animal and clinical studies. Particularly in rheumatic diseases, its clinical efficacy is well documented (51–54). Topical application of α-tocopherol inhibits anthralin dermatitis in humans (55) and reduces the enhancement of anthralin erythema produced by topical arachidonic acid (56). Tocopherol also inhibits the formation of persistent organic radicals in hairless mouse skin after topical treatment with anthralin (57). Systemic application of α-tocopherol causes a significant antiinflammatory effect in dextran-mediated rat paw edema (58). Croton acid dermatitis of the rabbit is inhibited by topical application of α-tocopherol (58). In an earlier study, topical treatment of croton acid dermatitis of the rabbit with tocopherol was not successful, in contrast to systemic therapy (59). Ayres and Mihan hypothesized that free radicals and lipid peroxidation participate in the pathophysiology of inflammation occurring in acne vulgaris and suggested that antioxidants, such as tocopherol, may be useful in the treatment of acne vulgaris (60).

Most probably, different mechanisms are responsible for the antiinflammatory effect of tocopherol. The effect of tocopherol on prostanoid metabolism is complex and strongly dependent on the animal species and the type of tissue (61). Tocopherol inhibits thromboxane and increases prostacyclin biosynthesis (62). In addition, α-tocopherol quinone has antiaggregatory activities. Lipid hydroperoxides can trigger prostaglandin synthesis (63), and tocopherol may modulate prostanoid metabolism by influencing steady-state concentrations of lipid hydroperoxides. Presumably, the generation and steady-state concentration of inflammatory mediators, such as prostanoids, lipid peroxidation products, and reactive oxygen species, are modulated by α-tocopherol.

Mesenchymal Metabolism-Modulating Effects

Topical application of *d*-α-tocopherol in rat skin before ultraviolet exposure inhibits long-term UVA- and UVB-induced hydroxyproline loss but does not influence the UVA-induced decrease in collagen solubility. It was suggested that tocopherol is involved in the regulation of collagen metabolism following long-term exposure of hairless mouse skin to ultraviolet light (64). Tocopherol deficiency in rat skin results in an increase in soluble collagen, probably via a defect of the inter- and intramolecular collagen cross-linking (65). A single subcutaneous injection of tocopherol in rats inhibits mesenchyme metabolism in dermis, aorta, and cartilage for many weeks (inhibition of sulfate insertion rate) (66). The growth of foreign body granulomas and the unspecific mesenchymal reaction (increased sulfate incorporation into mucopolysaccharides) induced by intraperitoneal injection of staphylococcus toxin is inhibited by tocopherol (66).

In rats, tocopherol inhibits collagen synthesis and delays wound healing (67). In rats exposed to ionizing radiation, however, an improvement in wound healing by tocopherol is observed (68), which may be caused by stimulation of epithelial regeneration. Indeed, Melkumian et al. reported that epithelial regeneration of burn wounds is improved by topical therapy with tocopherol (69). The data indicate that tocopherol is essentially involved in mesenchymal metabolism. In animals, high doses may cause inhibition of mesenchymal metabolism but low (physiological) concentrations are required for well-balanced collagen synthesis.

PHARMACOKINETIC PROPERTIES IN SKIN

Cutaneous Permeation

Autoradiographic investigations in human skin revealed that α-tocopherol acetate permeates rapidly into epidermis after topical application. Two permeation pathways have been characterized: (1) via stratum corneum, and (2) via hair follicles. The apocrine and eccrine glands probably do not play a significant role in permeation. A high affinity of tocopherol acetate for the small dermal vessels and the septs in the subcutis was observed (70). Quantitative penetration studies in rat skin showed that after 1 h 7%, and after 6 h 39% of the topically applied dose (5% tocopherol solution in ethanol) was detected in skin layers below the stratum corneum, which is the penetration barrier (71). The vehicle significantly influences skin penetration of d-α-tocopherol; penetration is better from an oil-water microemulsion than from a water-oil emulsion (72). Penetration of topical α-tocopherol into epidermis was also documented by Ohazawa et al. (43). An increase in serum α-tocopherol following topical application of α-tocopherol was not observed (73). However, Martini et al. reported an increase in tissue tocopherol concentration (liver, kidney, muscle, and fat tissue), starting 6 h after topical application of radioactively labeled tocopherol (72).

Distribution and Metabolism

The total tocopherol tissue pool of an healthy adult varies between 4 and 8 g. The tissue concentration is highest in the adrenal gland and in fat tissue. The adrenal gland is highly vulnerable to oxidative injury because of its high content of oxygenases, and fat tissue is readily susceptible to oxidative attack as a result of its high content of polyunsaturated fatty acids. Tocopherol is readily mobilized from the plasma and liver pool, whereas mobilization is intermediate from skeletal and heart muscle and low from fat tissue.

The d,l-α-tocopherol was given orally (600 IU/day) to atopic dermatitis patients. Their plasma tocopherol concentration reached a maximum (2-fold increase) followed by a plateau during weeks 6–8 of therapy (74). Oral administration of 600 mg d-α-tocopherol in young healthy male adults resulted in a 2.5 to 3-fold increase in plasma levels, and red and white blood cells, with a maximum occurring in week 4. The maximum was reached in thrombocytes after 8 weeks and in oral mucous membrane epithelia after 12 weeks. There was no further increase after the maximum was reached, although oral intake of tocopherol was continued. Within 1 week after termination of therapy, tissue concentrations declined to basal values. This indicates that after saturation of the membranes either resorption is reduced or elimination is increased (75). Supplementation of young healthy men with 200 or 400 mg *d,l*-α-tocopherol acetate resulted in an increase in the plasma tocopherol concentration that was statistically not significantly different. There was, however, a tendency that the group treated with 400 mg had higher plasma levels than the men taking 200 mg (76). Supplementation of volunteers with *all-rac*-α-tocopherol doses of 150, 225, 800, and 1000 IU/day over 3 weeks results in increased oxidation resistance of their low-density lipoprotein (LDL) of 1.17, 1.56, 1.35, and 1.75, respectively (relative values) (77). This indicates that oral supplementation with increasing tocopherol doses results in increased protection from oxidative injury.

Natural *d*-α-tocopherol is better retained in the tissues than synthetic *d,l*-α-tocopherol (78–80). Ingold et al. investigated how quickly tocopherol enters different tissues in the absence of metabolic stress. The biological half-life of α-tocopherol has been determined in different tissues of the rat after oral feeding with deuterated *d*-α-tocopherol acetate. The physiological tocopherol turnover is dependent on the type of tissue. Tocopherol half-life in rat skin (23.4 days) is between the minimal value, for lung (7.6 days), and the maximal value, for spinal cord (76.3 days). The half-life in plasma is 10.9 days and in erythrocytes, 12.5 days (80). From these results it may be extrapolated to the human condition that after oral tocopherol therapy changes in skin tocopherol content are expected only after several weeks (under the assumption of a physiological tocopherol turnover). *d*-α-Tocopherol acetate has a tissue uptake rate in humans similar to that of *d*-α-tocopherol (81).

Oral application of vitamin E (300 mg/day) for 10 days does not result in a tocopherol increase in human skin lipids (73). Hairless mice were fed a diet containing 10,000 IU α-tocopherol per kg (for a 30 g mouse consuming 5 g diet per day, this is about 1700 IU/kg body weight per day). The vitamin E content of the skin increased from 6.38 nmol/g skin (wet) to 56 nmol/g after 15 days and a maximum of 83 nmol/g after 30 days (82). Oral application of 300 or 900 mg *d*-α-tocopherol per day to humans increases heart tissue concentration after 2 weeks from 60 nmol/g (wet weight) to 120 nmol/g. No significant differences in tissue concentration were found in the groups treated with 300 or 900 mg tocopherol (81). After intravenous application of *d*-α-tocopherol, remarkably higher tissue levels (liver, kidney, heart, lung, and fatty tissue) are achieved in sheep in comparison to intramuscular or oral application (83). Shiratori reported that intravenous application of radioactive α-tocopherol in rats resulted 2 h after injection in a 10% increase and 20 days after injection in a 40% accumulation of the applied dose in skin (84). A preferential uptake of the radioactivity in skin and subsequent transepidermal elimination were noted (84).

Tocopherol is subject to oxidative metabolism; however, it is uncertain whether certain tissue metabolites (tocopherol quinone and tocopherol oligiomers) represent isolation artifacts. In skin of hairless mice, bioconversion of the tocopherol ester *d,l*-α-tocopherol L ascorbate phosphodiester to α-tocopherol and ascorbate was reported (85). In human skin an unspecific ali-esterase activity is found, which probably induces activation of the prodrug. The pharmacodynamic activity of α-tocopherol esters in skin depends on this enzyme-induced bioactivation.

TOCOPHEROL AS A DRUG IN DERMATOLOGY

The therapeutic uses of tocopherol in medicine have been reviewed (86). Vitamin E is used clinically to correct a deficiency state and its associated pathophysiological disturbances. In contrast to the characteristic sequela of tocopherol deficiency in certain animal species and premature infants, most vitamin E deficiency symptoms in humans are rather unspecific. Examples of specific deficiency diseases are respiratory distress syndrome and hemolytic anemia of low-birth-weight infants. In adults, tocopherol hypovitaminosis is observed only very rarely and is then mainly a consequence of malabsorption and alipoproteinemia. The reduced erythrocyte life span in patients with cystic fibrosis, the hyperaggregatability of platelets in patients with biliary atresia, and the myopathy, cerebellar dysfunction, and retinitis pigmentosa in patients with hereditary abetalipoproteinemia can clinically be treated well with adequate tocopherol substitution (87).

Vitamin E has also been reported to improve pathological conditions mediated by a deficiency of physiological antioxidants and/or increased formation of reactive oxidants. Examples include retrolental fibroplasia and bronchopulmonary dysplasia in premature infants exposed to prolonged oxygen administration in the treatment of respiratory distress syndrome. Vitamin E was employed successfully in the treatment of patients with hereditary hemolytic anemia due to deficiencies in glutathione synthase and glucose-6-phosphate dehydrogenase and leukocyte dysfunction in patients with a deficiency in glutathione synthase (87). Several clinical studies indicate that α-tocopherol may also be useful in the prevention of arteriosclerosis and cataract, in cancer prevention, and in the adjuvant treatment of Parkinson's disease and for improvement of the immune response (88).

The designation of vitamin E as a fertility vitamin initially led to applications for the treatment of male fertility disturbances. According to the clinical experience of many authors (89–98), the use of vitamin E in the treatment of male subfertility is considered justified as a therapeutic trial. The therapeutical benefit of male subfertility treatment with α-tocopherol is scientifically not clearly established, however. Recently it has been recognized that oxidative damage of spermatozoa may contribute to male infertility disorders (99,100), and attempts are being made to investigate the clinical efficacy of antioxidants in male infertility (101).

Parallel to the use of vitamin E in the treatment of male subfertility, the substance was also used as a drug for the treatment of skin diseases. Literature reviews (59,90,102–108) of clinical trials with vitamin E in dermatology indicate a broad spectrum of skin diseases in which clinical improvement has been reached with varying success rates. The broad clinical application spectrum reflects an enthusiastic therapeutic approach and a lack of knowledge with regard to the pharmacokinetic and pharmacodynamic properties of the vitamin. Consequently, the therapeutic expectations were not always fulfilled.

Only recently was a comprehensive concept of oxidative injury in experimental and clinical dermatopathology introduced (109). This concept is based on the reports of Niwa in Japan (110) and Meffert in Germany indicating the involvement of reactive oxygen species and free radicals in dermatopathobiochemical events. The clinical application of tocopherol in dermatological therapy is not based on the hypothetical concept of hypovitaminosis; tocopherol deficiency dermatoses in the human are unknown. However, tocopherol deficiency in epidermal or dermal microcompartments may be a consequence of endogenously or exogenously induced oxidative injury in skin. Dermatoses with significant involvement of reactive oxygen species in their pathogenesis include skin conditions mediated by electromagnetic radiation, inflammatory cells, prooxidant xenobiotics, and lipid peroxidation products. Furthermore, increased cellular metabolism during hyperproliferation, induction of prooxidant skin enzymes, and inhibition of skin antioxidants also contribute to prooxidant conditions. Dermatological disorders with significant involvement of oxidative damage may benefit from local and systemic antioxidant therapy. Hence, tocopherol may be of potential use in the treatment of distinct dermatoses, such as those associated with vasculitis or tissue infiltration with inflammatory cells, and those aggravated or induced by ultraviolet light. (See Table 1.)

Tocopherol Serum Concentrations in Dermatological Patients

To our knowledge there are no reports on tocopherol skin concentration in different dermatoses. Although it is not known whether there is a direct correlation of tocopherol serum con-

TABLE 1

Neutrophilic vasculitis	Behçet's disease, pyoderma gangrenosum, erythema elevatum et diutinum, urticaria vasculitis, vasculitis allergica, nodular fasciitis, granuloma faciale
Lymphocytic vasculitis	Erythema multiforme, systemic lupus erythematosus, dermatomyositis, rheumatic vasculitis, pityriasis lichenoides, purpura pigmentosa et progressiva, venous stasis dermatitis
Granulomatous vasculitis	Granuloma annulare, rheumatoid nodule, necrobiosis lipoidica, Wegener's granulomatosis
Eosinophilic vasculitis	Churg-Strauss syndrome
Tissue neutrophilia	Psoriasis vulgaris, Sweet's syndrome, dermatitis herpetiformis Duhring
Tissue eosinophilia	Bullous pemphigoid, pemphigus herpetiformis
Tissue lymphocytosis	Atopic dermatitis
Idiopathic photodermatoses	Polymorphous light eruption, solar urticaria, actinic reticuloid, hydroa vacciniforme
Chemical photosensitivity	Phototoxicity, photoallergy, persistent light reaction
Metabolic and genetic	Porphyrias
Photodermatoses	Disorders of tryptophan metabolism, melanin deficiency syndromes, skin diseases with increased cellular susceptibility (not all diseases have abnormal reaction to ultraviolet)

centration with tocopherol tissue content, particularly in skin, serum concentrations have been analyzed in several skin diseases.

Lupus erythematosus

Welsh reported that the average tocopherol blood concentration in a group of patients with lupus erythematosus was 567 μg per 100 ml; the blood level of a control group was 2370 μg per 100 ml. The blood concentration of β-carotene was also decreased in the lupus patients. However, in this study the standard deviation was not calculated (111). After oral application of different tocopherol derivatives (*d,l*-α-, *d*-α-, and solubilized *d*-α-tocopherol acetate), the tocopherol blood concentration was measured in these patients. Oral doses of *d*- and *d,l*-α-tocopherol acetate and solubilized tocopherol (500–2000 mg daily) resulted in an increase in blood tocopherol of 2500–5800 μg per 100 ml for *d*- and *d,l*-α-tocopherol acetate and 3100–7100 μg per 100 ml for the solubilized vitamin. In this study, the bioavailability of the natural tocopherol derivative (*d*-α-tocopherol acetate) was not higher than that of the synthetic tocopherol derivative (*d,l*-α-tocopherol acetate) (111).

Keratosis Follicularis

In Indian children diagnosed with keratosis follicularis who belonged to a low socioeconomic population group, the serum tocopherol concentration was significantly lower (330 μg per 100 ml) than in healthy children (660 μg per 100 ml) of a comparable socioeconomic group (112). Oral supplementation of these children with keratosis follicularis with daily doses of 300 mg *d,l*-α-tocopherol acetate over 4 weeks resulted in normalization of the serum tocopherol concentration (1090 μg per 100 ml) and to significant improvement in the skin condition. In this study hypovitaminosis E was evident (< 500 μg per 100 ml = 5 mg/L). The average plasma concentration of α-tocopherol in a healthy adult varies from 6 to 20 mg/L (10 mg/L = 22 μmol/L); γ-tocopherol is found at concentrations of 1–2 mg/L. Other tocopherol stereoisomers and tocotrienols are present in only trace amounts. It was suggested that in addition to hypovitaminosis E, hypovitaminosis A may be present in these children and that tocopherol inhibits the oxidative inactivation of vitamin A and thereby improves keratosis follicularis via an indirect sparing effect on vitamin A.

Miscellaneous Skin Diseases

Determination of serum tocopherol levels in dermatology patients (acne vulgaris, neurodermitis, and other eczema rosacea, psoriasis, scleroderma, ulcus cruris, and urticaria) showed a mean value of 850 ± 162 μg per 100 ml. This value is below the mean value of healthy persons of about 1000 μg per 100 ml (104). The variability in the physiological serum concentration is large (600–2000 μg per 100 ml), however, so that conclusions should be drawn from this study with caution. Fegeler could not demonstrate a clear correlation of tocopherol plasma concentration with severity of the skin diseases. In a different study (neurodermitis, other eczema, ichthyosis vulgaris, pityriasis rubra pilaris, dermatitis herpetiformis Duhring and others) no significant differences in tocopherol serum concentrations were noted in comparison to healthy controls (113). A high variability in serum tocopherol concentration (530–1560 μg per 100 ml) was also evident in this study.

Treatment of Lupus Erythematosus

Probably the most comprehensive clinical data with regard to the dermatological use of tocopherol are available for the treatment of lupus erythematosus. Lupus erythematosus is a chronic inflammatory autoimmune disorder that may affect the skin as well as internal organs. Clinically, three forms of lupus erythematosus have been described. Cutaneous diskoid lupus erythematosus (CDLE) is at the benign end of the spectrum, systemic lupus erythematosus (SLE) at other end, and there is an intermediate variant. Cutaneous lupus erythematosus is considered a special monosymptomatic form of SLE of the skin only (114,115). Pathological alterations of blood vessels and immunological disturbances are less expressed in chronic cutaneous lupus erythematosus than in SLE. A polygenetic or exogen-induced defect of cellular microdebridement has been suggested to be involved in the pathogenesis of SLE (116). Consequently, a reduced clearance of immune complexes results, which triggers the inflammatory reaction.

Oxidative Injury

Clinical (117–119) and animal studies (120,121) indicate involvement of reactive oxidants in the pathogenesis of lupus erythematosus, in particular with reference to vascular injury and photosensitivity. Tocopherol can intervene as an antioxidant in the pathophysiological scenario and may inhibit oxidative injury by scavenging reactive oxidants and prevent the formation of clastogenic factors.

Vascular Injury. Reactive oxygen species may participate in the pathogenesis of various inflammatory skin diseases with endothelial damage and/or circulating immune complexes (122). Immune complexes were found in the serum of patients with SLE (123,124). They activate neutrophil granulocytes of diseased as well as healthy patients, which results in the elevated production of reactive oxygen species (110,125). These mediators may ultimately be responsible for tissue and endothelial damage.

Clastogenic Factors. SLE patients have increased numbers of chromosome breaks (126). This correlates with a low-molecular-weight chromosomal damaging agent that is released from patient lymphocytes into the serum. The clastogenic factor was detected in plasma of SLE but not in CDLE patients (Emerit, personal communication). The presence of clastogenic products in plasma is not specific for lupus erythematosus. Clastogenic products were first described in plasma from patients irradiated with ionizing radiation and are also found in patients with spontaneous chromosomal instability, as well as chronic inflammatory and autoimmune diseases, such as progressive systemic sclerosis, dermatomyositis, periarteritis nodosa, rheumatoid arthritis, and Crohn's disease (127).

Lymphocytes from SLE patients have abnormal sensitivity to UVA, and ultraviolet-induced formation of clastogenic products was suggested to be involved in photosensitivity (128). Generation of reactive oxidants in the medium of lymphocyte cultures either by photoreduction of flavins or by the xanthine-xanthine oxidase system results in the production of chromosome breakage factors (clastogenic material), which induces chromosome breaks and sister chromatid exchanges (129). It was suggested that this clastogenic material may be composed of lipid peroxidation products. Clastogenic factor isolated from lymphocyte cultures contains thiobarbituric acid-reactive substances and conjugated dienes (129,130). Chromosome breaks are produced in normal lymphocytes when they are incubated in the presence of (1) SLE patient serum, (2) SLE patient lymphocytes, or (3) purified clastogenic factor. Reactive oxygen species have a short half-life, and their hazardous potential is limited to the immediate vicinity in which they arise. Secondary, longer lived clastogenic material, which is produced in reactions with short-lived highly reactive oxidants, can exert their clastogenic effect again via free-radical intermediates at a site distant from their formation. This mechanism may contribute to the chronicity of the inflammatory reaction (131).

Analysis of Clinical Trials

Burgess first reported promising results with the treatment of lupus erythematosus (CDLE) by oral or systemic application of vitamin E (3,4). His reports were later confirmed by other authors (102,111,132–137). We summarize the clinical success rates in these reports here.

  In 67 patients, d,l-α-tocopherol acetate, 1000–2000 mg orally per day, and pantothenic acid, duration of therapy several months, significant clinical improvement in most cases (111)

  In 47 patients, vitamin E, 300–900 mg orally per day, duration of therapy 1–6 months, significant clinical improvement in 41 cases (135)

  In 25 patients, tocopherol 100–300 (600) mg orally per day and 250–2000 mg intramuscularly one or two times per week, duration of therapy several weeks, significant clinical improvement in most cases (3,4)

  In 25 patients, d,l-α-tocopherol, 150 mg orally per day, duration of therapy several months, 13 patients cleared of skin lesions, significant clinical improvement in 2 cases, no significant clinical improvement in 10 cases (102)

  In 4 patients, vitamin E, 900–1600 IU orally per day, duration of therapy several months, significant clinical improvement in all cases (137)

  In 3 patients, vitamin E, 150 mg orally per day, and vitamin E, 100 mg intramuscularly daily, in all cases significant clinical improvement (134)

Less successful therapy results in the treatment of lupus erythematosus (CDLE) with vitamin E was reported by several other authors (47,138–142). Ayres and colleagues suggested that tocopherol therapy failures in lupus erythematosus may be due to a low-dose tocopherol regimen (107,137,143). Kaminsky et al. attributed a large proportion of therapy failures to insufficiency of the dosage utilized (140) and Welsh concluded that megadose therapy with tocopherol is justified to achieve a rapid therapeutic success in patients (111).

  In 45 patients, vitamin E, 600 mg orally per day, and vitamin E, 400 mg intramuscularly two times per week, duration of therapy 2–9 months, significant clinical improvement in 5 cases, in 1 patient transient therapy success, in 39 patients no therapeutic success (139)

  In 45 patients, d,l-α-tocopherol acetate, 200 mg orally per day, and vitamin E, 100–400 mg intramuscularly two times weekly, duration of therapy 2–9 months, in 5 patients significant clinical improvement, in 40 patients no therapy success (142)

In 22 patients, vitamin E, 200–300 mg orally per day, duration of therapy 10 weeks to several months, significant clinical improvement in 4 patients, no clinical improvement in 11 patients, in 7 patients deterioration of the clinical condition (138)

In 10 patients, vitamin E, 300–1000 mg orally per day, and vitamin E, 300–1500 mg intramuscularly daily, duration of therapy 10 days intramuscular and 90 days oral, no significant clinical improvement of skin lesions in all cases, but significant slight protection in most cases (47)

Although several investigators have used tocopherol for the treatment of lupus erythematosus, most of these studies have not yielded consistent results. Some clinical investigators have shown beneficial effects; others suggest that tocopherol was more or less ineffective. These controversial results could partially be explained by clinical subsets of the disease, the lack of adequate disease classification, and the heterogeneity of patient populations with regard to duration, severity, and dynamics of progression or regression. Multicenter prospective, long-term therapeutic trials with tocopherol in carefully selected cohorts of lupus erythematosus patients with comparable disease activity and duration are needed to resolve these differences.

Treatment of Skin Disorders Associated with Sclerosis and Fibrosis

The second largest patient group reported to benefit from tocopherol therapy includes those suffering from disorders associated with skin sclerosis and fibrosis. Sclerosis is characterized by excessive deposition of connective tissue in a mesenchymal organ, and fibrosis is defined by pronounced accumulation of fibroblasts in connective tissue. Frequently, however, the terms "sclerosis" and "fibrosis" are used synonymously. Sclerosis of the skin occurs in the clinically heterogeneous group of "connective tissue diseases": progressive systemic scleroderma, dermatomyositis, systemic lupus erythematosus, mixed connective tissue disease, and overlapping syndromes. Furthermore, sclerosis is frequently seen in a variety of dermatopathological conditions distinct from these autoimmune diseases: localized scleroderma, lichen sclerosus, hypertrophic scars, keloids, dermatosclerosis due to venous insufficiency, phenylketonuria, porphyria cutanea tarda, carcinoid syndrome, juvenile onset diabetes mellitus, and occupational (chemically induced) scleroderma. Generally, in chronic inflammatory processes and after hyperoxygenation and radiation therapy, an enhanced production of collagen is observed that may be mediated, at least in part, by reactive oxygen species.

Oxidative Injury and Skin Sclerosis or Fibrosis

The effects of reactive oxygen species on mesenchymal metabolism are quite complex. Regulation of collagen biosynthesis by reactive oxygen species may occur at different cellular levels, for example at the gene level or in the posttranslational phase. Reactive oxygen species can degrade and depolymerize collagen by attacking critical amino acid residues but may also induce cross-linking of collagen molecules and stimulate fibroblasts to increase collagen production. Fibroblast collagen synthesis is stimulated by elevated steady-state concentrations of reactive oxygen species, for example in hyperoxic periodontal tissue culture of mice (144), hyperoxic rat lung organ culture (145), rat lung exposed to ozone (146), and paraquat-exposed organ cultures (147). The superoxide anion radical stimulates collagen synthesis in vitro, probably by mediating ascorbate-dependent enzymic proline hydroxylation. The superoxide anion radical activates procollagen proline hydroxylase, which catalyzes the hydroxylation of proline (145). Prolyl hydroxylation is considered the rate-limiting step of collagen formation. High steady-state concentrations of reactive oxygen species cause cellular damage of cultured human fibroblasts. Cell density is reduced and thymidine incorporation inhibited. Low steady-state concentrations stimulate thymidine incorporation and increase cell density. Reactive oxygen species may provide a specific and sensitive trigger for fibroblast proliferation; prolonged stimulation may result in fibrosis (148). In conclusion, the reports clearly indicate that reactive oxygen species affect fibroblast metabolism. However, it is important to emphasize that fibroblast function and collagen gene expression are also modulated by various cytokines (e.g., transforming growth factor and interleukin-1). Silica-induced skin sclerosis is thought to be mediated by lymphokines and monokines (149).

Tocopherol (150) and retinoic acid or retinol (151), which both inhibit lipid peroxidation, also inhibit collagen synthesis in human dermal fibroblasts. Tocopherol deficiency promotes lipid peroxidation in rat skin and accelerates the cross-linking of dermal collagen (33). Several reports indicate that antioxidants may have some antifibrotic activity. Antioxidants were suggested to be of therapeutic value in the prevention (148) and treatment of fibrotic processes (152). To explain the profibrotic potential of reactive oxygen species and the clinical antifibrotic activity of such antioxidants as tocopherol, it seems plausible to suggest that the initiation and growth of fibrotic tissue results from the ability of reactive oxygen species to produce cross-linking of mesenchymal proteins and other macromolecules. In addition, a direct effect on the induction of collagen synthesis contributes to the profibrotic potential of reactive oxygen species. If the continuing effect of polymerization and increased de novo synthesis is stopped by an antioxidant, physiological depolymerization processes (collagenase and elastase) may predominate and result in clearing of the fibrotic lesion (153).

Analysis of Clinical Trials

Tocopherol and other antioxidants, such as superoxide dismutase, are used clinically with varying success rates in the treatment of various fibrotic skin diseases, such as induratio penis plastica, keloids, and hypertrophic scars. Tocopherol is particularly used in the treatment of fibromatosis palmoplantaris, induratio penis plastica, keloids and hypertrophic scars, scleroderma and lichen sclerosus. In the following we analyze the success rates of the clinical reports.

I. Fibromatosis palmaris et plantaris
   A. In 14 patients (also suffering from induratio penis plastica), vitamin E, 300 mg orally per day, and fractionated radiation therapy, duration of vitamin E therapy 1 year, no significant clinical improvement in any case (154).
   B. In 7 patients, vitamin E, 300 mg orally/day, duration of therapy 42–98 days, significant clinical improvement in 5 cases (155).
   C. In 6 patients, vitamin E, 60–100 mg orally per day, duration of therapy several weeks, significant clinical improvement in most cases (156).
II. Induratio penis plastica
   A. In 53 patients, vitamin E, 600–800 mg orally per day, fractionated radiation therapy and intrafocal injection of hyaluronidase, triamcinolone, and xylocaine, duration of vitamin E therapy several months, significant clinical improvement in most cases (157).
   B. In 21 patients, mixture of α-, β-, and γ-tocopherol, 250 mg orally per day, duration of therapy 4–12 months, significant clinical improvement in 10 cases, slight or no significant clinical improvement in 11 cases (158).
   C. In 20 patients, vitamin E, 300 mg orally per day, and fractionated radiation therapy, duration of vitamin E therapy 12 months. A total of 14 patients were treated within 4 months after the disease started: 5 patients without skin lesions after therapy, significant clinical improvement in 9 cases; 6 patients were treated 6 months after the disease became clinically apparent: significant clinical improvement in 1 patient, no clinical improvement in 5 patients (154).

    D. In 10 patients, vitamin E, 300 mg orally per day, duration of therapy several months, 2 patients without skin lesions, significant clinical improvement in 6 cases, no clinical improvement in 2 patients (159).

    E. In 4 patients, vitamin E, 300–400 mg orally or parenterally per day, duration of therapy several months, significant clinical improvement in 3 cases, no significant clinical improvement in 1 patient (160).

    F. In 1 patient, vitamin E, 300–600 mg orally per day, duration of therapy several months, significant clinical improvement (161).

III. Keloids. In 2 patients, *d,l*-α-tocopherol acetate, 200 mg orally per day, and *d,l*-α-tocopherol, 50–100 mg intramuscular injection two times per week, duration of therapy several months, only minor clinical improvement (142).

IV. Lichen sclerosus et atrophicus

    A. In 10 female patients, *d*-α-tocopherol acetate, 300–1200 IU orally per day, and *d*-α-tocopherol acetate cream or solution (30–50 IU/g), significant clinical improvement in 7 patients, only slight clinical improvement in 3 cases (107).

    B. In 3 female patients (kraurosis vulvae), vitamin E, 100 mg orally per day, duration of therapy 13 days, clinical improvement of pruritus in all cases (162).

    C. In 2 patients, *d*-α-tocopherol acetate, 300–600 IU orally per day, significant clifnical improvement in 1 patient, only minor improvement in 1 case (143).

    D. In 1 patient, *d,l*-α-tocopherol acetate, 200 mg orally per day, and *d,l*-α-tocopherol, 50–100 mg intramuscular injection two times per week, duration of therapy several months, no clinical improvement (142).

V. Scleroderma, systemic or localized

    A. In 6 patients, *d*-α-tocopherol acetate, 800–1200 mg orally per day, duration of therapy several months, significant clinical improvement in most cases (137,143).

    B. In 2 patients, *d,l*-α-tocopherol acetate, 200 mg orally per day, duration of therapy several months, no significant clinical improvement in all cases (142).

Treatment of Miscellaneous Skin Diseases

Other skin diseases reported to benefit from tocopherol therapy are those associated with vasculitis (acrodermatitis chronica atrophicans, necrobiosis lipoidica, and granuloma annulare), other types of skin inflammation (acne vulgaris and psoriasis vulgaris), and radiation and ultraviolet damage. In the following we analyze the clinical treatment results.

I. Acne vulgaris

    A. In 29 patients, sodium selenite, 0.4 mg, and tocopherol succinate, 20 mg orally per day, duration of therapy 6–12 weeks, significant clinical improvement in most cases (particularly in cases of pustular acne vulgaris) (163).

    B. In 23 female patients (premenstrual exacerbated acne papulopustulosa), *d,l*-α-tocopherol, 30–100 mg orally per day, 1–3 weeks per months, duration of therapy 3–24 weeks, significant clinical improvement in most cases (164).

II. Acrodermatitis chronica atrophicans

    A. In 11 patients, vitamin E, 10–100 mg orally per day, significant clinical improvement in all cases (135).

    B. In 8 patients, vitamin E, 100–250 mg orally per day, duration of therapy 2 months, significant clinical improvement in most cases (165).

III. Atopic dermatitis. In 20 patients, *d,l*-α-tocopherol acetate, 600 IU orally per day, and 600 μg selenium orally per day, duration of therapy 12 weeks, no clinical improvement in any case (74).

IV. Epidermolysis bullosa dystrophicans
   A. In 3 patients, *d,l*-α-tocopherol acetate, 300 mg orally per day, duration of therapy 12 weeks, significant clinical improvement in all cases (166).
   B. In 3 patients, vitamin E, 100–600 IU orally per day, duration of therapy several months, 2 patients without skin lesions, 1 patient no clinical improvement (167).
   C. In 2 patients, vitamin E, 1200–2000 mg orally per day, significant clinical improvement in all cases (168).
   D. In 2 patients, controlled double-blind crossover study, vitamin E, 1600 U orally per day, duration of therapy 8 weeks, significant clinical improvement in both cases (169).
   E. In 1 patient, α-tocopherol and α-tocopherol succinate, 1600–6000 IU orally per day, and tocopherol ointment application daily, duration of therapy 1 year, significant clinical improvement (170).
   F. In 1 patient, *d,l*-α-tocopherol succinate, 1600–3200 IU orally per day, and daily application of tocopherol ointment, 30 IU/g, significant clinical improvement (Price, 1962, unpublished).

V. Epidermolysis bullosa simplex Weber-Cockayne
   A. In 11 patients, vitamin E, 1200–2000 mg orally per day, significant clinical improvement in most cases (168).
   B. In 1 patient, vitamin E, 60–200 IU orally per day, duration of therapy 2 years, significant clinical improvement (171).

VI. Granuloma annulare
   A. In 13 patients, α-tocopherol, 150–600 mg orally per day, duration of therapy 2–9 weeks, 9 patients without skin lesions, significant clinical improvement in 1 case, questionable improvement in 2 cases, no improvement in 1 case (172).
   B. In 2 patients, *d*-α-tocopherol acetate, 200–300 IU orally per day, duration of therapy several months, significant clinical improvement (143).
   C. In 1 patient, vitamin E, 140 mg orally per day, and vitamin A, 60,000 IU orally per day, duration of therapy several weeks, significant clinical improvement (173).
   D. In 1 patient, mixture of α-, β-, and γ-tocopherol, 300 mg orally per day, significant clinical improvement (174).

VII. Keratosis follicularis
   A. In 56 children, 4 randomized patient groups, oral therapy with (1) placebo, (2) *d,l*-α-tocopherol acetate, 300 mg/day, (3) *d,l*-α-tocopherol acetate, 300 mg/day plus vitamin B complex, (4) vitamin B complex, duration of therapy 4 weeks. Group 1: in 15% of cases some clinical improvement, in 85% of cases no improvement. Group 2: in 36% of cases patients without skin symptoms, in 15% only minor clinical improvement. Group 3: in 77% of cases patients without skin symptoms, and only minor improvement in 23% of cases. Group 4: in 75% of cases partial clinical improvement, in 25% of cases no improvement (112).
   B. In 1 patient, oral vitamin A (200,000 IU) therapy over 5 years without skin improvement, *d*-α-tocopherol acetate, 1200 IU orally per day, duration of therapy 11 months, significant clinical improvement (175).

VIII. Lichen ruber. 12 patients mixture of α-, β-, γ-, and δ-tocopherol, 150 mg orally per day, duration of therapy 3–10 weeks, significant clinical improvement in most cases with skin and mucous membrane involvement, clinical success rates: lichen ruber planus > lichen ruber exanthematicus > lichen ruber verrucosus (103).

IX. Necrobiosis lipoidica diabeticorum
   A. In 1 patient, *d,l*-α-tocopherol acetate, 200 mg orally per day, duration of therapy several months, no significant clinical improvement (142).

B. In 1 patient, *d*-α-tocopherol acetate, 100 IU orally/day, and *d*-α-tocopherol cream (30–100 IU/g), duration of therapy several months, significant clinical improvement (143).

X. Neuralgia, postzoster. In 13 patients, vitamin E, 400–1600 IU orally per day, duration of therapy several weeks, 9 patients without symptoms, 2 patients significant improvement, 2 patients only minor improvement (176).

XI. Pemphigus benignus familiaris. In 3 patients, *d*-α-tocopherol acetate, 800–1200 IU orally per day, significant clinical improvement in all cases (107).

XII. Pigment disturbances (chloasma and postinflammatory hyperpigmentation. In 178 patients, multicenter double-blind study, three randomized groups. Group 1: *d,l*-α-tocopherol acetate, 150 mg, and ascorbate, 300 mg orally per day. Group 2: *d,l*-α-tocopherol acetate, 150 mg orally per dya. Group 3: ascorbate, 300 mg orally per day, duration of therapy 4–12 weeks, significant clinical improvement in group 1 in comparison to group 2 + 3 (177).

XIII. Porphyria, hepatic and erythropoietic
A. In 5 patients (4 porphyria acuta intermittens and 1 porphyria variegata), *d,l*-α-tocopherol, 100–200 (400) mg orally per day, duration of therapy 4 weeks, no significant increase in porphyrin excretion and no improvement of clinical symsptoms (178).
B. In 4 patients (2 porphyria cutanea tarda and 2 porphyria acuta intermittens), water-soluble vitamin E, 100 mg orally per day, duration of therapy 4 weeks, significant increase in porphyrin excretion and clinical improvement in 3 patients, no improvement in 1 patient (179).
C. In 1 patient (porphyria cutanea tarda), vitamin E, 100 mg orally per day, duration of therapy 1 month, significant increase in porphyrin excretion and patient without skin lesions (180).
D. In 1 patient (erythropoietic protoporphyria), vitamin E, 1500 IE orally per day, plus cholestyramine and prednisolone, significant clinical improvement without improvement of light hypersensitivity (181).

XIV. Pseudoxanthoma elasticum
A. In 3 patients, vitamin E, 90 mg orally/day, vitamin C, and anabolic steroids, significant clinical improvement in all cases (182).
B. In 2 patients, mixture of vitamin E homologs, 300 mg orally per day, and vitamin E acetate, 100 mg intramuscular injection once per week, significant clinical improvement of skin lesions in 1 patient and clearing of neurological symptoms in the other case (183).
C. In 1 patient, mixture of vitamin E homologs, 300 mg orally per day, and vitamin E acetate, 100 mg intramuscular injection once per week, duration of therapy several months, 90% clinical improvement (171).
D. In 1 patient, mixture of vitamin E homologs, 300 mg orally per day, and vitamin E acetate, 100 mg intramuscular injection once per week, duration of intramuscular therapy 3 months, duration of oral therapy 5 months, significant clinical improvement of skin and eye lesions (184).

XV. Psoriasis vulgaris
A. In 50 patients, vitamin E, 60 mg orally per day, vitamin D_2 and linolen- and linoleic acid, duration of therapy 4–6 weeks, in most cases the therapeutic efficiency of this regimen was comparable to that with steroid or anthralin treatment (185).
B. In 40 patients, vitamin E, 300 mg orally per day, duration of therapy several weeks, no significant clinical improvement in most cases (103).

XVI. Purpura, different etiopathologies
A. In 17 patients, *d,l*-α-tocopherol acetate, 200–400 mg orally per day, duration of therapy 4–23 days, significant clinical improvement in most cases (186).

    B. In 7 patients, α-tocopherol, 400–600 mg orally per day, duration of therapy 5–21 days, significant clinical improvement in 6 cases (187).
    C. Tocopherol, 50–150 mg orally or parenterally, significant clinical improvement in most cases (188).
XVII. Radiodermatitis and radiation ulcers
    A. In 3 patients, vitamin E, 300 mg orally per day, duration of therapy several months, significant clinical improvement in all cases (189).
    B. In 2 patients, vitamin E, 300 mg orally per day, duration of therapy several weeks, significant clinical improvement in most cases (190).
    C. In 1 patient (radiation ulcer resistant to previous therapy), tocopherol, 100 mg orally per day, duration of therapy 5 months, almost complete healing (191).
    D. In 1 patient, d,l-α-tocopherol acetate, 200 mg orally per day, d,l-α-tocopherol, 50–100 mg intramuscularly two times per week, local therapy with α-tocopherol ointment, duration of therapy several months, complete healing (142).
XVIII. Subcorneal pustulosis Sneddon Wilkinson. 1 patient (no improvement with steroids, sulfones, or antibiotics), d-α-tocopherol acetate, 100–400 IU orally per day, duration of therapy 2 years, patient without skin lesions (192).
XIX. Ulcera cruris
    A. In 18 patients, oral tocopherol, significant clinical improvement in 16 cases (193).
    B. Tocopherol local therapy, significant clinical improvement in most cases (1).
    C. Oral and local tocopherol, significant clinical improvement in most cases (2).
    D. Tocopherol, 200–300 mg orally per day, significant clinical improvement in most cases (194).
XX. Yellow nail syndrome
    A. In 3 patients, vitamin E, 800–1200 IU orally per day, duration of therapy 5–14 months, 1 patient total clearance of all clinical symptoms, 1 patient significant clinical improvement of nail lesions, no improvement in 1 patient (195).
    B. In 1 patient, α-tocopherol acetate, 400–800 IU orally per day, duration of therapy 2 years, total clearance of nail lesions, significant improvement of bronchitis and sinusitis (176).
    C. In 1 patient, vitamin E, 1200 IU orally per day, significant improvement of nail lesions (196).

THERAPEUTIC RISKS ASSOCIATED WITH TOCOPHEROL THERAPY

Local Side Reactions

Skin sensitivity to vitamin E has been judged in several hundreds of patients, and not a single case of a sensitization was reported (143). In 194 patients with skin diseases and 56 healthy humans, a positive test reaction (intra- and epicutaneous) to vitamin E was reported in 3 subjects (197). In 3 cases a contact dermatitis after application of vitamin E as an aerosol deodorant was reported (198). The appearance of an eruption like erythema multiforme after a topical application of vitamin E was documented in 2 cases (199). After high oral doses of vitamin E, yellow spots on the tooth may become apparent in a few cases (200). In a few other case reports (201–204), contact sensitization to vitamin E was reported. In conclusion, the topical sensitizing potential of vitamin E seems to be very low compared to that of other topical drugs and natural compounds.

Systemic Side Reactions

In contrast to the other fat-soluble vitamins—A, D, and K—the toxicity of vitamin E is extremely low. An excellent overview of the toxicological aspects of tocopherol was given by

Kästner and Kappus (205). Roberts published several reports about the presumed side effects of megadose (>800 IU) tocopherol therapy. Mainly nonspecific side effects, such as headache, dizziness, muscle weakness, and gastrointestinal discomfort, were noted (206). The side effects reported in the older literature (207)—thrombophlebitis, hypertension, myopathia, serum increase in cholesterin, triglycerides, and uric acid—have not been confirmed in placebo-controlled double-blind studies. Even with the intake of very high doses (3200 mg daily), only very few side effects have been observed (208). Salkeld analyzed multiple reports on side effects (209) and found a side effect rate of 0.8% in 10,068 cases (tocopherol application, 200–3000 mg/day over 11 years). This very low percentage is significantly increased with other medications and is also reported after administration of placebos. Patients receiving anticoagulant therapy with vitamin K antagonists or who are vitamin K deficient should not be treated with tocopherol to avoid exacerbation of the coagulation defect. In individuals who are not vitamin K deficient, tocopherol does not produce coagulation abnormalities.

Safety and Recommended Allowances

In adults the daily average tocopherol intake with food probably varies from 6 to 12 mg *d-α*-tocopherol. The World Health Organization recommends an accepted daily intake (ADI) of 0.15–2.0 mg α-tocopherol per kg body weight, and the U.S. Food and Drug Administration recommends a dietary allowance of 10 mg. However, many U.S. citizens probably have a daily intake of about 100–400 mg α-tocopherol. Breast-fed babies have a daily intake of about 2 mg tocopherol per kg body weight, which corresponds to a daily dose of 140 mg for a 70 kg adult. α-Tocopherol doses up to 700 mg/day are considered to be without side effects (210). For therapeutic purposes, α-tocopherol doses varying from 400 to a maximum of 3000 mg/day for an adult may be justified.

Future Prospects

Biochemical and pharmacological studies indicate that α-tocopherol is a highly effective antioxidant with antiinflammatory, antifibrotic, photoprotective, and other properties. However, evaluation of the clinical case studies and reports presented here on the treatment of various dermatoses with tocopherol do not yet provide convincing evidence of its clinical efficiency in several cases. This discrepancy may be explained by false indications for drug therapy, insufficient dosage and/or duration of therapy, inadequate study protocols, low numbers of patients, heterogeneity of patient populations investigated, and distinct clinical subsets of skin diseases. Nonresponders and remissions caused by self-healing are further complicating factors in the analysis. The uncritical and widespread use of tocopherol in the treatment of various skin diseases in the late 1940s to 1060s reflected a lack of detailed knowledge of the pharmacodynamic properties of tocopherol and the pathophysiology of the distinct skin disease. Furthermore, it was the result of a therapeutic dilemma in a preglucocorticosteroid and retinoid epoch.

Today tocopherol plays only a minor role in conventional dermatological therapy. The developments of the last decade, however, made it possible to understand the pharmacokinetics and pharmacodynamics of the substance and to learn about oxidative injury mechanisms in dermatopathology. Now is the time to use tocopherol logically and selectively in dermatological therapy. Prospective, long-term, multi-centered, double-blind studies in carefully selected patients are needed to unequivocally demonstrate the clinical efficacy of the vitamin in distinct skin diseases. It may be expected that tocopherol, especially with regard to its good local and systemic acceptance, will probably experience a renaissance in dermatological therapy.

Tocopherol has an extremely low acute, subacute, and chronic toxicity and lacks mutagenic, carcinogenic, and embryotoxic effects. Changes in skin tocopherol concentration and saturation are expected several weeks after initiation of therapy; hence therapeutic results occur only after a long time period. Therefore patients should be advised to take α-tocopherol over several months. The natural *RRR*-α-tocopherol (e.g., Malton or Spondyvit), which has better tissue retention than synthetic *all-rac*-α-tocopherol, is preferred, and therapeutic doses should be above 400 and below 3000 mg/day. We usually start a therapeutic trial in selected patients (e.g., epidermolysis bullosa dystrophicans, porphyria cutanea tarda, lupus erythematosus, scleroderma, induratio penis plastica, palmoplantar fibrosis, keloids, and hypertrophic scars), administering about 800 mg *RRR*-α-tocopherol per day orally over at least 3 months. After this time period the clinical status is assessed and either tocopherol therapy is continued (improvement) or the dosage is increased (no improvement or deterioration). Thus far we have had not one single patient report any side effects. A therapeutic trial with tocopherol seems to be justified in patients with inflammatory skin diseases, photodermatoses, and fibrotic skin diseases. Tocopherol is not a wonder drug but an innocous, natural compound and may be of significant clinical value for monotherapy and/or adjuvant therapy of distinct dermatological disorders.

REFERENCES

1. Reffy F. Über die Bedeutung der Vitamin in der Wundheilung. Verh 9 Int Kong Dermatol 1936; 2:528-30.
2. Leranth G, Laszlo F. Dermatologische Bedeutung des "E" Vitamins. Orv Hetil 1936; 778-9.
3. Burgess JF, Pritchard JE. Tocopherols (vitamin E). Treatment of lupus erythematosus: preliminary report. Arch Dermatol Syphilol 1948; 57:953-64.
4. Burgess JF. Vitamin E (tocopherols) in the collagenoses. Lancet 1948; 2:215-7.
5. Weiser H, Vecchi M. Stereoisomers of α-tocopheryl acetate. II. Biopotency of all eight stereoisomers, individually or in mixtures, as determined by rat resorption gestation test. Int J Vitam Nutr Res 1982; 52:351-70.
6. Burton W, Cheeseman KH, Doba T, Ingold KU, Slater F. Vitamin E as an antioxidant in vitro and in vivo. In: Biology of vitamin E. London: Pitman. Ciba Foundation Symposium 101, 1983; 4-14.
7. Burton GW, Ingold KU. Vitamin E: application of the principles of physical organic chemistry to the exploration of its structure and function. Acc Chem Res 1986; 19:194-201.
8. Stoyanovsky DA, Kagan VE, Packer L. Iron binding to alpha-tocopherol containing phospholipid liposomes. Biochem Biophys Res Commun 1989; 160:834-8.
9. Rousseau-Richard C, Richard C, Martin R. Kinetics of bimolecular decay of α-tocopherol free radicals studied by ESR. FEBS Lett 1988; 233:307-10.
10. Reddy CC, Scholz RW, Thomas CE, Massaro EJ. Vitamin E dependent reduced glutathione inhibition of rat liver microsomal lipid peroxidation. Life Sci 1982; 31:571-6.
11. Haenen GRMM, Bast A. Protection against lipid peroxidation by a microsomal glutathione dependent labile factor. FEBS Lett 1983; 159:24-8.
12. Hill KE, Burk RF. Influence of vitamin E and selenium on glutathione-dependent protection against microsomal lipid peroxidation. Biochem Pharmacol 1984; 33:1065-8.
13. McCay PB, Lai EK, Powell SR, Breugmann G. Vitamin E functions as an electron shuttle for glutathione dependent "free radical reductase" activity in biological membranes. Fed Proc 1986; 45:451.
14. Maguire JJ, Wilson DS, Packer L. Mitochondrial electron transport linked tocopheroxyl radical reduction. J Biol Chem 1989; 264:21462-5.
15. Cillard J, Cillard P, Cormier M. Effect of experimental factors on the prooxidant behavior of α-tocopherol. J Am Oil Chem Soc 1980; 57:255-61.
16. Husain SR, Cillard J, Cillard P. α-Tocopherol prooxidant effect and malondialdehyde production. J Am Oil Chem Soc 1987; 64:109-11.
17. Khrapova NG. Kinetic aspects of the action of tocopherols as antioxidants. Biofizika 1977; 22: 436-442.

18. Hicks M, Gebicki JM. Inhibition of peroxidation in linoleic acid membranes by nitroxide radicals, butylated hydroxytoluene and α-tocopherol. Arch Biochem Biophys 1981; 210:56-63.

19. Lai MZ, Düzgünes N, Szoka C. Effects of replacement of the hydroxyl group of cholesterol and tocopherol on the thermotropic behavior of phospholipid membranes. Biochemistry 1985; 24: 1646-1663.

20. Urano S, Yano K, Matsuo M. Membrane stabilizing effect of vitamin E: effect of α-tocopherol and its model compounds on fluidity of lecithin liposomes. Biochem Biophys Res Commun 1988; 10:469-75.

21. Stillwell W, Bryant L. Membrane permeability changes with vitamin A/vitamin E mixed bilayers. Biochim Biophys Acta 1983; 731:483-6.

22. Ernst E, Matrai A, Einfluss von alpha-Tokopherol (Vitamin E) auf die FließBeigenschaften des Blutes. Therapiewoche 1985; 35:5701-2.

23. Uerimov RF, Goloshchapov AN, Burlakova EB, Dzhafarov AI. Effect of the functional groups of tocopherol molecules on lipid viscosity of mitochondrial lipids. Bull Eksp Biol Med (USSR) 1987; 103:540-3.

24. Gutterridge JMC. The membrane effects of vitamin E, cholesterol and their acetates on peroxidative permeability. Res Commun Chem Pathol Pharmacol 1978; 22:563-72.

25. Ohki K, Takamura T, Nozawa Y. Effect of α-tocopherol on lipid peroxidation and acyl chain mobility of liver microsomes from vitamin E deficient rat. J Nutr Sci Vitaminol 1984; 30:221-34.

26. Massey JB, She HS, Pownall HJ. Interaction of vitamin E with saturated phospholipid bilayers. Biochem Biophys Res Commun 1982; 106:842-7.

27. Yoshioka A, Miyachi Y, Imamura S. Mechanisms of reactive oxygen species induced skin erythema and superoxide dismutase activities in guinea pigs. J Dermatol 1987; 14:569-75.

28. Fuchs J, Milbradt R, Zimmer G. Dihydrolipoate inhibits reactive oxygen species mediated skin inflammation. J Invest Dermatol 1990; 94:526.

29. Waravdekar VS, Saslaw LD, Jones WA, Kuhns JG. Skin changes induced by UV irradiated linolenic acid extract. Arch Pathol 1965; 80:91-5.

30. Tanaka T. Skin damage and its prevention from lipoperoxide. Vitamins (Kyoto) 1979; 53:577-86.

31. Ogura R. Cellular damage of epidermis exposed to ultraviolet light. Kurume Med J 1981; 5:279-301.

32. Ohsawa K, Watanabe T, Matsukawa R, Yoshimura Y, Imameda K. The possible role of squalene and its peroxide of the sebum in the occurence of sunburn and protection from the damage caused by u.v. irradiation. J Toxicol Sci 1984; 9:151-9.

33. Igarashi A, Uzuka M, Nakajima K. The effects of vitamin E deficiency on rat skin. Br J Dermatol 1989; 121:43-9.

34. Yamamoto A, Nagai T, Shimizu M, Shitara A, Sato Y, Inomata N. Skin irritation and cytotoxicity by squalene peroxides. XVI International Congress of Dermatology, Tokyo, 1982.

35. Saint-Leger D, Bague A, Cohen E, Chivot M. A possible role for squalene in the pathogenesis of acne. I. In vivo study of squalene oxidation. Br J Dermatol 1986; 114:535-42.

36. Werninghaus K, Handjani RM, Gilchrest B. Protective effect of alpha-tocopherol in carrier liposomes on ultraviolet light mediated epidermal cell damage in vitro. J Invest Dermatol 1990; 94: 590.

37. Kralli A, Moss SH. The sensitivity of an actinic reticuloid cell strain to near ultraviolet radiation and its modification by Trolox-C, a vitamin E analogue. Br J Dermatol 1987; 116:761-72.

38. Goldstein BD, Harber LC. Erythropoietic protoporphyria: lipid peroxidation and red cell membrane damage associated with photohemolysis. J Clin Invest 1972; 51:892-902.

39. Potokar M, Holtmann W, Werner-Busse A. Protective effect of vitamin E against ultraviolet light: comparative testing of natural tocopherol homologues on the skin of hairless mice. Society for Free Radical Research Winter Meeting, Free radicals in medicine. Current status of antioxidant therapy. Paris, December 9-10, 1988.

40. Pugliese PT, Saylor GB, Salter DC. The effects of *dl*-alpha-tocopherol-acetate on ultraviolet light induced epidermal ornithine decarboxylase activity. Annual Meeting of the SCC, New York, December 1-2, 1983.

41. Khettab N, Amory MC, Briand G, et al. Photoprotective effect of vitamins A and E on polyamine and oxygenated free radical metabolism in hairless mouse epidermis. Biochimie 1988; 70:1709-13.

42. Roshchupkin DI, Pitsov MY, Potapenko AY. Inhibition of ultraviolet light induced erythema by antioxidants. Arch Dermatol Res 1979; 266:91-4.

43. Ohzawa S, Arai M, Takeda Y. Vitamin E in topical preparations. Nippon Koshohin Kakakkaishi (BJJAEV) 1984; 8:18-27.

44. Potapenko AY, Abiev GA, Pliquett F. Inhibition of erythema of the skin photosensitized with 8-methoxypsoralene by a-tocopherol. Bull Exp Biol Med 1980; 89:611-5.

45. Potapenko AY, Abijev GA, Pitsov MY, et al. PUVA induced erythema and changes in mechano-electrical properties of skin. Inhibition by tocopherols. Arch Dermatol Res 1984; 276:12-6.

46. Namigata A. A study on the vitamin E and skin. Jpn J Dermatol 1960; 70:507-529.

47. Jordan P, Wulf K. Erythematodes Behandlung mit Vitamin E. Hautarzt 1950; 1:233-4.

48. Snyder DS. Cutaneous effects of topical indomethacin, an inhibitor of prostaglandin synthesis, on UV-damaged skin. J Invest Dermatol 1975; 64:322-5.

49. Kaidbey KH, Kurban AK. The influence of corticosteroids and topical indomethacin on sunburn erythema. J Invest Dermatol 1976; 66:153-6.

50. Msika P, Cesarini JP, Poelman MC. Antioxidants and aggressions from ultraviolet radiation in man. Society for Free Radical Research Winter Meeting, Free radicals in medicine: current status of antioxidant therapy. Paris, December 8-9, 1988.

51. Machtey I, Quaknine L. Tocopherol in osteoarthritis: a controlled pilot study. J Am Geriatr Soc 1978; 26:328-30.

52. Blankenhorn G, Fischer I, Seuss J. Möglichkeiten und Grenzen der Antioxidans-Therapie entzünd-licher rheumatischer Erkrankungen mit hochdosiertem Vitamin E. Akt Rheumatol 1985; 10:125-8.

53. Blankenhorn G. Klinische Wirksamkeit von Spondyvit (Vitamin E) bei aktivierten Arthrosen. Z Orthop 1986; 124:340-3.

54. Klein KG, Blankenhorn G. Vergleich der klinischen Wirksamkeit von Vitamin E und Diclofenac-Natrium bei Spondylitis ankylosans (Morbus Bechterew). VitaMinSpur 1987; 2:133-8.

55. Finnen MJ, Lawrence CM, Shuster S. Inhibition of dithranol inflammation by free radical scav-engers. Lancet 1984; 2:1129-30.

56. Lawrence CM, Shuster S. Effect of arachidonic acid on anthralin inflammation. Br J Clin Pharmacol 1987; 24:125-31.

57. Fuchs J, Packer L. Investigations on anthralin free radicals in model systems and in skin of hair-less mice. J Invest Dermatol 1989; 92:677-82.

58. Kamimura M. Antiinflammatory activity of vitamin E. J Vitaminol 1972; 18:204-9.

59. Yano S. Experimental and clinical studies on vitamin E from the dermatological point of view. Sapporo Med J 1959; 16:449-58.

60. Ayres S, Mihan R. Acne vulgaris and lipid peroxidation: new concepts in pathogenesis and treat-ment. Int J Dermatol 1978; 17:305-7.

61. Anonymous. Effect of vitamin E on prostanoid biosynthesis. Nutr Rev 1981; 39:317-20.

62. Panganamala RV, Cornwell DG. The effects of vitamin E on arachidonic acid metabolism. Ann N Y Acad Sci 1982; 393:376-90.

63. Hemler ME, Cook HW, Lands WEM. Prostaglandin biosynthesis can be triggered by lipid perox-ides. Arch Biochem Biophys 1979; 193:340-5.

64. Trautinger F, Grunewald C, Trenz A, Pittermann W, Kokoschka EM. Influence of natural vitamin E and UV radiation on dermal collagen content in hairless mice. XVIII Jahrestagung der Arbeits-gemeinschaft Dermatologische Forschung. Mannheim, November 9-11, 1990.

65. Brown RG, Button GM, Smith JT. Effect of vitamin E deficiency on collagen metabolism in the rat's skin. J Nutr 1967; 91:99-106.

66. Rave O, Wagner H, Junge-Hülsing G, Hauss WH. Zur Wirkung von Vitamin E auf den Mesen-chymstoffwechsel. Z Rheumaforsch 1971; 30:266-277.

67. Ehrlich HP, Tarver H, Hunt T. Inhibitory effects of vitamin E on collagen synthesis and wound repair. Ann Surg 1972; 17:235-40.

68. Taren DL, Chvapil M, Weber CW. Increasing the breaking strength of wounds exposed to pre-operative irradiation using vitamin E. Int J Vitam Nutr 1987; 57:133-7.

69. Melkumian AS, Tumanian EL, Aghajanov MI. Histologic estimation of regenerative ability of skin in treatment of burn traumas with the ointment, prepared on the basis of vitamin E. Zh Eksp Klin Med 1978; 18:52-3.

70. Kamimura M, Matsuzawa T. Percutaneous absorption of α-tocopherol acetate. J Vitaminol 1968; 14:150-9.

71. Djerassi D, Machlin LJ, Nocka C. Vitamin E: biochemical function & its role in cosmetics. Drug Cosmet 1986; January:29-34.

72. Martini MC. Role des microemulsions dans l'absorption percutaneous de l'alpha-tocopherole. J Pharm Belg 1984; 39:348-54.

73. Stüttgen G, Betzler H, Witte W. Zur Erfassung des Vitamin E im Serum und Hautfett mit Phosphormolybdänsäure. Arzneimittelforschung/Drug Research 1957; 7:407-8.

74. Fairris GM, Perkins P, Lloyd B, Hinks L, Clayton BE. The effect on atopic dermatitis of supplementation with selenium and vitamin E. Acta Derm Venereol (Stockh) 1989; 69:359-62.

75. Kitagawa M, Mino M. Effects of elevated *d*-alpha (*RRR*)-tocopherol dosage in man. J Nutr Sci Vitaminol 1989; 35:133-42.

76. Böhlers H. Resorptionsdynamik unterschiedlicher Vitamin E Mengen bei gesunden jungen Männern. VitaMinSpur 1988; 3:134-6.

77. Esterbauer H, Dieber-Rotheneder M, Waeg G, Puhl H, Tatzber F. Endogenous antioxidants and lipoprotein oxidation. Biochem Soc Trans 1990; 18:1059-61.

78. Hung SSO, Moon TW, Hilton JW, Slinger SJ. Uptake, transport and distribution of *d,l*-alpha-tocopherylacetate compared to *d*-alpha-tocopherol in rainbow trout. J Nutr 1982; 112:1590-6.

79. Cheng SC, Burton GW, Ingold KU, Foster DO. Chiral discrimination in the exchange of α-tocopherol stereoisomers between plasma and red blood cells. Lipids 1987; 22:469-73.

80. Ingold KU, Burton GW, Foster DO, Hughes L, Lindsay DA, Webb A. Biokinetics of and discrimination between dietary *RRR*- and *SRR*-α-tocopherols in the male rat. Lipids 1987; 22:163-72.

81. Ingold KU. Uptake and loss of vitamin E in animals and in man. German international workshop on α-tocopherol. Kluvensiek, Germany, January 31–February 5, 1991.

82. Witt E, Han D, Packer L. Protection of the skin of hairless mice against UV AB oxidative damage by oral supplementation with vitamin E. In Preparation.

83. Hidiroglou M. Vitamin E response in sheep to various modes of administration. Int J Vitam Nutr Res 1986; 56:27-252.

84. Shiratori T. Uptake, storage and excretion of chylomicra-bound ³H-alpha-tocopherol by the skin of the rat. Life Sci 1974; 14:929-35.

85. Tojo K, Lee AC. Bioconversion of a provitamin to vitamins C and E in skin. J Soc Cosmet Chem 1987; 38:333-9.

86. Horwitt MK. Therapeutic uses of vitamin E in medicine. Nutr Rev 1980; 38:105-13.

87. Oski FA. Vitamin E—a radical defense. N Engl J Med 1980; 303:454-5.

88. Sies H. Vitamin E. Dtsch Arztebl 1989; 86:B1475-7.

89. Nikolowski W. Die Bedeutung des Vitamin E für die Behandlung männlicher Fertilitätsstörungen. Ther Gegenw 1950; 89:329-32.

90. Nikolowski W. Vitamin E und seine Indikation für die Behandlung von Hautkrankheiten. Med Klin Wochenschr 1960; 55:415-8.

91. Nikolowski W. Inwieweit kann sich die Vitamin E-Therapie der Subfertilität des Mannes auf gesicherte Ergebnisse stützen? Derm Wochenschr 1962; 145:185-92.

92. Nikolowski W. Beschwerden während der sogenannten Wechseljahre des Mannes und Möglichkeiten ihrer Behandlung. Z Gerontol 1970; 3:404-7.

93. Nikolowski W. Das sogenannte climacterium virile. Dtsch Arztebl 1974; 5:299-302.

94. Lindner E. Therapeutische Bedeutung des Vitamin E bei Störungen der Spermiogenese. Int Z Vitaminforsch 1958; 20:33-40.

95. Bayer R. Ergebnisse der Vitamin E Vorbehandlung zur Beeinflussung der primären und sekundären essentiellen Infertilität (Bericht über 100 Ehen). Wien Med Wochenschr 1959; 109:271-5.

96. Milbradt R. Neuere Erkenntnisse in der Ätiologie, Therapie und Prophylaxe männlicher Fertilitätsstörungen. Hautarzt 1966; 17:97-100.

97. Bartak V. Zur Behandlung der Fertilitätsstörungen des Mannes mit Vitamin E und Methyltestosteron. Z Haut Geschl Kr 1972; 47:919-24.

98. Glander HJ, Liebsch F. Genügen Vitamin E-Kapseln mit 0.1 g alpha-Tokopherol für die Belange der Fertilitätstherapie? Dermatol Monatsschr 1973; 159:1048-50.

99.   Aitken RJ, Clarkson JS. Cellular basis of defective sperm function and its association with the genesis of reactive oxygen species by human spermatozoa. J Reprod Fertil 1987; 81:459-69.

100. Mann T, Jones R, Sherins R. Oxygen damage, lipid peroxidation and motility of spermatozoa. In: Steinberger A, Steinberger E, eds. Testicular development, structure and function. New York: Raven Press, 1980; 497-501.

101. Ochsendorf FR, Fuchs J, Milbradt R. Reaktive Oxidantien, Lipidperoxidation und Antioxidantien: Übersicht über ihre Bedeutung für die Spermatozoenfunktion. Dreiländertagung Fertilität und Sterilität, Goslar, May 29–June 1, 1991.

102. Sterzi G. La Vitamina E, con speciale riferimento alla dermatologia. Arch Ital Derm Sif Venerol 1950; 23:257-85.

103. Lohel H. Vitamin E in der Dermatologie. Dtsch Gesundh 1951; 6:984-7.

104. Fegeler F, Beckmann R. Über den Vitamin E Gehalt in Serum bei verschiedenen Hautkrankheiten, mit Bemerkungen über den Wirkungsmechanismus und die klinische Anwendung des Vitamin E. Derm Wochenschr 1955; 131:433-40.

105. Nikolowski W. Die Indikation für die Behandlung mit Vitamin E in der Dermatologie. Therapiewoche 1959; 10:136-7.

106. Wulf K. Vitamin E. In: Marchionini A, ed. Handbuch der Haut- und Geschlechtskrankheiten. Ergänzungswerk, Vol. 5. Berlin: Springer-Verlag, 1962; 447-69.

107. Ayres S, Mihan R. Vitamin E and dermatology. Cutis 1975; 16:1017-21.

108. Nikolowski W. Vitamin E in der Dermatologie. Apotheker J 1985; 7:53-6.

109. Fuchs J, ed. Oxidative injury in dermatopathology. Heidelberg: Springer-Verlag, in press, 1992.

110. Niwa Y, Sakane T, Shingu M, Miyachi Y. Role of stimulated neutrophils from patients with systemic lupus erythematosus in tissue injury, with special reference to serum factors and increased active oxygen species generated by neutrophils. Inflammation 1985; 9:163-72.

111. Welsh AL. Lupus erythematosus. Arch Dermatol 1954; 70:181-98.

112. Nadiger HA. Role of vitamin E in the aetiology of phrynoderma (follicular hyperkeratosis) and its interrelationship with B-complex vitamins. Br J Nutr 1980; 211-4.

113. Rietschel L. Die Bestimmung des Vitamin A und E im Serum von Hautkranken. Dermatol Wochenschr 1963; 25:635-8.

114. Gilliam JN, Sontheimer RD. Distinctive cutaneous subsets in the spectrum of lupus erythematosus. J Am Acad Dermatol 1981; 4:471-5.

115. Meigel W. Einteilung und Klinik des Lupus erythematodes. In: Holzmann H, Altmeyer P, Marsch WC, Vogel HG, eds. Dermatologie und Rheuma. Berlin: Springer-Verlag, 1987; 269-77.

116. Stone OJ. Lupus erythematosus and rheumatoid arthritis—groups of defects in microdebridement (polygenetic defects exceeding the fault tolerance threshold—a consequence of natural defense mechanisms). Med Hypoth 1988; 27:327-32.

117. Emerit I, Michelson AM. Chromosome instability in human and murine autoimmune disease: anticlastogenic effect of superoxide dismutase. Acta Physiol Scand (Suppl) 1980; 492:59-65.

118. Emerit I. Oxidative reactions and connective tissue diseases. In: Chow CK, ed. Cellular antioxidant defense mechanism, Vol. III. Boca Raton, FL: CRC Press, 1988; 111-21.

119. Emerit I. Chromosomal instability in collagen diseases. J Rheumatol 1989; 39:84-9.

120. Harman D. Free radical theory of aging: beneficial effect of antioxidants on the life span of male NZB mice; role of free radical reactions in the deterioration of the immune system with age and in the pathogenesis of systemic lupus erythematosus. Age 1980; 3:64-73.

121. Emerit I, Levy A, De Vaux Saint Cyr C. Chromosome damaging agent of low molecular weight in the serum of New Zealand Black mice. Cytogenet Cell Genet 1980; 26:41-7.

122. Miyachi Y, Yoshioka A, Imamura S, Niwa Y. Polymorphonuclear leucocyte derived reactive oxygen species in inflammatory skin diseases. In: Hayaishi O, Imamura S, Miyachi Y, eds. The biological role of reactive oxygen species in skin. New York: Elsevier, 1987; 135-40.

123. Shingu M, Oribe M, Todoroki T, et al. Serum factors from patients with systemic lupus erythematosus enhancing superoxide generation by normal neutrophils. J Invest Dermatol 1983; 81:212-5.

124. Yancey KB, Lawley TJ. Circulating immune complexes: their immunochemistry, biology, and detection in selected dermatologic and systemic disease. J Am Acad Dermatol 1984; 10:711-31.

125. D'Onofrio C, Maly FE, Fischer H, Maas D. Differential generation of chemiluminescence-detectable oxygen radicals by polymorphonuclear leukocytes challenged with sera from systemic lupus erythematosus and rheumatoid arthritis patients. Klin Wochenschr 1984; 62:710-6.

126. Tuschl H, Kovac R, Wolf A, Smolen JS. SCE frequencies in lymphocytes of systemic lupus erythematosus patients. Mutat Res 1984; 128:167-72.

127. Emerit I. Chromosome breakage factors: origin and possible significance. Prog Mutat Res 1982; 4:61-72.

128. Emerit I, Michelson AM. Mechanism of photosensitivity in systemic lupus erythematosus patients. Proc Natl Acad Sci U S A 1981; 78:2537-40.

129. Emerit I, Khan SH, Cerutti PA. Treatment of lymphocyte cultures with a hypoxanthine xanthine oxidase system induces the formation of transferable clastogenic material. Free Radic Biol Med 1985; 1:51-7.

130. Khan SH, Emerit I. Lipid peroxidation products and clastogenic material in culture media of human leukocytes exposed to the tumor promotor phorbol myristate acetate. J Free Radic Biol Med 1985; 1:443-9.

131. Emerit I. Properties and action mechanism of clastogenic factors. Lymphokines 1983; 8:413-24.

132. Bazex A. La vitamin E dans le traitment des lupus erythemateux chroniques. Bull Soc Fr Derm 1949; 56:56-7.

133. Doucas C. Vitamin E in case of lupus erythematosus with poikiloderma and in Meleda disease: successful therapy. Bull Soc Fr Dermatol Syph 1949; 56:449-54.

134. Silvers SH. Chronic lupus erythematosus successfully treated with vitamin E. Arch Dermatol Syph Suppl. 1950; 62:163.

135. Grubb E. Our experiences of vitamin E treatment. Acta Derm Venereol (Stockh) 1952; 32:256-8.

136. Shinskii GE, Telegina KA, Shephovtsova VV. Experience with vitamin E in the treatment of erythrematosus. Vestn Dermatol Venerol 1962; 36:64-8.

137. Ayres S, Mihan R. Is vitamin E involved in the autoimmune mechanism? Cutis 1978; 21:321-5.

138. Sweet RD. Vitamin E in Collagenoses. Lancet 1948; 2:310-1.

139. Sawicky HH. Therapy of lupus erythematosus. Arch Dermatol Syphilol Suppl. 1950; 61:906-8.

140. Kaminsky A, Kaplan H, Viglioglia P. Lupus eritematoso y tocoferoles. In: Separata de sesiones dermatologicas en homenjae al profesor Luis E. Pierini, Buenos Aires, November 1949. Buenos Aires: Lopez & Etchegoyen SRL, 1950.

141. Morgan J. A note on the treatment of lupus erythematosus with Vitamin E. Br J Dermatol 1951; 63:224-7.

142. Pascher F, Sawicky HH, Mabel MD, Silverberg MG, Braitman M, Kanof NB. Therapeutic assays of the skin and cancer unit of the New York University hospitals. Assay V. Tocopherols (vitamin E) for discoid lupus erythematosus and other dermatoses. J Invest Dermatol 1951; 17:261-3.

143. Ayres S, Mihan R. Vitamin E (tocopherol)—a reappraisal of its value in dermatoses of mesodermal tissues. Cutis 1971; 7:35-45.

144. Yen EHK, Sodek J, Melcher AH. The effect of oxygen partial pressure on protein synthesis and collagen hydroxylation by mature peridontal tissues maintained in organ cultures. Biochem J 1979; 178:605-12.

145. Bhatnagar RS, Liu TZ. Evidence for free radical involvement in the hydroxylation of proline: Inhibition by nitro blue tetrazolium. FEBS Lett 1972; 26:32-4.

146. Last JA, Greenberg DB. Ozone induced alterations in collagen metabolism of rat lungs. II. Long term exposures. Toxicol Appl Pharmacol 1980; 55:108-14.

147. Hussain MZ, Bhatnagar RS. Involvement of superoxide in the paraquat-induced enhancement of lung collagen synthesis in organ culture. Biochem Biophys Res Commun 1979; 89:71-6.

148. Murrell GAC, Francis MJO, Bromley L. Modulation of fibroblast proliferation by oxygen free radicals. Free Radic Biol Med 1990; 9(Suppl. 1):109.

149. Haustein UF, Ziegler V, Herrmann K, Mehlhorn J, Schmidt C. Silica induced scleroderma. J Am Acad Dermatol 1990; 22:444-8.

150. Geesin JC, Hendricks L, Gordon JS, Berg RA. Lipid peroxidation stimulates collagen synthesis in human dermal fibroblasts. Free Radic Biol Med 1990; 9(Suppl. 1):4.

151. Geesin JC, Gordon JS, Berg RA. Retinoids inhibit collagen synthesis through inhibition of ascorbate stimulated lipid peroxidation. J Invest Dermatol 1989; 92:433.

152. Borel JP. Radicaux libres oxygenes, tissu conjonctif et inflammation. Pathol Biol (Paris) 1983; 31:8-10.

153. Baillet F, Housset M, Michelson AM, Puget K. Treatment of radiofibrosis with liposomal superoxide dismutase. Preliminary results of 50 cases. Free Radic Res Commun 1986; 1:387-94.

154. Pambor M, Schmidt W, Wiesner M, Jahr U. Induratio penis plastica—Ergebnisse nach kombinierter Behandlung mit Röntgenbestrahlung und Tokopherol. Z Klin Welt 1985; 40:1425-7.

155. Reifferscheid M, Matis P. Das Vitamin E in der Behandlung von Durchblutungschäden Dupuytren-Kontraktur und Thrombosen. Med Welt 1951; 1168-72.

156. Hanfstaengel E. Die Behandlung der Dupuytrenschen Palmarfascienkontraktur mit Vitamin E. Med Klin Wochenschr 1951; 373-5.

157. Menzel I. Induratio penis plastica. Z Allgemeinmed 1977; 53:324-6.

158. Gartmann H, Sieler H. Zur Vitamin E-Behandlung der Induratio penis plastica. Derm Wochenschr 1953; 128:1213-9.

159. Kunstmann H. Zur Behandlung der Induratio penis plastica mit Vitamin E. Medizinische 1954 125-6.

160. Pult H. Erfahrungen bei der Behandlung der Kollagenosen mit Vitamin E. Dtsch Med Wochenschr 1954; 471-2, 481-3.

161. Marazzini Z. Induratio penis plastica e vitamina E. Urologia (Treviso) 1955; 22:352-5.

162. Martin HAP. Uber die Behandlung des Pruritus und der Kraurosis vulvae mit Vitamin E. Med Klin Wochenschr 1952; 1603-5.

163. Michaelsson G, Edquist LE. Erythrocyte glutathione peroxidase activity in acne vulgaris and the effect of selenium and vitamin E treatment. Acta Derm Venereol (Stockh) 1984; 64:9-14.

164. Dainow PI. La vitamin E dans le traitment de l'acne. Dermatologica 1953; 106:197-200.

165. Hagerman G. On the use of vitamin E in dermatologic practice. Acta Derm Venereol (Stockh) 1951; 31:225-6.

166. Sehgal VN, Vadiraj SN, Rege VL, Beohar PC. Dystrophic epidermolysis bullosa in a family. Dermatologica 1972; 144:27-34.

167. Michaelson JD, Schmidt JD, Dresden MH, Duncan C. Vitamin E treatment of epidermolysis bullosa. Arch Dermatol 1974; 109:67-9.

168. Degreef H, Fluor M. Epidermolysis bullosa: treatment with vitamin E, preliminary results. Arch Belges Derm 1974; 30:83-7.

169. Smith EB, Michener WM. Vitamin E treatment of dermolytic bullous dermatosis. Arch Dermatol 1973; 108:254-6.

170. Wilson HD. Treatment of epidermolysis bullosa dystrophica by alpha-tocopherol. Can Med Assoc J 1964; 90:1315-6.

171. Ayres S, Mihan R. Pseudoxanthoma elasticum and epidermolysis bullosa—response to vitamin E (tocopherol). Cutis 1969; 5:287-94.

172. Cochrane T. Granuloma annulare: treatment with vitamin E. Br J Dermatol Syph 1950; 62:316-8.

173. Dainow I. Granuloma annulaire traite par les vitamins E et A. Dermatologica 1958; 116:367-9.

174. Cornbleet T, Cohen D, Bielinski S. Granuloma annulare generalized. Arch Dermatol Syphilol 1951; 64:249.

175. Ayres S, Mihan R. Keratosis follicularis. Arch Dermatol 1972; 106:909-10.

176. Ayres S, Mihan R. Yellow nail syndrome. Arch Dermatol 1973; 108:267-8.

177. Hayakawa R, Ueda H, Nozaki T, et al. Effects of combination treatment with vitamins E and C on chloasma and pigmented contact dermatitis. A double blind controlled clinical trial. Acta Vitaminol Enzymol 1981; 3:31-8.

178. Mustajoki P. Vitamin E in porphyria. JAMA 1972; 221:714-5.

179. Nair PP, Mezey E, Murty HS, Quartner J, Mendeloff AI. Vitamin E and porphyrin metabolism in man. Arch Intern Med 1971; 128:411-5.

180. Murty HS, Pinelli A, Nair RP, Mendeloff AI. Porphyria cutanea tarda: therapeutic response to vitamin E. Clin Res 1969; 17:474.

181. Conley CL, Chisholm JJ. Ausheilung einer Leber-Dekompensation bei einer Protoporphyrie (EPP oder erythropoetische Protoporphyrie). Johns Hopkins Med J 1979; 145:237-40.

182. Kahan A, Kahan IL, Tapaszto I, Bencze G. Therapie Versuche beim Pseudoxanthoma elasticum Syndrom. Klin Monatsbl Augenheilk 1963; 142:1047-58.

183. Hamilton-Gibbs JS. Pseudoxanthoma elasticum. Aust J Dermatol 1964; 7:156-63.

184. Stout OM. Pseudoxanthoma elasticum with retinal angioid streaking, decidedly improved on tocopherol therapy. Arch Dermatol Syph 1951; 63:510-1.

185. Schade W. Zur Therapie der Psoriasis vulgaris. Hautarzt 1952; 3:373-4.

186. Gerloczy F, Lancos F, Szabo J. Wirkung des Vitamin E auf die Kapillarresistenz bei Purpura im Kindesalter. Acta Paediatr Acad Sci Hung 1966; 7:363-7.

187. Fujii T. The clinical effects of vitamin E on purpuras due to vascular defects. J Vitaminol 1972; 18:125-30.

188. Comi G, Nesi G. Contributo allo studio delle azioni esercitate dalla vitamina E. Riv Clin Med 1950; 50:214-26.

189. Frey JG. Über die Kombinationsbehandlung von Röntgenspätschäden der Haut mit Kurzwellen und Vitamin E. Strahlentherapie 1954; 95:440-3.

190. Hütter H. Neue Anwendungsgebiete für die Vitamin E Behandlung. Dtsch Gesundh 1954; 277-9.

191. Kemmer-Görlitz CH. Vitamin E Behandlung des Röntgenulkus. Derm Wochenschr 1952; 126: 1209-10.

192. Ayres S, Mihan R. Subcorneal pustular dermatosis controlled by vitamin E. Arch Dermatol 1974; 109:914.

193. Bijdendijk A, Noordhoeck FJ. Die Behandlung von Ulcus cruris mit hohen Dosen Vitamin E. Nederl Tijdschr Geneesk 1951; 1039-42.

194. Siedentopf H, Krüger A. Die Wirkung hoher Vitamin E (alpha-Tocopherol) Gaben auf Gefässerkrankungen, im besonderen auf das Ulcur cruris. Med Klin Wochenschr 1949; 1060-2.

195. Norton L. Further observations on the yellow nail syndrom with therapeutic effects of oral alpha-Tocopherol. Cutis 1985; 457-62.

196. Hazelrigg DE, McElroy RJ. The yellow nail syndrom. J Assoc Milit Dermatol 1980; 6:1-21.

197. Rust S. Über allergische Reaktionen bei Vitamintherapie. Z Haut Geschl Kr 1954; 17:317.

198. Aeling JL, Panagotacos PJ, Andreozzi RJ. Allergic contact dermatitis to vitamin E aerosol deodorant. Arch Dermatol 1973; 108:579-80.

199. Saperstein H, Rapaport M, Rietschel RL. Topical vitamin E as a cause of erythema multiforme like eruption. Arch Dermatol 1984; 120:906-8.

200. Bork K. Medikamentöse Zahnverfärbungen. In: Bork K, ed. Kutane Arzneimittelnebenworkungen. Schattauer Verlag, Stuttgart, 1985; 297-302.

201. Brodkin RH, Bleiberg J. Sensitivity to topically applied vitamin E. Arch Dermatol 1965; 92:76-7.

202. Minkin W, Cohen HJ, Frank SB. Contact dermatitis from deodorants. Arch Dermatol 1973; 107:774-5.

203. Kassen B, Mitchell JC. Contact urticaria from a vitamin E preparation in two siblings. Contact Dermatitis News Lett 1974; 16:482.

204. Roed-Petersen J, Hjorth N. Patch test sensitization from *dl*-alpha-tocopherol (vitamin E). Contact Dermatitis 1975; 1:391.

205. Kästner W, Kappus H. Sicherheit bei der Einnahme von Vitamin E. Toxikologische Aspekte und Verträglichkeit bei oraler Aufnahme. VERIS Vitamin E Res Information Service, 1990; 1:1-24.

206. Roberts HJ. Perspective on vitamin E as therapy. JAMA 1981; 26:129-31.

207. Roberts HJ. Thrombophlebitis associated with vitamin E therapy. With commentary on other medical side effects. Angiology 1979; 30:169-77.

208. Bendich A, Machlin LJ. Safety of oral intake of vitamin E. Am J Clin Nutr 1988; 48:612-9.

209. Salkeld RM. Safety and tolerance of high dose vitamin E administration in man: a review of the literature. Fed Reg 1979; 44:16172.

210. WHO Food Additives Series. 21. Toxicological evaluation of certain food additives and contaminants. Prepared by the 30th meeting of joint FAO/WHO expert committee on food additives, Rome, June 2-11, 1986. Cambridge University Press, 1987; 55-69.

53

Vitamin E and Coenzyme Q_{10} in Normal Human Skin and in Basal Cell Epitheliomas

Luigi Rusciani, Giuseppina Petrelli, and Silvio Lippa

Catholic University of the Sacred Heart, Rome, Italy

VITAMIN E AND COENZYME Q_{10} AS INTRACELLULAR ANTIOXIDANTS

Vitamin E seems to be the intracellular antioxidant par excellence because, in addition to its own activity, it is able to influence the activity of other antioxidants. Recent studies have shown that glutathione and ascorbic acid are capable of inhibiting lipid peroxidation only in the presence of normal levels of vitamin E. The presence of this vitamin is thus fundamental for the other two substances to exert their protective effect against lipid peroxidation (1). Certain investigators believe that vitamin E carries out its antioxidant activity by blocking the peroxyl radical and thus interrupting lipid peroxidation (2). Others have hypothesized that the methyl groups of the vitamin E molecule react with the "pockets" formed by the double hydrophobic bonds of the arachidonic acid of membrane phospholipids (3). At this point, it is possible to imagine that α-tocopherol located at the level of the cell membrane plays a role not only as an antioxidant by also as a structural regulator of the cell itself (3,4).

Coenzyme Q_{10}, or ubiquinone, functions as an electron carrier in the respiratory chain at the level of the internal mitochondrial membrane, where oxidative phosphorylation occurs (5–8). Recent studies have demonstrated that this molecule exerts an antioxidant activity as well as vitamin E (9–11) and it is able of furnish protection against lipid peroxidation not only in vitro but also in vivo (12). The in vivo antioxidant activity of CoQ_{10} has been demonstrated by the fact that this molecule, when administered systemically, is capable of curing syndromes of vitamin E deficiency (9).

Various ubiquinones can react in vitro with superoxide radicals enzymatically generated by the xanthine-xanthine oxidase system (9). It has also been shown that short-chain ubiquinones, which are more hydrosoluble than CoQ_{10}, significantly block the formation of superoxide radicals, CoQ_{10} could also exert an important protective effect in vivo on membrane phospholipids (9). More recent studies have shown how coenzyme Q is capable of exerting its antioxidant activity not only by reacting with free oxygen radicals, like vitamin E, to render them incapable of damaging cellular structures, but also by interacting with the lipid radicals of the membrane. In these experiments an oxidized form of CoQ_3 was used. Unlike the reduced form of α-tocopherol, this form does not possess a phenol group and thus cannot donate a hydrogen atom to a free radical to interrupt the autocatalytic peroxidation reactions.

Under these conditions the antioxidant effect of the coenzyme is due to its ability to modify the order of the other molecules of the membrane, increasing the fluidity of the double lipid layer. It is possible that this effect facilitates the interaction between the quinone and the lipid radicals, thus impeding the propagation of peroxidation (13). As far as CoQ_{10} is concerned, it is important to remember that this ubiquinone, when forced to play its role as an antioxidant because of increased oxidative threat or because of a deficit of vitamin E, cannot function in the respiratory chain for the production of energy (12).

The study of these two antioxidants in the skin is especially interesting because the skin is constantly exposed to external physical and chemical agents capable of generating free radicals. These radicals are probably involved in the processes of premature aging and cutaneous carcinogenesis (14–18). It is likely that physiological aging is also a consequence of free-radical damage (14), and the appearance of tumors is caused by the effects of these radicals on DNA (15,16,18–20).

EXPERIMENTAL STUDY OF VITAMIN E AND COENZYME Q_{10} LEVELS IN NORMAL HUMAN SKIN AND IN EPIDERMAL TUMORS

The aim of the present study was to measure the content of both vitamin E and coenzyme Q_{10} in normal human skin that was not exposed to the sun and to compare them to those found in normal exposed skin and in basal cell epithelioma specimens.

Vitamin E and CoQ_{10} content was measured in 30 skin biopsy specimens taken from 30 different subjects. These specimens were divided into three groups:

Group A), 10 samples of non–sun-exposed skin taken from the backs of 10 healthy subjects

Group B), 10 samples of sun-exposed skin taken from the faces, arms, or chests of 10 other healthy subjects

Group C), 10 lesional samples from 10 patients suffering from facial basal cell epitheliomas

In all these cases, the tumor type was histologically diagnosed. The samples used for our study were taken with a biopsy punch 6 mm in diameter during the initial excision of the tumors.

The ages of the healthy subjects in groups A and B ranged from 24 to 55 years and those of group C varied from 44 to 57 years.

The method described by Lippa et al. (21) was used to measure the coenzyme Q content in these skin samples. Briefly, specimens were homogenized for 30 s at 0°C in 0.01 M phosphate buffer (pH 7.4) in an Ultra-Turrax, and then coenzyme Q_8 was added as an internal standard and coenzyme Q_{10} was extracted three times with 5 ml acetone. The combined extracts were dried under nitrogen, and the residue, reconstituted with 4 ml hexane, was purified in a 15 × 0.5 cm silica gel (80–230 mesh) column.

After initial washing with 7 ml of a hexane-diethyl ether (98:2) mixture, coenzyme Q was eluted with 12 ml of a hexane-ethyl ether-dichloromethane (48:50:2) mixture. The eluate was dried under nitrogen and reconstituted in 50 μl ethanol. High-performance liquid chromatography (HPLC) was then carried out on 20 μl of this solution using a Beckman HPLC chromatograph under the following operating conditions:

Column: ultrasphere optical density 250 × 4 mm
Flow rate: 1 ml/ml
Eluent: ethanol-methanol, 70:30
Detector photometric at = 280 nm.

The mean recovery for the internal standard was 80 ± 5%. Vitamin E content was measured as described by Erikson and Soerensen (22), and proteins were determined according to Lowry's method (23). Coenzyme Q_{10} and vitamin E values were expressed as ng/mg protein.

TABLE 1 Vitamin E and Coenzyme Q_{10} Levels in Exposed and Nonexposed Normal Human Skin and Basal Cell Epitheliomas

	Normal nonexposed skin (ng/mg protein)	Normal sun-exposed skin (ng/mg protein)	Basal cell epitheliomas (ng/mg protein)
Vitamin E	181 ± 50	675 ± 273	2437 ± 107
CoQ_{10}	211 ± 64	637 ± 397	587 ± 211

The contents of vitamin E and CoQ_{10} in the three groups of skin biopsy specimens, as determined by HPLC, are reported in Table 1. Examination of these data reveals that the contents of both coenzyme Q_{10} and vitamin E vary widely from one subject to another. In fact, the standard deviations from the mean are very high. However, statistical analysis performed using Student's *t*-test shows that normal, sun-protected skin specimens contained significantly lower levels of both vitamin E ($p < 0.025$) and coenzyme Q_{10} ($p < 0.001$) than samples of normal, sun-exposed skin (Fig. 1). When nonexposed, normal skin values are

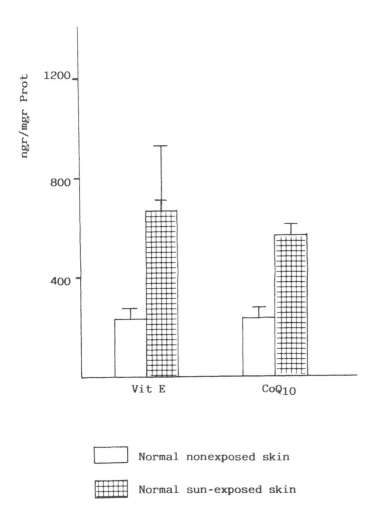

FIGURE 1 Vitamin E and coenzyme Q_{10} levels in sun-exposed and sun-protected normal skin.

compared to those from the basal cell epithelioma specimens, markedly higher values, which were statistically significant, for both vitamin E ($p < 0105$) and CoQ_{10} ($p < 0.025$) are seen in the neoplastic samples (Fig. 2). No significant differences were found between the content of CoQ_{10} in normal sun-exposed skin and that in basal cell epithelioma specimens, although vitamin E levels were significantly higher ($P < 0.001$) in the neoplastic samples (Fig. 3).

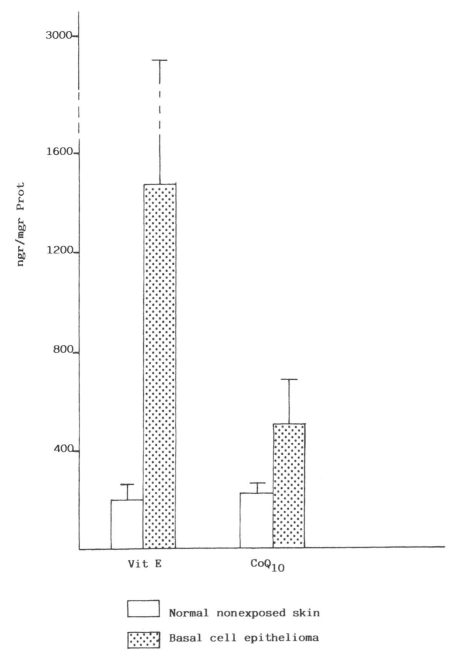

FIGURE 2   Vitamin E and coenzyme Q_{10} levels in nonexposed normal skin and basal cell epitheliomas.

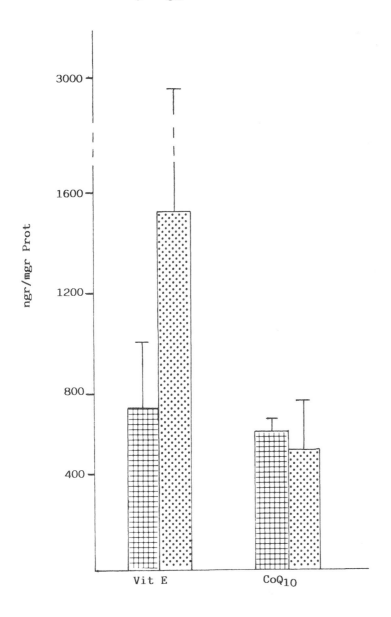

FIGURE 3   Vitamin E and coenzyme Q$_{10}$ levels in sun-exposed normal skin and basal cell epitheliomas.

POSSIBLE INTERPRETATIONS OF THE INCREASED VALUES OF VITAMIN E AND COENZYME Q_{10} IN BASAL CELL EPITHELIOMAS

Exposure to solar radiation is capable of provoking a series of alterations ranging from erythema, the characteristic immediate response of the skin to an excessive absorption of ultraviolet (UV) rays, to actinic keratoses and tumors, which reflect chronic and prolonged exposure. To clarify the mechanisms activated by the skin to defend itself from UV irradiation-induced damage, a series of studies have been conducted that show that the systemic administration of a mixture of antioxidant substances, involving vitamin E, is able to protect hairless mice from the erythematous reaction induced by UV irradiation exposure (24–27). The same mixture has also been found to be capable of reducing not only the number and the severity of lesions produced in mice by chronic UV irradiation exposure, but also the frequency of squamous cell carcinoma (25).

In humans, then, the differences in vitamin E and CoQ_{10} contents found between exposed and protected skin may reflect the different levels of oxidative damage risk to which these two types of tissues are subjected, the exposed skin requiring a higher level of defense than that needed for sun-protected skin. The increased concentrations of these two substances in cutaneous specimens from exposed areas would represent a *defense measure* against the production of free radicals provoked by UV exposure.

As far as neoplastic tissues are concerned, experimental studies conducted in animals have shown that the following:

1. There is a decreased level of lipid peroxidation in the microsomes of neoplastic cells. This is probably related to reduced contents of arachidonic acid and total lipids, as well as to an increased cholesterol-fatty acid ratio (17,28) and to a reduced level of cytochrome P_{450}, which acts as a free-radical producer during hydroxylation reactions, thus stimulating lipid peroxidation (17,28).
2. Total antioxidant activity relative to total lipids is much higher in the tumor extracts (17,28).
3. α-Tocopherol is responsible for the majority of all antioxidant activity carried out in tumor cells (28).

The increased values of vitamin E that we found in basal cell epithelioma specimens also confirm results obtained by other investigators who found increased levels of this antioxidant in experimentally induced hepatic tumor in animals. The higher concentrations of CoQ_{10} found in basal cell epitheliomas than in normal, nonexposed skin are probably related to the antioxidant role played by this substance together with vitamin E. The increased contents of the two antioxidants in tumor tissues could be due to (1) an accumulation of these substances caused by the relative lack in tumor cells of polyunsaturated fatty acids available to bind them. This reduction in polyunsaturated fatty acids is also responsible for the reduced level of lipid peroxidation observed in tumor cells (17). It is important to remember that lipid peroxidation may play a protective role against carcinogenesis. During this process, in fact, not only free radicals are produced but also toxic aldehydes, such as 4-hydroxyalkenals-ali, which interfere with synthesis of DNA, thus slowing tumor cell multiplication (17,28). Two such aldehydes in particular, 4-hydroxypentanal and to an even greater extent 4-hydroxynonenal, have been found to produce therapeutic effects in animal tumors (28). The reduction in lipid peroxidation in neoplastic cells, then, and the consequent reduction in aldehydes of this type, could facilitate the systemic dissemination of tumor tissue throughout the body.

Alternatively, some authors observed a reduction in mitochondrial superoxide dismutase (SOD) in two experimental tumors, Morris's hepatoma 3924A, and ascites cells from Ehrlich's tumor (29). Other investigators later reported reduced levels of SOD, glutathione peroxidase, and scavenger enzymes in general in cutaneous epitheliomas (20,30). The decrease in cyto-plasmic SOD would even seem to be directly proportional to the degree of malignancy of the tumor: in fact, the lowest levels of this enzyme were found in squamous cell epitheliomas, which are known for their highly invasive character (20). In light of these findings, the increased levels of vitamin E and of coenzyme Q$_{10}$ that we observed in our basal cell epithelioma specimens might reflect an attempt by these cells to compensate for the reduction in their scavenger enzymes.

Note that the 10 lesional samples, all histologically confirmed as basal cell epitheliomas, came from subjects who were, on the average, older than those from whom specimens of nonneoplastic tissues were taken. However, we do not believe that our findings can be interpreted as a function of subject age, since recent studies indicated that at least one of the antioxidants, CoQ$_{10}$, tends to be lower, rather than higher, in tissues from elderly subjects than younger ones (31).

That vitamin E levels in basal cell epithelioma were significantly higher than those in normal sun-exposed skin and that the differences between the levels of CoQ$_{10}$ in these types of specimens were not significant may reflect the fact that the former substance is the primary intracellular antioxidant and CoQ$_{10}$ exerts its antioxidant activity only when there is a deficiency of vitamin E or when there is a massive peroxidation attack.

Additional studies are in progress that will, it is hoped, add to our knowledge of the relationship between vitamin E, coenzyme Q$_{10}$, and peroxidative insult to tissues.

SUMMARY

Free oxygen-derived radicals have lately been implicated in the pathogenesis of a number of conditions, including aging of the skin and cutaneous carcinogenesis. These highly reactive substances, which are normal by-products of oxidative cellular metabolism, are capable of causing damage to numerous structural elements of the cell itself as well as to DNA. The cell's defense against these free radicals includes protection from antioxidant substances, which interact with the oxygen radicals to prevent their damaging interaction with cellular structures.

To clarify the protective roles played by two of these antioxidant substances, vitamin E and coenzyme Q$_{10}$, we measured the contents of these two substances using high-performance liquid chromatography in three types of skin: (1) normal skin from sun-exposed areas, (2) normal skin from non–sun-exposed areas, and (3) lesional skin from patients with basal cell epithelioma. The results obtained showed that quantities of both antioxidants in nonexposed skin were significantly lower than those in exposed skin. Basal cell epithelioma specimens also contained markedly higher levels of both substances than specimens from nonexposed areas. No significant differences were found between the content of coenzyme Q$_{10}$ in normal sun-exposed skin and that in basal cell epithelioma, although vitamin E levels were found to be significantly higher in the neoplastic samples.

REFERENCES

1. Wefers H, Sies H. The protection by ascorbate and glutathione against microsomal lipid peroxidation is dependent on vitamin E. Eur J Biochem 1988; 174(2):353-7.
2. Burton GW, Cheeseman KH, Doba T, Ingold KV, Slater TF. Vitamin E as an antioxidant in vitro and in vivo. In: Biology of vitamin E. Ciba Foundation Symposium 101. London: Pitman Books, 1983; 4-18.
3. Diplock AT. The role of vitamin E in biological membranes. In: Biology of vitamin E. Ciba Foundation Symposium 101. London: Pitman Books, 1983; 45-55.

4. Urano S, Iida M, Otani I, Matsuo M. Membrane stabilization of vitamin E: interactions of alpha tocopherol with phospholipids in bilayer liposomes. Biochem Biophys Res Commun 1987; 146(3): 1413-8.

5. Leboulanger J. Le Vitamine (Rindi G, Italian ed.). Milan: Mondadori Publishers, 1987; 62.

6. Lehninger AL. Buiochimica. Bologna: Zanichelli, 1979; 443-5.

7. Monesi V. Istologia. Padova: Piccin Editore, 1980; 157-66.

8. Moruzzi G, Rossi CA, Rabbi A. Principi di Chimica Biologica. Bologna: Libreria Universitaria Tinarelli, 1979; 341-9.

9. Littarru GP, Lippa S. Coenzyme Q_{10} and antioxidant activity: facts and perspectives. Drugs Exp Clin Res 1984; 10(7):491-6.

10. Landi L, Sechi AM, Pasquali P, et al. Antioxidative mechanism of CoQ incorporated in liposomes. Cardiologia 1986; 31(7):493-6.

11. Littarru GP, De Sole P, Lippa S, Oradei A. Study of quenching of singlet oxygen by coenzyme Q in a system of human leucocytes. In: Folkers K, Yamamura Y, eds. Biochemical and clinical aspects of coenzyme Q, Vol. 4. 201-8.

12. Freeman BA. Biological sites and mechanism of free radical production. In: Armstrong D, et al., eds. Free radicals in molecular biology. Aging and diseases. New York: Raven Press, 1984; 43-52.

13. Landi L, Sechi AM, Pasquali P, et al. Antioxidative mechanism of CoQ_{10} incorporated in liposomes. Cardiologia 1986; 31(7):493-6.

14. Harman D. Free radicals and the origination, evolution and present status of free radical theory of aging. In: Armstrong D, et al., eds. Free radicals in molecular biology. Aging and diseases. New York: Raven Press, 1984; 1-11.

15. Southorn PA, Powis G. Free radicals in medicine. II. Involvement in human disease. Mayo Clin Proc 1988; 63(4):390-408.

16. Borek C. Radiation and chemically induced transformation: free radicals, antioxidants and cancer. Br J Cancer (Suppl) 1987; 8:74-86.

17. Slater TF, Cheeseman KH, Proudfoot K. Free radicals, lipid peroxidation and cancer. In: Armstrong D, et al., eds. Free radicals in molecular biology. Aging and diseases. New York: Raven Press, 1984; 293-305.

18. Ward JF, Evans JW, Limoli CL, Calabro-Jones PM. Radiation and hydrogen peroxide induced free radicals damage to DNA. Br J Cancer (Suppl) 1987; 8:105-12.

19. Halliwell B. Oxygen radicals: a commonsense look at their nature and medical importance. Med Biol 1984; 62:71-7.

20. Galeotti T, Borrello S, Seccia A, Farallo E, Bartoli GM, Serri F. Superoxide dismutase content in human epidermis and squamous cell epithelioma. Arch Dermatol Res 1980; 267:83-6.

21. Lippa S, Littarru GP, Oradei A. Determinazione routinaria del Coenzima Q mediante HPLC in campioni biologici. Abstracts I Conf Naz, La Cromatografia liquida ad alta risoluzione in analitica clinica: situazione attuale e prospettive, Verona, 1985; 51.

22. Erikson T, Soerensen B. High performance liquid chromatography of vitamin E. Acta Pharmac Suec 1977; 14:478-84.

23. Lowry OM, Rosebrough Y, Farr L, Randall RJ. Protein measurement with the folinphenol reagent. J Biol Chem 1951; 193:265-73.

24. Black HS. Effects of dietary antioxidants on actinic tumor induction. Res Commun Chem Pathol Pharmacol 1974; 7(4):783-6.

25. Black HS, Chan JT. Suppression of UV light-induced tumor formation by dietary antioxidants. J Invest Dermatol 1975; 65:412-4.

26. Chan JT, Ford JO, Rudolph AH, Black HS. Physiological changes in hairless mice maintained on an antioxidant supplemented diet. Experientia 1977; 33(1):41-2.

27. De Rios G, Chan JT, Black HS, Rudolph AH, Knox JM. Systemic protection by antioxidants against UVL-induced erythema. J Invest Dermatol 1978; 70:123-5.

28. Burton GW, Cheeseman KH, Ingold KV, Slater TF. Lipid antioxidants and products of lipid peroxidation as potential tumor protective agents. Biochem Soc Trans 1983; 11:261-2.

29. Dionisi O, Galeotti T, Terranova T, Azzi A. Superoxide radicals and hydrogen peroxide formation in mitochondria from normal and neoplastic tissues. Biochim Biophys Acta 1975; 403:292-300.

30. Oberley LW, Buettner GR. Role of superoxide dismutase in cancer: a review. Cancer Res 1979; 39:1141-9.

31. Di Palma A. Coenzyme Q: a probable longevity marker. Abstracts VI international sympsoium on biomedical and clinical aspects of coenzyme Q, Rome. 1990; 72-3.

54

Vitamin E in Skin: Antioxidant and Prooxidant Balance

Eric Witt, Valerian E. Kagan, and Lester Packer

University of California, Berkeley, California

INTRODUCTION

The skin is a unique organ in that it must constantly withstand direct exposure to environmental insults. One such insult is solar radiation, which contains light in the UV (ultraviolet) B (290–320 nm), UVA (320–400 nm), and visible (400–700 nm) regions of the spectrum. Chronic exposure to such radiation has been linked to a number of types of skin damage, such as phototoxic injury and photoallergy (1), skin aging (2), and melanoma and nonmelanoma skin cancer (3–7).

One likely hypothesis for the genesis of skin pathologies due to exposure to solar light is that such exposure causes the formation of free radicals and that these radicals can overwhelm the skin's antioxidant defenses, causing damage to lipids, proteins, and DNA. Such damage could lead to a variety of pathogenic conditions. A consequence of this theory is that antioxidant supplementation of the skin, by dietary or topical means, should prevent such damage to some degree. Since vitamin E is the major lipid peroxidation chain-breaking antioxidant in skin, it seems a logical choice for such supplementation. However, vitamin E also absorbs light in the solar UV region and may itself become a free radical (the tocopheroxyl radical) under solar irradiation; thus, in the special environment of the skin, vitamin E may act in two conflicting roles, as an antioxidant and as a prooxidant.

In this chapter we briefly review some of the evidence for the free-radical theory of photodamage to skin and some of the work that has been done on the porotection of skin by antioxidant supplementation involving vitamin E. We then present some of our work that illustrates the prooxidant and antioxidant actions of the vitamin E analog, α-C6 chromanol, in skin.

EVIDENCE FOR FREE-RADICAL PROCESSES IN SKIN DAMAGE

If it is true that photodamage to skin is induced, at least in part, by free radicals generated by solar light, then in UV light-exposed skin one would expect to find generation of free radicals, depletion of antioxidants, and damage of the type caused by free radicals. Indeed, all these have been observed.

Free Radicals Form in Skin in Response to Irradiation with UV Light

Several endogenous metabolites in skin, such as NADPH and flavins, have been shown to be photosensitizers whose exposure to light in the solar radiation spectrum results in the formation of the superoxide anion radical (8,9). ESR (electron spin resonance) measurements have also revealed the presence of free radicals in excised patches of human skin exposed to UV light (10,11). One such free radical has been identified as that of ascorbate (the ascorbyl radical), which has been detected in nonirradiated and irradiated skin; its ESR signal increases in irradiated compared to nonirradiated skin (12).

Skin Antioxidant Defenses Depleted by UV Irradiation

Antioxidant inhibition in response to UV irradiation has been demonstrated in skin. UVB irradiation causes inhibition of superoxide dismutase (SOD), glutathione peroxidase, and glutathione reductase (13). A single exposure of mouse skin to UV irradiation results in a significant decrease in SOD activity 24–48 h after exposure, with a return to normal levels 72 h after irradiation (14). Other studies confirm the impairment of enzymic and nonenzymic antioxidants by exposure to ultraviolet light. Exposure of excised mouse skin to UVB light caused a decrease in the concentration of reduced and total glutathione (15), and both UVA and UVB exposure caused a loss of catalase, glutathione reductase, α-tocopherol, ubiquinol 9, and ubiquinone 9 (15,16). Other studies confirm the loss of nonenzymatic skin antioxidants by acute exposure to ultraviolet light (17,18).

Free-Radical–Mediated Alterations of Skin Cellular Components Due to UV Irradiation

Most attention has been paid to lipid peroxidation, a process initiated by free radicals (19) that is thought to be involved in carcinogenesis (20). Lipid epoxides form via lipid peroxidation reactions (21), and the concentration of one such epoxide, $5\alpha,6\alpha$-cholesterol epoxide (thought to be carcinogenic), increases dramatically in hairless mice chronically irradiated with UV light (22). Other products of lipid peroxidation increase in human skin after UV irradiation (23) and in UV-irradiated hairless mice (24,25).

Free radicals also appear to be involved in photoinduced changes in DNA expression. Heme oxygenase, a stress protein, is induced in human skin fibroblasts by UVA radiation and hydrogen peroxide, and its induction is prevented by iron chelators, desferrioxamine, or *o*-phenanthroline, suggesting a UV-induced free-radical–mediated process (26–28). Ornithine decarboxylase (ODC) in epidermis is also induced by UV irradiation (29) and is thought to be associated with tumor promotion. Butylated hydroxytoluene (BHT), a lipophilic antioxidant, reduces the UV-induced induction of ODC, but other antioxidants, such as β-carotene (30) and vanillin (31), do not; thus it is not clear what free-radical processes mediate the induction of ODC by UV light.

VITAMIN E AND OTHER ANTIOXIDANTS IN THE PREVENTION OF SKIN PHOTODAMAGE

From the preceding evidence it can be predicted that antioxidant and/or dietary therapy could affect the course of photodamage to skin: decreasing antioxidant content or increasing polyunsaturated fat content in the diet (and thus, presumarly, in skin cell membranes) should increase damage; increasing antioxidant content or decreasing dietary polyunsaturated fat should decrease skin damage. We do not exhaustively review such studies here, but the evidence points to both these effects, although supplementation with vitamin E alone has not had clear-cut results.

Vitamin E-deficient rats experience an increase in lipid peroxide and insoluble collagen content of the skin compared to vitamin E-sufficient rats, and when skin of vitamin E-deficient rats was irradiated with UV light, the lipid peroxide content became even greater compared to vitamin E-sufficient rats (32). In hairless mice fed diets containing either saturated or polyunsaturated fatty acids and chronically UV irradiated, the mice receiving the saturated fat diet had a longer tumor latency time than mice receiving the unsaturated fat diet (33); hairless mice receiving diets with increasing content of polyunsaturated lipids and chronically UV irradiated had increased levels of skin tumor multiplicity strongly correlated with the increasing content of polyunsaturated lipid in the diet (24). These results are consistent with free-radical processes playing a role in photocarcinogenesis.

Many approaches involving antioxidants have been used in an attempt to protect against photodamage. In the hairless mouse, diets containing a mixture of antioxidants, including vitamin E, have been found to be protective against UV light-induced erythema (34) and against UV-induced skin damage, including cancer (24,35). Single antioxidants other than vitamin E have also been found to have protective effects on skin, but the evidence for vitamin E alone is mixed. Ascorbic acid (36,37) and BHT and vanillin (31) only have been found to reduce carcinogenesis in chronically irradiated hairless mice, but vitamin E alone has no effect (37). Dietary vitamin E alone also fails to prevent UV-induced erythema (34). On the other hand, topical application of vitamin E to hairless mice before UV irradiation reduced lipid peroxidation and polyamine synthesis measured after irradiation (25). Also, in contrast to dietary supplementation with vitamin E, topical application of vitamin E to rabbit skin before irradiation prevents erythema (38).

Thus, although vitamin E is considered the major lipid peroxidation chain-breaking antioxidant in membranes and free-radical processes, especially lipid peroxidation, almost certainly occur in photoinduced skin damage, the protective effects of E are not at all clear-cut. One possible explanation for this discrepancy is offered here.

POSSIBLE PROOXIDANT AND ANTIOXIDANT EFFECTS OF VITAMIN E IN SKIN

It is possible that vitamin E in skin absorbs UV light, generating the tocopheroxyl radical, and, by the mechanism of its own recycling, depletes other antioxidants, in this way acting as a prooxidant. Vitamin E in the skin is exposed to UV irradiation from about 295 nm and above, and the vitamin E absorbance spectrum (maximum at 295 nm) extends well into this solar spectrum (Fig. 1) (39). Thus, vitamin E may absorb solar UV light and become a free radical (the tocopheroxyl radical). Tocopheroxyl radicals can be reduced (recycled) back to tocopherol by reductive antioxidants, such as ascorbate, thiols, and ubiquinols (40–43), producing intermediate radicals and final oxidized forms of these antioxidants, thus depleting them. The antioxidant function of vitamin E in membranes is well known. It is considered the major chain-breaking antioxidant in membranes (44,45). Hence, vitamin E in skin may act in two conflicting manners: as a peroxyl radical scavenger, and as an endogenous photosensitizer enhancing light-induced oxidative damage.

EXPERIMENTAL

To test the hypothesis that the vitamin E radical may be directly formed by illumination of skin with solar UV light and that it may deplete other skin antioxidants, we have illuminated skin homogenates enriched with the vitamin E homolog, α-C6 chromanol, and followed the formation of the α-C6 chromanoxyl and ascorbyl radicals by electron spin resonance (ESR). α-C6 chromanol has the same chromanol nucleus as vitamin E, but its hydrocarbon tail is

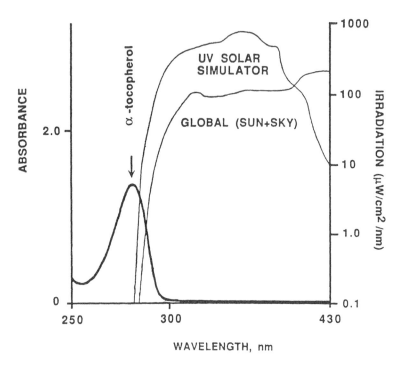

FIGURE 1 Comparison of UV absorbance spectrum of an ethanolic solution of α-tocopherol with the spectrum of global (sun emission + sky) overhead sunlight and of the light emitted by the solar UV simulator. α-Tocopherol concentration was 0.4 mM.

much shorter. It is thus expected to interact with light in the same way as vitamin E. We found previously (43) that α-C6 chromanol gives a stronger ESR signal than vitamin E when incorporated into membrane systems, and so we used it rather than vitamin E to enrich skin homogenates.

Methods

Hairless mice (Simonsen Laboratories, Emeryville, CA; strain SKH-1) were used as a skin source. Skin samples were removed immediately after sacrifice of the animal by CO_2 asphyxiation. Skin homogenates were prepared as described previously (15). To enrich the homogenate with α-C6 chromanol, 3 μl of 80 mM α-C6 chromanol in ethanol was added to 60 μl homogenate. ESR measurements of chromanoxyl and ascorbyl radicals were made on a Varian E 109E spectrometer at room temperature in gas-permeable Teflon tubing (0.8 mm internal diameter, 0.013 mm thickness; Zeus Industrial Products, Raritan, NJ). The permeable tube (approximately 8 cm in length) was filled with the sample, folded into quarters, and placed in an open 3.0 mm internal diameter ESR quartz tube such that all of the sample was within the effective microwave irradiation area. Spectra were recorded at 100 mW power, 2.5 G modulation, and 25 G/minute scan time. ESR spectra were registered at room temperature under aerobic conditions by flowing oxygen gas through the ESR cavity. Illumination was applied directly to the ESR cavity from a UV source (Solar Light Co., Philadelphia, PA), which emits light in the wavelength range from 295 to 400 nm in a pattern similar to natural sunlight (Fig. 1). The power density of the light at the sample surface in the spectral region 310–400 nm was 1.5 mW/cm² and dropped to 10% of this value at 295 nm. Chromanoxyl radical ESR signals were determined at 3245 G magnetic field strength and the power strength of 100 mW, modulation amplitude 2.5 G, scan range 100 G, and time constant 0.064 s.

Results

In freshly prepared homogenates of hairless mouse skin, the ESR signal of the ascorbyl radical was observed (Fig. 2A). The identity of the ascorbyl free-radical signal was confirmed by comparing the intrinsic signal with that obtained after addition of (1) pure sodium ascorbate to the skin homogenate (the magnitude of the signal increased) or (2) ascorbic acid oxidase (the signal disappeared). In the dark, the intensity of the signal decreased with time and eventually disappeared (Fig. 2B).

In skin homogenate samples to which no chromanol was added, illumination caused the ascorbyl radical signal to decay more quickly than in the dark; there was no detectable chromanoxyl signal (Fig. 2C). In samples enriched with α-tocopherol, the tocopheroxyl radical ESR signal was barely discernible after the ascorbyl radical disappeared, whereas in skin samples

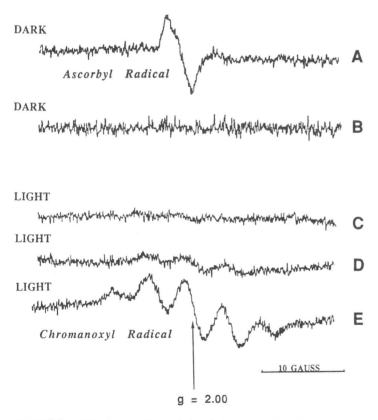

FIGURE 2 ESR spectra of ascorbyl and chromanoxyl radicals in skin homogenate from SKH—1 hairless mice. (A) In the dark, the ascorbyl radical signal is apparent. (B) The ascorbyl radical signal disappears with time in the dark. (C) In illuminated, nonenriched homogenates, the ascorbyl radical disappears more quickly, and no chromanoxyl radical signal is apparent. (D) In illuminated, α-tocopherol-enriched homogenates, the ascorbyl radical signal is replaced by a faint α-tocopheroxyl signal. (E) In illuminated, α-C6-chromanol-enriched homogenates, the ascorbyl radical signal is replaced by a strong α-C6-chromanoxyl radical signal.

enriched with α-C6 chromanol, a strong chromanoxyl radical could be seen after the ascorbyl radical signal disappeared (Fig. 2D and E, respectively). Because the α-C6 chromanoxyl radical signal was so much stronger, α-C6 chromanol was used to enrich skin homogenates rather than α-tocopherol.

The kinetics of ESR signal appearance and disappearance in illuminated skin homogenates is shown in Figure 3. In homogenates not enriched with α-C6 chromanol, the ascorbyl radical signal gradually disappeared over 30 minutes; no chromanoxyl radical signal was evident. In illuminated α-C6 chromanol-enriched homogenates, the disappearance of the ascorbyl signal is much more rapid, and as the ascorbyl radical signal disappears, the chromanoxyl radical signal appears, reaching a maximum at the time the ascorbyl radical signal has completely disappeared.

Discussion

It has been known for some time that the vitamin E radical can be reduced back to vitamin E in vitro by the action of ascorbate (40,41). In the skin, vitamin E is constantly exposed to UVA and B radiation, and the absorbance spectrum of E overlaps with the solar emission spectrum (Fig. 1). Thus, it is not surprising that we observed the α-tocoperoxyl radical signal and the much stronger α-C6 chromanoxyl radical signal in illuminated, enriched skin samples (Fig. 2D and E). That the chromanoxyl radical signals did not reach a maximum until the ascorbyl radical disappeared indicates that chromanols (α-C6 chromanol and presumably vitamin E) are recycled in skin by ascorbate in a manner similar to that in more defined in vitro chemical systems. The more rapid disappearance of the ascorbyl radical signal in α-C6

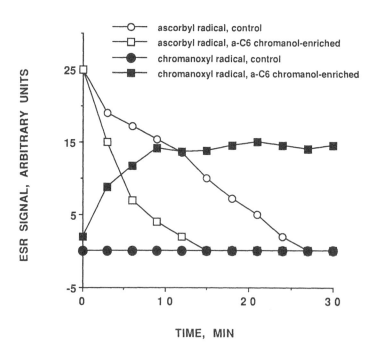

FIGURE 3   Time course of UV-induced ESR signals of chromanoxyl (closed symbols) and ascorbyl (open symbols) radicals in illuminated homogenates from control (circles) and α-C6 chromanol-enriched (squares) skin homogenates of hairless SKH-1 mice.

chromanol-enriched skin homogenates is consistent with such recycling. It is also thought that other antioxidants, such as glutathione, may serve to recycle the ascorbyl radical formed when ascorbate recycles vitamin E (unpublished observations). Hence, in the special environment of the skin, in which the vitamin E radical may be constantly directly formed as a result of UVA and B irradiation, vitamin E may act not only in its normal lipid peroxidation chain-breaking role but also as an agent of antioxidant depletion. The results we obtained here are consistent with such a dual role. It is yet to be determined where the balance lies in vivo.

CONCLUSIONS

In skin it appears that solar light may be directly absorbed by the vitamin E molecule and cause the formation of the vitamin E radical. This has at least two implications for research involving vitamin E supplementation, either dietary or topical, on UV-induced cutaneous pathologies, such as carcinogenesis and skin aging. The first is that vitamin E in topical preparations may act as a sunscreen in addition to its antioxidant effects; if such preparations are applied in excess or do not penetrate beneath the outer epidermal cell layer, they may exert a beneficial effect simply as a result of the UV-absorbing properties of vitamin E. Thus, a beneficial effect of topical vitamin E application on UV-induced skin disorders does not necessarily imply that such disorders are mediated by free-radical formation. The second is that vitamin E appears to act as a photosensitizer, and this effect is antagonistic to its antioxidant effect. The balance of the opposing effects is not known at present, nor is it known how supplementation would affect this balance. However, it may ultimately prove more fruitful to bolster cutaneous membrane antioxidant defenses indirectly, through increasing concentrations of substances that recycle vitamin E (such as ascorbate), or even substances that recycle ascorbate (such as dihydrolipoic acid), rather than by enhancing the concentration of vitamin E itself.

REFERENCES

1. Morliere P. Drug-induced photosensitivity: phototoxic and photoallergic reactions—a few molecular aspects. Biochemie 1986; 68:849-55.
2. Gilchrest BA, Szabo G, Flynn E, Goldwyn RM. Chronologic and actinically induced aging in human facial skin. J Invest Dermatol 1983; 80:81s-5s.
3. Rundel RD, Nachtwey DS. Skin cancer and ultraviolet radiation. Photochem Photobiol 1978; 28: 345-56.
4. Epstein JH. Photocarcinogenesis, skin cancer, and aging. J Am Acad Dermatol 1983; 9:487-502.
5. Sober AJ. Solar exposure in the etiology of cutaneous melanoma. Photodermatol 1987; 4:23-31.
6. Swerdlow AJ, English JS, MacKie RM, et al. Fluorescent lights, ultraviolet lamps, and risk of cutaneous melanoma. Br Med J 1988; 297:647-50.
7. Roza L, Baan RA, van der Leun JC, Kligman L, Young AR. UVA hazards in skin associated with the use of tanning equipment. J Photochem Photobiol [B] 1989; 3:281-7.
8. Cunningham ML, Krinsky NI, Giovanazzi SM, Peak MJ. Superoxide anion is generated from cellular metabolites by solar radiation and its components. J Free Radic Biol Med 1985; 1:381-5.
9. Ballou D, Palmer G, Massey V. Direct demonstration of superoxide anion production during the oxidation of reduced flavin and its catalytic decomposition by erythrocuprein. Biochem Biophys Res Commun 1969; 36:898-904.
10. Norrins AL. Free radical formation in the skin following exposure to ultraviolet light. J Invest Dermatol 1962; 39:445-8.
11. Pathak MA, Stratton K. Free radicals in human skin before and after exposure to light. Arch Biochem Biophys 1968; 123:468-76.

12. Buettner GR, Motten AG, Hall RD, Chignell CF. ESR detection of endogenous ascorbate free radical in mouse skin: enhancement of radical production during UV irradiation following application of chlorpromazine. Photochem Photobiol 1987; 46:161-4.

13. Maisuradze VN, Platonov AG, Gudz' TI, Goncharenko EN, Kudriashov IUB. Effect of ultraviolet rays on lipid peroxidation and various factors of its regulation in the rat skin (Russian). Biol Nauk 1987; 5:31-5.

14. Miyachi Y, Imamura S, Niwa Y. Decreased skin superoxide dismutase activity by a single exposure of ultraviolet radiation is reduced by liposomal superoxide dismutase pretreatment. J Invest Dermatol 1987; 89:111-2.

15. Fuchs J, Huflejt ME, Rothfuss LM, Wilson DS, Carcamo G, Packer L. Impairment of enzymic and nonenzymic antioxidants in skin by UVB irradiation. J Invest Dermatnol 1989; 93:769-73.

16. Fuchs J, Huflejt ME, Rothfuss LM, Wilson DS, Carcamo G, Packer L. Acute effects of near ultraviolet and visible light on the cutaneous antioxidant defense system. Photochem Photobiol 1989; 50:739-44.

17. Wheeler LA, Aswad A, Connor MJ, Lowe N. Depletion of cutaneous glutathione and the induction of inflammation by 8-methoxypsoralen plus UVA radiation. J Invest Dermatol 1986; 87:658-62.

18. Connor MJ, Wheeler LA. Depletion of cutaneous glutathione by ultraviolet radiation. Photochem Photobiol 1987; 46:239-45.

19. Halliwell B, Gutteridge JMC. Free radicals in biology and medicine. Oxford: Clarendon Press, 1989; 196.

20. Marnett LJ. Peroxyl free radicals: potential mediators of tumor initiation and promotion. Carcinogenesis 1987; 8:1365-73.

21. Smith LL, Kulig MJ. Sterol metabolism. XXXIV. On the derivation of carcinogenic sterols from cholesterol. Cancer Biochem Biophys 1975; 1:79-84.

22. Black HS, Douglas DR. Formation of a carcinogen of natural origin in the etiology of ultraviolet light-induced carcinogenesis. Cancer Res 1973; 33:2094-6.

23. Meffert H, Diezel W, Sonnichsen N. Stable lipid peroxidation products in human: detection, ultraviolet light-induced increase, pathogenic importance. Experientia 1976; 32:1397-8.

24. Black HS, Lenger WA, Gerguis J, Thornby JI. Relation of antioxidants and level of dietary lipid to epidermal lipid peroxidation and ultraviolet carcinogenesis. Cancer Res 1985; 45:6254-9.

25. Khettab N, Amory MC, Briand G, et al. Photoprotective effect of vitamins A and E on polyamine and oxygenated free radical metabolism in hairless mouse epidermis. Biochimie 1988; 70:1709-13.

26. Keyse SM, Tyrrell RM. Induction of the heme oxygenase gene in human skin fibroblasts by hydrogen peroxide and UVA (365 nm) radiation: evidence for the involvement of the hydroxyl radical. Carcinogenesis 1990; 11:787-91.

27. Keyse SM, Tyrrell RM. Heme oxygenase is the major 32-kDa stress protein induced in human skin fibroblasts by UVA radiation, hydrogen peroxide, and sodium arsenite. Proc Natl Acad Sci U S A 1989; 86:99-103.

28. Keyse SM, Tyrrell RM. Both near ultraviolet radiation and the oxidizing agent hydrogen peroxide induce a 32-kDa stress protein in normal human skin fibroblasts. J Biol Chem 1987; 262:14821-5.

29. Peterson AO, McCann V, Black HS. Dietary modification of UV-induced epidermal ornithine decarboxylase. J Invest Dermatol 1980; 75:408-10.

30. Mathews-Roth M, Breeding J, Lowe NJ. Effects of beta-carotene, canthazanthin and phytoen on ultraviolet light induced epidermal ornithine decarboxylase. Clin Res 1982; 30:597A.

31. Black HS, Tigges J. Evaluation of structurally-related phenols for anti-photocarcinogenic and photoprotective properties. Photochem Photobiol 1986; 43:403-8.

32. Igarashi A, Uzuka M, Nakajima K. The effects of vitamin E deficiency on rat skin. Br J Dermatnol 1989; 121:43-9.

33. Black HS, Lenger W, Phelps AW, Thornby JI. Influence of dietary lipid upon ultraviolet-light carcinogenesis. Nutr Cancer 1983; 5:59-68.

34. De Rios G, Chan JT, Black HS, Rudolph AH, Knox JM. Systemic protection by antioxidants against UVL-induced erythema. J Invest Dermatol 1978; 70:123-5.

35. Black HS. Effects of dietary antioxidants on active tumor induction. Res Commun Chem Pathol Pharmacol 1974; 7:783-6.

36. Dunham WB, Zuckerkandl E, Reynolds R, et al. Effects of intake of L-ascorbic acid on the incidence of dermal neoplasms induced by ultraviolet light. Proc Natl Acad Sci U S A 1982; 79:7532-6.
37. Pauling L, Willoughby R, Reynolds R, Blaisdell BE, Lawson S. Incidence of squamous cell carcinoma in hairless mice irradiated with ultraviolet light in relation to intake of ascorbic acid (vitamin C) and of D, L-alpha-tocopheryl acetate (vitamin E). Int J Vitam Nutr Res 1982; 23(suppl): 53-82.
38. Rochshupkin DI, Pitsov MY, Potapenko AY. Inhibition of ultraviolet light induced erythema by antioxidants. Arch Dermatol Res 1979; 266:91-4.
39. Baxter JG, Robeson CD, Taylor JD, Lehman RW. Natural α-, β-, and γ-tocopherols and certain esters of physiological interest. J Am Chem Soc 1943; 65:918-24.
40. Packer JE, Slater TF, Willson RL. Direct observation of a free radical interaction between vitamin E and vitamin C. Nature 1979; 278:737-8.
41. Niki E. Antioxidants in relation to lipid peroxidation. Chem Phys Lipids 1987; 44:227-53.
42. Packer L, Maguire JJ, Mehlhorn RJ, Serbinova E, Kagan VE. Mitochondria and microsomal membranes have a free radical reductase activity that prevents chromanoxyl radical accumulation. Biochem Biophys Res Commun 1989; 159:229-35.
43. Kagan VE, Serbinova EA, Packer L. Recycling and antioxidant activity of tocopherol homologs of differing hydrocarbon chain lengths in liver microsomes. Arch Biochem Biophys 1990; 282: 221-5.
44. Tappel AL. Vitamin E as the biological lipid antioxidant. Vitam Horm 1962; 20:493-510.
45. Burton GW, Ingold KU. Vitamin E—application of the principles of physical organic-chemistry to the exploration of its structure and function. Acc Chem Res 1986; 19:194-201.

B. Neurology

55

Vitamin E and Parkinson's Disease

Lawrence I. Golbe

University of Medicine and Dentistry of New Jersey-Robert Wood Johnson Medical School, New Brunswick, New Jersey

J. William Langston

California Parkinson's Foundation, San Jose, California

INTRODUCTION

Only 30 years ago Parkinson's disease (PD) languished on the long list of degenerative diseases of the brain for which our treatment was minimal and our pathophysiological understanding even less. The seminal events were Carlsson's 1959 proposal (1), confirmed in 1960 by Ehringer and Hornykiewicz (2), that the principal biochemical defect in PD is dopamine deficiency. Soon thereafter, small amounts of levodopa, a metabolic precursor of dopamine, were given without success to patients with PD. By 1967 Cotzias et al. (3) demonstrated that high levodopa dosages, approached gradually to minimize side effects, had a dramatic therapeutic benefit. This was the first successful use of a neurotransmitter replacement in the treatment of a neurodegenerative disease.

Levodopa remains the therapeutic mainstay against the outward symptoms of PD, but its antiparkinsonian efficacy wanes after several years. Furthermore, levodopa has no effect, or perhaps has a deleterious one, on the underlying neuronal degeneration in PD. Recent evidence of the role of oxidative mechanisms in the degeneration of dopaminergic neurons suggests that vitamin E may delay, decelerate, or even prevent premature loss of those brain cells. This chapter discusses the rationale for this hypothesis and the attempts to date to confirm that it may indeed provide a neuroprotective effect.

PARKINSON'S DISEASE

Epidemiology and Disease Course

PD is a relatively common illness of late middle age. It is present in 0.4–0.5% of the population over age 40 and in about 1–1.5% over age 60. There are approximately 400,000 people with PD in the United States. The lifetime risk of acquiring PD is approximately 2.5% (4).

The mean age at symptom onset is 60, but patients often note retrospectively that certain symptoms, such as constipation, loss of facial expression, a tendency to depression, or a general slowness of movement, were present for decades but were assigned no medical importance at the time. In fact, several lines of evidence suggest that 75–80% of the function of the substantia nigra, the principal dopaminergic nucleus of the brain, must be lost before symptoms become obvious (5,6). For these reasons, it seems likely that the disease process begins three or more decades before the appearance of symptoms diagnosable as PD (7). Unhappily, PD remains a diagnosis that for practical purposes can only be made by a careful history and physical examination, as there is no biomarker for the disease.

The initial overt motor symptoms are typically a 3–5 cycle/s tremor in one hand while the part is not in use (a "rest tremor"), a stiffness or slowness in the use of one hand, arm, or leg, or difficulty with truncal movements, such as arising from a deep chair. The symptoms always progress smoothly in intensity, and over the ensuing years the patient experiences a tendency to take short quick steps, loss of postural stability, deterioration in manual dexterity, softening of the voice, and, in a significant minority of cases, some degree of dementia. Untreated, the patient becomes chairbound within 5–7 years on average and dies 10–15 years after onset. With levodopa treatment, these intervals have roughly doubled.

After about 5 years of illness the majority of patients experience diminution of drug benefit and encounter neurological side effects, such as involuntary writhing movements (chorea), confusion, or hallucinations. These dictate lowering the levodopa dosage, allowing the underlying PD symptoms to reemerge.

Etiology

To date was have only a few promising clues to the cause of PD. An illness nearly identical to PD, along with loss of dopaminergic neurons in the zona compacta of the substantia nigra, as occurs in PD, is the toxic effect of 1-methyl-4-phenyl-1,2,3,6-tetrahydropyridine (MPTP). MPTP is a relatively simple compound, many analogs of which may be found in nature and industry (8). The initial observations of MPTP toxicity were of several intravenous drug users who unknowingly injected synthetic opiate analogs contaminated with MPTP and promptly displayed typical severe parkinsonian features that responded to treatment with levodopa (9,10). A cluster of such cases in California led to the tentative identification of MPTP as the actual offending agent. Since that time the MPTP animal model has offered unprecedented insight into the pathophysiology of the dopaminergic system and has provided clues to the cause of PD itself.

One such clue lies in the chemical resemblance of MPTP to some agricultural herbicides, such as paraquat (11,12). In fact, PD is more frequent among people with rural backgrounds and among those who have sprayed pesticides or herbicides (12–15). Although the pesticide rotenone is chemically unrelated to MPTP and is not known to cause PD, its mechanism of toxic activity at the mitochondrion is the complex I enzyme system (16) precisely the site of action of 1-methyl-4-phenylpyridinium ion (MPP$^+$), the metabolite responsible for the toxicity of MPTP (17,18).

Evidence against a primarily environmental cause of PD is the unchanging prevalence of the disease over time (19) and the absence thus far of confirmed geographical or occupational clusters (12).

OXIDATIVE MECHANISMS IN PD

Mechanism of Action of MPTP

The issue of vitamin E as an antiparkinson drug arises from the role of oxidative steps in the mechanism of toxic action of MPTP. MPTP is oxidized to MPP$^+$ (17,18) by monoamine oxidase type B (MAO-B) (20) in astrocytes and serotonergic neurons. MPP$^+$ is then taken into dopaminergic neurons via the dopamine reuptake system (21). There it inhibits MAD$^+$-

linked oxidation in mitochondrial complex I (16), impairing neuronal respiration and presumably killing the neuron. Activities of several subunits of complex I are in fact decreased in parkinsonian nigra (22), striatum (23), and platelets (24).

If the oxidation of MPTP to MPP$^+$ is prevented by pretreatment with the monoamine oxidase inhibitors deprenyl or pargyline, MPTP toxicity does not occur (25–27). If PD itself is the result of ongoing toxicity by MPTP or a toxic analog thereof, inhibition of MAO-B, which is the form of MAO present in human brain, may slow the rate of progression of the disease. Even if exposure to a culprit MPTP analog is brief, clinical parkinsonism may appear or worsen over the ensuing years, as it has in some of the exposed intravenous drug users (28).

Recent evidence suggests that MPP$^+$, like rotenone, has the additional toxic action of inducing NADH-dependent superoxide formation and lipid peroxidation in mitochondria (29). The potential role for antioxidants in the MPTP model of PD therefore extends to the submitochondrial level.

"Endogenous Toxin" Hypothesis

Toxic By-products of Dopamine Metabolism

Another potential role for oxidation reactions in the pathogenesis of PD lies in the metabolism of dopamine itself. MAO-B is the enzyme with primary responsibility for degrading dopamine in human brain (the other is catecholamine-*O*-methyltransferase). MAO-B oxidizes dopamine to 3,4-dihydroxyphenylacetaldehyde, ammonia, and 1 mol hydrogen peroxide. Also, nonenzymatic oxidation of dopamine produces cytotoxic by-products in the form of quinones and hydrogen peroxide (30). The normal brain scavenges hydrogen peroxide with glutathione peroxidase and catalase, preventing (or minimizing) the lipid peroxidation that can damage normal tissue.

In the aging brain, levels of MAO-B, for reasons still unclear, are actually elevated. In the patient with PD, dopamine is presumably being produced and degraded by nigral neurons at an increased rate to compensate for their reduced number. The supply of MAO-B is up to the task, but the paucity of glutathione peroxidase and catalase may allow the cytotoxic by-products of the MAO-B-catalyzed degradative reaction (and of nonenzymatic reactions) to accumulate (31–35). The result could be that the nigral neurons degenerate at an ever-increasing rate, in a "vicious cycle."

The presence of high levels of malondialdehyde, an indicator of lipid peroxidation, in parkinsonian substantia nigra supports the notion that excessive oxidative stress at least contributes to the pathogenesis of the disease (36). Furthermore, low levels in parkinsonian nigra of total glutathione, which is consumed by ongoing detoxification processes, raises the possibility that a neurotoxin continues to act throughout the clinical course of PD (34). The optimistic corollary is that a treatment that interferes with this process may slow or arrest the disease course even if it is initiated long after the process begins. For reasons already outlined, identification of persons at the earliest stages of the PD process is at present impossible, so the news that late prophylaxis may be better than none is welcome indeed. Even if the evidence of increased oxidation activity is a result, rather than a cause, of parkinsonian neuronal degeneration, it may prove useful as a biomarker.

Iron and PD

The Fenton reaction, detailed elsewhere in this volume, may have a major role in producing lipid peroxidation in the substantia nigra. The supply of ferrous iron is its rate-limiting step. Several lines of evidence suggest that it may be involved in the pathogenesis of PD.

First, the deficiencies of glutathione peroxidase and total glutathione in parkinsonian substantia nigra may render that pathway less available for the degradation of hydrogen peroxide produced as a by-products of dopamine metabolism. this offers an opportunity for the Fenton reaction to step into the breach, degrade excess hydrogen peroxide, and produce its cytotoxic hydroxyl radicals and their oxidative stress.

Second, the salient histological feature of the substantia nigra, and indeed of most of the catecholamine nuclei that degenerate in PD, is their neuromelanin content. Neuromelanin in normal catecholamine neurons may act as an efficient scavenger of the free radicals generated by catecholamine metabolism. However, neuromelanin also binds large amounts of iron and has the ability to reduce ferric to ferrous ion. Maximizing ferrous ion availability could drive the Fenton reaction in catecholamine neurons, producing more cytotoxic hydroxyl radicals (37).

Third, substantia nigra and globus pallidus in brains of patients with advanced PD contain excessive amounts of ferric iron (38,39). Magnetic resonance imaging of the brain using high field strength techniques (1.5 T) reveal areas of low signal intensity in those nuclei that correspond to known pattern of iron depositon (40).

STUDIES OF VITAMIN E SUPPLEMENTATION IN PD

Animal Models

Vitamin E deficiency, in the usual sense of a malabsorption state or as part of general malnutrition, does not produce PD. However, administering pharmacological doses to rodents attenuates glutathione loss and dopaminergic loss induced by MPTP (41,42). The same result could not be obtained, however, in a marmoset model (43). The literature on MPTP protection by other antioxidants reports variable results (44–47).

An interesting preliminary result, reported by Diamond et al. (48), suggests that a vitamin E-deficient diet produces a 50% loss of striatal dopamine in mice. However, mice with normal vitamin E levels suffered as much damage from MPTP as the vitamin E-deficient animals. The effect of supplementation with vitamin E on MPTP damage was not studied.

Vitamin E appears to confer partial protection against damage by intrastriatally injected 6-hydroxydopamine (49). This compound selectively damages catecholaminergic nerve cell bodies and is thought to act via an oxidative mechanism involving free radicals (50). A rat prepared in this manner rotates spontaneously in a direction ipsilateral to the side of the injected striatum and, given levodopa, rotates in the opposite direction. Until the advent of MPTP this was the leading animal model of PD.

Interventional Studies in Humans

Pilot Study

Fahn (51) reported the results of a semicontrolled pilot study in which high-dose antioxidants appear to have slowed the progression of symptoms in PD. He prescribed 3200 units vitamin E dialy and 3000 mg vitamin E daily to patients with early, mild PD. The experimental design was to follow the patients until such time as they required levodopa to treat emerging disability. This "end point" served as the primary dependent variable.

The sole available means of assessing the severity of PD and the need for levodopa is by physical and historical inventory of signs and symptoms, rather than by some direct measurement of substantia nigra function. It is therefore necessary to confine studies of PD prophylaxis to those few patients who neither require nor receive levodopa to control their symptoms. One way around this practical hurdle would be to study patients with more advanced disease who temporarily discontinue levodopa to allow baseline and end-point evaluations of their state unmodified by levodopa. However, such a design would be quite unsafe because a "drug holiday," even in moderately advanced patients with PD, can bring complications, such as falls, malignant hyperthermia, and dysphagia leading to aspiration.

Fahn compared his patients' "survival" without levodopa to that of a similar group of patients not receiving vitamin E or C followed by Tanner at another institution. Survival without levodopa for the patients receiving the vitamins exceeded that for the "control" group by a mean of 2.5 years (72 versus 40 months) for patients whose PD began before age 54 and by nearly 3.5 years (63 versus 24 months) for patients whose PD began at age 54 or later.

Fahn emphasizes that the patients were not randomized and that neither patients nor investigators were blinded to the treatment. Furthermore, the decision to begin levodopa for emerging disability relies on subjective judgments that could have differed systematically between the two treatment groups. Although Fahn and Tanner reported using the same policy regarding levodopa treatment, the study was far from truly controlled.

The DATATOP Study

A definitive examination of the issue of vitamin E prophylaxis against PD is currently underway as part of the multiinstitutional study on the deprenyl and tocopherol antioxidative therapy of parkinsonism (DATATOP). Vitamin E and deprenyl are being studied together in a 2 × 2 factorial parallel design. A total of 800 patients at 28 institutions in the United States and Canada were recruited in 1987 and 1988 and randomized into one of four treatment groups: deprenyl and vitamin E (D,L-α-tocopherol, 1000 units twice daily), deprenyl and placebo, placebo and vitamin E, or double placebo. All patients had PD that did not yet require levodopa or any other antisymptomatic therapy, and all received a standard daily multivitamin tablet containing 30 units vitamin E. They were given extensive baseline evaluations and were followed every 3 months with symptom histories, neurological examinations, including timed motor tasks, neuropsychological evaluations, videotaping, and an assessment of the degree to which their PD symptoms threaten their ability to perform their jobs and personal and household chores (52).

Patients in DATATOP were to have been treated until they required levodopa (end point) or for 2 years, whichever came sooner. However, a routine "peek" at the data by the independent safety monitoring committee 3 months after enrollment was completed revealed that deprenyl was producing an important delay in the end point (53). Only 97 of the 399 patients receiving deprenyl reached the end point during the abbreviated observation period, compared with 176 of the 401 patients not receiving deprenyl ($p < 10^{-8}$). This constitutes a reduction of 57% in the risk of requiring levodopa. All patients in DATATOP were offered the opportunity to take active deprenyl for the remainder of the study, and this transition was effected without breaking the blind. However, the results of vitamin E supplementation have not apparently been sufficiently dramatic to require interruption of the study. The final results were available in late 1992.

A study that served, in part, as a pilot for DATATOP was published shortly before DATATOP itself. Tetrud and Langston (54) administered deprenyl or placebo to 54 patients in a design identical to that of DATATOP except that vitamin E was not tested. They found that deprenyl slowed disease progression by 40–83% (on five separate assessment scales). Patients in the placebo group needed levodopa a mean of 312 days after randomization; those receiving deprenyl survived a mean of 549 days.

The results of Tetrud and Langston and those of the preliminary DATATOP analysis support the notion that MAO-B inhibition, which presumably reduces the oxidative stress imposed by the reactions it catalyzes, slows the rate of neuronal degeneration in PD. Whether vitamin E supplementation can have the same result remains to be determined upon completion of the DATATOP study.

Retrospective Epidemiological Studies

Even if vitamin E supplementation does not slow the progression of PD in interventional studies, the possibility remains that it may be efficacious if taken from the onset of dopaminergic degeneration early in life, long before loss of 80% of dopaminergic function produces symptoms that suggest the diagnosis. Several studies have examined this question.

Tanner et al. (55) assessed the premorbid use of vitamin supplements in 35 patients with PD and 70 controls matched for age and sex. Use of vitamin E or cod-liver oil was four times as common among controls as among patients with PD ($p < 0.05$). Also, use of a multivitamin supplement was about twice as common among controls as among patients with PD, although this difference did not reach statistical significance. To date, this study has been published only in abstract form.

Tanner et al. (56) subsequently calculated mean dietary and supplemental intake of vitamin E during the premorbid years in the same patients and controls. A slightly greater intake of vitamin E occurred in controls, but this failed to reach statistical significance. Tanner and colleagues performed the same assessment for 100 patients and 200 controls in the People's Republic of China and found no differences whatever. The authors hasten to add, however, that estimation of portion size is inexact despite their use of plastic food models during the interview. Furthermore, vitamin E content is even less precisely known for indigenous Chinese foods than it is for foods consumed in the United States.

A novel method of retrospectively assessing dietary habits was used by Golbe et al. (57) in comparing 81 married PD patients with their unaffected married same-sex siblings. They asked each respondent to compare him- or herself to his or her spouse with respect to preference for each of 17 food items during the premorbid years (or equivalent for siblings). Vitamin E intake was assessed in this relative manner rather than by asking respondents to recall absolute intake frequency or absolute portion size in multiple-decade retrospect. The list was formulated with the goal of identifying a dietary cause of PD, but no item emerged as a candidate etiological agent. Instead, the three foods with the greatest vitamin E content, nuts, salad dressing, and plums, emerged as possible protective factors. That is, patients with PD were more likely to exceed their spouses in consumption of these foods than was the case for controls.

The same authors (58) attempted to confirm and extend these findings using a different survey method. In this study, patients were compared with their spouses with respect to whether they considered their consumption of each food item greater, less than, or similar to that of most other people of their age and sex. The time interval under consideration was the years between age 18 and marriage. Again, "salad with dressing" had an apparently protective effective against PD (odds ratio 0.6, $p < 0.05$). "Nuts other than peanuts" and "peanuts or peanut butter" were also more likely to be preferred by spouse controls, but only among pairs in which the case was female and the control male (odds ratio 0.3 for each, $p < 0.05$).

A different approach used by Factor et al. (59) reached similar results. A group of eight patients with PD who had been self-administering supplemental vitamin E for a mean of 6.9 years (range 2–16 years) were compared with eight other patients with PD, matched for age, symptom duration, and daily levodopa requirement, who had not taken vitamin E supplements. The supplemented group had a lower (i.e., milder) mean total score on the unified Parkinson disability rating scale (26.3 ± 4.8 versus $37.1 \pm 4.2; p < 0.05$). Also yielding statistically significant differences between the two groups were a rating of complications of levodopa therapy, which generally worsens with overall disease progression (0.9 ± 0.6 versus 3.5 ± 1.3) and the Schwab and England subjective rating of the activities of daily living (91 ± 5 versus 79 ± 5).

A virtue of the methods of Factor et al. (59) is the comparison of PD patients among themselves. This avoids the confounding effect of the "premorbid Parkinson personality." Patients with PD are recalled by themselves and by family to have been relatively passive, cooperative, uncreative, anhedonic, morally conservative, and heedful of authority, even long before motor symptoms began (60–62). Nonsmoking is associated with some of these personality traits (63), and early-life nonsmoking was found to correlate with subsequent development of PD (64–67). Although cigarette combustion products may enhance dopaminergic function and carbon monoxide has antioxidant properties (68), there is no correlation among patients with PD between intensity of early-life cigarette use and severity of subsequent PD (69). This suggests that the abstemious premorbid PD personality produces nonsmoking, rather than that smoking protects against PD.

The difficulty with premorbid dietary studies is now apparent. What is the effect of the premorbid PD personality on dietary habits? One way to avoid having to assess this issue is to control for premorbid personality in any retrospective assessment of dietary factors in PD.

Another premorbid factor influencing dietary habits is the relative hyposmia in PD. Patients with PD were found by Doty et al. (70) to have markedly diminished olfactory function, relative to matched controls, using a smell identification test. The deficit was no worse in the more advanced patients and was present even in patients with very recent disease onset who were receiving no medication. This suggests that the abnormality may have been present long before motor symptoms appeared. Smell testing is now being evaluated as a presymptomatic test for PD.

What is the effect of asymptomatic smell dysfunction on one's dietary practices? Until this question is addressed, premorbid retrospective assessment of dietary factors in PD may give misleading results. Because there is relatively little overlap in olfactory function between persons with PD and controls (unlike the situation for personality), controlling for olfactory function in epidemiological surveys will be difficult. The time required to reliably test olfaction is another obstacle for large-scale survey studies.

CONCLUSIONS

The evidence that vitamin E attenuates the dopaminergic neuronal degeneration of PD remains circumstantial. Nevertheless, many neurologists specializing in movement disorders routinely prescribe vitamin E, usually 2000 IU daily, to all patients with PD from the time of diagnosis. The forthcoming DATATOP results will tell us whether this practice is of benefit. However, even if vitamin E fails to avert further progression in DATATOP subjects, who have already reached medical attention, the possibility will remain open for vitamin E to exert some effect at an earlier, presymptomatic, stage of PD. Can the PD process be halted or slowed before the vicious cytotoxic cycle (neuronal loss, additional demand on remaining neurons, additional free radical generation, and additional neuronal loss) has produced the critical 80% neuronal loss necessary for motor symptoms to appear?

The answer to this question may await the day when we can identify, treat, and follow patients long before the 80% loss stage has occurred. Until then, we must satisfy ourselves with retrospective epidemiological surveys and their biases. If the practice of treating all PD patients with vitamin E spreads to the primary care practitioners (who treat most patients with PD, especially early in the course), it may become more difficult for researchers to recruit patients who have not been exposed to high-dose supplemental vitamin E. Thus we could face the unhappy limbo of an unproven and perhaps unprovable prophylactic treatment that, because of its relative safety and low cost, will be widely prescribed until the future brings a new technology capable of assessing its true benefit.

Until then, one approach is to seek greater validity in our assessment of premorbid vitamin E intake by patients with PD. Epidemiological questionnaire assessments of lifelong dietary and supplemental vitamin E intake could be validated by comparing their results with adipose tissue vitamin E levels in a small sample of patients with PD (Tanner, personal communication). A perhaps more important issue is whether the parkinsonian brain itself has a relatively low concentration of vitamin E. The most german eissues, of course, are whether interventional measures taken before or after symptom appearance can alter those levels, and if they can, whether the course of the disease process is thereby modified. The forthcoming results of the DATATOP study will illuminate the issue for symptomatic patients, but for presymptomatic patients, an interventional study must await newer and more practical methods of assessing nigrostriatal function in vivo.

REFERENCES

1. Carlsson A. The occurrence, distribution and physiologic role of catecholamines in the nervous system. Pharmacol Rev 1959; 11:490-3.
2. Ehringer H, Hornykiewicz O. Verteilung von Noradrenalin und Dopamin (3-Hydroxytyramin) im Gehirn des Menschen und ihr Verhalten bei Erkrankungen des extrapyramidalen System. Wien Klin Wochenschr 1960; 38:1236-9.
3. Cotzias GC, Papavasiliou PS, Gellene R. Modification of parkinsonism—chronic treatment with L-dopa. N Engl J Med 1969; 28:337-45.
4. Kurland LT, Kurtzke JF, Goldberg ID. Epidemiology of neurologic and sense organ disorders. Chapter 3. Parkinsonism. Cambridge MA: Harvard University Press, 1973.
5. Riederer P, Wuketich S. Time course of nigrostriatal degeneration in Parkinson's disease. J Neural Transm 1976; 38:227-301.
6. Bernheimer H, Birkmayer W, Hornykiewicz O, Jellinger K, Seitelberger F. Brain dopamine and the syndromes of Parkinson and Huntington: clinical, morphological and neurochemical correlations. J Neurol Sci 1973; 20:415-55.
7. Agid Y, Ruberg M, Raisman R, Hirsch E, Javoy-Agid F. The biochemistry of Parkinson's disease. In: Stern G, ed. Parkinson's disease. London: Chapman and Hall, 1990; 99-125.
8. Maga JA. Pyridines in foods. J Agr Food Chem 1981; 29:895-8.
9. Davis GC, Williams AC, Markey SP, et al. Chronic parkinsonism secondary to intravenous injection of meperidine analogues. Psychiatry Res 1979; 1:249-54.
10. Langston JW, Ballard PA, Tetrud JW, Irwin I. Chronic parkinsonism in humans due to product of meperidine-analog synthesis. Science 1983; 219:979-80.
11. Solomon H, D'Amato RJ. Predicting Parkinson's disease. Nature 1985; 317:198-9.
12. Tanner CM, Chen B, Wang W-Z, et al. Environmental factors in the etiology of Parkinson's disease. Can J Neurol Sci 1987; 14:419-23.
13. Golbe LI, Farrell TM, Davis PH. Follow-up study of early-life protective and risk factors in Parkinson's disease. Mov Disord 1990; 5:66-70.
14. Ho SC, Woo J, Lee CM. Epidemiologic study of Parkinson's disease in Hong Kong. Neurology 1989; 39:1314-8.
15. Koller W, Vetere-Overfield B, Gray C, et al. Environmental risk factors for Parkinson's disease. Neurology 1990; 40:1218-21.
16. Nicklas WJ, Vyas I, Heikkila RE. Inhibition of NADH-linked oxidation in brain mitochondria by 1-methyl-4-phenyl-pyridine, a metabolite the neurotoxin 1-methyl,4-phenyl-1,2,3,6-tetrahydropyridine. Life Sci 1985; 36:2503-8.
17. Langston JW, Irwin I, Langston EB, Forno LS. 1-Methyl-4-phenyl-pyridinium ion (MPP$^+$): identification of a metabolite of MPTP, a toxin selective to the substantia nigra. Neurosci Lett 1984; 48:87-92.
18. Markey SP, Johannessen JN, Chiueh CC, Burns RS, Herkenham MA. Intraneuronal generation of a pyridinium metabolite may cause drug-induced parkinsonism. Nature 1984; 311:464-7.

19. Eldridge R, Rocca WA. The clinical syndrome of striatal dopamine deficiency: parkinsonism induced by MPTP. N Engl J Med 1985; 313:1159-60.

20. Chiba K, Trevor A, Castagnoli N. Metabolism of the neurotoxin tertiary amine MPTP by brain monoamine oxidase. Biochem Biophys Res Commun 1984; 120:574-8.

21. Javitch JA, D'Amato RJ, Strittmatter SM, Snyder SH. Parkinsonism-inducing neurotoxin, *n*-methyl-4-phenyl-1,2,3,6-tetrahydropyridine: uptake of the metabolite *n*-methyl-4-phenylpyridine by dopamine neurons explains selective toxicity. Proc Natl Acad Sci U S A 1985; 82:2173-7.

22. Schapira AHV, Cooper JM, Dexter D, Jenner P, Clark JB, Marsden CD. Mitochondrial complex I deficiency in Parkinson's disease (letter). Lancet 1989; 1:1269.

23. Mizuno Y, Ohta S, Tanaka M, et al. Deficiencies in complex I subunits of the respiratory chain in Parkinson's disease. Biochem Biophys Res Commun 1989; 163:1450-5.

24. Parker WD, Boyson SJ, Parks JR. Abnormalities of the electron transport chain in idiopathic Parkinson's disease. Ann Neurol 1989; 26:719-23.

25. Langston JW, Irwin I, Langston EB, Forno LS. Pargyline prevents MPTP-induced parkinsonism in primates. Science 1984; 225:1480-2.

26. Heikkila RE, Manzino L, Cabbat FS, Duvoisin RC. Protection against the dopaminergic neurotoxicity of 1-methyl,4-phenyl-1,2,3,6-tetrahydropyridine by monoamine oxidase inhibitors. Nature 1984; 311:467-9.

27. Cohen G, Pasik P, Cohen B, Leist A, Mytilineou C, Yahr MD. Pargyline and deprenyl prevent the neurotoxicity of 1-methyl-4-phenyl-1,2,3,6-tetrahydropyridine in the rat. Eur J Pharmacol 1984; 106:209-10.

28. Langston JW. MPTP: the promise of a new neurotoxin. In: Marsden CD, Fahn S, eds. Movement disorders 2. London: Butterworth Scientific, 1986; 73-90.

29. Hasegawa E, Takeshige K, Oishi T, Murai Y, Minakami S. 1-Methyl-4-phenylpyridinium (MPP$^+$) induces NADH-dependent superoxide formation and enhances NADH-dependent lipid peroxidation in bovine heart submitochondrial particles. Biochem Biophys Res Commun 1990; 3:1049-55.

30. Graham DG. Oxidative pathways for catecholamines in the genesis of neuromelanin and cytotoxic quinones. Mol Pharmacol 1978; 14:633-43.

31. Kish SJ, Morito C, Hornykiewicz O. Glutathione peroxidase activity in Parkinson's disease brain. Neurosci Lett 1985; 58:343-6.

32. Cote LJ, Kremzner LT. Biochemical changes in normal aging in human brain. Adv Neurol 1983; 38:19-30.

33. McGeer EG. Aging and neurotransmitter metabolism in the human brain. In: Katzman R, Terry RD, Bick KL, eds. Alzheimer's disease: senile dementia and related disorders. New York: Raven Press, 1978; 427-40.

34. Perry TL, Godin DV, Hansen S. Parkinson's disease: a disorder due to nigral glutathione deficiency? Neurosci Lett 1982; 33:305-10.

35. Ambani LM, Van Woert MH, Murphy S. Brain peroxidase and catalase in Parkinson's disease. Arch Neurol 1975; 32:114-8.

36. Dexter D, Carter C, Agid F, et al. Lipid peroxidation as cause of nigral cell death in Parkinson's disease. Lancet 1986; 2:639-40.

37. Youdim MBH, Ben-Shachar D, Yehuda S, Riederer P. The role of iron in the basal ganglion. In: Streifler MB, Korczyn AD, Melamed E, Youdim MBH, eds. Parkinson's disease: anatomy, pathology, and therapy. Advances in neurology, Vol. 53. New York: Raven Press, 1990; 155-62.

38. Dexter DT, Wells FR, Agid F, et al. Increased nigral iron content in postmortem parkinsonian brain. Lancet 1987; 2:1219-20.

39. Riederer P, Sofic E, Rausch WD, et al. Transition metals, ferritin, glutathione and ascorbic acid in parkinsonian brain. J Neurochem 1989; 52:515-20.

40. Drayer BP. Magnetic resonance imaging and brain iron: implications in the diagnosis and pathochemistry of movement disorders and dementia. BNI Q 1987; 3:15-30.

41. Perry TL, Yong VW, Clavier RM, et al. Partial protection from the dopaminergic neurotoxin *N*-methyl-4-phenyl-1,2,3,6-tetrahydropyridine by four different antioxidants in the mouse. Neurosci Lett 1985; 60:109-14.

42. Yong VW, Perry TL, Krisman AA. Depletion of glutathione in brainstem of mice caused by *N*-methyl-4-phenyl-1,2,3,6-tetrahydropyridine is prevented by antioxidant pretreatment. Neurosci Lett 1986; 63:56-60.

43. Perry TL, Yong VW, Hansen S, et al. Alpha-tocopherol and beta-carotene do not protect marmosets against the dopaminergic neurotoxicity of *N*-methyl-4-phenyl-1,2,3,6-tetrahydropyridine. J Neurol Sci 1987; 81:321-31.

44. Baldessarini RJ, Kula NS, Francoeur D, Finkelstein SP. Antioxidants fail to inhibit depletion of striatal dopamine by MPTP. Neurology 1986; 36:735.

45. Martinovits G, Melamed E, Cohen O, Rosenthal J, Uzzan A. Systemic administration of antioxidants does not protect mice against the dopaminergic neurotoxicity of 1-methyl-4-phenyl-1,2,3,6-tetrahydropyridine (MPTP). Neurosci Lett 1986; 69:192-7.

46. Sershen H, Reith MEA, Hashim A, Lajtha A. Protection against 1-methyl-4-phenyl-1,2,3,6-tetrahydropyridine neurotoxicity by the antioxidant ascorbic acid. Neuropharmacology 1985; 24:1257-9.

47. Wagner GC, Jarvis MF, Carelli RM. Ascorbic acid reduces the dopamine depletion induced by MPTP. Neuropharmacology 1985; 24:1261-2.

48. Diamond BI, Sethi K, Nguyen T, Nguyen H. Effect of vitamin E deficiency on MPTP toxicity in mice (abstract). Neurology 1989; 39(Suppl. 1):135.

49. Cadet JL, Katz M, Jackson-Lewis V, Fahn S. Vitamin E attenuates the toxic effects of intrastriatal injection of 6-hydroxydopamine (6-OH-DA) in rats: behavioral and biochemical evidence. Brain Res 1989; 476:10-5.

50. Heikkila RE, Cohen G. Inhibition of biogenic amine uptake by hydrogen peroxide: a mechanism for the toxic effects of 6-hydroxydopamine. Science 1971; 172:1257-8.

51. Fahn S. The endogenous toxin hypothesis of the etiology of Parkinson's disease and a pilot trial of high-dosage antioxidants in an attempt to slow the progression of the illness. Ann N Y Acad Sci 1989; 570:186-96.

52. Parkinson study group. DATATOP: a multicenter controlled clinical trial in early Parkinson's disease. Arch Neurol 1989; 46:1052-60.

53. Parkinson study group. Effect of deprenyl on the progression of disability in early Parkinson's disease. N Engl J Med 1989; 321:1364-71.

54. Tetrud JW, Langston JW. The effect of deprenyl (selegiline) on the natural history of Parkinson's disease. Science 1989; 245:519-22.

55. Tanner CM, Cohen JA, Summerville BC, Goetz CG. Vitamin use and Parkinson's disease (abstract). Ann Neurol 1988; 23:182.

56. Tanner CM, Chen B, Cohen JA, et al. Dietary anti-oxidant vitamins and the risk of developing Parkinson's disease (abstract). Neurology 1989; 39(Suppl 1):181.

57. Golbe LI, Farrell TM, Davis PH. Case-control study of early life dietary factors in Parkinson's disease. Arch Neurol 1988; 45:1350-3.

58. Golbe LI, Farrell TM, Davis PH. Follow-up study of early-life protective and risk factors in Parkinson's disease. Mov Disord 1990; 5:66-70.

59. Factor SA, Sanchez-Ramos JR, Weiner WJ. Vitamin E therapy in Parkinson's disease. In: Streifler MB, Korczyn AD, Melamed E, Youdim MBH, eds. Parkinson's disease: anatomy, pathology and therapy. Adv Neurol, Vol. 53.

60. Duvoisin RC, Eldridge R, Williams A, Nutt J, Calne D. Twin study of Parkinson's disease. Neurology 1981; 31:77-80.

61. Poewe W, Gerstenbrand F, Ransmayr G, Plorer S. Premorbid personality of parkinson patients. J Neual Transm Suppl 1983; 19:215-24.

62. Menza MA, Forman NE, Goldstein HS, Golbe LI. Parkinson's disease and dopamine-dependent personality characteristics. J Neuropsychiat Clin Neurosci 1990; 2:282-7.

63. Smith GM. Personality and smoking: a review of the empirical literature. In: Hunt WA, ed. Learning mechanisms in smoking. Chicago: Aldine, 1970; 42-61.

64. Nefzger MD, Quadfasel FA, Karl VC. A retrospective study of smoking in Parkinson's disease. Am J Epidemiol 1968; 88:149-58.

65. Marttila RJ, Rinne UK. Smoking and Parkinson's disease. Acta Neurol Scand 1989; 62:322-5.

66. Baumann RJ, Jameson HD, McKean HE, Haack DG, Weisberg LM. Cigarette smoking and Parkinson's disease. 1. A comparison of cases with matched neighbors. Neurology 1980; 30:839-43.

67. Godwin-Austein RB, Lee PN, Marmot MG, Stern GM. Smoking and Parkinson's disease. J Neurol Neurosurg Psychiatr 1982; 45:577-81.

68. Baron JA. Cigarette smoking and Parkinson's disease. Neurology 1986; 36:1490-6.

69. Golbe LI, Cody RA, Duvoisin RC. Smoking and Parkinson's disease: search for a dose-response relationship. Arch Neurol 1986; 43:774-8.

70. Doty RL, Stern MB, Pfeiffer C, Gollomp SM, Hurtig HI. Bilateral olfactory dysfunction in early stage medicated and unmedicated parkinsonism (abstract). Neurology 1990; 40(Suppl 1):221.

56

Vitamin E Therapy in Neurological Diseases

Tuomas Westermarck, Erkki Antila, and Faik Atroshi

University of Helsinki, Helsinki, Finland

FREE RADICALS IN THE BRAIN

Neurochemical aspects of vitamin E have been previously widely reviewed by Halliwell and Gutteridge (1). In their approach the therapeutic role of vitamin E in central nervous system (CNS) diseases is explained by its capacity to modulate oxygen toxicity.

The brain differs from many other tissues being a highly aerobic and totally oxygen-dependent tissue. Homeostatic mechanisms ensure in the first place vital functions of the CNS, which implies the supply of oxygen, glucose, vitamins, and trace elements. However, the brain levels of antioxidants are maintained even during a short period of lack of supply. Lipids account for 50% of the dry weight of the human brain. The structure of neurons, which have long and branching neurites, makes the surface-volume ratio extremely large. The cell membranes of neurons are especially rich in C20 and C22 polyunsaturated fatty acids (PUFAs). In the CNS the ratio of omega-3 to omega-6 fatty acids is 1. This ratio and the absolute amount of PUFAs is achieved gradually during fetal development and the perinatal period; they are then synthetized in situ from fatty acid precursors. The protection of PUFAs against peroxidation in the CNS is highly vital. As well as having a role in guaranteeing the fluidity of membranes, PUFAs function as precursors for the arachidonic acid cascade (2,3).

Lipid peroxidation causes a gradual loss of membrane fluidity and membrane potential and increases membrane permeability to ions. Radical attack may also destroy membrane-bound enzymes and receptors; for example, the binding of serotonin is decreased. Oxidative degradation and polymerization of lipids leads to the accumulation of lipofuscin, the age pigment. The presence of catalytic iron and copper complexes in human cerebrospinal fluid (CSF) and the high iron content of the brain suggest that these structures are very sensitive to oxygen radical generation (1). α-Tocopherol is the best recognized lipid-soluble secondary chain-breaking antioxidant. Cells utilizing molecular oxygen are, under normal conditions, adequately protected against the toxic intermediates of oxygen in both water and lipid phases.

In general the equilibrium in a cell seems to be regulated by balancing of the essential enzymes: superoxide dismutase (SOD), glutathione peroxidase (GSH-Px), and catalase. Sulfhydryl proteins, proteins containing thiol groups and controlling and utilizing the redox flow, also regulate the peroxide level. Nutrients cannot as such control the redox balance. The function of the blood-brain barrier is to ensure the proper levels of brain nutrients. For example, ascorbic acid is exceptionally effectively transported from plasma to CNS. The brain levels are well maintained even at very low plasma levels of ascorbic acid.

However, the brain tends to need radical reactions for the generation of physiological responses. Observations suggest that free-radical intermediates may be involved, for example in the coupling between depolarization of the plasma membrane, Ca^{2+} fluxes, and neurotransmitter release. Normally, cellular redox adjustments regulate the functional sulfhydryl groups of proteins. In addition to the conventional antioxidant processes, the brain is protected by specific or high endogenous levels of free-radical scavengers, such as dopamine, norepinephrine, catechol estrogens, taurine, and carnosine (4).

During aging, the pools of neurons become exhausted. The distinct hallmarks of aging and degeneration, such as ceroid and lipofuscin age pigments, paired helical filaments, and amyloids, can be experimentally induced by free-radical stress. Distortions of free-radical metabolism may take place as a result of hemodynamic changes caused by trauma, ischemia, or infection. Constitutive, inborn, or acquired pathognomonic errors of brain metabolism are frequently associated with free radicals. In animals, vitamin E deficiency causes intraneuronal lipofuscin accumulation; this can be avoided by treatment with vitamin E, selenium, and reduced glutathione. No evidence of vitamin E depletion has been observed in normal human aging, Alzheimer's disease, or Down's syndrome, characterized by mental retardation or dementia and increased lipid peroxidation (5).

However, neurological symptoms evolve in the human after prolonged vitamin E deficiency, such as spinocellular ataxia, including primary loss of deep tendon reflexes, truncal and limb ataxia, ophthalmoplegia, ptosis, muscle weakness, and dysarthria. There are diseases or conditions in which the decrease in the vitamin E level is secondary and that result in distinct neurological symptoms; these include abetalipoproteinemia (an inborn error of lipoprotein metabolism), chronic fat malabsorption as a result of liver disease, intestinal resection, and cystic fibrosis. In patients suffering from these diseases or from a severe deficiency of vitamin E without generalized malabsorption, the serum vitamin E level is low or not detectable. The clinical features of abetalipoproteinaemia are ataxic neuropathy and pigmentary retinopathy, which usually develop during the second decade of life. The neurological syndrome involves the central and peripheral nervous systems, as well as the retina and muscles. Muller (6) successfully cured these patients with large oral doses of vitamin E: 100 mg/kg body weight per day (the normal requirement is 10–30 mg/day).

The CNS antioxidant system is poorly understood. Vitamin E has the potential to scavenge radicals that initiate lipid peroxidation. The importance of vitamin E is its role as a chain-breaking membrane-bound antioxidant that tends to reduce the likelihood of hydroxyl free-radical generation as a result of Fenton-type reactions. Radical scavengers, such as vitamin E and coenzyme Q, may act by replacing a highly reactive alkyl radical with a phenolic radical that is more stable (2). Vitamin E prevents the random peroxidation of arachidonic acid and may protect the early steps of the cyclooxygenase pathway. Urano et al. (7) presented evidence that primary radicals of unsaturated fatty acids (L˙) may also form direct ether linkages with vitamin E.

Attempts have been made to relate vitamin E to different neurological disorders, such as epilepsy, intracranial hemorrhage, Parkinson's disease, Werdnig-Hoffman disease, neuronal ceroid lipofuscinoses (NCLs), multiple sclerosis, myotonic dystrophy, Down's syndrome (DS), and Alzheimer's disease (AD). It has also been suggested that oxygen-derived free-radical species may be involved in such neurological conditions as cerebral ischemia, tarditive dyskinesia, DS, Duchenne muscular dystrophy, and mental retardation. Free radicals are particularly involved in diseases with malplaced transition metal (Fe or Cu) complexes, which can stimulate the decomposition of lipid hydroperoxides to form peroxyl and alkoxyl radicals, which have sufficient reactivity to initiate further peroxidation of lipids in the nervous system. NCL diseases, the Hallervorden-Spatz syndrome, Parkinson's disease, Alzheimer's disease, and the Zellweger syndrome are also conditions with suspected iron involvement.

The relevant connection between vitamin E and the majority of these disorders rests on clinical observations during supplementation alone or in combination with other antioxidants (6,8,9).

VITAMIN E THERAPY

Vitamin E therapy has been tried in some neurological diseases with low serum vitamin E levels, such as Werdnig-Hoffman disease, a progressive degenerative disorder of the spinal and bulbar motor neurons that usually leads to death before the age of 2. However, vitamin E therapy did not result in any positive response in several treated infants (9).

In premature infants (with low birth weight and antioxidant stores), vitamin E has been used to correct a relative vitamin E deficiency and to prevent complications of periventricular hemorrhage and retinopathy. However, the results are conflicting. Most probably, the administration of vitamin E to newborn infants can reduce the severity but not the overall incidence of retinopathy (6).

Alzheimer's disease is a frequent cause of dementia in the elderly. Lassen et al. proposed in 1957 (10) that the decrease in cerebral blood flow, glucose utilization, and oxygen consumption common to many dementias results from abnormalities of brain structure with a high oxidative capacity. In dementia of the Alzheimer type brain blood flow and oxidative metabolism are reduced. This situation may lead to loss of balance between prooxidants and antioxidants.

To evaluate the peroxidative stress in dementias, autopsy brain samples should also be studied for GSH-Px, SOD, catalase, and selenium. A direct causal relationship between brain antioxidant defenses and dementia in aging and Alzheimer's disease is hard to demonstrate because of the extremely slow process. The decrease in activities of catalase and GSH-Px with aging in animals of short life cycle supports the free-radical theory of aging. Interestingly, a high proportion of Down's syndrome patients develop the neuropathological and clinical changes of AD, suggesting a close pathogenetic relationship between these disorders (4).

Increased primary gene products that may contribute to the pathology of Down's syndrome include cytoplasmic CuZn superoxide dismutase. Consistent with the gene dosage effect, SOD activity is increased by 50%, leading to noxious concentrations of H_2O_2; brain GSH-Px remains normal. The overall redox state in other tissues is corrected by an adaptive increase in GSH-Px activity. This means that the brain is especially susceptible to oxygen free-radical stress. Our primary survey of specific antioxidant therapy rests on this theory, which was recently reviewed (4).

Acceleration of oxidative processes in Down's syndrome (trisomy 21) is suggested by the clinical and biochemical signs of rapid aging, the early occurrence of clinical dementia with brain degeneration similar to that seen in Alzheimer's disease, and increased amounts of ceroid lipofuscin pigments in the neurons similar to that seen in NCL. A highly positive correlation has been reported between erythrocyte GSH-Px values and the intelligence quotient (IQ) in DS. Decreased serum zinc concentrations have been demonstrated in DS patients, and high doses of zinc were reported to correct some of the immunological defects. Annere'n et al. (11) demonstrated an increase in serum concentrations of IgG_2 and IgG_4 by selenium supplementations in children with Down's syndrome.

For the last 10 years, through intervention efforts, patients with DS have received different kinds of medical treatment, including antioxidative or megavitamin and mineral supplements. It has been known for a long time that vitamin E deficiency results in enhanced formation of ceroid lipofuscin in both humans and other animals. Therefore it is obvious that vitamin E should be included in these supplements. However, there are conflicting reports about the results of the megavitamin and mineral supplement trials. In 1981, Harrell et al. reported that

IQ scores in a heterogeneous group of mentally retarded children, four of these subjects with DS, improved after administration of megavitamin and mineral supplement. Smith et al. in 1984 reported that when they supplemented 56 school-aged DS children for 4 and 8 months with a daily cocktail including 600 IU d-α-tocopheryl succinate, no positive response could be seen in their cognitive intelligence test scores. The study was a double-blind two-group clinical trial. However, the megavitamin and mineral supplement also included some pro-oxidative substances (12).

In 1989, Metcalfe et al. (5) succeeded in measuring the concentration of vitamin E (α-tocopherol) in cortex samples from patients with Alzheimer's disease, fetuses with Down's syndrome, and a group of centenarians. The results showed that the normal aging processes, AD, and the increased in vitro lipid peroxidation reported in fetuses with DS did not indicate a gross lack of α-tocopherol or cause a significant depletion of vitamin E (5).

Free radicals and lipoperoxidation reactions seem to be involved in epileptic seizures developing after brain hemorrhage of different kinds. There is an association between hemosiderin deposition and posttraumatic epilepsy. Extravasation of blood and hemolysis of erythrocytes result in the decompartmentalization of free iron and accelerate the rates of lipoperoxidation and superoxide-dependent formation of OH˙ radicals, which are propagated by reperfusion and reoxygenation in postischemic tissue injury. Simultaneously, the activity of GSH-Px in the ischemic tissue is decreasing (13). Antioxidants have been observed to prevent synergistically the lipoperoxidation in animals and in humans (14).

Pretreatment of rats with vitamin E and selenium before iron injections have been shown to prevent the development of seizures to a high degree in a large percentage of experimental rats. There are also reports of the normalization of the electroencephalograph of patients with juvenile neuronal ceroid lipofuscinosis (JNCL) after vitamin E and sodium selenite supplementation. In addition, the onset of epilepsy occurs significantly earlier among JNCL patients not given this antioxidative therapy (11.1 years) compared with patients receiving antioxidant therapy (13.6 years) (15). Recently there was also a report of four children with intractable seizures, repeated infections, and intolerance to anticonvulsants who had evidence of GSH-Px deficiency. The clinical state of the children improved after discontinuation of anticonvulsant medication and selenium substitution.

The neuronal ceroid lipofuscinoses are a group of recessively inherited neurodegenerative lysosomal storage diseases. The diseases are characterized by deposits of ceroid and lipofuscin pigments in tissues, particularly in neural tissue, visual failure, and progressive mental retardation. A lesion in the degradative pathway of subunit c of mitochondrial ATP synthase may be one cause of juvenile NCL and infantile NCL (16).

Depending on the age of onset and clinical, electrophysiological, and neuropathological features, the NCLs can be subdivided into the infantile (INCL), the late infantile (LINCL), the juvenile (JNCL), and the adult type of NCL. The occurrence of the fluorescent pigments suggested the peroxidation of lipids in the etiology of NCL. It is likely that diseased tissue peroxidizes more rapidly than normal tissue, and cytotoxic end products of lipoperoxidation cause secondary damage. On a weight basis, the ceroid seen in JNCL patients binds five times more iron than the lipofuscin seen in normal elderly individuals. This finding is corroborated by the presence of complexable iron and copper in the cerebrospinal fluid of patients with NCL and other neurological disorders. When the pH value of the assay for iron was decreased, substantially more complexable iron was found in the CSF of NCL patients (17). It is well established that damaged tissue releases metals from protein-bound sites, and these metals stimulate peroxidative damage to lipids and other biomolecules (4). Interestingly, aluminum has been observed in CSF and in ceroid lipofuscin pigments of the brain of NCL patients. The increased levels of aluminum salts greatly enhance iron-dependent damage to membranes. The selenium-containing GSH-Px is one of the most essential enzymes counteracting lipoperoxidation. Previously, two independent reports demonstrated that erythrocyte GSH-Px activity was decreased in JNCL patients. This low GSH-Px activity was reversed to the normal level by selenium supplementation (18,19).

Apart from the low selenium status, very low vitamin E levels were also found in the serum of advanced and hospitalized NCL patients (19), probably chiefly as a result of an intake insufficient in selenium and vitamin E. However, this can also be explained by the recent finding of a pronounced reduction in apoprotein B as well as the whole fraction of very low density lipoprotein (VLDL) in JNCL patients (20).

In addition, some selenium has been reported to be associated with VLDL, low-density lipoprotein, and other specific plasma proteins. The reduction in circulating VLDL, the carrier of vitamin E and selenium, suggests that supplementation of these antioxidants is indicated in JNCL. In addition, there is also a metabolic relationship between vitamin E and selenium at high levels of selenium intake, and vitamin E may protect the selenium-treated patients from selenium toxicity. Vitamin E deficiency was demonstrated to decrease the amount of selenium needed to cause chronic selenium toxicity in animals. The high dosage of selenium was used in an attempt to maintain normal vision in NCL patients for as long as possible. Sodium selenite was shown to stimulate light sensitivity in animals, as judged by increases in the a and b waves of the electroretinogram (ERG).

Antioxidant therapy (vitamin E, selenium, and vitamins B_6 and B_2) has been used in the treatment of JNCL in Finland since 1970. Adverse reactions have been very few and of minor importance. Vitamin E as α-tocopherol acetate has been given (0.014–0.05 g/kg body weight), in practice 0.7–1.5 g/day in two or three divided doses. The recommended vitamin E serum concentration has been between 60 and 110 μmol/L, which has not always been easy to maintain. In recent years sodium selenite has been initiated with 0.5 mg/kg body weight as a single daily dose. The selenium serum concentration has been kept between 190 and 300 μg/L (2.4 and 3.7 μmol/L). Vitamin E as d,l-α-tocopherol acetate has been given (0.014–0.05 g/kg body weight), in practice 0.7–1.5 g/day divided in two or three doses. The recommended vitamin E serum concentration has been between 60 and 110 μmol/L, which has not always been easy to maintain. The daily dose of vitamin B_6 has been 20–40 mg and vitamin B_2, usually 3 mg (20).

Antioxidant treatments have indicated that they are not able to slow the course of INCL. However, antioxidants seemed to help prevent painful vasoconstriction in the extremities in the winter. Antioxidants have not been recommended for the treatment of LINCL patients, either, although a striking but transitory clinical improvement was seen in a 8-year-old boy with LINCL who had not been able to stand up or walk for 2 months and who began to walk after 2 weeks of treatment. This may have been prompted by adjustment of his phenytoin dosage to a proper level.

Santavuori et al. (15) studied 26 Finnish JNCL patients treated with antioxidants by using a JNCL disease-specific scoring system introduced by Kohlschütter et al. Scores were given for the problems of vision, intellect, language, motor function, and epilepsy and compared with the data of 17 German JNCL patients not treated with antioxidants. The study supported the theory that antioxidant treatment can slow the progress of JNCL. Compared with the unsupplemented German patients, epileptic seizures began in the Finnish group on the average about 2.5 years later and total blindness, about 4 years later. Similarly, the debut of mental retardation and loss of language and motor functions occurred later among the antioxidant-treated Finnish patients. The toxic range was considered 100 μmol/L for vitamin E and 4 μmol/L for selenium; it was only rarely exceeded. The clinical effect of the antioxidative treatment was so pronounced that the blind trial was stopped because it was considered unethical not to treat all the Finnish JNCL patients.

In a few other reports outside Finland, conflicting results have been reported on the antioxidant therapy of NCL patients. Rotteveel and Mullaart (21) found that application of the

Westermarck formula in three patients with JNCL resulted in a mitigated course of the disease, whereas Naidu et al. in 1988 reported an antioxidant trial involving only one INCL, one LINCL, and one JNCL patient, which resulted in no observable positive clinical response. Clausen et al. (18) documented in a few patients a low selenium status and low glutathione peroxidase activities of hematogenous cells; similarly to the Finnish biochemical studies, they reported that the biochemical parameters can be normalized by antioxidant treatment.

The benefits of antioxidative therapy are corroborated by the significant negative correlation of GSH-Px activity with neurological dysfunction of motor performance, balance, coordination, and speech. The mean age at death has been extended by 4 years compared to the beginning of the century. As the best responders to antioxidant therapy show no neurological dysfunction at the age of over 20 years, there is no doubt that the life expectancy of JNCL patients receiving antioxidants will be significantly prolonged in the future (1,15,20).

Another neurological use of vitamin E being investigated is the possible role of vitamin E therapy combined with selenium and vitamin C and balanced intake of omega-six and omega-three polyunsaturated fatty acids in the treatment of multiple sclerosis (MS). As early as in 1975, Mickel suggested that an increased peroxidation rate is a pathological factor in MS.

MS is a chronic disabling neurological disease that has a peak onset between the ages of 25 and 35 years. MS results at its worse in significant problems with vision, coordination, sensation, intellect, and sphincter control problems, resulting in death within a few years. However, in some cases MS may be diagnosed only accidentally or not at all. MS is considered to result from an immunological disorder involving the myelin or insulation of the nervous system, resulting in demyelination and scarring of the brain and spinal cord.

One of the most consistent epidemiological findings is the higher prevalence of the disease in populations living in countries with cold climates and consuming diets rich in animal fats containing saturated fatty acids (3). There is a high prevalence in many parts of the Nordic or Scandinavian countries. In a high-risk area in Finland, Wikström et al. (22) found the mean selenium content in human whole blood to be less than 50% of the internationally considered normal value. However, the values for both vitamin E and copper were within the international normal range. Low levels of polyenic acids are involved in the pathogenesis of both MS and JNCL. In 1972, Thompson found decreased levels of serum linoleate as well as unsaturated fatty acids of brain phospholipids in MS patients. It was also shown that supplementation with essential fatty acids may improve the clinical status of young MS patients diagnosed early (18).

As in NCL, selenium may activate GSH-Px (scavenger of organic peroxides), regulate the metabolic transformation of essential fatty acids, and biotransform these to prostaglandins, thromboxanes, and leukotrienes. Curiously, decreased GSH-Px activities in erythrocytes have been found in female but not in male MS patients. Blood selenium levels have been reported to be lower in MS patients than in healthy controls but have also been reported to be of equal levels. Impaired Se status has been found in MS largely in connection with severe protein-calorie malnutrition (18).

High-dose antioxidant supplementation has been recommended for MS patients. In a Danish study 18 MS patients were given 480 mg vitamin E, 6 mg sodium selenite equal to 2 mg selenium, and 2 g vitamin C per day for 5 weeks. It was concluded that the tested antioxidant treatment seemed to be safe and the biochemical abnormalities could be normalized, but the clinical response was not evaluated (23). During another trial in Denmark, MS patients were given daily 500 mg vitamin E, 2 g vitamin C, and 6 mg per 70 kg sodium selenite for 3 months to 2–3 years. A positive clinical response, such as more steady walking, was reported in 6 of 10 MS patients, and the condition of the others was stabilized.

Furthermore, controlled trials with the administration of omega-6 PUFAs and omega-3 PUFAs to MS patients with positive clinical results have been investigated in England. According to Bates (3), there was a trend suggesting that addition of omega-6 and omega-3 PUFAs to the diet of MS patients resulted in a reduction in the severity and frequency of relapses and in a mild overall benefit in a 2 year period. However, low levels of vitamin E were recently observed in some MS patients that were responsive to dietary advice. Thus the balanced intake of omega-3 and omega-6 PUFAs must take into consideration the increased requirement for vitamin E.

The Danish investigators reported that antioxidative therapy in MS patients can normalize the decreased linolic acid level of lymphocytes and erythrocytes. Treatment of MS with Se supplementation does not seem warranted in the absence of demonstrated deficiency. Thus in the selenium- and vitamin E-containing antioxidant treatments reported, the clinical benefit to the course of MS remains open to speculation (18).

In mitochondrial disorders, a group of neuromuscular disorders, the clinical manifestations are wide and varied. Many mitochondrial encephalomyopathies tend to show an episodic course. There is a vast literature on mitochondrial diseases classified as cytochrome c oxidase deficiencies involving defects of fatty acid oxidation, defects of pyruvate metabolism, and defects of the respiratory chain, for example the Kearn-Sayre syndrome (KSS), chronic progressive external ophthalmoplegia (CPEO), myoclonus epilepsy with ragged red fibers (MERRF), and mitochondrial encephalomyopathy with lactic acidosis and strokelike episodes (MELAS). The vitamins used in the therapy of mitochondrial diseases are mostly given in combinations, such as thiamine, biotin, riboflavin, lipoic acid, folate, nicotinic acid, pantothenate, pyridoxine, vitamin B_{12}, vitamin E, and vitamin K. The idea is that some of them are precursors of coenzymes or prosthetic groups of enzymes involved in the defect or distal from the defect. However, in no case has a favorable effect of this therapy been directly connected with any suspected or proven mitochondrial encephalopathy (24).

Parkinson's disease is a degenerative neurological disease with progressive degeneration of the pigmented nuclei in the brain stem with loss of the neurotransmitter dopamine in the basal ganglia. Clinical trials indicate that antioxidative treatment may postpone the need for levodopa antiparkinson treatment (6,9).

CONCLUSION

The primary form of vitamin E deficiency has provided substantial evidence that vitamin E is vital for the normal development and function of the CNS. In combination with other antioxidants, vitamin E seems to have a modulating and sparing function in tissue; it supports the antioxidative capacity of the tissue and prevents the side effects of other antioxidants. Further research is needed to determine the correct dosage and how vitamin E should be combined with other antioxidants to obtain an optimal response in different neurological disorders. However, the antioxidative therapy of disorders affecting the CNS should be evaluated more carefully.

REFERENCES

1. Halliwell B, Gutteridge JMC. Oxygen radicals and the nervous system. TINS 1985; 8:22-6.
2. Horrocks LA, VanRollins M, Yates AJ. Lipid changes in the ageing brain. In: Davison AN, Thompson RHS, eds. The molecular basis of neuropathology. London: Edward Arnold, 1981; 601-30.
3. Bates D. Dietary lipids and multiple sclerosis. Ups J Med Sci Suppl 1990; 48:173-87.
4. Antila E, Westermarck T. On the etiopathogenesis and therapy of Down syndrome. Int J Dev Biol 1989; 33:183-8.
5. Metcalfe T, Bowen DM, Muller DPR. Vitamin E concentrations in human brain of patients with Alzheimer's disease, fetuses with Down's syndrome, centenarians, and controls. Neurochem Res 1989; 14:1209-12.

6. Muller DPR. Antioxidant therapy in neurological disorders. Adv Exp Med Biol 1990; 264:475-84.

7. Urano S, Yamanoi S, Hattori Y, Matsuo M. Radical scavenging reaction of α-tocopherol. II. The reaction with some alkyl radicals. Liids 1977; 12:105-8.

8. Gutteridge JMC, Westermarck T, Halliwell B. Oxygen radical damage in biological systems. Modern Aging Res 1986; 8:99-139.

9. Sokol RJ. Vitamin E and neurologic function in man. Free Radic Biol Med 1989; 6:189-207.

10. Lassen NA, Munk O, Tottey ER. Mental function and cerebral oxygen consumption in organic dementia. Arch Neurol Psychiatr 1957; 77:126.

11. Annere'n G, Magnusson CGM, Nordwall SL. Increase in serum concentrations of IgG_2 and IgG_4 by selenium supplementation in children with Down's syndrome. Arch Dis Child 1990; 65:1353-5.

12. Smith GF, Spiker D, Peterson CP, Cicchetti D, Justine P. Use of megadoses of vitamins with minerals in Down syndrome. J Pediatr 1984; 1095:228-34.

13. Halliwell B, Gutteridge JMC. Free radicals in biology and medicine, 2nd ed. Oxford: Clarendon Press, 1989; 416-508.

14. Westermarck T, Tolonen M, Halme M, Sarna S, Keinonen M, Nordberg UR. Antioxidant supplementation for elderly living at a nursing home. A double-blind randomized one year clinical trial. In: Hurley LS, Keen CL, Lönnerdal B, Rucker RB, eds. TEMA 6. New York: Plenum Press, 1988; 325-6.

15. Santavuori P, Heiskala H, Autti T, Johansson E, Westermarck T. Comparison of the clinical courses in patients with juvenile neuronal ceroid lipofuscinosis receiving antioxidant treatment and those without antioxidant treatment. Adv Exp Med Biol 1989; 266:273-82.

16. Palmer DN, Fearnley IM, Medd SM, et al. Lysosomal storage of the DCCD reactive proteolipid subunit of mitochondrial ATP synthase in human and ovine ceroid lipofuscinoses. Adv Exp Med Biol 1989; 266:211-23.

17. Heiskala H, Gutteridge JMC, Westermarck T, Alanen T, Santavuori P. Complexable iron and copper in the cerebrospinal fluid of patients with neuronal ceroid lipofuscinosis. In: Rice-Evans C, Halliwell B, eds. Free radicals, methodology and concepts. London: Richelieu Press, 1988; 447-54.

18. Clausen J, Jensen GE, Nielsen SA. Selenium in chronic neurologic diseases. Biol Trace Elem Res 1988; 15:179-203.

19. Westermarck T. Selenium content of tissues in Finnish infants and adults with various diseases, and studies on the effects of selenium supplementation in neuronal ceroid lipofuscinosis patients. Acta Pharmacol Toxicol 1977; 41:121-8.

20. Santavuori P, Heiskala H, Westermarck T, Sainio K, Moren R. Experience over 17 years with antioxidant treatment in Spielmeyer-Sjögren disease. Am J Med Genet Suppl 5:265-74.

21. Rotteveel JJ, Mullaart RA. Anti-oxydatieve therapie bij de ceroid lipofuscinoses. Tijdschr Kindergeneeskd 1989; 57:181-6.

22. Wikström J, Westermarck T, Palo J. Selenium, vitamin E and copper in multiple sclerosis. Acta Neurol Scand 1976; 54:287-90.

23. Mai J, Sorensen PS, Hansen JC. High dose antioxidant supplementation to MS patients. Biol Trace Elem Res 1990; 24:109-17.

24. Przyrembel H. Therapy of mitochondrial disorders. J Inher Metab Dis 1987; 10:129-46.

57

Effects of Palm Oil Vitamin E on Atherogenesis

Erkki Antila, Faik Atroshi, and Tuomas Westermarck

University of Helsinki, Helsinki, Finland

INTRODUCTION

An increased risk of atherogenesis is attributed to a high level of serum cholesterol. However, an epidemiological risk factor is not causal unless proven by other means. It follows that strict logical criteria should be followed when correlations are explained. To prove such causality an exact mechanism that explains all the variables is needed. Experimental evidence will further confirm the causality. These premises have been thoroughly discussed by Stehbens (1).

The dominant sterols are nearly all ubiquitous end products derived from squalene. Although steroids derived from sterols are essential hormones or hormonelike compounds acting at the genomic level, sterols generally act as structural membrane constituents rather than through metabolism.

Eukaryotic membranes primarily contain free sterols, which means that the 3β-hydroxy group is in principle free. The OH group probably immerses itself in the polar milieu acting partially in the capacity of both a proton donor and a proton acceptor. Furthermore, cholesterol has the ability to condense, that is, restrict, the volume of phospholipid layers. This means that cholesterol decreases membrane fluidity, that is, increases the solidity of the membrane. Experimental data suggest that lecithin and cholesterol may form an equimolar complex in which lecithin phosphate and the sterol hydroxyl groups are juxtaposed.

In contrast to insects, which have developed specific ecological dependencies on plant sterols, all human cells possess the capacity of de novo synthesis of cholesterol. However, human cells accept and prefer exogenous and liver-derived cholesterol, available in plasma as low-density lipoprotein (LDL). Exogenous cholesterol is especially required at the site of cellular damage and repair, where the requirements are temporarily and rapidly increased.

Similar to cholesterol, α-tocopherol may be a constituent part of the cell membrane, is supplied by LDL to the cells, and is able to exchange between lipoproteins and erythrocytes.

LDL receptor-mediated cholesterol is internalized and metabolized in lysosomes. These processes are promoted by vitamin E. Preincubation of vitamin E with LDL or arterial wall increases cholesterol esterase activity. Conversely, cholesterol esterase activity is decreased in vitamin E deficiency. Stimulation of LDL metabolism may also result through vitamin E protection of peroxidative attack [reviewed by Saito and Morisaki (2)].

The control of cholesterol synthesis, as well as that of ubiquinone and dolichol, takes place through hydroxymethylglutaryl-coenzyme A reductase (HMG-CoA-reductase). This enzyme is regulated by negative feedback mechanisms via the end products and enzymatically through HMG-CoA-reductase kinase. Normally the uptake of LDL cholesterol and a nonsterol product derived from mevalonate keeps HMG-CoA-reductase suppressed.

In addition to real end products, such as ubiquinone, a variety of compounds, including lovastatin and tocotrienol, possess the ability to affect HMG-CoA-reductase. Qureshi et al. (3) suggested that unlike the former, tocotrienols do not inhibit enzyme activity but merely decrease the translational efficiency of HMG-CoA-reductase mRNA. However, the activity of avian hepatic HMG-CoA-reductase was increased by α-tocopherol treatment (4). The LDL cholesterol increase was precluded by a simultaneous increase in cholesterol 7α-hydroxylase activity. A high cholesterol level and a high 7α-hydroxylase activity may result in the accumulation of toxic levels of cholesterol oxides.

Lovastatin has been shown to significantly reduce by 60% the serum level of α-tocopherol; this was not prevented by α-tocopherol supplementation (5). This indicates that hepatic sterol metabolism is more complex than was first envisaged. The reduction in cholesterol in hepatocytes is somehow associated with endogenous lipoprotein production, resulting in a fall in circulating LDL (the serum compartment transporting vitamin E as well as cholesterol).

Manipulation of HMG-CoA-reductase poses the danger of affecting not only cholesterol but also dolichol and ubiquinone levels in an uncontrolled manner. Low cholesterol may alter the integrity of the cell membrane by affecting the wall lipoproteins and molecular cholesterol. Apart from the liver, cholesterol in the human body has a rather slow turnover rate, which is even slower in the brain. Dolichol is present in all cell membranes, particularly in lysosomes. The amount of dolichol increases with age in the human brain, especially in Alzheimer's and Batten-Mayou diseases.

In vitamin E-deficient rats, lovastatin treatment resulted in high mortality rates as a result of massive hepatic necrosis (0.5 g/kg of diet). Both serum cholesterol and hepatic dolichol levels were significantly elevated. Serum cholesterol in vitamin E-supplemented and lovastatin-treated rats did not significantly differ from those in controls (6).

High LDL cholesterol increases the risk of atherogenesis as a result of the special conditions prevailing at the site of cellular damage in the endothelial lining of arterial walls.

Continuous exposure to physical stress, microinjuries, and renewal of endothelial cells may occasionally result in cell death and inflammatory cascades. The possibility exists that these changes are secondary to events taking place in the adventitia, media, or even vasa vasorum. During initiative repair, LDL is concentrated in the extracellular space of the intima at the site of injury. A reduced number of receptors or an insufficient amount of LDL initiates homeostatic mechanisms that stimulate LDL production. Failure to provide enough material for the repair may result in increased disruption of the arterial wall and further potentiate the inflammatory reaction. Although almost nothing is known about the initiative events of LDL cholesterol oxidation, one can assume that inflamed conditions increase free-radical production and the risk of oxidative modification. Of the human plasma lipoproteins, LDL is the most susceptible to peroxidation (7). However, the polyunsaturated fatty acid side chains of membrane lipids are particularly sensitive to oxidation by molecular oxygen. Mechanical stresses, such as shear forces on the vascular endothelial cells, may promote lipid peroxidation followed by arachidonic acid release and prostanoid formation. Activation of phospholipase A_2 may further promote cell-specific prostanoid formation.

Monounsaturated fatty acids may block the desaturation-elongation reactions leading from linolenic to arachidonic acid. Linoleic acid reacts in vitro with molecular oxygen 20 times more rapidly than oleic acid. This indicates the necessity of monounsaturates in lipid bilayers, where they may even act in a manner similar in principle to vitamin E.

Lysophosphatidylcholine derived from the lecithin cholesterol acyltransferase reaction is a constituent of cell membranes and plasma that chemotactically attracts monocytes.

Complex formation of platelet-activating factor (PAF; 1-alkyl-2-acetyl-*sn*-glycerol-3-phosphocholine) is followed by rapid degradation catalyzed by a lipoprotein associated with PAF acetylhydrolase. Oxidized LDL promotes platelet aggregation more actively than native LDL. On the other hand, the formation of oxidatively modified LDL gives rise to high-uptake LDL, which does not use LDL receptors. In familial hypercholesterolemia, in which the cells of heterozygotes lack the proper amount of LDL receptors, the risk of oxidatively modified LDL is further increased.

It has been shown that only oxidized and denatured LDL is incorporated and metabolized in macrophages (8). Peroxidatively modified LDLs are known to be toxic to cultured endothelial cells (9), to attract macrophages from the blood, and to prevent their exit from tissues (10). The events taking place at the endothelial lining are stepwise or gradual modifications of LDL. Minimally modified LDL (MM-LDL) induces leukocyte adherence and monocyte transmigration and endothelially derived growth factors. These events have been experimentally verified in vitro, where they can also be prevented by such antioxidants as probucol and α-tocopherol. However, high-density lipoprotein (HDL) and HDL-derived phospholipids can also abolish the effects of MM-LDL on endothelial-smooth muscle cell cocultures.

There are several facts supporting the central role of oxidized LDL (oxLDL) in atherogenesis:

Favors foam cell formation
Is chemotactic to monocytes
Inhibits the motility of tissue macrophages
Is cytotoxic to endothelial cells
Stimulates the release of MCF-1 (endothelial chemotactic factor)
Stimulates the release of M-CSF (macrophage growth factor from endothelial cells)
Inhibits endothelially dependent relaxation of arteries
Is immunogenic

The capability of LDL particles to resist free-radical stress depends on the content of lipophilic antioxidants, such as tocopherols and tocotrienols. The tocopherol content is very rapidly depleted. In vitro incubations have shown a 97% drop within 6 h (11). However, most of the circulating vitamin E is found in the LDL fraction (12), and thus one of the functions of LDL is to provide vitamin E to the cells.

In addition to vitamin E, many lipophilic antioxidants may prevent the oxidative modification of LDL. Some flavonoids (morin) have a sparing effect on α-tocopherol consumption and thereby or directly protect LDL. In a cell-free system in which peroxidation of LDL was initiated by copper(II), even ascorbic acid seems to be more effective than α-tocopherol (13).

At present, the role of different plasma antioxidants in the recycling of LDL vitamin E from chromanoxyl radicals is an open question.

VITAMIN E THERAPY: PREVENTION OF ATHEROGENESIS

The aim of assessing the value of pharmacological or nutritional manipulation of a cholesterol-associated risk of atherogenesis is limited by many variables. Diet may vary to a significant extent, not only geographically, but also ethnically and individually. Dietary constituents that are important to control include fatty acids (PUFA, MFA, and SFA), cholesterol, antioxidant vitamins, trace elements, and fiber.

Sites where plasma cholesterol level can be controlled include:

1. The LDL receptor, which is a critical link between plasma and intracellular sterol pools
2. Interruption of the enterohepatic circulation of bile acids
3. The intestinal mucosa, which is limited in the number of sites available for sterol absorption
4. The enzymes responsible for cholesterol homeostasis: HMG-CoA-reductase, acyl-CoA-cholesterol acyltransferase, cholesterol 7α-hydroxylase (activates hepatocyte LDL receptor and extracts LDL cholesterol from the circulation)
5. Protection of LDL from oxidative attack, oxLDL goes to scavenger receptors in the macrophages

The aim of developing cholesterol-lowering drugs, such as lovastatin, has revealed many complex interactions. These dose-dependent interactions are clearly involved in vitamin E status and are reflected, in addition to cholesterol metabolism, in that of dolichol (6).

Proposals for the protective role in situ include ascorbate, vitamin E, ubiquinol, and flavonoids, and last but not least, cholesterol.

Natural vitamin E is a combination of tocopherols and tocotrienols. Palm oil, which is a rich source of vitamin E, predominantly tocotrienols, was recently shown to lower the levels of blood cholesterol and LDL cholesterol in humans and experimental animals (3,14–21).

Vitamin E is present in crude palm oil (600–1000 ppm) and in refined palm oil (470–670 ppm), together with β-carotene and vitamin A. The major components of palm oil vitamin E are α-tocopherol, α-tocotrienol, γ-tocotrienol, and δ-tocotrienol. Palm oil is commonly classified as a saturated fat, containing palmitic acid (42%), oleic acid (42%), linoleic acid (10.4%), and stearic acid (5%). The tocotrienol-rich fraction of palm oil recently became available in the form of gelatinous capsules, Palm vitee (Porim, Malaysia). One capsule contains 50 mg of the tocotrienol-rich fraction, including α-tocopherol (15–20%), α-tocotrienol (12–15%), γ-tocotrienol (35–40%), and δ-tocotrienol (25–30%).

There are several theoretical benefits of vitamin E supplementation in hypercholesterolemia and subsequent cardiovascular diseases. In addition to those involved in LDL cholesterol metabolism, the inhibitory effect of vitamin E on platelet aggregation, platelet adhesion (22), eicosanoid metabolism, aryl sulfatase B inhibition, and platelet-vascular interactions (23) may help retard the formation of arterial thrombosis.

However, it should be remembered that some of the effects of different forms of vitamin E may be diverse. Absorption, transport, and tissue uptake are specific for α-tocopherol. α-Tocopherol is the form of vitamin E that increases most in the serum after oral intake. Its molar concentrations in LDL can be increased in vitro from 9 to 30 mol/mol LDL. Calculated according to the estimated composition of LDL, 1 α-tocopherol molecule protects about 200 molecules of polyunsaturated fatty acid (PUFA) (24).

Holub et al. (25) demonstrated that 200 μM δ-tocotrienol causes in vitro an overall inhibition of platelet aggregation of 70% compared to 5 and 10% for α- and γ-tocotrienols, respectively.

Although γ-tocotrienol keeps the activity of HMG-CoA-reductase low, α-tocopherol apparently increases it (5).

If vitamin E was conclusively shown to raise HDL cholesterol in humans as it does in rats (26), this could further reduce the risk of atherosclerosis. Animal experiments suggest that vitamin E may reduce the incidence of epinephrine-induced arrhythmias. Finally, although vitamin E deficiency leads to a cardiomyopathy in primates, and in almost all other species studied, no such cardiac lesion has been described in vitamin E-deficient humans.

We believe that recommendations to lower cholesterol are not always the right way to reduce the risk of cardiovascular disease. To prevent the disease, the role of LDL oxidation and oxLDL internalization in the arterial wall and feedback regulation at the organ level in the etiopathogenesis of atherogenesis should be considered. Palm oil tocotrienol has thus far proven to be a beneficial vitamin E nutrient in this sense. Daily intake as crude palm oil could be replaced by refined palm oil capsules (Palmvitee), which contain tocopherols and tocotrienols in a palm superolein mixture. Its measurable effects may depend on the diet and on the LDL cholesterol status of the subject.

Palmvitee studies have been conducted with healthy adult volunteers in Malaysia (16,18, 20), the United States (3,15,19,21), and Finland (14,17). Double-blind follow-up studies have lasted from 4 to 6 weeks. The supplementation dose has been from 1 to 4 capsules/day, corresponding to 18–72 mg α-tocopherol and 42–168 mg α-, γ-, and δ-tocotrienols. Palmvitee supplementation was shown to (Fig. 1)

1. Reduce serum total cholesterol and LDL cholesterol
2. Increase serum α-tocopherol concentration
3. Decrease apo B concentration
4. Decrease thromboxane concentration
5. Decrease platelet factor 4 concentration

Palmvitee supplementation was shown not to:

1. Increase liver-derived serum enzymes
2. Result in subjective or objective side effects

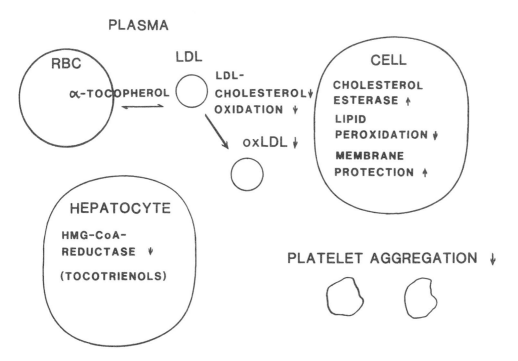

FIGURE 1 Basic compartments (hepatocyte, plasma, LDL, red blood cell (RBC), cell, and platelets) and events in which vitamin E (α-tocopherol and tocotrienols) controls the risk of atherogenesis.

Palmvitee supplementation may

1. Modulate beneficially relative proportions of PUFAs and MFAs in lipoproteins
2. Serve an active role as an inhibitor of tissue lipid peroxidation
3. Increase HDL cholesterol concentration

Although there are great individual and other variation depending on the experimental design, it appears that the cholesterol-lowering effect is relatively more effective when the cholesterol level at the onset is more than 8.0 mmol (drop of 35–40%, all 15–22%) (3). However, dietary palm oil and tocotrienol concentrate had no effect on total plasma cholesterol in rats when HDL cholesterol increased (26), which indicates the importance of species differences.

CONCLUSION

Dietary supplementation in humans with palm oil or Palmvitee could be of special interest because of several possibilities:

1. Fewer side effects compared to other medications.
2. When lowering LDL cholesterol there is no risk of depleting vitamin E (α-tocopherol).
3. It is effective enough to control the cholesterol-derived risk of atherogenesis.

REFERENCES

1. Stehbens WE. Basic precepts and the lipid hypothesis of atherogenesis. Med Hypoth 1990; 31: 105-13.
2. Saito Y, Morisaki N. Effect of vitamin E on low-density lipoprotein metabolism in the arterial wall. In: Miquel J, Quintanilha AT, Weber H, eds. CRC handbook of free radicals and antioxidants in biomedicine, Vol. II. Boca Raton, FL: CRC Press, 1989; 215-26.
3. Qureshi AA, Weber FE, Chaudhary VK, et al. Suppression of cholesterogenesis in hypercholesterolemic humans by tocotrienols of barley and palm oils. In: Packer L, Ames B, Smith M, Sies H, eds. Abstracts of antioxidants and degenerative diseases. Berkeley: University of California National Foundation for Cancer Research, 1990.
4. Qureshi AA, Peterson DM, Elson CE, Mangels AR, Din ZZ. Stimulation of avian cholesterol metabolism by alpha-tocopherol. Nutr Rep Int 1989; 40:993-1002.
5. Berry PH, MacDonald JS, Alberts AW, et al. Brain and optic system pathology in hypocholesterolemic dogs treated with a competitive inhibitor of 3-hydroxy-3-methylglutaryl coenzyme A reductase. Am J Pathol 1988; 132:427-43.
6. Porta EA, Monserrat AJ, Berra A, Rubio MC. Effects of lovastatin and leupeptin on ceroidogenesis of vitamin E-deficient and -supplemented young rats. Adv Exp Med Biol 1990; 226:169-90.
7. Szcczeklik A, Gryglewski RJ, Domagala B, et al. Serum lipoproteins, lipid peroxides and prostacyclin biosynthesis in patients with coronary heart disease. Prostaglandins 1981; 22:795-807.
8. Goldstein JL, Ho YK, Basu SK, Brown MS. Binding site on macrophages that mediates uptake and degradation of acetylated low density lipoprotein, producing massive cholesterol deposition. Proc Natl Acad Sci U S A 1984; 81:3883-7.
9. Hessler JR, Morel DW, Lewis LJ, Chisolm GM. Lipoprotein oxidation and lipoprotein-induced cytotoxicity. Arteriosclerosis 1983; 3:215-22.
10. Quinn MT, Parthasarthy S, Steinberg DA. Endothelial cell-derived chemotactic activity for mouse peritoneal macrophages and the effects of modified forms of low density lipoprotein. Proc Natl Acad Sci U S A 1985; 82:5949-53.
11. Esterbauer H, Jurgens G, Quehenberger O, Koller E. Autoxidation of human low density lipoprotein: loss of polyunsaturated fatty acids and vitamin E and generation of aldehydes. J Lipid Res 1987; 28:495-509.
12. Traber MG, Kayden HJ. Vitamin E is delivered to cells via the high affinity receptor for low-density lipoprotein. Am J Clin Nutr 1984; 40:747-51.

13. Jialal I, Vega GL, Grundy SM. Physiologic levels of ascorbic acid inhibit the oxidative modification of low density lipoprotein. Atherosclerosis 1990; 82:185-92.

14. Atroshi F, Westermarck T, Työppönen J, Rizzo A, Sankari S. Supplementation of palm oil vitamin E in human healthy adults; experimental and theoretical implications. In: Abstracts nutrition and health aspects of palm oil. 1989 Porim international palm oil development conference, Kuala Lumpur, 1989.

15. Qureshi AA, Qureshi N, Shen Z, et al. Lowering of serum cholesterol in hypercholesterolemic humans by Palmvitee. In: Abstracts nutrition and health aspects of palm oil. 1989 Porim international palm oil development conference, Kuala Lumpur, 1989.

16. Tan DTS, Khor HT. Studies on the hypocholesterolemic effect of palm oil vitamin E. In: Abstracts nutrition and health aspects of palm oil, 1989 Porim international palm oil development conference, Kuala Lumpur, 1989.

17. Atroshi F, Antila E, Westermarck T, Työppönen J, Sankari S, Saloniemi H. Effect of oral palm oil-vitamin E supplementation on human plasma lipid profiles. In: Packer L, Ames B, Smith M, Sies H, eds. Abstracts of antioxidants and degenerative diseases. Berkeley: University of California National Foundation for Cancer Research, 1990.

18. Tan DTS, Khor HT. The effect of palm oil vitamin E concentrate on serum lipid and lipoprotein levels in humans and in a hamster model. In: Packer L, Ames B, Smith M, Sies H, eds. Abstracts of antioxidants and degenerative diseases. Berkeley: University of California National Foundation for Cancer Research, 1990.

19. Qureshi AA, Qureshi N, Wright JJK, et al. Lowering of serum cholesterol in hypercholesterolemic humans by tocotrienols (palmvitee). Am J Clin Nutr 1991; 53:1021-26S.

20. Tan DTS, Khor HT, Low WHS, Ali A, Gapor A. Effects of a palm-oil-vitamin E concentrate on the serum and lipoprotein lipids in humans. Am J Clin Nutr 1991; 53:1027-30S.

21. Qureshi AA, Qureshi N, Hasler-Rapacz JO, et al. Dietary tocotrienols reduce concentrations of plasma cholesterol, apolipoprotein B, thromboxane B_2, and platelet factor 4 in pigs ;with inherited hyperlipidemias. Am J Clin Nutr 1991; 53:1042-6S.

22. Steiner M. Effect of alpha-tocopherol administration on platelet function in man. Thromb Hemost 1983; 49:73-7.

23. Betteridge J. Nutrition and platelet function in atherogenesis. Proc Nutr Soc 1987; 46:345-59.

24. Esterbauer H, Dieber-Rotheneder M, Striegl G, Waeg G. Role of vitamin E in preventing the oxidation of low-density lipoprotein. Am J Clin Nutr 1991; 53:314-21S.

25. Holub BJ, Sicilia F, Mahadevappa VG. Effect of tocotrienol derivatives on collagen- and ADP-induced human platelet aggregation. In: Abstracts nutrition and health aspects of palm oil, 1989 Porim international palm oil development conference, Kuala Lumpur, 1989.

26. Imaizumi K, Nagata JI, Sugano M, Maeda H, Hashimoto Y. Effects of dietary palm oil and tocotrienol concentrate on plasma lipids, eicosanoid productions and tissue fatty acid compositions in rats. Agr Biol Chem 1990; 54:965-72.

58

Vitamin E Deficiency and Neurological Disorders

Ronald J. Sokol

University of Colorado School of Medicine; Children's Hospital, Denver, Colorado

INTRODUCTION

Following its discovery in 1922 (1), vitamin E was found to have a diversity of actions in animals besides its role as a reproductive factor (2). However, no definitive human deficiency state was recognized until four decades later; thus, many ill-founded uses for vitamin E in the human were proposed, and the vitamin fell into disrepute among human nutrition scientists. In fact, vitamin E was even considered a "vitamin looking for a disease" (3). Despite the well-known neuromuscular deficits that developed in many experimental animals rendered vitamin E deficient (4), a similar role in human nutrition went largely unrecognized until recently. The neurological role of vitamin E was first reported in 1928, when suckling offspring of vitamin E-deficient mother rats developed paralysis (5). Degeneration of the brain in vitamin E-deficient chicks (nutritional encephalomalacia) (6) and muscular degeneration in vitamin E-deficient guinea pigs (7) were soon reported. Because of these and many other subsequent studies, vitamin E has been recognized as an important nutrient in agricultural and veterinary circles for decades. The similarity of animal disorders (nutritional muscular dystrophy) and human neuromuscular diseases (muscular dystrophy) led to trials of high-dose vitamin E (and other antioxidants) treatment in children with Duchenne's muscular dystrophy. It is now clear from the results of a number of studies (8–12) that vitamin E plays little role in the human muscular dystrophies and that vitamin E therapy is of no benefit.

The first clue that vitamin E deficiency affects the human nervous system was uncovered in patients with the very rare condition abetalipoproteinemia, in which the inability to secrete β lipoprotein results in severe malabsorption and deficiency of vitamin E from infancy. The ataxic neuropathy and retinal degeneration in this condition appeared in a pattern similar to abnormalities observed in vitamin E-deficient animals (4), leading Kayden et al. (13) to propose in 1965 that vitamin E deficiency may be the cause of neurological dysfunction in these patients. Subsequent long-term studies in patients with abetalipoproteinemia and other lipid malabsorption disorders conclusively led to the recognition that vitamin E deficiency is in a large part responsible for neurological and retinal deficits observed in patients with chronic steatorrhea (14). Moreover, the recently described inborn error in vitamin E metabolism, the isolated vitamin E deficiency syndrome in the absence of lipid malabsorption (15), and its associated neurological dysfunction, further verifies the neurological role of vitamin E in humans. In this chapter, pertinent neurochemical and neurobiological aspects of vitamin E are discussed, emphasizing recent developments in our understanding of the structure-function relationships of the α-tocopherol stereoisomers, transport of vitamin E from circulating lipoproteins to tissues, and the role of hepatic tocopherol binding protein. Clinical, electrophysiological, and histopathological features of the vitamin E deficiency neurological disorder are presented in detail, and treatment of the vitamin E deficiency in each of the underlying conditions is discussed.

NEUROCHEMISTRY AND NEUROBIOLOGY OF VITAMIN E

The four major forms of vitamin E (α-, β-, δ-, and γ-tocopherol) differ by the number and position of methyl group substitution on the chromanol ring (Fig. 1). Bioactivity is not uniform among the forms of vitamin E; the ranking of vitamin E activity is α > γ > β, δ (16). Esterification of the phenolic group on the chromanol ring with acetate, succinate, or nicotinate protects the tocopherol from oxidation; however, since the phenolic group is the active site for the antioxidant activity of vitamin E (scavenging of free radicals), hydrolysis of the ester is required before the compound is bioactive. In addition, tocopheryl esters are poorly absorbed by the intestine and must be hydrolyzed to the free tocopherol. Esters of tocopherol are uncommon in nature. Eight stereoisomers of vitamin E can be synthesized based on the rotational direction of the methyl groups at the 2-position on the chromanol ring and the 4'- and 8'-positions on the isoprenoid side chain. Only the *R-R-R* isomers (e.g., *d*-α-tocopherol) occur abundantly in nature. The *R*-rotation at the 2-position confers greater bioactivity to the molecule than the *S*-rotation. Recent studies employing the use of stable isotope technology indicate that the *R-R-R* isomer of α-tocopherol (the natural form) is preferentially taken up by the central nervous system compared to the *S-R-R* isomer and that the reverse may be the case in the liver (17). This implies that the liver may be involved in sequestering the synthetic isomers of vitamin E for degradation and removal from the body, by either metabolism or excretion in bile, thereby allowing the more bioactive *R-R-R*-α-tocopherol to be transported to the target neuromuscular tissues. The common synthetic form of vitamin E, *d,l*-α-tocopherol, more properly termed *all*-racemic α-tocopherol, is actually an equal mixture of the eight possible isomers and consequently is less bioactive than *d*-α-tocopherol.

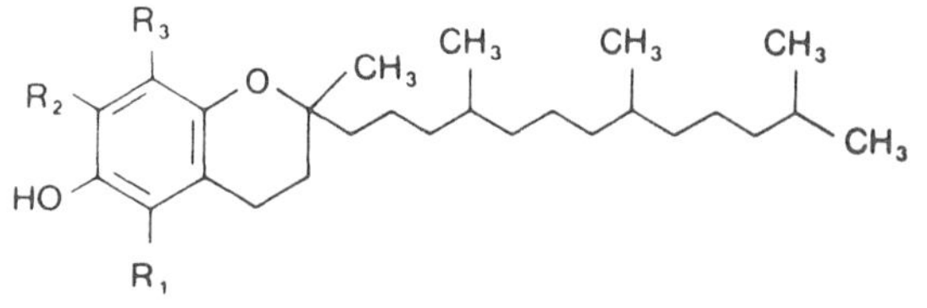

COMPOUND	R_1	R_2	R_3
α-tocopherol	CH_3	CH_3	CH_3
β-tocopherol	CH_3	H	CH_3
γ-tocopherol	H	CH_3	CH_3
δ-tocopherol	H	H	CH_3

FIGURE 1   Chemical structure of α-, β-, γ-, and δ-tocopherol.

In animals and humans, α-tocopherol is distributed almost exclusively in cellular membranes, the adipocyte fat globule, and the circulating lipoproteins. In the brain, vitamin E is localized primarily to the mitochondrial, microsomal, and synaptosomal subcellular fractions (18). An interaction between the phytyl side chain of the vitamin E molecule and polyunsaturated fatty acids present in membrane phospholipids has been suggested, thereby anchoring the vitamin E in an ideal position for free-radical scavenging by the hydroxyl group on the chromanol ring (19). This interaction has now been disputed and most likely is not operative in vivo.

The most widely accepted physiological function of vitamin E is its role as a scavenger of free radicals, preventing oxidant injury to polyunsaturated fatty acids, thiol-rich protein constituents of cellular membranes and the cytoskeleton, and nucleic acids, thus preserving the structure and functional integrity of subcellular organelles (20). Vitamin E is one component of the complex system of enzymatic and nonenzymatic antioxidants that detoxify reactive oxygen species and other free radicals. In experimental animal models of dietary vitamin E deficiency, diets rich in polyunsaturated fatty acids, the substrate for lipid perxoidation, worsen nervous system injury (21), whereas the addition of other antioxidants to the diet diminishes the neurological manifestations of vitamin E deficiency (22,23). In addition, the accumulation of breakdown products of peroxidized membrane lipids (ceroid or lipopigment) has been repeatedly observed in dorsal root ganglia, Schwann cell cytoplasm, and skeletal muscle in animal models and humans with the vitamin E deficiency neurological disorder (24–29). A primary structural role for α-tocopherol in the control of cell membrane permeability, stability, and fluidity, originally proposed by Lucy (30), is supported by studies in model membranes (31–34), however, has not been proven in isolated cellular membrane preparations.

There are several mechanisms by which alterations in membrane composition or structure caused by vitamin E deficiency may lead to neurological and muscle dysfunction and degeneration. Inasmuch as free-radical reactions are involved in the synthesis of neurotransmitters in central and peripheral nervous tissue, these tissues may be particularly susceptible to the lack of antioxidant protection during vitamin E deficiency. In the nervous system and muscle during vitamin E deficiency, peroxidation of mitochondrial membrane lipid constituents may interfere with mitochondrial energy production (35) or with ubiquinone's function in the cytochrome chain (36), the net result being an impairment in the production of ATP. Recent evidence indicates that axonal growth occurs in the growth cone, a terminal portion of the axon rich in unsaturated fatty acids. Growth cone activity is responsible for the establishment of complex connections and communications between neurons in the developing nervous system (37). It is possible that oxidation of the fatty acid components or critical proteins of the plasma membrane of the growth cone may alter the expression of membrane proteins and "secondary messengers" vital to these processes. Consistent with this proposed function of vitamin E is the observation that neurological dysfunction develops more rapidly in the more immature nervous system of young animals and human infants rendered vitamin E deficient compared to adults. Other possibly important functions of vitamin E in the nervous system include its participation in controlling brain prostaglandin and leukotriene synthesis (38) and regulating nucleic acid synthesis and gene expression (39).

Recent studies have focused on the role of impaired axonal transport in the pathogenesis of the vitamin E deficiency neuropathy. The morphological hallmark of vitamin E deficiency is swollen distal axons ("spheroids"), which contain densely packed smooth endoplasmic reticulum, mitochondria, lysosomes, and granular particles. Spheroids have been observed in other conditions in which axonal transport abnormalities have been implicated (40), the neuroaxonal dystrophies. The fast component of axonal transport carries small vesicles and organelles along microtubes in the axon in both the anterograde and retrograde directions and is dependent on the translocation protein, kinesin (40). Retrograde transport also delivers growth factors, viruses, and toxins to the cell body. Defects of fast axonal transport produce selective axonal degeneration of the "dying back" variety, the precise lesion found in vitamin E deficiency. Based on these observations, Goss-Sampson et al. (41) studied axonal fast transport of acetylcholinesterase in sciatic nerve of rats rendered vitamin E deficient by dietary restriction. After 52 weeks of vitamin E-deficient diet, a 25% reduction in anterograde accumulation and a 20% reduction in retrograde accumulation were found (Fig. 2). At earlier time points, axonal transport did not differ from that in the vitamin E-sufficient group. However, alterations in somatosensory evoked potential testing at 40–42 weeks of diet showed central but not peripheral pathway sensory conduction delays (42). Thus, abnormalities in peripheral axonal transport in this model may follow functional changes in central axonal pathways. Further studies of axonal transport and the pathobiology of the kinesin system in vitamin E deficiency may reveal the mechanism by which the lack of antioxidant leads to axonal membrane or protein alterations (presumably through oxidative modification), culminating in clinical neuromuscular degeneration.

In developed countries, vitamin E deficiency rarely occurs consequent to dietary insufficiency; thus, virtually all cases of symptomatic vitamin E deficiency have been observed in conditions in which dietary lipids and vitamin E are malabsorbed. Therefore, a review of the

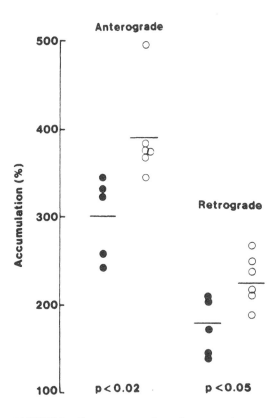

FIGURE 2   Fast anterograde and retrograde axonal transport of acetylcholinesterase in sciatic nerve of vitamin E-deficient (closed circles) and control (open circles) rats after 52 weeks of dietary therapy. Horizontal bar denotes mean. (Reproduced with permission from Muller DPR, Goss-Sampson MA, Ann N Y Acad Sci 1989; 570:146-55.)

normal absorption, transport, and metabolism of vitamin E is essential to understanding the pathogenesis of the human vitamin E deficiency state.

Dietary vitamin E is primarily in the form of α- and γ-tocopherol, of which approximately 20–40% is normally absorbed (43). Because of its hydrophobic nature, vitamin E must be solubilized into mixed micelles by bile acids secreted from the liver so that vitamin E can traverse the aqueous environment in the intestinal lumen and reach the surface of the absorptive enterocyte (Fig. 3) (44,45). Before intestinal absorption, the common esters of vitamin E are hydrolyzed by esterases either secreted by the pancreas or found in the intestinal mucosa (46). Vitamin E is absorbed into the intestinal mucosa by a nonsaturable, non–carrier-mediated, passive diffusion process (47). Recent evidence suggests that vitamin E may be absorbed within intact micelles that penetrate the mucosal cells (48). Inside the enterocyte, unesterified α- and γ-tocopherol are incorporated into chylomicrons and very low density lipoproteins (VLDL) with the other products of dietary lipid digestion and apolipoproteins, which are then transported via mesenteric lymphatics and the thoracic duct into the systemic circulation (Fig. 3). As circulating chylomicrons and VLDL undergo triglyceride hydrolysis by lipoprotein lipase and hepatic triglyceride lipase, vitamin E is transferred to some extent to tissues (49), although probably not the nervous system. The α- and γ-tocopherol remaining in chylomicron remnants are transported to the liver, the site for discrimination between α- and γ-tocopherol (Fig. 4). α-Tocopherol is resecreted as a component of hepatically derived VLDL and perhaps high-density lipoproteins (HDL) (50–52). γ-Tocopherol is not secreted substantially into hepatically derived lipoproteins but appears to be metabolized or excreted by the liver (52). The hepatic tocopherol binding protein (TBP), a 30,000–32,000 dalton cytosolic protein (53–56), may play a role in this discrimination process by preferentially binding to α-tocopherol and transferring it to endoplasmic reticulum or Golgi apparatus for incorporation into newly synthesized lipoproteins or by allowing the transfer of unbound γ-tocopherol to microsomes or lysosomes for degradation or excretion. Studies in progress using stable isotopes (deuterium) of the various tocopherols will be helpful in delineating the function of TBP and the mechanism of discrimination.

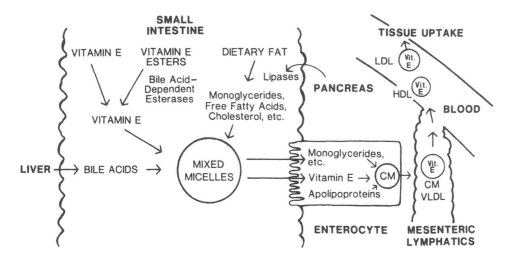

FIGURE 3 Processes involved in intestinal absorption of vitamin E in the human. (Reproduced with permission from *Clinical and Nutritional Aspects of Vitamin E*, Amsterdam: Elsevier, 1987; 169-81.)

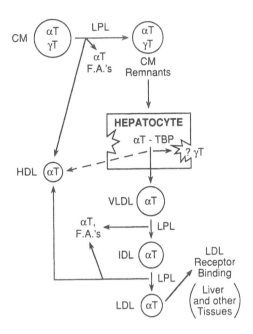

FIGURE 4 Proposed lipoprotein transport and delivery of vitamin E to liver and peripheral tissues. TBP, tocopherol binding protein; LPL, lipoprotein lipase; αT, α-tocopherol; γT, γ-tocopherol; F.A.'s, fatty acids. For other abbreviations, see text. (Reproduced with permission from Sokol RJ, Adv Pediatr 1990; 37:117-48.)

Most circulating vitamin E is found in LDL and HDL in the fasting state. The HDL α-tocopherol is derived from hepatic secretion as well as transfer from other lipoproteins. Other means of transfer of vitamin E to tissues include transfer during receptor binding (57), nonspecific binding of LDL (58) to various cells, and possibly by other uncharacterized mechanisms. The lipoprotein transport of α- and γ-tocopherol is summarized in Figure 3.

The liver functions as a rapid-turnover store of vitamin E, whereas adipose tissue may be slow to release vitamin E stored in the lipid droplet (59,60). Vitamin E is metabolized by oxidation and formation of quinones and other oxidative by-products. However, before the formation of quinones, the tocopheroxy free radical may be reduced by ascorbic acid (61–63) or by a glutathione-dependent pathway (63) back to the active tocopherol. This regeneration of α-tocopherol may help explain its rather small dietary requirement in comparison to its continuous utilization. An enterohepatic circulation of α-tocopherol was described in rats, with up to 14% of orally fed vitamin E appearing in bile as α-tocopherol and metabolites, whereas less than 1% is eliminated in the urine (64,65). The majority of excreted vitamin E appears in the stool either as nonabsorbed dietary vitamin E or as inactive biliary metabolites (66).

The precise mechanism by which vitamin E is transferred from circulating lipoproteins to the brain, spinal cord, peripheral nerves, and muscle and the factors regulating this transport have not been elucidated. Because of the development of neurological symptoms attributed to vitamin E deficiency in two conditions in which chylomicron transport of vitamin E appears to be normal but secretion of tocopherol into hepatically derived VLDL is greatly diminished [normotriglyceridemic abetalipoproteinemia (67) and isolated vitamin E deficiency syndrome (15)], it is suggested that the primary route of transfer of vitamin E to neuromuscular tissues is from hepatically derived lipoproteins (VLDL hydrolysis or transfer from LDL or HDL).

Compared to other tissues, the nervous system maintains higher vitamin E levels when vitamin E deficiency is produced by dietary exclusion in experimental rats (Fig. 5) (41). This suggests that the nervous system conserves vitamin E to a greater extent than other tissues during conditions that produce a deficiency state. After 52 weeks of deficiency in weanling rats, levels in the nervous system are less than 10% of control rats. The distribution of vitamin E in the central and peripheral nervous system in the vitamin E-sufficient rat is nonuniform (68). The medulla, spinal cord, and cerebellum, three sites that are affected during vitamin E deficiency, have significantly lower vitamin E content in young animals compared to other brain regions and compared with older animals (Fig. 6) (41,68,69). Surprisingly, intravenously injected radiolabeled α-tocopherol is taken up faster by the cerebellum compared to other nervous system tissues, suggesting that the cerebellum is particularly active in the metabolism or utilization of vitamin E (69). Thus, the susceptibility of certain brain regions to injury during vitamin E deficiency may relate to lower basal levels of vitamin E, higher utilization

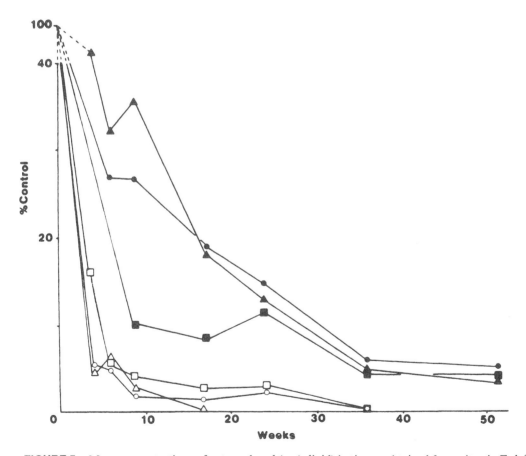

FIGURE 5 Mean concentrations of α-tocopherol (μg/g lipid) in tissues obtained from vitamin E-deficient rats from weanling to age 52 weeks; values expressed as a percentage of control values: (closed triangles) brain; (closed circles) spinal cord; (closed squares) nerve; (open triangles) serum; (open circles) liver; (open squares) adipose tissue. (Reproduced with permission from Goss-Sampson MA, et al. J Neurol Sci 1988; 87:25-35.)

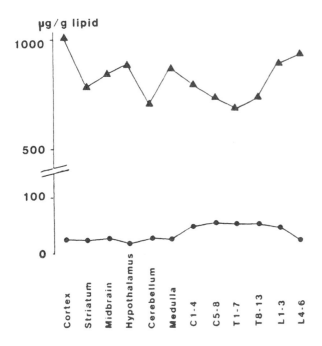

FIGURE 6 Mean concentrations of α-tocopherol content in regions of brain and spinal cord of control (triangles) and vitamin E-deficient (circles) rats after 55 weeks of dietary therapy. C, cervical; T, thoracic; L, lumbar regions of spinal cord. (Reproduced with permission from Muller DPR, Goss-Sampson MA, Ann N Y Acad Sci 1989; 570:146-55.)

rates of vitamin E, or differences in free-radical generation. In this regard, a recent report showed increased formation of oxygen free radicals in brain homogenate during vitamin E deficiency (70). Additional factors determining predisposition to injury may include local differences in sequestration of free radicals, concentrations of peroxidizable unsaturated fatty acids, and levels of other pro- or antioxidants. Interestingly, peripheral nerve, which also degenerates during vitamin E deficiency, actually contains more vitamin E than brain (69).

ETIOLOGY OF VITAMIN E DEFICIENCY

Vitamin E deficiency rarely occurs on a dietary basis in developed and most developing countries in older infants, children, and adults because of the ubiquitous distribution of tocopherol in grains, vegetable oils, and animal fat (71,72). Consequently, the most common cause of vitamin E deficiency is malabsorption of dietary vitamin E due to underlying gastrointestinal, pancreatic, and hepatic disorders that interfere with the digestion or absorption of lipid. Indeed, low serum concentrations of vitamin E have been observed in patients with a variety of steatorrheic disorders (73); an inverse relationship was evident between the serum vitamin E concentration and the severity of impairment of fat absorption. The major exception is the inborn error of metabolism, the isolated vitamin E deficiency syndrome, which leads to symptomatic vitamin E deficiency despite normal intestinal absorption of vitamin E (15). The other chronic conditions most commonly associated with a symptomatic vitamin E deficiency state include abetalipoproteinemia and other disorders of β lipoprotein secretion, chronic cholestatic hepatobiliary diseases, cystic fibrosis and other causes of pancreatic insufficiency, and short bowel syndrome.

Abetalipoproteinemia and Other Defects of β Lipoprotein Secretion

Abetalipoproteinemia is a rare autosomal recessively inherited defect of lipoprotein production and transport characterized by the failure of intestinal and hepatic secretion of chylomicrons and VLDL. The cause of abetalipoproteinemia is a failure to synthesize normal apolipoprotein B, a factor essential for the secretion of β lipoproteins. Thus, steatorrhea is present from birth because of the inability of the intestinal mucosal cell to transport absorbed dietary fat into the mesenteric lymphatic circulation (Fig. 2). The characteristic clinical features are steatorrhea, progressive ataxia, pigmented retinopathy, and acanthocytosis of red blood cells (74). The prolonged and severe steatorrhea and the absence of plasma β lipoproteins, the primary carriers for circulating vitamin E, combine to produce serum vitamin E levels that are almost undetectable and adipose tissue stores of tocopherol that are extremely low (75). The characteristic neurological symptoms begin in the first decade of life and evolve into a crippling ataxia with visual impairment by the second or third decades. The first clinical clue that vitamin E may play a role in human neurological function was the relationship between very low serum vitamin E concentrations and the progressive ataxia of abetalipoproteinemia (13). Long-term treatment trials with massive supplements of vitamin E have conclusively demonstrated that neurological function can be preserved if vitamin E deficiency is corrected at an early age (76–79).

Neurological findings identical to those observed in abetalipoproteinemia have been described in homozygous hypobetalipoproteinemia patients who also fail to synthesize adequate apolipoprotein B and consequently malabsorb lipid and vitamin E (80). This disorder is autosomal dominant; however, when the child of two heterozygotes receives both defective genes, the homozygous state mimics abetalipoproteinemia. Cases of apparent heterozygous hypobetalipoproteinemia have also been reported as suffering from vitamin E deficiency neuropathy and ataxia (81). Finally, other disorders of intestinal secretion of chylomicrons in which apolipoprotein B synthesis is intact but some other chylomicron-processing defect is present, such as chylomicron retention disease (Anderson's disease) (82,83), are associated with vitamin E malabsorption and deficiency and neurological degeneration.

Chronic Cholestatic Hepatobiliary Disorders

In 1981, a degenerative neurological disorder was linked to vitamin E deficiency in children with chronic cholestatic hepatobiliary disorders (24), including diseases of the liver (idiopathic neonatal hepatitis and familial cholestatic syndromes), of the intrahepatic bile ducts (Alagille syndrome and paucity of interlobular bile ducts), and of the extrahepatic bile duct (extrahepatic biliary atresia) (45). The reduction in bile flow from the liver to the intestine in cholestasis leads to diminished concentrations of bile acids in the intestinal lumen, causing failure of micellar solubilization and malabsorption of dietary lipids (45). Profound malabsorption of vitamin E in children with persisting, severe cholestasis results in biochemical (low serum vitamin E concentrations and ratio of serum E to total serum lipid concentration) and functional (elevated hydrogen peroxide hemolysis) evidence of vitamin E deficiency, as well as depleted tissue stores (45). Of children with these disorders, 50–75% show biochemical evidence of deficiency despite "routine" supplemental doses of vitamin E (84–87). Neurological abnormalities, which develop as early as the second year of life (84,85,88), may lead to irreversible, devastating consequences if the vitamin E deficiency is uncorrected into the second decade of life (24,45,89). Neurological symptoms occur much more commonly in vitamin E-deficient children with chronic cholestasis compared to those with cystic fibrosis

(CF). Biochemical evidence of vitamin E deficiency has also been described in approximately 20–50% of adults with chronic cholestasis due to primary biliary cirrhosis (PBC) (90–92), although a direct relationship to neurological symptoms was observed in only one patient (93). However, the majority of patients with PBC have severe malabsorption of vitamin E even when serum indices indicate vitamin E sufficiency (94).

Cystic Fibrosis

Cystic fibrosis is the most common lethal autosomal recessive disease among whites. In CF, increased viscosity of pancreatic secretions causes obstruction of pancreatic ducts, leading ultimately to destruction and fibrosis of the exocrine pancreas in 85% of patients. The resulting failure of secretion of pancreatic digestive enzymes causes steatorrhea and vitamin E malabsorption, even to some extent when pancreatic enzyme supplements are administered orally (95). In addition, a diminished bile acid pool size caused by increased fecal excretion of bile acids in CF may contribute to the impairment in vitamin E absorption. Although low serum vitamin E concentrations are very common in unsupplemented CF patients, overt neurological disease has been thought to be rare and to develop primarily in association with liver involvement by CF (96–98). It is not known if the liver disease renders the vitamin E deficiency more severe, results in malabsorption of another contributing nutrient, or imposes an oxidant stress that hastens the development of neurological dysfunction. Recent prospective evaluations of neurological function in CF patients identified a higher frequency of clinical (99), electrophysiological (99,100), and neurological abnormalities in vitamin E-deficient patients than previously reported. In addition, vitamin E deficiency causing neuropathy has been reported in a 4-year-old child with pancreatic insufficiency and sensorineural deafness [Johanson-Blizzard syndrome (101)] who responded to intramuscular injections of vitamin E (102).

Other Fat Malabsorption Disorders

In adults, chronic dysfunction or resection of the small bowel are the more common causes of acquired vitamin E deficiency (103–111). Lengthy intestinal resections (creating a short bowel syndrome) for treatment of Crohn's disease (103,104), mesenteric vascular thrombosis (104), or intestinal pseudoobstruction (103) have been associated with vitamin E deficiency and neurological degeneration. Chronic malabsorption of dietary vitamin E may not be overcome in these conditions unless adequate vitamin E supplementation is provided intravenously in the patient's parenteral nutrition infusates or as an oral supplement. Other medical conditions in adults causing chronic steatorrhea have led to vitamin E deficiency, including the blind loop syndrome (106), intestinal lymphangiectasia (107), celiac disease (108), and chronic pancreatitis leading to pancreatic insufficiency (109–111). In these diseases, vitamin E deficiency is also frequently associated with the deposition of lipofuscin pigment in smooth muscle cells of the digestive tract (brown bowel syndrome and intestinal ceroidosis) (109,112,113), presumably due to accumulation of peroxidized membrane lipid. Recent reports suggest that this pigment may actually impair smooth muscle function (114), possibly causing a mitochondrial myopathy (115,116) that may culminate in fatal intestinal pseudoobstruction (114). Serum vitamin E levels may fall within 1–2 years of acquired lipid malabsorption in adolescents and adults; however, a 10–20 year interval (as opposed to 1–2 years in young children) between the identification of biochemical vitamin E deficiency and the onset of neurological and ophthalmological symptomatology is generally observed in adults. This difference in rapidity of development of neurological injury is most likely related to the increased susceptibility of the young developing nervous system, the time required to deplete nervous tissue stores of vitamin E in the adult, and age-related differences in other antioxidant protective mechanisms.

Neurological symptoms occur in approximately 10% of cases of adult celiac disease (gluten-sensitive enteropathy and nontropical sprue) (117). Although biochemical evidence of vitamin E deficiency is commonly seen at diagnosis, its role in the neurological manifestations of celiac disease has been questioned (118,119), although several cases appear to exhibit the classic findings of vitamin E deficiency neuropathy (108).

Isolated Vitamin E Deficiency Syndrome

Over the past decade, a unique disorder causing a primary, isolated deficiency of vitamin E in the absence of lipid malabsorption has been described in nine patients (15,25,120–123). In most patients the onset of neurological deterioration (generally ataxia, head titubation, and clumsiness) occurred in the first decade of life; one patient was well until age 52 years (Table 1). Symptoms of gastrointestinal, liver, pancreatic, or lipoprotein disorders were absent, and thorough evaluation of these systems in all patients revealed normal function (15). Vitamin E deficiency was documented by very low serum vitamin E concentrations in all patients, and low serum vitamin E to total lipid ratios or vitamin E concentrations in nerve, muscle, or adipose tissue in those patients examined (15). The neurological features resembled those observed in cases of vitamin E deficiency secondary to lipid malabsorption; however, impairment of eye movements and retinal degeneration were very unusual (Table 1). Instead, decreased vibratory sensation, ataxia, head titubation, and muscle weakness predominated in the isolated vitamin E deficiency cases. These differences in symptomatology may be explained by adequate absorption of dietary lipid in this disorder in contrast to the lipid malabsorption diseases, resulting in differing tissue compositions of polyunsaturated fatty acids, the substrate for lipid peroxidation. Alternatively, another nutrient (e.g., retinol) that may interact with vitamin E deficiency in altering neurological function may become deficient in the malabsorption diseases. Available clinical, neurological, and treatment response data for the nine reported cases is summarized in Table 1.

The precise cause of this disorder is unclear; however, it appears to be autosomal recessive in inheritance in most cases. Indeed, consanguinity was present in the parents of 2 cases (25,122), and 3 affected siblings with normal parents were found in another family (15). Thus, an inborn error in vitamin E metabolism appears evident. Harding et al. (120) postulated that the biochemical defect responsible for this disorder was a selective impairment of intestinal absorption of vitamin E. In contrast, our investigations using deuterated α-tocopherol (a stable isotope) in physiological doses have demonstrated normal intestinal absorption and chylomicron transport of ingested vitamin E, but defective hepatic resecretion of absorbed tocopherol into hepatically derived lipoproteins (Fig. 6) (124). When given 15–20 mg deuterated α-tocopherol orally and deuterated tocopherol levels in plasma lipoprotein fractions were obtained serially, compared to normal controls the patients with this disorder showed similar levels of tocopherol in chylomicrons but lower levels in VLDL, LDL, and HDL. Thus, the absorption phase of vitamin E appeared to be intact; however, transport of absorbed vitamin E was impaired. The most logical explanation for these findings is a structurally or functionally abnormal hepatic TBP in the patients (15). Serum vitamin E concentrations normalize and neurological function improves in these patients in response to large oral doses of vitamin E, suggesting that α-tocopherol is transferred from circulating chylomicrons to other lipoproteins during chylomicron catabolism by lipoprotein lipase and is then transported to nervous system and muscle tissues by LDL or HDL. Whatever the etiology, this disorder may be more common than initially appreciated, obligating investigation of vitamin E status in patients with movement disorders, ataxia, or peripheral neuropathy, even in the absence of lipid malabsorption and with onset of symptoms at any age. Two surveys of patients with cerebellar ataxia (125,126) and one of patients with Friedreich's ataxia (127) failed

TABLE 1 Features of Nine Reported Cases of Isolated Vitamin E Deficiency Syndrome[a]

Reference	Sex	Family history	Age (years)		Serum vitamin E level (μg/ml)	Serum vitamin E to total lipid ratio (mg/g)	Neurological features	Histology		Treatment dose of vitamin E	Response
			At onset of neurological symptoms	At evaluation				Nerve lesion	Muscle lesion		
25	M	+	3	12	1.0	ND	Hypotonia, limb and truncal ataxia, areflexia, ↓ position and vibration	+	+	1500 mg/day	+
122	M	–	5	10	1.0	ND	Weakness, limb and truncal ataxia, areflexia, ↓ position and vibration, scoliosis	+	+	?	+
120	F	–	13	23	<1.0	ND	Hypotonia, weakness, limb and truncal ataxia, head titubation, areflexia, ↓ position and vibration	ND	ND	800–2000 mg/day	+
121	M	–	7	19	0.7	ND	Dysarthria, drooling, ophthalmoplegia, dystonia, scoliosis, pes cavus, areflexia, ↓ position and vibration, bedridden, truncal and limb ataxia previously	ND	+	1800 IU/day	+

26	M	−	52	62	1.1	ND	Dysarthria, limb and truncal ataxia, areflexia, ↓ pain and light touch, ↓ position and vibration	+	+	800 IU/day	+
15											
Patient 1	F	+	6	23	1.8	0.35	Limb and truncal ataxia, head titubation, ↓ vibration, ↓ vibration, pes cavus	−	+	800 IU/day	+
Patient 2	M	+	26	27	1.8	0.38	Intention tremor, ↓ vibration	ND	ND	800 IU/day	−
Patient 3	F	+	−	21	1.2	0.22	↓ vibration and position	ND	ND	800 IU/day	−
Patient 4	F	−	3	30	1.0	0.13	Dysarthria, weakness, areflexia, limb and truncal ataxia, head titubation, ↓ vibration and position, pes cavus, scoliosis, requires ambulatory assistance	−	±	800 IU/day	+

aND, not done; M, male; F, female; +, present; −, absent.

to demonstrate low serum vitamin E levels in 95 patients. Larger prospective evaluation of these types of disorders is needed to determine the frequency of isolated vitamin E deficiency.

NEUROLOGICAL DEFICITS IN VITAMIN E DEFICIENCY

For several decades, vitamin E deficiency has been known to cause neurological and muscular degeneration in a variety of vertebrate animal species. When raised on vitamin E-deficient diets from early in life, encephalomalacia and ataxia develop in growing chicks (4), ataxia and neuroaxonal degeneration in the brain stem, spinal cord, and peripheral nerves and pigmentary degeneration of the retina (128,129) develop in rats (130) and rhesus monkeys (131), and several species show clinical and histological evidence of skeletal and cardiac myopathy (4). These vitamin E deficiency-related disorders invariably lead to premature death of the affected animals. The relationship of vitamin E deficiency and neuromuscular abnormalities in man went unnoticed until the last 10–15 years; however, because of advances in the medical and surgical management of lipid malabsorption disorders, patients with conditions leading to prolonged malabsorption of vitamin E now survive for decades, allowing symptoms of vitamin E deficiency to become clinically evident. There is now a large body of compelling evidence proving that vitamin E deficiency causes human neuromuscular degeneration, including clinical, histological, and therapeutic response observations (132). Thus, the neurological role of vitamin E in humans is firmly established.

Clinical Features

The primary manifestations of a prolonged vitamin E deficiency include spinocerebellar ataxia, skeletal myopathy, and pigmented retinopathy (Fig. 7) (132). The development of the neurological syndrome is heralded by the loss of deep tendon reflexes, except in several reported cases of the isolated vitamin E deficiency syndrome in whom reflexes remained normal (15). Other common symptoms include truncal and limb ataxia manifested by a wide-based gait, positive Romberg sign, past-pointing on finger-to-nose examination, and intention tremor; markedly diminished proprioception and vibratory sensation; impairment of extraocular muscle function (ophthalmoplegia), particularly in upward eye movement; proximal muscle

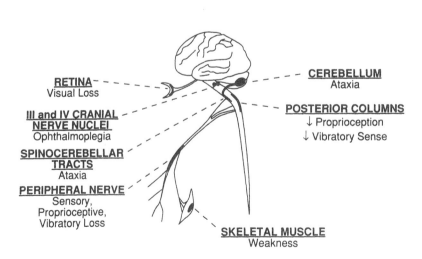

FIGURE 7 Neuromuscular lesions and associated clinical features in vitamin E deficiency.

weakness; ptosis; dysarthria; visual field loss; and in advanced cases, dystonic muscle contractions, shortened Achilles tendon, scoliosis, and pes cavus (95). These symptoms correspond to the underlying histopathological lesions (Fig. 7). Deterioration of visual function may also occur as a result of pigmented retinopathy (76,104), particularly in patients with abetalipoproteinemia. Not all patients show this entire spectrum of clinical findings. In fact, in adults biochemical deficiency may be detected yet neurological function may remain normal for years (104,111). It is now evident that distinct but overlapping patterns of neurological abnormalities correlate with the underlying cause of the vitamin E deficiency state (Table 2). For example, ophthalmoplegia and pigmented retinopathy are unusual in the isolated vitamin E deficiency syndrome, yet loss of vibratory sensation is marked and deep tendon reflexes may be normal. These differences, based on the cause of the vitamin E deficiency, may be due to concomitant deficiency of another malabsorbed nutrient, the effects of undernutrition, increased oxidative stress in certain conditions, or abnormalities of plasma transport and tissue uptake of vitamin E in addition to absorptive defects.

A distinct pattern of progression of neurological symptoms has been observed in children with vitamin E deficiency caused by chronic cholestatic hepatobiliary diseases (85). As illustrated in Figure 8, which shows the percentage of vitamin E-deficient children exhibiting each of five neurological signs at 2 year intervals, hypoflexia appears first at age 18–24 months. By age 3–4 years, the majority of vitamin E-deficient children show evidence of neurological dysfunction in several areas. Before the end of the first decade of life many untreated patients have a disabling combination of spinocerebellar ataxia, neuropathy, and ophthalmoplegia. Many children require ambulatory assistance devices or are even wheelchair bound (88,89, 133). Interestingly, the progression of the neurological symptoms appears to be slower in children with cystic fibrosis and abetalipoproteinemia, who presumably are also deficient in vitamin E from infancy. This suggests that other factors may be associated with chronic cholestatic liver disease, such as other antioxidant deficiencies or stimulation of oxidant stress by other mechanisms, which may accelerate the injury and dysfunction of the neuromuscular system. In adults, the latent period between onset of malabsorption and development of neurological symptoms may be as long as 10–20 years.

Accumulating evidence suggests that vitamin E deficiency may also have a direct effect on psychomotor, intellectual, and behavioral function. Lal et al. (134) demonstrated learning and memory deficits in rats maintained on a vitamin E-deficient diet for 14 months commencing

TABLE 2 Clinical Features of Vitamin E Deficiency Disorders[a]

	β Lipoprotein disorders	Chronic cholestasis	Other fat malabsorption disorders	Isolated vitamin E deficiency
Hyporeflexia or areflexia	+ +	+ +	+ +	±
Cerebellar ataxia	+ +	+ +	+ +	+ +
Loss of position sensation	+ +	+ +	+	±
Loss of vibratory sensation	+ +	+ +	+ +	+ +
Loss of touch or pain sensation	+	±	+	—
Ophthalmoplegia	+	+	+	—
Ptosis	+	+	±	—
Muscle weakness	+	+	+	+
Pigmented retinopathy	+ +	±	+	—
Dysarthria	+	±	+	±

[a]+ +, Always present; +, frequently present; ±, present inconsistently; —, absent.

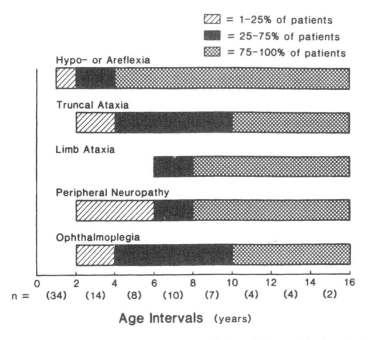

FIGURE 8   Percentage of vitamin E-deficient children with chronic cholestasis who show evidence of neurological abnormalities at 2 year age intervals. (Reproduced with permission from Sokol RJ, Guggenheim MA, Heubi JE, et al. Am J Dis Child 1985; 139:1211-5.)

in infancy. Sater and van der Linde (135) also demonstrated significant impairment in relearning ability of vitamin E-deprived rats. Satel and Riely (136) suggested that the high incidence of behavioral and personality disorders in their patients with chronic cholestatic liver disease may be similarly caused by vitamin E deficiency. Arria et al. (137) showed that vitamin E deficiency correlated with psychomotor impairment on a battery of neuropsychological tests in adults with advanced PBC who were being evaluated for liver transplantation. In addition, vitamin E deficiency appeared to correlate with cognitive and intellectual deficits in children with extrahepatic biliary atresia who were undergoing evaluation for liver transplantation (138,139). Despite the demonstration of significant correlations in these human studies, the possibilities of other confounding variables (such as the effect of prolonged hospitalizations on the development of a child) leaves the issue of vitamin E and psychobehavioral function in the human unresolved and deserving of further study.

Histopathology and Electrophysiology of Vitamin E Deficiency

Vitamin E deficiency in the rat, rhesus monkey, and human causes similar neuropathological lesions. Swollen, dystrophic axons (spheroids), the hallmark of vitamin E deficiency and of the aging process, have been observed in the gracile and cuneate nuclei of the brain stem in rats and monkeys and in humans with chronic cholestasis (24,140), cystic fibrosis (27), and short bowel syndrome (105). Interestingly, axonal spheroids are much less common in patients with abetalipoproteinemia, perhaps due to the limited amount of dietary fat that is absorbed or ingested, as reported in the dog (141). The primary axonapathy in vitamin E deficiency is one of axonal degeneration with secondary demyelination. Axonal dystrophy has been observed in the posterior columns of the spinal cord, Clarke's column, and dorsal and ventral

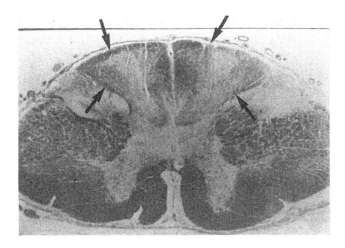

(a)

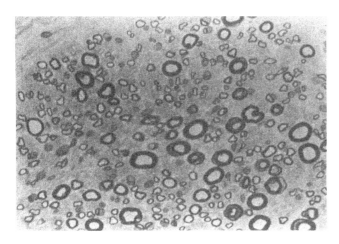

(b)

FIGURE 9 Central and peripheral nervous system lesions in human vitamin E deficiency. (a) Cervical spinal cord of 8-year-old boy with biliary atresia and vitamin E deficiency with a progressive neurological syndrome. Arrows denote loss of axons and demyelination from posterior columns. (Reproduced with permission from Nelson JS. In: *Biology of vitamin E*, London: Pitman, 1983; 92-105.) (b) Cross section of sural nerve biopsy removed from 14-year-old boy with abetalipoproteinemia, showing a decreased number of large-caliber myelinated axons. (c) Cross section of sural nerve biopsy obtained from 11-month-old girl with vitamin E deficiency due to chronic intrahepatic cholestasis, showing swollen large-caliber meylinated axons with attenuation of myelin sheaths (arrowheads). (d) Electron micrograph of swollen, large-caliber myelinated fiber from sural nerve of 25-month-old girl with vitamin E deficiency caused by arteriohepatic dysplasia. Neurofibrillar density is increased, ruffling and partial loss of myelin sheath are present, and axonal clefts are continuous with axolemma. ($\times$15,000.) (Reproduced with permission from Sokol RJ. Free Radic Biol Med 1989; 6:189-207.)

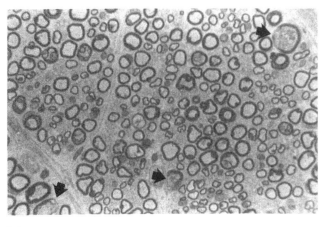

(c)

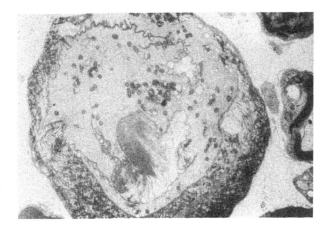

(d)

FIGURE 9

spinocerebellar tracts (Fig. 9a) (24). Mild atrophy of the cerebellar hemispheres and loss of nerve cell bodies in the nuclei of the third and fourth cranial nerves have also been observed at autopsy in children with chronic cholestasis and vitamin E deficiency (24). This constellation of lesions correlates excellently with the clinical manifestations. Loss of predominantly large-caliber, myelinated axons in peripheral sensory nerve has been an almost universal finding in patients with advanced vitamin E deficiency caused by cholestasis (24,45,142), abetalipoproteinemia (143), radiation enteritis (105), or the isolated vitamin E deficiency syndrome (Fig. 9b) (25). Swelling and ultrastructural evidence of axonal degeneration in large-caliber myelinated fibers has been observed before axonal loss in young children with cholestasis (Fig. 9c and d) (88).

Another characteristic finding in humans, rats, and monkeys has been the accumulation of lipofuscin in dorsal sensory neurons and peripheral nerve Schwann cell cytoplasm (24–27,29). Lipofuscin accumulation and axonal dystrophy are commonly observed after the third decade of life in normal people. Therefore, parallels have been drawn between vitamin E deficiency and the aging process, perhaps supporting the notion that vitamin E supplementation in normal people might retard aging. It remains unclear whether the deposition of lipofuscin

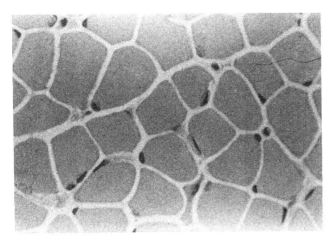

(a)

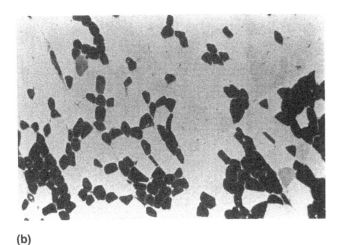

(b)

FIGURE 10 Muscle lesions in patients with vitamin E deficiency. (a) Variation in size of muscle fibers on cross section in 12-year-old child with vitamin E deficiency caused by chronic cholestasis. (b) Fiber type grouping of muscle fibers demonstrated on ATPase histochemical staining of muscle biopsy. (c) Presence of lipofuscin (arrowheads) in longitudinal sections of muscle biopsy from 14-year-old child with vitamin E deficiency and chronic cholestasis. (d) Acid phosphatase histochemical stain of muscle biopsy demonstrating small inclusions (arrows) scattered throughout numerous muscle fibers. (e) Electron micrograph of muscle demonstrating electron-dense, membrane-bound inclusion body lying between myofibrils from 12-year-old girl with vitamin E deficiency. (Reproduced with permission from Sokol RJ. Annu Rev Nutr 1988; 8:315-73; Free Radic Biol Med 1989; 6:189-207.)

in neurons and Schwann cells impairs function or merely is a marker for the presence of previous peroxidation of membrane lipids.

Two types of vitamin E deficiency-related muscle lesions have been described in patients with cholestasis (45,133,144,145), abetalipoproteinemia (146) or the isolated vitamin E deficiency syndrome (Fig. 10) (25,26,121,147). A neuropathic lesion, most likely representing a denervation-renervation process, consists of a variation in muscle fiber size and grouping of

(c)

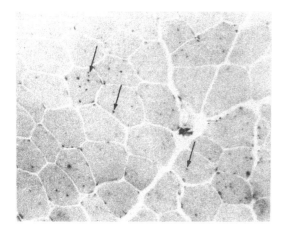

(d)

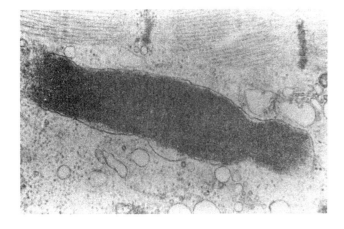

(e)

FIGURE 10

type I and type II fibers (detected by NADH or ATPase staining), without evidence of necrosis, inflammation, or fibrosis (Fig. 10a and b). This is in contrast to most other animal species, in which muscle necrosis and fibrosis predominate (5). The second type of lesion, a myopathic lesion, represents more intrinsic injury to the muscle fibers. Lipofuscin pigment can be observed on standard stains of muscle (Fig. 10c). In addition, small, grainlike basophilic deposits are scattered throughout many muscle fibers, which autofluoresce and stain positively for esterase and acid phosphatase on frozen section (144), suggesting a lysosomal nature (Fig. 10d). These deposits appear on electron microscopy to consist of densely osmiophilic membrane-bound cytosomes lying between myofibrils (Fig. 10e). It has been suggested that these ultrastructural deposits are secondary lysosomes containing lipofuscin pigment (25,26,123, 133,144). In addition, areas of Z-band streaming and autophagocytic vacuoles containing debris are also present. These latter lesions resemble those seen in vitamin E-deficient rats (148). It is uncertain which of these lesions is responsible for muscle weakness in advanced cases; however, the denervation injury may be the most important.

Pigmented retinopathy and photoreceptor degeneration have been observed in vitamin E-deficient rats and monkeys (128,129) and is made worse by concomitant vitamin A deficiency. The rich content of unsaturated fatty acids in the outer segment of the retinal rod predisposes this membrane to peroxidative injury and the formation of lipopigments. Older patients with abetalipoproteinemia (76) and some children with chronic cholestasis (149) develop this irreversible lesion, leading to significant visual impairment. As in animals, the synergistic role of vitamin A deficiency has been suggested in humans (149).

Thus, comparative neuropathological studies in humans and in experimental animals have demonstrated similar target organs (nervous system, muscle, and ocular retina) for vitamin E deficiency-induced injury across species. Unfortunately, determining the frequency and rate of progression of neuromuscular and ocular lesions in humans has been hampered by the limited amount of biopsy and autopsy tissue available for study. However, the development of sensitive electrophysiological techniques has made possible the noninvasive detection of dysfunction of the affected central and peripheral nerve pathways, muscle, and retina. Peripheral nerve conduction velocity testing has shown that diminished amplitudes of sensory nerve action potentials are common, whereas delayed conduction velocity (an indicator of demyelination) is unusual (103,123,142,149–151). These findings confirm that axonal degeneration rather than demyelination is the primary sensory nerve abnormality. Electromyographic studies reveal evidence of denervation injury of muscles (spontaneous fibrillation and positive sharp waves) in patients who have advanced vitamin E deficiency, consistent with the neuropathic muscle lesion. Similar findings have been demonstrated in the vitamin E-deficient rat (42). Somatosensory-evoked potential (SEP) testing quantitates the speed by which a peripheral sensory stimulus is conducted through peripheral nerve, spinal cord, and brain stem en route to the cerebral cortex, allowing noninvasive evaluation of these sensory pathways. A central delay in sensory conduction has been the most consistent finding in patients with vitamin E deficiency (26,103,105,150,151), as well as experimental rats (42), correlating with degeneration of the posterior columns in the spinal cord. Visually evoked potential testing and electroretinography have similarly revealed abnormalities in retinal function in vitamin E-deficient patients with chronic cholestasis (149,152), cystic fibrosis (153), short bowel syndrome (103,154), and isolated vitamin E deficiency syndrome (121). Electrophysiological studies not only allow sensitive detection of abnormalities but may also be used to monitor the progression or regression of neurological deficits during treatment.

ASSESSMENT OF VITAMIN E STATUS

Current means of evaluating vitamin E status can be divided into four categories: measuring vitamin E concentrations (and lipid concentrations), detecting the functional effect of vitamin

E deficiency (hemolysis, lipid peroxidation, or expired ethane), detecting clinical signs of deficiency, or obtaining histological documentation of lesions characteristic of vitamin E deficiency (155). Vitamin E can be measured accurately in plasma or serum by fluorometric (156) and high-performance liquid chromotography (HPLC) methods (157) and in red blood cells, adipose tissue, and other tissues by HPLC (75,157). The HPLC methodology, allowing separation and quantitation of α- and γ-tocopherol, is available in reference laboratories and is the method of choice. In the presence of normal circulating concentrations of triglycerides, cholesterol, and phospholipids, the sum of which equals the total serum lipids, a serum vitamin E concentration below the normal range reliably indicates vitamin E deficiency. Total serum lipids can conveniently be measured on a small volume of serum by a colorometric technique (158,159) available as a commercial kit (Total Lipids; Bio-Dynamics/bmc, Indianapolis, IN). The normal lower limit of serum vitamin E for children over age 12 years and adults is 5 μg/ml (160,161); for children aged 6 months to 12 years, the lower limit is somewhat lower, approximately 3–4 μg/ml (160), and for newborns and young infants normal levels may be even lower because of the low levels of circulating lipoproteins. Many authors consider the adult normal range of 5–20 μg/ml as the therapeutic goal when correcting vitamin E deficiency, even in infants. Since virtually all circulating vitamin E appears in the plasma lipoproteins, an elevation or lowering of serum lipoprotein levels causes a similar shift in serum vitamin E concentrations (161), without an actual alteration of vitamin E status. This is possibly as a result of partitioning of vitamin E between cellular membranes and the plasma lipid compartment (162). This effect is particularly important in chronic cholestatic liver disease in which the common finding of hyperlipoproteinemia (163) may elevate serum vitamin E concentrations into the normal range despite a deficiency state (89). For this reason, the ratio of serum vitamin E to total serum lipid concentration has been proposed to correct for hyperlipidemia and provide a better approximation of vitamin E status (89,161). The threshold for clinical deficiency in older children and adults is approximately below 0.8 mg total tocopherol per gram total lipid, and perhaps below 0.6 mg/g for children age 1–12 years (160,161). Although this ratio is particularly useful in states of hyperlipidemia, its role in states of hypolipidemia (e.g., abetalipoproteinemia) has not been established. It is clear, however, that lower serum lipid values are associated with diminished α-tocopherol levels in states of vitamin E sufficiency.

The measurement of vitamin E in tissues can be useful to confirm deficient vitamin E stores; however, it is impractical for everyday use. Adipose tissue levels have been a particularly popular index for estimating vitamin E status in patients with fat malabsorption (75, 80), especially in abetalipoproteinemia, in which serum tocopherol levels do not reflect nutritional status (75).

A lack of vitamin E may result in lipid peroxidation and destabilization of cell membranes upon exposure to an oxidant stress. The red blood cell hydrogen peroxide hemolysis test (164) and the red blood cell malondialdehyde (MDA) release test (165) take advantage of these membrane properties. Hemoglobin or MDA is released when vitamin E-deficient red blood cells lyse after exposure to hydrogen peroxide. There is an excellent correlation between hydrogen peroxide hemolysis testing in experienced laboratories and serum vitamin E concentrations; however, the relationship has not been studied during states of hyperlipoproteinemia as may occur during cholestasis. Other factors that influence tissue lipid peroxidation, such as selenium and polyunsaturated fatty acid status, may influence the results of these functional tests and should be taken into account when interpreting the results.

Serum lipid peroxide values have been measured (166) but may be difficult to interpret during hypo- or hyperlipidemia (167). Recent revisions in the methodology for measuring plasma MDA levels suggest that previous methods were inaccurate (168). Expired pentane or ethane gas, a sensitive measure of "total-body" peroxidation of polyunsaturated fatty acids, can be measured as a quantification of antioxidant deficiency (169,170). The previously tedious technology is undergoing modification and may soon be reasonable for clinical use (171), especially in patients with chylomicron secretion disorders who currently require adipose biopsies to assess vitamin E status (75).

Neurological evaluation of the patient and histological examination of nerve, muscle, and other tissues may be useful as well. The progressive nature of the clinical signs of vitamin E deficiency described previously are helpful in determining whether the vitamin E deficiency suggested by biochemical indices has led to tissue injury. Neurological injury undoubtedly occurs before clinical detection; however, it may be reversible at these subclinical early stages (88). Characteristic, but not necessarily unique, histological, histochemical, and ultrastructural lesions of sural nerve and muscle may be useful in confirming vitamin E deficiency. However, although these tissues may be relatively accessible, biopsies should be reserved for research purposes, for patients in whom other neurological diagnoses are considered, or when the role of vitamin E deficiency in the context of the neurological symptoms is in doubt.

TREATMENT OF VITAMIN E DEFICIENCY

Recognition of vitamin E deficiency in a patient with neurological disease is important because of the potential reversal, stabilization, or prevention of neurological deterioration upon correcting the deficiency. For this reason, a high index of suspicion is obligatory when evaluating patients with movement disorders, ataxia, neuropathy, or retinopathy. Because of the existence of the isolated vitamin E deficiency syndrome, even patients without historical evidence of gastrointestinal disease need to be considered as potentially vitamin E deficient and evaluated appropriately with a serum α-tocopherol concentration and total lipid level. However, the majority of vitamin E-deficient patients have underlying prolonged fat malabsorption. Therefore, because of the relationship between serum vitamin E concentrations and serum lipid levels, the specific manner of evaluating vitamin E status varies depending on the underlying gastrointestinal disease. Moreover, treatment programs are tailored to each underlying etiology because of differences in the severity of vitamin E malabsorption.

Abetalipoproteinemia and Other Disorders of β Lipoprotein Secretion

The failure of secretion of β lipoproteins, the major carrier of tocopherol, by the intestine and liver in abetalipoproteinemia leads to extremely low serum vitamin E levels irrespective of vitamin E status. Thus, serum vitamin E measurements do not accurately reflect vitamin E status in abetalipoproteinemia. Rather, adipose tissue vitamin E content (75) or erythrocyte hydrogen peroxide hemolysis (76) have been proposed as indicators of vitamin E status in this condition. Expired ethane measurements may one day be the least invasive, accurate measure of vitamin E status in this condition (171). All patients with abetalipoproteinemia [and homozygous hypobetalipoproteinemia (80)] should be assumed vitamin E deficient, unless they have received adequate α-tocopherol in parenteral nutrition infusates or by chronic intramuscular injection, necessitating therapy with 100–200 mg/kg/day of α-tocopherol or α-tocopheryl acetate in two or three divided oral doses. Pure α-tocopheryl acetate oil can be obtained in bulk from vitamin E manufacturers so that patients need not ingest scores of capsules each day. Inasmuch as serum vitamin E concentrations show little change with therapy in this condition, peroxide hemolysis testing (76), serial needle aspirate biopsies of adipose tissue for analysis of repletion of vitamin E stores (75), or serial electrophysiological studies (SEP or visual evoked response) should be used to monitor the efficacy of repletional therapy. In addition, most authorities recommend concomitant treatment with 15,000–20,000 IU/day of vitamin A (77–79) monitored by serum vitamin A concentrations that are not dependent on circulating lipoprotein levels. With this therapeutic regimen, neurological symptoms and electrophysiological abnormalities, as well as retinal dysfunction, have been stabilized in older patients and completely averted if therapy is started early in life (76–79). Patients have now been treated in this way for 15–22 years without development of significant visual impairment or appreciable neuropathy (172). Patients with chylomicron retention disease and other causes of hypobetalipoproteinemia may also require the "industrial-strength" dose (100–200 IU/kg/day) of vitamin E used in abetalipoproteinemia. An oral vitamin E tolerance test using 100 IU/kg body weight per dose as previously described (45) may help determine whether oral dosing with vitamin E is capable of raising serum tocopherol levels in these disorders.

Chronic Cholestasis

Hyperlipidemia, a consequence of regurgitation of biliary lipids into the systemic circulation (163), is common during childhood and adult cholestasis and can elevate serum vitamin E concentrations unrelated to underlying vitamin E status (89). Therefore, serum vitamin E concentrations alone do not accurately reflect vitamin E status during cholestasis and the ratio of serum vitamin E to total serum lipid concentration has been advocated to compensate for the hyperlipidemia of cholestasis (89). Serum vitamin E to total lipid ratios below 0.8 mg/g indicate biochemical deficiency, and ratios below 0.4 mg/g denote severe deficiency. A functional assay (hydrogen peroxide hemolysis or MDA release from red blood cells) can be obtained to confirm the deficiency state if there is a question; however, these assays have not been validated during states of hyperlipidemia.

A growing body of evidence proves that reversing the vitamin E deficiency state during cholestasis is beneficial. Tomasi initially reported improved muscle strength in a vitamin E-deficient, cholestatic child during oral vitamin E therapy (145). Subsequently, Guggenheim et al. (133) treated four cholestatic, vitamin E-deficient children with oral or intramuscular injections of vitamin E noting improved neurological function after 6–14 months of therapy. Five larger treatment trials of over 50 children with cholestasis have since been reported (151, 173–176). Vitamin E therapy for 1–4 years prevented neurological abnormalities in 25% of the patients, stabilized function in another 25%, and resulted in improvement in 50% of the patients. The magnitude of the response to vitamin E repletion was age dependent. Children under age 3 years showed complete reversal or prevention of development of symptoms, whereas older patients showed stabilization or more limited improvement; the most severely affected patients continued to have significant neurological handicaps. The better outcome in young children underscores the need to evaluate children at risk for vitamin E deficiency early in life and initiate repletional therapy when biochemical deficiency is identified.

Once vitamin E deficiency is detected, treatment should be started with large oral doses of available preparations of vitamin E, starting with 25–50 IU/kg/day and advancing by 50 IU/kg/day increments up to 150–200 IU/kg/day if the ratio of serum E to total lipids does not normalize (>0.8 mg/g) within 2–3 weeks. Attempts should be made to keep serum vitamin E below 25–30 μg/ml; however, in an occasional severely hyperlipidemic patient, it is necessary to allow the serum vitamin E concentration to rise above 30 μg/ml to maintain the ratio of serum vitamin E to total lipids above 0.8 mg/g. We prefer using α-tocopherol instead of esterified preparations because of the theoretical requirement of intraluminal bile acids to stimulate the mucosal tocopheryl ester hydrolytic enzyme (46) before absorption of tocopheryl esters. In addition, the *R-R-R* isomer (*d*-α-tocopherol) is more bioactive than the *all-rac* form (*d,l*-α-tocopherol) (177) and appears to confer optimal transport to and uptake by the nervous system. The vitamin E is given at the time of maximal bile flow, as a single morning dose with breakfast several hours before administration of any medication that may interfere with vitamin E absorption (e.g., cholestyramine, vitamin A, or ferrous sulfate). Dosage adjustments are dictated by the ratio of serum vitamin E to total lipids.

In patients with severe cholestasis, intraluminal bile acid concentrations are well below the critical micellar concentration resulting in failure of absorption of standard oral vitamin E preparations regardless of the size of the dose (45). These patients usually have serum direct bilirubin above 3–4 mg/dl and serum bile acid levels at least 10- to 20-fold elevated. However, patients may be anicteric (yet have very high serum bile acid concentrations) and be incapable of absorbing vitamin E (45). For this reason it is absolutely essential to monitor vitamin E status before and during therapy, not to assume that the recommended dose will be absorbed and correct the deficiency. If vitamin E status does not normalize after several months of the maximal oral dose and the child is over 6–9 months of age, therapy should be started with intramuscular injections of vitamin E to provide an average of 1.0–2.0 mg/kg/ day or, alternatively, the water-soluble oral form of vitamin E (see later). In the United States, parenteral *d,l*-α-tocopherol (50 mg/ml) is available from Hoffmann-LaRoche, Inc. (Nutley, NJ; Viprimol), but only as an investigational drug on a case-by-case basis. Elsewhere, parenteral *d,l*-α-tocopheryl acetate (100 mg/ml) is available from Hoffmann-LaRoche (Basel, Switzerland; Ephynal), without limitations on its use. Since both forms are manufactured in 1.0 ml ampules, we administer one ampule intramuscularly per dose and calculate the interval between doses to provide the recommended dose on average, starting with the lower dose (1.0 mg/kg/day). In small children, the ampule is divided into two 0.5 ml injections to prevent subcutaneous leakage that can cause erythema and induration (151). A French group has successfully used a slightly lower dose of Ephynal (173). The use of any other parenteral vitamin E preparation intramuscularly is discouraged unless vigorous safety testing in children and adults has been adequately performed and reviewed by the appropriate regulatory agencies.

Recent studies now indicate that a water-soluble ester of vitamin E, *d*-α-tocopheryl polyethylene glycol-1000 succinate (TPGS: Eastman Chemical Products, Inc., Kingsport, TN; Fig. 11), is absorbed after oral administration in children with severe cholestasis (178), corrects the biochemical vitamin E deficiency state at a dose of 15–30 IU/kg/day (176,178), appears to be nontoxic (175,176,178), and reverses and prevents neurological dysfunction (175,176). Inasmuch as a small portion of the polyethylene glycol present in TPGS is absorbed and then excreted in the urine (178), the potential is present for inducing a hyperosmolar state if the patient suffers from renal failure or dehydration. More than 50 children with severe cholestasis have now received TPGS in a multicentered trial (179), all children showing correction

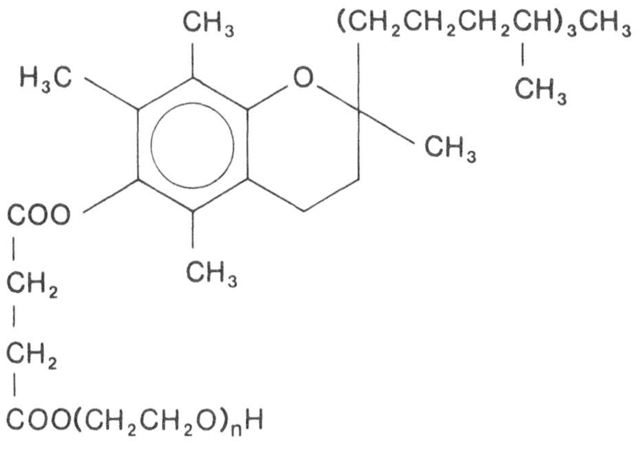

FIGURE 11   Structure of *d*-α-tocopheryl polyethylene glycol-1000 succinate (TPGS).

of vitamin E deficiency without adverse effects. A commercial product in the United States (Liqui-E; Twin Laboratories, Ronkonkoma, NY) that contains 26.6 IU TPGS per ml (as well as lecithin, glycerol, and sorbitol) is available for over-the-counter use.

Although biochemical vitamin E deficiency may be present in over 50% of adults with PBC (90–92), neurological dysfunction is rarer. Therefore, the aggressiveness with which vitamin E deficiency should be corrected in adults with cholestasis is controversial. If not for neurological indications, it has been recommended (91,180,181) that vitamin E therapy be started to prevent possible additional hepatic injury caused by copper accumulation (182, 183) and to correct possible impaired neutrophil function (184). Criteria for defining vitamin E deficiency and treatment modalities should be similar to those used in cholestatic children. The availability of TPGS increases the ease by which vitamin E deficiency can be corrected in these patients.

Cystic Fibrosis and Pancreatic Insufficiency

In general, patients with CF and other causes of pancreatic insufficiency receiving oral replacement of pancreatic enzymes respond to 5–10 IU/kg/day of oral vitamin E if there is no serious liver involvement. CF patients with pancreatic insufficiency who receive pancreatic enzyme supplements should be routinely treated with this dose of vitamin E even if not deficient because of the invariable development of vitamin E deficiency in nonsupplemented patients (185). The most inexpensive and readily available liquid or capsules of vitamin E or vitamin E esters should be utilized. CF or alcoholic patients with cholestatic liver disease or cirrhosis should be evaluated and treated in the same manner as other patients with chronic cholestasis. In the absence of hyperlipidemia, once treatment has been initiated, serum vitamin E concentrations above 5 μg/ml reflect successful vitamin E repletion and normal vitamin E status in patients with CF. Supplementation should continue for life. Serum vitamin E concentrations and neurological function should be monitored at least yearly to assure adequate absorption and compliance.

Other Fat Malabsorption Disorders

Evaluation of vitamin E status in children and adults with short bowel syndrome or other chronic fat malabsorption diseases is performed by serum vitamin E analyses, unless cholestatic liver disease, hyperlipidemia, or hypolipidemia is a complicating factor. Most patients with these illnesses who are found to be vitamin E deficient respond to large supplemental oral doses of vitamin E (200–3600 mg/day) (103–106). If parenteral nutrition is being administered chronically, the 15 IU/day of α-tocopheryl acetate in the standard intravenous multiple vitamin supplement usually prevents or reverses vitamin E deficiency. Rarely are intramuscular injections of vitamin E necessary. The true prevalence of vitamin E deficiency is unknown in adolescents and adults with long-standing Crohn's disease, gluten-sensitive enteropathy, radiation enteritis, and short bowel syndrome, primarily because 10–20 years of deficiency appear to be necessary before overt neurological symptoms develop, which would prompt evaluation of vitamin E status. Consequently, asymptomatic vitamin E deficiency may be more common in adults with these diseases than previously appreciated. Therefore, periodic serum vitamin E concentrations and neurological examinations should be incorporated into the standard clinical monitoring of patients with ongoing steatorrhea. The neurological response to vitamin E repletion has been dramatic in many of these patients (103,104).

Isolated Vitamin E Deficiency Syndrome

Patients with primary isolated vitamin E deficiency syndrome respond to 800–1200 mg/day of oral *all-rac-α*-tocopherol, *RRR-α*-tocopherol, or tocopheryl acetate, most patients requiring the lower dose (15,25,26,120–122). An occasional patient has required over 3000 mg/day. Patients with more advanced neurological deterioration before instituting therapy show a more limited neurological response. Identification of vitamin E deficiency in this disorder requires only a serum vitamin E concentration and total lipid level. Because patients with this disorder present with a movement disorder, poor balance, or impaired coordination without evidence of gastrointestinal symptoms, there are no clinical features to distinguish these patients from other people with ataxia or peripheral neuropathies of other etiologies. For this reason it is essential that all patients evaluated for movement disorders or neuropathies undergo screening for vitamin E deficiency by measurement of serum vitamin E concentrations and total serum lipid levels, lest this rare but treatable cause of neurological degeneration be overlooked. If vitamin E deficiency is identified, thorough evaluation for fat malabsorption, liver disease, and lipoprotein abnormalities should be performed and all first-degree relatives screened for vitamin E deficiency. Therapy should be initiated with 800–1200 IU/day of vitamin E in two divided doses. Asymptomatic relatives who have evidence of biochemical vitamin E deficiency should also be treated after a thorough neurological evaluation. Serum vitamin E levels should be monitored and doses adjusted accordingly. Neurological status should be serially evaluated during therapy as in all patients undergoing vitamin E repletion.

SUMMARY

Over the past decade it has become apparent that vitamin E is an essential nutrient for maintaining the structure and functional integrity of the human nervous system, skeletal muscle, and retina. The clinical and histological resemblance is striking between the human neuromuscular disorder associated with chronic fat and vitamin E malabsorption and that observed in experimental vitamin E-deficient animal models. Because of chronic malabsorption of vitamin E, patients with pancreatic insufficiency, chronic cholestasis, disorders of β lipoprotein secretion, short bowel syndrome, and other steatorrheic conditions are at risk for the development of neurological deficits. Correction of the vitamin E deficiency state prevents, reverses, or at least stabilizes the neurological dysfunction in susceptible individuals, generally improving the quality of the patient's life. Advances in stable isotope technology are now permitting study of the hepatic discrimination among the various stereoisomers and forms of vitamin E. Investigations into the cause of the primary form of vitamin E deficiency, the isolated vitamin E deficiency syndrome, promise to delineate the normal physiological processes involved in absorption, transport, and tissue delivery of vitamin E. Studies in progress are addressing the optimal route and form of vitamin E therapy to be used in each predisposing condition. One major task remaining is to better define the mechanism by which vitamin E deficiency leads to neurological injury.

ACKNOWLEDGMENTS

Supported in part by USPHS Grants RR00069 and RR00123 from the General Clinical Resources Centers Branch, Division of Research Resources, National Institutes of Health (NIH); NIH FIRST Award R 29 DK 38446; NIH Hepatobiliary Research Center (IP30AM34914); March of Dimes Basil O'Connor Award, Eastman Chemical Products, Inc.; and the Abby Bennett Liver Research Fund.

REFERENCES

1. Evans HM, Bishop KS. On the existence of a hitherto unrecognized dietary factor essential for reproduction. Science 1922; 56:650-1.

2. Wasserman RH, Taylor AN. Metabolic roles of fat soluble vitamin D, E, and K. Annu Rev Biochem 1972; 41:179-202.
3. Mason KE. The first two decades of vitamin E. Fed Proc 1977; 36:1906-10.
4. Nelson JS. Pathology of vitamin E deficiency. In: Machlin LJ, ed. Vitamin E—a comprehensive treatise. New York: Marcel Dekker, 1980; 397-428.
5. Evans HM, Burr GO. Development of paralysis in the suckling young of mothers deprived of vitamin E. J Biol Chem 1928; 76:273-97.
6. Pappenheimer AM, Goettsch M. A cerebellar disorder in chicks, apparently of nutritional origin. J Exp Med 1931; 54:145-65.
7. Goettsch M, Pappenheimer M. Nutritional muscular dystrophy in the guinea pig and rabbit. J Exp Med 1931; 54:145-65.
8. Berneske GM, Butson ARC, Gould EN, Levy D. Clinical trial of high dosage vitamin E in human muscular dystrophy. Can Med Assoc J 1960; 82:418-21.
9. Milhorat AT. Therapy in muscular dystrophy. Med Ann Dist Columbia 1954; 23:15-22.
10. Fenichel GM, Brooke MH, Griggs RC, et al. Clinical investigation in Duchenne muscular dystrophy: penicillamine and vitamin E. Muscle Nerve 1988; 11:1164-8.
11. Gamstorp I, Gustavsson KH, Hellstrom O, Nordgren B. A trial of selenium and vitamin E in boys with muscular dystrophy. J Child Neurol 1986; 1:211-4.
12. Backman E, Nylander E, Johansson I, Henriksson KG, Tagesson C. Selenium and vitamin E treatment of Duchenne muscular dystrophy: no effect on muscle function. Acta Neurol Scand 1988; 78:429-35.
13. Kayden HJ, Silber R, Kossman CE. The role of vitamin E deficiency in the abnormal autohemolysis of acanthocytosis. Trans Assoc Am Physicians 1965; 78:334-42.
14. Binder HJ, Solitare GB, Spiro HM. Neuromuscular disease in patients with steatorrhea. Gut 1967; 8:605-11.
15. Sokol RJ, Kayden HJ, Bettis DB, et al. Isolated vitamin E deficiency in the absence of fat malabsorption—familial and sporadic cases: characterization and investigations of causes. J Lab Clin Med 1988; 111:548-59.
16. Kasparek S. Chemistry of tocopherols and tocotrienols. In: Machlin LJ, ed. Vitamin E—a comprehensive treatise. New York: Marcel Dekker, 1980; 7-65.
17. Ingold KU, Burton GW, Foster DD, Hughes L, Lindsay DA, Webb A. Biokinetics of and discrimination between dietary *RRR* and *SRR*-alpha-tocopherols in the male rat. Lipids 1987; 22:163-72.
18. Vatassery GT, Angerhofer CK, Knox CA, Deshmukh DS. Concentrations of vitamin E in various neuroanatomical regions and subcellular fractions, and the uptake of vitamin E by specific areas, of rat brain. Biochim Biophys Acta 1984; 792:118-22.
19. Diplock AT, Lucy JA. The biochemical modes of action of vitamin E and selenium: a hypothesis. FEBS Lett 1973; 29:205-10.
20. Tappel AL. Vitamin E as the biological lipid oxidant. Vitam Horm 1962; 20:493-510.
21. Dam H. Toxicity of fractions of hog live fatty acids to chicks fed a vitamin E-deficient diet. J Nutr 1944; 28:297-302.
22. Nelson JS. Effects of free radical scavengers on the neuropathology of mammalian vitamin E deficiency. In: Hayaishi O, Mino M, eds. Clinical and nutritional aspects of vitamin E. Amsterdam: Elsevier, 1987; 157-9.
23. Singsen EP, Bunnell RH, Matterson LD, et al. Studies on encophalomalacia in the chick. 2. The protective action of diphenyl-*p*-phenylenediamine against encephalomalacia. Poultry Sci 1955; 34:262-71.
24. Rosenblum J, Keating JP, Prensky A, Nelson JS. A progressive neurologic syndrome in children with chronic liver disease. N Engl J Med 1981; 304:503-8.
25. Burck U, Goebel HH, Kuhlendahl HD, Meier C, Goebel KM. Neuromyopathy and vitamin E deficiency in man. Neuropediatrics 1981; 12:267-78.
26. Yokota T, Wada Y, Furakawa T, Tsukagoshi H, Uchihara T, Watabiki S. Adult onset spinocerebellar syndrome with idiopathic vitamin E deficiency. Ann Neurol 1987; 232:84-7.
27. Sung JH. Neuroaxonal dystrophy in mucoviscidosis. J Neuropathol Exp Neurol 1964; 23:567-83.
28. Hayes KC, Nelson SW, Rousseau JE. Vitamin E deficiency and fat stress in the dog. J Nutr 1969; 99:196-209.

29. Werlin SL, Harb JM, Swick H, Blank E. Neuromuscular dysfunction and ultrastructural pathology in children with chronic cholestasis and vitamin E deficiency. Ann Neurol 1983; 13:291-6.

30. Lucy JA. Functional and structural aspects of biological membranes: a suggested role for vitamin E in the control of membrane permeability and stability. Ann N Y Acad Sci 1972; 203:4-11.

31. Wassall SR, Thewalt JL, Wong L, Gorrissen H, Cushley RJ. Dueterium NMR study of the interaction of alpha-tocopherol with a phospholipid model membrane. Biochemistry 1986; 25:319-26.

32. Srivastava S, Phadke RS, Govil G, Rao CNR. Fluidity, permeability and antioidant behaviour of model membranes incorporated with alpha-tocopherol and vitamin E acetate. Biochim Biophys Acta 1983; 734:353-62.

33. Fukuzawa K, Chida H, Suzuki A. Fluorescence depolarization studies of phase transition and fluidity in lecithin liposomes containing alpha-tocopherol. J Nutr Sci Vitaminol 1980; 26:427-34.

34. Diplock AT, Lucy JA, Verrinder M, Zeileneiwski A. Alpha-tocopherol and the permeability to glucose and chromate of unsaturated lipsomes. FEBS Lett 1977; 82:341-4.

35. Heffron JJ, Chan AC, Gronert GA, Hegarty PV. Decreased phosphorylative capacity and respiratory rate of rabbit skeletal muscle mitochondria in vitamin E dystrophy. Int J Biochem 1978; 9:539-43.

36. Hornsby PJ. The role of vitamin E in cellular energy metabolism in cultured adrenocortical cells. J Cell Physiol 1982; 112:207-16.

37. Pfenninger KH. Of nerve growth cones, leukocytes, and memory: secondary messenger systems and growth-related proteins. Trends Neurosci 1986; 9:562-5.

38. Meydani M, Meydani SN, Macauley JB, Blumberg JR. Influence of dietary vitamin E and selenium in the ex-vivo synthesis of prostaglandin E_2 in brain regions of young and old rats. Prostaglandins Leuko Med 1985; 19:337-46.

39. Castignani GL. Role in nucleic acid and protein metabolism. In: Machlin LJ, ed. Vitamin E—a comprehensive treatise. New York: Marcel Dekker, 1980; 318-32.

40. Griffin JW, Watson DF. Axonal transport in neurological disease. Ann Neurol 1988; 23:3-13.

41. Goss-Sampson MA, MacEvilly CJ, Muller DPR. Longitudinal studies of the neurobilogy of vitamin E and other antioxidant systems, and neurological function in the vitamin E deficient rat. J Neurol Sci 1988; 87:25-35.

42. Goss-Sampson MA, Kriss A, Muddle JR, Thomas PK, Muller DPR. Lumbar and cortical somatosensory evoked potentials in rats with vitamin E deficiency. J Neurol Neurosurg Psychiatr 1988; 51:432-5.

43. Blomstrand R, Forsgren L. Labeled tocopherols in man. Int J Vitam Res 1968; 38:328-44.

44. Gallo-Torres HE. Obligatory role of bile acids for the intestinal absorption of vitamin E. Lipids 1978; 5:379-84.

45. Sokol RJ, Heubi JE, Iannaccone S, Bove KE, Balistreri WF. Mechanism causing vitamin E deficiency in children with chronic cholestasis. Gastroenterology 1983; 85:1172-82.

46. Muller DPR, Manning JA, Mathias PM, Harries JT. Studies on the intestinal hydrolysis of tocopheryl esters. Int J Vitam Nutr Res 1976; 46:207-10.

47. Hollander D, Rim E, Muralidhara KS. Mechanism and site of small intestinal absorption of alpha-tocopherol in the rat. Gastroenterology 1975; 68:1492-9.

48. Traber MG, Goldberg I, Davidson E, et al. Vitamin E uptake by human intestinal cells during lipolysis in vitro. Gastroenterology 1990; 98:96-103.

49. Traber MG, Olivecrona T, Kayden HJ. Bovine milk lipoprotein lipase transfers tocopherol to human fibroblasts during triglyceride hydrolysis in vitro. J Clin Invest 1985; 75:1729-34.

50. Bjorneboe A, Bjorneboe G-E, Hagen BF, Nossen JO, Drevon CA. Secretion of alpha-tocopherol from cultured rat hepatocytes. Biochim Biophys Acta 1987; 922:199-205.

51. Cohn W, Loechleiter F, Weber F. Alpha-tocopherol is secreted from rat liver in very low density lipoproteins. J Lipid Res 1988; 29:1359-66.

52. Traber MG, Kayden HJ. Preferential incorporation of alpha-tocopherol vs alpha-tocopherol in human lipoproteins. Am J Clin Nutr 1989; 49:517-26.

53. Catignani GL, Bieri JG. Rat liver alpha-tocopherol binding protein. Biochim Biophys Acta 1977; 497:349-57.

54. Murphy DJ, Mavis RD. Membrane transfer of alpha-tocopherol. Influence of soluble alpha-tocopherol binding factors from the liver, lung, heart, and brain of the rat. J Biol Chem 1981; 256: 10464-8.

55. Kaplowitz N, Yoshida H, Kuhlenkamp J, Slitsky B, Ren I, Stolz A. Tocopherol-binding proteins of hepatic cytosol. Ann N Y Acad Sci 1989; 570:85-94.

56. Verdon CP, Blumberg JB. Influence of dietary vitamin E on the intermembrane transfer of alpha-tocopherol as mediated by an alpha-tocopherol binding protein. Proc Soc Exp Biol Med 1988; 189: 52-60.

57. Traber MG, Kayden HJ. Vitamin E is delivered to cells via the high affinity receptor for lfow density lipoprotein. Am J Clin Nutr 1984; 40:747-51.

58. Cohn W, Kuhn H. The role of low density lipoprotein receptor for alpha tocopherol delivery to tissues. Ann N Y Acad Sci 1989; 570:61-71.

59. Machlin LJ, Keating J, Nelson J, Brin M, Filipski R, Miller ON. Availability of adipose tissue tocopherol in the guinea pig. J Nutr 1979; 109:105-9.

60. Traber MG, Kayden HJ. Tocopherol distribution and intracellular localization in human adipose tissue. Am J Clin Nutr 1987; 46:488-95.

61. Bendich A, Machlin LJ, Scandurra O, Burton GW, Wayner DN. The antioxidant role of vitamin C. Adv Free Radic Biol Med 1986; 2:419-44.

62. Bendich A, Diapolito P, Gabriel E, Machlin LJ. Interaction of dietary vitamin C and vitamin E in guinea pigs immune response to mitogens. J Nutr 1984; 114:1588-93.

63. Wefers H, Sies H. The protection of ascorbate and glutathione against microsomal lipid peroxidation is dependent on vitamin E. Eur J Biochem 1988; 174:353-7.

64. Bjorneboe A, Bjorneboe GE, Drevon CA. Serum half-life, distribution, hepatic uptake, and biliary excretion of alpha tocopherol in rats. Biochim Biophys Acta 1987; 921:175-81.

65. Lee-Kim YC, Meydani M, Kassarjian Z, Blumberg JB, Russell RM. Enterohepatic circulation of newly administered alpha-tocopherol in the rat. Int J Vitam Nutr Res 1988; 58:285-91.

66. Gallo-Torres HE. Transport and metabolism. In: Machlin LJ, ed. Vitamin E—a comprehensive treatise. New York: Marcel Dekker, 1980; 193-267.

67. Malloy MJ, Kane JP, Hardman DA, et al. Normotriglyceridemic abetalipoproteinemia. Absence of B-100 apilipoprotein. J Clin Invest 1981; 67:1441-50.

68. Vatassery GT, Angerhofer CK, Knox CA. Effect of age on vitamin E concentrations in various regions of the brain and a few selected peripheral tissues of the rat, and on the uptake of radioactive vitamin E by various regions of rat brain. J Neurochem 1984; 43:409-12.

69. Vatassery GT. Selected aspects of the neurochemistry of vitamin E. In: Hayaishi O, Mino M, eds. Clinical and nutritional aspects of vitamin E. Amsterdam: Elsevier, 1987; 147-55.

70. LeBel CP, Odunze IN, Adams JD, Bondy SC. Perturbations in cerebral oxygen radical formation and membrane order following vitamin E deficiency. Biochem Biophys Res Commun 1989; 163: 806-6.

71. Kutsky RJ. Vitamin E. In: Handbook of vitamins and hormones. New York: Van Nostrand-Reinhold, 1973; 23-31.

72. Bieri JG, Farrell PM. Vitamin E. Vitam Horm 1976; 34:31-74.

73. Muller DPR, Harries JT, Lloyd JK. The relative importance of the factors involved in the absorption of vitamin E in children. Gut 1974; 15:966-71.

74. Salt HB, Wolff OH, Lloyd JK, Fosbrooke AM, Cameron AH, Hubble DV. On having no betalipoprotein; a syndrome comprising abetaliproteinemia, acanthosis, and steatorrhea. Lancet 1960; 2:325-9.

75. Kayden HJ, Hatam LJ, Traber MG. The measurement of nanograms of tocopherol from needle aspiration biopsies of adipose tissue: normal and abetalipoprotenemic subjects. J Lipid Res 1983; 24:652-6.

76. Muller DPR, Lloyd JK, Wolff OH. Vitamin E and neurological function. Lancet 1983; 1:225-8.

77. Azizi E, Zaidman JL, Eschar J, Szeinberg A. Abetalipoproteinemia treated with parenteral and oral vitamin A and E and with medium chain tryglycerides. Acta Paediatr Scand 1978; 67:797-801.

78. Bishara S, Merin S, Cooper M, Azizi E, Delpre G, Deckelbaum RJ. Combined vitamin A and E therapy prevents retinal electrophysiological deterioration in abetalipoproteinemia. Br J Ophthalmol 1982; 66:767-70.

79. Muller DPR, Lloyd JK, Bird AC. Long-term management of abetalipoproteinemia. Possible role for vitamin E. Arch Dis Child 1977; 52:209-14.

80. Traber MG, Sokol RJ, Ringel SP, Neville HE, Thellman CA, Kayden HJ. Lack of tocopherol in peripheral nerves of vitamin E-deficient patients with peripheral neuropathy. N Engl J Med 1987; 317:262-5.

81. Griffiths RD, Taylor CJ, Isherwood PM, Jackson MJ. Fat malabsorption, vitamin E deficiency, scoliosis, and cataracts. J Inherited Metab Dis 1988; 11(suppl. 2):153-4.

82. Roy CC, Levy E, Green PHR, et al. Malabsorption, hypocholesterolemia, and fat-filled entero-cytes with increased intestinal apoprotein B: chylomicron retention disease. Gastroenterology 1987; 92:390-9.

83. Anderson CM, Townley RR, Freeman JP. Unusual causes of steatorrhea in infancy and child-hood. Med J Aust 1961; 11:617-21.

84. Guggenheim MA, Jackson V, Lilly JR, Silverman A. Vitamin E deficiency and neurologic dis-ease in children with cholestasis: a prospective study. J Pediatr 1983; 102:577-9.

85. Sokol RJ, Guggenheim MA, Heubi JE, et al. Frequency and clinical progression of the vitamin E deficiency neurologic disorder in children with prolonged neonatal cholestasis. Am J Dis Child 1985; 139:1211-5.

86. Tazawa Y, Nakagawa M, Yamada M, et al. Serum vitamin E levels in children with corrected biliary atresia. Am J Clin Nutr 1984; 40:246-50.

87. Alvarez F, Cresteil D, Lemonnier F, Lemonnier A, Alagille D. Plasma vitamin E levels in children with cholestasis. J Pediatr Gastroenterol Nutr 1984; 3:390-3.

88. Sokol RJ, Bove KE, Heubi JE, Iannaccone ST. Vitamin E deficiency during chronic cholestasis: presence of sural nerve lesion prior to 2½ years of age. J Pediatr 1983; 103:197-204.

89. Sokol RJ, Heubi JE, Iannaccone ST, Bove KE, Balistreri WF. Vitamin E deficiency with normal serum vitamin E concentrations in children with chronic cholestasis. N Engl J Med 1984; 310: 1209-12.

90. Sokol RJ, Balistreri WF, Hoofnagle JH, Jones EA. Vitamin E deficiency in adults with chronic liver disease. Am J Clin Nutr 1985; 41:66-72.

91. Jeffrey GP, Muller DRP, Burroughs AK, et al. Vitamin E deficiency and its clinical significance in adults with primary biliary cirrhosis and other forms of chronic liver disease. J Hepatol 1987; 4:307-17.

92. Munoz SJ, Heubi JE, Balistreri WF, et al. Vitamin E deficiency in primary biliary cirrhosis: gas-trointestinal malabsorption, frequency, and relationship to other lipid-soluble vitamins. Hepatol-ogy 1989; 9:525-31.

93. Knight RS, Bourne AJ, Newton M, Black A, Wilson P, Lawson MJ. Neurologic syndrome asso-ciated with low levels of vitamin E in primary biliary cirrhosis. Gastroenterology 1986; 91:209-11.

94. Sokol RJ, Kim YS, Hoffnagle JH, Heubi JE, Jones EA, Balistreri WF. Intestinal malabsorption of vitamin E in primary biliary cirrhosis. Gastroenterology 1989; 96:479-86.

95. Farrell PM. Deficiency states, pharmacological effects, and nutrient requirements. In: Machlin LJ, ed. Vitamin E—a comprehensive treatise. New York: Marcel Dekker, 1980; 520-620.

96. Bye AME, Muller DPR, Wilson J, Wright VM, Mearns MB. Symptomatic vitamin E deficiency in cystic fibrosis. Arch Dis Child 1985; 60:162-4.

97. Elias E, Muller DPR, Scott J. Association of spinocerebellar disorders with cystic fibrosis or chronic childhood cholestasis and very low serum vitamin E. Lancet 1981; 2:1319-21.

98. Willison HJ, Muller DPR, Matthews S, et al. A study of the relationship between neurological function and serum vitamin E concentrations in patients with cystic fibrosis. J Neurol Neurosurg Psychiatr 1985; 48:1097-102.

99. Kaplan PW, Rawal K, Erwin CW, et al. Visual and somatosensory evoked potentials in vitamin E deficiency with cystic fibrosis. Electroenceph Clin Neurophysiol 1988; 71:266-72.

100. Cynamon HA, Milov DE, Valenstein E, Wagner M. Effect of vitamin E deficiency on neurologic function in patients with cystic fibrosis. J Pediatr 1988; 113:637-40.

101. Johanson A, Blizzard R. A syndrome of congenital asplasia of the alae nasi, deafness, hypothy-roidism, dwarfism, absent permanent teeth, and malabsorption. J Pediatr 1971; 79:982-7.

102. Davidai G, Zakaria T, Goldstein R, Gilai A, Freier S. Hypovitaminosis E induced neuropathy in exocrine pancreatic failure. Arch Dis Child 1986; 61:901-3.

103. Harding AE, Muller DPR, Thomas PK, Willison HJ. Spinocerebellar degeneration secondary to chronic intestinal malabsorption: a vitamin E deficiency syndrome. Ann Neurol 1982; 12:419-24.

104. Satya-Murti S, Howard L, Krohel G, Wolf B. The spectrum of neurologic disorder from vitamin E deficiency. Neurology 1986; 36:917-21.

105. Weder B, Meienberg O, Wildi E, Meier C. Neurologic disorder of vitamin E deficiency in acquired intestinal malabsorption. Neurology 1984; 34:1561-5.

106. Brin MF, Fetell MR, Green PHA, et al. Blind loop syndrome, vitamin E malabsorption, and spinocerebellar degeneration. Neurology 1985; 35:338-42.

107. Gutmann L, Shockcor W, Gutmann L, Kien VCL. Vitamin E-deficient spinocerebellar syndrome due to intestinal lymphangiectasia. Neurology 1986; 36:554-6.

108. Ackerman Z, Eliashiv S, Reches A, Zimmerman J. Neurological manifestations in celiac disease and vitamin E deficiency. J Clin Gastroenterol 1989; 11:603-5.

109. Braunstein H. Tocopherol deficiency in adults with chronic pancreatitis. Gastroenterology 1961; 40:224-31.

110. Dutta SK, Bustin MP, Russel RM, Costa BS. Deficiency of fat-soluble vitamin in treated patients with pancreatic insufficiency. Ann Intern Med 1982; 97:549-52.

111. Yokota T, Tsuchiya K, Furukawa T, Tsukagoshi H, Miyakawa H, Hasumura Y. Vitamin E deficiency in acquired fat malabsorption. J Neurol 1990; 237:103-6.

112. Nye SW, Chittayasothorn K. Ceroid in the gastrointestinal smooth muscle of the Thai-Lao ethnic group. Am J Pathol 1967; 51:287-99.

113. Gallager RL. Intestinal ceroid deposition—"brown bowel syndrome." A light and electron microscopic study. Virchows Arch [A] 1980; 389:143-51.

114. Ruchti C, Eisele S, Kaufmann M. Fatal intestinal pseudo-obstruction in brown bowel syndrome. Arch Pathol Lab Med 1990; 114:76-80.

115. Foster CS. The brown bowel syndrome: a possible smooth muscle mitochondrial myopathy? Histopathology 1979; 3:1-17.

116. Lambert JR, Luk SC, Pritzker KPH. Brown bowel syndrome in Crohn's disease. Arch Pathol Lab Med 1980; 104:201-5.

117. Cooke WT, Smith WT. Neurological disorders associated with adult coeliac disease. Brain 1966; 89:683-722.

118. Ward ME, Murphy JT, Greenberg GB. Celiac disease and spinocerebellar degeneration with normal vitamin E status. Neurology 1985; 35:1199-201.

119. Tison F, Arne P, Henry P. Myoclonus and adult coeliac disease. J Neurol 1989; 236:307-8.

120. Harding AE, Matthews S, Jones S, Ellis CJ, Booth IW, Muller DP. Spinocerebellar degeneration associated with a selective defect of vitamin E absorption. N Engl J Med 1985; 313:32-5.

121. Krendel DA, Gilchrist JM, Johnson AO, Bossen EH. Isolated deficiency of vitamin E with progressive neurologic deterioration. Neurology 1987; 37:538-40.

122. LaPlante P, Vanasse M, Michaud J, Geoffroy G, Brochu P. A progressive neurological syndrome associated with an isolated vitamin E deficiency. Can J Neurol Sci 1984; 11:561-4.

123. Stumpf DA, Sokol RJ, Bettis D, et al. Friedreich's disease. V. Variant form with vitamin E deficiency and normal fat absorption. Neurology 1987; 37:68-74.

124. Traber MG, Sokol RJ, Burton GW, et al. Impaired ability of patients with familial isolated vitamin E deficiency to incorporate alpha-tocopherol into lipoproteins secreted by the liver. J Clin Invest 1990; 85:397-407.

125. Kay R, Teoh R, Woo J, Chin D, Mak YT. Serum vitamin E concentrations in adult-onset spinocerebellar degeneration. J Neurol Neurosurg Psychiatr 1989; 52:131-2.

126. Harding AE, Macevilly CJ, Muller DPR. Serum vitamin E concentrations in degenerative ataxias. J Neurol Neurosurg Psychiatr 1989; 52:132.

127. Muller DPR, Matthews S, Harding AE. Serum vitamin E concentrations are normal in Friedreich's ataxia. J Neurol Neurosurg Psychiatr 1987; 50:625-7.

128. Hayes KC. Retinal degeneration in monkeys induced by deficiencies of vitamin E or A. Invest Ophthalmol 1974; 13:499-510.

129. Robison WG Jr, Kuwabara T, Bieri JG. Deficiencies of vitamins E and A in the rat: retinal damage and lipofuscin accumulation. Invest Opthalmol Vis Sci 1980; 19:1030-7.

130. Pentschew A, Schwarz K. Systemic axonal dystrophy in vitamin E deficient adult rats: with implication in human neuropathology. Acta Neuropathol (Berl) 1962; 1:313-34.

131. Nelson JS, Fitch CD, Fischer VW, Brown GO, Chou AC. Progressive neuropathologic lesions in vitamin E-deficient rhesus monkeys. J Neuropathol Exp Neurol 1981; 40:166-86.

132. Sokol RJ. Vitamin E and neurologic function in man. Free Radic Biol Med 1989; 6:189-207.

133. Guggenheim MA, Ringel SP, Silverman A, Grabert BE. Progressive neuromuscular disease in children with chornic cholestasis and vitamin E deficiency: diagnosis and treatment with alpha tocopherol. J Pediatr 1982; 100:51-8.

134. Lal H, Pogacar S, Daly PR, Puri SK. Behavioral and neuropathological manifestations of nutritionally induced central nervous system "aging" in the rat. Prog Brain Res 1973; 40:129-40.

135. Sarter M, van der Linde A. Vitamin E deprivation in rats: some behavioral and histochemical observations. Neurobiol Aging 1987; 8:297-307.

136. Satel SL, Riely CA. Vitamin E deficiency and neurologic dysfunction in children. N Engl J Med 1986; 314:1389-90.

137. Arria AM, Tarter RE, Warty V, Van Thiel DM. Vitamin E deficiency and psychomotor performance in adults with primary biliary cirrhosis. Am J Clin Nutr 1990; 52:383-90.

138. Stewart SM, Uauy R, Waller DA, et al. Mental and motor development correlates in patients with end-stage biliary atresia awaiting liver transplantation. Pediatrics 1987; 79:882-8.

139. Stewart SM, Uauy R, Kennard BD, Waller DA, Benser M, Andrews WS. Mental development and growth in children with chronic liver disease of early and late onset. Pediatrics 1988; 82:167-72.

140. Sung JH, Stadlin EM. Neuroaxonal dystrophy in congenital biliary atresia. J Neuropath Exp Neurol 1966; 25:341-361.

141. Hayes KC, Nelson SW, Rousseau JE. Vitamin E deficiency and fat stress in the dog. J Nutr 1969; 99:196-209.

142. Landrieu P, Selva J, Alvarez F, Ropert A, Metral S. Peripheral nerve involvement in children with chronic cholestasis and vitamin E deficiency. A clinical, electrophysiological and morphological study. Neuropediatrics 1985; 16:194-201.

143. Wichman A, Buchthal F, Pezeshkpour GH, Gregg RE. Peripheral neuropathy in abetalipoproteinemia. Neurology 1985; 35:1279-89.

144. Neville HE, Ringel SP, Guggenheim MA, Wehling CA, Starcevich JM. Ultrastructural and histochemical abnormalities of skeletal muscle in patients with chronic vitamin E deficiency. Neurology 1983; 33:483-8.

145. Tomasi LG. Reversibility of human myopathy caused by vitamin E deficiency. Neurology 1979; 29:1182-6.

146. Lazaro RP, Dentinger MP, Rodichok LD, Barron KD, Satya-Murti S. Muscle pathology in Bassen-Kornzweig syndrome and vitamin E deficiency. Am J Clin Pathol 1986; 86:378-87.

147. Stumpf DA, Sokol RJ, Bettis D, et al. Friedreich's disease. V. Variant form with vitamin E deficiency and normal fat absorption. Neurology 1987; 37:68-74.

148. Lin CT, Chen LH. Ultrastructural and lysomal enzyme studies of skeletal muscle and myocardium in rats with long-term vitamin E deficiency. Pathology 1982; 14:375-82.

149. Alvarez F, Landrieu P, Laget P, Lemonnier F, Odievre M, Alagille D. Nervous and ocular disorders in children with cholestasis and vitamin A and E deficiencies. Hepatology 1983; 3:410-4.

150. Brin MF, Pedley TA, Lovelace RE, et al. Electrophysiologic features of abetalipoproteinemia: functional consequences of vitamin E deficiency. Neurology 1986; 36:669-73.

151. Sokol RJ, Guggenheim MA, Iannaccone ST, et al. Improved neurologic function following long-term correction of vitamin E deficiency in children with chronic cholestasis. N Engl J Med 1985; 313:1580-6.

152. Larsen PD, Mock DM, O'Connor PS. Vitamin E deficiency associated with vision loss and bulbar weakness. Ann Neurol 1985; 18:725-7.

153. Messenheimer JA, Greenwood RS, Tennison MB, Brickley JJ, Ball CJ. Reversible visual evoked potential abnormalities in vitamin E deficiency. Ann Neurol 1984; 15:499-501.

154. Howard L, Ovesen L, Satya-Murti S, Chu R. Reversible neurological symptoms caused by vitamin E deficiency in a patient with short bowel syndrome. Am J Clin Nutr 1982; 36:1243-9.

155. Sokol RJ. Assessing vitamin E status in childhood cholestasis. J Pediatr Gastroenterol Nutr 1987; 6:10-3.

156. Hansen LG, Warwick WJ. A fluorometric micromethod for serum vitamins A and E. Am J Clin Pathol 1969; 51:538-41.

157. Bieri JG, Tolliver TJ, Catignani GL. Simultaneous determination of alpha-tocopherol and retinol in plasma or red cells by high pressure liquid chromatography. Am J Clin Nutr 1979; 32:2143-9.

158. Zoellner N, Kirsch K. A micromethod for lipids using a sulphooxanillin reaction. Z Gesamte Exp Med 1962; 135:545-61.

159. Heubi JE, Sokol RJ, McGraw CA. Comparison of total serum lipids measured by two methods. J Pediatr Gastroenterol Nutr 1990; 10:468-72.

160. Farrell PM, Levine SL, Murphy D, Adams AJ. Plasma tocopherol levels and tocopherol-lipid relationship in a normal population of children as compared to healthy adults. Am J Clin Nutr 1978; 31:1720-6.

161. Horwitt MK, Harvey CC, Dahm CH, Searcy MT. Relationship between tocopherol and serum lipid levels for determination of nutritional adequacy. Ann N Y Acad Sci 1972; 203:223-36.

162. Bieri JG, Evarts RP, Thorp S. Factors affecting the exchange of tocopherol between red blood cells and plasma. Am J Clin Nutr 1977; 30:686-90.

163. Sabesin SM. Cholestatic lipoproteins—their pathogenesis and significance. Gastroenterology 1982; 83:704-9.

164. Gordon HH, Nitowsky HM, Cornblath M. Studies of tocopherol deficiency in infants and children. I. Hemolysis of erythrocytes in hydrogen peroxide. Am J Dis Child 1955; 90:669-81.

165. Cynamon HA, Isenberg JN, Nguyen CH. Erythrocyte malondialdehyde release in vitro: a functional measure of vitamin E status. Clin Chim Acta 1985; 151:169-76.

166. Satoh K. Serum lipid peroxide in cerebrovascular disorders determined by a new colorimetric method. Clin Chim Acta 1978; 90:37-43.

167. Lemonnier F, Cresteil D, Feneant M, et al. Plasma lipid peroxides in cholestatic hildren. Acta Paediatr Scand 1987; 76:928-34.

168. Lepage G, Munoz G, Champagne J, Roy CC. Preparative steps necessary for the accurate measurement of malondialdehyde by HPLC. Anal Biochem 1991; 197:227-283.

169. Tappel AL, Dillard CJ. In vivo lipid peroxidation: measurement via exhaled pentane and protection by vitamin E. Fed Proc 1981; 40:174-8.

170. Wispe JR, Bell EF, Roberts RJ. Assessment of lipid peroxidation in newborn infants and rabbits by measurements of expired ethane and pentane: influence of parenteral lipid infusion. Pediatr Res 1985; 19:374-9.

171. Refat M, Moore TJ, Perman JA, Schwarz KB. Utility of breath ethane as a noninvasive biomarker of vitamin E status in children with liver disease (abstract). Hepatology 1990; 12:930.

172. Lloyd JK. The importance of vitamin E in human nutrition. Acta Paediatr Scand 1990; 79:6-11.

173. Alvarez F, Landrieu P, Feo C, Lemonnier F, Bernard O, Alagille D. Vitamin E deficiency is responsible for neurologic abnormalities in cholestatic children. J Pediatr 1985; 107:422-5.

174. Perlmutter DH, Gross P, Jones HR, Fulton A, Grand RJ. Intramuscular vitamin E repletion in children with chronic cholestasis. Am J Dis Child 1987; 141:170-4.

175. Sokol RJ, Butler-Simon NA, Bettis D, Smith DJ, Silverman A. Tocopheryl polyethylene glycol 1000 succinate therapy for vitamin E deficiency during chronic childhood cholestasis: neurologic outcome. J Pediatr 1987; 111:830-6.

176. Sokol RJ, Butler-Simon N, Heubi JE, et al. Vitamin E deficiency neuropathy in children with fat malabsorption. Studies in cystic fibrosis and chronic cholestasis. Ann N Y Acad Sci 1989; 570:156-69.

177. Weiser H, Vecchi M. Stereoisomers and tocopheryl acetate. II. Biopotencies of all eight stereoisomers, individually or in mixtures, as determined by rat resorption-gestation tests. Int J Vitam Nutr Res 1982; 52:351-70.

178. Sokol RJ, Heubi JE, Butler-Simon N, McClung HJ, Lilly JR, Silverman A. Treatment of vitamin E deficiency during chronic childhood cholestasis with oral *d*-alpha tocopheryl polyethlene glycol-1000 succinate. Gastroenterology 1987; 93:975-85.

179. Sokol RJ, Butler-Simon N, Conner C, et al. Multi-center trial of d-alpha tocopherol polyethylene glycol-1000 succinate therapy for vitamin E deficiency in chronic childhood cholestasis (abstract). Gastroenterology 1991; 100:A799.

180. Sokol RJ, Kim YS, Hoffnagle JH, et al. Intestinal malabsorption of vitamin E in primary biliary cirrhosis. Gastroenterology 1989; 96:479-86.

181. Sokol RJ. The coming of age of vitamin E. Hepatology 1989; 9:649-653.
182. Sokol RJ, Devereaux MW, Traber MG, Shikes RH. Copper toxicity and lipid peroxidation in isolated rat hepatocytes: effect of vitamin E. Pediatr Res 1989; 25:55-62.
183. Sokol RJ, Devereaux MW, Mierau GW, Hambidge KM, Shikes RH. Oxidant injury to hepatic mitochondrial lipids in rats with dietary copper overload: modification by vitamin E deficiency. Gastroenterology 1990; 99:1061-71.
184. Sokol RJ, Harris RE, Heubi JE. Effect of vitamin E on neutrophil chemotaxis in chronic childhood cholestasis (abstract). Hepatology 1984; 4:1048.
185. Farrell PM, Bieri JG, Fratantoni JF, Wood RE, DiSant'Agnese PA. The occurrence and effects of human vitamin E deficiency: a study in patients with cystic fibrosis. J Clin Invest 1977; 60: 233-40.

59

Effects of Vitamin E and Its Derivative on Posttraumatic Epilepsy and Seizures

Akitane Mori, Isao Yokoi, and Hideaki Kabuto

Okayama University Medical School, Okayama, Japan

INTRODUCTION

In some experimental models of epilepsy, vitamin E (α-tocopherol) is known to reduce seizure activity. Jerrett et al. (1) showed that hyperbaric oxygen-induced seizures in rats could be prevented by prior administration of vitamin E: that is, convulsions occurred in 100% of vitamin E-deficient rats and in 50% of rats fed a normal diet but did not occur in any of the rats fed a diet containing a vitamin E supplement. Kryzhanovsky et al. (2) demonstrated that prior injection of vitamin E decreased the number of seizures recorded on the electrocortico-gram after penicillin application to the sensory motor cortex of rats. Shandra and Kryzhanovsky (3) observed that vitamin E inhibited the development of epileptic activity in pentylenetetrazol-kindled mice. In addition, vitamin E is reported to cause a delay in the onset of pentylenetetrazol-induced seizures (4) and to decrease the frequency of epileptiform seizures induced by hypoxia (5).

In a clinical evaluation, Ogunmekan (6) first reported in 1979 that children with grand mal convulsive disorders showed significantly lower plasma vitamin E levels compared with age-matched medically healthy children. Reduced plasma vitamin E levels were observed in seizure patients mainly receiving antiepileptic drugs (7–9). Kovalenko et al. (10) administered vitamin E (α-tocopherol, 600 mg once daily) to patients with various forms of epilepsy who had shown resistance to their routine anticonvulsant and psychotropic therapy and found that the inclusion of vitamin E in the multiple modality therapy of certain forms of epilepsy increased the efficacy of the treatment. Moreover, Ogunmekan and Hwang (11) treated epileptic children who were refractory to antiepileptic drugs with vitamin E (D-α-tocopheryl acetate, 400 IU/day) as adjunctive therapy to the usual antiepileptic therapy and showed a significant improvement in seizures compared to the double-blind placebo control group. These results suggest that vitamin E may play some role in anticonvulsant therapy, although the mechanism remains unclear.

Willmore et al. (12,13) first reported in 1978 that a single injection of ferrous or ferric chloride solution into the rat or cat brain resulted in epileptic seizures. Willmore and Rubin (14) then observed that vitamin E and selenium prevented the development of epileptic dis-charges induced by ferrous chloride injection into rat brain. This iron (ferric and ferrous)-induced epilepsy model is of interest because it may serve as a model of clinical epilepsy ap-pearing in persons who have suffered head trauma (15,16), as described later in this chapter.

Levy et al. (17) recently examined the anticonvulsant effects of vitamin E in a variety of animal seizure models, that is, the pentylenetetrazol threshold model, the maximal electroshock model, amygdala-kindled seizures, and the ferrous chloride model. They reported that only the onset of electroencephalographic seizures was significantly delayed in the intracerebral ferrous chloride model and that no effect of vitamin E was observed in the other animal models commonly used to screen for anticonvulsive drug activity. In this chapter we introduce our findings concerning the anticonvulsant effect of vitamin E and its derivative EPC-K$_1$ in the iron-induced epilepsy model.

POSSIBLE BIOCHEMICAL PATHOGENESIS OF POSTTRAUMATIC EPILEPSY

Posttraumatic epilepsy is characterized by epileptic seizures due to brain damage secondary to head trauma. From the standpoint of the first occurrence of seizures, neurosurgeons classify posttraumatic epilepsy into two types: early epilepsy (or early seizures) and late epilepsy. Early epilepsy is defined as a convulsion occurring within 1 week of the head trauma. Seizures occur more frequently during the first week after injury than during the next 7 weeks, and such cases have a high risk of the development of late epilepsy.

Head trauma is often followed by epilepsy and may be related to the breakdown of red blood cells and hemoglobin within the central nervous system. Injection of hemoglobin or iron salts into the rat cortex is known to induce a chronic epileptic focus.

A single injection of 5 or 10 μl of 100 mM ferrous or ferric chloride into the rat or cat sensorimotor complex resulted in chronic recurrent focal paroxysmal electroencephalographic discharges, as well as behavioral convulsions (12,13). These epileptic discharges continued for more than 5 months after the injection (14).

We observed the generation of superoxide anion (O_2^-) and hydroxyl radical ($OH^{\cdot}$) and also an accelerated formation of guanidino compounds, which are endogenous convulsants, after ferric chloride injection into the rat cerebral cortex (16). These results suggest that these

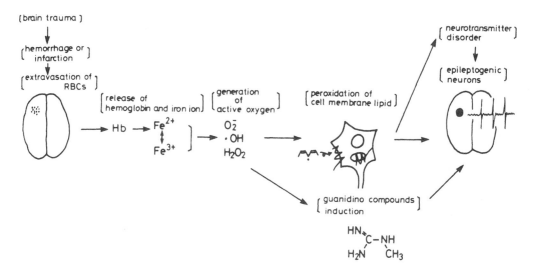

FIGURE 1   Possible pathogenesis of posttraumatic epilepsy.

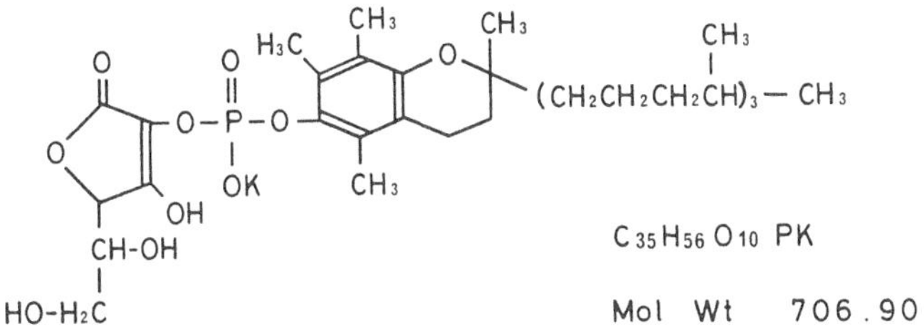

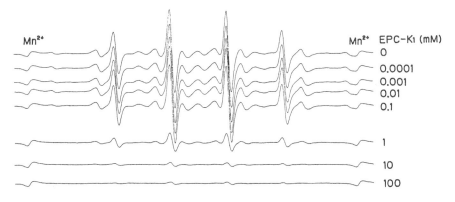

FIGURE 2   Chemical structure of EPC-K₁.

radicals, especially OH˙, may be reponsible for the initiation of lipid peroxidation in neuronal membranes and that the accelerated production of guanidino compounds in the brain may relate to the epileptogenesis.

Figure 1 illustrates a hypothesis for the pathogenesis of posttraumatic epilepsy (16). Head injury or hemorrhagic cortical infarction results in extravasation of blood and the breakdown of red blood cells and hemoglobin. Biological iron is normally protein bound in hemoglobin and transferrin. Iron liberated from hemoglobin is thought to be associated with the generation of active oxygen species. Moreover, hemoglobin itself may promote oxygen free-radical generation (18). Oxygen free radicals, especially OH˙, are responsible for the induction of peroxidation of neuronal lipids, that is, injury to the neuronal membranes (19). On the other hand, OH˙ accelerates the production of guanidino compounds, such as methylguanidine and guanidinoacetic acid, in the brain, where they are endogenous convulsants (16,20). These reactions may be followed by excitatory and inhibitory neurotransmitter disorders and may lead to the development of epileptic discharges in the epileptogenic focus.

EPC-K₁: A NEW HYDROXYL RADICAL SCAVENGER

L-Ascorbic and 2-[3,4-dihydro-2,5,7,8-tetramethyl-2(4,8,12-trimethyltridecyl)-2H-1-benz-pyrene-6-yl-hydrogen phosphate] potassium salt (EPC-K₁) is a new compound synthesized by Senju Pharmaceutical Co., Ltd. (Osaka) (21) and represents a phosphate diester linkage of vitamin E and vitamin C, as shown in Figure 2. EPC-K₁ has high water solubility, even

FIGURE 3   Effect of EPC-K₁ on the ESR spectra of the OH˙ radical generated by FeSO₄, H₂O₂, and DETAPAC. After 50 μl of 1.03 mM H₂O₂, 50 μl EPC-K₁ (0.1 μM to 100 mM), 10 μl DMPO, and 100 μl of 1 mM FeSO₄-DETAPAC were mixed in a test tube, DMPO-OH spin adducts were then analyzed exactly 60 s later by ESR spectrometry. HFS parameter of DMPO-OH: $A_N = A_H^\beta = 15.3G$.

though vitamin E itself is insoluble in water, and is known to have potent antioxidant (22) and antiinflammatory activities (23). We examined the effects of EPC-K$_1$ on oxygen free radicals using electron spin resonance spectrometry and the spin-trapping agent 5,5-dimethyl-1-pyrroline-1-oxide (24). EPC-K$_1$ was found to dose dependently scavenge hydroxyl radical (OH˙) generated by $FeCl_3 + H_2O_2$ + vitamin C or OH˙ generated by $FeSO_4 + H_2O_2$ + DETAPAC (Fig. 3), as well as OH˙ generated by *t*-butylhydroperoxide + ferric citrate, although almost no effect was observed on the O_2^- generated from the hypoxanthine-xanthine oxidase system. We also examined the effects of EPC-K$_1$ on thiobarbituric acid-reacting substances (TBARS) formation in the ferric citrate-induced epileptic focus of rat cerebral cortex and found that pretreatment with intraperitoneal EPC-K$_1$ (60 mg/kg IP 20 minutes before the ferric citrate injection) significantly inhibited the formation of TBARS in the ferric citrate-induced epileptic focus (24). It was suggested by these findings that EPC-K$_1$ may be useful as a clinical agent in the treatment and attenuation of progression of free-radical–induced degenerative diseases.

EFFECTS OF VITAMIN E AND ITS DERIVATIVES ON IRON-INDUCED SEIZURES

Effect of Vitamin E on Iron-Induced Seizures

Vitamin E inhibits the effects of oxidation in brain tissues and is a free-radical scavenger. Pretreatment vitamin E (dietary supplementation with α-tocopherol and with 2 ppm selenium for 5 days) is known to inhibit the development of epileptiform activity in 72% of animals (15) and to prevent the histopathological changes associated with iron injection (14).

On the other hand, we examined the effect of acute treatment with vitamin E on the development of epileptic seizures after the injection of iron into rat cerebral cortex.

Ea-0160-Z006 (water-solubilized vitamin E preparation; Eisai, Tokyo) had no ability to suppress the induction of iron-induced seizures when injected either intraperitoneally (250 mg/kg, 1 h before injection of 5 μl of 100 mM ferric chloride) or intravenously (250 mg/kg, simultaneously with the iron injection). Moreover, intracerebral injection of 125 μg Ea-0160-Z006 in 5 μl $FeCl_3$ solution (100 mM) was also found to have no effect.

Effect of EPC-K$_1$ on Iron-Induced Seizures

Experimental model rats were developed using our method (25). In brief, male Sprague-Dawley rats were immobilized with succinylcholine after all preparations for endotracheal intubation and artificial ventilation were completed under ether anesthesia. Four electrodes for an electrocorticogram (ECoG) were placed epidurally, as shown in Figure 4. A burr hole was made over the left sensorimotor cortex, 1 mm posterior and 1 mm lateral to the bregma for injection of solution.

In the rats given the ferric chloride (Fe, 500 nmol) injection into the sensorimotor cortex, spike discharges appeared on the ECoG 15–45 minutes after the Fe injection, and many spikes and/or seizure patterns were then observed 70–90 minutes after the injection (Fig. 4). When EPC-K$_1$ was injected with the Fe, the occurrence of epileptic discharges was prevented or delayed. As shown in Figure 4, the rat given 500 nmol $FeCl_3$ with 500 pmol EPC-K$_1$ exhibited no spike activity 3 h after the injection. Figure 5 shows the percentage occurrence of seizure patterns in each group of seven to nine rats. Seven control rats given $FeCl_3$ injection without EPC-K$_1$ exhibited seizure patterns 150 minutes after the injection, and the ECoG of two rats showed many spikes. In contrast, when 2500 pmol EPC-K$_1$ was injected with $FeCl_3$ (500 nmol), seizure patterns developed in the ECoG recorded in only one of seven rats up to 4 h after the injection. As shown in Figure 5, higher doses (500–5000 pmol) and even lower

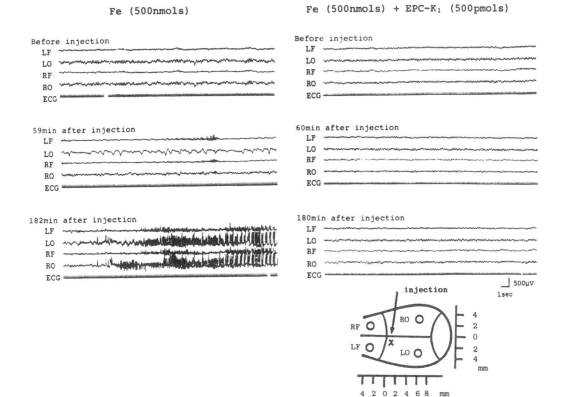

FIGURE 4 Effect of EPC-K₁ on the development of seizure activity induced by FeCl₃. Left: 5 μl FeCl₃ solution (100 mM) was injected into the sensorimotor cortex; spike bursts were seen in the LF 59 minutes after the injection, and a seizure pattern was seen in all leads 182 minutes after the injection. Right: 5 μl EPC-K₁ (100 μM) solution dissolved in FeCl₃ solution (100 mM) was injected into the sensorimotor cortex, and no spike activity was seen. The recordings were from four epidural electrodes: LF, unipolar recording from a left frontal electrode; LO, unipolar recording from a left occipital electrode; RF, unipolar recording from a right frontal electrode; and RO, unipolar recording from a right occipital electrode. ECG, electrocardiogram. Positions of each electrode and the injection site of the drugs are shown in the lower right corner.

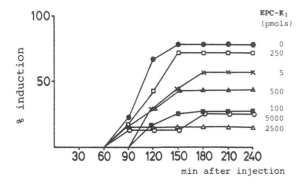

FIGURE 5 Occurrence of seizure pattern in electrocorticograms following FeCl₃ injection into the sensorimotor cortex of rat. Each rat was injected with 5 μl EPC-K₁ solution dissolved in FeCl₃ solution (100 mM) in the sensorimotor cortex. The 5 μl solution contained 0–5000 pmol EPC-K₁. Ordinate, percentage induction of ictal seizure patterns. Abscissa, time after injection. Each group consisted of seven to nine rats.

doses (5–100 pmol) of EPC-K₁ had a significant suppressing effect, but 250 pmol EPC-K₁ had almost no effect on the induction of seizure pattern; that is, five of seven rats developed seizure patterns. Such an interesting biphasic phenomenon might be related to multiactive sites acting as free-radical scavengers within the EPC-K₁ molecule.

Effect of EPC-K₁ on DOPAC and HVA Levels After Iron Injection into Rat Brain

A dialysis tube was inserted into the left caudate nucleus and the amounts of dihydroxyphenyl-acetic acid (DOPAC), homovanillic acid (HVA), and 5-hydroxyindoleacetic acid (5-HIAA) in the dialysate analyzed by our method (26) after injection of Fe into the right sensorimotor motor cortex of the rat to investigate the roles of dopaminergic and serotonergic neurotransmission in the development of Fe-induced seizures. Iron ions had no effect on the 5-HIAA levels in the dialysate samples but produced a significant increase in the DOPAC and HVA levels in the samples obtained after 180 minutes and thereafter (Fig. 6). Since dopamine released from the nerve endings is metabolized to HVA through DOPAC, the increases in the DOPAC and HVA contents in the dialysates indicates an increase in dopamine release in the

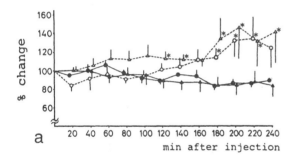

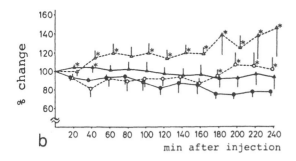

FIGURE 6   Time-dependent changes in the levels of DOPAC (a) and HVA (b) after injection of FeCl₃ and EPC-K₁ using a brain dialysis technique. The dialysis tube was inserted into the left caudate nucleus at the following coordinates: 2 mm anterior and 3 mm lateral to the bregma and 4 mm below the dura mater. The tube was perfused with physiological saline at 2 μl/minute and the dialysate was collected at 20 minute intervals. Data are expressed as percentage changes (mean ± SEM) from the value of the sample just before the injection: (closed circles) control, 5 μl of saline acidified by HCl to the same pH value as the FeCl₃ solution injected. (open circles) FeCl₃ (500 nmol); (closed triangle) EPC-K₁ (500 pmol); (open triangle) FeCl₃ (500 nmol) + EPC-K₁ (500 pmol). *$p < 0.05$ compared with control using the U test ($n = 5$–6).

caudate nucleus. On the other hand, a lack of change in the levels of 5-HIAA, which is the final product of serotonin released, indicated that iron ions did not alter the serotonergic neuroactivity. These data suggest that the development of an epileptic focus induced by iron ions accompanies the activation of dopaminergic neurons in the caudate nucleus.

There were no differences in the DOPAC, HVA, and 5-HIAA levels between the control group and the group given 500 pmol EPC-K$_1$ alone. When EPC-K$_1$ was injected with Fe, the increase in DOPAC and HVA contents began earlier, as they started to increase 120 and 40 minutes after the injection, respectively (Fig. 6). These results indicated that the activation of dopaminergic neurons in the caudate nucleus of rats injected with EPC-K$_1$ and Fe began earlier than in rats injected with Fe alone. Dopaminergic neuron are known to participate in an inhibitory system in the seizure mechanism (27,28). Therefore, EPC-K$_1$ may prevent and slow the occurrence of epileptic discharges induced by Fe in part through the activation of dopaminergic neurons.

CONCLUSION

Previous experimental studies have shown that vitamin E inhibited the development of seizure activity very effectively in some experimental animal models, such as hyperbaric oxygen-induced seizures (1). Such a potent anticonvulsant effect of vitamin E may be related to its antioxidant effect. Injection of iron salts into the rat cortex induced a chronic epileptic focus. In this case, iron is thought to be associated with the generation of active oxygen species. Vitamin E is a free-radical scavenger and inhibits the effects of oxidation in brain tissues. Chronic administration of vitamin E inhibited the development of an epileptic focus induced by iron ion (14,17). However, acute experiments with vitamin E have failed to suppress the induction of iron-induced seizure activity. The reason for a lack of an anticonvulsant effect in our acute experiment with vitamin E may relate to its insufficient delivery to the target neurons due to its less hydrophilic character. EPC-K$_1$ is a phosphate diester of vitamins E and C and is highly water soluble, as opposed to vitamin E itself. We found that EPC-K$_1$ was an effective hydroxyl radical scavenger and prevented the development of iron-induced seizure activity at a dose of 500–5000 pmol as well as at lower doses (5–100 pmol). Using a brain dialysis technique, we demonstrated that EPC-K$_1$ accelerated the dopaminergic nervous system. Dopamine participates as an inhibitory system in the seizure mechanism. Therefore, EPC-K$_1$ may prevent the occurrence of iron-induced epileptic discharges, not only through its antioxidant effect, but also in part through the activation of dopaminergic neurons. Such an inhibitory mechanism EPC-K$_1$ may be useful in explaining the anticonvulsant effects of vitamin E in some experimental models of epilepsy, which may not be closely related to a peroxidation procedure in the brain.

REFERENCES

1. Jerret SA, Jefferson D, Mengel CE. Seizure, H$_2$O$_2$ and lipid peroxides in brain during exposure to oxygen under high pressure. Aerospace Med 1973; 44:40-4.
2. Kryzhanovsky GN, Nikushkin EV, Braslavsky VE, Glebov RN. Lipoperoxidation in the hyperactive focus of rat cerebral cortex (Russian). Biull Eksp Biol Med 1980; 89:14-6.
3. Shandra AA, Kryzhanovsky GN. Antiepileptic effects of synthetic pharmacologic agents and vitamin preparations used in combination. (Russian). Zh Nevropatol Psikhiatr 1984; 84:889-903.
4. Sobaniec-Lotowska M, KvLak W. Cardiazol-induced seizures and the concentration of lipid peroxides in the brain of rats under the influence of valproic acid and vitamin E. Neuropathol Pol 1989; 27:229-36.
5. Kagan VE, Kopaladze RA, Prilipko LL, Savov VM. The role of lipid peroxidation in serotonin receptor, damage and appearance of epileptiform seizures during hypoxia. Biull Eksp Biol Med 1983; 96:16-8.

6. Ogunmekan AO. Vitamin E deficiency and seizures in animals and man. Can J Neurol Sci 1979; 6:43-5.

7. Ogunmenkan AO. Predicting serum vitamin E concentrations from the age of normal and anticonvulsant drug treated epileptic children using regression equations. Epilepsia 1979; 20:295-301.

8. Higashi A, Tamari H, Ikeda T, et al. Serum vitamin E concentration in patients with severe multiple handicaps treated with anticonvulsants. Pediatr Pharmacol (New York) 1980; 1:129-34.

9. Nagaura T, Mimaki T, Ono J, Yabuuchi H, Sugi K, Kamio M. Decreased serum vitamin E levels in epileptic children. Brain Dev 1985; 7:115.

10. Kovalenko NM, Kryzhanovsky GN, Pronina IG, Nikushkin EV. α-Tocopherol in the multiple modality therapy of some forms of epilepsy (Russian). Zh Nevropatol Psikhiatr 1984; 84:892-7.

11. Ogunmekan AO, Hwang PA. A randomized, double-blind, placebo-controlled, clinical trial of D-α-tocopheryl acetate (vitamin E), as add-on therapy, for epilepsy in children. Epilepsia 1989; 30: 84-9.

12. Willmore LJ, Sypert GW, Munson JB, Hurd RW. Chronic focal epileptiform discharges induced by injection of iron into rat and cat cortex. Science 1978; 200:150-3.

13. Willmore LJ, Hura RW, Sypert GW. Epileptiform activity initiated by pial iontophoresis of ferrous and ferric chloride on rat cerebral cortex. Brain Res 1978; 152:406-10.

14. Willmore LJ, Rubin JJ. Antiperoxidation pretreatment and iron-induced epileptiform discharges in the rat: histopathologic studies. Neurology 1981; 3:63-9.

15. Willmore J, Triggs WL, Gray JD. The role of iron induced hippocampal peroxidation in acute epileptogenesis. Brain Res 1986; 382:422-6.

16. Mori A, Hiramatsu M, Yokoi I, Edamatsu R. Biochemical pathogenesis of post-traumatic epilepsy. Pavlov J Biol Sci 1990; 25:54-62.

17. Levy SL, Burnham WM, Hwang PA. An evaluation of the anticonvulsant effects of vitamin E. Epilepsy Res 1990; 6:12-7.

18. Sadzadeh SMH, Panter SS, Hallaway PE, Eataon JW. Hemoglobin: a biological Fenton reagent. J Biol Chem 1984; 259:14354-6.

19. Mori A, Hiramatsu M, Edamatsu R, Kohno M. Possible involvement of oxygen free radicals in the pathogenesis of post-traumatic epilepsy. In: Hayashi O, Niki E, Kondo M, et al., eds. Medical, biochemical and chemical aspects of free radicals. Amsterdam: Elsevier, 1989; 1249-52.

20. Mori A. Biochemistry and neurotoxicology of guanidino compounds: history and recent advances. Pavlov J Biol Sci 1987; 22:85-94.

21. Ogata K. Phosphoric acid diesters or their salts and process for producing the same. U.S. Patent 4,564,686, 1986.

22. Seo K, Ogata K, Yoshida K, Uehara K, Tomita K. Antioxidant. Japanese Laid Patent Publication No. 139972, 1988 (Japanese).

23. Yamamoto I, Ogata K, Terashima K. Antiphlogistic drug. Japanese Laid Patent Publication No. 145019, 1987.

24. Mori A, Edamatsu R, Kohno M, Ohmori S. A new hydroxy radical scavenger: EPC-K$_1$. Neurosciences 1989; 15:371-6.

25. Yokoi I, Kabuto H, Akiyama K, Mori A, Ozaki M. Tannins inhibit the occurrence of epileptic focus induced by FeCl$_3$ injection in rats. Jpn J Psychiatr Neurol 1989; 43:552-3.

26. Kabuto H, Yokoi I, Okamura Y, Yamamoto M, Suyama K, Mori A. Effects of ferric ion and (−)-epigallocatechin-3-O-gallate on monoamine metabolite levels in rat striatum by brain dialysis. Neuroscience 1990; 16:377-82.

27. Kobayashi K, Mori A. Brain monoamine in seizure mechanism. Folia Psychiat Neurol Jpn 1977; 31:482-9.

28. Mori A. Alterations in biogenic amines in epileptic E1 mouse brain and in human epileptic focus. In: Suzuki, J, Seino M, Fukuyama Y, et al., eds. Art and science of epilepsy. Amsterdam: Excerpta Medica, 1989; 35-8.

C. Pediatrics

60

Vitamin E in the Newborn

Sunil Sinha and Malcolm Chiswick

Saint Mary's Hospital, Manchester, England

INTRODUCTION

Vitamin E is a major antioxidant, effective in stabilizing unsaturated lipids in cell membranes against autooxidation. It is also thought to be an important structural component of biological membranes, maintaining their integrity and function in a role different from its antioxidant properties. Furthermore, vitamin E scavenges free radicals produced by lipid peroxidation and by the normal activity of oxidative enzymes (1). Because of its position in cellular function, it is not surprising that diverse roles have been suggested for vitamin E in many biological reactions and in human disease.

Since its discovery in 1922 by Evans and Bishop (2) as a fat-soluble dietary factor necessary for normal reproduction in the rat, a variety of vitamin E-deficient syndromes have been produced experimentally in several animal species, showing effects on reproduction, neuromuscular, cardiovascular, and hematopoetic systems. Human clinical correlates have also been sought, but vitamin E deficiency does not readily occur in adult humans owing to resistance of their tissue stores to depletion and to the wide distribution of vitamin E in foods. However, low circulating levels of vitamin E in humans have been described in association with various malabsorption syndromes, such as cystic fibrosis, biliary atresia, and abetalipoproteinemia, and in neonates, especially those born preterm (3). This has led to a search for a corresponding vitamin E deficiency syndrome that might bridge the gap between experimental studies in laboratory animals and practical application of vitamin E therapy in human disease.

EVALUATION OF VITAMIN E STATUS

Various approaches have been used to evaluate vitamin E status, but there is still no widely agreed definition of vitamin E deficiency. On the basis of data obtained from normal and malnourished children and adults, Horwitt et al. (4) concluded that measurements of blood tocopherol concentrations alone may be misleading and cause erroneous interpretations. In their study, a substantial number of adults with plasma tocopherol levels below the normal range (1.05 mg/dl with a 95% confidence interval of 0.5–1.6 mg/dl) did not show any abnormality in peroxide-induced erythrocyte hemolysis. Their data also indicated that plasma or serum tocopherol concentrations were dependent on the levels of circulating lipids. Subsequent reports (5) have confirmed the advantages of using a blood lipid reference base when expressing the results of plasma tocopherol levels. This proposal that the tocopherol-lipid ratio be utilized to assess the vitamin E status corresponds to the mechanism of tocopherol transport in blood, since circulating vitamin E is carried with lipoproteins according to the fat concentration of various fractions. This is especially relevant to preterm newborn infants who initially have very low levels of circulating lipid because of the relatively ineffective transport of fat across the placenta. Dju et al. (6), in an earlier study, confirmed that in fetuses and the newborn there is a positive linear relationship between total-body tocopherol and lipid stores. A positive correlation between cord blood vitamin E levels, on the one hand, and b lipoprotein, cholesterol, and polyunsaturated fatty acids, on the other, has also been reported.

To a variable extent, investigators have also included a hydrogen peroxide hemolysis test to evaluate what is presumed to be a consequence of vitamin E deficiency. Circulating red cells are particularly susceptible to peroxidative damage because conditions that favor peroxidation are seemingly optimal in red cells. The membrane of these cells is rich in polyunsaturated fatty acids, the cells are continuously exposed to high oxygen tensions, and the cell contains hemoglobin, one of the most powerful catalysts for the initiation of peroxidative reactions. That normal red cells are protected from peroxidative damage in vivo can be attributed to efficient antioxidant mechanisms. This antioxidant protection, although partly a function of the structural integrity of each cellular constituent, is also a reflection of antioxidant systems in the cells. These include superoxide dismutase, glutathione peroxidase, catalase, and vitamin E. An impairment of any of these defense mechanisms may render the red cells more susceptible to peroxidative damage and eventually lead to its breakdown. Although this measure of peroxide-induced hemolysis has been considered to be only a crude test, a normal result probably implies an adequate level of intracellular antioxidant (5). Some authors have measured malondialdehyde formation—a secondary breakdown product of lipid peroxidation—by incubating red blood cells in vitro with hydrogen peroxide solution as a characterization of vitamin E status in an at-risk population (7).

PLASMA VITAMIN E LEVELS IN FETUS AND NEWBORN

A number of investigators (3) have reported blood tocopherol levels in premature infants that were consistently lower (Table 1) than the adult level, which averages 1.05 mg/dl. Dju et al. (6) measured the total-body store of vitamin E in human fetuses and neonates who died within 9 days of birth. In 22 fetuses between 2 and 6 months of gestational age, the mean total tocopherol content was 3.1 mg/kg body weight compared with 5.6 mg/kg in a 3570 g full-term stillborn infant and an estimated 70 mg/kg body weight in adults. Since adipose tissue was found to contain the greatest concentration of tocopherols and since the amount of adipose

TABLE 1 Plasma or Serum Tocopherol Levels (mg/dl) Reported in Premature Infant

Authors	Mean	Range
Owens and Owens (1949)	0.25	—[a]
Moyer (1950)	0.22	0.04–0.46
Mackenzie (1954)	0.20	0.06–0.39
Nitowsky et al. (1956)	0.26	—[a]
Hassan et al. (1966)	0.32	0.12–0.50
Oski and Barnes (1967)	0.22	0.09–0.32
Gross and Melhorn (1972)	0.25	—[a]
Gutcher et al. (1984)	0.27	0.13–0.64

[a]Range not reported.

TABLE 2 Plasma or Serum Vitamin E Levels (mg/dl) Reported in Mothers and Their Neonates

Mean maternal vitamin E (mg/dl)	Mean neonatal vitamin E (mg/dl)	Ratio of maternal to neonatal vitamin E levels	Source
1.70[a]	0.34[a]	5.7:1.0	Straumfjord and Quaife (1946)
1.92[b]	0.38[a]	5.1:1.0	Wright et al. (1951)
1.32[b]	0.24[c]	5.5:1.0	Nitowsky et al. (1956)
0.92[a]	0.24[a]	3.9:1.0	Leonard et al. (1972)
1.71[a]	0.41[a]	4.2:1.0	Mino and Nishino (1973)
2.53[d]	0.52[a]	4.9:1.0	Mino and Nishino (1973)
1.60[a]	0.61[a]	2.6:1.0	Tateno and Ohshima (1973)
1.20[a]	0.30[a]	4.0:1.0	Baker et al. (1975)
2.10[e]	0.42[a]	5.0:1.0	Haga and Lunde (1978)

[a]Blood samples obtained at delivery (cord blood samples).
[b]Blood samples obtained 2 days postpartum postnatally.
[c]Blood samples obtained 1 day postpartum postnatally.
[d]Women who took supplementary vitamin E acetate (600 mg/day).
[e]Maternal level estimated from graphic data.

tissue increases from less than 1% body weight during the first two trimesters to approximately 16% of body weight at term, Dju et al. concluded that tocopherol body stores were low in the fetuses. A number of studies have since also determined blood vitamin E levels in newborn infants and their mothers and have consistently noted that maternal levels at delivery or in the immediate postpartum period were higher than the neonatal levels. Mean maternal levels in these studies ranged from 0.92 to 2.53 mg/dl compared with mean levels ranging from 0.24 to 0.61 mg/dl in their babies (Table 2). Since the ratio of maternal to neonatal blood levels in mother-baby pairs ranged from 2.6:1 to 5.7:1, the data suggested that transfer of vitamin E from the mother to the fetus may be limited (3).

In a recent (8) study to assess the effect of vitamin E supplementation on plasma concentration, we enrolled 231 babies, ≤32 weeks of gestation, who were randomized to the vitamin E-supplemented or control group. The susceptibility of red blood cells to hydrogen peroxide hemolysis, plasma vitamin E concentration, and plasma total lipids was measured in samples of heparinized venous or arterial blood (1.5 ml) drawn from babies immediately before the first dose of vitamin E and at comparable times in the control babies (day 0). Measurements were repeated 24, 48, and 72 h later (days 1, 2, and 3). Plasma vitamin E concentration was measured as total tocopherol by a colorimetric method in which ferrous iron produced

TABLE 3 Values of Plasma Vitamin E, Lipid, and Vitamin E-Lipid Ratio at Birth (Day 0)

	Total tocopherol (mg/dl)	Total lipid (mg/dl)	Total tocopherol-lipid ratio (mg/dl)
Mean ± SD	0.43 ± 0.20	331.59 ± 80	1.39 ± 0.79
Median	0.40	337	1.23
Range	0.10–1.13	140–643	0.26–7.16
10th percentile	0.18	171	0.59
90th percentile	0.73	490	2.20

TABLE 4 Values of Plasma Tocopherol (mg/dl) and Tocopherol-Lipid Ratio (mg/g) According to Birth Weight Groups

Birth weight (g)	Total tocopherol (mean ± SD)	Tocopherol-lipid ratio (mean ± SD)
600–750 (15)[a]	0.51 ± 0.27	1.9 ± 1.57
751–1000 (41)	0.39 ± 0.16	1.32 ± 0.52
1001–1250 (62)	0.43 ± 0.21	1.35 ± 0.68
1251–1500 (51)	0.44 ± 0.21	1.36 ± 0.73
1501–1750 (38)	0.43 ± 0.20	1.35 ± 0.82
1751–2000 (16)	0.43 ± 0.13	1.34 ± 0.62

[a]Number in study group.

by the reduction of ferric iron by vitamin E was used as an index of the vitamin E concentration. A colorimetric method (sulfophosphovanillin reaction) was used to measure total plasma lipids. The value of plasma total tocopherol and the ratio of plasma tocopherols to lipid levels in 223 babies on day 0 (soon after birth) is shown in Table 3. There was no significant difference observed in the level of plasma tocopherol or its ratio to plasma lipids when these values were compared according to birth weight (Table 4) and their gestational age groups (Table 5). There was a statistically significant but weakly positive correlation between plasma tocopherol and lipid levels ($r = 0.18$, $p < 0.002$).

In addition to the neonatal supplementation study, the effect of maternal vitamin E supplementation on the vitamin E status of their preterm newborn babies was also assessed in a double-blind controlled design. A total of 60 mothers admitted to the delivery suite with threatened preterm delivery between 26 and 32 weeks of gestation were enrolled in this study after informed consent. These mothers were given either vitamin E capsules (α-tocopherol acetate, one capsule = 400 IU) or placebo up to a maximum of 16 capsules in divided doses at 4–6 h intervals until delivery. Those mothers ($n = 13$) who did not deliver within 24 h of taking the last capsule were excluded from the study and not reentered into the trial even if their delivery became imminent again. Paired blood samples were collected from each mother immediately before the administration of the capsule and at delivery. Blood samples from the newborn babies were collected at birth as part of the neonatal supplementation trial as already described.

Oral supplementation with vitamin E of mothers ($n = 28$) during labor resulted in a significantly higher mean ± standard deviation (SD) plasma vitamin E level at delivery (1.73 ± 0.70 mg/dl) compared to levels at a corresponding time in mothers ($n = 19$) who received placebo (0.97 ± 0.32 mg/dl; $p < 0.003$). The mean difference in the vitamin E levels between paired blood samples of supplemented mothers was 0.41 mg/dl compared to only 0.02 mg/dl

TABLE 5 Value of Plasma Tocopherol (mg/dl) and Tocopherol-Lipid Ratio (mg/g) According to Gestational Age Groups

Gestational age (weeks)	Total tocopherol (mean ± SD)	Tocopherol-lipid ratio (mean ± SD)
24–26 (38)[a]	0.40 ± 0.19	1.36 ± 0.58
27–29 (86)	0.42 ± 0.19	1.43 ± 0.89
30–32 (99)	0.45 ± 0.21	1.37 ± 0.76

[a]Number in study group.

in the placebo mothers. Despite a significant rise in plasma vitamin E levels in supplemented mothers, the mean ± SD vitamin E levels in their babies ($n = 30$, including two pairs of twins) at birth was only slightly higher (0.44 ± 0.18 mg/dl) than the vitamin E levels in babies ($n = 19$) whose mothers received placebo (0.37 ± 0.15 mg/dl), and this difference was not statistically significant (Fig. 1).

This finding is in keeping with there being a "barrier" to the passage of vitamin E from mothers to the fetus. Whether this threshold can be surpassed by supplementing mothers with parenteral preparation of vitamin E, which appears to have a superior pharmacokinetics in terms of achieving higher levels in a comparatively shorter time than the oral preparations, is under evaluation.

Although newborn babies have low plasma vitamin E concentration compared to adults, it is moot whether this represents true vitamin E deficiency. On the basis of results obtained in studies of adult patients, Horwitt et al. (4) concluded that a ratio above 0.8 mg tocopherol per gram total lipids in serum should be considered a sign of vitamin E nutrition. Subsequent studies (5), however, disputed the value of this index; they identified several infants and children with a ratio of 0.6–0.8 who did not show abnormal peroxide hemolysis results. Thus, additional data are needed to determine what value of the tocopherol-total lipid ratio should be established as the lower limit of normal, especially in newborn infants. Estimation of tocopherol content in adipose or other tissues, such as peripheral nerve, might provide better evidence of vitamin E deficiency, but in practice most investigators have relied on vitamin E in blood. Indeed, there is no widely accepted definition of vitamin E deficiency in adults. When newborn babies, especially those born preterm, are considered, the problem of defining vitamin E deficiency becomes even more difficult because the latter subset of babies may be considered abnormal by their very existence ex utero. Thus, the findings in such infants of low plasma vitamin E concentration compared to adults and older children raises the issue of whether the low levels represent normal age-adjusted values or whether they in fact represent a deficiency state.

The precise relationship between plasma levels and the biological activity of vitamin E in tissue is even more unclear. Although the plasma levels of vitamin E after supplementation,

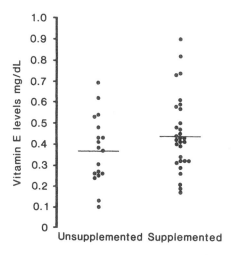

FIGURE 1   Vitamin E levels of babies at birth born to supplemented ($n = 30$) and placebo ($n = 19$) mothers.

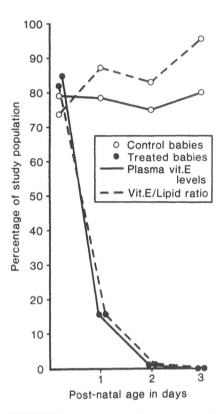

FIGURE 2   Proportion of supplemented and control babies with plasma vitamin E levels < 0.64 mg/dl and vitamin E-lipid ratio < 1.9 at birth and during the first 3 days of life.

noted in our study (Fig. 2) as well as other studies, were certainly close to the physiological range observed in adults, whether these enhanced levels indicate protective antioxidant levels is a question of semantics in view of the proposed role for tocopherol at tissue membrane levels (9). However, Gutcher et al. (5) evaluated the vitamin E status of 62 premature infants born between 27 and 34 weeks and weighing between 720 and 2240 g. Based on a mathematical and statistical modeling of the results of plasma tocopherol, its ratio to total lipids, and the percentage of in vitro hydrogen peroxide hemolysis test in their study, the authors concluded that the ex utero antioxidant protective role of vitamin E was best achieved at plasma concentrations of tocopherol very close to those observed in adults, especially when the total tocopherol was more than 0.64 mg/dl and a total tocopherol-lipid ratio of more than 1.9. Using these discriminant levels of plasma vitamin E and plasma vitamin E-lipid ratios as a measure of "normalcy" as suggested by Gutcher et al., it can be said that more than 75% of babies in our study were deficient at birth.

THERAPEUTIC USES OF VITAMIN E IN THE NEWBORN

The observation of low plasma vitamin E levels in preterm infants has led a number of investigators to assess the therapeutic role of vitamin E in a range of neonatal conditions. The evidence of its efficacy in many of these conditions (described here) has been a matter of intense controversy in recent years.

Anemia of Prematurity

Vitamin E has been used in premature newborns for the treatment or prevention of anemia. In this regard, two distinct entities must be distinguished. When a vitamin E deficiency exists with resultant hemolytic anemia associated with thrombocytosis, treatment with vitamin E is definitely beneficial. Oski and Barness (10) described the development of hemolytic anemia in chemically vitamin E-deficient preterm infants aged 6–10 weeks who were fed proprietary formulas. The anemia was characterized by marked reticulocytosis, abnormal red blood cell (RBC) morphology, and shortened RBC survival times. Administration of vitamin E to affected infants resulted in a prompt increase in the hemoglobin level and a fall in the reticulocyte count. Daily vitamin E supplements beginning on the third day of life also prevented or markedly decreased the severity of the hemolytic anemia. Subsequently, Ritchie et al. (11) reported that generalized edema, hemolytic anemia, reticulocytosis, and thrombocytosis accompanied vitamin E deficiency in premature infants aged 4–8 weeks who had received commercial formulas. Again, these abnormalites were corrected with vitamin E therapy. This vitamin E-responsive hemolytic anemia is the only true vitamin E deficiency state occurring naturally in preterm infants.

Whether the prophylactic use of vitamin E ameliorates the normochromic "physiological" anemia that occurs in premature neonates at 6–10 weeks is unclear and is the subject of current controversy. Most studies do not support a beneficial effect of supplemental vitamin E in premature infants fed breast milk or proprietary formulas (12).

Related to the question of vitamin E supplementation for the amelioration of the physiological normochromic anemia, several investigators evaluated the efficacy of the vitamin in reducing physiological jaundice in the newborn on the basis that the anemia might have a hemolytic component. However, in vivo hemolysis, as reflected by serum bilirubin concentration, has not been influenced by vitamin E therapy in any consistent way (13). In the latter controlled study of preterm infants weighing between 1000 and 2000 g, intramuscular administration of α-tocopherol, 50 mg/kg in divided doses on days 1, 2, and 3 of life, resulted in a reduction in serum bilirubin in the vitamin E-treated infants who weighed less than 1500 g. The differences were significant only on day 3 and for the peak bilirubin observed. There was also a significant decrease in duration of phototherapy in babies who received vitamin E supplements (48 versus 107 h). In infants weighing more than 1500 g, these differences were not observed.

Based on the same rationale, that vitamin E protects erythrocytes against hemolysis, its role has been assessed in infants with hemolytic uremic syndrome with encouraging results (14).

Retinopathy of Prematurity (ROP)

Retinopathy of prematurity is a condition that affects the growth of immature retinal blood vessels and may lead to blindness. Owen and Owen (15) first postulated that vitamin E deficiency might be linked to the development of retrolental fibroplasia and reported a lower incidence of this condition among 101 infants weighing 1360 g or less who received vitamin E (4.4 versus 21.8% in controls). Subsequently, interest in the role of vitamin E in the prevention of retrolental fibroplasia decreased when it was shown that the cause was clearly related to the level of oxygen therapy. The common use of assisted ventilation with high oxygen concentration in neonatal intensive care units has awakened the interest in the relation of vitamin E to retinal oxygen toxicity and has led to further investigations regarding its prevention with vitamin E supplementation during the immediate neonatal period.

In those studies that showed beneficial effects of vitamin E, the overall incidence of ROP was not affected by early supplementation, but the incidence of severe ROP was reduced

(16–18). In these studies vitamin E (α-tocopherol acetate) was given orally and/or intramuscularly from birth and continued until discharge, which ranged up to more than 10 postnatal weeks. Other controlled studies, however, have failed to demonstrate conclusively any benefit of vitamin E prophylactic supplementation on ROP (19). One randomized double-blind controlled study involving 287 newborns not only showed no protective effect of early vitamin E on the incidence and severity of ROP, but the treated group had a significantly higher incidence of retinal hemorrhages (20).

Bronchopulmomary Dysplasia (BPD)

Bronchopulmonary dysplasia is a form of chronic lung disease that occurs all too frequently in low-birth weight babies. The etiology of BPD has remained controversial and is probably multifactorial, including toxic effects of oxygen on the lungs.

However, the efficacy of vitamin E in the prevention of BPD is not convincing. Although animal studies have indicated that vitamin E deficiency enhances lung oxygen toxicity, its benefit to human newborns have not been demonstrated. An earlier preliminary study suggested a shorter requirement for respiratory support in newborns treated with intramuscular vitamin E (20 mg/kg IM) for 3 days (21). However, subsequent studies by the same authors were unable to confirm their preliminary observations (22). Therefore, the value of vitamin E in preventing neonatal oxidative lung injury has yet to be demonstrated.

Periventricular Hemorrhage

Serial ultrasound brain scanning in the newborn carried out by the cotside has shown that about 40% of preterm babies, 32 weeks of gestation, sustain bleeding in or around the lateral ventricles of the brain (periventricular hemorrhage, PVH). The hemorrhage commences in the floor of the ventricles (subependymal hemorrhage), where the primitive vascular bed is poorly supported in a matrix of loose connective tissue. The hemorrhage may be confined here or spread into the cavity of the lateral ventricles (IVH). Bleeding can also occur into the brain parenchyma adjacent to the lateral ventricles (parenchymal hemorrhage). Babies with subependymal or small intraventricular hemorrhages have a good prognosis for long-term development. Babies who have large intraventricular hemorrhages with ventricular dilatation and those with parenchymal hemorrhages have a much worse prognosis for neurodevelopmental outcome.

In a small study designed to assess the efficacy of intramuscular vitamin E in increasing plasma vitamin E levels and reducing hydrogen peroxide hemolysis in preterm newborn babies, Chiswick et al. (23) made the unexpected observation that treated babies have a reduced incidence of intraventricular hemorrhage at postmortem (2 of 14) compared with nontreated controls (9 of 21).

These findings were further explored in a large randomized controlled trial in which daily supplementation with 20 mg/kg of vitamin E intramuscularly during the first 3 days of life was associated with a very significant protective effect against intraventricular hemorrhage diagnosed by ultrasound in preterm babies (8). The incidence of this lesion was reduced from 34.3% (37 of 108) in nonsupplemented babies to 8.8% (9 of 102) in those who received vitamin E supplement ($p < 0.005$). The supplemented babies also had a lower combined frequency of intraventricular and parenchymal hemorrhages (referred to as PVH) than the controls (10.8 versus 40.7%; $p < 0.0001$; 95% confidence interval for difference, 18.9–41.0%). A multivariate analysis, adjusted for factors other than vitamin E that may influence the occurrence of PVH, showed that vitamin E had an independent protective effect against PVH ($p < 0.001$). The mean plasma vitamin E level of babies who did not develop PVH was higher than that of those who sustained hemorrhage throughout the first 3 days of life. In relation to the protective effect of vitamin E against PVH as observed in this study, it was speculated that vitamin E scavenges free radicals generated during ischemic injury of the subependymal region and thereby limits tissue damage and the extent of periventricular hemorrhage on reperfusion. This notion was supported by the finding that no protective effect of vitamin E was observed against subependymal hemorrhage but supplemented babies had a markedly reduced incidence of intraventricular and parenchymal hemorrhage. Although the frequency of ventricular enlargement was lower (20.2%; 21 of 104) in the supplemented group than in the controls (33.6%; 36 of 107; $p < 0.05$), vitamin E had no independent effect in influencing the occurrence of ventricular enlargement. The lower incidence of ventricular enlargement among vitamin E-supplemented babies was presumably secondary to reduction in the incidence of intraventricular hemorrhage. Nonetheless, the findings in the study (unpublished) of a reduced combined incidence of IVH and parenchymal hemorrhage in survivors with ventricular dilatation among vitamin E-supplemented babies (52.9%; 9 of 17) compared to control groups (79.2%; 19 of 24; $p < 0.05$) was relevant and may have potential advantage as the neurodevelopmental outcome among surviving babies with cerebral ventricular enlargement is thought to be related to the severity of associated intraventricular hemorrhage.

Two previous studies primarily designed to assess the role of vitamin E in preventing the retinopathy of prematurity have given conflicting results with respect to incidental protection against PVH. Phelps et al. (20) warned against the use of vitamin E in preterm babies and suggested that intravenous vitamin E was associated with a higher frequency of severe grades of PVH and death among babies weighing < 1 kg at birth. Variable vitamin E doses and diagnostic criteria of PVH were used in this study, however, and cranial ultrasound was neither part of the original study nor performed routinely. The increased moretality in treated babies was probably related to the toxicity of the preparation used in this study: intravenous tocopherol has been known to be associated with serious hepatic renal and hematopoietic toxicities. Although Phelps et al. monitored the plasma vitamin E levels in their study babies, these were performed only twice a week, which may not have detected the rapid elevation in plasma vitamin E levels seen to occur with intravenous preparation; this possibility is supported by their own findings of a very high concentration of tocopherol in the liver of one of their infants noted only 3.5 h after a single dose of vitamin E.

Speer et al. (25) reported a reduction in both the severity and incidence of PVH among babies < 1000 g who were given vitamin E supplementation (oral and IM) during the immediate neonatal period. This study, however, did not include referred babies who had an increased risk of developing IVH. More recently, Fish and associates (26), studying babies of birth weights of 501–1000 g, showed that vitamin E supplementation was associated with a significant reduction in the frequency of PVH, including the most severe grades among tiny babies who weighed 501–750 g at birth.

REFERENCES

1. Biology of vitamin E. Ciba Foundation Syumposium 101. London: Pitman, 1983.
2. Evans HM, Bishop KS. On the existence of a hitherto unrecognised dietary factor essential for reproduction. Science 1922; 56:650-1.
3. Ehrenkranz RA. Vitamin E and the neonate. Ann J Dis Child 1980; 134:1157-66.
4. Horwitt MK, Harvey CC, Dahm CH, et al. Relationship between tocopherol and serum lipid levels for determination of nutritional adequacy. Ann N Y Acad Sci 1972; 203:223-36.
5. Gutcher GR, Raynor WJ, Farrell PM. An evaluation of vitamin E status in premature infants. Ann J Clin Nutr 1984; 40:1078-89.
6. Dju MY, Mason KE, Filer LJ Jr. Vitamin E (tocopherol) in human fetuses and placenta. Etudes Neo-Natales 1952; 1:49-60.

7. Cynamon HA, Isenberg JN. Characterization of vitamin E status in cholestatic children by conventional laboratory standards and a new functional assay. J Pediatr Gastroenterol Nutr 1987; 6:46-50.

8. Sinha S, Davies J, Toner N, et al. Vitamin E supplementation reduces frequency of periventricular haemorrhage in very preterm babies. Lancet 1987; 1:466-71.

9. Molenaar I, Hulstaert CE, Hardonk MJ. Role in function and ultrastructure of cellular membranes. In: Machlin LJ, ed. Vitamin E: a comprehensive treatise. New York: Marcel Dekker, 1980; 372-89.

10. Oski FA, Barness LA. Vitamin E deficiency: a previously unrecognised cause of haemolytic anaemia in the premature infant. J Pediatr 1987; 70:211-20.

11. Ritchie JH, Fish MB, McMasters V, et al. Edema and haemolytic anaemia in preterm infants: a vitamin E deficiency syndrome. N Engl J Med 1968; 279:1185-90.

·12. Aranda JU, Chemtob S, Landigron N, et al. Frusemide and vitamin E. Pediatr Clin N Am 1986; 33:588-602.

13. Gross SJ. Vitamin E and neonatal bilirubinaemia. Pediatrics 1979; 64:321-3.

14. Powell HR, McCreedie DA, Taylor CM, et al. Vitamin E: treatment of haemolytic uraemic syndrome. Arch Dis Child 1984; 59:401-9.

15. Owen WE, Owen EV. Retrolental fibroplasia in premature infants. II. Studies on prophylaxis of disease. Use of alpha-tocopherol acetate. Ann J Ophthalmol 1949; 32:1631-7.

16. Johnson L, Schaffer D, Boggs TR Jr. The premature infant, vitamin E deficiency and retrolental fibroplasia. Ann J Clin Nutr 1974; 27:1158-72.

17. Hittner HM, Speer MR, Rudolph AJ, et al. Retrolental fibroplasia and vitamin E in preterm infant. Comparison of oral versus intramuscular and oral administration. Pediatrics 1984; 73:238-49.

18. Finer NN, Schindler RF, Peters KL, et al. Vitamin E and retrolental fibroplasia. Improved visual outcome with early vitamin E. Ophthalmology 1983; 90:428-35.

19. Puklin JE, Simon RM, Ehrenkranz RA. Influence on retrolental fibroplasia of intramuscular vitamin E during respiratory distress syndrome. Ophthalmology 1982; 89:96-9.

20. Phelps DL, Rosenbaum AL, Isenberg SJ. Tocopherol efficacy and safety for preventing retinopathy of prematurity: a randomised controlled, double masked trial. Pediatrics 1987; 79:489-99.

21. Ehrenkranz RA, Bonta BW, Ablow RC, et al. Amelioration of bronchopulmonary dysplasia after vitamin E administration. N Engl J Med 1978; 299:564-9.

22. Ehrenkranz RA, Ablow RC, Warshaw JB. Amelioration of bronchopulmonary dysplasia after vitamin E administration during the acute stages of respiratory distress syndrome. J Pediatr 1979; 95:873-8.

23. Chiswick ML, Johnson M, Woodhall C, et al. Protective effect of vitamin E (*dl*-alpha-tocopherol) against intraventricular haemorrhage in premature babies. Br Med J 1983; 287:81-4.

24. Chiswick M, Gladman G, Sinha S, et al. Prophylaxis of periventricular hemorrhage in preterm babies by vitamin E supplementation. Ann N Y Acad Sci 1989; 570:197-204.

25. Speer ME, Blifeld C, Rudolph AJ, et al. Intraventricular haemorrhage and vitamin E in the very low-birth-weight infant; evidence for efficacy of early intramuscular vitamin administration. Pediatrics 1984; 74:1107-12.

26. Fish WH, Cohen M, Franzek D, et al. Effect of intramuscular vitamin E on mortality and intracranial hemorrhage in neonates of 1000 grams or less. Pediatrics 1990; 85:578-84.

D. Ophthamology

61

Vitamin E and the Eye

John R. Trevithick, James McD. Robertson, and Kenneth Patrick Mitton

University of Western Ontario, London, Ontario, Canada

INTRODUCTION

Of the five senses, vision is perhaps the most valued. Loss of vision, for whatever reason, is a terribly debilitating and frustrating experience.

Yet the eye, uniquely of all organs, must not only develop by the normal mechanisms involving sequential turning on of genes in cells in the correct temporal order, but also must maintain the multiple functions necessary for vision (Fig. 1) throughout life. [For a more complete discussion the reader is referred to a basic ocular histology text (1).] These functions include (1) refraction of light, which occurs mainly at the cornea/air interface; (2) fine focusing as the light passes from the optically clear cornea and aqueous humor through the lens, which is continually deformed and moved by the ligaments attached to the lens capsule at the zonules; (3) regulation of the amount of light admitted to the lens and retina by the iris, which continuously responds to optimize visual acuity and sensitivity from bright light to very low light levels; and (4) detection of the light as an image, after passing through the fine-focusing lens and vitreous humor, at the retina. The retina is an extension of the brain with special cells located to permit particular functions: (1) the cones in the central area (the macula), densely packed to visualize high-resolution images in color, and (2) the rods located throughout the remainder of the retina, permitting images with very low levels of light to be perceived. In the imaging, stray light photons are absorbed by the retinal pigment epithelium (RPE) after passing through the neural retina and the rods and cones, which form the perceived image. This light absorption by the RPE prevents backscattering. In albino animals backscattering seriously lowers the resolving power for points, lines, and other small features of the image. Also, for imaging, the RPE in a diurnally (often light-activated) process digests and recycles the oldest portions of the rod and cone cells, permitting the reuse of both retinol and an essential fatty acid, docosahexaenoic (DHA) acid, both uniquely found in large concentrations in the photoreceptor cell membranes. These membranes gain fluidity by their presence in the membrane lipids.

SPECIAL RISK FACTORS FOR THE EYE

Like the skin, the cornea (Fig. 2) is in intimate contact with the external environment and has no blood circulation, although its posterior aspect is bathed by aqueous humor, formed in the ciliary body, with a composition closely approximating that of the blood plasma from

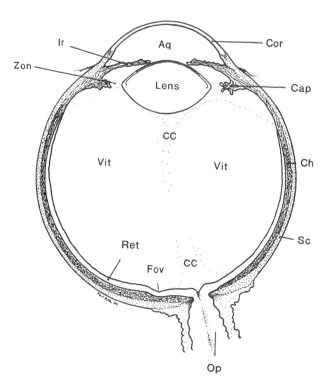

FIGURE 1 Horizontal meridional section of the left eye, top view: Aq, aqueous humor; Cap, capillary body; CC, canal of Cloquet; Ch, choroid; Cor, cornea; Fov, fovea centralis; Ir, iris; Op, optic nerve; Ret, retina; Sc, sclera; Vit, vitreous body; Zon, zonular fibers.

which it is derived. The cornea by its location and lack of circulation is protected only by a thin film of tear fluid and is thus in intimate contact with

1. Oxygen of the air
2. Noxious smoke-derived pollutants present in the air (2)
3. Elevated air temperatures, up to 50 or 60°C in desert and tropical sun
4. Chlorine and other oxidizing agents employed to maintain sterility in swimming pools
5. Many other stresses, including wind-borne allergens

The cornea can repair many types of damage, but if the damage is too severe, resulting in permanent, milky, light-scattering corneal scars, corneal transplantation may be necessary for restoration of normal vision. As we describe later, preservation of corneal grafts from oxidative damage before transplantation is thus an important current area of investigation.

The aqueous humor contains balancing levels of oxidizing hydrogen peroxide and of the antioxidants vitamin C (3–5) and vitamin E (probably bound to an α lipoprotein). Hydrogen peroxide at normal physiological levels very likely provides a useful bacteriostatic function in the aqueous by protecting against infections, just as lysozyme does in tears. Elevated levels of aqueous humor peroxide have been found by Spector's group (6) associated with lens damage in human senile cataract, particularly involving the sodium, potassium ATPase [Na,K-ATPase (7)].

The lens is particularly susceptible to a variety of stresses even at low levels, since it carries its entire life history with it, uniquely among all organs, including bone, building up layers

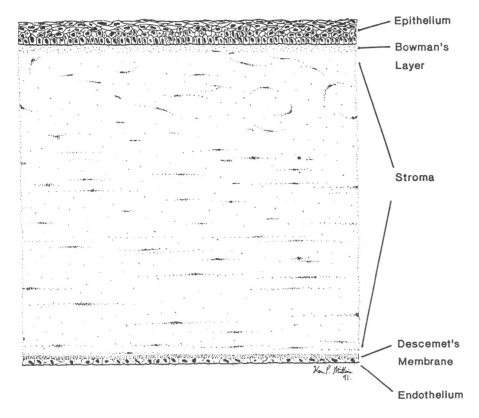

FIGURE 2 Section of typical cornea with the exterior surface of the eye at the top. The five major corneal layers are shown.

by differentiation of the lens epithelial cells located on the front hemispherical surface of the lens (1). These epithelial cells grow and move toward the equator of the lens, where they differentiate, elongating up to 500 times as they change to fiber cells expressing unique lens proteins β- and γ-crystallins. Thus the lens adds layers on the outside throughout life, resulting in a multilayered structure like an onion. This characteristic can result in a cataract. In this process, stress or damage suffered by an area while being formed is expressed many years later as a cataractous opacity within the lens located at the layer that was laid down at the time of the stress (8).

The vitreous humor is in contact with a basement membrane, Bruch's membrane, must posterior to the back of the lens and its posterior capsule. The lens posterior is not protected by epithelium, although in some microwave-induced cataracts (9,10) abnormal migration of epithelial cells to this area can occur, forming swollen cells termed bladder cells. Because of damage to the fiber cells, lens proteins can diffuse out of damaged regions of the posterior lens into the vitreous, where they accumulate because the vitreous turns over so slowly. Also in the vitreous are found breakdown products of the retina that may escape normal scavenging by the RPE. Such products may arise after such stresses as light, some genetic diseases, such as retinitis pigmentosa, diabetes, and age-related functional failure to phagocytose and digest the retinal photoreceptor cells. Both the lens proteins and the retinal breakdown products may be immunologically active. In addition to the formation of chain reaction free radicals from the docosahexaenoic (22:6) fatty acid, O'Keefe et al. (11) found that, during the peak of retinal degeneration, vitreous macrophage counts increased by over 100 times, which would result in additional oxidative stress by superoxide release. This vitreous stress could potentiate damage to both lens and retina. Peroxide is produced by dismutation of the superoxide produced by such macrophages. Further damage can be potentiated by the release of iron from damaged cells: the well-known Haber-Weiss and Fenton reactions (12) catalyzed by iron can produce very damaging hydroxyl radical from H_2O_2. The preceding examples show that damage or oxidative stress to one area of the eye can potentially result in damage to a spatially distant and distinct ocular component. For this reason, the overall oxidative defenses of the eye are likely of paramount importance in minimizing damage that could, in effect, spread to a secondary site after an initial primary site is damaged.

A particular concern is the accumulation with age of oxidative superoxidelike species capable of damaging the eye: this accumulation appears to occur in all parts of the eye studied (13), but its concentration is highest in the retina. Such an age-related increase in superoxidelike activity is potentially damaging. The increased oxidative stress is consistent with the age-dependent increase in the pigment lipofuscin in the retina, apparently as a result of lipid peroxidation (14). It seems possible that higher levels of dietary antioxidants are necessary to offset this increased oxidative stress in the aging organism.

THE ROLE OF VITAMIN E

Vitamin E is a lipid-soluble antioxidant, part of the antioxidant defenses of each cell, and functions as part of a team of antioxidants each cell must deploy to scavenge potentially damaging oxidative species: superoxide, hydrogen peroxide, singlet oxygen, hydroxyl radical, hypochlorite, and lipid peroxidase and peroxy free radicals, which form parts of the "oxygen cascade" (15).

There seems to be a preferred order in many cells in which the antioxidants are utilized in defending the tissue against oxidation. Vitamin E appears to be the last line of defense (16).

In some cases it has also been shown that such antioxidants as vitamin C can, by a coupled reaction, replenish the partially oxidized vitamin E even though E is fat soluble and C water soluble (17). A recent concept of utility in understanding the role of antioxidants in the process of oxidation was advanced by Tirmenstein and Reed (18). They suggest that the critical factor for most cells' antioxidative defense is the ratio of glutathione to vitamin E. If the glutathione level falls too low, the defense mechanisms dependent on it fail, that is, direct scavenging, peroxidation via the selenium-dependent enzyme, glutathione peroxidase, and free-radical chain breaking by the glutathione S-transferase. This places the entire burden of cellular defense on vitamin E, causing rapid depletion of E for which the cell cannot compensate. Of course, this concept ignores the other defense mechanisms, including water-soluble antioxidants, such as ascorbate, taurine (a hypochlorite scavenger), and uric acid; proteins, such as metallothionein (19); enzymes, such as catalase, peroxidases, and superoxide dismutase(s); and lipid-soluble antioxidants, such as α-, β-, and γ-tocopherols; bile pigments, such as bilirubin; and β-carotene and ubiquinol (ubiquinone). These also play roles either (1) indirectly, as antioxidant enzyme systems using antioxidants as substrates to detoxify component members of the oxygen cascade or lipoperoxy radicals, or (2) directly as scavengers of these or other oxidizing species, such as hypochlorite (taurine). Uniquely of all these antioxidants, tocopherol can act as a free-radical scavenger, breaking lipid peroxy radical chain reactions both as the free tocopherol and as the first oxidation product, the tocopheroxy radical. Thus one tocopherol can terminate two free-radical chain reactions (20). Because of its potency and lipid solubility, it has been viewed as the major lipid-soluble antioxidant of most tissues.

In the following we describe roles that have been indicated or suggested for vitamin E in various conditions affecting the different regions of the eye.

THE CORNEA

Vitamin E has been useful in medicinal applications for the cornea, especially in keratoplasty, or corneal transplantation. The transparent cornea is similar among various species (1), consisting of an outer multilayer of cells, the corneal epithelium; the largest and middle layer consisting of mostly acellular material, keratin, and collagen fiber bundles with specialized physically extended keratocytes; and an inner layer of cells termed the corneal endothelium that is in contact with the aqueous humor of the eye. The basement membrane of the endothelial cells is in contact with a collagenous band of material called Descemet's membrane, which lies between the endothelium and the posterior stroma. The anterior epithelium contacts the external environment of the fluid tear layer of glycoprotein and air (Fig. 2) (1).

The corneal endothelium has been shown to be absolutely necessary for maintaining the fluid balance and thickness of the cornea, which in turn are required for corneal transparency (21,22). Also, work involving eye or corneal preservation in the area of keratoplasty has revealed that lipid oxidation and loss of stromal fiber organization play a large role in the loss of corneal integrity (23–26). Most interesting is that α-tocopherol as well as some other lipid-soluble antioxidants has been shown effectively to decrease the rate of corneal autolysis (23,25,26). The use of antioxidants in organ preservation and transplantation to decrease tissue damage is not new and can decrease tissue damage (27). The European literature contains examples of antioxidant treatment in corneal preservation and keratoplasty, but the specific use of tocopherol as an antioxidant is not apparent in the 1980s literature concerning corneal and eye tissue.

A related possible clinical use of α-tocopherol may be in the conservative treatment of keratoconus, as reported in the Russian and German literature (28,29).

Keratoconus is a corneal dystrophy (Fig. 3), seen as a bilateral thinning and weakening of the cornea as a result of a progressive thinning of the corneal stromal layer (30–32). It is a major indication for the use of keratoplasty. The normal cornea has five distinct regions from the anterior (outside) to posterior surface: epithelium, Bowman's layer, stroma, Descemet's membrane, and endothelium (Fig. 2) (1,30).

Bowman's layer is a fibrous interweaving of collagen fibrils set in a matrix of proteoglycan. As one progresses further into the stroma, alternating layers or lamellae of parallel bundles of collagen fibrils are found. These alternating layers are arrayed with the fibrils roughly perpendicular to them (1,30). The thinning of the corneal stroma in keratoconus is physically manifested by the gradual loss of lamellae and thinning of the lamellae themselves. Also, interfibril distances increase (fewer collagen fibrils per unit volume), and therefore a weakened and thinner cornea results (32). The affected area tends to buckle outward from

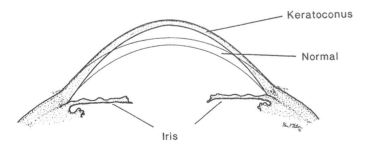

FIGURE 3 Effect of keratoconus on the cornea. The corneal thinning and conical projection of the cornea are shown relative to the normal spherical cornea.

the intraocular pressure of the eye. As the disease progresses the cornea develops a conical shape that eventually can be seen in the angulation of the lower eyelid as the eye is rotated downward, a symptom known as Munson's sign (Fig. 3). Eventually the base of the conical area is also enhanced by the appearance of Fleisher's ring, due to the disruption of the stromal lamellae and yellow or greenish coloration from deposition of hemosiderin (30).

Lipid Peroxidation in the Cornea

Puchkowskaya and Titarenko reported that treatment of keratoconus with α-tocopherol at various stages stabilized the progress of the condition, even at stage IV, over a 2 year period (29). Some improvement in visual acuity was found, especially in the earlier stages of the disease. The possible benefit of tocopherol treatment, especially in the earliest stages of keratoconus and in the preservation of corneal tissue, suggests a significant role for lipid peroxidation in these situations.

First, the study of healthy isolated corneal tissue provides evidence that the corneal potential for lipid peroxidation is a significant factor in the loss of endothelial viability. The survival time of endothelium in isolated cornea perfused with media of various antioxidant compositions has been reported (22). Lux-Neuwirth and Millar found that α-tocopherol inclusion in the perfusion medium (10 μg/ml) doubled the endothelial survival time of perfused rabbit cornea from 7.2 to 15.5 h. β-Carotene (1 μg/ml) had a similar effect. Conversely, the water-soluble antioxidants mannitol, sodium benzoate, and ascorbate had no significant effect. Superoxide dismutase or catalase also increased survival time to an extent similar to tocopherol, suggesting that the initial agents involved in the damage are superoxide and hydrogen peroxide (22). Electron microscopy of endothelial cells showed disruption of mitochondrial and endoplasmic reticular membranes in controls but not in vitamin E-treated corneal samples (22). Lux-Neuwirth and Millar utilized the method that Dikstein and Maurice used to show that the rabbit corneal endothelium is the site of fluid pumping out of the corneal stroma (21,33). This method involves swelling the deepithelialized cornea with salt solution, replacing the solution with oil, and monitoring the return toward normal thickness under the microscope (21,22,33). Dikstein and Maurice (21) also found that the duration of fluid pumping was greatly enhanced by the presence of glucose and adenosine in the perfusion medium, two components also used by Lux-Neuwirth and Millar in their media. From their work these authors made the important point that although the testing of antioxidants in order transplantation and preservation (kidney, heart, and liver) is common, serious evaluation of such antioxidants as tocopherol for use in corneal preservation is required. They also point out that studies with allopurinol (an inhibitor of xanthine oxidase) may be useful to determine the involvement of xanthine oxidase as a source of superoxide in the corneal endothelium (22).

The sensitivity of the corneal tissue to oxidative damage is hypothesized as the reason for the benefit of tocopherol treatment of keratoconus as well (28,29). Titarenko (28) evaluated tocopherol treatment reported in previous biochemical studies by Kim and Hassard (34). These authors found that corneal tissue from human cases of keratoconus displayed an almost complete deficiency in glucose-6-phosphate dehydrogenase (G6PDH) activity. Although not necessarily an initial cause of this condition, this change could have grave consequences for the normal cornea. Kim and Hassard (34) found that G6PDH activity was greatest in the corneal epithelium, which uses as much as 35% of metabolized glucose through the pentose phosphate pathway (35). Loss of substantial G6PDH activity could diminish the available supply of NADPH and pentose for synthesis and growth (34). Most important to the discussion of α-tocopherol is the suggestion that reduced NADPH may lead to a deficiency in reduced glutathione (GSH), thus impairing the neutralization of membrane-damaging lipid peroxides (34). Loss of the selective barrier and metabolic control of the epithelium would in turn affect the maintenance of a healthy stroma. The apparent benefit of tocopherol treatment found by Titarenko (28) certainly supports this hypothesis. It may supply the extra neutralization of lipid hydroperoxyl radicals required. Although not solving all the problems that may be present in keratoconus, the significant stabilization reported is worth further scrutiny.

Bron's review of the cornea (30) provides an excellent summary of careful investigations concerning collagen and proteoglycan changes in the stroma but does not consider in detail structural studies that suggest initial changes in the epithelial and Bowman's layer region. In light of the reports just discussed but not mentioned in Bron's review, the changes in keratoconus may be at least partly due to a compromised pentose phosphate pathway and therefore less NADPH and ribose for RNA and protein synthesis and lower glutathione levels. Future work will, it is hoped, clarify this.

Consistent with this hypothesis is an early study of vitamin E deficiency in turkey embryos, concentrating on lens development. The authors noted that a weakened keratoconus-like cornea with a loss of keratocytes in the central corneal region and a reduction in corneal thickness occurred in 27 day embryos (36).

LENS AND CATARACT

Human senile cataracts are usually divided into two types, depending on the area of the lens in which the opacity occurs (37). In *cortical* cataracts a milky opacity occurs in the outer layers or cortex of the lens; in *nuclear* cataracts the opacity is found in the lens nucleus, which corresponds to the embryonic lens over which the cortical layer of fiber cells was deposited during growth and maturation throughout life. Often mixed cataracts occur in which both nucleus and cortex show milky opacities. This results when a primary damage to the nucleus or cortex potentiates the complementary development of an opacity in the remaining portion of the lens. Elaborate classification schemes (38–40) have evolved to describe and catalog all the changes and the various areas of opacity within the lens. Although opacity is observed in both regions, nucleus and cortex, the causes of the opacity appear to differ strikingly for each site.

In the lens the long fiber cells are hexagonal in cross section and oriented parallel to each other with interdigitations at corner regions and ball and socket joints on opposing flat faces. These prevent the fiber cells sliding past one another when the lens is distorted in accommodation. In cortical opacities, cell death is accompanied by globule formation (41) as the cell cytoskeleton fails. Many small round globules bud from the cell surface (Fig. 4). After cell death the soluble crystallin proteins leak out of the cells, leaving behind globules, fibrous matrix, and insoluble matrix, which has a lower protein content. The scattering of light in the normal lens is minimized because of the correlation effect (42) of the parallel orientation of the lens fiber cells. This is lost in the area of globular degeneration because of the spherical globules and the randomly occurring refractive index differences in the area of dead and leaking cells compared with the higher refractive index of live cells. Such differences act as millions of small mirror surfaces, reflecting, refracting, and scattering the light to give a milky appearance to the area of globular degeneration (Fig. 4).

In the lens nucleus, by contrast, processes resulting in aggregation of the lens proteins by sulfhydryl oxidation and aggregation of low-molecular-weight proteins lead to very high molecular weight aggregates of size 35 million daltons or more (43,44). In the cataractous area the milkiness appears to be due to light scattering of these large aggregates. As the cataract matures, oxidation of methionines, tyrosine, and tryptophans, as well as membrane fatty acids, may occur, resulting in colored polymeric products that lead eventually to progression from deep yellow to brown (brunescent) to black (niger). Not all cataracts progress this far, but in Third World tropical areas of strong sunlight when cataracts are not operated on early (in contrast to Europe or North America), such colored progression of cataract

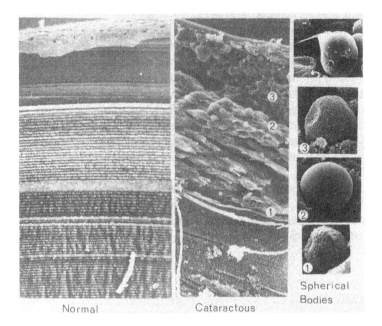

FIGURE 4 Normal morphology of human lens compared to posterior cortical cataract, with globules typically located at location numbered as shown enlarged at right. Note the thread of adhering cytoplasm still attached to globule located at 1.

development is often seen. In cortical cataracts, a portion of the opacity may also be due to aggregates formed by rapid phase separations, when for the protein concentration found in the lens cells, the temperature decreases below the critical point, as seen in cold cataract most dramatically. In both cortical and nuclear cataracts the increase in scattered light gives a much higher background of randomly scattered light at the retina, resulting in severe loss of image contrast. This in turn reduces visual acuity (41).

Although the mechanism of formation is different, both nuclear (43,44) and cortical cataracts (45) seem to involve some oxidative step or steps in their development. A large amount of work on animal models of a number of different causative agents for cortical cataracts (45–57) and nuclear cataracts (58) has indicated that antioxidants, such as vitamin E, typically delay or prevent the development of the opacity and, for cortical cataracts, the cellular changes.

Although vitamin E deficiency in the turkey (36) and rabbit (59) resulted in cataracts and Bunce and Hess showed an elevated incidence of cataract in vitamin E-deficient rats (60), very little work was done using supplementation. Evans recommended vitamin E supplementation for treatment of cataracts, but in combination with other supplemental vitamins (61,62).

Initially in studies performed in vitro, we used rat lenses incubated in tissue culture medium, with 10 times elevated glucose to simulate diabetic cataract (45). When vitamin E (2.43 μm) or glutathione (0.1 mM) was added to the incubation medium, development of the cataractous opacity and globular degeneration was prevented. Although the scientific community was somewhat skeptical of this effect, Bhuyans et al. (63) showed that vitamin E was effective in preventing cataracts induced in rabbits by 3-aminotriazole, which results in catalase inhibition and thus oxidative stress due to higher H_2O_2 concentration. Subsequent work in our laboratory showed that vitamin E and a number of other antioxidants were effective in maintaining lens clarity by preventing globular degeneration in vitro, without affecting the lens sorbitol concentration (47,48). This was surprising, since it was established much earlier (64) that the intracellular accumulation of sorbitol in the lens cells in hyperglycemic medium occurred as a result of the operation of the aldose reductase pathway:

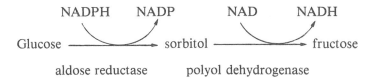

Kinoshita et al. hypothesized that the accumulation of sorbitol, which cannot pass out through the cell membrane, caused an osmotic stress on the lens cells, leading to the formation of swollen, distorted lens cells and the formation of lakes or vacuoles containing liquid in the diabetic lens (64). In vitro work showed that lenses (47) exposed to hyperglycemic conditions in the presence of vitamin E also swelled less in tissue culture medium and returned sooner to their original weight, but untreated sugar-stressed lenses never returned to the their original weight and became opaque, developing globular degeneration. This led us to a hypothesis that the sorbitol-induced osmotic stress resulted in turn in an oxidative stress, which was effectively counteracted by vitamin E or other antioxidants in the medium. Testing this hypothesis in vivo, we used rats in which diabetes was induced by streptozotocin (48). We found that in contrast to the 48 h required for opacity to develop in vitro, in vivo opaque lenses required 6–12 weeks to develop. Conversely, in rats in which vitamin E was supplemented by daily subcutaneous injection (later experiments used dietary treatment), no opacity was observed. Lenses from treated rats at 6 weeks showed morphological alterations typical of lenses removed after 4 days from untreated diabetic rats (Fig. 5). Even more surprising was the observation that lenses from treated animals had levels of sorbitol equal to or significantly higher than those of diabetic lenses, indicating that tocopherol treatment enabled the lenses to withstand the osmotic stress. The higher level of sorbitol in treated lenses probably was due to leakage of intracellular contents, including sorbitol, from the lens cells as they underwent globular degeneration in the untreated diabetic rat.

We developed a modified hypothesis, in which the steps of diabetic cataractogenesis included an oxidative damage step (Fig. 6) after the osmotic stress due to sorbitol (48).

If diabetic cataract involved an oxidative step, could an oxidative stress result in similar cataractous and histopathological changes? The best known example of such an oxidative stress is ionizing radiation (65–67). We treated lenses with ionizing radiation (51) to induce an oxidative stress and showed that γ-irradiation of rat lenses followed by incubation in vitro resulted in cortical globular degeneration similar to that observed in elevated glucose (Fig. 7). This was significantly reduced by adding vitamin E to the incubation medium. Reduction of the damage permitted visualization of "holes" in the cell surfaces along with globular degeneration. These results are consistent with Petkau's hypothesis that radiation cataractogenesis involves cell membranes (67), and they are consistent with the hypothesis that oxidative damage is involved in the pathogenesis of cataract.

Varma et al. (68), studying photoperoxidation of cultured rat lens lipids, corroborated the protective effect of vitamin E on the lens. Light-generated oxygen free radicals could cause cataract and inhibition of rubidium uptake by the lens (68). Vitamin E prevented lipid peroxidation, evaluated by thiobarbituric acid-reactive material, in the lens homogenate. Varma's group attributed this effect to the reaction of vitamin E with superoxide or other members of the oxygen cascade.

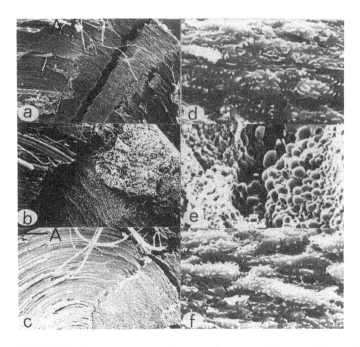

FIGURE 5 Morphology of rat lenses at low magnification (left) and higher magnification (right). Morphology of lenses: after 4 days of diabetes (a and d), 6 weeks of diabetes (b and e), and 6 weeks of diabetes with injection treatment by vitamin E (961 IU/kg) as described in Reference 48.

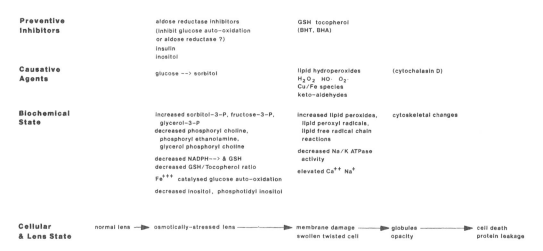

Preventive Inhibitors	aldose reductase inhibitors (inhibit glucose auto-oxidation or aldose reductase ?) insulin inositol	GSH tocopherol (BHT, BHA)	
Causative Agents	glucose --> sorbitol	lipid hydroperoxides H_2O_2 HO· O_2· Cu/Fe species keto-aldehydes	(cytochalasin D)
Biochemical State	increased sorbitol-3-P, fructose-3-P, glycerol-3-P decreased phosphoryl choline, phosphoryl ethanolamine, glycerol phosphoryl choline decreased NADPH--> & GSH decreased GSH/Tocopherol ratio Fe^{+++} catalysed glucose auto-oxidation decreased inositol, phosphotidyl inositol	increased lipid peroxides, lipid peroxyl radicals, lipid free radical chain reactions decreased Na/K ATPase activity elevated Ca^{++} Na^+	cytoskeletal changes
Cellular & Lens State	normal lens ➡ osmotically-stressed lens ➡	membrane damage ➡ swollen twisted cell	globules ➡ cell death opacity protein leakage

FIGURE 6 The many interacting events in cortical cataractogenesis, generally based on the diabetic model.

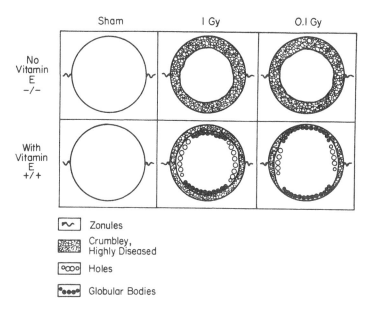

FIGURE 7 Damage after x-ray irradiation and/or vitamin E treatment in vitro. (From Ref. 51.)

Our further experiments indicated that oxidative stress appeared to be involved in several other types of stress capable of causing cataracts in vitro, which could be prevented by vitamin E, including heat [elevated temperature (53)], hygromycin B (50), Solu-Medrol (21-succinate sodium salt of 6α-methylprednisolone) (49), radiation-induced cataract in vitro (51) and in vivo (52), galactose-induced cataract in vitro (54,69), and diabetic cataract in vitro (47) and in vivo (48). Aurricchio's group (70) tested vitamin E in several systems and found that it did not affect galactose-induced cataract in vivo (69), confirming our results. They found that lysophosphatidylcholine damage to tissue-cultured lenses could be prevented by vitamin E (70), although the concentration used (10^{-3} M) was rather high. They suggested that the effect of vitamin E was due to stabilization of membranes rather than an antioxidant action. At this high concentration this suggestion appears likely since at lower concentrations (2.4 μm (45)] vitamin E is found to have a significant protective effect as an antioxidant.

Hattori et al. (71) used vitamin E-containing liposomes to prevent glucose-induced cataract in vitro. They concluded that the protective effect on cataract formation they observed was due both to the antioxidant role of vitamin E and its effect on membranes, although it significantly decreased malondialdehyde levels.

Bunce and Hess (60) found a significant reduction in cataract incidence after limitation of amino acid intake in rats during gestation and development. This may be due to a change in critical temperature of the proteins in the lens, which Eccarius and Clark showed to be reduced by vitamin E (72); this phase change effect is apparently not due to the antioxidant action of tocopherol.

All these examples involved acute stress acting over a rather short period of time. If an oxidative component is common to many cataracts, one might expect to see it as well in more slowly developing cataracts. Two such animal models exist: one, the Emory mouse model of senile cataract, and the other, the genetically determined retinal dystrophy found in the RCS (Royal College of Surgeons) rat strain, in which a retinal degeneration produces lipoperoxy free radicals from the docosahexaenoic acid of the degenerating photoreceptors (73–75).

In the Emory mouse cataract, two reports differ in the efficacy of vitamin E treatment: Varma et al. (58) observed a significant delay in the progression of the cataract by 10 mg/ week of intraperitoneal vitamin E acetate in coconut oil. He attributed this to an elevated level of lipid peroxidation. Consistent with this, Bhuyan et al. (76) showed elevated liver levels of the peroxidation product malondialdehyde in Emory strain (compared with other laboratory mice). Kuck and Kuck (77), who originated this Emory mouse senile cataract model, were able to prevent lens damage using the water-soluble antioxidant penicillamine, but vitamin E at 6 IU/g diet did not significantly delay the onset of cataract. The dietary intake expected from the Kucks' experiments, assuming a 20 g mouse eats 2 g food per day containing 12 IU vitamin E, is approximately 84 IU/week compared with the intraperitoneal injection level used by Varma et al. (10 mg/week = 10 IU/week). Thus it is uncertain in this model whether vitamin E may delay cataract. Unfortunately, no further studies have been done with this interesting model. Extrapolated to the human situation, the levels used by Varma et al. would be considered megadoses: this dose in a 35 g mouse corresponds to 2857 IU/day in a 70 kg human. When the levels given to rats (2850 IU/kg in food) in our studies are compared with those of the Varma study (58), they were even higher. They might be expected to result in levels of 2.850 IU × 3.5 = 9.98 (10.0 IU/day) in a 35 g mouse (or 2.85 × 20 = 57 IU/day in a 200 g rat; 70 IU/week), which corresponds to a human dose of 20,000 units/day.

For the human senile cataract, which develops over many years, with these dose levels in mind, the question that might be asked is, If antioxidants are effective at high levels in counteracting high stress levels over a short period of time, would the much lower stress levels encountered in human senile cataract be responsive to treatment with usually encountered supplemental vitamin E levels $\geq$ 300–400 IU/day. As discussed in detail subsequently, our collaborative case-control study with epidemiologists Robertson et al. (78) showed significant cataract risk reduction in persons who had been taking supplemental vitamin E for a minimum of 5 years. Complementing this, Jacques et al. (79) observed an association of low levels of at least two of three antioxidants (β-carotene, vitamin C, and vitamin E) with cataract, and more recently, Leske et al. (80), by correlating food levels of antioxidants with cataract grade, indicated significant risk reduction in persons with elevated vitamin E (see later).

These data are consistent with the observation of Varma et al. (58) of elevated malondialdehyde in human senile cataracts, approximately proportional to the severity of cataract grade and consistent with the Bhuyan group's (76) observation of increased malondialdehyde in animal cataracts. If the increased lipid peroxidation indicated by malondialdehyde were counteracted by vitamin E, the lowering of the oxidative stress would result in less tissue damage, less release of iron from the damaged tissue, and therefore a decrease in autocatalytic damage to the lens as a result of the iron-catalyzed Haber-Weiss and Fenton reactions. That such an effect is possible in human senile cataracts as well as animals was indicated by the three epidemiological studies, already discussed briefly and cited later.

Possible Mechanisms in Cataract Prevention or Risk Reduction

The best known function of vitamin E is probably its ability to react with free-radical intermediates of lipid peroxidation and with the peroxides (81). Vitamin E has been shown to react with hydroxyl (OH^-), perhydroxyl ($HO_2\cdot$), and superoxide ($O_2\cdot$) radicals (82). Although the hydroxyl radical is potentially the most efficient oxidant (81), the perhydroxy radicals can react directly with unsaturated lipids (83). The superoxide radical cannot oxidize lipids but may assist the degradative process by reacting with vitamin E (82).

Malondialdehyde is one of the products of lipid peroxidation, and levels of malondialdehyde have been shown to reflect the amount of lipid peroxidation occurring in a tissue (84). Vitamin E administration was shown to reduce the age-related increase in malondialdehyde levels in liver and serum of rats (85), whereas vitamin E deficiency in rats was shown to lead to increased malondialdehyde excretion (86). Bhuyan and Bhuyan (87) showed increased malondialdehyde in lenses as a result of many types of cataract, with vitamin E administration leading to a decrease in the malondialdehyde level. Therefore, one of the functions of vitamin E in the lens most probably is protection of lipids from oxidation.

Vitamin E has been shown to be capable of penetrating into monolayers of various phospholipids (88), liposomes (89), and sarcoplasmic reticulum of skeletal muscle (90). In 1982, Burlakova et al. (91) showed that, even though the chroman structure of vitamin E had the antiradical properties, the side chain was necessary to anchor the vitamin E in the membranes. The presence of vitamin E in membranes has been correlated with the fluidity of the membrane (91) and its permeability (89) and conductivity (92). The studies of Baig and Laidman (93) indicated a polar interaction between the phenolic head groups of α-tocopherol and the phosphate group of the phospholipid, possibly involving hydrogen bonding. Similarly, Erin et al. (90) suggested hydrogen bonding between the OH group of the dihydrobenzopyran nucleus of tocopherol and the carbonyl group of the fatty acid. They also reported possible hydrophobic interactions between the fatty acyl chains and the methyl group of the tocopherol nucleus. A proposed interaction between vitamin E and membrane phospholipids has been shown (94).

It appears from the preceding discussions, therefore, that vitamin E may have two functions in the membrane: it can prevent lipid peroxidation and physically stabilize the membrane. Mino and Sugita (95) demonstrated the incorporation of tocopherol acetate into membranes. They found that the vitamin E acetate so incorporated was not converted into the tocopherol form; therefore, vitamin E acetate used in vitro may offer some protection through stabilization of the membrane but is not able to prevent oxidative damage to the membrane.

Vitamin E has also been shown to be capable of repairing free radicals derived from the one-electron oxidation of the amino acids tryptophan and tyrosine (96) and methionine and histidine (97). This oxidation of amino acids may explain the oligomerization of membrane proteins that was shown to occur in retina and skeletal muscle in conjunction with lipid peroxidation (98). Kagan et al. (99) demonstrated membrane protein linking involving the Ca^{2+}-ATPase along with lipid peroxidation. The presence of vitamin E in membranes was shown to protect the Ca^{2+}-ATPase from this damage (100,101).

It is critical to the lens to maintain low internal Ca^{2+} levels, which is accomplished by an active Ca^{2+}-ATPase (102). Inhibition of this calcium pump was shown to lead to calcium accumulation and lens opacification (103). Protection from oxidation of the proteins of the calcium pump may therefore be another way in which vitamin E can reduce cataract formation.

Peroxidation of lipids has been shown also to lead to formation of DNA free radicals (104). Popov and Konev (105) established in Ehrlich ascites carcinoma cells that, as processes of lipid peroxidation were intensified, the rate of DNA synthesis decreased. The existence of a high-affinity tocopherol binding receptor protein in rat liver nuclei has been demonstrated (106,107). The presence of such a vitamin E binding protein may help to explain how vitamin E was able to prevent free-radical damage to DNA (104) and $K_2Cr_2O_7$-induced DNA degradation (108).

Vitamin E was also shown to be capable of inducing RNA synthesis (109). In the lens, the epithelial cells, but not the fiber cells, are actively involved in DNA synthesis; therefore, any protection against oxidation of nucleic acids vitamin E might offer would be of paramount importance for the integrity of the epithelial layer and the differentiation of these cells into fiber cells.

Lipid peroxides have been shown to modulate the biosynthesis of prostaglandins (110, 111). The major arachidonic acid pathway in rabbit lens has been shown to be through the lipoxygenase enzyme (112). Bovine epithelial lens cells in vitro have been shown to form leuko-triene B_4, which is a product of the lipoxygenase pathway (113). Although the lipoxygenase pathway may be the major arachidonate pathway in the lens, the lens was shown to be capable of synthesizing prostaglandins by way of the cyclooxygenase pathway (114). Ono et al. (115) found that administration of prostaglandin E_1 (PGE_1) to rats for 3 days encouraged phospholipid turnover, especially of phosphatidylcholine in the lens. The presence of PGE_1 (10^{-3} M) in the incubation medium of rabbit lens was shown to cause increased sodium and decreased potassium content of the lens with resultant lens opacification (116). Both vitamin E deficiency (117) and streptozotocin-induced diabetes (118) were shown to result in an unbalanced production ratio of thromboxane A_2 (TXA_2) and prostaglandin I_2 (PGI_2) in platelets and serum. Gilbert et al. (118) and Karpen et al. (119) reported the restoration of normal PGI_2/TXA_2 balance in platelets and serum by vitamin E. The addition of vitamin E to the diet may assist in preventing pathological changes in the lens as well (48).

Another way in which vitamin E may be able to protect against the hyperglycemia of streptozotocin diabetes is through the control of the enzymes involved in glucose utilization. Ulasevich et al. (120) reported that vitamin E (50 mg daily) increased the ability of rat erythrocytes to use glucose 2.4-fold. There was simultaneous activation of aldolase and phospho-hexosisomerase by 34 and 48%, respectively. A decreased level of glucose in all tissues, including the lens, resulting from the presence of high levels of vitamin E, may result in less stress on the lens with a resultant decrease in cataract formation.

Vitamin E may also have an indirect affect on the integrity of the lens through its interaction with other protective compounds. Vitamin E has been shown to modulate tissue retinol levels in vitro and in vivo by influencing retinal palmitate hydrolysis (121). The incorporation together into liposomes of vitamin A and E was shown to lead to an increase in the antioxidant effects over those of vitamin E alone (122). The ability of vitamin E to act as an antioxidant is greatly affected by the levels of vitamin C (123). Vitamin E was also shown to have an effect on the enzymes involved in the synthesis and metabolism of glutathione (124–128).

From the preceding data there appear to be many ways in which vitamin E may protect the lens from the damage that may lead to cataract formation. That oxidative damage seems to be a major factor in cataract is indicated by the variety of in vitro cataract models in which we found vitamin E to be effective in preventing lens damage: hygromycin B cataract in vitro (50), x-ray-induced cataract in vitro (51) and in vivo (52), galactose-induced cataract in vitro (54), steroid-induced cataract in vitro (49), diabetic cataract in vitro (47) and in vivo (48), and cataract induced by elevated temperature in vitro (53).

The actual mechanisms operating in the prevention of cataract remain to be worked out. Recent work on diabetic cataract by Wolff's group (129) suggests that protein glycation may induce peroxide formation, which would result in increased oxidative stress. In response to oxidative stress, one of the most attractive recent ideas, mentioned earlier, is the importance of the glutathione-vitamin E ratio in maintaining cell health (18); for this reason it will be important in future studies of cataract to focus on the entire antioxidant complement of the lens. Such studies should pinpoint the lens epithelium for particular studies on the pivotal role of epithelial cells in protecting against incipient damage that might become autocatalytic, with the release of iron from damaged tissues.

Epidemiological Studies

The studies of experimentally induced cataracts just described demonstrated protective effects of antioxidant substances both in vivo and in vitro. The key question is whether these findings may be extrapolated with any certainty to naturally occurring human senile cataracts. The results of three recent epidemiological studies (78–80,130) suggest a positive answer to this question, whereas a fourth study (131) does not.

Jacques et al. (79) compared the nutritional status, measured by both nutrient intake and biochemical markers, of 112 cataract patients with that of 35 cataract-free control subjects. Significantly increased risks of cataract were found in subjects who consumed less than 3.5 servings of fruit or vegetables per day and in those with low vitamin C intake (below the 20th percentile). Conversely, low carotene and vitamin E intakes were not found to be associated with elevated risks. Subjects with low plasma carotenoids showed a 7-fold increase in the risk of cortical cataracts ($p < 0.05$). Those with low plasma vitamin C, had an 11-fold increase ($p < 0.10$) in the risk of posterior subcapsular cataracts. Plasma vitamin E status was unrelated to either type of cataract. The association between both the nutrient intake data and the plasma vitamin C levels and freedom from posterior subcapsular cataracts suggests a protective effect for this antioxidant. On the other hand, the contradictory findings for carotene intake and plasma carotenoids and the lack of an association with intake or plasma levels of vitamin E argue against a role for these substances. However, the sample size in this study was small, particularly in the control group, and the estimation of nutrient intake tends to be inexact at best. The interindividual variability of the estimates of nutrient intake and of the biochemical measurements of nutritional status in a small sample probably accounted, at least partially, for the inconsistencies in these findings.

A case-control study by Robertson et al. (78) compared the consumption of supplementary vitamins C and E of 175 cataract cases with that of 175 age- and sex-matched cataract-free control subjects. After adjusting for potential confounding variables in a conditional multiple logistic regression analysis, they found that the cases were significantly less likely to have taken regular supplements of vitamin C or vitamin E during the previous 5 years (Table 1). Subjects who took 300–600 mg/day of vitamin C experienced a 70% reduction in cataract risk. Those taking 400 IU vitamin E daily decreased their risk by 54% (see Tables 2 through 5). In addition, a dose-response relationship, which was significant in the case of vitamin C was evident. The authors concluded that, provided dietary vitamin intake was unrelated to vitamin supplementation in these subjects, the consumption of supplementary vitamins C and E reduced their risks of cataracts significantly.

Leske et al. (80) studied 1380 ophthalmology outpatients aged 40–79 years, 915 with cataracts and 435 who were cataract free. Polychotomous logistic regression analyses revealed a negative association (odds ratio OR = 0.64) between all cataract types and the use of multiple vitamin supplements. Low dietary intakes of vitamins A, C, and E were found to be associated with cortical, nuclear, and mixed cataracts (OR = 0.48–0.56). These odds ratios for dietary intake were similar to those found by Robertson et al. (78) for supplementary vitamin consumption.

Among other potential risk factors for cataract, Mariani et al. (131) investigated the role of nutritional variables by dietary interview or laboratory analysis. In this clinic-based case-control study of 1008 cases and 469 controls aged 45–79 years, the authors found no associations between any of the nutritional factors and freedom from cataracts.

TABLE 1 Matched-Pairs, Multiple Logistic Regression Analysis of the Associations Between Vitamins C and E and Cataracts

Variable	Case-control pairs	Coefficient ± SEM	Adjusted odds ratio	95% Confidence limits	Significance level p
Vitamin C	40	−1.19 ± 0.46	0.30	0.12–0.75	0.01
Vitamin E	76	−0.82 ± 0.29	0.44	0.24–0.77	0.004
Vitamins C and E	25	−1.12 ± 0.58	0.32	0.11–0.99	0.05

Source: From Reference 78.

The lack of association between nutritional factors and cataracts in the preceding study is difficult to explain in light of the results of the other three epidemiological studies, which revealed tentative evidence of antioxidant prevention of human cataracts. Although some inconsistencies were evident, the results indicated statistical associations between low levels of vitamins C and E and increased risk of cataracts. That the associations were found in three different population groups using different methods of assessing vitamin C and E status serves to strengthen the evidence. However, statistical associations between vitamins C and E and senile cataracts do not necessarily connote causation. Only direct experiment can settle that question. Lacking such evidence, the potential causality of these epidemiological associations may be assessed by considering the following factors: (1) the strength of the association; (2) evidence of a dose-response relationship; (3) the consistency of the association; (4) the time sequence; and (5) the biological plausibility. How well do these studies fulfill these criteria?

One study (79) showed 7- and 11-fold increases in cataract risk. The odds ratios found in the other two (78,80) were similar, varying from 0.30 to 0.64. These findings suggest a strong association. A significant dose-response relationship was observed for vitamin C in one study (78). The results of the three studies were reasonably consistent in finding associations between low intake and/or low plasma levels of antioxidant vitamins and cataracts. Definite evidence of exposure preceding effect was found in one (78). The association is biologically plausible and is supported by the experimental data reviewed in the previous section. Given the foregoing, the results of these studies appear to fulfill the preceding criteria reasonably well, providing fair to good support for a causal association.

In summary, the mounting evidence from in vitro and in vivo experiments, coupled with the epidemiological data, suggests that vitamins C and E may have a role in cataract prevention.

RETINA

Nonretinopathy of Prematurity Studies

The major retinal problems in which vitamin E may influence, with the exception of retinopathy of prematurity (ROP), appear to involve lipid peroxidation, often associated with the deposition of the fluorescent insoluble pigment lipofuscin in the retinal tissue, particularly the retinal pigmented epithelium. For this reason, we discuss ROP in a separate section.

The remainder of the studies can be subdivided conveniently into studies of hereditary retinal dystrophies and retinitis pigmentosa, on the one hand, and, on the other, normal diet- and age-related changes in the retina.

Retinal Dystrophies and Retinitis Pigmentosa

When the retinal photoreceptor cells degenerate and autolyse as in the rCS rat hereditary failure (Fig. 8), their polyunsaturated fatty acids (PUFAs), in particular docosahexaenoic acid (22:6), lose the normal defense of intracellular antioxidants. At the same time the oxidative stress is increased by the release of iron from damaged cells and mitochondria. At the beginning of this autolytic process, if some means of protecting the cell membrane integrity were available, the damage to the membrane might be prevented, delaying the eventual leakage of intracellular contents associated with retinal photoreceptor cell death. Hess et al. (73–75) used the RCS rat model to provide evidence for a number of conditions that can reduce the damage to the retina. This is evaluated by a delay or prevention of the train of biochemical events leading to transient cataract formation in the posterior region of the RCS lens at 8–12 weeks after weaning; this transient cataract goes on to form a mature cataract approximately one-third of the time. Thus in this model, the formation of a cataract is related to the stress

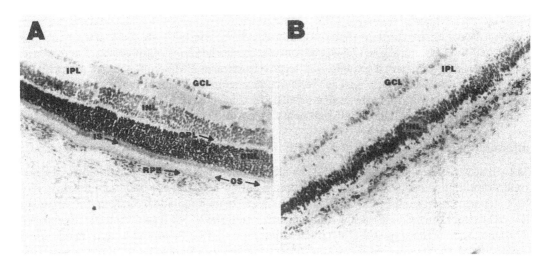

FIGURE 8 (A) Section of control normal retina for the rat strain RCS N − rdy + p. (B) Similar section of retina from the dystrophic strain RCS D. Eyes were from 12-week-old animals. Note loss of photoreceptors (ONL, IS, and OS). The RPE region is quite disordered. GCL, ganglion cell layer; IPL, inner plexiform layer; INL, inner nuclear layer; OPL, outer plexiform layer; ONL, outer nuclear layer; IS, inner segments; OS, outer segments; RPE, retinal pigment epithelium. The choroidal and scleral tissue are below the RPE. (Frozen sections, toluidine blue staining.) (From Mitton KP, Dzialoszynski T, and Trevithick JR.)

resulting from the retinal degeneration, due apparently to failure of the RPE cells to phagocytose the photoreceptors (132). Although Hess and her collaborators did not emphasize the retinal changes, they indicated a number of protective regimens that reduce or eliminate cataract formation: prevention of exposure to light by dark rearing results in normal retinal morphology at the age when degeneration is usually maximum; addition of β-carotene and butylated hydroxytoluene to the diet; provision of a defined diet of purified components; and addition of vitamin E to the diet (73–75). Our studies have provided some additional information (133). We investigated RCS degeneration using additives added singly or in combination to the diet of RCS rats after weaning. Protein patterns obtained by sodium dodecyl sulfate (SDS) from dissected retinas of treated rats were compared with those from untreated RCS as positive controls or, as a negative control, Wistar rats whose retinal protein patterns we found to mimic closely the untreated RCS strain lacking the genetic defect (rdy^-). Prevention of the retinal changes by the particular additive or mixture resulted in a "normal" protein SDS-PAGE (SDS–polyacrylamide gel electrophoresis) pattern (133). Simultaneous monitoring of cataract by slit lamp examination was performed. Cataract was prevented most effectively by dietary vitamin A (12,500 IU/kg), slightly less effectively by vitamin E (2.85 g/kg) and vitamin C (3 g/kg), and not at all by the combination of all three (ACE). Correlating with this, vitamin A treatment resulted in normal retinal protein SDS-PAGE patterns, and the retinas from vitamin E- or C-treated rats showed patterns intermediate between those of untreated RCS and controls. The ACE combined treatment pattern was almost identical to that of the control untreated RCS rat. Heat shock proteins are often induced by oxidative stress (134) as well as elevated temperatures. For this reason we investigated the presence of HSP70, the main heat stress protein (135), during the period of development of the RCS degeneration (6–12 weeks postpartum) while monitoring cataract incidence and the presence of a normal retinal protein, S antigen (136). The results showed a significant correlation between the appearance of the stress protein and cataract, consistent with the peroxidative changes postulated by Hess and Zigler (73) as responsible for both effects. Conversely, the presence of S antigen, a normal retinal protein, was negatively correlated with both cataract and HSP70, indicating that the loss of S antigen as the retina degenerated was correlated with the expected cataracts and also with the appearance of HSP70. Clarke et al. (133) also found protection of the 100 kD stress protein by all treatments except the combined A, C, and E.

Interactions Between Vitamins A and E: Formation of Lipofuscin

Dilley and McConnell (137) showed that bovine rod outer segments (ROS) contained a higher concentration of α-tocopherol. The ROS contain 4 mol rhodopsin per mol tocopherol. In air, ROS preparations consume vitamin E (VE) and produce malondialdehyde from the peroxidation of docosahexaenoic acid (22:6). Farnsworth and Dratz (138) found that the VE content of bovine ROS membranes was much higher than that of other membranes, and furthermore noted that for in vitro ROS preparations, a seasonal dietary change made it necessary to supplement the ROS preparations with tocopherol. That tocopherol actually had a role in vivo in maintaining retinal function was suggested by Hayes (139), who observed degeneration of the macular photoreceptor (PR) outer segments (OS) of monkeys after 2 years of vitamin E deficiency. Antioxidant deficiency (including VE) was later (140) shown to result in massive accumulations of an autofluorescent pigment (lipofuscin) in the RPE, although no apparent neural retinal damage was observed. These authors observed that antioxidant deficiency appeared to mimic aging and that this could be artificially induced by diets high in polyunsaturated fatty acids but deficient in vitamin E, selenium, sulfur-containing amino acids, and chromium (140). The age-dependent increase in lipofuscin is consistent with increased age-dependent lipid peroxidation. Ephimova (141) observed lipid peroxidation to be increased when retinas of older rats were compared with those of younger rats.

Oxidative Stress

Yoshida, using the oxidative stress-inducing chemical alloxan, showed that vitamin E could restore retinal anaerobic glycolysis decreased by alloxan, both in vivo and in vitro (142).

Robison et al. (143) looked into relationships between dietary vitamin A and E and their effect on the retina. In rats fed a vitamin E-deficient diet, they found damage to photoreceptor cells (equally to rods and cones) and both ultrastructural changes and a fivefold increase in lipofuscin deposition in the RPE. The changes initially mimicked those found for rats deficient in vitamin A, which also causes damage to the photoreceptor cells, but later during the E deficiency the changes differed significantly. The changes noted early were swelling of the distal portions of the outer segments with disorganization of the disk membranes and vesiculation. Later, unlike vitamin A deficiency, no extensive formation of small vesicles was observed, nor was the loss of nuclei different in the central compared to the peripheral retina, and cone nuclei were not preferentially spared as observed in vitamin A deficiency.

Although Hayes reported preferential damage in the cone-rich macula of E-deficient monkeys (139), Robison et al. (142) tentatively ascribed the different results they found in the rat to the diurnal-nocturnal differences in the experimental animal used.

Complementary to this study, Farnsworth et al. studied the effect of vitamin E deficiency in the fatty acid composition of the rat retina, using combined vitamin E and selenium deficiency, which would be expected to affect both major detoxification routes for radicals in the retina: vitamin E and glutathione peroxidase (144). They noted major decreases in PUFA in the RPE, although the levels for the whole rat did not change, and concluded that the RPE appeared particularly sensitive to lipid peroxidation. In the outer segment, deficiency resulted in a decrease in total PUFA; in the RPE, the deficient rats had only half the normal levels of docosahexaenoic fatty acid but showed an increase in linoleic acid (18:2ω6) during phago-cytosis. Associated with the deficiency, a buildup of the fluorescent pigment was detected in the RPE. These changes were attributed to inactivation of the phagocytic cells of the selenium-deficient rats due to a buildup of lipid hydroperoxide (LPO) during phagocytosis. Serfass and Ganther (145) found that superoxide dismutase (SOD) is high in the retina outer segments; along with ascorbate and glutathione these form the other major antioxidant systems of the retina (146).

Further work by Robison's group (147) explored the relationship of retinol, tocopherol, and lipofuscin deposition in RPE cells. Since vitamin A (retinol) is subject to oxidation, they tested the hypothesis that protection by vitamin E might be important in maintaining vitamin A levels using three diets (high A, 80.5 μmol/kg diet; normal A; and deficient A, retinoic acid instead of vitamin A). Vitamin E deficiency resulted in a reduction in plasma vitamin A levels (Table 2A). The RPE retinal ester content of both E^+ and E^- groups showed decreases only in vitamin A deficiency (Table 2B). Unlike the case in plasma, vitamin E deficiency did not reduce retinal vitamin A. In the A groups, vitamin E intake had no influence on RPE retinal ester content, but in both A^+ and A^- groups, vitamin E deficiency caused an elevation in the RPE retinyl ester content and an elevation in RPE lipofuscin (Table 2C). Using quan-tiative morphometry for both E^- and E^+ diets, retinol deficiency led to a decrease in RPE lipofuscin (Table 2D). For high A, no differences were detected. Similar results were obtained by fluorescence (Table 2C). Vitamin A deficiency led to a decrease in fluorescence intensity in both E^- and E^+ groups (Table 2D), consistent with the morphometry results. For high vitamin A, no effect of dietary vitamin E on fluorescence was observed. Photoreceptor cell densities were affected similarly at all levels of retinol in the diet; after 26 weeks, the E^- groups had a decrease of 19% in PR cell density.

TABLE 2A Plasma Vitamin A Level Changes (%)

	High A	Normal	A
E^-/E^+	-33	-45	-66

TABLE 2B Retinal Ester Content of RPE Changes Compared to A^+ (%)

	High A	$A-$
E^+	NS	-90 ($p < 0.005$)
E^-	NS	-75

TABLE 2C Elevation in RPE Lipofuscin

	High A	A^+	A^-
E^-/E^+	4.5	3.9	6
E^-/E^+	2.8	2.3	3

TABLE 2D Vitamin A Deficiency Effect on Lipofuscin in E^- and E^+ Dietary Regimens

	E^+ (%)	E^- (%)
A^-/A^+	-50	-22
A^-/A^+	-56	-44

Since the photoreceptor cell density was less in the A$^+$, E$^-$ groups, the increased lipofuscin deposition could not be attributed to PR cell density. The vitamin A deficiency also lowered PR cell density while increasing RPE fluorescence and lipofuscin granules, consistent with an explanation involving simply numbers of photoreceptor cells, although explanations involving increased turnover rates of PR cells in the E deficiency may still be possible.

The formation of lipofuscin in the RPE has long been thought to occur by autooxidation of PUFA, perhaps in combination with vitamin A: such autooxidation should be increased in vitamin E deficiency. Since vitamin E$^+$- and vitamin A$^+$-treated rats deposited 50% less RPF lipofuscin than vitamin E$^+$, vitamin A$^-$ but vitamin E$^-$, vitamin A$^+$ rats deposited only 22% less RPE lipofuscin (Table 5), the retinol effect was lower in vitamin E-deficient rats; this is not consistent with an autooxidative mechanism, but when the retinyl ester content of RPE was compared, a 90% reduction occurred in E-supplemented rats but only a 75% reduction was found for E-deficient rats (Table 2B). Because of this, for the vitamin A-deficient regimen, the E$^-$ group would have more vitamin A available for enhancing lipofuscin formation than the E$^+$. These authors believed that the results suggested a more complex interaction between vitamin E and A than simply an antioxidant-prooxidant relationship (147). Indeed, the later (see subsequent discussion) purification of the interphotoreceptor matrix retinol binding protein (IRBP) by the Bridges group (148) and Chader's group (149) opened a new possibility for such interactions since IRBP can bind retinol, vitamin E and retinoic acid and may thus play a role in vitamin E transport in the retina. Thus the vitamin A deficiency could permit better transport of vitamin E, resulting in the lower lipofuscin levels recorded in the A$^-$ versus A$^+$ or (A high) for either E$^-$ or E$^+$ regimens.

Hermann and colleagues (150,151) also found that retinol deficiency reduced the amount of lipofuscin that accumulated in the choroid and extraocular muscle in response to vitamin E deficiency. Although this is consistent with these results, the involvement of some other protein than IRBP, a retinal protein, in transport, may have to be invoked.

In subsequent work, Katz and Robison and their colleagues went on to explore the role of light and IRBP along with vitamin E in the deposition of lipofuscin in the RPE as the retina ages (152). They began by studying the effects of light and aging on the vitamin content of the retina in pigmented rats. Contrary to expectations that light stress might lower the vitamin content, RPE and choroid sclera showed much higher vitamin E levels in light-adapted than dark-adapted animals. Furthermore, this effect was most pronounced in the oldest age group (1429 versus 1051 pmol per eye) at 32 months of the three ages examined (12, 22, and 32 months). Further dietary experiments (152) explored the influence of dietary vitamin A and vitamin E on retinol ester composition and content of RPE using the same three levels of vitamin A described earlier (147), with a view to determining the mechanisms by which lipofuscin accumulation is accelerated by vitamin E deficiency and suppressed by vitamin A deficiency. Vitamin A taken up from the diet is destined for the photoreceptor cells, where its derivatives play a central role in visual processes (153). To reach that destination it is carried by a plasma retinol binding protein but must be taken up by the RPE, which lies between the capillary bed and the photoreceptor cells. The RPE cells have receptors for plasma retinol binding protein along their lateral and basal surfaces (154) and rapidly esterify any retinol taken up into the form of stearate or palmitate. IRBP may be involved in the transport of these retinol esters from the RPE to the photoreceptor cells (148). In vitamin E deficiency, the retinol stearate levels were little affected, but as a result of the increased RPE retinol palmitate, the palmitate-stearate ratio was also increased. This suggested that vitamin E does not act as a regulator of vitamin A levels as a result of its antioxidant function but may actually compete with vitamin A for transport by IRBP and act as an inhibitor of vitamin A uptake and/or storage. Consistent with this idea of competition for a site in the RPE cells were the observations that the vitamin E$^-$ diet led to increased RPE vitamin A and also retarded the loss of

RPE vitamin A in rats on the vitamin A$^-$ diet, indicating that vitamin E may displace vitamin A in competition for the same site in the RPE. In the converse experiment, the vitamin A$^-$ diet led to an increase in neural retina vitamin E. An alternative possibility to explain these data is that there is differential sequestration of vitamin A esters into the RPE lipofuscin (LPF) granules, so that the ratio of retinol palmitate (RP) to retinal stearate (RS) correlated with the RPE LPF formed. For instance, in the vitamin A$^-$ diet, the RP/RS ratio decreases as does LPF, suggesting that RP may be sequestered into the LPF. These authors noted that the RPE retains its retinol very closely, despite depletion in other tissues: alternatively, the fatty acid esterification process may additionally regulate the retinol and retinol ester content since stearate is more labile than palmitate.

These studies of the normal retinal metabolism of vitamins E and A are complemented by a number of studies of stress-induced changes in the retina or its blood supply.

Stress-Related Retinal Damage and the Role of Vitamin E in Its Prevention

Persad et al. (155) found that retinal cell damage by chlorpromazine could be monitored in tissue cultures by release of ^{51}Cr from the damaged cells. Both in the dark and on irradiation, tocopherol alone was capable of preventing the damage, although several other antioxidants tested were not effective (oxygen radical scavengers such as glutathione, β-carotene, mannitol, and D-penicillamine, as well as such enzymes as superoxide dismutase and catalase). Phenothiazines such as chlorpromazine are widely used for treatment of psychiatric disorders (156), and can result in ocular lesions in treated patients who are exposed to sunlight (157–159). The possibility of preventing the damage by tocopherol deserves further study: although the in vitro concentration of tocopherol most effective in preventing the damage was rather high (500 μM), in vivo a lower level of vitamin E may be effective in preventing damage.

Neural Retina

The neural retina is the first stage in the nervous system's visual processing of the image focused on the retina, which is an extension of the brain. The well-known damaging effects on neural tissue from vitamin E deficiency and genetic conditions, such as abetalipoproteinemia, result in changes in both the neural retina and the optic nerve.

Complementary to the work by Katz et al. (152), Kagan et al. (160) showed that vitamin E-deficient rats develop increased levels of lipid peroxidation, as estimated by the content of conjugated dienes (spectrophotometrically) hydroperoxides and dialkyl lipoperoxides (polarographically) and by the characteristic fluorescence of intermolecular cross-links (λ_{ex} 360 nm). Associated with free-radical lipid oxidation, which increased with time on the vitamin E-free diet, they observed damage to the photoreceptor cells, with the degeneration mainly confined to the rod outer segments. These workers suggested a mechanism linking peroxidative damage to cell degeneration. The presence of polar peroxidized lipids in the ROS membranes leads to weakening of lipid-lipid and protein-lipid interactions, resulting in (1) increased ease of extraction of rhodopsin and (2) ROS membrane disorganization. Such changes in the ROS membranes would labilize them, making them more accessible to enzyme digestion by both phospholipases and proteases, such as cathepsin D, which is present in the RPE. Other stresses to which the retina is exposed include normal exposure to light (161), sensitization due to fluorescent dyes used together with strong illumination in fluorescein angiography (162), and combined physiological and psychological stress like that found in the emotional and painful stress which is known to induce lipoperoxide formation in various tissues (including brain, heart, and muscles) (163). Chemical injury to the retina by injection of iodoacetate results in an experimental retinal degenerationf, which results in a significant increase in malondialdehyde (MDA) levels (164). All these conditions have relevance to vitamin E. Vitamin E treatment of fluorescent dye damage, resulting in oligomer formation by rhodopsin, reduces the oligomer conversion of rhodopsin and also MDA levels (162). Stress lipid peroxidation results in large reductions in electroretinogram (ERG) amplitude for both a and b waves and a corresponding fall of 35% in retinal vitamin E. The ERG change was prevented by either fat- or water-soluble antioxidants belonging to the hydroxypyridine group and by 4-methyl-2,6-di-*tert*-butyl phenol (MBTB).

Experimental degeneration induced by monoiodoacetic acid or oxygen poisoning in the rabbit increased the MDA level in the retina. This was decreased by approximately 30% by injection of vitamin E (163). Electrical stimulation of rabbit visual cortex was used by Dzhafarov et al. (165) to stimulate lipid peroxidation (LPO). Vitamin E injection blocked the inhibitory effect on ERG, reducing the accumulation of MDA in the retina. These peroxidation-inducing stresses either affected vitamin E levels in the retina or had effects on the ERG, or effects on rhodopsin oligomerization, which could be prevented by vitamin E supplementation.

The best method to monitor retinal function is simply to measure its response to light, for which it has evolved in the visual process. Akopyan et al. (161) using isolated frog and turtle retinas to test the photic response. The ERGs, both single and rhythmic, were recorded and remained high until challenged by a $FeSO_4$-ascorbate solution, which induced lipoperoxide formation in the retina and reduced the ERG amplitude. Tocopherol was ineffective in a 10^{-3} M solution in preventing the reduction in ERG amplitude, but 0.01% M sodium selenite was effective. The level of vitamin E used was probably not high enough to protect significantly, since previous authors showed effects on the retina using higher vitamin E levels (10^{-3} M) (155). If the lack of an effect is observed when other similar experiments have shown the effect, as in this case, this suggests that further experiments may demonstrate the effect, perhaps at a higher concentration or using a different preparation of vitamin E. These experiments used the ERG, a measure of the neural retina physiological response to light. The best evidence regarding the mechanism of the damage that vitamin E prevented came from an older paper mentioned earlier (162). In this study the methylene blue and fluorescent light, which initiated the production of MDA, oligomerization of rhodopins, and reduction of ERG in the frog retina, produced singlet oxygen (generated by the light excitation of methylene blue). Singlet oxygen was postulated to initiate lipid peroxidation according to the following series of reactions:

$$S \xrightarrow{h\nu} S^*$$
$$S^* + O_2 \rightarrow S + {}^*O_2$$
$${}^*O_2 + RH \rightarrow ROOH$$

where S and S* represent the basic and excited dye states, $^*O_2$ the singlet excited state of O_2, and RH and ROOH the exogeneous substrates (such as DHA) and their peroxidation product, respectively.

MDA generation and its prevention by vitamin E and other antioxidants is consistent with a mechanism in which vitamin E acts as a scavenger of singlet oxygen and lipoperoxides by the following reactions (166,167). Singlet oxygen:

$$^*O_2 + T-OH \rightarrow O_2 + T-OH$$

Peroxyradical chain breaking:

$$ROO^{\cdot} + T-OH \rightarrow ROOH + T-O^{\cdot}$$

Human Deficiency and Abetalipoproteinemia and Vitamin E

Vitamin E deficiency and the genetic disease abetalipoproteinemia both result in neuropathological changes (168–170), the latter disease because the deficiency of β lipoprotein, the normal plasma carrier protein for vitamin E, results in a de facto tissue deficiency of vitamin E. Presumably nervous tissue is especially sensitive because, as for the stresses already described, lipid peroxidation results in disorganization of the cell membranes with an associated dysfunction of nerve conduction. Associated with this, one might expect to find functional changes in the retina, both electrophysiological (electroretinogram amplitude changes) and perceptual (visualization of images), and pathological changes in the optic nerve.

Abetalipoproteinemia was reported initially to cause decreased night vision, with one patient exhibiting eventual progression (171) to bilateral optic atrophy, retinitis pigmentosa, and seriously decreased visual acuity. There were large bilateral central scotoma and moderate constrictions of the peripheral visual fields, although external eye movements were normal. No further deterioration of this condition occurred when the patient began taking 2 g supplemental vitamin E per day and 10 g triglycerides per day. Muller's group (172) reported on eight patients with abetalipoproteinemia who received early treatment with supplemental vitamin E (100 mg/kg/day). In six of the eight no retinal abnormalities developed; in a seventh, the vitamin E treatment at 8 years arrested a mild pigmentary retinopathy that developed at 5 years of age. An eighth patient, displaying abnormal retinal function, pigmentary retinopathy, marked ataxia, absent tendon reflexes, and decreased motor nerve conduction velocities, began treatment at age 10. Vitamin E supplementation resulted in marked improvement within 2 years and normal retinal function and nerve conduction velocities after 13 years of treatment. Other investigators have observed similar beneficial effects of large doses of vitamin E, sometimes in combination with vitamin A (173), in abetalipoproteinemia. Azizi et al. (174) noted improvement in scotopic vision of an 11-year-old girl treated with combined vitamins A and E; Miller et al. (175) described the use of vitamin E supplementation to halt deterioration in neurological and retinal function in a 22 year old. Herbert et al. (176) reported on three patients over 30 years old who received long-term oral vitamin E and did not develop either retinal or neurological complications.

A more detailed study of retinal function using electroretinography, and electroautography when indicated (173), showed no evidence of further progression of any defect in visual function once therapy with combined vitamin A (50,000 IU twice per week) and vitamin E (100 mg/kg/day) was begun. Thus it appears that vitamin E (along with A) can effectively treat these degenerative changes due to the failure to transport vitamin E to ocular tissues.

Several other conditions that result in vitamin E deficiency have also been shown to result in similar ocular degenerative changes. These include cystic fibrosis (177), small bowel resection resulting in vitamin E deficiency (178), and congenital biliary atresia (179). Saty-Murti et al. (180) found that changes persisted after treatment. They concluded that since treatment may prevent further progression of degenerative changes, such as pigmentary retinopathy and abnormal ERGs, it may be best to detect and treat the deficiency of vitamin E at an asymptomatic stage. Other authors confirm that early treatment can prevent further progression of the degenerative changes, which include impaired night vision, bilateral ring scotomas, and progressive ophthalmoparesis (178) with progressively increasing adduction weakness, exotropia, and ptosis.

Complementary to these studies, the study of the effect of the combined A and E supplement Rovigon (Hoffman-LaRoche & Co.) on vision in elder atherosclerotic patients indicated that a decrease in visual acuity in old age could be significantly improved (181). Of a total of 192 patients, 169 were evaluated critically in four groups: 92 cases of atherosclerotic retinopathy, 12 of diabetic retinopathy, 39 of myopic chorioretinopathy, and 11 others. Improvement was noted in 3 old central vein thromboses and 1 partial optic atrophy. In myopic chorioretinopathy of 76 eyes, only 3 became worse, and 40 either showed clear improvement (8 eyes) or slightly measurable improvement (32 eyes). In diabetic retinopathy of 24 eyes, 2 declined but 10 showed no change and 12 either slightly or markedly improved. With miscellaneous conditions, of 4 patients studied, improvements in visual acuity cited were (1) 5/30 to 5/15 after treatment, (2) 1/20 to 1/10, (3) 5/30 to 5/10, and (4) 5/50 to 5/15, respectively. This study indicates that supplemental vitamins may be of utility in improving vision in a variety of elderly patients. Together with the cataract studies (78–80), this suggests that the elderly may benefit from sharper vision if supplementation with vitamin E and A is taken.

Spielmeyer-Vogt syndrome (Batten's disease, juvenile lipofuscinosis) also seems to involve pathogenetic peroxidative processes, characterized by blindness (4–6 years of age), epilepsy (8–10 years of age), and dementia (10–20 years of age) (182). Biochemically this is caused by a low cellular glutathione peroxidase activity and decreased blood selenium concentration. The blindness seems to be due to retinal lipofuscinosis as a result of autooxidation of retinaldehyde and linolenic acid derivatives. Clausen (182) recommends antioxidant treatment for such patients based on evidence that the juvenile onset (but not the infantile type) seems to benefit from a therapy combining vitamin E, sodium selenite, vitamin C, and vitamin B_6 (183,184).

Retinal pigments, such as melanin, found in the RPE, appear to have an antioxidative effect on lipid peroxidation (185). Nonpigmented RPE from rabbits was more sensitive to peroxidation. Conversely, although glutathione peroxidase (GSH-Px) and superoxide dismutase were practically equal in the RPE of pigmented and nonpigmented rabbits, albino RPE contains a much higher α-tocopherol concentration, apparently to compensate for the lower melanin concentration.

The effects of combined deficiency of vitamin E and selenium on rat ocular capillaries was studied by Ameniya (186). A deficiency of vitamin E led to striking differences in both retinal capillaries and rectus muscle capillaries that could contribute to the increased peroxidative changes in vitamin E deficiency. The retinal capillaries in the outer plexiform layer show increased basement membrane thickening with age, enlarged mitochondria, a high ribosome content in the endothelial cytoplasm, and frequently a very narrow luminal space. Basement membrane thickening is also observed in diabetic kidney basement membrane (187). It is associated in the kidney with reduced filtration efficiency, resulting in larger molecules being able to pass such thickened but more loosely cross-linked basement membranes. This has been thought to be due in part to lysyl oxidase inactivation, resulting in fewer collagen cross-links and a looser collagen structure in the basement membrane. Perhaps related to this, in the rectus muscle capillaries, the vitamin E-deficient diet resulted in significantly thinner basement membranes and hemorrhaging in which occasional endothelial cell connections were broken (186). Stress due to oxygen can cause retinal cell degeneration in the rabbit (188) that is not prevented by administration of DL-α-tocopherol acetate. Subsequent experiments with rats, however, have shown that it is possible by dietary treatment using vitamin E alone or vitamin E in combination with selenium to prevent changes in the ERG a and b wave amplitude usually seen in rats exposed to hyperbaric oxygen (189). This holds significant promise because hyperbaric oxygen treatment for a variety of conditions is becoming more common (182), and prevention of complications from this therapy should be a high priority.

Retinopathy of Prematurity

Retinopathy of prematurity appears to be a disease caused by modern medical technology, since as has been shown (190,191) it appears to be due to the interaction between the survival of smaller and smaller premature infants and the use of oxygen to prevent hypoxia. As more infants of younger gestational age survive, the problem is how to complete developmental stages that should occur in utero but are forced to take place in the incubators of a neonatal intensive care unit. Complicating the process at present is the necessity of using oxygen to counter the lower efficiency of the lungs if the infant is suffering respiratory distress, previously known as hyaline membrane disease. A major factor in such respiratory distress is the lack of lung surfactant at these early gestational ages, resulting in damage to the lungs due to an inappropriate surface tension at the air/liquid alveolar interface.

The history of the association of vitamin E in treating retinopathy of prematurity is a long one. As Silverman (192) describes, during the first 40–50 years of this century, premature infants were cared for in premature units, based on the ideas of Budin of Paris and his associate Couney. The strict isolation measures discouraged traffic in glass-enclosed rooms, and patients and even physicians were discouraged from touching the infants too often. Imagine the alarm in the late 1940s that spread when the frequency of retrolental fibroplasia (RLF), not observed until 1942, reached epidemic proportions (193). Even more alarming was the observation that it occurred most frequently in premature nurseries with the most organized and "high-tech" programs for care.

By the early 1950s the occurrence of RLF rose sharply first in the United States and then in other developed countries and was regarded as the major cause of infant blindness (194). An early study in 1949 by Owens and Owens (195) on the treatment of premature infants with vitamin E showed promise in significantly reducing the incidence of RLF, and for a while Kinsey and Chisholm incorporated this treatment with success into evolving treatment regimens (196). An early study found a correlation between RLF and (1) the use of iron, (2) water-soluble vitamins and vitamin preparations, and (3) the use of oxygen, but they noted a "less striking" correlation with oxygen use than for iron and vitamins. In the Owens's study two alternative treatments were used: 150 mg/day of synthetic vitamin E, or no treatment. During the 10 month trial of 11 infants receiving vitamin E, none developed RLF, but 5 of the 15 untreated developed RLF. This early result so impressed Friedenwald and Woods, ophthalmologists at Johns Hopkins Hospital, that they persuaded Owens and Owens to abort the trial and begin treating all infants. Unfortunately, nine other groups informally using a similar treatment regimen were unable to confirm the effectiveness of vitamin E for RLF, although as noted, Kinsey and Chisholm found a positive effect (196), particularly in a reduction in the number of the most severely affected cases. The early vascular changes that were initially taken as a sign of RLF were subsequently reported by Owens and Owens (197) to regress spontaneously in three-quarters of cases. Silverman (192) tested adrenocorticotropic hormone (ACTH) after some encouraging early results but could not confirm a beneficial effect when a controlled clinical trial was performed. Then in Birmingham, England, Crosse (198) found that in the period 1944–1950, when high oxygen was used, 3 of 13 infants developed RLF; in 1951, using a limited oxygen regimen (in which the shortest possible exposure was offered, with just enough oxygen to prevent cyanosis and return to normal air gradually but as soon as possible), 0 of 6 infants developed RLF.

This prompted several further studies in the United States (described in Silverman, Ref. 192) and eventually issued in a National Cooperative Study, beginning July 1, 1953 to June 30, 1954 involving 18 hospitals (189). This study showed a striking difference between the percentage of infants developing RLF after routine oxygen (70%) or curtailed oxygen (31%) and a dramatic decrease in *scarring RLF* (SRLF) when routine oxygen (17% SRLF) was compared to curtailed oxygen (5% SRLF). For multiple births, the figures were even more striking: routine O_2 (83% RLF) versus curtailed O_2 (42% RLF) and scarring RLF-normal oxygen (67% SRLF) versus curtailed (14% SRLF).

From this point on oxygen practices at most premature nurseries were adjusted, in particular to respond to the observation that it was not the concentration of oxygen administered but the duration of exposure, particularly for the first few days. The risk increased with exposure for up to 2 weeks for multiple-birth infants. Furthermore, any concentration of oxygen above that in room air increased the risk of RLF, but the rate of withdrawal of infants from oxygen did not appear to be important in RLF development. This study resulted in major changes in the practice of oxygen administration in premature nurseries and curtailing of the exposure of premature infants to high levels of oxygen with the result that the incidence of RLF dropped dramatically over the succeeding years.

As treatment of premature infants improved over the next 20 years, many advances in the medical care of premature nurseries, along with improvements in oxygen monitoring and delivery, have permitted the survival of smaller and smaller infants, who are at higher risk for ROP. Thus the number at risk for ROP increased, and increasing numbers of premature infants developed blinding ROP. In the early 1970s, Johnson and coworkers began a series of studies reinvestigating the role of vitamin E in the prevention and treatment of ROP. Their 1972–1974 study (199) substantially reduced the incidence of RLF in premature infants < 2 kg birth weight and was marginally significant for < 1500 g birth weight infants ($p < 0.06$). Soon this was supported by proof of vitamin E efficacy in animal studies by Phelps and Rosenbaum using the kitten model of ROP (200,201). Several other centers undertook clinical studies with encouraging results.

Helen Hittner and colleagues reported a significant reduction in severity but not in the incidence of ROP (202) in infants less than 1500 g birth weight. A number of other studies reported varying effects (203–207), mostly protective: however, Puklin et al. (204) did not find any difference between treated and placebo groups < 1500 g birth weight using intramuscular (IM) vitamin E in the first 24 h, nor did Milner et al. (207). Finer et al. (203) reported that although the incidence of ROP was not affected by early IM vitamin E, it significantly reduced the severity of retinal damage. Since then a number of studies have appeared, with differing results. A more recent trial was designed very carefully as a controlled, randomized, masked trial (208). In this report, there was no significant difference in the effect of vitamin E-treated and placebo-injected infants. Unfortunately, a detailed examination of the composition of the placebo injection reveals that 1% benzyl alcohol was included: a molar amount of benzyl alcohol was present, equal to 40 mol % of the molar quantity of vitamin E present in the treatment injection. Recently, Testa (209) proposed benzyl alcohol as an anticataract agent, suggesting, that it acts as an antioxidant. If, like mannitol, it does have antioxidant or free radical scavenging activity, since the molar quantity of benzyl alcohol is so close to that of the tocopherol, both treatment and placebo groups might be above any threshold for a preventive effect on ROP. Although this does not invalidate the placebo used, it certainly casts serious doubt on this study's conclusions. For this reason, it should be given special attention when considering whether it offers guidance in treatment of ROP.

Proposed Mechanism

Initially the mechanism believed to function (210) in inducing the retinopathy of prematurity involved vasoconstriction (in response to oxygen) of the retinal capillaries, followed by destruction of the retinal endothelial cells by oxygen. To understand the mechanism by which this could occur, it is necessary to know how the retinal circulation develops.

Normal Vascularization

Foos (211) outlined the steps of retinal vascularization. Usually vascularization, which begins at the fourth lunar month of gestation, is fully developed at birth, with the mature capillary system in place, covering most of the retina except a 1–3 mm zone behind the ora serrata that has no vessels.

ROP can only occur in a vascular system that is not fully developed or mature, so the details of growth of the vascular system are important for an understanding of how ROP can arise. The first event is a wave of vasculogenic cells following from the optic disk in the inner retina toward the periphery of the retina (211). The wave of vasculogenic cells has two active zones: (1) the vanguard (anterior) has primitive spindle-shaped cells, of mesenchymal origin, which differentiate from angioblast precursors (212); and (2) the rearguard consists of mesenchymal cells that differentiate into endothelial cells, which express factor VIII. These aggregate into cords that later lumenize and then become primordial capillaries. Anterior to the vanguard the retina is avascular (in its superficial aspect) and contains cystic spaces that intercommunicate. These appear as superficial blebs bordered by Mueller cells (211). Since the optic nerve is located eccentrically in the retina, more to the nasal side, tissue forming vessels must travel farther temporally. Because of this, in the temporal quadrants the capillaries mature later than in the nasal quadrants. Remodeling of these capillaries into their adult form requires 2–3 months.

Stages of Retinopathy

Stage 1. In retinopathy, in stage 1 of the process the hyperoxic injury occurs, probably due to elevated levels of free radicals that may induce oxidative damage and lipid peroxidation, which vitamin E is known to prevent or minimize. This damage presumably leads to a proliferative response. Macroscopically the normal advancing line of vasculature becomes thickened and exaggerated, with marked vasodilation behind the line. Microscopically, at this advancing line can be seen hyperplasia of the spindle-shaped mesenchymal cells of the vanguard. Most retinal damage is found in the temporal region of the retina since in the nasal region the vasculature matures earlier and vascularization is complete in the nasal region before premature birth, leaving further development to occur temporally.

Stage 2. At stage 2 of retinopathy, in response to continuing injury, macroscopically, the line becomes thickened and a thin white line of approximately equal width is seen behind the vanguard, as a result of proliferation of the rearguard cells. Progressive thickening occurs at this ridge, and by fluorescein angiography, an arteriovenous shunt can often be seen in this area. This stage usually arrests and spontaneously regresses with no residual scars.

Stage 3. Stage 3 retinopathy involves extraretinal vascularization (ERV), which occurs in three forms: (1) placoid (circumferential orientation); (2) polyploid (isolated hemispherical mounds on the rear surface, behind the ridge); and (3) pedunculated (stalk and palm frondlike projections into the vitreous body). Microscopically ERV occurs from the rearguard region (the rear of the ridge), leading to either proliferation of cords of endothelial cells not containing any lumina or, more commonly, the growth of delicate, thin-walled vessels through the surface of the retina into the vitreous cavity.

Retinal buckling eventually occurs as the first stage of detachment due to processes occurring in the vitreous:

1. Synchysis: usually over arrested lesions are found large pockets, over the ridge. Usually these pockets are caused by lysis of the vitreous by products released from the incompetent vessels in the retina or vitreous.
2. Condensation is seen as sheets and strands extending from the retina into the vitreous toward the lens, associated with hyaluronate depolymerization and collapse of the vitreous collagen into visible structures (210). This process usually exerts force on the retina, leading to retinal buckling that precedes actual detachment.

Stage 4. At stage 4 of ROP, the characteristic feature is retinal detachment, the classification features of which have been published (213). The detachment may be classified as exudative or tractional and may spatially be characterized as segmental or circumferential as well as macular or extramacular. This stage of cicatricial damage is not reversible and is usually associated with blindness.

Hypotheses

Vasoconstriction.   The original hypothesis was that endothelial cell death occurs following capillary contraction after oxygen toxicity, leading to vasobliteration and reproliferation of rearguard cells. This was disproved by Flower (212). He showed in the puppy model that vasoconstriction appeared to be a protective effect, since aspirin, which reduced vasoconstriction, resulted in more severe retinopathy than found in untreated animals. He suggests, as an alternative to the direct cytotoxic effect of oxygen on endothelial cells (214), that normally at birth vasoconstriction occurs, leading to a rise in arterial blood pressure. In the retinal arteries such vasoconstriction would result in excessively high blood pressure, possibly producing retinal capillary hemorrhage, an effect that could be independent of or work together with direct oxygen cytotoxicity (212). Aspirin was shown to cause vasoconstriction (212), and alternatively to that caused by aspirin, a similar vasoconstriction occurred in puppies exposed to high CO_2/O_2 (10% CO_2, resulting in focal hemorrhage in puppies. In the data of Baieer and Widmayer (215), the best predictor of ROP was the highest PCO_2 measured, especially if PCO_2 elevation was associated with elevated PO_2.

Suggested Mechanism Involving Spindle Cells.   Why, then, are premature infants at 27–28 weeks of gestation at such risk for ROP? Kretzer and Hittner suggested a mechanism (216–218) that explained the observation that vitamin E appears to have an effect in preventing severe ROP in infants of 1 kg birth weight (gestational age 28 weeks) but is not effective on those of 27 weeks or less gestational age. Kretzer's explanation is based on animal experiments and ultrastructural analysis of the retinas of premature infants who died. His observation is that the spindle cells, which are the precursors of the capillary endothelium of the inner retina, have their mesenchymal origin in the adventitia of the myeloid artery. The uterus provides a hypoxic environment in which these cells are in an undifferentiated state, termed trend 1 by Hittner and Kretzer (218), occupying a minimum volume of the nerve fiber layer as an anastomosing circumferential homogeneous apron in which the cells lying adjacent to each other have few gap junctions. Following administration of oxygen after as little as 4 days, activation of the spindle cells occurs, with differentiation to a stage Kretzer termed trend 2 in all infants less than 32 weeks of gestational age. A striking feature of this stage is a large increase in gap junctions, a major route of intercommunication between adjacent spindle cells. Vitamin E supplementation in infants to a level $\geq$ 1.2 mg per 100 ml but $\leq$ 3.5 mg per 100 ml vitamin E, which is viewed as nontoxic as described by Hittner and Kretzer (217), inhibits this gap junction formation, except in the very young (27 week gestational age or less). The proliferation of these spindle cells is a precursor of RLF, since they participate in abnormal retinal vascularization at the boundary between the vascular and avascular retina, as described earlier. In addition to the effect of vitamin E on spindle cell gap junction formation and activation, Johnson et al. (219) indicated that late supplementation with vitamin E may induce regression or reduce damage in some cases of severe RLF, which Phelps and Rosenbaum (200) also found effective in a related but inexact animal model, the kitten.

Controversy in the Use of Vitamin E to Treat ROP

The provision of a rational explanation for the effect of vitamin E in inhibiting spindle cell differentiation was very useful, but subsequent events have resulted in a "benign neglect" of vitamin E therapy for ROP in recent years.

In late 1983 and early 1984, a preparation of vitamin E was marketed for intramuscular injection, apparently without appropriate testing. It resulted in a syndrome characterized by unexplained thrombocytopenia, renal dysfunction, hepatomegaly, cholestasis, ascites, hypotension, and metabolic acidosis (220–223). The toxicity of the preparations has been attributed to the high polysorbate levels used as a solubilizing agent rather than to the vitamin E acetate (222,223).

Several toxic side effects of vitamin E administration were noted. These included intraventricular hemorrhage (IVH) (224–228), which some groups reported that vitamin E reduces in severity but not in overall incidence (225), but others reported reduced in both incidence and severity (224,226,228). Even the same group at Baylor found no effect in one study of oral supplementation (202), but another study involving oral vitamin E supplemented with intramuscular vitamin E showed a reduction in both incidence and severity of IVH (218). In contrast, the Pennsylvania study (205) showed no effect and the University of California–Los Angeles (UCLA) study showed increased IVH incidence for vitamin E treatment, comparing more severe grades of damage (3 and 4) in infants < 1000 g at birth, but no differences for less severe IVH (grades 1 and 2) or for larger infants (229). When this was reviewed by a committee of the U.S. National Academy of Medicine, the committee believed that no firm conclusion could be reached regarding the efficacy of vitamin E for treating IVH (228).

The other major medical problems reported to be associated with vitamin E administration were necrotizing enterocolitis (NEC) and neonatal sepsis (NS). Finer and colleagues, in a retrospective study (228,230), found an increased incidence of NEC after oral but not intramuscular administration of vitamin E, especially in infants ⩽ 1250 g birth weight. Johnson and colleagues (205,231), treating to achieve high blood levels of vitamin E (5 mg/dl), observed an increase in the incidence of NEC and neonatal sepsis (231).

Keith, Kitchen and colleagues (240,241) found no significant benefit from vitamin E treatment of ROP.

Risk-Benefit Analysis

Phelps (232) first raised the issue of vitamin E used to prevent severe ROP compared to the possible harmful effects of toxicity to IVH or NEC and sepsis. Later she was asked by the American Academy of Pediatrics to serve as a consultant for the Committee on Fetus and Newborn report entitled Vitamin E and the Prevention of Retinopathy of Prematurity. The committee considered several controlled clinical trials: (1) in one trial, a decrease in the severity of ROP after prophylactic oral administration of vitamin E, and (2) in three other controlled trials a lower incidence of severe ROP in vitamin E-treated groups. The committee concluded, regarding these trials, that "none of these differences was statistically significant" (233).

The report goes on to note that in the United States 22,000 infants of birth weight < 1500 g would have to be treated annually to prevent severe ROP in approximately 2000. It observes that this would be acceptable if it were certain that the administration of vitamin E was completely safe, or at least that the benefits of its use outweighed the risks by a substantial margin. The report cites two papers as preliminary reports indicating that complications may arise from the administration of pharmacological doses of vitamin E. One refers to the E-Ferol preparation (234), which of course was taken off the market and so is not now relevant to considerations of complications, although it was relevant at the time. The second, a report by Sobel et al. (235), refers to data submitted to the U.S. Food and Drug Administration (FDA) in support of a parenteral vitamin E preparation that indicated a higher incidence of NEC in vitamin E-treated infants: at blood levels > 3.5 mg/dl, NEC developed in "30% of all treated infants"; at ⩽ 3.5% mg/dl, only 4% of treated infants developed NEC. Citing Finer et al. (230) and Johnson's group (231) to show that NEC and sepsis occur only infrequently in the neonatal population, the committee concluded that studies to date "lacked sufficient sample size to appropriately evaluate a potential increase of such problems or to be sufficiently reassuring that unexpected complications of therapy are unlikely to occur." The committee also concluded that it could not recommend that high doses of vitamin E be given routinely to infants weighing less than 1500 g even if such infants require the use of supplemental oxygen (233).

Analyzing a slightly larger group of six studies 2 years later, the Institute of Medicine of the U.S. National Academy of Sciences reported that the "data from the prospective studies are conflicting and do not warrant a recommendation at the present time for the routine use of vitamin E to prevent or modify ROP" (229).

Metaanalysis Using the Mantel-Haenzel-Peto Technique

Since the studies used as a basis for these reports, several new studies have appeared that deal with efficacy of vitamin E treatment of ROP as well as possible complications, such as IVH, NEC, and sepsis. In the interim, too, novel statistical techniques of metaanalysis to compare and combine the results of clinical trials that share common major features (even though the detailed studies are not identical) have been developed by Peto's group (236,237).

Using this method we decided to take a second look at the data, which was available in the many published reports designed to study the effect of supplemental vitamin E on ROP, as well as on IVH, NEC, and sepsis.

In choosing the studies we decided to take two approaches: (1) include all studies for which control groups, either consecutive (when previous experience in the same facility provided a baseline) or concurrent, which is now the accepted norm; and (2) choose a defined group acceptable for study as previously defined and reanalyze the data: the Institute of Medicine for its 1986 report chose six studies that were amenable to such an analysis, as had Schaffer et al. (238).

All Studies.   Using all the studies that could be arranged into an appropriate form for analysis by the Peto technique (Appendix Table 1) we identified a number of studies, some of which used shared controls. When this was the case, the data for occurrence of the ROP and unaffected cases were summed for treatment groups, matching with the appropriate single set of controls (Table 3). Because of the large number of studies, it was thought that the use of a procedure of this type for 1 or 2 groups in 21 would not have a large effect on the results. Omission of the groups tested can easily be done simply by subtracting their individual $O - E$ values from the grand total of $O - E$ values and their variance from the total variance, which did not have a large effect on the result. The results of the analysis indicated a reduction in the odds ratio for retinopathy of prematurity in treated or untreated infants was 0.69 with confidence limits 0.85 and 0.57. This odds ratio reduction of 31% is highly significant ($p < 0.0005$).

Selected Groups.   For the data for incidence used in the Institute of Medicine study (Appendix Table 2A) (229), the odds ratio (treated-untreated) is 0.79 with confidence limits 1.018 and 0.616, resulting in a marginal significance ($p < 0.07$). For the same study, severe ROP (Appendix Table 2B) treated with vitamin E (Table 8) had a odds ratio of 0.51 with confidence limtis 0.87–0.30, and this was highly significant ($p < 0.01$).

For the incidence of cicatricial complications (Appendix Table 2C; Table 4A–C) four studies only were included because this was not monitored in all six chosen for the incidence study. The odds ratio (treated-untreated, Table 2C) was 0.52 with confidence limits 0.23–1.18 due to the small numbers, and this was not significant ($p < 0.2$).

Schaffer et al. (238) assembled a group of studies (Appendix Table 3; Table 5) in which the effect of vitamin E on ROP was compared for severe retinopathy only. Six studies, when combined by the Peto technique, gave an odds ratio (treated-untreated) of 0.67 ($p < 0.01$) with 95% confidence limits of 0.20–0.78.

Intraventricular Hemorrhage: All six studies of IVH were combined (Appendix 4A) to give odds ratio (treated-control) of 0.52 ($p < 0.001$), which is highly significant, with 95% confidence limits of 0.35–0.77. If two recent studies [Chiswick's group study (227) and Fish

TABLE 3 Summary of Statistical Analysis of ROP and Other Conditions Affected by Vitamin E

Condition	Appendix table	Total N	O − E	Variance	No. studies	Odds ratio	Confidence limits
ROP, all studies	A.1	2249	−34.64	94.58	20	0.69 (p < 0.0005)	0.57–0.85
ROP, NIM report	A.2	973	−14.22	60.79	6	0.79 (p < 0.1)	0.62–1.02
Severe ROP, NIM report	A.3	973	−9.33	13.44	6	0.51 (p < 0.01)	0.30–0.86
Cicatricial damage	A.4	657	−3.71	5.71	4	0.52 (p < 0.2)	0.23–1.18
Severe ROP, Schaffer (1988) (238)	A.5	1077	−7.9	8.51	6	0.395 (p < 0.01)	0.20–0.78
Intraventricular hemorrhage	A.6	653	−15.42	24.89	6	0.52 (p < 0.001)	0.35–0.77
	A.6	882	−25.81	34.11	7	0.46 (p < 0.0001)	0.33–0.64
	A.6	1019	−29.42	42.10	8	0.50 (p < 0.0001)	0.37–0.67
Necrotizing enterocolitis and sepsis	A.7	677	−0.21	11.39	3	0.98 (NS)	0.92–1.05
	A.7	1152	14.86	23.37	4	1.86 (p < 0.005)	1.24–2.77
	A.7	822	−1.62	12.57	4	0 (NS)	0.45–1.9
Necrotizing enterocolitis (NEC)	A.8	860	−0.47	10.29	5	0.96 (NS)	0.51–1.75
	A.8	1338	8.29	21.56	6	1.46 (p < 0.1)	0.963–2.24
	A.8	1005	−1.94	12.41	6	0.58 (p < 0.1)	0.22–1.01
	A.8	1483	6.82	23.41	7	1.22 (NS)	0.9–1.99

TABLE 4A   Vitamin E Concentration in Normal Animals

Reference	Species	Diet or condition	Retina	Choroid	Vitreous
136	Bovine	—	0.52–0.96 nmol/mg ROS protein		
273	Kitten	—	40 ± 7.5 µg/g protein	50 µg/g protein	9 µg/g protein
276	Human neonate	—	0.33 ± 0.16 µg/ protein	0.25 ± 0.16 µg/g protein	0.21 ± 0.041 µg/g protein
277	Rat,				
	18–20 days	Cyclic light	26 ± 0.03 nmol/ retina		
	30–40 days	Cyclic light	43 ± 0.05 nmol/ retina		
	44–60 days	Cyclic light	47 ± 0.05 nmol/ retina		
	18–20 days	Dark reared	20 ± 0.01 nmol/ retina		
	30–40 days	Dark reared	38 ± 0.06 nmol/ retina		
	44–60 days	Dark reared	44 ± 0.03 nmol/ retina		
278	Human	Donor eye	13.7 ± 5.1 nmol/ eye	Includes RPE 15.2 ± 5.9	
			1.1 ± 0.5 nmol/cm²	1.2 ± 1.5 nmol/cm²	
			2.8 ± 1.2 mg/100 g	3.3 ± 1.5	
	Cattle		0.72 nmol/mg protein		
275	Human	Infant	51 ± 57 µg/g protein		
283	Cat	Adult	108 ± 10 µg/g dry weight	21 ± 6 µg/g dry weight	1 ± 0.03 µg/g dry weight
	Rabbit		116 ± 11 µg/g dry weight	35 ± 6 µg/g dry weight	1 ± 0.4 µg/g dry weight
	Rat		83 ± 9 µg/g dry weight	114 ± 12 µg/g dry weight	1 ± 0.05 µg/g dry weight

TABLE 4B   Vitamin E in Lens

Reference	Species	Diet or condition	Lens	Anterior capsule	Posterior capsule
278	Rabbit	VE, 30 mg/ kg/day	1.97 ± 0.15		
280	Human 60–80 years		373 ± 40 ng/lens (n = 5)		
	Cataract		338 ± 30 ng/lens (n = 5)		
283	Cat		2 ± 0.1 µg/g dry weight	2 ± 0.1 µg/g dry weight	1 ± 0.05 µg/g dry weight
	Rabbit		1 ± 0.05 µg/g dry weight	2 ± 0.1 µg/g dry weight	1 ± 0.06 µg/g dry weight
	Rat		2 ± 0.03 µg/g dry weight	1 ± 0.05 µg/g dry weight	1 ± 0.04 µg/g dry weight

TABLE 4C Vitamin E Levels After Supplementation

Reference	Species	Diet or condition	Retina	Choroid	Vitreous
273	Kitten IM	N 20 mg 4 h	$y = 48(x)^{0.24}$ $x =$ time (h) $y = \mu g/g$ protein $r = 0.36, p < 0.05$	$y = 24.3(x)^{0.006}$ $x =$ time (h) $y = \mu g/g$ protein $r = 0.56, p < 0.001$	$y = 113(x)^{0.005}$ $x =$ time (h) $y = \mu g/g$ protein $r = 0.58, p < 0.001$
	IV Oral	Supplement	NS $y = 23(x)^{0.41}$ $713 \pm 581 \ \mu g/g$ protein x, y as above $r = 0.66, p < 0.001$	NS	NS
275	Human neonate	Supplement 50 mg/ kg/day (oral)	$713 \pm 581 \ \mu g/g$ protein	$956 \pm 586 \ (n = 5)$ $\mu g/g$ protein	$139 \pm 69 \ (n = 5)$ $\mu g/g$ protein

et al. (see Ref. 281)] were included, the odds ratio was 0.50 ($p < 0.001$) with 95% confidence limits of 0.37–0.67.

Necrotizing Enterocolitis and Sepsis: If the study of Johnson et al. (231) is included, analysis of four studies gives an odds ratio (treated-control, Appendix Table 4B) of 1.86 ($p < 0.005$, Table 6) with 95% confidence limits of 1.24–2.77. Since other groups did not exhibit such a high incidence of sepsis and NEC, they were compared without including the data of Johnson et al. (231); the odds ratio for treated to control was 0.98 ($p < 0.95$) with confidence limits of 0.92–1.04.

For necrotizing enterocolitis alone, comparison of all studies (five in total, Appendix Table 4C) gave an odds ratio (treated-control) of 1.46 (NS) and confidence limits of 0.96–2.22; excluding the data of Johnson's group (231) led to an odds ratio (treated-control) of 0.96 (NS) and confidence limits of 0.57–1.75.

TABLE 5 Concentration of Tocopherol in Eye as a Function of Age

Age	Number	α-Tocopherol (nmol per eye)	γ-Tocopherol (nmol per eye)	Total $\alpha + \gamma$ (nmol per eye)
23–30	2	23.4	2.5	35.9
30–40	3	22.6	3.7	26.3
40–50	4	21.9	7.9	29.8
50–60	8	25.8	5.9	31.7
60–70	10	29.8	6.1	35.4
70–80	8	30.5	5.2	35.7
80–90	2	25.8	1.5	27.3
90–100	1	17.9	5.0	22.9

Source: From Reference 258.

TABLE 6   Estimated Benefits of Vitamin E Treatment Calculated by Method One

Condition	(O − E)/N Percent	Number of infants[1]
R. O. P. (Table 2A)	− 1.46	− 289
Severe R. O. P. (Table 2B)	− 0.96	− 209
I.V.H. (Table 4A)	− 2.42	− 531
N.E.C. (Table 4B)	+ 0.95	+ 208
Benefits versus risks:		
All R. O. P./Severe R. O. P.	− 498	− 209
I.V.H.	− 531	− 531
N.E.C.	+ 208	+ 208
Net gain	− 821	− 532
Percent of 22,000	3.73	2.42

[1]22,000[(O − E)/N].

TABLE 7A   Odds Ratios and Incidence Rates (per 100) Used in Calculations for Method Two

Condition	Odds ratios			Incidence rates		
	LCL[1]	Mean	UCL[2]	Low	Mean	High
R.O.P. (2a)	0.57	0.69	0.85	19.5	23.6	29.1
Severe R.O.P.	0.20	0.40	0.78	1.0	2.7	3.6
I.V.H. (4A)	0.35	0.52	0.67	9.4	13.9	17.9
N.E.C. (4B)	1.24	1.86	2.77	10.8	16.2	24.2

[1]95% Lower confidence limit.
[2]95% Upper confidence limit.

TABLE 7B   Estimated Benefits of Vitamin E Treatment Calculated by Method 2

Condition	Mean	Number of infants		
		High/High[1]	Low/Low[2]	Low/High[3]
R. O. P.	− 2332	− 3234	− 1122	− 1122
I. V. H.	− 2838	− 3828	− 1958	− 1958
N. E. C.	+ 1650	+ 3410	+ 462	+ 3410
Net gain	− 3520	− 3652	− 2618	+ 330
Percent of 22,000	16.0	16.6	11.9	1.5
Severe R.O.P.	− 396	− 770	− 198	− 198
I.V.H.	− 2838	− 3828	− 1958	− 1958
N.E.C.	+ 1650	+ 3410	+ 462	+ 3410
Net gain	− 1584	− 1188	− 1694	+ 1254
Percent of 22,000	7.2	5.4	7.7	5.7

[1]High estimates for R.O.P., I.V.H., and N.E.C.
[2]Low estimates for R.O.P., I.V.H., and N.E.C.
[3]Low estimates for R.O.P. and I.V.H., high estimate for N.E.C.

Calculation of Benefits versus Risks

In order to assess the net benefit of vitamin E treatment of premature infants, two methods were used based on the 22,000 infants estimated to be eligible for treatment (233).

In the first method, the risk reduction, $O - E/N$, was calculated as a percent for the conditions which were treated with vitamin E. This figure was applied to 22,000 to estimate the reduction or increase in numbers of affected infants. In the case of R. O. P. and I. V. H., the calculations which produced the least reduction in cases were used (see Appendix Tables 2A and 4A). Conversely, the calculations which produced the largest increase in cases were used in the case of N. E. C. and neonatal sepsis (Appendix Table 4B). The figures, reported in Table 6 constitute a worst case scenario. The estimated benefit of vitamin E treatment was a reduction of R. O. P. and I. V. H. cases by 1,029 at a cost of increasing N. E. C. and neonatal sepsis by 208 cases. In the case of severe R. O. P. and I. V. H., the estimated reduction was 740.

In the second method, the weighted average incidence rates of the control sample for each condition were multiplied by the odds ratios and their upper and lower 95 percent confidence limits to produce low, mean and high estimates of incidence with vitamin E treatment (Table 7A). These incidence rates were then applied to 22,000 to estimate the number of infants who would be affected. The differences between these numbers and the number of infants affected in the control sample are reported in Table 7B. With the exception of the low estimates for prevention of R. O. P. and I. V. H. versus the high estimate for N. E. C. and neonatal sepsis, the results suggest sizeable reductions in the numbers of affected infants.

EVOLUTION OF CURRENT TREATMENT PRACTICES

When Owens and Owens began their work, it was customary to use artificial formulas with both supplemental polyunsaturated fatty acids and iron for premature infants (245), but it may not have been appreciated, as Katz and Robison (246) point out, that the PUFAs are highly susceptible to autooxidation if insufficient antioxidants are given along with them. Coupled with the fact that in most infants vitamin E levels are quite low, the increased oxidative stress due to the load of PUFA probably placed a peroxidative stress on the premature newborn, which was exacerbated by oxygen, along with the PUFA and the iron supplements believed necessary to nutritionally supplement the premature infant (245,247).

Normal breast-fed infants receive the majority of their new vitamin E in the colostrum and through the mother's milk (248). In 1958 a recommendation was made (249) that premature infants receive vitamin E supplementation. A condition called vitamin E deficiency anemia was identified (247,250–253) in the late 1960s. In 1962 iron was identified as an initiator of linoleate peroxidation (253).

No deleterious effect was found for iron supplements, although Kinsey and Chisholm (196) thought of them as possibly contributing to ROP. Kinsey's group showed that removing them had no effect on ROP induction. Following the discovery of the role of oxygen in ROP, oxygen use was drastically curtailed, which led to a concomitant increase in premature mortality (254). The "cautious" use of oxygen (255) resulted in an increase in ROP, so that currently some 4000 children per year develop active ROP (245), about 1300 suffering irreversible damage (256). The discovery of high tocopherol levels in bovine retinal outer segments in 1970 (137) gave new impetus, in light of the increasing incidence of ROP, to the investigations already described, which were not thought by the two professional committees to justify adequately the clinical use of tocopherol supplements. Nevertheless, practice has reflected the realization of the premature infant's low vitamin E levels (205,249), which is usually obtained from the mother's milk (245,248,249). Colostrum in particular contains a relatively large amount of α-tocopherol relative to PUFA (248,257). This low vitamin E level may place the premature infant at risk for hemolytic anemia and perhaps for ROP and IVH.

Both infant formula and parenteral vitamin preparations now contain larger amounts of vitamin E (258). Parenteral vitamin preparations [Rorer; MVI Pediatric (MVIP) or Lymphomed in Canada] given the newborn premature infant now include 7 IU vitamin E; these are mixed into 100–150 ml intravenous fluid to be given parenterally.

The composition of MVIP was formulated to be based on the recommended dietary allowances (RDA) for children (259), which, however, specifically exclude individuals with special needs, such as premature babies. The Statement on Multivitamin Preparations for Parenteral Use by the Nutritional Advisory Group (NAG) of the Department of Foods and Nutrition, American Medical Association (260), portions of which appeared in 1979 (261), used these RDAs as a foundation for their recommendations. The NAG recommendation for children less than 10 kg weight is 10% of a vial per kg. For the Canadian package insert the suggested dose for infants < 3 kg is 0.65 vial/day; in the United States for infants < 1 kg the recommendation is not >30% of a vial/day. Several groups (262–267) have evaluated the blood levels following MVIP parenteral administration (see later).

Nevertheless, it appears to be the case that many premature nurseries may be using the entire vial when the intravenous fluid is prepared for administration to the infant. Thus for a newly born 1 kg infant, it can be calculated roughly that he or she may receive up to 7 IU/day of D-α-tocopherol, which would correspond roughly to 490 IU/day in an adult of 70 kg body weight. At 0.65 vial, the infant would receive 4.55 IU/day, corresponding to a level of 320 IU/day in a 70 kg adult. Since over 300–400 IU/day is considered a normal supplemental level (78), these infants are definitely receiving supplemental levels of vitamin E. Etches and Koo (262) confirm that these preparations (in Canada, marketed by Rorer or Lymphomed) contained sufficient vitamin E to raise plasma levels to an adequate level of vitamin E (≥1.2 mg/dl and ≤3.5 mg/dl), if given in a regimen of 0.6 vial MVIP per day for the first 24 h followed by 0.2 vial/kg/day. Greene et al. (263) showed that 65% of a vial of MVIP daily achieved plasma tocopherol concentrations of 1.47 ± 0.2 mg/dl in premature infants not receiving additional oral supplemental vitamin E. Potentially toxic levels were reached in 3 weeks in infants receiving additional oral vitamin E. Subsequent work by these investigators (264) showed that 30% of a vial of MVIP may be unable to maintain levels of 1 mg/dl in infants who weighed less than 1 kg. The 1 mg of IV vitamin E daily, however, can maintain naturally occurring serum tocopherol (0.5 mg/dl) (261), but 4.5 mg vitamin E per day as found in MVIP can result in a small percentage of very small preterm infants developing potentially toxic levels of tocopherol (265).

If the whole vial is given in 100 ml of IV fluid, it can be calculated that parenteral levels would be approximately 7 IU per 100 ml, which would dilute into plasma (volume ≃ 50 ml) at a rate 1 per 24 h, so the infants would be receiving approximately 0.61 IU/h/dl.

In the event that the infant is receiving formula, the commonly used premature formula in Canada, Wyeth's Premie SMA, contains supplemental levels of vitamin E (0.23 IU per ounce formula or 0.65 IU per dl formula. In 1985 the vitamin E level in the United States was higher (248): for Enfamil Premature, 16 IU/L, for SMA Preemie, 15 IU/L, and the Similac Special Care, 30 IU/L.

Thus it appears that current practice avoids the development of any vitamin E deficiency such as occurred rather frequently in the 1970s and early 1980s, when hospitals began monitoring vitamin E levels in preterm infants and when ROP began to receive attention again. The levels of vitamin E recommended in such documents as the Guidelines for Perinatal Care of the American Academy of Pediatrics are 5–10 IU/day (268), which as suggested earlier would provide an adequate level, even for premature infant of 1000 g birth weight.

TREATMENT OF STAGE 3 ROP BY HIGH DOSES OF VITAMIN E

Recently, Johnson's group, continuing to test the hypothesis that the progression of ROP and its long-term outcome are dependent on the balance between the antioxidant defenses and the antioxidant stress, have tested high doses of vitamin E on infants exhibiting severe ROP (282). They treated infants with bilateral severe ROP with vitamin E administered intravenously, intramuscularly, and/or orally to maintain vitamin E levels of 3–5 mg/dl until ROP was clearly regressing or cicatricial ROP developed.

Using the Hoffman-LaRoche parenteral *d,l*-α-tocopherol preparation, they found no complications and no apparent decrease in resistance to infection until completion of retinal vascularization. They ascribed this to the shorter duration of administration of the vitamin E 2–4 weeks), combined with the fact that the host defenses are more fully developed by 2 months after birth, when severe ROP usually begins. In their study only 24% of infants developed ≥ grade 3 cicatricial ROP or worse, and 48.5% had a normal posterior pole and little or no peripheral scarring (≤ grade 1 CIC). By comparison, the Cryo Trial (282) reported an unfavorable outcome (≥ grade 3 CIC ROP) in 23% of eyes randomized to cryotherapy versus 48% for eyes assigned to treatment.

VITAMIN E TOXICITY

As Hittner (269,270) points out, two main problems with the use of vitamin E have occurred.

1. At levels in plasma ≥ 5 mg/dl, there was a greater incidence of necrotizing enterocolitis and sepsis (231). Johnson and her group believed this was due to inhibition of macrophage activity.
2. Any other toxic effects of vitamin E preparations have been attributed to diluents or additives used in the preparations: for oral Aquasol E, necrotizing enterocolitis resulting from either hyperosmolality and/or toxicity from propylene glycol (230), for intravenous (MVI) Pediatric, infants weighing less than 1 kg at birth require a reduction in dose due to toxicity of polysorbate 80 and 20 (220) for intramuscular, E-Vicotrat, transient calcification of the thigh muscle resulting from the toxicity of polysorbate 80 (271,272); and as mentioned previously, for intravenous E-Ferol, hepatic failure, renal failure, and death apparently resulted from toxicity of polysorbates 80 and 20.

VITAMIN E CONCENTRATIONS AND TRANSPORT IN OCULAR TISSUES

Several studies have investigated the vitamin E concentration in different ocular tissues (273–279,283). Unfortunately, except for Stephens et al. (283, see later), no comprehensive analysis of major ocular tissues were done in the same studies. The highest concentration of vitamin E is in the retina (Tables 7–9) (137). This is rapidly depleted in vitamin E deficiency. Of the two fluids of the eye, the aqueous and vitreous, the aqueous must be involved in any transport of tocopherol to the avascular tissues, lens and corneal endothelium, and the tear fluid may play the same function for the avascular corneal epithelium. The previously noted treatment of keratoconus with vitamin E implies that topical tocopherol applied to the cornea (28) is absorbed from tear fluid.

The vitamin E content of ocular tissues is outlined in Tables 7 and 8. The levels of tocopherol fall from retinal tissue to lens and cornea, and in the ocular fluids, the vitreous contains more than the aqueous. The vitreous is a relatively stable jellylike fluid compared to the more liquid aqueous. This probably accounts for the longer period of accumulation of esterified α-tocopherol in the vitreous compared to the serum and choroid (275). Injection intramuscularly of α-tocopherol acetate raised concentrations in retina, choroid, and vitreous body (276). Hunt and coworkers (277) explored the effect of light on vitamin E levels in retina and ROS. In the developing rat, the molar ratios of tocopherol to rhodopsin were 0.25, 0.29, and -0.27 in cyclic light reared for 18–20, 30–40, and 44–60 days, respectively; for dark-reared rats these values were 0.2, 0.22, and 0.22, respectively. The values for dark-reared rats were significantly lower than for cyclic light-reared animals for the ROS. Age-related trends (278), although not statistically significant (Table 10), indicated that the total γ- + α-tocopherol levels were remarkably similar throughout the age span examined. Although a slight decline occurred in the last two decades examined, the numbers are not significant. Such a decline would be consistent with an increased requirement for tocopherol in the elderly, similar to that postulated by Robertson et al. (78) to result in reduced cataract risk, and consistent with the low levels reported for vitamin E in cataractous human lenses (280).

Stephens et al. (283) documented the levels of vitamin E in the various areas of the eye for the rat, rabbit, and cat (Tables 7 and 8). They tested the effect of vitamin E supplementation or depletion in the rat: many ocular tissues (cornea, lens, anterior and posterior lens capsules, vitreous, and sclera) showed only small changes in their levels of vitamin E whether supplemented or depleted rats were tested, but 20- to 60-fold differences in VE content were found for ciliary body, iris, and retina.

It is of some interest, in the light of Reed's hypothesis that the ratio of vitamin E to glutathione is important, to note that Castigliola et al. (279) observed that vitamin E could prevent the loss of glutathione from tissues and cells subjected to oxidative stress. This offers a possible mechanism for preserving and thus positively influencing the antioxidant defenses of most cells and tissues.

APPENDIX TABLES

TABLE 1A All Studies ROP

Reference	Treated ROP a	Treated No b	Control ROP c	Control No d	Total N	Observed O	Expected E	Observed–expected O – E	Variance V
199	9	32	15	25	81	9	12.14	−3.14	4.27
199	6	10	12	5	33	6	8.72	−2.72	2.10
199	11	54	15	45	125	11	13.52	−2.52	5.18
239	46	21	33	18	118	46	44.86	+1.14	6.46
239 combined	83	16	17	2	118	83	83.90	−0.90	3.81
241	17	71	20	66	174	17	18.71	−1.71	7.81
239	48	20	46	21	135	48	47.39	+0.65	7.18
239	21	6	19	3	49	21	22.04	−1.04	3.98
219 (pooled controls)	11	38	30	57	136	11	14.77	−3.77	6.65
242 pool experimental	73	46	33	18	170	73	74.2	−1.42	8.42
195	1	22	17	61	101	1	4.10	−3.10	2.63
203	4	44	10	41	99	4	6.79	−2.79	1.92
243	0	44	9	14	67	0	5.9	−5.9	1.63
243	12	36	17	34	99	12	14.06	−2.06	3.92
204	9	28	8	29	74	9	8.5	0.5	3.32

TABLE 1A Continued

Reference	Treated ROP a	Treated No b	Control ROP c	Control No d	Total N	Observed O	Expected E	Observed − expected O − E	Variance V
196	5	41	10	28	84	5	8.21	− 3.21	3.04
196	8	38	33	130	247	8	7.46	+ 0.36	3.44
208	25	72	28	71	196	25	26.23	− 1.23	9.72
208 < 1 kg	12	6	10	8	36	12	11	1	2.2
207	13	98	19	95	225	13	15.78	− 2.78	6.9
Totals			401	771	2249			− 34.64	94.58
Odds ratio 0.69;								− 34.64	94.58
95% confidence								3.56 SD	SD =
limits 0.57–0.85								from 0	9.73
								$p < 0.0005$	

TABLE 2A 1986 National Institute of Medicine Report 10M-86-02 (Derived from Data Given in Report)

Reference	Treated ROP a	Treated No b	Control ROP c	Control No d	Total N	Observed O	Expected E	Observed− expected O − E	Variance[a] V
202	32	18	33	18	101	32	32.18	− 0.18	5.85
207	13	98	19	95	215	13	16.5	− 3.5	7.89
203	9	39	12	39	99	9	10.18	− 1.18	16.08
206	25	72	28	71	196	25	26.23	− 1.23	9.72
205 (1 day)	56	85	76	71	288	56	64.63	− 8.63	17.93
204	9	28	8	29	74	9	8.5	+ 0.5	3.32
Totals					973			− 14.22	60.79
Odds ratio 0.79								1.8238 SD	SD =
95% confidence								from 0	7.80
limits 0.62–1.02								$p < 0.1$	

[a]Cancer.

TABLE 2B Severe Retinopathy

Reference	Treated ROP a	Treated No b	Control ROP a	Control No b	Total N	Observed O	Expected E	Observed− expected O − E	Variance[a] V
202	0	50	5	45	101	0	2.47	− 2.47	1.164
207	3	108	5	109	215	3	8.15	− 5.15	4.380
203	2	46	4	47	99	2	2.91	− 0.91	1.42
206	11	86	8	91	196	11	9.40	− 1.60	4.31
205 (1 day)	3	138	7	140	288	3	4.90	− 1.90	2.42
204	0	37	1	36	74	0	0.5	+ 0.5	0.25
Totals					973	19		− 9.33	13.94
Odds ratio 0.512;								2.50	SD =
95% confidence								SD from 0;	3.73
limits 0.30–0.86								$p < 0.01$	

[a]Cancer.

TABLE 2C   Cicatricial Damage

| Reference | Treated | | Control | | Total | Observed | Expected | Observed −expected | Variance |
	ROP a	No b	ROP c	No d	N	O	E	O − E	V
203	3	45	5	46	99	3	3.88	−0.88	1.85
206	3	44	1	98	196	1	1.98	−0.98	2.66
205	4	137	7	140	288	4	5.39	−1.39	2.66
204	0	34	1	36	74	0	0.45	−0.45	0.22
Totals					657			−3.71	5.71
Odds ratio 0.522;								1.55	SD =
95% confidence							SD from 0;		2.38
limits 0.23–1.18								$p < 0.2$	

[a]Cancer.

TABLE 3   Severe Retinopathy

| Study | Treated | | Control | | Total | Observed | Expected | Observed−expected | Variance |
	ROP a	No b	ROP c	No d	N	O	E	O − E	V
202	0	50	5	46	101	0	2.47	−2.47	1.20
207	3	108	5	109	215	3	4.13	−0.13	2.22
238	0	22	1	19	42	0	.52	−0.52	0.025
203	2	46	4	47	99	2	2.90	−0.90	1.42
244	1	96	1	98	196	1	0.99	+0.01	0.50
205	1	205	1	207	424	3	5.89	+0.01	0.50
Totals					25	973		−7.9	8.51
Odds ratio 0.395;								2.7	SD =
95% confidence							SD from 0;		2.92
limits 0.20–0.78								$p < 0.01$	

Source: From Reference 238.

TABLE 4A   Possible Complications of Vitamin E in Premature Infants

| Reference | Treated | | Control | | Total | Observed | Expected | Observed −expected | Variance |
	ROP a	No b	ROP c	No d	N	O	E	O − E	V
1. 239	43	58	9	10	120	43	43.76	−0.76	3.96
2. 204	16	11	21	18	66	16	15.14	0.86	3.99
3. 223	3	13	9	7	32	3	6	−3	1.94
4. 225	9	50	37	71	210	9	16.25	−7.25	7.66
5. 226	10	54	24	46	134	10	16.24	−6.24	6.37
6. 208	2	43	2	44	91	2	1.98	0.2	0.97
7. 227	16	98	37	71	229	16	26.38	−10.38	9.22
8. 281	24	44	35	34	137	24	27.67	−3.67	7.99
Totals (1–6)	83	229	102	196	653			−15.42	24.89
Totals (1–7)	99	327	139	267	882			−25.81	34.11
Totals (1–8)	123	371	174	301	1019			−29.48	42.10
Odds ratio (6)							Omitting	3.28	SD =
0.52;							Ref. 228	SD from 0;	4.98
95% confidence								$p <$	
limits 0.33–0.77								0.001	

TABLE 4A Continued

Reference	Treated		Control		Total N	Observed O	Expected E	Observed − expected O − E	Variance V
	ROP a	No b	ROP c	No d					
Odds ratio (7) 0.46; 95% confidence limits 0.33–0.64							Including all trials	4.58 SD from 0; p < 0.0001	SD = 5.84
Odds ratio (8) 0.50; 95% confidence limits 0.37–0.67								4.54 SD from 0; p < 0.0001	SD = 6.49

TABLE 4B Necrotizing Enterocolitis (NEC) and Sepsis

Reference	Treated		Control		Total N	Observed O	Expected E	Observed − expected O − E	Variance V
	ROP a	No b	ROP c	No d					
1. 218	39	215	7	44	305	39	37.48	1.52	5.34
2. 232	43	185	24	223	475	43	27.9	15.07	12.48
3. 208	8	132	10	137	287	8	8.78	− 0.78	4.22
4. 208	3	39	5	38	85	3	3.95	− 0.95	1.83
5. 281	3	69	6	67	145	3	4.47	− 1.47	2.12
Totals (4)	93	571	46	442	1152			14.86	23.87
Totals (1–5)	96	640	52	509	1297			12.29	25.99
Odds ratio (1–4) 1.86 (p < 0.005); 95% confidence limits 1.24–2.77								3.04 SD from 0;	SD = 4.88
Odds ratio (1–4) 0.98 PNS; 95% confidence limits 0.92–1.05 (excluding 2)					677			− 0.21 SD from 0	11.39 SD = 3.37
Odds ratio (1–5) 1.6; 95% confidence limits 1.24–2.77								2.41 SD from 0	SD = 5.1
Totals 1–5 excluding 2)	53	455	28	286	822			− 1.62	12.57
Odds ratio (1, 3–5) 0.92; 95% confidence limits 0.45–1.9								0.6 SD from 0	SD = 2.68

TABLE 4C Necrotizing Enterocolitis (NEC)

Reference	Treated ROP a	Treated No b	Control ROP c	Control No d	Total N	Observed O	Expected E	Observed – expected O – E	Variance V
1. 218	9	245	2	44	305	9	9.16	−0.16	1.48
2. 228	7	41	6	45	99	7	6.3	0.7	2.85
3. 232	42	185	28	223	478	42	33.24	8.76	14.92
4. 208 (all)	6	141	6	141	287	6	6.15	−0.15	3.1
5. 208 <1 kg)	3	39	4	39	85	3	3.45	−0.45	1.62
6. 228	10	47	6	21	84	10	10.86	0.86	2.86
7. 281	3	69	6	67	145	3	4.47	−1.47	2.12
Totals (1–6); odds ratio 1.46 ($p < 0.1$); 0.96–2.24	77	698	52	513	1338			8.29 1.78 SD from 0	21.56 SD = 4.64
Totals (1–6) excluding 3; odds ratio 0.96 NS; 0.57–1.75	35	513	24	290	860			−47 0.15 SD from 0	10.29 SD =
Totals (1–7) odds ratio 1.22; 95% confidence limits 0.9–1.99	80	767	58	580	1483			+6.82 1.40 SD from 0	23.68 SD =
Totals (1–7) excluding 3; odds ratio 0.58; 95% confidence limits 0.22–1.01	38	582	30	357	1005			−1.94 0.55 SD from 0	12.41 SD =

REFERENCES

1. Fine BS, Yanoff M. Ocular histology, 2nd ed. New York: Harper and Row, 1979.
2. Pryor WA, Prier DG, Church DF. Electron-spin resonance study of mainstream and sidestream cigarette smoke: nature of the free radicals in gas-phase smoke and in cigarette tar. Environ Health Perspect 1983; 47:345-55.
3. Green K, Nye RA, Nelson E, Checks L. Role of eicosanoids in the ocular response to intracameral hydrogen peroxide. Lens Eye Toxicity Res 1989; 7:79-101.
4. Riley MV. A role of glutathione and glutathione reductase in control of corneal hydration. Exp Eye Res 1984; 39:751-8.
5. Giblin FJ, McCready JP, Kodama T, Reddy VN. A direct correlation between the levels of ascorbic acid and H_2O_2 in aqueous humor. Exp Eye Res 1984; 38:87-94.
6. Spector A, Garner WH. Peroxide and human cataract. Exp Eye Res 1981; 33:673-82.
7. Garner WH, Spector A. Hydrolysis kinetics by Na K-ATPase in cataract. Exp Eye Res 1986; 42: 339-48.
8. Hayes RP, Fisher RF. Influence of a prolonged period of low dosage x-rays on the optic and ultrastructural appearances of cataract of the human lens. Br J Ophthalmol 1979; 63:457-64.
9. Carpenter RL, van Ummersen CA. The action of microwave irradiation on the eye. J Microwave Power 1968; 3:3-19.
10. Van Ummerson CA, Cogan FC. Effects of microwave irradiation on the lens epithelium in the rabbit eye. AMA Arch Ophthalmol 1976; 94:828-34.

11. O'Keefe TL, Hess HH, Zigler JS Jr, Kuwabara T, Knapka JJ. Prevention of cataracts in pink-eyed RCS rats by dark rearing. Exp Eye Res 1990; 51:509-17.

12. Gutteridge JMC. Lipid peroxidation: some problems and concepts. In: Halliwell B ed. Oxygen radicals and tissue injury. Fed Am Soc Exp Biol 19??; 9-19.

13. Trevithick JR, Dzialoszynski T. Possible posterior segment influences on oxidative damage to the lens. Proc. 33rd annual meeting Canadian Federation of Biological Societies, Ottawa, Canada, 1990; 265.

14. Katz ML, Eldred GE. Retinal light damage reduces antifluorescent pigment deposition in the retinal pigment epithelium. Invest Ophthalmol Vis Sci 1989; 30:37-43.

15. Fisher AB. Intracellular production of oxygen-derived free radicals. In: Halliwell B, ed. Oxygen radicals and tissue injury. American Society for Experimental Biology; 34-42.

16. Machlin LJ, Bendich A. Free radical tissue damage: protective role of antioxidant nutrients. FASEB J 1987; 1:441-5.

17. Niki E, Saito T, Kawakami A, Kamiya Y. Inhibition of oxidation of methyl linoleate in solution by vitamin E and vitamin C. J Biol Chem 1984; 259:4177-82.

18. Tirmenstein M, Reed DJ. Effects of glutathione on the α-tocopherol-dependent inhibition of nuclear lipid peroxidation. J Lipid Res 1989; 30:959-65.

19. Wilson J. Personal communication, 1990.

20. Yamauchi R, Matsui T, Veno Y. Reaction products of alpha-tocopherol with methyl linoleate peroxy radicals. Lipids 1990; 23:152-8.

21. Dikstein S, Maurice DM. The metabolic basis to the fluid pump in the cornea. J Physiol (Lond) 1972; 221:29-41.

22. Lux-Neuwirth O, Millar TJ. Lipid soluble antioxidants preserve rabbit corneal cell function. Curr Eye Res 1990; 9:103-8.

23. Travkin AG, Gundorowa RA, Bordjugowa GG, Derewjanko WP, Zypin AB. Erforschung des zustandes der zellmembran und der mechanismen der hornhautautolyse bei der konservierung. Klin Monatobl Augenheilkd 1976; 169:500-4.

24. Travkin AG, Derevyanko VP. Restoration of deranged autoregulated oxidation of the lipid structural elements of the donor's cornea in the course of its transplantation. Vestn Oftalmol 1978; 2: 61-5.

25. Travkin AG, Derevyanko VP, Tsypin AB. Supramolecular changes in preserved corneal tissue during treatment with α-tocopherol. Biull Eksp Biol Med 1975; 80:38-41.

26. Travkin AG, Vinetskaya MI, Slepukhina LV. Effects of antioxidants with different anti-radical activity on the preservation of the cornea. Biofizika 1976; 21:1064-6.

27. Hernandez LA, Granger DN. Role of antioxidants in organ preservation and transplantation. Crit Care Med 1988; 16:543-9.

28. Titarenko ZD. The action of vitamin E on the cornea in keratoconus. Oftalmol Zh 1985; 35:163-5.

29. Puchkowskaya NA, Titarenko ZD. Neues in der bahandlung des keratoconus. Klin Monatsbl Augenheilkd 1986; 189:11-4.

30. Bron AJ. Keratoconus. Cornea 1988; 7:163-9.

31. Hensyl WR, ed. Steadman's Medical Dictionary, 25th ed. Baltimore: Williams & Wilkins, 1990.

32. Patey A, Savoldelli M, Pouliquen Y. Keratoconus and normal cornea: a comparative study of the collagenous fibres of the corneal stroma by image analysis. Cornea 1984; 3:119-24.

33. Maurice DM. The location of the fluid pump in the cornea. J Physiol (Lond) 1972; 221:43-54.

34. Kim JO, Hassard DTR. On the enzymology of the cornea: a new enzyme deficiency in keratoconus. Can J Ophthalmol 1972; 7:176-80.

35. Kinoshita JH, Masurat T, Helfant M. Pathways of glucose metabolism in corneal epithelium. Science 1955; 122:72-3.

36. Ferguson TM, Rigdon RH, Couch JR. Cataracts in vitamin E deficiency. Arch Ophthalmol 1956; 55:346-55.

37. Bellows JG, Bellows RJ. Presenile and senile cataract. In: Bellows JG, ed. Cataract and abnormalities of the lens. New York: Grune & Stratton, 1975; 303-13.

38. Chylack LT. Classification of human cataracts. Arch Ophthalmol 1978; 96:838-92.

39. Sparrow JM, Bron AJ, Brown NAP, Aycliffe W, Hill AR. The Oxford clinical cataract classification and grading system. Int Ophthalmol 1986; 9:207-25.

40. Taylor HR. Use of photographic techniques to grade nuclear cataracts. Invest Ophthalmol Vis Sci 1988; 29:73-7.

41. Creighton MO, Trevithick JR, Mousa GY, et al. Globular bodies: a primary cause of the opacity in senile and diabetic posterior cortical subcapsular cataracts? Can J Ophthalmol 1978; 13:166-81.

42. Benedek GB. Theory of transparency of the eye. Appl Optics 1971; 10:459-73.

43. Bessems GJH, deMan B, Hoenders JH, Wollensak J. Alpha-tocopherol in the normal and nuclear cataractous human lens. Lens Res 1984; 2:233-42.

44. Augusteyn RC. Protein modification in cataract: possible oxidative mechanisms. In: Duncan G, ed. Mechanisms of cataract formation in the human lens. New York: Academic Press, 1981; 71-115.

45. Creighton MO, Trevithick JR. Cortical cataract formation prevented by vitamin E and glutathione. Exp Eye Res 1979; 29:689-93.

46. Creighton MO, Stewart-DeHaan PJ, Ross WM, Sanwal M, Trevithick JR. Modelling cortical cataractogenesis. 1. In vitro effects of glucose, sorbitol and fructose on intact rat lenses in medium 199. Can J Ophthalmol 1980; 15:183-8.

47. Trevithick JR, Creighton MO, Ross WM, Stewart-DeHaan PJ, Sanwal M. Modelling cortical cataractogenesis. 2. In vitro effects on the lens of agents preventing glucose- and sorbitol-induced cataracts. Can J Ophthalmol 1981; 16:32-8.

48. Ross WM, Creighton MO, Stewart-DeHaan PJ, Sanwal M, Hirst M, Trevithick JR. Modelling cortical cataractogenesis. 3. In vivo effects of vitamin E on cataractogenesis in diabetic rats. Can J Ophthalmol 1982; 17:61-6.

49. Creighton MO, Sanwal M, Stewart-DeHaan PJ, Trevithick JR. Modelling cortical cataractogenesis. 5. Steroid cataracts induced by solumedrol partially prevented by vitamin E in vitro. Exp Eye Res 1983; 37:65-75.

50. Creighton MO, Trevithick JR, Sanford SE, Dukes TW. Modelling cortical cataractogenesis. 4. Cataracts induced by Hygromycin B partially reversed by vitamin E. Exp Eye Res 1982; 34:467-76.

51. Ross WM, Creighton MO, Inch WR, Trevithick JR. Radiation cataract diminished by vitamin E in rat lenses in vitro. Exp Eye Res 1983; 36:645-53.

52. Ross WM, Creighton MO, Trevithick JR. Radiation cataractogenesis induced by neutron or gamma irradiation in the rat lens is reduced by vitamin E. Scanning Microsc 1990; 4:641-50.

53. Stewart-DeHaan PJ, Creighton MO, Sanwal M, Ross WM, Trevithick JR. Effects of vitamin E on cotrical cataractogenesis induced by elevated temperature in intact rat lenses in medium 199. Exp Eye Res 1981; 32:54-60.

54. Creighton MO, Ross WM, Stewart-DeHaan PJ, Sanwal M, Trevithick JR. Modelling cortical cataractogenesis. 7. Effects of vitamin E treatment on galactose-induced cataracts. Exp Eye Res 1985; 40:213-22.

55. Ross WM, Creighton MO, Trevithick JR, Stewart-DeHaan PJ, Sanwal M. Modelling cortical cataractogenesis. 6. Induction by glucose in vitro or in diabetic rats: prevention and reversal by glutathine. Exp Eye Res 1983; 37:559-73.

56. Trevithick JR, Linklater HA, Mittton KP, Dzialoszynski T, Sanford SE. Modelling cortical cataractogenesis. IX. Activity of vitamin E and esters in preventing cataracts and γ-crystallin leakage from lenses in diabetic rats. Ann N Y Acad Sci 1989; 570:358-71.

57. Linklater HA, Dzialoszynski T, McLeod HL, Sanford SE, Trevithick JR. Modelling cortical cataractogenesis. X. Vitamin C reduces γ-crystallin leakage from lenses in diabetic rats. Exp Eye Res 1990; 51:241-7.

58. Varma SD, Chand D, Sharma YR, Kuck JF, Richards RD. Oxidative stress on lens and cataract formation: role of light and oxygen. Curr Eye Res 1984; 3:35-57.

59. Devi A, Raina PL, Singh A. Abnormal protein and nucleic acid metabolism as a cause of cataract formation induced by nutritional deficiency in rabbits. Br J Ophthalmol 1965; 49:271-5.

60. Bunce GE, Hess JL. Lenticular opacities in young rats as consequence of maternal diets low in tryptophan and/or vitamin E. J Nutr 1976; 106:222-9.

61. Evans SC. Nutrition in eye health and disease. London: Roberts Publications, 1978; 28-34.

62. Evans SC. The significance of nutrition in cataract. 4. Optician 1979; 178:41-3.

63. Bhuyan KC, Bhuyan DK, Podos SM. The role of vitamin E in therapy of cataract in animals. Ann N Y Acad Sci 1982; 393:169-71.

64. Kinoshita JH, Merola LO, Dikmak E. Osmotic changes in experimental galactose cataracts. Exp Eye Res 1962; 1:405-10.

65. Prasad KN, ed. CRC handbook of radiobiology. Boca Raton, FL: CRC Press, 1984; 39-40, 234-6.

66. Giblin FJ, Chakrapani B, Reddy VN. The effects of x-irradiation on lens reducing systems. Invest Ophthalmol Vis Sci 1979; 18:468-75.

67. Petkau A. Radiation carcinogenesis from a membrane perspective. Acta Physiol Scand 1980; 492: 81-90.

68. Varma SD, Kuman S, Richards RD. Light-induced damage to ocular lens cation pump: prevention by vitamin C. Proc Natl Acad Sci U S A 1979; 76:3504-6.

69. Libondi T, Menzione M, Lulcano G, Della Corte M, Lane F, Auricchio G. Changes in some biochemical parameters of the lens in galactose-treated weaned rats with and without vitamin E therapy. Ophthalmic Res 1985; 17:42-8.

70. Libondi T, Menzione M, Aurriccho G. In vitro effect of α-tocopherol on lysoposphatidylcholine-induced lens damage. Exp Eye Res 1985; 40:661-6.

71. Hattori H, Majima Y, Nagamura Y, Ishiguro I. Effect of vitamin-E containing liposomes on experimental sugar cataract. Nippon Ganka Gakkai Zasshi 1989; 93:97-102.

72. Eccarius S, Clark JI. Effect of aspirin and vitamin E on phase separation in calf lens homogenate. Ophthalmic Res 1987; 19:65-71.

73. Zigler JS, Hess HH. Cataracts in the Royal College of Surgeons rat: evidence for initiation by peroxidation products. Exp Eye Res 1985; 41:67-76.

74. Hess HH, Kuwabara T, Zigler S, Westney IV. Posterior subcapsular cataracts in the Royal College of Surgeons (RCS) rat: light as a factor in development and maturation. Invest Ophthalmol Vis Sci (Suppl) 1986; 27:203.

75. Hess HH, Knapka JJ, Newsome DA, Westney IV, Wartofsky L. Dietary prevention of cataracts in the pink-eyed RCS rat. Lab Anim Sci 1985; 35:47-53.

76. Bhuyan KC, Bhuyan DK, Kuck JFR, Kuck KD, Kern HL. Increased lipid peroxidation and altered membrane functions in Emory mouse cataract. Curr Eye Res 1983; 2:597-606.

77. Kuck JFR, Kuck KD. The Emory mouse cataract: the effects on cataractogenesis of α-tocopherol, penicillamine, triethylenetetraamine and mercaptopropionylglycine. J Ocul Pharmacol 1988; 4: 243-51.

78. Robertson JM, Donner AP, Trevithick JR. Vitamin E intake and risk of cataracts in humans. Ann N Y Acad Sci 1989; 570:372-82.

79. Jacques PF, Chylack LT, McGandy RB, Harts SC. Antioxidant status in persons with and without senile cataract. Arch Ophthalmol 1988; 108:337-40.

80. Leske MC, Chylack LT Jr, Wu SY, the LOCS research group. The lens opacities case-control study. Proceedings of the 9th International Congress of Eye Research. Int Soc Eye Res. New York, N.Y. 1990; 140.

81. Tappel AL. Vitamin E as the biological lipid antioxidant. Vitam Horm 1962; 20:493-510.

82. Fukuzawa K, Gebicki JM. Oxidation of α-tocopherol in micelles and liposomes by the hydroxyl, perhydroxyl, and superoxide free radicals. Arch Biochem Biophys 1983; 266:242-51.

83. Gebicki JM, Bielski BHJ. Comparison of the capacities of the perhydroxyl and superoxide radicals to initiate chain oxidation of linoleic acid. J Am Chem Soc 1981; 103:7020-2.

84. Willson RL. Free radicals and tissue damage: mechanistic evidence from radiation studies. In: C. de Duve and O. Hayashi, eds. Biochemical mechanisms of liver injury. Elsevier, Amsterdam. 1978; 233-245.

85. Takeuchi N, Iritani N, Fukuda E, Tanaka F. Effects of long-term administration and deficiency of alpha-tocopherol on lipid metabolism of rats. In: de Duve C, Hayaishi O, eds. Tocopherol, oxygen and biomembranes. Amsterdam: Elsevier/North Holland Biomedical Press, 1978; 257-72.

86. Draper HH, Polensek L, Hadley M, McGirr LG. Urinary malondialdehyde as an indicator of lipid peroxidation in the diet and in the tissues. Lipids 1984; 19:836-43.

87. Bhuyan KC, Bhuyan DK. Molecular mechanism of cataractogenesis. II. Evidence of lipid perioxidation and membrane damage. In: Greenwald RA, Chohen G, eds. Oxy radicals and their scavenger systems, Vol. II. Cellular and medical aspects. Amsterdam: Elsevier/North Holland, 1983; 349-56.

88. Maggio B, Diplock AT, Lucy JA. Interactions of tocopherols and ubiquinones with monolayers of phospholipids. Biochem J 1977; 161:111-21.

89. Diplock AT, Lucy JA, Verrinder M, Zieleniewski A. α-Tocopherol and the permeability to glucose and chromate of unsaturated liposomes. FEBS Lett 1977; 82:341-4.

90. Erin AN, Spirin MM, Tabixze LV, Kagan VE. Formation of complexes of α-tocopherol with fatty acids. The possible mechanism of stabilization of biomembranes by vitamin E. Biokhimiya 1983; 48:1855-61.

91. Burlakova EB, Kukhtina EN, Sarycheva IK, Khrapova NG, Aristarkohova. Effect of phytyl side chain of tocopherols on oxidative reactions occurring in lipids. Biokhimiya 1982; 47:987-92.

92. Agadzhanov MI, Badzhinyan SA, Karagesyan KG, Mkhitaryan VG. The regulating action of α-tocopherol on the conductivity of bilayer phospholipid membranes constructed from rat brain and liver phospholipids when the organism is under stress. Dokl Akad Nauk SSSR 1979; 244: 1496-9.

93. Baig MMA, Laidman DL. Spectrophotometric evidence for a polar interaction between α-tocopherol and phospholipids: the effects of different salts and pH. Biochem Soc Trans 1983; 11: 601-2.

94. Diplock AT, Lucy JA. The biochemical modes of action of vitamin E and selenium: a hypothesis. FEBS Lett 1973; 29:205-10.

95. Mino M, Sugita K. The membrane action of alpha-tocopherol upon oxidative damage in erythrocytes. In: de Duve C, Hayashi O, eds. Tocopherol, oxygen and biomembranes. Amsterdam: Elsevier/North Holland Biomedical Press, 1978; 71-81.

96. Hoey BM, Butler J. The repair of oxidized amino acids by antioxidants. Biochim Biophys Acta 1984; 791:212-8.

97. Bisby RH, Ahmed S, Cundall RB. Repair of amino acid radicals by a vitamin E analog. Biochem Biophys Res Commun 1984; 119:245-51.

98. Korchagin VP, Bratkovskaya LB, Shvedova AA, Arkhipenko YV, Kagan VE, Shukolyukov SA. Oligomerization of integral membrane proteins during lipid peroxidation. Biokhimiya 1980; 45: 1767-72.

99. Kagan VE, Archipebnko YV, Kozlov YP. Modification of the enzyme system of Ca^{++} transport in the sarcoplasmic reticulum in peroxide oxidation of lipids. Change in the chemical composition and ultrastructural organization of the membranes. Biokhimiya 1983; 48:158-66.

100. Ananieva LK, Inanov II, Tabixze LV, Kagan VE. Mechanism of the stabilization of Ca^{++}-ATPase of the sarcoplasmic reticulum by tocopherol against thermal denaturation activated by fatty acids. Biokhimiya 1984; 49:60-6.

101. Tabixse LV, Ritov VB, Kagan VE, Koslov YP. Protection of sarcoplasmic reticular membranes against damage by free fatty acids by vitamin E. Bull Exp Biol Med 1983; 96:1548-50.

102. Hightower KR, Leverenz V, Reddy VN. Calcium transport in the lens. Invest Ophthalmol 1980; 19:1059-66.

103. Hightower KR, Reddy VN. Metabolic studies on calcium transport in mammalian lens. Curr Eye Res 1981; 1:197-207.

104. Nakayama T, Kodama M, Nagata C. Free radical formation in DNA by lipid peroxidation. Agr Biol Chem 1984; 48:571-2.

105. Popov GA, Konev VV. Influence of lipid peroxidation on DNA synthesis in Ehrlich ascites carcinoma cells. Biokhimiya 1984; 49:1199-202.

106. Nair PP, Patnaik RM, Hanswirth JW. Cellular transport and binding of *d*-α-tocopherol. In: de Duve C, Hayaishi O, eds. Tocopherol, oxygen and biomembranes. Holland: Elsevier/North Holland Biomedical Press, 1978; 121-30.

107. Rajaram OV, Fatterpaker P, Sreenivasan A. Occurence of α-tocopherol binding protein in rat liver cell sap. Biochem Biophys Res Commun 1973; 52:459-65.

108. Kalinina LM, Minseitova SR. Mutagenic effects and DNA degradation in *E. coli* cells treated with $K_2Cr_2O_7$. Genetika 1983; 19:1941-7.
109. Hauswirth JW, Nair PP. Some aspects of vitamin E in the expression of biological information. Ann N Y Acad Sci 1972; 203:111-22.
110. Samuelsson B, Granstrom E, Hammberg M. On the mechanism of the biosynthesis of prostaglandins. In: Bergstrom S, Samuelsson B, eds. Prostaglandins. Almqvist & Wiksell, Interscience Publishers, New York, N.Y. 1967; 31-4.
111. Dakhil T, Vogt W. Hydroperoxides, the active principles of polyunsaturated fatty-acid preparations which cause intestinal muscle to contract. J Physiol (Lond) 1962; 106:21p-2p.
112. Guivernau M, Terragno A, Dunn MW, Terragno NA. Estrogens induces lipoxygenase derivative formation in rabbit lens. Invest Ophthalmol 1982; 23:214-7.
113. Longhampt MO, Bonne C, Regnault F, Masse JP, Conquelet C, Sincholle D. Evidence of leukotriene B4 biosynthesis in epithelial lens cells. Prostaglandins Leukot Med 1983; 10:381-7.
114. Fleisher LN, McGahan MC. Endotoxin-induced ocular inflammation increases prostaglandin E_2 synthesis by rabbit lens. Exp Eye Res 1985; 40:711-9.
115. Ono S, Obara Y, Hatano M. The effect of prostaglandin E_1 on the phospholipid metabolism of the lens. Ophthalmic Res 1973; 4:281-3.
116. Paterson CA, Eck BA. Prostaglandin E_1 and the ocular lens. Ophthalmic Res 1971; 2:246-9.
117. Falanga A, Doni MG, Delaini F, et al. Unbalanced plasma control of TxA_2 and PGI_2 synthesis in vitamin E-deficient rats. Am J Physiol 1983; 245(14):H867-70.
118. Gilbert VA, Zebrowski EJ, Chan AC. Differential effects of megavitamin E on prostacyclin and thomboxane synthesis in streptozotocin-induced diabetic rats. Horm Metab Res 1983; 15:320-5.
119. Karpen CW, Pritchard KA Jr, Arnold JH, Cornwell DG, Panganamala RV. Restoration of prostacyclin/thromboxane A_2 balance in the diabetic rat. Influence of dietary vitamin E. Diabetes 1982; 31:947-51.
120. Ulasevich JJ, Grozina AA, Vorobyova EN. Effect of vitamin E on glycolytic processes in erythrocytes. Vopr Med Khim 1984; 30:118-20.
121. Napoli JL, McCormick AM, O'Meara B, Dratz EA. Vitamin A metabolism: alpha-tocopherol modulates tissue retinol levels in vivo and retinol plamitate hydrolysis in vitro. Arch Biochem Biophys 1984; 230:194-202.
122. Bascetta E, Gunstone FD, Walton JC. Electron spin resonance study of the role of vitamin E and vitamin C in the inhibition of fatty acid oxidation in a model membrane. Chem Phys Lipids 1983; 33:207-10.
123. Niki E, Tsuchiya J, Tanimura R, Kamiya Y. Regeneration of vitamin E from α-chromanoxyl radical by glutathione and vitamin C. Chem Lett 1982; 789-92.
124. Harison WH, Gander JE, Blakley ER, Boyer PD. Interconversions of α-tocopherol and its oxidation products. Biochim Biophys Acta 1956; 21:150-8.
125. Rousseau C, Richard C, Martin R. Synergistic effect of glutathione on the inhibiting power of vitamin E during oxidation of methyl linolenate in solution. J Chim Phys Phys-Chim Biol 1984; 81:137-8.
126. Yang NY, Desari TD. Glutathione peroxidase and vitamin E relationship. In: de Duve C, Hayashi O, eds. Tocopherol, oxygen and biomembranes. Amsterdam: Elsevier, 1978; 233-45.
127. Kruglikova AA, Donchenko GV, Shvachko OP, Karpov AV. Synthesis of S-adenosylmethionine and S-adenosylcysteine in rat liver during E-avitaminosis. Biokhimiya 1983; 48:639-44.
128. Lankin VZ, Tikhaze AK, Rakita DR, Pomoinetskii VD, Vikhert AM. Influence of α-tocopherol on the superoxide dismutase and glutathione lipoperoxidase activities of mouse liver cytoxol and mitochondria. Biokhimiya 1983; 48:1555-9.
129. Jiang Z-Y, Woolard ACS, Wolff SP. Hydrogen peroxide production during experimental protein glycation. FEBS Lett 1990; 268:69-71.
130. Robertson JM, Donner AP, Trevithick JR. A possible role for vitamins C and E in cataract prevention. Am J Clin Nutr 1991; 53:3465-515.
131. Maraini G, Pasquini P, Sperduto RD. Risk factors in age-related cataract formation: Parma experience. Proc 9th International Congress of Eye Research Int. Soc. Eye Res., New York, N.Y. 1990; 140.

132. LaVail MM, Batelle BA. Influence of eye pigmentation and lipid deprivation on inherited retinal dystrophy in the rat. Exp Eye Res 1975; 21:167-92.

133. Clarke IS, Dzialoszynski T, Sanford SE, Chevendra V, Trevithick JR. Dietary prevention of damage to retinal proteins in RCS rats. In: Lerman S, Tripathi R, eds. Ocular toxicity. New York: Marcel Dekker, 1989; 253-72.

134. Keyse SM, Tyrrell RM. Both near ultraviolet radiation and the oxidizing agent hydrogen peroxide induce a 32 kDa stress protein in normal human skin fibroblasts. J Biol Chem 1987; 262:14821-5.

135. Barbe MF, Tyrell M, Gower DJ, Welch WJ. Hyperthermia protects against light damage in the rat retina. Science 1988; 241:1817-20.

136. Clarke IS, Dzialoszynski T, Sanford SE, Trevithick JR. A possible relationship between increased levels of the major heat shock protein HSP 70 and decreased levels of S-antigen in the retina of the RCS rat. Exp Eye Res 1991; 53:545-48.

137. Dilley RA, McConnell DG. Alpha-tocopherol in the retinal outer segment of bovine eyes. J Membr Biol 1970; 2:317-23.

138. Farnsworth CC, Dratz EA. Oxidative damage of retinal outer segment membranes and the role of vitamin E. Biochim Biophys Acta 1976; 433:556-70.

139. Hayes KC. Pathophysiology of vitamin E deficiency in monkeys. Am J Clin Nutr 1974; 27:1130-40.

140. Katz ML, Sonte WL, Dratz EA. Fluorescent pigment accumulation in retinal pigment epithelium of antioxidant-deficient rats. Invest Ophthalmol Vis Sci 1978; 17:1049-58.

141. Ephimova MG. The system of lipid peroxidation in retina and brain during the early postnatal period of life. Zh Evol Biokhim Fiziol 1990; 26:130-3.

142. Yoshida N. Influence of vitamin E on the tissue metabolism of the normal nd alloxan-added rat retina. Nippon Gahka Gakkai Zasshi 1964; 68:238-42.

143. Robison WG, Kuwabara T, Bieri JG. Vitamin E deficiency and the retina: photoreceptor and pigment epithelial changes. Invest Ophthalmol Vis Sci 1979; 18:683-90.

144. Farnsworth CC, Stone WL, Dratz EA. Effects of vitamin E and selenium deficiency on the fatty acid composition of rat retinal tissues. Biochim Biophys Acta 1979; 552:281-93.

145. Serfass RE, Ganther HE. Effects of dietary selenium and tocopherol on glutathione peroxide and superoxide dismutase activities in rat phagocytes. Life Sci 1976; 19:1139-44.

146. Hall M, Hall D. Superoxide dismutase of bovine and frog outer segments. Biochem Biophys Res Commun 1975; 67:1199-204.

147. Katz ML, Drea CM, Robison WG. Relationship between dietary retinol and lipofuscin in the retinal pigment epithelium. Mech Ageing Dev 1986; 35:291-305.

148. Fong SL, Liou GI, Landers RA, Alvarez RA, Bridges CD. Purification and characterization of a retinol-binding glycoprotein synthesized and secreted by bovine neural retina. J Biol Chem 1984; 259:6534-41.

149. Pfeffer B, Wiggert B, Lee L, Zonnenberg B, Nesome N, Chader G. The presence of a soluble interphotoreceptor retinol-binding protein (IRBP) in the retinal interphotoreceptor space. J Cell Physiol 1983; 117:334-41.

150. Hermann RK, Robison WG, Bieri JG. Deficiencies of vitamins E and A in the rat: lipofuscin accumulation in the choroid. Invest Ophthalmol Vis Sci 1984; 25:429-33.

151. Hermann RK, Robison WG, Bieri JG, Spitznas M. Lipofuscin accumulation in extra ocular muscle of rats deficient in vitamins E and A. Graefes Arch Clin Exp Ophthalmol 1985; 223:272-7.

152. Katz ML, Robison WG. Light and aging effects on vitamin E in the retina and retinal pigment epithelium. Vision Res 1987; 27:1875-9.

153. Wald G. Molecular basis of visual excitation. Science 1968; 162:230-9.

154. Bok D. Retinal photoreceptor-pigment epithelium interactions (Friedenwald lecture). Invest Ophthalmol Vis Sci 1985; 26:1659-94.

155. Persad S, Menon IA, Basu PK, Carre F. Phototoxicity of chlorpromazine on retinal pigment epithelial cells. Curr Eye Res 1988; 7:1-9.

156. Cohen H, Schoenfeld D, Wolter J. Randomized trial of chlorpromazine caffeine and methyl-CCNU in disseminated melanome. Cancer Treat Rep 1980; 64:151-3.

157. Greener AC, Berry K. Skin pigmentation and corneal and lens opacities with prolonged chlor-promazine therapy. Can Med Assoc J 1964; 90:663-5.
158. Prien RF, Delong SL, Cole JO, Levine J. Ocular changes occuring with prolonged high dose chlorpromazine therapy. Arch Gen Psychiatr 1970; 23:464-8.
159. Silverman HI. The adverse effects of commonly used systemic drugs on the human eye. II. Am J Optom 1972; 49:335-62.
160. Kagan VE, Barybina GV, Novikov KN. Peroxidation of lipids and degeneration of photoreceptors in the retina of rats with avitaminosis E. Zh Exp Biol Med 1977; 83:473-6.
161. Akopyan GK, Tagiev SK, Szhafarov AI. Time course of the electroretinogram of isolated frog and turtle retinas exposed to repetitive photic simulation and induced lipid peroxidation. Bull Exp Biol Med 1984; 97:406-9.
162. Shvedova AA, Kagan VE, Kuliev IY, Vekshina OM. Mechanisms of the harmful action of fluor-escent dyes on the retina. Bull Exp Biol Med 1983; 96:1085-1089.
163. Shvedova AA, Kagan VE, Kuliev IY, et al. Lipid peroxiation and retinal injury in stress. Bull Exp Biol Med 1982; 93:408-11.
164. Magomedov NM, Neiman-Zade NK, Kulieva EM, Dzhafarov AI, Akhmedli GT, Guseinova ES. Lipid peroxidation in experimental degeneration of the retina. Bull Exp Biol Med 1983; 95:296-8.
165. Dzhafarov AI, Gadzhieva NA, Mamedkhanly TA, Kul'gavin LE, Dagkesamanskaya DN, Alieva NI. Effect of visual cortex stimulation on lipid peroxidation in the rabbit retina. Bull Exp Biol Med 1987; 103:318-20.
166. Packer L, Landvik S. Vitamin E: introduction to biochemistry and health benefits. Ann N Y Acad Sci 1989; 570:1-6.
167. Burton GW, Ingold KU. Vitamin E as in vitro and in vivo antioxidant. Ann N Y Acad Sci 1989; 570:7-22.
168. Sokol RJ, Butler-Simon N, Heubi JE, et al. Vitamin E deficiency neuropathy in children with fat malabsorption. Ann N Y Acad Sci 1989; 570:156-69.
169. Kayden H, Traber MG. Neuropathies in adults with or without fat malabsorption. Ann N Y Acad Sci 1989; 570:170-5.
170. Muller DPR, Lloyd JK, Wolff OH. Vitamin E and neurological function: abetalipoproteinemia and other disorders of fat absorption. In: R. Porter and J. Whelan, eds. (CIBA Foundation Sym-posium 101) Biology of vitamin E. London: Pitman Books 1983; 106-21.
171. Yuill GM, Scholz C, Lascelles RG. Abetalipoproteinemia. A case report with pathological stud-ies. Postgrad Med J 1976; 52:713-20.
172. Runge P, Muller DRP, McAlister J, Calver D, Lloyd JK, Taylor D. Oral vitamin E supplements can prevent the retinopathy of abetalipoproteinemia. Br J Ophthalmol 1986; 70:166-73.
173. Bishara S, Merin S, Cooper M, Szizi E, Delpre G, Deckelbaum RJ. Combined vitamin A and E therapy prevents retinal electrophysiological deterioration in abetalipoproteinemia. Br J Ophthal-mol 1982; 66:767-70.
174. Azizi E, Zardman JL, Eshchar J, Szeinberg A. Abetalipoproteinemia treated with parenteral and oral vitamins A and E and with medium chain triglycerides. Acta Pediatr Scand 1978; 67: 797-801.
175. Miller RG, Davis CJF, Illingworth DR, Bradley W. The neuropathy of abetalipoproteinemia. Neurology 1980; 30:1286-91.
176. Herbert PN, Assmann G, Gottoam J, Frederickson DS. Familial lipoprotein deficiency: abeta-lipoproteinemia hypobetalipoproteinemia and Tangier disease. In: Stanbury JB, Wyngaarden JB, Frederickson DS, Goldstein JL, Brown MS, eds. Metabolic basis of inherited disease, 4th ed. New York: McGraw-Hill, 1983; 589-621.
177. Sitrin MD, Lieberman F, Jensen WE, Noronha A, Milburn C, Addington W. Vitamin E defi-ciency and neurologic disease in adults with cystic fibrosis. Ann Intern Med 1987; 107:51-4.
178. Bertoni JM, Abraham FA, Falls HF, Itabashi HH. Small bowel resection with vitamin E defi-ciency and progressive spinocerebellar syndrom. Neurology 1984; 34:1046-52.
179. Nelson JS. Neuropathological studies of chronic vitamin E deficiency in mammals including hu-mans. In: R Porter and J Whelan, eds. Ciba Foundation Symposium 101. Biology of vitamin E. London: Pitman, 1983; 92-105.

180. Satya-Murti S, Howard L, Krohel G, Wolfe B. The spectrum of neurologic disorder from vitamin E deficiency. Neurology 1986; 36:917-21.

181. Kager S. Langzertbehandlung arteriosklerotisch bedingter durchblutungsstorungen der netz und aderhaut mit den vitaminen A and E sowie einem nikotinsaurederivat. Klin Monatsbl Augenheilkd 1968; 153:571-7.

182. Clausen J. Dementical syndromes and lipid metabolism. Acta Neurol Scand 1984; 70:345-55.

183. Westmarck T Santaviori P, Marklund S, Polya P, Salmi A. In: Armstrong D, Koppag N, Rider JA, eds. Ceroid-lipofuscinosis (Batten's disease). Amsterdam: Elsevier, 1982; Chap. 32.

184. Santavuori P, Moren P. Experience of antioxidant treatment in neuronal ceroid-liporuscinosis of Spielmeyer-Sjogren type. Neuropadiatics 1977; 8:333-4.

185. Ostrovsky MA, Sakina NL, Dontsov AE. An antioxidative role of ocular screening pigments. Vision Res 1987; 27:893-9.

186. Ameniya T. Effects of vitamin E and selenium deficiencies on rat capillaries. Int J Vitam Nutr Res 1989; 59:122-6.

187. Friedman EA. Overview of diabetic nephropathy. In: Keen H, Legrain M, eds. Prevention and treatment of diabetic nephropathy. Boston: MTP Press, 1983; 3-19.

188. Bresnick GH. Oxygen induced visual cell degeneration in the rabbit. Invest Ophthalmol Vis Sci 1970; 9:372-87.

189. Stone WL, Henderson RA, Howard GH Jr, Hollis AL, Payne PH, Scott RL. The role of antioxidant nutrients in preventing hyperbaric oxygen damage to the retina. Free Radic Biol Med 1989; 6:505-12.

190. Kinsey VE, Arnold HJ, Kalina RE, et al. PaO₂ levels and retrolental fibroplasia. A report of the cooperative study. Pediatrics 1977; 60:655-67.

191. Lucey JF, Dangman B. A re-examination of the role of oxygen in retrolental fibroplasia. Pediatrics 1984; 73:82-96.

192. Silverman WA. Retrolental fibroplasia. A modern parable. In: Oliver TK, ed. Monographs in neonataology. New York: Grune & Stratton, 1980; 9-58.

193. Terry TL. Extreme prematurity and fibroblastic overgrowth of persistent vascular sheath behind each crystalline lends. Preliminary report. Am J Ophthalmol 1942; 25:203-4.

194. Ashton N. Some aspects of the comparative pathology of oxygen toxicity in the retina (Donders lecture). Br J Ophthalmol 1968; 52:505-31.

195. Owens WC, Owens EU. Retrolental fibroplasia in premature infants. II. Studies on the prophyllaxis of the disease: the use of alpha tocopherol acetate. Am J Ophthalmol 1949; 32:1631-4.

196. Kinsey VE,, Chisholm JF. Retrolental fibroplasmia. Evaluation of several changes in dietary supplements of premature infants with respect to the incidence of the disease. Am J Ophthalmol 1951; 34:1259-69.

197. Owens WC. Spontaneous regression in retrolental fibroplasia. Trans Am Ophthalmol 1953; 51: 555-79.

198. Crosse VM. The problem of retrolental fibroplasia in the city of Birmingham. Brans Ophthalmol Soc UK 1951; 71:609-12.

199. Johnson L, Schaffer D, Boggs TR. The premature infant, vitamin E deficiency and retrolental fibroplasia. Am J Clin Nutr 1974; 27:1158-73.

200. Phelps DL, Rosenbaum A. The role of tocopherol in oxygen induced retinopathy: kitten model. Pediatrics 1977; 59:998-1005.

201. Phelps DL, Rosenbaum AL. Vitamin E in kitten oxygen-induced retinopathy. II. Blockage of vitreal neovascularization. Arch Ophthalmol 1979; 97:1522-6.

202. Hittner HM, Godio LB, Rudolph AJ. Retrolental fibroplasia: efficacy of vitamin E in a double-blind clinical study of preterm infants. N Engl J Med 1981; 305:1356-71.

203. Finer NN, Schindler RF, Grant G. Effect of intramuscular vitamin E on frequency and severity of retrolental fibroplasia: A controlled trial. Lancet 1982; 1:1087-91.

204. Puklin JE, Simon RM, Ehrenkranz RA. Influence on retrolental fibroplasia of intramuscular vitamin E administration during respiratory distress syndrom. Ophthalmology 1982; 89:97-103.

205. Johnson L, Schaffer DB, Quinn GE. Pennsylvania Hospital, Philadelphia. The effect of sustained pharmacologic vitamin E serum levels on the incidence and severity of retinopathy of prematurity

(ROP): a controlled clinical trial. Cited in report of Institute of Medicine, U.S. Natl. Acad. Sci. report 10M-86-02. Washington, DC: National Academy Press, 1986.

206. Phelps DL, Rosenbaum AL, Isenberg SJ. University of Rochester School of Medicine, Rochester, NY. Safety and efficacy of tocopherol for preventing retinopathy of prematurity: a randomised, controlled, double-masked trail. Cited in report of Institute of medicine, U.S. Natl. Acad. Sci. report 10M-86-02. Washington, DC: National Academy Press, 1986.

207. Milner RA, Watts JL, Paes B. RLF in 1500 gram neonates: part of a randomized clinical trial of the effectiveness of vitamin E. In: Retinopathy of prematurity conference syllabus. Conference sponsored by Ross Laboratories, Columbus, OH 1981.

208. Phelps DL, Rosenbaum AL, Isenberg SJ, Leake RD, Dorey FJ. Tocopherol efficacy and safety for preventing retinopathy of prematurity: a randomized controlled, double-masked trail. Pediatrics 1987; 79:489-500.

209. Testa M. Personal communication, 1990.

210. Ashton N. Oxygen and the growth and development of retinal vessels; in vivi and in vitro studies. Am J Ophthalmol 1966; 62:412-35.

211. Foos RY. Pathologic features of retinopathy of prematurity. Birth Defects 1988; 24:73-85.

212. Flower RW. Physiology of the developing retinal vasculature. Birth Defects 1988; 24:129-46.

213. Schaffer DB, Johnson L, Quinn GE. Classification of retrolental fibroplasia to evaluate vitamin E therapy. Ophthalmology 1979; 86:1749-60.

214. Ashton N, Pedler C. Studies on developing retinal vessels. IX. Reaction of endothelial cells to oxygen. Br J Ophthalmol 1962; 46:257-76.

215. Bauer CR, Widmayer SM. A relationship between PaO$_2$ and retrolental fibroplasia. Pediatr Res 1981; 15:1236A.

216. Kretzer FL, Hittner HM. Spindle cells and retinopathy of prematurity: interpretations and predictions. Birth Defects 1988; 24:147-68.

217. Kretzer FL, Hittner HM. Retinopathy of prematurity: clinical implications of retinal development. Arch Dis Child 1988; 63:1151-67.

218. Hittner HM, Kretzer FL. Vitamin E and retrolental fibroplasia: ultrastructural mechanism of clinical efficacy. In: R Porter and J. Whelen, eds. Biology of vitamin E. Ciba Foundation Symposium 101. London: Pitman Books, 1983; 165-85.

219. Johnson L, Schaffer D, Quinn G, et al. Vitamin E supplementation and the retinopathy of prematurity. Ann N Y Acad Sci 1982; 393:473-95.

220. Bove KE, Kosmetatos N, Wedig KE, et al. Vasculopathic hepatotoxicity associated with E-Ferol syndrome in low-birth-weight infants. JAMA 1985; 254:2422-30.

221. Lorch V, Murphy D, Hoersten LR, et al. Unusual syndrome among premature infants: association with a new intravenous vitamin E products. Pediatrics 1985; 75:598-602.

222. Reyniers JP, Machado EA, Farkas WR. Hepatotoxicity of chemically-defined diets containing polysorbate-80 in germ free and conventional mice. In: BS Woostman and VR Plesants, eds. Germfree research: Microflora control and its application to the biomedical sciences, vol. 181: Progress in Clin Biol Res. New York: Alan R. Liss, 1985; 91-4.

223. Alade SL, Brown RE, Paguet A. Polysorbate 80 and E-ferol toxicity. Pediatrics 1986; 77:593-7.

224. Chiswick ML, Johnson M, Woodhall C, et al. Protective effect of vitamin E (DL-alpha-tocopherol) against intraventricular haemorrhage in premature babies. Br Med J 1983; 287:81-4.

225. Sinha S, Davies J, Toner N, Bogle S, Chiswick M. Vitamin E supplementation reduces frequency of preventricular haemorrhage in very preterm babies. Lancet 1987; :466-70.

226. Speer ME, Blifeld C, Rudolph AJ, et al. Intraventricular haemorrhage and vitamin E in the very low-birth-weight infant: evidence for efficacy of early intramuscular vitamin E administration. Pediatrics 1984; 74:1107-12.

227. Chiswick M, Gladman G, Sinhar S, Toner N, Davies J. Prophyllaxis of periventricular haemorrhage in preterm babies of vitamin E supplementation. Ann N Y Acad Sci 1989; 570:197-204.

228. Finer NN, Peters KL, Schindler RF, Grant GD. Vitamin E and retrolental fibroplasia: prevention of serious ocular sequellae. In: Porter R, Whelan J, eds. Biology of vitamin E. Ciba Foundation Symposium 101. London: Pitman Books, 1983; 147-64.

229. Oliver TK, Avery ME, Feinstein AR, et al. Vitamin E and retinopathy of prematurity. Report 10M-86-02. Washington, DC: Institute of Medicine of U.S. National Academy of Sciences, 1986.

230. Finer NN, Peters K, Hayek L, et al. Vitamin E and necrotizing enterocolitis. Pediatrics 1984; 73: 387-93.

231. Johnson L, Bowen FW, Abasi S, et al. The relationship of prolonged pharmacologic serum vitamin E levels to incidence of sepsis and necrotizing enterocolitis in infants with birth weight 1500 grams or less. J Pediatr 1985; 75:619-37.

232. Phelps DL. Vitamin E and retrolental fibroplasia in 1982. Pediatrics 1982; 70:420-5.

233. Little GA, Freeman JM, Kattwinkel J, et al. Vitamin E and the prevention of retinopathy of prematurity. Pediatrics 1985; 76:315-6.

234. Centers for Disease Control. Unusual syndrome with fatalities among premature infants: association with a new intravenous vitamin E product. MMWR 1984; 33:198.

235. Sobel S, Gueriguian J, Trendle G. Letter to the editor. N Engl J Med 1982; 306:867.

236. Yusuf S, Peto R, Lewis J, Collins R, Sleight P. Beta blockade during and after myocardial infarction: an overview of the randomized trials. Prog Cardiovasc Dis 1985; 27:335-71.

237. Acheson J, APT collaboration. Secondary prevention of vascular disease by prolonged antiplatelet treatment. Br Med J 1988; 296:320-31.

238. Schaffer D, Johnson L, Quinn GE, Abbasi S, Otis C, Bowen FW. Vitamin E and retinopathy of prematurity: the ophthalmologist's perspective. Birth Defects 1988; 24:219-35.

239. Hittner HM, Rudolph AJ, Kretser FL. Suppression of severe retinopathy of prematurity with vitamin E supplementation. Ophthalmology 1984; 91:1512-23.

240. Keith CG, Smith ST, Landsell BJ. Retrolental fibroplasia: a study of the incidence and aetiological factors 1977–79. Med J Aust 1981; :589-92.

241. Keith CG, Kitchen WH. Retinopathy of prematurity in extremely low birthweight infants. Med J Aust 1984; :225-7.

242. Kretzer FL. Vitamin E and retrolental fibroplasia: ultrastructural support of clinical efficacy. Ann N Y Acad Sci 1982; 393:145-66.

243. Finer NN, Schindler RF, Peters KL, Grant GG. Vitamin E and retinolental fibroplasia. Ophthalmology 1983; 90:428-35.

244. Phelps DL. Vitamin E and retinopathy of prematurity. In: Silverman WA, Flynn JT, eds. Contemporary issues in fetal and neonatal medicine, Vol. 2. Retinopathy of prematurity. Boston: Blackwell Scientific, 1985; 185-205.

245. Karp WA, Robertson AF. Vitamin E in neonatalology. Adv Pediatr 1986; 33:127-48.

246. Katz ML, Robinson WG Jr. Autooxidative damage to the retina: potential role in retinopathy of prematurity. Birth Defects 1988; 24:237-48.

247. Williams ML, Shott RJ, O'Neal PL, et al. Role of dietary iron and fat on vitamin E deficiency anemia of infancy. N Engl J Med 1975; 292:887-90.

248. Jansson L, Holmberg L. Vitamin E and fatty acid composition of human milk. Am J Clin Nutr 1981; 34:8-13.

249. Gordon HH, Nitowsky HM, Tildon JT, Levin S. Studies of tocopherol deficiency in infants and children. V. An interim summary. Pediatrics 1958; 21:673-81.

250. Hassan H, Hashim SA, Van Itallie TB, et al. Syndrome in premature infants associated with low plasma vitamin E levels and high polyunsaturated fatty acid diet. Am J Clin Nutr 1966; 19:147-57.

251. Oski FA, Barness LA. Vitamin E deficiency: a previously unrecognized cause of hemolytic anemia in the premature infant. J Pediatr 1967; 70:211-20.

252. Ritchie JH, Fish MB, McMasters V, et al. Edema and hemolytic anemia in premature infants. N Engl J Med 1968; 279:1185-90.

253. Smith GJ, Dunkley WL. Initiation of lipid peroxidation by a reduced metal ion. Arch Biochem Biophys 1962; 98:46-8.

254. Avery ME, Oppenheimer EH. Recent increase in mortality from hyaline membrane disease. J Pediatr 1960; 57:553-9.

255. DeLeon AS, Eliott JH, Jones DB. The resurgence of retrolental fibroplasia. Pediatr Clin North Am 1970; 17:309-22.

256. Patz A. Symposium on retrolental fibroplasia. Summary. Ophthalmology 1979; 86:1761-3.

257. Gutcher GR, Raynor WJ, Farrell PM. An evaluation of vitamin E status in premature infants. Am J Clin Nutr 1984; 40:1078-84.

258. Mauer AM, Dweck HS, Finberg L, et al. Nutritional needs of low birth-weight infants. Pediatrics 1985; 75:976-86.

259. National Research Council, Food and Nutrition Board. Recommended dietary allowances, 9th ed. Washington, DC: National Academy of Sciencies, 1980.

260. Nutritional advisory group. Guidelines for multivitamin preparation for parenteral use. Chicago: American Medical Association, 1975.

261. American Medical Association department of foods and nutrition. 1975: multivitamin preparations for parenteral use. A statement by the nutrition advisory group. J Parenter Enter Nutr 1979; 3:258.

262. Etches PC, Koo WWK. Parentreal vitamins A, D and E for premature infants. J Perinatol 1988; 8:93-5.

263. Greene, HL, Courtrey-Moore ME, Phillips B, et al. Evaluation of a pediatric multiple vitamin preparation for total parenteral nutrition. II. Blood levels of vitamins A, D and E. Pediatrics 1986; 77:539-47.

264. Greene HL, Phillips BL. Vitamin dosage in premature infants (letter). Pediatrics 1987; 79:655.

265. Huston RK, Brenda GI, Carlson CV, et al. Selenium and vitamin E sufficiency in premature infants requiring total parenteral nutrition. J Parenter Enter Nutr 1982; 6:507-10.

266. DeVito V, Reynolds JW, Benda GI, et al. Serum vitamin E levels in very low-birth weight infants receiving vitamin E in parenteral nutrition solutions. J Parenter Enter Nutr 1986; 10:63-5.

267. Moore ML, Greene HL, Phillips B, et al. Evaluation of a pediatric multiple vitamin preparation for total parenteral nutrition in infants and children. (1) Blood levels of water-soluble vitamins. Pediatrics 1986; 77:530-8.

268. American Academy of Pediatrics. Maternal and newborn nutrition. In: FD Frigoletti, GA Little, eds. Guidelines for perinatal care, 2nd ed. Elk Grove Village, IL: American Academy of Pediatrics, 1980; 203.

269. Hittner H-M. Discussion. Ann N Y Acad Sci 1989; 570:205-6.

270. Hittner HM, Kretzer FL. Toxicity of vitamin E in preterm infants. In: McPherson AR, Hittner HM, Kretzer FL, eds. Retinopathy of prematurity: current concepts and controversies, vol. 11. Toronto: B.C. Decker, 1986; 111-6.

271. Schroder H, Schulz M, Aissen K. Muscular calcification following injection of vitamin E in newborn infants. Eur J Pediatr 1984; 142:145-6.

272. Smith IJ, Buchanan MFG, Goss I, Congdon PJ. Correspondence: vitamin E in retrolental fibroplasia. N Engl J Med 1983; 309:669.

273. Bhat R, Raju T, Barrada A, Evans M. Disposition of vitamin E in the eye. Pediatr Res 1987; 22: 16-20.

274. Nishiyama J, Ellison EC, Mizuno GR, Chipault JR. Microdetermination of α-tocopherol in tissue lipids. J Nutr Sci Vitaminol 1975; 21:355-61.

275. Bhat R. Serum, retinal, choroidal, vitreal vitamin E concentrations in human infants. Pediatrics 1986; 78:866-70.

276. Nishida A, Togari H. Effect of vitamin E administration on α-tocopherol concentrations in the retina, choroid, and vitreous body of human neonates. J Pediatr 1986; 108:150-3.

277. Hunt DF, Organisciak DT, Wang HM, Wu RLC. α-Tocopherol in the developing rat retina: a high pressure liquid chromatographic analysis. Curr Eye Res 1984; 3:1281-8.

278. Alvarez RA, Liou GI, Fong S-L, Bridges CDB. Levels of α- and γ-tocopherol in human eyes: evaluation of the possible role of IRBP in intraocular α-tocopherol transport. Am J Clin Nutr 1987; 46:481-7.

279. Costagliola C, Italiano G, Menzione M, Rinaldi E, Vitro P, Auricchio G. Effect of vitamin E on glutathione in red blood cells, aqueous humor and lens of humans and other species. Exp Eye Res 1986; 43:905-14.

280. Dillon J, Mehlman B, Ponticorvo L, Spector A. The state of neutral lipids in normal and cataractous human lenses. Exp Eye Res 1983; 37:91-8.

281. Fish WH, Cohen M, Franzik D, Williams JM, Lemons JA. Effect of intramuscular vitamin E on mortality and intracranial hemorrhage in neonates of 1000 grams or less. Pediatrics 1990; 85: 578-84.
282. Johnson L, Quinn GE, Abassi S, Delevoria-Papadopoulos M, Peckham G, Bowen FW Jr. Bilateral stage 3-plus retinopathy of prematurity (ROP). Effect of treatment (Rx) with high-dose vitamin E. Ann N Y Acad Sci 1989; 570:464-6.
283. Stephens RJ, Negi DS, Short SM, Van Kulik FJGM, Dratz EA, Thomas DW. Vitamin E distribution in ocular tissues following long-term dietary depletion and supplementation as determined by microdissection and gas chromatography-mass spectrometry. Exp Eye Res 1988; 47:237-45.

E. Internal Medicine

62

Vitamin E in Gastroenterology

Toshikazu Yoshikawa, Yuji Naito, and Motoharu Kondo

Kyoto Prefectural University of Medicine, Kyoto, Japan

INTRODUCTION

The implication of active oxygens and free radicals in the pathogenesis of mucosal injury in the gastrointestinal tract has been reported in different models (1–5). Much evidence suggests the possible contribution of these active species derived from molecular oxygen in tissue injury, such as gastric (2,4) and duodenal (6) mucosal injuries, inflammatory bowel diseases (7), and ischemic colitis (8). Furthermore, lipid peroxidation mediated by free radicals is believed to be one of the important causes of biological membrane destruction and cell damage (9). We have already reported that lipid peroxidation plays a significant role in the pathogenesis of gastric mucosal injury induced by water-immersion restraint stress (10), burn stress (11), treatment with platelet-activating factor (12) and compound 48/80 (13), and ischemia-reperfusion (4).

Vitamin E functions as a potent chain-breaking lipid-soluble antioxidant in living systems by scavenging oxygen radicals and terminating free-radical chain reaction (9,14). Therefore, it is important to determine the change in vitamin E in serum and tissue in gastrointestinal mucosal injury to elucidate the mechanism of the effect and the role of vitamin E. In the present study, we deal with some further aspects of the chronological changes in vitamin E and lipid peroxides in serum and the gastric mucosa after ischemia and reperfusion and the effect of vitamin E-deficient and E-supplemented conditions on these changes.

GASTRIC MUCOSAL INJURY INDUCED BY ISCHEMIA-REPERFUSION

It has been reported that gastric mucosal injury induced by ischemia-reperfusion can be prevented by administering superoxide dismutase (SOD) (2,4), catalase, dimethylsulfoxide (DMSO) (15), and other antioxidants (16) in several experimental models. Itoh and Guth (2) first demonstrated that SOD could significantly protect against gastric mucosal injury induced by a withdrawal of the blood for 20 minutes and retransfusion in rats. Currently, we have made a new model of gastric mucosal injury induced by ischemia-reperfusion (4), which was prepared by clamping the celiac artery for a certain period with a hemostat (ischemia) and then releasing the artery (reperfusion). By the clamping the celiac artery, the gastric mucosal blood flow decreased to 10% of that measured before clamping and recovered to the normal range by subsequent reperfusion (4). After more than 45 minutes of clamping, the gastric mucosal blood flow could not recover to the normal range. The total area of erosion, a morphological index of gastric injury, did not increase after 30 minutes of ischemia and significantly increased

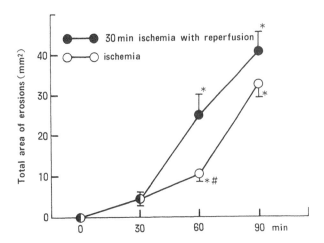

FIGURE 1 Changes in total area of gastric erosion after ischemia or ischemia-reperfusion in rats. Each value indicates the mean ± SEM of 5–12 rats. *$p < 0.001$, difference from values for rats before clamping celiac artery. #$p < 0.05$, difference from values for rats 30 minutes after reperfusion following 30 minutes of ischemia.

after more than 60 minutes of ischemia and 30 minutes of ischemia with 30 and 60 minutes of reperfusion (Fig. 1). However, the increase in the total area of erosion after 30 minutes of ischemia with 30 minutes of reperfusion was significantly higher than that after 60 minutes of ischemia only. Therefore, these results are consistent with the view that the injury produced by ischemia-reperfusion is more severe than that produced by ischemia per se.

LIPID PEROXIDATION AND VITAMIN E IN GASTRIC MUCOSAL INJURY

To investigate the role of lipid peroxidation in the pathogenesis of gastric mucosal injury induced by ischemia-reperfusin, thiobarbituric acid (TBA)-reactive substances (17,18) and α-tocopherol (19) in gastric mucosa and serum were measured as an index of lipid peroxidation. Since all vitamin E is carried in plasma lipoprotein, there is a linear relationship between serum vitamin E and the total serum lipid concentration. For this reason, the ratio of serum α-tocopherol to total serum cholesterol was utilized to better approximate α-tocopherol status. TBA-reactive substances in the gastric mucosa barely increased after 30 minutes of ischemia and increased remarkably and significantly after following reperfusion in parallel with the increase in the total area of gastric erosion (Table 1). The increases in both TBA-reactive substances and total area of erosion in the gastric mucosa after ischemia-reperfusion were significantly inhibited by pretreatment with SOD plus catalase (4). However, there was no significant change in the levels of TBA-reactive substances in serum. The level of α-tocopherol in the gastric mucosa tended to decrease during ischemia and decreased further during early reperfusion. The α-tocopherol-cholesterol ratio also decreased after 30 and 60 minutes of reperfusion. These results suggest that lipid peroxidation or lipid peroxides may play an important role in the formation of gastric mucosal injury induced by ischemia-reperfusion. Therefore, vitamin E, which reacts with free radicals and terminates the free-radical–mediated chain reaction, is speculated to be consumed in serum and gastric mucosa to prevent the development of tissue damage. Conversely, the decrease in vitamin E suggests the implication of free radicals in gastric mucosal injury induced by ischemia-reperfusion.

TABLE 1 Changes in the Total Area of Erosions, TBA-Reactive Substances, and α-Tocopherol in Gastric Mucosa After Ischemia-Reperfusion[a]

	Total area of erosion (mm²)		TBA-reactive substances (nmol/g wet weight)		α-Tocopherol (μg/mg protein)	
Before ischemia	0.0 ± 0.0	(6)	46.6 ± 2.3	(6)	0.16 ± 0.07	(8)
30 minutes ischemia	4.6 ± 0.9	(10)	53.9 ± 8.1	(7)	0.15 ± 0.08	(5)
30 minutes ischemia + 30 minutes reperfusion	25.0 ± 5.3^b	(7)	87.6 ± 6.8^b	(16)	0.13 ± 0.05^c	(8)
30 minutes ischemia + 60	41.5 ± 4.7^b	(8)	96.2 ± 10.2^b	(10)	0.11 ± 0.09^c	(10)

[a]Each value is the mean $\pm$ SEM followed by number of rats in parentheses. Indicated values are significantly different from values before ischemia.
[b]$p < 0.001$.
[c]$p < 0.01$.

EFFECTS OF VITAMIN E ON GASTRIC MUCOSAL INJURY

We studied the degree of gastric mucosal injury induced by ischemia-reperfusion in rats raised on vitamin E-deficient, E-sufficient, and E-supplemented diets. Sprague-Dawley rats at 4 weeks of age were maintained on a vitamin E-deficient, E-sufficient, or E-supplemented diet for 8 weeks. The total tocopherol content of the basal vitamin E-deficient diet (Eisai Co., Ltd., Tokyo) is less than 0.1 mg per 100 g diet, and the vitamin E-sufficient and E-supplemented diets are prepared by adding 2.0 mg (2.0 IU) d,l-α-tocopheryl acetate and 58.5 mg (50 IU) d,l-α-tocopheryl nicotinate, respectively, to a 100 g basal vitamin E-deficient diet. As shown in Table 2, the levels of α-tocopherol in serum and gastric mucosa of the vitamin E-deficient rats decreased to 17.4 and 16.4%, respectively, compared to that of vitamin E-sufficient rats after 8 weeks of feeding (20). Vitamin E deficiency, which is an inadequate content for scavenging free radicals generated in ischemia-reperfusion, significantly aggravated gastric mucosal injury induced by ischemia-reperfusion and also significantly increased TBA-reactive substances in the gastric mucosa compared with vitamin E-sufficient rats (Table 3) (20). These results are compatible with the assumption that the cellular damage caused by gastric ischemia-reperfusion can be explained by a free-radical reaction process during ischemia, especially reperfusion, and suggest that α-tocopherol plays an important role as an endogenous antioxidant to protect tissue from oxidative stress during ischemia-reperfusion.

TABLE 2 Levels of α-Tocopherol in Serum and Gastric Mucosa of Rats Fed Vitamin E-Deficient, E-Sufficient, and E-Supplemented Diets for 8 Weeks[a]

Group	Serum (μg/ml)		Gastric mucosa (ng/mg protein)	
Deficient	0.50 ± 0.05	(6)	21.0 ± 1.6	(7)
Sufficient	2.88 ± 0.39	(7)	128.0 ± 6.2	(8)
Supplemented	10.11 ± 1.44	(7)	338.0 ± 105	(7)

[a]Each value indicates the mean $\pm$ SEM followed by number of rats in parentheses.

TABLE 3   Effects of Vitamin E on Gastric Mucosal Injury and TBA-Reactive Substances in Gastric Mucosa After Ischemia-Reperfusion in Rats[a]

Group	Total area of erosions (mm²)		TBA reactants (nmol/mg protein)	
Deficient	52.9 ± 9.6^b	(10)	1.79 ± 0.45^b	(8)
Sufficient	23.9 ± 5.0	(7)	1.15 ± 0.36	(7)
Supplemented	20.7 ± 3.8	(10)	1.07 ± 0.13	(8)

[a]Each value indicates the mean $\pm$ SEM followed by number of rats in parentheses.
[b]$p < 0.05$, significant difference from values for rats in the vitamin E-sufficient group.

There are many reports that the administration of α-tocopherol has a protective effect on damage by ischemia and reperfusion in several organs, such as liver (21), kidney (22), and heart (23); however, a significant reduction in gastric mucosal injury induced by ischemia-reperfusion could not be obtained by treatment with d,l-α-tocopheryl nicotinate. The mechanism of α-tocopherol inhibition of peroxidative damage to membrane lipids in vitro has been suggested to be scavenging of lipid peroxyl radicals. It has been proposed that hydroxyl radicals generated from the Haber-Weiss reaction, which is catalyzed by transition metal ions, induce cellular damage under oxidative stress through a direct action on biological molecules, especially lipids, and lipid peroxidation has been explained as a late event occurring only at the point of cell death. As shown in the present study, lipid peroxides accumulated in the gastric mucosa in parallel with the development of gastric mucosal injury. Therefore, the ineffectiveness of α-tocopherol treatment is considered reasonable in gastric mucosal injury induced by ischemia-reperfusion, however, further studies are needed.

REFERENCES

1. Granger DN, Rutili G, McCord JM. Superoxide radicals in feline intestinal ischemia. Gastroenterology 1981; 81:22-9.
2. Itoh M, Guth PH. Role of oxygen-derived free radicals in hemorrhagic shock-induced gastric lesions in the rat. Gastroenterology 1985; 88:1162-7.
3. Parks DA, Bulkley GB, Granger DN. Ischemic injury in the cat small intestine: role of superoxide radicals. Gastroenterology 1982; 82:9-15.
4. Yoshikawa T, Ueda S, Naito Y, et al. Role of oxygen-derived free radicals in gastric mucosal injury induced by ischemia or ischemia-reperfusion in rats. Free Radic Res Commun 1989; 7:285-91.
5. Parks DA, Bulkley GB, Granger DN. Role of oxygen-derived free radical in digestive tract diseases. Surgery 1983; 94:415-22.
6. Salim AS. Role of oxygen-derived free radicals in mechanism of acute and chronic duodenal ulceration in the rat. Dig Dis Sci 1990; 35:73-9.
7. Keshavarzian A, Morgan G, Sedghi S, Gordon JH, Doria M. Role of reactive oxygen metabolites in experimental colitis. Gut 1990; 31:786-90.
8. Miller MJS, McNeill H, Mullane KM, Caravella SJ, Clark DA. SOD prevents damage and attenuates eicosanoid release in a rabbit model necrotizing enterocolitis. Am J Physiol 1988; 255:G558-65.
9. Niki E. Antioxidants in relation to lipid peroxidation. Chem Phys Lipids 1987; 44:227-53.
10. Yoshikawa T, Miyagawa H, Yoshida N, Sugino S, Kondo M. Increase in lipid peroxidation in rat gastric mucosal lesions induced by water-immersion restraint stress. J Clin Biochem Nutr 1986; 1:271-7.

11. Yoshikawa T, Yoshida N, Miyagawa H, et al. Role of lipid peroxidation in gastric mucosal lesions induced by burn shock in rats. J Clin Biochem Nutr 1987; 2:163-70.

12. Yoshida N, Yoshikawa T, Ando T, et al. Pathogenesis of platelet-activating factor-induced gastric mucosal damage in rats. Scand J Gastroenterol 1980; 24(suppl. 162):210-4.

13. Takemura T, Yoshikawa T, Yoshida N, et al. Role of lipid peroxidation and oxygen radicals in compound 48/80-induced gastric mucosal injury in rats. Scand J Gastroenterol 1989; 24(suppl 162):51-4.

14. Packer L, Landvik S. Vitamin E: introduction to biochemistry and health benefits. In: Diplock AT, Machlin LJ, Packer L, Pryor WA, eds. Vitamin E: biochemistry and health implications. New York: New York Academy of Sciences, 1989; 1-6.

15. Perry MA, Wadhwa S, Parks DA, Pickard W, Granger DN. Role of oxygen radicals in ischemia-induced lesions in the cat stomach. Gastroenterology 1986; 90:362-7.

16. Smith SM, Grisham MB, Manci EA, Granger DN, Kvietys PR. Gastric mucosal injury in the rat. Role of iron and xanthine oxidase. Gastroenterology 1987; 92:950-6.

17. Yagi K. A simple fluorometric assay for lipoperoxide in blood plasma. Biochem Med 1976; 15: 212-6.

18. Ohkawa H, Ohnishi N, Yagi K. Assay for lipid peroxide for animal tissues by thiobarbituric reactions. Anal Biochem 1979; 95:351-8.

19. Abe K, Yuguchi Y, Katsui G. Quantitative determination of tocopherols by high-speed liquid chromatography. J Nutr Sci Vitaminol 1975; 21:183-8.

20. Yoshikawa T, Yasuda M, Ueda S, et al. Role of vitamin E in gastric mucosal injury induced by ischemia-reperfusion. Am J Clin Nutr 1991; in press.

21. Marubayashi S, Dohi K, Ochi K, Kawasaki K. Role of free radicals in ischemic rat liver cell injury. Prevention of damage by α-tocopherol administration. Surgery 1986; 99:184-91.

22. Takenaka M, Tatsukawa Y, Dohi K, Ezaki H, Kawasaki T. Protective effects of α-tocopherol and coenzyme Q_{10} on warm ischemic damages of the rat kidney. Transplantation 1981; 32:137-41.

23. Guarnieri C, Ferrari R, Visioli O, Caldarera CM, Nayler WG. Effect of α-tocopherol on hypoxic-perfused and reoxygenated rabbit heart muscle. J Mol Cell Cardiol 1978; 10:893-906.

63

Vitamin E Disturbances During Alcohol Intoxication

Roger Nordmann and Hélène Rouach

Université René-Descartes, Paris, France

INTRODUCTION

The hypothesis that ethanol-induced liver injury could be linked to the formation of lipo-peroxides and an alteration in the antioxidant balance of the hepatic cell was initially proposed in 1963 by Di Luzio (1). This hypothesis resulted from the finding that the acute ethanol-induced fatty liver could be prevented by the administration of some antioxidants (1). The effects of antioxidants on fatty liver following acute and chronic ethanol administration in rats were extensively studied by Di Luzio et al. (2,3) to support this hypothesis. These authors reported that the administration of the lipid antioxidant N-N'-diphenyl-p-phenylenediamine (DPPD) prevented the ethanol-induced fatty liver as well as the increased lipid peroxide generation assessed by the formation of thiobarbituric acid-reactive substances. Di Luzio and Hartman (3) also reported that the liver triglyceride concentration following an acute oral ethanol load (6 g/kg body weight) was 36% lower in α-tocopherol-pretreated rats than that of the ethanol control group (3). This result indicated that the prevention of the ethanol-induced fatty liver is not limited to a specific class of antioxidants. It was furthermore reported that the administration of various antioxidants before an acute ethanol load is able to prevent ethanol-induced hypertriglyceridemia in rats (3). Antioxidant administration simultaneous with ethanol also inhibited hepatic triglyceride accumulation as well as ethanol-corn oil–induced hypertriglyceridemia (3). Furthermore, Di Luzio and Hartman (3) reported that addition of ethanol to normal rat liver homogenates resulted in an increase in lipid peroxidation that could be prevented by the simultaneous addition of antioxidants, such as DPPD (3). Studying the prevention of the chronic ethanol-induced fatty liver by antioxidant administration, they reported that rats maintained for 3 weeks on an alcohol diet and treated with DPPD had hepatic triglyceride levels that were not different from those in sucrose-fed rats (3). The daily intake of ethanol in these experiments ranged from 9.6 to 15.9 g/kg.

In contrast to these observations, Porta et al. (4,5) and Lieber and De Carli (6) failed to inhibit the fatty liver induced by chronic alcohol ingestion by the oral administration of a variety of antioxidants. These negative findings could be due to the limited absorption of orally administered antioxidants (3). They point out the importance of the dose, pattern, and mode of antioxidant administration, as well as the antioxidant itself, in antioxidant modification of hepatic injury (3). The dose and mode of ethanol administration is obviously also of primary importance. The same holds true for the composition of the diet, especially concerning its fat and vitamin E content.

Despite these controversial results concerning the effects of antioxidants on ethanol-induced fatty liver, it can be considered that the hypothesis initially proposed by Di Luzio more than 25 years ago (1) and linking ethanol-induced liver injury to free-radical mechanism(s) has been supported by many experimental data. The observation by Porta et al. (5) that ethanol-induced hepatic cell ultrastructural changes, especially at the mitochondrial level, are effectively inhibited in antioxidant-treated rats shows that antioxidants are able to protect the hepatocyte against the manifestations of ethanol toxicity different from triglyceride accumulation. The finding that ethanol administration could, under certain experimental conditions, lead to enhanced liver peroxide levels suggested that ethanol can induce an oxidative stress in the liver by increasing the generation of prooxidants and/or by reducing the antioxidant level. The protective action of antioxidants would then probably be due to an inhibition of free-radical–induced chain reactions with the resulting prevention of peroxidative deterioration of structural lipids in membranous organelles (3).

This concept is still a subject of debate. As a matter of fact, many controversial results concerning the onset and the putative importance of hepatic lipid peroxidation during acute or chronic ethanol intoxication have been reported. They were summarized in 1985 by Dianzani (7). However, the recent use by Müller and Sies (8) of noninvasive methods, such as low-level chemiluminescence assay or determination of alkane generation, clearly demonstrated that lipid peroxidation occurs during the metabolism of ethanol by perfused rat livers. Moreover, the use of electron spin resonance (ESR) spectroscopy allowed the detection, in the liver of ethanol-fed rats receiving a spin-trapping agent, of ESR signals that appear to correspond most likely to lipid carbon-centered radicals resulting from peroxidative events (9). It therefore appears that ethanol administration, either acute or chronic, may elicit oxidative stress in the liver, at least under certain experimental conditions. That Slater devoted to the hepatotoxic effects of alcohol a whole chapter of his book *Free Radical Mechanisms in Tissue Injury*, published in 1972 (10), therefore today seems greatly justified.

It should furthermore be pointed out that, in contradiction to the initial suggestion made by Di Luzio and Hartman (3), ethanol administration may induce an oxidative stress in some extrahepatic tissues, and not only in the liver, where ethanol is actively oxidized. We recently summarized the experimental data concerning ethanol-induced lipid peroxidation and oxidative stress in extrahepatic tissues (11).

All these considerations show the importance of studying the relationship between alcohol intake and vitamin E, the most important natural lipid chain-breaking antioxidant. In the present review article we shall consider first the effects of acute and chronic ethanol administration on vitamin E levels in the liver and various extrahepatic tissues in the rat. Some experiments using isolated hepatocytes to clarify the mechanisms responsible for the changes in vitamin E in the liver are considered at the same time. After discussing these experimental data obtained mainly in rats, we consider the vitamin E changes induced by ethanol in the human, most of the results dealing with changes in the blood. The last section is devoted to the effects of vitamin E administration on alcohol toxicity in experimental animals and human patients.

EFFECTS OF ETHANOL ADMINISTRATION ON VITAMIN E LEVELS IN VARIOUS RAT TISSUES

Effects of Ethanol Administration on Vitamin E in Rat Liver

Acute Models of Ethanol Administration

Hannah and Soares (12) studied the effects of giving orally 1.5 ml of a 50% solution of ethanol per 100 g body weight on liver α-tocopherol in rats fed essentially vitamin E-free or vitamin E-supplemented diets. Whereas the vitamin E content of the liver was directly related to the dietary vitamin E level, there was no effect of ethanol treatment on this parameter. This contradicted the ethanol effect on thiobarbituric acid-reactive substances, which were increased even in the high vitamin E group.

Bjørneboe et al. (13) also reported that acute ethanol treatment (4 g/kg body weight injected intraperitoneally, IP) had no effect on hepatic α-tocopherol concentrations in rats.

Using a lower ethanol dose (50 mmol/kg body weight IP), we observed a significant decrease in the α-tocopherol level of liver mitochondria isolated from ethanol-treated rats (14). The susceptibility of these organelles to lipid peroxidation induced by a xanthine-xanthine oxidase system was increased at the same time. We further observed that chronic iron overload obtained by feeding rats during 7 weeks a diet containing 5% elemental (carbonyl) iron also resulted in a decreased liver mitochondrial vitamin E level and an increased mitochondrial susceptibility to lipid peroxidation. An additive effect on these parameters was furthermore observed when ethanol was acutely administered to iron-overloaded rats (14). Such an additive effect may result from an expanded low-molecular-weight chelatable iron pool due partly to the iron overload (15) and partly to the increase in this iron pool induced by acute ethanol administration (16). These results suggest that ethanol-induced iron mobilization could at least partly be responsible for the changes in vitamin E and susceptibility to lipid peroxidation observed at the liver mitochondrial level following acute ethanol administration.

In connection with these in vivo experiments concerning acute ethanol administration, it may be added that primary cultures of rat hepatocytes exposed to ethanol (60 mM) during incubation secrete a significant lower amount of α-tocopherol than control hepatocytes (17). Since the secretion of very low density lipoproteins (VLDL) is inhibited in the same order of magnitude, the authors suggest that the effects of acute ethanol exposure on α-tocopherol secretion may be mediated via an inhibition of VLDL secretion (17). This effect could partially counteract the lowering of the vitamin E level that would result from increased vitamin E cellular demand secondary to the ethanol-induced enhancement of free-radical reactions.

Chronic Models of Ethanol Administration

Hannah and Soares (12) fed rats over 6 weeks different diets containing 36% ethanol based on calorie content and providing various levels of α-tocopherol acetate. They reported that the liver concentration of vitamin E was higher in the ethanol-treated groups receiving low dietary vitamin E diet than in controls; this trend was reversed in the two high vitamin E supplementation diet groups, however. They suggest that supplementation with large amounts of vitamin E to animals receiving ethanol treatment may result in enhanced vitamin E incorporation into lipoproteins because such lipoproteins are being synthesized at an increased rate. Contrary to findings concerning acute ethanol intoxication, chronic ingestion of ethanol did not result in changes in liver lipid peroxidation in these experiments (12).

Bjørneboe et al. (13,18,19) further studied the effects of chronic ethanol administration on α-tocopherol content and distribution in rat liver. Their first results (13,18) showed that, in rats fed ethanol (35% of total energy) for 5–6 weeks, the concentration of α-tocopherol was reduced by 25% compared to pair-fed controls. This reduction was significant in the parenchymal cells but not in nonparenchymal cells. Mitochondrial α-tocopherol was reduced by 55% in the ethanol-treated rats, whereas no significant difference was observed in microsomes, light mitochondria, or cytosol. However, different results were later reported by the same group (19). In this study a reduction by 33% in the α-tocopherol content was reported in the liver light mitochondrial fraction of the ethanol-fed group, whereas no significant difference from the control group was observed in the mitochondrial group. As stated by the authors (19), the difference in the effect of ethanol on the subcellular α-tocopherol content compared to their previous result might be explained by a somewhat different fractionation procedure and/or differences in the size and age of the rats at the start of ethanol feeding. They also reported that long-term administration of ethanol promotes an enrichment (178%) of α-tocopherol in the Golgi apparatus (19), possibly due to reduced secretion of VLDL-associated α-tocopherol. Such a reduced secretion was in fact observed by the same group when studying the effects of chronic ethanol exposure on primary cultures of rat hepatocytes (17).

Kawase et al. (20) recently reconsidered the effects of chronic ethanol feeding on rat liver lipid peroxidation and antioxidant defense systems. Feeding rats a liquid diet containing 36% of energy as ethanol during 22 days, they observed that chronic ethanol feeding elicited a decrease in liver α-tocopherol content in rats given a low-vitamin E diet (containing 2 IU α-tocopherol acetate equivalent per liter).

The same authors (20) reported that chronic ethanol feeding had no significant effect on vitamin E in liver mitochondria. On the contrary, it resulted in a significant increase in the α-tocopheryl quinone (α-TQ) level in microsomes without a significant effect on the microsomal α-tocopherol (α-T) content. The resulting increase in the microsomal ratio α-TQ/α-T was especially apparent in rats fed low-vitamin E diets and ethanol, both factors being synergistic.

Considering the mechanism(s) of the chronic ethanol effects on the α-tocopherol level in the liver, one should agree that we are still in a state of uncertainty. According to Bjørneboe et al. (18), the reduction in hepatic α-tocopherol content is probably not due to decreased absorption of this vitamin, since measurements of α-tocopherol in other tissues and serum revealed no significant changes, except for skeletal muscle, the ethanol-fed rats having significantly higher muscle levels of α-tocopherol than the controls. The same authors reported that the hepatic uptake of α-[^{3}H]tocopherol was not significantly affected in the ethanol-fed rats. Their results indicate some delay in the initial uptake of α-tocopherol to the liver after long-term ethanol feeding without major effects on the total hepatic uptake. The same group, studying the secretion of α-tocopherol from primary cultures of rat hepatocytes, reported that this secretion is reduced in hepatocytes from chronic ethanol-fed rats (17). Taken together, their observations suggest an increased turnover of α-tocopherol during chronic ethanol intake. The reduced content of α-tocopherol they observed in the hepatocytes of ethanol-fed rats could be linked to the increased demand and utilization of α-tocopherol due to enhanced generation of free radicals and lipid peroxidation (17).

Such an enhanced generation of free radicals is also likely responsible for the increase in the α-TQ/α-T ratio reported by Kawase et al. (20), since the generation of α-TQ from α-T, which is irreversible under physiological conditions, is mediated by free-radical mechanisms. It appears important to emphasize that an increased α-TQ/α-T ratio is especially apparent in rats fed a low-vitamin E diet, a group in which hepatic lipid peroxidation is apparent after chronic ethanol feeding, whereas no peroxidation is evident in rats receiving an adequate vitamin E diet (20).

Rikans and Gonzalez (21) studied the influence of chronic ethanol inhalation on the antioxidant protection systems of rat liver and lung. They reported that exposing rats to ethanol vapors for 35 days lowered the hepatic vitamin E level (expressed per g wet weight) by 33% without enhancing lipid peroxidation. The decrease in vitamin E was no longer apparent, however, when hepatic vitamin E concentration was expressed per mg protein.

Using milder conditions of chronic ethanol intoxication, Työppönen and Lindros (22) did not observe any ethanol effect on the liver vitamin E content in rats given ethanol chronically in drinking water (10% wt/vol for 2 weeks). The same lack of effect was reported both with vitamin E-adequate and vitamin E-deficient diets.

To conclude, it appears that the effects of chronic ethanol administration on rat liver vitamin E depend largely on the dose administered and the vitamin E content of the diet. Although no constant changes are apparent, the more consistent results show that ethanol administrated chronically can lower the α-tocopherol level in some liver fractions. Furthermore, the experimental results favor the hypothesis that free-radical mechanisms are involved in these disturbances and suggest that a poor vitamin E status may potentiate the toxic effects of ethanol on the liver (22).

Effects of Ethanol Administration on Vitamin E in Rat Extraheptic Tissues

Numerous studies suggest that ethanol administration elicits in the rat, under certain experimental conditions, oxidative stress in various extrahepatic tissues, such as heart, testis, and cerebellum (11). Only a few of these studies, however, include the effects of ethanol on the vitamin E status in these tissues. They are mainly devoted to the cerebellum, an organ that has special characteristics in terms of vitamin E metabolism. Vatassery et al. (23) indeed reported that cerebellar gray matter altogether contains a low level of α-tocopherol and takes up intravenously labeled α-tocopherol faster than other nervous system tissues. These findings suggest that cerebellum is particularly active in the metabolism or utilization of vitamin E. Meydani et al. (24) also reported that cerebellum and brain stem are more susceptible to challenge by peroxidative agents than other regions and may have higher requirements for α-tocopherol. Furthermore, the ability to produce lipid peroxides is higher in the cerebellum than in other brain areas (25). It is therefore not surprising that cerebellum appears particularly sensitive to α-tocopherol deficiency (26).

All these considerations prompted our group to study whether ethanol administration would induce both changes in the α-tocopherol level and other disturbances suggesting the occurrence of an oxidative stress in the cerebellum. We thus observed that an acute ethanol load (50 mmol/kg body weight IP) elicits a significant decrease in the levels of α-tocopherol and ascorbate, together with an enhanced lipid peroxidation in rat cerebellum (27). The changes in α-tocopherol and ascorbate were interpreted as indicative of a consumption of these antioxidants in quenching free radicals (28). Further studies showed that an acute ethanol load also induced at the cerebellar level a decrease in selenium (29), a trace element involved in antioxidant defense. Furthermore, it elicited an increase in the cerebellar cytosolic low-molecular-weight chelatable iron content (LMW-Fe) (30), which by favoring the biosynthesis of reactive free radicals may contribute to lipid peroxidation (31). Allopurinol, but not desferrioxamine, administration appeared effective in preventing the ethanol-induced cerebellar decrease in α-tocopherol concentration (32). Chronic ethanol administration performed by giving rats during 4 weeks as sole drinking fluid an aqueous solution of ethanol (10% vol/vol) also resulted in a significant decrease in the cerebellar level of α-tocopherol and selenium as well as an increase in cytosolic LMW-Fe (33). These findings suggest that chronic ethanol feeding may induce an oxidative stress in rat cerebellum even when the experimental conditions result only in a relatively mild intoxication.

Whereas the cerebellar α-tocopherol level thus appears disturbed following chronic ethanol feeding to rats, no changes in the vitamin E content were reported in the lung after chronic ethanol inhalation, which did not produce a significant degree of oxidative stress in rat lung (21).

EFFECTS OF ETHANOL INTAKE ON THE HUMAN VITAMIN E STATUS

The first clinical data concerning disturbances in serum levels of vitamin E in alcohol abusers were extensively reviewed by Bonjour in 1981 (34). Most of these data show significant lower than normal mean circulating levels of α-tocopherol in alcoholics, especially in those with liver cirrhosis or chronic alcoholic pancreatitis. Using high-performance liquid chromatography to study plasma α-tocopherol concentrations, Majumdar et al. (35) more recently observed that 30% of chronic alcoholics had reduced plasma vitamin E levels on admission. They showed furthermore that 20% of these patients were still deficient while undergoing conventional detoxification and receiving a routine hospital diet. Tanner et al. (36) compared the vitamin E level in two groups of alcoholics, one with established alcoholic liver disease and the other comprising alcoholic patients without clinical and biological evidence of liver disease. They observed that the serum vitamin E level was lowered to a similar extent in both groups. No correlation between the serum α-tocopherol level and the nutritional status of the patients could be established. Bjørneboe et al. (37,38) reported that about half the alcoholic patients they considered had a serum concentration of α-tocopherol below the lower limit of reference (14 μmol/L). Furthermore, the mean serum concentration of α-tocopherol was reduced by 37% compared to controls. Ward et al. (39) further reported that the mean value of plasma α-tocopherol was significantly reduced in patients with alcoholic myopathy compared to alcoholic patients with normal skeletal muscle histology. The same group, studying the relationship between plasma vitamin E and liver damage, observed that the mean concentration of α-tocopherol was within the reference range (12.5–25 μmol/L) in both the fatty and fibrotic groups of patients, whereas only in the cirrhotic group did the concentration of this vitamin decrease significantly (40). The subdivision in different groups was undertaken in this study by liver histology, each patient having a liver biopsy to determine the degree of damage. This may explain the discrepancy with previous studies, in which the alcoholic abusers were not classified by liver histology. Moscarella et al. (41) reported that the circulating vitamin E levels as well as the ratio of vitamin E to total lipids were depressed to a similar extent in patients with both alcoholic and nonalcoholic liver disease compared to controls. In contrast to the results of Ward et al. (40), Girre et al. (42) recently reported that the vitamin E levels were significantly depressed in both plasma and erythrocytes in chronic alcoholic patients without liver cirrhosis. The plasma vitamin E levels were furthermore still low after 14 days of abstinence and not significantly different from the levels measured before abstinence. Whereas most reported data show a decreased plasma α-tocopherol level in alcoholic patients, at least in those who suffer from severe liver alcoholic disease, a few reports show either no significant difference from healthy controls (43) or even a higher level in drinkers than in control subjects (44).

Different mechanisms likely contribute to the changes in the plasma α-tocopherol level in alcoholic patients. Bjørneboe et al. (38) reported that the vitamin E intake was reduced by 62% during the month preceding their plasma α-tocopherol determination among the alcoholics compared to the controls. This reduction in α-tocopherol intake was particularly marked during hard-drinking periods compared to moderate-drinking and abstinence periods (37). Malabsorption of vitamin E (45,46), partly related to pancreatic insufficiency, could also be involved. A decreased secretion of VLDL by the liver may furthermore be a contributing factor (40), since an important part of α-tocopherol is transported in the plasma by VLDL. Finally, an increased consumption of α-tocopherol linked to the quenching of free radicals produced in excess in some tissues of alcoholic patients could play an important role (36,39–41). This would explain why lowered plasma vitamin E levels are particularly encountered in alcoholics with severe liver disease or myopathy.

If this holds true, the changes in plasma vitamin E would reflect the oxidative stress induced in some tissues, mainly in the liver, by chronic alcohol abuse. Direct measurements of vitamin E in the liver of alcoholics have been reported by few authors. In a limited group of patients, a lowered hepatic vitamin E level expressed per mg total lipids was apparent in alcoholic as well as in nonalcoholic liver diseases (41). A lowered vitamin E-total lipid ratio was also reported in 35 of 39 liver biopsy specimens from alcoholic subjects. However, the decrease in this ratio was unrelated to the severity of the cellular disturbances as judged by histological criteria (47). Besides the changes in α-tocopherol, an increase in conjugated dienes and a decrease in glutathione were reported in this study (47). This was in accordance with the previous findings of Suematsu et al. (48) and Shaw et al. (49), who suggested that lipid peroxidation may be an important mechanism in the pathogenesis of alcoholic liver disease.

EFFECTS OF VITAMIN E ADMINISTRATION ON ETHANOL TOXICITY

As stated before, the finding that α-tocopherol administered to rats before an acute ethanol load partly prevented the enhancement in the liver triglyceride concentration (3) played a fundamental role in the consideration of the role of vitamin E and other antioxidants in alcohol-induced liver injury. We now focus attention on data concerned with the increase in ethanol toxicity in animals fed a vitamin E-deficient diet and the preventive effects of vitamin E supplementation against such toxicity.

Litov et al. (50) considered the effects of various conditions of exposure to ethanol on pentane expiration in the rat. Although pentane is a minor product of autooxidized ω-6-unsaturated fatty acids, the fact that it is biologically inert and can be detected in picomole amounts allowed them to consider that measuring pentane expiration represented a powerful technique to study in vivo lipid peroxidation. They reported that an acute oral ethanol dose (6 g/kg body weight PO) given to rats fed during 14 weeks a vitamin E-deficient diet together with an 18% ethanol solution replacing the drinking water dramatically increased pentane production above basal levels in some of the animals. More recently, Eskelson et al. (51) observed that an increase in ethane exhalation was apparent in rats fed during 28 days the diet proposed and recently modified by Lieber and De Carli (52). Supplementing the basal content of this diet with an additional 1.42 IU α-tocopherol per kcal significantly reduced ethane exhalation in ethanol-fed rats. Eskelson et al. therefore suggest feeding rats a diet enriched with vitamin E to meet the increased α-tocopherol requirement resulting from ethanol administration (51).

A reduction in lipid peroxidation and ethanol toxicity by α-tocopherol supplementation was also reported when studying the effects of ethanol administration on some rat tissues. We already mentioned the data obtained by Kawase et al. (20) indicating that dietary vitamin E is an important determinant of hepatic lipid peroxidation induced by chronic ethanol feeding. Somewhat similar results were obtained by Nadiger et al. (53) when considering the effects of vitamin E deficiency and supplementation on lipid peroxidation and ethanol toxicity in rat brain. Whereas the level of thiobarbituric acid-reactive substances was increased in cerebellum and cerebral cortex following the administration of a single large dose of ethanol in rats maintained on normal vitamin E diets, no such increase was observed in rats fed an α-tocopherol-supplemented diet.

Administration of vitamin E has furthermore been shown by several authors (54–56) to reduce the cardiotoxic effects of ethanol. These effects of administered α-tocopherol were apparent on both the changes in the cardiac phospholipid composition of ethanol-dependent mice (54) and the increase in myocardial lipid peroxidation in rats after chronic alcohol ingestion (56). The beneficial role of vitamin E on these parameters could be related, at least partly, to the recently described modulation exerted by α-tocopherol on the ethanol-induced disturbances of phosphatidylcholine metabolism in the heart (57).

A report concerning the effects on brain fetal dysfunction of vitamin E supplementation given together with ethanol during pregnancy did not show any improvement related to the supplementation (58).

Since, besides this isolated observation concerning the fetal alcohol syndrome, most experimental data show that ethanol toxicity to various tissues may be potentiated by vitamin E-deficient diets and, at least partly, be reduced by α-tocopherol supplementation, it is not surprising that such supplementation has been recommended by many authors for alcoholic patients. We recall that Majumdar et al. (35) showed that the plasma α-tocopherol level of alcoholics fed a routine hospital diet was still often reduced after 6 days of abstinence. They therefore suggest that chronic alcoholics should be treated routinely with vitamin E along with other polyvitamins during detoxification in alcoholic units. Kawase et al. (20) also suggest that, if the data they reported in rats can be extrapolated to humans, alcoholics might have increased vitamin E requirements compared to controls. Although most of these considerations suggest the usefulness of vitamin E supplementation in alcoholics, further well-controlled clinical trials should be performed to assess the possible beneficial effects of such supplementation. Since vitamin E deficiency in the absence of fat malabsorption is able to induce human neurological degeneration (59), special attention should be given to the possible effects of α-tocopherol supplementation on neurological disturbances in alcoholic patients.

CONCLUSION

Since the pioneering experiments undertaken more than 25 years ago showing that the administration of α-tocopherol or other lipid-soluble antioxidants is able to partly prevent fatty liver induction by acute ethanol, the interest in possible links between α-tocopherol disturbances and alcohol toxicity has always been strong.

The numerous studies concerning the changes in hepatic α-tocopherol distribution in chronically ethanol-intoxicated rats showed no constant abnormalities. It appears, however, that the liver α-tocopherol level is often reduced and that this decrease is particularly evident in animals fed a vitamin E-deficient diet. Controversial results have been published about the subcellular localization of disturbances of α-tocopherol concentration in the liver. The mechanisms involved in these disturbances seem multifactorial. An increased demand for α-tocopherol, likely related to a free-radical–mediated oxidative stress, appears to represent an important contributing factor. Experimental data also show that acute as well as chronic ethanol administration to rats induces a decrease in the α-tocopherol level in the cerebellum, an organ characterized by an important α-tocopherol turnover and a high susceptibility to vitamin E deficiency.

Clinical studies have shown that the plasma vitamin E level is lowered in cirrhotic alcoholic patients. Controversial results have been reported concerning similar changes in alcoholics with milder forms of liver disease or even without overt signs of liver dysfunction. Lowering of the α-tocopherol content was also observed in liver biopsies of some alcoholic patients.

Various experimental data show that α-tocopherol supplementation above the amount present in "vitamin E-adequate" diets may reduce the intensity of some of the toxic effects of ethanol on liver, heart, and brain. This would justify supplementation of vitamin E for alcoholic patients, especially for those suffering from liver dysfunction or myopathy. Further well-controlled clinical trials, however, are needed to clearly demonstrate that such a supplementation is beneficial to these patients.

REFERENCES

1. Di Luzio NR. Prevention of the acute ethanol-induced fatty liver by antioxidants. Physiologist 1963; 6:169-73.
2. Di Luzio NR, Costales F. Inhibition of the ethanol and carbon tetrachloride induced fatty liver by antioxidants. Exp Mol Pathol 1964; 4:141-54.

3. Di Luzio NR, Hartman AD. Role of lipid peroxidation in the pathogenesis of ethanol-induced fatty liver. Fed Proc 1967; 26:1436-42.

4. Porta E, Hartroft W, de la Iglesia F. Hepatic changes associated with chronic alcoholism in rats. Lab Invest 1965; 14:1437-55.

5. Porta E, Hartroft W, de la Iglesia F. Structural and ultrastructural hepatic lesions associated with acute and chronic alcoholism in man and experimental animals. In: Maickel R, ed. Biochemical factors in alcoholism. New York: Pergamon Press, 1967; 201-33.

6. Lieber CS, De Carli LM. Study of agents for the prevention of the fatty liver produced by prolonged alcohol intake. Gastroenterology 1966; 50:316-22.

7. Dianzani MU. Lipid peroxidation in ethanol poisoning: a critical reconsideration. Alcohol Alcohol 1985; 20:161-73.

8. Müller A, Sies H. Alcohol, aldehydes and lipid peroxidation: current notions. Alcohol Alcohol 1987; Suppl 1:67-74.

9. Reinke LA, Lai EK, Du Bose CM, McCay PB. Reactive free radical generation in vivo in heart and liver of ethanol-fed rats: correlation with radical formation in vitro. Proc Natl Acad Sci U S A 1987; 84:9223-7.

10. Slater TF. Hepatotoxic effects of alcohol. In: Slater TF, ed. Free radical mechanisms in tissue injury. London: Pion, 1972; 171-97.

11. Nordmann R, Ribière C, Rouach H. Ethanol-induced lipid peroxidation and oxidative stress in extrahepatic tissues. Alcohol Alcohol 1990; 25:231-7.

12. Hannah JS, Soares JH. The effects of vitamin E on the ethanol metabolizing liver in the rat. Nutr Rep Int 1979; 19:733-44.

13. Bjørneboe GEA, Bjørneboe A, Hagen BF, Mørland J, Drevon CA. Reduced hepatic α-tocopherol content after long-term administration of ethanol to rats. Biochim Biophys Acta 1987; 918:236-41.

14. Rouach H, Park MK, Orfanelli MT, et al. Effects of ethanol on hepatic and cerebellar lipid peroxidation and endogenous antioxidants in naive and chronic iron-overloaded rats. In: Nordmann R, Ribière C, Rouach H, eds. Alcohol toxicity and free radical mechanisms. Advances in the biosciences, Vol. 71. Oxford: Pergamon Press, 1988; 49-54.

15. Bacon BR, Tavill AS, Recknagel RO. Lipid peroxidation and iron overload. In: Rice-Evans C, ed. Free radicals, oxidant stress and drug action. London: Richelieu Press, 1987; 259-75.

16. Rouach H, Houzé P, Park MK, Orfanelli MT, Nordmann R. Role of iron in the hepatic and cerebellar oxidative stress after acute ethanol administration to rats. In: Kuriyama K, Takada A, Ishii H, eds. Biomedical and social aspects of alcohol and alcoholism. Amsterdam: Excerpta Medica, 1988; 675-80.

17. Bjórneboe A, Bjórneboe GEA, Hagen BF, Drevon CA. Acute and chronic effects of ethanol on secretion of alpha-tocopherol from primary cultures of rat hepatocytes. Biochim Biophys Acta 1987; 922:357-63.

18. Bjørneboe A, Bjørneboe GEA, Drevon CA. Kinetics of α-tocopherol in control and alcohol-fed rats. In: Nordmann R, Ribière C, Rouach H, eds. Alcohol toxicity and free radical mechanisms. Advances in the biosciences, Vol. 71. Oxford: Pergamon Press, 1988; 87-91.

19. Hagen BF, Bjørneboe A, Bjørneboe GEA, Drevon CA. Effect of chronic ethanol consumption on the content of α-tocopherol in subcellular fractions of rat liver. Alcohol Clin Exp Res 1989; 13: 246-51.

20. Kawase T Kato S, Lieber CS. Lipid peroxidation and antioxidant defense systems in rat liver after chronic ethanol feeding. Hepatology 1989; 10:815-21.

21. Rikans LE, Gonzalez LP. Antioxidant protection systems of rat lung after chronic ethanol inhalation. Alcohol Clin Exp Res 1990; 14:872-7.

22. Työppönen JT, Lindros KO, Combined vitamin E deficiency and ethanol pretreatment: liver glutathione and enzyme changes. Int J Vitam Nutr Res 1986; 56:241-5.

23. Vatassery GT, Angerhofer CK, Knox CA, Deshmukh DS. Concentrations of vitamin E in various neuroanatomical regions and subcellular fractions, and the uptake of vitamin E by specific areas, of rat brain. Biochim Biophys Acta 1984; 792:118-22.

24. Meydani M, Macauley JB, Blumberg JB. Effect of dietary vitamin E and selenium on susceptibility of brain regions to lipid peroxidation. Lipids 1988; 23:405-9.

25. Zaleska M, Floyd RA. Regional lipid peroxidation in rat brain in vitro: possible role of endogenous iron. Neurochem Res 1985; 10:397-410.

26. Le Bel CP, Odunze IN, Adams JD, Bondy SC. Perturbations in cerebral oxygen radical formation and membrane order following vitamin E deficiency. Biochem Biophys Res Commun 1989; 163: 860-6.

27. Rouach H, Ribière C, Park MK, Saffar C, Nordmann R. Lipid peroxidation and brain mitochondrial damage induced by ethanol. Bioelectrochem Bioenerg 1987; 18:211-7.

28. Rouach H, Park MK, Orfanelli MT, Janvier B, Nordmann R. Ethanol-induced oxidative stress in the rat cerebellum. Alcohol Alcohol 1987; Suppl 1:207-11.

29. Houzé P, Rouach H, Gentil M, Orfanelli MT, Nordmann R. Effect of allopurinol on the hepatic and cerebellar iron, selenium, zinc and copper status following acute ethanol administration to rats. Free Radic Res Commun 1991; 12-13:663-8.

30. Rouach H, Houzé P, Orfanelli MT, Gentil M, Bourdon R, Nordmann R. Effect of acute ethanol administration on the subcellular distribution of iron in rat liver and cerebellum. Biochem Pharmacol 1990; 39:1095-100.

31. Ferrali M, Ciccoli L, Comporti M. Allyl alcohol-induced hemolysis and its relation to iron release and lipid peroxidation. Biochem Pharmacol 1989; 38:1819-25.

32. Park MK, Rouach H, Orfanelli MT, Janvier B, Nordmann R. Influence of allopurinol and desferrioxamine on the ethanol-induced oxidative stress in rat cerebellum. In: Nordmann R, Ribière C, Rouach H, eds. Alcohol toxicity and free radical mechanisms. Advances in the biosciences, Vol. 71. Oxford: Pergamon Press, 1988; 135-40.

33. Rouach H, Houzé P, Orfanelli MT, Gentil M, Nordmann R. Effects of chronic ethanol intake on some anti- and pro-oxidants in rat cerebellum. Alcohol Alcohol 1991; 26:257.

34. Bonjour JP. Vitamins and alcoholism. X. Vitamin D. XI. Vitamin E. XII. Vitamin K. Int J Vitam Nutr Res 1981; 51:307-18.

35. Majumdar SK, Shaw GK, Thomson AD. Plasma vitamin E status in chronic alcoholic patients. Drug Alcohol Depend 1983; 12:269-72.

36. Tanner AR, Bantock I, Hinks L, Lloyd B, Turner NR, Wright R. Depressed selenium and vitamin E levels in an alcoholic population: possible relationship to hepatic injury through increased lipid peroxidation. Dig Dis Sci 1986; 31:1307-12.

37. Bjørneboe GEA, Johnsen J, Bjørneboe A, Bache-Wiig JE, Mørland J, Drevon CA. Diminished serum concentration of vitamin E in alcoholics. Ann Nutr Metab 1988; 32:56-61.

38. Bjørneboe GEA, Johnsen J, Bjørneboe A, Mørland J, Drevon CA. Effect of heavy alcohol consumption on serum concentrations of fat-soluble vitamins and selenium. Alcohol Alcohol 1987; Suppl 1:533-7.

39. Ward RJ, Duane PD, Peters TJ. Nutritional, selenium and α-tocopherol status of alcohol abusers with and without chronic skeletal muscle myopathy. In: Nordmann R, Ribière C, Rouach H, eds. Alcohol toxicity and free radical mechanisms. Advances in the biosciences, Vol. 71. Oxford: Pergamon Press, 1988; 93-6.

40. Ward RJ, Jutla J, Peters TJ. Antioxidant status in alcoholic liver disease. In: Poli G, Cheeseman KH, Dianzani MU, Slater TF, eds. Free radicals in the pathogenesis of liver injury. Advances in the biosciences, Vol. 76. Oxford: Pergamon Press, 1989; 343-51.

41. Moscarella S, Mazzanti R, Gentilini P. The role of lipid peroxidation and antioxidants in alcoholic liver disease in man. In: Poli G, Cheeseman KH, Dianzani MU, Slater TF, eds. Free radicals in the pathogenesis of liver injury. Advances in the biosciences, Vol. 76. Oxford: Pergamon Press, 1989; 275-80.

42. Girre C, Hispard E, Therond P, Guedj S, Bourdon R, Dally S. Effect of abstinence from alcohol on the depression of glutathione peroxidase activity and selenium and vitamin E levels in chronic alcoholic patients. Alcohol Clin Exp Res 1990; 14:909-12.

43. Mézes M, Par A, Németh P, Javor T. Studies of the blood lipid peroxide status and vitamin E levels in patients with chronic active hepatitis and alcoholic liver disease. Int J Clin Pharmacol Res 1986; 6:336-8.

44. Herbeth B, Didelot-Barthelemy L, Lemoine A, Le Devehat C. Plasma fat-soluble vitamins and alcohol consumption. Am J Clin Nutr 1988; 47:343-4.

45. Leevy CM, Baker H. Nutritional deficiencies in liver disease. Med Clin North Am 1970; 54:467-77.
46. Kalvaria I, Labadarios D, Shephard GS, Visser L, Marks IN. Biochemical vitamin E deficiency in chronic pancreatitis. Internat J Pancreatol 1986; 1:119-28.
47. Situnayake RD, Crump BJ, Thurnham DI, Davies JA, Gearty J, Davies M. Lipid peroxidation and hepatic antioxidants in alcoholic liver disease. Gut 1990; 31:1311-7.
48. Suematsu T, Matsumura T, Sato N, et al. Lipid peroxidation in alcoholic liver disease in humans. Alcohol Clin Exp Res 1981; 5:427-30.
49. Shaw S, Rubin KP, Lieber CS. Depressed hepatic glutathione and increased diene conjugates in alcoholic liver disease. Evidence of lipid peroxidation. Dig Dis Sci 1983; 28:585-9.
50. Litov RE, Gee DL, Downey JE, Tappel AL. The role of lipid peroxidation during chronic and acute exposure to ethanol as determined by pentane expiration in the rat. Lipids 1981; 16:52-63.
51. Eskelson CD, Odeleye OE, Watson RR, Earnest D, Chvapil M. Is the Lieber-De Carli liquid ethanol diet adequate in vitamin E? Alcohol Alcohol 1990; 25:433-4.
52. Lieber CS, De Carli L. Liquid diet technique of ethanol administration: 1989 update. Alcohol Alcohol 1989; 24:197-211.
53. Nadiger HA, Marcus SR, Chandrakala MV. Lipid peroxidation and ethanol toxicity in rat brain: effect of vitamin E deficiency and supplementation. Med Sci Res 1988; 16:1273-4.
54. Abu Murad C, Littleton JM. Cardiac phospholipid composition during continuous administration of ethanol to mice: effect of vitamin E. Br J Pharmacol 1978; 63.
55. Redetzki JE, Griswold KE, Nopajaroonsri C, Redetzki HM. Amelioration of cardiotoxic effects of alcohol by vitamin E. J Toxicol Clin Toxicol 1983; 20:319-31.
56. Edes I, Toszegi A, Csanady M, Bozoky B. Myocardial lipid peroxidation in rats after chronic alcohol ingestion and the effects of different antioxidants. Cardiovasc Res 1986; 20:542-8.
57. Choy PC, O K, Man RYK, Chan AC. Phosphatidylcholine metabolism in isolated rat heart: modulation by ethanol and vitamin E. Biochim Biophys Acta 1989; 1005:225-32.
58. Tanaka H, Iwasaki S, Nakazawa K, Inomata K. Fetal alcohol syndrome in rats: conditions for improvement of ethanol effects on fetal cerebral development with supplementary agents. Biol Neonate 1988; 54:320-9.
59. Sokol RL. Vitamin E and neurologic function in man. Free Radic Biol Med 1989; 6:189-207.

64

Interaction Between Vitamin E and Selenium Under Clinical Conditions

Bodo Kuklinski and Eckhard Koepcke

Klinikum Südstadt, Rostock, Germany

INTRODUCTION

During the recent decades a considerable increase in knowledge has been achieved in biochemical basic research that requires its application to prophylactic and curative medicine. Acute or chronic oxidative stress (latent and cumulative) has an essential influence on numerous pathogenetic processes on the molecular and biological levels. In the future, all medical fields can be expected to be significantly influenced by taking them into consideration. Investigations into the complex correlations between the individual nutrition, and the environment have led to conclusions that human beings in highlyindustrialized countries with technologically dense areas are increasingly exposed to oxidative stress. The imbalance between oxidants and antioxidants has increased (Table 1).

On the one hand, the nutritive supply of antioxidants is marginally low or insufficient, but the load due to direct or indirect substances generating radicals or reactive O_2^- species has increased. The doctor engaged in caring for patients can never cover all antioxidative or oxidative parameters even when making full use of the available laboratory capacity, which would also be uneconomical. In decision making he or she must start from the following facts:

It is an illusion to make the attempt to eliminate all oxidative stress factors in an industrialized society.

Oxidative stress factors will continue having an influence on human beings in a chronic or acute form.

Any therapeutic relevance may result in an enhancement of the antioxidative potential.

Laboratory and chemical parameters must be applied, which on the basis of a broad population allow screening to be made of oxidative stress in vivo.

As antioxidants such as vitamin E, vitamin C, β-carotenoids, ubiquinone Q_{10}, and the selenoenzymes glutathione peroxidase (GSH-Px) and PLH-GPx are in close correlation the doctor may only rely on some parameters. Vitamin E and selenium cannot be regarded separately because their interrelations through lipolytic and hydrophilic compartments are very close.

There is no ideal marker of oxidative stress in the human at present. It is feasible to determine the concentrations of H_2O_2 excretion (urine and expiratory air), lipofuscin in blood, malondialdehyde, chemiluminescence, DNA adducts, daily urine excretion of DNA base fragments, and others.

TABLE 1 Imbalance Between Nutritive Antioxidants, Vitamin
E, Vitamin C, β-Carotenoids, Selenium, and Oxidants

Reactive and O_2^- species
  Carcinogenic (chemical, physical, biological)
  Chronic inflammation
  Chronic hypoxia
  Increased O_2 metabolism and O_2 supply
  Medications
Increased demand for antioxidants
  Incorporation of heavy metals (Hg, Pb, Cd)
  Nitrate load,, increased endogenous formation of nitrite
  PUFA-rich nutrition (linoleic, linolenic, eicosapentaenic, and
    docosaheaxaenic acids)
Exogenous radical-generating sources
  Smoking
  Ozone and O_2 supply
  Alcohol consumption
  UV light

The skin test with a 7.5% H_2O_2 solution is suitable as a simple screening method. Lacking a skin reaction may indicate a deficient capacity of antioxidants or a disturbed protein SH/SS ratio (1). For the purpose of ambulatory and stationary treatment we used the determination of malondialdehyde (MDA). In healthy individuals about one-third of MDA is from thrombocyte aggregation, the rest from peroxidation of polyunsaturated fatty acids with more than three double bindings. Any increase in MDA in vivo is only possible after exhaustion of the pool of vitamin E and other antioxidants.

Transverse and follow-up studies revealed how different the status of antioxidants may be in closely adjacent regions and, above all, how rapidly and measurably it may react. In this chapter clinical experience is discussed.

METHODS OF DETERMINATION

The determination of laboratory and chemical methods was always made after 14 h starvation in fasting healthy individuals or patients between 8 and 9 a.m. from uncongested veins, after the first 5 ml blood was rejected after puncturing the cubital vein.

Thiobarbituric acid-reactive substance (MDA) was determined according to the method of Yagi (2) or Uchiyama and Michara (3) modified according to Andreeva (4). Determining vitamin E in serum was according to Peters et al. (5), and selenium was determined by means of the AAS-3 method (Carl Zeiss Jena GmbH, Germany) according to Peters and Dittrick (6).

STATUS OF ANTIOXIDANTS IN HEALTHY INDIVIDUALS

In healthy individuals ($n = 518$), age- and sex-dependent determinations of selnium, vitamin E, and malondialdehyde concentrations resulted in marginally low levels of vitamin E in men between 30 and 50, with a shift of 10 years in women between 40 and 60, the most active period of life. It cannot be ruled out that the higher values of vitamin E found in elderly individuals are a result of selection (premature death due to chronically latent deficiency of vitamin E). Malondialdehyde, which showed lower values in elderly people, had exactly the same behavior; however it was inverse correlated with vitamin E. Concentrations of selenium showed a falling tendency with growing age (Fig. 1) (7).

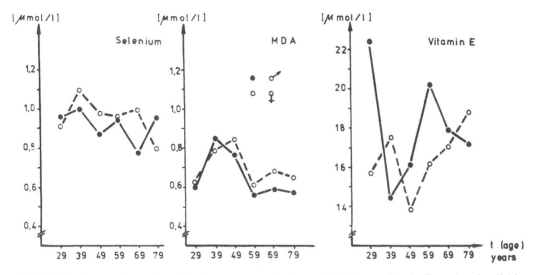

FIGURE 1   Age- and sex-dependent serum concentrations of selenium, vitamin E, and malondialdehyde (MDA) in healthy individuals (each age and sex group $n = 40$).

How different in regional respects serum selenium concentrations may be in an area of 450 km can be shown by the results of investigations made in (East) Germany with regard to age and sex. Healthy test subjects selected from the area along the Baltic Sea coast (Rostock) revealed excessively low values of selenium. In men the concentrations were around 0.59 μmol/L, in women it was below 0.7 μmol/L. This coastal area showed a top position in (East) Germany as far as the incidence of myocardial infarction and mortality was concerned.

TABLE 2   Concentrations of MDA in Healthy Individuals

Healthy individuals	n	MDA (μmol/L)
Men		
20–29 years	28	8.7 ± 2.0
30–39 years	31	8.6 ± 3.3
40–49 years	25	8.8 ± 4.6
50–59 years	34	7.6 ± 3.4
Women without contraceptives		
20–29 years	42	7.7 ± 3.2
30–39 years	35	7.3 ± 1.9
40–49 years	38	7.0 ± 2.4
50–59 years	40	8.4 ± 2.6
Women without contraceptives		
20–39 years	50	9.8 ± 3.7[a]
Geriatric patients	111	13.2 ± 4.5[a]
Pregnant women before delivery	46	7.5 ± 4.2
1 day after solarium	12	24.6 ± 7.9[a]

[a]Significance $p < 0.05$ by Mann-Whitney U test.

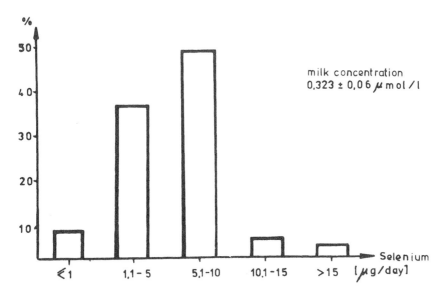

FIGURE 2 Distribution of selenium supply over mother's milk in babies from day 4 postpartum as a percentage (*n* = 123).

Moderate alcohol consumption correlated inversely with selenium concentrations (0.49 μmol/L) in men, as well as such noxious substances as car exhaust gases, asbestos, and formaldehyde (1).

The results of MDA transverse investigations in healthy individuals are presented in Table 2. Examinations of pregnant women and mother's milk, as well as daily nutritive supply with selenium, showed an insufficient uptake of selenium in breast-fed infants. Only 10% of the babies book up more than 10 μg selenium per day. Because in the first weeks of lactation colostrum and mother's milk are particularly abundant in vitamin E and selenium (Fig. 2), a deficiency of selenium in that period of life, with a high cell replication rate, may be regarded as a serious indication for a deficiency of antioxidants.

We do not know whether the MDA values in healthy individuals are normally because a moderate supplementation with antioxidants decreased MDA in serum more than 30% (Table 3).

TABLE 3 Behavior of Malondialdehyde Concentrations During Antioxidative Medication in Healthy Individuals

		Malondialdehyde (μmol/L)		
Therapy groups	*n*	Before treatment	After 10 days	*p* (Wilcoxon)
Control group	10	8.6 ± 2.6	9.1 ± 3.1	NS
Vitamin E, 300 mg/day	12	9.4 ± 3.6	6.2 ± 2.4	<0.05
Sodium selenite, 300 μg/day	17	9.0 ± 3.7	5.9 ± 2.8	<0.05
Ginkgo biloba	21	8.8 ± 2.7	5.3 ± 2.5	<0.05
Ubiquinone (Q_{10}), 100 mg/day	11	8.4 ± 4.2	4.8 ± 3.4	<0.05
β-Carotenoids, 15 mg/day	12	9.2 ± 3.6	8.8 ± 4.2	NS

In a follow-up study made in healthy blood donors ($n = 360$) patients with high and very low malondialdehyde values were noted. We believe both to be risk groups, but at present no significant statements can be made about dignity. From the group of test subjects with MDA < 6 μmol/L ($n = 24$) a carcinoma could be diagnosed in three female patients after 1 year. Four men with values > 12 μmol/L died (myocardial infarction, acute pancreatitis, and alcohol intoxication).

CHANGES IN ANTIOXIDANT STATUS UNDER PREMORBID AND MORBID CONDITIONS

Transverse examinations of MDA concentrations yielded the values in Table 4. All MDA increases indicate that perpetuation of the arachidonic acid cascade owing to lipid peroxidation must have taken place, with numerous releasing noxious substances taking part in it. Among other things, this could be inflammation with activation of leukocytes and macrophages, hypoxia, or postaggression metabolism. Transverse examinations do not reflect the true extent of oxidative stress. Follow-up investigations point to serious deviations that, at least medically, require adjuvant therapy.

Oxidative Stress in the Treatment of Hyper- and Dyslipoproteinemia

Hyperlipidemia

We pointed out that any dietetic change in antilipemic, antiatherogenic, cholesterol-poor nutrition will lead to a serious deficiency of antioxidants (7). An analysis carried out in 1000 kinds of food and foodstuffs proved that only cholesterol-rich foodstuffs, such as meat, meat products, entrails, and egg yol, contained selenium. The consumption of these foodstuffs must be strongly limited in hypercholesterolemia and hyperlipoproteinemia. Cereals, vegetables, and fruits from Central European countries showed traces of selenium (7). Also, industrially manufactured baby food was selenium poor. It was only after adding meat or liver that its amount increased to minimal amounts (Table 5).

By changing the diet an excessive drop in serum selenium and GSH-Px activities could be proved. The prescribed plant oils, which are known to be rich in vitamin E, also could not prevent vitamin E concentrations from decreasing. We assume that the drop in vitamin E was a consequence of the selenium deficiency with ensuing decreased vitamin E radical-regenerating capacity.

The changed relations of quotients Se + vitamin E to MDA are shown in various patients with a risk of atherosclerosis or even manifest disease (Table 6).

The alterations of parameters caused only by changing the diet, with sport, and later by therapy is represented in Figure 3. In addition to this picture we also observed an apparent cardiomyopathy (7) that was reversible with vitamin E and selenium supplementation and the appearance of hyperthyrosis in one patient. Since selenium is also essential for 5'-dejodase type I (significant in converting T4 to T3) a connection cannot be excluded by eventual stimulation of thyrotropin releasing hormone (TRH) or thyroid stimulating hormone (TSH).

Figure 4 illustrates additional factors contributing to accelerated antioxidant deficiency:

Cholesterol-poor diet (decrease in vitamin E and selenium)
Sport activity (increased demand for vitamin E and selenium)
Ultraviolet light strain by solarium or sunshine (increased demand for vitamin E and selenium)
Cholestyramine therapy (daily irreversible loss of selenium amounting to about 30 μg via the enterohepatic circulation).

TABLE 4 Concentrations of Malondialdehyde in Premorbid and Morbid Patients Compared to Healthy Individuals

Investigated groups	n	MDA (μmol/L)	p (Mann-Whitney)
Male smokers	64	8.4 ± 4.6	NS
Consumption of alcohol			
Moderate (<30 g/day)	33	10.7 ± 3.8	<0.05
Alcoholics (⩾60 g/day)	23	15.6 ± 6.8	<0.05
Delirium tremens	16	19.4 ± 7.8	<0.01
Alcoholic-toxic hepatitis (second day)	15	30.6 ± 13.4	<0.01
Diabetes mellitus			
Insulin independent	22	10.5 ± 4.2	<0.05
Insulin dependent	25	7.5 ± 2.8	NS
Hyper-, dyslipoproteinemias			
With antilipemic diet (week 12)	45	11.4 ± 3.7	<0.05
With HMG-CoA-reductase inhibitors and diet second week)	7	10.8 ± 3.8	NS
Tangier disease (cholesterol-poor diet only)	1	89.9	
Arteriosclerotic disease			
Acute myocardial infarction (first day)			
Without diabetes mellitus	22	7.6 ± 2.9	NS
With diabetes mellitus	18	12.4 ± 4.3	<0.05
Apoplectic stroke			
Female patients	24	13.0 ± 3.7	<0.05
Male patients	16	9.4 ± 3.1	NS
Acute pancreatitis (first day)	41		
Mild course	20	12.5 ± 4.3	<0.05
Necrotizing course	21	20.7 ± 11.6	<0.01
Rheumatoid arthritis			
Moderate activity	28	8.6 ± 3.7	NS
High activity	32	12.7 ± 3.5	<0.05
Postoperation (fourth day)			
Uncomplicated	22	8.3 ± 2.7	NS
Multiple trauma	15	17.6 ± 8.4	<0.05
Sepsis			
Compensated	4	14.3 ± 6.6	<0.05
Decompensated	3	6.2 ± 1.5	NS
Attacks of bronchial asthma (first day)	18	63.4 ± 22.8	<0.01
Coma			
Coma diabeticum			
Ketoacidotic	7	17.3 ± 8.7	<0.05
Hyperosmolaric	8	12.7 ± 5.8	<0.05
Coma hepaticum	11	78.8 ± 10.3	<0.01
Left heart insufficiency			
Latent	23	12.1 ± 4.5	<0.05
Decompensated	12	25.3 ± 11.9	<0.01
Patients with positive Hg mobilization test resulting from dental amalgam fillings (ratio of Se to Hg, Cd, Pb in serum < 0.05)	7	13.4 ± 3.3	<0.05

TABLE 5 Serum Concentrations of Selenium, Vitamin E, and Glutathione Peroxidase GSHPx) During a 2 Year Antilipemic Diet Regimen

	Dietary Period		
Group[a]	Before diet	After 1 year	After 2 years
Control group (n = 16)			
Selenium, μmol/L	0.92 ± 0.31	0.36 ± 0.04[b]	0.38 ± 0.01[b]
Vitamin E, μmol/L	28.5 ± 12.2	8.2 ± 17.6[b]	14.7 ± 2.9[b]
GSHPx, U/mmol Hb	4.3 ± 2.2	2.9 ± 1.7	1.2 ± 0.6[b]
Sunflower oil group (n = 37)			
Selenium	0.86 ± 0.29	0.38 ± 0.07[b]	0.42 ± 0.11[b]
Vitamin E	27.4 ± 18.4	8.7 ± 16.5[b]	16.1 ± 4.8[b]
GSHPx	4.7 ± 3.8	3.3 ± 1.3	1.8 ± 0.7[b]
Linseed oil group (n = 33)			
Selenium	0.91 ± 0.34	0.41 ± 0.03[b]	0.42 ± 0.10[b]
Vitamin E	26.5 ± 12.2	7.3 ± 15.1[b]	13.8 ± 7.6[b]
GSHPx	4.8 ± 3.4	3.2 ± 1.5	1.8 ± 0.8[b]

[a]Control group, cholesterol-poor diet only; sunflower oil group, diet and 40 ml sunflower oil per day; linseed oil group, diet and 40 ml linseed oil per day.
[b]Significance $p < 0.001$ by Mann-Whitney U test compared to values before diet.

The addition of prescription of HMG-CoA inhibitors could lead to a significant enhancement of oxidative stress because we proved increases in MDA concentration after its intake. Thus, the occurrence of cataracts and myopathies could be explained if inhibitors of cholesterol synthesis are prescribed, yielding insufficient antioxidants.

From this the conclusion can be drawn that any antilipemic and antisclerotic therapy must take into account the metabolic background. On the strength of the most recent results obtained by Esterbauer et al. (8), it must also be postulated that the decrease in plasma lipids without antioxidant protection will increase their atherogenicity, thus exchanging the lipid by an enhanced atherogenous potency of blood lipids. Vitamin E prevents the development of atherogenic oxidized low-density lipoprotein (LDL) and, in addition, inhibits vascular

TABLE 6 Relation of Vitamin E and Selenium to Malondialdehyde in Antilipemic and Antiatherogenic Cholesterol-Poor Nutrition

		Ratio vitamin E × selenium/malondialdehyde		
Group	n	Before diet	After 1 month	After 1 year
Hyperlipemias	42	88 ± 19	66 ± 15	18 ± 17[a]
Hyperlipemias + arteriosclerosis + smoking	38	49 ± 11	37 ± 22	4 ± 4[a]

[a]Significance $p < 0.001$ by Wilcoxon test compared to values before diet.

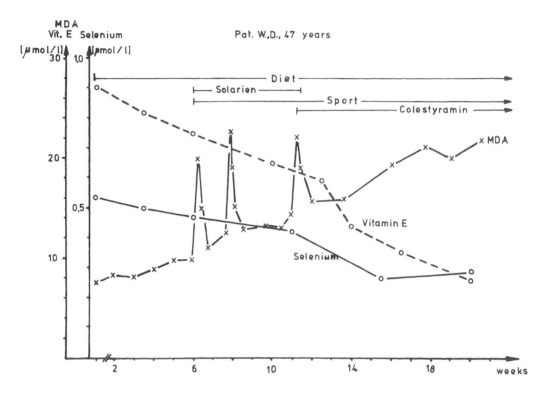

FIGURE 3 Progressive oxidative stress under antilipemic therapy with diet, sport activity, and cholestyramine in a 47-year-old male patient with hypercholesterolemia.

smooth muscle cell proliferation and protein kinase C activity induced by platelet-derived growth factor (PDGF) or endothelin (9).

Lipid hydroperoxide levels, but not those of thiobarbituric acid-reactive substances (TBARs) were correlated with LDL cholesterol and triglyceride levels. The administration of vitamin E at a daily dosage of 300 mg for 1 month depressed elevated hydroperoxides by 33% and TBARs by 44% on average (10). High levels of serum lipid peroxides were associated with low serum high-density lipoprotein (HDL$_2$) cholesterol concentrations and low HDL$_2$/HDL$_3$ cholesterol and apolipoprotein AI/AII ratios. Low levels of LDL cholesterol and apolipoprotein B and low apolipoprotein B/AI ratios were typical of subjects with high serum glutathione peroxidase activity (11). In addition it was shown that selenium supplementation increased GSH-Px activity and the HDL cholesterol-total cholesterol ratio in healthy subjects (12). It is well established that the effects of oxidized LDL are inhibited by high-density lipoprotein.

Tangier disease

Previous surveys deal only with the diagnostics of the lipoprotein and apolipoprotein situation. A causal therapy was not possible. The high risk of carcinoma involved in these patients with very low total cholesterol, HDL cholesterol, and apolipoprotein A values has been repeatedly referred to. This can be prevented by taking the antioxidant situation into consideration. Two factors might be essential for reducing antioxidants in patients:

High-density lipoproteins are known to have an antioxidative effect. Their deficiency leads to a further loss of antioxidative capacity.

Due to the very low LDL cholesterol there is no carrier for vitamin E in the blood.

Figure 4 illustrates the efficiency of a vitamin E and selenium substitution as a protective measure.

In 2 week infusions of 2 g HDL, each carried out in patients with hyperlipoproteinemia, dyslipoproteinemia, and Tangier disease, we could identify this antioxidative HDL effect by increasing vitamin E and selenium concentrations and decreasing malondialdehyde (Fig. 5).

This was accompanied by extraordinarily good and complex antiarteriosclerotic and anti-arrhythmic effects (e.g., lowering of blood pressure, thrombocyte aggregation, erythrocyte aggregation, decrease in cholesterol and triglycerides, and improved cardiac function and contractility) (13–15).

Because antioxidant supplementation leads to an increase in HDL, this therapy aimed at improving the endogenous reverse cholesterol transport in hypercholesterolemia would be preferable to a physiological cholesterol decrease. There are signs indicating that antioxidants may inhibit accelerated hepatic HDL apo AI and apo AII catabolism, possibly by inhibiting the premature loss of *N*-acetylneuraminic acid. Asial HDL is known to be preferably degraded and eliminated through the liver. In addition, the model of Tangier disease indicates that low cholesterol values may also enhance oxidative stress.

Oxidative Stress in Other Clinical Diseases

Acute Pancreatitis

Acute pancreatitis may be considered a genuine free-radical disease. Initially, a generally increased status of peroxidation can be expected to suddenly activate PLA_2 lipase activities and the cascade of arachidonic acid by such simple causes as ethanol or fried meals. Table 7 shows that parameters of the pancreatic juice differed significantly from those of the blood and showed higher activities of leukocytes and macrophages, as well as greater oxidative stress.

MDA increases above 20 μmol/L, as well as the extent of hypocalcemia, are a signal for a serious prognosis (1). In a randomized clinical study in 18 patients with acute necrotizing

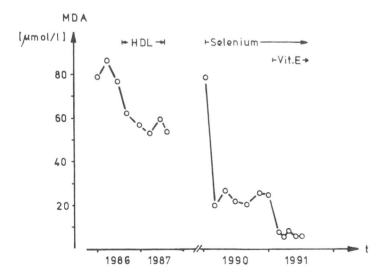

FIGURE 4 Concentrations of malondialdehyde (MDA) in a 26-year-old male patient with Tangier disease with infusions of HDL (1986–1987) and supplementation of sodium selenite and vitamin E.

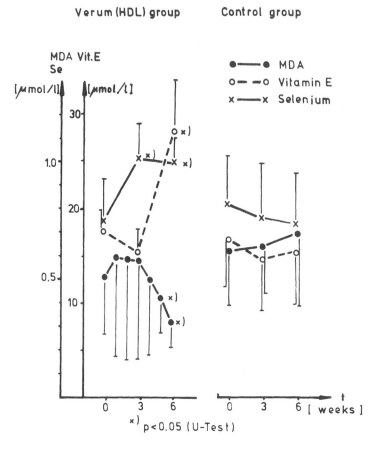

FIGURE 5   Increase in vitamin E and selenium concentration with 2 week infusions of 2 g high-density lipoproteins with a consecutive decrease in malondialdehyde in serum. MDA was estimated according to Yagi (Chem Phys Lipids 1987; 45:327-51).

pancreatitis (Fig. 6), the lethal outcome in the antioxidative therapy group could be lowered to 0 ($n = 8$) solely by using sodium selenite (Na_2SeO_3), whereas it amounted to 89% in the control group (16).

Markedly milder forms of the disease were observed in edematous-serous acute pancreatitis, and its duration was shortened. Potentially, this disease might be avoided by improved antioxidant supplementation with vitamin E and selenium.

TABLE 7   Parameters of Inflammation and Antioxidants in Serum and Pancreatic Juice on the First Day of Disease ($n = 3$)

	Serum	Pancreatic juice
Malondialdehyde, μmol/L	17.3 ± 5.4	52.5 ± 22.8
Elastase, μg/L	62.7 ± 12.9	212.8 ± 33.7
Vitamin E, μmol/L	14.5 ± 5.6	0
Selenium, μmol/L	0.36 ± 0.16	0.74 ± 0.10

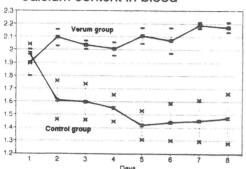
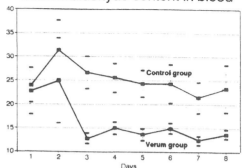

FIGURE 6 Concentrations of serum total calcium and malondialdehyde in 17 patients with necrotizing pancreatitis. Experimental group ($n = 8$) receiving sodium selenite therapy (Selenase pro injectione, Stuttgart, Germany); control group $n = 9$. Concentrations of MDA in μmol/L and calcium in mmol/L.

Alcoholic Hepatitis

In disease we measured the highest MDA values at 170 μmol/L. Even on the first day of treatment high MDA increases could be identified. The MDA levels correlated positively with the degree of seriousness of the disease. With the antioxidant status being neglected, values > 40 μmol/L signal a lethal outcome owing to multiorgan failure. In pilot investigations therapy with selenium alone had no effect and pointed to a complex disturbance of the liver as the most important radical scavenger organ, with hepatic glutathione content playing an important role in this connection. In chronic alcohol abuse the liver has lowered levels of selenium, vitamin E, Zn, β-carotenoids, glutathione, and other antioxidants. More complex measures of therapy with vitamin E, sodium selenite, and *N*-acetylcysteine (a glutathione precursor) could essentially improve the patient's prognosis. An efficiency of adjuvant therapy could be proved in male patients with compensated alcohol-toxic liver damage. The daily administration of 1.2–1.5 g vitamin E, 600 mg *N*-acetylcysteine, and 300 μg sodium selenite prevented a serious course and the lethal outcome (Fig. 7).

Similar results were obtained by Lang et al. (17) in a 1 month double-blind clinical trial. They described the hepatoprotective and immunomodulatory effects of an antioxidative treatment of patients suffering from compensated alcoholic cirrhosis of liver.

Chronic Ischemic Heart Disease

We described 6 years ago the effect of sodium selenite on experimental myocardial infarction in 47 domestic pigs, with major emphasis on the hemodynamic anti

Rheumatoid Arthritis

The clinical activity of rheumatoid arthritis correlates positively with parameters of inflammation and MDA concentration in serum. Compared with healthy individuals, there were nearly 20–50% higher MDA values in serum. In the punctates of inflammable synovial fluid, however, there were none. The latter showed MDA concentrations nearly 50% lower than those in serum. Depending on the activity of inflammation higher values of elastase could be identified (Table 8).

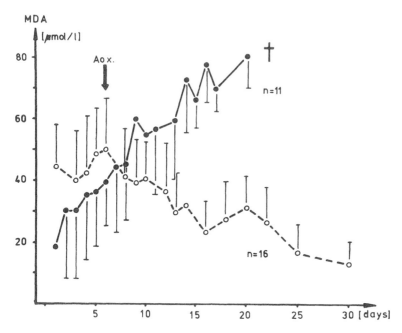

FIGURE 7 Concentrations of MDA in patients with severe alcoholic hepatitis without ($n = 11$) and with antioxidative therapy (Aox., $n = 16$). Underlying oxidative stress can cause hepatic decompensation.

Increased leukocyte and macrophage activity predominantly generates hypochloric acid (HOCl) in synovial fluid through myeloperoxidase; this cannot be covered by means of TBARs reaction. Thus, therapy with vitamin E can be concluded to have no or only small effects only, and an antioxidative effect of therapy can only be expected by applying amino acids in higher concentrations.

Intensive Care Patients

Stroke.   After cerebral trauma or ischemic events in the skull, the highest MDA values could be found in patients immediately after the acute event in postaggression metabolism of the first 2 or 3 days of treatment with a subsequent fall. Serum MDA values for 32 patients are shown in Table 9. These high serum levels may be connected to basic metabolism (post-aggression metabolism). The cerebral vessel spasm beginning 6 h after the acute event may

TABLE 8 Concentrations of Malondialdehyde and Elastase in Serum and Synovial Fluid in Patients with Rheumatoid Arthritis ($n = 5$)

	Serum	Synovial fluid
Malondialdehyde, μmol/L)	12.4 $\pm$ 3.5	7.8 $\pm$ 3.5
Elastase, μg/L	54.8 $\pm$ 9.9	243.7 $\pm$ 64.4

TABLE 9 Concentrations of Malondialdehyde (μmol/L) in Serum and Liquor in the First Week

	Days				
	1	2	3	5	6
Serum	14.8 ± 8	16.4 ± 6	12.6 ± 4	8.4 ± 3.8	9.2 ± 2
Liquor	1.4 ± 0.2	—	1.1 ± 0.3		0.9 ± 0.2

be favored by high lipid peroxidation rates in vivo, especially in synapses and in areas with a higher content of polyunsaturated fatty acids favored by the high oxygen demand of the cerebrum. Table 10 shows MDA, lactate, and pH values in blood and liquor.

Surgery and Multiple Trauma Patients. Postoperatively, increased MDA concentrations appeared from the third to the fourth day, which correlated negatively with vitamin E and C concentrations. The more serious the operation or polytraumatization, the more distinct are the changes (18). In uncomplicated operations with rapid healing and mobilization of patients, the MDA values amount to < 10 μmol/L. Longer parenteral phases of nutrition and/or periods of artificial respiration lead in the first week to MDA values > 15 μmol/L postoperation in patients 65 years old or more, whereas younger pagients showed further increases only after the second week or later. It can be concluded that elderly people have a weaker antioxidant potential.

During septic complications MDA levels above 20 μmol/L correlated positively with other inflammatory parameters. The decrease in MDA observed in concentrations < 6 μmol/L can be seen as an early symptom of exhaustion in the cellular immunodefense system. This serious prognostic sign can be identifieed before the decrease in leukocyte elastase.

From our results the conclusion can be drawn that in those operations that can be scheduled in advance supplementation with vitamin E, vitamin C, and selenium should be made 14 days before the operation.

Pathological Pregnancy

In the last trimester of normal pregnancy there are decreased values of serum selenium without increases in MDA. In the amniotic fluid MDA is below the limit of detection. Patho-

TABLE 10 Concentrations of Malondialdehyde and Lactate and pH in Serum and Liquor in Patients with Different Cerebral Diseases ($n = 62$)

	MDA (μmol/L)		Lactate (μmol/L)		pH	
	Serum	Liquor	Serum	Liquor	Serum	Liquor
Male patients	9.41	1.05	2.15	2.96	7.40	7.03
Female patients	13.0	1.14	2.40	3.04	7.39	7.65
≤60 years	12.17	0.80	1.96	2.25	7.40	7.68
>60 years	9.23	1.26	1.58	3.15	7.41	7.57
Alzheimer's disease	10.9	0.75	1.70	2.97	7.38	7.67
TIA,[a] stroke	9.43	0.86	1.78	2.96	7.40	7.68
Hepatic encephalopathy	20.24	1.18	2.16	2.33	7.42	7.73
Died	13.94	1.64	2.11	3.29	7.41	7.81

[a]Transient ischemic attack.

logical pregnancy with hypertension showed MDA concentrations in serum 200–300% higher. They did not correlate with the diastolic or systolic blood pressure, or with concentrations of serum selenium and vitamin E (19). This pathological process is poorly understood; nevertheless it seems that moderate antioxidative therapy with vitamin E and selenium is indicated because O_2 radicals inhibit the EDRF. A biological preponderance of endothelin with its vasoconstrictive effect and the ensuing hypertension may thus be explained.

Malignant Tumors

We showed in a double-blind placebo-controlled trial in healthy subjects a significantly increase in $CD4^-$ T lymphocytes (20) by application of sodium selenite. This effect could be important to treatment of malignant diseases.

In tumor diseases we noted MDA concentrations to be lower the more undifferentiated and invasive the tumor growth. Even surgical extirpation caused no essential changes. The MDA levels were between 2 and 6 μmol/L and must be regarded as an expression of generalized cellular immunosuppression.

Various measures of therapy with x-rays, cytostatics, and interferon-α coincided with MDA increases. Because the increased radical levels inhibit cell proliferation and various tumors reveal high antioxidant capacities, we performed no adjuvant antioxidative therapy.

Because basic research at present can give no general statement about the significance of SH/SS status, radical generation, and radical inhibition, it would be premature in our view to draw any conclusions in this respect. It cannot be ruled out, however, that the protection of cellular compartments (macrophages and leukocytes) against free radicals, for example by applying selenium or vitamin E, eventually encourages natural killing activity, not the tumor itself.

CONCLUSIONS

Increases in malondialdehyde concentrations reflect peroxidative damage to membrane PUFA with three or more double bonds. Then they appear in vivo if there is exhaustion of the radical scavenger capacity of vitamin E and/or tocopherol radical generation caused by a deficiency of selenium and vitamin C.

In endemic areas of selenium deficiency a higher demand for vitamin E is expected if no general supplementation of selenium has been made. The antioxidant status may markedly deteriorate even by applying an antilipemic and antiatherogenic diet.

Transverse and longitudinal section examinations carried out in healthy, premorbid, and morbid individuals resulted in risk groups for whom adjuvant antioxidant supplementation is indicated to yield sufficient SH/SS status.

In healthy individuals these risk groups include breast-fed infants, pregnant women, geriatric patients, smokers, and individuals with regular moderate alcohol consumption, chronic strains due to occupational noxious substances (heavy metals, herbicides, pesticides, and nitrates), and ultraviolet light.

In premorbid and morbid individuals antioxidant supplementation serves to protect enzymes containing SH groups. The following indication groups for antioxidant therapy were found here on the basis of preliminary results:

Hyperlipoproteinemia, dyslipoproteinemia under lipid-lowering therapy
Arteriosclerotic diseases, including chronic heart disease, heart failure, and rhythmic disturbances
All diseases with postaggression metabolism, parenteral nutrition, and artificial respiration, especially in higher O_2 application
Diabetes mellitus type II (insulin independent)

In the future additional studies of intervention will be necessary to document the demand for an efficient antioxidant supplementation for preventive and curative medicine. Until now these are known in the treatment of hyperlipoproteinemia, chronic ischemic heart disease, acute necrotizing pancreatitis, and alcohol-toxic liver damage.

In particular, studies in those fields are necessary that confirm the increase in resistance against carcinogenic substances by measuring the DNA repair capacity and unspecific protection parameters of the mucosa (mucin, sIgA, cellular immune barriers), as well as arteriosclerotic diseases.

REFERENCES

1. Kuklinski B, Vorberg B, Koepcke E, et al. Latent deficiency of antioxidants (selenium and vitamin E) in the population of the German Democratic Republic. Reasons and clinical relevance. Arztl Lab 1990; 36:288-94.
2. Yagi K. Lipid peroxides and human diseases. Chem Phys Lipids 1987; 45:327-51.
3. Uchiyama M, Michara M. Determination of malondialdehyde precursor in tissues by thiobarbituric acid test. Anal Biochem 1978; 86:271-8.
4. Andreeva LJ. A modified thiobarbituric acid test for measuring lipid peroxidation products. Lab Delo 1988; 11:41-3.
5. Peters HJ, Vorberg B, Höhler H. Eine modifizierte Halbmikromethode zur Vitamin E—Bestimmung im Serum. Z Gesante Hyg 1983; 29:441-3.
6. Peters HJ, Dittrich K. Bestimmung von Selen in biologischen Materilien mittels der Hydridtechnik. Z Med Lab Diagn 1981; 22:237-8.
7. Kuklinski B, Vorberg B, Rühlmann C, et al. Latenter Antioxidantienmangel in der DDR-Population. I. Mitteilung. Z Gesamte Inn Med 1990; 45:33-7.
8. Esterbauer H, Dieber-Rothender M, Waeg G, Striegl G, Jürgens G. Biochemical, structural, and functional properties of oxidized low-density lipoprotein. Chem Res Toxicol 1990; 3:77-92.
9. Boscoboinik D, Szewczyk A, Azzi A. Alpha-tocopherol regulates vascular smooth muscle cell proliferation and protein kinase C activity. Arch Biochem Biophys 1991; 286:264-9.
10. Domagala B, Hartwich J, Szczeklil A. Indices of lipid peroxidation in patients with hypercholesterolemia and hypertriglyceridemia. Wien Klin Wochenschr 1989; 101:425-8.
11. Luoma PV, Stengard J, Korpela H, et al. Lipid peroxides, glutathione peroxidase, high density lipoprotein subfraction and apolipoproteins in young adults. J Intern Med 1990; 227:287-9.
12. Luoma PV, Sotaniemi EA, Korpela H, Kumpulainen J. Serum selenium, glutathione peroxidase activity and high density lipoprotein cholesterol—effect of selenium supplementation. Res Commun Chem Pathol Pharmacol 1984; 46:469-72.
13. Kuklinski B, Perfionov AS, Zimmermann R, et al. In vivo-Wirkungen High Density Lipoprotein-reicher Serumfraktionen beim Menschen. I. Mitteilung. Z Gesamte Inn Med 1989; 44:413-20.
14. Kuklinski B, Perfionov AS, Rühlmann C, et al. In vivo—Wirkungen High Density Lipoprotein-reicher Serumfraktionen beim Menschen. Z Gesamte Inn Med 1989; 44:421-6.
15. Beitz J, Kuklinski B, Beitz A, et al. Influence of high density lipoprotein (HDL), prepared from human blood, on prostanoid formation, serum and tissue lipids and development of arteriosclerosis in cholesterol rich fed fed rabbits. Prostaglandins Leuko Essent Fatty Acids 1990; 40:211-5.
16. Kuklinski b, Buchner M, Schweder R, Nagel R. Acute pancreatitis—a "free radical disease." Decrease of lethality by sodium selenite (Na_2SeO_3). Z Gesamte Inn Med 1991; 46:7-11.
17. Lang I, Nekam K, Deak G, et al. Immunomodulatory and hepatoprotective effects of in vivo treatment with free radical scavengers. Ital J Gastroenterol 1990; 22:283-8.
18. Kreinhoff U, Elmadfa I, Salomon F, Weidler B. Untersuchungen zum Antioxidantienstatus nach operativem Stress. Infusionstherapie 1990; 17:261-7.
19. Peiker G, Kretschmar M, Dawczynski H, Müller B. Lipid peroxidation in pathologic pregnancy: pregnancy-induced hypertension. Zentralbl Gynakol 1991; 113:183-8.
20. Herzfeld A, Kuklinski B, Köhler H, Peters HJ, Vorberg B. Behavior of selected antioxidant and immune parameters in selenium-supplemented healthy subjects. Arztl Lab 1990; 36:295-8.

F. Obstetrics and Gynecology

65

Fetomaternal Vitamin E Status

Makoto Mino

Osaka Medical College, Takatsuki Osaka, Japan

INTRODUCTION

Historically static indices have been used more commonly than functional indices in assessing vitamin E nutritional status. Static indices used in human subjects involve tocopherol levels in plasma or serum, erythrocytes, platelets, leukocytes, and buccal mucosal cells. Although measurement of the plasma or serum tocopherol level is easiest, the level is altered by lipemic status. Since tocopherol in blood does not have a specific carrier protein but travels with the different lipoprotein fractions, the tocopherol-lipid ratio in plasma was used for the assessment in some cases. Horwitt et al. (1) proposed that plasma tocopherol expressed in terms of total lipids (tocopherol-lipid ratio) is a very reliable index of vitamin E status. Studies of tocopherol levels in platelets, leukocytes, and buccal mucosal cells as biological samples obtained from humans are limited. Tocopherol can also be measured in adipose tissue stores (2), but this method is not practical except in a research setting. On the other hand, the assessment of vitamin E is considered under a functional systems classification, the role of vitamin E in the structural integrity of cells and homeostasis. In the classic hemolysis test, diluted hydrogen peroxide was used to estimate vitamin E deficiency, but the hemolysis was unreliable because of its poor reproducibility (3). More recently, ethane and pentane in breath have been proposed in clinical approaches (4,5). Having undergone further studies, these breath tests have been confirmed as better measurements of vitamin E status as compared with static indices. With this view, studies on the nutritional status of vitamin E in pregnant women and human fetuses will be discussed here.

PREGNANT WOMEN

Classically it is well known that plasma tocopherol levels in women are elevated throughout gestation (6,7). Thus, the problem of whether the rise in plasma tocopherol during gestation reflects the physiological demands on both the fetus and mother in an effort to increase the placental and mammary transfer of tocopherol to the fetus and infants, remains unsolved. However, with respect to the relationship between plasma lipids and tocopherol, the increased plasma tocopherol levels during pregnancy are proposed to be due to elevated β lipoproteins, and the tocopherol-β lipoproteins ratio remains unchanged during pregnancy (8,9). As shown in Figure 1, the plasma tocopherol level increased to twice the basal level at the end of gestation, in parallel with the change in lipids. The increase in plasma tocopherol is due to the lipid changes, as shown by the finding that the tocopherol-lipid ratio was almost constant throughout

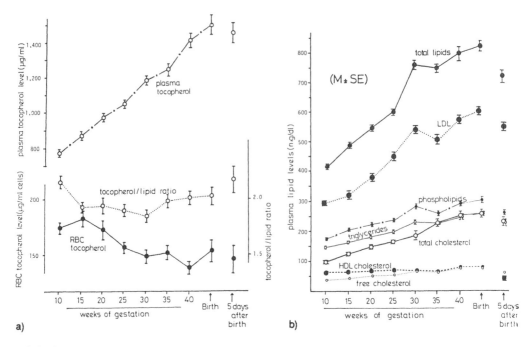

FIGURE 1 Changes in tocopherol concentrations in plasma and erythrocytes and the tocopherol-lipid ratios throughtout gestation (a) compared with gestational changes in plasma lipids and lipoproteins (b). Each bar represents the mean ± SEM.

pregnancy. The erythrocyte tocopherol level, which is another index of vitamin E status (10–12), also seems to remain unchanged throughout gestation, although it decreased during the third trimester of gestation in Figure 1 (13). These findings indicate an unchanged vitamin E status during pregnancy. As a functional index for the assessment of vitamin E status in pregnant women, oxygen consumption in erythrocyte ghosts of pregnant women at delivery has been examined and reported (14). This study was performed using erythrocyte ghosts in the kinetic study of peroxidation in relation to membrane tocopherol as a defense mechanism against peroxidation initiated by an azo compound, AAPH [2,2'-azobis-(2-amidinopropane)dihydrochloride]. AAPH is known to generate free radicals by its thermal decomposition (15). The rate of generation of free radicals from AAPH can be easily controlled and measured by adjusting the concentration of the compound. When erythrocyte ghosts are reacted with AAPH solutions, oxygen uptake patterns occur at a slow rate in the first phase and a faster rate during the second phase, as shown in Figure 2. Tocopherol in these ghosts is constantly consumed during the first phase, and after tocopherol decreases to an undetectable level, the second phase, characterized by increased oxygen consumption, commences (16). The second phase of oxygen consumption suggests extensive propagation of an oxidative chain reaction in membrane lipids, whereas slow oxygen uptake in the first phase suggests that the chain reaction is constantly inhibited by tocopherol remaining in the membrane. The oxygen consumption rate in phase 1 is termed R_{inh} and that in phase 2, R_p (mole oxygen consumption per second); the length of the first phase is termed t_{inh} (second). In this study, the erythrocyte tocopherol content in pregnant women was slightly lower, R_p was larger, and the kinetic chain length (KCL) was longer, but t_{inh} and R_{inh} were comparable, compared with a normal population of nonpregnant adults, as shown in Table 1. The KCL is defined as the ratio of oxygen

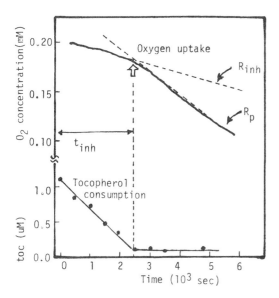

FIGURE 2 Oxygen uptake and tocopherol consumption during oxidation of RBC ghosts. t_{inh}, length of phase 1 (s); R, rate of oxygen consumption (decrease of O_2, mol/s); R_{inh}, rate of oxygen consumption in phase 1; R_p, rate of oxygen consumption in phase 2.

TABLE 1 Relationship Between Peroxidizability of Cord, Pregnant Mother, and Adult[a]

	Cord ($n = 11$)	Pregnant women ($n = 7$)	Adults ($n = 10$)
Tocopherol, μM			
Total	1.67 ± 0.47	1.37 ± 0.26	1.67 ± 0.18
α	1.54 ± 0.47	1.25 ± 0.23	1.49 ± 0.17
γ	0.13 ± 0.13	0.12 ± 0.04	0.18 ± 0.04
t_{inh}, s	2021 ± 542	1927 ± 386	2089 ± 492
R_{inh}, $10^{-9}/s$	27.7 ± 7.1	24.6 ± 6.6	24.7 ± 4.4
R_p, $10^{-9}/s$	56.6 ± 7.1[b]	49.9 ± 8.9	46.9 ± 7.1
Active H/protein	7246 ± 1528[b]	5602 ± 2754	3640 ± 788
KCL in R_{inh}	13.5 ± 2.4	11.2 ± 3.2	12.1 ± 2.1
KCL in R_p	27.5 ± 3.5[b]	24.2 ± 4.3	22.7 ± 3.7

[a]Erythrocyte ghosts reacted with AAPH. t_{inh}, length of phase 1 (s); R, rate of oxygen consumption (decrease in O_2, mol/s); R_{inh}, rate of oxygen consumption in phase 1; R_p, rate of oxygen consumption in phase 2; KCL, kinetic chain length, defined as the ratio of oxygen uptake rate to basic formation rate. KCL (R_{inh} of R_p) = R_{inh} or R_p/R initiation. R initiation is calculated from the rate of tocopherol consumption (2× tocopherol) in ghosts reacted with AAPH.
[b]$p < 0.05$ (versus adults).

uptake rate to radical formation rate, indicating that when one radical is generated from oxidants, given numbers of fatty acid molecules are oxidized. This indicates greater numbers of KCL as chain propagation is prolonged. In addition, fatty acid analysis in these ghosts shows an increase in linoleic acid and arachidonic acid in the erythrocyte ghosts of pregnant women. Thus, the quantities of active bisallylic hydrogen (active hydrogen) increased in pregnant women rather than normal adults (14). It is well known that the dietary vitamin E requirement increases as polyunsaturated fatty acid (PUFA) composition increases in situ (17). The amounts of active H numbers in the ghosts have been reported to closely correlate with R_p but poorly with R_{inh}. This also indicates that the rate of suppression of chain propagation by the remaining tocopherol in the membranes is not greatly affected by the amount of active H, but the propagation rate is proportional to the amount of active H. Since R_p in pregnant women's ghost membranes with AAPH was significantly greater than that in control adult ghosts (Table 1), slightly higher peroxidizability is assumed in pregnant women's ghosts and is probably attributable to greater active H numbers in pregnant women's membranes. These kinetic findings indicate a slight increase in susceptibility of the ghost membrane in pregnant women to oxidant stress compared with healthy adults. Therefore, the vitamin E status in pregnant women at the terminal stage of gestation seems to be poorer than that in normal adults with respect to the tocopherol concentration in ghost membranes and its susceptibility to oxidant stress, despite an exceptionally high plasma tocopherol concentration (18).

In addition to this evidence, to our knowledge there has been no further precise information on vitamin E status in pregnant women during gestation.

FETUSES

In the past there have been many reports with respect to the relationship in tocopherol levels between cord and maternal serum to assess fetal vitamin E status (7,19–21). All the reports show that the tocopherol level in cord blood is very low compared with the mother's level, indicating impaired transport of tocopherol across the placenta. Although, some investigators have reported that there is a direct relationship between the plasma or serum tocopherol levels in infants at birth and that of the mother (19–21). Even in the latter report (20), the highest tocopherol level in mother's blood, which was induced by administration of vitamin E preparations to the pregnant mother, did not lead to an adequate increase in the level in cord blood.

In a recent study of the nutritional status of vitamin E assessed in very low birth weight infants with a birth weight of less than 1500 g, the erythrocyte tocopherol level was within the normal range of that in healthy adults even in the extremely low birth weight infants, less than 1000 g at birth, but their plasma tocopherol levels were much lower (22). The lowest level of plasma tocopherol in newborn infants is attributable to lower plasma lipid levels (23–26). However, the erythrocyte tocopherol level in infants weighing less than 1000 g at birth decreased below the normal range during the 4–10 weeks of life after birth, possibly because of immaturity of intestinal absorption of tocopherol, as shown in Figure 3. This result confuses our understanding of the vitamin E status of newborn infants, because in the classic investigations, it was believed that newborn and premature infants were marginally vitamin E deficient on the basis of findings of low plasma tocopherol levels and the increased susceptibility of their erythrocytes to hydrogen peroxide. Thus, the finding of a normal tocopherol content in erythrocytes of very low birth weight infants conflicts with the increased susceptibility of their erythrocytes to oxidant stress with hydrogen peroxide. To resolve this question, the peroxidizability of erythrocytes in newborn infants (using cord blood) was examined in a kinetic study of oxygen consumption in cord blood ghosts with AAPH, as described for pregnant women's erythrocyte ghosts (14). The results are also shown in Table 1. The R_p was greater

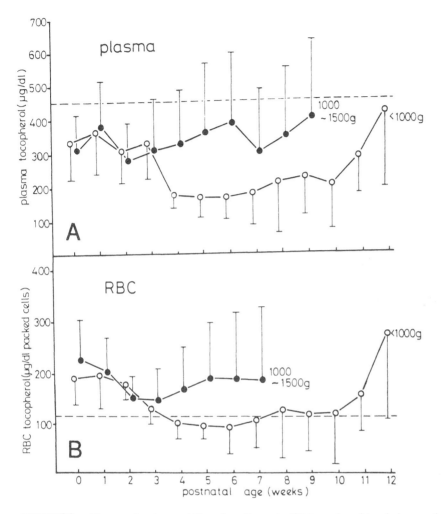

FIGURE 3 Changes in plasma (A) and erythrocyte (B) tocopherol levels in very low birth weight infants after birth. The values are represented as the mean ± SD. (Open circles) values in the infants less than 1000 g birth weight; (closed circles) infants 1000–1500 g birth weight.

in the cord ghosts than in the mother's and healthy adult control ghosts, despite similar concentrations of tocopherol in the ghost membranes between the three groups. Fatty acid analysis of the ghost membranes showed the greatest numbers of active hydrogen due to the abundant content of arachidonic acid in cord ghosts as compared to mother's and adult control ghosts. Therefore, the ghost membranes in cord erythrocytes are highly susceptible to oxidant stress compared with mother's and adult control ghosts, indicating a relative vitamin E deficiency in newborn membranes even with a normal content of tocopherol in membranes. The transportation of a relatively large amount of tocopherol to erythrocytes is probably due to a low level of plasma lipoproteins, because the tocopherol concentration in plasma lipids (indicated as the tocopherol-lipid ratio) seems to be relatively high. Traber and Kayden (26) emphasized that tocopherol delivery to tissues is mainly based on a lipoprotein receptor mechanism (especially specifically on LDL receptors) rather than a collision mechanism, the latter playing a role in the delivery of tocopherol in red blood cell (RBC) membranes without low-density lipoprotein (LDL) receptors (27).

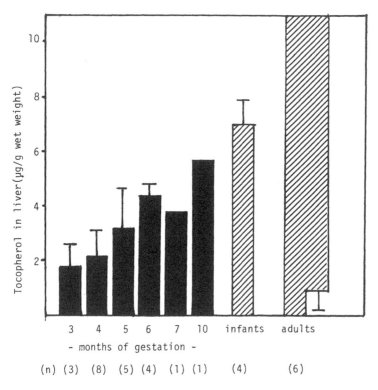

FIGURE 4 Tocopherol levels in human liver in fetuses, infants, and adults. Each bar represents the mean ± SD.

There have been no recent reports regarding tocopherol changes during gestation in human fetuses, but by colorimetric analysis accompanied by fluoridine column and thin-layer separation, changes in tocopherol concentrations in the livers of human fetuses and infants after birth were examined using autopsy samples (29). The result is shown in Figure 4. A steady increase in the levels of tocopherol is found throughout fetal life: the younger the fetus, the lower the tissue tocopherol levels. In this study, the level in fetal liver at the end of gestation

TABLE 2 α-Tocopherol Concentrations in Plasma, Erythrocytes, Platelets, Leukocytes, and Buccal Mucosal Cells in Neonates and Children[a]

	Neonates		Children	
Plasma, μg per 100 ml	402 ± 155	(81)	769 ± 99	(65)
Tocopherol-lipid ratio	1.43 ± 0.34	(67)	1.58 ± 0.34	(55)
Erythrocyte, μg per 100 ml packed cells	203 ± 48	(29)	241 ± 47	(55)
Platelets, μg/mg protein	0.03 ± 0.02	(29)	0.12 ± 0.05	(52)
Leukocytes, μg per 10^9 cells				
PMN	2.59 ± 1.33	(24)	4.35 ± 1.23	(25)
MN	4.06 ± 1.02	(40)	8.47 ± 2.66	(51)
Buccal mucosal cells, μg/mg protein	25.4 ± 7.6	(25)	47.8 ± 15.8	(30)

[a]Mean ± SD. MN, mononuclear cells; PMN, polymorphonuclear cells. Numbers in parentheses.

TABLE 3 % α-Tocopherol Levels Below Lower Limit in Neonates[a]

Age (days)	Lower limit	PL, 450 g/dl	RBC, 115 μg/dl%	PLT, 0.01 μg/mg protein	MN, 4.0 μg per 10⁹%	PMN, 2.5 μg per 10⁹%	BMC, 20.0 ng/mg protein
0		100.0	0	9.1	60.0	44.4	44.4
		(10)	(10)	(11)	(15)	(9)	(9)
1–3		66.7	0	16.7	50.0	62.5	66.7
		(12)	(12)	(12)	(16)	(8)	(9)
4–7		50.0	0	33.3	22.2	71.4	14.3
		(6)	(7)	(6)	(9)	(7)	(7)
Total		75.0	0	17.2	47.5	50.0	44.0
		(28)	(29)	(29)	(40)	(24)	(25)

[a]Number in parentheses. PL, plasma; RBC, erythrocyte; PLT, platelet; MN, mononuclear cell; PMN, polymorphonuclear cell; BMC, buccal musocal cell.

was one-third or one-fifth of adult levels. The tocopherol status in fetal liver conflicts with that of erythrocytes in very low birth weight infants at birth, comparable to that in adult erythrocytes (23). The discrepancy that may occur because of the different transportation mechanism of tocopherol between erythrocytes (27) and other tissues (26), requires further investigation on the static vitamin E status of the human tissues and cells excluding erythrocytes. Recently, tocopherol concentrations in biological samples in human newborns were examined (28,30, 31). During the neonatal period, the lowest levels were documented in the all blood fractions excluding erythrocytes and also in buccal mucosal cells (BMC) (30). In this study, tocopherol in platelets, leukocytes, and buccal mucosal cells in newborns amounted to only about one-half or one-third of older child values, as shown in Table 2. In this report, stable tocopherol levels were documented after 2 years of life in all the biological samples. Then, from the distribution of the levels in the biological samples of children older than 2, the lowest levels of normal ranges were first arbitrarily proposed in respective biological samples. On the basis of the lowest normal levels, the incidence of the tocopherol levels lower than the normal ranges in neonatal biological samples were calculated. Table 3 shows the percentage of tocopherol values below the lower limit in neonatal samples of the first 7 days of life. The tocopherol concentration in erythrocytes did not fall below the lowest normal range, but the incidence of values lower than the normal lowest range of tocopherol concentrations was 17.2, 47.5, 50.0, and 44.0% in platelets, mononuclear leukocytes, polymorphonuclear leukocytes, and buccal mucosal cells, respectively (29). With respect to the static vitamin E status, there may be a lower level of vitamin E in all fetal tissues and cells, excluding erythrocytes. In addition, the peroxidizability of membrane lipids increases even with a normal content of tocopherol, as shown by the AAPH reaction on cord erythrocyte ghosts.

As another functional index, ethane and pentane in breath as volatile products of unsaturated fatty acid peroxidation, have been measured in human newborns (5). In this study, newborn infants excrete 16 pmol ethane per kg body weight per minute and 15 pmol pentane per kg body weight per minute. This compares with 1.4 pmpol ethane per kg body weight per minute and 1.3 pmol pentane per kg body weight per minute in healthy adult men.

The majority of these findings, except the tocopherol-lipid ratio and erythrocyte tocopherol concentrations, support the belief that human fetuses are in vitamin E deficiency states.

REFERENCES

1. Horwitt MK, Harvey CC, Dahm CH Jr, Searey MT. Relationship between tocopherol and serum lipid levels for determination of nutritional adequacy. Ann N Y Acad Sci 1972; 203:223-36.

2. Traber MG, Kayden HJ. Tocopherol distribution and intracellular localization in human adipose tissue. Am J Clin Nutr 1987; 46:488-95.

3. Mino M, Nishida Y, Murata K, Takegawa M, Katsui G, Yuguchi Y. Studies on the factors influencing the hydrogen peroxide hemolysis test. J Nutr Sci Vitaminol 1978; 24:383-95.

4. Leyone M, Van Gossum A, Kurian R, Ostro M, Axler J, Jeejeebhoy KN. Breath pentane analysis as an index of lipid peroxidation: a functional test of vitamin E status. Am J Clin Nutr 1987; 46:267-72.

5. Wisple JR, Bell EF, Roberts RJ. Assessment of lipid peroxidation in newborn infants and rabbits by measurements of expired ethane and penthane: influence of parenteral lipid infusion. Pediatr Res 1985; 19:374-9.

6. Fergson ME, Bridgeforth E, Quife ML, et al. The Vanderbilt cooperative study of maternal and infant nutrition. Tocopherol in relation to pregnancy. J Nutr 1954; 54:305-21.

7. Straumfjord JV, Quife ML. Vitamin E levels in maternal and fetal blood plasma. Proc Soc Exp Biol Med 1946; 61:369-74.

8. Takahashi Y, Shitara H, Uruno K, Kimura S. Vitamin E and lipoprotein levels in the sera of pregnant women. J Nutr Sci Vitaminol 1978; 24:471-6.

9. Haga P, Johan EK, Kran S. Plasma tocopherol levels and vitamin E/beta-lipoprotein relationships during pregnancy and in cord blood. Am J Clin Nutr 1982; 36:1200-4.

10. Kitagawa M, Nakagawa S, Mino M. Influence of plasma lipids and adiposity on red blood cell tocopherol level. Eur J Pediatr 1983; 140:238-43.

11. Mino M, Kitagawa M, Nakagawa S. Red blood cell tocopherol concentrations in a normal population of Japanese children and premature infants in relation to the assessment of vitamin E status. Am J Clin Nutr 1985; 41:631-8.

12. Mino M, Kasugai O, Shimizu T. Red blood cell tocopherol and liver tocopherol in hyperlipemic rats as compared with plasma tocopherol. Lipids 1985; 20:488-91.

13. Mino M, Nagamatsu M. Evaluation of nutritional status of vitamin E in pregnant women with respect to red blood cell tocopherol level. Int J Vitam Nutr Res 1986; 56:149-53.

14. Mino M, Miki M, Miyake M. Vitamin E status in neonatal erythrocytes under oxidative stress. In: Hayaishi O, Niki E, Kondo M, Yoshikawa T, eds. Medical, biochemical and chemical aspects of free radicals. Amsterdam: Elsevier, 1989; 251-9.

15. Niki E. Antioxidants in relation to lipid peroxidation. Arch Biochem Biophys 1987; 258:373-80.

16. Yamamoto Y, Niki E, Eguchi J, Kamiya K, Shimazaki H. Oxidation of biological membranes and its inhibition. Free radical chain oxidation of erythrocyte ghost membranes by oxygen. Biochim Biophys Acta 1985; 819:29-36.

17. Hassan H, Hashim SA, Van Itallie TB, Sebell WH. Syndrome in premature infants associated with low plasma vitamin E levels and high polyunsaturated fatty acid diet. Am J Clin Nutr 1966; 19:147-57.

18. Mino M, Miki M, Miyake M, Ogihara T. Nutritional assessment of vitamin E in oxidative stress. Ann N Y Acad Sci 1989; 570:296-310.

19. Leonard PJ, Doyle E, Harrington W. Levels of vitamin E in the plasma of newborn infants and the mothers. Am J Clin Nutr 1972; 25:480-4.

20. Mino M, Nishino H. Fetal and maternal relationship in serum vitamin E level. J Nutr Sci Vitaminol 1973; 19:475-82.

21. Ibeziako PA, Ette SI. Vitamin E in pregnant Nigerian women and newborn. J Trop Med Hyg 1982; 85:256-68.

22. Tanaka H, Mino M, Takeuchi T. A nutritional evaluation of vitamin E status in very low birth weight infants with respect to changes in plasma and red blood cell tocopherol levels. J Nutr Sci Vitaminol 1988; 34:293-307.

23. Martinez PE, Concalves AL, Jorge SM, Desai ID. Vitamin E, placental blood and its interrelationship to maternal and newborn level of vitamin E. J Pediatr 1981; 99:298-300.

24. Desai ID, Swann MA, Garcia Traveres ML, Dutra de Oliveria BS, Pharm B, Duarte de Oliveria JE. Vitamin E status of agricultural migrant workers in southern Brazil. Am J Clin Nutr 1980; 33:2669-73.

25. Ogihara T, Miki M, Kitagawa M, Mino M. Distribution of tocopherol among human plasma lipoproteins. Clin Chim Acta 1988; 174:299-306.

26. Traber MG, Kayden HJ. Vitamin E is delivered to cells via the high affinity receptor for low-density lipoprotein. Am J Clin Nutr 1984; 40:747-51.

27. Tanaka H, Mino M. Cell to cell transfer of tocopherol. J Nutr Sci Vitaminol 1986; 32:463-74.

28. Ogihara T, Miyake M, Kawamura N, Tamai H, Kitagawa M, Mino M. Tocopherol concentrations of leukocytes in neonates. Ann N Y Acad Sci 1989; 579:487-90.

29. Mino M, Nishino H, Ymaguchi T, Hayashi M. Tocopherol level in human fetal and infant liver. J Nutr Sci Vitaminol 1977; 23:63-9.

30. Yokota K, Tamai H, Mino M. Clinical evaluation of alpha-tocopherol in buccal mucosal cells of children. J Nutr Sci Vitaminol 1990; 36:365-75.

31. Kaempf D, Ogihara T, Mino M. Unpublished data. 1992.

VI
FUTURE PERSPECTIVES

66

Vitamin E: Biological Activity and Health Benefits: Overview

Lester Packer

University of California, Berkeley, California

THE VITAMIN E PARADOXES

Vitamin E is the major, if not the only, chain-breaking antioxidant in membranes, but its membrane concentration is very low, usually equal to or less than 0.5–0.1 nmol/mg protein (less than 1 per 1000–2000 membrane phospholipids). Despite this incredibly low molar concentration, the rate of lipid radical generation in membranes can be very high, about 1–5 nmol/mg protein per minute. Nevertheless, under normal conditions "rancidification," that is, oxidation of membrane lipids and proteins, does not occur. Moreover, it is very difficult to render animals deficient in vitamin E, and vitamin E deficiency is seldom found in adult humans. Hence, there must exist a remarkably efficient mechanism for permitting low concentrations of vitamin E to have such high efficiency in protecting membranes against damage and in supporting normal biological activity.

THE VITAMIN E CYCLE

We hypothesize that these remarkable properties of vitamin E may be explained by its ability to be efficiently re-reduced from its free-radical form, the form it takes after quenching free radicals, to its native state.

   This occurs by the actions between water- and lipid-soluble substances and by nonenzymatic and enzymatic mechanisms for regenerating the vitamin E radical (tocotrienoxyl and tocopheroxyl radical back to tocotrienol and tocopherol, respectively). If true, this means that vitamin E interacts with water- and lipid-soluble substances either through chemical reactions or enzymatic pathways to drain away or accept their reducing power. Under conditions in which these auxiliary systems act synergistically to keep the steady-state concentration of vitamin E radicals low, the loss or consumption of vitamin E is prevented.

   This means that vitamin E is lost only when these backup systems in either the aqueous or membrane domain become rate limiting. The consequence is that vitamin E consumption will commence. The loss of vitamin E is accompanied by increased rates of lipid and protein oxidation and the destruction of membrane function and inactivation of membrane enzymes and receptors. Thus, vitamin E not only has an antioxidant action but also acts as a biological response modifier by modulating membrane-associated enzyme systems whose effects are amplified by the production of low-molecular-weight substances, like secondary messengers and the products of the arachidonic acid cascade, which have profound effects at low concentrations in cell regulation and proliferation.

Oxidative damage never occurs while vitamin E is present. Thus, accurate and sensitive measurements of vitamin E are of paramount importance because it is only when vitamin E levels begin to fall and it disappears that the point is reached at which oxidative damage commences. The index of potential damage is high only when vitamin E is gone.

The concentrations of various other antioxidants that act in the cytosol to quench radical reactions or serve to donate their reducing power to vitamin E may fluctuate. The same may be true for membrane-associated redox couples, such as ubiquinol and ubiquinones or oxidized and reduced cytochrome c, which lowers the steady-state concentrations of vitamin E radicals. Fluctuations in their concentrations are not as important as the concentration changes in vitamin E.

REGULATORY EFFECTS OF VITAMIN E: AMPLIFYING THE MESSAGE

Intracellular Signaling and Secondary Messenger Formation

Protein kinase C (PKC) (1,2) activity and cell proliferation are downregulated in a concentration-dependent manner by α-tocopherol. Trolox, phytol, and α-tocopherol esters were without effect. These results may be relevant to the anticancer effects of vitamin E.

Production of Cell Mediators by the Arachidonic Acid Cascade

Phospholipase A_2 (3,4) preferentially hydrolyzes peroxidized fatty acid esters of phospholipid membranes. Since lipid peroxides activate phospholipase A_2, a decrease in lipid peroxides by vitamin E decreases phospholipase A_2 activity. Vitamin E inhibits platelet phospholipase A_2. Vitamin E modulates arachidonic acid release from membrane phospholipids and therefore arachidonic acid metabolism.

Vitamin E inhibits plant and mammalian 5- and 15-lipoxygenases (5,6). Fatty acid hydroperoxides necessary since activation of lipoxygenase can overcome this inhibition.

Prostaglandin and HETE (hydroxyeicosatetraenoic acid) production from arachidonic acid in bovine seminal vesicles and kidney was inhibited by α-tocopherol (7); HETE production was inhibited less than that of prostaglandin. It appears that vitamin E influences both the cyclooxygenase and lipoxygenase pathways; this modulation of arachidonic acid oxidation may have important in vivo implications.

CLINICAL APPLICATIONS OF VITAMIN E

I. Clearly defined need
  A. Malabsorption
    1. Chlolestatic liver disease
    2. Cystic fibrosis
    3. Abetalipoproteinemia
    4. Others (celiac disease, sprue, and pancreatitis)
  B. Familial deficiency
  C. Prolonged total parenteral nutrition
II. Some evidence, but not completely accepted
  A. Premature infants (reduced risk of intraventricular hemorrhage and reduced severity of retrolental fibroplasia)
  B. Intermittent claudication

    C. Peyronie's disease

    D. Hemolytic anemias (sickle cell, glucose-6-phosphate dehydrogenase deficiency, and thalassemia)

  III. Adult respiratory distress syndrome (ARDS) (8): low plasma E levels in ARDS suggest a need for parenteral E administration, which may delay the onset of acute respiratory failure.

  IV. Epilepsy (9): vitamin E supplementation in addition to antiepileptic drug therapy resulted in a reduction in seizures in a majority of epileptic children refractory to antiepileptic drugs.

   V. Tardive dyskinesia (10): after vitamin E administration, scores on the abnormal involuntary movement scale were lower in subjects with persistent tardive dyskinesia.

  VI. Cancer therapy

    A. Radiotherapy (11): plasma E and β-carotene levels were decreased in patients during radiotherapy before bone marrow transplantation. Antioxidant loss must be considered a possible cause of early posttransplant organ toxicity.

    B. Chemotherapy (12): heart damage due to Adriamycin therapy was decreased by vitamins A and E.

 VII. Burns (13): vitamin E supplementation stimulated T-helper cells to nearly normal levels in patients after burn injury (20–64% of body surface area).

VITAMIN E BENEFITS IN CHRONIC DISEASES AND AGING

Aging

Free-radical damage accumulates during the aging process. Evidence is increasing that lipid peroxidation may be an important factor in making aging less than the long and healthy process it should be.

Cancer

Reactive oxygen species have been implicated in the process of cancer initiation and promotion. Vitamin E and the other antioxidants alter cancer incidence and tumor growth by functioning as anticarcinogens, quenching free radicals or reacting with their products. Epidemiological studies indicate that vitamin E, alone or in combination with other antioxidants, decreases the incidence of certain forms of cancer.

    Skin cancer is the most common form of human cancer and the most common malignant cancer. The incidence of ultraviolet irradiation on the earth correlates with increased malignant and nonmalignant melanoma. Alarmingly, this disease has increased at the rate of about 100% in the United States in the past decade. Thus, it is one of the fastest developing human cancers. There is increasing evidence that antioxidants retard the multistep process of carcinogenesis and that molecular damage due to ultraviolet irradiation may be reduced by vitamin E.

    Anticancer agents based upon redox cycling that generate reactive oxygen species damage normal tissue. Vitamin E is beneficial in protecting the skin and the heart in studies involving Adriamycin (doxorubicin) therapy for cancer.

Arthritis

Increased free-radical production has been observed in animal and human studies on arthritis. Vitamin E has beneficial effects on the symptoms of arthritis. The results of studies in patients with osteoarthritis indicate that vitamin E therapy is effective in relieving pain and also results in greater improvement of mobility.

Ischemic Heart Disease

Animal and human studies show that, following myocardial ischemia-reoxygenation, the heart vitamin E content falls. Human trials are underway to demonstrate if vitamin E therapy is beneficial in reducing ischemic heart disease. Epidemiological studies of ischemic heart disease mortality (Monica study) show a lower risk for populations having higher plasma vitamin E and A levels.

Circulatory Conditions

Excessive platelet aggregation speeds the development of atherosclerosis. Vitamin E supplementation significantly decreases platelet aggregation in healthy adults and reduces elevated platelet aggregation rates in patients with high blood lipid levels and in oral contraceptive users. In coronary artery disease patients undergoing bypass surgery, free-radical concentrations did not increase significantly during or after bypass surgery even if patients were pretreated with 2000 IU vitamin E 12 h before surgery.

Cataracts

Photooxidative mechanisms and ROS (reactive oxygen species) are important in cataractogenesis. Vitamin E delays or minimizes cataract development in isolated animal lenses. High plasma antioxidant concentrations correlate with reduced cataract risk in adults.

Exercise

Strenuous physical exercise is associated with an increased rate of lipid and protein oxidation. Animal studies have demonstrated that vitamin E is consumed by tissues during periods of increased physical exercise. In a study of mountain climbers, 400 IU/day of vitamin E improved physical performance and decreased breath pentane output associated with exercise and prolonged exposure to high altitudes.

Air Pollution

Vitamin E is an important component of the lung's defense against the injurious effects of smog, smoke, and smoking.

REQUIREMENTS FOR VITAMIN E

Vitamin E Paradox

How do we reconcile the great efficiency of minute quantities of vitamin E in membranes with the beneficial effects of vitamin E supplementation on human health seen in chronic diseases and acute clinical conditions?

A local, but vitally important deficiency in the membrane content of vitamin E can arise rapidly. Exhaustion of vitamin E could occur when oxidative stress overloads the antioxidant defenses, which back up (recycling mechanisms) to protect vitamin E from being lost. Under these conditions, vitamin E decreases to a point at which molecular damage to lipids, proteins, and nucleic acids can be expected. Examples of such acute situations are ischemia-reperfusion injury—myocardial infarction or stroke and subsequent reoxygenation—hemorrhaghic and other forms of shock, exposure to extremes of environmental pollution or irradiation, or antineoplastic drugs.

Replenishment of this localized vitamin E deficiency in membranes often occurs slowly, even in tissues in which vitamin E turnover is rapid. Since considerable time is required before the vitamin E content can be reequilibrated, the conventional method of vitamin E loading by dietary supplementation may not be successful in a time frame required to prevent inflicting tissue injury.

It usually takes days to weeks to substantially increase the vitamin E content of membranes. Water-soluble substances that recycle vitamin E may be helpful in increasing the antioxidant deficiency of the residual vitamin E in membranes or lipoproteins. The use of water-soluble recyclers or water-soluble forms of vitamin E that are more rapidly distributed, such as vitamin E phosphate or vitamin E succinate, may also be helpful.

Vitamin E Requirements

Determination of requirements for vitamin E are complicated by variations in the susceptibility of dietary and tissue fatty acids to peroxidation and the difficulty in demonstrating vitamin E inadequacies in healthy adults. Vitamin E requirements may vary fivefold, depending on dietary intake and/or tissue composition; high polyunsaturated fat intakes increase vitamin E requirements because of the increased peroxidative potential of body tissues.

Free-radical–mediated damage has been implicated in cellular and extracellular changes that occur over time in the aging process and in the development of chronic diseases. Vitamin E and other antioxidants prevent or minimize the oxidative damage in biological systems. How adequate the antioxidant defense should be to protect the body from the high free-radical concentrations that are unavoidable at the present time is one of the many new horizons for vitamin E research.

Health status, life-style, and environment also exert a marked influence on the requirements for vitamin E. The following is the text of a public statement prepared by Packer, Pryor, and Diplock (October 29, 1990):

> An increasing number of studies link vitamin E to lowered risk of many chronic and acute health problems. Vitamin E's role in maintaining the integrity of the nervous and cardiovascular systems and its role in age-related health problems, such as cancer, cataracts, and the maintenance of immunity, is becoming ever more evident.
>
> Moreover, it is known that
>
> 1. Humans are exposed to high levels of oxidant stress from the environment and as a result of current life-style patterns.
> 2. The U.S. recommended daily allowance (RDA) levels of vitamin E (15 IU), although adequate to prevent deficiency syndromes, do not supply one-tenth of the vitamin E necessary for optimal antioxidant protection against the health problems just listed.
> 3. Daily intake of vitamin E in amounts up to 800 IU is safe for normal humans.

There is not yet a consensus in the research community about the exact daily intake of vitamin E for optimal protection, but there is compelling evidence of health benefits by the increased intake of vitamin E above RDA levels.

In light of the evidence of the benefits of vitamin E intake at higher levels to prevent oxidant damage and because such intake of the vitamin is known to be safe, those who are most knowledgeable about the benefits of antioxidant nutrients, researchers in the field of free-radical damage who are the leading authorities in the field of antioxidants and degenerative diseases, should recommend higher intakes of vitamin E now. To postpone such a recommendation is not in the best interests of the public, who look to us for guidance in matters of health and nutrition.

REFERENCES

1. Mahoney CW, Azzi A. Vitamin E inhibits protein kinase C activity. Biochem Biophys Res Commun 1988; 154(2):694-7.
2. Boscoboinik D, Szewczyk A, Hensey C, Azzi A. Inhibition of cell proliferation by alpha-tocopherol. Role of protein kinase C. J Biol Chem 1991; 226(10):6188-94.
3. Douglas CE, Chan AC, Choy PC. Vitamin E inhibits platelet phospholipase-A_2. Biochim Biophys Acta 1986; 876(3):639-45.
4. van Kuijk FJGM, Sevanian A, Handelman GJ, Dratz EA. A new role for phospholipase-A_2—protection of membranes from lipid-peroxidation damage. Trends Biochem Sci 1988; 12(1):31-4.
5. Reddanna P, Rao MK, Reddy CC. Inhibition of 5-lipoxygenase by vitamin E. FEBS Lett 1985; 193(1):39-43.
6. Baklova RA, Nekrasov AS, Lankin VZ, Kagan VE, Stoichev TS. A mechanism of the inhibitory effect of alpha-tocopherol and its synthetic analogs on the oxidation of linoleic acid catalyzed by lipoxygenase from reticulocytes. Dokl Akad Nauk SSSR 1988; 299(4):1008-11.
7. Halevy O, Sklan D. Inhibition of arachidonic acid oxidation by beta-carotene, retinol and alpha-tocopherol. Biochim Biophys Acta 1987; 918(3):304-7.
8. Richard C, Lemonnier F, Thibault M, Couturier M, Auzepy P. Vitamin E deficiency and lipoperoxidation during adult respiratory distress syndrome. Crit Care Med 1990; 18(1):4-9.
9. Ogunmekan AO, Huang PA. A randomized, double-blind, placebo-controlled, clinical-trial of *d*-alpha-tocopheryl acetate (vitamin E), as add-on therapy, for epilepsy in children. Epilepsia 1989; 30(1):84-9.
10. Elkashef AM, Ruskin PE, Bacher N, Barrett D. Vitamin E in the treatment of tarditive dyskinesia. Am J Psychiatry 1990; 147(4):505-6.
11. Clemens MR, Ladner C, Ehninger G, et al. Plasma vitamin E and beta-carotene concentrations during radiochemotherapy preceding bone marrow transplantation. Am J Clin Nutr 1990; 51(2): 216-9.
12. Milei J, Boveris A, Llesuy S, et al. Amelioration of adriamycin-induced cardiotoxicity in rabbits by prenylamine and vitamin A and vitamin E. Am Heart J 1986; 111(1):95-102.
13. Haberal M, Hamaloglu E, Bora S, Oner G, Bilgin N. The effects of vitamin E on immune regulation after thermal injury. Burns Including Thermal Injury 1988; 14(5):388-93.

67
Outlook for the Future

Lester Packer

University of California, Berkeley, California

Sharon V. Landvik

Vitamin E Research and Information Service, Edina, Minnesota

It is well-accepted that vitamin E is the major lipid-soluble, chain-breaking antioxidant in biological systems. Studies using deuterated vitamin E have provided estimates of α-tocopherol turnover in body tissues in vitamin E-sufficient animals. Additional animal research is required to determine the effects of vitamin E deficiency and age on vitamin E utilization.

The use of radiolabeled tocopherols in human studies should provide more definitive evidence on absorption, transport, utilization, and retention of vitamin E in various body organs and tissues. Stable isotope studies will, it is hoped, lead to a better understanding of vitamin E requirements and status of individuals of varying ages, health conditions, dietary habits, and environments. It is known that life-style and such influences as low-fat diets, increased polyunsaturated fat intake, alcohol consumption, cigarette smoking, and air pollution have an impact on vitamin E requirements and status, but their quantitative effects are unknown at present.

A number of studies are currently underway to evaluate the efficacy of vitamin E therapy in various clinical conditions. Recent documentation of vitamin E deficiency resulting in a progressive neurological syndrome in patients with chronic malabsorption has resulted in research interest on the role of vitamin E treatment of neurological disorders. Studies are in progress to evaluate the effectiveness of vitamin E as add-on therapy or before initiation of drug therapy in such neurological conditions as Parkinson's disease, Huntington's disease, epilepsy, and tardive dyskinesia. Other clinical areas in which preliminary research suggests a beneficial role for vitamin E therapy include intermittent claudication, arthritis, burns, and ischemia-reperfusion injury. Future research is expected to provide substantiation for the efficacy of vitamin E in these clinical applications and more precisely define the optimal dosage range.

Studies continue on clinical applications of vitamin E therapy, but there is increasing research emphasis on the role of vitamin E and the other antioxidants in the prevention of free-radical–mediated degenerative diseases and conditions, including cancer and heart disease. Results from the majority of epidemiological studies on blood antioxidant levels and cancer risk have suggested that vitamin E and the other dietary antioxidants decrease the incidence of certain cancers. Recent epidemiological evidence in Europe has demonstrated an association between low plasma vitamin E levels and the increased mortality rate of ischemic heart disease. Preliminary research data also suggest that vitamin E and other antioxidants protect low-density lipoprotein from oxidation, which is believed to be an important event in the development of atherosclerosis. Free radicals are also implicated in the development of other degenerative conditions, such as premature aging and cataracts, and there is research evidence to suggest that the antioxidants may help delay or prevent their development. Based on the results to date, vitamin E and the other antioxidants merit continued active research evaluation in the future, including intervention trials, to more precisely define their protective roles, alone or in combination, in the prevention of major degenerative diseases.

As discussed in previous chapters, an increasing number of studies link vitamin E to a lowered risk of a number of chronic and acute health problems. There is not yet a consensus in the research community about the exact daily intake of vitamin E for optimal protection, but there is compelling evidence of health benefits by the increased intake of vitamin E above levels required to prevent vitamin E deficiency. As we continue to be exposed to high levels of oxidant stress from the environment and as a result of current life-style patterns, determination of the optimal vitamin E intake to prevent peroxidative damage in individuals over time is essential.

Index

About the Editors

LESTER PACKER is Professor of Molecular and Cell Biology and Director, Membrane Bioenergetics Group, University of California at Berkeley, Berkeley, California. Winner of the 1992 Vitamin E Research and Information Service (VERIS) Award for significant contributions in clinical applications of Vitamin E, he is the author of over 400 published articles and a member of the American Society for Biochemistry and Molecular Biology, the American Physiological Society, and the Society for Investigative Dermatology, among others. Dr. Packer received the B.S. (1951) and M.S. (1952) degrees in biology and chemistry from Brooklyn College, Brooklyn, New York, and the Ph.D. degree (1956) in microbiology and biochemistry from Yale University, New Haven, Connecticut.

JÜRGEN FUCHS is Research Associate and clinical faculty member, Department of Dermatology, Johann Wolfgang Goethe University, Frankfurt, Germany. The author of approximately 100 published articles, he is a member of the Society for Investigative Dermatology and the Society for Electron Resonance Paramagnetic Tomography in Biomedicine. Dr. Fuchs received the Ph.D. degree (1985) in biology and the M.D. degree (1986) from the University of Frankfurt, Frankfurt, Germany.